Methods in Cell Biology

VOLUME 74

Development of Sea Urchins, Ascidians, and Other Invertebrate Deuterostomes: Experimental Approaches

Series Editors

Leslie Wilson

Department of Biological Sciences
University of California
Santa Barbara, California

Paul Matsudaira

Whitehead Institute for Biomedical Research
Department of Biology
Massachusetts Institute of Technology
Cambridge, Massachusetts

Methods in Cell Biology

Prepared under the Auspices of the American Society for Cell Biology

VOLUME 74

Development of Sea Urchins, Ascidians, and Other Invertebrate Deuterostomes: Experimental Approaches

Edited by

Charles A. Ettensohn

Department of Biological Sciences
Carnegie Mellon University
Pittsburgh, Pennsylvania

Gregory A. Wray

Department of Biology
Duke University
Durham, North Carolina

Gary M. Wessel

Department of Molecular and Cell Biology and Biochemistry
Brown University
Providence, Rhode Island

ELSEVIER
ACADEMIC
PRESS

AMSTERDAM • BOSTON • HEIDELBERG • LONDON
NEW YORK • OXFORD • PARIS • SAN DIEGO
SAN FRANCISCO • SINGAPORE • SYDNEY • TOKYO

Elsevier Academic Press
525 B Street, Suite 1900, San Diego, California 92101-4495, USA
84 Theobald's Road, London WC1X 8RR, UK

This book is printed on acid-free paper. ♾

For all information on all Academic Press publications visit our Web site at www.academicpress.com

ISBN: 0-12-480278-8 Case
ISBN: 0-12-480279-6 Paperback

PRINTED IN THE UNITED STATES OF AMERICA
04 05 06 07 08 9 8 7 6 5 4 3 2 1

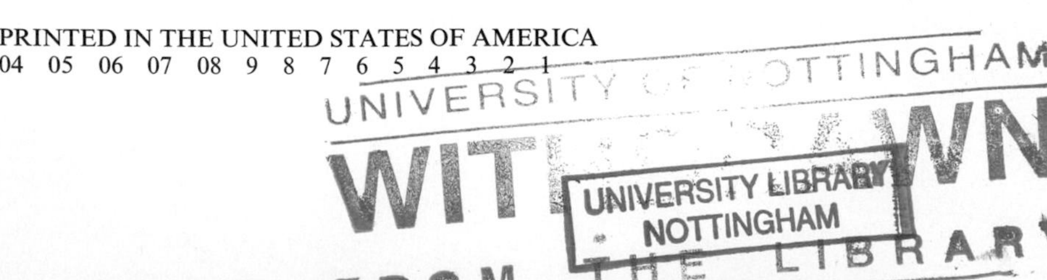

CONTENTS

PART II Embryological Approaches

PART IV Molecular Biological Approaches

PART V Genomics

PART VI Echinoderm Eggs and Embryos in the Teaching Lab

CONTRIBUTORS

Numbers in parentheses indicate the pages on which authors' contributions begin.

Nikki L. Adams (39), Department of Biological Sciences, California Polytechnic State University, San Luis Obispo, California 93407

Shonan Amemiya (243), Department of Integrated Biosciences, Graduate School of Frontier Sciences, University of Tokyo, Kashiwa, Chiba 277-8562, Japan

Lynne M. Angerer (699), Department of Biology, University of Rochester, Rochester, New York 14627

Robert C. Angerer (699), Department of Biology, University of Rochester, Rochester, New York 14627

Maria I. Arnone (621), Stazione Zoologica Anton Dohrn, Villa Comunale, 80121 Napoli, Italy

Carmen Beltrán (545), Department of Developmental Genetics and Molecular Physiology, Institute of Biotechnology, Universidad Nacional Autónoma de México, Cuernavaca, Morelos, México

Stephen C. Benson (273), Department of Biological Sciences, California State University, Hayward, Hayward, California 94542

S. Anne Böttger (17), Department of Zoology and Marine Biology, Biomedical Research Group, Durham, New Hampshire 03824

Bruce P. Brandhorst (579), Department of Molecular Biology and Biochemistry, Simon Fraser University, Burnaby, British Columbia, V5A 1S6, Canada

C. Titus Brown (733), Division of Biology and the Center for Computational Regulatory Genomics, Beckman Institute, California Institute of Technology, Pasadena, California 91125

Jacqueline M. Brooks (87), Department of Molecular and Cell Biology and Biochemistry, Brown University, Providence, Rhode Island 02912

David Burgess (371), Department of Biology, Boston College, Chestnut Hill, Massachusetts 02167

Robert D. Burke (411), Departments of Biology and Biochemistry/Microbiology, University of Victoria, Victoria, British Columbia V8W 3P6, Canada

R. Andrew Cameron (733), Division of Biology and the Center for Computational Regulatory Genomics, Beckman Institute, California Institute of Technology, Pasadena, California 91125

Melani S. Cheers (287), Department of Biological Sciences, Carnegie Mellon University, Pittsburgh, Pennsylvania 15213

Lionel Christiaen (143), INRA Junior Group, UPR 2197 DEPSN, CNRS, Institut de Neurobiologie A. Fessard, Gif-sur-Yvette, France

James A. Coffman (653), Stowers Institute for Medical Research, Kansas City, Missouri 64110

Bruce J. Crawford (411), Department of Anatomy and Cell Biology, University of British Columbia, Vancouver, British Columbia V6T 1Z4, Canada

Alberto Darszon (545), Department of Developmental Genetics and Molecular Physiology, Institute of Biotechnology, Universidad Nacional Autónoma de México, Cuernavaca, Morelos, México

Eric H. Davidson (775), Division of Biology, California Institute of Technology, Pasadena, California 91125

Karine Deschet (143), Neuroscience Research Institute and Department of Molecular, Cellular, and Developmental Biology, University of California, Santa Barbara, California 93106

Ivan J. Dmochowski (621), Department of Chemistry, University of Pennsylvania, Philadelphia, Pennsylvania 19104

Jan Ellenberg (371), European Molecular Biological Lab, Heidelberg, Germany

David Epel (797), Hopkins Marine Station of Stanford University, Pacific Grove, California 93950

Charles A. Ettensohn (1, 287), Department of Biological Sciences, Carnegie Mellon University, Pittsburgh, Pennsylvania 15213

Vera Lynn Flowers (333), Department of Cell Biology and Anatomy, Louisiana State University Health Sciences Center, New Orleans, Louisiana 70112

Victoria Foe (371), Friday Harbor Laboratories, University of Washington, Seattle, Washington 98195

Kathy R. Foltz (39), Department of Molecular, Cellular, and Developmental Biology, and the Marine Science Institute, University of California, Santa Barbara, California 93106

Christian Gache (621, 677), Laboratory of Developmental Biology, CNRS-Université Pierre et Marie Curie (Paris VI), Observatoire Océanologique, 06230 Villefranche-sur-Mer, France

John Gerhart (171), Department of Molecular and Cellular Biology, University of California Berkeley, Berkeley, California 94720

Julia Gorelik (545), Division of Medicine, Imperial College London, MRC Clinical Sciences Centre, London W12 0NN, United Kingdom

Carolyn Hendrickson (143), INRA Junior Group, UPR 2197 DEPSN, CNRS Institut de Neurobiologie A. Fessard, Gif-sur-Yvette, France

Jonathan Henry (243), Department of Cell and Structural Biology, University of Illinois, Urbana, Illinois 61801

Noritaka Hirohashi (523), Department of Biology, Ochanomizu University, Tokyo, 112-8610 Japan

Linda Z. Holland (195), Marine Biology Research Division, Scripps Institution of Oceanography, University of California San Diego, La Jolla, California 92093

Tom Humphreys (171), Kewalo Marine Laboratory, Honolulu, Hawaii 96813

Laurinda A. Jaffe (219, 469), Department of Cell Biology, University of Connecticut Health Center, Farmington, Connecticut 06032

Di Jiang (143), Neuroscience Research Institute and Department of Molecular, Cellular, and Developmental Biology, University of California, Santa Barbara, California 93106

Jean-Stéphane Joly (143), INRA Junior Group, UPR 2197 DEPSN, CNRS, Institut de Neurobiologie A. Fessard, Gif-sur-Yvette, France

Marc Kirschner (171), Systems Biology, Harvard Medical School, Boston, Massachusetts 02115

Chisato Kitazawa, (75), Department of Biology, Duke University, Durham, North Carolina 27708

Masato Kiyomoto (243), Tateyama Marine Laboratory, Ochanomizu University, Tateyama, Chiba 294-0301, Japan

Yuri E. Korchev (545), Division of Medicine, Imperial College London, MRC Clinical Sciences Centre, London W12 0NN, United Kingdom

Ritsu Kuraishi (243), Marine Biological Station, Graduate School of Science, Tohoku University, Aomori 039-3501, Japan

Laurent Legendre (143), INRA Junior Group, UPR 2197 DEPSN, CNRS, Institut de Neurobiologie A. Fessard, Gif-sur-Yvette, France

Peter Lenart (371), European Molecular Biological Lab, Heidelberg, Germany

Thierry Lepage (677), Laboratory of Developmental Biology, CNRS-Université Pierre et Marie Curie (Paris VI), Observatoire Océanologique, 06230 Villefranche-sur-Mer, France

Christopher J. Lowe (171), Department of Organismal Biology and Anatomy, University of Chicago, Chicago, Illinois 60612 and Department of Molecular and Cellular Biology, University of California Berkeley, Berkeley, California 94720

David R. McClay (243, 311), Department of Biology, Duke University, Durham, North Carolina 27708

Pam Miller (797), Seaside High School, Seaside, California 93955

Benjamin Miner (75), Department of Zoology, University of Florida, Gainesville, Florida 32611

Takuya Minokawa (243), Division of Biology, California Institute of Technology, Pasadena, California 91125

Yuki Nakatani (143), Neuroscience Research Institute and Department of Molecular, Cellular, and Developmental Biology, University of California, Santa Barbara, California 93106

Hiroki Nishida (243), Department of Biological Sciences, Tokyo Institute of Technology, Nagatsuta, Midori-ku, Yokohama 226-8501, Japan

Takuya Nishigaki (545), Department of Developmental Genetics and Molecular Physiology, Institute of Biotechnology, Universidad Nacional Autónoma de México, Cuernavaca, Morelos, México

Paola Oliveri (775), Division of Biology, California Institute of Technology, Pasadena, California 91125

Chris Patton (797), Hopkins Marine Station of Stanford University, Pacific Grove, California 93950

Margaret Peeler (797), Department of Biology, Susquehanna University, Selinsgrove, Pennsylvania 17870

Carmen Pepicelli (333), Curis, Inc., Cambridge, Massachusetts 02138

Andrew Ransick (243, 601), Division of Biology, California Institute of Technology, Pasadena, California 91125

Jonathan P. Rast (731), Division of Biology and the Center for Computational Regulatory Genomics, Beckman Institute, California Institute of Technology, Pasadena, California 91125

Esmeralda Rodríguez (545), Department of Developmental Genetics and Molecular Physiology, Institute of Biotechnology, Universidad Nacional Autónoma de México, Cuernavaca, Morelos, México

Linda L. Runft (39), Department of Molecular, Cellular, and Developmental Biology, and the Marine Science Institute, University of California, Santa Barbara, California 93106

Daniel Sánchez (545), Division of Medicine, Imperial College London, MRC Clinical Sciences Centre, London W12 0NN, United Kingdom

Nori Satoh (759), Department of Zoology, Graduate School of Science, Kyoto University, Kyoto 606-8502, Japan

William C. Smith (143), Neuroscience Research Institute and Department of Molecular, Cellular, and Developmental Biology, University of California, Santa Barbara, California 93106

Laila Strickland (371), Department of Biology, Boston College, Chestnut Hill, Massachusetts 02167

Billie J. Swalla (115), Biology Department and Friday Harbor Laboratories, University of Washington, Seattle, Washington 98195

Hyla Sweet (243), Department of Biological Sciences, College of Science, Rochester Institute of Technology, Rochester, New York 14623

Kuni Tagawa (171), Kewalo Marine Laboratory, Honolulu, Hawaii 96813

Mark Terasaki (219, 469), Department of Cell Biology, University of Connecticut Health Center, Farmington, Connecticut 06032

Jason Tresser (143), Neuroscience Research Institute and Department of Molecular, Cellular, and Developmental Biology, University of California, Santa Barbara, California 93106

Tatsuya Unuma (17), National Research Institute of Aquaculture, Nansei, Mie 516-0193, Japan

Victor D. Vacquier (491, 523, 797), Center for Marine Biotechnology and Biomedicine, Scripps Institution of Oceanography, University of California, San Diego, La Jolla, California 92093

Judith M. Venuti (333), Department of Cell Biology and Anatomy, Louisiana State University Health Sciences Center, New Orleans, Louisiana 70112

George von Dassow (371), Friday Harbor Laboratories, University of Washington, Seattle, Washington 98195

Ekaterina Voronina (87), Department of Molecular and Cell Biology and Biochemistry, Brown University, Providence, Rhode Island 02912

Charles W. Walker (17), Department of Zoology and Marine Biology, Biomedical Research Group, Durham, New Hampshire 03824

Gary M. Wessel (1, 87, 491), Department of Molecular and Cell Biology and Biochemistry, Brown University, Providence, Rhode Island 02912

Michael Whitaker (443), School of Cell & Molecular Biosciences, Faculty of Medical Sciences, University of Newcastle upon Tyne, Framlington Place, NE2 4HH, United Kingdom

Athula Wikramanayake (243), Department of Zoology, University of Hawaii at Manoa, Honolulu, Hawaii 96822

Fred H. Wilt (273), Department of Molecular Cell Biology, University of California, Berkeley, California 94720

Christopher Wood (545), Department of Developmental Genetics and Molecular Physiology, Institute of Biotechnology, Universidad Nacional Autónoma de México, Cuernavaca, Morelos, México

Gregory A. Wray (1, 75), Department of Biology, Duke University, Durham, North Carolina 27708

Ju-Ka Yu (195), Marine Biology Research Division, Scripps Institution of Oceanography University of California, San Diego, La Jolla, California 92093

Chiou-Hwa Yuh (653), Division of Biology 156-29, California Institute of Technology, Pasadena, California 91125

Robert W. Zeller (713), Department of Biology, San Diego State University, San Diego, California 92182

PREFACE

The past decade has been one of the most exciting periods in the history of developmental biology. Numerous genomes have been sequenced, powerful new experimental approaches are rapidly being developed, and we are now able to visualize processes in living embryos in ways that that were inconceivable a short while ago. From these powerful new experimental tools has emerged a much richer view of developmental processes across scales of biological organization, as well as a growing appreciation for the shared genetic and molecular underpinnings of development in diverse species.

Within this broad research effort, the invertebrate deuterostomes are playing an important role. As the closest living relatives of vertebrates, the invertebrate deuterostomes occupy a uniquely informative phylogenetic position, providing a crucial perspective on the organization and evolution of vertebrate genomes and clues concerning origin of the vertebrate body plan. As experimental subjects, the embryos of these animals provide a singular and powerful combination of advantages: they are simple, optically transparent, develop externally, are robust to physical manipulation, and can be raised in large quantities. Invertebrate deuterostomes have genomes that are similar to those of vertebrates in many respects, but contain fewer gene duplications.

With the completion of an ascidian genome project in 2003 and the imminent completion of a sea urchin genome project in 2004, we are entering the "post-genome" era for the invertebrate deuterostomes. As the pace of research accelerates, there is a clear need for accessible protocols to guide experimental studies with the embryos of these animals. The most recent compendium of experimental methods was published 18 years ago. "Echinoderm Gametes and Embryos," edited by Thomas Schroeder, has been an invaluable resource at the bench, but many essential new methods have been developed since that book was published and an updated collection of methods is clearly needed. Moreover, while the focus of the earlier volume was on echinoderms, a growing number of researchers are working with other invertebrate deuterostomes: urochordates, hemichordates, and cephalochordates. Our aim with the present volume is to bring together a set of protocols for rearing and working with these embryos.

This book would not have been possible without the collective efforts of the research community to which we belong: the scientists who study the development of invertebrate deuterostomes. We would like to extend our thanks and appreciation to the many people who developed and tested the methods described in the following chapters, and who have inspired us by applying these methods to a wide range of exciting problems in developmental biology. We also thank our many colleagues who wrote, reviewed, and otherwise contributed to the chapters in this book. To the extent that this book proves useful, it will be thanks to their generous efforts to distill years of experience into brief, practical guidelines. Finally, we extend our special thanks to Mica Haley and Kristi Savino at Elsevier, who patiently guided us through the process of assembling this book.

Charles A. Ettensohn
Gregory A. Wray
Gary M. Wessel

The editors dedicate this volume to their scientific mentor,
Dr. David R. McClay,
the Arthur S. Pearse Professor of Biology at Duke University.

CHAPTER 1

The Invertebrate Deuterostomes: An Introduction to Their Phylogeny, Reproduction, Development, and Genomics

Charles A. Ettensohn,* Gary M. Wessel,[‡] and Gregory A. Wray[†]

*Department of Biological Sciences
Carnegie Mellon University
Pittsburgh, Pennsylvania 15213

[†]Department of Biology
Duke University
Durham, North Carolina 27708

[‡]Department of Molecular Biology, Cell Biology, and Biochemistry
Brown University
Providence, Rhode Island 02912

I. Introduction

The invertebrate deuterostomes comprise approximately 10,000 species of marine animals distributed throughout the oceans of the world. Early studies on the eggs and embryos of these animals played a critical historical role in embryology,

0091-679X/04 $35.00

cell biology, and genetics. They have continued to be widely used as model organisms for developmental and cell biological studies for more than a century. In recent years, the traditional strengths of these systems for developmental studies have been supplemented by a variety of powerful methodologies for analyzing and perturbing gene expression and function. Moreover, with the advent of modern genomics and the re-emergence of evolutionary–developmental biology, this group of organisms promises to yield critical new insights in many areas. Analysis of early embryonic patterning and gene regulatory networks is advancing rapidly. The invertebrate deuterostomes are also proving to be a rich resource for analysis of the evolution of developmental programs. There are two major reasons for this: (1) embryos of a diverse collection of species can be obtained and studied with relative ease, and (2) there are examples of remarkable variations in life cycles, developmental patterns, and adult body plans within the invertebrate deuterostomes, notwithstanding the many shared features of their development. Finally, the invertebrate deuterostomes are the closest living relatives of the vertebrates, and analysis of their development promises to shed light on the origins of the vertebrates, the chordates, and the entire deuterostome lineage.

II. Phylogeny

Recent molecular phylogenetic studies suggest that the deuterostomes include only the echinoderms, hemichordates, and chordates (Adoutte *et al.*, 2000; Aguinaldo *et al.*, 1997) (Fig. 1). Certain groups that were previously considered to be deuterostomes, notably, the chaetognaths and lophophorates, are instead likely to be protostomes. The invertebrate deuterostomes consist of all deuterostomes outside Subphylum Vertebrata; i.e., the echinoderms (Phylum Echinodermata), hemichordates (Phylum Hemichordata), and the members of two invertebrate subphyla within the chordates, Urochordata and Cephalochordata (Fig. 1). Among the invertebrate deuterostomes, morphological and molecular data indicate that echinoderms and hemichordates are closely related to one another and represent a distinct clade, that the urochordates are monophyletic and could be considered a separate phylum, and that the cephalochordates are most closely related to vertebrates (Bromham and Degnan, 1999; Cameron *et al.*, 2000; Swalla *et al.*, 2000; Fig. 1). A close relationship between hemichordates and echinoderms is also supported by recent studies showing similarities in patterns of gene expression during early embryogenesis (Gross and McClay, 2001; Harada *et al.*, 2002; Nakajima *et al.*, 2004; Shoguchi *et al.*, 1999; Tagawa *et al.*, 1998).

Most present-day species of invertebrate deuterostomes are members of Phylum Echinodermata (~7000 species). Adult echinoderms are bottom-dwelling, radially symmetrical animals, but their larvae are free-swimming and bilaterally symmetrical (Fig. 2A, B, Fig. 3I). Modern echinoderms are grouped into five classes: Echinodea (sea urchins and sand dollars), Holothuroidea (sea cucumbers), Asteroidea (starfish), Ophiuroidea (brittle stars), and Crinoidea (feathers stars and

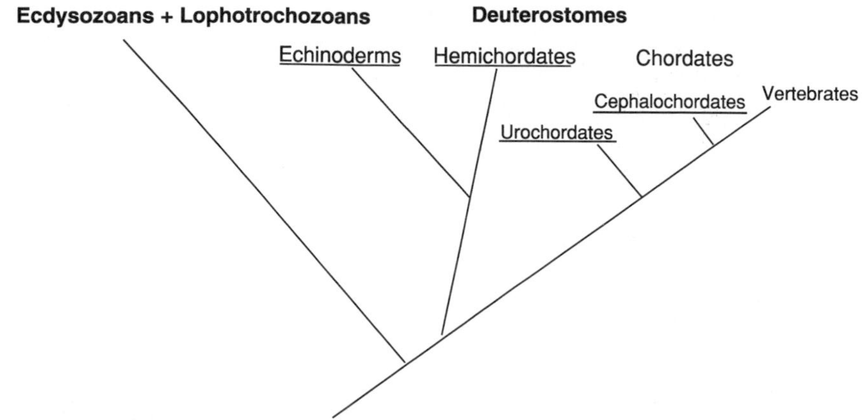

Fig. 1 A phylogenetic tree of the bilaterally symmetrical animals (Bilateria). These are usually grouped into three multiphyletic clades: Ecdysozoa, Lophotrochozoa, and Deuterostomia. Taxa that comprise the invertebrate deuterostomes (the organisms considered in this book) are underlined. They consist of Phylum Echinodermata and Phylum Hemichordata, and two chordate subphyla (Subphylum Urochordata and Subphylum Cephalochordata).

sea lilies). Among the echinoderms, sea urchins have been most widely used for developmental studies. Indeed, the developmental biology of sea urchins is better understood than that of any other invertebrate deuterostome. Starfish have also been widely used for studies of gametogenesis and embryonic development, but far less attention has been paid to members of the remaining three classes of echinoderms.

The urochordates, or tunicates (Subphylum Urochordata), are represented by some 3000 extant species. They are usually grouped into three classes: Ascidiacea (ascidians), Thaliacea, and Larvacea. The ascidians, the most numerous group (~2300 species) have been the most thoroughly studied from a developmental perspective. They are benthic, sessile, filter feeders that develop via a bilaterally symmetrical, nonfeeding tadpole larva (Figs. 2C, Fig. 3K). Some live as individuals (solitary, or simple, ascidians) while others form colonies (colonial, or compound, ascidians). The solitary ascidians have been the most widely used for developmental studies (Satoh, 1994). Unlike ascidians, thalaceans and larvaceans remain free-swimming throughout their life cycle (Bone, 1998).

The hemichordates (Phylum Hemichordata) are represented by relatively few (~85) extant species and are grouped into the classes Enteropneusta, Pterobranchia, and Planctosphaeroidea. The enteropneusts (acorn worms) are likely to be basal within the group (Cameron *et al.*, 2000) and are solitary, sedentary, burrowing worms (Fig. 2D). The pterobranchs are colonial, sessile organisms that live in secreted tubes. The monotypic Planctosphaeroidea form spherical, transparent larvae, the adult forms of which have not yet been described (see Cameron *et al.*, 2000).

Fig. 2 Representative adult invertebrate deuterostomes. (A) Sea urchin (*Lytechinus variegatus*, Phylum Echinodermata). (B) Starfish (*Linkia lavigata*, Phylum Echinodermata). (C) Ascidians (*Boltenia villosa* and *Styela gibbsii*, Subphylum Urochordata). Photograph courtesy of Billie Swalla. (D) Hemichordate (*Saccoglossus kowalevskii*, Phylum Hemichordata). Photograph courtesy of Chris Lowe. (E) Amphioxus (*Branchiostoma floridae*, Subphylum Cephalochordata). Photograph courtesy of Linda Holland. (See Color Insert.)

The cephalochordates, commonly known as amphioxus (or the lancelet), include ~28 different species, grouped into two genera. They are the closest living relatives of the vertebrates and share a dorsal hollow nerve cord, notochord, and phayngeal gill slits, but lack paired limbs, paired ears, and image-forming eyes (Fig. 2E). They are relatively sedentary, sand-dwelling filter feeders. Most developmental studies have been carried out on three species of the genus *Branchiostoma* (*B. lanceolatum*, *B. floridae*, and *B. belcheri*) (Holland and Yu, Chapter 9).

III. Reproduction

Most species of invertebrate deuterostomes have separate sexes. The major exceptions are the ascidians, which are hermaphrodites. Sexual reproduction is by far the most common mode within the group. Most species have a distinct reproductive season during which sperm and eggs are released under appropriate environmental conditions. Invertebrate deuterostomes are typically broadcast spawners, and fertilization and development occur externally. Colonial ascidians and some echinoderms, however, brood their eggs within body cavities of the adult. Although sexual reproduction is predominant, reproduction by asexual budding occurs in a number of groups that have colonial forms, including pterobranch hemichordates and the colonial ascidians (Cameron *et al.*, 2000; Satoh, 1994). In addition, among the echinoderms, some solitary forms are capable of asexual reproduction through the splitting of adult animals (McGovern, 2003) or larvae (Bosch *et al.*, 1989; Eaves and Palmer, 2003).

Many invertebrate deuterostomes develop through a free-swimming, feeding larval stage that bears little anatomical resemblance to the adult form (maximal indirect development) (Fig. 3). There is a complex nomenclature associated with these larval forms, particularly those of hemichordates and echinoderms, which reflects variations in certain aspects of their morphology, such as the distribution of cilia and the organization of calcified skeletal elements. For example, the larva of enteropneust hemichordates is known as a tornaria, the sea urchin larva is called a pluteus, and larvae of asteroids, holothuroids, ophiuroids, and crinoids are known as bipinneria, auricularia, ophiopluteus, and doliolaria larvae, respectively. These various ciliated, feeding larvae often undergo dramatic morphological changes at metamorphosis to give rise to the adult. For example, the radially symmetrical, adult sea urchin is formed from a rudiment (the echinus rudiment) derived from the left coelomic pouch (hydrocoel) and the overlying ectoderm (vestibule) of the bilaterally symmetrical larva, while most of the other larval tissues degenerate (Burke, 1989). Metamorphosis in indirect developing enteropneust hemichordates results in less pronounced anatomical changes, and the body plan of the tornaria is more directly inherited by the adult (Lowe *et al.*, Chapter 8).

Indirect development was clearly the ancestral mode of development within the echinoderms (Nakano *et al.*, 2003; Peterson *et al.*, 1999; Raff, 1987), and morphological similarities between the larvae of echinoderms and hemichordates suggest that this was the ancestral mode within the echinoderm + hemichordate clade. Direct development (i.e., the loss of a feeding larval stage) has arisen independently many times, however, within the group (Raff, 1987; Strathmann, 1978). In its

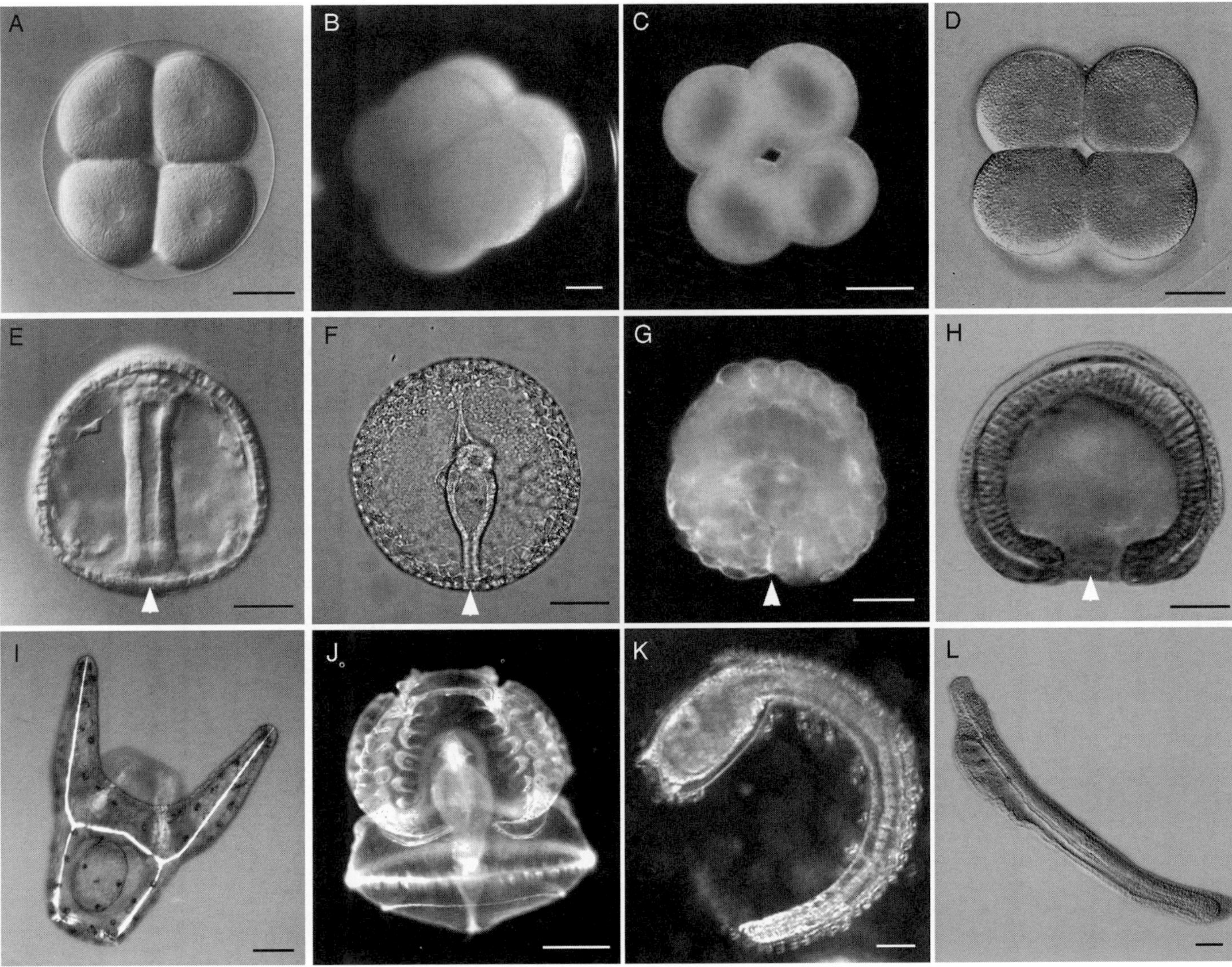
A
B
C
D
E
F
G
H
I
J
K
L

extreme form, direct development is associated with the formation of a late-stage embryo that has few morphological features of the feeding larva and develops rapidly and directly into a juvenile adult. As others have noted, however, many species of echinoderms show intermediate modes of development that challenge a simple "indirect" vs "direct" classification scheme (Chia *et al.*, 1993; McEdward and Janies, 1997). These include larvae that are morphologically very similar to feeding larvae but do not feed (or feed facultatively) as well as various late embryonic forms with morphologies intermediate between those of extreme indirect and direct developing forms (Chia *et al.*, 1993; Raff, 1987). Direct development has also arisen within the enteropneust hemichordates. Again, however, there are examples of intermediate forms, including species with swimming, tornaria-like larvae that do not feed (Barrington, 1965).

A simple categorization of urochordates and cephalochordates as either direct or indirect developers is also somewhat problematic, given the specific characteristics of their larval forms. In cephalochordates, embryogenesis gives rise to a ciliated, pelagic larva that can swim and feed for many weeks prior to metamorphosis. In these organisms, the morphological changes at metamorphosis are considerably less pronounced than in echinoderms, however. They include the generation of a second row of gill slits, movement of gills slits and the mouth, and changes in the digestive tract (Holland and Yu, Chapter 9). Most ascidian species develop via a swimming tadpole larva and are considered to undergo indirect development. Ascidian tadpole larvae differ from the larvae of other invertebrate deuterostomes, however, in that they are uniformly nonfeeding and swim by muscular contraction rather than ciliary beating. Ascidian larvae usually swim for only minutes or a few hours before settling and undergoing metamorphosis. Extensive anatomical changes take place at that time, including resorption of the tail, loss of the outer layer of tunic and some organs, and the repositioning of visceral organs (Satoh, 1994). This indirect mode of development is ancestral within the ascidians, but in so-called "tailless" or "aneural" species, it has given way to direct development, in which various features of the tadpole, such as differentiated tail muscle cells, notochord, and the neural sensory organ, have been lost (Jeffery and Swalla, 1992; Jeffery *et al.*, 1999).

Fig. 3 Representative embryonic and larval stages of selected invertebrate deuterostomes. (A, E, I) Sea urchin (*Lytechinus variegatus*). (B, F, J) Enteropneust hemichordate (B, *Saccoglossus kowalevskii*; F, J, *Ptychodera flava*). (C, G, K) Ascidian (C, G, *Ciona intestinalis*; K, *Boltenia villosa*). (D, H, L) Amphioxus (*Branchiostoma floridae*). The upper panels (A–D) show early cleavage stage embryos (4–8 cell stage) and illustrate the holoblastic pattern of cleavage common to the group. The middle panels (E–H) show gastrula stage embryos. Endoderm and mesoderm are internalized through a roughly circular blastopore (arrowhead) positioned at the vegetal pole of the spherical gastrula. The lower panels (I–L) show larval stages (1–3 d larvae except for J, which is much older). There is considerable diversity in larval morphology within the group. Scale bars = 50 μm except for J (300 μm). Photographs courtesy of Chris Lowe (B, J), Nori Satoh (C, G), Linda Holland (D, H, L), Jonathan Henry (F), Rachel Fink (I), and Billie Swalla (K).

These variations in developmental modes and larval and adult morphologies among the extant, invertebrate deuterostomes raise several questions: Did the ancestral chordate and the ancestral deuterostome develop in an "indirect" fashion, via a free-swimming, feeding larva? If so, what did the ancestral larva look like? How different was its anatomy from that of the adult? These questions are obviously difficult to answer in light of the many independent evolutionary changes that have taken place in the various lineages since they first diverged. One hypothesis is that the ancestral deuterostome developed via a ciliated, pelagic, feeding larva that may have given rise to a sessile, wormlike adult. Garstang (1928) suggested that vertebrates arose through neoteny, the sexual maturation of a swimming larva that may have resembled the ascidian tadpole larva (see Barrington, 1965; Gee, 1996).

IV. Development

A. Historical Contributions

Experimental work with invertebrate deuterostomes has played an important role in developmental biology, cell biology, and genetics. Descriptive and experimental studies on the embryonic development of these organisms date back to the mid-nineteenth century. The earliest papers describing *in vitro* fertilization and embryonic development in sea urchins were published by Derbes and von Baer in 1847 (see Hörstadius, 1973). The blastomere separation experiments of Chabry (1887), who worked with the ascidian *Ciona intestinalis*, are often considered the first work in the field of experimental embryology (cited in Fischer, 1992). These were soon followed by the pioneering studies of Driesch on sea urchin embryos, which led to the discovery of regulative development (Driesch, 1892). Other early work with sea urchins pointed to the central role of the nucleus in development and heredity (Hertwig, 1875) and revealed the nonequivalence of chromosomes (Boveri, 1907). Several decades later, the famous cell isolation and transplantation studies of Sven Hörstadius provided early evidence of inductive interactions among blastomeres and helped define the notion of developmental gradients (reviewed by Hörstadius, 1939). Early experimental studies on ascidian embryos were central in revealing cytoplasmic regionalization of the egg, the role of ooplasmic constituents in regulating cell fates, and the relationship between the egg and larval body axes (Castle, 1896; Conklin, 1905; van Beneden and Julin, 1884). Observations on hemichordate and cephalochordate embryos also date back to the mid-nineteenth century (Bateson, 1884; Kowalevsky, 1866; Müller, 1841).

B. Experimental Characteristics

The characteristics of invertebrate deuterostomes that attracted many early experimental embryologists remain extremely useful today. The ease with which gametes can be collected and the external fertilization and development of the embryos originally made it possible to study their early development. Today, the ability to produce vast numbers of synchronously developing embryos (using methods that are best developed with sea urchins) facilitates a wide variety of biochemical and molecular biological approaches, from protein purification to cDNA library construction. The unsurpassed transparency of many invertebrate deuterostome eggs and embryos originally made it possible to directly observe developmental processes using only a light microscope. The clarity of these embryos now facilitates the application of a wide variety of modern light optical technologies, including fluorescence-based methods for monitoring gene expression, protein localization, protein–protein interactions, and biochemical activity. The embryos of many invertebrate deuterostomes are highly amenable to physical manipulations, including embryo dissociation, cell isolation, and cell transplantation. These approaches continue to be invaluable in analyzing cellular interactions that pattern the embryo. Molecular biological approaches including expression of exogenous mRNAs, injection of reporter DNA constructs for cis-regulatory analysis, and injection of morpholino antisense oligonucleotides, make it possible to dissect gene regulation and function. The development of genomics-based resources, including the assembly of draft genomic sequences of several organisms, is spurring work of all kinds (see Section V).

C. Shared Features of Development

Several features of early development are shared among the invertebrate deuterostomes, as might be expected from their close phylogenetic relationships. The eggs of invertebrate deuterostomes are isolecithal (i.e., they have sparse, evenly distributed yolk). They are usually 100 to 200 μm in diameter, but can range up to 4 mm, with larger, more yolky eggs typical of direct-developing species (Chia *et al.*, 1993; Raff, 1987). Primary oocytes typically undergo germinal vesicle breakdown shortly before spawning. Depending upon the organism, at the time of fertilization the oocyte may be in metaphase of the first meiotic division (starfish and ascidians), metaphase of the second meiotic division (amphioxus), or may have completed meiosis (sea urchins). The unfertilized egg is polarized along a single maternal axis; the animal–vegetal axis. In ascidians, other axes are established immediately after fertilization, as reflected by movements of the cytoplasm and sperm aster (Satoh, 1994). In sea urchins, a secondary axis (the oral–aboral axis) is probably entrained soon after fertilization, although the initial cues have not been identified (Duboc *et al.*, 2004; Ettensohn and Sweet, 2000).

Cleavage is almost always holoblastic (i.e., the cleavage furrows pass completely through the egg) and generally radial (Fig. 3A–D). Urochordates, however, exhibit bilateral cleavage, which produces an early embryo with mirror-image right and left halves. Early cleavages are also typically equal (or nearly equal), with the best-known exception being the unequal divisions of the vegetal blastomeres of the 8-cell stage sea urchin embryo, which generate the micromeres. Gastrulation begins 5 to 10 hours after fertilization in warm water species, when the embryo consists of between 100 (ascidians) and 800 cells (amphioxus and sea urchins). As deuterostomes, all these organisms gastrulate through a vegetally positioned blastopore, which subsequently becomes the anus (Fig. 3E–H). The mouth opening forms secondarily, and body cavities form from mesodermal pouches that balloon outward from the primitive gut (enterocoely). Gastrulation involves a process of invagination, by which mesoderm and endoderm move inward through a circular blastopore. Typically, there is some involution of cells near the margins of the blastopore and limited epibolic spreading of animal cells. The animal region of the embryo (opposite the blastopore) invariably gives rise to the ectoderm and the mesoderm and endoderm are derived from the vegetal half of the egg. In amphioxus and ascidians, the cells at the vegetal pole give rise to endoderm, and the mesoderm arises from more equatorial cells. In echinoderms, however, the orientation of these two prospective territories is reversed. Embryonic development is often extremely rapid, particularly in warm water species, and the swimming, feeding larva may form in less than a day in some species. If the ascidian tadpole and sea urchin pluteus are representative of the group, the swimming larva contains 2000 to 3000 cells and some 10 to 15 basic cell types, including pigment cells, muscle cells, neurons, other mesenchyme cells, and gut cells.

V. Genomics

Because of the utility of invertebrate deuterostomes for developmental analysis and their close phylogenetic relationship to vertebrates, genomic resources are being developed from a variety of these organisms. The draft genomic sequences of two invertebrate deuterostomes, the related ascidians *C. intestinalis* and *C. savignyi*, have been released (see Cameron *et al.*, Chapter 30; Satoh, Chapter 31. The assembly of the *C. intestinalis* genome is considerably more complete, although in neither species is it yet possible to assign scaffolds of assembled genomic sequence to specific chromosomes. The draft genomic sequence of a sea urchin, *Strongylocentrotus purpuratus*, is nearing completion, with a release expected in late 2004. The genome of *S. purpuratus* is large (~800 megabases, or about 25% the size of the human genome), while the genomes of *C. intestinalis* and *C. savignyi* are considerably smaller (150–180 megabases). Gene density in the ascidians is therefore likely to be higher. It seems almost assured that as these three initial genome sequencing projects are completed, efforts will be made to sequence the complete genomes of other invertebrate deuterostomes.

In addition to the whole genome sequencing projects, many other genomic resources are being developed. Vast EST databases have been generated for *C. intestinalis*, including EST collections from many developmental stages and several adult tissues (Satoh, Chapter 31). Considerable progress has also been made in this organism in determining full-length cDNA sequences, developing DNA microarrays, and in carrying out large-scale, whole mount *in situ* hybridization analyses to determine the spatial patterns of expressions of genes. In the sea urchin, *S. purpuratus*, extensive EST databases have been developed for whole embryos of various developmental stages and for specific embryonic cell types (Cameron *et al.*, Chapter 30; Poustka *et al.*, 2003; Zhu *et al.*, 2001). Robotically arrayed cDNA and genomic libraries have also been developed for a number of other sea urchin species and these are proving to be extremely useful for comparative studies. Computational tools have been developed that facilitate the use of these genomic resources; for example, the FamilyRelations program facilitates comparisons of genomic sequences for the purpose of identifying conserved regions that may function as cis-regulatory elements (Cameron *et al.*, Chapter 30). A large EST database is currently being developed for the direct developing, enteropneust hemichordate, *Saccoglossus kowalevskii* (Lowe *et al.*, 2003).

A useful feature of invertebrate deuterostome genomes is that many genes appear to be present in single copies that are found in multiple copies in vertebrate genomes. A frequently cited example is the Hox gene cluster, which is found in four copies in mammals but in only one copy in cephalochordates and sea urchins (Arenas-Mena *et al.*, 2000; Bailey *et al.*, 1997). Models of genome evolution within the vertebrate lineage posit sequence duplications on scales ranging from individual genes to multiple duplications of entire genomes (Durand, 2003; Holland, 1999, 2003; Panopoulou *et al.*, 2003; Wagner, 2001). The "simplified" nature of invertebrate deuterostome genomes may prove very useful in dissecting developmental pathways, gene networks, and the functions of conserved gene products that are more challenging to study in vertebrates because of the proliferation of the genes involved. Undoubtedly, analysis of the genomes of invertebrate deuterostomes will also reveal genes and pathways that are specific to particular lineages of deuterostomes. This may provide insight into the genetic changes that led to the diversification of the deuterostomes, including the innovations that gave rise to the vertebrates. The prospects are bright, indeed, that this group of organisms, which played such a central role in the emergence of developmental biology, will yield important and exciting new insights in the future.

References

Adoutte, A., Balavoine, G., Lartillot, N., Lespinet, O., Prud'homme, B., and de Rosa, R. (2000). The new animal phylogeny: Reliability and implications. *Proc. Natl. Acad. Sci. USA* **97,** 4453–4456.

Aguinaldo, A. M., Turbeville, J. M., Linford, L. S., Rivera, M. C., Garey, J. R., Raff, R. A., and Lake, J. A. (1997). Evidence for a clade of nematodes, arthropods, and other moulting animals. *Nature* **387,** 489–493.

Arenas-Mena, C., Cameron, A. R., and Davidson, E. H. (2000). Spatial expression of Hox cluster genes in the ontogeny of a sea urchin. *Development* **127,** 4631–4643.

Bailey, W. J., Kim, J., Wagner, G. P., and Ruddle, F. H. (1997). Phylogenetic reconstruction of vertebrate Hox cluster duplications. *Mol. Biol. Evol.* **14,** 843–853.

Barrington, E. J. (1965). "The Biology of Hemichordata and Protochordata." W. H. Freeman and Co., San Francisco.

Bateson, W. (1884). The early stages in the development of Balanoglossus (sp. uncert.). *Quart. J. Micros. Sci.* **24,** 209–235.

Bone, Q. (1998). "The Biology of Pelagic Tunicates." Oxford University Press, Oxford, UK.
Bosch, I., Rivkin, R. B., and Alexander, S. P. (1989). A sexual reproduction by oceanic planktotrophic echinoderm larvae. *Nature* **337,** 169–170.
Boveri, T. (1907). Zellenstudien VI: Die Entwicklung dispermer Seeigelier. Ein Beitrag zur Befruchtungslehre und zur Theorie des Kernes. *Jena Zeit. Naturw.* **43,** 1–292.
Bromham, L. D., and Degnan, B. M. (1999). Hemichordates and deuterostome evolution: Robust molecular phylogenetic support for a hemichordate + echinoderm clade. *Evol. Dev.* **1,** 166–171.
Burke, R. D. (1989). Echinoderm metamorphosis: Comparative aspects of the change in form. *In* "Echinoderm Studies" (M. Jangoux and J. Lawrence, eds.), pp. 81–108. A. A. Balkema Pub., Rotterdam, Netherlands.
Cameron, C. B., Garey, J. R., and Swalla, B. J. (2000). Evolution of the chordate body plan: New insights from phylogenetic analyses of deuterostome phyla. *Proc. Natl. Acad. Sci. USA* **97,** 4469–4474.
Castle, W. E. (1896). The early embryology of *Ciona intestinalis* Flemming (L.). *Bull. Mus. Comp. Zool.* **27,** 203–280.
Chia, F. S., Oguro, C., and Komatsu, M. (1993). Sea-star (asteroid) development. *Oceanogr. Mar. Biol. Annu. Rev.* **31,** 223–257.
Conklin, E. G. (1905). The organization and cell lineage of the ascidian egg. *J. Acad. Nat. Sci. Phila.* **13,** 1–119.
Driesch, H. (1892). The potency of the first two cleavage cells in echinoderm development. Experimental production of partial and double formations. *In* "Foundations of Experimental Embryology" (B. H. Willier and J. M. Oppenheimer, eds.). Hafner, New York, 1974.
Duboc, V., Rottinger, E., Besnardeau, L., and Lepage, T. (2004). Nodal and BMP2/4 signaling organizes the oral–aboral axis of the sea urchin embryo. *Dev. Cell.* **6,** 397–410.
Durand, D. (2003). Vertebrate evolution: Doubling and shuffling with a full deck. *Trends Genet.* **19,** 2–5.
Eaves, A. A., and Palmer, A. R. (2003). Reproduction: Widespread cloning in echinoderm larvae. *Nature* **425,** 146.
Ettensohn, C. A., and Sweet, H. (2000). Patterning the early sea urchin embryo. *Curr. Top. Dev. Biol.* **50,** 1–44.
Fischer, J. L. (1992). The embryological oeuvre of Laurent Chabry. *Roux's Arch. Dev. Biol.* **201,** 125–127.
Garstang, W. J. (1928). The morphology of the Tunicata, and its bearing on the phylogeny of the Chordata. *Quart. J. Micros. Sci.* **72,** 51–187.
Gee, H. (1996). "Before the Backbone: Views on the Origins of the Vertebrates." Chapman and Hall, London, UK.
Gross, J. M., and McClay, D. R. (2001). The role of Brachyury (T) during gastrulation movements in the sea urchin *Lytechinus variegatus*. *Dev. Biol.* **239,** 132–147.
Harada, Y., Shoguchi, E., Taguchi, S., Okai, N., Humphreys, T., Tagawa, K., and Satoh, N. (2002). Conserved expression pattern of BMP-2/4 in hemichordate acorn worm and echinoderm sea cucumber embryos. *Zoolog. Sci.* **19,** 1113–1121.
Hertwig, O. (1875). Beiträge zur Kenntnis der Bildung, Befruchtung, und Teilung des tierischen Eies, I. Morphologische Jahrbuch 1.
Holland, P. W. (1999). Gene duplication: Past, present, and future. *Semin. Cell Dev. Biol.* **10,** 541–547.
Holland, P. W. (2003). More genes in vertebrates? *J. Struct. Funct. Genomics* **3,** 75–84.
Hörstadius, S. (1939). The mechanics of sea urchin development, studied by operative methods. *Biol. Rev.* **14,** 132–179.
Hörstadius, S. (1973). "Experimental Embryology of Echinoderms." Clarendon Press, Oxford, UK.
Jeffery, W. R., and Swalla, B. J. (1992). Evolution of alternate modes of development in ascidians. *Bioessays* **14,** 219–226.
Jeffery, W. R., Swalla, B. J., Ewing, N., and Kusakabe, T. (1999). Evolution of the ascidian anural larva: Evidence from embryos and molecules. *Mol. Biol. Evol.* **16,** 646–654.

Kowalevsky, A. (1866). Entwicklungsgeschichte des Amphioxus lanceolatus. *Mémoires de l'Académie des Sciences de St. Pétersbourg* **7**(11), No. 4, 1–17.

Lowe, C. J., Wu, M., Salic, A., Evans, L., Lander, E., Stange-Thomann, N., Gruber, C. E., Gerhart, J., and Kirschner, M. (2003). Anteroposterior patterning in hemichordates and the origins of the chordate nervous system. *Cell* **113,** 853–865.

McEdward, L. R., and Janies, D. A. (1997). Relationships among development, ecology, and morphology in the evolution of echinoderm larvae and life cycles. *Biol. J. Linnean Soc.* **60,** 381–400.

McGovern, T. M. (2003). Plastic reproductive strategies in a clonal marine invertebrate. *Proc. R. Soc. Lond. B Biol. Sci.* **270,** 2517–2522.

Müller, J. (1841). Mikroskopische Untersuchungen über den Bau und die Lebenserscheinungen des Branchiostoma lubricum Costa, Amphioxus lanceolatus Yarrell, pp. 396–411. Ber. Preuss. Akad. Wissensch. Berlin, Germany.

Nakajima, Y., Humphreys, T., Kaneko, H., and Tagawa, K. (2004). Development and neural organization of the tornaria larva of the Hawaiian hemichordate, Ptychodera flava. *Zoolog. Sci.* **21,** 69–78.

Nakano, H., Hibino, T., Oji, T., Hara, Y., and Amemiya, S. (2003). Larval stages of a living sea lily (stalked crinoid echinoderm). *Nature* **421,** 158–160.

Panopoulou, G., Hennig, S., Groth, D., Krause, A., Poustka, A. J., Herwig, R., Vingron, M., and Lehrach, H. (2003). New evidence for genome-wide duplications at the origin of vertebrates using an amphioxus gene set and completed animal genomes. *Genome Res.* **13,** 1056–1066.

Peterson, K. J., Harada, Y., Cameron, R. A., and Davidson, E. H. (1999). Expression pattern of Brachyury and Not in the sea urchin: Comparative implications for the origins of mesoderm in the basal deuterostomes. *Dev. Biol.* **207,** 419–431.

Poustka, A. J., Groth, D., Hennig, S., Thamm, S., Cameron, A., Beck, A., Reinhardt, R., Herwig, R., Panopoulou, G., and Lehrach, H. (2003). Generation, annotation, evolutionary analysis, and database integration of 20,000 unique sea urchin EST clusters. *Genome Res.* **13,** 2736–2746.

Raff, R. A. (1987). Constraint, flexibility, and phylogenetic history in the evolution of direct development in sea urchins. *Dev. Biol.* **119,** 6–19.

Satoh, N. (1994). "Developmental Biology of Ascidians." Cambridge University Press.

Shoguchi, E., Satoh, N., and Maruyama, Y. K. (1999). Pattern of Brachyury gene expression in starfish embryos resembles that of hemichordate embryos but not of sea urchin embryos. *Mech. Dev.* **82,** 185–189.

Strathmann, R. R. (1978). The evolution and loss of feeding larval stages of marine invertebrates. *Evolution* **32,** 894–906.

Swalla, B. J., Cameron, C. B., Corley, L. S., and Garey, J. R. (2000). Urochordates are monophyletic within the deuterostomes. *Syst. Biol.* **49,** 52–64.

Tagawa, K., Humphreys, T., and Satoh, N. (1998). Novel pattern of Brachyury gene expression in hemichordate embryos. *Mech. Dev.* **75,** 139–143.

van Beneden, E., and Julin, C. H. (1884). La segmentation chez les ascidens dans ses rapportes avec l'organization de la larve. *Arch. Biol.* **5,** 111–126.

Wagner, A. (2001). Birth and death of duplicated genes in completely sequenced eukaryotes. *Trends Genet.* **17,** 237–239.

Zhu, X., Mahairas, G., Illies, M., Cameron, R. A., Davidson, E. H., and Ettensohn, C. A. (2001). A large-scale analysis of mRNAs expressed by primary mesenchyme cells of the sea urchin embryo. *Development* **128,** 2615–2627.

PART I

Procurement, Maintenance & Culture of Oocytes, Embryos, Larvae, and Adults

CHAPTER 2

Care and Maintenance of Adult Echinoderms

S. Anne Böttger,* Charles W. Walker,* and Tatsuya Unuma[†]

*Department of Zoology and Marine Biology
Biomedical Research Group
Durham, New Hampshire 03824

[†]National Research Institute of Aquaculture
Nansei, Mie 516-0193

METHODS IN CELL BIOLOGY, VOL. 74

0091-679X/04 $35.00

I. Overview

This chapter addresses the care and maintenance of adult echinoderms that produce gametes and embryos commonly used for molecular, cellular, and developmental biology (see Chapter 1). Emphasis is placed on sea urchin and sea star species from North and South America, Europe, and Japan. Sea cucumbers, brittle stars, and crinoids are not discussed because they are less amenable to fertilization in the laboratory and are thus less widely used for experimental manipulation of their eggs and embryos. Specific topics addressed in this chapter are: (a) adult echinoderm models: their reproductive cycles and gametogenesis; (b) methods for obtaining adult echinoderms; (c) maintenance of adult echinoderms in land-based systems; and (d) methods for care and handling of adult echinoderms.

II. Introduction

Sea urchins and sea stars have been employed by generations of molecular, cellular, and developmental biologists as models in studies of gametogenesis, egg and sperm interaction and activation, fertilization, and early development (see Chapter 1). Echinoderms that produce large numbers of gametes do so using annual cycles of reproduction in which the germinal epithelium may contain amitotic germ-line stem cells (oogonia or spermatogonia) for one to several months. This period is followed by a gametogenic phase lasting several months during which germ-line stem cells initiate mitosis, increase in numbers in both sexes to form resulting gonocytes (primary and secondary oocytes or spermatocytes) that complete gametogenesis, and produce gametes (fully mature ova in sea urchins, primary oocytes in sea stars, and spermatozoa in both). Depending upon the stage of gametogenesis and the degree of spawning, gonads in these echinoderms may contain few or copious numbers of gametes. An understanding of the cell biology of gametogenesis in sea urchins and sea stars is vital in determining when one can expect to obtain "ripe" eggs (fully mature ova in female sea urchins and primary oocytes in female sea stars) that can be successfully fertilized to yield embryos for experimentation. An understanding of gametogenesis in these two echinoderm groups can be matched with available data on the reproductive cycles for particular species in different parts of the world to determine when each will be most useful in experimental applications (Tables I and II).

III. Adult Echinoderm Models: Their Reproductive Cycles and Gametogenesis

Annually, within the gonads of both sexes of most echinoderms, millions of gonial cells (oogonia and spermatogonia) originate by mitosis, produce gonocytes (primary and secondary) that undergo gametogenesis and result in gametes that

Table I
Summary of Species Information Relating to Collection of Echinoids (Sea Urchins) and Their Breeding Season[a]

Species	Size	Depth	Substrate	Distribution	Reproductive season
Anthocidaris crassispina	<7 cm	0–70 m	rock	Japan, Southern China	June–August
Arbacia punctulata	<5.6 cm	0–230 m	rock, shell	Cape Cod–Florida, Texas, Yucatan, Cuba, Jamaica, West Indies	May–September
Echinus esculentus	15–16 cm	0–40 m	rock	British Isles	February–June
Hemicentrotus pulcherrimus	<5 cm	0–40 m	rock, coarse gravel	Japan, Northern China, Korea	January–March
Lytechinus variegatus	<7.6 cm	0–55 m	sand, gravel	North Carolina–Florida, Bahamas, West Indies	December–July
Loxechinus albus	<7.5 cm	0–15 m	rock, gravel sand, seagrass	Chile, Peru	July–September
Paracentrotus lividus	<7 cm	<3 m	rock	Mediterranean, Western Atlantic	April–May
Psammechinus miliaris	<5.5 cm	0–10 m	coarse gravel	Britain & Ireland	February–November
Pseudocentrotus depressus	<8 cm	0–50 m	rock	Japan	October–December
Strongylocentrotus droebachiensis	<8.3 cm	0–1160 m	rock	Arctic-New Jersey, Alaska–Puget Sound	January–April
Strongylocentrotus fransciscanus	<12.7 cm	0–91 m	rock	Alaska–Baja California	February–August
Strongylocentrotus intermedius	<8 cm	0–40 m	rock	Pacific coasts of Asia and Siberia	August–October
Strongylocentrotus nudus	<8 cm	0–180 m	rock	Pacific coasts of Asia and Siberia	September–November
Strongylocentrotus purpuratus	<10.2 cm	0–160 m	rock	Alaska–Baja California	November–June
Tripneustes gratilla	<12 cm	0–30 m	hard substrate	Circumtropical extending to tropical areas in Indian and West Pacific Ocean	July–September (in Japan)

[a]The reproductive seasons have been compiled for sea urchins from different populations. In different geographical areas, breeding seasons can therefore vary slightly but will be within the reported range. Further information on British species can be found at http://www.marlin.ac.uk/.

are stored. Such prolific and orderly gametogenesis depends upon the subdivision of the germinal epithelium into small groups of interrelated somatic and germinal cells (Walker, 1979, 1982; Walker *et al.*, 2001). Subdivisions increase the surface area in the interior of an otherwise sac-like gonadal lumen and provide a mechanism for supplying nutrients to smaller groups of germinal cells among the

Table II
Summary of Species Information Relating to Collection of Asteroids (Sea Stars) and Their Reproductive Season[a]

Species	Size	Depth	Substrate	Distribution	Reproductive season
Asterias amurensis	<40 cm	5–200 m	sand	South of Alaskan Peninsula, British Columbia, Canada, Japan, invading Australia & Tasmania	February–April (in Japan)
Asterias forbesi	<13 cm	0–49 m	rock, gravel, sand	Gulf of Maine–Texas (South of Cape Cod)	May–July
Asterias rubens	10–30 cm	30–100 m	gravel rock		February–April
Asterias vulgaris	<20 cm	0–349 m	rock, gravel	Labrador–Cape Hatteras (north of Cape Cod)	May–July
Asterias miniata	<10.2 cm	0–293 m	rock	Alaska–Baja California	January–November
Asterina pectinifera	<20 cm	0–100 m	rock, gravel, sand	Japan	May–July

[a]The reproductive seasons have been compiled for sea stars from different populations. In different geographical areas breeding seasons can therefore vary slightly but will be within the reported range. Further information on British species can be found at http://www.marlin.ac.uk/.

immense numbers produced within each gonad. The nature of these subdivisions differs substantially between sea urchins and sea stars, particularly in the ovaries (Walker, 1979, 1982; Walker *et al.*, 2001). Histological examination of the gametogenic stages in echinoderms is traditionally accomplished using paraffin-embedded gonads. Conspicuous changes within the gonads can be appreciated using sections of these tissues, but resolution is greatly enhanced in 1 μm sections of plastic-embedded gonads. Echinoderm germ cells are very small (5–10 μm) and events such as germ-line stem cell mitosis and meiotic divisions cannot be seen without the use of plastic sections. Appropriate methods for preparation of echinoderm gonads for histological examination in plastic sections are provided in Section C.

A. Sea Urchin Models from North and South America, Europe, and Japan

Throughout the world, adult sea urchin species are relatively easy to obtain and maintain (Table I) and their gametes are readily available in large numbers. In obtaining gametes from sea urchins, it is important to recognize that gonad size in both sexes does not directly relate only to the progress of gametogenesis or numbers of gametes. Gametogenesis and intragonadal nutrient storage and utilization are linked processes in the sea urchin (Walker *et al.*,

2001). Uniquely among the echinoderms, sea urchin gonads grow in size not only because gametogenesis increases the size and/or numbers of germinal cells present but also because somatic cells within the germinal epithelium, the nutritive phagocytes (NP), store extensive nutrient reserves before gametogenesis begins (see the green sea urchin Web page, *http://zoology.unh.edu/faculty/walker/urchin/gametogenesis. html*). In sea urchins of both sexes, a glycoprotein termed "major yolk protein" (MYP) is the principal nutrient stored in NP (Brooks and Wessel, 2002; Unuma *et al.*, 2001). This protein was originally identified as the most abundant component of yolk granules in sea urchin eggs (Yokota and Sappington, 2002), but is predominantly found prior to gametogenesis in the NPs of both ovaries and testes (Unuma *et al.*, 2003). MYP is mobilized from NP during gametogenesis, transported in an unknown way to gametogenic cells, and used in the synthesis of new protein, nucleic acids, and other constituents of ova and spermatozoa (Unuma *et al.*, 2003).

Nutritive phagocytes are surrounded by oogonia, oocytes, and subsequent gametogenic stages. During oogenesis, follicle cells are not present and ovarian follicles do not exist in sea urchins as they do in sea stars (Walker, 1982; Walker *et al.*, 2001). Meiotic divisions of sea urchin post-vitellogenic primary oocytes occur as each is released into the ovarian lumen for storage. As a result, meiosis in sea urchin primary oocytes does not occur at one discrete time (as it does during spawning of the primary oocytes of sea stars), but rather continues during several months as successive groups of primary oocytes become meiotically competent (Shirai and Walker, 1988; Smiley, 1990; Walker *et al.*, 2001). Following oogenesis, fully mature sea urchin ova are stored within the ovarian lumen for periods varying from days to weeks prior to their release at spawning (Pearse and Cameron, 1991; Walker, 1982; Walker and Lesser, 1998; Walker *et al.*, 2001). Stages of spermatogenesis are similarly associated with NP as they progress from spermatogonial stem cell clusters that are located near the testicular wall to the lumen where spermatozoa are stored prior to spawning (Pearse and Cameron, 1991; Walker, 1982; Walker and Lesser, 1998; Walker *et al.*, 2001). (Fig. 1).

B. Sea Star Models from North America, Europe, and Japan

Adult sea star model organisms from around the world (Table II) have been instrumental in understanding the initiation of meiosis in primary oocytes and in determining the biochemistry and physiology of molecules that are involved. Such studies depend upon the nature of oogenesis in sea stars that differs substantially from that of sea urchins. During oogenesis in sea stars, individual primary oocytes are enveloped by a single layer of follicle cells, resulting in ovarian follicles. During vitellogenesis, each ovarian follicle supplies nutrients derived from extragonadal sources to primary oocytes (Schroeder *et al.*, 1979). A single meiotically arrested primary oocyte is contained within each ovarian follicle. Fully developed ovarian follicles are stored within the ovarian lumen until ovulation, when primary oocytes synchronously complete meiosis (Kishimoto, 1998, 1999). Three interrelated chemical messengers have been implicated in this annual event. These are

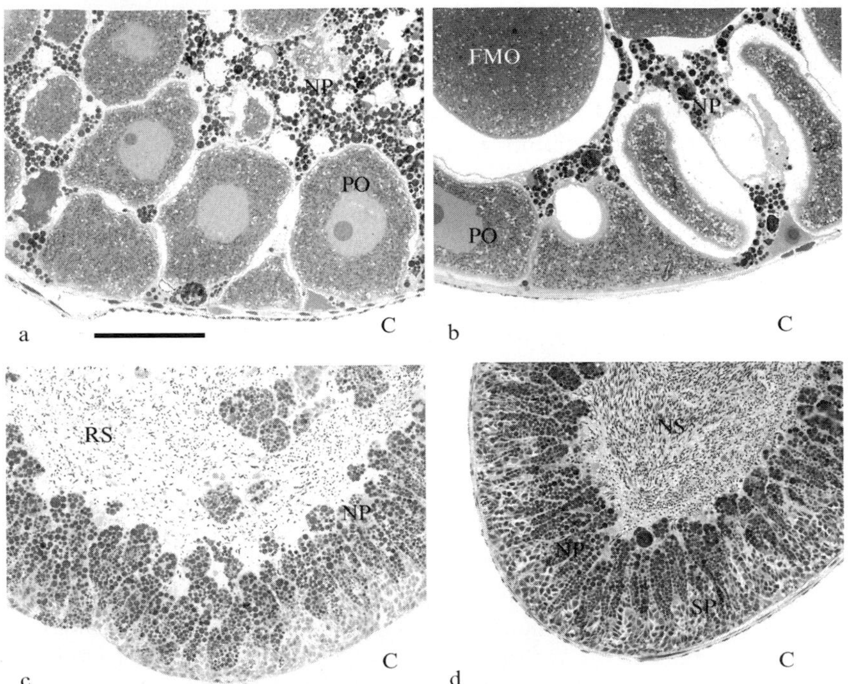

Fig. 1 Gametogenesis in sea urchins: (1a) Ovary of a sea urchin after the initiation of oogenesis filled with large nutritive phagocytes (NP) and vitellogenic primary oocytes (PO); (1b) Ovary of a sea urchin later in oogenesis, showing primary oocytes peripherally (PO), diminished nutritive phagocytes (NP), and fully mature ova (FMO) in the lumen; (1c) Testis of a sea urchin before the initiation of spermatogenesis containing large nutritive phagocytes containing dark granules (NP) and residual spermatozoa from the spermatogenesis in the previous year (RS); (1d) Testis of a sea urchin after mobilization of nutrients from nutritive phagocytes and mitosis of spermatogonia (SP). Notice new spermatozoa in the lumen (NS). C = coelom in la–d. Scale bar = 100 μm.

gonad-stimulating substance (GSS from the radial nerves); maturation-inducing substance (MIS; 1-methyl adenine from the follicle cells), and maturation-promoting factor (MPF; generated within the cytoplasm of each primary oocyte). Immediately before spawning, these messengers sequentially lead to germinal vesicle and follicular envelope breakdown (GVBD and FEBD) and to meiosis of primary oöcytes (Shirai and Walker, 1988; Smiley, 1990). Resulting fully mature ova and any remaining meiotically competent, primary oocytes (which complete meiosis during ovulation or shortly thereafter) are spawned immediately. Stages of spermatogenesis in sea stars are associated with somatic accessory cells that are similar to the NP of both sexes of sea urchins, except that they do not store nutrients. Spermatogenic cells progress along discrete spermatogenic columns, each associated with a somatic accessory cell, from stem cell clusters located near the testicular wall to the lumen where spermatozoa are stored prior to spawning (Walker, 1980, 1982) (Fig. 2). Unlike the situation in sea urchins, where gonad size

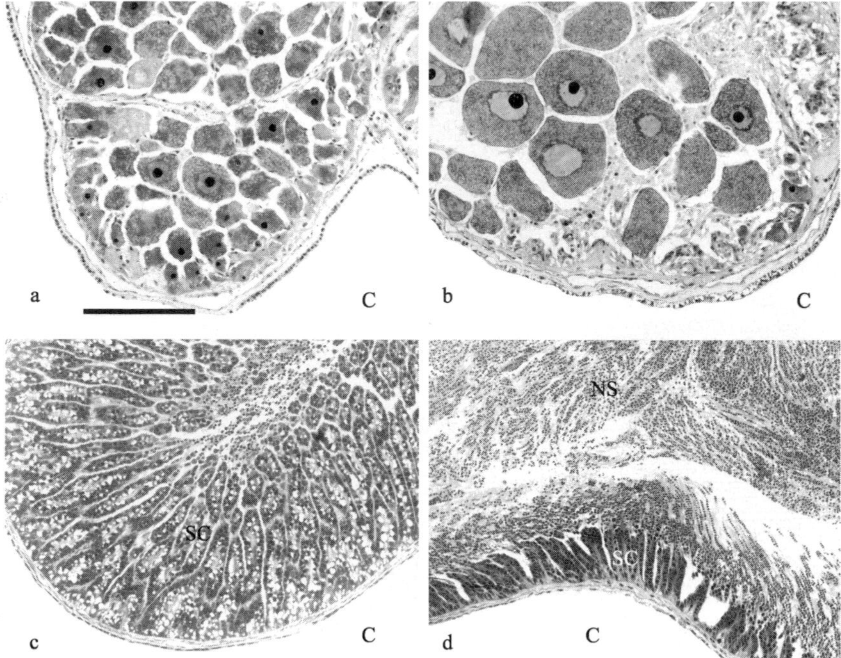

Fig. 2 Gametogenesis in sea stars: (2a) Ovary of a sea star early during vitellogenesis, showing ovarian follicles containing primary oocytes in various stages of growth (notice black follicle cell nuclei on the surface of the primary oocytes); (2b) Ovary of a sea star later in vitellogenesis with maturing ovarian follicles released into the lumen and still surrounded by follicle cells; (2c) Testis of a sea star prior to the initiation of meiosis in primary spermatocytes which are arranged along the axes of axial somatic cells forming separate spermatogenic columns (SC); (2d) Testis of a sea star near the completion of annual spermatogenesis, showing reduced spermatogenic columns (SC) and new spermatozoa stored in the testicular lumen (NS). C = coelom in 2a–d. Scale bar = 100 μm.

reflects the growth of NP as well as the progress of gametogenesis, the size of gonads in both sexes of sea stars does indicate the progress of gametogenesis. In sea stars, nutrients are mobilized from remote sources like the pyloric caeca and are transported to the gonads during gametogenesis.

C. Echinoderm Gametogenesis in Plastic Sections for Light and Electron Microscopy

1. Quick Fixation Method

Small pieces of gonadal tissue are placed in 2.5% glutaraldehyde in 0.45 μm filtered H_2O at room temperature for 1 hour to overnight.

Fixative is removed and gonads are rinsed (×2) in 2.5% $NaHCO_3$ (pH 7.2) for 10 min and finally brought to distilled H_2O prior to embedding. To avoid damage, do not move gonads from one fluid to another during this process; instead, remove fluids from the tissues by decanting or aspiration.

2. Extended Fixation Method

Small pieces of gonadal tissue are placed for 2 hr in 3% glutaraldehyde in 0.2 M sodium cacodylate buffer (pH7) containing a balanced salt solution (NaCl [30 mg/ml] and $CaCl_2$ [2 μg/ml] at 4 °C. This is followed by washing for 10 min in 0.2 M sodium cacodylate buffer (pH 7) with salts added in the same proportions as in the fixative, then by four 10-min rinses in buffer with the NaCl concentration decreased each time (Walker, 1979) until only buffer (plus $CaCl_2$) remains followed by post fixation for 3 hr in 1% OsO_4 (minus salts) and two rinses in 0.2 M sodium cacodylate buffer (pH 7) (minus salts).

3. Embedding in Plastic

Fixed gonads can be dehydrated in an ethanol series to 100% and embedded in (a) Bioacryl polymerized at 4 °C under UV light, (b) JB-4 polymerized under vacuum, or (c) Epon/Araldite polymerized with heat and followed in each case by preparation of one μm plastic sections. Resulting sections are mounted on poly-L-lysine coated cover slips or slides (Walker, 1979).

4. Sectioning and Staining

One micrometer sections of gonadal tissues embedded in Bioacryl or JB-4 can be observed by light microscopy after application of stains also used for paraffin histology (Burns and Bretschneider, 1981). Those embedded in the Epon/Araldite plastic can be observed using either light microscopy following staining with 0.2% buffered Azure B (Lane and Europa, 1965). The latter can also be prepared on copper grids for electron microscopy and contrasting with 2% uranyl acetate and Reynolds lead citrate (Walker, 1979).

IV. Obtaining Adult Echinoderms

A. Collection and Transport of Adult Echinoderms

Echinoderms are highly sensitive to changes in their environment and must be handled very carefully during collection. Gloves decrease the risk of infection through disruption of their external epithelia (Scheibling and Stephenson, 1984). Echinoderms collected by SCUBA will survive longer than those collected by mechanical dredging. Subtropical echinoderms are easier to transport than are those from temperate climates because changes in temperature are less likely to affect their survival. However, these echinoderms must be more heavily aerated because of the lower oxygen concentration of warmer water. Subtropical species like *Lytechinus variegatus* survive well in containers (30- to 36-quart coolers available from a variety of manufacturers) supplied with battery-operated minomizers (Hypark Speciality Corporation Inc., Minnetonka, Minnesota) that cool and aerate seawater. During transport, temperate species like *Strongylocentrotus droebachiensis* may be placed on a bed of macroalgae that is cooled with bags of ice or gel coolants (e.g., North America–*Laminaria saccharina*; Europe–*Laminaria saccharina, Fucus vesiculosus, Ulva* ssp.; Japan–*Sargassum* ssp.). When macroalgae are not available, moistened towels or newspapers can be substituted (Yokota, 2002). Some sea stars, like *Asterias vulgaris*, have relatively flaccid bodies that leak coelomic fluid and must be transported in seawater. Others like *Asterina pectinifera* can be treated like sea urchins. If echinoderms are ready to spawn, it is advisable to keep each one in separate Zip-Loc bags, since a single individual can trigger mass spawning. Following transport and prior to feeding and experimentation, echinoderms should be acclimated to the tank conditions at their destination for 48 to 72 hours.

Commercial harvesting of sea urchins is usually tightly regulated by local governments. It is essential to contact local Departments of Marine Resources to ascertain local regulations before attempting to collect these species or you could be prosecuted!

B. Commercial Suppliers

Sea urchins are harvested worldwide as a source of the sushi termed "uni" (Andrew *et al*., 2002; Keesing and Hall, 1998; Williams, 2001) and can usually be purchased from local fisherman along the coast. Problems associated with utilizing such urchins depend upon the methods used in their collection. Such individuals may be severely compromised if collected by dredging. Some commercial suppliers (e.g., Connecticut Valley Biological and Carolina Biological) will collect animals by hand, a method that is less efficient, so prices may be higher. It is also important to point out that by using field-collected echinoderms, it may be difficult to obtain individuals of similar quality, primarily because of seasonal and environmental variability at different collection sites. Before using them in experiments, it may be necessary to maintain field-collected animals for a period of time under similar conditions, including feeding them a similar diet.

In Japan, there are many semi-governmental hatcheries that produce seedlings of sea urchins (Hagen, 1996). When surplus stocks of sea urchins are available, they can be purchased at reasonable prices. The benefit of buying the young sea urchins from these facilities is that they are of the same age and size. The disadvantage of buying sea urchins from the hatcheries is that they must be cultured before they reach sexual maturity. Buying adult sea urchins from aquaculturists would be most convenient. However, in Japan and the United States, sea urchin aquaculture ventures are in their infancy and elsewhere in the world they do not exist.

V. Maintenance of Adult Echinoderms in Land-Based Systems

A. Culture in Small Aquaria

Successful maintenance of sea urchins in small-scale, closed systems where water is recirculated has been described for different species of sea urchins (Cellario and Fenaux, 1990; Fridberger *et al.*, 1979; Hinegardner, 1969; Leahy, 1986). Such closed systems can easily be compromised by overcrowding, low oxygen concentrations, and inadequate water circulation and filtration.

Crowding can be avoided by maintaining only a limited number of echinoderms in each aquarium. In general, each medium-sized animal of 5 to 10 cm should occupy one liter of seawater (Leahy, 1986) and increasing the number of animals may lead to spine loss, bacterial infections, and ultimately mortality.

Oxygen consumption will increase in animals that have been stressed during transport or that experience changing environmental conditions during culture. When maintaining a number of aquaria, a large air pump (Aquatic Biosystems) should provide filtered air. Long air stones crossing the length of the aquarium should be used to maximize the dispersal of oxygen. Crowding will also be minimized since echinoderms tend to cluster around areas of high water and airflow.

Water circulation depends on ambient environmental conditions experienced by particular echinoderm species. Species collected in active surf environments will require higher air and water circulation to remain healthy than those collected sub-tidally. However, for most species, water circulation should be restricted to 100 to 150 ml seawater/minute so that animals are not detached. Repeated detachment can lead to damage of tube feet and resulting stress. In a recirculating system, water should enter the aquarium at the surface to increase dissolved oxygen. Water filtration is important to remove ammonia and pollutants, such as nitrogen and phosphorous. It is also essential to monitor ammonia (Lawrence *et al.*, 2003) and nitrate concentrations.

B. Large-Scale, Land-Based Facilities

Two designs for land-based sea urchin aquaculture systems have been successfully employed and discussed in the literature. These include a toboggan system (Grosjean *et al.*, 1998; Le Gall, 1990) and a trough system (Devin, 2001). The toboggan system used by Grosjean *et al.* (1998) consists of a series of stacked shallow tanks (Fig. 3a). Each tank is sloped, encouraging water flow. Flow in the top tank is maintained from a reservoir, while subsequent tanks are supplied from those directly above. The water finally returns either to the reservoir in a zigzag fashion or may be discharged. For this system, densities of sea urchins are 250 animals meter^{-1}. Sea urchins have been successfully maintained from fertilization to sexual maturity using this system.

The trough design (Fig. 3b) uses modified rain gutters, with a drainage channel along the length of each trough and a drainage sump at the discharge of each

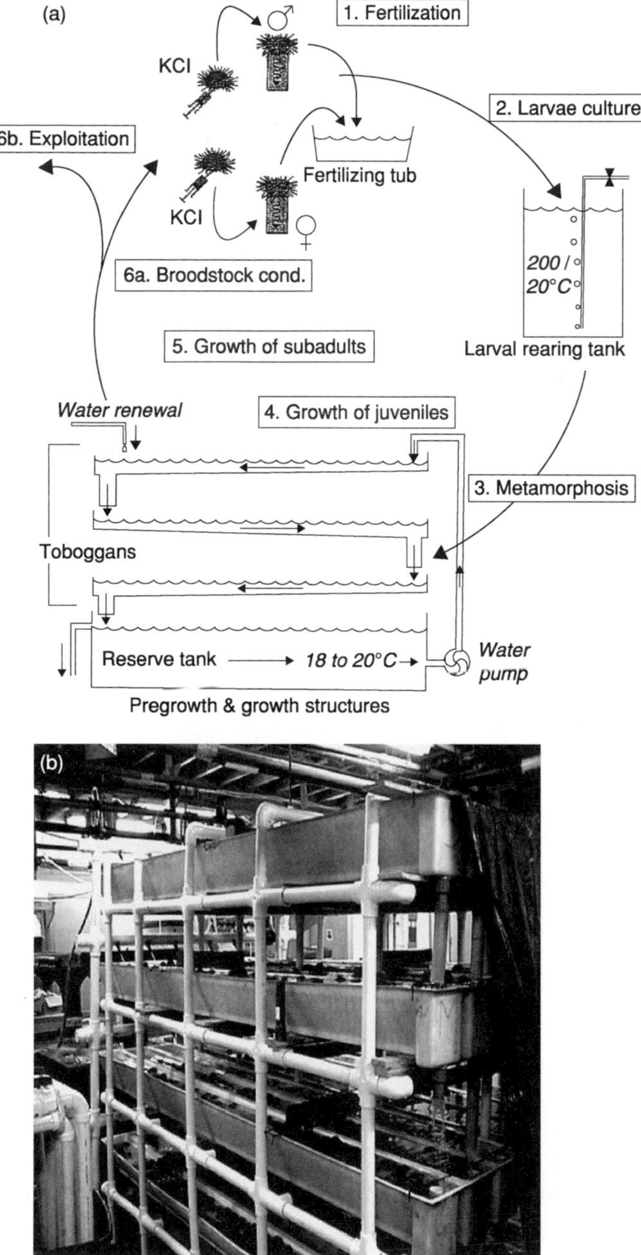

Fig. 3 Large-scale, land-based facilities: (3a) Toboggan system for maintaining large numbers of sea urchins from fertilization to sexual maturity (from Grosjean *et al.*, 1998); (3b) Trough system for maintaining sea urchins throughout their life cycle (from Devin, 2001).

trough (Devin, 2001). The drainage channel, which collects feces and uneaten food, is a very anoxic environment and should be narrow enough (3.1 cm) to prevent access to cultured sea urchins. The troughs may be attached in sets of three or more, with each trough connected to separate water intakes.

Land-based aquaculture facilities for a variety of organisms commonly use tanks in which animals are held at high densities for specific periods of time while they grow to some predetermined, marketable size (e.g., shrimp). Although sea urchins and sea stars are commonly found in high densities in the wild, bringing these animals into a land-based facility presents some unique animal husbandry issues that need to be resolved. High densities of animals require sufficient and rapid water flow to provide continuous replenishment of oxygen, elimination of nitrogenous wastes, and consistent temperatures. Both sea urchins and sea stars are more susceptible to bacterial and protozoan diseases upon exposure to elevated temperatures. Problems associated with the production and elimination of feces and the removal and disposal of unused food must be carefully considered in order to avoid high organic loading and the creation of hypoxic or anoxic sites within the tanks. It is advisable to clean the bottom of the tank of feces and uneaten food at regular intervals, normally once a week.

Flow-through seawater systems directly connected to the ocean are the most successful in echinoderm husbandry and can be coupled with the trough system already described. As long as continuous, external water flow is provided, less attention can be paid to salinity, dissolved oxygen, ammonia, etc. Unavoidably, water supplies may sometimes stop abruptly because of mechanical malfunction or blockage of water flow (wood fragments, sea weed, biofouling, etc.). Such problems can lead to a rapid deterioration of water quality and can cause immediate and high mortality. Tanks should always be equipped with emergency or backup aeration systems. If the tank is set up outdoors it should be covered to limit UV radiation, which may inhibit feeding.

VI. Care and Handling of Adult Echinoderms

Successful maintenance of echinoderms is dependent on continuous monitoring of their environmental conditions in both small aquaria and large land-based facilities. Physical and chemical factors can be monitored with the help of readily available instruments. This can be accomplished simply by using thermometers and refractometers purchased from Aquatic Biosystems or continuously using light, salinity, and temperature probes from LI-COR, Lincoln, NE; Rickly Hydrological Co, Columbus, OH; and Onset Computer, Bourne, MA, as well as test kits for certain pollutants. Biological factors, such as diseases, are more difficult to monitor and may depend on the health of the individuals at the time of collection. Individuals should therefore be collected from pristine environments that have not suffered any recent environmental perturbations or disease outbreaks.

A. Physical and Chemical Factors

1. Temperature

In combination with other physical, chemical, and biological factors, temperature can be important in the husbandry of adult sea urchins. The optimum temperature for most echinoderms depends upon the time of year and should approximate the ambient temperature at which they were collected as closely as possible. Temperate echinoderms (e.g., *Strongylocentrotus droebachiensis*) show a negative correlation between feeding and temperature (Garrido and Barber, 2001). Subtropical species (e.g., *Lytechinus variegatus*) can acclimate to elevated temperatures (Klinger *et al.*, 1986), while other species (e.g., *Eucidaris tribuloides, Strongylocentrotus franciscanus*) (Lares and McClintock, 1991; McBride *et al.*, 1997) seem to be unaffected by mild changes in temperature and their feeding rates and absorption efficiencies remain unchanged. Most echinoderms, however, show a positive relationship between feeding, absorption rates, and temperature. Thus, gross feeding and assimilation rates in such echinoderms in their natural environment are highest in spring and summer and decline in the winter (Fernandez and Boudouresque, 2000).

2. Photoperiod

Although exceptions occur (Bay-Schmith and Pearse, 1987; Cochran and Engelmann, 1975; Ito *et al.*, 1989; Sakairi *et al.*, 1989; Spirlet *et al.*, 1998; Yamamoto *et al.*, 1988), changing photoperiod can be correlated with the simultaneous initiation of gametogenesis and the mobilization of nutrients from NP in both sexes of a number of species of sea urchins and with mobilization of nutrients from pyloric caeca and the subsequent initiation of gametogenesis in sea stars (Pearse *et al.*, 1986a,b; Walker and Lesser, 1998; Walker *et al.*, 2001). In the Gulf of Maine, the onset of shortening day length in the fall results in the initiation of gametogenesis in the northern sea star, *Asterias vulgaris*, and the green sea urchin, *S. droebachiensis* (Pearse and Walker, 1986; Walker and Lesser, 1998). This occurs at photosynthetically active irradiances (PAR: 400–700 nm) found at approximately 10 m of depth (Walker and Lesser, 1998). Fall photoperiod is so influential in the green sea urchin that one can manipulate the ambient light regime and promote out-of-season gametogenesis in urchins maintained in the laboratory (Walker and Lesser, 1998). Green sea urchins were brought into the laboratory (in March, following natural spawning), fed the Lawrence pelletized diet *ad libitum* (Lawrence *et al.*, 2001), and subjected to a photoperiod advanced by four months. During this study, temperatures and salinities for experimental urchins mirrored those recorded at the collection site. Experimental urchins had a significantly higher gonad index (GI) in March, April, and May (18 $\pm$ 6%) compared with field urchins (11 $\pm$ 3%). Subsequently, experimental urchins had a mean monthly GI of 25 to 30%, while the mean GI for field urchins was 11 to 13%. Germ-line stem cell mitosis and gametogenesis began in June in experimental male and female urchins, 4 to 5 months earlier than in field urchins.

Studies of the effects of photoperiod on gametogenesis in other organisms may help in understanding its effect in echinoderms. Vertebrates with seasonal patterns of gametogenesis are affected by photoperiod in differing ways. Despite the often conflicting and incomplete explanations of the effects of photoperiod on seasonal gametogenesis, one pattern is apparent. Among those species studied that have seasonal reproduction, animals respond either positively (like echinoderms—Pearse and Walker, 1986; Pearse *et al.*, 1986a,b; and fish – Miura *et al.*, 1991; Nagahama, 1994) or negatively (other vertebrates—Chandola *et al.*, 1976; Heideman *et al.*, 1992; Paniagua *et al.*, 1990) to shortening fall photoperiod. Light also affects feeding and continuous low levels of light result in increased feeding rates (Grosjean *et al.*, 1998).

3. Salinity

Echinoderms are stenohaline osmoconformers, so salinity is an important variable to check regularly in their husbandry. Some populations of echinoderms can live in hyper- or hyposaline environments (Stickle and Diehl, 1987), but most species require an ambient salinity of 28 to 33 ppt (Lawrence, 1996). Salinity below 25 and above 40 ppt will result in significant mortality (Lawrence, 1975a, 1996; Roller and Stickle, 1993). Less dramatic changes in salinity may result in a decrease in activity (Ellington and Lawrence, 1974; Lawrence, 1975a; Sabourin and Stickle, 1981). In most cases, normal activity should return after adjusting the salinity (Ellington and Lawrence, 1974). If unfavorable saline conditions are maintained for an extended period of time, a decrease in activity will result in decreased feeding (Arnn, 1963; Glynn, 1974; Shirley and Stickle, 1982; Yamaguchi, 1974). The resulting lack of energy input will ultimately affect gametogenesis as well as somatic body growth and maintenance (more significant in sea stars than sea urchins in which the gonads contain somatic NP). Changes in salinity can affect ionic (Binyon, 1972; Diehl and Lawrence, 1985; Ellington and Lawrence, 1974; Stephens and Virkar, 1966; Stickle and Diehl, 1987) and amino acid concentrations in tissues (Stickle and Diehl, 1987). Such changes could also directly influence biochemical pathways associated with nutrient storage and resulting gamete production.

Ambient salinity at the collection site should be monitored with commercially available refractometers (Aquatic Biosystems) and matched in closed or recirculating aquarium systems. Measurements taken from the surface water may differ from those at depth. Recirculating systems should be adjusted to ambient environmental salinity and checked twice a week for variations. In flow-through, near-shore seawater systems, it is not possible to change salinity. Transient changes in salinity may result from freshwater runoff. As a result, it is advisable to pump seawater from deeper sources that are not subject to major river drainage. Alternatively, echinoderms could be housed and fed in pens in deeper water to avoid major changes in salinity.

4. Pollutants

Another important factor to consider in the maintenance of sea urchins is the accumulation of undesirable compounds in the culture system. In closed seawater systems, where water is recirculated, excretion products such as ammonia and phosphates may become a problem. Despite the use of a biofiltration device, not all of these compounds may be eliminated. Testing for harmful compounds should occur regularly. Phosphates are a chronic problem in recirculating seawater systems, since they may leak from dietary pellets and can also be produced by the animals themselves. Accumulation of phosphates (Böttger *et al.*, 2001) or crude oil (Temara *et al.*, 1999) can decrease feeding, which would result in smaller nutritive phagocytes and subsequently the production of fewer or lower quality gametes (Fig. 4). Aquaculture in near-shore facilities presents an even larger problem resulting from the effects of anthropogenic pollutants. Phosphates (Böttger and McClintock, 2002), cadmium (Au *et al.*, 2001a,b; Gnezdilova *et al.*, 1985, 1987; Vashchenko *et al.*, 1993), and oil (Krause, 1994) are known inhibitors of gametogenesis. It may therefore be necessary to limit near-shore aquaculture to areas that are strictly monitored for environmental contaminants or to design land-based running seawater systems that reduce the amounts of pollutants that animals may encounter.

B. Biological Factors

1. Diet

The quantity and quality of food provided to echinoderms can substantially alter their morphology, physiology, and reproduction (Ebert, 1980; Fernandez and Boudouresque, 1997; George, 1996; Gonor, 1973; Levitan, 1991; Walker and

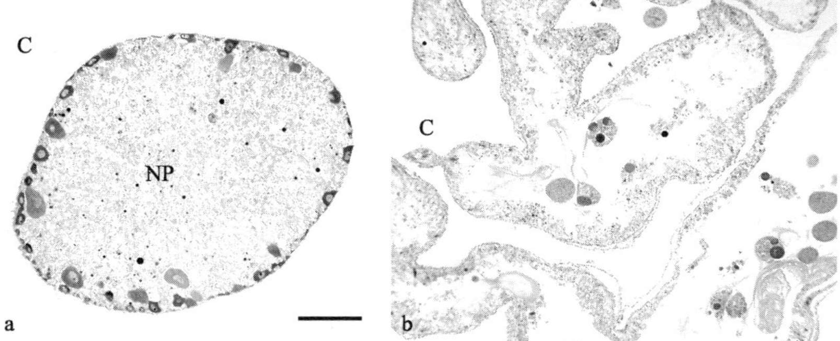

Fig. 4 The effects of chronic phosphate exposure on the ovaries of *Lytechinus variegatus* (Lamarck): (4a) Well-developed nutritive phagocytes (NP) and young vitellogenic oocytes can be seen in the ovary of animals maintained in artificial seawater; (4b) An ovary of the animals maintained in an organic phosphate has poorly developed nutritive phagocytes (NP) and is devoid of oocytes. C = coelom in 4a–b. Scale bar = 200 μm.

Lesser, 1998). Any change in the biochemical composition of the food may result in different rates of gonadal and somatic growth and could impact overall health. If animals are utilized within a week or two, it is not necessary to feed them; however, if left unfed for longer periods of time, gametogenesis will be affected.

The use of marine macroalgae (fresh or dry) as part of a sea urchin diet is popular, especially where it is easily accessible in coastal areas. Macroalgae may be dried in the sun before use as food for urchins. Dry algae can then be stored for more than half a year in plastic bags in a cool room. Alternatively, dry seaweeds can be purchased commercially from Asian markets. When introduced into culture tanks, dry macroalgae decay much faster than live algae. Since the rate of ingestion of algae by urchins depends on food preferences, it is important to use dried algae that are normally ingested by particular species (Anderson and Velimirov, 1982; De Ridder and Lawrence, 1982; Frantzis and Gremare, 1992; Vadas, 1977). This is true even when algae are used to prepare artificial diets (Klinger and Lawrence, 1984). Algae show significant seasonal variation in their biochemical composition and may therefore not be a reliable year-round food source to promote gonadal growth. For sea stars, a diet comprised of mollusks would result in maximum growth. Mollusks can be acquired directly from fishermen or seafood stores. Unfortunately, this is an expensive diet and may not be appropriate for prolonged maintenance.

Because of their commercial value, many sea urchin diets have been developed (Lawrence *et al.*, 2001). These are usually extruded as food pellets (Klinger, 1982). Since echinoderms manipulate their food with tubefeet, spines, and, in the case of the regular echinoids, the teeth of the Aristotle's lantern, food pellets need to be of small size and soft to medium-hard texture. Also, such pellets should retain their shape and not disintegrate between scheduled feedings. An example is the Lawrence pelletized diet available from Wenger International, Inc. (Kansas City, KS), which is administered at 4 grams every other day. Its components are approved by the Food and Drug Administration and the U.S. Department of Agriculture. For all of the diets described in this section, long-term storage can be a problem. The most successful strategy for long-term storage involves packaging the diet in small quantities (sufficient for 1 month's feeding) under vacuum (FoodSaver Delux II, Tilia International) and freezing.

Urchins fed a low protein diet increase their food intake in an attempt to compensate for the low protein in prepared vegetable diets (Fernandez and Boudouresque, 2000). Vegetable diets also contain high amounts of insoluble carbohydrates (structural carbohydrates originating from plant cell walls), which are poorly absorbed in many echinoderms when compared to soluble sugars and protein (Lawrence, 1975b, 1976, 1982). These diets result in decreased absorption rates, which will eventually lead to decelerated growth.

Prepared animal feeds containing fish or chicken meal have high protein and low carbohydrate levels (Fernandez and Boudouresque, 2000). The high amount of protein in these diets results in decreased feeding rates but also in increased absorption efficiencies. Animal-based diets will result in excellent somatic and gonadal growth for most echinoderms.

2. Diseases

A variety of pathogens and parasites from protozoans to chordates (Jangoux, 1987a,b, 1990) can infect echinoderms. Perhaps most interesting to molecular and cellular biologists is the lack of any evidence for diseases involving cellular proliferation (e.g., cancer) in any echinoderm (Wellings, 1969). Ultimately, investigation of this remarkable feature of the biology of echinoderms may yield information on the progress and prevention of cancer in other organisms.

Sea urchins from a variety of geographical areas suffer from the spectacular sea-urchin balding disease (Jangoux, 1990; Scheibling and Stephenson, 1984). The disease progresses from discoloration of the epidermis surrounding the spines, to loss of spines and other appendages followed by loss of epidermis and superficial dermal tissues, and finally to destruction of the skeleton and appearance of lesions on the aboral surface (Maes and Jangoux, 1984; Maes *et al.*, 1986; Scheibling and Stephenson, 1984). This communicable disease is caused by the marine bacteria, *Vibrio anguillarum* and *Aeromonas salmonicida* (Gilles and Pearse, 1986; Maes and Jangoux, 1985) and can result in mass mortalities affecting 10 to 90% of sea urchin populations in nature and in culture (Bourdouresque *et al.*, 1980, 1981; Pearse *et al.*, 1977). Once infected, sea urchins do not seem to recover, and there is no known cure for this communicable disease.

It is not likely that small-scale aquaculture will have a problem with diseases, unless infected individuals are brought in from the field. Infections may be avoided by treating individuals for 1 to 2 hours with Gentamycin (10 mg l^{-1}), Neomycin (30 mg l^{-1}), Novobiocin (30 mg l^{-1}), or Sulfisoxalole (0.25 mg l^{-1}) (Böttger and McClintock, in prep.). There is a major problem with bacterial infections in mass aquaculture, where millions of juvenile organisms (seed stock) have been lost from disease in culture (Tajima and Lawrence, 2001), with no treatment known at present.

Acknowledgments

The methods described in this chapter were developed with support by USDA NRI-Competitive Grant (2002-35206-11631) and Sea Grant (R/FMD-146; R/FMD-166) to CWW.

References

Anderson, R. J., and Velimirov, B. (1982). An experimental investigation of palatability of kelp bed algae to the sea urchin *Parechinus angulosus* Leske. *PSZNI Mar. Ecol.* **3,** 357–373.

Andrew, N. L., Agatsuma, Y., Ballesteros, E., Bazhin, A. G., Creaser, E. P., Barnes, D. K. A., Botsford, L. W., Bradbury, A., Campbell, A., Dixon, J. D., Einarsson, S., Gerring, P., Hebert, K., Hunter, M., Hur, S. B., Johnson, C. R., Juinio-Menez, M. A., Kalvass, P., Miller, R. J.,

Moreno, C. A., Palleiro, J. S., Rivas, D., Robinson, S. M. L., Schroeter, S. C., Steneck, R. S., Vadas, R. I., Woodby, D. A., and Xiaoqi, Z. (2002). Status and management of world sea urchin fisheries. *Oceanogr. Mar. Biol. Ann. Rev.* **40,** 343–425.

Arnn, B. L. (1963). Feeding rates of the sea star, *Asterias forbesi* (Desor), at reduced salinities. M. S. thesis, Duke University, Durham, NC.

Au, D. W. T., Lee, C. Y., Chan, K. L., and Wu, R. S. S. (2001a). Reproductive impairment of sea urchins upon chronic exposure to cadmium. Part I: Effects of gamete quality. *Env. Poll.* **11,** 1–9.

Au, D. W. T., Reunov, A. A., and Wu, R. S. S. (2001b). Reproductive impairment of sea urchins upon chronic exposure to cadmium. Part II: Effects on sperm development. *Env. Poll.* **11,** 11–20.

Bay-Schmith, E., and Pearse, J. S. (1987). Effect of fixed daylengths on the photoperiodic regulation of gametogenesis in the sea urchin *Strongylocentrotus purpuratus. Int. J. Invert. Repro. Develop.* **11,** 287–294.

Binyon, J. (1972). Physiology of Echinoderms, pp. 1–263. Pergamon Press, New York, NY.

Böttger, S. A., McClintock, J. B., and Klinger, T. S. (2001). Effects of inorganic and organic phosphates on feeding, feeding absorption, nutrient allocation, growth, and righting responses of the sea urchin *Lytechinus variegatus. Mar. Biol.* **138,** 741–751.

Böttger, S. A., and McClintock, J. B. (2002). Effects of inorganic and organic phosphate exposure on aspects of reproduction in the common sea urchin *Lytechinus variegatus* (Echinodermata: Echinoidea). *J. Exp. Zool.* **292,** 660–671.

Böttger S. A., and McClintock, J. B. (in prep.). The effects of inorganic and organic phosphate exposure on bactericidal activity of the coelomic fluid of the sea urchin *Lytechinus variegatus* (Lamarck) (Echinodermata: Echinoidea). Prepared for *Environmental Toxicology*.

Bourdouresque, C. F., Nedelec, H., and Shepherd, S. A. (1980). The decline of a population of the sea urchin *Paracentrotus lividus* in the bay of Port-Cros (Var, France). *Trav. Scient. Parc Nat. Port-Cros.* **6,** 243–251.

Bourdouresque, C. F., Nedelec, H., and Shepherd, S. A. (1981). The decline of a population of the sea urchin *Paracentrotus lividus* in the Bay of Port-Cros (Var, France). Rapp. P.v.Reun. *Commn. Int. Explor. Scient. Mer. Mediterr.* **27,** 223–224.

Brooks, J. M., and Wessel, G. M. (2002). The major yolk protein in sea urchins is a transferrin-like, iron-binding protein. *Develop. Biol.* **245,** 1–12.

Burns, W. A., and Bretschneider, A. (1981). Thin Is In: Plastic Embedding of Tissues for Light Microscopy, p. 58. American Society of Clinical Pathologists, Chicago.

Cellario, C., and Fenaux, L. (1990). *Paracentrotus lividus* (Lamarck) in culture (larval and benthic phases): Parameters of growth observed during 2 years following metamorphosis. *Aquaculture* **84,** 173–188.

Chandola, A., Singh, R., and Thapliyal, J. P. (1976). Evidence for a circadian oscillation in the gonadal response of the tropical weaver bird (*Ploceus phillippinus*) to programmed photoperiod. *Chronobiologia* **3,** 219–227.

Cochran, R. C., and Engelman, F. (1975). Environmental regulation of the annual season of *Strongylocentrotus purpuratus* (Stimpson). *Biol. Bull.* **148,** 393–401.

De Ridder, C., and Lawrence, J. M. (1982). Food and feeding mechanisms. *In* "Echinoderm Nutrition" (M. Jangoux and J. M. Lawrence, eds.), pp. 57–92. Balkema, Rotterdam, The Netherlands.

Devin, M. G. (2001). Land-based echinoculture: A novel system to culture adult sea urchins. *In* "The Sea Urchin: From Basic Biology to Aquaculture" (Y. Tokota, V. Matranga, and Z. Smolenicka, eds.), pp. 145–159. Balkema, Rotterdam, The Netherlands.

Diehl, W. J., and Lawrence, J. M. (1985). Effect of salinity on the intracellular osmolytes in the pyloric caeca and tube feet of *Luidia clathrata* (Say) (Echinodermata: Asteroidea). *Cop. Biochem. Physiol.* **82A,** 559–566.

Ebert, T. A. (1980). Relative growth of sea urchin jaws: An example of resource allocation. *Bull. Mar. Sci.* **30,** 467–474.

Ellington, W. R., and Lawrence, J. M. (1974). Coelomic fluid volume regulation and isosmotic intracellular regulation by *Luidia clathrata* (Echinodermata: Asteroidea) in response to hyposmotic stress. *Biol. Bull.* **146,** 20–31.

Fernandez, C., and Boudouresque, C. F. (1997). Phenotypic plasticity of *Paracentrotus lividus* (Echinoidea: Echinodermata) in a lagoonal environment. *Mar. Ecol. Prog. Ser.* **152,** 145–154.

Fernandez, C., and Boudouresque, C. F. (2000). Nutrition of the sea urchin *Paracentrotus lividus* (Echinodermata: Echinoidea) fed different artificial food. *Mar. Ecol. Prog. Ser.* **204,** 131–141.

Frantzis, A., and Gremare, A. (1992). Ingestion, absorption, and growth rate of *Paracentrotus lividus* (Echinodermata: Echinoidea) fed different macrophytes. *Mar. Ecol. Prog. Ser.* **95,** 169–183.

Fridberger, A., Fridberger, T., and Lundin, L. G. (1979). Cultivation of sea urchins of five different species under strict artificial conditions. *Zoon.* **7,** 149–151.

Garrido, C. L., and Barber, B. J. (2001). Effects of temperature and food ratio on gonad growth and oogenesis of the green sea urchin *Strongylocentrotus droebachiensis*. *Mar. Biol.* **138,** 447–456.

George, S. B. (1996). Echinoderm egg and larval quality as a function of adult nutritional state. *Oceanol. Acta* **19,** 297–308.

Gilles, K. W., and Pearse, J. S. (1986). Disease in sea urchins *Strongylocentrotus purpuratus*: Experimental infection and bacterial virulence. *Dis. Aquat. Org.* **1,** 105–114.

Glynn, P. W. (1974). The impact of *Acanthaster* on corals and coral reefs in the eastern Pacific. *Environ. Conserv.* **1,** 255–304.

Gnezdilova, S. M., Khristoforova, N. K., and Lipina, I. G. (1985). Gonadotoxic and embryotoxic effects of cadmium in sea urchins. *Symp. Biol. Hungarica* **29,** 239–251.

Gnezdilova, S. M., Lipina, I. G., and Khristoforova, N. K. (1987). Morphological changes in the ovaries of the sea urchins *Strongylocentrotus intermedius* exposed to cadmium. *Dis. Aquat. Org.* **2,** 127–133.

Gonor, J. J. (1973). Reproductive cycles in Oregon populations of the echinoid *Strongylocentrotus purpuratus* (Stimpson). Annual gonad growth and ovarian gametogenic cycles. *J. Exp. Mar. Biol. Ecol.* **12,** 45–64.

Grosjean, P., Spirlet, C., Gosselin, P., Vaitilingon, D., and Jangoux, M. (1998). Land-based, closed-cycle echinoculture of *Paracentrotus lividus* (Lamarck) (Echinoidea: Echinodermata): A long-term experiment at a pilot scale. *J. Shellfish Res.* **17,** 1523–1532.

Hagen, N. T. (1996). Echinoculture: From fishery enhancement to closed-cycle cultivation. *World Aquacult.* Dec: 7–19.

Heideman, P. D., Deoraj, P., and Bronso, F. H. (1992). Seasonal reproduction of a tropical bat, *Anoura geoffroyi*, in relation photoperiod. *J. Reprod. Fertil.* **96,** 765–773.

Hinegardner, R. T. (1969). Growth and development of the laboratory cultured sea urchin. *Biol. Bull.* **137,** 465–475.

Ito, S., Shibayama, M., Kobayakawa, A., and Tani, Y. (1989). Promotion of maturation and spawning of sea urchin *Hemicentrotus pulcherrimus* by regulating water temperature. *Nippon Suisan Gakkaishi* **55,** 757–763.

Jangoux, M. (1987a). Diseases of Echinodermata. II. Agents metazoans (Mesozoa to Bryozoa). *Dis. Aquat. Org.* **2,** 205–234.

Jangoux, M. (1987b). Diseases of Echinodermata. III. Agents metazoans (Annelida to Pisces). *Dis. Aquat. Org.* **3,** 59–83.

Jangoux, M. (1990). Diseases of Echinodermata. *In* "Diseases of Marine Animals" (O. Kinne, ed.), Vol. 3, pp. 439–567. Biologische Anstalt Helgoland, Germany.

Keesing, J. K., and Hall, K. C. (1998). Review of harvests and status of world sea urchin fisheries points to opportunities for aquaculture. *J. Shellfish Res.* **17,** 1597–1604.

Kishimoto, T. (1998). Cell cycle arrest and release in starfish oocytes and eggs. *Seminars Cell Develop.* **9,** 549–557.

Kishimoto, T. (1999). Activation of MPF at meiosis reinitiation in starfish oocytes. *Develop. Biol.* **214,** 1–8.

Klinger, T. S. (1982). Feeding rate of *Lytechinus variegatus* Lamarck (Echinodermata: Echinoidea) on differing physiognomies of an artificial food of uniform composition. *In* "Echinoderms: Proceedings of the International Conference, Tampa Bay" (J.M. Lawrence, ed.), pp. 29–32. Balkema, Rotterdam, The Netherlands.

Klinger, T. S., and Lawrence, J. M. (1984). Phagostimulation of *Lytechinus variegatus* (Lamarck) (Echinodermata: Echinoidea). *Mar. Behav. Physiol.* **11,** 49–67.

Klinger, T. S., Hsieh, H. L., Pangallo, A., Chen, C. P., and Lawrence, J. M. (1986). The effects of temperature on feeding, digestion, and absorption of *Lytechinus variegatus* (Lamarck) (Echinodermata: Echinoidea). *Physiol. Zool.* **59,** 332–336.

Krause, P. R. (1994). Effects of an oil production effluent on gametogenesis and gamete performance in the purple sea urchins (*Strongylocentrotus purpuratus* Stimpson). *Env. Tox. Chem.* **13,** 1153–1161.

Lane, B. P., and Europa, D. L. (1965). Differential staining of Epon-embedded tissues for light microscopy. *J. Histochem. Cytochem.* **12,** 579–582.

Lares, M. T., and McClintock, J. B. (1991). The effect of temperature on the survival, organismal activity, nutrition, growth, and reproduction of the carnivorous *Eucidaris tribuloides. Mar. Behav. Physiol.* **19,** 75–96.

Lawrence, J. M. (1975a). The effect of temperature–salinity combinations on the functional well-being of adult *Lytechinus variegatus* (Echinodermata: Echinoidea). *J. Exp. Mar. Biol. Ecol.* **18,** 271–275.

Lawrence, J. M. (1975b). On relationships between marine plants and sea-urchins. *Oceanogr. Mar. Biol. Ann. Rev.* **13,** 213–286.

Lawrence, J. M. (1976). Absorption efficiencies of four species of tropical echinoids fed *Thalassia testudinum. Thalassia Jugosl.* **12,** 201–205.

Lawrence, J. M. (1982). Digestion. *In* "Echinoderm Nutrition" (M. Jangoux and J.M. Lawrence, eds.), pp. 283–316. Balkema, Rotterdam, The Netherlands.

Lawrence, J. M. (1996). Mass mortality of echinoderms from abiotic factors. *In* "Echinoderm Studies" (J. M. Lawrence, ed.), Vol. 5, pp. 103–137. Balkema Press, Rotterdam, The Netherlands.

Lawrence, J. M., Lawrence, A. L., McBride, S. C., George, S. B., Watts, S. A., and Plank, L. R. (2001). Developments in the use of prepared feeds in sea-urchin aquaculture. *World Aquaculture* **32**(3), 34–39.

Lawrence, J. M., McBride, S. C., Plank, L. R., and Shpigel, M. (2003). Ammonia tolerance of the sea urchins *Lytechinus variegates, Strongylocentrotus franciscanus, and Paracentrotus lividus. In* "Echinoderm Research, 2001" (J.-P. Feral and B. David, eds.), pp. 233–236. Swets & Zeitlinger, Lisse, The Netherlands.

Leahy, P. S. (1986). Laboratory culture of *Strongylocentrotus purpuratus* adults, embryos, and larvae. *In* "Methods in Cell Biology" (T.S. Schroeder, ed.), Vol. 27, pp. 1–13. Academic Press.

Le Gall, P. (1990). Culture of echinoderms. *In* "Aquaculture" (G. Barnabe, ed.), Vol. 1, pp. 443–462. Ellis Horwood, NY.

Levitan, D. R. (1991). Skeletal changes in the test and jaws of the sea urchin *Diadema antillarum* in response to food limitation. *Mar. Biol.* **11,** 431–435.

Maes, P., and Jangoux, M. (1984). The bald-sea-urchin disease: A biopathological approach. *Helgoländer Meeresunters.* **37,** 217–224.

Maes, P., and Jangoux, M. (1985). The bald-sea-urchin disease: A bacterial infection. *In* "Proceedings of the International Echinoderm Conference, Galway" (B. F. Keegan and B. D. O'Connor, eds.), pp. 313–314. Balkema, Rotterdam, The Netherlands.

Maes, P., Jangoux, M., and Fenaux, L. (1986). La maladie de l'oursin chauve: Ultrastructure des lesions et caracterisation de leur pigmentation. *Annls. Inst. Oceanogr., Monaco (N.S.)* **62,** 37–45.

McBride, S. C., Pinnix, W. D., Lawrence, J. M., Lawrence, A. L., and Mulligan, T. M. (1997). The effect of temperature on production of gonads by the sea urchin *Strongylocentrotus franciscanus* fed natural and prepared diets. *J. World Aquat. Soc.* **28,** 357–365.

Miura, T., Yamaguchi, K., Takahashi, H., and Nagahama, Y. (1991). Hormonal induction of all stages of spermatogenesis *in vitro* in the male Japanese eel (*Anguilla japonica*). *PNAS* **88,** 5774–5778.

Nagahama, Y. (1994). Endocrine regulation of gametogenesis in fish. *Int. J. Dev. Biol.* **38,** 217–229.

Paniagua, R., Fraile, B., and Sáez, F. J. (1990). Effects of photoperiod and temperature on testicular function in amphibians. *Histol. Histopathol.* **5,** 365–378.

Pearse, J. S., and Walker, C. W. (1986). Photoperiodic control of gametogenesis in the North Atlantic sea star, *Asterias vulgaris. Int. J. Invert. Repro. Dev.* **9,** 71–77.

Pearse, J. S., and Cameron, R. A. (1991). Echinodermata: Echinoidea. *In* "Reproduction of Marine Invertebrates" (A. C. Giese, J. S. Pearse, and V. B. Pearse, eds.), Vol. VI, pp. 514–662. Echinoderms and Lophophorates. The Boxwood Press, Pacific Grove, CA.

Pearse, J. S., Sosta, D. P., Yellin, M. B., and Agegian, C. R. (1977). Localized mass mortality of red sea urchin, *Strongylocentrotus purpuratus*, near Santa Cruz, California. *Fish. Bull. U. S.* **53,** 645–648.

Pearse, J. S., Eernisse, D. J., Pearse, V. B., and Beauchamp, K. A. (1986a). Photoperiodic regulation of gametogenesis in sea stars, with evidence for an annual calendar independent of fixed daylength. *Amer. Zool.* **26,** 417–431.

Pearse, J. S., Pearse, V. B., and Davis, K. K. (1986b). Photoperiodic regulation of gametogenesis and growth in the sea urchin, *Strongylocentrotus purpuratus. J. Exp. Zool.* **237,** 107–118.

Roller, R. A., and Stickle, W. B. (1993). Effects of temperature and salinity acclimation of adults on larval survival, physiology, and early development of *Lytechinus variegatus* (Echinodermata: Echinoidea). *Mar. Biol.* **116,** 583–591.

Sabourin, T. D., and Stickle, W. B. (1981). Effects of salinity on respiration and nitrogen excretion in two species of echinoderms. *Mar. Biol.* **65,** 91–99.

Sakairi, K., Yamamoto, M., Ohtsu, K., and Yoshida, M. (1989). Environmental control of gonadal maturation in laboratory-reared sea urchins, *Anthocidaris crassispina* and *Hemicentrotus pulcherrimus. Zoo. Sci.* **6,** 721–730.

Scheibling, R. E., and Stephenson, R. L. (1984). Mass mortality of *Strongylocentrotus droebachiensis* (Echinodermata: Echinoidea) off Nova Scotia, Canada. *Mar. Biol.* **78,** 153–164.

Schroeder, P. C., Larsen, J. H., and Waldo, A. E. (1979). Oocyte-follicle cell relationships in a starfish. *Cell Tiss. Research* **203,** 249–256.

Shirai, H., and Walker, C. W. (1988). Chemical control of asexual and sexual reproduction in echinoderms. *In* "Endocrinology of Selected Invertebrate Types" (H. Laufer and R. G. H. Downer, eds.), pp. 453–476. Alan Liss, Inc., New York.

Shirley, T. C., and Stickle, W. B. (1982). Responses of *Leptasteris hexactis* (Echinodermata: Asteroidea) to low salinity. I. Survival, activity, feeding, growth, and absorption efficiency. *Mar. Biol.* **69,** 147–154.

Smiley, S. (1990). A review of echinoderm oogenesis. *J. Electron Micros. Tech.* **16,** 93–114.

Spirlet, C., Grosjean, P., and Jangoux, M. (1998). Reproductive cycle of the echinoid *Paracentrotus lividus*: Analysis by means of maturity index. *Invert. Reprod. Dev.* **34,** 69–81.

Stephens, G. C., and Virkar, R. A. (1966). Uptake of organic material by aquatic invertebrates. IV. The influence of salinity on the uptake of amino acids by the brittle star, *Ophiactis arenosa. Biol. Bull.* **131,** 172–185.

Stickle, W. B., and Diehl, W. J. (1987). Effects of salinity on echinoderms. *In* "Echinoderm Studies" (M. Jangoux and J. M. Lawrence, eds.), Vol. 2, pp. 235–285. Balkema, Rotterdam, The Netherlands.

Tajima, K., and Lawrence, J. M. (2001). Disease in edible sea urchins. *In* "Edible Sea Urchins" (J. M. Lawrence, ed.), pp. 139–148. Elsevier Science B.V.

Temara, A., Gulec, I., and Holdway, D. A. (1999). Oil-induced disruption of foraging behavior of the asteroid keystone predator, *Coscinasterias muricata* (Echinodermata). *Mar. Biol.* **133,** 501–507.

Unuma, T., Okamoto, H., Konishi, K., Ohta, H., and Mori, K. (2001). Cloning of cDNA encoding vitellogenin and its expression in red sea urchin *Pseudocentrotus depressus. Zool. Sci.* **18,** 559–565.

Unuma, T., Yamamoto, T., Akiyama, T., Shiraishi, M., and Ohta, H. (2003). Quantitative changes in yolk protein and other components in the ovary and testis of the sea urchin *Pseudocentrotus depressus. J. Exp. Biol.* **206,** 365–372.

Vadas, R. L. (1977). Preferential feeding: An optimization strategy in sea urchins. *Ecol. Monogr.* **47,** 337–371.

Vashchenko, M. A., Zhadan, P. M., Karaseva, E. M., and Luk'yanova, O. N. (1993). Disturbance of reproductive function of the sea urchin *Strongylocentrotus intermedius* in polluted areas of the Peter the Great Bay, Sea of Japan. *Russ. J. Mar. Biol.* **19,** 35–41.

Walker, C. W. (1979). Ultrastructure of the somatic portion of the gonads in asteroids, with emphasis on flagellated-collar cells and nutrient transport. *J. Morph.* **162,** 127–161.

Walker, C. W. (1980). Spermatogenic columns, somatic cells, and the micro-environment of germinal cells in the testes of asteroids. *J. Morph.* **166,** 81–107.

Walker, C. W. (1982). Nutrition of gametes. *In* "Echinoderm Nutrition" (J. M. Lawrence and M. Jangoux, eds.), pp. 449–468. Balkema, Rotterdam, The Netherlands.

Walker, C. W., and Lesser, M. P. (1998). Manipulation of food and photoperiod promotes out-of-season gametogenesis in the green sea urchin, *Strongylocentrotus droebachiensis*: Implications for aquaculture. *Mar. Biol.* **132,** 663–676.

Walker, C. W., Unuma, T., McGinn, N. A., Harrington, L. M., and Lesser, M. P. (2001). Reproduction of sea urchins. *In* "Edible Sea Urchins: Biology and Ecology" (J. M. Lawrence, ed.), pp. 5–26. Elsevier Science.

Wellings, S. R. (2001). Neoplasia and primitive vertebrate phylogeny: Echinoderms, prevertebrates, and fishes—A review. *In* "Neoplasms and related disorders of invertebrates and lower vertebrate animals" (C. J. Dawe and J. C. Harshbarger, eds.), Natn. Cancer Inst. Monogr.

Williams, H. (2001). Sea urchin fisheries of the world: A review of their status, management strategies, and biology of the principal studies. Report for the Department of Primary Industries, Water, and Environment, Tasmania. p. 27.

Yamaguchi, M. (1974). Growth of juvenile *Acanthaster planci* (L.) in the laboratory. *Pacific Sci.* **28,** 123–138.

Yamamoto, M., Ishine, M., and Yoshida, M. (1988). Gonadal maturation independent of photic conditions in laboratory-reared sea urchins, *Pseudocentrotus depressus* and *Hemicentrotus pulcherrimus*. *Zool. Sci.* **5,** 979–988.

Yokota, Y. (2002). Introduction to the sea urchin biology. *In* "The Sea Urchin: From Basic Biology to Aquaculture" (Y. Yokota, V. Matranga, and Z. Smolenicka, eds.), pp. 1–10. Balkema, Rotterdam, The Netherlands.

Yokota, Y., and Sappington, T. W. (2002). Vitellogen and vitellogenin in echinoderms. *In* "Reproductive Biology of Invertebrates: Progress in Vitellogenesis" (A. S. Raikhel and Sappington, eds.), Vol. XII, pp. 201–221. Science Publishers, New Hampshire.

CHAPTER 3

Echinoderm Eggs and Embryos: Procurement and Culture

Kathy R. Foltz,* Nikki L. Adams,† and Linda L. Runft*

*Department of Molecular, Cellular, and Developmental Biology
and the Marine Science Institute
University of California
Santa Barbara, California 93106

†Department of Biological Sciences
California Polytechnic State University
San Luis Obispo, California 93407

METHODS IN CELL BIOLOGY, VOL. 74

0091-679X/04 $35.00

I. Introduction

A. Background

Echinoderms in general and sea urchins in particular have a rich history as model systems for the study of oogenesis, fertilization, and early embryogenesis (Chapter 1). Since the mid-nineteenth century, work in the sea urchin embryo has lead to many contributions in cell and developmental biology (Monroy, 1986; see Chapter 1). The ease of collecting and maintaining adult animals (Chapter 2), as well as in obtaining gametes and culturing large quantities of synchronous embryos, is complemented by the ability to do biochemistry, reverse genetics, and embryo manipulations, all in an organism with unparalleled transparency in its early developmental stages. More recently, studies in sea urchin embryos have provided detailed analyses of gene regulatory networks (Chapter 32) and echinoderms have been viewed as a key group of organisms in terms of formulating ideas about the evolutionary aspects of developmental biology (Chapter 1).

B. Purpose of Approaches

Here, we provide detailed methods for obtaining gametes, achieving synchronous *in vitro* fertilization, and culturing embryos through early larval stages for several echinoderm species. There are many excellent books providing comprehensive reviews of echinoderm reproductive and developmental biology (cf. Giese *et al.*, 1991; Hyman, 1955; Strathmann, 1987a) as well as published protocols for specific culturing conditions (cf. Leahy, 1986; Lowe and Wray, 2000).[1] There is also an extensive literature on natural spawning and fertilization in the field, which we do not cover in any detail here (cf. Levitan *et al.*, 1991; Starr *et al.*, 1990; Strathmann, 1987a). We refer the reader to published literature throughout the methods sections when appropriate. The intent of this abbreviated chapter is to provide specific examples of protocols for obtaining gametes and culturing embryos of a selected number of species for experimental analysis of their development. In most of the procedures outlined here, we refer to the use of filtered

[1]We thank our many colleagues who provided advice and comments, especially Shane Anderson, Gordon Hendler, Laurinda Jaffe, Chris Lowe, and Gary Wessel. We also acknowledge the extensive use and referencing of Megumi Strathmann's *Reproduction and Development of Marine Invertebrates of the Northern Pacific Coast* as well as Volume VI of *Reproduction of Marine Invertebrates*, edited by Arthur Giese, John Pearse, and Vicki Pearse. We encourage all readers, especially students, to consult and enjoy these excellent books.

seawater (FSW). In many cases, high quality, natural seawater may not be available and thus, artificial seawaters (ASW) must be used. Table I lists recipes for these artificial seawaters as well as for specific seawaters that are buffered or lacking certain ionic components.

C. Specific Examples

Several key species of echinoderms representing four classes (Echinoidea, Asteroidea, Ophiuroidea, and Holothuroidea) are covered here. The species were chosen to provide breadth across the phylum Echinodermata, as well as to provide practical guidelines for handling some of the more commonly studied species. For each species, we highlight specific advantages and special note is made of key issues to consider when handling adults, collecting gametes, or setting and maintaining embryo cultures. Finally, information regarding interspecific crosses is provided.

II. Method for Sea Urchins (Class Echinoidea)

A. General Explanation

Sea urchins spawn eggs that have completed meiosis and are in an arrested stage ("G_o") of the first mitotic cell cycle. However, there are methods for obtaining immature oocytes from sea urchins (Chapter 5). Millions of eggs can typically be spawned by a single female. In the wild, male and female urchins release their gametes into the seawater in response to environmental cues that are as yet unknown. There are many hypotheses about spawning cues, but little agreement about whether there is one specific cue or synergy among various physical parameters. Regardless, fertilization naturally occurs externally and involves a series of species-preferential interactions between the gametes (reviewed in Vacquier, 1986; Vacquier *et al.*, 1995). Following fertilization, embryos of most species develop by cleaving in a radial holoblastic fashion to form a blastula that ciliates, hatches out of the fertilization membrane, and becomes free swimming. The blastula then undergoes gastrulation to form the archenteron, the tip of which grows to one side to eventually form the ventral surface of the prism stage. Development of the larval gut and a completed internal skeleton comprised of triradiate primary spicules (see Chapter 12) and supporting extension rods mark the echinopluteus (or pluteus) stage. After several weeks (depending on the species), the feeding larva eventually attaches to a substrate and undergoes metamorphosis to create a juvenile sea urchin (Pearse and Cameron, 1991; Strathmann, 1987b). See Table II for a description of the timing of development and various stages for different species.

This Section focuses on how to procure eggs and sperm, perform *in vitro* fertilization, and culture embryos from the California purple sea urchin (*Strongylocentrotus purpuratus*) and from both the Pacific and Atlantic *Lytechinus*

Table I
Recipes for Commonly used Artificial Seawaters[a]

Chemical	g/L	Concentration (mM)
A. Marine Biological Laboratory Artificial Seawater (ASW) (for Calcium-Free Seawater, CaFASW, leave out the $CaCl_2$)		
NaCl	25.5	437
KCl	0.67	9
$MgCl_2$ $6H_2O$	4.7	22.9
$MgSO_4$ $7H_2O$	6.3	25.5
$CaCl_2$ $2H_2O$	1.35	9.3
$NaHCO_3$	0.18	2.1
B. Calcium–Magnesium–Free Seawater (CaMgFASW; Keller *et al.*, 1999)		
NaCl	30.7	526
KCl	0.745	10
EGTA	0.76	2
$NaHCO_3$	0.21	2.5
C. Calcium–Magnesium–Free Seawater (CaMgFASW; Strathmann, 1987a)		
NaCl	26.22	449
KCl	0.67	9
Na_2SO_4	4.62	33
$NaHCO_3$	0.21	2.5
Na_2EDTA	0.37	1
D. Low Sodium–Choline Substituted Seawater (Jaffe, 1980; Schuel and Schuel, 1981)		
Choline Cl[b,c]	67.2	484
KCl	0.76	10
$MgSO_4$ $7H_2O$	7.2	29
$MgCl_2$ $6H_2O$	5.6	27
$NaHCO_3$	0.2	2.4
$CaCl_2$ $2H_2O$	1.6	11
E. Sodium–Free Seawater (Keller *et al.*, 1999; NaFASW)[c]		
Glycerol	84.7	920
KCl	0.745	10
$CaCl_2$ $2H_2O$	1.62	11
$MgCl_2$ $6H_2O$	5.29	26
$MgSO_4$ $7H_2O$	7.15	29
$KHCO_3$	0.2	2

[a]Note that there are several different formulations for some of the artificial seawaters. Prepare the artificial seawater at room temperature using deionized, distilled H_2O. Add components to a volume of H_2O that is less than the final, desired volume and allow them to dissolve completely. Then chill or warm the artificial seawater to the desired temperature and, only then, adjust the pH to 8.0 with NaOH, followed by a final adjustment of the volume with H_2O.

[b]Choline Chloride has a limited shelf life. It is normally a white crystal and hydroscopic; if discolored, do not use. It can be purchased from SIGMA, catalog #C-1879. For the choline-substituted seawater, adjust the pH to 8.0 using 1 M NaOH before adding the $CaCl_2$. Then add the $CaCl_2$, bring the solution up to volume, and readjust pH to 8.0 using 1 M NaOH.

[c]For both the choline-substituted and sodium-free ASW, keep track of how much NaOH is used so that the total amount of sodium being added can be calculated. Alternatively, use NH_4OH to adjust the pH.

Table II
Culture Temperatures and Developmental Time Courses for Selected Echinoderm Species[a]

Time after insemination	Developmental stage
S. purpuratus raised at 12 °C	
2–2.5 h	1st cleavage
4 h	2nd cleavage
5.5 h	3rd cleavage
6.5 h	4th cleavage
27 h	Hatching of blastula from FE
30 h	Gastrulation
3 days	Early prism larva
4 days	Early pluteus larva
40–80 days	Metamorphosis
L. variegatus raised at 25 °C (Schoenwolf, 1995)	
1 h	1st cleavage
1.5 h	2nd cleavage
2 h 40 min	32 cell stage
7.5 h	Hatching of blastula from FE
11 h	Gastrulation
15 h	Early prism larva
20 h	Early pluteus larva
L. pictus raised at 18 °C (Sea Urchin Embryology website)	
1.5 h	1st cleavage
2.5 h	2nd cleavage
24 h	Blastula
2 days	Gastrula
5 days	Pluteus
A. miniata raised at 15 °C (see Strathmann, 1987c)	
2.5 h	1st cleavage
9 h	Blastula
18 h	Hatching of blastula from FE
2 days	Gastrulation is complete
3–4 days	Bipinnaria larva
1 month	Brachiolaria larva
2 months	Metamorphosis
2 months and 1 day	Juvenile seastar
Ophiopholis aculeata raised at 8 °C (see Hendler, 1991; Strathmann and Rumrill, 1987)	
3 h	1st cleavage
24–36 h	Blastula
1.5–2 days	Gastrula
3 days	2-arm stage
10.5 days	6-arm stage
17 days	8-arm stage (ophiopluteus)
30 days	Incipient metamorphosis
83–216 days	Complete metamorphosis

(*continues*)

Table II *(continued)*

Time after insemination	Developmental stage
Ophiothrix angulata raised at 24–27 °C (see Hendler, 1991)	
0.5 day	Gastrula
1 day	2-arm stage
2.5 days	6-arm stage
7 days	8-arm stage
>19 days	Complete metamorphosis
Ophiothrix fragilis (temperature not specified; see Hendler, 1991)	
24 h	Blastula
1.5 days	Gastrula
1.5–2 days	2-arm stage
3 days	4-arm stage
4 days	6-arm stage
7–10 days	8-arm stage
16 days	Incipient metamorphosis
26 days	Complete metamorphosis
P. chitonoides raised at 11 °C (see McEuen, 1987)	
2 h	1st cleavage
8 h	32-cell stage
18 h	Blastula
40 h	Gastrula
60–70 h	Early larva
80–90 h	Early doliolaria
11–12 days	Settlement
26–32 days	Ossicle plates form
P. californicus raised at 11 °C (see McEuen, 1987)	
2.75 h	1st cleavage
10.75 h	32-cell stage
19 h	Blastula
27 h	Gastrula
3.25–4.25 days	Early larva
52 days	Early doliolaria
60–61 days	Settlement
61–65 days	Ossicle plates form

[a]See Strathmann (1987a) and Giese *et al.* (1991) for a more extensive list of species.

spp, L. pictus and *L. variegatus*, respectively. All three of these species are indirect developers with plankotrophic larvae, but it should be noted that some species are direct developing and others have slight variations on the larval developmental plan (reviewed in Pearse and Cameron, 1991; Strathmann, 1987b). Purple urchins are commonly found in the Pacific intertidal and shallow subtidal zones and their natural spawning season spans mid-November through March (Appendix). *Lytechinus spp* typically are gravid in the summer months, from May through September. Kept in tanks and fed regularly on a diet of *Macrocystis* (kelp), the adults can remain gravid for several months longer, if not year round (Chapter 2).

Appendix
Breeding Seasons for Various Echinoderm Species[a]

Echinoderm species	Breeding season
S. purpuratus	mid Nov–March
L. variegatus	mid May–Sept
L. pictus	mid May–Sept
A. miniata	mid May–August
Ophioderma brevispinum	May–August
Ophiopholis aculeata	April–September[b]
Ophiothrix angulata	March–September
Ophiothrix fragilis	June–September
P. chitonoides	mid March–mid May
P. californicus	mid May–mid July

[a]See Strathmann (1987a) and Giese *et al.* (1991) for a more extensive list of species.

[b]This species of brittle star has been reported to spawn in the lab January–April, July, and October–November (Hendler, 1991; Strathmann and Rumrill, 1987).

These species have been important model systems for studying fertilization (reviewed in Jaffe *et al.*, 2001) and for both the molecular mechanisms (cf. Angerer and Angerer, 2000, 2003) and morphogenesis (cf. Ettensohn, 1999; Ettensohn and Sweet, 2000) aspects of early embryogenesis. Different species have different optimal temperatures for both gametogenesis and culture of embryos and larva (see note II.C.1; Table II) and care must be taken to ensure that the gametes and embryos are kept at the appropriate temperature. In general, warmer temperatures contribute to polyspermy and arrested development. Most of the Strongylocentrotids prefer 12 to 17 °C (although *S. droebachiensis*, a trans-polar species, requires even colder temperatures, 4 to 10 °C). *Lytechinus spp.* tend to require warmer temperatures (17–25 °C). The methods described in this Section can be applied readily, with minor modifications, to other sea urchin species such as *Arbacia punctulata*, as well as to sand dollars and heart urchins (for specific protocols and seasonality for other sea urchin species, see Pearse and Cameron, 1991; Strathmann, 1987b). Readers are also directed to excellent protocols provided in Lowe and Wray (2000).

B. Detailed Method

1. Materials Needed

1. Gravid sea urchins. See note II.C.2 for commercial vendors and see Appendix for breeding seasons.
2. Filtered seawater (FSW), artificial seawater (ASW), and, depending on the procedure, calcium-free artificial seawater (CaFASW; see note II.C.3). See Table I for recipes.

3. 0.55 M KCl
4. Small plastic paddles for resuspending eggs and embryos
5. 3 cc syringe and 23–28 gauge hypodermic needles for injecting KCl intracoelomically
6. Antibiotics for long-term culturing (gentamicin or penicillin-streptomycin; In Vitrogen; catalog numbers 15750-060 and 15070-063, respectively)
7. Food for larvae; see Chapter 4.
8. Glassware: various size glass beakers, glass finger bowls, Pyrex baking dishes (see note II.C.4)
9. 1.5 ml microfuge tubes
10. Pasteur pipets and pipet bulbs
11. Ice bucket
12. Glass slides
13. Nylon mesh for filtering gametes and larvae and for dejellying eggs; see note II.C.5 and Table III for mesh sizes used for various procedures for a given species.
14. Compound light microscope with 10× objective
15. ATAZ (3-amino-1,2,4-triazole; Sigma, cat # A8056; see note II.C.6) to prevent hardening of the fertilization envelope, if necessary).

2. Obtaining Gametes

1a. Most species of sea urchin are not possible to sex by visual inspection. Generally, spawning must be induced and then viewing of the gametes easily distinguishes male from female; eggs are yellow-orange, sperm are white. Vigorous shaking of ripe animals sometimes induces enough spawning to determine the sex. Obtain the adult urchin and rinse it free of any debris using seawater. The five gonopores are situated on the aboral surface (the top of the animal). Spawning of ripe urchins can be triggered by injecting <1 mL of 0.55 M KCl directly into the coelomic cavity through the perioral membrane surrounding the lantern on the oral surface using a 23 to 27 gauge needle. Injecting $\sim$1 ml 0.55 M KCl in different spots around the mouth will often cause the animal to completely spawn its store of gametes (see note II.C.7). Thoroughly rinse the needle and syringe before using for another animal to avoid transferring sperm.

1b. Alternatively, a small electrical shock can be applied to the adult animal using a 9V battery and small metal paddles (with insulated handles) wired to the battery. Place the animal oral side down on a glass petri dish in a volume of FSW that just covers the dish bottom. Apply a paddle attached to the battery to each side of the animal for approximately 15 to 20 s. Current flow is indicated by vigorous movement of the animal's spines. Within a minute or two, a small amount of gametes will be released from the gonopores. Pipette sperm directly

Table III
Average Egg Sizes and Nylon Mesh Sizes to be Used for Various Procedures[a]

Echinoderm species	Average egg diameter	Size mesh for filtering eggs	Size mesh for stripping jelly	Size mesh for stripping FEs
S. purpuratus	80 μm	120 μm	120 μm	64 μm
L. variegatus	120 μm	210 μm	210 μm	80 μm
L. pictus	120 μm	210 μm	210 μm	100 μm
A. miniata	180 μm	265 μm	265 μm	N/A
Ophiopholis aculeata (coral-orange)	105 μm	210 μm	N/A	N/A
Ophiura sarsii (orange-pink)	100 μm	210 μm	N/A	N/A
Ophioderma brevispinum (olive green or orange-yellow)	300 μm	400 μm	N/A	N/A
Ophiothrix angulata (white)	100 μm	210 μm	N/A	N/A
Ophiothrix fragilis (yellow)	90 μm	120 μm	N/A	N/A
P. chitonoides	625 μm	1000 μm	N/A	N/A
P. californicus	204 μm	310 μm	N/A	N/A

[a]See Strathmann (1987a) and Giese *et al.* (1991) for a more extensive list of egg sizes for various species. The average egg diameters listed refer to the egg proper and do not include extracellular coats or jelly. Egg color is indicated for the Ophiuroids, as color often can be viewed through the epidermis on the oral side and serves as an indicator of sex. In general, when using the nylon mesh to clean up negatively buoyant eggs or to strip jelly layers, use a mesh size that is 1.5× the diameter of the egg so the eggs can pass through without damage. To "strip" fertilization envelopes (FEs), use a mesh size slightly smaller than the egg diameter. In all cases, use care in filtering and never force the eggs through the mesh—allow gravity to do the job. To clean up positively buoyant (floating) eggs or motile embryo cultures, use a mesh size that is substantially smaller than the egg or embryo diameter so that eggs or embryos will collect on the mesh, which can then be rinsed into fresh culture media. For example, to clean sea urchin or sea star embryo cultures, use a 30 μm mesh for early cultures, a 100 μm mesh for early larval cultures, and a 200 μm mesh for later larval cultures (N/A = Not Applicable, as no standard protocols have been developed). Nylon mesh is available from Sefar America (BriarCliff Manor, NY; tel. 1-800-995-0531; www.sefaramerica.com) or from Barnstead International/Labline, Inc (Melrose Park, IL; tel. 708-450-2600; FAX 708-450-0943).

from the gonopores into a microfuge tube on ice. Invert females over small beakers of FSW to collect eggs. This method is desirable when sex determination is the goal or when only small amounts of gametes are desired.

1c. Finally, to obtain maximal quantities of gametes, the adult can be cut open with a pair of scissors, rinsed, and the gonads carefully dissected to remove all of the gametes by aspiration. This typically is not required, but has been used successfully. Use care in this procedure, as ripe gonads may be easily ruptured and thus the gametes may be contaminated with coelomic fluid or gut contents. Use special precautions to avoid unwanted exposure of eggs to sperm (e.g., rinse scissors and other tools in between animals). See the excellent detailed protocol for large-scale preparations using KCl-induced spawning (up to 1000 adult animals at a time) by Coffman and Leahy (2000). Another variation on this for collecting eggs is to remove the lantern from the female urchin, place animal upside down on a beaker of FSW, and then fill the body cavity with 0.55 M KCl.

2. Place spawning male urchins right side up (oral side down) on a tray of ice. Collect sperm "dry" by pipetting it directly from the gonopores as the semen is exuded. Use a P200 pipettor with a yellow tip that has had the very end cut off to make it wider. Alternatively, place the male urchin upside down on a petri dish on ice and allow semen to collect in the dish. Transfer the sperm to microfuge tubes and keep on ice. Sperm will be good for the rest of the day if kept cold. Do not mix sperm from different individuals. Do not dilute sperm until just before use; dilution activates the sperm, which have a limited life span.

3. For collection of eggs, invert females onto a small glass beaker (note II.C.4) of FSW at the appropriate temperature for the species being used (note II.C.1). Make sure the gonopores are fully submerged so that the eggs are shed directly into the seawater. Allow the eggs to fall into the beaker and settle. View a sample of the eggs microscopically (a 10× objective lens is sufficient). Mature sea urchin eggs have completed meiosis, so there is no germinal vesicle visible and the eggs will be of a uniform size. No lysis or blebbing of the egg surface should be visible. The jelly coat is not visible by light microscopy unless stained with India ink or toluidine blue (use a 5% solution of dye in ASW to stain eggs), but jelly-intact eggs are characterized by a lack of clumping. Mature *S. purpurtus* eggs typically are ~80 μm in diameter while *L. variegatus* and *L. pictus* eggs are ~100 to 130 μm in diameter (Table III). See Strathmann (1987b) and Pearse and Cameron (1991) for sizes of mature eggs from various other urchin species. Note the volume of settled eggs when the female has spawned completely. Transfer (by carefully pouring) the suspension of eggs gently into a large glass beaker and add a large (~500×) volume of FSW. Resuspend the eggs gently by using a small paddle. Allow the eggs to settle by gravity. Decant the FSW and repeat. Do this twice more. Do not allow the eggs to sit overly long—ideally, no longer than 4 to 5 h. The quicker the washes, the better the success at fertilizing. If eggs are to be transferred or manipulated using pipettes, use care to pre-wet the pipette and to ensure that the tip bore is larger than the egg diameter. Finally, do not mix eggs from different individuals.

4a. For some procedures, it is desirable to remove the jellycoat of the egg, which naturally disperses in seawater over time. If the jellycoat is to remain intact, skip to step II.B.3.1. To dejelly the eggs, resuspend the eggs in FSW as a 10% vol–vol suspension. Gently pass the egg suspension once through nylon mesh (120 μm for *S. purpuratus*; 210 μm for *L. pictus* and *L. variegatus*; see note II.C.5 and Table III) that has been pre-wetted with FSW. After pouring the egg suspension through the mesh into a glass beaker, rinse the mesh with FSW and pass the egg suspension through the mesh 7 to 9 more times. After a total of 8 to 10 gentle passages through the mesh, allow the eggs to settle in the glass beaker. Check microscopically that there is no lysis of or damage to the eggs. After the eggs have settled by gravity, collect (by pipetting) the jellywater (which will be cloudy) and save it, chilled to the species-appropriate temperature. It can be used later to fertilize the eggs. Wash the dejellied eggs 2 to 3 times by gravity settling in a large volume of FSW.

4b. Another method for dejellying is a short exposure to acidified buffered seawater. However, this method, of which there are several variations, tends to result in less reliable fertilizability of the dejellied eggs and can cause more damage to the eggs as compared with mechanical removal (personal observations). Briefly resuspend the washed eggs in ASW as a 10% vol–vol suspension. Rapidly acidify the solution to pH 4.5 by adding, dropwise, 0.1 N HCl while monitoring the pH. As soon as pH 4.5 is reached, quickly add 2 volumes of Trisbuffered ASW (ASW with 1M Tris, pH 8.0) to bring the pH back to ~6. Then add 1N NaOH, dropwise, while monitoring the pH with constant, gentle stirring, to restore the solution to ~pH 7.9. Check microscopically that there is no lysis of or damage to the eggs. Wash the dejellied eggs 2 to 3 times by gravity settling in FSW.

3. Fertilizing the Eggs

1a. *Fertilization of jelly-intact eggs.* After the final wash, resuspend the eggs in FSW as a final 10% (volume/volume) suspension. For each 100 mL of egg suspension, prepare 5 mL of seawater in a disposable glass test tube and add 25 uL of dry sperm to this tube. Mix well with a paddle to make a uniform suspension of sperm. Add the diluted sperm to the egg suspension and mix gently with a paddle (final concentration of sperm is ~1:4000). If eggs are healthy, it is actually quite difficult to force polyspermy in FSW, so excess amounts of sperm can be used in order to achieve complete and synchronous fertilization. The actual amount of sperm required may vary depending on the individual batch of sperm or eggs. More dilute sperm suspensions (even as low as 1:100,000) often also work fine, as long as gametes are healthy. It may be necessary to conduct a few small scale "test fertilizations" to determine the appropriate dilution of sperm. Precise sperm counts can be achieved by making small dilutions of sperm, fixing in an equal volume of 2% glutaraldehyde in FSW for 10 min, and then using a hemacytometer to count the sperm.

1b. *Fertilization of dejellied eggs.* The jellycoat contains components necessary to trigger the acrosome reaction in sperm (Darszon *et al.*, 1996; Vacquier, 1986; Vacquier and Moy, 1997; Vilela-Silva *et al.*, 1999, 2002). Therefore, sperm must be exposed to the jellywater that was collected in the dejellying step (II.B.2.4a) in order to become fertilization-competent. For each 100 mL of egg suspension, prepare 50 mL of jellywater + 50 mL FSW in a beaker. Dilute 50 μL of dry sperm into this, and quickly but gently mix with a small plastic paddle to disperse the sperm. Pour this gently into the egg suspension and mix gently with a paddle (the final concentration of sperm is ~1:4000). Sometimes, a higher concentration of sperm is required to fertilize dejellied eggs compared to jelly-intact eggs, so small test fertilizations may be necessary (II.B.3.1a). When diluted into jellywater, sperm may appear to clump. This is a transient aggregation and healthy sperm will disperse quickly. Using the larger volume of diluted jellywater for the initial sperm dilution helps minimize clumping.

1c. In some cases, it may be desirable to fertilize eggs under conditions that prevent hardening of the fertilization envelope (FE). If so, then after dejellying, wash the eggs once in artificial seawater (ASW) and then once in ASW supplemented with 1 mM ATAZ (see note II.C.6). Resuspend the eggs in ASW/ATAZ as a 10% volume:volume suspension and then fertilize as indicated above in II.B.3.1b. The FE will elevate and will be visible and can serve as an indicator of successful fertilization, but will not "harden." The fertilized eggs can then be stripped of the "soft" FE by passage through nylon mesh (Table III). Be aware that the sticky hyaline layer of the blastomeres will mediate clumping of embryos that lack a fertilization envelope. While culturing the embryos in CaFASW can reduce clumping, it also induces blastomeres to dissociate. It may be helpful to perform the fertilization step under more dilute egg suspension conditions in order to minimize clumping. Because of these complications, careful consideration of the necessity for removing the FE should be exercised.

2. Immediately view a drop of inseminated eggs under the microscope; elevation of the FE initiates within 30 s of insemination, from the point of sperm entry, and proceeds to completion very quickly. Once elevated FEs are observed for the majority of the eggs, immediately dilute the culture in a large volume of FSW. If less than 95% fertilization is observed, or if polyspermy is apparent, discard and start over. You may want to try sperm as well as eggs from different individuals. Polyspermy can be detected by the presence of multiple fertilization cones (small projections on the egg surface) within about 10 min of insemination, although these may be difficult to observe in some species. Usually, polyspermic eggs will tend to have abnormal-looking cytoplasm (patchy-looking) or produce small membrane blebs after an hour or so. Finally, polyspermy is characterized by inability of the zygote to cleave properly (zygotes may look like soap bubbles) and death ensues after several hours.

3. Wash the fertilized eggs twice with large volumes of seawater using gravity settling. This is very important as it removes excess sperm.

4. Embryo Culture

1. Resuspend the washed, fertilized eggs in a final volume of FSW to give a 0.5 to 1% (volume:volume) suspension.

2. Add penicillin–streptomycin (or gentamicin) to keep down bacterial growth. Typically, a 1:1000 dilution of a stock of 5U/mL is sufficient.

3. Gently transfer the zygotes to the culture dish. This can be done by gently pouring the suspension from beaker to dish or by using a large bore pipette (like a turkey baster). Standing cultures should contain no more than a single layer of zygotes resting on the bottom, as overcrowding will slow development, produce abnormalities, and potentially cause the culture to crash. The ideal density is to set the culture at $\geq$1 mL/embryo. Early developmental stages are usually less affected by high density, so early cultures can be maintained at higher densities and then thinned later. Use a shallow vessel to culture the embryos in order to maximize the surface exposure. Small volume cultures do well in small finger bowls and standing larger

cultures do well in large Pyrex baking dishes (covered loosely with plastic wrap). A small amount of water motion seems to be advantageous for developing cultures. This can be achieved by a small rocking movement or by gentle stirring using paddles (see Lowe and Wray, 2000, for details on setting up continuous stirring apparati). Cultures of *S. purpuratus* should be raised at 12 to 15 °C (note II.C.1; for appropriate temperatures if another species is being cultured, see Table II).

4. Refer to Chapter 4 for details on long-term culturing of larval stages through metamorphosis. The FSW of standing cultures must initially be changed daily to remove debris and dead embryos. For small cultures, the embryos can be gently transferred to a new dish containing FSW using a pipette with an opening that is 3 mm in diameter. For large cultures, gently decant the culture into a 200 ml plastic beaker which has had the bottom removed and covered with a nylon mesh (the mesh can be attached to the beaker with a glue gun; note II.C.5). The nylon mesh beaker should be partially immersed in a bowl of FSW positioned in a sink (Chapter 4; Lowe and Wray, 2000; Strathmann, 1987b). The old FSW from the culture will overflow into the sink while the embryos remain on the mesh (for early cultures, use a 30 μm mesh; for early larva, use a 100 μm mesh; and for late larva, use a 200 μm mesh). Gently wash the embryos with several rinses of FSW. Gently pour the contents of the mesh-bottom beaker (or transfer the embryos with a turkey baster) into a clean culture dish with fresh FSW. Once feeding larva have developed, the water only needs to be changed every 2 to 3 days with the fresh FSW now containing food (see Chapter 4, and following text).

5. As an alternative to standing cultures, embryos may be raised in mesh-bottomed plastic beakers that are suspended in filtered flowing or recirculating seawater (see Strathmann, 1987b). The mesh size used must be less than the diameter of the eggs (see Table III).

6. Monitor the development of the culture microscopically. A healthy culture should develop fairly synchronously. For most urchin species, first cell division will occur within 90 to 120 min of insemination and then every 30 min thereafter up to the blastula stage. As has been noted, if polyspermy has occurred, the first cleavage may be irregular and the embryos will develop abnormally. By 3 to 4 days after fertilization, feeding pluteus larva will have developed. Larva can be fed *Rhodomonas lens* algae (Chapter 4). See Table II for the approximate times at which developmental stages occur in *S. purpuratus, L. pictus*, and *L. variegatus* embryos.

C. Notes

1. See Table II for appropriate temperatures for maintenance and culture of various sea urchin species. It is important that gametes are not exposed to a temperature shock and that insemination and culturing of embryos be conducted at the species-appropriate temperature.

2. See Appendix for breeding seasons to determine when particular species are gravid. Adult sea urchins can be obtained from the following suppliers:

Marinus Scientific
11771 St. Mark St.
Garden Grove, CA, 92845
Tel. 714-901-9700 (info@marinusscientific.com)

Pacific BioMarine Laboratories, Inc.
PO Box 536
Venice, CA 90291
Tel. 310-677-1056

Marine Biological Laboratory Resources Center
7 MBL Street
Woods Hole, MA, 02543-1015
Tel. 508-289-7700

Carolina Biological Supply Co.
2700 York Road
Burlington, NC 27215
Tel. 800-334-5551

Beaufort Biological
135 Duke Marine Lab Road
Beaufort, NC 28516
Tel. 252-504-7567, fax 252-504-7648

Sue Decker
141400 SW 22nd Place
Davie, FL 33325
Tel. 954-424-2620

Gulf Specimen Marine Laboratories, Inc.
222 Clark Ave., PO Box 237
Panacea, FL 32346
Tel. 850-984-5297 (gspecimen@sprintmail.com, www.gulfspecimen.org).

3. For fertilization and embryo culture experiments, it is usually sufficient to use natural seawater that has been filtered through a glass fiber filter and then a 0.2 or 0.45 micron filter (Millipore, Inc., Bedford, MA). Let filtered seawater (FSW) reabsorb O_2 for 12 to 24 hrs. before it is used. If natural seawater is not available, artificial seawater (ASW) may be used (Table I). Seawater should be pre-chilled or pre-warmed to the desired temperature before use (avoid exposing eggs or embryos to temperature shocks).

4. All glassware should be designated for use with gametes and embryos only. Many chemical reagents, including soaps and detergents, will cling to glassware. Glassware should be kept clean using water and a paper towel to wipe the inside. It is also a good idea to periodically treat the glassware with mild acid (dilute nitric acid works well) and then rinse extensively in deionized and then

distilled water. Finally, rinse the glassware in seawater (of the desired temperature) just prior to use. See Lutz and Inoué (1986) for more details on treating glassware.

5. Nylon meshes are useful for removing debris from gamete and embryo culture suspensions as well as for removing extracellular coats from eggs. Table III lists the average diameter of the eggs of various echinoderm species along with the size of nylon mesh that is recommended for use in a given procedure. A convenient way to make the mesh filtering apparatus is to cut the bottom out of a small plastic beaker or a 50 mL conical tube and affix a piece of the nylon mesh to the bottom using a glue gun. Alternatively, cut the mesh slightly larger than the opening and use a rubber band to secure it across the opening of the beaker or tube. Always pre-wet the nylon mesh in FSW prior to use. Be sure to backwash and thoroughly rinse the mesh after use.

6. ATAZ (a peroxidase inhibitor) should be prepared fresh as 0.5 M stock solution in ASW. It may be necessary to prepare the stock at room temperature, using a gentle rocking motion to dissolve the chemical completely. Alternatively, the ATAZ can be dissolved in distilled water. Dilute the stock solution of ATAZ into the desired volume of ASW at the appropriate temperature for a final concentration of 1 mM. Avoid using ATAZ with natural seawaters, as the activity will vary widely.

7. If not too much 0.5 mM KCl has been injected ($\leq$1 ml), the animals can recover after spawning. Place them in running seawater, in a bucket or tank that does not contain other urchins (to avoid group spawning) for a day and then transfer to a permanent tank. This is a good way to keep a stock of females and males separated. If urchins are fed a regular diet of *Macrocystis* (kelp; available from most urchin suppliers; see note II.C.2) they can be kept tanked for several years (Leahy, 1986; see also Chapter 2).

III. Method for Sea Stars (Class Asteroidea)

A. General Explanation

As is the case with sea urchins, sea stars have many advantages as a model system for studying oocyte maturation, fertilization, embryology, and the evolutionary aspects of development. These advantages include optically clear eggs and embryos, the ease of *in vitro* maturation and fertilization, the ability to obtain large quantities of gametes (one female sea star can provide millions of eggs), ease of oocyte microinjection, rapid rate of development, and a potentially informative evolutionary position as an invertebrate deuterostome. In the wild, male and female sea stars release gametes through gonopores into the seawater, where fertilization occurs externally. Development then proceeds according to one of several modes including: (i) pelagic development with feeding larvae; (ii) pelagic development with nonfeeding larvae; or (iii) brooding of nonfeeding larvae. In addition, at least one species of seastar (*Pteraster tesselatus*) undergoes direct development (McEdward, 1992).

This Section focuses on how to procure and mature oocytes, perform *in vitro* fertilization, and culture embryos from the sea star *Asterina miniata* (formally known as *Patiria miniata*). These colorful "vermilion" sea stars are commonly found in the Pacific intertidal and subtidal zones from Alaska to Baja California (Ricketts *et al.*, 1985). They are omnivorous scavengers, and due to "webbing" between their short triangular arms, are commonly called "bat stars." Their natural spawning season typically spans mid-May through August (Appendix). Kept in tanks and fed regularly on a diet of mussels and other invertebrates, the adults can remain gravid for several months longer. Following fertilization, *A. miniata* embryos develop by cleaving radially in typical deuterostome fashion to form a swimming blastula. The blastula then undergoes gastrulation to form a gut and create a feeding larva called a bipinnaria. After several weeks, the bipinnaria larva takes on a more elaborate form called a brachiolaria larva, which eventually attaches to a substrate and undergoes metamorphosis to create a junvenile seastar (Brusca and Brusca, 1990; Chia and Walker, 1991; Strathmann, 1987c). *A. miniata* has been an important model system for studying oocyte maturation (Chapter 5; Kanatani *et al.*, 1969; Kishimoto, 1998) and fertilization (reviewed in Jaffe *et al.*, 2001; Runft *et al.*, 2002; Stricker, 1999) and has also been used in embryological studies (cf. Hoegh-Guldberg and Manahan, 1995, and references in Strathmann, 1987c). The methods described in this section can be generally applied with minor modifications to other sea star species such as *Asterias forbesi* (Atlantic sea star) and *Pisaster ochraceus* (Pacific sea star; for specific protocols and seasonality for other sea star species, see Chia and Walker, 1991; Strathmann, 1987c). We also refer the reader to Lowe and Wray (2000) for a detailed protocol for culturing *P. ochraceus* embryos and larvae.

B. Detailed Method

1. Materials Needed

1. Gravid *A. miniata* sea stars (note III.C.1)
2. Filtered seawater (FSW) or artificial seawater (ASW) and calcium-free artificial seawater (CaFASW) (note II.C.3 in Method for Sea Urchins; Table I)
3. 3 mm sample corer (Fine Science Tools, Foster City, CA, model #18035-03)
4. Dissecting scissors (Noyes scissors are nice; Fine Science Tools, model #15012-12)
5. Curved forceps (Graefe forceps from Fine Science Tools, model #11051-10)
6. Razor blades
7. 1-methyladenine (Sigma, St. Louis, MO, catalog #M-4876; note III.C.2)
8. Antibiotics for long-term culturing (gentamicin or penicillin–streptomycin; In Vitrogen; catalog numbers 15750-060 and 15070-063, respectively)

9. Food for larvae (Chapter 4)
10. Glassware: 30 ml beakers, glass finger bowls, Pyrex baking dishes (note II.C.4 in Method for Sea Urchins)
11. 1.5 ml microfuge tubes
12. Pasteur pipets and pipet bulbs
13. Ice bucket
14. Glass slides and coverslips
15. Microcentrifuge
16. 200 ml plastic beaker
17. Nylon mesh (Table III)
18. Small plastic paddles for resuspending eggs and cultures
19. Compound light microscope with 10× objective

2. Collecting Immature Oocytes and Sperm (see note III.C.3)

1. Gravid sea stars usually have a "full" or plump look. Males and females typically cannot be distinguished externally, so the sex has to be determined by whether the animal contains ovary or testis. A pair of gonads runs along the length of each arm of the animal. To obtain a small amount of ovary or testis, push a 3 mm sample corer through the upper dorsal arm of the sea star. Sometimes this requires quite a bit of force (using a grinding motion often helps) to pop the sample corer through the sea star surface. See step III.B.2.4 for an alternative method for obtaining larger amounts of gonad. Finally, Section III.B.4.3 provides a method for obtaining a large amount of matured oocytes.

2. Using curved forceps, pull out a bit of gonad through the hole made in the arm using the sample corer. If the sea star is quite gravid, gonad material should spill out of the small hole. Ovary will be bright orange (occasionally yellow, light brown, or green) and testis will be white. If a greyish-brown, slick, soft substance is pulled out, this is probably part of the gut. Stuff the gut back into the seastar and probe with the forceps to find gonad. The gonad will generally have a plump, lobed appearance (gonad that is not healthy or is immature will look grainy and condensed). Squeezing the sea star will sometimes help force the gonad out of the hole made by the sample corer. Place pieces of ovary in filtered seawater (FSW) in a tall, narrow 30 to 50 mL glass beaker. Place testis fragments in a 1.5 ml microfuge tube on ice. This testis or "dry sperm" will be good all day on ice. Take care to wash the coring tool and forceps before coring a new animal to prevent the transfer of sperm.

3. Return the sea star to 15 °C seawater. Push down on the surface of the seastar to force out any air that may have entered through the sample corer hole. The sea star will heal within a few days and can be cored again many times in untouched arms or in arms that have completely healed.

4. To obtain a large amount of ovary or testis, use a razor blade to cut a slit between the arms of the sea star and retrieve the gonad lobes out of the arms using forceps. Place the ovary pieces in FSW in a glass beaker and put the testis in

microfuge tubes on ice. For some species of sea star, such as *Asterias forbesi*, the surface of the animal is rather brittle and weak, so using a coring tool will not work. Instead, a convenient way to obtain gonad material is to remove an arm from the sea star and dissect out the gonad. If a large number of matured oocytes (i.e., eggs) are desired for culturing or biochemical analysis, skip to Section III.B.4.3 in the following text.

5. Clip the ovary into small bits in FSW using small dissecting scissors. This will release the oocytes from the ovary. Using a Pasteur pipette, gently pick up some of the oocytes that have settled at the bottom of the beaker (often, damaged oocytes remain floating) and inspect them microscopically (10× objective) to ascertain their quality. Full grown, healthy oocytes are ~180 μm in diameter and are arrested in prophase I of meiosis I. As *A. miniata* oocytes are optically clear, a germinal vesicle (i.e., nucleus) of ~80 μm in diameter is clearly visible. Within the nucleus can be seen a clear round vesicle, which is the nucleolus. Healthy oocytes should have a smooth perimeter (no blebbing or lysing) and a homogenous, slightly grainy cytoplasm (no large dark spots or aggregates). Lying on top of the oocyte jelly (the jelly has a striated appearance) are flat, thin follicle cells. They interact with the oocyte plasma membrane via a projection (not visible at 10×) that extends through the jelly to the oocyte surface. It is through these projections that the follicle cells are believed to secrete 1-methyladenine (1-MA) onto the surface of the oocyte (Schroeder, 1981; Schroeder *et al.*, 1979). 1-MA is the "maturation hormone" that induces the oocyte to reenter the cell cycle and proceed through meiosis (Kanatani *et al.*, 1969). See Chapter 5 for a more detailed description of echinoderm oocyte maturation.

3. Defolliculation of Oocytes

1. In order to prevent the oocytes from spontaneously maturing (i.e., entering meiosis) due to the release of 1-MA by the follicle cells, the follicle cells can be removed using calcium-free artificial seawater (CaFASW). After mincing the ovary fragments and checking the quality of the oocytes, swirl the beaker of FSW they are in, let the big ovary chunks settle to the bottom of the beaker, and then decant the suspension of oocytes surrounded by follicle cells into a fresh 30 ml beaker (this is to separate the oocytes from large ovary fragments that may secrete 1-MA). Let the follicle-enclosed oocytes settle to the bottom of the new beaker, decant the FSW, and then add ice-cold CaFASW.

2. Place the beaker of CaFASW and oocytes on ice. The CaFASW will cause the follicle cells to slough off. Every few minutes decant the CaFASW, add more ice-cold CaFASW, and inspect the state of the oocytes using the microscope. When only 1/4 of the oocyte surface is covered by follicle cells, pour off the CaFASW and replace with room temperature FSW. Depending on the individual animal, the time in CaFASW may vary from just a few minutes to as long as 30 min (maximum length of time oocytes should be left in CaFASW). It is critical to get the oocytes out of the CaFASW as soon as most of the follicle cells have dispersed, because prolonged

exposure causes damage to the oocyte surface. If you know from previous experience that a particular sea star has good ovary, you can put the ovary fragment directly into ice cold CaFASW and begin the defolliculation process right away.

3. After the CaFASW treatment, wash the oocytes 3 times with room tempertaure FSW (any remaining follicle cells are generally moved by these FSW washes). Store the defolliculated oocytes in FSW at 15 to 18 °C. They can be used for experiments all day.

4. Maturation of Oocytes

1. In order for sea star oocytes to become fertilization-competent eggs, the oocytes must be exposed to the maturation hormone 1-methyladenine (1-MA; note III.C.2). This hormone stimulates the oocytes to enter meiosis and undergo other changes that are collectively referred to as "oocyte maturation." An obvious sign that an oocyte has been exposed to 1-MA and has begun oocyte maturation is germinal vesicle breakdown (GVBD). See Chapter 5 for details on echinoderm oocyte maturation.

2. To induce defolliculated oocytes to undergo oocyte maturation, replace the FSW they are in with a solution of 1 μM 1-MA made in ASW (note III.C.2). GVBD will take place 20 to 40 min after exposure to 1-MA (the time to GVBD is temperature dependent and will vary from animal to animal). GVBD refers to the moment when the perimeter of the nucleus appears to disintegrate. Soon after GVBD, the remainder of the nucleus and the nucleolus will disappear. Within 10 to 15 min after GVBD, the stage of metaphase I is reached. At metaphase I, the oocyte can now be fertilized and is referred to as a mature oocyte or an egg. Note that the egg remains at metaphase I for only a few minutes as it then continues on in the meiotic cycle regardless of whether or not it is fertilized. If the egg is fertilized, it will finish meiosis and then begin the mitotic cleavages of embryogenesis. If an egg is not fertilized, it will finish meiosis and then undergo apoptosis at roughly 10 h after 1-MA exposure (Sasaki and Chiba, 2001; Yuce and Sadler, 2001).

3. An alternative way to obtain mature oocytes (especially when a large number of eggs is desired) is to collect the ovary in FSW, clip any large pieces using small scissors, and apply the 1 μM 1-MA solution (note III.C.2) directly to the ovary fragments (skipping the CaFASW defolliculation step). This will cause the ovary to ovulate defolliculated, maturing oocytes within about 30 to 60 min. Monitor the ovary closely for the release of the eggs and for GVBD. Transfer the released eggs (using a wide orifice pipette) into a beaker of FSW. Mature oocytes can also be obtained by inducing adult sea stars to spawn *in vivo* matured eggs.[2] Sea stars can be spawned either by coelomic injection near the central disc of 1 ml of radial nerve extract (RNE) or 1 ml of 100 μM 1-methyladenine for each 100 ml of estimated body volume. Spawning commences 20 to 30 min or sooner after injection. Spermatozoa should

[2]We thank our colleagues S. A. Böttger, C. W. Walker, and T. Unuma for providing the procedures for preparing RNE and *in vivo* spawning.

be collected "dry" either by pipetting into a collection tube on ice or by inverting males over a chilled biologically clean container. Females can be maintained in small glass dishes filled with filtered seawater and eggs should be collected either by gentle pipetting or by filtering seawater through a 100 μm mesh (for detailed spawning protocols, see Lowe and Wray, 2000; Strathmann, 1987c; also see note III.C.2).

5. Fertilization

1. For optimal fertilization, insemination should be performed at metaphase I (10–15 min after GVBD) though eggs can be fertilized at any time during meiosis I. Eggs finish meiosis I at about 1 hr after GVBD when the first polar body is extruded. Once eggs proceed beyond the first meiotic division, they are referred to as "over mature" because they often become polyspermic if inseminated.

2. If eggs were in 1-MA, decant the 1-MA solution and add fresh FSW for a 10% (volume/volume) suspension in a glass beaker. Clip the "dry" testis (in a microfuge tube on ice) with small dissecting scissors. Spin the testis for 30 s at top speed in a microcentrifuge to separate sperm from testis tissue. Right before insemination, dilute the sperm 1:1000 in FSW. Check that the diluted sperm have good motility by placing a drop on a slide and observing the sperm under a 10× or 20× objective. Then add the appropriate volume of 1:1000 sperm to the eggs so that the final concentration of sperm is 1:10,000. Insemination can be performed at 15 °C up to room temperature. Mix the sperm–egg mixture gently for a few minutes using a small paddle. Then let the eggs settle and gently wash the eggs twice with large volumes of FSW to remove excess sperm. As is the case with sea urchins (Section II.B.3.1a), it may be necessary to perform small-scale test fertilizations to optimize the sperm dilution.

3. Almost immediately after adding sperm, put a drop of inseminated eggs on a slide for viewing under a microscope with a 10× objective. Within 30 s of adding sperm, the majority of eggs should begin to raise a fertilization envelope (FE). The fertilization envelope of *A. miniata* eggs elevates slowly and is less visible compared to sea urchin eggs and will not be fully elevated until several minutes after fertilization. Roughly 5 to 10 min after the FE has fully elevated, the fertilization cone making the site of sperm entry may become visible. The fertilization cone will look like a thin cytoplasmic strand reaching out from the egg surface to the edge of the elevated fertilization envelope. Multiple fertilization cones are an indication of polyspermy.

4. If fewer than 95% of the eggs raise fertilization envelopes by 5 min after insemination or if the majority of the eggs are polyspermic, throw out this particular insemination mix and try again using new eggs and/or sperm from different animals.

6. Embryo Cultures

1. Follow the same procedure described in Section II.B.4, Method for Sea Urchins. Cultures of *A. miniata* should be raised at 15 to 18 °C (Table II).

2. Monitor the development of the culture by looking at embryos under a 10× objective. A healthy culture should develop fairly synchronously. In *A. miniata*, the first cleavage occurs at about 2 to 2.5 h after fertilization at 15 to 18 °C. If polyspermy has occurred, the first cleavage may be irregular and the embryos will develop abnormally. By 3 to 4 days after fertilization, feeding bipinnaria larva will have developed. Larva can be fed *Rhodomonas lens* algae (Chapter 4). Table II summarizes the approximate times at which developmental stages occur in *A. miniata* embryos raised at 15 °C (Strathmann, 1987c).

C. Notes

1. Most sea urchin suppliers (see note II.C.2) can also provide sea stars. Adult *A. miniata* generally have full grown oocytes (~180 μm diameter) between the beginning of May through the end of October. Adults can be kept in an aquarium at ~15 °C with a flow-through seawater system or a recirculating seawater system. Ideally, the temperature and salinity of the seawater should be similar to what the animals are used to in the field. Note that animals may spawn if warmed or if handled roughly. Remove spawning animals from the aquarium to prevent spawning in the other sea stars. *A. miniata* should be fed periodically to maintain ripe gonads (mussels or pieces of thawed seafood from a frozen seafood mix found in grocery stores works well). For more information on aquaculture of adult sea stars, see Chapter 2 and Strathmann (1987c).

2. For oocyte maturation, prepare a 10 μM solution of 1-methyladenine in ASW. Make small (100 μL) aliquots and store at −20 °C. Stock aliquots usually retain activity for over 1 year if not thawed and refrozen. Use 1-MA at a final concentration of 1 μM. Higher concentrations (2–10 μM) can be used, but typically are not necessary. For spawning, use 1-MA as a stock of 100 μM in distilled water. Radial nerve extract (RNE) can also be used to induce oocyte maturation or spawning. Methods for preparing radial nerve extract are described in Strathmann (1987c) and in Strathmann and Sato (1969). In brief, radial nerve extract can be prepared by relaxing an adult sea star in 8% $MgCl_2$ in distilled water. In this condition, the tube feet, which occur on both sides of the ambulacra, can be reflexed to either side of the ambulacrum, thus exposing the nerve as a white or cream-colored line. Grasp an individual nerve with tweezers at the tip of each ray and gently pull along the length of the ray toward the disc. Immediately freeze the isolated nerves (5 nerves can be isolated per adult sea star). Radial nerve extract can then be prepared by any one of the following three methods and can be stored at minus 15 °C for one year.

   a. Macerate nerves in 1 ml distilled water/nerve, and allow them to lyse for 30 to 60 min. Centrifuge the tissue, collect the supernatant, and dilute with an equal volume of 1 M NaCl or sterile double strength seawater. Aliquot this RNE and store frozen.

b. Heat the isolated nerves in filtered seawater at 76 °C for 2 min, then centrifuge, and collect the supernatant as RNE.

c. Wash the isolated nerves in chilled seawater, lyophilize, and then pulverize using a mortar and pestle. Lyophilized nerve extract should be dissolved at 5 mg/100 ml sterile seawater prior to use.

3. Much of the information for procuring, defolliculating, and maturing sea star oocytes was taken with permission from L. A. Jaffe from the on-line microinjection manual located at: http://terasaki.uchc.edu/panda/injection/ (see also Chapter 10).

IV. Method for Brittle Stars (Class Ophiuroidea, Order Ophiurida)

A. General Explanation

The class Ophiuroidea is divided into two orders: Ophiurida, the brittle stars, and Phrynophiurida, the basket stars. This Section focuses on the brittle stars, as very little work has examined acquisition of gametes from the basket stars. Brittle stars arguably are the most diverse group of echinoderms, exhibiting an extensive array of distinct reproductive and developmental modes, which makes them difficult to characterize in general terms (reviewed in Byrne and Selvakumaraswamy, 2002; Hendler, 1991; Strathmann and Rumrill, 1987). These organisms are notoriously difficult ones from which to isolate gametes and there is not one simple protocol that holds for all species. This Section describes the various modes of development and the methods for handling gametes from species that are most reliable for cellular and developmental biology studies.

The ophiuroids can reproduce asexually by a process called fissiparity (splitting across the plane of the disc; cf. Mladenov and Emson, 1984; Mladenov *et al.*, 1983), but most species reproduce sexually. Although most species are dioecious, some are hermaphrodites. As is the case with many of the Echinodermata, development may be direct or indirect and some species brood. McEdward and Miner (2001) have reorganized and described in detail the seven basic patterns of reproduction and development exhibited by the ophiuroids, originally outlined by Strathmann and Rumrill (1987). Indirect developing phiuroids have three primary larval types: ophiopluteus, vitellaria, and doliolaria. In addition, some ophiuriods undergo direct development (reviewed in Hendler, 1991; Strathmann and Rumrill, 1987). The four general patterns of larval development include: (i) pelagic with planktotrophic larvae; (ii) pelagic or demersal development with lecithotrophic larvae; (iii) benthic lecithotrophic development; or (iv) direct development.

Ophiuroid ova undergo germinal vesicle breakdown (GVBD) prior to or soon after spawning in most species, regardless of developmental mode (reviewed in Hendler, 1991; Strathmann and Rumrill, 1987; cf. Fenaux 1970; Holland, 1979; Yamashita, 1985). Species with pelagic, plankotrophic larvae tend to shed thousands of small, oligolecithal eggs (70–200 μm diameter) and undergo external fertilization. They develop in the plankton into ophioplutei and have a pelagic period of about 20 to 90 days, depending on the water temperature (Byrne and Selvakumaraswamy, 2002; Hendler, 1975, 1991; Strathmann and Rumrill, 1987). Other species with pelagic larval stages are lecithotrophic. These species tend to shed several thousand medium- to large-sized eggs (130–420 μm diameter) which are nutrient rich and opaque. Development into nonfeeding larvae with pelagic periods of roughly 1 week or less occurs in the water column or near the bottom (Byrne and Selvakumaraswamy, 2002). Interestingly, there are species that exhibit benthic lecithotrophic development. After spawning, the several thousand medium eggs (250–300 μm) adhere to firm surfaces. A reduced pluteus develops that then progresses to the vitellariae stage before hatching, or develops directly to a modified, nonfeeding vitellariae (Hendler, 1991). Species with direct development exhibit both external and internal brooding (reviewed in Strathmann and Rumrill, 1987; Hendler, 1991).

As evidenced even from this extremely brief overview, this class of Echinodermata is quite diverse. Here, we focus on methods for obtaining gametes and culturing larvae from fairly common, free-spawning species of ophiuroids with pelagic planktotrophic or lecithotrophic larvae, specifically, *Ophioderma brevispinum, Ophiopholis aculeata, Ophiothrix angulata*, and *Ophiothris fragilis*. (Byrne and Selvakumaraswamy, 2000, 2002; Selvakumaraswamy and Byrne, 1995, 2000) have published dctailed, successful protocols for working with the Australian species *Ophiactis resiliens, Ophiothrix caespitosa*, and *Ophiothrix spongicola*, which may apply to other species as well, and we refer to their methods where appropriate. We also refer the reader to more extensive reviews (Hendler, 1991; Strathmann and Rumrill, 1987) for handling other species of brittle stars.

B. Detailed Methods

1. Materials Needed

1. Gravid brittle stars. See note IV.C.1 and Appendix for gravid seasons.
2. Filtered seawater (FSW), artificial seawater (ASW), and CaFASW (note II.C.3)
3. Antibiotics for long-term culturing (gentamicin or penicillin–streptomycin; In Vitrogen; catalog numbers 15750-060 and 15070-063, respectively)
4. Food for larvae (Chapter 4)
5. Glassware: 30 ml beakers, glass finger bowls, Pyrex baking dishes, small petri dishes (note II.C.4)
6. 1.5 ml microfuge tubes
7. Pasteur pipets and pipet bulbs
8. Ice bucket
9. Glass slides and coverslips

10. Microcentrifuge
11. Nylon mesh (see Table III)
12. Small plastic paddles for resuspension of eggs and cultures
13. Compound light microscope with 10× objective

2. Obtaining Gametes

1a. Gravid individuals can be recognized by swelling of the gonads or bursae. Unlike the Asteroidea, the gonads of Opiuroids are typically located in the disc, not the rays. Identification of males and females is possible in some species, based on the color of the gametes and gonads, which often are visible through the epidermis on the oral side. Testes are typically white, with some exceptions (yellow, pink, blood red) and the color of ova varies widely across species (see Hendler, 1991, for a comprehensive list). Sexual dimorphism is rare, but has been described as variation in shape of genital papillae size. In many cases, spawning is the only way to distinguish the two sexes. Recently collected, gravid animals may spawn naturally when tanked, and thus should be closely monitored. It is important to note that most females may not spawn in isolation (Selvakumaraswamy and Byrne, 2000). A range of stimulus tactics to artificially induce spawning in gravid ophiuroids have been attempted, but there are few accounts of successful, controlled spawning and it is unlikely that spawning or successful fertilization will be achieved during times when the animals are not in season (Lowe, personal communication). The most successful spawning techniques reported are to place individuals in a small bowl of FSW and expose them to stressors, including temperature, light, or mechanical shock, alone or in combination. Sperm can be isolated from dissected, ripe gonads (see Section V.B.2.1b in the following text regarding sea cucumber dissections), but fertilizable oocytes typically cannot be obtained by this route.

1b. *Light shock.* Despite the assumption that most ophiuroids are nocturnal spawners, some species have been induced to spawn by exposure to high intensity light for brief periods and then moving them to the dark. Place the animal oral side up in a small bowl of FSW. Shine intense light (a 75-watt bulb held about 1 foot away from the animal) on the oral surface of the animal for 15 to 30 min, then transfer to the dark for 15 to 30 min. Monitor closely for gamete release. Repeat as necessary for 2 to 3 h.

1c. *Temperature shock.* One of the most reliable spawning techniques reported has been to elevate the seawater temperature for brief periods, either alone or in combination with light shock. Temperate species have been induced to spawn when they are exposed to indirect daylight and placed in seawater 10 °C above ambient seawater temperature for 5 to 10 min, and then transferred to indoor darkened tanks with briskly flowing seawater for 30 min (reviewed by Strathmann and Rumrill, 1987).

1d. *Combination of stressors.* Selvakumaraswamy and Byrne (2000) report success in inducing spawning by transferring multiple individuals from ambient temperature seawater 19 to 23 °C to 34 °C (>10 °C increase) repeatedly for a few seconds then holding the animals in the dark for 30 min in contact with aerated seawater at ambient temperature. Then, use a combination of exposing the brittle stars to artificial light and disturbance by shaking. Repeat this routine every 30 min for 2 to 3 h or until spawning is observed.

2. Invert spawning animals in a bowl, oral side down. The gametes are shed through bursal slits on the oral surface. Sperm typically are released as a dense concentration in ribbons from the bursae and become motile immediately upon release in seawater. The sperm are active for several hours (Hendler, 1977). Alternatively, sperm may be collected "dry" by pipetting and kept on ice for later use. Oocytes are released in loose ribbons that break up soon after release. Once released, they sink or float, depending on the species. Because females may not spawn in isolation, the numbers of eggs spawned per female often must be estimated. Transfer the eggs to a small beaker of FSW and wash several times. Visualize under the microscope to ensure that GVBD has occurred and that the eggs are generally healthy (no obvious lysing or damage).

3. Fertilization

1. As with many echinoderms, the optimal time for insemination is very soon after GVBD but before first polar body formation. As has been noted, spawned oocytes usually have undergone GVBD, so eggs should be fertilized as soon as possible after spawning (Hendler, 1991; Yamashita, 1985). The temperature may influence the rate and success of fertilization; higher temperatures may facilitate polyspermy (see Table II for temperatures at which various species should be inseminated).

2. Follow the insemination protocol for Asteroidea provided in Section III.B.5.

3. Brittle star eggs undergo a cortical reaction when fertilized, and the elevation of the fertilization envelope should be monitored microscopically. In general, the brittle stars exhibit less-synchronous fertilization than do other echinoderms.

4. Culturing Embryos

1. For species with feeding larvae, see Lowe and Wray (2000) and Chapter 4. Follow the basic procedures as outlined in Section II.B.4 for sea urchin embryos. See note IV.C.2. for details about developmental stages.

2. For nonfeeding (lecithotrophic) larval species, culture in glass baking dishes or large beakers, changing the FSW every day. See Chapter 4 for details.

C. Notes

1. Intertidal ophiuroids are typically found under rocks on sand or mud, in kelp hold-fasts, and in eelgrass root mats. Subtidal species can be collected from sandy bottoms by diving or dredging. Animals must be handled as gently as possible to prevent damage to fragile arms. Transfer adults to seawater tanks, preferably with a sandy or rocky substrate, at the ambient temperature where collected. Although brittle stars can be obtained commercially (most sea urchin and sea star suppliers can also provide brittle stars; see note II.C.2), it is important to realize that gravid animals may spawn easily during shipping due to stress.

2. Table II provides the optimal temperature for culturing embryos of various species and also a timetable of development. As with other echinoderms, once gametes are collected, avoid temperature shocks to gametes and to the embryos.

V. Method for Sea Cucumbers (Class Holothuroidea)

A. General Explanation

Sea cucumbers have gained importance recently as yet another example of echinoderm diversity in larval morphology and life history (McEdward, 1995) and in comparative studies of the evolution of the metazoan body plan (Lowe *et al.*, 2002; Wray, 2000). Aquaculture of Holothuroideans is of economic importance in many countries as a commercial enterprise. In particular, a number of aquaculture techniques and larval rearing studies have been published for commercially important species (cf. Ramofafia *et al.*, 1997). In general, adults of most sea cucumber species do well in flow-through seawater systems when provided with natural substrata. Sea cucumber species tend to have separate sexes, though they are not easily distinguished by external morphology (cf. McEuen, 1987). There are a few externally brooding species (most notably, in a few *Cucumaria spp.*), and even internal brooders (e.g., *Leptosynapta clarki*), but most sea cucumbers broadcast spawn and have free-swimming larva. Many have nonfeeding (lecithotrophic) larva, but a few species have feeding larva (McEuen, 1987). Smiley *et al.* (1991) provide an excellent overview of the reproductive biology and larval development of this class of Echinodermata.

In general, there is no reliable way to induce spawning or successful oocyte maturation and fertilization in sea cucumbers unless they are in season (Lowe, personal communication; McEuen, 1987; Smiley, 1988). Indeed, in some species, spawning cannot be induced even in gravid animals at the peak of the season and gonads must be dissected in order to obtain gametes. The protocol below is based on working with two subtidal North American Pacific species, *Psolus chitonoides* and *Parastichopus californicus. P. chitonoides* tends to spawn in the early morning hours in response to light during their gravid period (mid-March to mid-May; McEuen, 1987) and to form lecithotrophic (nonfeeding) larvae, which are easily cultured. *P. californicus* typically cannot be induced to spawn in the laboratory, but instead must be dissected in order to obtain gametes. Care must be taken when handling *P. californicus* as it will self-eviscerate in response to stressful conditions such as rough handling, warm temperatures, or overcrowding. *P. californicus* tends to be gravid from mid-May to mid-July (depending on latitude; McEuen, 1987) and has feeding larvae. In both *P. chitonoides*, and *P. californicus*, metamorphosis begins when the larval body decreases in size and forms a barrel-shaped form called a doliolaria. Eventually, the doliolaria will form tentacles (the pentactula stage) and settle on a substrate to complete metamorphosis. Most species will settle and finish metamorphosis in glass bowls without the need of natural substrata (see McEuen, 1987, for exceptions). The protocols described here can be extrapolated to other Holothuroidea species, keeping in mind that successful fertilization and culturing may be achieved only for a few weeks out of the year, during the gravid period. A few species (especially *Cucumaria spp.*) spawn their sperm in long strands or bundles and may require patience in diluting. Finally, some species have negatively, and others, positively, buoyant eggs. If eggs float, wash by filtering through nylon mesh. If eggs settle, they can be easily washed by gravity settling and decanting, as with sea urchin and sea star eggs (Sections II.B and III.B).

B. Detailed Methods

1. Materials Needed

1. Gravid sea cucumbers. See note V.C.1 for sources and reproductive seasons.
2. Filtered seawater (FSW), artificial seawater (ASW), and CaFASW (Note II.C.3)
3. Dissecting scissors (Noyes scissors; Fine Science Tools, model 15012-12)
4. Curved forceps (Graefe forceps; Fine Science Tools, model 11051-10)
5. Razor blades
6. Antibiotics for long-term culturing (gentamicin or penicillin-streptomycin; In Vitrogen; catalog numbers 15750-060 and 15070-063, respectively)
7. Food for larvae (Chapter 4)
8. Glassware: 30 ml beakers, glass finger bowls, Pyrex baking dishes, small petri dishes (note II.C.4)
9. 1.5 ml microfuge tubes
10. Pasteur pipets and pipet bulbs
11. Ice bucket
12. Glass slides and coverslips

13. Microcentrifuge
14. Nylon mesh (see Table III)
15. Small plastic paddles for resuspending eggs and cultures
16. Compound light microscope with 10× objective
17. Radial Nerve Factor/Extract (RNE) for inducing maturation (III.C.2 and Chapter 5).

2. Collecting Gametes

1a. For species that spawn in response to light, individuals should be collected late in the day during the gravid season (note V.C.1) and placed in seawater tables or shallow tanks in the dark. The next morning, allow sunlight to fall on the tanks and monitor gamete release closely. Gamete release typically occurs within ~45 min. Warming the water by shutting off water flow may enhance spawning and may even induce spawning during afternoon/evening hours (McEuen, 1987). Keep females and males separate (in small dishes) as soon as spawning initiates. Collect the gametes directly from the spawning animals. *P. chitonoides* eggs are ~625 μm in diameter and will appear brick red in color and should be collected into a beaker. The eggs will be maturing as they are spawned. Monitor them for the absence of a germinal vesicle and the absence of follicle cells. Filter the eggs through nylon mesh (Table III) to remove remaining follicle cells and debris. Sperm should be pipetted directly from the gonopores of spawning males into plastic microfuge tubes and kept on ice. See note V.C.2 for information about temperature optima.

1b. For species that cannot be induced to spawn, the adults must be dissected and the large, ripe ovaries (orange in color) or testes (milky white in color) removed. Cut the animal open lengthwise, away from the mid-dorsal axis (to avoid cutting the gonaduct), using a razor blade or scalpel and use a small pair of scissors to dissect the gonad. Remove the entire gonad using forceps. Place the testis in a small petri dish on ice. Concentrated sperm can be obtained by easy rupture of the testis wall with a pipet tip, transferred to a microfuge tube, and kept on ice. If necessary, centrifuge the sperm suspension at top speed in microfuge for a few seconds to pellet any testis tissue debris, leaving the sperm in suspension. If the animal is very ripe, the ovary appears turgid and glassy (McEuen, 1987). Place ovary pieces in a small beaker of CaFASW at room temperature. Some oocytes may be released spontaneously. Fully grown *P. californicus* oocytes are ~204 μm in diameter.

3. Maturation of Oocytes

1. For species (such as *P. californicus*) that require dissection to obtain gametes, the oocytes must be matured, as they are arrested in meiotic prophase I in the ovary (Chapter 5; Smiley, 1988). Rinse the pieces of dissected ovary in CaFASW and transfer to a small glass beaker of ASW.

2. *In vivo*, holothuroid follicle cells are stimulated upon spawning to produce a low molecular weight substance that induces oocyte maturation. However, the identity of this "maturation substance" has not been determined (see McEuen, 1987). To mature holothuroid oocytes *in vitro*, radial nerve factor or extract (RNE) prepared from a sea star can be used to stimulate maturation. Prepare the RNF extract as detailed in Section III.C.2, Chapter 5, and in Strathmann and Sato (1969) and incubate the ovary pieces in the extract (0.2–0.3 mg/mL RNE) at room temperature. Mature oocytes will be naturally shed from the ovary. Collect the oocytes and inspect them for general health (no lysis or membrane blebbing) and maturation (germinal vesicle breakdown). It may take several hours for GVBD to occur. After GVBD, transfer the oocytes to a small beaker of FSW and wash several times in FSW.

4. Fertilization

1. As with many echinoderms, the optimal time for insemination is very soon after GVBD, but before first polar body formation. The temperature may influence the rate and success of fertilization; higher temperatures may facilitate polyspermy (see Table II for temperatures at which various species should be inseminated). *P. chitonoides*, and *P. californicus* should be inseminated and early cultures raised at ~10 to 12 °C.

2. Follow the fertilization protocol for Asteroidea, Section III.B.5.

5. Culturing Embryos

1. For species with feeding larvae (such as *P. californicus*), see Lowe and Wray (2000) and Chapter 4. Follow the basic procedures as outlined in Section II.B.4 for sea urchin embryos. See note V.C.2. for details about developmental stages.

2. For nonfeeding (lecithotrophic) larval species, culture in glass baking dishes or large beakers, changing the FSW every day. See Chapter 4 for details.

C. Notes

1. Most sea urchin suppliers can also supply sea cucumbers. Refer to note II.C.2 for vendors. See Appendix for seasonal information.

2. Table II provides the optimal temperature for culturing embryos of various holothuroidean species and also a timetable of development. As with other echinoderms, avoid temperature shocks to gametes and to embryos.

VI. Method for Interspecific Crosses

A. General Explanation

Most hybridization studies that have been performed in echinoderms have been carried out in sea urchins. Boveri (1889) used such crosses in some of the initial studies that revealed the differential contribution of the maternal and paternal genomes to the developing embryo. In addition to studying the regulation of gene expression during embryogenesis in terms of the contribution of maternal vs zygotic information to developmental events, crossing two different species of sea urchin in an attempt to create a hybrid has been used to study a variety of biological problems. These include analyses of sperm–egg interactions, the evolution of reproductive barriers, and the evolutionary relationship between different species (Brandhorst and Davenport, 2001; Brandhorst *et al.*, 1991; Chen and Baltzer, 1975; Giudice, 1973; Lessios and Cunningham, 1990; Palumbi, 1992, 1996; Palumbi and Metz, 1991; Palumbi *et al.*, 1997; Raff *et al.*, 1999; Rahman *et al.*, 2001; Uehara *et al.*, 1990). Cross-fertilization between different sea urchin species is not common, although some natural cross-fertilization between congeners (such as *S. purpuratus* and *S. franciscanus*) can occur naturally (Strathmann, 1987b). In general, intraclass crosses are dramatically improved when the egg vitelline layer is first modified by proteases (see Section VI.B.3.a), suggesting that this extracellular layer contributes to the extremely species-specific gamete interactions observed in these free-spawning organisms (see Summers and Hylander, 1975). There are some published examples of crosses between different classes of echinoderms, but these rarely progress beyond the initial sperm–egg interaction, although there are exceptions (cf. Osani and Kyozuka, 1982). Giudice (1973) provides an excellent and thorough review of sea urchin intra- and interclass hybrids, both naturally occurring and experimentally induced. This Section briefly describes how to set up a reciprocal hybrid cross of *S. purpuratus* x *L. pictus*. See Table IV for references that describe methods for creating interspecific hybrids using other sea urchin species. In some cases, the cross will work only in one direction.

B. Detailed Method

1. Materials Needed

1. Gravid *S. purpuratus* and *L. pictus* animals (note II.C.1)
2. Materials listed in Section II.B.1, Method for Sea Urchins
3. Trypsin (Sigma, catalog #T4665)

2. Obtaining Gametes

1. Procure eggs and sperm as described in Section II.B.2, Method for Sea Urchins.

Table IV
Interspecific Crosses of Echinodermata[a]

Interspecies cross	Developmental stage reached by hybrid	Reference
S. purpuratus X *L. pictus* (and reciprocal)	Pluteus larva	Brandhorst and Davenport, 2001; Conlon *et al.*, 1987
S. purpuratus X *S. franciscanus* (and reciprocal)	Pluteus larva	Giudice, 1973; Moore, 1943
S. purpuratus eggs X *S. droebachiensis* sperm	Blastula[b]	Crain and Bushman, 1983
S. droebachiensis eggs X *S. purpuratus* sperm	Prism larva	
L. variegatus eggs X *S. purpuratus* sperm	Prism larva	Crain and Bushman, 1983
S. purpuratus eggs X *L. variegatus* sperm	Gastrula	
S. pallidus X *S. droebachiensis* (and reciprocal)	Pluteus larva	Giudice, 1973; Strathmann, 1981
S. purpuratus eggs X *Dendraster excentricus* sperm	Pluteus larva	Giudice, 1973
Paracentrotus lividus eggs X *Arbacia lixula* sperm	Gastrula	Giudice, 1973
Arbacia punctulata X *L. variegatus* (and reciprocal)	Pluteus larva	Giudice, 1973
Arbacia punctulata eggs X *S. droebachiensis* sperm	Pluteus larva	Giudice, 1973
Asterias forbesii eggs X *Arbacia punctulata* sperm	Gastrula	Giudice, 1973
Heliocidaris erythrogramma eggs X *H. tuberculata* sperm	Juvenile adults	Raff *et al.*, 1999

[a]See Giudice (1973) and Strathmann (1987b) for comprehensive reviews and further information on interspecific and interclass crosses.

[b]This cross also has been reported to result in development to pluteus stage larvae (Tufaro and Brandhorst, 1982).

3. Fertilization

1. Removing or modifying the egg surface (jelly layer and vitelline layer) by trypsin treatment or exposure to acid seawater (see note VI.C.1) will often facilitate fertilization by heterologous sperm (Brandhorst and Davenport, 2001). In addition, exposing the sperm to conspecific jellywater prior to their addition to the heterospecific eggs may facilitate the interaction as well (see Section II.B.3.1b). For a cross of *S. purpuratus* x *L. pictus*, treat *S. purpuratus* eggs with 125 μg/ml trypsin in ASW for 30 s at 10 to 12 °C and *L. pictus* eggs with 250 μg/ml trypsin in ASW for 5 min at 16 °C. Following the trypsin treatment, dilute the egg suspension two-fold with FSW. Let the eggs settle, decant the FSW, and then wash the eggs 2× more with large volumes of FSW.

2. In general, if overwhelming amounts of sperm are used to inseminate eggs, different sea urchin species can be induced to fertilize one another (Strathmann, 1987b). For a cross of *S. purpuratus* (eggs or sperm) x *L. pictus* (eggs or sperm), inseminate eggs so that the final concentration of sperm is 1:500 (Brandhorst and

Davenport, 2001). Incubate the sperm and eggs in suspension for 30 min with gentle rocking or stirring at 10 to 12 °C for *S. purpuratus* eggs and at 14 to 16 °C for *L. pictus* eggs. After 30 min, wash the fertilized eggs twice with large volumes of seawater, using gravity settling to remove excess sperm.

4. Embryo Culture

1. Set up cultures as described in Section II.B.4, Method for Sea Urchins. Typically, the embryos are best cultured at the temperature optimal for the egg-derived species (note VI.C.2). For example, for a cross in which *S. purpuratus* eggs were used, grow cultures at 10 to 12 °C and for a cross in which *L. pictus* eggs were used, grow cultures at 14 to 18 °C.

2. In many sea urchin interspecific crosses, development initiates but later arrests in early developmental stages or at the larval stage (Strathmann, 1987b). In the case of an *S. purpuratus* x *L. pictus* cross (reciprocal), plutei larva can be obtained. Interestingly, studies have indicated that expression of many *L. pictus* genes is reduced in embryos obtained from an *S. purpuratus* × *L. pictus* cross, regardless of whether the *L. pictus* genome comes from the egg or the sperm (Brandhorst and Davenport, 2001; Brandhorst *et al.*, 1991).

C. Notes

1. One protocol recommends treating sea urchin eggs with FSW or ASW acidified to pH 5.0 with glacial acetic acid (Raff *et al.*, 1999). Incubate sea urchin eggs 45 s in the acidified seawater, then rinse with normal FSW.

2. See Table II for optimal temperatures at which to culture embryos of various sea urchin species.

VII. Summary

The protocols outlined here hopefully will provide researchers with healthy, beautiful echinoderm oocytes, eggs, and embryos for experimental use. The large size of echinoderm oocytes and eggs, the ease with which they can be manipulated, and (in many species) their optical clarity, make them an ideal model system for studying not only the events specific to oocyte maturation and fertilization, but also for investigating more general questions regarding cell cycle regulation in an *in vivo* system. The quick rate at which development proceeds after fertilization to produce transparent embryos and larva makes the echinoderm an advantageous organism for studying deuterostome embryogenesis. Continued use of the echinoderms as model systems will undoubtedly uncover exciting answers to questions regarding fertilization, cell cycle regulation, morphogenesis, and how developmental events are controlled.

References

Angerer, L. M., and Angerer, R. C. (2000). Animal–vegetal axis patterning mechanisms in the early sea urchin embryo. *Dev. Biol.* **218,** 1–12.

Angerer, L. M., and Angerer, R. C. (2003). Patterning the sea urchin embryo: Gene regulatory networks, signaling pathways, and cellular interactions. *Curr. Top. Dev. Biol.* **53,** 159–198.

Boveri, T. (1889). Ein geschlechtich erzeugter Organismus ohne mütterliche Eigenschaften. Sitz. Ges. Morph. Phys. München 5. Translated by T. H. Morgan 1893. An organism produced sexually without characteristics of the mother. *Am. Nat.* **27,** 222–232.

Brandhorst, B. P., and Davenport, R. (2001). Skeletogenesis in sea urchin interordinal embryos. *Cell Tissue Res.* **305,** 159–167.

Brandhorst, B. P., Filion, M., Nisson, P. E., and Crain, W. R., Jr. (1991). Restricted expression of the *Lytechinus pictus* Spec 1 gene homologue in reciprocal hybrid embryos with *Strongylocentrotus purpuratus*. *Dev. Biol.* **144,** 405–411.

Brusca, R. C., and Brusca, G. J. (1990). "Invertebrates." Sinauer Associates, Inc. Publishers, Sunderland, MA.

Byrne, M., and Selvakumaraswamy, P. (2000). Vestigal ophiopluteal structures in the lecithotrophic larvae of *Ophionereis schayeri*. *Biol. Bull.* **198,** 379–386.

Byrne, M., and Selvakumaraswamy, P. (2002). Phylum Echinodermata: Ophiuroidea. *In* "Atlas of Marine Invertebrate Larvae" (C. M. Young, M. A. Sewell, and M. E. Rice, eds.), pp. 483–512. Academic Press, San Diego, C.

Chen, P. S., and Baltzer, F. (1975). Morphology and biochemistry of diploid and androgenetic (merogonic) hybrids. *In* "The Sea Urchin Embryo: Biochemistry and Morphogenesis" (G. Czihak, ed.), pp. 424–472. Springer-Verlag, Berlin.

Chia, F. S., and Walker, C. W. (1991). Echinodermata: Asteroidea *In* "Reproduction of Marine Invertebrates" (A. C. Giese, J. S. Pearse, and V. B. Pearse, eds.), Vol. VI, pp. 301–353. Boxwood Press, Pacific Grove, CA

Coffman, J. A., and Leahy, P. S. (2000). Large-scale culture and preparation of sea urchin embryos for isolation of transcriptional regulatory proteins. *In* "Developmental Biology Protocols" (Rocky S. Tuan and Cecilia W. Lo, eds.), Vol. 1, pp. 17–23. Humana Press, Totowa, NJ.

Conlon, R. A., Tufaro, F., and Brandhorst, B. P. (1987). Post-transcriptional restriction of gene expression in sea urchin interspecies hybrid embryos. *Genes Dev.* **1,** 337–346.

Crain, W. R., Jr., and Bushman, F. D. (1983). Transcripts of paternal and maternal actin gene alleles are present in interspecific sea urchin embryo hybrids. *Dev. Biol.* **100,** 190–196.

Darszon, A., Lievano, A., and Beltran, C. (1996). Ion channels: Key elements in gamete signaling. *Curr. Top. Dev. Biol.* **34,** 117–167.

Ettensohn, C. A. (1999). Cell movements in the sea urchin embryo. *Curr. Opin. Genet. Dev.* **9,** 461–465.

Ettensohn, C. A., and Sweet, H. C. (2000). Patterning the early sea urchin embryo. *Curr. Top. Dev. Biol.* **50,** 1–44.

Fenaux, L. (1970). Maturation of the gonads and seasonal cycle of the planktonic larvae of the ophiuroid *Amphiura chiajei* Forbes. *Biol. Bull.* **138,** 262–271.

Giese, A. C., Pearse, J. S., and Pearse, V. B. (1991). *In* "Reproduction of Marine Invertebrates" (A. C. Giese, J. S. Pearse, and V. B. Pearse, eds.), Vol. VI, Boxwood Press, Pacific Grove, CA.

Giudice, G. (1973). Hybrids. *In* "Developmental Biology of the Sea Urchin Embryo," pp. 469. Academic Press, NY, NY.

Hendler, G. (1975). Adaptational significance of the patterns of ophiuroid development. *Am. Zool.* **15,** 691–715.

Hendler, G. (1977). Development of *Amphioplus abditus* (Verrill) (Echinodermata: Ophiuroidea). I. Larval Biology. *Biol. Bull.* **152,** 51–63.

Hendler, G. (1991). Echinodermata: Ophiuroidea. *In* "Reproduction of Marine Invertebrates" (A. C. Giese, J. S. Pearse, and V. B. Pearse, eds.), Vol. VI, pp. 355–511. Boxwood Press, Pacific Grove, CA.

Hoegh-Guldberg, O., and Manahan, D. (1995). Coulometric measurement of oxygen consumption during development of marine invertebrate embryos and larvae. *J. Exp. Biol.* **198**(Pt. 1), 19–30.

Holland, N. D. (1979). Electron microscope study of the cortical reaction of an ophiuroid echinoderm. *Tiss. Cell* **11,** 445–455.

Hyman, L. H. (1955). *In* "The Invertebrates: Echinodermata," Vol. IV, pp. 763. McGraw-Hill, New York, NY.

Jaffe, L. A. (1980). Electrical polyspermy block in sea urchins: Nicotine and low sodium experiments. *Dev. Growth Differ.* **22,** 503–507.

Jaffe, L. A., Giusti, A. F., Carroll, D. J., and Foltz, K. R. (2001). Ca^{2+} signalling during fertilization of echinoderm eggs. *Semin. Cell Dev. Biol.* **12,** 45–51.

Kanatani, H., Shirai, H., Nakanishi, K., and Kurokawa, T. (1969). Isolation and identification of meiosis inducing substance in starfish *Asterias amurensis*. *Nature* **221,** 273–274.

Keller, L. R., Evans, J. H., and Keller, T. C. S. (1999). *In* "Experimental Developmental Biology: A Laboratory Manual," pp. 114. Academic Press, San Diego, CA.

Kishimoto, T. (1998). Cell cycle arrest and release in starfish oocytes and eggs. *Sem. Cell Dev. Biol.* **9,** 549–557.

Leahy, P. (1986). Laboratory culture of *Strongylocentrotus purpuratus* adults, embryos, and larvae. *In* "Methods in Cell Biology: Echinoderm Gametes and Embryos" (T. E. Schroeder, ed.), Vol. 27, pp. 1–13. Academic Press Inc., New York, NY.

Lessios, H. A., and Cunningham, C. W. (1990). Gametic incompatibility between species of the sea urchin *Echinometra* on the two sides of the Isthmus of Panama. *Evolution* **44,** 933–941.

Levitan, D. R., Sewell, M. A., and Chia, F.-S. (1991). Kinetics of fertilization in the sea urchin *Strongylocentrotus franciscanus*: Interaction of gamete dilution, age, and contact time. *Biol. Bull.* **191,** 371–378.

Lowe, C. J., Issel-Tarver, L., and Wray, G. A. (2002). Gene expression and larval evolution: Changing roles of *distal-less* and *orthodenticle* in echinoderm larvae. *Evol. Dev.* **4,** 111–123.

Lowe, C. J., and Wray, G. A. (2000). Rearing larvae of sea urchins and sea stars for developmental studies. *In* "Developmental Biology Protocols" (Rocky S. Tuan and Cecilia W. Lo, eds.), Vol. 1, pp. 9–15. Humana Press, Totowa, NJ.

Lutz, D. A., and Inoué, S. (1986). Techniques for observing living gametes and embryos. *In* "Methods in Cell Biology: Echinoderm Gametes and Embyos" (T. E. Schroeder, ed.), Vol. 27, pp. 89–110. Academic Press Inc., New York, NY.

McEdward, L. R. (1992). Morphology and development of a unique type of pelagic larva in the starfish *Pteraster tesselatus* (Echinodermata: Asteroidea). *Biol. Bull.* **182,** 177–187.

McEdward, L. R. (ed.) (1995). "Ecology of Marine Invertebrate Larvae." CRC Press, Boca Raton, FL.

McEdward, L. R., and Miner, B. G. (2001). Larval and life-cycle patterns in echinoderms. *Can. Journ. Zool.* **79,** 1125–1170.

McEuen, F. S. (1987). Phylum Echinodermata, Class Holothuroidea. *In* "Reproduction and Development of Marine Invertebrates of the Northern Pacific Coast" (M. F. Strathmann, ed.), pp. 574–596. The University of Washington Press, Seattle, WA.

Mladenov, P. V., Emson, R. H., Colpit, L. V., and Wilkie, I. C. (1983). Asexual reproduction in the West Indian britle star *Ophiocomella ophiactoides* (H. L. Clark) (Echinodermata: Ophiuroidea). *J. Exp. Mar. Biol. Ecol.* **72,** 1–24.

Mladenov, P. V., and Emson, R. H. (1984). Divide and broadcast: Sexual reproduction in the West Indian britle star *Ophiocomella ophiactoides* and its relationship to fissiparity. *Mar. Biol.* **81,** 273–282.

Monroy, A. (1986). A centennial debt of developmental biology to the sea urchin. *Biol. Bull.* **171,** 509–519.

Moore, A. R. (1943). Maternal and paternal inheritance in the plutei of hybrids of the sea urchins *S. purpuratus* and *S. franciscanus*. *J. Exp. Zool.* **94,** 211–228.

Osani, K., and Kyozuka, K. (1982). Cross fertilization between sea urchin eggs and oyster spermatozoa. *Gamete Res.* **5,** 49–60.

Palumbi, S. R. (1992). Marine speciation on a small planet. *Trends Ecol. Evol.* **7,** 114–118.

Palumbi, S. R. (1996). What can molecular genetics contribute to marine biogeography? An urchin's tale. *J. Exp. Mar. Biol. Ecol.* **203,** 75–92.

Palumbi, S. R., Grabowsky, G., Duda, T., Geyer, L., and Tachino, N. (1997). Speciation and population genetic structure in tropical Pacific sea urchins. *Evolution* **51,** 1506–1517.

Palumbi, S. R., and Metz, E. (1991). Strong reproductive isolation between closely related tropical sea urchins (genus *Echinometra*). *Mol. Biol. Evol.* **8,** 227–239.

Pearse, J. S., and Cameron, R. A. (1991). Echinodermata: Echinoidea. *In* "Reproduction of Marine Invertebrates" (A. C. Giese, J. S. Pearse, and V. B. Pearse, eds.), Vol. VI, pp. 513–662. Boxwood Press, Pacific Grove, CA.

Raff, E. C., Popodi, E. M., Sly, B. J., Turner, F. R., Villinski, J. T., and Raff, R. A. (1999). A novel ontogenetic pathway in hybrid embryos between species with different modes of development. *Development* **126,** 1937–1945.

Rahman, M. A., Uehara, T., and Pearse, J. S. (2001). Hybrids of two closely related tropical sea urchins (genus *Echinometra*): Evidence against postzygotic isolating mechanisms. *Biol. Bull.* **200,** 97–106.

Ramofafia, C., Foyle, T. P., and Bell, J. D. (1997). Growth of juvenile *Actinopyga mauritiana* (Holothuroidea) in captivity. *Aquaculture* **152,** 119–128.

Ricketts, E. F., Calvin, J., and Hedgpeth, J. W. (1985). *In* "Between Pacific Tides," pp. 652. Stanford University Press, Stanford, CA.

Runft, L. L., Jaffe, L. A., and Mehlmann, L. M. (2002). Egg activation at fertilization: Where it all begins. *Dev. Biol.* **245,** 237–254.

Sasaki, K., and Chiba, K. (2001). Fertilization blocks apoptosis of starfish by inactivation of the MAP kinase pathway. *Dev. Biol.* **237,** 18–28.

Schoenwolf, G. C. (1995). Embryology of the sea urchin. *In* "Laboratory Studies of Vetebrate and Invertebrate Embryos" (Sheri Snavely, ed.), pp. 125–143. Prentice Hall, Englewood Cliffs, NJ.

Schroeder, P. C., Larsen, J. H., and Waldo, A. E. (1979). Oocyte-follicle cell relationships in a starfish. *Cell Tissue Res.* **203,** 249–256.

Schroeder, T. E. (1981). Microfilament-mediated surface change in starfish oocytes in response to 1-methyladenine: Implications for identifying the pathway and receptor sites for maturation-inducing hormone. *J. Cell Biol.* **90,** 362–371.

Schuel, H., and Schuel, R. (1981). A rapid, sodium-dependent block to polyspermy in sea urchin eggs. *Dev. Biol.* **87,** 249–258.

Sea Urchin Embryology Website (1997). *URL:* www.stanford.edu/group/Urchin/dev.htm

Selvakumaraswamy, P., and Byrne, M. (1995). Reproductive cycle of two populations of *Ophionereis schayeri* (Ophiuroidea) in New South Wales. *Mar. Biol.* **124,** 85–97.

Selvakumaraswamy, P., and Byrne, M. (2000). Reproduction, spawning, and development of 5 ophiuroids from Australia and New Zealand. *Invert. Biol.* **119,** 394–402.

Smiley, S. (1988). The dynamics of oogenesis in *Stichopus californicus* and its annual ovarian cycle. *Biol. Bull.* **175,** 79–93.

Smiley, S., McEuen, F. S., Chaffee, C., and Krishnan, S. (1991). Echinodermata: Holothuroidea. *In* "Reproduction of Marine Invertebrates" (A. C. Giese, J. S. Pearse, and V. B. Pearse, eds.), Vol. VI, pp. 663–750. Boxwood Press, Pacific Grove, CA.

Starr, M., Himmelman, J. H., and Therriault, J.-C. (1990). Direct coupling of marine invertebrate spawning with phytoplankton blooms. *Science* **247,** 1071–1074.

Strathmann, M. F. (1987a). *In* "Reproduction and Development of Marine Invertebrates of the Northern Pacific Coast" (M. F. Strathmann, ed.). The University of Washington Press, Seattle, WA.

Strathmann, M. F. (1987b). Phylum Echinodermata, class Echinoidea. *In* "Reproduction and Development of Marine Invertebrates of the Northern Pacific Coast" (M. F. Strathmann, ed.), pp. 511–534. The University of Washington Press, Seattle, WA.

Strathmann, M. F. (1987c). Phylum Echinodermata, Class Asteroidea. *In* "Reproduction and Development of Marine Invertebrates of the Northern Pacific Coast" (M. F. Strathmann, ed.), pp. 535–555. The University of Washington Press, Seattle, WA.

Strathmann, M. F., and Rumrill, S. S. (1987). Phylum Echinodermata, Class Ophiuroidea. *In* "Reproduction and Development of Marine Invertebrates of the Northern Pacific Coast" (M. F. Strathmann, ed.), pp. 556–573. The University of Washington Press, Seattle, WA.

Strathmann, R. R. (1981). On barriers to hybridization between *Strongylocentrotus droebachiensis* (O.F. Müller) and *S. pallidus* (G. O. Sars). *J. Exp. Mar. Biol. Ecol.* **55,** 39–48.

Strathmann, R. R., and Sato, H. (1969). Increased germinal vesicle breakdown in oocytes of the sea cucumber *Parastichopus californicus* induced by starfish radial nerve extract. *Exp. Cell Res.* **54,** 127–129.

Stricker, S. A. (1999). Comparative biology of calcium signaling during fertilization and egg activation in animals. *Dev. Biol.* **211,** 157–176.

Summers, R. G., and Hylander, B. L. (1975). Species-specificity of acrosome reaction and primary gamete binding in echinoids. *Exp. Cell Res.* **96,** 63–68.

Tufaro, F., and Brandhorst, B. P. (1982). Restricted expression of paternal genes in sea urchin interspecific hybrids. *Dev. Biol.* **92,** 209–220.

Uehara, T., Asakura, H., and Arakaki, Y. (1990). Fertilization blockage and hybridization among species of sea urchins. *In* "Advances in Invertebrate Reproduction" (M. Hoshi and O. Yamashita, eds.), pp. 305–310. Elsevier Science Publishers, Amsterdam, The Netherlands.

Vacquier, V. D. (1986). Activation of sea urchin spermatozoa during fertilization. *Trends Biochem. Sci.* **11,** 77–81.

Vacquier, V. D., and Moy, G. W. (1997). The fucose sulfate polymer of egg jelly binds to sperm REJ and is the inducer of the sea urchin sperm acrosome reaction. *Dev. Biol.* **192,** 125–135.

Vacquier, V. D., Swanson, W. J., and Hellberg, M. E. (1995). What have we learned about sea urchin sperm binding? *Develop. Growth Differ.* **37,** 1–10.

Vilela-Silva, A.-S. E. S., Alves, A.-P., Valente, A.-P., Vacquier, V. D., and Mourão, P. A. S. (1999). Structure of the sulfated α-*L*-fucan from the egg jelly coat of the sea urchin *Strongylocentrotus franciscanus*: Patterns of preferential 2-*O*- and 4-*O*-sulfation determine sperm cell recognition. *Glycobiol.* **9,** 927–933.

Vilela-Silva, A.-S., Castro, M. O., Valente, A.-P., Biermann, C. H., and Mourão, P. A. S. (2002). Sulfated fucans from the egg jellies of the closely related sea urchins *Strongylocentrotus droebachiensis* and *Strongylocentrotus pallidus* ensure species-specific fertilization. *J. Biol. Chem.* **277,** 379–387.

Wray, G. A. (2000). The evolution of embryonic patterning mechanisms in animals. *Sem. Cell Dev. Biol.* **11,** 385–393.

Yamashita, M. (1985). Embryonic development of the brittle-star *Amphipholis kochii* in laboratory culture. *Biol. Bull.* **169,** 131–142.

Yuce, O., and Sadler, K. C. (2001). Postmeiotic unfertilized starfish eggs die by apoptosis. *Dev. Biol.* **237,** 29–44.

CHAPTER 4

Culture of Echinoderm Larvae through Metamorphosis

Gregory A. Wray,* Chisato Kitazawa,* and Benjamin Miner[†]

* Department of Biology
Duke University
Durham, North Carolina 27708

† Department of Zoology
University of Florida
Gainesville, Florida 32611

I. Introduction

Although developmental studies of echinoderms have focused largely on early phases of the life cycle, there are many reasons to study post-embryonic developmental processes. Many cell types, such as neurons, differentiate during larval life, and interesting morphogenetic events, such skeletogenesis and formation of the coeloms, unfold primarily during larval stages (Chapter 1). Perhaps most significantly, the radial symmetry of the adult body is patterned late in larval life in most species of echinoderms. Analyzing these post-embryonic events has led to interesting insights about developmental mechanisms (e.g., Arenas-Mena *et al.*, 2000; McCain and McClay, 1994; Ransick *et al.*, 1996). Echinoderm larvae also exhibit fascinating developmental processes, including a surprising degree of

0091-679X/04 $35.00

phenotypic plasticity (Hart and Strathmann, 1994; Strathmann *et al.*, 1992) and remarkable forms of larval cloning (Balser, 1998; Jaeckle, 1994). In addition, echinoderm larvae are a mainstay in studies of evolutionary developmental biology (e.g., Lowe and Wray, 1997; Raff *et al.*, 1999) and common in studies of larval ecology, physiology, and biomechanics (e.g., Emlet, 1982; McEdward, 1985; Strathmann, 1971).

Almost all echinoderm species exhibit a markedly indirect mode of development via a tiny, planktonic, bilaterally symmetrical larva that bears no resemblance to the adult. In the majority of echinoderm species, larvae are planktotrophic ("plankton-eating"). Such larvae must feed for an extended interval, typically for weeks to months, in order to reach metamorphosis. Developmental biologists primarily study echinoderms with planktotrophic larvae, which is almost certainly the ancestral condition for the phylum (Nielsen, 1998; Strathmann, 1988). A number of echinoderm species have lecithotrophic ("yolk-eating") larvae, a condition that has evolved many times independently within the phylum (McEdward and Miner, 2001; Wray, 1996). Lecithotrophic larvae derive their energy from maternally provisioned macromolecules and do not need to feed in order to complete metamorphosis; as a consequence, feeding structures are reduced or lost in lecithotrophic larvae.

This chapter outlines general methods for culturing larval food (unicellular algae), for caring and feeding larvae, and for inducing metamorphosis. Additional sources of information on culturing echinoderm larvae are Hinegardner (1969), Cameron and Hinegardner (1974), Leahy (1986), Strathmann (1987), and Lowe and Wray (2000). For general information about the reproductive biology of echinoderms, see Giese and Pearse (1991).

II. Materials for Culturing Algae and Larvae

Most of the equipment needed for culturing echinoderm larvae is relatively inexpensive or part of a standard developmental biology lab.

Culture vessels. For culturing algae, Ehrlenmeyer flasks are convenient for most needs and 5-gal glass water containers work well for very large cultures. For culturing larvae, ~3 L glass food storage jars make excellent and inexpensive culture vessels (restaurant supply companies commonly sell them as pickle jars in the United States). Alternatively, 4-L Pyrex beakers can be used. Scrupulously avoid contaminating culture vessels (or anything else that contacts cultures) with soap, heavy metals, toxins, or fixatives. It is helpful to mark this glassware as "E" (for embryologically clean) with permanent ink. Wash culture vessels with tap water, rinse with distilled or deionized water, dry, cover with foil, and autoclave to sterilize. Alternatively, dilute household bleach (~1% by volume in distilled water) can be used to sterilize glassware, followed by rinsing thoroughly with distilled water. To clean culture vessels of accumulated protein and lipid, rinse with dilute acetic acid, followed by several rinses with tap water. For stubborn protein or algal accumulation, rinse with 2 to 4 mL of 0.3 to 0.5% potassium dichromate in concentrated sulfuric acid and rinse well with tap water. Glassware suspected of contamination should be discarded or retired into general lab use (and the "E" crossed out).

Additional equipment. One or more aquarium air pumps, along with flexible plastic tubing, valves, and connectors are needed to bubble air through algal and larval cultures (all are available from pet stores). Access to a microscope is critical in order to monitor health and development of larvae and to quantify algal density. Since saltwater is corrosive, it is convenient to keep an inexpensive microscope near algal and larval cultures for this purpose. A hemocytometer is needed for counting algal cells. Access to a low-speed centrifuge and an autoclave is also desirable, although alternatives are possible and discussed at the appropriate points in this chapter. Larvae of some species must be cultured below room temperature (see Appendix; a cold room, incubator, sea table, or water chiller is needed in these cases). Strainers are needed for changing seawater in larval cultures (see "Changing Water" under "Culturing Echinoderm Larvae" in the following text), as well as some mechanism for stirring cultures (see "Stirring" under "Culturing Echinoderm Larvae").

Seawater. Although natural seawater (SW) is preferable, algae and larvae can be reared reliably with ASW supplemented with trace elements (see following text) or using commercial mixes for marine aquaria that contain trace elements (e.g., CoralLife™ or Instant Ocean™, available from pet stores). Check ingredients lists for commercial salt mixes, as not all include trace elements.

To prepare natural SW, collect from an unpolluted source as far from shore as convenient. Pass the SW through a coarse strainer to remove debris and planktonic organisms. If preparing SW from salt mixes or by formula, use only double-distilled or deionized water. Stir overnight to ensure that all salts have gone into solution. Note that natural SW has a pH of about 7.6 and that some salts will not dissolve at lower pH. Salinity, which can be measured using a refractometer, should be 1.025. Sterilize SW using one of the methods described here. A widely used defined formulation of ASW suitable for rearing larvae is provided:

MBL trace solution seawater formula. From Cavanaugh (1975). Prepare standard MBL ASW as described in Appendix but add only 8.27 mL of 1 M KCl. Then add H_3BO_3 0.0024 g/L, KBr 0.089 g/L, NaF 0.003 g/L, $SrCl_2 \cdot 6H_2O$ 0.037 g/L.

Whether SW is natural or prepared from salt mix or reagents, it must be sterilized to avoid contaminating cultures. Several methods are useful. (1) *Vacuum filtration*. Pass SW through a 0.45 or 0.22 μm cellulose nitrate filter (Millipore, Gelman, or Corning). Although disposable vacuum filtration units are convenient, a much more cost-effective solution is to use a device that holds standard 47 mm filter disks (e.g., Nalgene DS0315-0047 or Fisher 09-753-1G) with a side-arm Ehrlenmeyer flask. Before filtering SW, pass hot deionized or distilled water through the filter to remove detergents that are added as wetting agents and discard. (2) *Brief autoclaving*. 15 min on liquid cycle is adequate; longer autoclaving will cause some salts to precipitate. (3) *Pasteurization*. Heat to 60 °C for 20 min on a hot plate.

Following microsurgery or microinjection, embryos sometimes develop bacterial infections. If this becomes a problem, antibiotics can be added to sterilized SW: streptomycin (final concentration: 50 μg/ml), penicillin (final concentration: 100 units/ml), or ampicillin (final concentration: 100 units/ml).

III. Establishing Algal Cultures

Planktotrophic echinoderm larvae are herbivorous and, in a natural situation, consume a variety of species of unicellular algae. Culturing planktotrophic larvae in the lab therefore requires establishing reliable cultures of algae. Since algal cultures take several days to establish, they should be set up at least 4 days before larvae are ready to feed.

Algal stocks. Several species of algae are routinely used for feeding echinoderm larvae (Table I). Stocks of most of these species and strains can be obtained from UTEX (www.bio.utexas.edu/research/utex) or commercial sources (e.g., Carolina Biological Supply Co.; www.carosci.com) in the US. A useful source in Japan is the National Research Institute of Aquaculture, Fisheries Research Agency (www.nria.affrc.go.jp).

Some algal species can be stored at −20 °C in 50% SW/50% glycerol (vol/vol). Survivorship is generally low, but backup stocks can be useful in the event algal cultures become contaminated.

Appropriate species. Echinoderm larvae differ in their food preferences and requirements. Various species of diatoms and green and golden brown algae have

Table I
Unicellular Algae Useful for Feeding Echinoderm Larvae

Species	Group	Max. temp.[a]	Notes[b]
Dunaliella tertiolecta	Chlorophyceae	28°	hardy; protein-rich; U
Chaetoceros gracilis	Bacillariophyceae	19°	can be used alone; U
Isochrysis glabana[c]	Prymnesiophyceae	26°	use in mixed diet; E, A, H
Pavlova lutheri	Prymnesiophyceae	?	lipid- and protein-rich; A, H
Phaeodactylum tricornutum[d]	Bacillariophyceae	?	lipid-rich; E, A, H
Rhodomonas lens	Cryptophyceae	?	slow-growing; lipid-rich; U
Thalassiosira weissflogii[e]	Bacillariophyceae	?	use in mixed diet; E, A

[a]Above this temperature, algae stop swimming or die.
[b]Abbreviations; A, E, H, O, U = has proven useful for rearing larvae of some asteroids, echinoids, holothuroids, ophiuroids, or all of the above, respectively.
[c]Strain T-ISO, the so-called "Tahiti strain," has a higher thermal tolerance.
[d]Formerly known as *Nitzchia closterium*.
[e]Clone "Actin" is recommended.

been used successfully (Table I). Although it is possible to rear some echinoderm larvae through metamorphosis on a single species of alga, better results are often obtained with two species. A good general starting point is a 1:1 mixture of *Rhodomonas lens* and *Dunaliella tertiolecta*. This mixture works well for larvae of phylogenetically diverse sea urchins and seastars. Less is known about good food sources for larvae from other classes of echinoderms and some experimentation may be necessary.

Culture media. Commercial formulations are available, including Guillard's Medium F/2 (sold as F/2 Algae Food, Fritz Industries, Dallas, TX, USA) and Alga Grow® (Carolina Biological: www.carosci.com). Alternatively, stock solutions for Medium F/2 can be prepared from standard reagents (for formula, see Guillard (1983) or Strathmann (1987). To prepare Medium F/2 for use, add 387 mL each of solution A and solution B to 3 L of sterilized ASW or SW.

Culture methods. Use basic sterile procedures to handle algal cultures. Glassware should be sterilized either by autoclaving or by rinsing with dilute household bleach (~1% by volume in distlled water) followed by several rinses with distilled water. Fill a culture vessel about 2/3 full with culture medium and inoculate with actively growing algae. Insert an air line to bubble cultures (e.g., a plastic 1 mL volumetric pipette or glass Pasteur pipette) connected by plastic tubing to a small aquarium pump (Fig. 1A). Avoid using lab outlets for compressed air as these can introduce toxins into cultures. A single air pump can usually aerate several algal cultures with a gang valve. Cover the opening of the culture flask with a stopper or plastic wrap to prevent contamination. Place the flask in indirect sunlight or under fluorescent lights (ideally, bulbs with a spectrum suitable for growing plants). Replace water that evaporates every couple of days (using distillied or deionized water, not SW).

Algal stock cultures are typically shipped as a suspension of cells in culture medium. Upon arrival, pour the suspension of cells into a sterilized 125 ml Ehrlenmeyer flask containing ~50 ml of algal culture medium or filtered natural SW (see Preceding text). The culture will darken as cells multiply. This should be noticeable within a few days. Increase the volume of the culture as it darkens by adding more culture medium.

It is important to keep algae in log-phase growth, as they lose nutritional value if senescent or nutrient-limited (Caperon and Meyer, 1972). Shake culture bottles once a day to prevent cells from accumulating on the bottom. Maintaining appropriate algal concentration is important for other reasons: too few cells result in slow growth while too many encourage contamination. To maintain active growth, passage dense cultures by simply pouring off about one-half to three-quarters of the culture and replacing with culture medium. As a rule of thumb, cultures are actively growing if you can see a difference in color from day to day. Pay attention to culture color and use it as an indication of when to passage cultures. This interval differs among algal species and is sensitive to light, temperature, and initial culture density.

Feeding larvae. Algae need to be concentrated and rinsed of culture medium before they are fed to larvae. Pour algal culture into centrifuge tubes and spin for 1 min at ~2300 rpm (~3000 g) to pellet the cells. Discard the medium and

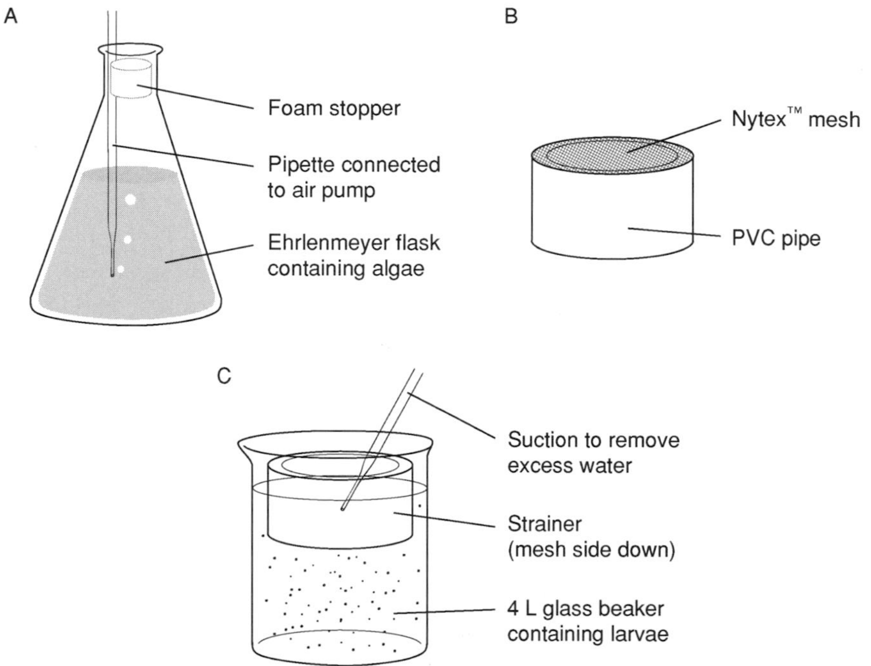

Fig. 1 Basic culture methods. (A) Algal culture. Place flasks in sunlight or under direct artifical illumination. Multiple flasks can be aerated from a single pump. (B) Construction of a sturdy sieve. Saw 8 cm diameter plastic commercial pipe (polyvinyl chloride; PVC) into ~7 cm lengths. Attach nylon screen (e.g., Nytex™) using a nontoxic glue. Construct several sieves with different mesh sizes (see text). (C) Reverse-filtration method of changing seawater in larval cultures. Attach a catchment flask to the vacuum line if using a pump. Seawater can be changed in many culture vessels quickly using a long vacuum line.

resuspend the pellet in about one-quarter the original volume of sterilized SW. Use a hemocytometer to quantify algal cell density (once you can estimate algal culture density by eye, this is no longer necessary). Add algal cell suspension directly to larval cultures to a final concentration of approximately 8 to 10 $\times$ 10^5 cells/ml. If using two species of algae, each should be added to cultures at this concentration. Well-fed larvae contain 1 to 2 algal cells in their stomach. Empty stomachs may signal too low a concentration of algal cells, too little mixing of SW in culture vessels, or that a different food species should be tried.

The volume of algal culture that needs to be maintained will depend on how many echinoderm larvae are being raised and the growth rate of the alga. To protect against accidents and contamination, it is a good idea to maintain at least two separate cultures of each species of alga.

Wild algae. If culturing echinoderm larvae at a marine lab or at sea, the most convenient approach may simply be to harvest algae directly from the ocean. This approach has the virtue of providing larvae with the species of phytoplankton they would encounter in the wild; the primary disadvantage is that algae must be harvested each time they are needed. Collect SW as far from shore as conveniently possible. Upon return to the lab, filter out detritus and planktonic organisms by passing the SW through a coarse (~500 μm) filter. Pellet the remainder, which will mostly consist of phytoplankton, in a centrifuge as has been described. Resuspend, estimate cell density, and add to a final concentration of $\sim 20 \times 10^5$ cells/ml.

IV. Culturing Echinoderm Larvae

Success in culturing echinoderm larvae is largely a matter of consistent attention to a few key parameters. Start cultures as described in Chapter 3 and switch to the methods described here when larvae form a mouth (e.g., prism stage for echinoids). Special considerations for culturing lecithotrophic larvae are described at the end of this section.

Density. A critical component of success in rearing echinoderm larvae is low larval density. Although embryos can be cultured at relatively high densities (Chapter 3), larvae are very sensitive to crowding. Dense cultures are more prone to fatal infestation by bacteria and protozoa. Importantly, even when healthy, larvae in dense cultures have slower growth and development rates. Culture feeding larvae at a density of no more than 1 larva per ml of SW. Lower densities (1 larva per 10 ml) are even better. While such low densities may seem excessive, experience has shown that dilute cultures have faster development times and much higher survivorship.

Changing water. Another critical component of success in rearing echinoderm larvae is keeping cultures free of contamination. The most common problems are molds, bacteria, and ciliate protozoans. Contaminated cultures will quickly "crash" (i.e., experience 100% mortality). To lower the risks of contamination, culture larvae at low density, do not overfeed with algae, make sure algal cultures are not themselves contaminated prior to feeding, and change SW every 2 or 3 days (more often to optimize developmental rates, less often to save labor). Change SW immediately if cultures become contaminated and at least once per day thereafter for several days.

Because larvae are fragile, changing SW requires care. This task is most easily accomplished using strainers of nylon mesh (e.g., Nytex™, Tetko Inc., Elmsford, NY). A simple strainer can be constructed by stretching nylon mesh across one end of a plastic beaker whose bottom has been removed and securing the mesh with a rubber band. Sturdier strainers can be constructed from 8 cm diameter plastic pipe (commercial plumbing pipe made of polyvinyl chloride, PVC, is cheap and works well) cut to ~7 cm lengths using a hacksaw. Attach the nylon mesh using a hot glue gun (not solvent-based glue), as illustrated in Fig. 1B. Some larvae are sensitive to chemicals that leach from PVC pipe, so use plastic beakers if this is a problem. Several mesh sizes are useful: 30 to 50 μm for straining embryos and very early larval stages, 100 μm for intermediate larvae, 200 μm for late larvae, and

500 μm for straining debris from natural seawater. Methods for changing SW are discussed in the following text.

Reverse filtration method: Hold a strainer in a culture vessel and draw off most of the seawater from inside (Fig. 1C), either by hand or using a Pasteur pipette connected to a vacuum line with a catchment flask. Resuspend the larvae in a few hundred mL of sterilized SW, drain, and repeat.

High-throughput reverse filtration method: Fabricate strainers from 50 ml polypropylene centrifuge tubes (Falcon™ tubes or similar) by first cutting away the flat part of the cap, then stretching nylon mesh over the opening and replacing the cap. Next, cut away the pointed end of the tube and attach flexible tubing (Tygon™ or rubber tubing) in a watertight seal with a hot glue gun (not solvent-based glue). To use, place the filter into a culture vessel with the cap down and either siphon SW into a bucket or sink or attach the tubing to a vacuum line with a catchment flask. This method has the virtue that several culture vessels can be filtered simultaneously with minimal handling.

Overflow method: Set a strainer of the appropriate mesh size into an "E" shallow dish with the mesh down and place the dish in a sink. Next, slowly pour the contents of a culture vessel into the strainer. The SW will flow through the strainer and over the edges of the dish. Decant most of the SW from the dish, taking care to keep larvae suspended, then pour a few hundred mL of sterilized SW through the mesh to rinse larvae. Repeat twice.

Following SW changes, transfer larvae to clean culture vessels to minimize the risk of contamination. In most cases, simply rinse an empty culture vessel with very hot tap water, followed by a swirl of sterilized SW. (If cultures are contaminated, use a sterilized culture vessel.) Gently transfer the rinsed larvae back into the clean culture vessel and add sterilized SW to restore culture density.

Food. Feed larvae right after changing SW, ideally every two days. Appropriate algal density and species are described under "Establishing algal cultures." Do not overfeed, as too many algae increases the likelihood of contamination. Note that echinoderm larvae capture unicellular algae from sea water at substantial rates. For instance, each late-stage larva of *Dendraster excentricus* can clear algae from several ml of SW per day (Hart and Strathmann, 1994). Thus, it is important to feed larvae often, but to avoid adding too many algae at any one time.

Stirring. Some larvae develop more reliably when cultures are gently stirred. A simple solution is to bubble air through the culture vessel (not using an aquarium airstone, just an open-ended tube). For some species, bubbling does not induce enough water circulation and larvae may become trapped on the bottom of culture vessels. Small cultures can be stirred by placing larvae in tissue culture plates on a slow rotary shaker table or in small bottles on a slow rotator (8 rpm; Amemiya, 1996). Strathmann (1987) describes simple devices for stirring cultures, including a multiplexed, swinging paddle system that can be built from inexpensive materials. Magnetic stir bars don't work well for stirring cultures.

Temperature and light. Most echinoderm larvae tolerate a range of temperatures, but this range differs considerably among species (Chapter 3). Development will be faster at higher temperatures, but be careful of pushing temperatures too high; the difference between rapid development and death is a matter of only a few degrees. SW temperatures where reproductive adults live provide a good guide to optimal culture temperatures. Larvae develop more reliably when on a light cycle that approximates the time of year they are in the plankton.

Development rates. Developmental rates at a given temperature can differ considerably among species. In general, the larvae of most echinoids develop relatively quickly, while those of ophiuroids, asteroids, holothuroids, and cidaroid echinoids develop more slowly. Some tropical sea urchins can be cultured through metamorphosis in as few as 5 days (e.g., *Leodia sexiesperforata*), but 3 weeks to 3 months is more typical of echinoderm larvae in general. Time to metamorphosis is strongly influenced by food level and quality, culture density, temperature, and presentation of cues that induce metamorphosis.

Variation among cultures. The development of planktotrophic echinoderm larvae is not "hard wired." Batch-to-batch variation in developmental rates and in the sequence of developmental events is common, even among larvae from the same cross that are maintained under apparently uniform conditions. Maternal nutrition (George *et al.*, 1990) and phenotypic plasticity in response to food levels experienced by larvae (Boidron-Metairon, 1988; Strathmann *et al.*, 1992) are further complications and can produce dramatic developmental consequences. For instance, differences in food level will shift the onset of distal-less protein expression in larvae of *Strongylocentrotus droebachiensis* by a matter of days and at different stages of development relative to morphogenetic events (L. Issel-Tarver and G. Wray, unpublished). Developmental differences can be minimized by keeping culture conditions as uniform as possible, but some difference in developmental rates between culture vessels is difficult to avoid and rearing replicate cultures is a good general practice. A practical consequence of variation among cultures and of phenotypic plasticity is that it is not possible to define precise times at which developmental events occur in larvae.

Culturing small numbers of larvae. Some experiments, such as microsurgery or microinjection, require culturing individual larvae or small numbers of larvae separately. Individual larvae can be reared through metamorphosis in 6- or 24-well plastic tissue culture dishes. Place dishes on a shaker platform and rotate at 40 to 60 rpm. Alternatively, a single larva or small numbers of larvae can be cultured in 15 or 50 mL plastic centrifuge tubes (Amemiya, 1996). Leave a small air space before closing and rotate obliquely on a rotor at 2 to 8 rpm (e.g., TAITEC model RT-5 or an inexpensive rotisserie for home cooking). In either case, keep the density of larvae below 1/mL (see Previous text).

Culturing lecithotrophic larvae or brooded embryos. Many lecithotrophic larvae can be reared with methods similar to those that have been outlined, although feeding is, of course, not necessary. Lecithotrophic larvae typically reach metamorphosis much sooner than their planktotrophic relatives, in some cases within just a few days. As with planktotrophic larvae, it is usually necessary to provide appropriate settlement cues (see following text). Lecithotrophic larvae are generally lipid-rich and relatively fragile. Dead larvae can create an oil "slick" at the water surface that promotes bacterial growth. Remove any oil and debris that accumulate with a pipette and change SW every day. Most lecithotrophic larvae are positively buoyant and do not require stirring.

V. Metamorphosis and Beyond

While many species of echinoderms have been reared through metamorphosis (Mortensen, 1921; Strathmann, 1987), less practical information is available to guide investigators than is the case for culturing larvae.

Induction of metamorphosis. Few echinoderm larvae will readily undergo settlement and metamorphosis in a clean glass culture jar, even when fully competent to do so. A variety of methods have been used to induce metamorphosis (Burke and Gibson, 1986). A simple and effective approach is to take advantage of the fact that larvae often use the presence of adults or food as a cue for settlement (Amemiya, 1996; Burke, 1980; Cameron and Hinegardner, 1974; Gosselin and Jangoux, 1995). Transfer into the culture vessel a small quantity of sand, gravel, or shell fragments from an aquarium or natural habitat where the adult of the same species lives. Alternatively, beakers or petri dishes can be left in aquaria containing adults for at least a week to accumulate a biofilm, and then retrieved and filled with cultures of competent larvae. Other methods are sometimes effective, including elevated K^+ concentration (Yool *et al.*, 1986), electrical stimulation (Burke, 1983), or thyroxine (Hodin *et al.*, 2001; Johnson and Carwright, 1996). Settlement and metamorphosis take less than an hour, but individual larvae will respond to suitable cues over a period of many hours or even days.

Culturing juveniles. Because of their minute size, juveniles are usually cultured in small petri dishes or beakers (Leahy, 1986; Strathmann, 1987). Although little is known about the diet of juvenile echinoderms in the wild, many can be reared in the laboratory on diatoms and other organisms that live in biofilms. Cultured diatoms such as *Phaeodactylum tricornutum* can be used (Hinegardner, 1969; Leahy, 1986), but simply placing culture dishes in aquaria inhabited by adult animals is usually sufficient to establish a suitable biofilm. Some juvenile sea urchins will eat macroalgae such as the green alga *Ulva* (Amemiya, 1996; Leahy, 1986). To avoid contamination, soak a small piece of alga in tap water for 1 h before adding it to a dish containing juveniles. Try different species of algae and note which are eaten. As juveniles grow, sand or gravel from aquaria or habitats where adults live can be added. Once juveniles reach about 1 cm in diameter, they can be cultured in aquaria like adults (see Chapter 2).

Acknowledgments

Thanks to Ina Arnone and Christian Gache for sharing information and to Gary Wessell and an anonymous reviewer for helpful comments.

References

Amemiya, S. (1996). Complete regulation of development throughout metamorphosis of sea urchin embryos deprived of macromeres. *Devel. Growth Differ.* **38,** 465–476.

Arenas-Mena, C., Cameron, R. A., and Davidson, E. H. (2000). Spatial expression of *Hox* cluster genes in the ontogeny of a sea urchin. *Development* **127,** 4631–4643.

Balser, E. J. (1998). Cloning by ophiuroid echinoderm larvae. *Biol. Bull.* **194,** 187–193.

Boidron-Metairon, I. (1988). Morphological plasticity in laboratory-reared echinoplutei of *Dendraster excentricus* (Eschscholtz) and *Lytechinus variegatus* (Lamarck) in response to food conditions. *J. Exp. Mar. Biol. Ecol.* **119,** 31–41.

Burke, R. (1980). Podial sensory receptors and the induction of metamorphosis in echinoids. *J. Exp. Mar. Biol. Ecol.* **47,** 223–234.

Burke, R. (1983). Neural control of metamorphosis in the sand dollar, *Dendraster excentricus. Biol. Bull.* **164,** 176–188.

Burke, R. D., and Gibson, A. W. (1986). Cytological techniques for the study of larval echinoids with notes for inducing metamorphosis. *Meth. Cell Biol.* **27,** 295–308.

Cameron, R. A., and Hinegardner, R. T. (1974). Induction of metamorphosis in laboratory cultured sea urchins. *Biol. Bull.* **146,** 335–342.

Caperon, J., and Meyer, J. (1972). Nitrogen limited growth of marine phytoplankton. I. Changes in population characteristics with steady state growth rate. *Deep Sea Res.* **19,** 601–618.

Cavanaugh, G. M. (ed.) (1975). "Formulae and Methods of the Marine Biological Laboratory Chemical Room," 6th ed. Marine Biological Laboratory, Woods Hole, MA.

Emlet, R. B. (1982). Echinoderm calcite: A mechanical analysis from larval spicules. *Biol. Bull.* **163,** 264–275.

George, S. B., Cellario, C., and Fenaux, L. (1990). Population differences in egg quality of *Arbacia lixula* (Echinodermata, Echinoidea)—Proximate composition of eggs and larval development. *J. Exp. Mar. Biol. Ecol.* **141,** 107–118.

Giese, A. C., and Pearse, J. S. (eds.) (1991). "Reproduction of Marine Invertebrates," Vol. 6. Palo Alto CA, Boxwood Press.

Gosselin, P., and Jangoux, M. (1995). Induction of metamorphosis in *Paracentrotus lividus* larvae (Echinodermata, Echinoidea). *Oceanol. Acta* **19,** 292–296.

Guillard, R. L. (1983). Culture of phytoplankton for feeding marine invertebrates. *In* "Culture of Marine Invertebrates: Selected Readings" (C. J. Berg, Jr., ed.), pp. 108–132. Hutchinon Ross Publ. Co.

Hart, M. W., and Strathmann, R. R. (1994). Functional consequences of phenotypic plasticity in echinoid larvae. *Biol. Bull.* **186,** 291–299.

Hinegardner, R. T. (1969). Growth and development of the laboratory cultured sea urchin. *Biol. Bull.* **137,** 465–475.

Hodin, J., Hoffman, J. R., Miner, B. G., and Davidson, B. J. (2001). Thyroxine and the evolution of lecithotrophic development in echinoids. *In* "Echinoderms 2000" (M. Barker, ed.), pp. 447–452. Swets and Zeitlinger, Lisse, The Netherlands.

Jaeckle, W. B. (1994). Multiple modes of asexual reproduction by tropical and subtropical sea star larvae—An unusual adaptation for genet dispersal and survival. *Biol. Bull.* **186,** 62–71.

Johnson, L. G., and Cartwright, C. M. (1996). Thyroxine-accelerated larval development in the crown-of-thorns starfish, *Acanthaster planic. Biol. Bull.* **190,** 299–301.

Leahy, P. S. (1986). Laboratory culture of *Strongylocentrotus purpuratus* adults, embryos, and larvae. *Methods in Cell Biology* **27,** 1–13.

Lowe, C. J., and Wray, G. A. (1997). Radical alterations in the roles of homeobox genes during echinoderm evolution. *Nature* **389,** 718–721.

Lowe, C. J., and Wray, G. A. (2000). Rearing larvae of sea urchins and sea stars for developmental studies. *In* "Developmental Biology Protocols" (R. S. Tuan and C. W. Lo, eds.), pp. 9–15. Humana Press, Totowa, NJ.

McCain, E. R., and McClay, D. R. (1994). The establishment of bilateral asymmetry in sea urchin embryos. *Development* **120,** 395–404.

McEdward, L. R. (1985). Effects of temperature on the body form, growth, electron-transport system activity and development rate of an echinopluteus. *J. Exp. Mar. Biol. Ecol.* **93,** 169–181.

McEdward, L. R., and Miner, B. G. (2001). Larval and life-cycle patterns in echinoderms. *Can. J. Zool.* **79,** 1125–1170.

Mortensen, T. (1921). "Studies of the Development and Larval Forms of Echinoderms." GEC Gad, Copenhagen, Denmark.

Nielsen, C. (1998). Origin and evolution of animal life cycles. *Biol. Rev.* **73,** 125–155.

Raff, E. C., Popodi, E. M., Sly, B. J., Turner, F. R., Villinski, J. T., and Raff, R. A. (1999). A novel ontogenetic pathway in hybrid embryos between species with different modes of development. *Development* **126,** 1937–1945.

Ransick, A., Cameron, R. A., and Davidson, E. H. (1996). Postembryonic segregation of the germ line in sea urchins in relation to indirect development. *Proc. Natl. Acad. Sci. USA* **93,** 6759–6763.

Strathmann, M. F. (1987). "Reproduction and Development of Marine Invertebrates of the Northern Pacific Coast." Univ. Washington Press, Seattle, WA.

Strathmann, R. R. (1971). The feeding behavior of planktotrophic echinoderm larvae: Mechanisms, regulation, and rates of suspension feeding. *J. Exp. Mar. Biol. Ecol.* **6,** 109–160.

Strathmann, R. R. (1988). Functional requirements and the evolution of developmental patterns. *In* "Echinoderm Biology" (R. D. Burke, P. V. Mladenor, P. Lambert, and R. L. Parsley, eds.), pp. 55–61. Balkema Press, Amsterdam, The Netherlands.

Strathmann, R. R., Fenaux, L., and Strathmann, M. F. (1992). Heterochronic developmental plasticity in larval sea urchins and its implications for evolution of nonfeeding larvae. *Evolution* **46,** 972–986.

Wray, G. A. (1996). Parallel evolution of nonfeeding larvae in echinoids. *Sys. Biol.* **45,** 308–322.

Yool, A. J., Grau, S. M., Hadfield, M. G., Jensen, R. A., Markell, D. A., and Morse, D. E. (1986). Excess potassium induces larval metamorphosis in four marine invertebrate species. *Biol. Bull.* **170,** 255–266.

CHAPTER 5

Obtaining and Handling Echinoderm Oocytes

Gary M. Wessel, Ekaterina Voronina, and Jacqueline M. Brooks

Department of Molecular and
Cell Biology and Biochemistry
Brown University
Providence, Rhode Island 02912

0091-679X/04 $35.00

I. Introduction

The sea urchin egg is a favorite for studies of the molecular biology and physiology of fertilization and early development, yet we know sparingly little of its oocytes and the mechanisms of oogenesis in these cells. In contrast, the starfish oocyte has been studied intensively and has been instrumental in our understanding of meiotic mechanisms. But much less is known about the molecular mechanisms of embryogenesis in starfish. With the recent advances in genome technology and information, it will be easier to take advantage of the rich diversity of echinoderms at all stages of development and this chapter is intended to integrate methods of working with different echinoderm oocytes.

The process of oogenesis in most echinoderms is asynchronous so each ovary lobe has hundreds of oocytes at all stages of development (Fig. 1). At the beginning of oogenesis, the oocyte is about 10 μm in diameter. During the vitellogenic phase of oogenesis (about a 1-month period), the oocytes accumulate yolk proteins and grow to ten times their original size—to 80 to 100 μm in sea urchins, 150 microns in starfish, and even larger in sea cucumbers. The oocyte is apparent with its large nucleus, the germinal vesicle, containing a prominent nucleolus. Echinoid (sea urchins, sand dollars) and Holothurian (sea cucumber) oocytes complete meiotic maturation prior to fertilization, distinct from other echinoderms and almost all other animals (except Cnidarians). These eggs may then be stored for weeks to months within the female before they are spawned, and the proportion of eggs in the ovary increases from early to late season, as the numbers of oocytes decline.

This chapter describes the basic procedures for handling echinoderm oocytes, with particular emphasis on sea urchin oocytes. Other echinoderm oocytes, especially starfish, have received a great deal of attention and such references will be provided along with a brief description of handling procedures.

II. Experimental Preparation

Oocytes are far more vulnerable to mechanical, chemical, and bacterial demise than the resultant eggs or embryos. Some species, such as the sea urchin *Lytechinus variegatus*, are particularly sensitive, and extra care is required for their use. Here are some basic considerations for handling:

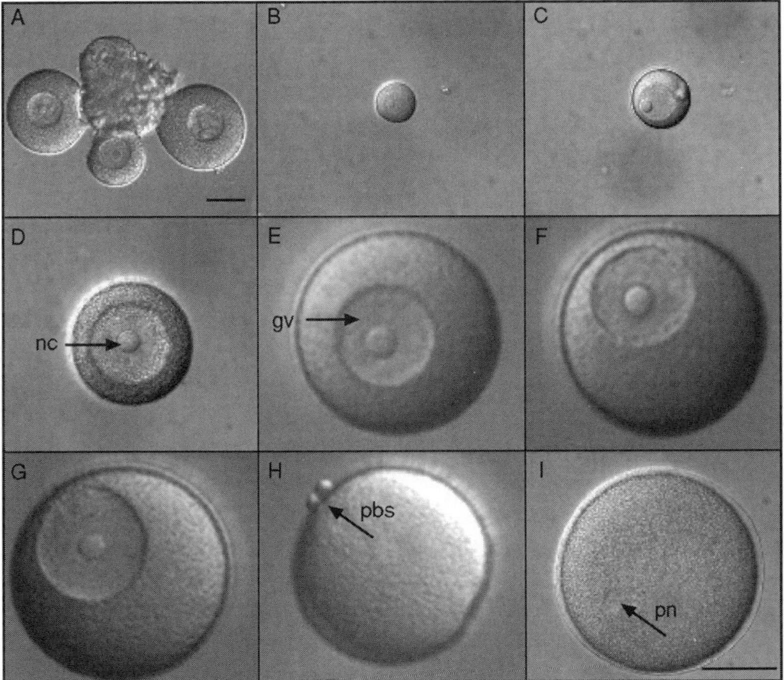

Fig. 1 Oogenesis. These samples were shed and dissected from the ovary of *L. variegatus* and viewed with a light microscope. (A) Oocytes attached to the somatic cells of the ovary. (B) Previtellogenic primary oocytes. (C–E) Vitellogenic growth phase of oogenesis. (E–I) Full-grown oocytes complete meiosis with the large germinal vesicle moving asymetrically to the cell periphery, where it breaks down and extrudes two polar bodies to produce a haploid mature egg that is now ready for fertilization. Scale bar, 50 μm; nc, nucleolus; gv, germinal vesicle; pbs, polar bodies; pn, pronucleus. Figure adapted from Brooks and Wessel (2003a).

- Glass and plasticware should be dedicated to sea urchin use and should never be washed with soap or other detergents. We mark our "urchin-ware" with an X using nail polish so it does not get mixed with general glassware. We also use a sink that is soap-free and dedicated to urchin use.
- Avoid oocyte contact with sperm. The oocytes are not equipped to encounter sperm and will readily die upon premature insemination.
- Prepare a humid chamber for oocyte culturing so the seawater doesn't evaporate. We use shallow plastic containers with hinged lids. Line the bottom of the container with Whatman paper and moisten with deionized water when the chamber is in use. For incubating slides, we line the bottom of the container with an additional support (glass or plastic rods) to keep the slides suspended.
- Oocytes stick to glass surfaces, and their removal usually damages them mechanically. Glass surfaces can be coated by use of "Safetycoat" nontoxic coating (J. T. Baker, Phillipsburg, NJ, thru VWR cat JT4017-01) to minimize this problem.

A. Making Micropipettes for Mouth Pipetting

Oocytes can be sorted from eggs and manipulated by use of a mouth micropipette (Fig. 2). This takes practice and making a mouthpipette that is comfortable is paramount. We have two basic designs that we use in our lab. One is made from a 9″ glass Pasteur pipette and the other is made from a small glass capillary. Both types of pipettes require a mouthpiece and flexible rubber tubing.

- plastic mouthpiece
  HPI Hospital Products, Altamonte Springs, FL
  red plastic mouthpiece cat #1501P (B4036-2)
  white plastic mouthpiece cat #1506P
- yellow rubber tubing
  HPI Hospital Products, Altamonte Springs, FL
  cat #1503P (B4036-1)

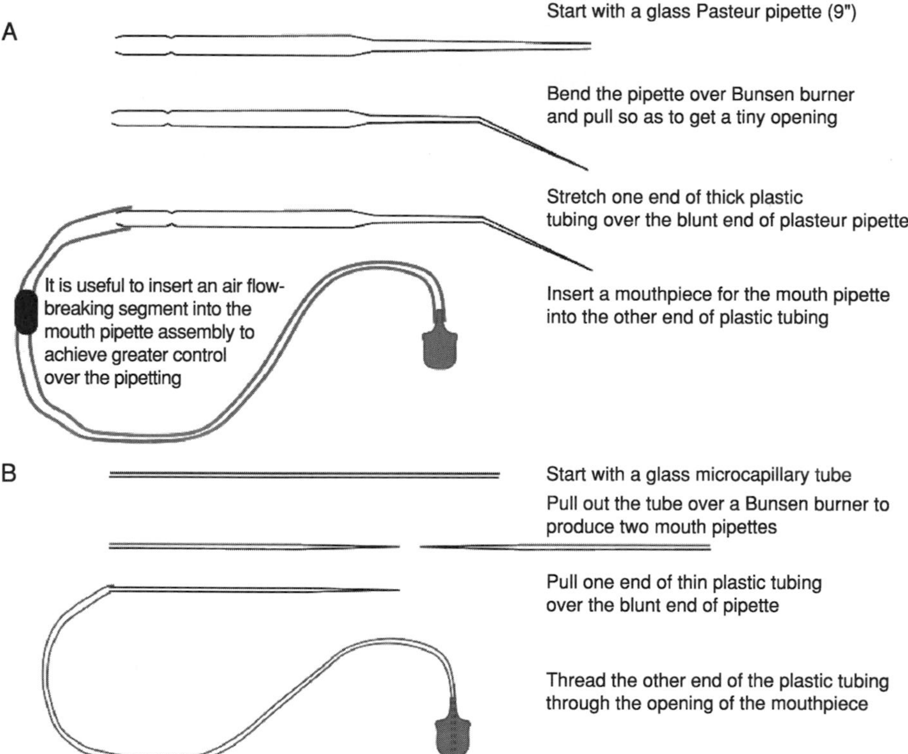
A
Start with a glass Pasteur pipette (9")
Bend the pipette over Bunsen burner and pull so as to get a tiny opening
Stretch one end of thick plastic tubing over the blunt end of plasteur pipette
It is useful to insert an air flow-breaking segment into the mouth pipette assembly to achieve greater control over the pipetting
Insert a mouthpiece for the mouth pipette into the other end of plastic tubing
B
Start with a glass microcapillary tube
Pull out the tube over a Bunsen burner to produce two mouth pipettes
Pull one end of thin plastic tubing over the blunt end of pipette
Thread the other end of the plastic tubing through the opening of the mouthpiece

Fig. 2 Construction of mouth pipette. (A) Starting from a 9″ glass Pasteur pipette. (B) Starting from a glass microcapillary tube. (See Color Insert.)

- thin clear intramedic polyethylene tubing
  Becton Dickinson, Oxnard, CA
  cat #427415
- custom glass capillary tubes
  Drummond Scientific Company thru VWR Scientific Products
  cat #9-00-1061-6
  Pyrex custom glass tubing; length 6.0 inches
  OD 0.8 mm/ID 0.6 mm
- glass Pasteur pipette; 9 inch
  VWR Scientific Products, West Chester, PA
  cat #14672-380

To begin construction of your pipette, you will need a Bunsen burner and either a glass Pasteur pipette or a glass capillary tube. If you are using a glass Pasteur pipette, do the following:

1. Heat the narrow portion of your Pasteur pipette over a Bunsen burner about 0.5 inches from the thick part. When it gets soft, remove it from the flame and bend a 90° angle.
2. Allow it to cool for a few seconds.
3. Hold the narrow tip of the bent pipette in one hand and the thick end in the other. Heat the narrow portion of the pipette again, about halfway between the bent angle and the narrow tip end. When it gets soft, remove it from the flame and immediately pull the two ends away from each other. You will end up with one functional micropipette that has a 90° angle and a small piece of glass for disposal.

If you are using microcapillaries to make your micropipette, do the following:

1. Hold each end of the microcapillary.
2. Place the capillary over the flame and heat the center.
3. When it gets soft, remove it from the flame and pull the ends away from each other. You will end up with two functional mouth pipettes.

Keep in mind that the micropipettes you have made are probably heat sealed so score the pipette lightly with a diamond scribe and use your fingers or a forceps to snap the tip open. Minimizing rough ends is important to reduce damage to the oocytes, so briefly flame the snapped tip to polish it. For the Pasteur mouth pipette, use the thicker yellow tubing and connect one end to your pulled Pasteur pipette and the other end to your mouthpiece. The length of tubing used is arbitrary and depends on personal preference and individual mouth control. For a microcapillary mouth pipette, the thin clear polyethylene tubing is used to connect the mouthpiece to the pulled microcapillary. Some users enjoy additional control over their mouth pipette by inserting a filtered micropipette tip along the length of the yellow rubber tubing. Before mouth pipetting oocytes, coat the pipette by drawing Safetycoat (see supplies) through the glass portion and then rinse in seawater.

B. Culturing Material

- Pyrex glass spot plates
  VWR Scientific Products, West Chester, PA
  3 well spot plate #53631-001
  9 well spot plate cat #53632-002
- Falcon 3911 assay plates (PVC)
  Becton Dickinson, Oxnard, CA
  96U-bottom wells
  VWR cat #62406-220
- Falcon 3913 Microtest III flexible lid
  VWR cat #62406-263
- Plastic disposable transfer pipettes
  (Samco brand distributed through VWR cat #14670-205)

III. Methods of Oocyte Collection

A. Methods of Sea Urchin Oocyte Collection

Our favorite sea urchin is *Lytechinus variegatus*, found along the east coast of the United States, which produces oocytes from December to June in Florida, and from May to September in North Carolina. The choice of *L. variegatus* is also based on large (100 micron), clear (nonpigmented) eggs and oocytes that can be successfully cultured at room temperature. Furthermore, a plethora of molecular and biochemical information is already available for this species, and even more can be inferred from *S. purpuratus*, and its newly available genome information (for a morphological comparison of the oocytes from these two species, see Fig. 3).

Often, you will find some oocytes in the female shed and the percentage depends on the season of the animal. In gravid females, the gonads contain a majority of mature eggs but early in the gravid season, and periodically during the season in regenerating ovaries, oocytes will constitute several percent of the shed gametes. An injection of 0.5 M KCl usually procures a steady stream of gametes within 1 to 2 min of injection. For these animals, eggs and sperm are easy to discern by color since eggs are yellow/orange and sperm are white. However, if the animal has immature gametes, the shedding process may require more KCl and may take several minutes longer. When a shed contains numerous oocytes, it will look lighter, will glisten, and can be mistaken for sperm. This is because immature

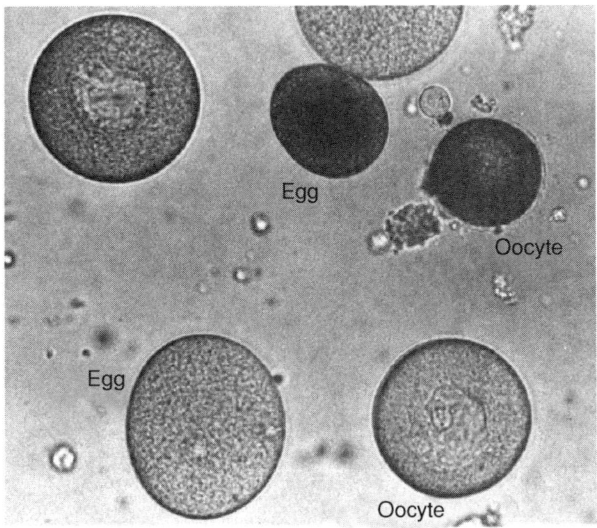

Fig. 3 Eggs and oocytes from two different species of sea urchin. The *Lytechinus variegatus* full-grown oocytes and eggs are large (100 microns) and not as heavily pigmented as the *Strongylocentrotus purpuratus* eggs and mature full-grown oocytes (80 microns).

oocytes do not contain as much yolk as mature eggs. Check the shed under a dissecting scope if you are not certain. If there are too few (for your purposes) or no oocytes in the shed, you will need to dissect the ovary (Fig. 4), but first ridding the gonad of eggs will enrich the oocyte population of gametes in the ovary.

1. Shedding Sea Urchin Oocytes

During certain times of the gravid season, the ovary is full of oocytes, and these can often be shed from the female with repeated KCl injection.

Procedure

1. Use a small gauge needle to inject 0.1 to 1 ml of 0.5 M KCl solution through the peristomal membrane surrounding the mouth into the body cavity.
2. Wait 1 to 2 minutes to see if the animal will release gametes. If you don't see anything, you may try to inject with KCl again.
3. Place females (gonadopore side down) onto a seawater-filled beaker whose diameter is smaller than that of the animal.
4. Usually a steady stream of gametes is observed from the gonadopore and they settle on the bottom of the beaker. Allow the animal to finish shedding, remove it from the beaker, and allow it to recover in a beaker of seawater for 10 to 15 minutes. We then inject KCl a second time to see if the animal will shed again. Often these "supershed" females will release gametes enriched with the most full-grown oocytes. These are good quality oocytes and will mature more readily when cultured.

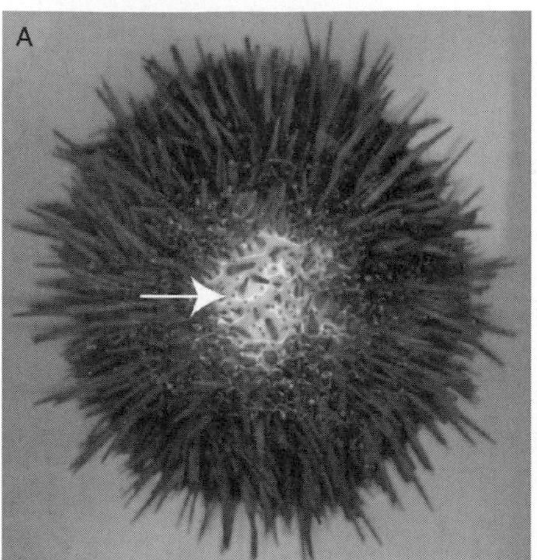

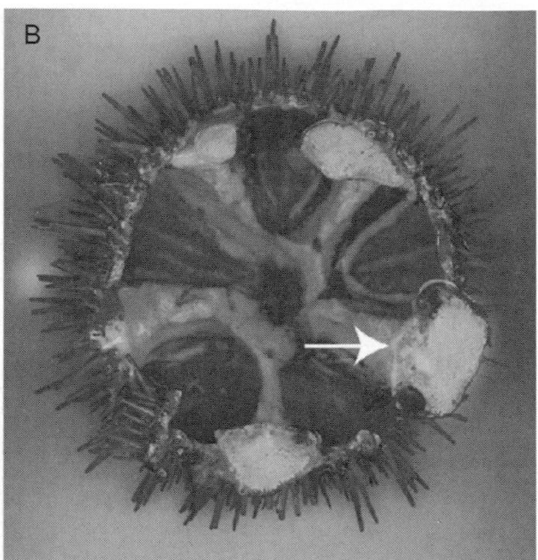

Fig. 4 Panel A shows the top side of a female *Strongylocentrotus purpuratus* sea urchin shedding eggs (arrow) from its five gonadopores. Panel B shows a bottom-side view of this same urchin that has been cut away along the periphery of the test surrounding the mouth (Aristotle's lantern) to show the five ovaries (arrow) that fill the body cavity. (See Color Insert.)

5. Expelled gametes should be thoroughly washed to remove debris and any toxic secretions. This is accomplished by allowing the gametes to settle to the bottom of the beaker by gravity, decanting the seawater above, and resuspending the gametes in fresh seawater. Repeat this process 2 to 3 times. If there are large pieces of debris, filter the gametes 1× through cheesecloth or 100-micron Nitex into another beaker.

6. Collect a small sample of shed from the bottom of the beaker using a plastic transfer pipette and examine for oocyte content using a microscope. Usually, this determination can be made readily at 50× magnification on a dissecting microscope.

2. Dissecting Sea Urchin Oocytes from the Ovary

The ovary may have to be dissected if the adult does not shed oocytes. Dissection is also of utility to isolate smaller and more numerous oocyte populations. Isolation of the gonad (Fig. 4) may also be of utility in studying oocytes *in situ*, in ovary cultures (see following text).

Procedure

1. Begin by shedding adult females. This will release a majority of the mature eggs from the ovary.

2. Open the animal with a pair of scissors by inserting one of the tips of the scissors into the soft peristomal membrane around the mouth and then proceed to cut the rest of the animal tangentially.

3. Five orange ovaries line the body cavity (Fig. 4). Use forceps to dissect the ovaries by gently pulling the lobes away from the body cavity. Rinse the ovaries in seawater and then place them in a glass culture dish with sufficient water to cover the tissue. Using two scalpels, mince the ovaries into very small pieces.

4. Remove the ovarian fragments by passing the dissected ovary in seawater through cheesecloth and into a new beaker. Alternatively, allow the minced tissue to settle for a few minutes in a 50 ml tube—the tissue chunks will settle rapidly, within a few minutes, whereas the oocytes, eggs, and debris will remain longer in suspension.

5. Rinse oocytes by allowing them to settle, decanting the seawater above, and adding fresh seawater. The number of times you do this depends on the condition of the seawater. If you choose to hand centrifuge, use great caution not to pack the oocytes. They readily stick to other cells and are often damaged during resuspension. Sometimes, minced ovaries are relatively clean, while at other times, the somatic cells make the culture cloudy and hard to work with.

Notes

- If you allow the dissected tissue to sit for 30 min, contractions of the ovarian pieces force more oocytes out into the seawater.
- Dissected oocytes can be in many different shapes that, due to stickiness, can be more fragile for mouth pipetting. We often allow the oocytes to sit for 30 min to allow them to "round up" and become easier to work with.
- A maturation-inducing substance in sea urchins is not known.

Additional procedure for Strongylocentrotus purpuratus

Often for *S. purpuratus*, simple dissection of the ovaries does not liberate many oocytes, and they remain attached to somatic cells (Fig. 5). Collagenase digestion of the ovary is then necessary:

1. Prepare a working solution of collagenase type I (Sigma Scientific cat #C-0130) at a concentration of 10 μg per ml freshly made in calcium-free seawater and chilled to 4 °C. Follow the dissection procedure and proceed with collagenase treatment.

2. Place the ovarian tissue pieces in a 50-ml conical tube and rinse eggs and debris off with ASW. Wash by inverting, allow the larger ovary bits to settle, and then pour off the supernatant. Repeat this step several times.

3. Add fresh collagenase solution to the washed ovary pieces.

4. Place the conical tube containing the minced ovary fragments at 4 °C for 1 to 2 h on a rocker. Periodically remove some ASW to check and see if oocytes are being liberated.

5. When maximum oocytes have been released, remove the collagenase with repeated washing of the oocyte suspension by settling and decanting off the supernatant. Additional mincing of ovaries with a small scalpel may help to liberate more oocytes as the ovarian tissue has been partially digested by the collagenase.

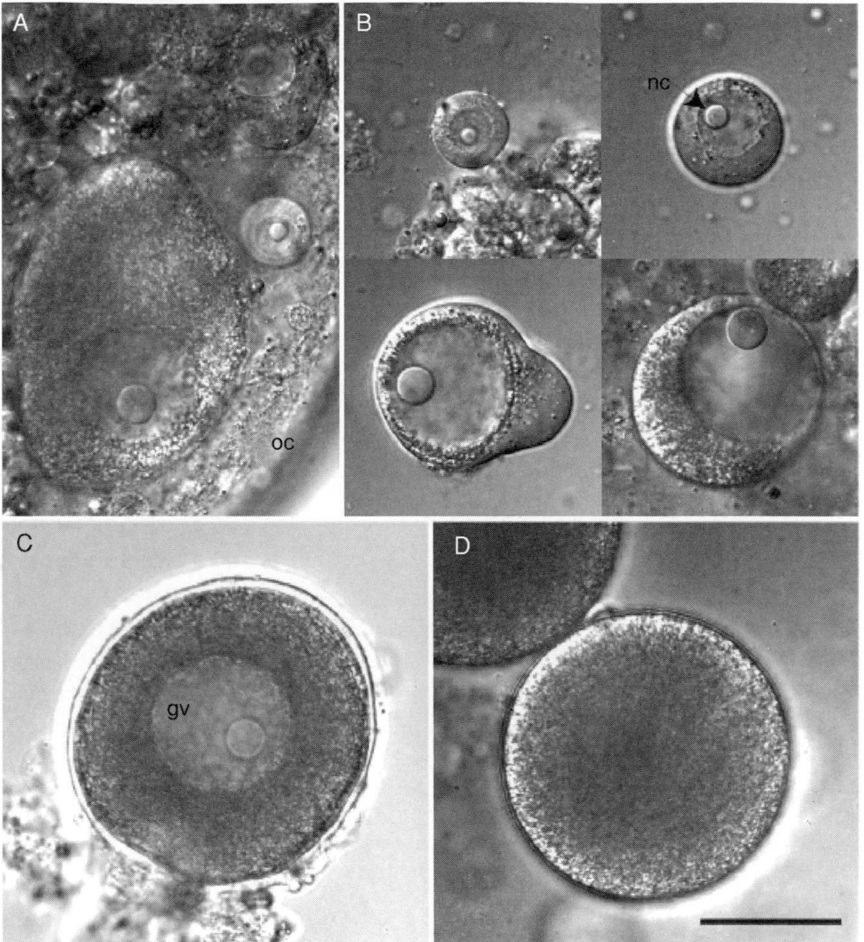

Fig. 5 *Strongylocentrotus purpuratus* oogenesis. These samples were dissected from the ovary of *S. purpuratus* and viewed with a light microscope. (A) Oocytes embedded in the somatic tissue of the ovary surrounded by the ovarian capsule. (B) Previtellogenic primary oocytes begin their transition through the growth phase of oogenesis. (C) Full-grown oocyte. (D) Egg. Scale bar, 40 μm; OC, ovarian capsule; gv, germinal vesicle; nc, nucleolus.

Readers are also referred to classic treatments of this subject in Harvey (1956), Costello *et al.* (1957), and Czihak (1975).

B. Obtaining and Handling Starfish Oocytes

See the excellent online manual from Laurinda Jaffe at http://egg.uchc.edu/injection for both microinjection and oocyte-handling methods, and Chapter 10 in this book by Terasaki and Jaffe for additional oocyte experimental methods.

Female and male starfish cannot be distinguished externally. Sex can be determined by taking a sample with a 3 mm sample corer from Fine Science Tools (Foster City, CA) or a small cork borer. Push the sample corer through the dorsal surface of the arm and save the small tissue plug, as this can be replaced into the arm to minimize infection in the adult. Alternatively, you can cut a slit between the arms with a razor blade. Ovaries are orange and testes are white and both are located along the length of each arm. Since the animal may be returned to the tank and used multiple times, it is best to start from the distal ends of each arm and work proximal over time. It is also important to feed the animals that may be stored and used over long periods of time (more than a few weeks; see Chapter 2 in this volume by Böttger *et al.* on maintenance of adults) as the health of the oocytes is critically dependent upon nutrition, and animals will often resorb the oocytes in times of nutritional depletion.

Pull out a piece of ovary with a forceps through the hole in the arm. Put the ovary piece into a beaker of cold calcium-free seawater (kept in an ice bucket). Mince the ovary into small bits, using a fine scissors, as this releases oocytes surrounded by follicle cells. After a few minutes, pour off the calcium-free seawater, replace with the same, let the ovary fragments settle out, and decant the suspension of oocyte/follicle cell complexes into a second beaker. Pick up some of the oocytes that settle on the bottom with a Pasteur pipette (often, the damaged oocytes remain floating). Transfer the oocytes to a microscope slide and inspect them with a microscope (10× objective) to see the state of the follicle cells. The follicle cells are a layer of cells surrounding the oocyte and the calcium-free seawater causes them to slough off. Check a sample every few minutes. Periodically decant the calcium-free seawater and add more (ice cold). When only about 1/4 of the oocyte surface is still covered by follicle cells, pour off most of the calcium-free seawater and replace with room temperature seawater. Depending on the individual animal, the time in calcium-free seawater may vary from just a few minutes to as much as 30 min. It is important to get the oocytes out of calcium-free seawater as soon as most of the follicle cells have dispersed, because prolonged exposure causes damage to the oocyte surface. Let the oocytes settle in the seawater, then wash them once or twice more, again by pouring off most of the water in the beaker and replacing it with fresh seawater. It is best to store the oocytes at a cool temperature; 15 to 18 °C is preferable. They can be used for experiments all day.

To stimulate the oocytes to resume meiosis, and to undergo other changes necessary for normal fertilization ("oocyte maturation"), use the hormone 1-methyladenine (1-MA; Sigma R-751138). From a frozen 10 μM stock solution in seawater, make a 1 μM solution of 1-MA also in seawater, and add the oocytes to it. This will cause nuclear envelope breakdown (germinal vesicle breakdown = GVBD) to occur about 20 to 30 min later (at 18–22 °C). Approximately 10 to 15 minutes following GVBD, the oocytes will have reached first meiotic metaphase, where they remain until about 60 min after GVBD, when the first meiotic division occurs. Oocytes at first metaphase are referred to as "mature eggs." For optimal fertilization, insemination should be performed at this stage. After the oocytes pass the first meiotic division, they will continue on to complete both meiotic divisions. Oocytes that have proceeded beyond the first meiotic division are referred to as "overmature," and often become polyspermic if inseminated, or will apoptose in a few hours if not fertilized. An alternative way to obtain mature eggs is to collect the ovary in natural seawater, clip it in pieces, and apply 1-MA directly to the ovary fragments. This will cause the ovary to spawn eggs at first metaphase, within about 30 to 60 min.

C. Obtaining and Handling Oocytes from Sand Dollars

Eggs from the common sand dollars can be obtained by 0.5 M KCl injection, as in sea urchins. In this case, though, the needle must be inserted through the mouth as parallel as possible to the oral surface. A single injection of approximately 0.5 mls for an animal of about 30 mls total will induce shedding of nearly all the ripe eggs and sperm. Oocytes, however, are infrequent in these sheds.

To obtain oocytes of different stages, the calcareous plates must be cracked open along the center of the fused "arms" apparent in the pentagonal symmetry of the calcareous sieve plate slits. The gonads are then accessible and can be placed into filtered seawater. Oocytes can be dissected out of the ovaries by mincing in calcium-free seawater, as for starfish. See starfish applications already discussed for removal of the follicle cell layer.

Note

- The maturation-inducing substance in sand dollars is not known.

D. Obtaining and Handling Oocytes from Sea Cucumbers

Females are dissected laterally and ovaries placed into filtered seawater. Oocytes are removed by mincing the ovary and shaking them gently. The oocytes settle rapidly and are then pipetted into calcium-free seawater to remove the follicle cells as for starfish and sand dollars. Following removal, the oocytes are placed into filtered seawater and used immediately. Oocytes of the sea cucumber can be induced to undergo GVBD as a dissection artifact; others may undergo GVBD when exposed to seawater.

A very useful, detailed description of sea cucumber oogenesis can be found in Smiley (1990). The maturation-inducing substance in sea cucumbers is not known, although some reducing reagents can artificially stimulate GVBD very efficiently in certain holothurians (e.g., Maruyama, 1980).

IV. Sea Urchin Oocytes Cultured *In Vitro*

Sea urchin oocytes removed from their ovarian environment are viable *in vitro* and can be cultured for up to several days. These cells are fragile and are best used for experiments right away, with the least handling possible. The most important consideration for culture of sea urchin oocytes is a sterile environment. Oocytes are not equipped to counter microbial invaders, and are easily overtaken by bacteria and other microorganisms in the seawater. To prevent contamination of the oocyte culture, it is essential to use filter-sterilized ASW, supplemented with 100 ug/ml ampicillin. For prolonged culture, it is critical to change the oocyte incubation medium once every 12 h. The best containers for *in vitro* culture are either silicon-coated glass Pyrex multiwell spot plates, or Falcon plastic 96-well microtiter plates (see experimental preparation, Section II). These containers need to be stored in a humid chamber, to prevent excessive evaporation of seawater. The oocytes are usually cultured at the temperatures preferred by their embryos (e.g., *L. variegates*: 20 °C; *S. purpuratus*: 16 °C).

A. *In Vitro* Maturation

Animal oocytes are arrested at the prophase of the first meiotic division during their growth period with an enlarged nucleus, called a germinal vesicle (GV). The signal for meiotic maturation is different for different animals (see Voronina and Wessel, 2003): progesterone for frog, 1-MA for starfish, but for most animals, including sea urchin, this signal is not known. Sea urchin oocytes will, however, spontaneously and asynchronously enter maturation when removed from the ovary. The predominant fraction of oocytes entering meiosis is close to 100% when large oocytes are used, though small oocytes will seldom mature. *In vitro* maturation recapitulates normal maturation and can be used to study processes and changes at maturation (Fig. 6; Berg and Wessel, 1997). The first indication of meiotic resumption in sea urchin oocytes is migration of the germinal vesicle to a peripheral location. By definition, this site is referred to as the animal pole. As maturation progresses, the GV breaks down (GVBD), and the chromosomes arrange in pairs in the middle of the forming meiotic spindle during metaphase I. Separation of the paired homologous chromosomes is followed by the first polar body formation, at about 3 to 4 h post-GVBD. The chromosomes remaining in the oocyte then are arranged on a second meiotic spindle at metaphase II. With the second meiotic division, chromatids separate and the second polar body is formed, at about 7 to 8 h post-GVBD. Finally, the chromatids remaining in the oocyte decondense and a haploid pronucleus forms. This whole process takes about 8 to 9 h in *L. variegates*, from GVBD of the oocyte to reformation of the haploid pronucleus in the mature egg (Berg and Wessel, 1997).

When removed from the ovary, the oocytes begin maturation asynchronously, so that an increase in the number of total GVBD events occurs linearly over time (Berg and Wessel, 1997). The rate of oocytes entering maturation depends greatly on the seasonality and physiological state of the oocyte donor. For example, 80 to 99% of oocytes obtained from a healthy donor at the beginning of egg production cycle will resume meiotic maturation after 24 h incubation *in vitro*, in contrast to only about 40 to 50% of the oocytes obtained from an "end-of-season" female. Similarly, if the donor female has been kept in captivity without food, the health and maturation rates of the oocytes also decline.

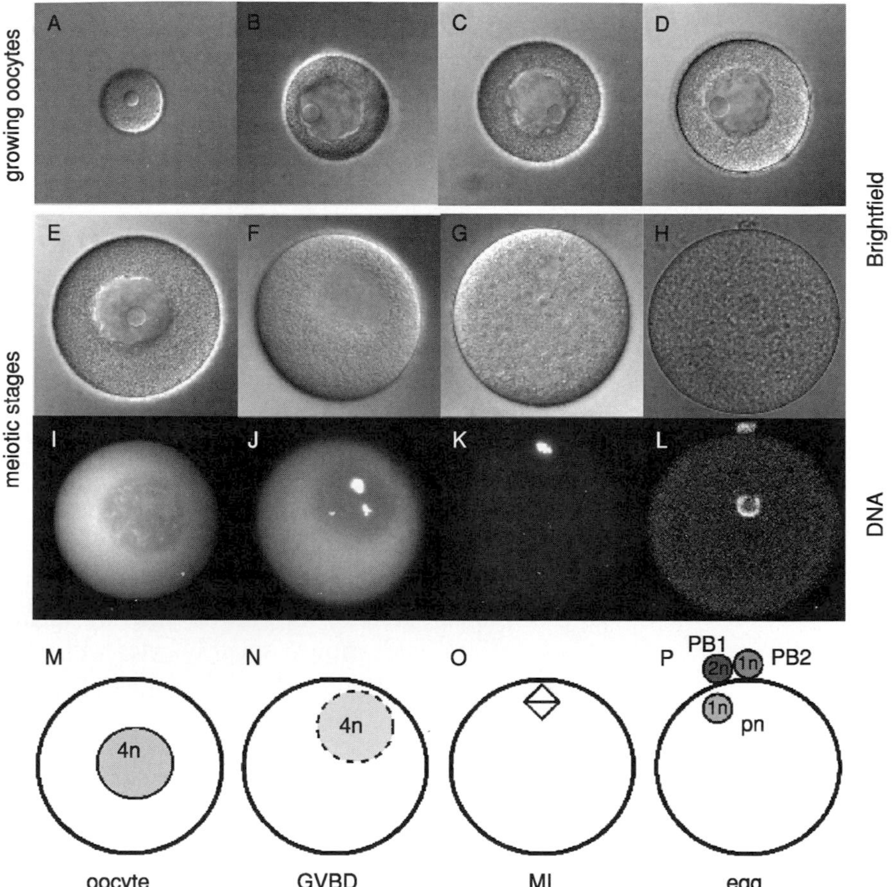

Fig. 6 Growth and meiotic maturation of *L. variegatus* oocytes. (A–D) growing oocytes; brightfield images. (E–P) stages of meiotic maturation, (F–H) brightfield images; (I–L) DNA staining (Hoechst); (M–P) schematic representations of the stages. (E, I, M) full grown oocyte, decondensed DNA in the nucleus; (F, J, N) germinal vesicle breakdown. DNA condensation; (G, K, O) first metaphase of meiotic divisions; (H, L, P) mature haploid egg with two polar bodies.

Only a limited number of ways are currently known to control maturation of the oocytes (mostly, by preventing it from happening). One of these is incubation of the oocytes with reagents that artificially increase cytoplasmic cAMP concentration: dbcAMP and theophylline at 100 μg/ml (Wessel *et al.*, 2002). These treatments do not interfere with oocyte viability, but can be used experimentally to effectively delay spontaneous maturation until necessary. The oocytes will mature when removed from the inhibitors and placed into fresh ASW. We have not yet found a treatment that will artificially induce entry of sea urchin oocytes into meiotic divisions.

B. Culture of Ovarian Explants *In Vitro*

We have found that ovarian explants will remain viable in culture for several days, enabling experiments on oocytes and their associated somatic cells *in situ*. To remove ovarian tissue, first open up the animal with a pair of scissors by inserting one of the tips of the scissors into the soft peristomal membrane around the mouth, then proceed to cut a circle around this area and drain the coelomic fluid. Locate the ovaries (Fig. 4b) and then carefully cut the test of the animal in half being careful not to sever the ovaries. Use forceps to remove the ovaries by gently pulling the lobes away from the body cavity and keeping them intact. Rinse the ovaries in seawater several times and then place them in a glass culture dish with seawater containing 100 μg/ml ampicillin (Sigma, St. Louis, MO). Place them at the appropriate temperature for your assay.

C. Culture of Whole Animals Following Introduction of Exogenous Macromolecules

All of the major organ systems, including the ovaries, are in direct contact with the circulatory coelomic fluid. Therefore, one can test the function or effect of molecules on the gonads or guts of adults simply by injecting it into the coelom. The adults will remain viable for weeks following such treatment. To first sex the individual, the adults can be partially shed with small volumes of KCl, light electrical shock, or even rigorous shaking, and then returned to an aquarium where they will remain viable. Care must be taken, however, to ensure the animal has stopped shedding before exposing it to the general population so as to not stimulate others to shed. We were interested in examining vitellogenesis *in vivo* in female sea urchins. We fluorescently labeled the major yolk protein (see labeling of oocyte components; endocytosis section) that normally circulates in the coelomic fluid and injected it into the coelomic cavity of adult sea urchins using a microfine 1 cc insulin syringe (Becton Dickinson, Oxnard, CA) to gently pierce the peristomal membrane surrounding Aristotle's lantern. Injected animals were incubated for 1 to 12 days in circulating water tables, their eggs and oocytes were collected by KCl injection, and then their ovaries were dissected. Fluorescently labeled MYP was specifically targeted to the ovary and packaged in the yolk platelets of the developing oocytes in these animals (Brooks and Wessel, 2003b).

V. Labeling of Oocyte Components

A. Live Whole Mount Immunolocalization of Oocyte Surface Molecules

This technique is used to analyze antibody surface labeling on unfixed, living cells. Be gentle with your oocytes during this procedure. It is fine if you have some eggs mixed in as you are going to look at individual cells with the microscope.

1. Collect oocytes and wash 2× in ASW using cheesecloth to remove jelly. For each wash, allow oocytes to settle and then aspirate off the supernatant.

2. Next, acid treat the oocytes by washing in ASW pH 5.2 for 2 to 5 minutes; this will ensure that the jelly is completely removed. It is important that you pH the ASW by using HCl, as acetic acid activates the eggs. The presence of jelly results in high background nonspecific antibody binding.

3. Wash the cells in ASW (normal pH 8.0) 2x.

4. Aliquot cells to snap cap Eppendorf tubes and allow the cells to settle on ice.

5. Dilute the primary antibody in ASW and keep on ice until ready to use.

6. Remove the supernatant from the samples and add the primary antibody. Incubate the samples at 4 °C using a rocking table to keep the cells gently resuspended. Don't forget your pre-immune control. It is essential that the live samples are kept cold to prevent or slow the endocytosis of the antibodies and some investigators simply put the tubes on ice and resuspend occasionally, to minimize damage to the cells.

7. Wash samples 3× in cold ASW, allowing cells to settle.

8. Dilute the secondary antibody in ASW and keep on ice. Add the secondary antibody to the samples and incubate at 4°C using a rocking table to keep the cells gently resuspended, or periodically resuspend if kept on ice.

9. Wash 3× in cold ASW, allowing the cells to settle.

10. Mount the samples on slides and visualize. As the samples are still alive, they may begin to endocytose the antibodies, so keep the samples on ice as long as possible and work quickly, or fix the samples following the last wash.

B. Oocyte Fixation (Whole Mounts)

For basic information on fixation techniques, see the fixation of whole mount embryos protocols (this volume). Below are notes and recommendations from our experience with oocytes and eggs. You can use 4% formaldehyde, 4% paraformaldehyde, methanol, or a succession of 4% paraformaldehyde and methanol. The optimal fixation and staining protocol is empirical and depends on the epitope, so be prepared for trial and error.

Four percent paraformaldehyde is the fixative of choice for labeling microfilaments and works well for detection of secretory granules. The stock of paraformaldehyde is made at 10% in dH_2O.

- Warm 10 ml dH_2O in a glass bottle on a heated stirplate, in the chemical hood
- Add 1 gm paraformaldehyde to the warm water while stirring
- Add 0.1 N NaOH slowly to clear (you need about 80 microliters).

This stock is best fresh, but can be stored frozen at −20 °C, in single-use (~1 ml) aliquots.

For fixation, use 4% paraformaldehyde in ASW made from 10% paraformaldehyde.

2 × ASW or 2 × calcium-free ASW	500 μl	1.25 ml
10% paraformaldehyde	400 μl	1.0 ml
*dH_2O	100 μl	0.25 ml
	total 1 ml	**2.5 ml**

Note

Instead of using dH_2O, one may add 0.5 M EGTA pH 8.0. EGTA chelates calcium and prevents formation of precipitate. Also, note that 2× calcium-free ASW seems to be best for fixation of fine membrane components.

1. Concentrate oocytes by settling and decant supernatant.
2. Add about 10 volumes of 4% paraformaldehyde and gently resuspend oocytes. One may do this on ice or at room temperature.
3. Incubate cells with regular suspensions for at least 30 min.
4. Allow the oocytes to settle and decant the fixative and wash several times by settling in ASW. Wash the cells thoroughly after fixation or else the residual fixative will cross-link the antibody to the sample.
5. Either store cells at 4 °C and add sodium azide at 0.01% or proceed immediately to immunolabeling.

Formaldehyde fixation works well for visualizing endocytosis of fluorescently labeled molecules (such as fixable dextrans or fluorescently labeled proteins). These procedures do not require permeablization for immunolabeling, since formalin solutions routinely contain 10% methanol, and care should be taken if your protein of interest is sensitive to such organics. Fixation with formaldehyde allows excellent immunolabeling of yolk.

1. Concentrate oocytes by settling and decant supernatant.
2. Add about 10 volumes of 3.7% formaldehyde (1:10 dilution of commercial formalin stock in ASW) and gently resuspend oocytes.
3. Incubate cells on ice with regular suspensions for at least 1 h.
4. Allow the oocytes to settle, decant the fixative, and wash several times by settling in ASW.
5. Either store cells at 4 °C and add sodium azide at 0.01% or proceed immediately to immunolabeling.

Methanol fixation is preferred for labeling microtubules. However, as sea urchin oocytes are much more fragile than embryos and eggs, they often get damaged during methanol fixation.

1. Concentrate oocytes by settling and decant or aspirate as much of the supernatant as possible.

2. Add 10 volumes (or more) of –20 °C methanol (or 90% methanol/50 mM EGTA, pH 6.0) to the sample.

3. Incubate the cells in the freezer for 10 to 20 min. At this point, samples can be stored for extended periods (at −20 °C, in a closed and Parafilmed container to avoid evaporation of the methanol).

4. Decant the fixative, and wash/rehydrate the sample with ASW.

5. Proceed immediately to immunolabeling.

Immunolabeling Notes:

• If you are going to do immunolocalizations following fixations in formaldehyde or paraformaldehyde, you must use 0.05% Tween-20 in your wash solutions. The fixed oocytes need to equilibrate with Tween-20 for at least 10 min to permeabilize the membranes and allow the antibodies to enter the samples, and to prevent nonspecific antibody binding. Whichever wash solution you choose, stick with it for the whole immunolocalization procedure to keep the cells osmotically happy.

• Methanol extraction can be helpful after paraformaldehyde fixation when labeling with antibodies, since it effectively removes the lipids. The procedure for methanol extraction is exactly the same as for methanol fixation. As the cells have already been crosslinked by paraformaldehyde, the structural damage introduced by methanol is minimal. Methanol extraction is not required after formaldehyde fixation, since commercial formaldehyde or formalin (non-EM grade) contains 6 to 15% methanol.

• One may also immobilize the cells on poly-L-lysine-coated coverslips and then immerse into cold methanol. Subsequent processing of the cells is then also performed on the coverslips.

• If the antibody gives high background staining, you might need to add filtered Blotto, 1% BSA, or 10% fetal calf serum to your buffer for all incubation steps.

C. Oocyte Stratification

Organelles can be stratified according to their density *in situ*. Isolated oocytes are resuspended in 28% sucrose (an iso-osmotic concentration in deionized water; Harvey 1956) to give a final concentration of 16% sucrose. The cells are placed in a flat-bottom microfuge tube and then spun in an Eppendorf microcentrifuge at varying speeds, depending on the application. For example, $1500 \times g$ for 10 min will partially displace cortical granules, or $5000 \times g$ for 10 min will completely displace the cortical granules (Wessel *et al.*, 2002) or yolk (Fig. 7; Brooks and Wessel, 2003b) to the centrifugal pole. Significantly higher speeds or times will cause the oocytes to readily fragment. Eggs withstand these same speeds easily with their robust microfilament layer in the cortex. To measure cortical granule or other organelle displacement, cells are often fixed immediately after centrifugation and prepared for immunolabeling (see Preceding text).

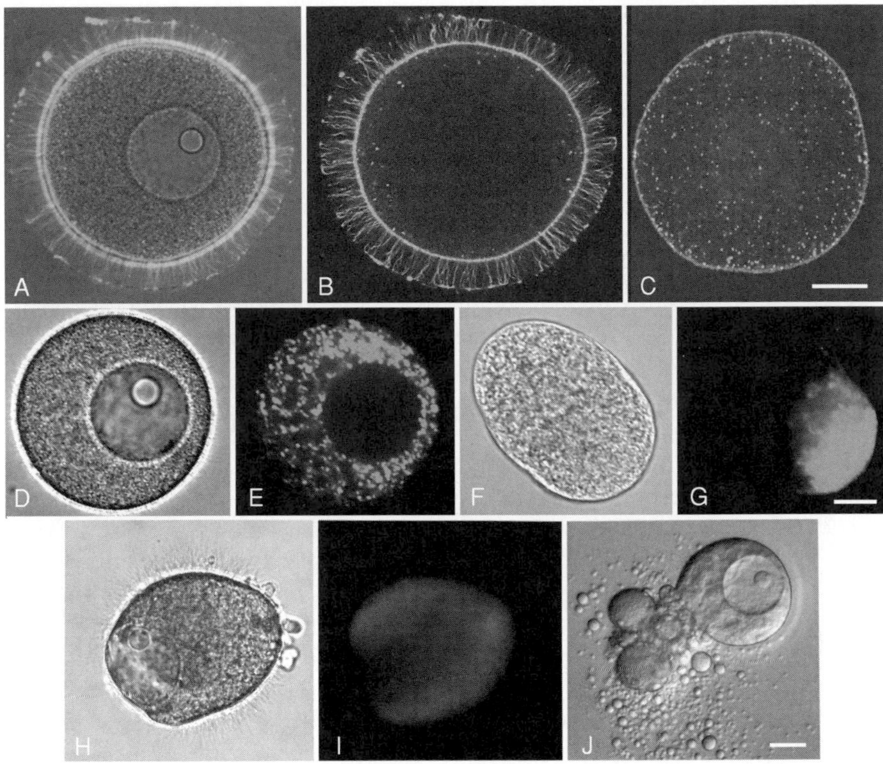

Fig. 7 Oocyte labeling and manipulation. (A, B) Oocyte labeled with FM1-43 to delineate the plasma membrane and fine membrane extensions emanating from the surface. (A) is FM1-43 overlain with DIC (B) is fluorescence alone. (C) An oocyte fixed and labeled with antibodies to a cortical granule content protein, hyalin (green) and phalloidin to label the dense actin ring at the cortex (red). (D–G) Oocytes that have endocytically packaged the major yolk protein MYP that was fluorescently labeled. (E) shows the distribution of the newly packaged and fluorescent MYP in yolk platelets. (G) shows an oocyte as in (D) and (E) that was centrifuged to stratify its organelles, including the yolk platelets. These are readily seen concentrating to the centrifugal or heavy end of the elongated oocyte by virtue of its endocytosed, fluorochrome-labeled MYP. (H–J) Oocyte apoptosis. (H) and (I) shows an oocyte stimulated to apoptose by staurosporine and then labeled for caspase activity with FITC-VAD-fmk. (J) shows the characteristic blebbing phenotype of an advanced apoptotic oocyte. Each bar represents 25 microns. (See Color Insert.)

D. Endocytosis Assays

Oocytes are capable of endocytosing and packaging massive amounts of protein, lipids, and carbohydrates. Both fluid phase and receptor-mediated endocytosis are active transport mechanisms and are temperature sensitive: reduced temperatures (<4 °C) block internalization. The earliest characteristics of internalization can be observed in cells after warming. You can stain cells at 0°C, warm them to room temperature, and follow the dye transport over time with fluorescence microscopy. Washing away the free dye followed by warming the cells allows for pulse/chase experiments whereby you can track internalization of your dye over time. After just several minutes at room temperature, discrete fluorescent vesicles appear within the cytoplasm. We see a rapid drop in this endocytosis when the oocytes mature, indicating that some endocytic mechanisms change during meiosis.

To assay membrane dynamics and endocytosis, we routinely use two types of markers available by Molecular Probes, which have previously been used in sea urchin eggs and embryos. These include FM1-43 as a marker for the fate of internalized plasma membrane and rhodaminated dextran as a fluid phase marker (Molecular Probes, Eugene, OR; Whalley *et al.*, 1995).

FM1-43 (cat #T-3163) is a lipophilic, nontoxic, water-soluble dye used to label plasma membranes and their derivatives, including sea urchin eggs (Whalley *et al.*, 1995; Fig. 7). This dye is virtually nonfluorescent in aqueous medium, but upon insertion in the outer leaflet of the plasma membrane, it becomes intensely fluorescent. This dye is membrane impermeable and is used to view plasma membrane dynamics including the internalization of cell membranes during endocytosis. We incubated isolated eggs and oocytes of various sizes in seawater with FM1-43 and recorded endocytic activity using a confocal microscope. The results indicate that oocytes were far more endocytically active than were eggs (Brooks and Wessel, 2003b). The amount of FM1-43 endocytosed for individual cells clustered into three distinct groups. Eggs were the least endocytically active, small oocytes were more active, and larger oocytes were most active. In our assays, FM1-43 was resuspended in methanol at 1 mg/ml, then diluted in ASW to give a working concentration of 1 μg/ml. To evaluate endocytosis, eggs and oocytes were transferred to the FM1-43 in ASW.

Note

FM1-43 is a vital dye that cannot be fixed, and its absorbance peak overlaps the spectra of the red dyes so care should be taken when designing double-label experiments.

Dextran Conjugates: A multitude of dextran conjugates are available from Molecular Probes. These dextrans are nontoxic and water-soluble. We use these dextrans for both endocytosis assays and molecular loading during microinjection. We mainly use the 10,000 MW dextrans that have lysine residues incoporated into them so that the dyes are aldehyde fixable. The amount of dextran used is empirical, depending on your experiment. We routinely use a 2.5 milligram per milliliter concentration of the rhodamine dextran and 0.25 milligram per milliliter of the Alexa dextran.

- Cat #D-3312 tetramethylrhodamine dextran (mini-ruby)
- Cat #D-22910 Alexa Fluor dextran

Protein Labeling Kits: Molecular Probes and other vendors also offer kits that are convenient to label proteins including antibodies. The reactive dyes in the kit label primary amines and form stable dye–protein conjugates. We have successfully used these kits to label the major yolk protein from coelomic fluid and other proteins. To fluorescent label MYP, coelomic fluid enriched with MYP was prepared as described (Brooks and Wessel, 2003b). This protein was conjugated to fluorochromes using both the FluoReporter Oregon-green 488 (Cat #F-6153) and Texas-red-X protein labeling kits (Cat #F-6162). The integrity of the protein was evaluated by SDS-PAGE and the degree of fluorochrome labeling was determined according to the manufacturer's instruction.

Note
Molecular Probes also offers anti-fluorescent dye antibodies.

E. Microfilament Labeling and Cytoskeletal Inhibitors

1. Labeling Microfilaments with Phalloidin

This protocol can be used with oocytes, eggs, and embryos. Fixation in paraformaldehyde (vs formaldehyde) is much better for fixing microfilaments in oocytes. In addition, do not use methanol (commercial formaldehyde or formalin contains 6 to 15% methanol). Staining is best performed according to the manufacturer's protocol.

1. Dissolve contents of 1 vial of Texas Red-X Phalloidin (Molecular Probes; cat #T-7471; Eugene, OR) in 1.5 ml methanol. The final concentration is 200 U/mL. (Be sure to check unit size on manufacturer's information sheet.)
2. Place 3 μL dissolved Texas-Red phalloidin in microfuge tube. Allow all methanol to evaporate.
3. Staining.
   a. Add 100 μL washed, paraformaldehyde-fixed oocytes/eggs/embryos resuspended in ASW with 0.05% Tween 20 (ASWT). Avoid using nonosmotic buffers that will damage the cells.
   b. Incubate at least 15 min.
   c. Wash 2× in ASWT.
4. Mount samples on slides.

Note
You can also use this staining method in conjunction with antibody labeling. If double labeling samples with an antibody, follow the phalloidin stain with the primary antibody incubation (also in ASWT) for an hour, wash, then incubate with the secondary antibody for an hour and wash. We have also performed the phalloidin label successfully after the antibody labeling, but we feel it is important to first assess the phalloidin labeling before committing the cells to immunolabeling.

2. Cytoskeletal Inhibitors

The following inhibitors have been used successfully in sea urchin oocytes: Microtubule inhibitors: nocodazole and colchicines; Microfilament inhibitors: latrunculin A, cytochalasin B, and cytochalasin D.

We use the cytoskeletal inhibitors at the following concentrations: nocodazole, 10 μg/ml; colchicine, 10 μM; cytochalasin B, 1 μg/ml; cytochalasin D, 1 μg/ml; 2 μM latrunculin A (Calbiochem, San Diego, CA; Schatten *et al.*, 1986), all in ASW with 100 μg/ml ampicillin. In cases where DMSO (Sigma Chemicals, St. Louis, MO) was used as the solution vehicle, DMSO was tested alone at the same dilution and was found to have no effect in the assays (Brooks and Wessel, 2003b; Wessel *et al.*, 2002). The cytoskeletal inhibitor concentrations we used effectively disrupt the specific cytoskeletal element but do not damage oocytes, as ascertained by recovery of the cells following treatment. In these experiments, oocytes were isolated and incubated for 15 min either in the presence or absence of inhibitors and then given the reagent to be tested.

F. Labeling of DNA

A variety of fluorescent dyes for nucleic acids detection are presently available. Sea urchin oocytes and eggs contain significant amounts of maternal mRNA stores, making it critical to use dyes that preferentially stain DNA vs RNA, such as DAPI or Hoechst. In contrast, the SYTO dye series from Molecular Probes (Eugene, OR) is less selective for DNA, and can be effectively used only following RNase A treatment of the fixed samples (in our hands, it mostly stains cytoplasm of the live oocytes). Note that DNA in the oocyte is spread throughout a large nucleus (and not concentrated in a small volume as in embryos), which makes it difficult to visualize (Fig. 6).

Hoechst 33342 (Molecular Probes, Eugene, OR) can be effectively used for DNA labeling in live cells as well as in fixed samples, when diluted to 0.1 to 0.2 micrograms/ml in ASW. Note that higher concentrations of Hoechst in seawater may precipitate. Fixed samples generally require a 10-min incubation in Hoechst, followed by a brief wash. We have noticed that the quality of Hoechst labeling deteriorates quickly in the stored fixed samples. Addition of TE to inhibit nucleases in the stored samples is suggested to increase the storage period to about 1 week.

G. Apoptosis Assays

Sea urchin oocytes and eggs possess apoptotic machinery, and can undergo regulated cell death (Voronina and Wessel, 2001). The onset of apoptosis in sea urchin oocytes can be ascribed by several criteria, including cell blebbing and caspase activation. Oocyte blebbing morphology is obvious (Fig. 7) as multiple cytoplasmic fragments will form at the periphery of the cell. In contrast to many other cell types, sea urchin oocytes do not exhibit DNA condensation and degradation during apoptosis, and are, on occasion, seen expelling intact nuclei from a mass of blebs. Therefore, the apoptosing oocytes are usually TUNEL-negative (TUNEL assay detects chromatin fragmentation during apoptosis).

Caspase activation in the oocytes can be easily detected by fluorescence caspase marker, FITC-VAD-fmk (FVf; Promega, Madison, WI; Fig. 7). Staining with FVf is performed according to manufacturer's protocol:

1. Transfer the cells into a 10 μM solution of FVf in ASW in a well of a microtiter plate.
2. Incubate at room temperature for 20 min.
3. Wash the cells twice with ASW; mount on a slide (wet mount).
4. Observe the fluorescence under the appropriate (FITC) filter

H. Mounting the Samples

For visualization under the microscope, the labeled oocyte whole mounts are mounted onto slides. For most applications, wet mount with the wash solutions used for immunolocalizations (TBST/PBST/ASWT). Put 30 μl of your wash buffer on a 22 mm square coverslip with a sparing amount of silicone grease around the edges. Under the dissecting scope, carefully mouth-pipette the oocytes into the center of the drop. Next, invert a glass slide over the coverslip; the water tension will hold the coverslip to the glass slide. Invert the mounted sample coverslip up, and nail polish around the edges to prevent water loss and/or detachment of the coverslip from the slide. Such slides have to be stored horizontally, as the cells are not attached to either the slide or the coverslip surface, and they will fall to the border if the slide is stored vertically.

If fading of fluorochrome is a concern, an antifade solution should be used. We find that commercially available antifade solutions such as VectaShield (Vector Laboratories, Burlingame, CA) do not perform well for the oocyte whole mounts, as they draw the water out of the cells and cause oocytes to "shrivel." An excellent "homemade" alternative is elvanol mounting medium:

Elvanol recipe

0.2 M Tris buffer, pH 6.5 (12 ml)

Polyvinyl alcohol (2.4 g; average MW 30–70 $\times$ 10^3)

Glycerol (6.0 g)

H_2O (6.0 ml)

DABCO (1,4-Diazabicyclo[2.2.2]octane; Sigma cat# D2522) (0.8 g)—anti-fade reagent

Elvanol stock preparation

1. Combine the Tris, polyvinyl alcohol, glycerol, and water. Stir thoroughly and incubate for 1 h at room temperature.

2. Incubate in a water bath at 50 °C and continue stirring until all ingredients are dissolved.
3. Clarify the solution by centrifugation at 5000 g at room temperature for 15 min.
4. Collect the clarified gel-supernatant, add DABCO to 2.5%, and mix into solution.
5. Store as 0.5 ml aliquots at −20 °C.

Mounting. If labeling the cells adhered to the coverslip, invert the coverslip over a small drop of elvanol on the slide. If labeling whole mounts in suspension, gently mouth-pipette the cells into a minimal amount of wash buffer on a small drop of elvanol on a coverslip, and invert the slide over it, as described earlier for the wet mount. When left at room temperature (in the dark), elvanol polymerizes, so nail polishing of the slides is not required. In addition, after polymerization, the embedded whole mounts do not slide to one side when the slide is stored vertically.

VI. Introduction of Experimental Substances into the Oocytes

See the excellent online manual from Laurinda Jaffe at http://egg.uchc.edu/injection for additional detail on both microinjection and oocyte-handling methods and the chapter in this book by Terasaki and Jaffe for additional oocyte experimental methods.

A. Microinjection Procedures: Oocyte Specific Supplies

- Scotch double-coated tape linerless: 1 roll 1/2 inch × 1296 in (36 yd); 12.7 mm × 32.9 m
- Dimethypolysiloxane (oil for loading capillary): Sigma cat #DMPS-1C
- Fluorinert (oil for microinjector): Sigma cat #F-9880
- High vacuum grease (for sealing chambers): Dow Corning, Midland, MI
- Safety Coat nontoxic coating (to treat coverslips and glassware to prevent sticking of oocytes to glass): J. T. Baker, distributed by VWR Scientific Products cat #4017-01.

Microinjection in the Kiehart chamber. The basic procedures for microinjection are described elsewhere (Jaffe: http://egg.uchc.edu/injection; Kiehart, 1982; see also Fig. 8). Here, we only mention the additions to the protocols useful when injecting sea urchin oocytes. The coverslips used for microinjecting the oocytes need to be thoroughly cleaned, as detailed in the general microinjection protocol. In addition, the coverslips used for the experiments with oocytes need to be pretreated so that the oocytes do not stick to them.

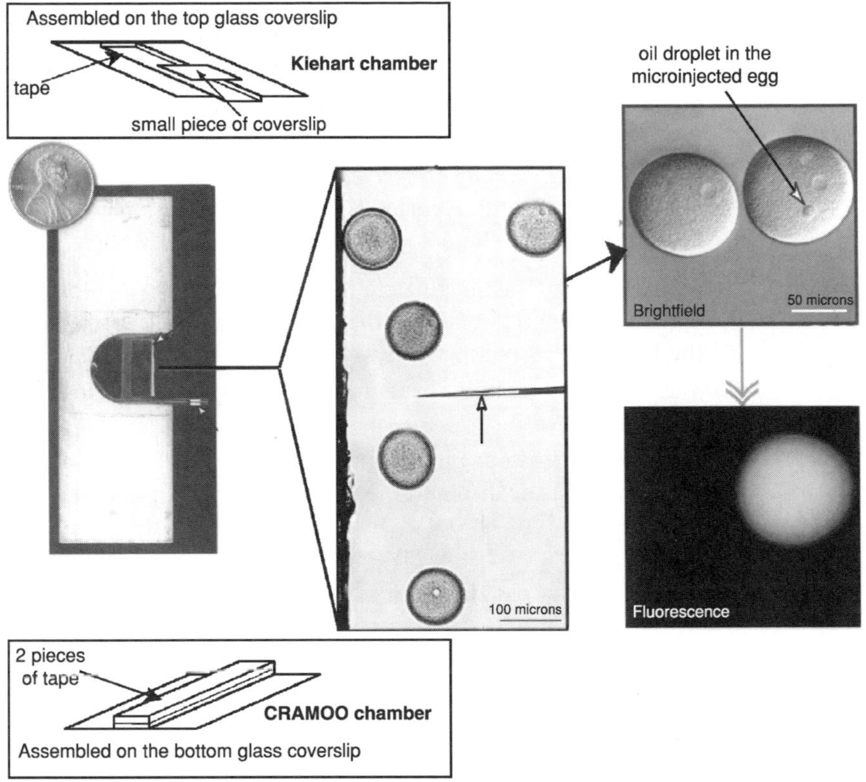

Fig. 8 Microinjection setup. The cells to be microinjected are loaded (left; arrow) in a Kienart chamber or a CRAMOO chamber. The microinjection needle is then brought into the chamber for injection (middle; arrow). The cell injected with a solution of fluorescent dextran contains an oil droplet (top right; arrow) and exhibits fluorescence (bottom right).

Prepare 50 ml conical tube filled with dH_2O with a couple of drops of Safety Coat. Individually drop the coverslips into the solution with forceps, shake; incubate for 30 minutes–1 hour, shaking occasionally. Rinse well with dH_2O (3 times). Store coverslips as usual, under 80% ethanol.

We have recently developed a modification to the basic Kiehart chamber technique, which we call CRAMOO (*c*hamberless *ra*pid *m*icroinjection *o*f *o*ocytes). Instead of assembling a chamber on a coverslip, a double layer of a double-stick tape forms a barrier on the bottom coverslip. This coverslip is attached at the bottom of the support slide, and another coverslip is attached at the top. The assembly is filled with ASW. Then the oocytes are gently loaded onto the bottom coverslip with a mouth pipette. The oocytes sit on this coverslip and can be injected as is, or pushed up against the tape barrier for injection (Fig. 8). CRAMOO is very useful, in that it takes less time to make the chamber, and it is much easier to load, as sea urchin oocytes are quite fragile and are easily torn apart when loaded into a standard Kiehart chamber by capillary action. The disadvantage of CRAMOO is that it is comparatively more difficult to change the incubation media, so for prolonged incubations (over 24 h), the oocytes should be transferred to another container (for example, into wells of a multiwell dish). Microinjected oocytes should be incubated in the humid chamber.

Microinjection in an angled injector setup. To microinject oocytes with an angled injector setup, Andy Ransick (California Institute of Technology, Pasadena, CA) uses agarose tunnels to support cells for microinjection.

1. Pull out Pasteur pipettes to get capillaries ~100 to 150 microns in diameter. With a diamond pencil, cut the capillaries into short (~1 cm) pieces; lay them on the bottom of small petri dish parallel to one another.
2. Make 10 ml of 0.5% agarose in dH_2O, microwave to melt, cool down to ~40 to 50 °C.
3. With a Pasteur pipette, pour enough agarose into the petri dish to submerge the bottom of the dish and the capillaries. Then, immediately aspirate all the liquid agarose. The bottom of the dish is now coated with the agarose, and the capillaries are embedded into the agarose. Allow the agarose to solidify.
4. Carefully cut the agarose plugs from the ends of capillaries, push the capillaries out, and save for later reuse. You are left with agarose tunnels on the bottom of the petri dish.
5. Fill the petri dish with ASW. Mouth-pipette the oocytes into the tunnels by aspirating the oocytes in a limiting volume of ASW, inserting the end of the mouth pipette into a tunnel, and gently expelling the oocytes into it.
6. Proceed to inject cells through the agarose as usual.

This technique allows rapid microinjection of large numbers of cells, immobilizes the injected cells such that they can always be uniquely identified, and still allows observation of the samples under the fluorescent microscope (inverted). Furthermore, the exchange of incubation media is very easy. If the injected cells must be fixed for further analysis, they can be easily flushed out of tunnels with a mouth pipette.

Following the injected cells. The cells microinjected in Kiehart chamber (or CRAMOO) can be easily identified by the oil droplet introduced by microinjection. An alternative method especially helpful when planning to immunolabel the cells or using continuous-flow microinjector is to co-inject fluorescently labeled dextran with the solution of interest (Fig. 8). Usually, injection of 0.1 mg/ml of fluorescent dextran (labeled with rhodamine or Oregon Green; Molecular Probes) provides a sufficient signal for detection.

RNA Microinjection:

The artificial expression of a defined protein fused with green fluorescent protein (GFP) has been applied to a wide variety of cells. We find that the presence of a poly(A) tail is necessary for the translation of GFP RNA injected into sea urchin oocytes. When capped GFP RNA is *in vitro* transcribed using the Message

Machine kit (Ambion, Austin, TX) and injected into the cytoplasm of oocytes, GFP fluorescence could not be detected. The integrity of the transcript was verified by its subsequent injection into fertilized eggs where GFP fluorescence could be detected after several hours. Subsequently, when this same transcript was *in vitro* polyadenylated with poly (A) polymerase (Ambion, Austin, TX) and injected into oocytes, fluorescence was detected after several hours, as seen in the fertilized eggs. Therefore, it seems that sea urchin oocytes require capped-GFP-poly(A) RNA for GFP activity (Brooks and Wessel, unpublished results). Recently, it has also been shown that mouse oocytes show increased GFP expression when the injected RNA is added with a long poly(A) tail (Aida *et al.*, 2001).

B. Endocytosis

Sea urchin oocytes are endocytically very active. This feature allows one to introduce certain experimental substances into the oocytes by uptake from the media. We were successful in using this route to treat the oocytes with antisense oligonucleotides (Voronina *et al.*, 2003) and for uptake of labeled major yolk protein (Brooks and Wessel, 2003b). In our experiments, 10 μM antisense cyclin B oligonucleotide solution in ASW was sufficient to interfere with the progression of oocyte maturation. The cells took up the oligonucleotides from the medium, as was also seen in mammalian tissue culture (reviewed in Dokka and Rojanasakul, 2000). In contrast, sea urchin eggs and embryos appeared resistant to such treatments, suggesting that the efficiency of oligonucleotide uptake is greatly reduced in these stages. The eggs are much less active in endocytosis, and the embryos have a robust extracellular matrix that may hinder entry of macromolecules.

VII. Concluding Remarks/Outlook

Sea urchin eggs have been a champion for many experimental questions over the past century, but little is known about the regulation and development of its oocytes. In contrast, starfish oocytes have been used for the past 30 years, especially in studies of meiotic maturation. The technical applications of its meiotic reactivation is well documented so that, at will, an investigator can stimulate a population of (manipulated) oocytes to begin meiosis and become fertilization competent. What is becoming feasible now is to perform many of the ocoyte experiments in sea urchins, and thereby take advantage of the vast resources of genome information, antibody and cDNA reagents, and seasonality. Perhaps more importantly, it enables investigators to use the rich diversity of echinoderms to maximize the experimental model, and to use comparison biology to learn what features are generally conserved. Surprisingly, the mechanisms of meiotic maturation are greatly diverse among species of a family. Perhaps this is best seen in echinoderms, where Echinoids represent one of only two animal groups that complete meiosis prior to fertilization, whereas Asteroids more closely resemble some vertebrates that are fertilized during the meiotic process. These differences are, at the same time, complicating in practice, but enriching in our biological understanding. Hopefully, this chapter will encourage additional venture into a broader array of oocyte models and applications.

Acknowledgments

We are grateful to the following individuals for their contributions to this work: Linnea Berg for attention to detail in the early development of the sea urchin oocyte techniques; Rindy Jaffe for excellent documentation and dissemination of oocyte handling and microinjection procedures; Bradley Schnackenberg for introducing us to his Elvanol recipe; Sheila Haley for advances in the whole mount immunolabeling protocol; Mariana Leguia for protocol development; Andy Ransick for sharing his agarose tunnel microinjection approach; and Sean Conner, for among other things, acronyms. We gratefully acknowledge the support for work in this laboratory from the National Institutes of Health and the National Science Foundation.

References

Aida, T., Oda, S., Awaji, T., Yoshida, K., and Miyazaki, S. (2001). Expression of a green fluorescent protein variant in mouse oocytes by injection of RNA with an added long poly(A) tail. *Mol. Hum. Reprod.* **11,** 1039–1046.

Berg, L., and Wessel, G. M. (1997). Cortical granules of the sea urchin translocate early in oocyte maturation. *Development* **124,** 1845–1850.

Brooks, J. M., and Wessel, G. M. (2003a). A diversity of yolk protein dynamics and function. *Recent Devel. Cell Res.* **1,** 1–30.

Brooks, J. M., and Wessel, G. M. (2003b). Selective transport and packaging of the major yolk protein in the sea urchin. *Dev. Biol.* **261,** 353–370.

Costello, D. P., Davidson, M. E., Egger, A., Fox, M. H., and Henley, C. (1957). "Methods for Obtaining and Handling Marine Eggs and Embryos." Lancaster Press Inc., Lancaster, PA.

Czihak, G. (1975). "The Sea Urchin Embryo." Springer-Verlag, Berlin, Germany.

Dokka, S., and Rojanasakul, Y. (2000). Novel non-endocytic delivery of antisense oligonucleotides. *Adv. Drug Deliv. Rev.* **44,** 35–49.

Harvey, E. B. (1956). "The American Arbacia and Other Sea Urchins." Princeton University Press.

Jaffe, L. Microinjection manual. http://egg.uchc.edu/injection

Kiehart, D. (1982). Microinjection of echinoderm eggs: Apparatus and procedures. *Meth. Cell Biol.* **25,** 13–31.

Maruyama, Y. K. (1980). Artificial induction of oocyte maturation and development in the sea cucumbers *Holothuria leucospilota* and *Holothuria pardalis*. *Bio. Bull.* **158,** 339–348.

Schatten, H., Schatten, G., Mazia, D., Balczon, R., and Simerly, C. (1986). Behavior of centrosomes during fertilization and cell division in mouse oocytes and in sea urchin eggs. *Proc. Natl. Acad. Sci. USA* **83,** 105–109.

Smiley, S. (1990). A review of Echinoderm oogenesis. *J. Electron Microscopy* **16,** 93–114.

Voronina, E., and Wessel, G. M. (2001). Apoptosis in oocytes, eggs, and early embryos of the sea urchin. *Mol. Reprod. Dev.* **60,** 553–561.

Voronina, E., Marzluff, W. F., and Wessel, G. M. (2003). Cyclin B synthesis is required for sea urchin oocyte maturation. *Dev. Biol.* **256,** 258–275.

Voronina, E., and Wessel, G. M. (2003) The regulation of oocyte maturation *In* "Current Topics in Developmental Biology" (G. Schatten, ed.), pp. 53–110, Vol. 58. Academic Press.

Wessel, G. M., Berg, L., and Conner, S. D. (2002). Cortical granule translocation is linked to meiotic maturation in the sea urchin oocyte. *Development* **129,** 4315–4325.

Whalley, T., Terasaki, M., Cho, M. S., and Vogel, S. S. (1995). Direct membrane retrieval into large vesicles after exocytosis in sea urchin eggs. *J. Cell Biol.* **131,** 1183–1192.

CHAPTER 6

Procurement and Culture of Ascidian Embryos

Billie J. Swalla

Biology Department and Friday Harbor Laboratories
University of Washington
Seattle, Washington 98195

METHODS IN CELL BIOLOGY, VOL. 74

0091-679X/04 $35.00

I. Overview

Ascidians are marine invertebrate chordates, with a tadpole larva that metamorphoses into a sessile adult. Ascidian embryos have been used for embryological experiments for over 100 years, and are still an important molecular model system for understanding chordate evolution and development. Many ascidian species are easy to harvest from docks and ropes hanging in the water for anyone interested in watching their development. Gametes are readily obtained from gravid adults, development is rapid, and the eggs and embryos can be experimentally manipulated. Cleavage is bilateral, determinant, and invariant for solitary species, and the cell lineage of solitary ascidians is known. Hatching occurs after 12 to 36 h of development and metamorphosis can be induced and observed easily. Ascidian cell lineages can be followed with simple enzymatic staining, muscle cells can be visualized with acetylcholinesterase and endoderm followed with alkaline phosphatase that students can accomplish on the first try. Cell separation and cleavage arrest experiments can be carried out in addition to histological staining for understanding mosaic development. Gene expression can be observed with *in situ* hybridization and functional experiments can be carried out by overexpressing and/or knocking out specific genes. Furthermore, there are closely related solitary and colonial species as well as tailed and tailless ascidian species, making them excellent choices to study the evolution of larval morphology.

II. Ascidian Development and Metamorphosis

In the late 1800s, Laurent Chabry, working in France, separated the first two blastomeres of a *Ciona intestinalis* ascidian embryo, and began the field of experimental embryology (Chabry, 1887). Since then, ascidian embryos have continued to be an excellent model system for developmental studies. In 1905, Edwin G. Conklin published the cell fate map for ascidian embryos by carefully mapping the localization of the colored cytoplasms in another ascidian, *Cynthia partita* (now known as *Styela canopus*) at the Marine Biological Laboratory in Woods Hole, MA (Conklin, 1905). Recently, the genomes of two related species, *Ciona intestinalis* and *Ciona savignyi*, have been sequenced, allowing even greater molecular access for developmental biologists to study ascidian embryos (Dehal *et al.*, 2002; http://www2.bioinformatics.tll.org.sg/ciona_savignyi). For a good overview of ascidian development and metamorphosis, see Satoh (1994) and Jeffery and Swalla (1997). The purpose of this chapter is to allow even the novice student to be able to work with ascidian embryos, culturing and metamorphosing them while performing experiments to understand their development and evolution. Ascidians are filter feeders as adults and thus prefer fresh seawater from the oceans. They are best studied at marine labs near the ocean, so breathe in the salt air, eat seafood, and enjoy yourself!

A. Obtaining and Maintaining Ascidians

Ascidians are relatively easy to collect and maintain in running seawater. See Table I for a list of widely used ascidians and how and where they may be obtained. Taxonomic keys that may be especially helpful include Van Name (1945), Berrill (1950), Abbott (1975), Hayward and Ryland (1990), Kott (1985, 1990, 1992), Monniot and Monniot (1996), and Monniot *et al.* (1991, 2001).

B. Gamete Collection and Fertilization

Ascidian embryos are an excellent model system for developmental studies for a number of reasons. First, gametes can be collected throughout the year in large numbers in many species. Second, synchronous development and high fertilization rates are easy to achieve with appropriate handling of gametes. Third, ascidian cleavage is invariant and determinant in all solitary species that have been studied, allowing comparison of cell lineages across species. The phylogenetic relationship of various ascidian species within the deuterostomes is shown in Fig. 1 (Cameron *et al.*, 2000; Swalla *et al.*, 2000). Ascidians are part of the phylum Tunicata, which is considered a sister group to the vertebrata and cephalochordata. There are three orders of ascidians—phlebobranchs, stolidobranchs, and aplousobranchs—although phylogenetic analysis suggests that the Molgulidae family should also be considered a separate order (Fig. 1; Swalla *et al.*, 2000). The relationship of the aplousobranchs to the other orders of ascidians has not yet been resolved. Several stolidobranchs have colored cytoplasms, allowing students and researchers to follow specific lineages with natural pigments during development (Fig. 2), and many phlebobranch species have crystal clear eggs, making them excellent for immunocytochemistry and gene expression studies.

All ascidians are hermaphrodites, producing both eggs and sperm when gravid. However, most species of ascidians are self-sterile. Because the gonad structure and architecture differ in ascidians, each order will be treated separately. The advantages and disadvantages of different species for various studies will be discussed throughout this chapter.

1. Phlebobranchs

Ripe eggs and sperm accumulate in ovarian and sperm ducts in phlebobranch ascidians, allowing easy gamete collection. None of the widely used phlebobranchs have colored cytoplasms, but some of them have crystal clear embryos, making them excellent species for immunological studies and experiments utilizing fluorescent dyes (Sardet *et al.*, 1989, 1992; Zalokar, 1974). *Ciona* gametes can be obtained by natural spawning or, more reliably, by cutting through the mantle and squeezing the gametes out through the gonoducts. In gravid animals, the two gonoducts are visible through the mantle; one is white (sperm duct) and the other is pinkish, yellowish, or brownish (ovarian duct). To obtain eggs and sperm, cut

Timetable for *Ciona intestinalis* Development (18 °C)

Stage	Time after fertilization
2-cell	60 min
4-cell	90 min
8-cell	120 min
16-cell	160 min
32-cell	210 min
64-cell	250 min
Early gastrula	5 h
Neurula	7 h
Early tailbud	9 h
Otolith pigmentation	12 h
Ocellus pigmentation	15 h
Tadpole	18 h

through the tunic and underlying body wall of a *Ciona* adult. Be careful not to cut into the internal organs. Spread the body, cut the ovarian duct first, and gently squeeze the unfertilized (matured) eggs into a dish with seawater. Alternatively, blot the exposed gonoducts with a Kimwipe, puncture the oviduct with a needle held parallel with the long axis, and pick up the eggs with a Pasteur pipette that has been moistened with seawater. After collection, wash the eggs once or twice by gently settling, pouring off the water, and then resuspending in filtered seawater. Sperm are obtained by cutting the sperm duct. Dilution of sperm is recommended, although polyspermy is rare in ascidians. Avoid overcrowding of eggs and keep them cool, to match the local seawater temperature. Other phlebobranchs such as *Phallusia*, *Ascidiella*, and *Ascidia* are even easier to dissect. Simply remove the tunic with forceps and collect the eggs from the exposed oviduct, which is visible as soon as the tunic is removed. Drying the oviduct with a Kimwipe before puncture will allow the eggs to stick to the outside of a Pasteur pipette; it will also help to avoid losing eggs that are washed away.

a. *Ciona intestinalis*—*Ciona intestinalis* has been used for years for elegant embryological experiments done originally by Reverberi and Ortolani at the Stazione Zoologica in Naples, Italy. Rosario de Santis and Roberto Di Lauro continue to study ascidian embryos at this historical marine lab with large research groups. *Ciona intestinalis* has been developed for genetic studies in Naples (see Part I, 7). If you love ascidians, you really should consider traveling to Naples and spending some time working in this historical lab. *Ciona intestinalis* has been sequenced, making it an excellent choice for developmental studies (Dehal *et al.*, 2002).

b. *Ciona sayvigni*—*Ciona sayvigni* has also been sequenced (http://www2.bioinformatics.tll.org.sg./ciona_savignyi) and has been used to generate tissue-specific monoclonal antibodies and for cell lineage studies by Noriyuki Satoh's group and Hiroki Nishida's laboratory in Japan (Hirano and Nishida, 1997, 2000; Nishida, 1987; Nishida and Satoh, 1983, 1985) Recently, William Smith has

Table I
Places to Obtain Different Ascidian Species

MBL Woods Hole, MA	HFHL Friday Harbor, WA	Stazione Zoologica Naples, Italy	Station Biologique Roscoff, France	Japan	Hopkins Marine Station, Pacific Grove, CA
Ciona intestinalis	*Ascidia columbiana*	*Ciona intestinalis*	*Ciona intestinalis*	*Ciona savignyi*	*Ciona intestinalis*
	A. paratropa		*Phallusia mammillata*	*Ciona intestinalis*	*Ascidia ceratodes*
Styela clava	*Boltenia villosa*		*Styela clava*	*Halocynthia roretzi*	*Styela clava*
				Styela plicata	*Styela plicata*
				Herdmania pallida	
Botryllus schlosseri		*Botryllus schlosseri*	*Botryllus schlosseri*	*Botrylloides violaceus*	*B. violaceus*
					B. schlosseri
Botrylloides violaceus	*Botrylloides violaceus*			*Botryllus schlosseri*	
Molgula manhattensis	*Molgula pugetiensis*		*Molgula oculata*	*Molgula tectiformis*	
M. provisionalis			*M. occulta*		
M. citrina			*M. citrina (echinosiphonica)*		
			M. socialis		

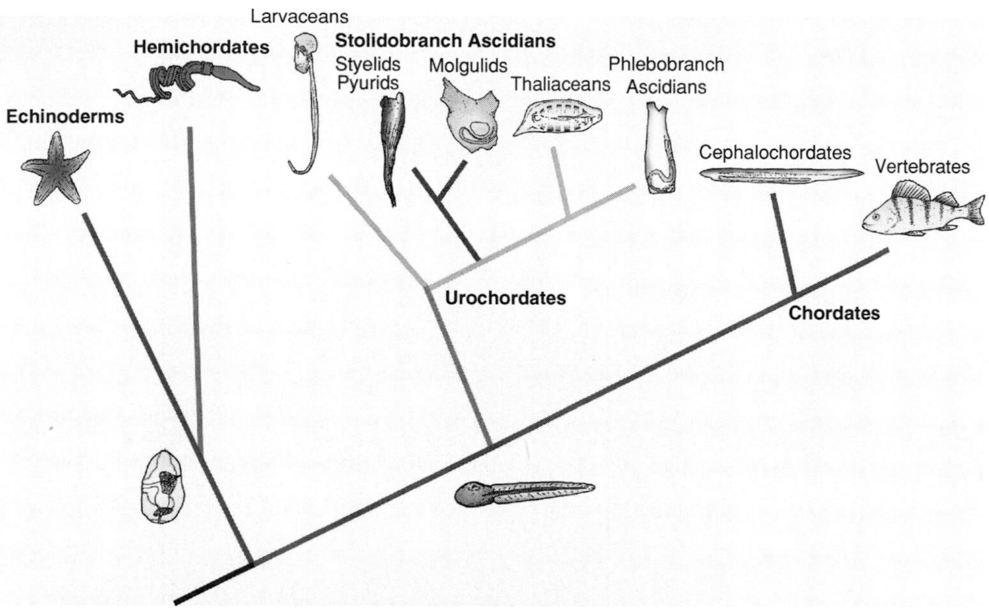

Fig. 1 Deuterostome Evolution—The Tunicata is the sister group of vertebrates and Cephalochordata = Acrania (Lancelets). Echinoderms and hemichordates are sister groups, and together form an outgroup to the chordates. The Tunicata have at least four distinct clades, the sessile ascidians, the planktonic Thaliacea (including Pyrosomes, salps, and doliolids), and the planktonic Appendicularia (formerly called larvaceans). Modified from Swalla, 2001.

developed genetics with *Ciona savignyi* at the University of California at Santa Barbara (see Section IV.A).

c. *Phallusia mammillata*—*Phallusia mammillata* has been used by Christian Sardet's research group at Villefranche in France (Roegiers *et al.*, 1999; Sardet *et al.*, 1992) and by Thomas Honegger's group in Switzerland (Koyanagi and Honegger, 2003). This species has crystal clear eggs, making it excellent for antibody studies and also for studies of early fertilization. It has the advantage of a sperm which is much larger than that of *Ciona*.

2. Stolidobranchs—Pyurids, Styelids, and Molgulids

Gametes in stolidobranchs are obtained by dissection of the gonad. The tunics of styelids and pyurids can be very tough, so take care not to cut yourself! The tunic can be cut with scissors or a razor blade by cutting along a plane through the siphons. Be very careful not to cut into the internal organs; this may release body fluids and reduce the yield of fertilizable eggs. Spread the body halves out, flatten them in a dish and carefully dissect the branchial sac away from the gonads. In most species, the testes (white) and the ovaries (colored) are adjacent and run parallel to each other. Excise the gonads and place in a Syracuse

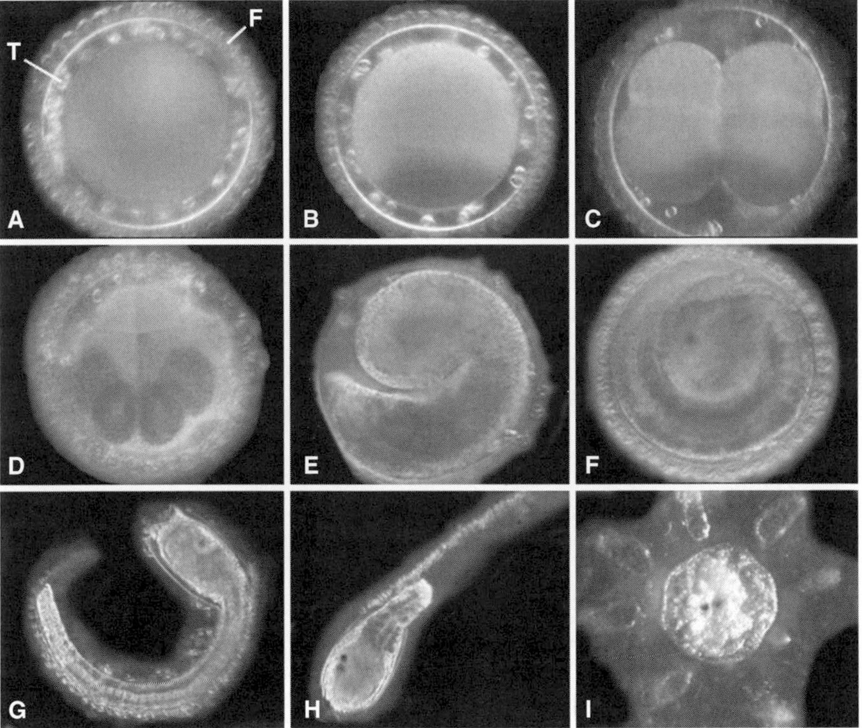

Fig. 2 *Boltenia villosa* development. In *Boltenia villosa*, the myoplasm is colored a dark orange by pigment granules in the egg as in other pyurids and some styelids. (A) An unfertilized egg. The orange myoplasm is seen throughout the cortex of the egg. The test cells (T) float around the egg inside of the thick chorion. Attached to the chorion are a number of follicle cells (F). Most ascidian eggs have both test cells and follicle cells. (B) After fertilization, the myoplasm contracts toward the vegetal pole (bottom); this location will become the dorsal side of the embryo. The myoplasm marks where gastrulation will begin at the vegetal pole. Next, the myoplasm moves to the future posterior of the embryo (not shown). (C) At the 4-cell stage, the polarity of the embryos is visible by the bright orange myoplasm at the posterior (anterior is up). (D) A 16-cell embryo shows the distinct bilateral cleavage in an ascidian embryo. The plane of bilateral symmetry is in the center, anterior is up. At this stage, there are 4 myoplasm-containing cells, seen by the orange blastomeres in this vegetal view of the embryo. (E) After gastrulation, the tailbud embryo has a white head and the orange muscle cells surround the notochord in the posterior. (F) The tadpole larva just before hatching has undergone extensive convergence and extension, so now the tail wraps around the head and the white notochord cells are visible in the center of the tail. (G) A freshly hatched tadpole, with test cells still clinging to it. Note the palps (adhesive papillae) at the anterior of the larva. (H) After the larva swims for a period of time, the tail retracts during the process of metamorphosis. (I) In *Boltenia villosa*, the metamorphosing juvenile flattens down and makes a number of ectodermal ampullae that radiate out from the larva. (See Color Insert.)

dish containing seawater. Eggs should be washed with Nytex mesh before fertilizing. For cross-fertilization, mix sperm suspensions from two or more individuals; the suspension should be murky with sperm. Usually, enough sperm is present in gonad mixtures for fertilization. No pyurids are self-fertile. Some styelids are self-fertile but better fertilization will result from cross-fertilization. Therefore, you should routinely cross-fertilize stolidobranchs.

a. *Halocynthia roretzi*—*Halocynthia roretzi*, an ascidian found in Japan, has been used extensively for experimental and molecular biological studies because of the large size of the egg and the ease of obtaining adults (Hirano and Nishida, 1997, 2000), which are grown by mariculture. A number of EST studies have been published for *Halocynthia roretzi* (MAJEST: Maboya Gene Expression patterns and Sequence Tags; http://www.genome.ad.jp/magest/). This species is completely self-sterile so fertilization does not occur until you cross-fertilize.

b. *Styela clava*—*Styela clava* is available worldwide in temperate regions, primarily because it tends to be an invasive species in ports. In some populations, it has a yellow myoplasm in the eggs and embryos. Fertilizable eggs can be obtained by dissection from the gonads. Place a piece of 300 μ Nytex over a 100 ml beaker of seawater. Cut a specimen through the siphons, remove the branchial basket, and place the gonads on top of the Nytex. Mince the gonads with fine scissors and gently mash the gonads through the Nytex with your forceps. Now with a Plexiglass cylinder slightly smaller than the beaker with a 100 μ Nytex glued on the bottom surface, wash the sperm away by withdrawing seawater from the cylinder with a 5 ml pipette, leaving the eggs behind. Refill the beaker with fresh seawater.

c. *Styela plicata*—*Styela plicata* also has a yellow crescent in the embryo. This species has been used for molecular studies and for studies of blood cells. It is an easy species to spawn reliably in the summer with a light–dark cycle (West and Lambert, 1976).

d. *Botryllus spp., Botrylloides spp.*—The botryllid ascidians are all colonial. They are easy to grow in culture and have been used for several decades for genetic and ecological studies. The entire colony will be at the same life history stage. Therefore, not every colony will have the stage you are interested in, so you may have to dissect several colonies.

e. *Molgulids*—The Molgulidae are an interesting family of ascidians because they have almost equal numbers of tailed and tailless species (Fig. 3; Huber *et al.*, 2000) Many of the Molgulidae are self-fertile, making them an excellent choice of study for students who have not had a lot of experience in fertilization. *Molgula occidentalis* is a commercially available tailed species that yields 100% fertilization from a single animal. *M. occidentalis* is tolerant to ambient temperatures and can be used for middle school and high school labs. This species never fails to delight students with the large number of tadpoles that develop and metamorphose on the lab bench in filtered seawater. *M. occidentalis* can be ordered from Gulf Specimen Co., Panacea, FL.

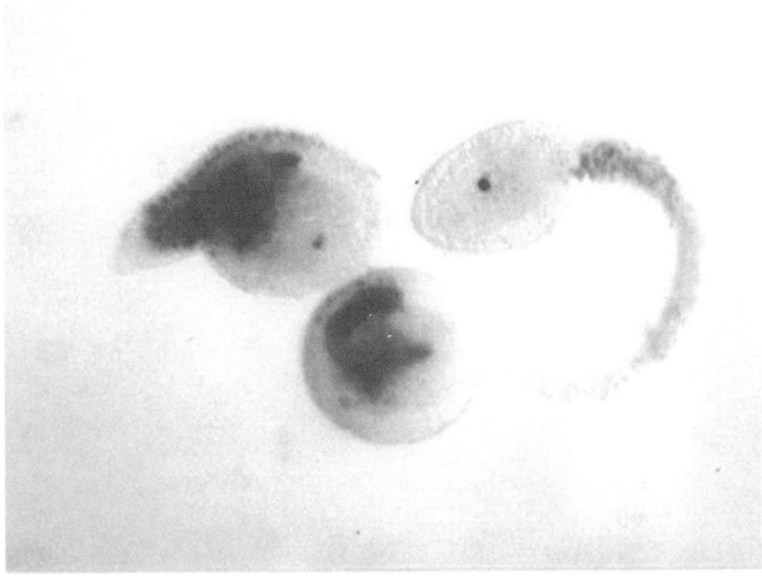

Fig. 3 Three *Molgula* larvae after hatching that have been stained with acetylcholinesterase to visualize the muscle lineage cells. At the bottom is *Molgula occulta*, a tailless ascidian that is found off the coast of Roscoff, France. This species does not have a pigmented otolith or muscle genes expressed in the larva, but the muscle lineage cells attain dark brown coloration with the acetylcholinesterase stain. The larva to the right is a closely related tailed species *Molgula oculata*; the pigmented otolith is seen in the head and acetylcholinesterase stains the muscle cells brown in the tail. The hybrid larva to the left is a result of a tailless *Molgula occulta* egg fertilized by a tailed *Molgula oculata* sperm. A short tail has developed which contains 20 notochord cells instead of 40. The hybrid also has a pigmented otolith in the head.

3. Aplousobranchs—Didemnidae, Polyclinidae, Polycitoridae

All aplousobranch ascidians are colonial (Van Name, 1945). Colonial ascidians have been less well studied developmentally than have solitary ascidians because their embryos are brooded in brood chambers and must be dissected from the brood pouch. However, colonial ascidians also have a large range of larval phenotypes and so are interesting evolutionarily (Satoh, 1994; Davidson *et al.*, 2004). One of the best reviews on the development of colonial ascidians is Berrill (1935, 1936).

C. Larval Development

1. Fertilization

Eggs from every species of solitary ascidian have an outer layer of follicle cells surrounding a thick chorion (Fig. 2A). A number of test cells are located between the chorion and the egg (Fig. 2A). These extra-embryonic layers are important to prevent self-fertilization and polyspermy. Any experimental manipulation of embryos requires that they be dechorionated first. Protocols for dechorionation are included in

Section IV.B. Pronase E and sodium thioglycolate are mixed in particular proportions. For each species, the concentration of pronase E can be adjusted depending on the length of time that it takes to dechorionate the eggs. Increase the pronase concentration if your species takes longer than 10 min to dechorionate the eggs or embryos and decrease it if they go faster than 10 min.

2. Ooplasmic segregation

Following fertilization, the myoplasm in the cortex contracts towards the vegetal pole (Fig. 2B). The place of contraction defines the future site of gastrulation at the vegetal pole and determines the dorsal side of the embryo. The cytoplasm of unfertilized *Styela* or *Boltenia* eggs has colored pigment granules in the myoplasm, found in the egg periphery (Fig. 2A). After fertilization, the myoplasm undergoes two distinct movements, setting up the embryonic axes. The first cytoplasmic movement occurs while the egg is finishing its maturation division. It is initiated by sperm entry and consists of the myoplasm contracting toward the vegetal pole by the contraction of microfilaments in the cortex. At the end of the first movement, the myoplasm is collected in a cap at the vegetal pole, marking the point of invagination during gastrulation (Fig. 2B). This will be the future dorsal side of the larva. The second phase of ooplasmic segregation follows sperm aster formation and the movement of the male pronucleus toward the animal hemisphere, where it will eventually fuse with the female pronucleus. The myoplasmic pigment granules move upward along the vegetal periphery of the egg, pulled by the sperm aster, until they reach what will be the future posterior region of the embryo. At this location, they spread out to form the yellow or orange myoplasmic crescent. At the end of ooplasmic segregation, the three major axes of the embryo are established. The myoplasm will be distributed to the larval tail muscle and mesenchyme cells during embryogenesis.

3. Cleavage

Ascidian cleavage is determinate and invariant in all solitary species that have been examined. Ascidians have bilateral cleavage, with the first cleavage furrow separating the embryo (and later, the larva) into a left and right half. The posterior can be discerned in some species by the presence of the colored myoplasm (Fig. 2). The second cleavage furrow divides the embryo into an animal half and vegetal half. The animal–vegetal axis demarcates the future dorsal–ventral axis, with the vegetal half becoming the future dorsal side of the larva. After the third cleavage, the 8-cell embryo consists of 4 animal blastomeres and 4 vegetal blastomeres. Each pair of anterior and posterior blastomeres have specific cell fates, as observed if isolated at this stage and cultured (Section IV.C; Deno *et al.*, 1985; Nishida, 1992).

4. Gastrulation

For a recent review of gastrulation, see Swalla (2004). The ascidian embryo begins gastrulation by invagination of the yolky endodermal cells at the vegetal pole. The endodermal cells move toward the animal pole; at the mid-gastrula stage, the notochord cells are found anterior of the blastopore and the muscle cells are found at the posterior. The 10 notochord precursor cells divide twice to make a total of 40 notochord cells that converge and extend, elongating the embryo along the anterior–posterior axis.

5. Tailbud

The tailbud embryo is formed with three rows of muscle cells surrounding the notochord in the tail and the endoderm located in the head (Fig. 2E). Muscle cells are colored in some species or can be visualized by acetylcholinesterase staining (see Section III.A.1). The endoderm in the head can be visualized by simple alkaline phosphatase histochemistry (see Section III.A.2). In most species, the embryo continues to converge and extend until the tail wraps around the head in the chorion and the pigmented cells of the ocellus and otolith develop (Fig. 2F). Then the embryo hatches from the chorion (Fig. 2G) and the nonfeeding tadpole swims for a short period of time before settling (Fig. 2H).

D. Metamorphosis

After swimming for several hours to a few days, ascidian larvae undergo metamorphosis (Cloney, 1978; Davidson *et al.*, 2003). Larvae become competent to settle and secrete adhesives at the anterior, where they will stick to a suitable substrate by the adhesive papillae (Cloney, 1978). The tail is pulled in and undergoes apoptosis, then the organ rudiment rotates 90° so that the siphons are pointing upward. The swimming larvae initiate new transcription in the process of becoming competent (Davidson and Swalla, 2001). In stolidobranch species, competent larvae can be induced to settle by the addition of 50mM KCl to the seawater (Degnan *et al.*, 1997).

Swimming larvae can be transferred to a multiwell dish with 50 to 100 larvae put into each well. Several hours after hatching, 50mM KCl can be added to the dish and the number of larvae that retract their tails within 30 min can be recorded. Once 50% of the larvae retract their tails within 30 min of KCl addition, they are considered competent. Ascidian larvae can be synchronized by adding KCl and then plating out on dishes or on coverslips to observe the later stages of metamorphosis.

Metamorphosis involves rapid morphogenetic changes during settling (Cloney, 1978, 1982; Davidson *et al.*, 2003) including (1) secretion of adhesives by the papillae or the epidermis; (2) eversion and retraction of the anterior papillae;

(3) resorption of the tail; (4) loss of the outer larval tunic; (5) emigration of blood cells or pigment cells into the tunic; (6) rotation of visceral organs through an arc of about 90°, expansion of the branchial basket, and elongation of the oozooid or juvenile; (7) expansion, elongation, or reciprocation of ampullae, reorientation of test vesicles, and expansion of the tunic; (8) retraction of the sensory vesicle; (9) phagocytosis of visceral ganglion, sensory organs, and cells of the axial complex; (10) release of organ rudiments from an arrested state of development (Cloney, 1978, 1982). Some events of metamorphosis may be completed in seconds or minutes (papillary eversion and tail resorption), and others may take hours or days (rotation, ampullar outgrowth, phagocytosis of the axial complex). You can observe these events by watching the larvae each day and/or by setting up a time-lapse of the process.

1. Metamorphosis of Colonial Ascidian Larvae

All colonial ascidians brood their larvae, a form of development that may be truly viviparous in a few species such as *Botrylloides violaceus*, but in most is ovoviviparous; the developing embryo is brooded but not nourished. Colonial species sequester large eggs in brood pouches, usually found on the bottom of the colony. They release large larvae (at dawn) rather than eggs. Larvae of the colonial ascidians *Botrylloides diegensis* (a species limited to southern California) or *Botrylloides violaceus* can be observed metamorphosing in real time if you float adult colonies in plastic colanders in the sea tables; the large tadpoles will be trapped and can be collected swimming around the dish in the morning. To examine embryos, you must dissect these colonies by flipping them over and gently pulling the colony apart. The larvae are quite opaque but they may start to undergo metamorphosis quickly (within 20 min) and are extremely hardy. Watch the process of larval tail resorption under the microscope. Isolate some of the larvae into clean seawater dishes with coverslips or slides (*before* they metamorphose) and you will find small zooids in your dishes. Many colonial larvae are released with 2 or more buds already developed, so they might have 4 to 5 individuals immediately after settling.

III. Larval Tissue Specification

A. Histochemical Methods

1. Acetylcholinesterase (Larval Muscle and Adult Nervous System)

Acetylcholinesterase is used to visualize muscle cells in ascidian larvae or partial embryos (Whittaker, 1973; Whittaker and Meedel, 1989). Fix embryos in 5% formalin in ASW for 30 min. The embryos then need to be washed extensively with phosphate buffer pH 6. If the embryos are not washed extensively, a dull brown precipitate can form on the eggs and embryos, making it difficult to see the true staining pattern. The pH is also very important; if it is too high or too low, the reaction will not occur.

Make up fresh staining medium:

1. 10 mg of acetylthiocholine iodide in 6.5 ml of 0.1 M sodium phosphate pH 6.
2. Add in order the following and be sure to stir after each addition:
   (1) 0.5 ml 0.1 M sodium citrate
   (2) 1 ml 30 mM cupric sulfate
   (3) 1 ml water
   (4) 1 ml 5 mM potassium ferricyanide

(final color should be clear and green and mixture is stable for only a few hours).

3. Rinse embryos 4 to 5× in 0.1 M phosphate buffer pH 6.5 to remove formalin.
4. Incubate embryos for 2 to 4 h at 37 °C, checking the reaction periodically. The brown reaction will develop over time. If you are getting an unsightly precipitate, it is likely that you did not rinse enough times in the phosphate buffer. (For controls, incubate embryos in the absence of substrate.)
5. Stop the reaction in distilled water. This also clears the embryos a bit.

The samples should then be dehydrated and can be stored in 70% ethanol for several weeks. Permanent slides can be prepared by dehydrating samples in a graded series of ethanols (30%, 50%, 70%, 80%, 90%, 100%, 100%), clearing in xylene or toluene, and mounting in Permount.

2. Alkaline Phosphatase (Larval and Adult Endoderm)

Alkaline phosphatase (AP) can be used to stain the presumptive endoderm in embryos, larvae, or adults of ascidians (Whittaker, 1990). This simple procedure allows identification of the endoderm in intact, partial, or cleavage-arrested embryos (Whittaker and Meedel, 1989). In order to examine the tissues expressing alkaline phosphatase, first fix specimens with 4% paraformaldehyde in seawater for 15 to 30 min on ice. After fixation, rinse twice in Buffer 3 (100 mM Tris-HCI, pH 9.5, 100 mM NaCl, 5 mM $MgCl_2$), then incubate in AP Detection Buffer for 30 to 90 min until the desired blue stain is achieved. We normally use the detection solution from *in situ* hybridization kits, but it is also possible to buy alkaline phosphatase reagent from Sigma. It is called Nitro-BT Stock Solution: Dissolve 4-nitroblue tetrazolium chloride (Sigma N6876) in 50 mg/ml of 70% DMF (N,N-dimethyl formamide). The AP Detection Buffer will turn bright blue in areas where endogenous alkaline phosphatase is located. The samples should then be dehydrated and can be stored in 70% ethanol for several weeks. Permanent slides can be prepared by dehydrating samples in a graded series of ethanols (30%, 50%, 70%, 80%, 90%, 100%, 100%), clearing in xylene or toluene, and mounting in Permount.

B. *In Situ* Hybridization

Embryos should be dechorionated and fixed in 4% paraformaldehyde in 0.5 M NaCl, 0.1 M MOPS (pH 7.5) at room temperature (RT) for 90 min and then washed in PBS.

DIG-Labeled RNA Probes:

The first step in making a probe from a cloned cDNA in a plasmid is to linearize the cDNA. This can be accomplished either by appropriate restriction enzyme linearization or by PCR with the appropriate primers. To make DIG-labeled RNA probes, the Roche DIG RNA Labeling Kit (SP6/T7) is a simple one to use for students in a course. For plasmids with T3 and T3 Polymerase, T3 10X transcription buffer are used instead of the T7 buffer. Before either the T7/T3 Polymerase is added, samples should be spun briefly at high speed in a microfuge. After probe synthesis, 10% of the total sample volume of 3M Na Acetate (pH 5.8) is added (5 μl for a 50 μl reaction). Two and two-tenths times the total sample volume of 100% ethanol should be added (121 μl for a 50 μl reaction) and left overnight at −20 °C to precipitate.

Probes are then spun down at 15,000 rpm for 15 min. Remove the ethanol and add 80% EtOH. This step is repeated using 100% EtOH, spun down at 15,000 rpm for 15 min, then the EtOH is removed and the probe is dried. The samples are then rehydrated, depending on the size of the pellet of RNA. For a small pellet, 5 μl DEP water is added; for a medium pellet, 8 μl DEP water is added; and for a large pellet, 10 μl DEP water is added.

Digoxigenin (DIG) Detection:

To check whether labeled RNA probes have been properly prepared, 1/10, 1/50, 1/100, and 1/500 serial dilutions are made using the resuspended probes. Serial dilutions of the control-labeled RNA from the DIG Nucleic Acid Detection Kit (Roche Molecular Biochemicals) are also prepared. On a hybridization membrane (Hyband-N; Amersham) 1 μl of each dilution is blotted on a grid. The membrane can then be placed into a 50 ml tube for detection and follow the "immunological detection" procedure from the DIG Nucleic Acid Detection Kit (Roche Molecular Biochemicals).

C. Immunocytochemistry

There are several different protocols for immunocytochemistry, and the one that works best will depend on the species of ascidian, the antibody that you are using, and the protein that you are trying to detect.

We routinely use a protocol from Mita-Miyazawa *et al.* (1987) for staining with monoclonal antibodies. In this protocol, embryos are fixed for 10 min in ice-cold methanol, followed by 10 min in ice-cold ethanol.

A second widely used fixative is 4% paraformaldehyde for 30 min at room temperature (RT) or at 4 °C for a longer period of time. For cytoskeletal proteins, we use a short paraformaldehyde fix, followed by extracting for 10 min in acetone.

1. fix embryos in the fixative of choice.
2. wash embryos 3 to 4× in Phosphate Buffered Saline (PBS)
3. immerse specimens in 100 μl of primary antibody for 1 hr at RT
4. wash with PBS (10 min × 3)
5. incubate for 30 min with 100 μl of secondary antibody diluted 1:40 in PBS, conjugated with a fluorescent dye
6. wash samples well in PBS (10 min × 3)
7. mount in 80% glycerol, and then observe with a fluorescence microscope.

IV. Experimental Techniques

A. Cleavage Arrest

Ascidian development has classically been called "mosaic," because many of the embryonic tissues will develop normally, even if cleavage is inhibited or the blastomeres are separated from the rest of the embryo (Zalokar, 1974). Classical experimental embryological experiments are satisfying to perform on ascidian embryos. Autonomous development was initially discovered in ascidians through blastomere isolations, which can be difficult to perform. A different, but still informative, approach is to cleavage arrest ascidian embryos with cytochalasin at various stages and examine tissue development by specific stains and/or antibody staining. Cytochalasin B or D inhibits cleavage by inhibiting microfilament assembly (Zalokar, 1974). Early embryos are arrested at the 1-, 2-, 4-, 8-, 16-, 32-, 64-, or 128-cell stage by adding 2 μg/ml cytochalasin B or D in FSW. Cytochalasin-treated embryos are then cultured until the controls have made tailbud embryos, and arrested embryos are stained for a particular tissue. By looking at the cleavage-arrested lineages, one can see what blastomeres had that particular fate at the time of cleavage arrest.

After cytochalasin treatment, stain for acetylcholinesterase in tailbud embryos (control) and embryos that have been cleavage arrested. You can also split your samples and stain half with alkaline phosphatase; both histological stains work equally well to show that muscle lineages and endodermal lineages continue to differentiate in the absence of cell division. You may also observe epidermal, muscle, or notochord differentiation in cleavage-arrested embryos with tissue-specific antibodies (Nishikata *et al.*, 1987). The ability to differentiate without cleavage has classically been known as mosaic development. We now know that these results suggest that ascidian eggs have a number of maternal factors that are localized in the cytoplasm of the egg (Swalla, 2004).

B. Dechorionation

Dechorionation can be performed manually with sharpened tungsten needles or number 2 insect needles. The needles can be melted into a glass Pasteur pipet with a Bunsen burner. The needles are sharpened on a sharpening stone with a little oil rubbed on the stone. Then, gently stab the chorion on one side of the egg with one needle and tear the chorion with the second needle. The egg can then be slowly extruded through the hole torn in the chorion.

Chemical dechorionation with protease allows large batches of dechorionated eggs to be prepared at once. Dechorionated live eggs and embryos stick and lyse when they contact glass or plastic. This includes Pasteur pipettes, slides, and cover slips. The following treatment prevents dechorionated eggs from sticking (Sardet *et al.*, 1989).

1. Soak briefly all glassware, plasticware, coverslips, slides, and pipettes in 0.1% gelatin, 0.1% formaldehyde in distilled H_2O.
2. Drain thoroughly.
3. Air dry.
4. Rinse in tap water, then distilled water.
5. Dry and store until needed. Mark GF (do not use this on E-marked glassware).

For dechorionation of Molgula species, make up the dechorionation solution and leave on ice. Seawater containing 1% sodium thioglycolate and 0.05% protease (freshly made), with the pH of the solution adjusted to about 11.0 by addition of drops of 1N NaOH. Gently pipetting the eggs removes the chorion within 10 to 20 min at room temperature (18–23 °C). The dechorionated eggs should be washed several times with filtered seawater, transferred into plastic petri dishes coated with 1 to 2% agar. Dechorionated eggs or isolated blastomeres can reared in Millipore-filtered seawater containing 1 μg/ml streptomycin. *Phallusia mammillatta* chorions can be removed with trypsin in seawater without the use of thioglycolate. Dechorionated eggs and embryos tend to stick to glass and plastic, so either coat your dishes as described in Sardet (1989) or culture the embryos in 1% agar in seawater.

C. Blastomere Isolation and Recombination

For blastomere isolation and recombination, sharpened tungsten needles may be used. Alternatively, you can prepare fresh glass needles by pulling Pasteur pipettes on a Bunsen burner. One can also use a #2 insect pin that is mounted in a glass pipet by melting with a Bunsen burner. The pin is then sharpened with a little oil on a sharpening stone. Dechorionated eggs are cultured in 1% agar-coated Falcon petri dishes until the 2-, 4-, or 8-cell stage. At the appropriate stage, put a glass needle gently on the boundary of the blastomeres, and push and pull the needle. If you have skillful hands, try separation of the 4 cell-pairs of 8-cell embryos. The location of polar bodies, the configurations of the blastomeres, and the distribution of pigments can be used as landmarks for orientation of the embryos. Isolated blastomeres should be cultured separately in 1% agar-coated Falcon 24-well multiwell dishes for about 15 hr before fixation for histochemistry, as has been described.

Fusion of Eggs, Blastomeres, or Egg Fragments

1. Two specimens (eggs, blastomeres, or egg fragments) are placed in a hole made in the 1% agar-coated dish that is filled with Millipore-filtered seawater (MFSW). The specimens are brought into close contact with each other with the aid of tungsten needles.

2. The specimens are overlaid with 40% (w/v) PEG, which makes them adhere firmly to each other.

3. Adhering specimens are rinsed once with fresh MFSW and washed three times with the fusion medium (0.77 M D-Mannitol in 0.25% Ca^{2+} and Mg^{2+}-free artificial seawater).[1]

4. The fusion is triggered by passing one or more pulses of electricity (about 400 V/cm^2) over them.

5. Fused specimens are washed two times with fresh MFSW.[3]

D. Microinjection

Microinjection of recombinant DNAs into fertilized *Ciona* eggs can be performed through the intact chorion. Dissolve plasmid DNAs in 1 mM Tris-HCl, 0.1 mM EDTA, pH 8.0. Microinjection can be carried out with the aid of a holding pipette and an injection pipette, held by micromanipulators under a stereo-microscope. Micropipettes should be made on a horizontal puller from 1.2 mm fiber-filled glass capillary tubing (Microcaps; Drummond Sci. Co., Broomall, PA) Each tip should be approximately 4 μm in diameter. Micropipettes are siliconized and then sterilized. The DNA solution can be injected into the cytoplasm of fertilized eggs under pressure to aid in microinjection.

E. Summary

In summary, ascidians are chordate marine invertebrates that are easy to use for developmental studies in laboratory settings. The gametes are easy to obtain, development is rapid, and the eggs and embryos are easy to experimentally manipulate. Cleavage is bilateral, determinant, and invariant, and the cell lineage is completely known. Ascidian cell lineages can be followed with simple enzymatic stains, and gene expression can be observed with *in situ* hybridization and immunocytochemistry. Functional experiments can be carried out by overexpression of RNA constructs, and knockout studies are possible with morpholinos. Furthermore, ascidians have evolved a number of different larval types, making them excellent species with which to study evolution of larval morphology.

[1]When specimens with MFSW are transferred into fusion medium, specimens tend to move upward because of the high density of fusion medium. Take care that specimens do not lyse at the upper surface of medium.

[2]This value may need to be varied depending on species that are used.

[3]In the case of unfertilized egg or egg fragments, insemination is carried out several minutes after fusion.

V. Protocols

A. Fertilization of Ascidian Eggs

Glassware: All glassware and pipettes must either be new, unused plasticware or properly cleaned laboratory glassware. To clean glassware: soak in liquid detergent solution diluted according to manufacturer's specifications, brush well with a clean laboratory brush, and rinse 10 times in VERY hot tap water. The hot tap water removes all traces of detergent. Then rinse 2× with distilled water and air dry. This glassware is suitable for fertilization, development, and metamorphosis.

1. Fill labeled 100 to 200 ml "E" ware containers with filtered seawater (cooled to around 12 °C).
2. Slice adults longitudinally (through both siphons) with a razor blade.
3. Remove gonads with "E" forceps and push through 200 μm Nytex mesh gently with forceps (don't use gloves) into seawater in the beakers, one adult/beaker. Note: rinse forceps and razor with ddH_2O in between adults to avoid early fertilization.
4. Allow to stand 10 to 20 minutes; pour off sperm (top half of beaker) from all beakers into separate labeled 15 ml falcon tubes. Avoid getting eggs in with the sperm if possible. Keep sperm on ice.
5. Rinse eggs. Suction off remaining SW through mesh filter to avoid removing eggs (leave about 30–50 ml, enough to cover bottom of dish) and refill with fresh seawater.
6. Rinse remaining sperm from eggs by allowing eggs to settle (takes 10–15 min), pouring off seawater (leave about 30–50 ml) and refilling with fresh seawater. Continue to do this until the seawater is no longer cloudy from sperm (about 3 times).
7. Add 1 to 2 drops of pH 9 Tris to each tube of sperm to activate. Mix one min. Test to see that it is pH 9. Check sperm for swimming if you are having difficulties.
8. Pour off seawater from settled eggs to leave 30 to 50 ml in beaker. Add about 1 ml of each tube of sperm to each beaker of eggs. Do not add self sperm to the beaker.
9. Check for fertilization after 30 min, if successful:
   a. Fill beaker to the top with seawater to dilute sperm, let eggs settle.
   b. Rinse three times as follows: allow eggs to settle 10 to 20 min, pour off top half and then refill.
   c. Settle eggs, pour off top half again, and pour settled eggs into petri dish; place at 12 °C and allow to develop.
10. If necessary, refertilize by repeating step 9 with more sperm. If fertilization is abnormal, you may need to dilute sperm more.

11. At F+6 to 8 hrs (after gastrulation), transfer to seawater w/50 μg/ml ampicillin or kanomycin (1:1000 with FSW) if desired.
Keep eggs/embryos between 10 and 16 °C at all times!!!!!

B. ACHE Histochemistry

1. Let embryos develop to the desired stage.
2. Fix embryos in 5% formalin in ASW for 30 min.
3. Make up fresh staining medium as follows:
   (a) 10 mg of acetylthiocholine iodide in 6.5 ml of 0.1 M sodium phosphate pH 6.
   (b) Add in order the following and be sure to stir between each addition:
      (1) 0.5 ml 0.1 M sodium citrate.
      (2) 1 m 30 mM cupric sulfate.
      (3) 1 ml water.
      (4) 1 ml 5 mM potassium ferricyanide.

(Final color should be clear and green and mixture is stable for only a few hours.)

4. Rinse 4 to 5× in 0.1 M phosphate buffer pH 6.5 to remove formalin.
5. Incubate embryos for 2 to 4 h at 37°C, checking the reaction periodically. The brown reaction will develop over time. If you are getting an unsightly precipitate, it is likely that you did not rinse enough times in the phosphate buffer. (For controls, incubate embryos in the absence of substrate.)
6. Rinse in ddH_2O.
7. Dehydrate in ethanol.
8. Clear in toluene.
9. Mount in Permount or embed.

Na Phosphate Buffer
Make stocks at 0.2 M. Dilute 1:1 in distilled H_2O to make 0.1 M.

pH	Monobasic	Dibasic
pH 7.2	280 ml	720 ml
pH 7	390 ml	610 ml
pH 6.5	200 ml	100 ml

C. Whole Mount Alkaline Phosphate Staining Protocol for Ascidians

1. Sample Fixation and Storage:

4% paraformaldehyde in PBS (pH 7.2)	12–24 h @ 4 °C
Transfer to PBS @ 4 °C	
Store @ 4 °C	

1. Rinse in Buffer 3 (100 mM Tris-HCI, pH 9.5, 100 mM NaCl, 5 mM $MgCl_2$)
   Wipe excess buffer off of slide
2. Transfer to AP Detection Solution:
   5 ml Buffer 3 (pH 9.5)
   22 μl NBT, 16.5 μl BCIP
   light sensitive—Keep in the dark!
   leave in color solution until the desired darkness of stain comes up.
3. Stop reaction in PBS, wash 2×
   can leave in PBS for short period of time after staining
4. Mount in glycerol
5. Photograph.

OR

4. Dehydrate through a series of ethanol 30%, 50%, 80%, 90%, 100%, 100%
   10 min each
5. Clear in benzyl alcohol: Benzyl benzoate 1:2. Mix thoroughly. Change 1×
6. Photograph.

D. Dechorionation of *Boltenia villosa* for *In Situ* Hybridization or Antibody Staining

1. Let eggs develop to desired stage in FSW (filtered seawater).
2. Collect 3 ml of eggs into a small Corning tube.
3. Make up the dechorionation solution fresh each time:

Dechorionation solution	5×
500 μl of 4% Na thioglycolate in FSW	2.5 ml
400 μl of 2% pronase E	2.0 ml
80 μl of 1 M NaOH	400 μl
.98 ml total	4.9 ml total

4. Add 1 ml of dechorionation solution to each tube containing 3 ml of eggs. Mix gently. Continue to mix gently every 2 to 3 min.

5. Watch a sample (100 μl) under the dissection scope for chorion or test cell removal. The test cells will come off first, but be patient; watch for the chorion to break open and the fertilized egg to drop out.

6. Dechorionation should take about 10 min; label a silanized DEP tube on the top and side with:

Species
F + # hrs
Date

7. When finished, spin the tubes gently in a hand centrifuge, pour off the dechorionation solution and add 5 ml fresh filtered SW. Rinse 3×.

8. Transfer the embryos to a silanized, labelled DEP tube. Spin down.

9. Add 500 μl *In situ* hybridization fix solution (4% paraformaldehyde in MOPS, pH 7.5)

4% paraformaldehyde in 0.5 M NaCl, 0.1 M MOPS pH 7.5
10 ml 16% ultrapure formaldehyde (Polysciences) stored @ 4 °C
20 ml, 0.2 M MOPS pH 7.5
5 ml, 4 M NaCl
5 ml DEPC-treated water
40 ml total in sterile disposable 50 ml tube

10. Leave at 4 °C for at least 12 h.

11. Spin samples in a picofuge for 10 s. Take off fix and put into fix waste container.

12. Add 50% ethanol and leave at 4 °C for 30 min.

13. Add 80% ethanol and leave at −20 °C until use.

E. Whole Mount *In Situ* Hybridization Protocol

(all steps done in siliconized 1.5 ml microcentrifuge tubes)

Volume of each solution:

PBT for wash	500 μl
Hybridization Solutions	50 μl
Others	200 μl

Sample fixation and storage:

4% paraformaldehyde in 0.5 M NaCl, 0.1 M MOPS (pH 7.5);	12 h @ 4 °C
50% EtOH	30 min
80% EtOH	30 min
store @ −20 °C	

Hybridization: (room temperature unless stated otherwise)

1. Rehydration of sample: wash with PBT	5 min × 4
2. Proteinase K treatment (10 μg/ml Proteinase K in PBT)	10 min @ 37 °C
3. Stop reaction in 2 mg/ml glycine in PBT	10 min
4. Wash with PBT	5 min × 4
5. Post-fix in 4% paraformaldehyde in PBS; pH 7.5	60 min room temp.

6. Wash with PBT — 5 min × 4
   - 0.1 M triethanolamine (pH 8.0, adjust w/HCl); — 5 min × 2
   - 0.25% anhydrous acetic acid in 0.1 M triethanolamine (pH 8.0) prepare just before use; — 10 min
7. Wash with PBT — 5 min × 4
8. PBT: prehybridization solution 1:1 — 10 min
9. Hybridization solution — 10 min @ 45 °C
10. Fresh hybridization solution — 1 h @ 45 °C
11. Hybridization solution with probe added — 16 h @ 45 °
    46 μl of hybridization buffer and add DEP ddH2O and probe for 50 ul/ sample.
    ___μl of hybridization buffer
    ___μl of_____probe
    (Usually 1–4 μl of RNA probe per 50 μl.)

Wash

1. 4 × SSC, 50% formamide, 0.1% Tween 20 — 15 min × 2, 45 °C
2. 2 × SSC, 50% formamide, 0.1% Tween 20 — 15 min × 2, 45 °C
3. Solution A — 10 min × 3, 37 °C
4. 20 μg/ml RNase A in solution A — 20 min 37 °C
5. Solution A — 15 min 37 °C
6. 2 × SSC, 50% formamide, 0.1% Tween 20 — 20 min 45 °C
7. 1 × SSC, 50% formamide, 0.1% Tween 20 — 15 min × 2, 45 °C
8. 1 × SSC, 50% formamide, 0.1% Tween 20: PBT → 3:1; 10 min Room Temp
9. 1 × SSC, 50% formamide, 0.1% Tween 20: PBT → 1:1; 10 min RT
10. 1 × SSC, 50% formamide, 0.1% Tween 20: PBT → 1:3; 10 min RT

Detection

1. Wash with PBT — 5 min × 4 RT
2. 0.1% blocking reagent in PBT — 30 min RT
3. 1/2000 anti DIG-AP in PBT (can stop and leave overnight in antibody) — 1 h RT
4. Wash with PBT — 5 min × 4
5. Wash with Buffer 3 (pH 8.0) — 5 min × 2
6. Wash with Buffer 3 (pH 9.5) — 10 min × 2 RT
7. Transfer to AP Detection Solution:
   5 ml of Buffer 3 (pH 9.5)
   50 μl of 200 mM levamisole
   100μl of NBT/BCIP from the kit
   - light sensitive—Keep in the dark!
   leave in color solution until desired darkness of stain comes up.

8. Stop reaction in PBS, wash 2× (can leave in PBS for short period of time)
9. Dehydrate through a series of ethanol 30%, 50%, 80%, 90%, 100%, 100% 10 min each
10. Clear in Benzyl alcohol: Benzyl benzoate 1:2. Mix thoroughly. Change 1×

Solutions:

4% paraformaldehyde in 0.5 M NaCl, 0.1 M MOPS pH 7.5

---

10 ml 16% ultrapure formaldehyde (Polysciences) stored @ 4 °C
20 ml, 0.2 M MOPS pH 7.5
5 ml, 4 M NaCl
5 ml DEPC-treated water

---

40 ml total in sterile disposable 50 ml tube
- PBT ⇒ 0.1% Tween 20 in PBS
- Solution A ⇒ 0.5 M NaCl, 10 mM Tris-Cl (pH 8.0), 5 mM EDTA, 0.1% Tween 20
- Buffer 3 (pH 8.0) ⇒ 100 mM Tris-Cl (pH 8.0), 100 mM NaCl, 50 mM MgCl2
- Buffer 3 (pH 9.5) ⇒ 100 mM Tris-Cl (pH 9.5), 100 mM NaCl, 50 mM MgCl2
- 4% paraformaldehyde in PBS ⇒ 250 ul of 16% paraformaldehyde 750 ul of PBS adjust to pH 7.5 with 2 ul 1N NaOH

Salt solutions:

20 × SSC = 175.3 g NaCl + 88.2 g NaCit (this is diluted to make 4 × SSC, 2 × SSC, and 1 × SSC solutions)

Hybridization buffer:

- 50% formamide
- 6 × SSC
- 5 × Denhardt's (contains no BSA)
- 100 μg/ml salmon sperm DNA
- 0.1% Tween 20
- (1 μg/μl BSA)

F. HRP Whole Mount Antibody Protocol Modified for Ascidian Staining (Panganiban *et al.*, 1997)

Fix Buffer (Stock Solution), filter and sterilize for up to 1 year
1.33 × PBS and 67 mM EGTA

Fix
10 × PBS, pH 7.0

(200 mM KPO4, 140 mM NaCl)

500 mM EGTA pH 7.4

37% formaldehyde (Sigma)

Block (PBT, 1 × PBS, 0.1% Triton X-100, 2% BSA) make up in 15 ml tube store at −20 °C

10 × PBS, pH 7.0

Triton X-100

BSA

Wash (PT) make up in 1-liter volumes, filter sterilize OR use sterile 10 × PBS and put into sterile container

10 × PBS

Triton X-100

Developer
Metal-enhanced DAB solutions (Pierce)

Other

Methanol

30% H_2O_2 (Sigma)

goat anti-rabbit peroxidase (Jackson) or goat anti-rabbit biotin and Elite ABC reagents (Vector)

Part 1—Fixation (note: before fixation, you must dechorionate embryos with pronase E or, after fixation, manually dechorionate with needles)

1. Mix 3 parts Fix buffer stock solution with 1 part 37% formaldehyde JUST prior to use.
   Fix embryos for 2 to 8 hours in 1.5 ml flip-top tubes or 15 ml tubes.
2. Wash embryos 3 to 4 ×, each wash 5 min; use picofuge to collect embryos in bottom of tube with MeOH to remove fix.
3. Treat for 4 min with 3% H_2O_2 in MeOH to inactivate endogenous peroxidases.
4. Wash away H_2O_2 with 4 to 5 changes of MeOH, each wash 5 min.
5. Store fixed embryos indefinitely in MeOH at −20 °C.

Part 2—Incubation with primary antibody

1. Rehydrate embryos in 50% 1 × PBS, pH 7.0, 50% MeOH, then 100% 1 × PBS.
2. Treat embryos with acetone by first removing 1 × PBS, then adding 1 ml of acetone for 5 min.
3. Wash 2 × 100% 1 × PBS.
4. Block embryos in PBT (1 × PBS pH 7.0, 0.1% Triton X-100, 2% BSA) for 1 to 2 h (longer OK) at 4 °C.

5. Incubate embryos overnight at 4 °C with 5 μg/ml anti-Dll AB in PBT (no DMSO!) Stain approx. 20 μl of packed ascidian embryos with 20 μl of AB. Add 10 μl, mix gently, then add the final 10 μl, making sure to get all embryos that may be on sides of tube. This can be done with gentle shaking. (STOCK AB soln is 200 μg/ml).

Part 3—Incubation with secondary antibody

1. Wash embryos 10 × over 2 h with PT (1 × PBS pH 7.0, 0.1% Triton X-100).
2. Incubate embryos with 1:400 goat anti-rabbit peroxidase (or goat anti-rabbit biotin), 50 to 60 μl for 2 to 4 h (longer OK) at 4 °C.
3. Wash embryos 10 × over 2 h with PT.

Part 4—Developing

1. Remove embryos from tubes and place into 24-well plate (culture plate).
2. Develop in Pierce DAB solns (according to manufacturer's instructions). First mix:
   9 parts DAB buffer
   1 part metal-enhanced DAB

Add 1 ml per well to embryos. This step may go rapidly (fastest to date with ascidian embryos with chorions ON-10 min). Be prepared to wash with PT. You can slow the reaction by using less developer or by doing this on ice.

3. Wash embryos with PBT, whole-mounts put into depression slides with glycerol. Store in 4 °C.

References

Abbott, D. P. (1975). Phylum Chordata: Introduction and Urochordata. *In* "Light's Manual: Intertidal Invertebrates of the Central California Coast" (R. I. Smith and J. T. Carlton, eds.), pp. 638–655. University of California Press, Berkeley, CA.

Berrill, N. J. (1935). Studies in tunicate development. Part 3: Differential retardation and acceleration. *Philosophical Transactions of The Royal Society of London Series B Biological Sciences* **225,** 255–326.

Berrill, N. J. (1936). Studies in tunicate development. Part 5: The evolution and classification of ascidians. *Philosophical Transactions of The Royal Society of London Series B Biological Sciences* **226,** 43–70.

Berrill, N. J. (1950). "The Tunicata with an Account of the British Species." The Ray Society, London, UK.

Cameron, C. B., Garey, J. R., and Swalla, B. J. (2000). Evolution of the chordate body plan: New insights from phylogenetic analyses of deuterostome phyla. *Proc. Natl. Acad. Sci.* **97,** 4469–4474.

Chabry, L. (1887). Embryologie normale et tératologique des Ascidies simples. *J. Anat. Physiol. (Paris)* **23,** 167–319.

Cloney, R. A. (1978). Ascidian metamorphosis: Review and analysis. *In* "Settlement and metamorphosis of marine invertebrate larvae" (F. S. Chia and M. E. Rice, eds.). Elsevier, New York.

Cloney, R. A. (1982). Ascidian larvae and the events of metamorphosis. *Am. Zool.* **22,** 817–826.

Conklin, E. G. (1905). The organization and cell lineage of the ascdian egg. *J. Acad. Nat. Sci. Phil.* **13,** 1–119.

Davidson, B., and Swalla, B. J. (2001). Isolation of genes involved in ascidian metamorphosis: Epidermal growth factor signaling and metamorphic competence. *Dev. Genes Evol.* **211,** 190–194.

Davidson, B., Jacobs, M., and Swalla, B. J. (2004). The individual as a module: Metazoan evolution and coloniality. *In* "Modularity in Development and Evolution" (G. Schlosser and G. Wagner, eds.). University of Chicago Press, pp. 443–465.

Davidson, B., Smith Wallace, S. E., Howsmon, R. A., and Swalla, B. J. (2003). A morphological and genetic characterization of metamorphosis in the ascidian *Boltenia villosa. Develop. Genes & Evol.* **213,** 601–611.

Degnan, B. M., Souter, D., Degnan, S. M., and Long, S. C. (1997). Induction of metamorphosis with potassium ions requires development of competence and an anterior signaling centre in the ascidian *Herdmania momus. Dev. Genes Evol.* **206,** 370–376.

Dehal, P., Satou, Y., Campbell, R. K., Chapman, J., Degnan, B., De Tomaso, A., Davidson, B., Di Gregorio, A., Gelpke, M., Goodstein, D. M., Harafuji, N., Hastings, K. E., Ho, I., Hotta, K., Huang, W., Kawashima, T., Lemaire, P., Martinez, D., Meinertzhagen, I. A., Necula, S., Nonaka, M., Putnam, N., Rash, S., Saiga, H., Satake, M., Terry, A., Yamada, L., Wang, H. G., Awazu, S., Azumi, K., Boore, J., Branno, M., Chin-Bow, S., DeSantis, R., Doyle, S., Francino, P., Keys, D. N., Haga, S., Hayashi, H., Hino, K., Imai, K. S., Inaba, K., Kano, S., Kobayashi, K., Kobayashi, M., Lee, B. I., Makabe, K. W., Manohar, C., Matassi, G., Medina, M., Mochizuki, Y., Mount, S., Morishita, T., Miura, S., Nakayama, A., Nishizaka, S., Nomoto, H., Ohta, F., Oishi, K., Rigoutsos, I., Sano, M., Sasaki, A., Sasakura, Y., Shoguchi, E., Shin-i, T., Spagnuolo, A., Stainier, D., Suzuki, M. M., Tassy, O., Takatori, N., Tokuoka, M., Yagi, K., Yoshizaki, F., Wada, S., Zhang, C., Hyatt, P. D., Larimer, F., Detter, C., Doggett, N., Glavina, T., Hawkins, T., Richardson, P., Lucas, S., Kohara, Y., Levine, M., Satoh, N., and Rokhsar, D. S. (2002). The draft genome of *Ciona intestinalis*: Insights into chordate and vertebrate origins. *Science* **298,** 2157–2167.

Deno, T., Nishida, H., and Satoh, N. (1985). Histospecific acetylcholinesterase development in quarter ascidian embryos derived from each blastomere pair of the eight-cell stage. *Biol. Bull.* **168,** 239–248.

Hayward, P. J., and Ryland, J. S. (eds.) (1990). Chapt. 16: Hemichordata and Urochordata, pp. 872–904. *In* "The Marine Fauna of the British Isles and North-West Europe." Vol. 2—Molluscs to Chordates. Oxford Univ. Press.

Hirano, T., and Nishida, H. (1997). Developmental fates of larval tissues after metamorphosis in ascidian *Halocynthia roretzi.* I. Origin of mesodermal tissues of the juvenile. *Dev. Biol.* **192,** 199–210.

Hirano, T., and Nishida, H. (2000). Developmental fates of larval tissues after metamorphosis in the ascidian, *Halocynthia roretzi.* II. Origin of endodermal tissues of the juvenile. *Dev. Genes Evol.* **210,** 55–63.

Huber, J. L., Burke da Silva, K., Bates, W. R., and Swalla, B. J. (2000). The evolution of anural larvae in molgulid ascidians. *Sem. in Develop. Biol.* **11,** 419–426.

Jeffery, W. R., and Swalla, B. J. (1997). Embryology of the Tunicates. *In* "Embryology: Constructing the Organism" (S. Gilbert, ed.), pp. 331–364. Sinauer, Sunderland, MA.

Kott, P. (1985). The Australian Ascidiacea. Part 1, Phlebobranchia and Stolidobranchia. *Mem. Queensland Mus.* **23,** 1–440.

Kott, P. (1990). The Australian Ascidiacea. Part 2. Aplousobranchia (1). *Memoirs of the Queensland Museum.* **29,** 1–266.

Kott, P. (1992). The Australian Ascidiacea, supplement 2. *Memoirs of the Queensland Museum* **32,** 621–655.

Koyanagi, R., and Honegger, T. G. (2003). Molecular cloning and sequence analysis of an ascidian egg β-N-acetylhexosaminidase with a potential role in fertilization. *Develop. Growth Differ.* **45,** 209–218.

Mita-Miyazawa, I., Nishikata, T., and Satoh, N. (1987). Cell- and tissue-specific monoclonal antibodies in eggs and embryos of the ascidian *Halocynthia roretzi. Development* **99,** 155–162.

Monniot, C., and Monniot, F. (1996). New collections of ascidians from the western Pacific and southeastern Asia. *Micronesica* **29,** 133–279.

Monniot, C., Monniot, F., and Laboute, P. (1991). "Coral Reef Ascidians of New Caledonia," 247Orstom, Paris.

Monniot, C., Monniot, F., Griffiths, C. L., and Schleyer, M. (2001). South African ascidians. *Annals of the South African Museum* **108,** 1–141.

Nishida, H. (1987). Cell lineage analysis in ascidian embryos by intracellular injection of a tracer enzyme. III. Up to the tissue restricted stage. *Dev. Biol.* **121,** 526–541.

Nishida, H. (1992). Regionality of egg cytoplasm that promotes muscle differentiation in embryo of ascidian, *Halocynthia roretzi. Development* **116,** 521–529.

Nishida, H., and Satoh, N. (1983). Cell lineage analysis in ascidian embryos by intracellular injection of a tracer enzyme. 1. Up to the eight-cell stage. *Dev. Biol.* **99,** 382–394.

Nishida, H., and Satoh, N. (1985). Cell lineage analysis in ascidian embryos by intracellular injection of a tracer enzyme. II. The 16- and 32-cell stages. *Dev. Biol.* **110,** 440–454.

Nishikata, T., Mita-Miyazawa, I., Deno, T., and Satoh, N. (1987). Muscle cell differentiation in ascidian embryos analyzed with a tissue-specific monoclonal antibody. *Development* **99,** 163–171.

Panganiban, G., Irvine, S. M., Lowe, C., Roehl, H., Corley, L. S., Sherbon, B., Grenier, J., Fallon, J. F., Kimble, J., Walker, M., Wray, G., Swalla, B. J., Martindale, M. Q., and Carroll, S. B. (1997). The origin and evolution of animal appendages. *Proc. Natl. Acad. Sci.* **94,** 5162–5166.

Roegier, F., Djediat, C., Dumollard, R., Rouviére, C., and Sardet, C. (1999). Phases of cytoplasmic and cortical reorganizations of the ascidian zygote between fertilization and first division. *Development* **126,** 3101–3117.

Sardet, C., Speksnijder, J. E., Inoué, S., and Jaffe, L. (1989). Fertilization and ooplasmic movements in the ascidian egg. *Development* **105,** 237–249.

Sardet, C., Speksnijder, J. E., Terasaki, M., and Chang, P. (1992). Polarity of the ascidian egg cortex before fertilization. *Development* **115,** 221–237.

Satoh, N. (1994). "Developmental Biology of Ascidians." Cambridge Univ. Press, New York.

Swalla, B. J. (2001). Phylogeny of the Urochordates: Implications for Chordate Evolution. *In* "The Biology of Ascidians" (H. Sawada, H. Yokosawa, and C. Lamberts, eds.), pp. 24–29. Springer-Verlag, Tokyo.

Swalla, B. J. (2004). "Protochordate Gastrulation: Lancelets and Ascidians" *In* "Gastrulation" (C. Stern, ed.). Cold Spring Harbor Press, NY.

Swalla, B. J., Cameron, C. B., Corley, L. S., and Garey, J. R. (2000). Urochordates are monophyletic within the deuterostomes. *Systematic Biology* **49,** 122–134.

Van Name, W. G. (1945). The North and South American ascidians. *Bull. Am. Mus. Nat. Hist.* **84,** 1–476.

West, A. B., and Lambert, C. C. (1976). Control of spawning in the tunicate *Styela plicata* by variations in a natural light regime. *J. Exp. Zool.* **195,** 263–270.

Whittaker, J. R. (1973). Segregation during ascidian embryogenesis of egg cytoplasmic information for tissue-specific enzyme development. *Proc. Natl. Acad. Sci. USA* **70,** 2096–2100.

Whittaker, J. R. (1990). Determination of alkaline phosphatase expression in endodermal cell lineages of an ascidian embryo. *Biol. Bull.* **178,** 222–230.

Whittaker, J. R., and Meedel, T. H. (1989). Two histospecific enzyme expressions in the same cleavage-arrested one-called ascidian embryos. *J. Exp. Zool.* **205,** 168–175.

Zalokar, M. (1974). Effect of colchicine and cytochalasin B on ooplasmic segregation in ascidian eggs. *Wilheim Roux's Arch. Dev. Biol.* **175,** 243–248.

CHAPTER 7

Culture of Adult Ascidians and Ascidian Genetics

Carolyn Hendrickson,* Lionel Christiaen,† Karine Deschet,* Di Jiang,* Jean-Stéphane Joly,† Laurent Legendre,† Yuki Nakatani,* Jason Tresser,* and William C. Smith*

*Neuroscience Research Institute and
Department of Molecular, Cellular, and Developmental Biology
University of California
Santa Barbara, California 93106

†INRA Junior Group, UPR 2197 DEPSN, CNRS
Institut de Neurobiologie A. Fessard
Gif-sur-Yvette, France

I. Overview

Ascidians are primitive chordates that are used extensively in experimental biology. This chapter describes procedures used to culture ascidians, to isolate and propagate mutant lines, and to identify markers for genomic loci of interest.

METHODS IN CELL BIOLOGY, VOL. 74

0091-679X/04 $35.00

Ascidian culturing methods described here include an open seawater system in use in Santa Barbara and a closed seawater system in use in Gif-sur-Yvette. Many procedures used for ascidian genetics are still in the early stages of development. Preliminary results with chemical mutagenesis are discussed as well as the prospects for positional cloning of mutant genes.

II. Introduction

Ascidians were one of the earliest groups of model organisms used for experimental embryology. Interest in these primitive chordates has increased steadily in recent years as new experimental techniques and resources have become available. For most experimental embryological procedures that are commonly performed in ascidians, such as cis-regulatory element analysis (e.g. Erives *et al.*, 1998), there is no need for culturing ascidians past the larval stage. However, our laboratory, along with several others, is intrigued by the possibilities presented by ascidians for classical genetic analysis (Moody *et al.*, 1999; Nakatani *et al.*, 1999; Sordino *et al.*, 2000). Some of the earliest work using ascidians for genetic characterization of traits and linkage analysis was done with the polymorphic fusibility/histocompatibility locus in the colonial ascidian *Botryllus schlosseri* (De Tomaso *et al.*, 1998). Experimental embryology has focused primarily on the solitary ascidians, which produce more gametes and are more amenable than colonial ascidians to experimental manipulation. Ascidian species are numbered in the thousands and many of them are used for experimental embryology. Each experimental species has its own unique advantages, such as large, transparent embryos (e.g., *Halocynthia roretzi*), or closely related species with morphological variations (e.g., the *Molgulid* ascidians). Efforts at developing an ascidian model for developmental genetics have focused on the genus *Ciona*. Members of this genus, in particular, *Ciona intestinalis* and *Ciona savignyi*, have proven to be well suited for developmental genetics. Like one of the most widely used invertebrate genetic model organisms, *C. elegans, Ciona* are hermaphrodites with a capacity for self-fertilization. This trait of *Ciona*, as we will detail later in this chapter, is particularly useful for the identification and characterization of recessive mutations. Also like *C. elegans*, and unlike their vertebrate cousins, ascidians have an invariant cell lineage up through the larval stage (Satoh, 1994). These traits, together with their relatively simple embryology and genomic organization in comparison to vertebrates, make *Ciona* ideal genetic model chordates.

One of the requirements for forward genetics, as well as a number of other techniques such as stable transgenesis (Deschet *et al.*, 2003), is the reliable culturing of lines through multiple generations. Culturing marine organisms in general, and filter feeders such as ascidians in particular, is challenging. A reliable laboratory culturing system has been developed for *Botryllus schlosseri* (Boyd *et al.*, 1986).

However, the larger solitary ascidians, like *Ciona*, are more challenging to culture. Our research group in Santa Barbara, located at a seaside campus with extensive marine biology facilities, has been in a unique position to explore *Ciona* species as genetic model organisms. One key resource that is not likely to be found outside of marine laboratories is plumbed running seawater. Having access to running seawater has made the culturing of thousands of *Ciona* through multiple generations very easy. It has allowed us to focus our efforts on development of techniques for the identification, characterization, and mapping of developmental mutations rather than on animal husbandry. The largest obstacle to the more widespread use of *Ciona* for classical genetics is the difficulty in culturing animals in closed systems. However, as we will describe here, progress is being made in this area as well.

III. Culturing Ascidians

A. Settling of Larvae and Metamorphosis

Chapter 6 of this volume details procedures for obtaining ascidian adults, collecting gametes, and laboratory fertilization to generate embryos and larvae. The procedures used for generating larvae for genetic analysis are very similar, although, as we will detail, controlled matings between individuals of specific genotypes are critical. If one wishes to grow the animals past the larval stage, the next step is to allow and, with some species, to induce, the larvae to attach to a substrate, after which most species will spontaneously undergo metamorphosis. The settlement of *B. schlosseri* on glass slides for laboratory culture has been described previously (Boyd *et al.*, 1986).

For both *C. savignyi* and *C. intestinalis*, the majority of larvae will hatch within 18 h of fertilization at 18 °C. *Ciona* larvae are about 1 mm in length and, unlike the larvae of some colonial ascidians, they lack circulatory and digestive systems at this stage. The *Ciona* larva is a very transient form that subsists on maternally deposited yolk stores and appears to function only for dispersal and selecting a location to settle. When grown in the laboratory, we find that *Ciona* will swim for an average of about 6 h before attaching via their adhesive palps and starting metamorphosis.

For settlement of larvae, the embryos are first fertilized and grown in untreated petri dishes until they hatch. The tadpole larvae are then washed through a 70 μm Nitex filter to remove the remains of the chorion and any other debris. We have found the most useful tool for washing and transferring embryos and larvae is a "transfer filter." The transfer filter is made from a 50 ml conical tube with both the cone-shaped bottom and the center of the lid cut off, leaving just the threaded ring for a top. An approximately 4-cm square piece of Nitex is then placed on the threaded end of the tube, and the cut top is screwed on over the filter to hold it in place. Larvae to be washed are placed in the transfer filter and moved through two

100-ml beakers of seawater, allowing the seawater to drain between each beaker. To recover the larvae, the transfer filter is inverted above a new dish of seawater and the larvae washed out from the bottom using a wash bottle of seawater. We have found that the larvae attach more readily to charged surfaces, such as is found in tissue culture dishes. As tissue-culture dishes are expensive, we use regular plastic petri dishes coated with poly-L-lysine. Dishes are soaked or filled with 1 mg/ml of poly-L-lysine in water, overnight at room temperature. Coated dishes are then washed several times with water and dried.

Depending on the number of adults one wishes to grow, the larvae are settled on either 10 or 15 cm dishes containing seawater with 50 μg/ml each streptomycin and kanamycin (SW/Ab). A 15 cm dish will accommodate about 50 adults, and at least twice that number of larvae can be placed in the dish, as many of them will not settle or will fail to grow to adults. During the first few hours of the larval stage, *Ciona* larvae have a natural tendency to swim upwards, against the force of gravity. As a consequence, in the shallow culture dishes, many of the swimming larvae would become trapped at the air/water interface. In fact, the larvae perceive the interface as a solid surface and will attempt to attach to it and start metamorphosis. To overcome this problem, the larvae are sandwiched in SW/Ab between the inside of the lid and the outside of the bottom of a culture dish (Fig. 1). We observe that larvae preferentially attach to the tops, but also to the bottoms and sides of the dishes.

Once the larvae are attached, they quickly begin tail resorption, and metamorphosis is well underway within two days of fertilization. After the larvae are well attached, the SW/Ab should be changed daily. Juveniles are cultured sandwiched in the dishes for 7 to 10 days in an 18 °C incubator. At 5 to 6 days after fertilization, most of the adult organs are developed and the beating heart should be apparent. A mass of brown (*C. intestinalis*) or orange (*C. savignyi*) yolk will persist near the stalk for the first week or longer. The juveniles can be fed microalgae at this stage (Cirino *et al.*, 2002), but we have found it to be unnecessary. Once the juveniles have completed metamorphosis, they are moved to the feeding tanks/aquaria. We have found that the seawater in the tanks has a tendency to degas, which can result in bubbles forming on the plates and inside the animals. To

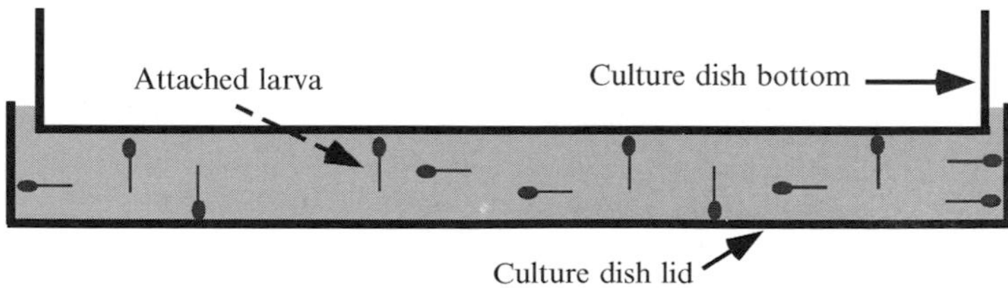

Fig. 1 *Ciona* larvae attaching to a culture dish. Swimming tadpole larvae in seawater are sandwiched between the bottom of a culture dish and an inverted lid.

ensure that the dishes remain submerged, we attach a small weight to the bottom of the dishes.

B. Raising Juveniles to Adults

Given that a sufficient diet and clean seawater are provided, *Ciona* can grow to be reproductive adults in 2 to 3 months, while *B. schlosseri* take less than 2 months. Ascidians are, with a few exceptions, indiscriminate filter feeders, and thus their diet consists of an assortment of microalgae and zooplankton. For raising the juveniles to maturity, one must either provide the cultured ascidians with an artificial diet or, as in our case in Santa Barbara, use running unfiltered seawater. The growth conditions and diet for laboratory culture *B. schlosseri* are well established and described (Boyd *et al.*, 1986). We focus our attention here on the culture of solitary ascidians, and describe our experience with an open system in Santa Barbara and with a closed culture system in Gif-sur-Yvette. A third system, developed in Naples, that uses running filtered seawater supplemented with an artificial diet has also been described (Cirino *et al.*, 2002).

Because juveniles settled on petri dishes will stay firmly attached as they grow to adults (Fig. 2A), keeping track of animals of specific genotype is simple. Animals are tracked by labeling the back of the dishes with the appropriate information. In addition, because the animals stay attached to their plates, animals of various genotypes can be grown together in the same culturing tank with no danger of mix-up. However, for screening and studying individuals, the adults have to be gently peeled from the plates so that gametes can then be collected. Culturing and tracking loose individual adults is more difficult. While the loose adults will eventually reattach, we have found it takes several weeks for the attachment to become firm. To address this problem, we have designed cages that can be labeled to hold the adults as they reattach. For adult *C. intestinalis*, which are much larger than *C. savignyi* (10–15 cm vs 5–8 cm), the cages are made from plastic "strawberry baskets" (Fig. 2B). The *C. savignyi* cages are smaller (about 5 cm on a side) and are made from a stiff plastic mesh with ≈3 mm openings (Fig. 2C). Both types of cages are weighted at the bottom and suspended in the culturing tanks. The perforations in the sides of the cages allows for water flow and for normal light–dark cycles. After several weeks, the adult will have firmly reattached, and the lids can be opened.

1. Open Seawater Systems

Researchers working at marine laboratories have the convenience of using running seawater for culturing ascidians. Most marine facilities provide plumbed running seawater that is collected from an offshore intake pipe. Marine laboratories typically filter the incoming seawater through a sand filter with the objective of removing particles and plankton before it is distributed to the laboratories. Such filtered seawater is too nutrient-poor for raising ascidians, and if that is

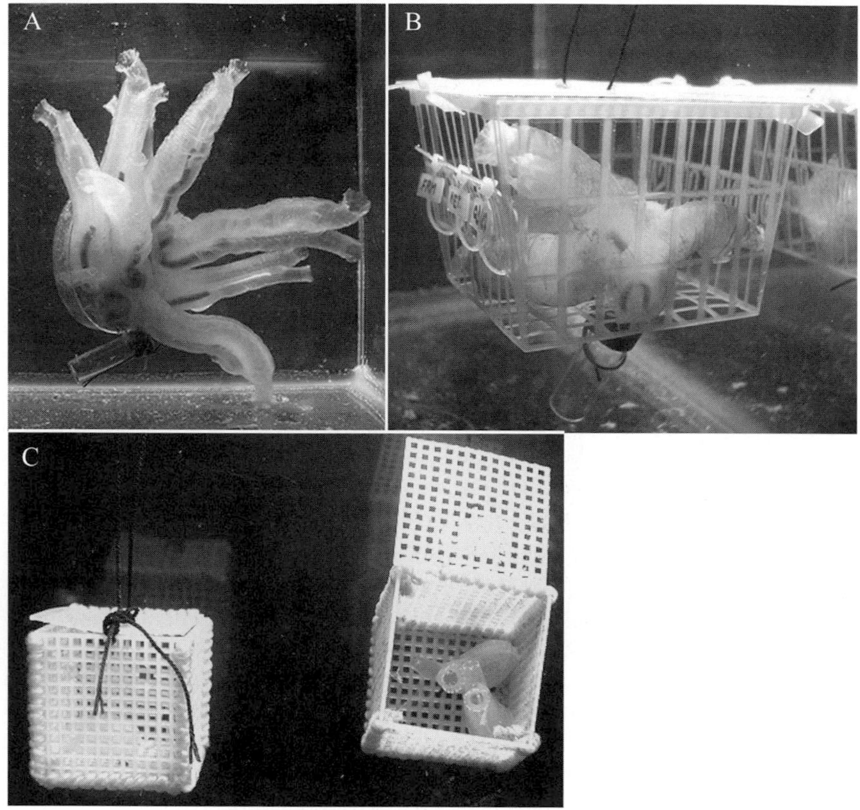

Fig. 2 (A) Adult *Ciona intestinalis* cultured in a 10-cm petri dish. (B) Typical "cage" used for holding *C. intestinalis*. (C) Cages used for holding *Ciona savignyi*. (See Color Insert.)

all that is available, it will need to be supplemented. In Santa Barbara, we initially experimented with filtered seawater, and found that addition of cultured microalgae could provide an adequate diet. However, we also experimented with unsupplemented raw (unfiltered) seawater as an alternative for culturing ascidians. We found that our initial concerns about using unfiltered seawater, particularly the potential for wild *Ciona* coming in with the seawater and settling on our culture plates, to be unfounded. While we did find some organisms, such as barnacle larvae, coming in with the seawater and settling in the tanks, we found that a weekly washing of the tanks with an abrasive pad kept this problem under control. The situation in Santa Barbara may be somewhat unique in that the intake pipe is located in approximately 20 meters of water approximately 500 meters from shore on a sandy bottom. Thus, the absence of *Ciona* in the intake pipe may be due to the fact that the intake pipe is not located in a favorable habitat for *Ciona*. Anyone wishing to culture *Ciona* at a marine laboratory should certainly explore the possibility of using unfiltered seawater. Even if the raw seawater were found to have significant *Ciona* larvae, this could be overcome by passing the water through a commercial UV sterilizer, which probably would not decrease the nutritional content, but would, as we have found, kill ascidian larvae. The ease, economy, and reliability of using raw seawater has more than outweighed any shortcomings. Furthermore, the culturing capacity of a raw seawater facility is very easily expanded, and our current facility allows us to culture over 10,000 adult *Ciona* with very little maintenance.

For the Santa Barbara system, dedicated duplicate 4-inch seawater PVC lines were installed, which branch from the main intake line before the sand filter (Fig. 3). The two 4-inch lines branch into parallel sets of 2-inch "feeder lines" that distribute water to the culturing tanks. The duplicate lines are alternated each month so that when one is pressurized, the other is dry. Alternating the lines in this way prevents fouling organisms from building up and occluding the lines. Seawater entering the culturing tanks is first passed though a degassing column consisting of a 3-foot piece of 8-inch diameter PVC pipe filled with 1.5-inch "bioballs." The bioballs are available from any aquaculture supply dealer, and serve to increase the surface area as the incoming water equilibrates with atmospheric pressure. In the absence of the degassing column, the entering water, which is supersaturated with gas at sea level due to the depth from which it is collected, will seed out bubbles inside the ascidian pharyngeal baskets and kill them. The culturing tanks themselves hold approximately 450 liters and are constructed of polypropylene or fiberglass, and stacked two-high on fiberglass A-frame racks. Incoming seawater enters each of the top tanks at a rate of approximately 20 liters per minute and then drains to the lower tanks through a 2-inch connecting pipe. The outflow from the lower tank is passed through a 160-watt commercial UV sterilizer before being returned to the ocean.

The Santa Barbara facility currently has 18 culturing tanks. Each tank can accommodate up to 600 full-sized adult *Ciona* (and many more juveniles than

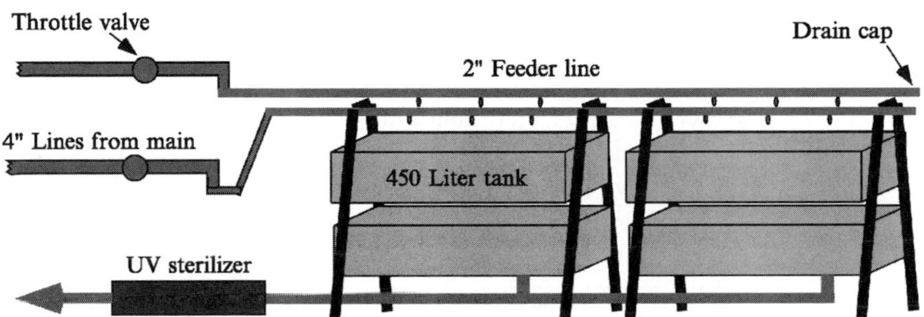

Fig. 3 *Ciona* culturing tanks used in Santa Barbara. Unfiltered seawater from an offshore intake pipe is passed through the tanks at approximately 20 liters per min. Efflux seawater is passed through a UV sterilizer before being discharged.

this). We have found that the raw seawater alone has sufficient nutrients to support the robust growth of the animals. To assess the nutritional content of the incoming seawater, we measure the chlorophyll content of the incoming seawater monthly. We find the chlorophyll content of the incoming water varies widely, from 12.5 mg/liter to 0.4 mg/liter. By comparing the chlorophyll content of the incoming and outgoing water, we are able to assess whether the density of animals growing in the tanks is too high. On average, we find that *C. savignyi* reach sexual maturity at 10 to 12 weeks, while *C. intestinalis* are mature at 12 weeks. The sexual maturity of the animals can be monitored by inspecting the gonoducts: the mature sperm (white) and eggs (orange in *Cs*; brown in *Ci*) are found in two conspicuous ducts that lead to the atrial siphon. The generation time in the Santa Barbara facility is somewhat slower than is reported for other systems (Cirino *et al.*, 2002), which we attribute to the lower temperature of the seawater in our facility, which averages $\approx$15 °C.

The maintenance of the Santa Barbara facility is relatively simple. The bottoms of the tanks are vacuumed with a siphon twice a week to remove debris. The incoming raw seawater is mostly contaminated by mollusks, sponges, and nudibranchs that represent potential competitors for food. All these contaminants, which tend to cover the walls of the tanks, are scraped off once a week with a sponge. We have occasionally observed the presence of other ascidian species on the culture plates. These species are mostly colonial and they are presumably transported through the intake pipe or brought in along with wild-caught *Ciona*. Contaminating ascidians have to be removed from the plates manually or with a sponge on a regular basis. Less frequently (once every 6 months), the tanks are emptied and filled with fresh water to get rid of particularly tough contaminating organisms. The only other maintenance is an annual changing of the bulbs in the UV sterilizer.

2. Closed Seawater Systems

Inland breeding is an important challenge that must be met by laboratories using *Ciona* species in order to facilitate a more widespread use of this model organism. The availability of such systems would serve to complement and back up larger facilities at marine laboratories, and would help advance the current growth of the ascidian scientific community around the world. We have developed a closed and self-contained culturing system for *Ciona* at the CNRS laboratory at Gif-sur-Yvette, near Paris. The closed nature of the system should help buffer the animals against changing seawater conditions and potential pathogens. A manuscript providing a more detailed description of the culturing system and of our results is in preparation.

The culturing system for adults is based on 1000-liter tanks that are divided into two compartments (Fig. 4A). One compartment is for growing the ascidians, and the other compartment is designed for sustainable photosynthetic production for

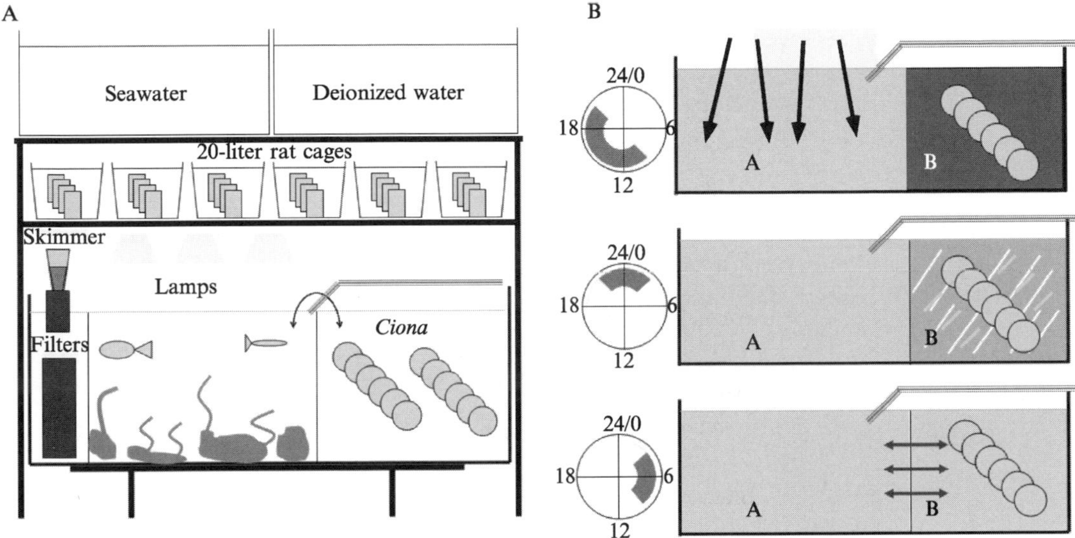

Fig. 4 (A) *Ciona* culturing tanks used in Gif-sur-Yvette. The entire culturing module consists of a three-compartment 1000-liter tank at the bottom, an upper row of 20-liter rat cages for juvenile rearing, and on top, two tanks containing deionized water and fresh artificial seawater (Instant Ocean, Crystal Reef). Water levels in the tanks are maintained by electromagnetic valves driven by switches sensing water in levels in the tanks. Evaporation in the rearing tank is compensated automatically with deionized water in the upper tank. To maintain algal growth in the bottom tank, three powerful halogen lamps are placed above the central compartment. This results in a 2 to 3 °C increase in tank temperature, compared to room temperature, which is held at 15 °C. Water mixing is achieved in the central compartment with three 1000 1/h pumps (Turbelle, France) activated alternatively by a "wave generator" (Wavemaster Pro, Red Sea). The rightmost compartment is kept dark by a removable plastic cover and moderate mixing is achieved by a 500 1/h pump. The leftmost compartment contains a skimmer (Tunze) designed to eliminate floating algae or organic particles. An external filtration device (Eheim 2250) eliminates large particles, resulting from the degradation of dead macrophytes. (B) Three daily phases of illumination and mixing of the tanks. During Phase 1 (9 AM–9 PM), the central compartment is illuminated and no water flow is allowed, with the right compartment containing the ascidians. During Phase 1, the ascidians are fed by the addition of 15 to 20 ml of a concentrated diatom preparation (*skeletonema costatum*; 10^9 cells per gram; SATMAR, France) obtained by centrifugation of large-scale cultures. During Phase 2 (9 PM–3 AM), the right compartment is kept dark while it is cleaned of uneaten food with a small external fine-meshed filter (cleaned weekly together with the walls of the right compartment). During Phase 3 (3 AM–9 AM), water is circulated between the two compartments with use of a pump. The conditions in the reconstituted ecosystem, in particular, the denitrification capacity of the coral sand at the bottom of the central compartment, allow for the natural elimination of excess food.

feeding the ascidians. We reasoned that seawater conditions in larger tanks would be slower to change, and easier to monitor and control, than conditions in multiple small aquaria. Additionally, our tanks are based on two leading aquaculture principles: the use of natural and productive conditions, and the inclusion and separation of several environmental zones. To generate equilibrated and self-sustaining conditions in the "production tanks," we attempted to reconstitute a basic ecological cycle including multitudes of marine microorganisms together with "living" porous limestone (collected at the bottom of calcareous D-Day cliffs in Arromanche, Normandie), macrophytes placed under strong illumination, herbivorous animals (nonfiltering molluscs) that control algal growth, carnivorous fish, and detritivorous invertebrates (crabs, shrimp). To minimize the need for water changes, a complete nitrogen cycle is expected to take place in this type of tank; while macrophytes do contribute to denitration, the denitrification step is mainly performed by a layer of coral sand placed on a grid at the bottom of the tanks, following a previously described process (Jaubert and Gatusso, 1989). To generate good productivity in tanks, strong mixing of the seawater to promote optimum oxygenation was achieved by powerful pumps driven alternatively by "wave generators."

We reasoned that the conditions in the production tanks, while optimal for controlled nutrient growth, were not ideal for *Ciona*, which prefer quiet, dark, and eutrophic water. We thus decided to breed *Ciona* juveniles in separate dark compartments which are allowed moderate water mixing, are easy to inspect and clean, and are free of potential predators. Exchanges of nutritive substances and conditioning of the seawater occurs by alternative circulation between the two compartments (Fig. 4B).

While the details of long-term, closed system culturing facilities are being worked out, it is necessary to import ascidians from a coastal supplier. Divers collect the animals each Monday and ship them to Gif-sur-Yvette on Wednesday. Sorted mature adults are then placed in 100-liter tanks to recover. On Friday, they are placed under constant illumination to promote gamete production and used during the following week. Alternatively, we raised *Ciona savignyi* juveniles imported from Santa Barbara. Crosses of *Ciona intestinalis* adults collected at the Marine Roscoff station and breeding of larvae in culture plates were performed, as described later in the chapter.

About one week after fertilization, a maximum of six plates containing 200 to 500 juveniles are placed vertically in 20-liter rat cages. The cages were filled with one-third volume of seawater coming from our production tanks (as described) and two-thirds fresh seawater. One air-pulsed internal filter filled with lava gravel (pre-equilibrated in the tank for 2 weeks before use) is added in each cage. Air exits the filter via a bent tube placed at the surface, thereby resulting in a moderate water mixing in the cages. One milliliter of *skeletonema costatum* preparation is added daily to each cage (see Fig. 4).

After 4 weeks, juveniles have grown to about 2 to 5 mm and are then moved to the 1000-liter tanks for further culturing. When the largest animals have grown to about 3 cm, smaller ones are removed from the plates, leaving only $\approx$100 adults per plate. The plates are then transferred to a second 1000-liter tank containing mature animals. The plates are checked regularly for potential contamination.

Our tanks were installed over 2 years ago, and we have observed no deviation in water quality. Most animals brought from the seashore are still alive. Fucus and other macrophytes are still growing actively. No obvious pathology affecting *Ciona* has occurred. During the first year, we were able to support the growth of three generations of *Ciona savignyi* that reached a normal adult size but produced a moderate amount of gametes (about 100–500 eggs). These animals grew to adulthood in about 4 months and survived for about 9 months. However, it became clear that the growth of the animals was slower than in the wild. We observed that, with time, the productivity of the tanks declined to the point that they were not able to sustain the growth of a high number of *Ciona*. To compensate for the low productivity, live diatoms were then added (Pr. Tsuda, personal communication). In addition, we have begun to feed them with flagellate cultures (*Isochrysis*). This regimen has accelerated the growth rate of *Ciona*, and gametes can now be obtained after 2 months of culture. Using a mathematical analysis of our results from the 18 rat cages available, these methods were tested systematically (manuscript in preparation).

Although more experience is needed with these breeding strategies, the modules described in this chapter open the route to *Ciona* inland breeding. When a means for inland culturing of *Ciona* at large scale is developed and viable at long term, it will offer laboratories worldwide, regardless of geographical location, the opportunity to work with species such as *Ciona savignyi*.

IV. Induced Developmental Mutants and Natural Variants/Mutants

While molecular genetic techniques have been applied to ascidians for many years, serious efforts at applying classical genetic techniques to ascidians are very recent, and were developed in parallel with laboratory culturing techniques. For classical developmental genetics, the goal is to identify and propagate ascidian lines carrying mutations in developmentally important genes. We have previously reviewed techniques and strategies for isolating *Ciona* mutants at projects based in Santa Barbara and Naples (Sordino *et al.*, 2000). However, our techniques have been significantly modified and improved since this first review. In Santa Barbara, we have taken two approaches for collecting mutants lines: chemical mutagenesis and screening wild populations for spontaneous mutations. While a number of interesting mutants have been isolated in chemical mutagenesis screens (Section III.B.2), screening for spontaneous mutations has been surprisingly successful (Section III.B.1).

A. General Techniques

For the selfing and controlled outcrossing techniques described here, it is essential to either dispose of or wash all materials (beakers, dissecting tools, transfer filters, pipets, etc.) with hot fresh water (or isopropanol) after they are used to collect gametes and before proceeding to the next animal. *Ciona* are broadcast spawners and make an enormous number of gametes. It requires only a small volume of sperm to result in an inadvertent cross-fertilization.

1. Screening by Selfing

In our screening, we have taken advantage of the fact that ascidians are hermaphroditic. Despite reports that *Ciona intestinalis* are not self-fertile (Byrd *et al.*, 2000), we have found that high percentages of both *C. savignyi* and *C. intestinalis* individuals (roughly 90 and 80%, respectively) self-fertilize at various efficiencies (5 to 100% of eggs in one clutch will fertilize with self-sperm). Because *C. savignyi* and *C. intestinalis* produce so many eggs, 5% fertilization of a large spawn can provide enough embryos to score for phenotypes. Self-fertilization offers a rapid way to generate homozygous animals for screening recessive mutations. We routinely use two methods to self-fertilize animals: dissection and light-induced spawning. In cases where high-throughput screening is necessary, or there is no further need for the parent, we dissect the adults and recover the gametes from the gonoducts. Alternatively, the adults can be induced to spawn (Section III.A.1.b). This allows us to keep the parent for repeated spawning or for further genetic study. For either method, the animals are first allowed to accumulate gametes by keeping them under constant fluorescent lighting for 2 to 4 days in running, filtered seawater.

a. Dissection

After the animal has been kept in constant light for several days, we dissect it under a dissection microscope and recover the gametes. First, the tunic is removed with scissors by gently cutting along the side opposite the gonoducts. This decreases the chance of prematurely tearing the genital ducts, which lie on the dorsal side of the animal near the atrial (outcurrent) siphon. Once the tunic is removed, we use a pair of fine forceps to expose the genital ducts. This is done by carefully grasping and tearing the body wall muscle with two forceps, making sure not to tear the genital ducts. This may have to be done several times along the body to expose the entire length of the ducts. Eggs should be removed first by tearing the oviduct and gently pressing along the length of the duct to release the eggs. This can easily be done without tearing the sperm duct. Eggs are then collected with a Pasteur pipet and transferred to 5 cm plastic petri dishes with SW/Ab. Isolated eggs are viable for up to one day at 18 °C. Next, the sperm duct is torn with forceps and sperm is quickly collected with a Pasteur pipet and transfer to a 1.5 ml microfuge tube on ice. Isolated sperm is viable at 4 °C for up to one week. For fertilization, several drops of sperm are added to the isolated eggs. After approximately 5 min, a transfer filter can be used to wash off excess sperm. This is important because excess sperm can result in abnormal embryos, which could complicate the scoring of phenotypes. The Petri dishes with the fertilized eggs are then kept in an incubator at 15 to 18 °C. If the fertilization is successful, the first cleavage should be visible within an hour.

b. Light-Induced Spawning

C. savignyi: After being allowed to accumulate gametes, the adults are placed in individual plastic cups containing 300 ml of seawater. The cups containing the adults are first placed in a dark 18 °C incubator for 5 to 8 h. The cups are then strongly illuminated with fluorescent lights. We have found that most of the adults (≈75%) will spawn within 15 min of being illuminated, and once spawning has started, sperm and eggs are shed simultaneously over a period of several minutes. Embryos are, therefore, highly synchronous in their development. Of the remaining 25% of the adults that do not spawn on cue, we find a variable fraction of them will have already spawned in the dark, and some will not spawn. Those that have not spawned can often be induced to spawn by a second dark–light cycle. Once the spawning is complete, the adult is removed and, if it is to be kept, it is transferred back to a feeding tank for recovery for at least 2 weeks before spawning again. The eggs in the beakers are collected using a transfer filter and moved to petri dishes with SW/Ab.

C. intestinalis: Ciona intestinalis adults are also kept in constant light for several days to accumulate gametes. However, animals spawn gametes in the dark over a period of roughly 1 to 2 h after the lights are turned off. Animals spawn their gametes over a longer time period than *C. savignyi*, so highly synchronous embryos are more difficult to obtain. Eggs are collected as has been described. As with *C. savignyi*, individuals differ in their spawning response, and some animals will spawn when the lights are turned on.

2. Controlled Outcrossing and Complementation Screening

The principles for crossing individuals either for propagation of recessive mutants or for complementation testing are the same as have been described for self-fertilization, with the exception that gametes from two individuals are mixed. A controlled cross-fertilization is most easily achieved by collecting the gametes by dissection. However, it is also possible to use spawning for outcrosses. Sperm that has been spawned into the seawater will remain viable for approximately 60 min and, after the spawned eggs have been removed by filtration, the sperm-containing seawater can be used to fertilize eggs collected from a different individual. This technique is particularly useful for outcrossing or complementation analysis when one does not wish to sacrifice the adult. It is also possible to spawn several adults together in a beaker, but the result will consist of a collection of selfed and outcrossed progeny in unknown proportions.

While self-fertilization is a rapid and easy way to screen for recessive mutations, we have found the progeny of self-fertilization rarely grow to healthy adults. As a consequence, our usual procedure when screening for recessive mutations is to use only a fraction of the sperm to mix with self-eggs. The remainder of the sperm is stored at 4 °C until the phenotype of the self-fertilized embryos is known. Sperm from individuals showing potentially interesting phenotypes in their progeny at the expected Mendelian ratios is then used to fertilize eggs from a wild-type individual. The resulting "outcrossed" progeny are then cultured to adults and rescreened for the phenotype by self-fertilization.

There are several important considerations when outcrossing. Because efforts to generate inbred laboratory strains of *Ciona* have met with limited success (Kano *et al.*, 2001), all wild-type animals are literally that: they are collected from the wild population. Because any wild-caught *Ciona* could potentially carry a lethal recessive mutation, it is important to screen the animals for preexisting mutations by self-fertilization before they are used. Typically, a large number of wild *Ciona* are collected, spawned in cups, and the resulting embryos examined for obvious mutations. Adults that give wild type-appearing progeny are saved, allowed to recover for 2 weeks in the culturing tanks, and then sacrificed for egg collection by dissection. When collecting the wild-type eggs for outcross, it is also important to ensure that the eggs have not inadvertently been contaminated with self-sperm during the dissection. To control for this possibility, the dissected eggs are allowed to remain at 18 °C for one hour. If no cleavage is observed, the eggs can then be used for the outcross.

3. Maintaining Mutant (and Transgenic) Lines

Both *C. savignyi* and *C. intestinalis* will live for approximately one year in culture. For mutant lines that are currently being studied, we maintain a pool of adults for repeated spawning. New adults are generated by outcrossing approximately every 6 months. To protect against the loss of a mutant line, and to store lines that are not being actively studied, samples of sperm are stored in liquid nitrogen.

a. Freezing sperm

Animals are first kept for several days in constant light to accumulate sperm. The animal is then dissected as has been described, but extra care is taken to blot up all seawater before tearing the sperm duct so that the sperm is not diluted. The sperm is then quickly collected into a microfuge tube on ice. Twenty-μl aliquots of the concentrated sperm is added to 500 μl cryogenic vials and quickly mixed with 80 μl of a chilled solution containing 10% DMSO in seawater. Samples are frozen by placing the cryotubes on a floating styrofoam raft in liquid nitrogen. After 10 to 20 min, the tubes can be stored by suspending in liquid nitrogen. One can typically collect 100 μl or more of concentrated sperm from a large adult, and it is best to make as many aliquots as possible. After a few days, it is a good idea to test one aliquot of the sperm because, for unknown reasons, some batches of frozen sperm are unable to fertilize eggs.

b. Fertilization with Frozen Sperm

Dissect eggs from several healthy wild-type animals previously screened for preexisting mutations. Ensure eggs have not been self-fertilized during dissection by keeping eggs in 18 °C seawater for one hour, monitoring for cleavage. After removing the sperm aliquot from liquid nitrogen, thaw it quickly by submerging the tube in a 30 °C water bath. Once the sperm solution begins to thaw, add 200 μl of "thawing solution" (0.4 mM NaOH in seawater) directly to sperm and mix by pipetting. The slightly higher pH of this solution ($\approx$pH 8.6) appears to activate the sperm. Quickly add the thawed sperm to the eggs and mix.

B. Screens for Developmental Mutants

1. Spontaneous Mutants

We have screened a large number of *C. savignyi* and *C. intestinalis* collected from the Santa Barbara Yacht Harbor for preexisting recessive mutations by self-fertilization. Although this work is still underway, we estimate that $\approx$20% of the wild-caught animals carry recessive mutations that can be easily detected. We are screening self-fertilized progeny only at the hatched tadpole stage. At this stage, the chordate body plan is evident, but many of the organ systems have not yet formed. As a consequence, the true percentage of animals carrying lethal recessive mutations is probably higher than 20%. A complete complementation analysis of spontaneous mutants has not been done, but we have isolated several non-complementing mutants, suggesting that either certain loci are "hot spots" for mutations, or certain mutant alleles are widespread within the population.

The success of the screens for spontaneous mutants is attributable not only to the apparent high frequency of preexisting recessive mutations, but also to the enormous number of *Ciona* that can easily be collected and the ease with which recessive mutations can be detected in these animals by self-fertilization. In most locations, both boat owners and marine ecologists view *Ciona* as a nuisance and/or an invading species. We have almost always found harbor staff more than willing to allow us to collect them by the bucketful. However, be advised to check with your local government before collecting any marine specimen and acquire any necessary permits.

Although the various molecular lesions causing the spontaneous mutations are not known (point mutations, inversions, etc.), we have found that spontaneous mutant lines are easily propagated through multiple generations, and these mutations behave as single loci. We are currently culturing and studying several spontaneous mutant lines of *C. intestinalis* and *C. savignyi* that carry recessive mutations that profoundly disrupt early development. These include the *C. savignyi* mutants *draemong* that fails to form the pharyngeal gill slits and atrial siphon at metamorphosis, and *chobiesque* that has disrupted notochord morphogenesis, and the *C. intestinalis* mutant *frimousse* that fails to form anterior ectoderm derivatives (manuscripts in preparation).

2. Induced Mutagenesis

Screens for induced mutations have been an effective way to isolate genes involved in the development of *C. elegans, Drosophila*, zebrafish, and mouse (Anderson and Brenner, 1995; Driever *et al.*, 1996; Haffter *et al.*, 1996; Justice, 2000; Nusslein-Volhard and Wieschaus, 1980). Ascidians are ideal for genetic studies due to their ease of maintenance, relatively short generation time, small genome, and phylogenetic relationship to vertebrates. Several induced mutagenesis screens have been performed in *Ciona* (Moody *et al.*, 1999; Sordino *et al.*, 2001, 2000). All of these screens have utilized the chemical mutagen N-ethyl-N-nitrosourea (ENU), which has been shown to be an effective point mutagen in zebrafish and other animals (Driever *et al.*, 1996; Haffter *et al.*, 1996; Justice *et al.*, 1999; Mullins *et al.*, 1994). While a number of mutants have been isolated in ENU screens, ENU does not appear to be as robust a mutagen in ascidians as it is in other animals. Table I gives a summary of several ENU screens. The low frequency of recovered mutants in *Ciona* after ENU treatment differs significantly from the high frequencies in other organisms (Driever *et al.*, 1996; Haffter *et al.*, 1996). There are several possibilities to account for the differences. First, the potency of ENU may be reduced in seawater due to its high salinity. To our knowledge, ascidians are the first marine animals in which ENU-mutagenesis has been attempted. More experiments to address the stability of ENU and other mutagens in seawater must be done. Second, we do not have detailed knowledge of spermatogenesis in ascidians, especially the time-course of spermatogenesis and the number of spermatogonia. Both of these factors can strongly influence the efficiency of mutagenesis.

We have included a detailed ENU mutagenesis protocol in this chapter. However, along with other investigators, we are currently exploring other mutagenesis strategies, including insertional mutagenesis. The final, preferred strategy is likely to be different from the ENU protocol outlined here. In fact, because ENU is a point mutagen, and the wild populations that are currently being used are so polymorphic, ENU is probably not the mutagen of choice if one wishes to identify the exact genetic lesion involved in developmental abnormalities. Nonetheless, the development of the ENU mutagenesis protocol is a useful starting point, and

Table I
Frequency of Induced Mutations

	Species	# of F1 screened	# of potential mutants	Potential mutant frequency (%)	# of potential mutants	References
1	*Ciona savignyi*	80	25	31	25	(Moody *et al.*, 1999)
2	*Ciona intestinalis*			10–20		(Sordino *et al.*, 2000)
3	*Ciona savignyi*	256	21	8	21	unpublished

will help guide future investigations. Insertional mutagenesis and possibly other mutagens such as trimethylpsoralen (TMP), which in other species makes ≈ 1-kilobase deletions (Gengyo-Ando and Mitani, 2000), may prove to be more appropriate for ascidians. A classical method to test for the efficacy of a mutagen is to assay for new induced mutations that fail to complement known mutations (Mullins *et al.*, 1994). We currently have several nonlethal recessive mutants that give obvious phenotypes (e.g., no pigment or short tails) and that can be cultured as homozygotes. Complementation experiments to test the ability of various mutagenesis protocols to induce new mutations at these nonlethal loci are underway.

ENU mutagenesis involves treating the whole animal, with the target being the spermatogonia. The entire ENU treatment procedure takes several months. At present, because the starting animals for mutagenesis are wild-caught rather than laboratory strains, it is essential to ensure that the animals to be mutagenized do not have obvious preexisting mutations. Thus, the first step is to collect adults and screen them for mutations by self-fertilization (Section III.A.1). One should identify at least 20 mutation-free animals, as the ENU-treatment regime will kill 50 to 75% of the animals. The selected animals are then allowed to recover for 2 weeks in culture tanks before starting the ENU treatment. The ENU treatment involves injecting the animals and then soaking them in ENU-seawater. The treatments are repeated three times, once a week, following the protocol which will be detailed.

Procedure:

Note that ENU is a powerful mutagen and precautions must be taken to avoid both contact with the skin and inhalation. All materials that come into contact with ENU should be decontaminated with 25% sodium thiosulfate, pH 10, including all seawater.

1. Start with a 1-gram sealed isopec bottle of ENU (Sigma, N-3385). Remove from the freezer and equilibrate in the dark to room temperature for 15 min in a fume hood.

2. Inject 85.4 ml of 10 mM sodium phosphate, pH 6.5, into the sealed ENU vial, to make a 100 mM solution. The ENU will dissolve very slowly. To help the ENU dissolve, cover the vial with aluminum foil (to keep it dark), and attach it to a vortex mixer in fume hood. Allow the vial to mix for 2 hours.

3. The animals are treated in 2-liter plastic tubs. Fill the tub with 998 ml filtered sea water and add 2 ml 0.5 M $NaPO_4$ pH 6.0. The final pH of the sea water should be 6.5. Hold the temperature of solution in the tub to 18–20 °C by putting it in a water bath.

4. Draw up 30 ml of the dissolved ENU into a large syringe and add to the phosphate-buffered seawater in the tub (3 mM final concentration).

5. A "mock-treatment" tub can also be set up, in which 30 ml of 10 mM $NaPO_4$, pH 6.5 is added in place of the ENU.

6. The animals are first injected with ENU before being placed in the ENU-seawater. Take the first animal to be treated and inject with 50 μl of the 100 mM ENU solution using a 25 gauge needle and a 1 cc syringe. We aim the injection at the gonads, near the base of the ascidian. Place the injected animal into the tub of ENU-seawater. Repeat this with all of the animals to be ENU treated. Mock-treated animals should receive 50 μl 10 mM $NaPO_4$ pH 6.0.

7. Leave animals in ENU- or mock-tubs for 1 h at room temperature.

8. At the end of the treatment, place the animals in a new 2-liter tub of seawater to wash out the ENU. Give the animals several changes of seawater over the course of 24 h. Be sure to decontaminate the seawater before disposing of it.

9. Return the animals to the culturing tanks and allow them one week to recover before the next treatment.

10. After the last of the three treatments, the animals are allowed to recover for several weeks.

11. Sperm from the ENU-treated, or mock animals, is collected by dissection and used to fertilize wild-type eggs (Section III.A.2). (Make sure to prescreen the egg-donor animals for preexisting mutations). The resulting larvae are settled on plates and cultured to adults (Sections II.A, II.B), at which time they are screened for recessive mutations by self-fertilization (Section III.A.1).

V. Linkage Analysis and Mapping Genes in Ascidians

For the ascidian projects focusing on both developmental genetics and allorecognition, the long-term goal is to identify the genes responsible for observed phenotypes. Until an insertional mutagenesis procedure is developed in ascidians, efforts will focus on screening candidate genes and/or positional cloning. The *C. intestinalis* genome is about 5% the size of the human genome and it contains approximately 16,000 predicted genes. Both the small genome size of ascidians and the relatively low number of predicted genes relative to vertebrates are tremendous advantages. Furthermore, the completed and partially assembled genome sequences of both *C. intestinalis* (Dehal *et al.*, 2002) and *C. savignyi* (manuscript in preparation) are immeasurably important resources for all gene identification efforts. For positional cloning in *Ciona*, the extremely high fecundity of solitary ascidians means that hundreds of meioses can easily be analyzed from a single individual. One potential obstacle with positional cloning in *Ciona* is the lack of inbred laboratory strains. The traditional route for linkage analysis has been to rely on genetic maps consisting of various markers such as Amplified Fragment Length Polymorphisms (AFLPs), microsatellites, Random Amplified Polymorphic DNAs (RAPDs), or cloned genes. Because of the absence of inbred laboratory strains of solitary ascidians, and especially due to the highly polymorphic nature of ascidian genomes, genetic maps based on anonymous markers such as RAPDs will vary enormously between individuals and, consequently, may be of little use for positional cloning. Given this challenge, and the high degree of polymorphism present in ascidian genomes, bulked segregant analysis (BSA) is proving to be a valuable tool for positionally cloning genes in both *B. schlosseri* (De Tomaso *et al.*, 1998) and *Ciona*. The BSA technique allows one to identify markers such as AFLPs that show tight linkage to the locus of interest. Linked AFLP fragments can be readily sequenced and constitute a starting point for higher resolution linkage analysis with single tadpoles (Section IV.B.2), for candidate gene identification, and for rescue experiments using genomic fragments.

A. Mapping the Fu/HC Locus in *B. schlosseri*

Botryllus schlosseri, a colonial ascidian, is a model system for studying allorecognition and histocompatibility, stem cells, and programmed cell death. Many of the *Botryllus* studies have focused on the three levels of allorecognition: colonial fusion and rejection outcomes, chimeric resorption, and somatic and germ cell parasitism (Rinkevich, 2002). The Fu/HC locus is the single gene locus that governs the fusion or rejection outcome of contact between neighboring colonies. If the two colonies share a Fu/HC allele, then they will fuse and form a chimera. If they do not share a common allele at the Fu/HC locus, then a rejection pathway is initiated, resulting in inflammation, hemorrhage, cytolysis, and scarring, ensuring that no future contacts occur at that particular site. The evolutionary antecedents of the vertebrate immune system may be rooted in this allorecognition system. The rapid fusion or rejection response resembles graft acceptance or rejection in vertebrates. It has been speculated that this system could be the basis for the immune recognition found in vertebrate T-cell recognition/corecognition of MHC (Weissman *et al.*, 1990). The techniques used for the genetic characterization of traits and linkage analysis of the highly polymorphic fusibility/histocompatibility (Fu/HC) locus are particularly relevant to our genetic mapping efforts. Microsatellites have been used to genetically identify the origins of somatic and gametic tissues within chimeras (Stoner *et al.*, 1999). Restriction Fragment Length Polymorphisms (RFLPs) have been used for linkage analysis of heat shock protein gene loci and Fu/HC alleles (Fagan and Weissman, 1998). De Tomaso *et al.* (1998) elected a genomic approach in isolating the Fu/HC locus by making a genetic map, isolating tightly linked markers, and using these as starting points for a genomic walk. AFLP was used to identify DNA polymorphisms as molecular markers for this genetic map. BSA was used to evaluate the meiotic assortment of certain AFLP markers with respect to predetermined Fu/HC alleles. Two AFLP markers that flanked the Fu/HC locus were identified and the pooling strategy was reiterated by making new DNA pools from animals that showed crossovers at either of the two flanking loci. Any marker found to show linkage in both the original and the new crossover DNA pools was determined to be proximal to the

Fu/HC locus. Without the advantage of a highly inbred laboratory strain, this procedure was used to quickly narrow down the location of the Fu/HC locus to less than 6 cM within the uncharacterized *Botryllus* genome. AFLP served to significantly reduce the amount of physical mapping required to identify the Fu/HC locus.

B. Pilot Studies for Positional Cloning in *Ciona*

1. BSA

AFLP is a cost-efficient, reliable, and robust DNA fingerprinting method that uses the Polymerase Chain Reaction (PCR) to selectively amplify a subset of genomic restriction fragments (Vos *et al.*, 1995). AFLP requires no prior knowledge of genomic DNA sequence and can be used on any DNA, regardless of origin or complexity. We use AFLP to identify markers linked to mutant loci starting with pools of DNA from sibling larva that had been segregated into mutant and wild type phenotypes. We typically make triplicate pools, each containing DNA from 15 larva. Markers linked to the mutation are differentially present in the wild-type and mutant pools. Conversely, unlinked markers are present in both pools because of crossing over between distant loci on the same chromosome and the independent assortment of loci on different chromosomes during meiosis (for a complete description of the technique, see Ransom and Zon, 1999; Vos *et al.*, 1995). The AFLP products are resolved on polyacrylamide gels, and one looks for AFLP bands present in the phenotypically wild-type pools and absent in the mutant pools. Because we work on recessive mutations, the pools of phenotypically wild-type tadpoles will contain heterozygotes, and thus will contain both the wild-type and mutant alleles. The mutant pools contain only the mutant allele. Therefore, bands found in all of the wild-type pools and none of the mutant pools represent a molecular marker associated with the wild-type allele of the affected gene. The goal is to identify several linked markers that are clustered in a particular region of the genome.

Although the procedures and materials of AFLP are relatively simple, we have chosen to use a commercial kit from Life Technologies^{TM} (AFLP Analysis System II), with a few modifications to adapt it to ascidians. Most of the techniques we describe here are based on those developed in zebrafish (Ransom and Zon, 1999). As in zebrafish, the BSA technique is most easily performed on embryos or larvae. However, *Ciona* larvae have only about 2500 cells, and expected yield of DNA from one larva is about 1 ng. Consequently, the reaction volumes used in a few steps in the AFLP procedure have been scaled down.

Ascidian larvae present one additional challenge for linkage analysis. There is a group of maternally contributed cells called "test cells" that are found between the chorion and the developing embryo. At the tailbud stage, these cells adhere to the surface of the embryo and contribute to the formation of the larval and juvenile test/tunic (Sato *et al.*, 1997). Thus, a DNA preparation made from whole larvae would be potentially contaminated with maternal DNA. Fortunately, the test cells

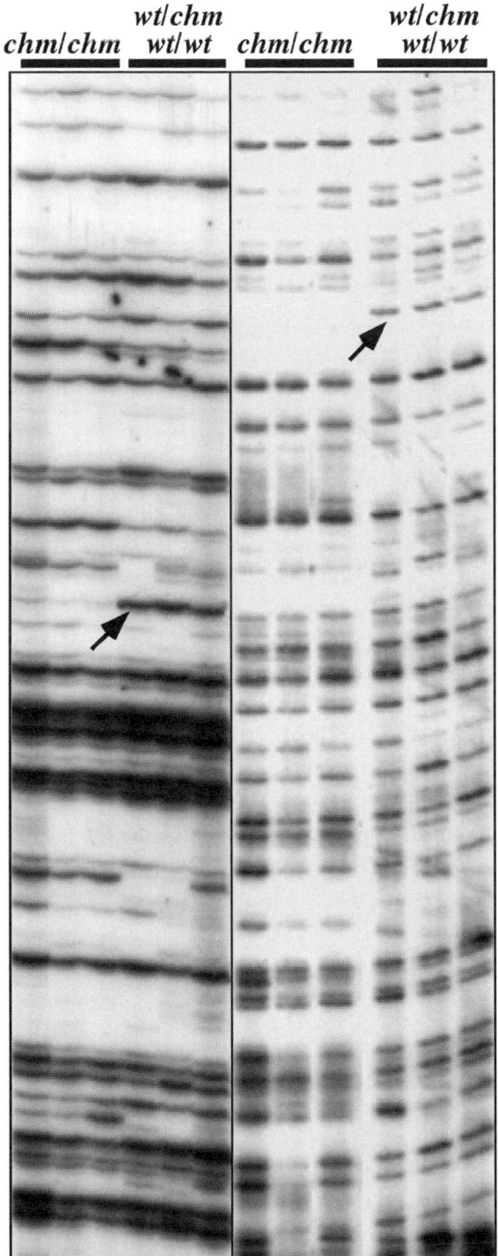

Fig. 5 Bulked segregant analysis of *chongmague* (*chm*) mutant. Genomic DNA was prepared from triplicate pools of 15 tadpole larvae from crossed *chm/wt* adults. The pools either had tadpoles showing the short-tailed *chm* phenotype (*chm/chm*) or were wild type in appearance (*wt/wt* or *chm/wt*). The pools of DNA were assayed for linked markers by AFLP analysis. The results from two different selective primer sets are shown. Putative linked bands are indicated with arrows.

can be removed by dechorionating the embryos at the one-cell stage. The absence of the test cells does not noticeably affect embryonic development, although such animals undergo abnormal metamorphosis (Sato and Morisawa, 1999).

In Santa Barbara, we are currently using BSA to find linked markers for several recessive mutations. One of the *Ciona savignyi* mutants being investigated is the notochord mutant *chongmague* (*chm*) (Nakatani *et al.*, 1999). We have begun to map the *chm* mutation using AFLP. Examples of AFLP gels for *chm* are shown in Fig. 5. AFLP involves amplifying a unique subset of markers using "selective primers." Several sets of selective primers are tested individually in order to examine a large number of potentially polymorphic markers. We have examined the products of 50 selective amplifications and have identified four markers that show strong linkage to the *chm* locus. Based on our sampling of the highly polymorphic wild population of *C. savignyi* from which this mutant was propagated, we estimate that we have examined about 11,000 polymorphic markers to identify these four potentially linked markers. The sequences of the markers have allowed us to identify genomic regions from the sequenced *Ciona* genomes. New markers for SSCP analysis were designed from genes within these genomic regions and are being tested in single tadpoles for higher-resolution mapping (Section IV.B.2). All indications are that the *C. intestinalis* and *C. savignyi* genomes have a very high degree of synteny, and we have taken advantage of both genome assemblies. Although the genomes of both *C. savignyi* and *C. intestinalis* are only partially assembled, we have been able to place two of the *chm* markers on the same 353,958 bp *C. intestinalis* scaffold. Within this scaffold is at least one strong candidate that is currently being examined.

a. Extraction of Tadpole Larva DNA for AFLP

1. Self-fertilize an adult heterozygous for the recessive mutation of interest (or cross two heterozygous siblings). The embryos need to be dechorionated (Chapter 6) before the first cleavage. Grow the embryos to the late-tailbud stage.
2. Using a transfer filter (Section II.A), wash embryos through 3 beakers of 100 ml distilled water.
3. Using distilled water, wash larvae onto a petri dish coated with 1% agarose/distilled water.
4. Using a pipette coated with bovine serum albumin, transfer the mutant and wild-type embryos separately to new dishes of distilled water, carrying over as little debris as possible.
5. Make triplicate pools of 15 mutant and 15 wild-type embryos in 1.5 ml microfuge tubes (six tubes total).
6. Centrifuge the tubes at max speed for 5 min.
7. Remove supernatant, making sure not to lose larvae (they do not pellet well).
8. Freeze pellets on dry ice.

9. Thaw pellets quickly at 37 °C. Add 100 μl of embryo lysis buffer and pipet up and down several times to homogenize larvae. Add 2 μl of 10 mg/ml proteinase K. Vortex, then incubate at 50 °C for 3 or more hours.

10. Extract the homogenate once with an equal volume of phenol, then once with an equal volume of phenol/chloroform.

11. Recover the aqueous phase, about 100 μl, to a new tube. Add 10 μl of 3M NaOAc and 1 μl of Pellet Paint (Novagen). Vortex to mix solution. Add 250 μl of ethanol and mix again. Incubate on ice for 15 min.

12. Centrifuge at max speed for 10 min. Remove supernatant and wash pellet with 100 μl of 70% ethanol. Remove as much fluid as possible from the pellet and allow to air dry. Resuspend pellet in 9 μl of water.

Embryo Lysis Buffer:
100 mM NaCl
20 mM Tris pH 7.8
10 mM EDTA
1% SDS

b. AFLP

Protocols for AFLP have been published previously (Ransom and Zon, 1999; Vos *et al.*, 1995). We followed the protocol supplied with the Life Technologies™ AFLP Analysis System II and AFLP Small Genome Primer Kit exactly, with the exception that the entire 9 μl of pooled tadpole DNA (see previous text) was used in half-volume reactions for the restriction digest and adapter ligation steps.

c. Gel Electrophoresis and AFLP Band Recovery

AFLP products are resolved in 6% polyacrylymide/urea gels as described previously (Ransom and Zon, 1999; Vos *et al.*, 1995). After the gel has been run and dried, before the x-ray film is exposed, mark the perimeter of the Whatman paper with glow-in-the-dark tape, paint, or marker. These markings are used for band recovery from the gel. Linked bands will be those that appear in the pooled wild-type lanes but not in the mutant lanes (Fig. 5).

To recover AFLP bands:

1. Overlay dried gel with x-ray film. Line up the glow-in-the-dark markings. Trace band of interest by punching holes through the film, into the gel, using a needle. It is important to outline the band of interest as tightly as possible, carefully avoiding isolation of extraneous areas of the gel.

2. Cut band out of Whatman paper with dried gel using a razor to connect the punched holes.

3. Put the slip of paper with the dried gel into a microcentrifuge tube. Add 500 μl of elution buffer (0.5 mM sodium acetate, 1 mM EDTA pH 8.00, 2% SDS). Incubate at 55 °C for 1 h.

4. Spin down paper at 14K rpm for 5 min. Remove 250 μl of supernatant and re-spin supernatant to remove all debris.
5. Transfer re-spun supernatant to a clean 1.5 ml microcentrifuge tube and add 250 μl of ethanol. Place on ice for 30 min.
6. Spin at 14K rpm for 15 min at 4 °C.
7. Wash twice with 1 ml of 70% ethanol.
8. Allow pellet to dry by leaving tube uncapped at room temperature.
9. Resuspend in 50 μl TE.

Use 2 μl of recovered DNA per 20 μl PCR reaction with 20 pmol of AFLP primers to reamplify the band. If in doubt, confirm that the correct band has been isolated by running a sample of a 1:50 dilution of the reamplified DNA next to the original AFLP reactions that generated the band. The reamplified band can then be subcloned and sequenced or sequenced directly.

2. Linkage Analysis on Single *Ciona* Tadpoles by SSCP

Although the combination of BSA and AFLP allows one to simultaneously assay hundred of markers for potential linkage, higher resolution mapping within a candidate genomic region, or testing for linkage of a candidate gene, will require examining many individual meioses. Doing such higher-resolution mapping will, in most cases, require developing new polymorphic markers within the gene or genomic region of interest. Due to the extremely polymorphic nature of wild *Ciona* genomes, we have had good success identifying new polymorphic markers using Single-Strand Conformation Polymorphism (SSCP) assays (Orita *et al.*, 1989). SSCP takes advantage of the fact that even single nucleotide changes in short single-stranded DNA fragments alter the intramolecular secondary structure of these molecules. When DNA fragments are denatured at a high temperature then quickly cooled, renaturation into double-stranded DNA is inhibited. Each single strand assumes its most favored conformation, corresponding to its lowest free-energy state. The particular conformation it assumes depends on the nucleotide sequence of the DNA. Under the appropriate conditions, the differences in secondary structures, generated by differences in nucleotide sequence, result in detectable shifts in the gel migration of the DNA fragments. SSCP requires PCR amplification of short DNA fragments, ideally 150 to 350 bp. Primers should anneal to conserved regions of DNA to ensure reliable amplification. It is also important that the primers amplify a region with a polymorphism that changes the secondary structure of the single-stranded DNA, generating a gel shift. We have obtained successful SSCP results using primers that anneal to exon sequences flanking small introns (Fig. 6). Because introns, unlike exons, are not subject to the stringent selective pressure to conserve nucleotide sequence, they are a rich source of polymorphisms.

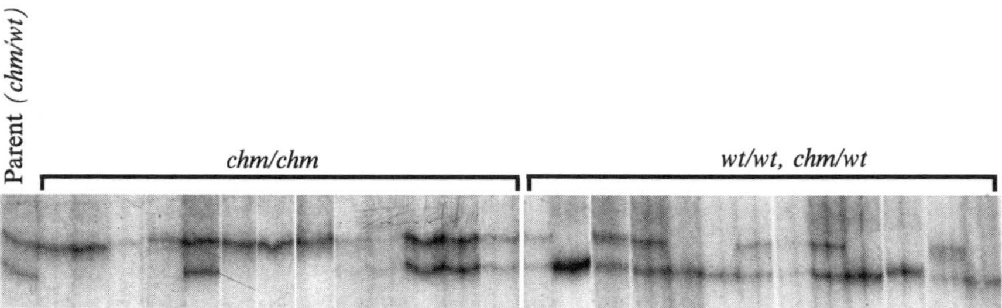

Fig. 6 Linkage analysis with DNA from single *Ciona savignyi* tadpoles. Shown are the results from a PCR-amplified genomic region thought to be linked to the *chongmague (chm)* locus. The PCR products were analyzed by the single-strand conformation polymorphism technique. Note that this genomic region appears to show some degree of linkage as the upper allele is always found in *chm/chm* tadpoles, and none are homozygous for the lower allele.

a. Single **Ciona** *Tadpole DNA Preparation*

Ascidian larvae are tiny (≈1 mm), have very few cells, and consequently, little DNA. It is necessary to be careful when preparing DNA from single tadpoles. The following protocol is the most simple method available for obtaining genomic DNA from single tadpoles that can be used as templates for PCR reactions. If a cleaner DNA prep is needed, it is also possible to perform a phenol chloroform extraction after the tadpole is digested, followed by an ethanol precipitation step.

1. Start with a plate of dechorionated tadpoles. To remove debris, gently wash tadpoles through three agarose/seawater-coated dishes of Millipore seawater, using a pipet coated with bovine serum albumin. Using a bovine serum albumin-coated pipet, transfer dechorionated tadpoles to an agarose-coated petri dish containing distilled water. Minimize the volume of seawater transferred.
2. Set pipetman to 2.5 μl. Using filter tips coated with bovine serum albumin, collect a single tadpole and transfer it to a 1.5 m clear microcentrifuge tube containing 2.5 μl of 2X Single Tadpole Lysis Buffer.
3. Centrifuge tubes briefly to bring larvae to the bottom of the tubes. Using the dissection microscope, check each tube to confirm that one tadpole has successfully been transferred.
4. Freeze fracture the tadpoles by incubating them for at least 1 h in the −80 °C freezer, then incubating tubes at 65 °C for at least 1 h.
5. Under dissection microscope, check to be sure tadpoles are digested. Denature proteinase K by incubating tubes at 70 °C for 10 min.
6. Bring volume of each tadpole prep up to 100 μl with DI water.

In the first round of nested PCR, use 16 μl of this solution for each 25 μl PCR reaction.

- *2X Single Tadpole Lysis Buffer:*

- 100 mM KCl
- 20 mM Tris HCl pH 8.3
- 5.0 mM $MgCl_2$
- 0.9% NP40
- 0.9% Tween
- 0.01% gelatin
- 400 μg/ml proteinase K (add just before use)

b. SSCP

SSCP was performed as described previously (Foernzler *et al.*, 1999), with a few modifications. Only one primer was labeled and an equal amount of the second primer was added to the primer mix after the kinase was denatured. We did not find it necessary to quantify the DNA recovered from the tadpoles before following the nested PCR procedure described earlier (Section IV.B.1.a).

3. Prospects for Positional Cloning in Ascidians

Positional cloning efforts in ascidians are at an early stage of development, but methods adapted from other species, such as BSA, appear to work well in ascidians. Efforts to map the Fu/HC locus in *B. schlosseri* have probably progressed the furthest. Positional cloning in *Ciona* will be aided immensely by the sequenced *C. intestinalis* and *C. savignyi* genomes. Hopefully, a more finished genome will be available for future studies. Currently, the *C. intestinalis* assembly contains 116.7 million base pairs of nonrepetitive sequence in 2501 scaffolds greater than 3 kb. Half of this (60 Mbp) is assembled into 117 scaffolds longer than 190 Kbp; 85% of the assembly (104.1 Mbp) is found in 905 scaffolds longer than 20 kb (http://genomic.jgi-psf.org/ciona4/ciona4.home.html). Until a more complete assembly is available, it is likely that, for positional cloning, one will have to create a more finished sequence in specific regions of interest. These efforts should be aided by arrayed bacterial artificial chromosome (BAC) libraries that are available for both *Ciona* species. Depending upon the particular genetic lesion, identifying the specific mutant locus may be challenging. Point mutations may be difficult to find within the highly polymorphic background of the *Ciona* genome. However, the potential to rescue mutants with cloned BAC DNA should allow one to narrow the locus to a single gene with a high degree of certainty.

Acknowledgments

We thank Lisa Belluzzi and Dave Johnson for their comments on this manuscript. This work was supported by grants from the NIH and NSF to W.C.S and from the CNRS to J.S.J.

References

Anderson, P., and Brenner, S. (1995). Mutagenesis. *Methods Cell Biol.* **48,** 31–58.

Boyd, H. C., Brown, S. K., Harp, J. A., and Weissman, I. L. (1986). Growth and sexual maturation of laboratory-cultured Monterey *Botryllus schlosseri. Biol. Bull.* **170,** 91–109.

Byrd, J., and Lambert, C. C. (2000). Mechanism of the block to hybridization and selfing between the sympatric ascidians *Ciona intestinalis and Ciona savignyi. Mol. Reprod. Dev.* **55,** 109–116.

Cirino, P., Toscano, A., Caramiello, D., Macina, A., Miraglia, V., and Monte, A. (2002). Laboratory culture of the ascidian *Ciona intestinalis* (L.): A model system for molecular developmental biology research. *Mar. Mod. Elec. Rec.* [serial online]http://www.mbl.edu/html/BB/MMER/CIR/Cir-Tit.html, Nov. 27.

De Tomaso, A. W., Saito, Y., Ishizuka, K. J., Palmeri, K. J., and Weissman, I. L. (1998). Mapping the genome of a model protochordate. I. A low resolution genetic map encompassing the fusion/histocompatibility (Fu/HC) locus of *Botryllus schlosseri. Genetics* **149,** 277–287.

Dehal, P., *et al.* (2002). The draft genome of *Ciona intestinalis*: Insights into chordate and vertebrate origins. *Science* **298,** 2157–2167.

Deschet, K., Nakatani, Y., and Smith, W. C. (2003). Generation of *Ci-Brachyury*-GFP stable transgenic lines in the ascidian *Ciona savignyi. Genesis* **35,** 248–259.

Driever, W., Solnica-Krezel, L., Schier, A. F., Neuhauss, S. C., Malicki, J., Stemple, D. L., Stainier, D. Y., Zwartkruis, F., Abdelilah, S., Rangini, Z., Belak, J., and Boggs, C. (1996). A genetic screen for mutations affecting embryogenesis in zebrafish. *Development* **123,** 37–46.

Erives, A., Corbo, J. C., and Levine, M. (1998). Lineage-specific regulation of the *Ciona* snail gene in the embryonic mesoderm and neuroectoderm. *Dev. Biol.* **194,** 213–225.

Fagan, M. B., and Weissman, I. L. (1998). Linkage analysis of HSP70 genes and historecognition locus in *Botryllus schlosseri. Immunogenetics* **47,** 468–476.

Foernzler, D. and Beier, D. R. (1999). Gene mapping in zebrafish using single-strand conformation polymorphism analysis. *In* "Methods in Cell Biology" (W. H. Detrich M. Westerfield, and L. I. Zon, eds.). vol. 60, pp. 185–193. Academic Press, San Diego.

Gengyo-Ando, K., and Mitani, S. (2000). Characterization of mutations induced by ethyl methanesulfonate, UV, and trimethylpsoralen in the nematode *Caenorhabditis elegans. Biochem. Biophys. Res. Commun.* **269,** 64–69.

Haffter, P., Granato, M., Brand, M., Mullins, M. C., Hammerschmidt, M., Kane, D. A., Odenthal, J., van Eeden, F. J., Jiang, Y. J., Heisenberg, C. P., Kelsh, R. N., Furutani-Seiki, M., Vogelsang, E., Beuchle, D., Schach, U., Fabian, C., and Nusslein-Volhard, C. (1996). The identification of genes with unique and essential functions in the development of the zebrafish, *Danio rerio. Development* **123,** 1–36.

Jaubert, J., and Gatusso, J. P. (1989). An integrated nitrifying–denitrifying biological system capable of purifying seawater in a closed circuit system. *Bulletin de l'Institut Oceanographique* **5** (special), 101–106.

Justice, M. J. (2000). Capitalization on large-scale mouse mutagenesis screens. *Nature Reviews Genetics* **1,** 109–115.

Justice, M. J., Noveroske, J. K., Weber, J. S., Zheng, B., and Bradley, A. (1999). Mouse ENU mutagenesis. *Hum. Mol. Genet.* **8,** 1955–1963.

Kano, S., Chiba, S., and Satoh, N. (2001). Genetic relatedness and variability in inbred and wild populations of the solitary ascidian *Ciona intestinalis* revealed by arbitrarily primed polymerase chain reaction. *Marine Biotechnology* **3,** 58–67.

Moody, R., Davis, S. W., Cubas, F., and Smith, W. C. (1999). Isolation of developmental mutants of the ascidian *Ciona savignyi. Molecular and General Genetics* **262,** 199–206.

Mullins, M. C., Hammerschmidt, M., Haffter, P., and Nusslein-Volhard, C. (1994). Large-scale mutagenesis in the zebrafish: In search of genes controlling development in a vertebrate. *Curr. Biol.* **4,** 189–202.

Nakatani, Y., Moody, R., and Smith, W. C. (1999). Mutations affecting tail and notochord development in the ascidian *Ciona savignyi*. *Development* **126,** 3293–3301.

Nusslein-Volhard, C., and Wieschaus, E. (1980). Mutations affecting segment number and polarity in Drosophila. *Nature* **287,** 795–801.

Orita, M., Suzuki, Y., Sekiya, T., and Hayashi, K. (1989). Rapid and sensitive detection of point mutations and DNA polymorphisms using the polymerase chain reaction. *Genomics* **5,** 874–879.

Ransom, D. G., and Zon, L. I. (1999). Mapping zebrafish mutations by AFLP. *In* "Methods in Cell Biology" (W. H. Detrich, M. Westerfield, and L. I. Zon, eds.). vol. 60, pp. 195–211. Academic Press, San Diego.

Rinkevich, B. (2002). The colonial urochordate *Botryllus schlosseri*: From stem cells and natural tissue transplantation to issues in evolutionary ecology. *Bioessays* **24,** 730–740.

Sato, Y., and Morisawa, M. (1999). Loss of test cells leads to the formation of new tunic surface cells and abnormal metamorphosis in larvae of *Ciona intestinalis* (Chordata, Ascidiacea). *Dev. Genes Evol.* **209,** 592–600.

Sato, Y., Terakado, K., and Morisawa, M. (1997). Test cell migration and tunic formation during post-hatching development of the larva of the ascidian, *Ciona intestinalis*. *Develop. Growth Differ.* **39,** 117–126.

Satoh, N. (1994). "Developmental Biology of Ascidians." Cambridge University Press, Cambridge.

Sordino, P., Belluzzi, L., De Santis, R., and Smith, W. C. (2001). Developmental genetics in primitive chordates. *Philos. Trans. R. Soc. Lond. B. Biol. Sci.* **356,** 1573–1582.

Sordino, P., Heisenberg, C. P., Cirino, P., Toscano, A., Giuliano, P., Marino, R., Rosario Pinto, M., and De Santis, R. (2000). A mutational approach to the study of development of the protochordate *Ciona intestinalis* (Tunicata, Chordata). *Sarsia* **85,** 173–176.

Stoner, D. S., Rinkevich, B., and Weissman, I. L. (1999). Heritable germ and somatic cell lineage competitions in chimeric colonial protochordates. *Proc. Natl. Acad. Sci. USA* **96,** 9148–9153.

Vos, P., Hogers, R., Bleeker, M., Reijans, M., van de Lee, T., Hornes, M., Frijters, A., Pot, J., Peleman, J., and Kuiper, M. (1995). AFLP: A new technique for DNA fingerprinting. *Nucleic Acids Res.* **23,** 4407–4414.

Weissman, I. L., Saito, Y., and Rinkevich, B. (1990). Allorecognition histocompatibility in a protochordate species: Is the relationship to MHC somatic or structural? *Immunol. Rev.* **113,** 227–241.

CHAPTER 8

Hemichordate Embryos: Procurement, Culture, and Basic Methods

Christopher J. Lowe,[*,†] Kuni Tagawa,[§] Tom Humphreys,[§] Marc Kirschner,[‡] and John Gerhart[†]

[*]Department of Organismal Biology and Anatomy
University of Chicago
Chicago, Illinois 60612

[†]Department of Molecular and Cellular Biology
University of California Berkeley
Berkeley, California 94720

[‡]Systems Biology
Harvard Medical School
Boston, Massachusetts 02115

[§]Kewalo Marine Laboratory
Honolulu, Hawaii 96813

METHODS IN CELL BIOLOGY, VOL. 74

0091-679X/04 $35.00

I. Introduction

The phylum Hemichordata is composed of two extant classes: the solitary burrowing marine worms, or enteropneusts, and the pterobranchs, which are colonial sessile organisms. Bateson originally placed this group within the chordates, due to many of the adults morphological affinities between the two groups (Bateson, 1885) and it was not until much later that they were reclassified into their own phylum (Hyman, 1959). Current molecular phylogenies place them as the sister group of the echinoderms (Adoutte *et al.*, 2000; Cameron *et al.*, 2000; Turbeville *et al.*, 1994; Wada and Satoh, 1994). Despite the close relationship of these two groups, their respective adult body plans are highly divergent, and even gross axial comparisons between the groups are problematic. The bilateral symmetry, gill slits, stomochord, post-anal tail, and hollow dorsal nerve cord of hemichordates provide strong morphological affinities with the chordates, making them a particularly interesting group for testing hypotheses of chordate origins (Lowe *et al.*, 2003; Nubler-Jung and Arendt, 1999; Tagawa *et al.*, 2001). For any consideration of deuterostome evolution, hemichordates are a critical component: their key phylogenetic placement and proposed morphological affinities with the chordates make them an intriguing but poorly described group. Developmental studies on this phylum are beginning to generate a substantial amount of gene expression data for conserved developmental regulatory genes (Harada *et al.*, 2000, 2001, 2002; Lowe *et al.*, 2003; Ogasawara *et al.*, 1999; Okai *et al.*, 2000; Peterson *et al.*, 1999; Tagawa *et al.*, 1998a,b, 2000, 2001; Taguchi *et al.*, 2000, 2002; Takacs *et al.*, 2002) allowing a more comprehensive phylogenetic sampling of developmental data within the deuterostome lineage. These results are beginning to provide critical insights into early deuterostome evolution. Only by a comprehensive understanding of the evolution of all three phyla can we hope to reconstruct the early evolutionary history of the deuterostomes and understand the transitions that led to the unique and unusual body plans of this group.

Within the enteropneusts, there are two contrasting developmental strategies, one indirect and the other direct. Indirect-developing species, the Ptychoderids and Spengellidae, produce a larva that swims and feeds in the plankton for months before transforming into the adult worm and settling. Direct-developing species (the Harramanids) bypass the larval stage and develop to the adult worm from the egg. Comparative studies of these contrasting developmental strategies is particularly interesting as current diverse hypotheses of chordate origins alternatively propose emergence of the dorsal nervous system of the chordate body plan either from progenitor larval structures (Garstang, 1928; Nielsen, 1999; Tagawa *et al.*, 2000) or from the basic invertebrate adult nervous system (Arendt and Nubler-Jung, 1996; DeRobertis and Sasai, 1996; Lowe *et al.*, 2003). Since extant chordates lack larval stages, investigation of both life history modes will probably be necessary to comprehensively analyze the implications that various molecular observations have for these hypotheses.

This chapter focuses on practical considerations for collection of hemichordate adults and procurement of gametes for carrying out developmental studies. We describe methods for collecting, spawning, and rearing embryos and larvae from two species of hemichordate—*Saccoglossus kowalevskii* and *Ptychodera flava*, the latter an indirect developer producing a true larval form and the former a direct-developing species. In addition, we present two protocols for whole mount *in situ* hybridization.

II. Procurement, Spawning, and Culture of *S. kowalevskii*

A. Background

Despite the general recognition of hemichordates as a phylogenetically key group, very few studies have focused on their development. In large part, this is due to some of the technical challenges associated with collecting adults and reliably obtaining viable gametes in large enough quantities for molecular studies. Artificial triggers for completion of meiosis and germinal vesicle breakdown have not been determined for hemichordates thus far, and experimental manipulation of gravid animals to stimulate natural spawning is required to induce oocyte maturation (Colwin and Colwin, 1962).

S. kowalevskii is found along the eastern seaboard of the United States from South Carolina to New Hampshire. It is an intertidal species found mainly in sheltered bays and forms its burrows in coarse organic sand, often in clam beds. The worms are approximately 6 to 9 inches long and form U-shaped burrows. Figure 1A shows a diagram of this species with major body regions labeled. Figure 1B,C shows pictures of a gravid adult male and female.

B. Method

1. Reproductive Season

Our experience is limited to animals collected on Cape Cod close to the Marine Biological Laboratory at Woods Hole, MA. The animal collection staff of the Marine Resources Center at the Marine Biological Laboratory has extensive experience in both collecting and shipping these animals (www.mbl.edu). From late August to the end of September, ripe animals of both sexes are abundant in the area. The peak of spawning tends to be toward the middle of September, with few ripe animals being found by the beginning of October. In addition to the main spawning season in the fall, mature individuals of both sexes have been reliably collected in May, though we have not thoroughly investigated this period of reproductive activity.

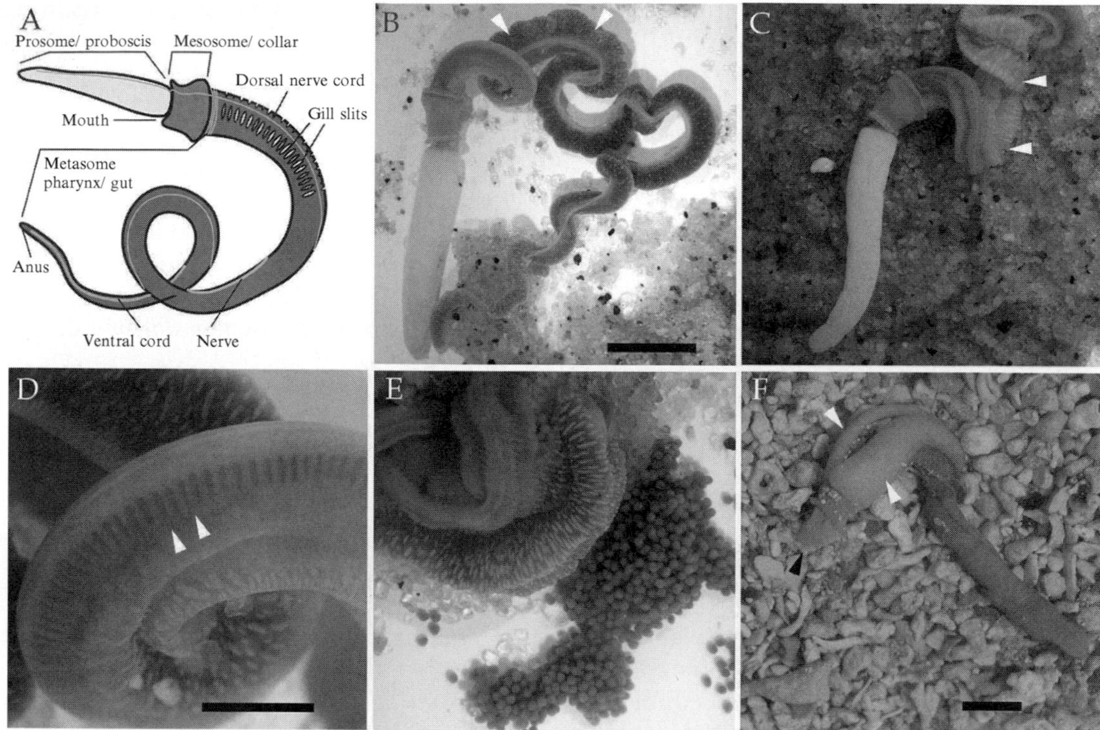

Fig. 1 Adult enteropneust hemichordates, *Saccoglossus kowalevskii* and *Ptychodera flava*. (A) Model of an acorn worm adult based on *S. kowalevskii* outlining some of the characteristic features of the adult morphology. (B) Adult *S. kowalevskii* female. White arrowheads indicate the position of the green ovary. Black scale bar = 3 cm. (C) Adult *S. kowalevskii* male. White arrowheads indicate the position of the orange/white testes. (D) High magnification of branchial region of metasome on *S. kowalevskii*. Lateral view. White arrowheads indicate the position of two of the multiple pairs of gill slits on the dorsal side of the worm. Scale bar = 0.5 cm. (E) Spawning female of *S. kowalevskii*. Oocytes are green/blue-colored and are spawned along the length of the gonad. (F) Adult *P. flava*. Black arrowhead indicates the position of the proboscis. White arrowheads indicate the position of the genital wings. Scale bar = 1 cm. (See Color Insert.)

2. Adult Collection

a. Their populations are patchy and can be found most commonly in sheltered bays. Populations seem to be most common in brackish water and are present in the mid to low intertidal zone.

b. The burrows of the adults are most easily identified by the presence of fragile, spiral fecal castings, 3 to 4 cm in diameter, which are very obvious in calm water, but are easily disturbed by water flow.

c. Adults are collected by digging vertically with a shovel close to the burrows, which are often in groups, at low tide, carefully transferring the sand into a large sieve (mesh diameter of approximately 0.5 cm). Both digging and sieving are best done in shallow water, which aids in locating the burrows and also allows sand to easily pass through the sieve, causing less damage to the fragile adults. The sieve should remain in the water to avoid entangling the animals in the sieve and damaging the epidermis. The worms are transferred into widemouth, screw-top plastic bottles by placing the mouth of the bottle next to worm on the mesh and submerging the sieve, washing the worm into the vessel. Avoid washing in additional debris and ensure there is no sharp shell debris or algae in the vessel with the worm.

3. Transportation

Adult animals are extremely fragile and are easily damaged in transport. Fine, clean sand should be added to the vial, just enough to cover the animal. Once collected, the animals should be kept cool and not left in the sun to heat up above 25 °C. Animals can be shipped by FeDEX®, after filling the vials with fresh seawater aerated with oxygen for a few seconds before sealing the vials. Add 20 μg/ml gentamycin if warm. Twenty or so vials of animals can be shipped in one box by double-packing the vials with large plastic bags, with the inner one filled with seawater. Add two cool packs to the box before shipping. It is imperative that the sand be free of sharp shells or filamentous algae to avoid damaging the animals during shipment.

4. Adult Maintenance in the Lab

Ideally, the adults should be kept in running seawater aquaria at a marine station, but a recirculating aquarium would also be adequate. Animals remain healthier when entirely covered by sand and allowed to establish burrows. However, this makes egg collection impractical, so maintaining the adults in a small amount of clean sand that is changed daily is the best compromise to ensure animals can be effectively monitored for spawning. If mucus accumulates around the collar of the animals, it should be removed with forceps as such "nooses" can slice through the epithelium of the collar within a day or two, leading to a rapid decline in health. We have not attempted to maintain a long-term population of worms in the lab. For short-term storage of several days to two weeks, animals may be kept in their shipping vessels after replacing the seawater with filtered seawater and gentamycin (20 μg/ml) and kept at 4 °C.

The next section describes the treatment of animals following either collection from the field or receipt of a shipment.

a. Animals can be tipped with their sand into a much larger container filled with fresh seawater to expose the worm, and transferred into a small glass dish using a turkey baster or spoon.

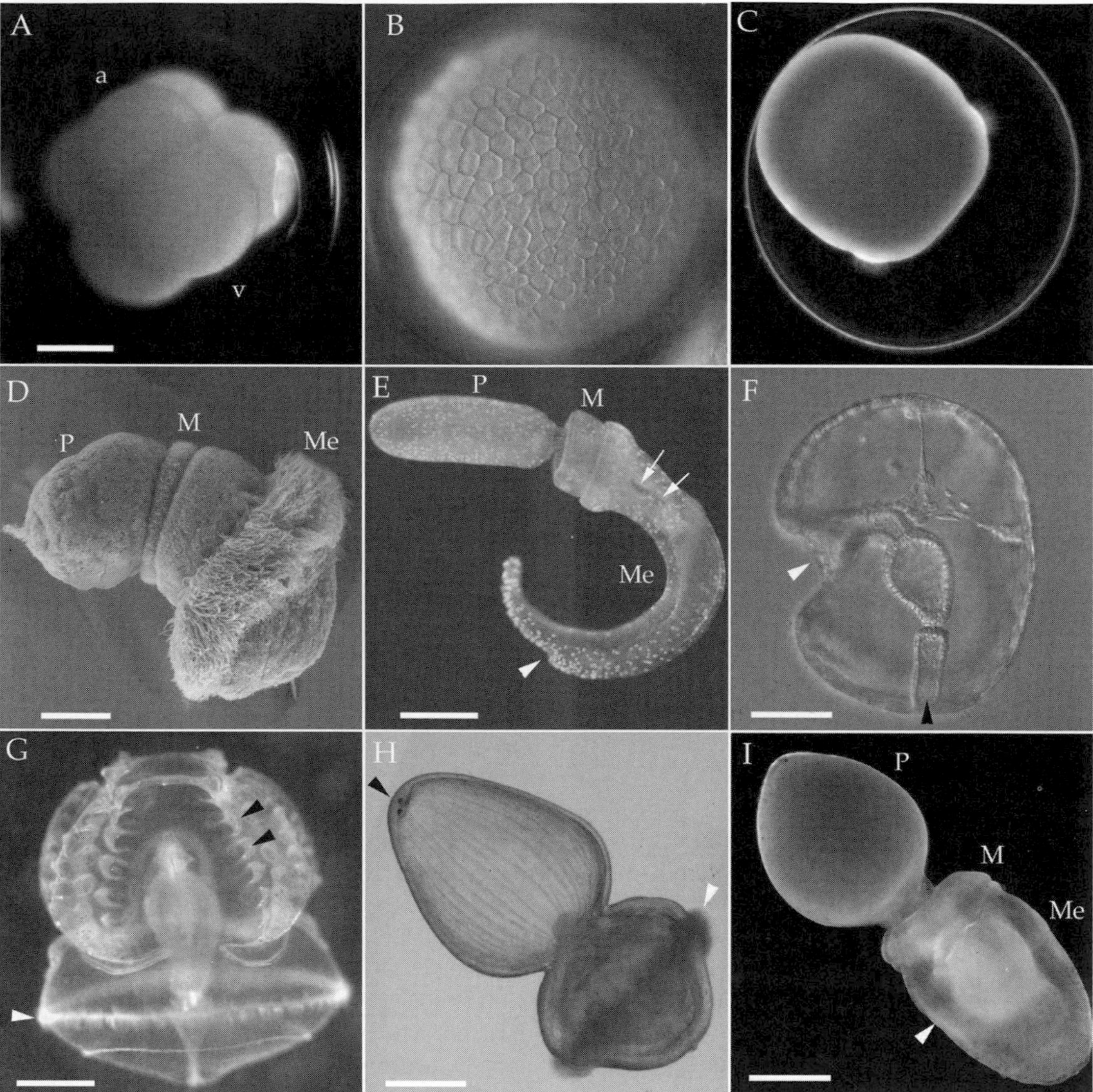

Fig. 2 Embryological, larval, and juvenile stages of *S. kowalevskii* and *P. flava*. A through E reflect stages of *S. kowalevskii* development. F through I reflect stages of *P. flava* development. (A) Early cleavage stage within the vitelline envelope. Animal blastomeres are slightly smaller than vegetal ones. a = animal pole, v = vegetal pole. Scale bar = 100 μm. (B) Late blastula. (C) Post-gastrula with prominent telotroch at the posterior end of the embryo swimming within the vitelline envelope. (D) SEM micrograph of midstage embryonic development (day 4), lateral view. The tuft of cilia at the far left of the panel is the apical tuft. The ectoderm is evenly ciliated except for the dense telotroch at the posterior of the embryo. Scale bar = 75 μm. (E) Late juvenile 14 days old, lateral view. All three body regions clearly visible. White arrowhead indicates the position of the anus and, posterior to

b. Examine the animal under a dissecting scope to check for damage. The long gut should be amputated with fine scissors just anterior to any region where there is damage, and we routinely remove large stretches of post-branchial gut, as the animals tend to tie themselves in tight knots without large amounts of sand to burrow in. If the proboscis is damaged at the tip, then a clean cut should be made with a scalpel posterior to the damage. Much of the proboscis can be removed without compromising the health of the animal significantly in the short term or inhibiting gamete production. If the entire proboscis is removed, females are unable to spawn or regenerate new tissue. Ovary that is damaged during shipping or storage rapidly turns orange and should be amputated immediately anterior to the damage to avoid degeneration in animal health. Removal of this tissue does not compromise the spawning ability of the animal from the remaining ovary. Gently rinse each animal free of residual sand from the collection site as sperm are sometimes contained in it.

c. Transfer animals into small, white plastic containers: we have found disposable plastic weigh boats of approximately 10 cm diameter to be useful for this purpose. The white background helps to identify spawned oocytes that are hard to see among sand grains. The worm should only just be covered with clean sand, free of any sharp shell debris or algae, and placed into a running seawater aquarium. Adult males and females can be distinguished only when reproductive by the color of their gonads (Fig. 1B, C). Males have a milky white gonad visible through the ectoderm, whereas females have a green/blue color to their gonad. Oocytes can clearly be seen through the ectoderm during their reproductive season (Fig. 1B, D, E). Males and females should be kept in separate tanks. Animals should be checked every day for general health and body knots, and mucus strands removed from their collars. Sand should be changed every other day to avoid buildup of mucus secreted during their feeding and burrowing, which can attract ciliates and bacteria.

5. Spawning

All further steps of the protocol should be carried out in dedicated embryological glassware that should be washed in water only with no detergent. **Any glassware that has been exposed to fixative, detergent, or heavy metals should not be used to culture embryos.**

---

that, the ventral post-anal tail. White arrows indicate the position of the first two gill slits. Scale bar = 150 μm. (F) Lateral view, DIC micrograph of 3-day-old tornaria larva. White arrowhead indicates the larval mouth and black arrowhead, the anus. Scale bar = 50 μm. (G) Krohn stage larva. Black arrowheads indicate the folds, or epaulets, in the dorsal ciliated bands, and white arrowhead marks the telotroch. Scale bar = 300 μm. (H) Larva during transformation with clearly formed proboscis. The black arrowhead marks the position of the eyespots, and white arrowhead, the telotroch. Scale bar = 300 μm. (I) Juvenile worm one week after the initiation of transformation. White arrowhead indicates the band that persists after the cilia of the telotroch are shed upon settlement. Scale bar = 500 μm. P = prosome, M = mesosome, and Me = metasome. (See Color Insert.)

a. Active sperm can be obtained by rupturing the testes of a ripe male with fine forceps. Concentrated sperm will ooze out of the wound and can be collected with a Pasteur pipette.

b. Oocytes are arrested in meiosis in the female gonad and, in order to achieve successful fertilization, the females must be induced to spawn. This can be achieved by a heat treatment in seawater to 29 °C for 8 h, followed by reimersion in the seawater at ambient temperature (22 °C) (Colwin and Colwin, 1962). Do not extend the heat treatment for any longer than 8 h. Often, females will start spawning 6 to 8 h after heating and continue for 3 to 4 h, so we recommend heating overnight to avoid animals spawning late at night. At peak spawning season, females may spawn unpredictably in the seawater tables several hours after collection, purely from the stress of transport, without any heat stimulus. Regular animal inspections are advised during these peak times, both in the holding aquaria and during the heat treatment, to ensure spawning females are identified quickly. Egg collection is most effectively carried out using a Pasteur pipet to separate eggs from sand. Oocytes are released along the length of the ovary and often line the burrow in the sand. If the sand is kept to a minimum in the containers, then these burrows are easily cracked open to reveal the oocytes. It is not advised to heat worms without any sand as they will become entangled in mucus threads, leading to rapid decline in health.

6. Fertilization

One drop of concentrated sperm solution should be diluted in 100 ml of seawater before fertilization of oocytes. If oocytes are not immediately fertilized, then they can be stored for up to 12 h at 10 °C (and possibly longer) before fertilization. The oocytes should be washed with several rinses of filtered seawater to remove the mucus and debris carried over during collection. Approximately 3 ml of sperm suspension prepared as has been described should be added to washed oocytes and stirred in 100 ml of filtered seawater. Fertilization success can be monitored by observation of the zygotes under a dissecting scope. A thick fertilization envelope should begin to rise 1 to 2 minutes following the addition of sperm if successful fertilization occurs.

7. Culture of Embryos

Figure 2A–E shows a range of developmental stages of *S. kowalevskii* from cleavage (Fig. 2A) through to juvenile (Fig. 2E). At 22 °C, first cleavage begins at approximately 120 min, and then every 30 min subsequently (Bateson, 1884; Colwin and Colwin, 1953). Gastrulation begins at approximately 18 h and is completed by 36 h. At this stage, the telotroch (ciliated band) forms and the embryo begins to swim within the vitelline (Fig. 2C). One ectodermal, circumferential groove forms at the boundary of the prospective proboscis and collar, and is closely followed by a second groove marking the boundary between the collar and metasome. At this stage, approximately 3 to 4 days postfertilization (Fig. 2D), the telotroch is very prominent and all three body regions are clearly visible. Embryos hatch at 7 days and swim for up to 24 h before searching on the bottom of the glass dish for a settlement site. At this stage, the juveniles begin to develop their post-anal ventral tail (Fig. 2E) and actively burrow in fine sand. The following section describes the basic protocol required for culturing embryos and juveniles.

a. Embryos are cultured in small glass dishes at room temperature (22 °C). Do not crowd the dishes with too many embryos, as this will result in slower development and asynchrony of the culture. There should be no more than a loose monolayer of embryos on the bottom of the dish. Examine cultures at advanced cleavage stages: any damaged or irregular embryos and unfertilized eggs should be discarded from the culture. Some embryos will have small yolk particles in the perivitelline space which have no effect on the viability of the embryos. Post-gastrula embryos can be kept at 4 °C for several days to slow down development without adverse effects; however, earlier embryonic stages tolerate temperatures lower than 16 °C poorly. Transfer the embryos once they begin to hatch into a larger glass dish with a thin layer of fine reptile sand coating the base of the dish. The sand is very important as the worms begin to secrete mucus at this stage and will become entangled in mucus threads if cultured without a substrate to burrow in.

b. Two days following hatching, juveniles should be fed by adding a suspension of phytoplankton (see Chapter 4 by Greg Wray or Lowe and Wray, 2000). We have used *Dunaliella terectiola, Rhodomonas lens, and Isochrysis galbana.* These can be prepared ahead of time, spun down to a concentrated pellet, aliquoted, and frozen at −20 °C, or added live as a suspension. Filtered seawater should be replaced every day, and the cultures should be very gently aerated at this stage with an air pump to keep water circulating in the culture dish. Juveniles have been successfully cultured for at least 20 days after hatching. Harvesting the juveniles out of the sand before fixation can be problematic, but if MgSW is added to anaesthetize the juveniles and the sand disturbed by squirting forcefully with a turkey baster, then the juveniles become dislodged. When the culture is swirled inside the dish, juveniles will move into the center on top of the sand. This procedure should be repeated several times to recover the majority of animals.

III. Procurement, Spawning, and Culture of *Ptychodera flava*

A. Background

Ptychodera flava was the first hemichordate described (Eschscholtz, 1825), although the original classification, based only on adult structure, placed it within the holothuroids. The species has an extensive range throughout the IndoPacific. The worms are approximately 10 to 25 cm long and 1 to 2 cm in diameter. Fig. 1F is a photograph of a gravid adult female. Like all the ptychoderids, *P. flava*, is an indirect developer. Females spawn large numbers of small oocytes that develop into tornaria larvae. The larvae remain in the water column for several months, feeding and growing until they become competent to transform into the juvenile worm. Unlike echinoderm metamorphosis, the larval body plan of the tornaria is directly inherited by the adult without the loss of larval tissue or cells (Agassiz, 1873). This suggests the possibility of comparing meaningfully gene expression patterns among embryonic, larval, young juvenile, and adult, stages that all preserve the bilateral body plan laid down during gastrulation. This body plan is also directly comparable with the vertebrate body plan.

B. Method

1. Adult Collection

Ptychodera flava adults are abundant in sand flats and are easily collected by snorkeling in shallow water around the islands of Hawaii. We have collected primarily on the island of Oahu at either the large sandbar across the entrance to Kaneohe Bay on the windward (east) coast or at the shallow reef that extends off Paiko on the leeward (south) coast. These habitats are strikingly different, the former constituted by deep, coarse, clean sand while at Piko, the animals are found in shallow, small collections of muddier, fine sand in pockets in the reef. At Piko, when the sand layer is very thin, the animals may even be found exposed. The animals do not seem to maintain fixed burrows but to move continuously in the top few inches of the sand.

In both locations, animals are found by using a vigorous wave of the hand in the water to stir up the top one or two inches of sand. The exposed animals are caught in the hand and placed into Ziplock bags of seawater. During collection, we often place 10 to 20 animals in a sandwich bag full of seawater and transport them in this manner for the 2 or 3 h required to reach the lab.

Ripe animals may be recognized by fuller, more distended genital wings which are a bit brighter yellow in color. The genital wings are a yellowish tan, reflecting the color of the underlying gonad, when compared with the tan color of the rest of the animal. Generally, there is not a great distinction between the females and males, but the males at Piko exhibit a darker, greenish gray color in their genital wings and usually can be distinguished from females.

Gravid animals can be collected from late October through January in Hawaii. The adults can be maintained in a healthy state for several months in running seawater tanks with 3 to 4 cm sand from their native habitat. We often keep the animals in small groups placed in 10×15 cm or 20×20 cm glass baking dishes with a few cm of sand stacked in the running seawater aquaria. In the Kewalo Marine Lab seawater system, artificial sources of food are not required if the running seawater is not filtered. Tanks should be kept clean and algal growth kept to a minimum.

2. Spawning

Oocytes are stored in the ovary arrested in meiosis with prominent germinal vesicles. Attempts to artifically induce *P. flava* oocyte maturation after dissection from the ovary have been unsuccessful (Tagawa *et al.*, 1998a). Unlike

S. kowalevskii, active sperm cannot be dissected from the ripe testes, and natural spawning must be induced in both males and females to achieve fertilization (Tagawa *et al.*, 1998a). Spontaneous spawning of individuals generally occurs in the evening after sundown, suggesting that spawning in *P. flava* may be stimulated by onset of darkness (Tagawa *et al.*, 1998a). Ripe individuals can be kept in holding tanks for at least 2 weeks without loss of fecundity if they do not spawn spontaneously, which is most frequent a day or two following collection.

Spawning is reliably induced at about sundown by placing individual worms into glass or plastic containers with sand covering the bottom, raising the seawater temperature from ambient temperature of 22 to 26 °C and maintaining the containers in a dark room. The room must be kept dark until the animals spawn. Spawning usually occurs within 2 h of the temperature increase. Unlike *S. kowalevskii*, spawning occurs during, rather than following, the heat treatment. Gametes are shed through the epidermis along the length of the genital wings. Females usually produce 0.05 to 0.2 ml of eggs, but some females shed up to 1 ml of eggs.

3. Fertilization

a. The oocytes of *P. flava* are translucent yellow, 120 μm in diameter, and surrounded by a thick, sticky jelly coat that makes them difficult to collect with a simple Pasteur pipette. We find a 5 or 10 ml Komagome Pipette (Iwaki Glass Co. Ltd, Tokyo) works best because you can suck up the eggs with a constant flowing stream of water. Transfer the oocytes to a 15 ml centrifuge tube and pellet the eggs using a hand centrifuge. Wash the eggs several times with filtered seawater. Concentrated sperm can be collected by Pasteur pipette from the genital wing area of spawning males or from the bottom of the vessel where the males have spawned.

b. Transfer the eggs into a petri dish. A few drops of semen should be diluted in approximately 100 ml of filtered seawater and 2 to 3 ml of this suspension should be swirled with the washed eggs. Fertilization success can be assayed by the appearance of the first polar body between 30 and 90 min later; first cleavage begins approximately 3 h post-fertilization.

4. Culture of Embryos and Larvae

a. Fertilized eggs are transferred into glass beakers up to a density that creates a monolayer on the bottom of the beaker. Wash the zygotes with filtered seawater several times and allow them to develop at room temperature (22 to 24 °C). Cultures should be washed daily with several rinses of fresh, filtered seawater containing streptomycin (50 mg/l). Before hatching, the embryos will settle to the bottom of the container and the water can be decanted easily.

b. Gastrulation begins at 18 h post-fertilization and is completed before hatching occurs at 44 h post-fertilization. After hatching, we culture the larvae at a density of about 1 per ml of seawater. Following hatching, the ciliated embryos and larvae disperse throughout the seawater and further seawater changes must be carried out by an alternative procedure using 100 μm Nitex® mesh. Do not collect the larvae on the mesh. We remove the bottom of a plastic beaker and replace it with the Nitex® mesh. (A hot glue gun works well to stick the Nitex® to the base; solvent-based adhesives are not recommended.) We immerse this modified beaker in a slightly larger beaker into which the culture has been transferred and aspirate as much old medium as possible. Add fresh filtered seawater, repeat once or twice, and transfer the culture to a clean culture container.

c. We have not successfully fed *P. flava* larvae. We have attempted to feed tornaria with cultured algal strains of *Tetraselmis, Nanochlorophsis, Nanochloris, Isochrysis, and Rhodomonas*. Young tornaria after 72 h ingest and swallow all of them; the algal cells can be observed to enter the stomach. They also seem to be able to digest all of them except *Tetraselmis*, which remains intact and even appears viable in the stomach. However, the tornaria do not seem to thrive when fed with any one or a mixture of these strains.

d. We have been successful culturing larvae for 2 months, from early tornaria through the Metschnikoff stage tornaria (Fig. 2F) (Stiansny-Wijnhoff and Stiasny, 1926; 1927) by standing cultures in a sunlit window. However, attempts to culture the tornaria through to late larvae that will transform to young juvenile worms have so far proven unsuccessful.

Larvae competent to transform to juvenile worms are easily collected in plankton tows during the months of April to July. Figure 2G shows a Krohn stage tornaria, the stage after the Metschikoff stage, collected in a plankton tow (Stiansny-Wijnhoff and Stiasny, 1926, 1927). Collected animals soon begin to transform as shown in Fig. 2H. When they are placed in a petri dish with a single layer of sand to provide an attractive substrate, the transforming larvae shed the cilia of their telotroch and burrow in the sand as young juvenile worms, as shown in Fig. 2I. At this point, the two pigmented eye spots of the larva are still evident at the tip of the proboscis and one or two pairs of gill slits are present in the pharynx of the young worm. By 2 weeks, the trunk region has extended considerably, more than doubling the length of the animal, and four or five gill slits are evident.

IV. Removal of Vitelline Envelope in *S. kowalevskii*

A. Background

Following fertilization, a thick vitelline envelope is raised around the embryo. This envelope can be removed manually with fine forceps post-gastrulation, but is very difficult to remove manually without damaging the embryo before the onset of gastrulation. We have developed a protocol to remove the vitelline chemically in *S. kowalevskii*. Removal of the membrane has been crucial for carrying out embryological manipulations and for microinjection. Denuded oocytes and embryos should be handled with extreme care following this procedure, as they become extremely fragile and sticky. Embryos are cultured in small petri dishes coated with agar, and seawater should be filtered and supplemented with antibiotic and calcium. The blastomeres tend to disaggregate without the vitelline in place.

B. Method

1. Obtain unfertilized eggs, as described in Section II. The egg's color is dark grey/green to cream colored, depending on the female. The unfertilized egg's diameter is 0.35 to 0.40 mm. Fertilize eggs as described in Section II. Add filtered SW at 22 °C. After fertilization, the fertilization envelope lifts from the surface, at least doubling the diameter within 30 min. The transparent fluid space between the egg surface and envelope is readily apparent.
2. Dejelly eggs at 60 min post-fertilization (second polar body forms by 50 min; first cleavage at about 120 min), as follows:
3. Prepare a small dish of 3.5 ml DTT-NaOH-FSW as described in Section VII.8. The pH will be approximately 8.0. Add eggs (up to 100) with minimal seawater carryover, pipet up and down gently with a Pasteur pipet at 15-s intervals to get uniform exposure and reduce sticking of the eggs to the glass surface. Within 3 min, the fertilization envelope dissolves. Materials on the egg surface (hyalin-like?) are also removed. Take care not to break the bare eggs by passing them through the meniscus.
4. Transfer bare eggs to 10 ml FSW (fresh seawater) in a dish and rinse by slow swirling. Do a second FSW rinse if there are many eggs (>30). Transfer eggs to a holding dish with thin agar coating on the surface (that is, rinse the dish surface with 1% agarose and decant the excess) in 10 ml FSW to which 10 mM $CaCl_2$ has been added.

V. Whole-Mount *In Situ* Hybridization

A. *S. kowalevskii*

In situ hybridization is a critical technique for determining the expression of developmental genes. The deployment of conserved developmental genes during the development of the hemichordate body plan can potentially give key insights into axial similarities and differences between the deuterostome taxa. The protocol for *S. kowalevskii* is based largely on the *Xenopus* protocol developed initially by Harland (1991) and modified by Salic *et al.* (1997) and Lowe *et al.* (2003). Some of the steps have been changed due to the smaller size of the embryos. The signal-to-noise ratio varies between different genes but, in general, is excellent. Probe quality is particularly important and probe should be free of all unincorporated nucleotides. We do not remove the vitelline for this protocol and have not noticed an attenuation of signal or increase in background as a result of its presence.

1. Fixation
   a. Harvest embryos or juveniles and rinse in filtered seawater, then incubate for 10 min in MgSW to anaesthetize in a 50 ml falcon tube (only embryos after day 2 of development).
   b. Draw off most of the seawater and pipet the anaesthetized specimens into a 50 ml falcon tube. Add 20 ml of fixative A and incubate on a shaking stage at low speed for 1 h at room temperature. The vitellines will stick to the wall of the tube if the embryos are allowed to settle.
   c. Drain the fixative and wash embryos 3× for 10 min in buffer of fixative A (without fixative). **Do not use PBST to wash the embryos as embryos become rapidly deformed**.
   d. Dehydrate the fixed embryos by adding 100% EtOH directly without stepping through an ethanol series. The tube should be continuously agitated during the first few minutes following addition of EtOH to avoid the vitellines sticking to the side of the tube. Incubate for 10 min and repeat 2× before storing the embryos in 100% EtOH −20 °C.
2. Rehydration

Transfer embryos into 1 gill screw to glass vials and rehydrate through an ethanol series into PBST. Five minutes for each wash

75% EtOH/25% H_2O

50% EtOH/50% H_2O

25% EtOH/75% H_2O

3X 100 % PBST

3. Proteinase K Treatment
   a. Incubate samples for 5 min in a 1 μg/ml solution of Proteinase K in PBST at 37 °C.
   b. Rinse 2×, 5 min each in 0.1 M Triethanolamine.
4. Acetic anhydride treatment
   a. Incubate for 10 min in 0.1 M Triethanolamine and acetic anhydride (250 μl in 100 ml of 0.1 M Triethanolamine).
   b. Rinse 2×, 5 min each in PBST.
5. Refix in formaldehyde
   a. 10 min in 4% formaldehyde in PBST.
   b. Rinse 3× for 5 minutes each in PBST.

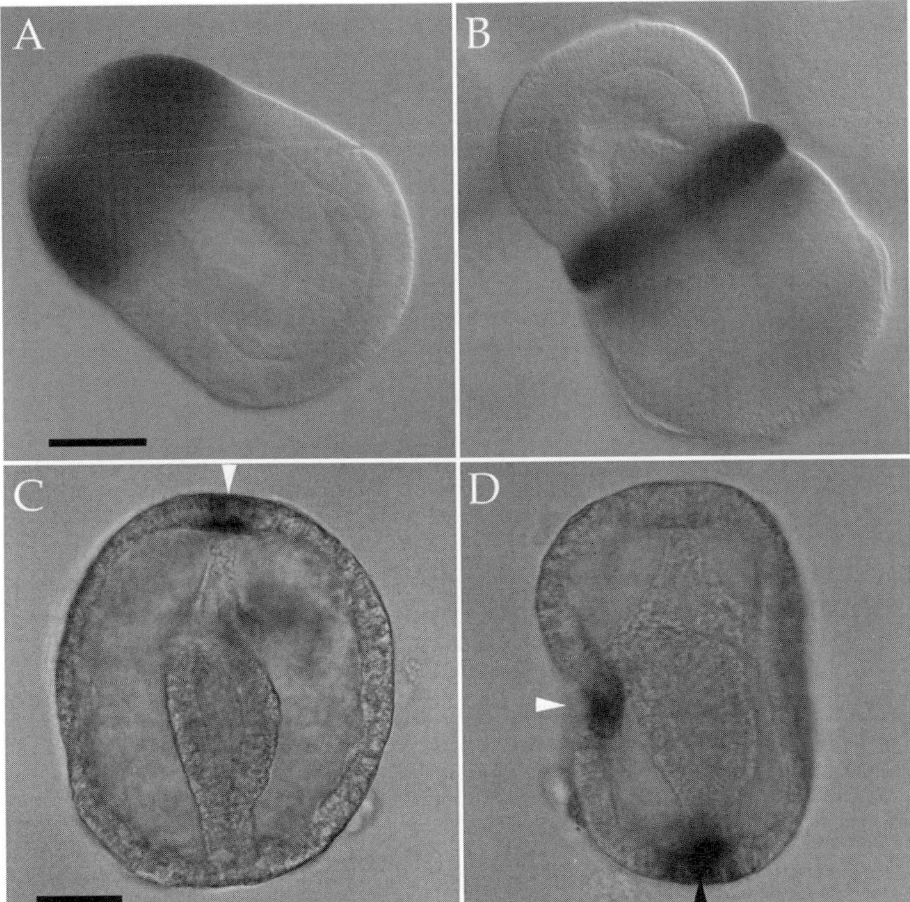

Fig. 3 Examples of patterns of expression in hemichordate embryos of genes implicated in specifying vertebrate structures. Whole mount *in situ* hybridization in *S. kowalevskii* and *P. flava*. A and B show *S. kowalevskii*, C and D show *P. flava*. (A) Retinal homeobox expression in the developing proboscis ectoderm of a 2-day-old early embryo. Optical frontal section of a cleared specimen. Scale bar = 150 μm. (B) Expression of BarH, a neural determining gene, in the mesosome ectoderm of a 3-day-old early embryo. Frontal optical section of a cleared embryo. (C) Expression of the forebrain gene, T-brain in a 3-day-old tornaria larva at the apical organ, as marked by a white arrowhead. Scale bar = 50 μm. (D) Expression of Brachyury, a gene important in notochord development, in a 2-day-old tornaria larva. White arrowhead indicates the position of the mouth and black, the position of the anus. (See Color Insert.)

6. Prehybridization.
   a. Wash embryos with small amount of hybridization A solution for 10 min.
   b. Replace solution in each vial with approximately 1 ml of fresh hybridization buffer A.
   c. Incubate for 1 to 3 h at 60 °C in a shaking water bath.
7. Hybridization
   a. Replace solution in vials with 1 ml of hybridization A solution with probe diluted to approximately 1 μg/ml. Riboprobes should be prepared using standard protocols and purified using G50 columns (Ambion).
   b. Hybridize overnight at 60 °C in shaking water bath.
8. Post-hybridization washing
   a. Remove hybridization solution with probe and discard. Replace with hybrization buffer A and fill vial with 2× SSC.
   b. Invert each vial into one Netwell sample vial in a 12-well tissue culture plate. Fill the reservoirs with 2× SSC and rinse 3×, 20 min each at 60 °C.
   c. Rinse 2× in 0.2× SSC, 30 min each at 60 °C.
9. Blocking

Transfer the Netwell vials into a fresh 12-well culture plate. Add 1× MAB with 2% Roche blocking agent for 1 h at room temperature.

10. Antibody incubation
    a. Replace blocking solution with the same solution containing a 1:1000 dilution of anti-digoxigenin, alkaline phosphatase-conjugated (Roche), or 1:2000 dilution of an antifluoroscein, alkaline phosphatase-conjugated antibody (Roche).
    b. Incubate for 4 h at room temperature.
11. Washes

Wash 4×, 30 min each in 1× MAB at room temperature, and an additional 60 min at room temperature or O/N at 4 °C.

12. Chromogenic reaction
    a. Rinse 2×, 5 min each in alkaline phosphatase buffer.
    b. Remove embryos from the Netwells and transfer into the reservoirs of the 12-well plate directly. Remove as much of the buffer as possible and add 1 ml of BM purple.
    c. Allow to develop in the dark and monitor every 30 min for staining intensity.
13. Post fixation
    a. Once embryos have reached the desired level of staining, they should be post-fixed overnight in Bouins fixative.

b. Following fixation, embryos should be repeatedly rinsed in 80% EtOH and 0.1M Tris pH 8.0 until all yellow stain from the embryos has disappeared.

14. Clearing

a. Further dehydrate samples in sequential rinses into 100% EtOH and 3× 5 min in 100%MeOH.

b. Transfer embryos to a glass depression slide and remove all but a trace amount of MeOH.

c. Immediately add several drops of a 1:2 mix of benzyl alcohol and benzyl benzoate. Allow embryos to become transparent.

d. Make permanent mounts of embryos by replacing clearing agent with Permount and covering specimens with a cover slip with thin clay feet on each corner.

B. General Considerations

The critical parameters in the protocol are: (1) Proteinase K digestion should be very light as the embryos will not tolerate long incubations or high concentrations. Adequate results can be achieved by eliminating the step altogether albeit with a slightly higher background staining. The collar of late juveniles (2–3 gill slits) is particularly sensitive to this treatment. (2) Probe quality: we use full-length probes and generally have not found it necessary to hydrolyze. We also routinely use fluoroscein-labeled probe rather than digoxigenin, as background is generally slightly lower with fluoroscein. We use G-50 columns from Ambion® to remove unincorporated nucleotides and check probe quality on an agarose gel. The fluorosceinated probes glow green/blue under UV illumination and give a rough indication of fluoroscein incorporation in the probe.

We use Netwell vials®, which are small plastic vials with a Nytex mesh that fit into the wells of a 12-well tissue culture dish. These vials ensure minimal fluid carryover between each wash step, and minimize the loss of embryos through accidental uptake during aspiration of the post-hybridization washes. They are not critical to the procedure and washes may be carried out in the hybridization vials. The fixation in Bouins is only necessary if the embryos are to be cleared in benzyl benzoate/benzyl alcohol. The blue precipitate from the AP reaction is unstable in this clearing agent and will rapidly disperse without the post-fixation in Bouins.

C. *P. flava*

The *in situ* hybridization protocol for *P. flava* was developed largely from the protocol established for ascidian embryos and larvae (Tagawa *et al.*, 1998a; Yasuo and Satoh, 1994). The level of background varies with different genes but, in general, is quite low. We do not remove the vitelline membrane from prehatching stages because we have not noticed an attenuation of signal nor an increase in background as a result of its presence. After hatching, the handling of tornaria larva requires more care because the larvae settle slowly in the reaction tube and it is easy to lose the sample during changes of solutions.

To reduce background, we use a preparation of powered adult hemichordate tissue prepared and used much as the one developed for Amphioxus (Holland *et al.*, 1996). The hybridization solution becomes viscose during hybridization because of the mucus from young juveniles. To reduce background, it should be removed when you start washing the hybridized preparations.

1. Fixation
   a. Embryos, larvae, and juveniles are fixed in freshly made 4% formaldehyde in Fixation buffer B at 4 °C overnight or for 1 h at room temperature in a 1.5 ml Eppendorf tube. (The formaldehyde fixative is made from 4% paraformaldehyde powder that is dissolved in the fixation buffer by heating.)
   b. Fixative is removed and embryos washed 3× for 10 min each in buffer B.
   c. Samples are dehydrated in an ethanol series up to 80%, 30 min each step.
      50% EtOH/50% H_2O
      70% EtOH/30% H_2O
      80% EtOH/20% H_2O
   d. Embryos in 80% EtOH can be stored for over a year at −20 °C.
2. Rehydration

Transfer embryos into transparent 1.5 ml tube (Fukae Kasei Co., Ltd) and rehydrate through EtOH into PBST, 5 min for each wash.

75% EtOH/25% H_2O
50% EtOH/50% H_2O
25% EtOH/75% H_2O
3X 100% PBST

3. Proteinase K Treatment
   a. Incubate in a solution of Proteinase K in PBST at 37 °C.
      Embryos: 2 μg/ml for 30 min
      Tornaria larva: 2 μg/ml for 10 min
      Transforming larva: 5 μg/ml for 30 min
4. Refix in paraformaldehyde
   a. 1 h in 4% paraformaldehyde in PBST
   b. Rinse 3× for 5 min each in PBST.
5. Acetic anhydride treatment
   a. Incubate for 5 min in 0.1 M Triethanolamine and acetic anhydride (250 μl in 100 ml of 0.1 M Triethanolamine).
   b. Rinse 4× for 10 min each in PBST.

6. Prehybridization
   a. Wash embryos with equal parts of hybridization B solution and PBST for 10 min.
   b. Replace solution in each vial with approximately 1 ml of fresh hybridization buffer B and incubate at room temperature for 10 min.
   c. Replace with fresh hybridization buffer and incubate for 1 h at 42 °C for embryos and tornaria larvae and at 60 °C for metamorphosing larva and young juveniles.
7. Hybridization
   a. Replace hybridization solution with diluted probe between 0.1 μg/ml – 1 μg/ml in hybridization buffer.
   b. Hybridize overnight (16 h) at 42 °C for embryos and tornaria larvae and at 60 °C for metamorphosing larva and young juveniles in a shaking water bath.
8. Post-hybridization washing

Post-hybridization washes at 42 °C for embryos and larvae, and 65 °C for metamorphosis larvae

   a. 4× SSC, 50% FA, 0.1% Tween 20 for 20 min, 2×
   b. 2× SSC, 50% FA, 0.1% Tween 20 for 20 min, 2×
   c. Solution A at 37 °C for 10 min, 3×
   d. 20 μg/ml Rnase A in Solution A at 37 °C for 30 min.
   e. Solution A at 37 °C for 15 min.
   f. 1× SSC, 50% FA, 0.1% Tween 20 at 42 or 65 °C for 20 min, 4×
   g. 1× SSC, 50% FA, 0.1% Tween 20: PBST @ 3:1 at RT for 10 min
   h. 1× SSC, 50% FA, 0.1% Tween 20: PBST @ 1:1 at RT for 10 min
   i. 1× SSC, 50% FA, 0.1% Tween 20: PBST @ 1:3 at RT for 10 min
   j. 4× PBST at RT for 20 min.
9. Blocking
   a. 0.5% Roche blocking agent in PBT at 4 °C for 20 min.
   b. 0.5% Roche blocking agent in PBT at 4 °C overnight.
10. Antibody incubation
   a. For embryos and early larvae, place in 1/2000 anti-digoxigenin-alkaline phosphatase in 0.5% blocking reagent in PBST and incubate at RT for 1 to 1.5 h.
   b. For transforming larvae juvenile worms, dilute the preparation of alkaline phosphatase-coupled anti-digoxigenin antibody pre-absorbed with adult powder (see Section VII) 1/3000 with 0.5% blocking reagent in PBST. Incubate at 4 °C overnight.

11. Washes
    a. Wash in PBST for 15 to 20 min 6× at RT. (During the first washes for juveniles when adult powder has been used to block the antisera, be sure to adequately swirl the larvae to disperse and subsequently remove the mucus that forms around them.)
    b. 2× in AP buffer (pH 8.0).
    c. 2× in AP buffer (pH 9.5).
12. Chromogenic reaction
    a. Develop color reaction in NBT 0.9 μl/ml/×-phos 0.7 μl in 200 μl AP buffer (pH 9.5) for 15 min to several hours until the desired level of color is achieved.
    b. Stop reaction by washing in PBST several times for 5 min each. Postifix with 4% paraformaldehyde in PBT for 1 h at RT. Then wash 2 to 3 times with PBST for 5 to 10 min.

VI. Preparation of Blocking Reagent for *P. flava*

A. Background

Anti-digoxigenin antisera used to immunolocalize the hybrids tends to react nonspecifically to the older *P. flava* developmental stages. To reduce background in the reactions, we have resorted to the tactic of blocking such nonspecific reactions with extracts of *P. flava* tissue. Our approach follows closely the procedure developed for Amphioxus (Holland *et al.*, 1996) of mixing an aqueous preparation from an acetone powder of animal tissue with the antisera before the incubation with the experimental preparations.

B. Protocol to Prepare Pre-Blocked Antisera

1. Preparation of acetone extracted tissue powder.
    a. Grind frozen adult tissue in a cold glass pestle with the smallest possible volume of PBS.
    b. Add 4 volumes ice-cold acetone.
    c. Incubate on ice for 30 min.
    d. Transfer to a centrifuge tube and centrifuge at 10,000g.
    e. Collect pellet and wash twice in acetone.
    f. Air dry in a mortar and grind further with the pestle.
    g. Store dry at minus 20 °C.
2. Preparation of pre-blocked antiserum
    a. Weigh out 1.5 mg hemichordate powder.
    b. Add 400 μl PBS/0.1% Triton ×-100.

c. Heat at 70 °C for 30 min.
d. Add 50 μl of 20 mg/ml bovine serum albumen.
e. Add 50 μl pretreated sheep serum.
f. Add 0.5 μl anti-digoxigenein antibody.
g. Mix and incubate on shaker for 1 h or more at room temperature or overnight at 4 °C.
h. Add 1 ml PBS containing 0.1% Triton ×-100.
i. Add 2 mg/ml bovine serum albumen powder.
j. Add 50 μl sheep serum.
k. Store frozen at −20 °C in 200 μl aliquots.
l. This solution can safely be reused in a second reaction.

VII. Materials and Reagents

A. Solutions for Fixation and *in Situ* Hybridization

1. Magnesium seawater (MgSW): 3.5% $MgCl_2$ in filtered seawater
2. Fixation Buffer A
   0.1 M MOPS pH 7.5
   0.5 N NaCl
   2 mM EGTA pH 8.0
   1 mM $MgCl_2$
   1X PBS

   Fixation buffer B
   0.1 M MOPS, pH 7.5
   0.5 M NaCl
3. PBST: 1XPBS, 0.1% Tween 20
4. Hybridization buffer A

Reagent	*Conc.*
Heparin	100 μg/ml
Denhardt's solution	1X
CHAPS	0.1%
Tween 20	0.1%
20× SSC	5×
EDTA	5 mM
Formamide	50%
yeast tRNA	1 mg/ml
Water to volume	
Store at −20 °C	

Hybridization buffer B

Reagent	*Conc.*
Denhardt's solution	5X
SSC	6X
Tween 20	0.1%
yeast tRNA	100 μg/ml
formamide	50%
Water to volume	
Store at −20 °C	
Store at −20 °C	

5. 100X Denhardts

Reagent	*Supplier*
2% nuclease-free BSA	Calbiochem cat #126609
2% Ficoll 400	Calbiochem cat #341691
2%PVP-40	Calbiochem cat #5295

Make up in DEPC-treated water and store in aliquots at −20 °C.

Maleic Acid Buffer (MAB) 5X Stock: 500 mM maleic acid; 750 mM NaCl; pH with solid NaOH to 7.5

Alkaline phosphate buffer (AP buffer): 0.1 M Tris pH 9.5 or 8.0; 50 mM MgC12; 0.1 M NaCl; 0.1% Tween 20.

Solution A: 0.5 M NaCl, 10 mM Tris (pH8.0), 5 mM EDTA, 0.1% Tween 20.

Bouins fixative: For 100 ml; 75 ml of saturated picric acid, 25 ml of 40% aqueous formaldehyde, 5 ml of glacial acetic acid. **Attention: Powdered picric acid is highly explosive**.

DTT, NaOH-FSW; Prepare 3.5 ml DTT-NaOH-FSW by adding to FSW a volume of 30 μl 1.0 M DTT (made fresh), swirled to mix, and then adding and immediately mixing 20 μl 1N NaOH to partially ionize the DTT and raise the pH to 8.

B. Materials

1. Corning Netwells® plates 74 μm mesh size; Fisher Scientific cat #07-200-211
2. BM purple AP Substrate®; Roche CAT #1442074
3. Anti-Digoxigenin FAB fragments; Roche #1 093 274
4. Anti-Fluoroscein FAB fragments; Roche #1426338
5. Algal F/2 Algae Food (Fritz Industries, Dallas, TX) or Alga grow® (Carolina Biologicals)
6. BCIP; Roche #1 383 221
7. NBT; Roche #1 383 313
8. 1.5 ml tube; Fukae Kasei Co., Ltd., #131-415C
9. 10 ml Nissui Spitz tube; P. Nissui Pharmaceutical Co., Ltd., #302064700
10. 5 ml Komagome pipette; Iwaki Glass Co., Ltd., Tokyo, http://www.tech-jam.com/items/2051-1a.phtml
11. Acrodisc®, 0.2 μm; Gelman Sciences, AnnArbor, MI
12. Triethanolamine; Sigma T-1377
13. Roche blocking reagent; Roche #10057177

References

Adoutte, A., Balavoine, G., Lartillot, N., Lespinet, O., Prud'homme, B., and de Rosa, R. (2000). The new animal phylogeny: Reliability and implications. *Proc. Nat. Acad. Sci.* **97,** 4453–4456.

Agassiz, A. (1873). The history of *Balanoglossus* and Tornaria. *Mem. Amer. Acad. Arts & Sciences* **9,** 421–436.

Arendt, D., and Nubler-Jung, K. (1996). Common ground plans in early brain development in mice and flies. *Bioessays* **18,** 255–259.

Bateson, W. (1884). The early stages in the development of Balanoglossus (sp. Incert.). *Quarterly Journal of Microscopic Science* **24,** 208–234.

Bateson, W. (1885). The later stages in the development of *Balanoglossus kowalevskii*, with a suggestion as to the affinities of the enteropneusta. *Quart. J. Microscop. Sci. B*, 81–128.

Cameron, C. B., Garey, J. R., and Swalla, B. J. (2000). Evolution of the chordate body plan: New insights from phylogenetic analyses of deuterostome phyla. *Proc. Nat. Acad. Sci.* **97,** 4469–4474.

Colwin, A. L., and Colwin, L. H. (1953). The normal embryology of *Saccoglossus kowalevskii. J. Morphol.* **92,** 401–453.

Colwin, L. H., and Colwin, A. L. (1962). Induction of spawning in *Saccoglossus kowalevskii* (Enteropneusta) at Woods Hole. *Biol. Bull.* **123,** 493.

De Robertis, E. M., and Sasai, Y. (1996). A common plan for dorsoventral patterning in bilateria. *Nature* **380,** 37–40.

Eschscholtz. (1825). Bericht über die zoologische Ausbeuteder Reise von Kronstadt bis St. Peter und paul. *Oken's Isis.*

Garstang, W. (1928). The morphology of the Tunicata. *Quarterly Journal of Microscopic Science* **72,** 51–189.

Gee, H. (1996). "Before the Backbone: Views on the Origin of the Vertebrates." Chapman & Hall, London, UK.

Harada, Y., Okai, N., Taguchi, S., Tagawa, K., Humphreys, T., and Satoh, N. (2000). Developmental expression of the hemichordate *otx* ortholog. *Mech. Dev.* **91,** 337–339.

Harada, Y., Shoguchi, E., Okai, N., Taguchi, S., Tagawa, K., Humphreys, T., and Satoh, N. (2001). Embryonic expression of a hemichordate *distal-less* gene. *Zool. Sci.* **18,** 57–61.

Harada, Y., Shoguchi, E., Shunsuke, T., Okai, N., Humphreys, T., Tagawa, K., and Satoh, N. (2002). Conserved and divergent aspects of expression pattern of *BMP-2/4* in hemichordate acorn worm and echinoderm sea cucumber embryos. *Zool. Sci.* **19,** 1113–1121.

Harland, R. (1991). *In situ* hybridization: An improved whole-mount method for *Xenopus* embryos *In* "Methods in Cell Biology," (B. K. Kay and H. B. Peug, eds.). Vol. 36, pp. 685–695.

Henry, J. Q., Tagawa, K., and Martindale, M. Q. (2001). Deuterostome evolution: Early development in the enteropneust hemichordate *Ptychodera flava. Evol. Dev.* **3,** 375–390.

Holland, L. Z., Holland, P. W. H., and Holland, N. D. (1996). Whole mount *in situ* hybridization applicable to Amphioxus and other small larvae. *In* "Molecular Zoology, Advances, Strategies, and Protocols" (J. D. Ferraris and S. R. Palumbi, eds.), pp. 473–483. Wiley-Liss Inc.

Hyman, L. H. (1959). "The Invertebrates V: Smaller Coelomate Groups: Chaetognatha, Hemichordata, Pogonophora, Phoronida, Ectoprocta, Brachiopoda, Sipunculida, The Coelomate Bilateria." McGraw-Hill, New York.

Lowe, C. J., and Wray, G. A. (1997). Radical alterations in the roles of homeobox genes during echinoderm evolution. *Nature* **389,** 718–721.

Lowe, C. J., and Wray, G. A. (2000). Rearing larvae of sea urchins and sea stars for developmental studies. *In* "Methods in Molecular Biology," (C. Lo and R. Tuan, eds.). Vol. 135, pp. 9–15.

Lowe, C. J., Wu, M., Salic, A., Evans, L., Lander, E., Strange-Thomann, N., Gruber, C. E., Gerhart, J., and Kirschner, M. (2003). Anteroposterior patterning in hemichordates and the origins of the chordate nervous system. *Cell* **113,** 853–865.

Nielsen, C. (1999). Origin of the chordate central nervous system and the origin of chordates. *Dev. Genes. Evol.* **209,** 198–205.

Nübler-Jung, K., and Arendt, D. (1999). Dorsoventral axis inversion: Enteropneust anatomy links invertebrates to chordates turned upside down. *J. Zool. Systematics & Evol. Res.* **37,** 93–100.

Ogasawara, M., Wada, H., Peters, H., and Satoh, N. (1999). Developmental expression of *Pax1/9* genes in urochordate and hemichordate gills: Insight into function and evolution of the pharyngeal epithelium. *Development* **126,** 2539–2550.

Okai, N., Tagawa, K., Humphreys, T., Satoh, N., and Ogasawara, M. (2000). Characterization of gill-specific genes of the acorn worm *Ptychodera flava. Dev. Dyn.* **217,** 309–319.

Peterson, K. J., Cameron, R. A., Tagawa, K., Satoh, N., and Davidson, E. H. (1999). A comparative molecular approach to mesodermal patterning in basal deuterostomes: The expression pattern of Brachyury in the enteropneust hemichordate *Ptychodera flava. Development* **126,** 85–95.

Peterson, K. J., and Eernisse, D. J. (2001). Animal phylogeny and the ancestry of bilaterians: Inferences from morphology and 18S rDNA gene sequences. *Evolution & Development* **3,** 170–205.

Salic, A. N., Kroll, K. L., Evans, L. M., and Kirschner, M. W. (1997). *Sizzled*: A secreted *Xwnt8* antagonist expressed in the ventral marginal zone of Xenopus embryos. *Development* **124,** 4739–4748.

Stiasny-Wijnhoff, G., and Stiasny, G. (1926). Uber Tornarien-Typen und ihre Beziehung zur Systematik der Enteropneusten. *Zool. Anz.* **68,** 159–165.

Stiasny-Wijnhoff, G., and Stiasny, G. (1927). Die Tornarien. *Ergebn. Fortschr. Zool.* **7,** 38–208.

Tagawa, K., Nishino, A., Humphreys, T., and Satoh, N. (1998a). The spawning and early development of the Hawaiian acorn worm (Hemichordate), *Ptychodera flava. Zool. Sci.* **15,** 85–91.

Tagawa, K., Humphreys, T., and Satoh, N. (1998b). Novel pattern of *Brachyury* gene expression in hemichordate embryos. *Mech. Dev.* **75,** 139–143.

Tagawa, K., Humphreys, T., and Satoh, N. (2000). *T-Brain* expression in the apical organ of Hemichordate Tornaria larvae suggests its evolutionary link to the vertebrate forebrain. *J. Exp. Zool. (Mol. & Dev. Evol.)* **288,** 23–31.

Tagawa, K., Satoh, N., and Humphreys, T. (2001). Molecular studies of hemichordate development: A key to understanding the evolution of bilateral animals and chordates. *Evol. Dev.* **3,** 443–454.

Taguchi, S., Tagawa, K., Humphreys, T., Nishino, A., Satoh, N., and Harada, Y. (2000). Characterization of a hemichordate *fork head/HNF-3* gene expression. *Dev. Genes & Evol.* **210,** 11–17.

Taguchi, S., Tagawa, K., Humphreys, T., and Satoh, N. (2002). *Group B Sox* genes that contribute to specifications of the vertebrate brain are expressed in the apical organ and ciliary bands of hemichordate larvae. *Zool. Sci.* **19,** 57–66.

Takacs, C. M., Moy, V. N., and Peterson, K. J. (2002). Testing putative hemichordate homologues of the chordate dorsal nervous system and endostyle: Expression of NK2.1 (TTF-1) in the acorn worm *Ptychodera flava* (Hemichordata, Ptychoderidae). *Evol. Dev.* **4,** 405–417.

Turbeville, J. M., Schulz, J. R., and Raff, R. A. (1994). Deuterostome phylogeny and the sister group of the chordates: Evidence from molecules and morphology. *Mol. Biol. Evol.* **11,** 448–455.

Wada, H., and Satoh, N. (1994). Details of the evolutionary history from invertebrates to vertebrates, as deduced from the sequences of 18s rDNA. *Proc. Natl. Acad. Sci.* **91,** 1801–1804.

Yasuo, H., and Satoh, N. (1994). An ascidian homolog of the mouse *Brachyury* (*T*) gene is expressed exclusively in notochord cells at the fate restricted stage. *Dev. Growth Differ.* **36,** 9–18.

CHAPTER 9

Cephalochordate (Amphioxus) Embryos: Procurement, Culture, and Basic Methods

Linda Z. Holland and Ju-Ka Yu

Marine Biology Research Division
Scripps Institution of Oceanography
University of California San Diego
La Jolla, California 92093

I. Overview

Early development in amphioxus, the closest living invertebrate relative of the vertebrates, resembles that of echinoderms, while later development is close to that of vertebrates. In natural populations of the Florida amphioxus *Branchiostoma floridae*, each female spawns about once every 2 weeks during summer. A given local population will typically spawn-out over 2 to 3 days. However, local populations may be out of phase with one another. On the afternoon of the day of

METHODS IN CELL BIOLOGY, VOL. 74

0091-679X/04 $35.00

spawning, oocytes undergo meiotic maturation and arrest at second meiotic metaphase. Spawning occurs about 30 min after sunset, but can be delayed if animals collected from the field in the afternoon are maintained in the light. Spawning can then be induced either by a mild electric shock or by putting animals in the dark. Since methods for inducing meiotic maturation have not been developed, at present, it is necessary to rely on natural populations for ripe adults. Eggs are highly resistant to polyspermy. They are most conveniently raised in petri dishes without aeration. Antibiotics are not essential if excess sperm and dead eggs are removed. The seawater should be changed daily. Embryos develop normally from 22 to 32 °C. Development is rapid. At 30 °C, neurulae hatch at 6.5 h after insemination. Feeding is critical and should commence at 30 h after insemination at 30 °C. The diet consists of unicellular algae from 10 to 30 μm in diameter. The larger algae should not exceed more than 20% of the total. If food is not constantly present, the larvae will cease eating and will not resume even if food is supplied. Foreign DNA and/or RNA can be introduced into embryos by microinjection of unfertilized eggs. LacZ reporter constructs are preferable to those with green fluorescent protein, since the eggs and embryos fluoresce green when illuminated with blue light.

II. Introduction

The invertebrate chordate amphioxus, also called the lancelet, is the closest living invertebrate relative of the vertebrates. Amphioxus is similar to vertebrates in many respects, but simpler. Thus, both have a dorsal hollow nerve cord, notochord, and pharyngeal gill slits, but unlike vertebrates, amphioxus does not form paired limbs, paired ears, or image-forming eyes (Figs. 1 and 2). In addition, the amphioxus genome has representatives of most of the vertebrate gene families, but it has not undergone the extensive gene duplications characteristic of vertebrates. For example, amphioxus has a single *Hox* cluster with fourteen co-linear genes while mammals have four *Hox* clusters that evidently arose by duplication of a single *Hox* cluster with thirteen co-linear genes (Ferrier *et al.*, 2000; Garcia-Fernàndez and Holland, 1994). As a result of this vertebrate-like simplicity and its close phylogenetic relationship to vertebrates, amphioxus has long been used in studies of evolution and development as a stand-in for the vertebrate ancestor.

There are two genera of amphioxus, *Branchiostoma* and *Epigonichthys*. Three of the approximately 22 species of *Branchiostoma* (Poss and Boschung, 1996) have been used extensively for developmental studies. These are *B. lanceolatum* (Atlantic and Mediterranean coasts of Europe), *B. floridae* (gulf coast of Florida, USA) and *B. belcheri* (East Asia, especially on the China coast at Xiamen and Qingdao; a population in Japan has recently been identified) (Kubokawa *et al.*, 1998). In Xiamen, conservation areas for *B. belcheri* were designated in 1991 see http://ois.xmu.edu.cn/cbcm/english/resource/save/save04.htm. Unfortunately, before these areas were designated, most of the amphioxus populations

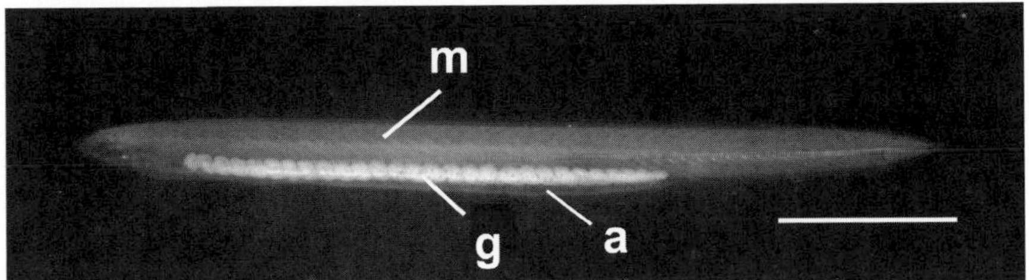

Fig. 1 Live adult specimen of *Branchiostoma floridae*. Anterior is at left. Scale line = 1 cm; m = segmented axial muscles; g = gonads; a = atrium.

in Xiamen, which once supported a fishery, had disappeared. There are about 6 species of *Epigonichthys*; however, its development has not been studied.

Amphioxus has a maximum length of 3 to 7 cm, depending on the species (Fig. 1). It is a filter-feeder that lives in relatively shallow marine habitats, burrowing tail-first in the sand. Population densities can be over $1000/m^2$ (Stokes, 1996b). The choice of species for study has been dictated by the accessibility and density of populations. Depending on the depth of the water, adults are collected either from a boat with a small dredge or with a shovel and sieve. Although *B. lanceolatum* was the first species used for embryological studies (Garcia-Fernàndez and Holland, 1994), the highly accessible populations of this species in southern Italy have largely disappeared due to habitat destruction. Today, most developmental studies concern *B. floridae* and *B. belcheri.* All three species of *Branchiostoma* commonly studied have similar reproductive habits and development (Cerfontaine, 1906; Conklin, 1932; Hirakow and Kajita, 1990; Müller, 1841; Stokes and Holland, 1995; Wilson, 1893; Wu *et al.*, 1994). However, the frequency of spawning and the rate of development depend on the environmental temperature. Those from warmer waters reproduce more often and have faster development than those from cold water (Table I) (Courtney, 1975; Fang *et al.*, 1991; Lin *et al.*, 1996; Sager and Gosselick, 1986; Stokes, 1996a,b; Stokes and Holland, 1996).

At present, no species of amphioxus is in continuous culture in the laboratory. This is chiefly because a concerted effort to develop breeding colonies in the laboratory has not yet been made. Adults, embryos, and larvae of *B. floridae* live well in the laboratory, and the time from fertilization to sexual maturity is only 6 to 8 weeks. However, to maintain gonad growth, adults must have moderate concentrations of particulate food present at all times. In an uncontrolled experiment, specimens of *B. belcheri* that had spawned out were kept for one year in aquaria in a room with large windows admitting ambient light. They were provided with sand for burrowing and fed well on a culture of mixed unicellular algae. All the animals spawned over 2 days in summer when animals in the field were expected to spawn naturally (Zhang *et al.*, 1999, 2001). Thus, it seems likely that breeding colonies could be established by carefully controlling light, temperature, and food.

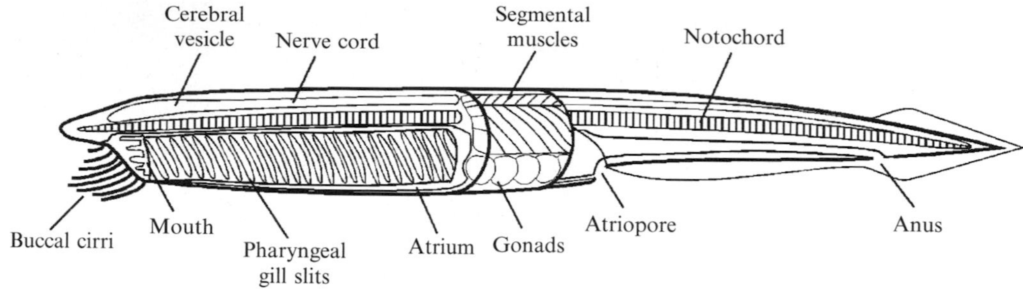

Fig. 2 Diagram of the major anatomical features of amphioxus. Anterior is at left. The swollen anterior part of the nerve cord, the cerebral vesicle, is homologous to the vertebrate diencephalon.

At present, however, it is necessary to depend on natural populations for ripe adults. Unfortunately, although it is known that a long day/short night light cycle and a water temperature of 30 °C are required for spawning to occur in *B. floridae*, the environmental cues that trigger meiotic maturation and ovulation are unknown. There is no correlation with phases of the moon or tides (Stokes, 1996a,b). Moreover, no means of experimentally inducing meiotic maturation of large primary oocytes has been discovered. A wide range of substances known to induce meiotic maturation in oocytes of other deuterostomes has failed to induce meiosis in oocytes of both *B. floridae* and *B. belcheri* (Watanabe *et al.*, 1999; our unpublished data). These included gonadotropin-releasing hormone (GnRH) of several species (chick, lamprey, teleost, tunicate). Although there is immunocytochemical evidence that amphioxus secretes GnRH (Fang, 1998; Yin *et al.*, 1994), attempts to purify and/or clone it have been unsuccessful (Kubokawa *et al.*, 2002; N. Anderson, personal communication). Cloning is particularly difficult because the active GnRH peptide, which tends to be fairly conserved among species, is only 10 amino acids long. However, libraries for genome sequencing have recently been constructed (see Section IV) and it should ultimately be possible to obtain the sequence of amphioxus GnRH from the whole genome sequence.

The most accessible populations of amphioxus are of *B. floridae*, which is common in intertidal and shallow subtidal waters along the Gulf Coast of Florida. It is particularly abundant in Old Tampa Bay, Florida, where it can compose up to 70% of the infaunal biomass (Stokes and Holland, 1996). The size at sexual maturity is 23 mm. *B. floridae* breeds in late spring and summer, when the water temperature exceeds 27 °C. The optimum temperature for reproduction is 30 to 32 °C (Table I). Each individual spawns every 10 to 15 days, depending on temperature—the higher the temperature, the more frequently animals spawn. Thus, over a 3-month breeding season, each individual may spawn a maximum of 6 to 8 times. However, since a given population may spawn out over 2 or 3 successive nights and since local populations within Old Tampa Bay may be

Table I
Schedule of Development for *Branchiostoma floridae*

Stage	Time at 25 °C	Time at 30 °C
Insemination	0 h	
Fertilization envelope begins to elevate	30 sec	
Second polar body	10 min	
Fertilization envelope completes elevation	20 min	
2-cell	45 min	30 min
4-cell	60 min	50 min
8-cell	90 min	60 min
16-cell	2 h	75 min
32-cell	2.25 h	90 min
64-cell	2.5	105 min
128-cell	3 h	2 h
Blastula	4 h	2.5 h
Onset of gastrulation	4.5 h	3.5 h
Capped-shaped gastrula	5 h	4 h
Cupped-shaped gastrula	6 h	4.5 h
Late gastrula/early neurula	6.5 h	5 h
Neurula, ciliated begins rotating inside fertilization envelope	8.5 h	6 h
Hatching	9.5 h	6.5 h
Anterior somites visible	10.5 h	7.5 h
Late neurula; onset of muscular movement	20 h	12 h
Mouth and first gill slit	30–32 h	23–24 h
Anus open	36 h	28 h
2 gill slits	3 days	36 h
3 gill slits	6 days	3 days
4 gill slits	15 days	7 days
6 gill slits	23 days	14 days
8 gill slits	29 days	19 days
11 gill slits	37 days	25 days
Metamorphosis	41–49 days	26–32 days

out of phase with one another, there may be considerably more days during the summer on which it is possible to obtain ripe gametes.

Sexes are separate in amphioxus. *B. floridae* has 26 gonads on each side of the animal. Each female spawns from 500 to several thousand eggs, depending on the size of the animal. Since the animals are transparent, the sex can easily be determined visually. Sperm are typically white and eggs yellow, although some eggs are very pale. The female gametes, which are up to about 140 μm in diameter, are stored in the gonads as primary oocytes. In the early afternoon of the day of spawning, oocytes undergo meiotic maturation, arresting at second meiotic metaphase, and at the same time, are ovulated from the thin layer of follicle cells surrounding them. The mature eggs remain within the gonads as long as light levels remain high. About 30 min after sunset, animals emerge from the sand to spawn. The control of spawning by light levels is advantageous, since it allows spawning to be controlled in the laboratory.

Until recently, experiments with amphioxus eggs and embryos were largely limited to exposing embryos to reagents such as retinoic acid (Escriva *et al.*, 2002; Holland and Holland, 1996) that readily penetrate them when added to the seawater. The ease of removing the fertilization envelopes has allowed blastomere recombination and some limited transplantation and cell lineage experiments (Tung *et al.*, 1958, 1960, 1962a,b; Wilson, 1893). However, the second polar body, which marks the animal pole, is generally lost when the fertilization envelope is removed, making it difficult to distinguish animal and vegetal blastomeres. Moreover, blastomeres do not adhere well together until the mid-blastula stage, complicating blastomere recombination experiments. In addition, the small size of the embryos makes experiments like transplantation of part of the blastoporal lip into another organism quite difficult. Consequently, such transplantation experiments have not been repeated in the over 40 years since they were initially performed. In recent years, although the limited number of nights on which embryos are available has restricted experimentation, there have been a few studies in which specific groups of cells have been labeled with fluorescent dyes (Holland and Yu, 2002; Zhang *et al.*, 1997). In addition, in 2002, we developed methods for introducing foreign DNA, morpholino-oligonucleotides, RNA, or dye into embryos. Of several methods tried, including liposomes and electroporation, only microinjection was successful. This success has opened a wide range of possibilities for manipulating signaling pathways and studying gene regulation.

The present chapter concerns methods for obtaining, raising, and manipulating living embryos of *B. floridae*. Previously unpublished methods for expressing reporter constructs and for manipulating levels of gene expression are included. Methods for *in situ* hybridization and for fixing and embedding embryos for light and electron microscopy have previously been published and will not be included here. Since development of *B. lanceolatum* and *B. belcheri* is virtually identical to that of *B. floridae*, the methods described here should be applicable to all species of *Branchiostoma*, with adjustments made for differences in the optimum temperature for development.

III. Obtaining Gametes of *Branchiostoma floridae*

A. Collecting Adults

Without any available method to induce meiotic maturation of large oocytes, it is at present only possible to obtain fertilizable gametes from amphioxus on the days when they would normally spawn in the field. Adults should be collected in the afternoon after females have ovulated and the eggs have arrested at second meiotic metaphase. If they are collected in the morning, they must be kept at field temperature or they will not spawn. Even so, fewer will spawn than if collected in the afternoon. Presumably, this is because the trauma of collection and/or a temporary drop in temperature inhibit the trigger for the meiotic divisions and ovulation. In Old Tampa Bay, adult amphioxus are found wherever the bottom is sandy. In most locations within the bay, the sand in which they burrow is rather fine. To separate adult amphioxus, the sand is shoveled into a 12″ × 18″ sieve made from 1″ × 4″ boards. Nylon window screening is nailed to the bottom of the frame and supported by 1/4-inch wire mesh (hardware cloth). Handles available from any hardware store facilitate sieving. (Fig. 3). The animals are sieved out of the sand and transferred to a plastic container. Care should be taken that the water in the container does not cool off or heat substantially during collection. To transport the collected animals to a nearby laboratory, the lid can be placed on the container of amphioxus if dusk is not imminent. Presumably, there is a terminal hormone responsible for spawning that is synthesized during the late afternoon, and the animals cannot be induced to spawn until there is sufficient hormone built up. However, if it is late in the day and the animals are particularly ripe, they should be kept in the light and kept warm (28–30 °C) until gametes are needed. Close examination of ripe females under a dissecting microscope will show whether ovulation and the meiotic divisions have occurred. If germinal vesicles are present, the animals will not spawn that day. Germinal vesicles, which in very large oocytes form a clear or light grey area at the animal pole, are easiest to see in the most anterior and most posterior gonads. If no germinal vesicles are present and if the

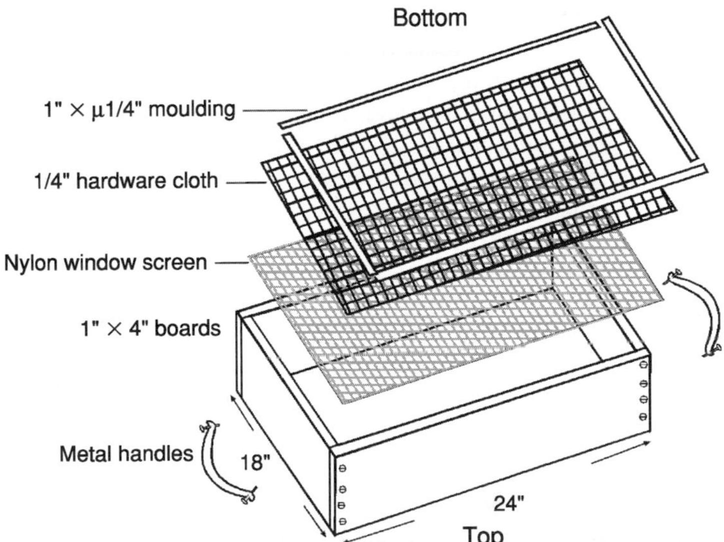

Fig. 3 Construction of a sieve for collecting amphioxus in areas where the sand is fine. The sieve is made of 1″ × 4″ boards held together with screws. An 18″ × 24″ piece of nylon window screening is nailed over the bottom of the frame (a staple gun is useful) and topped with an 18″ × 24″ piece of hardware cloth. Strips of 1″ × $\frac{1}{4}$″ molding are nailed over the edges of the frame. Galvanized handles available at any hardware store are added for ease of handling.

eggs seem to be rounder and more loosely packed within the gonads, it is likely that they are ovulated.

B. Spawning Adults

Spawning can be prevented if animals collected from the field on the afternoon of the day on which they would normally spawn are brought into the laboratory and kept in the light. The ordinary room light in a typical laboratory is sufficient to prevent spawning. The time of spawning can then be controlled, allowing the investigator to obtain freshly spawned eggs throughout the night. Unfortunately, the percentage of normally developing embryos begins to decrease toward dawn the next day. Even if spawned eggs are kept with antibiotics to inhibit bacterial growth (see following text), the percentage of normally developing embryos typically becomes quite low by 24 h after ovulation. Spawned eggs will not develop if refrigerated.

Spawning of animals with fully mature gametes can be induced either by placing the males or females in the dark for 30 min or by applying 10 msec pulses of 50 V DC with two platinum electrodes connected to an electrical stimulator (e.g., Grass model SD9; Grass-Telefactor division of Astro-Med Inc., Astro-Med Industrial Park, 600 East Greenwich Avenue, West Warwick, RI 02893; Tel: 401-828-4000, Fax: 401-822-2430, Tollfree: 1-877-472-7779 (USA & Canada only), e-mail: grass-telefactor@astromed.com, website: www.grass-telefactor.com). Two or three ripe adults of the same sex are placed in a beaker (plastic disposable cocktail glasses work well) in about 1 cm of seawater. Current pulses are applied for 1 to 3 sec. Spawning will typically occur within a few minutes. At spawning, the gametes emerge from the atriopore and should be collected with a pipette. If all the gonads do not spawn out, additional application of current generally suffices. Spawned eggs should be transferred to seawater filtered through a Whatman no. 1 filter to remove contaminating organisms such as copepods. It is generally not necessary to use Millipore-filtered seawater. Sperm shed from the atriopore should be collected before they become diluted and begin to swim. They can be stored on ice for at least 24 h. Fully grown sperm do not become capable of swimming when diluted into seawater until about a day before natural spawning. Motility of somewhat immature sperm can be increased by squeezing sperm from males into freshly prepared seawater with 10 mM NH_4Cl. Add drops of 1 N NaOH to adjust pH to 8.0. However, this method is only successful if the males are within a day or two of a natural spawning.

IV. Raising Embryos

A. Fertilization

To fertilize eggs, add several drops of concentrated sperm to several thousand eggs in a 5- to 9-cm petri dish. Since the eggs are not prone to polyspermy, it is better to use more, rather than fewer sperm, particularly if the males are not maximally ripe and it has been necessary to use NH_4Cl to increase motility. The eggs begin to undergo a cortical reaction within 20 to 30 s of insemination (Holland and Holland, 1989). It does not progress as a wave around the embryo but is initiated from many points. Even though the cortical reaction is completed by 1.5 min after insemination, the fertilization envelope continues to rise for about 20 min. During this time, there is continuing secretion of material from the egg into the perivitelline space. If the eggs are crowded in the petri dish, they should be dispersed as the fertilization envelope begins to rise, since they will stick together, with the result that the fertilization envelope will not rise fully and development will be abnormal.

B. Raising Embryos and Schedule of Development

1. General Care

Embryos are easily raised in petri dishes. No aeration is needed because of the large surface/volume ratio. Before hatching, the embryos should be kept at a concentration that results in no more than a monolayer on the bottom of the dish. Antibiotics are not essential; however, a 1:1000 dilution of penicillin G/streptomycin sulfate (100 mg/ml each penicillin G, sodium salt (>1400 units/mg) and streptomycin sulfate) will help prevent bacterial growth, which can be a problem if there are many dead eggs or if excess sperm has not been removed. Embryos should be disturbed no more than necessary. The fertilization membrane can readily be lost and/or the blastomeres dissociated if the embryos are mechanically agitated. At hatching, the neurulae, which are uniformly ciliated, swim up to the surface of the water and toward the light. They can be easily concentrated in this way and pipetted from the surface of the seawater into a dish with clean seawater, thus removing them from abnormal embryos, which tend to remain on the bottom, as well as dead eggs or sperm and the discarded fertilization envelopes. This change of water should be done before about 20 h of development, when muscular movements begin. Shortly thereafter, the embryos move to the bottom of the dish where they spend most of their time crawling on the bottom via their ectodermal cilia. However, by 3 days, if the larvae are growing well, they will hover in the water column while feeding (Stokes and Holland, 1995). The seawater should be changed once a day by transferring embryos with a drawn-out Pasteur pipette to a clean dish. Petri dishes for embryo culture can be reused, but should be washed with fresh water (no soap) to kill the bacterial film. As the larvae grow, their numbers per dish should be reduced or growth will be hindered. For maximal growth, by the three to four gill-slit stage, there should be only a few hundred larvae per 25 cm petri dish.

Table I shows a schedule of development for *B. floridae* at both room temperature (24 °C) and field temperature (30 °C). If raising the embryos in petri dishes, the temperature can be controlled by either putting the dishes in an incubator set to the appropriate temperature or by placing the dishes under an incandescent light. *B. floridae* does not develop normally below 20 °C or above 32 °C.

Metamorphosis is initiated at the 9 to 10 gill slit stage. A second row of gill slits appears on the right side above the first row. The first row moves around the ventral side of the larva to become the left gill slits. Subsequently, each gill slit becomes divided in two by a branchial bar growing from the dorsal side. The mouth rotates from the left side to the anterior end of the larva. Atrial folds grow out from above the gill slits and extend ventrally around the gill slits and gonads to fuse in the ventral midline. Finally, a digestive diverticulum, called the liver, grows out from the anterior part of the gut behind the pharynx (Conklin, 1932).

Late larvae can also be obtained by pulling a 150 μm mesh plankton net by hand through the bay after sunset. A 10 to 15 min tow generally suffices. Newly metamorphosed larvae that have recently moved into the sand can be collected with a shovel and a fine sieve made from 150 μm plankton netting substituted for window-screening.

2. Feeding Embryos

Starting at 30 to 36 h of development, the embryos must be fed. The mouth opens and begins to enlarge around 30 h of development and the anus is opened by 36 h. The larvae are ciliary filter feeders and must have food constantly available from about 34 h. If feeding is delayed beyond 36 h, the larvae shut down the ciliary feeding mechanism and do not restart it even if food becomes available. They will die within a few days. Similarly, if food is withdrawn for more than a few hours at any time before the digestive diverticulum forms, the larvae will stop eating and will not recommence even if food is subsequently provided. The appropriate diet consists of mixed unicellular motile algae ranging in size from 10 to 30 μm. Suitable algae that are easy to raise include species of *Dunaliella, Isochrysis, Monochrysis, Platymonas, Thallosiosira, Ellipsoidion, Tetraselmis*, and *Rhodomonas*. Use cultures that are 2 weeks to 1 month old to feed embryos. Do not use the dead algae that sink to the bottom of the algal culture vessels. To remove the culture medium, which is toxic, aliquots of the algal cultures should be pelleted in a tabletop centrifuge (about 1500X G) and resuspended in clean seawater before feeding. The final concentration should be approximately 1×10^6 cells/ml. The water should be visibly colored with the algae. Concentrations sufficient for feeding sea urchin larvae are too low for feeding amphioxus. A little trial and error will reveal the amount of food that is enough but not too much. The diet should include a mixture with a small percentage (10–20%) of some of the larger (~30 μm diameter) algae. Inclusion of the larger algae keeps the feeding mechanism operating. However, a diet including a large percentage of larger algae can prove fatal since the larvae will feed beyond the point of satiation and their guts will burst.

3. Sources of Micro-Algae

There are several sources of unicellular algae. In addition to those listed in the following text, some varieties of live algae can be obtained from aquaculture suppliers.

In the United States:
Provasoli-Guillard National Center for Culture of Marine Phytoplankton (CCMP) http://ccmp.bigelow.org
In Europe:
(1) Scandinavian Culture Center for Algae and Protozoa. Botanical Inst. Univ., Copenhagen. Øster Far Farimagsgade 2D, DK-1353, Copenhagen K, Denmark; Curator at SCCAP: Niels H. Larsen http://www.sccap.bot.ku.dk
(2) Culture Collection of Marine Micro-Algae, Instituto de Ciencias Marinas de Andalucia (ICMAN), Cadiz, Spain; http://www.icman.csic.es/servic/servCmicroalgas en.htm
In Australia:
CSIRO collection of microalgae. Orders and enquiries to Ms. Cathy Johnston, CSIRO Marine Research, GPO Box 1538, Hobart, Tasmania 7001, Australia; Telephone: +61 (3) 6232 5316 (international); (03) 6232 5316 (Australia); Facsimile: +61 (3) 6232 5000 (international); (03) 6232 5000 (Australia); Email: microalgae @marine.csiro.au; http://www.marine.csiro.au/microalgae/strainlist/strains_pl.html

4. GPM Algal Culture Medium

[modified from Loeblich (1975)]

1. *PII Trace minerals*; for 1 l

Na_2EDTA	6.0 g
$FeCl_3.6H2O$	0.29 g
$MnCl_2.4H2O$	0.86 g
$ZnCl_2$	0.06 g
$CoCl_2.6H2O$	0.26 g
H_3BO_3	6.84 g

2. *GPM medium*

Seawater	750 ml
Distilled water	225 ml
1 M KNO_3	2 ml
PII trace metals	5 ml
Vitamin B 12 (100 μg/ml)	10 μl
Thamine-HCl (10 mg/ml)	10 μl
Biotin (100 μg/ml)	20 μl

3. *Method*

Autoclave medium without the KPO_4 buffer. Autoclave the 0.5 M KPO_4 buffer separately. After the medium has cooled to room temperature, add 0.4 ml of cool 0.5 M KPO_4 buffer pH 7.4. Keep small algal cultures (about 100–150 ml) under constant illumination from a fluorescent light at 15 to 20 °C, depending on the optimum temperature for the alga. Turn over cultures about once a month. Before the breeding season, raise up larger cultures.

V. Manipulating Embryos

A. Removal of the Fertilization Envelope

Removal of the fertilization envelope is necessary for spotting dye on the surface of embryos or for doing blastomere manipulations. After the fertilization envelope is fully elevated but before first cleavage, the fertilization envelope can be removed mechanically by gently sucking the eggs into a Pasteur pipette with the tip drawn out in a flame and cut off evenly with a diamond scriber. The diameter of the tip of the pipette should be slightly larger than the egg and the drawn-out end of the pipette should be as short as possible. The fertilized, uncleaved eggs can be gently sucked into and expelled from the pipette. If the diameter and length of the tip of the pipette are suitable, most of the eggs will survive this treatment. If the fertilization envelope is removed before it is fully elevated, the material the eggs are secreting will stick them together. Although mechanical removal is cleaner, faster, and less damaging to the eggs, the fertilization envelope can also be removed enzymatically in a solution of 0.2 g Na thioglycolate and 0.01 g protease (Pronase E, Sigma no. 5147; Sigma-Aldrich Co., P.O. Box 14508, St. Louis MO 63178, USA. http://www.Sigma-Aldrich.com/order, Tel: 1-800-325-3010; Fax: 800-325-5052) in 20 ml seawater brought to pH >10 with about 20 drops of 1.0 M NaOH. After enzymatic removal of the fertilization envelope, embryos should be thoroughly rinsed in seawater. Because it is necessary to wait to begin removal of the fertilization envelope until it is fully elevated or nearly so, it is difficult to completely remove it enzymatically before the onset of first cleavage. Embryos with the fertilization envelope removed should be raised in petri dishes coated with a thin layer of 1% agar in seawater and provided with antibiotics. Otherwise, they will stick to the bottom. All manipulations of fertilized eggs lacking the fertilization envelope should be completed before first cleavage. Embryos should be raised to the mid-blastula stage without being disturbed. Until the mid-blastula stage, the blastomeres do not adhere well to one another. If separated before the third cleavage, individual blastomeres can develop into entire embryos. However, since there is a pole plasm that is segregated into a single blastomere at each cleavage, it is possible that only one of the embryos resulting from the separated blastomeres of a given embryo will contain germ cells. Embryos lacking the fertilization envelope that touch each other will fuse, creating monsters.

B. Microinjection of Unfertilized Eggs

Unfertilized eggs can be readily microinjected. Embryos cannot be microinjected after the fertilization envelope has elevated since the embryos are not fastened within the fertilization envelope and rotate out from under the microelectrode. If the fertilization envelope is removed, the embryos are excessively fragile and the percentage of normal development is low. Typically, more than 50% of the injected eggs will survive injection. Most of those that survive injection will develop normally if the particular batch of eggs is good. In one night of spawning, it is possible to inject 500 or more eggs.

1. Solutions and Supplies

Sterile 50–100% glycerol

Sterile deionized water (DEPC-treated for RNA injections)

50 mg/ml Texas Red dextran (Molecular Probes, Inc., P.O. Box 22010, Eugene, OR 97402-0414; Tel: 1-800-438-2209)

XGal, 40 mg/ml in dimethylformamide

5-cm petri dishes

Circular plasmid DNA of lacZ reporter construct at a concentration of 0.5 μg/μl. 200–500 μM antisense morpholino oligonucleotide (Gene Tools, Inc.; http://www.gene-tools.com/. One Summerton Way, Philomath, OR 97370; Tel: (541) 929-7840; (541) 929-7841

Polylysine, mol. wt. 30,000–70,000, 0.25 mg/ml in distilled water

Capped mRNA (The mMESSAGE mMACHINE kit from Ambion (2130 Woodward St., Austin, TX 78744-1882; Tel: 512-651-0200; Fax: 512-651-0201; www.ambion.com) works well).

PBST; 0.9% NaCl, 0.1% Tween 20, 20 mM KPO_4 buffer pH 7.4

2. Equipment

Dissecting microscope

Fluorescence microscope

Picospritzer, single-channel with foot pedal (General Valve Division Parker Hannifin, 19 Gloria Lane, Fairfield NJ 07004; Tel: 973-575-4844; Fax: 973-575-4011)

Micromanipulator with fine control

Capillary tubing for microelectrodes; borosilicate glass; Omega Dot fiber for rapid fill. 1.0 O.D. by 0.75 mm I.D. catalog no. 30-30-0 FHC, http://www.fh-co.com/; Bowdoinham, ME 800-326-2905

Horizontal microelectrode puller

3. Protocol

DNA preparation: For amphioxus, do not use reporter constructs with green fluorescent protein. Amphioxus eggs and embryos fluoresce brightly green when illuminated with blue light. Plasmids larger than about 12 kb should be purified on a cesium chloride gradient (Sambrook *et al.*, 1989) or on an ion exchange column (e.g., Qiagen Plasmid Maxi Kit, catalog no. 12162; Qiagen Inc., 28159 Avenue Stanford, Valencia, CA 91355; www.Qiagen.com). For smaller plasmids, a Qiagen miniprep spin column works well. Add an extra wash in 80% ethanol before eluting the DNA. Circular DNA works well. It is not necessary to linearize

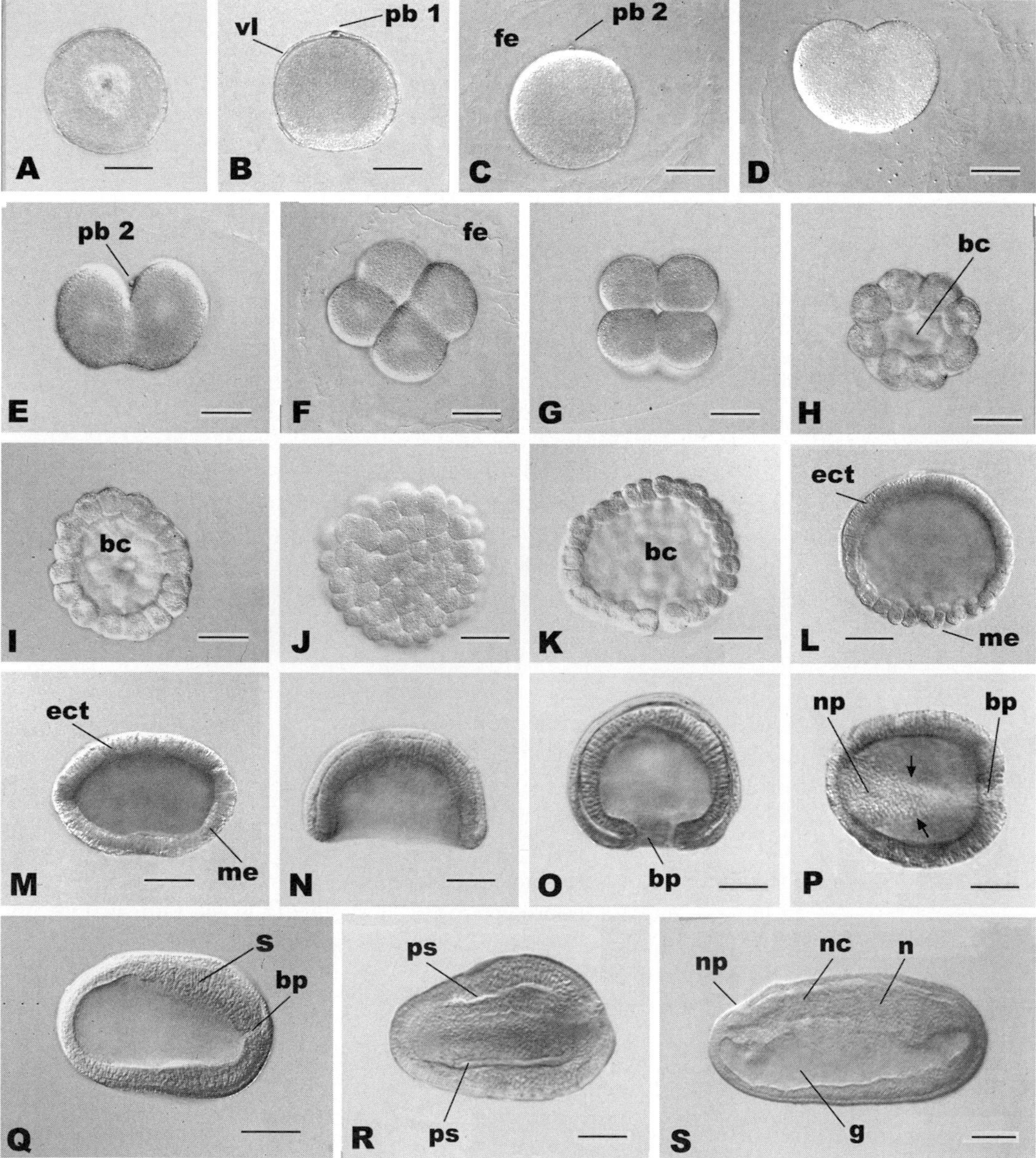

Fig. 4 Light micrographs of early development of *Branchiostoma floridae*. (A) Large primary oocyte with central nucleus and nucleus. The vitelline layer is not visible. (B) Spawned egg arrested at second meiotic metaphase. The vitelline layer (vl) is slightly elevated from the egg surface over the first polar body (pb1). (C) Zygote 20 min after insemination. The second polar body (pb2) is visible at the animal pole. The fertilization envelope (fe) has maximally elevated, increasing the diameter of the zygote to about 450 μm, but has somewhat collapsed in this fixed specimen. (D) An embryo at the onset of first cleavage, which begins at the animal pole. (E) Two-cell stage after the completion of karyokinesis but

the plasmid. A stock solution of DNA at 500 ng/μl in deionized water should be kept frozen. The optimal concentration for microinjection may range from 20 to 100 ng/μl.

1. Capped mRNA. The mRNA should be precipitated with ethanol and stored as a pellet in 70% ethanol until just before use. Addition of a long polyA tail can enhance translation.

2. Coat dishes with polylysine solution. Let polylysine solution remain in dish for 2 to 5 min. Pour out into another dish. Let dishes dry upside down. Do not rinse. Dishes should be used within a day or two of coating.

3. Pull microelectrodes. These should have a tip about 1.25 to 1.5 cm long. The tip should not curl. If it does, reduce the heat. Electrodes should be used within 3 to 4 days of pulling. Make injection solution: 1 μl Texas Red dextran (50 mg/ml), enough glycerol for a final concentration of 10 to 30% plus sufficient DNA, capped mRNA, or antisense morpholino for final concentrations of 50 ng/μl DNA, 200 to 500 nM morpholino-oligonucleotide or up to 1 μg/μl capped mRNA. Although Rhodamine dextran has been used for microinjections, it can inhibit development, perhaps due to contaminants in some lots.

4. Backfill electrodes with 2 to 3 μl of injection solution using a microloader pipette tip (Eppendorf Microloaders, pipette tip for filling microinjection capillaries, pkg. of 200, catalog number 930001007; http://www.eppendorfsci.com. Also available in the United States from Fisher Scientific).

---

before the completion of cytokinesis. The second polar body (pb2) is in the cleavage furrow. The two nuclei are visible as clear areas in the center of each blastomere. (F) Four-cell stage viewed from the animal pole. Fe = fertilization envelope. (G) Eight-cell stage viewed from the animal pole. (H) Early blastula; bc = blastocoel. (I) Mid-blastula. All the blastomeres are approximately the same size; bc = blastocoel. (J) Surface view of late blastula. (K) Optical section through late blastula. The cells at the vegetal pole are slightly larger and more loosely adherent than elsewhere; bc = blastocoel. (L) Optical section through very late blastula. The future ectoderm (ect) is distinguished from the future mesendoderm (me) by tighter packing of the cells. (M) Early gastrula. The blastula has slightly flattened at the equator, which is the future blastoporal lip. The mesendoderm has begun to invaginate with the center of the mesendoderm (me) slightly indented; ect = ectoderm. (N) Mid-gastrula. Invagination is complete and the blastopore is wide open. (O) Late gastrula. The blastopore (bp) has nearly closed. (P) Dorsal view of early neurula. Anterior is at left. The blastopore (bp) marking the posterior of the embryo is at the right. In amphioxus, the non-neural ectoderm (arrows) detaches from the edges of the neural plate (np) and migrates over it to fuse in the dorsal midline. Then the neural plate rounds up. (Q) Side view of early neurula. The blastopore (bp) is posterior and is covered by the non-neural ectoderm that has migrated over it. At this stage, the three most anterior somites (s) have begun to form. (R) Dorsal view of early neurula. Optical section through the forming somites. The anteriormost somites pinch off from presomitic grooves (ps) in the dorsolateral edges of the archenteron. Anterior is at left. (S) Side view of neurula. Anterior at left. The neuropore (np) is open at the anterior end of the nerve cord (nc). The notochord (n) has pinched off from the dorsal medial mesoderm; g = gut. All scale bars = 50 μm.

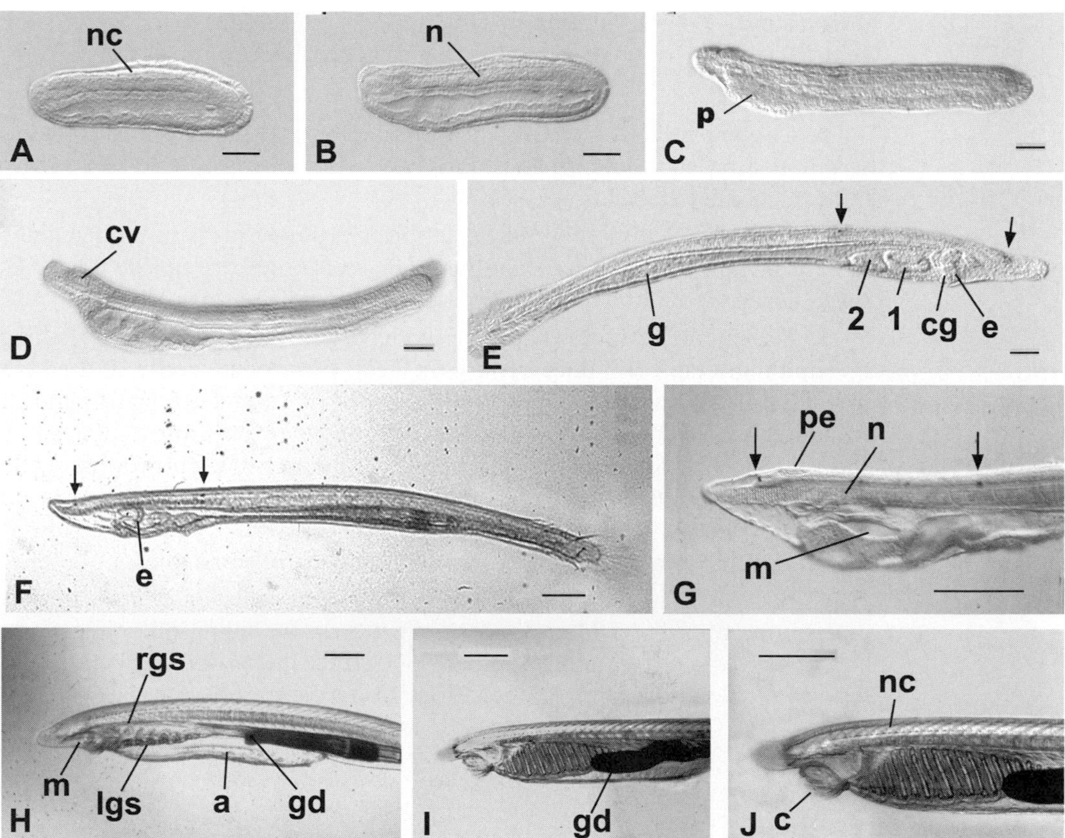

Fig. 5 Later development of amphioxus. Side views. Anterior at left in A–D, F–J, anterior at right in E. (A) Mid-neurula (18 h); nerve cord (nc). (B) Mid-neurula (20 h). Muscular movements begin at this stage; n = notochord. (C) Late neurula (24 h); p = pharynx. (D) Early larva (36 h). The cerebral vesicle (cv) is clearly visible at the anterior end of the nerve cord. In the pharynx, the mouth has opened on the left (not visible). (E) Six-day larva viewed from the right side. The first two gill slits (1, 2) have penetrated on the right behind the club-shaped gland (cg), an enigmatic structure, which is just behind the endostyle (e), the homolog of the thyroid. Arrows show the first two pigment spots associated with photoreceptors in the nerve cord. The first of these, the more posterior one, first appears at the mid-neurula stage (15 h). The anteriormost one, associated with the frontal eye, perhaps homologous to the vertebrate paired eyes, appears at 2.5 d; g = gut. (F) Photograph of a live, 10-day larva with 3 gill slits. E = endostyle visible through the transparent pharynx. Arrows show the first two pigment spots in the nerve cord. The dark material in the gut is food. The gut cells turn green from the algal pigments. (G) Anterior end of a 10-day larva. Arrows show the first two pigment spots in the nerve cord. N = notochord. In this fixed specimen, the pineal eye (pe) is visible in the cerebral vesicle; m = mouth, which in larvae is on the left side. (H) Side view of a living larva at the onset of metamorphosis (about 3 weeks). At the 9 gill-slit stage, the first row of gill slits to form on the right migrates around the ventral side of the larva to become the left gill-slits (lgs). Then a new row of gill slits (rgs) forms above them on the right. The mouth is still on the left, but has begun to migrate anteriorly. Atrial folds grow out and downward from above the gill slits to fuse in the ventral midline, forming the atrium (a). A diverticulum (gd) has begun to grow anteriorly from the gut. The gut cells are dark from algal pigments taken up by

5. The quality of eggs is very important for successful injection. Choose a batch of unfertilized eggs from a single female that are nicely round, of a uniform size, with vitelline layers that are not partially elevated, and that does not include dead eggs. Test to make sure that 100% of the eggs will fertilize.

6. Make 5 to 10 parallel scratches on the bottom of a polylysine-coated petri dish with a Pasteur pipette or diamond scriber for orientation. Transfer a few hundred eggs to the center of the dish. A Pasteur pipette with the tip pulled out to a slightly larger diameter than that of the eggs and used with a mouthpiece gives fine control. Wait several minutes to let the eggs settle down and attach to the dish. Then add filtered seawater to a depth about one-half the height of the dish.

7. Insert the microelectrode into its holder. Break the needle tip. For the long electrodes recommended here, the most effective method is to use sharp forceps to break the tip in air under a dissecting microscope. Test the size of the tip and the injection volume by inserting the tip into a 3 cm petri dish containing mineral or vegetable oil. With the injection pressure set to 20 psi and a duration of 300 ms, inject a drop into the oil. Adjust the duration of the injection to obtain a drop diameter about $\frac{1}{4}$ to $\frac{1}{5}$ of the diameter of an egg. The optimum injection duration is generally between 100 and 200 msec. If the injection time required to obtain an appropriate drop size is longer than 300 msec, the tip may be so fine that it will clog rapidly. If less than 100 msec is required for the appropriate drop size, the tip is probably too large; more eggs will die when injected. After confirming the drop size, move needle to the dish with attached eggs as quickly as possible.

8. Inject the eggs under a dissecting microscope with fiber optics. Because eggs of *Branchiostoma floridae* develop well from 24 to 30 °C, it is not necessary to control the temperature. Starting at one edge of the group of eggs, inject them in order, using the scratches on the dish as guidelines. The microelectrode should approach the upper side of the egg at a 45° angle. Dent the egg with the tip of the microelectrode and advance it with the fine and/or medium controls of the micromanipulator until the dent comes out, indicating that the egg has been penetrated. Pushing the injection pedal as the egg is being dented may help to determine when the electrode has penetrated into the egg. If no dye comes out, the needle has clogged. Break the tip larger or use a new microelectrode. Injected dye can be seen to persist within the egg for a second or two before it disperses. Withdraw the electrode after injection and move on to the next egg. Depending on the type of micromanipulator, use a combination of moving the dish with one

---

the gut cells. (I) A newly metamorphosed adult. The gut diverticulum (gd) extends anteriorly. (J) Higher magnification of the newly metamorphosed adult in I. Numerous pigment spots associated with photoreceptor cells (the organs of Hesse) have formed in the nerve cord (nc). The mouth has moved anteriorly, and the buccal cirri (c) have formed. These keep particles that are too large to pass through the digestive tract from entering the mouth. Each of the gill slits has become divided into two by downward growth of a medial bar. Additional gill slits form as the animal grows larger. Scale bars A–E = 50 μm; F, G = 100 μm; H–J = 500 μm.

hand and the micromanipulator controls with the other to inject the eggs as quickly as possible. After all the eggs in the dish are injected, the microelectrode, if not clogged, should be placed into the dish of oil while the next dish of eggs is prepared.

9. Fertilize the injected eggs with a drop or two of sperm. When the fertilization envelopes are visibly elevated, add antibiotics (Section III.B.1), cover the dish, and allow to develop at 30 °C.

10. Use fluorescent microscope to ensure that injections were successful. The co-injected Texas Red will emit red fluorescence under a rhodamine filter (green light) and orange fluorescence under a fluorescein filter (blue light).

11. Fix embryos at specific stage of interest. The fluorescence of Texas Red dextran persists at least several hours after fixing. Fixation is in 1% glutaraldehyde in seawater for lacZ detection or, for subsequent *in situ* hybridizations or antibody labeling, in 4% paraformaldehyde in 0.1 M MOPS buffer, 0.5 M NaCl, 1 mM EGTA, 2 mM $MgSO_4$, pH 7.4. Subsequent washes in buffer only (the above MOPS buffer for antisense morpholinos or PBST for lacZ staining) will decrease the fluorescence and increase the background fluorescence. It is particularly important when injecting antisense morpholino olignucleotides to remove any embryos that were not successfully injected while they are still in fixative.

12. Staining of embryos for lacZ detection. Wash fixed embryos twice in PBST. Stain in freshly made 1 mM $MgCl_2$, 3 mM K ferrocyanide, 3 mM K ferricyanide in PBST with a 1:100 dilution of 40 mg/ml XGal. When stained, wash in PBST, store in 80% glycerol in PBS.

13. Methods for *in situ* hybridization are in Holland *et al.*, 1996.

VI. Amphioxus Resources Available

In the 15 years since amphioxus was first spawned in the laboratory, a number of DNA resources have been generated and made freely available. The mitochondrial genomes of *B. floridae, B. lanceolatum,* and *B. belcheri* have been sequenced (Boore *et al.*, 1999; Spruyt *et al.*, 1998). These sequences are available at http://www.ebi.ac.uk/cgi-bin/genomes/genomes.cgi?genomes=organelles or http://megasun.bch.umontreal.ca/ogmp/projects/other/mt_list.html. In addition, gridded cDNA libraries of late neurula (26 h) and adults, as well as two cosmid libraries of *B. floridae*, are available at cost from the RZPD in Berlin (http://www.RZPD.de). The 26-h library has been subject to EST analysis (Panopoulou *et al.*, 2003). A genome library in pCYPAC7 has been constructed by Chris Amemiya, Virginia Mason Research Center Benaroya Research Institute, Seattle, WA 98101. This library was made from 6 individuals and has an average insert size of 90 kb. It has approximately 5-fold coverage of the genome and is freely available (http://www.vmresearch.org/lab_research/amemiya/default.htm). A large-insert BAC library of *B. floridae* made from a single ripe male has recently been constructed by the laboratory of P. de Jong, CHORI, Oakland, CA, and will be available at cost from the BACPAC resource center at CHORI (http://www.chori.org/bacpac/).

VII. Concluding Remarks

Amphioxus, as the closest living invertebrate relative of the vertebrates, has given considerable insight into the evolution of the vertebrate body plan. For much of the first 10 years since *B. floridae* was first spawned in the laboratory, research focused on using gene expression patterns to identify homologies between amphioxus and vertebrate embryos. This research showed that amphioxus is an appropriate simplified model for understanding the evolution of the vertebrate body plan. The recent development of methods for microinjection of amphioxus eggs has greatly expanded the potential of amphioxus as a model for studying the evolution of developmental mechanisms. What remains is the development of continuous laboratory cultures and sequencing of the amphioxus genome. It seems likely that these goals will be achieved in the near future, adding amphioxus to the short list of "model developmental systems."

Acknowledgments

Laboratory space in Florida over the years has been generously provided by Prof. John Lawrence and the Biology Department of the University of South Florida, Tampa, FL, USA. This work was supported by National Science Foundation grant no. IBN00-78599 and NASA grants NAG-1376 and NAG 2-1585.

References

Boore, J. L., Daehler, L. L., and Brown, W. M. (1999). Complete sequence, gene arrangement, and genetic code of mitochondrial DNA of the cephalochordate *Branchiostoma floridae*. *Mol. Biol. Evol.* **16,** 410–418.

Cerfontaine, P. (1906). Recherches sur le développement de l'amphioxus. *Arch. Biol. Liège* **22,** 229–418 + pl. XII–XXII.

Conklin, E. G. (1932). The embryology of amphioxus. *J. Morphol.* **54,** 69–150.

Courtney, W. A. M. (1975). The temperature relationships and age-structure of North Sea and Mediterranean populations of *Branchiostoma lanceolatum*. *Symp. Zool. Soc. Lond.* **36,** 213–233.

Escriva, H., Holland, N. D., Gronemeyer, H., Laudet, V., and Holland, L. Z. (2002). The retinoic acid signaling pathway regulates anterior/posterior patterning in the nerve cord and pharynx of amphioxus, a chordate lacking neural crest. *Development* **129,** 2905–2916.

Fang, Y., Qi, X., Liang, P., and Hong, G. (1991). Annual change of gonadal development of the amphioxus in Xiamen. *Acta Oceanol. Sin.* **10,** 477–479.

Fang, Y. Q. (1998). Position in the evolution of reproductive endocrine of amphioxus, *Branchiostoma belcheri*. *Chinese Sci. Bull.* **43,** 177–185.

Ferrier, D. E. K., Minguillón, C., Holland, P. W. H., and Garcia-Fernández, J. (2000). The amphioxus Hox cluster: Deuterostome posterior flexibility and *Hox14*. *Evol. Dev.* **2,** 284–293.

Garcia-Fernández, J., and Holland, P. W. H. (1994). Archtypal organization of the amphioxus *Hox* gene cluster. *Nature* **370,** 563–566.

Hirakow, R., and Kajita, N. (1990). An electron microscopic study of the development of amphioxus, *Branchiostoma belcheri tsingtauense*: Cleavage. *J. Morph.* **203,** 331–334.

Holland, L. Z., and Holland, N. D. (1996). Expression of *AmphiHox-1* and *AmphiPax-1* in amphioxus embryos treated with retinoic acid: Insights into evolution and patterning of the chordate nerve cord and pharynx. *Development* **122,** 1829–1838.

Holland, L. Z., Holland, P. W. H., and Holland, N. D. (1996). Revealing homologies between body parts of distantly related animals by *in situ* hybridization to developmental genes: Amphioxus versus vertebrates. *In* "Molecular Zoology: Advances, Strategies, and Protocols" (J. D. Ferraris and S. R. Palumbi, eds.), pp. 267–282. Wiley-Liss, New York.

Holland, N. D., and Holland, L. Z. (1989). Fine structural study of the cortical reaction and formation of the egg coats in a lancelet (=amphioxus), *Branchiostoma floridae* (Phylum Chordata: Subphylum Cephalochordata = Acrania). *Biol. Bull.* **176,** 111–112.

Holland, N. D., and Yu, J.-K. (2002). Epidermal receptor development and sensory pathways in vitally stained amphioxus (*Branchiostoma floridae*). *Acta Zool.* **83,** 309–319.

Kubokawa, K., Azuma, N., and Tomiyama, M. (1998). A new population of the amphioxus (Branchiostoma belcheri) in the Enshu-Nada Sea in Japan. *Zool. Sci.* **15,** 799–803.

Kubokawa, K., Okuno, T., Terakado, K., and Nozaki, M. (2002). Survey of pituitary hormone genes expressed in Hatschek's pit of amphioxus. *In* "Perspective in Comp. Endocrinol.: Unity and Diversity," pp. 809–812. Monduzzi Editore, Bologna, Italy.

Lin, J. H., Fang, Y. Q., Iiu, J., and Lin, Q. M. (1996). Effects of different temperatures on gonadal development in lancelet, *Branchiostoma belcheri*. *J. Oceanogr. Taiwan Strait* **15,** 170–173.

Loeblich, III, A. R. (1975). A seawater medium for dinoflagellates and the nutrition of *Cachonina niei*. *J. Phycology* **11,** 80–86.

Müller, J. (1841). Mikroskopische Untersuchungen über den Bau und die Lebenserscheinungen des Branchiostoma lubricum Costa, Amphioxus lanceolatus Yarrell. *Ber. Preuss. Akad. Wissensch. Berlin* **1841,** 396–411.

Panopoulou, G., Hennig, S., Grotu, D., Krause, A., Poustka, A. J., Herwig, R., Vingron, M., and Lehrach, H. (2003). New evidence for genomic-wide duplications at the origin of vertebrates using an amphioxus gene set and completed animal genomes. *Genome Res.* **13,** 1056–1066.

Poss, S. G., and Boschung, H. T. (1996). Lancelets (Cephalochordata: Branchiostomatidae): How many species are valid? *Israel J. Zool.* **42 Suppl.,** 13–66.

Sager, G., and Gosselick, F. (1986). Investigation into seasonal growth of *Branchiostoma-lanceolatum* off Heligoland, according to data by Courtney (1975). *Int. Rev. Ges. Hydrobiol.* **71,** 701–707.

Sambrook, J., Fritsch, E. F., and Maniatis, T. (1989). "Molecular Cloning. A Laboratory Manual." Cold Spring Harbor Laboratory Press, Cold Spring Harbor, New York.

Spruyt, N., Delarbre, C., Gachelin, G., and Laudet, V. (1998). Complete sequence of the amphioxus (Branchiostoma lanceolatum) mitochondrial genome-relations to vertebrates. *Nucleic Acids Res.* **26,** 3279–3285.

Stokes, M. D. (1996a). Larval settlement, post-settlement growth, and secondary production of the Florida lancelet (=amphioxus) *Branchiostoma floridae*. *Mar. Ecol. Prog. Ser.* **130,** 71–84.

Stokes, M. D. (1996b). Reproduction of the Florida lancelet (*Branchiostoma floridae*): Spawning patterns and fluctuations in gonad indexes and nutritional reserves. *Invert. Biol.* **115,** 349–359.

Stokes, M. D., and Holland, N. D. (1995). Ciliary hovering in larval lancelets (= amphioxus). *Biol. Bull.* **188,** 231–233.

Stokes, M. D., and Holland, N. D. (1996). Life-history characteristics of the Florida lancelet, Branchiostoma floridae: Some factors affecting population dynamics in Tampa Bay. *Israel J. Zool.* **42 Suppl.,** 67–86.

Tung, T. C., Wu, S. C., and Tung, Y. F. Y. (1958). The development of isolated blastomeres of amphioxus. *Scientia Sinica* **7,** 1280–1320.

Tung, T. C., Wu, S. C., and Tung, Y. Y. F. (1960). The developmental potencies of the blastomere layers in amphioxus egg at the 32-cell stage. *Scientia Sinica* **9,** 119–141.

Tung, T. C., Wu, S. C., and Tung, Y. Y. F. (1962a). The presumptive areas of the egg of amphioxus. *Scientia Sinica* **11,** 629–644.

Tung, T. C., Wu, S. C., and Tung, Y. Y. F. (1962b). Experimental studies on the neural induction in amphioxus. *Scientia Sinica* **11,** 805–820.

Watanabe, T., Yoshida, M., and Shirai, H. (1999). Effect of light on the time of spawning in the amphioxus *Branchiostoma belcheri tsingtauense*. *Contributions Ushimado Marine Lab. Okayama Univ.* **37,** 1–7.

Wilson, E. B. (1893). Amphioxus, and the mosaic theory of development. *J. Morph.* **8,** 579–683 + plates XXIX–XXXVIII.

Wu, X. H., Zhang, S. C., Wang, Y. Y., Zhang, B. L., Qu, Y. M., and Jiang, X. J. (1994). Laboratory observations on spawning, fecundity, and larval development of amphioxus (*Branchiostoma belcheri tsingtauense*). *Chin. J. Oceanol. Limnol.* **12,** 289–294.

Yin, H., Zhang, C.-L., Wang, H., and Shen, W.-B. (1994). Studies on amphioxus GnRH during breeding season. *Acta Zool. Sinica* **40,** 63–68.

Zhang, S. C., Holland, N. D., and Holland, L. Z. (1997). Topographic changes in nascent and early mesoderm in amphioxus embryos studied by DiI labeling and by *in situ* hybridization for a *brachyury* gene. *Dev. Genes. Evol.* **206,** 532–535.

Zhang, S. C., Zhu, J. T., Jia, C. H., and Li, G. R. (1999). Production of fertile eggs and sperms in laboratory-reared amphioxus *Branchiostoma belcheri tsintauense*. *Chin. J. Oceanol. Limnol.* **17,** 53–56.

Zhang, S. C., Zhu, J. T., Li, G. R., and Wang, R. (2001). Reproduction of the laboratory-maintained lancelet *Branchiostoma belcheri tsingtaunese*. *Ophelia* **54,** 115–118.

PART II

Embroyological Approaches

CHAPTER 10

Quantitative Microinjection of Oocytes, Eggs, and Embryos

Laurinda A. Jaffe and Mark Terasaki
Department of Cell Biology
University of Connecticut Health Center
Farmington, Connecticut 06032

I. Introduction

In contrast to other techniques for manipulating the intracellular environment, such as membrane permeabilization, patch pipet dialysis, or use of a cell homogenate, microinjection offers the unique advantage that precisely defined amounts of test substances can be introduced into the cytoplasm without altering other components within the cell. This chapter describes a method for microinjection of oocytes, eggs, and embryos using quantitative techniques initially used by Pierre de Fonbrune in the 1930s to inject amoeba (see de Fonbrune, 1949). These techniques were developed and applied to echinoderm eggs by Yukio Hiramoto (see Hiramoto, 1962), and have been used and further developed by many investigators (in particular, see Kiehart, 1982). This chapter describes the version used in our laboratories.

0091-679X/04 $35.00

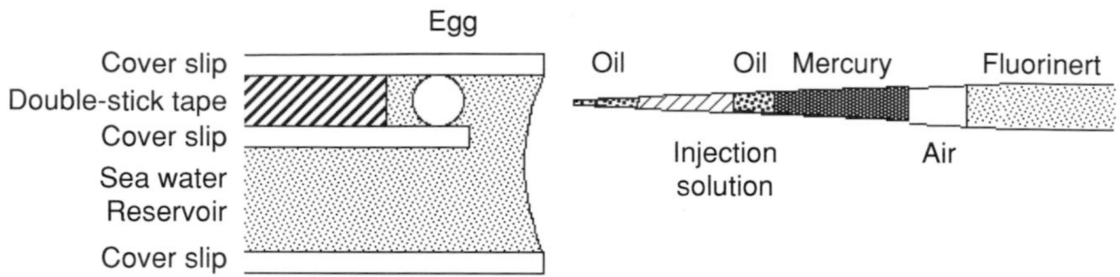

Fig. 1 A side view of an egg in a microinjection slide, with a loaded injection pipet coming in horizontally.

In this method, a screw-controlled syringe is used to draw solutions in and out of the tip of the micropipet. Small turns of the syringe cause displacements of fractions of a milliliter; to transform these displacements into picoliter volumes for injection, a small droplet of mercury is back-loaded, then pushed to the tip of the micropipet. Behind the mercury is an air space and an oil-filled tube leading to the syringe. Adjusting the syringe applies pressure to the air, which, in turn, applies pressure to the mercury. Because of the very high surface tension of mercury, large displacements of the syringe volume (on the order of 100 μl) result in very small displacements of the volume in front of the mercury (on the order of 10 pl). This allows precise control of the injection, and as will be described, precise quantitation.

To view the egg clearly during microinjection, it is held between two parallel coverslips separated by a spacer (Kiehart, 1982); for most sea urchin or starfish eggs, the spacer can be made of one or two layers of double-stick tape (Fig. 1). The injection slide is held on the stage of an upright compound microscope, and the microinjection pipet is brought in horizontally (Fig. 2). With some modification, the technique can be used to inject eggs of a variety of species, including ascidians (Runft and Jaffe, 2000), frogs (Runft *et al.*, 1999), and mammals (Mehlmann and Kline, 1994; Mehlmann *et al.*, 2002), and can also be used to inject other large cells (Terasaki *et al.*, 1995).

Many of the components needed for this method of injection are specialty items, and a major purpose of this chapter is to provide a guide to where they can be obtained. Updated information about suppliers, as well as additional technical details, information pertaining to other species of eggs, and new developments of methods will be available at http://egg.uchc.edu/injection.

II. Methods

A. Preliminary Preparations

1. Assembling the Apparatus

Sources of the equipment and supplies needed for these microinjection methods are listed in Section III. The costly items are an upright microscope, a micromanipulator, and a horizontal micropipet puller. You also need a number of

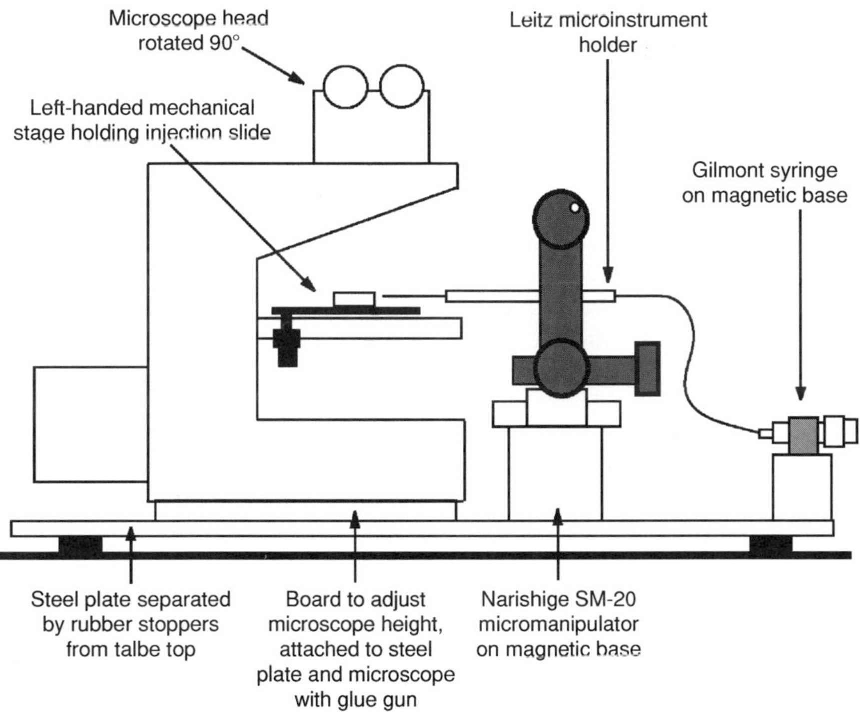

Fig. 2 Arrangement of the microscope, micromanipulator, injection syringe, and microinstrument holder.

inexpensive small parts including a microinjection syringe and micropipet holder, attached to each other with a piece of Teflon tubing. This assembly is filled with a fluorocarbon oil (Fluorinert).

The microinjection apparatus should be set up in a quiet location, out of the way of traffic, vibration, and distractions. A comfortable chair of the right height helps. Since many echinoderm species require cool temperatures for development, a small room with an air conditioner is ideal (see Section III.A.4).

To avoid vibration, the microscope as well as the micromanipulator should be firmly attached to a heavy steel plate. The plate should be positioned on a solid table or bench; vibration can be greatly reduced by placing 1 to 2″ diameter rubber stoppers between the plate and the table at each corner (Fig. 2). The microscope can be attached to the plate with a glue gun, which does no harm to the microscope. Apply the glue to the solid metal of the microscope, not the rubber feet, which should be removed. Usually, it is necessary to adjust the height of the microscope to match the height of the micromanipulator; this can be accomplished by using the glue gun to attach a Plexiglas or wood board between the microscope and the steel plate. The micromanipulator and the microinjection syringe can be attached to the plate with magnetic bases (Fig. 2).

The microscope should be mounted with the front of the stage facing to the right, facing the micromanipulator. The head of the microscope (the part with the eyepieces) should be rotated 90°, such that it faces the front of the table (Fig. 2; see details in Section III).

Finally, check that everything is level, using a small "bubble" level. Check the table, the steel plate, and the microscope stage. Sometimes you will find that even with the steel plate level, the microscope stage is not. This is a common problem with Zeiss Axioskops. To fix this, you can insert a shim where the stage is mounted to the microscope body. Alternatively, for a "quick fix," you can tilt the microinjection slide slightly, by lifting its front edge (a few tenths of a mm) such that it is not resting flat on the stage but is still held securely by the stage slide holder.

2. Cleaning Coverslips

Coverslips to be used for constructing microinjection chambers must be thoroughly cleaned. New, uncleaned coverslips are highly toxic to echinoderm eggs, which you can see for yourself by attempting to fertilize eggs sitting on such a coverslip. Failure to wash coverslips is a major cause of bad results with microinjection. The following procedure should be used:

a. Assemble a "coverslip washer" (see Section III.B.16).

b. Prepare a dilute solution of detergent in hot water (a pinch of Alconox in ~200 ml).

c. Using fine forceps, drop individual coverslips (#1 1/2, 22 × 22 mm square) into the detergent solution. Put in half of one box of coverslips. Let sit 15 to 30 min.

d. Attach the coverslip washer to the deionized water line and fill with water. Remove the lid and, using a fine forceps, transfer the individual coverslips into the water in the washing tube. Put the lid on, start the water running, and shake gently to redistribute the coverslips.

e. Let deionized water run through the washer for 1 h, shaking occasionally.

f. Let 1 to 2 gallons of milliQ water run through, while shaking gently.

g. Using fine forceps, transfer the coverslips individually to a jar containing 85% EtOH, 15% H_2O for storage.

h. Before use, wipe the EtOH from the coverslip using a Kimwipe.

3. Making Micropipets

Using a double-stage horizontal puller, pull a micropipet. (See Sections III.A.3 and III.C.10 for sources of pullers and glass tubing, as well as details on pullers and tubing dimensions.) From the shoulder to the tip, the length should be ~1 cm. It should be straight, not curved. If it is curved or too long or too short, adjust the micropipet puller. Test the pipet in the injection set-up. If it works well, you may want to prepare a supply of ~10 or more pipets (keeping the same settings on the pipet puller) before continuing with the series of injections. Store pipets by attaching them to the sticky side of a pair of foam weatherstripping strips attached with double-stick tape to the surface of a 150 mm petri dish. Alternatively, use a micropipet storage jar. See Section III.C.11 for further details about micropipet storage.

Using a 10 μl Hamilton syringe inserted from the back of the pipet, deposit ~1 μl of Hg so that it forms a bead in the region where the pipet begins to taper. You will later push the Hg to the tip of the pipet under microscopic observation.

B. Preparation of the Microinjection Slide

1. Initial Assembly

a. Apply silicon grease to both sides of the U-shaped cutout of the plastic support slide (see Begg and Ellis, 1979; Kiehart, 1982; and Section III.B.9), so that coverslips can be attached to both sides. Attach the bottom coverslip, flush with the front of the support slide (Fig. 3A).

b. Prepare the top coverslip that will hold the eggs (Figs. 3B and C). This involves cutting a small piece of coverslip using a diamond pencil, and attaching the small piece of coverslip to the larger piece of coverslip (or an uncut coverslip), using double-stick tape to form a spacer. The distance between the tape and the front of the larger coverslip piece should be ~3 to 5 mm. The eggs go between the 2 coverslip pieces. For different sizes of eggs, different spacers may be used (Mylar of various thicknesses; see Kiehart, 1982), or coverslips may be assembled to form a wedge-shaped space (Kishimoto, 1986; Lutz and Inoué, 1986).

A single piece of double-stick tape produces a space about 100 μm thick (good for most sea urchin or sand dollar eggs); two pieces of tape produce a space about 200 μm thick (good for many starfish eggs). The distance between the overhanging edge of the small coverslip piece and the tape can be varied, depending on the particular use. If sperm are to be added, it is best to keep the egg chamber shallow (~200–300 μm for sea urchin eggs), since sperm do not swim well between the coverslips. Several chambers may be assembled and stored. The plastic strips from slide boxes make convenient storage racks, or use plastic petri dishes.

2. Loading Eggs into the Chambers

Let the eggs settle to the bottom of a beaker or test tube. The seawater in the tube should be at 18 to 22°C, not cold, as this will cause air bubbles to form. With a mouth pipet, pick up about 10 μl of a dense suspension of eggs. Deposit the suspension on the edge of the small coverslip, such that the eggs enter the chamber by capillary action. Observe by holding the coverslip up to the light or use a stereoscope. Touching a Kimwipe to the opposite end of the chamber may help to draw in eggs.

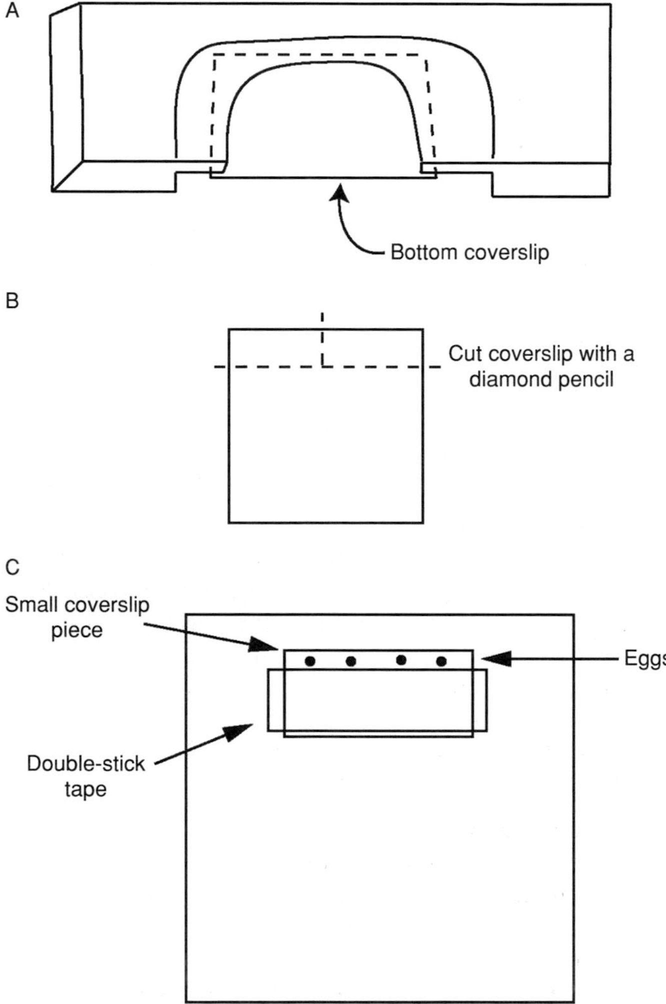

Fig. 3 Assembly of the microinjection slide. (A) The plastic support slide with the bottom coverslip attached. (B) Cutting a small coverslip piece, for assembling the top coverslip. (C) The top coverslip, with a piece of double-stick tape and the small coverslip piece to hold the eggs.

3. Final Assembly

As soon as possible after loading the eggs into the coverslip chamber, invert the coverslip and attach it to the plastic support slide by pressing it against the silicon grease. Fill the chamber with seawater. For better optics, it helps to push the top coverslip back about 1 to 2 mm with respect to the bottom coverslip, to eliminate the vertical air–water interface. A drop of seawater should occasionally be added to the front of the slide, since slow evaporation will occur.

4. Observation of the Preparation

Check to be sure that the eggs are in good shape and not overly flattened (diameter should not be more than about 10% bigger than that of an egg outside the chamber).

5. Preparation of the Loading Capillary

Using a diamond pencil, cut a piece of glass tubing ~2 cm long (see Section III.C.10 for tubing dimensions and ordering information). Make a clean break on the end of the capillary. This is accomplished by making a scratch on the glass and then bending the capillary back until it snaps at the scratch. Wipe the capillary glass with a Kimwipe so that it is clean. Using a P-20 pipetman, pick up ~1 μl of silicon oil. Touch the pipet tip to the end of the loading capillary and eject the oil. The oil should stay at the tip of the loading capillary. After loading the oil, load ~1 μl of the solution to be injected (as little as 0.6 μl can be used). Then load 1 μl more of oil. See Fig. 4A.

Now apply a soft wax ("valap"; see Section III.C.9) to the other end of the capillary to seal it. Push the end of the capillary into a dish containing the valap. Be careful not to expel the injection solution and oil. After this, wipe the oil off the outside of the capillary with a Kimwipe. Apply a dab of valap to the center of the capillary and attach it to one side of the top coverslip of the injection slide (Fig. 4B).

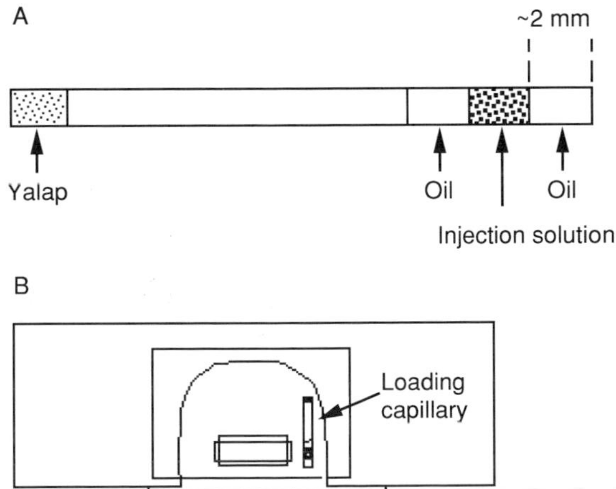

Fig. 4 The loading capillary (A) and assembled microinjection slide (B).

C. Injection

1. Mount the Micropipet in the Microinstrument Holder

First, expel any air that may be trapped in the instrument holder, by turning the micrometer syringe until oil drips out the front. Then, loosen the cap of the holder, and gently push a pipet containing a bead of mercury (Section II.A.3) through the silicon tubing. See Fig. 5.

If the pipet breaks in the holder, unscrew the cap, remove the front brass collar, then pull out the silicon tubing and back brass collar using a syringe needle (~20 gauge). Clean off all glass fragments carefully, and flush through some oil to clean inside the tube. If the silicon tubing is frayed, put in a new piece. To do this, first insert the back brass collar. Then insert a long piece of silicon tubing (a few cm long) and cut off flush with the front of the holder. Insert the front brass collar into the cap, and screw on the cap.

Check that the pipet is parallel to the surface of the microscope stage. If not, adjust the micromanipulator.

2. Break Off the Tip of the Micropipet

Observe with a 10 or 20x objective. Back the injection slide out of the field of view. Bring the micropipet into view and focus on the tip, using the micromanipulator vertical control. Now, back the pipet away from the field of view, and move the injection slide into view. Using the microscope focus control, focus on the edge of the loading capillary. Slowly bring the micropipet tip into the field of view, being careful not to hit the loading capillary. Then very delicately touch the tip of the micropipet to the edge of the loading capillary, to break the tip very slightly. The tip size should be about 1 to 3 μm (with a 10x lens, each small division in the eyepiece micrometer is 10 μm; 5 μm for a 20x lens). The optimal tip size varies for different egg species and materials being injected. Too large a tip causes damage,

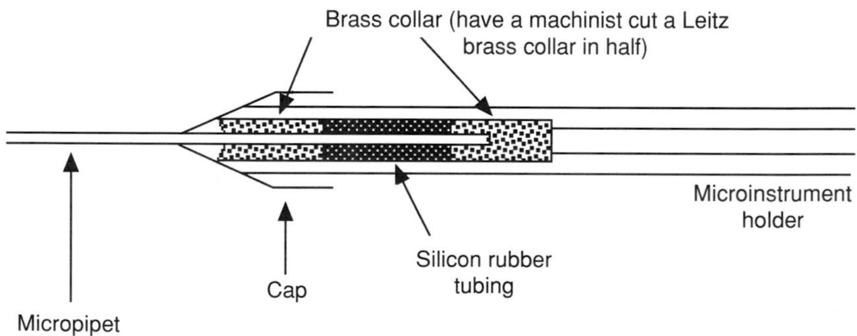

Fig. 5 A cross-sectional view of the Leitz microinstrument holder, holding a micropipet.

while too small a tip results in difficulty in loading and clogging during injection. Starfish oocytes are particularly nice for injection because tips up to 5 μm can be used without damage.

Note that movement and focusing for the injection chamber is done with the microscope stage controls (using your left hand), while movement and focusing for the micropipet is done with the micromanipulator (using your right hand).

3. Bring the Mercury to the Tip of the Pipet

While watching the micropipet through the microscope, advance the micrometer syringe until mercury starts to move toward the tip of the micropipet. The tip may move forward slightly during this process, so it should be moved away from the loading capillary. If the mercury doesn't move and the syringe pressure becomes great, the tip will start to move forward, which is the last stage before the pipet shoots out of the holder. Back off on the syringe pressure and break the tip some more by touching it again to the loading capillary edge. When the mercury does move forward, you want it to end up within a few μm of the tip; it is not necessary to bring it all the way to the end.

4. Calibrate the Micropipet

Manuever the pipet tip into the oil cap of the loading capillary. Using the reticle as a measuring tool, draw up a column of oil of a known length (for instance, with a 10X objective, 50 divisions = 500 μm). To determine the volume of this column of oil, you will expel it into seawater and measure the diameter of the resulting oil drop.

First, you need to bring the pipet into the chamber, between the coverslips. Move the injection pipet away from the slide. Focus on the edge of an egg in the chamber. Then move the slide out of the field of view. Move the pipet into view and focus on the tip. It is now necessary to lower the pipet tip ~60 μm, compensating for the difference in focusing through glass and water (to see the egg) and focusing through air (to see the pipet tip). With a Narishige SM-20 micromanipulator, this is accomplished by turning the height control knob ~1 to 1/4 turns clockwise. Now move the pipet out of the field of view, and bring the egg into view. Slowly advance the pipet toward the egg. When both pipet and egg are in view under the coverslip, adjust the vertical control on the micromanipulator to bring the pipet tip to the same focus as the edge of the egg.

Now that the pipet is in the chamber, expel oil. Measure the diameter of the oil drop and determine the volume (see Table I). This should be 1 to 5% of the volume of the egg. If the oil drop volume is not as desired, go back to the loading capillary and try again with a different length of oil. For each injection to be done with this pipet, draw up the solution to the value of h determined in this calibration (Fig. 6).

Table I
Calibration Table for Determining Injection Volumes by Measuring the Diameter of an Oil Drop

Diam (μm)	Vol (pl)	Diam (μm)	Vol (pl)	Diam (μm)	Vol (pl)	Diam (μm)	Vol (pl)
10	0.5	36	24	62	125	88	360
12	0.9	38	29	64	140	90	380
14	1.4	40	34	66	150	100	520
16	2.1	42	39	68	160	110	700
18	3.1	44	45	70	180	120	900
20	4.2	46	51	72	200	130	1150
22	5.6	48	56	74	210	140	1400
24	7.2	50	65	76	230	150	1800
26	9.2	52	74	78	250	160	2100
28	11.5	54	82	80	270	170	2600
30	14	56	92	82	290	180	3100
32	17	58	102	84	310	190	3600
34	20	60	113	86	330	200	4200

Note: Volumes were calculated using the following equation:

$$V = \frac{4}{3}\pi r^3 = \frac{4}{3}\pi\left(\frac{d}{2}\right)^3$$

Example: The volume of an *Asterina miniata* oocyte (d ~ 180 μm) is ~ 3000 pl. Therefore, an injection of 30 pl is 1% of the oocyte volume; an injection of 150 pl is 5%. Injections greater than 5% require special techniques (see Kishimoto, 1986).

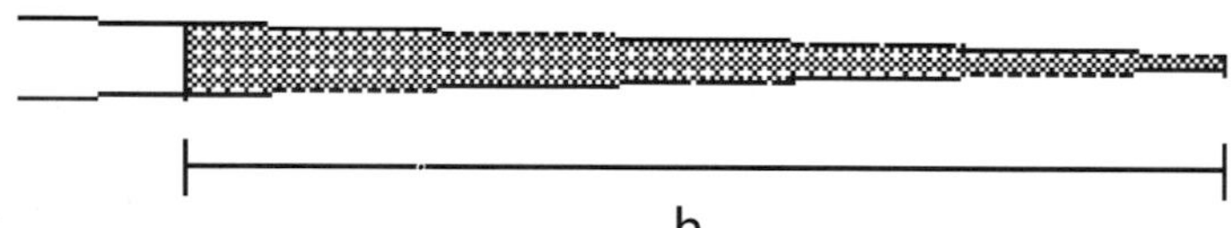

Fig. 6 Calibration of the micropipet. h = the length of solution drawn into the micropipet in order to have a defined picoliter volume.

5. Load the Injection Solution and Oil into the Pipet

First, draw up some oil into the pipet; this oil will separate the mercury from the injection solution. Focus on the interface between the oil and the injection solution in the loading capillary, and move the tip of the pipet into the solution. Apply suction to draw solution into the pipet. The column of solution should have the same length as determined during the calibration. Now move the loading capillary so that the pipet tip is in the oil. Draw up a cap of oil (~10–30 divisions long), and then adjust the micrometer syringe so that there is no excess pressure or suction. Often, the oil cap is difficult to draw up. Apply extra suction and wait 30 s. It sometimes helps to move the tip back and forth within the oil. If this doesn't work, reduce the excess suction and break the tip some more.

6. Bring the Pipet Tip Close to the Egg

Follow the procedure described in Section II.C.4 to bring the pipet into the chamber. Be prepared to adjust the micrometer syringe if the liquid in the pipet moves up the pipet when the pipet enters the seawater (a result of excess suction being left on the pipet). Bring the pipet next to the egg to be injected. Adjust the focus of the microscope to see the periphery of the egg sharply, and adjust the focus of the micromanipulator to see the tip of the pipet sharply.

7. Inject

Push the pipet tip into the egg. Sometimes, the coverslips hold the egg sufficiently tightly that the egg doesn't move. Other times, it is necessary to push the egg back against the double-stick tape, pushing with the micropipet. The egg will dimple when the pipet is pushed against it, and then the tip will go in. The tip should be positioned near the center of the egg to avoid damage to the plasma membrane when solution is expelled. Once the pipet is in the cytoplasm, turn the micrometer syringe to expel the oil cap and injection solution. You will see the injection solution displace the cytoplasm as it moves in; if not, you may not have the tip in the egg. Stop the injection at the interface of the injection solution and the back volume of oil. It is OK if a little bit of the back volume of oil enters the egg. Then withdraw the pipet from the egg.

Observe the egg to be sure that it is not damaged, and to determine whether the injected substance caused any response, such as fertilization envelope elevation. Vacuoles in the cytoplasm or lysis at the injection site indicate damage. Do a control injection (e.g., buffer only) to be sure that the injection procedure itself is not having an effect on the egg. Sometimes, a control injection will cause a local elevation of a fertilization envelope at the injection site. If this happens, try using a smaller pipet tip and performing the injection more delicately.

Except for highly concentrated protein solutions, the same pipet can be used multiple times. With practice, you can expect a series of injections to require about 2 min per injection.

8. Culture of Eggs after Injection

The microinjection slide can be used as a culture and observation chamber until the embryos reach a stage where they swim away. For culturing, keep the slide in a moist environment (a petri dish with wet Kimwipes). Alternatively, the injected eggs can be swept out of the coverslip shelf, using the micropipet, and then collected using a mouth pipet and transferred to a separate container for culturing.

Note that the oil drop left in the egg after injection does not inhibit fertilization or development, although it is better to keep it small if the embryo is to be followed through cleavage stages.

III. Equipment and Supplies (Prices are as of December, 2002)

A. Major Equipment

1. Upright Microscope

A Zeiss Standard, Axioskop, or Axiostar, or a similar microscope made by another company works well. In general, the body of the microscope shouldn't be so large that it is difficult to bring in the micromanipulator (see Fig. 2). The stage should move up and down to focus (vs a fixed-stage design). There must be a mechanical stage for X–Y movement. Ideally, the stage should have the X–Y controls on the left side. (Microscope stage controls are usually on the right side, but left-hand stages are available, or the stage can be changed from right to left by a machinist. In principle, if you are left-handed, you might want to set up the apparatus with a left-hand micromanipulator and a right-hand stage. We've never tried this, but it should work.) A 10 or 20× objective should be used. A Zeiss Axiostar microscope, equipped with a left-hand stage and a 10× objective costs about $2000 and is an excellent microscope for this purpose. Recommended components for a typical setup are listed here; these should be modified, depending on the particular application.

Carl Zeiss, Inc.
One Zeiss Drive
Thornwood, NY 10594
Phone: 1-800-233-2343
Fax: 1-914-681-7453
Website: http://www.zeiss.com/

Basic Components:
1169150 Axiostar Plus Microscope Stand; specify left-hand stage
1029023 Condenser 09/1.25 F/Axiostar
440930 CP-Achromat 10×/0.25 objective
452928 Binocular Tube 45/20 ICS
4442329902 Eyepieces (2), E-PL 10×/20 FOC/26 DIA
Other small parts as recommended by Zeiss dealer and needed for your particular use.

2. Micromanipulator

A Narishige SM-20 works well. A Leitz manipulator is also very good. The manipulator must have fine controls for X, Y, and Z movements. It must be mounted

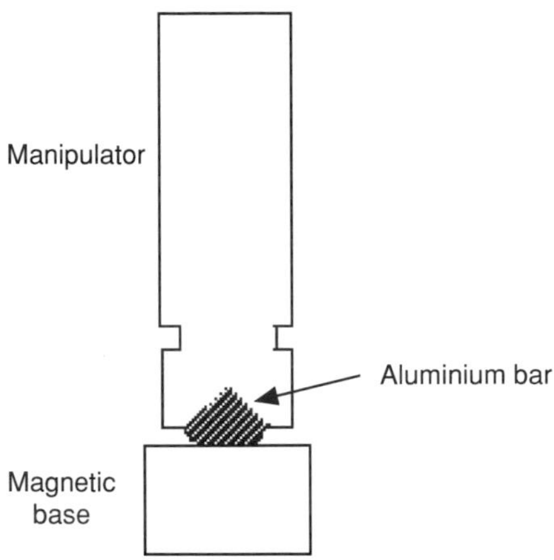

Fig. 7 Bar for attaching the SM-20 micromanipulator to a magnetic base.

securely, preferably with a magnetic base. The magnetic base from Flexbar works well; to attach the manipulator to the magnetic base, have a machinist make a small aluminum bar (Fig. 7).

The manipulator can be ordered directly from Japan:

Narishige Scientific Instrument Laboratory
27-9 Minamikarasuyama 4-chome
Setagaya-ku, Tokyo, 157 Japan
Fax: 81-3-3308-2005
Email: sales@narishige.co.jp
Website: http://www.narishige.co.jp

Model SM-20: $4400 including shipping. Specify for right-hand use (unless you prefer left). (Right- and left-handed manipulators are easily interconvertible: loosen the screw under the plate on the upper X–Y control. Slide the plate off and rotate 180°. Slide the plate back on and turn the knob until the screw comes up into the groove. Tighten the screw.

This manipulator is much more expensive if ordered from a United States distributor. The Japanese company provides excellent service; you can ship back the manipulator if service is needed.

The magnetic base can be ordered from:

Flexbar Machine Corp.
250 Gibbs Road
Islandia, NY 11722
Phone: 800-879-7575
Website: http://www.flexbar.com
Model 11002: $86

3. Micropipette Puller

Horizontal pullers are generally best for pulling micropipets with a long taper, as needed for inserting the pipet between 2 parallel coverslips. Using a wide ribbon filament (~6 mm) is important for obtaining a long taper. It is also possible to use a vertical puller (see Sluder *et al.*, 1999).

Horizonal pullers are currently sold by:

Narishige Scientific Instrument Laboratory
PN-30 Glass microelectrode puller, \$2950
PN-30H Heater for PN-30, 6 mm platinum plate, pkg. of 3, \$162
(Recommended settings: heater = 80/81, main magnet = 78, submagnet = minimum; start here, but "fine tuning" will likely be needed.)

Sutter Instrument Company
51 Digital Drive
Novato, CA 94949
Phone: 415-883-0128
Website: www.sutter.com
P-97 Flaming/Brown type micropipet puller, \$6700
For advice on the use of this puller, contact Adair Oesterle (adair@sutter.com).

Two other horizontal pullers, that are no longer sold, are often available in neurophysiology labs, and are often no longer in use:

Industrial Science Associates
Model M1

Narishige Scientific Instrument Laboratory
Model PN-3

These older model pullers make particularly nice pipets for this application and are somewhat easier to use than the current models.

An alternative to buying a puller is to use one in another lab and make a large supply of pipets. Pulled pipets can be stored for years, with or without mercury in them (see Section III.C.11 for a convenient storage device).

4. Air Conditioner

Many species of echinoderm embryos do not develop normally at room temperature (22–24 °C) and require temperatures below 20 °C. Also, in general, injection damage problems are fewer at cooler temperatures. For experiments that require cooling, it is possible either to cool the entire room, or to blow cool air over the microscope stage and use an incubator for storing the eggs and embryos. To cool the room or to blow cool air over the stage, a Koldwave water-cooled air conditioner is convenient.

Often, a good solution is to construct a small "microinjection room" around a Koldwave; if the room is constructed in a corner, 2″ thick styrofoam insulation boards can be used to make the other two walls and ceiling. A $4' \times 5'$ space is sufficient, and for this small size, 2″ aluminum adhesive tape is sufficient to assemble the styrofoam boards. Styrofoam boards and aluminum tape are available from Home Depot. A curtain is adequate for a door. For a larger room, additional supports for the styrofoam should be used. Construct a frame using 2×4 s, or incorporate the backs of bookshelves to make one of the walls.

Another way to direct cool air over the microscope stage is to connect a piece of dryer duct tubing from an air conditioner that may be built in to the room. Lead the dryer duct tubing to a position near the stage.

For a supplier of Koldwave air conditioners, contact:

Koldwave Division
Heat Exchangers, Inc.
8100 N. Monticello
Skokie, IL 60076
Phone: 708-679-0300

Cost is about $1300–2000, depending on the model.

To monitor temperature inside the injection room or on the microscope stage, use a small electronic thermometer. Inexpensive electronic thermometers are sold by:

Thomas Scientific
P.O. Box 99
Swedesboro, NJ 08085
Phone: 800-345-2100
Website: www.thomassci.com

Cat.#9327-L19 ($17)

High temperature is another major cause of microinjection failure! Equipment in a small lab can heat the room well above normal room temperature.

5. Stereoscope with a reticle

This is not essential, particularly if you are under 40, but is very useful for assembling microinjection chambers.

B. Other Equipment

1. Eyepiece Micrometer Reticle

For the Zeiss Axiostar, you need a 26 mm diameter disc. (Other microscopes may require slightly larger or smaller discs.) The reticle should have a 10 to 12 mm scale (12 mm is better), divided into 100 to 120 divisions. Reticles can be ordered from:

Klarmann Rulings Inc.
480 Charles Bancroft Hwy.

Litchfield, NH 03052
Phone: 603-424-2401
Fax: 800-252-2401
Website: www.reticles.com
KR-221 Microscope reticle, 12 mm in 120 parts, specify diameter of disc, $62 each

2. Micrometer syringe

The Gilmont model S-1200 works well.

Barnant Co.
Gilmont Instruments Div.
28W092 Commercial Ave.
Barrington, IL 60010
Phone: 800-962-7142
Website: www.barnant.com
Cat. # GS-1200 ($100)

3. Magnetic Stand for Holding the Micrometer Syringe

Flexbar Machine Corp.
(address in III.A.2)
Cat. # 11004 ($30)

The stand comes with a post, which should be removed and replaced with a bolt, and a clamp made from a strip of 1/2″ metal braid with a metal tab should be soldered on the end. Metal braid is available from electronics suppliers such as Newark. See Fig. 8.

Another convenient way to attach the syringe to the magnetic base is with a metal hose clamp (a metal ring that tightens down with a screwdriver). Hose

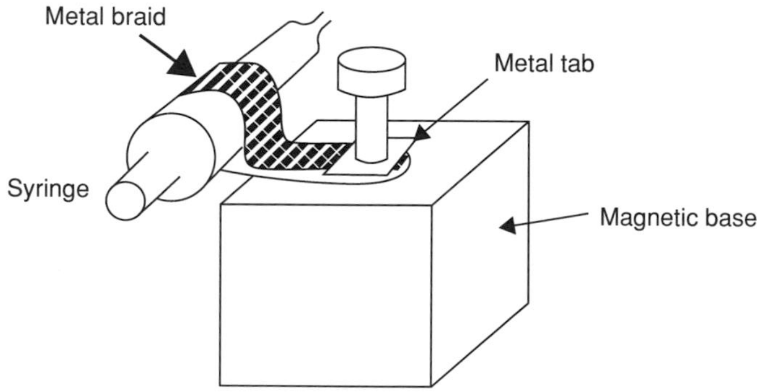

Fig. 8 Assembly for attaching the Gilmont syringe to a magnetic base.

clamps are available in hardware stores. An aluminum clamp made for holding a broom handle to the wall can also be used.

4. Leitz Microinstrument Holder

Kramer Scientific Corp.
5 Westchester Plaza
Elmsford, N.Y. 10523
Phone: 845-267-5050

or

Leica Inc.
24 Link Dr.
Rockleigh, NJ 07647
Phone: 201-767-8304

Cat. # 520145
Set of 3 ($157)

Also, order the following replacement parts for the microinstrument holder:
W832688 Tubing replacement* $10/yard
026-350-012-007 Brass collar replacement (pressure piece) $6 each

*Specify silicon rubber tubing. Leica also supplies Tygon tubing as a replacement, but this does not work well. The silicon rubber tubing may have to be special-ordered from Leica in Germany.

To hold the microinstrument holder in the SM-20 micromanipulator, have a machinist make a metal collar.

5. Syringe Needle Adapter

PGC Scientifics Corp.
7311 Governors Way
Frederick, MD 21704
Phone: 800-424-3300
Website: www.pgcscientifics.com
Cat. # 79-4162-02 ($15)

Have a machinist attach the syringe needle adapter to the microinstrument holder.

You can do this yourself using silver solder (4% silver, 96% tin) and stainless steel soldering flux, available from:

Ed Mar/Freed
706-710 Sansom Street
Philadelphia, PA 19106
Phone: 800-346-7614

Cat. # 54.452 Staybrite low temperature silver bearing solder and flux kit ($7) Kester Solder (Chicago, IL) also makes this type of solder. Ordinary solder for electronics will not work for soldering brass and stainless steel. Before soldering, cut off the back of the microinstrument holder and file it down to fit in the syringe needle adapter. Apply flux, heat the metal pieces with a soldering iron, and apply solder.

6. Teflon Tubing

With a CTFE hub on both ends, 1 to 3 ft. long (2 ft. is best, but 1–3 ft is OK), 20 gauge.

Scientific Commodities, Inc.
P.O. Box 2458
Lake Havasu City, AZ 86405
Phone: 800-331-7724
Website: www.scicominc.com

Cat. # BB635-02, pkg. of 3 = $33; specify on order: 20 gauge, 24″ length (custom order, 2–3 weeks)

or

Hamilton Company
P.O. Box 10030
Reno, Nevada 89520
Phone: 800-648-5950
Website: www.hamiltoncorp.com

Cat. # 86510, $16 each (min. order of 3); specify on order: 20 gauge, 24″ length.

You can also use polyethylene or Teflon tubing, with cut-off syringe needles pushed into the ends, for making the connections to the syringe and microinstrument holder.

7. Steel Plate

24″ × 18″ × 1/4″ is good; larger is OK, too. Steel plates can be obtained from a machine shop; a plate this size costs about $50. It is supported on rubber stoppers to prevent vibration. The plate should sit on a solid table.

8. Plexiglas or Plywood Boards

These are often needed to raise the microscope to the level of the micromanipulator. The height difference is typically ~1/4″ to 1″.

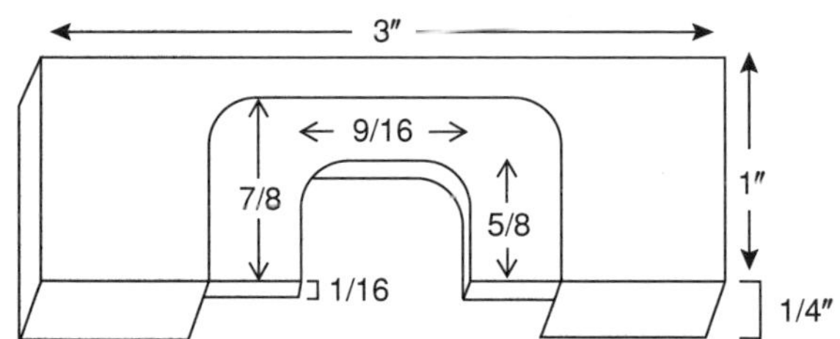

Fig. 9 Microinjection slide dimensions.

9. Microinjection Chamber Support Slide

This is a variation of a support slide described by Kiehart (1982). It can be machined from a 3″ × 1″ × 1/4″ piece of Plexiglas (Fig. 9).

The machinist at the Marine Biological Lab at Woods Hole has made these slides for us. Contact:

Rick Langill
Marine Biological Lab
7 MBL Street
Woods Hole, MA 02543
Phone: 508-289-7237

Cost ~$10–20 each, depending on the number ordered.

See Kishimoto (1986) for an alternative design that also works well. For some microscope condensers, it is better to use a thinner slide. Slides can also be made of stainless steel or aluminum.

10. Black Plexiglas Sheet

(6″ × 12″ × 1/4″) for a work surface; available from a machinist.

11. Diamond Pencil

(Glass marker, diamond-tipped)
Thomas Scientific

12. Pipetmen

(P20 and P10)

13. Fine forceps (2)

Fine Science Tools
373-G Vintage Park Drive
Foster City, CA 94404

Phone: 800-521-2109
Website: www.finescience.com

14. Small Scissors

Fine Science Tools.

15. Flexible Clear Plastic Ruler (6″)

Thomas Scientific.

16. Coverslip Washer

Made from a 50 ml polypropylene centrifuge tube, a 3′ long piece of Tygon tubing (3/8″ O.D., 1/4″ I.D.) and a tubing connector to attach the washer to a water line. See Fig. 10.

17. Glass or Polyethylene Jar

For storing coverslips in 85% EtOH.

18. Mouth Pipet

Mouthpieces for mouth pipets can be ordered from:

MEDTECH International/HPI Hospital
P.O. Box 162992

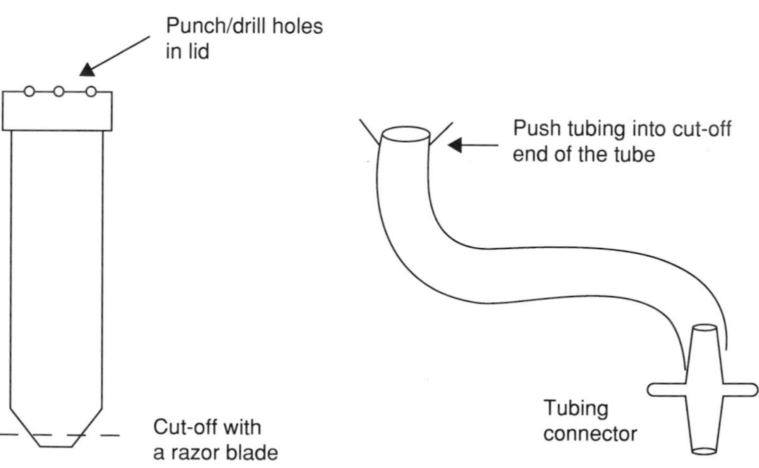

Fig. 10 Coverslip washer.

Altamonte Springs, FL 32716
Phone: 407-880-6904

Cat. # 1506P, white mouthpiece, pkg. of 144, $100
Unfortunately, the minimum number of pieces sold is 144.

The mouth pipet can be assembled by attaching a mouthpiece to a piece of intramedic polyethylene tubing (Clay Adams, PE-60, from Thomas Scientific). Attach the tubing to the mouthpiece, as shown in Fig. 11.

Attach a piece of glass tubing (0.8 mm O.D., 0.6 mm I.D., 100 mm length, see Section III.C.10) to the end of the polyethylene tubing.

Another style of mouth pipet, assembled and ready to use, can be purchased from Sigma.

A5177 Aspirator tube assembly for microcapillary pipets, pkg. of 5, $8.50.

Although we prefer the design we have described, the Sigma mouth pipets also work.

For some uses, it is possible to transfer eggs using a pipetman rather than a mouth pipet. A 20 μl pipetman set on 2 to 3 μl works well, although a mouth pipet allows better control of the number of eggs in the chamber.

19. Hamilton Syringe, 10 μl

Hamilton #80300, Fisher Scientific 14-824, $20.

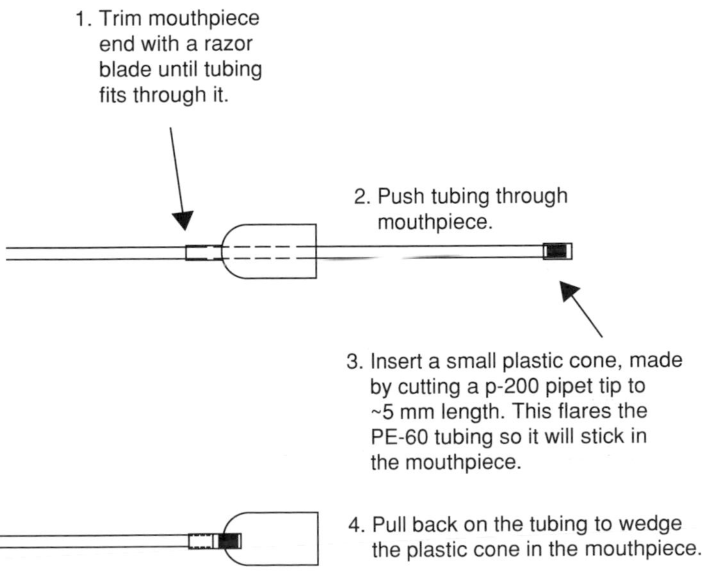

1. Trim mouthpiece end with a razor blade until tubing fits through it.
2. Push tubing through mouthpiece.
3. Insert a small plastic cone, made by cutting a p-200 pipet tip to ~5 mm length. This flares the PE-60 tubing so it will stick in the mouthpiece.
4. Pull back on the tubing to wedge the plastic cone in the mouthpiece.

Fig. 11 Connection of polyethylene tubing to a mouthpiece to make a mouth pipet.

20. Glue Gun

Hardware store (~\$20).

21. Bubble Level

Hardware store (~\$5).

C. Supplies

1. Mercury

Aldrich Chemical Co., #21,545-7, 100 g, \$32. Another good source of clean mercury is to break a glass thermometer.

2. Silicon Oil

(100 centistokes) (From Sigma: dimethylpolysiloxane, DMPS-1C, 100 g, \$20). Don't use mineral oil in the micropipet; it is toxic to eggs.

3. Fluorinert Oil

Type FC-70 (Sigma F9880, 100 ml, \$147) for filling the tubing to the syringe. Silicon oil can be used instead and is 7X cheaper. (But don't mix with Fluorinert!) Silicon oil is more viscous, so it tends to be messier.

4. Silicon High Vacuum Grease

(for assembling the injection slide). Fisher, 14-635-5D, \$22.

5. Alconox Detergent

(for cleaning coverslips). Fisher, 04-322-4, \$14.

6. 85% EtOH

(for storing coverslips).

7. Coverslips

(No. 1/1/2 thickness, 22 mm square).

8. Double-Stick Scotch Tape

Drugstore.

9. Valap

(a 1:1:1 mixture of beeswax, Vaseline, and lanolin, melted together). Beeswax is sold by Fisher. Vaseline and lanolin can be purchased at the drugstore.

10. Glass Tubing

For injection pipets, use Pyrex glass with an outer diameter of 1.0 mm. The glass should have a thin wall (~0.1 mm) and should not contain a filament (commonly present in glass used for electrophysiology). Fifty-microliter "microcaps" from Drummond work very well. These are 1.0 mm in diameter.

Drummond Scientific Company
P.O. Box 700
Broomall, PA 19008
Phone: 800-523-7480
Website: www.drummondsci.com

Cat. # 1-000-0500, $9 per vial of 100

Note that it is best to order directly from Drummond; ordering through Fisher can result in a delay of several weeks.

For mouth pipets and loading capillaries, get some slightly smaller tubing.

Cat # 9-000-1061-100 Custom Pyrex Tubing, 100 pieces per vial, O.D. 0.8 mm, I.D. 0.6 mm, length 100 mm, $13 per vial of 100 (from Drummond), minimum order of 2 vials.

11. Sticky Foam Weatherstripping Tape

Can be purchased at the hardware store. Mount 2 strips of the weatherstripping tape with the sticky side up, with double-stick tape to attach it to a 150 mm petri dish surface. The 2 strips should be parallel to each other and ~1 cm apart. Lay pulled pipets across the 2 strips for storage; ~40 pipets can be stored in one dish. For holding the pipets securely during transport, tape a piece of foam rubber inside the lid of the petri dish and tape the lid over the dish of pipets. You can also purchase a micropipet storage jar:

Cat. # E210 micropipet storage jar, micropipet O.D. 1.0 mm, $35 from:

World Precision Instruments
175 Sarasota Center Blvd.
Sarasota, Florida 34240
Phone: 941-371-1003
Website: www.wpiinc.com

Acknowledgments

We thank the many colleagues who passed on to us their expertise in microinjection, and who contributed to optimizing these methods. Much of the "bioengineering" described here is a legacy of Ray Kado; this chapter is dedicated to the memory of Ray.

References

Begg, D. A., and Ellis, G. W. (1979). Micromanipulation studies of chromosome movement. I. Chromosome-spindle attachment and the mechanical properties of chromosomal spindle fibers. *J. Cell Biol.* **82,** 528–541.

de Fonbrune, P. (1949). Technique de Micromanipulation. Masson, Paris, France.

Hiramoto, Y. (1962). Microinjection of the live spermatozoa into sea urchin eggs. *Exp. Cell Res.* **27,** 416–426.

Kiehart, D. P. (1982). Microinjection of echinoderm eggs: Apparatus and procedures. *Meth. Cell Biol.* **25,** 13–31.

Kishimoto, T. (1986). Microinjection and cytoplasmic transfer in starfish oocytes. *Meth. Cell Biol.* **27,** 379–394.

Lutz, D. A., and Inoué, S. (1986). Techniques for observing living gametes and embryos. *Meth. Cell Biol.* **27,** 89–110.

Mehlmann, L. M., and Kline, D. (1994). Regulation of intracellular calcium in the mouse egg: Calcium release in response to sperm or inositol trisphosphate is enhanced after meiotic maturation. *Biol. Reprod.* **51,** 1088–1098.

Mehlmann, L. M., Jones, T. L. Z., and Jaffe, L. A. (2002). Meiotic arrest in the mouse follicle maintained by a G_s protein in the oocyte. *Science* **297,** 1343–1345.

Runft, L. L., and Jaffe, L. A. (2000). Sperm extract injection into ascidian eggs signals Ca^{2+} release by the same pathway as fertilization. *Development* **127,** 3227–3236.

Runft, L. L., Watras, J., and Jaffe, L. A. (1999). Calcium release at fertilization of *Xenopus* eggs requires type I IP_3 receptors, but not SH2 domain-mediated activation of PLCγ or G_q-mediated activation of PLCβ. *Dev. Biol.* **214,** 399–411.

Sluder, G., Miller, F. J., and Hinchcliffe, E. H. (1999). Using sea urchin gametes for the study of mitosis. *Meth. Cell Biol.* **61,** 439–472.

Terasaki, M., Schmidek, A., Galbraith, J. A., Gallant, P. E., and Reese, T. S. (1995). Transport of cytoskeletal elements in the squid giant axon. *Proc. Natl. Acad. Sci. USA* **92,** 11500–11503.

CHAPTER 11

Blastomere Isolation and Transplantation

Hyla Sweet,[*] Shonan Amemiya,[†] Andrew Ransick,[‡] Takuya Minokawa,[‡] David R. McClay,[§] Athula Wikramanayake,[||] Ritsu Kuraishi,[¶] Masato Kiyomoto,[#] Hiroki Nishida,[**] and Jonathan Henry[††]

[*]Department of Biological Sciences
College of Science
Rochester Institute of Technology
Rochester, New York 14623

[†]Department of Integrated Biosciences
Graduate School of Frontier Sciences
University of Tokyo
Kashiwa, Chiba 277-8562, Japan

[‡]Division of Biology
Calfornia Institute of Technology
Pasadena, California 91125

[§]Department of Biology
Duke University
Durham, North Carolina 27708

[||]Department of Zoology
University of Hawaii at Manoa
Honolulu, Hawaii 96822

[¶]Marine Biological Station
Graduate School of Science
Tohoku University
Aomori 039-3501, Japan

[#]Tateyama Marine Laboratory
Ochanomizu University
Tateyama, Chiba 294-0301, Japan

[**]Department of Biological Sciences
Tokyo Institute of Technology
Nagatsuta, Midori-ku
Yokohama 226-8501, Japan

[††]Department of Cell and Structural Biology
University of Illinois
Urbana, Illinois 61801

METHODS IN CELL BIOLOGY, VOL. 74

0091-679X/04 $35.00

I. Introduction

Blastomere isolation and transplantation experiments in sea urchin embryos had been performed in the period from 1920s to 1960s, mainly by Hörstadius who produced extensive experimental results using his skillful techniques (Hörstadius, 1973). Operation of sea urchin embryos can be difficult because of small egg size, the diameter being less than 0.1 mm. Thus, the results obtained by Hörstadius were rarely reexamined by other researchers. These types of experiments have been tried again in the 1990s, mainly because of the requirement of combining techniques of molecular biology and experimental embryology. Many of Hörstadius' experiments that had been considered to be almost impossible have been successfully performed in various laboratories with the improved optical instruments and micromanipulators now available.

Although blastomere isolation and transplantation experiments have been mainly performed on the sea urchin embryo among nonvertebrate deuterostomes, a considerable amount of experimental data have also been accumulated for starfish and ascidians. Relatively few blastomere isolation experiments have been reported from amphioxus and hemichordates.

In the beginning of this chapter, methods for setting up and preparation of instruments and tools needed for embryo operation are described. The following sections provide the methods for isolation and recombination of blastomeres in sea urchin embryos. In the latter sections, the methods for surgical operation on starfish, amphioxus, ascidians, and hemichordates are described.

II. Preparation of Mouth Pipettes and Needles

A. General Explanation

Mouth pipettes, needles, and other tools are used for handling embryos during blastomere isolation and transplantation. Mouth pipettes are used to gently pick up and move embryos, for example, from the operating dish to the recovery dish. Hand-pulled or machine-pulled glass needles are used as knives to slice through an embryo. They can also be used to manipulate transplanted cells into their final position in the host embryo.

B. Preparation Methods

1. Preparation of Mouth Pipettes

The tips of a long Pasteur pipette are held in each hand and the thin end is held over the flame of an alcohol lamp, Bunsen burner, or micro-burner (18-gauge disposable needle inserted into flexible tubing connected to a gas source and clamped in a ringstand). As the glass begins to melt, the ends of the pipette are pulled apart while taking the pipette out of the flame. The pipette can be pulled straight or at an angle. Any excess glass from the pulled pipette can be clipped with fingernails to adjust the length and diameter of the tip according to the application and the size of the eggs or embryos. The pulled pipette is connected to tubing and a mouthpiece can be connected (HPI Hospital Products, Alamonte Springs, FL, Catalog # 1501PK, for use with their rubber latex tubing, Catalog # 1503PK). Another option is to add a "brake" for fine control. A brake consists of a constricted region within the tubing, such as a fine gauge needle or capillary (Fig. 1A). As an alternative to Pasteur pipettes, glass capillaries can be pulled over a flame or in a micropipette puller, and attached to thin tubing (Fig. 1B).

To use the mouth pipette, the open end (or the end with the mouthpiece) is placed in the mouth and secured with the teeth. The pulled pipette is held with the writing hand as if it were a pen and the hand is stabilized on the microscope stage. The dish of embryos is held with the other hand. While looking through the microscope, the opening of the pipette is positioned next to embryos. One may gently suck or blow for coarse control, or inhale or exhale around the opening for fine control.

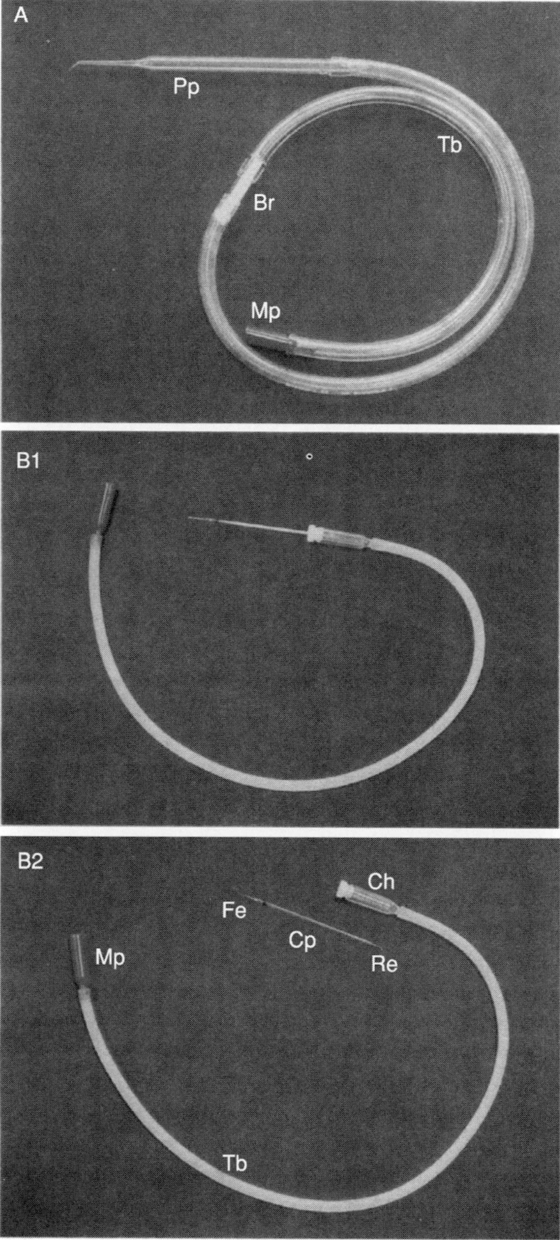

Fig. 1 Mouth pipettes. (A) A mouth pipette composed of a pulled Pasteur pipette, a braking tube, a mouthpiece, and tubing made of rubber or plastic. (B) A mouth pipette composed of a pulled glass capillary, a capillary holder, a mouthpiece, and tubing. A completed pipette (B1) and a pipette in which the glass capillary is detached from the capillary holder (B2) are shown separately. The rear end of the glass capillary is pulled to make the open end thinner. The thin rear end works as a brake for fine control. Pp, Pasteur pipette; Br, braking tube; Mp, mouthpiece; Tb, tubing; Ch, capillary holder; Cp, capillary; Fe, front end of the capillary; Re, rear end of the capillary. (See Color Insert.)

2. Preparation of Needles

Glass rods, glass capillaries, or long Pasteur pipettes can be used to make dissecting needles. The tips of two glass rods with the diameter of 5 to 7 mm are heated over a flame to make a slender tip with a diameter of about 1 mm. The slender tips of the two rods are then touched and pulled apart to make a sharp needle (Fig. 2). Similarly, the tips of the glass capillaries or the thin ends of long Pasteur pipettes are touched over a flame and then pulled. New needles can be dipped into molten agar to reduce stickiness. The smaller glass needles can be mounted into a holder such as a Starret pin vise (L.S Starrett Co., Anthol Mass.) or Narishige needle holder (H-1).

An alternative to hand-pulled glass needles is to use a horizontal micropipette puller (Flaming/Brown micropipette puller from Sutter Instrument Company or Narishige PN-3) or a vertical puller (Narishige PP-83), following the manufacturer's instructions. Machine-pulled glass needles are made by pulling glass rods (Narishige P-1000 or equivalent) of about 1 mm in diameter and 9 cm in length. The horizontal pullers may be better than the vertical puller for making relatively long needles. These needles can also be used with a needle holder for microsurgery by hand (Fig. 2), or they can be connected to a micromanipulator.

C. Cautions

Hand-pulled glass needles can be difficult to make and a great deal of practice is needed before the embryologist has enough experience to judge a good needle. Good needles are critical for success. The needle should have a fine tip of about 2 to 6 mm to slice between cells without damaging the rest of the embryo. The tip has to be long enough so that the thick part of the needle does not obscure the

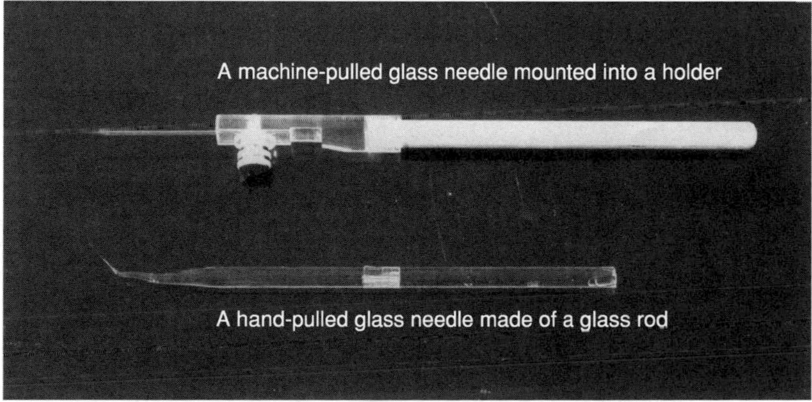

Fig. 2 A machine-pulled glass needle mounted into a holder and a hand-pulled glass needle made of a glass rod.

view of the embryo through the microscope; but if it is too long and thin, the user will have poor control over the needle under seawater. Hand-pulled glass needles vary in shape and they must be carefully selected. An advantage to hand-pulled glass needles is that they can be angled, which may make it easier to cut the embryos. An advantage to machine-pulled needles is that they are highly reproducible. They are suitable for blastomere recombination, but work less well for blastomere isolations.

III. Removal of Fertilization Envelope of Sea Urchin Eggs

A. General Explanation

Removal of the fertilization envelope is usually necessary for blastomere isolation and transplantation. In some species of sea urchin (*Lytechinus variegatus* and others), the fertilization envelope hardens within a few minutes after fertilization and it can be removed mechanically within 10 to 15 min after fertilization. In other species (*Hemicentrotus pulcherimus* and others), the eggs should be fertilized in the presence of a compound that prevents the hardening of the fertilization envelope. The envelope can then be removed mechanically at a later time. Eggs and embryos without their fertilization envelopes can be very sticky. They should be placed in a dish that is coated with 1 to 2% agar or agarose, or horse serum.

B. Procedure to Remove the Fertilization Envelope

Eggs are fertilized in a fresh solution of 0.5 to 1.0 mM 3-amino-1,2,4-trizole (ATA) in filtered seawater (FSW) or 1 to 10 mM *p*-aminobenzoic acid (PABA) in FSW. Some species, such as *Anthocidaris crassisipina*, are resistant to these reagents. After about 10 to 15 min, the eggs are transferred to fresh filtered seawater to remove the ATA or PABA and excess sperm. The softened fertilization envelopes can be removed mechanically at the desired time by shaking the embryos or by passing them through a pulled pipette or fine mesh.

In the first method, 5 ml of egg suspension is added to a 10 ml test tube. The opening of the test tube is covered with a thumb and the test tube is vigorously shaken horizontally one or two times along the longitudinal axis of the tube with an amplitude of 30 to 40 cm. Additional shaking will damage the embryos. The fertilization envelope is stripped from the eggs as they rub against each other. Therefore, the removal of the fertilization envelope is inefficient if there are not enough eggs. In the second method, embryos are drawn into a hand-pulled Pasteur pipette or a pipette that has been flamed to narrow the tip diameter. In both cases, the inner diameter of the pipette is smaller than the fertilization envelope. The eggs are drawn in while watching through a dissecting microscope to ensure debris does not clog the pipette tip. In the third method, embryos are passed through Nitex nylon mesh (Sefar America, Inc.) that has a pore diameter slightly smaller than the fertilization envelope.

C. Cautions

If used improperly, each of these methods may result in mechanical damage to the embryos. Pipetting too many times, repeated passing of embryos through mesh, and overshaking can all damage the embryos. Removing the fertilization envelope using Nitex mesh is better for large quantities of embryos. All methods are suitable for smaller quantities of embryos. Early removal of the fertilization envelope occasionally results in abnormal cleavage patterns. Thus, removal immediately before microsurgery may be a more effective method than removal right after fertilization.

IV. Isolation and Recombination of Blastomeres in Sea Urchin Embryos

A. General Explanation

Surgical isolation and recombination of blastomeres by hand (Section IV.B.1) includes techniques that were originally used by Hörstadius on sea urchin embryos (Hörstadius, 1973). These techniques are very powerful and some embryologists continue to use them with only minor modifications. A disadvantage to these techniques is that they require much patience and practice. Other techniques for blastomeres isolation and transplantation have been developed to overcome some of these difficulties. These techniques include similar surgical experiments using micromanipulators to control the dissection needles and bulk methods for isolation and recombination of blastomeres from a population of embryos.

B. Isolation and Recombination

1. Surgical Isolation and Recombination of Blastomeres by Hand

The following techniques were used with embryos of the sea urchin species *Lytechinus variegatus* (Sweet *et al.*, 1999, 2002), *Hemicentrotus pulcherimus* and *Scaphechinus mirabilis* (Amemiya, 1996), and *Strongylocentrotus purpuratus* (Ransick and Davidson, 1993, 1995).

a. Preparations

Materials required for microsurgery by hand include operating dishes, recovery dishes, dissecting needles, mouth pipette, filtered seawater (FSW), hyaline extraction medium (HEM; Fink and McClay, 1985; see recipe), calcium-free seawater (CFSW; see recipe), and filtered seawater with antibiotics (100 μg streptomycin and 100U penicillin per ml FSW; SPFSW). A staining solution such as rhodamine isothiocyanate (RITC) should be prepared for labeling donor cells for blastomeres transplantations. Operating dishes should be coated with agar, agarose, or horse serum. For the former two, 1 to 2% agar or agarose in distilled water is melted in a microwave. About 1 mm of the bottom of a 35 mm or 60 mm plastic petri dish or glass dish is coated with the molten solution. The agar in the operating dishes should be equilibrated with HEM or CFSW by soaking the dishes in fresh solution three times, 10 min each. A backstop may be made in the operating dish to stabilize the embryos during microsurgery. The tip of the mouth pipette is lightly dragged across the center of the agar-coated dish to form a shallow groove. To prepare dishes for recovery, agar-coated dishes like those already described are equilibrated with FSW.

Fertilizations can be staggered every 30 min to maximize the time for microsurgery. The fertilization envelope is removed and the embryos are raised in FSW in agar-coated dishes. A few embryos are transferred using a mouth pipette to an operating dish filled with HEM or CFSW and the operating dish is placed under a dissecting microscope. It is best to transfer the embryos into the operating dish immediately prior to blastomeres isolation, because the embryos occasionally divide abnormally in CFSW or HEM.

The dissecting needle is held as if it were a pen in the writing hand and this hand is stabilized on the microscope stage. The operating dish is stabilized with the other hand. The side of the needle is used to position an embryo against the backstop in the operating dish. The side of the glass needle is then used as a knife to slice the embryo along cell boundaries. Slicing the embryo is easiest right after and right before cell divisions. Blastomere isolation becomes easier when the embryos are left several minutes in CFSW. However, embryos left too long in CFSW are apt to dissociate into single blastomeres. Occasionally, embryos are resistant to dissection because of a tough hyaline layer. The tough hyaline layer can be dissected by vibrating or sawing the needle along the boundary between the blastomeres. Dissection of embryos using the distal part of the needle causes less damage to the blastomeres, although it is more time consuming.

b. Animal Cap Isolation

For animal cap isolation at the 8-cell stage, it is important to know the orientation of the animal–vegetal axis before proceeding. In the case of *L. variegatus*, the four animal cells are slightly larger than the four vegetal cells. The embryos are positioned on their side and a cut is made to separate the two halves. To verify that the animal half has been isolated, the halves can be examined after the following cleavage to ensure that the four cells divide into eight cells of equal size. To isolate animal caps at the 16-cell stage, the micromeres serve as a marker for the vegetal pole. The embryos are placed on their sides and the glass needle is pressed along the equatorial plane between the mesomeres and macromeres.

c. Micromere Removal and Isolation

To remove the micromeres, a 16-cell stage embryo is left in CFSW for several minutes to loosen the adhesion of blastomeres. When a quartet of micromeres in the embryo becomes rounded and projects from the vegetal pole, the micromeres are removed from the embryos by pressing a needle along the boundary between the micromeres and macromeres.

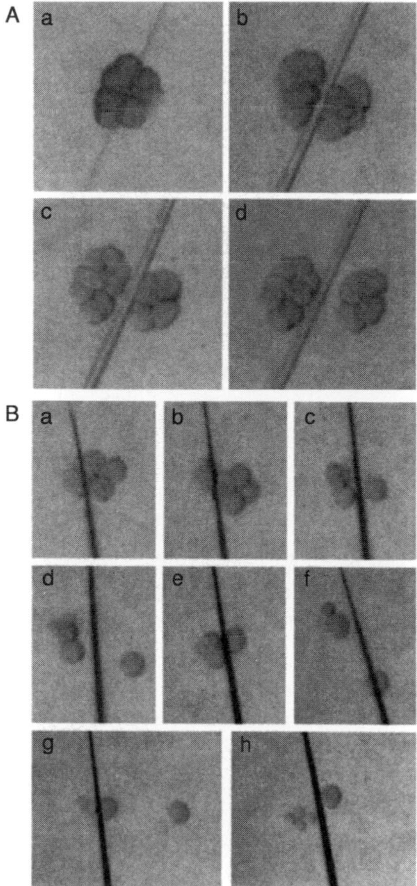

Fig. 3 Sequence photographs on the processes for isolation of an animal cap and a vegetal half (A) and of a quartet of micromeres (B) from a 16-cell stage embryo (Photos by Dr. S. Amemiya). (See Color Insert.)

The micromeres can be isolated, as has been described. An alternative tactic is to dissect a 16-cell stage embryo along the equatorial plane (Fig. 3A) and then remove the macromeres one by one (Fig. 3B). The large micromeres can be isolated from the small micromeres after the subsequent cleavage by removing all small micromeres one by one from an aggregate of a quartet of large and small micromeres that is isolated from a 32-cell stage embryo.

d. Micromere Transplantation to the Animal Pole and Animal Cap

The following technique for micromere transplantation to the animal pole is derived from Ransick and Davidson (1993) using the embryos of *S. purpuratus*. As this is a coldwater species, all solutions must be chilled and the microsurgeries are performed in an operating dish within an icewater bath. Donor embryos are labeled at an earlier stage by immersion in rhodamine B isothiocyanate (RITC; after Ettensohn, 1990). Donor embryos at the 16-cell stage are incubated in HEM until the blastomeres are rounded (2–5 min). The exposure to HEM should be limited to 10 min. The embryos are dissected to obtain an intact set of four micromeres. These isolated micromeres are carefully transferred by mouth pipette to a dish containing CFSW. The animal pole side of an unlabeled 8-cell (or later) stage host embryo, whose cell contacts had been loosened by incubation in CFSW, is then maneuvered into the proximity of the micromeres and pressed against the micromere tier by gentle pressure with the shank of the glass needle. Perfusion of these constructs with normal seawater promotes adhesion between cells and, within minutes, the chimeric embryos can be transferred to FSW for long-term culture.

The following technique for micromere transplantation to animal caps is derived from Amemiya (1996) using the embryos of *H. pulcherimus* and *S. mirabilis*. At the 8-cell stage, embryos with an equatorial third cleavage plane are selected and transferred to FSW in a petri dish coated with agar. At the 16-cell stage, the embryos are transferred to CFSW in a petri dish coated with horse serum. The animal cap is dissected (Fig. 3A). Quartets of micromeres are isolated from RITC-labeled 16-cell stage embryos (Fig. 3B). The animal cap and micromere quartet are transferred to SPFSW in a small petri dish coated with agar and are recombined (Fig. 4). A glass needle is used to position the animal cap toward a quartet of micromeres and to press the blastocoelic side of an animal cap on top of the quartet of micromeres.

e. *Postsurgery Procedures*

As soon as possible after microsurgery, the experimental embryos are very gently transferred from the operating dish to a recovery dish with filtered seawater supplemented with antibiotics (SPFSW). The transfer should be watched through the microscope to ensure that the embryo is not lost or does not fall apart. Long exposure to HEM or CFSW (greater than 10 min) and rough treatment with the needle or mouth pipette will increase the probability that the embryo will fall apart, especially around the time of cell division. Subsequent cleavages should be examined to ensure that the experimental embryos continue to divide properly and that transplanted cells adhere to the host. Any embryos that are visibly disorganized or damaged by the manipulations should be discarded. Embryos should be enclosed within a chamber to protect the swimming stages from being destroyed by the surface tension at the interface between the air and seawater.

In one method, agar tunnels are made by pouring molten 2% Noble agar in distilled water over a glass filament (1–2 cm long, 100–200 μm diameter) positioned in a plastic petri dish and pouring off the excess ($\times 2$); after the agar has gelled into a thin layer, the glass is extracted from the side with fine forceps to create a hollow chamber with a thin and transparent covering. Such agar-coated dishes require a 10-min equilibration in FSW. Once inserted into such a tunnel, the

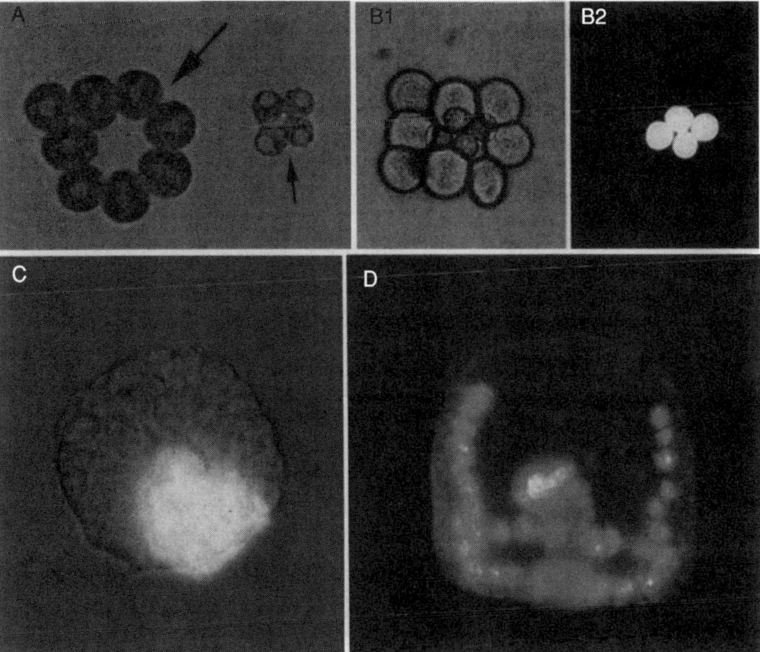

Fig. 4 Development of a chimeric embryo derived from an animal cap recombined with a quartet of rhodaminated micromeres isolated from 16-cell stage embryos. (A) An animal cap (large arrow) and a quartet of micromeres (small arrow) just after isolation. (B) An animal cap recombined with a quartet of rhodaminated micromeres just after recombination viewed under a light microscope (B1) and an epifluorescence microscope (B2). (C) A morula stage embryo derived from an animal cap recombined with a quartet of rhodaminated micromeres. (D) A middle gastrula derived from an animal cap recombined with a quartet of rhodaminated micromeres (Photos by Dr. S. Amemiya). (See Color Insert.)

embryos remain corralled, but there is ample room for each to swim and grow into pluteus larvae. These preparations are ideal for using inverted compound microscopes to make quick and detailed observations on cultures containing from a dozen up to hundreds of embryos.

In a second method, individual wells from a flexible 96-well plate (Falcon Flexible Plate, 96 well, U bottom without lid, Becton Dickinson) are inverted over the experimental embryos. Individual wells are cut from the plate and a thin layer of vacuum grease or Vaseline is applied to the edges of the well. The experimental embryos are clustered in the center of the recovery dish; the well is filled with FSW, gently inverted using forceps over the embryos, and pressed into place. Slow movements are essential; otherwise, water currents will cause the embryos to move away from their position. The cover of the petri dish should be thoroughly secured with a strip of Parafilm. The recovery dish is placed in a water bath at the appropriate temperature.

In a third method, embryos can be raised in a glass bottle that is fully filled with SPFSW. The bottle should be capped in a manner such that no air bubbles are present.

f. Examining Embryos

For transplantations, the resulting larvae are examined with a microscope equipped with epifluorescence to visualize the RITC-labeled descendants of the transplanted cells (Fig. 4). A whole-mount microscope slide is made using clay or Vaseline to prop up the coverslip to prevent squashing of the larvae. To immobilize the larvae, gentle pressure is applied to lower the coverslip. Alternatively, the larvae may be immobilized by pretreating the microscope slide or coverslip with 0.1% poly-L-lysine or 1% protamine sulfate for 1 min, followed by several washes in distilled water, and dried. After examination, the coverslip can be removed with fine forceps while watching under a dissecting microscope. The living embryos can be quickly rescued with a mouth pipette and used for further analysis.

2. Isolation of Cells Derived from Recombined Blastomeres

The method to produce the partial embryos derived from mesomeres that had been under the inductive influence of micromeres for a limited period has been reported by Minokawa and Amemiya (1999). In this section, a brief explanation of the practical method of the manipulation is described, as well as several tips for using this procedure.

a. Microscopes and Micromanipulators

Required equipment includes an epifluorescence microscope (Nikon Optiphot or equivalent), a set of fine and coarse micromanipulators, a microinjector, and a stereomicroscope for handling of the embryos. The stereomicroscope equipped with the fluorescence unit (Leica MZ FLIII or equivalent) can be a substitution for the epifluorescence microscope. As a fine micromanipulator, a three-axis joystick micromanipulator (Narishige MO-102 or equivalent) is strongly recommended because of its ability to provide both three-dimensional and smooth movements. The fine micromanipulator is mounted on the coarse micromanipulator (Narishige MN-4 or equivalent). The micromanipulators must be attached to the microscope such that the micropipette approaches the embryo perpendicular to the optical axis of the microscope (that is, parallel to the stage surface) (Fig. 5). Using this setup, the embryos and the tip of the micropipette will be in the same focal plane (Fig. 6).

b. Micropipettes

Glass micropipettes are pulled from a glass capillary (outer diameter 1 mm, with microfilament; Narishige GDC-1 or equivalent) using a micropipette puller (Narishige PC-10 or equivalent). The micropipette is back-filled with seawater, and a small amount of silicon oil is put in the end of the pipette and attached to a

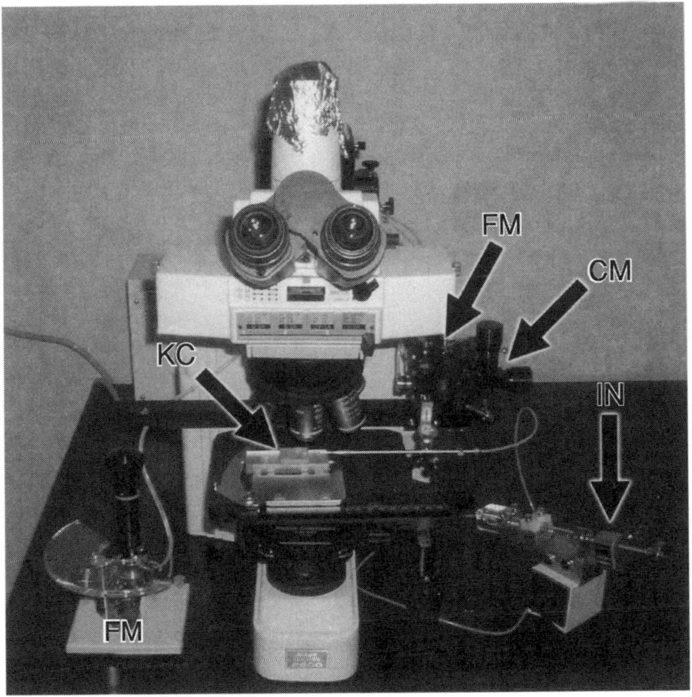

Fig. 5 The micromanipulation setup. This system includes an epifluorescence microscope, a set of fine and coarse micromanipulators (FM and CM, respectively), and a microinjector (IN). KC, the Kiehart micromanipulation chamber (Photo by Drs. Minoru Ijima and T. Minokawa).

microinjector (Narishige IM-5A or equivalent) filled with deionized water through a micropipette holder (Narishige HI-7 or equivalent).

c. The Handling of the Recombinant Embryos

The recombinant embryos are assembled by the method of Amemiya (1996), and consist of RITC-labeled cells and unlabeled cells. For species with tough extracellular matrix, a brief pretreatment with CFSW, or calcium-deficient seawater (seawater: CFSW=1:19) is sometimes helpful to weaken the adhesion between cells. The pretreatment is not necessary for the sand dollar *Scaphechinus mirabilis*.

Addition to seawater of trace amounts of protamine sulfate (<0.001%) helps with the handling of the embryos after hatching blastula stage. This treatment causes the cilia to stick together and prevents the swimming of the embryos. The embryos recover the swimming ability after extensive washes.

Single recombinant embryos are inserted into the narrow space of a Kiehart micromanipulation chamber (Kiehart, 1982). Because the Kiehart chamber keeps the embryo in the limited space between two coverslips, both the embryo and the tip of the micropipette can be observed in the same focal plane. The Kiehart

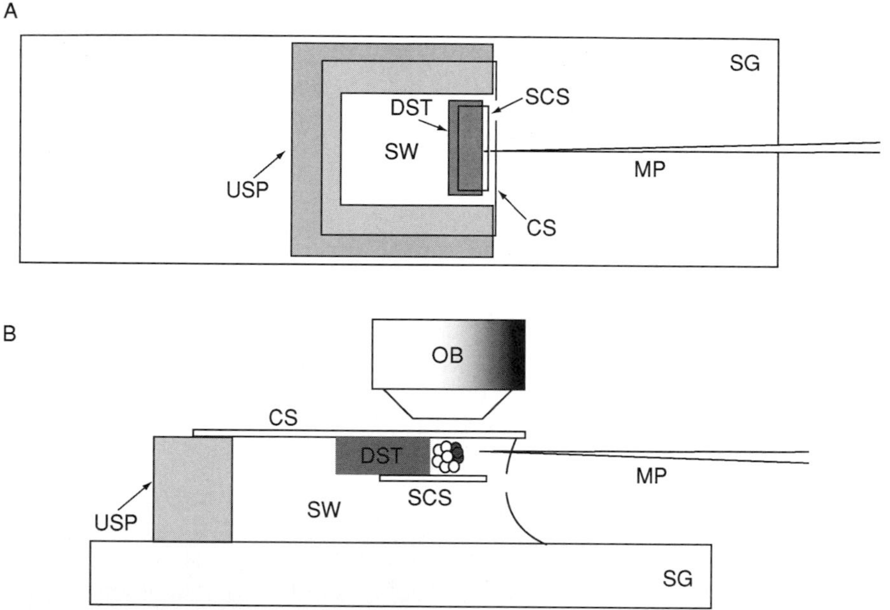

Fig. 6 The Kiehart micromanipulation chamber (see Kiehart (1982) for the original description). (A) Top view. (B) Side view. CS, cover slip; DST, double-stick tape; MP, micropipette; OB, objective; SCS, small fragment of cover slip; SG, slide glass; SW, seawater; USP, U-shaped silicone rubber plate (Illustration produced by Dr. T. Minokawa).

chamber and a mouth pipette should be coated with BSA solution or serum to prevent sticking of the embryo. A drop of BSA or serum is put on the space of the chamber; after several minutes, it is withdrawn, and the chamber is washed several times with seawater. This coating of the chamber is necessary if protamine sulfate is used to prevent the swimming of the embryos.

d. Brief Description of the Micromanipulation

All of the microsurgical processes described in this section must be performed under the combined illumination of fluorescent and low-intensity bright field on an epifluorescence microscope. This mixed illumination enables visualization of both the shape of the embryo and the position of labeled cells.

A continuous stream of seawater is blown on the boundary between the RITC-labeled and unlabeled cells in a recombinant embryo to loosen adhesion between the cells. After adhesion between the cells is significantly reduced, the labeled cells are blown away by the stream. Remaining labeled cells can be trimmed or sucked off with the micropipette tip. The embryo is then transferred to SPFSW for long-term culture.

A possible alternative of this protocol is the photoablation method reported by Ettensohn (1990). In this method, RITC-labeled donor cells are selectively killed by irradiating the embryos with excitation light for the RITC dye.

3. Isolation and Recombination of Blastomeres by Micromanipulator

a. Transplantations with the Use of Micromanipulators

Hörstadius accomplished a remarkable series of cell transplantations and embryo manipulations using a compound microscope and handheld glass needles. Those same manipulations and many more are now possible using a dissecting microscope and two micromanipulators. This section describes the apparatus and the kinds of experiments that can be performed with these technical improvements.

b. Microscope Apparatus and Dissection Chamber

Figure 7 shows a diagram of the dissection chamber and the approach for micromanipulation. Two joystick micromanipulators (Narishige) are mounted on the column of a Leica MZIII fluorescent dissecting microscope or equivalent. The micromanipulators are mounted with brackets so the glass needles for dissection (A and B on Fig. 7) enter the chamber at right angles to one another. This involves special mounts for the coarse and fine micromanipulator controls, which are available from a good microscope representative. The microscope is equipped with a 1.6 plan apo objective. A floating stage enables the operator to gently move the stage without shaking the embryos in the chamber. The fine focus mechanism allows an excellent 3D perspective of the embryo, which helps for fine dissections. A continuous zoom lens is valuable for both the close-in look as well as examination of the field to locate cells or embryos during a transplant procedure. Fluorescence capability enables the operator to unambiguously distinguish between fluorescent and nonfluorescent cells in transplantation operations.

The chamber holding sea urchin embryos is a modified Kiehart chamber (Fig. 7). The chamber is roughly the dimension of a 2.5 cm × 6 cm microscope slide but made out of aluminum. The modified chamber contains a U-shaped cutout in the center of the slide, and a shelf surrounding the U milled to a thickness of 1 mm. This shelf is used to separate two coverslips, one mounted with Vaseline on the top of the Kiehart chamber and the other mounted on the shelf beneath the U. Half of the slide is cut to allow the second needle to approach at a 90° angle to the first.

Glass needles are pulled using custom glass capillary tubes purchased from Drummond Scientific Co. (OD. 0.85 mm; ID. 0.55 mm; Catalog item 9-000-1000). One needle is mounted without breaking (needle A in Fig. 7) and is used as a manipulating and holding pipette, or as a knife. The second needle is broken and used as a suction needle. The suction needle initially is broken on the stage of a dissection microscope using a razor blade. It is then further broken, if necessary, by smashing it end-on into the edge of the coverslip of the Kiehart chamber. With practice, one can obtain a needle with a blunt tip and an inner diameter (ID) just

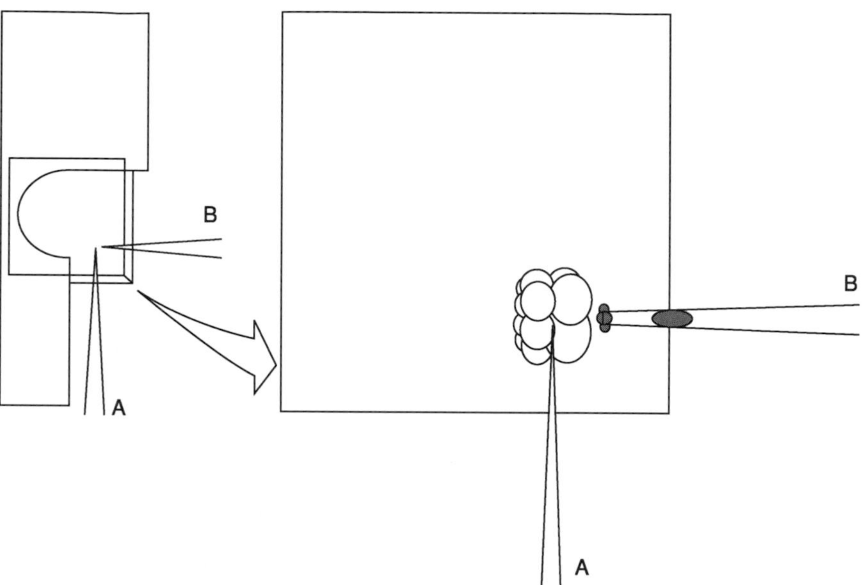

Fig. 7 Modified Kiehart chamber. An aluminum slide of the shape illustrated is milled so two coverslips are separated by 1 mm. The coverslips form the dissection chamber by sealing the coverslips to the slide with Vaseline. Two micromanipulators are mounted to the dissection microscope to hold glass needles that approach the Kiehart chamber as illustrated. Needle A is unbroken and serves as either a knife or a "thumb" to maneuver the embryos. Needle B is broken with a blunt opening of a diameter and this needle is attached to a mouth pipette at the other end. When the suction needle is introduced into the chamber, a small amount of seawater backfills the needle by capillary action. Once the flow stops, one has very sensitive control over the suction, using this needle as a micro-mouth pipette. To the right of the Kiehart chamber is a blowup of the corner of the chamber where the surgeries are conducted. Shown is an embryo at an approximate magnification as seen during the surgery (Illustration produced by Dr. D. McClay). (See Color Insert.)

smaller than the cells to be suctioned. The suction pipette is attached to Intramedic tubing with an ID of 0.80 mm, flared slightly to accommodate the outer diameter (OD) of the glass needle, and at the other end of the tubing, a pipette tip normally used for a P10 micropipettor to serve as a mouth pipette suction device.

The Kiehart chamber is mounted on the stage of the dissection microscope and held in place with Paraplast attached to the stage in the shape of an L to provide a defined mounting location on the stage. The glass needles are mounted on the micromanipulators to approach the chamber at 90° to each other and at only a slight angle out of parallel with the stage. This allows both needles to enter between the coverslips and into the chamber up to several mm distance without smashing into either the top or the bottom coverslip. The chamber holds modified seawater adjusted for optimal results for each species. The seawater solution should be changed every hour or so due to evaporative increase in salinity.

c. Cell Transplantations and Embryo Operations

To perform transplantations, a second dissection microscope is necessary for loading and unloading the Kiehart chamber. The chamber is filled with seawater modified for optimal surgical transfer of cells, and the embryos to be manipulated are transferred into the chamber with a mouth pipette. The chamber is then placed onto the transplantation microscope and the surgery performed. If the surgery is to be on post-hatched embryos, coverslips containing protamine sulfate are used (see Section IV.B.1.*f*).

Several kinds of manipulations are easily performed with the Kiehart chamber and micromanipulator apparatus (McClay *et al.*, 2000). One can use the holding pipette as a knife to slice through the embryo at any location. In this operation, the suction pipette is used as a holding pipette. More frequently, the suction pipette is used to remove single cells into the tip of the pipette and transfer them to host embryos (as illustrated in the blowup of Fig. 7).

Figure 8 shows a series of transplantations that allow one to perform Hörstadius transplantations and a mosaic analysis. At the stage desired for surgery, labeled donor embryos are placed into the Kiehart chamber along with unlabeled host embryos of the same stage. The holding pipette is used to rotate the embryo until one can see the tiers of blastomeres aligned. One or two cells of the host are removed from the tier in question, and then one or two donor cells are suctioned from the same tier and transplanted into the vacant site on the host embryo. The suction pipette should have an ID slightly smaller than the cell to be transplanted so the donor cell is shaped like a sausage while in the pipette. This allows the donor cells to slip into the host site easily. Care must be taken to place the cells in the plane of the monolayer. If the cells are attached outside the hyaline layer, they will not participate in morphogenesis of the host, and if the cells are dropped through the basal lamina, they float in the blastocoel and fail to participate in normal morphogenetic events. After surgery, the transplant combinations are removed from the chamber and placed into FSW for culture. Many operations of the sorts illustrated in Fig. 8 are possible with the manipulator apparatus. Tiers can be removed or replaced. Embryo halves can be recombined. Single tiers can be substituted. Or, later in development, the suction pipette can remove whole sections of ectoderm or endoderm with that hole then filled with a plug of donor tissue. The advantage to the apparatus is the precision gained over manual manipulation and an increase in the variety of experiments that can be performed.

The mosaic analysis illustrated in Figure 8 (I–L) allows one to determine whether a molecular perturbation modifies the donor cells in each tier. In addition to a lineage tracer, eggs can be injected with RNA or morpholino to perturb specific gene products. If there is a general perturbation, the donor cells will develop abnormally no matter which tier is examined. With examination of cells carrying the lineage tracer at the end of the analysis, one can determine whether there is an autonomous or a nonautonomous response due to the manipulated, donor cells of each tier. In this way, the analysis is identical to mosaic analyses in

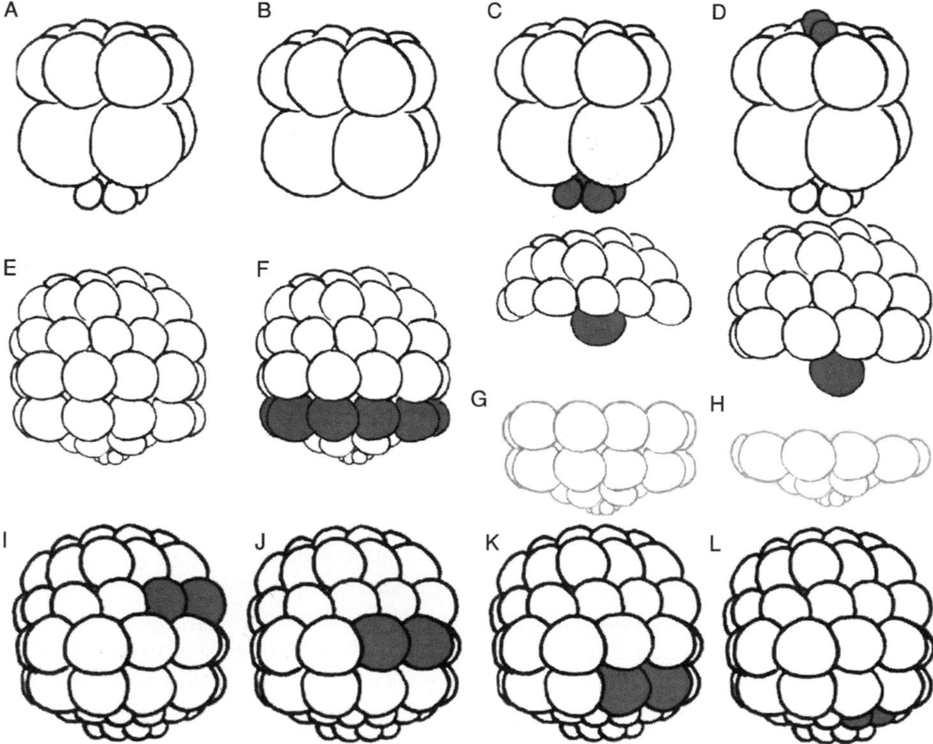

Fig. 8 Hörstadius transplants and Mosaic analysis. Donor cells from embryos injected with RNA, a morpholino, or other perturbation are coinjected with rhodamine dextran. Examples of transplant combinations are shown and are described in the text. (A–H) Hörstadius transplants. (I–L) Mosaic analysis (Illustration produced by Dr. D. McClay). (See Color Insert.)

Drosophila, with the one exception being that the construct is introduced and begins expression at fertilization and therefore expression of the perturbed molecule is not temporally controlled as it is in fly mosaic analysis. The mosaic analysis has been useful for detection of abnormal signal transduction responses and abnormal fate decisions, such that expected cell migrations fail in the transplanted cells of one tier only.

4. Bulk Methods for Isolation and Reassociation of Blastomeres

This method is used to isolate large numbers of intact animal halves or vegetal halves from 8-cell sea urchin embryos. A method is also described to recombine relatively large numbers of animal and vegetal halves following the isolation protocol.

Denuded 8-cell stage embryos are placed in about 30 ml of HEM in 50 ml plastic conical tubes for 5 min and then settled and resuspended in 30 ml of CFSW. The embryos are then partially dissociated by tipping the tube upside down repeatedly for 1 to 2 min. The degree of dissociation can be assessed during this period by examining a sample of the culture with a microscope. Ideally, the culture should have a population of embryos in various stages of dissociation and it is not necessary to have all embryos undergo dissociation. A small volume of the partially dissociated culture (~5 ml) is poured into a 100 × 15 mm agar-coated petri dish containing FSW. Using a mouth pipette, quartets of blastomeres are collected individually and placed in a small agar-coated petri dish (35 × 10 mm or 60 × 15 mm) containing cooled seawater. This can be achieved by placing a petri dish lid on ice, placing several layers of paper tissue on top of the lid, and then placing the small petri dish with the seawater on top of the paper tissue. The rest of the dissociated culture can be kept on ice for coldwater species such as *S. purpuratus*, or at 15 °C (for warm water species such as *L. variegatus*). Undissociated embryos and blastomeres that have not separated as a quartet should be avoided.

Once the desired number of blastomere quartets has been collected, the petri dish containing the isolated quartets is placed under the stereo microscope stage and observed until the quartets undergo the fourth cleavage. These quartets will divide in three easily recognized patterns. Two patterns are generated by an equatorial separation of the 8-cell embryo and one pattern is generated by a separation of the 8-cell embryos along the animal–vegetal axis. Isolated animal halves will divide equally to produce a ring of equal-sized mesomeres. Isolated vegetal halves will have four large macromeres and four small micromeres. Those embryos that were separated along the animal–vegetal axis will typically have two pairs of mesomeres and two pairs of the macromere/micromere combination. The proportion of embryos that divide along the animal–vegetal axis and along the equatorial plane can vary from batch to batch.

Once the quartets of blastomeres have undergone the fourth cleavage, animal halves and vegetal halves can be placed in separate dishes and either cultured in the presence of antibiotics or harvested for further analysis (Fig. 9).

If the investigator simply needs to isolate mesomeres and/or macromere/micromere pairs and it is not necessary to isolate whole animal or vegetal halves, an easier method to collect these cells is to completely dissociate embryos at the 8-cell stage. Blastomeres can be observed as they go though the fourth cleavage and mesomere pairs and the macromere/micromere pairs can be easily distinguished based on the cleavage patterns. To achieve complete dissociation of embryos, one should follow the same protocol that has been outlined for obtaining animal and vegetal halves. After the embryos are placed in CFSW, they are gently aspirated and expelled through a 10-ml syringe (without the needle) several times. An aliquot of the culture should be checked under a microscope to determine the desired degree of dissociation. Overaspirating the culture will cause blastomere

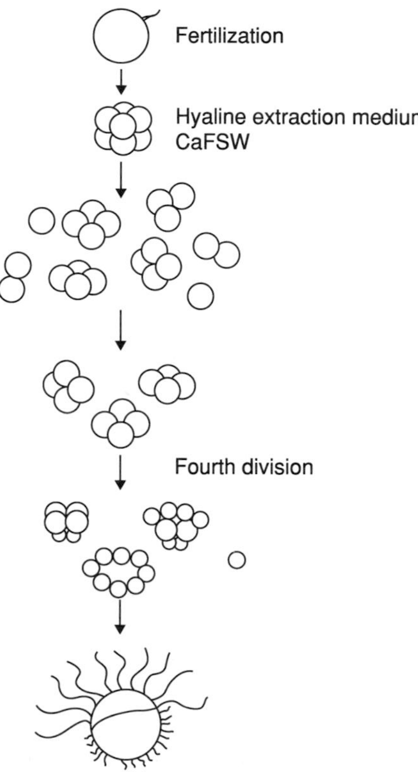

Fig. 9 A process to isolate quartets of blastomeres by bulk method (Illustration produced by Dr. A. Wikramanayake).

lysis. After this treatment, some blastomeres will appear elongated; however, these will quickly become spherical and will divide normally. Single blastomeres are collected and placed in a petri dish with cooled seawater until sufficient numbers of blastomeres have been collected. Isolated blastomeres are then placed under a stereo microscope and allowed to undergo the fourth cleavage.

Relatively large numbers of isolated animal and vegetal halves can be recombined using the following protocol. A thick layer of 1% agarose is placed at the bottom of the wells of a 72-well plate with 20 μl well volume (Nunc, Denmark). Once the agarose has solidified, a small hole is made in the agarose in each well using a sharp needle. The plates are allowed to equilibrate in seawater before adding any cells. An animal half and a vegetal half are placed in a hole in one of the wells and brought into contact using a mouth pipette and a glass needle made from a drawn-out glass capillary needle. After a few hours, the halves fuse and develop into normal-appearing embryos. Those halves that do not fuse should be removed from the dish using a mouth pipette.

C. Cautions

Each of the techniques described here has strengths and weaknesses. One of the strengths of blastomeres isolation and transplantation by hand is that, with practice, very precise types of microsurgery can be performed. Also, the setup for this technique is inexpensive, requiring only simple tools and a good dissecting microscope. A significant weakness to this technique is that it requires much patience, practice, and a steady hand. If these are lacking, one should try blastomeres isolation and transplantation using a micromanipulator. This technique also requires some practice. Bulk methods for blastomeres isolation and reassociation are, perhaps, the quickest and easiest methods. However, the significant disadvantage is the lack of precision when recombining host and donor embryos.

V. Blastomere Isolation and Transplantation in Starfish Embryos

A. General Explanation

Since early development of starfish is similar to that of sea urchins, embryos and larvae of starfish can be handled in practically the same way as those of sea urchins. However, there are some differences between these two classes. These differences bring advantages in some experimental analysis. At the same time, they make it necessary to employ some specific techniques.

Larger size, simpler shape of larvae, and absence of larval skeleton may make some manipulative operations easy. The position of the germinal vesicle in immature oocytes and that of polar bodies in fertilized eggs and cleavage stage embryos are good markers for the animal pole. However, since cleavages are almost equal, it is difficult to identify blastomeres according to their size.

B. Fertilization and Reconstruction

1. Removal of Fertilization Envelope

The methods for handling gametes, induction of oocyte maturation, and insemination are described in Chapter 3. To remove the fertilization envelope, eggs or embryos are treated with 1% sodium thioglycolate in seawater (pH10) for 3 min at any time after fertilization (Maruyama *et al.*, 1986). Repeated pipetting of fertilized eggs through a fine glass pipette with a tip diameter about 80% of the fertilization envelope is effective for removal of the fertilization envelope without chemical treatment (Maruyama and Shinoda, 1990). Prolonged chemical treatment of eggs and embryos may cause a delay or arrest in cleavage. After removal of the fertilization envelope, eggs and blastomeres are placed in an agar-coated dish to prevent their attachment to the bottom of the dish.

Denuded eggs and blastomeres of starfish are physically weaker than those of ordinary sea urchins because of their larger size and the absence of a thick apical extracellular matrix (hyaline layer). Rough pipetting using a thin pipette or strong centrifugal force may cause fragmentation of eggs and blastomeres.

2. Blastomere Isolation and Reaggregation (Reconstruction)

Unlike sea urchin embryos, cleavage of starfish embryos is equal, producing blastomeres of the same size. Another difference from sea urchin embryos is that following removal of the fertilization envelope, cleavage-stage embryos of starfish easily dissociate into single blastomeres so that they spread into a sheet on the substrate. These characteristics of the starfish embryos make it difficult to identify individual blastomeres. In order to resolve this difficulty, several methods can be used to identify specific blastomeres in the starfish embryos. Among them, vital staining of specific blastomeres with the dyes that pass through the fertilization envelope is commonly used. The fertilization envelope is then broken, and the stained blastomeres are isolated. Alternatively, the blastomeres of animal and vegetal halves in 8-cell stage embryos are isolated without vital staining.

a. Blastomere Isolation after Staining of Specific Blastomeres

Partial Vital Staining of Eggs and Embryos. A fine glass needle is prepared with tip diameter of 2 to 4 μm and loaded with 0.05% Nile blue sulfate in half-strength seawater. The needle is attached to a microinjection apparatus fixed on a microscope. The tip of the needle is pushed against egg or embryo surface and the dye is gradually ejected to mark the cell (Maruyama and Shinoda, 1990).

Partial vital staining of larger quantities of eggs or embryos is achieved by placing eggs or embryos on a 1% agar plate containing 0.1% Nile blue sulfate for 0.5 to 3 min (Kuraishi and Osanai, 1989). They are then washed with seawater, and those with the appropriate marking pattern are selected using a stereomicroscope.

Nile blue sulfate itself is fluorescent; thus, it may interfere with signals from other fluorescent dyes. Alternatives to staining with Nile blue sulfate include intracellular injection of marker dyes and membrane labeling with DiI.

Blastomere Isolation. Because no rigid structure holds blastomeres together in cleavage-stage starfish embryos, the blastomeres become dissociated into single cells when the fertilization envelope is removed. The blastomeres are easily isolated from a mass of the dissociated single blastomeres using a mouth pipette.

b. Isolation of Blastomeres without Vital Staining

Four-cell stage embryos are treated with 0.002% Actinase E (Kaken Pharmaceutical Co., Ltd.); by the next cleavage or 8-cell stage, embryos are treated with sodium thioglycolate for 1 min to weaken the fertilization envelope. Embryos with a correct blastomere distribution of animal and vegetal layers are collected in an agar-coated dish. The position of the polar bodies is checked and the animal and vegetal quartets are separated within the softened fertilization envelope with a fine glass needle. Each blastomere is taken out separately through a tear in the envelope using another glass needle (Kiyomoto and Shirai, 1993).

c. Reaggregation of Isolated Blastomeres

A cleavage stage, denuded embryo easily dissociates into a single layer on the substrate, as has been described. These spread blastomeres retain the ability to develop into a spherical blastula. If the sheets of blastomeres derived from different embryos are cultured in close contact with each other, they may fuse and develop into a chimeric embryo (Dan-Sohkawa, 1977). The same procedure can be used to reaggregate isolated blastomeres from different embryos. Because the cell–cell adhesion is weak during cleavage stages, the position of blastomeres can be changed easily. Thus, during reaggregation experiments, the culture dish must not be disturbed. Also, the dish must not be coated with agar because the sticky substrate is helpful to keep the blastomeres in the original position.

d. Reconstruction of Embryo within Fertilization Envelope

Single blastomeres or a group of blastomeres inside the fertilization envelope can be rotated along the animal–vegetal axis or the axis perpendicular to it. Eggs are washed with CFSW containing 1 mM EGTA several times soon after fertilization and incubated in the same media for 5 to 15 min. An embryo is held in a chamber for micromanipulation and a fine glass needle attached to a micromanipulator is inserted into the perivitelline space. The orientation of blastomeres is changed by pushing and dragging the surface of blastomeres using the needle.

Treatment with CFSW containing 1 mM EGTA destroys fibrous material in the perivitelline space, facilitating the operation. Embryos can develop into spherical blastulae as long as the apico–basal polarity is preserved (Kuraishi and Osanai, 1994), while they form separate or constricted blastulae if the polarity is disturbed (Kuraishi and Osanai, 1989).

VI. Blastomere Isolation and Transplantation in Amphioxus

Methods for handling adults and embryos of amphioxus are described in Chapter 9 of this volume. The development of isolated blastomeres from 2- through 16-cell stage embryos of amphioxus has been examined by gently shaking (Wilson, 1893; Morgan, 1896) or by expelling the embryos through a pipette (Conklin, 1933). The embryos typically fall apart within the spacious fertilization envelope. These methods, however, can cause individual blastomeres to fragment. Also, as the blastomeres are still within the fertilization envelope, they may adhere to each other in abnormal orientations. These problems can obscure the meaning of the results.

The most recent studies for blastomeres isolation and transplantation of amphioxus embryos were performed by Tung *et al.* (1958, 1960a,b, 1962, 1965), using the species *Branchiostoma belcheri tsingtaoense*. A convenient feature of these embryos is that the poles can be distinguished during cleavage stages even though the blastomeres are similar in size. The animal pole of the 8-cell stage embryo can be distinguished by a more transparent cytoplasm. At the 32-cell stage, the size and timing of formation of openings at the poles can be used to determine the orientation of the animal/vegetal axis. The opening at the vegetal pole is larger and lasts longer than the opening at the animal pole.

Because the fertilization envelope of these embryos is several times the diameter of the egg, these researchers were able to remove the fertilization envelope with fine forceps. Their techniques for microsurgery are similar to those used by Hörstadius (1973), which are described above in Section IV.B.1. They isolated and transplanted blastomeres and tiers of blastomeres with a glass needle immediately following cleavage, when the blastomeres were loosely connected to each other. When doing blastomeres transplantations, they found that doubling the amount of calcium in the seawater resulted in greater adhesion of donor and host cells.

VII. Blastomere Isolation and Transplantation in Ascidians

A. General Explanation

Described here are methods of microsurgery in commonly used ascidian species, *Halocynthia roretzi, Ciona intestinalis*, and *C. savignyi*. The diameter of eggs is 280 μm in *Halocynthia* and 130 μm in *Ciona*. *Halocynthia* is more suitable for micromanipulative approaches because of its larger size of the egg and the translucency, although its habitat is restricted to Asian countries. One can find either *C. intestinalis* or *C. savignyi* in most countries.

B. Manipulation of Blastomeres

1. Removal of Vitelline Membrane

Unfertilized eggs already have a perivitelline space. The vitelline membrane can be manually removed before and after fertilization with sharpened tungsten needles. The vitelline membrane of *Halocynthia* can be removed at any developmental stage because of its wider perivitelline space. Denuded embryos are reared in 0.9% agar-coated plastic dishes filled with FSW containing 50 μg/ml streptomycin sulfate and 50 μg/ml kanamycin sulfate.

For chemical devitellination, eggs just after fertilization are treated for 5 to 15 min with seawater containing 0.05% actinase E (Crude proteinase from bacteria, Kaken Co. Ltd., Tokyo, Japan) and 1% sodium thioglycolate (pH 10 adjusted with 1–2 N NaOH) (Mita-Miyazawa *et al.*, 1985). Softening of the vitelline membrane should be monitored by touching with a tungsten needle. The dissolving membrane is removed by washing the eggs five times just before the membrane is completely dissolved in order to avoid damage to the eggs.

In *Halocynthia*, naked embryos often fail to close their neural tubes in the anterior region during the latter half of the spawning season. This abnormality is avoidable by rearing the embryos in the supernatant of a homogenate of the embryos (Nishida and Satoh, 1985). Cleaving eggs are homogenized in seawater at the concentration of about 700 embryos/ml and this homogenate is centrifuged for 30 min at 200*g*. The supernatant thus obtained is used as a culture medium of naked embryos. The medium can be stored at −80 °C without losing its activity.

2. Dissociation of Blastomeres

Blastomeres of vitellinated embryos at cleavage stage can be dissociated at any time by treatment with devitellination medium described above (Nishida, 1992). After the membrane is dissolved, gentle pipetting will help the dissociation of loosened blastomeres. If one wants to continuously dissociate cleaving blastomeres, dissociated cells are placed into CFSW containing 0.2 mM ethylene-bis(oxyethylenenitrilo)-tetraacetic acid (EGTA). Dissociation should be monitored at frequent intervals and is facilitated by frequent agitation or gentle pipetting.

3. Isolation of Identified Blastomeres

The cell lineage is invariant in ascidians and each blastomere is numbered up to the gastrula stage (Conklin, 1905; Nishida, 1987). A very fine glass needle is used to isolate blastomeres. Blastomeres are identified and isolated with a glass needle under a good dissecting stereomicroscope (e.g., Olympus SZX). In *Halocynthia*, each blastomere can be isolated at least up to the 110-cell stage (Nishida, 1990, 1992), and in *Ciona*, up to the 64-cell stage (Sebastien Darras, personal communication).

4. Recombination of Blastomeres

Two isolated blastomeres are transferred into a small hole made in an agar layer that coats the bottom of a plastic dish. Blastomeres are allowed to establish close contact. A 30% solution (w/v) of polyethylene glycol (PEG 6000; Wako Ltd., Osaka, Japan) in distilled water, which has higher specific gravity than seawater, is placed in the hole. After a 30 s incubation, the blastomeres, which are now firmly adhered to each other, are immediately transferred several times to dishes filled with seawater (Nakatani and Nishida, 1994).

VIII. Lineage Tracing, Blastomere Isolation, and Transplantation in Hemichordates

A. Hemichordates

1. Lineage Tracing in Hemichordates by Iontophoretic Labeling with DiI

Previous lineage tracing studies in hemichordate embryos include those of Colwin and Colwin (1951, 1953, *Saccoglossus kowalevskii*) and Henry *et al.* (2001, *Ptychodera flava*). The vitelline envelopes are difficult to remove and denuded embryos do not survive well. However, blastomeres can be labeled with DiI directly through the vitelline envelope via iontophoresis (Henry *et al.*, 2001). Hoder and Ettensohn (1998) describe this technique and the construction of an inexpensive power supply. Quicker labeling is achieved by lowering the value of the current-limiting resistor to 100,000 or 10,000 ohms.

A standard mechanical puller is used to pull glass microelectrodes with a tip length of about 4 to 5 mm. Microelectrodes are back-filled with DiI-EtOH, as described by Hoder and Ettensohn (1998) and Henry *et al.* (2001; 25 mg/ml DiI, cat. no. D-282, Molecular Probes Inc., Eugene, OR, dissolved in 100% EtOH). If the tip of the needle breaks, DiI will crystallize and no labeling will take place. The needle is held using standard microelectrode holders. The positive platinum electrode wire is threaded into the back of the glass microelectrode so that it contacts the DiI-EtOH solution. The negative reference electrode (platinum wire) is immersed in FSW. Using a micromanipulator, firm contact is established between the tip of the needle and the surface of the embryo. Pressure is applied such that the embryo surface is visibly indented. It is not necessary to actually penetrate the envelope. To prevent the embryos from rolling away, a small groove is engraved in the bottom of the plastic petri dish. Current is applied until the desired amount of fluorescence labeling is achieved (15 s to a few min). Labeling is monitored with an epifluorescence dissecting microscope. If the needle becomes coated with extracellular matrix or cell debris, it must be cleaned by wiping the tip gently against the bottom of the operating dish, or replaced with a fresh one. Specimens can be fixed at any stage in 0.1 to 2.0% formalin in FSW containing 100 mM EDTA to preserve DiI fluorescence (Henry *et al.*, 2001).

2. Blastomere Isolation in Hemichordate Embryos

Two studies describe the isolation of blastomeres in hemichordate embryos (Colwin and Colwin, 1950, *Saccoglossus kowalevskii*; Henry *et al.*, 2001, *Ptychodera flava*). As there is currently no viable approach to remove the vitelline envelope, it is difficult to isolate blastomeres, particularly at later stages of development. Furthermore, grafting of cells is not possible. Of course, these problems may not be encountered for all species of hemichordates. Four approaches are successful for isolating blastomeres in these embryos. The last two approaches offer the only means of isolating corresponding halves of embryos so that their development may be compared directly.

The first method involves the use of fine glass or tungsten needles to ablate specific cells within the vitelline envelope (Henry *et al.*, 2001). Glass needles are pulled by hand using a gas microburner (Hamburger, 1960). Alternatively, sharpened tungsten needles may be used. Tungsten wire (diameter of 0.004 inches) may be sharpened in a saturated solution of either sodium hydroxide, or potassium or sodium nitrite (Conrad *et al.*, 1993; Hubel, 1957) or by burning the tungsten wire using a small microburner (Dossel, 1958; Hamburger, 1960). These needles can be mounted in small needle holders or melted glass rods. To ablate specific blastomeres, firm pressure is applied so that the needle penetrates the vitelline envelope and lyses the blastomere. The presence of residual cell debris does not interfere with the development of the remaining live cells (Henry *et al.*, 2001). To avoid killing sister cells, one must wait until each division is complete.

Individual blastomeres may also be ablated using the technique of "blastomere electrocution" (Henry *et al.*, 2001). This is accomplished by passing a prolonged electric current through selected cells using the same device used to label blastomeres with DiI, as has been described (Hoder and Ettensohn, 1998). The circuit is modified by replacing the single current-limiting resistor with a lower value of 10,000 or 1000 ohms. Needles are back-filled with DiI-EtOH or 100% EtOH and firm contact is made with the desired blastomere. Constant current is applied for 30 s to a few minutes. Cell death is accompanied by a clearly visible change in its appearance.

Calcium-low or calcium-free seawater (CFSW) has been used to separate blastomeres within the looser-fitting vitelline envelopes of *S. kowalevskii* (Colwin and Colwin, 1950). Approximately 45 min prior to first cleavage, the fertilized eggs are washed five times in CFSW. The Colwins used Rulon's (1941) modification of Herbst's CFSW (see also Cavanaugh, 1956). After completion of the desired cleavage stage, the embryos are washed three times in FSW. Due to the presence of the vitelline envelope, some blastomeres may not be entirely separated or they may fuse at later stages.

Early and late cleavage stage blastomeres/embryos may also be separated with ligatures. Very fine nylon ligatures, which are approximately 20 μm in diameter, may be teased from dental floss (POH, "Personal Oral Hygene," unwaxed dental floss, Oral Health Products, Inc., P.O. Box 470623, Tulsa, OK 74147) or from nylon stockings. Short pieces (approximately 2 mm) are pre-tied into single overhand knots. The loop is placed around the embryo and tightened using watchmaker's forceps. One must be careful not to allow the nylon thread to contact the air–water interface, as the surface tension may damage the embryo. A glass or tungsten needle can be used to pre-crease the embryo to aid in the proper placement of the ligature. The ligature is left in place until the desired stage of development is reached.

Recipes

Hyaline Extraction Medium (HEM; Fink and McClay, 1985):

0.3 M glycine, 0.3 M NaCl; 0.01 M KCl; 0.01 M $MgSO_4 \bullet 7H_20$, 0.01 M Tris, (pH 8.0) and 0.002 M EGTA

Calcium-Free Sea Water (CFSW) from Pacific coast of Japan (Misaki: Prescription by K. Dan):

Per Liter, 27.6 g NaCl, 0.7 g KCl, 4.6 g $MgCl_2 \bullet 6H_2O$, 0.5 g $NaHCO_3$

CFSW from East coast of United States (Woods Hole: From G. M. Cavanaugh):

Per Liter, 24.72 g NaCl, 0.67 g KCl, 1.50 g $K_2C_2O_4$, 4.66 g $MgCl_2{\bullet}6H_2O$, 6.29 g $MgSO_4{\bullet}7H_2O$, 0.18 g $NaHCO_3$

References

Amemiya, S. (1996). Complete regulation of development throughout metamorphosis of sea urchin embryos devoid of macromeres. *Dev. Growth Differ.* **38,** 465–476.

Cavanaugh, G. M. (1956). *In* "Formulae and Methods VI of the Marine Biological Laboratory Chemical Room," p. 84. Marine Biological Laboratory, Woods Hole, MA.

Colwin, A. L., and Colwin, L. H. (1950). The developmental capacities of separated early blastomeres of an enteropneust *Saccoglossus kowalevskii. J. Exp. Zool.* **115,** 263–295.

Colwin, A. L., and Colwin, L. H. (1951). Relationships between the egg and larva of *Saccoglossus kowalevskii* (Enteropneusta): Axes and planes; general prospective significance of the early blastomeres. *J. Exp. Zool.* **117,** 111–138.

Colwin, A. L., and Colwin, L. H. (1953). The normal embryology of *Saccoglossus kowalevskii* (Enteropneusta). *J. Morphol.* **92,** 401–453.

Conklin, E. G. (1905). The organization and cell lineage of the ascidian egg. *J. Acad. Nat. Sci. (Philadelphia)* **13,** 1–119.

Conklin, E. G. (1933). The development of isolated and partially separated blastomeres of *Amphioxus. J. Exp. Zool.* **64,** 303–375.

Conrad, G. W., Bee, J. A., Roche, S. M., and Teillet, M. A. (1993). Fabrication of microscalpels by electrolysis of tungsten wire in a meniscus. *J. Neurosci. Methods* **50,** 123–127.

Dan-Sohkawa, M. (1977). Formation of joined larvae in the starfish, *Asterina pectinifera. Dev. Growth Differ.* **19,** 233–239.

Dossel, W. E. (1958). Preparation of tungsten micro-needles for use in embryology research. *Lab. Invest.* **7,** 171–173.

Fink, R. D., and McClay, D. R. (1985). Three cell recognition changes accompany the ingression of sea urchin primary mesenchyme cells. *Dev. Biol.* **107,** 66–74.

Ettensohn, C. A. (1990). Cell interactions in the sea urchin embryo studied by fluorescence photoablation. *Science* **248,** 1115–1118.

Hamburger, V. (1960). *In* "A Manual of Experimental Embryology," revised ed., p. 221. Univ. of Chicago Press.

Henry, J. Q., Tagawa, K., and Martindale, M. Q. (2001). Deuterostome evolution: Early development in the enteropneust hemichordate *Ptychodera flava. Evolution & Development* **3,** 375–390.

Hoder, P. G., and Ettensohn, C. A. (1998). The dynamics and regulation of mesenchymal cell fusion in the sea urchin embryo. *Dev. Biol.* **199,** 111–163.

Hörstadius, S. (1973). *In* "Experimental Embryology of Echinoderms," pp. 1–192. Clarendon Press, Oxford, UK.

Hubel, D. H. (1957). Tungsten microelectrode for recording from single units. *Science* **125,** 549–550.

Kiehart, D. P. (1982). Microinjection of echinoderm eggs: Apparatus and procedures. *Methods in Cell Biology.* **25,** 13–31.

Kiyomoto, M., and Shirai, H. (1993). The determinant for archenteron formation in starfish: Co-culture of an animal egg fragment-derived cell cluster and a selected blastomere. *Dev. Growth Differ.* **35,** 99–105.

Kuraishi, R., and Osanai, K. (1989). Structural and functional polarity of starfish blastomeres. *Dev. Biol.* **136,** 304–310.

Kuraishi, R., and Osanai, K. (1994). Contribution of maternal factors and cellular interaction to determination of archenteron in the starfish embryo. *Development* **120,** 2619–2628.

McClay, D. R., Peterson, R. E., Range, R. C., Winter-Vann, A. M., and Ferkowicz, M. J. (2000). A micromere induction signal is activated by β-catenin and acts through notch to initiate specification of secondary mesenchyme cells in the sea urchin embryo. *Development* **127,** 5113–5122.

Maruyama, Y. K., Yamamoto, K., Mita-Miyazawa, I., Kominami, T., and Nemoto, S. (1986). Manipulative methods for analyzing embryogenesis. *In* "Methods in Cell Biology" (T. Schroeder, ed.), Vol. 27, pp. 326–343. Academic Press, New York.

Maruyama, Y. K., and Shinoda, M. (1990). Archenteron-forming capacity in blastomeres isolated from eight-cell stage embryos of the starfish, *Asterina pectinifera*. *Dev. Growth Differ.* **32,** 72–84.

Minokawa, T., and Amemiya, S. (1999). Timing of the potential of micromere-descendants in echinoid embryos to induce endoderm differentiation of mesomere-descendants. *Dev. Growth Differ.* **41,** 535–547.

Mita-Miyazawa, I., Ikegami, S., and Satoh, N. (1985). Histospecific acetylcholinesterase development in the presumptive muscle cells isolated from 16-cell-stage ascidian embryos with respect to the number of DNA replication. *J. Embryol. Exp. Morphol.* **87,** 1–12.

Morgan, T. H. (1896). The number of cells in larvae from isolated blastomeres of *Amphioxus*. *Wilhelm Roux' Archiv für Entwicklungsmechanik der Organismen.* **3,** 269–294.

Nakatani, Y., and Nishida, H. (1994). Induction of notochord during ascidian embryogenesis. *Dev. Biol.* **166,** 289–299.

Nishida, H., and Satoh, N. (1985). Cell lineage analysis in ascidian embryos by intracellular injection of a tracer enzyme. II. The 16- and 32-cell stages. *Dev. Biol.* **110,** 440–454.

Nishida, H. (1987). Cell lineage analysis in ascidian embryos by intracellular injection of a tracer enzyme. III. Up to the tissue-restricted stage. *Dev. Biol.* **121,** 526–541.

Nishida, H. (1990). Determinative mechanisms in secondary muscle lineages of ascidian embryos: Development of muscle-specific features in isolated muscle progenitor cells. *Development* **108,** 559–568.

Nishida, H. (1992). Developmental potential for tissue differentiation of fully dissociated cells of the ascidian embryo. *Roux's Arch. Dev. Biol.* **201,** 81–87.

Ransick, A., and Davidson, E. H. (1993). A complete second gut induced by transplanted micromeres in the sea urchin embryo. *Science* **259,** 1134–1138.

Ransick, A., and Davidson, E. H. (1995). Micromeres are required for normal vegetal plate specification in sea urchin embryos. *Development* **121,** 3215–3222.

Rulon, O. (1941). Modification of development in the sand dollar by NaCNS and Ca-free seawater. *Phys. Zool.* **14,** 305–315.

Sweet, H. C., Hodor, P. G., and Ettensohn, C. A. (1999). The role of micromere signaling in Notch activation and mesoderm specification during sea urchin embryogenesis. *Development* **126,** 5245–5254.

Sweet, H. C., Gehring, M., and Ettensohn, C. A. (2002). LvDelta is a mesoderm-inducing signal in the sea urchin embryo and can endow blastomeres with organizer-like properties. *Development* **129,** 1945–1955.

Tung, T. C., Wu, S. C., and Tung, Y. Y. F. (1958). The development of isolated blastomeres of *Amphioxus*. *Scientia Sinica* **7,** 1280–1320.

Tung, T. C., Wu, S. C., and Tung, Y. Y. F. (1960a). The developmental potencies of the blastomeres layers in *Amphioxus* egg at the 32-cell stage. *Scientia Sinica* **9,** 119–141.

Tung, T. C., Wu, S. C., and Tung, Y. Y. F. (1960b). Rotation of the animal blastomeres in *Amphioxus* egg at the 8-cell stage. *Science Record* **4,** 389–394.

Tung, T. C., Wu, S. C., and Tung, Y. Y. F. (1962). The presumptive areas of the egg of *Amphioxus*. *Scientia Sinica* **11,** 639–644.

Tung, T. C., Wu, S. C., and Tung, Y. Y. F. (1965). Differentiation of the prospective ectodermal and entodermal cells after transplantation to new surroundings in *Amphioxus*. *Scientia Sinica* **14,** 1785–1794.

Wilson, E. B. (1893). *Amphioxus*, and the mosaic theory of development. *J. Morphol.* **8,** 579–683 plus plates XXIX–XXXVI.

CHAPTER 12

Isolation and Culture of Micromeres and Primary Mesenchyme Cells

Fred H. Wilt* **and Stephen C. Benson**†

*Department of Molecular Cell Biology
University of California
Berkeley, California 94720

†Department of Biological Sciences
California State University, Hayward
Hayward, California 94542

I. Introduction

A. Background

The sea urchin embryo became a favorite of experimental zoologists during the late nineteenth century because its regular, rapid development was relatively easy to observe with contemporary microscopes. In the indirectly developing species

METHODS IN CELL BIOLOGY, VOL. 74

0091-679X/04 $35.00

that were examined, especially *Paracentrotus lividus* from the Mediterranean and *Psammechinus miliaris* from the Baltic, the fourth cleavage division was very asymmetric. A quartet of small micromeres arise by oblique equatorial division at the vegetal pole of the zygote; their sister cells are four large macromeres. The eight animal mesomeres arise by meridional cleavage. The micromeres received considerable attention by experimental embryologists, and were studied intensively by Boveri, von Ubisch, and Horstadius, among others (reviewed by Horstadius, 1973). Direct observation of the fate of the descendants of these four micromeres showed clearly that they gave rise to the primary mesenchyme cells, which differentiate into the endoskeletal spicules of the pluteus larva. Transplantation experiments showed that the micromeres played a powerful role in the specification of the overall pattern of the developing embryo. In the vocabulary of those halcyon days of experimental embryology, the sea urchin embryo was considered to be "regulative" and micromeres and their descendants were supposed to exert a potent "vegetalizing" influence. Thus, the isolation, manipulation, and culture of the micromeres and their descendants became an important experimental system. A breakthrough in technique was announced in a paper by Kayo Okazaki (1975), in which she described the isolation, en masse, of micromeres and their subsequent culture, *in vitro*, during which they differentiated and formed calcareous spicules.

B. Purpose of Approaches

The isolation and culture of micromeres and their descendants, the primary mesenchyme cells (PMC), have been a useful experimental tool for studies of morphogenesis and terminal differentiation of elements of the larval skeleton. Isolation of large numbers of micromeres has also been instrumental in construction of EST libraries from this cell type (Zhu *et al.*, 2001), and previously was used to isolate or characterize specific mRNAs involved in the terminal differentiation of these cells (Harkey *et al.*, 1988). The interaction between mesenchyme cells and other cell types has been primarily studied by transplantation of micromeres or mesenchyme cells, subjects which are covered in Chapters 11 and 13.

C. Specific Examples

1. Morphogenesis

The primary mesenchyme that arises from micromere progeny shows a complex program of changes in cell behavior, and Okazaki (1975) first noticed that the changes that occurred *in vitro* mimicked those that occur in the intact embryo. A number of workers, notably Michael Solursh and his colleagues, studied the behavior of isolated micromeres *in vitro*, focusing on migration, putative chemotaxis, and adhesion (reviewed in Solursh, 1986). Kitajima and Okazaki (1980) also used this approach to demonstrate that micromeres formed precociously at the

3rd cell division have the same potential for skeletogenesis as the conventional micromeres. Khaner and Wilt (1991) used a variant of the technique to show that the two daughter cells of 16 cell micromeres, the so-called large and small micromeres that form in the 5th division cycle, have different developmental potentials. Only the larger daughters of the 16 cell micromeres form PMCs; Pehrson and Cohen (1986) showed that the fate of the small micromeres is to contribute to the coelomic pouches.

2. Isolation and Characterization of RNA

Susan Ernst first demonstrated that the RNA population of micromeres was substantially different from what is found in the whole embryo at the 16 cell stage (Ernst *et al.*, 1980). Michael Harkey isolated polyadenylated RNA from cultured micromeres and was able to characterize several PMC specific cDNA clones (Harkey *et al.*, 1988). One of these, PM27, is an occluded mineral matrix protein. Zhu *et al.* (2001) employed micromere isolation to create a library of sequences expressed by primary mesenchyme cells (EST library), thereby allowing characterization of a large number of mRNAs that could be involved in differentiation of PMCs.

3. Biomineralization

Several workers have exploited the use of micromere cultures to study the differentiation of PMCs and secretion of calcareous spicules. Though the overall pattern of the skeletal elements formed in culture do not duplicate the elaborate morphology of the intact skeleton, the timing and, insofar as is known, the construction of the skeletal spicule is the same in culture as in the intact embryo. This area has been reviewed extensively (Decker and Lennarz, 1988; Wilt, 1999, 2002).

II. Isolation and Culture of Micromeres

A. General Explanation

The general approach used for isolation and culture of micromeres is to dissociate the embryo into single cells at the 16-cell stage, a stage when the difference in size between micromeres and other blastomeres is maximal. The dissociated cells of different sizes are separated by the sedimentation at $1 \times g$ through a sucrose gradient, and the resultant purified micromeres are cultured on plastic Petri plates in seawater containing low concentrations of horse serum. There are three essential considerations: (1) a suitable species of sea urchin must be used, (2) undamaged micromeres should be quickly isolated, and (3) the culture medium must contain 1 to 4% horse serum.

B. Detailed Method

1. Fertilization and Demembranation

The species most often used in North America is *Strongylocentrotus purpuratus*; in Japan, *Hemicentrotus pulcherrimus* is often used. Eggs are collected, washed, and fertilized in the presence of agents that interfere with hardening of the fertilization membrane, as discussed in Chapter 3. We find that a sea water solution of 3 to 5 mM p-aminobenzoate, sodium salt (stock solution adjusted to pH 8 with NaOH) effectively keeps the fertilization membrane from hardening. Cultures should be allowed to develop between 12 and 18 °C at an approximate concentration of 1 ml of washed eggs (egg volume is measured after sedimentation of dejellied eggs at 200 × g) per 100 ml of seawater. One hundred ml of such a 1% culture is a convenient amount, which results in four culture plates (100 mm diameter). All developmental times mentioned in this chapter are based on cultures maintained at 15°. Seawater, whether artificial (ASW) or natural, should be passed through a 0.45 μm Millipore filter before use (MPFSW).

At any time between 10 and 30 minutes post-fertilization, the fertilization membrane should be removed by passing the culture ($\leq$1% v/v) through nylon mesh of suitable size ($\sim$55 μm for *S. purpuratus*). We do this by placing a piece of the mesh over the cutoff end of a 50 ml plastic syringe, keeping the mesh in place with a rubber band. The zygotes are poured into the barrel of the cutoff syringe and flow through the mesh into a beaker. One, or sometimes two, passes of the culture through the mesh will disrupt the fertilization membranes. The zygotes are allowed to settle at 1 × g and the seawater, containing the ruptured fertilization envelopes, is aspirated. The culture is then allowed to develop with stirring at 15 °C. We stir the cultures by using 60 rpm clock motors, mounted on Plexiglas with plastic rods—fixed to the motor—to turn plastic paddles.

2. Weakening the Hyaline Membrane

We have found it useful to culture the zygotes in seawater with a reduced concentration of calcium in order to weaken the forming hyaline membrane that invests the blastomeres and keeps them together. This step is not essential, and sometimes the lowering of calcium favors the formation of subequatorial third cleavages. The treatment does not, however, alter the 4th cleavage plane. If the investigator chooses to impose this treatment, the concentration of calcium should be reduced from the usual 10 mM to 3.3 mM. We find it convenient to mix two volumes of calcium-free seawater (CFSW) with one volume of standard seawater (MPFSW) for this step. The treatment weakens the forming hyaline layer, thereby facilitating the dissociation of the blastomeres after the 4th cell division.

After the second cell division is completed, at about 160 to 180 min. After fertilization, the embryos are allowed to settle, low calcium seawater is aspirated, and the embryos are suspended in CFSW. This settling and resuspension is repeated once more. The culture is now allowed to continue to develop with stirring until the 4th cleavage division is complete, about 5 h after fertilization.

3. Embryo Dissociation

The micromeres complete cytokinesis about 10 to 15 min after the mesomeres divide. When the vegetal blastomeres of the 8-cell stage begin to divide, embryos are settled at 1 × g. This should take no more than 10 min. CFSW is aspirated and the embryos transferred to 50 ml plastic centrifuge tubes with conical bottoms (e.g., Falcon conical polypropylene tubes). Embryos are centrifuged at 400 to 500 × *g* for 2 min and the CFSW is aspirated.

The embryos are dissociated by repeated suspension and centrifugation in CFSW and calcium/magnesium free seawater (CMFSW). The embryos from ~1 ml of packed eggs are a suitable amount for 50 ml of suspension and centrifugation. Between centrifugations, the tubes containing the suspended embryos are kept on ice. Embryos in a pellet after centrifugation and aspiration are resuspended by adding CFSW (or other solutions) up to the 40 ml mark, then gently inverting the capped tube several times until all the embryos are suspended. The procedure of centrifugation and resuspension is carried out at least three times. Following the first resuspension in CFSW, the second is carried out with CMFSW, and then subsequent resuspensions are carried out with CFSW. The state of the embryos is monitored under the microscope. As the procedure is carried out, a gradual loosening of the blastomeres and, finally, extensive dissociation takes place. When most of the embryos are partially or completely dissociated, the suspension is diluted so that 1 ml of embryos is suspended in 10 ml of CFSW. Then the suspension is sucked up and down through the orifice of a 10 ml plastic pipette 1 to 3 times. This dense suspension of dissociated and partially dissociated embryos is layered over a CFSW–sucrose gradient for micromere collection.

4. Purification of Micromeres on A Sucrose Gradient

The linear sucrose gradient for micromere purification is constructed from mixtures of CFSW and 0.75 M sucrose in distilled water. Both parent solutions contain 10 μg/ml of gentamycin. The lighter component is 5% 0.75 M sucrose/95% CFSW; the heavier component is 25% 0.75 M sucrose/75% CFSW. A 100 ml cushion of the heavier solution (25% sucrose/75% CFSW) is placed in a 400 ml glass beaker. A 180 ml gradient (90 ml of light and 90 ml of heavier solution) is layered over the cushion with a peristaltic pump. It should take 20 to 40 min to make one gradient. The gradients are assembled at ambient laboratory temperatures, approximately 20 °C. We make the gradients while the cultures are developing. They are stored at 15 °C, or sometimes at 4 °C, until use. The gradients may be used any time within 7 h of construction.

The 10 ml suspension of dissociated embryos is layered over the gradient with a 10 ml pipette. The effluent from the pipette is allowed to flow slowly over a baffle floating on top of the gradient. The floating baffle is approximately 2 cm square and cut from a plastic weighing dish. Use of the baffle helps maintain a sharp interface between the embryo suspension and the gradient. The cells are allowed to settle into the gradient at 1 × g. The temperature should be approximately 15 °C.

About 25 to 35 min are required for the micromeres to be resolved about 5 to 10 mm below the gradient surface. (Fig. 1). There is often a thin opaque layer of lysed cells on top of the gradient. Mesomeres, macromeres, and undissociated embryos sediment below the micromere layer. The micromere layer may not be visible if the dissociation has been incomplete because the micromeres have little pigment. Even if the band is not visible, removal of solution from this area of the gradient can produce good cultures.

The cells are removed by suction. A Pasteur pipette is inserted, tip inward, into clear plastic tubing (3–4 mm. i.d.). The large end of the pipette is used as a kind of "vacuum cleaner" to scour the micromere area of the gradient. The other end of the plastic tubing is attached to a 50 ml plastic syringe barrel. Remove 35 to 40 ml of solution from this area of the gradient with the use of the syringe and transfer the cells to a plastic tube. One hundredth of a volume of 1.0 M $CaCl_2$ is added to the micromeres, mixed well, and the cells deposited on plastic petri plates at a ratio of 10 ml of collected micromeres for a 100 mm diameter plate. When different size plates are used, the surface-to-volume ratios should be maintained approximately. Bacteriological grade plastic plates are preferred, though the tissue culture variety work almost as well.

The cells are allowed to attach to the plate surface for 30 to 60 min, after which the fluid is carefully and slowly aspirated. Ten ml of filtered seawater is then added to the plate. It is removed after 15 min with careful aspiration. Finally, 10 ml of MPFSW containing 10 μg/ml of gentamycin is added to the plate and the cultures are maintained at 15 °C.

5. The Culture Medium

The medium is MPFSW containing gentamycin with the addition of 2 to 4% (v/v) horse serum. The horse serum is previously heated at 55 °C for 30 min to destroy complement and dialyzed against seawater. It can be stored frozen at −20 °C

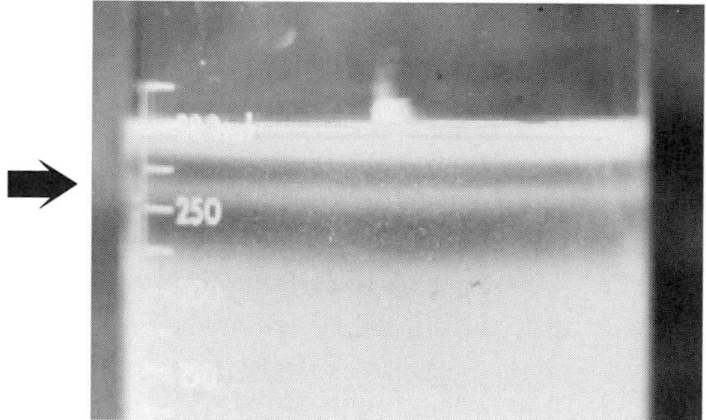

Fig. 1 Sedimentation of blastomeres of 16-cell stage embryos at 1× g. Dissociated embryos were layered over the linear sucrose gradient, as discussed in the text. After 30 min of sedimentation, a clear micromere layer can be seen just above the 250 ml mark on the beaker (arrow). The illumination of the beaker was from the side and the background was black paper. (See Color Insert.)

before use. We have found, based on the results of Page and Benson (1992), that the serum needs to be present for at least 10 h between 30 and 40 h post-fertilization. We find the differentiation of spicules is better if micromeres are cultured without serum for the first 24 h. After approximately 24 h have elapsed, the seawater is gently aspirated and then serum containing seawater is added on this second day of culture; no further changes of medium are necessary. The micromeres form PMCs, which then differentiate the calcareous spicules, at about the same tempo as intact embryos (Fig. 2). After 4 days of incubation, the cultures start to become moribund, probably because of nutritional deficiency.

We have examined several types of mammalian sera as well as chicken serum, and find horse serum is superior. Some lots of horse serum are better than others, but they all support differentiation. Benson and Chuppa (1990) have shown that the extracellular matrix from EHS cells (sold as Matrigel) will support the differentiation of micromeres from *S. purpuratus* in the absence of serum.

It is possible to place plastic or glass coverslips in the petri dishes and the micromeres will adhere to the coverslips; this provides good material for light microscopy, cytochemistry, or immunocytochemistry. Cells do not adhere well to coverslips that are easily sectioned for electron microscopy. We have found that Aclar plastic can be used to support micromere differentiation. However, in order to obtain good attachment of micromeres and PMCs to this surface, the Aclar should be "conditioned" by a previous culture of micromeres. After the first use, the Aclar is soaked in distilled water, then air dried, and used for a second round of culture of micromeres. PMCs that differentiate in culture on the conditioned coverslips can be fixed, *in situ*, after removal of the culture medium, and then flat embedded (Ingersoll *et al.*, 2003) for sectioning for electron microscopy.

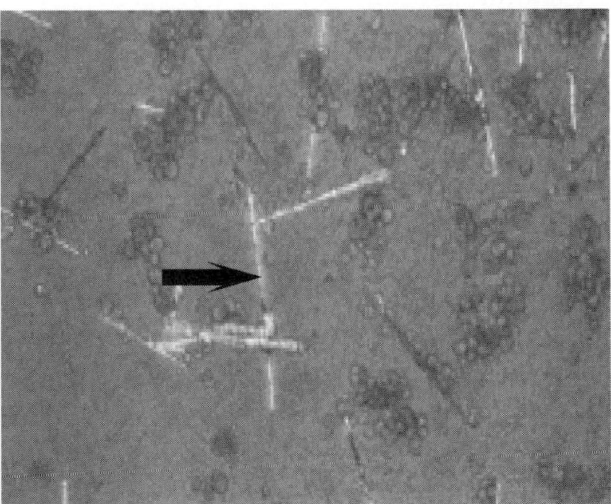

Fig. 2 The appearance of a micromere culture after 72 h. Micromeres were cultured on coverslips, as discussed in the text. The coverslips were inverted and observed at a magnification of 200× using differential interference contrast microscopy. An arrow points to a spicule. (See Color Insert.)

C. Cautions and Other Considerations

1. Dissociation

While embryos of many species can be dissociated by the techniques described here, some are more difficult to dissociate than others. In particular, *Arbacia punctulata* requires much harsher treatment. During the washing and resuspension steps a glycine-EDTA (Harkey and Whiteley, 1980) solution can be used to hasten dissociation, though exposure to this solution should be kept to a minimum. Furthermore, exposure to CMFSW should also be kept to a minimum. The absence of magnesium is very toxic.

2. Gradient Resolution

We were not able to separate mesomeres from macromeres using the described techniques. However, if the gradients are constructed and maintained at 4 °C, and the sedimentation of cells is carried out at 4 °C., there is often a sharp separation of mesomeres from macromeres (Norris Armstrong, personal communication). Thus, these cell types can then be used for various experimental purposes.

3. Species

Though micromeres can be isolated from species of *Lytechinus* and subsequently cultured using these techniques, they resolutely refuse to differentiate spicules. We have employed the method and obtained spicule formation using *S. purpuratus, S. franciscanus, P. depressus, H pulcherrimus, A. punctulata*, and *P. lividus*. Micromeres collected from animals at the beginning or end of their normal season of fecundity are not as robust in spicule formation as in the middle of their season.

4. Yield and Purity

The yield of micromeres under the usual conditions is about 20% according to Wilt *et al.* (2000). The purity of cultures was also estimated to be in the range of 85 to 95%. There is definitely variation in the different clutches of eggs in several respects: ease of dissociation, ability of micromeres to form spicules, absolute size of the micromeres, and exact position of the 3rd and 4th cleavage planes.

5. Miscellaneous

Radioactive amino acids and nucleosides can be added to the culture medium and are readily incorporated into protein or nucleic acids. This is a definite advantage because PMCs are difficult to label isotopically in the intact embryo; this is presumably because the surface epithelium of the blastula and gastrula is a barrier to transport of compounds into the blastocoel fluid.

Cells and spicules can be harvested from the culture plates by scraping the surface with a plastic policeman and decanting the released cells and other materials into a centrifuge tube. Many of the cells are very loosely attached to the surface, and CFSW containing 20 mM EGTA will release most of the cells without scraping (and without imposing the attendant damage to the cells). However, use of EGTA will begin to dissolve the mineral in the spicules, if that is an experimental consideration. The collected material can then be sedimented by centrifugation and subjected to techniques of cell biology and biochemistry.

D. Buffers and Reagents

1. Artificial seawater (ASW)

Concentration	g/L
0.48 M NaCl	28.32
0.01 M KCl	0.77
0.027 Mg Cl_2	5.48 of $MgCl_2 \cdot 6\ H_20$
0.03 M $MgSO_4$	7.39 of $MgSO_4 \cdot 7\ H_20$
0.01 M $CaCl_2$	1.11 anhyd.

Dissolve all components in 1 liter of water, add 0.2 gm $NaHCO_3$, and check that pH is 7.8 to 8.4. Tris buffer can be added to 0.01 M to stabilize pH.

2. Calcium-free seawater (CFSW)

   Prepare as above but omit the $CaCl_2$.

3. Calcium–magnesium free seawater (CMFSW)

   Prepare just as for ASW, but omit $MgCl_2$, $MgSO_4$, and $CaCl_2$. Adjust tonicity by adding Na_2SO_4 to 0.06 M. (8.52 gm/lit).

4. DM (after Harkey and Whiteley, 1980)

   1.0 M glycine
   100 μM EDTA
   adjusted to pH 8 with NaOH

5. Sources of:

   a. 60 rpm stirring motors

      Synchron Motors, 50Hz, 110 V, 60 rpm, type "C" mount, 600 series A-C timing motor Hansen Corporation, 901 S. First St., Princeton, IN 47670–2369
      Ph: 812–385–3415

   b. Nylon mesh (50, 75, or 100 μm mesh size)

      PGC Scientific, 7311 Governors Way, Frederick, MD 21704
      Ph: 800–424–3300

   c. Wheat Germ Agglutinin (WGA)

      ICN Biochemicals, #790164
      www.icnbiomed.com

Sigma-Aldrich, #61767
www.sigmaaldrich.com
d. FITC-WGA
Vector Labs, #FL-1021
www.vectorlabs.com

E. Alternatives

There are some alternatives to isolation of micromeres by sedimentation at $1 \times g$ through a sucrose gradient. Kiyomoto and Tsukuhara (1991) have published a technique for removal of micromeres from individual embryos of *Hemicentrotus pulcherrimus*. The micromeres, which are then cultured in small volumes of seawater containing blastocoel fluid, form spicules.

The Angerer laboratory (Nasir *et al.*, 1992) has devised a procedure using elutriation rotors to collect large numbers of micromeres, mesomeres, and macromeres from 16-cell stage embryos.

Finally, as mentioned previously, Benson and Chuppa (1990) have shown that the commercially available extracellular matrix derived from EHS tumor cell cultures (Matrigel) can substitute for horse serum. Presumably, both the Matrigel and horse serum can supply an unknown growth factor(s) or ligand(s) that is required for PMCs to initiate spicule formation.

III. Isolation and Culture of Primary Mesenchyme Cells

A. General Explanation

The experimental aims of some investigators may require the isolation of primary mesenchyme cells. Questions regarding PMC migration, aggregation, ECM interaction, and gene expression may be more conveniently answered by studying PMCs rather than their micromere precursors. The approach described in the following text has its origins in observations by Spiegel and Burger (1982) that wheat germ agglutinin (WGA) binds specifically to PMCs. This led to the methodology paper by Ettensohn and McClay (1987) that describes a panning method utilizing wheat germ agglutinin-coated dishes. Our modification of this method is described.

B. Detailed Method

1. Embryo Dissociation

Eggs are collected and fertilized as has been described, Embryos are cultured at 15 °C as a 1% (v/v) suspension in MFSW until the hatching blastula stage (18 h for *S. purpuratus* and 16 h for *L. pictus*). Embryos are then collected by centrifugation at $400 \times g$, 3 min, and resuspended in ASW containing 3.3 mM $CaCl_2$. At the mesenchyme blastula stage, embryos are collected by centrifugation using a 50 ml conical tube and washed three times with ice-cold CFSW containing 0.2 mM EGTA. After each centrifugation, the embryos were resuspended in at least 25 volumes of the CFSW-EGTA. Following the third resuspension, the embryos were continuously agitated by inversion and the extent of dissociation monitored by microscopy. Periodic trituration with a serological pipette can sometimes facilitate dissociation but care must be taken to avoid excessive shear since this can damage the cells. Again, the time involved to achieve dissociation should be determined empirically for each species. In our hands, this process takes about 15 to 20 min for *S. purpuratus* and 20 to 30 min for *L. pictus*. At a point when 90% of the embryos appear dissociated, the culture is poured through a 64 μm mesh Nitex, secured over the end of a cutoff syringe barrel, as described for micromere isolation. This traps undissociated embryos and partially dissociated clumps and results in a single-cell suspension. Cells are then counted with a hemocytometer and the suspension adjusted to 1×10^6 cells/ml.

2. Coating of Dishes with WGA

Polystyrene cell culture dishes are preferred for this protocol. Bacteriological plastic yields unsatisfactory results and should not be used. We have used 100×20 mm or 60×10 mm dishes, depending on the quantity of PMCs to be isolated. Wheat germ agglutinin (WGA) is dissolved in distilled water at 1 mg/ml and poured over the plates until just covering the surface and allowed to stand for 30 min at room temperature. Following the coating period, the WGA solution is removed and stored at 4 °C for subsequent use. The dishes are washed five times with distilled water, blocked by incubating for 30 min at room temperature in 100% fetal bovine serum or calf serum, and rinsed again with distilled water. Following the distilled water rinse, the plates are rinsed three times with calcium-free artificial seawater (CFASW) and used for panning. Plates can be prepared ahead of time and stored at 4 °C for at least three days prior to use. If the plates are stored, it is recommended that the CFASW step be omitted and the plates stored moist in the refrigerator after the distilled water rinse.

3. WGA Panning

Following dissociation and counting, the cells were plated on the WGA-coated dishes at 1×10^6 cell/cm^2. The cells are allowed to settle and attach at 4 °C. The time of settling and attachment is somewhat variable. We have found that 30 min is sufficient for *S. purpuratus* and 45 min is sufficient for *L. pictus*. The investigator should determine the optimal time for each species. Following attachment, the unbound cells are removed by decantation, and the culture plate rinsed gently three times with ASW. Attached PMCs are then dislodged by a shear stream of ASW from a serological or Pasteur pipette. The detached PMCs can be collected by centrifugation at $1000 \times g$ for 2 min. To investigate spicule formation, the cells are plated in ASW containing 4% horse serum, as described for isolated micromeres.

Typical purity varies but generally exceeds 90% as determined by FITC-WGA staining.

4. Alternative Method, Plastic Panning

Although we have found the WGA method for PMC "panning" to be the best in terms of the purity of the resulting population, another simpler approach has been published. Carson *et al.* (1985) and Venkatasubramanian and Solursh (1984) reported that dissociated embryos could be added to tissue culture plastic plates in the presence of 4% horse serum, previously dialyzed against ASW. The cell suspension is allowed to attach for 1 to 2 days and the unattached cells and aggregates poured off and replaced with fresh ASW containing 4% serum. Although the authors claim a purity of 90 to 95% spicule forming cells, no evidence is provided. In our hands, we cannot achieve this level of purity and the method is not useful for the establishment of PMC-enriched cultures early in their development.

References

Benson, S., and Chuppa, S. (1990). Differentiation *in vitro* of sea urchin micromeres on extracellular matrix in the absence of serum. *J. Exp. Zool.* **256,** 222–226.

Carson, D. D., Farach, M. C., Earles, D. S., Decker, G. L., and Lennarz, W. J. (1985). A monoclonal antibody inhibits calcium accumulation and skeleton formation in cultured embryonic cells of the sea urchin. *Cell* **41,** 639–648.

Decker, G. L., and Lennarz, W. J. (1988). Skeletogenesis in the sea urchin embryo. *Develop.* **103,** 231–247.

Ernst, S. G., Hough-Evans, B. R., Britten, R. J., and Davidson, E. H. (1980). Limited complexity of the RNA in micromeres of 16-cell sea urchin embryos. *Devel. Biol.* **24,** 119–127.

Ettensohn, C. E., and McClay, D. R. (1987). A new method for isolating primary mesenchyme cells of the sea urchin embryo. *Exp. Cell Res.* **168,** 431–438.

Harkey, M. A., and Whitely, A. H. (1980). Isolation, culture, and differentiation of echinoid primary mesenchyme cells. *Wilhemlm Roux Arch. Devel. Biol.* **189,** 111–122.

Harkey, M. A., Whiteley, H. R., and Whiteley, A. H. (1988). Coordinate accumulation of five transcripts in the primary mesenchyme during skeletogenesis in the sea urchin embryo. *Devel. Biol.* **125,** 381–395.

Horstadius, S. (1973). "Experimental Embryology of Echinoderms." Clarendon Press, Oxford, UK.

Ingersoll, E., Wilt, F., and MacDonald, K. (2003). Ultrastructural localization of spicule matrix proteins in normal and metalloproteinase inhibitor-treated sea urchin primary mesenchyme cells. *J. Exp. Zool.* **300,** 101–112.

Khaner, O., and Wilt, F. (1991). Interactions of different vegetal cells with mesomeres during early stages of sea urchin development. *Develop.* **112,** 881–890.

Kitajima, T., and Okazaki, K. (1980). Spicule formation *in vitro* by the descendants of precocious micromere formed at the 8-cell stage of sea urchin embryo. *Dev. Growth Differ.* **22,** 265–279.

Kiyomoto, M., and Tsukuhara, J. (1991). Spicule formation inducing substance in the sea urchin embryo. *Dev. Growth Differ.* **33,** 443–450.

Nasir, A., Reynolds, S. D., Keng, P. C., Angerer, L. M., and Angerer, R. C. (1992). Centrifugal elutriation of large fragile cells: Isolation of RNA from fixed embryonic blastomeres. *Anal. Biochem.* **203,** 22–26.

Okazaki, K. (1975). Spicule formation by isolated micromeres of the sea urchin embryo. *Am. Zool.* **15,** 567–581.

Page, L., and Benson, S. (1992). Analysis of competence in cultured sea urchin micromeres. *Exp. Cell Res.* **203,** 305–311.

Pehrson, J. R., and Cohen, L. H. (1986). The fate of the small micromeres in sea urchin development. *Dev. Biol.* **113,** 522–526.

Solursh, M. (1986). Migration of sea urchin primary mesenchyme cells. *In* "Developmental Biology: A Comprehensive Synthesis. The Cellular Basis of Morphogenesis" (L. W. Browder, ed.), Vol. II, pp. 391–431. Plenum Press, New York.

Spiegel, M., and Burger, M. M. (1982). Cell adhesion during gastrulation. A new approach. *Exp. Cell Res.* **139,** 377–382.

Venkatasubramanian, K., and Solursh, M. (1984). Adhesive and migratory behavior of normal and sulfate-deficient sea urchin cell *in vitro*. *Exp. Cell Res.* **154,** 421–431.

Wilt, F. H. (1999). Matrix and mineral in the sea urchin larval skeleton. *J. Struct. Biol.* **126,** 216–226.

Wilt, F. H. (2002). Biomineralization of the spicules of sea urchin embryos. *Zool. Sci.* **19,** 253–261.

Wilt, F. H., Hamilton, P., Benson, S., and Kitajima, T. (2000). Culture of micromeres of *Strongylocentrotus purpuratus in vitro*. *Dev. Biol.* **225,** 212–213.

Zhu, X., Mahairas, G., Illies, M., Cameron, R. A., Davidson, E. H., and Ettensohn, C. A. (2001). A large-scale analysis of mRNAs expressed by primary mesenchyme cells of the sea urchin embryo. *Develop.* **128,** 2615–2627.

CHAPTER 13

Rapid Microinjection of Fertilized Eggs

Melani S. Cheers and Charles A. Ettensohn

Department of Biological Sciences
Carnegie Mellon University
Pittsburgh, Pennsylvania 15213

I. Introduction

Microinjection of reagents into eggs and early embryos has been a critical tool in analyzing many cellular and developmental processes. This approach has been essential in studying cell lineages and cell fate specification, gene regulation, gene function, and biochemical pathways. Although other methods have been used to deliver substances into eggs pressure-controlled microinjection continues to be the most commonly used approach. Currently, this method is widely used to introduce DNA, mRNA, fluorescent probes, and morpholino antisense oligonucleotides into eggs and blastomeres of sea urchins and other invertebrate deuterostomes (in this

0091-679X/04 $35.00

volume, see Angerer and Angerer, Chapter 28; Arnone *et al.*, Chapter 25; Holland and Yu, Chapter 9; Lepage and Gache, Chapter 27). Many other molecules and cellular components can also be introduced into oocytes and zygotes by microinjection, including sperm (Hiramoto, 1962), cytoplasm (Kishimoto, 1986), calcium (Kiehart, 1981), aequorin (Steinhardt *et al.*, 1977), phosphatases (Kumano *et al.*, 2001), peptides (Conner and Wessel, 2000), and antibodies (Chui *et al.*, 2000; Rogers *et al.*, 2000).

The advantages of pressure-based microinjection as a delivery method are that it allows precise control over (1) the amount of reagent introduced into the cell, (2) the time of introduction, and (3) the specific cell(s) into which the reagent is introduced. By co-injecting a visible or fluorescent dye, successful delivery of the reagent can easily be confirmed. Almost any molecule or component can be injected, including membrane-impermeant substances, provided the material is in a solution or suspension that does not clog the microinjection pipette. The major disadvantage of pressure-based microinjection is that it is more time-consuming than approaches that can introduce molecules into large numbers of eggs simultaneously, such as electroporation and biolistics. Although electroporation is an effective means of introducing molecules into ascidian eggs (see Zeller, Chapter 29), this method has thus far proven too damaging to the more fragile eggs of echinoderms to be useful with those organisms. DNA constructs have also been successfully delivered into sea urchin eggs using a particle gun (Kurita *et al.*, 2003), but this method has not yet gained widespread use.

Current methods for microinjecting oocytes and embryos of echinoderms and other invertebrate deuterostomes are based on a pressure-injection method described by Hiramoto (1956, 1962). This approach involves immobilizing eggs and microinjecting them using an injection pipette controlled by a micromanipulator. Hiramoto's original method used an injection pipette which was pressurized with a fluid-filled micrometer syringe and which released a continuous stream of injection solution. Modifications of this basic pressure injection method remain widely used. Several variations have been described which differ with respect to the microscope (stereo or compound), micromanipulator, microinjection chamber (semi-enclosed chamber or open dish), and the system used to generate constant or intermittent pressure (micrometer syringe or picospritzer). In an earlier volume of this series, Colin (1986) and Kishimoto (1986) described useful methods for microinjecting starfish and sea urchin eggs. In Chapter 10 of this volume, Jaffe and Terasaki outline methods for quantitative microinjection using a fluid-filled micrometer syringe and a semi-enclosed microinjection chamber.

In this chapter, we focus on methods for using a picospritzer to inject large numbers of eggs that have been immobilized in open plastic dishes. This particular approach is especially useful for rapid, repetitive microinjection of reagents such as morpholino antisense oligonucleotides, mRNAs, and DNA constructs. Using these methods, a practiced hand can inject several hundred eggs in a typical session lasting 2 to 3 h (approximately 15 eggs/min). We describe specific methods currently used in our laboratory for microinjecting fertilized eggs of

Strongylocentrotus purpuratus and *Lytechinus variegatus*, two species of sea urchins used commonly for developmental studies in our own laboratory and throughout the United States. These methods are applicable to the eggs of many species of echinoderms and other invertebrate deuterostomes (e.g., Holland and Yu, Chapter 9). With slight modifications, the same methods are also useful for microinjecting molecules into specific blastomeres of early cleavage stage embryos (Ruffins and Ettensohn, 1996).

II. Equipment

A. Microinjection Apparatus

The microinjection apparatus consists of the following items (Figs. 1 and 2):

1. Magnetic stand
2. Micromanipulator (Leitz, Inc.)
3. Microscope, fixed stage, equipped for transmitted light and epifluorescence optics
4. Picospritzer II (Parker Instrumentation, General Valve Division, Fairfield, NJ)
5. Vibration isolation table

1. General Description of the Apparatus

We microinject using an inverted, fixed-stage microscope equipped for epifluorescence optics. An open plastic petri dish, coated with protamine sulfate and with a row of immobilized eggs adhering to its surface, is placed on the stage of the microscope. A freestanding micromanipulator is positioned next to the microscope and is used to position the microinjection pipette. The pipette is connected to a picospritzer which is attached to a tank of compressed air. The picospritzer delivers controlled, timed pulses of pressurized air when activated by a hand switch or foot pedal. The XY stage controller of the microscope is used to move the eggs relative to the injection pipette. To control vibration, the entire apparatus is placed on a heavy metal breadboard that rests on natural rubber foam pads or another vibration-absorbing material. A better, but more expensive, alternative is to place the apparatus on a vibration isolation unit that uses compressed air-filled pistons. These are available as freestanding tables or tabletop units from several manufacturers (e.g., Technical Manufacturing Corporation, Peabody, MA, USA).

2. Other Possible Configurations

A variety of configurations other than those shown in Figs. 1 and 2 are possible. For example, instead of a freestanding micromanipulator, several manufacturers sell high-quality micromanipulators with pipette holders that mount directly onto

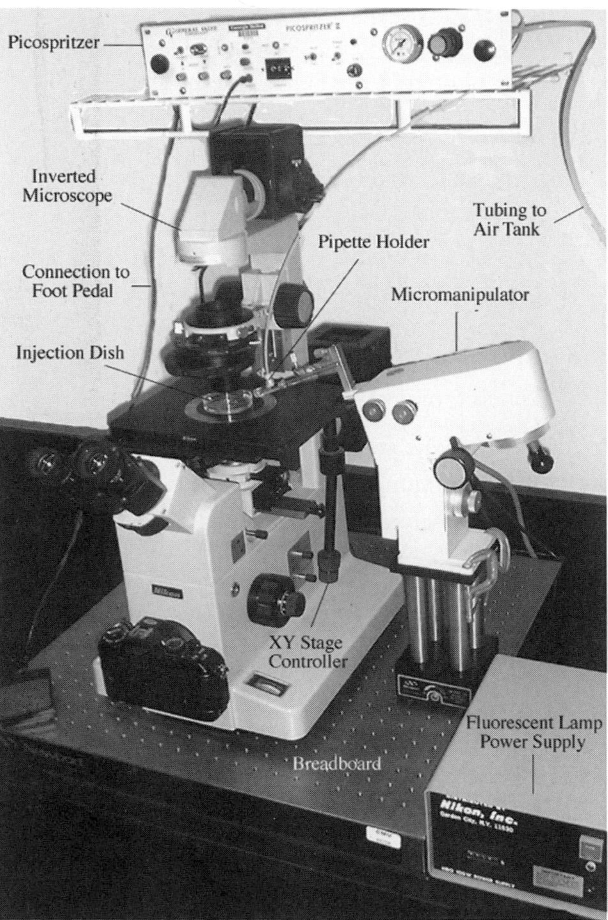

Fig. 1 The microinjection apparatus. An open plastic injection dish with a row of immobilized eggs adhering to its surface is placed on the stage of an inverted, fixed-stage microscope. The microscope is equipped for transmitted light and epifluorescence optics. A freestanding micromanipulator is positioned on the right side of the microscope and used to move the microinjection pipette. The pipette is connected to a picospritzer, which is attached to a tank of compressed air. The picospritzer delivers controlled, timed pulses of pressurized air when activated by a foot pedal (note that the foot pedal normally rests on the floor, not on the tabletop as shown in the photograph). The XY stage controller of the microscope is used to move the eggs relative to the injection pipette. To control vibration, the entire apparatus is placed on a heavy metal breadboard that rests on natural rubber foam pads or other vibration-absorbing material. (See Color Insert.)

the microscope stage. Such a micromanipulator eliminates the need for a fixed-stage microscope, because the position of the injection pipette relative to the specimen remains constant even when the stage moves during focusing. Other microscope configurations are also possible. We use an epifluorescence microscope

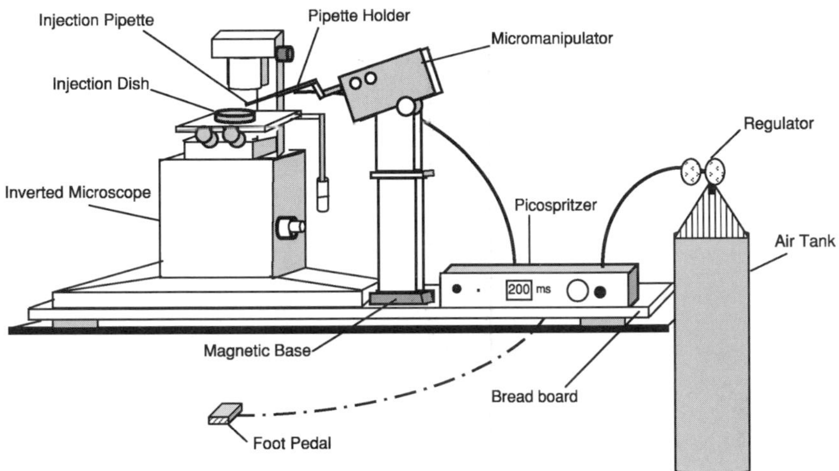

Fig. 2 Schematic diagram of the microinjection apparatus shown in Fig. 1.

because we frequently inject solutions that contain fluorescent marker dyes. As will be described, such fluorescent dyes are useful for several reasons: (1) The fluorescent marker makes it easy to determine whether reagent has been successfully delivered into the cytoplasm. (2) If not all eggs on the dish are to be injected, the marker dye makes it possible to distinguish injected from uninjected eggs at a later time. (3) The intensity of the marker dye fluorescence can provide a qualitative (or quantitative) indication of the amount of reagent injected. (4) Use of a fluorescent injection solution allows one to determine whether backflow of medium into the pipette tip is occuring after each pressure pulse (see following text). Although for these reasons we highly recommend equipping the injection microscope with epifluorescence optics, this is an added expense. As will be described (Section III.F.2.b), one can also carry out injections of glycerol-containing solutions under conventional brightfield illumination. Finally, although we use a compound microscope for injections, typically only low magnification is required (~100× total magnification). Thus, dissecting (stereo) microscopes can also be used effectively (see Colin, 1986). Most microscope manufacturers sell dissecting microscopes that can be used for epifluorescence (e.g., the Zeiss M^2BIO microscope).

B. Additional Equipment

1. Centrifuge, clinical (tabletop)
2. Stereomicroscope
3. Eyepiece micrometer reticle
4. Horizontal micropipette puller (Sutter Instrument Co., Novato, CA, USA)

C. Reagents and Supplies

1. Agarose
2. Bunsen burner
3. Capillary tubes, borosilicate glass (Kwik-Fil™, World Precision Instruments, M1B100F-6)
4. Cell culture dishes (Corning Inc., Polystyrene Cell Culture Dishes, 60 mm, #430166)
5. Centrifuge tubes, glass, 15 ml conical
6. Citric acid
7. Coverslips, glass (#1½, 22 × 50 mm)
8. Diamond pen
9. Fluorescent dextran (Molecular Probes, Inc., Eugene, OR, USA)
10. Forceps, Dumont (watchmaker's)
11. Glycerol (sterilized by autoclaving)
12. Ice bucket
13. Microcentrifuge tubes
14. Microscope slides, glass
15. Mouthpiece for mouth pipette
16. Pasteur pipettes, glass (6 and 9″)
17. Plastic petri dishes, 100 mm
18. Potassium chloride solution (0.5 M)
19. Protamine sulfate (Sigma, #P-4020)
20. Scissors, small
21. Seawater (see following text)
22. Stopcock grease
23. Syringe, plastic, 60 cc
24. Syringe filter, 0.2 μm (Nalgene #1909920)
25. Tape, double-sided (Scotch 3M #665)
26. Tubing, plastic, ID = 3/16″, OD = 5/16″ (Tygon)
27. Vacuum flask

III. Experimental Protocols

A. General Notes on Glassware and Seawater

Our laboratory keeps separate all glassware and plasticware that will come into contact with eggs or embryos, or that will be used to make up solutions that will come into contact with eggs and embryos. This "embryological grade" plasticware and glassware is never exposed to detergents, fixatives, or other reagents that might be toxic to live embryos. After each use, it is washed thoroughly with hot water only, rinsed with deionized water, and allowed to dry.

We primarily use natural seawater, collected from the Marine Biological Laboratory at Woods Hole, MA. The seawater is sterile-filtered (0.2 μm pore size) just before use. We have also used a variety of artificial seawater formulations with success, although marginal batches of embryos do best in natural seawater. Of the synthetic formulations, we typically use Instant Ocean (Aquarium Systems, Inc., Mentor, OH, USA). Instant Ocean is available at almost all aquarium stores in the United States.

B. Preparation of Protamine Sulfate-Coated Injection Dishes

Injection dishes are prepared by coating plastic cell culture dishes with a 1% solution of protamine sulfate. The protamine sulfate creates a positively charged substrate to which the negatively charged surfaces of the eggs adhere. Immobilizing the eggs in this fashion facilitates rapid injection without adversely affecting development. Embryos are released naturally from the surface of the dish when they hatch from the fertilization envelope at the blastula stage. If necessary, embryos can be manually detached from the substrate prior to hatching by using a microinjection pipette to tear the fertilization envelope, thereby releasing the embryo.

Prepare a 1% solution of protamine sulfate by adding 0.4 g protamine sulfate to 40 ml distilled water in a sterile 50 ml conical tube. Incubate the solution on a tilter until the protamine sulfate is fully dissolved (>1 h). This solution can be used for at least 3 months if stored at 4 °C. Some protamine sulfate will precipitate after storage at 4 °C, and each time plates are prepared, the solution should be warmed to room temperature and the protamine sulfate fully redissolved. Note that this may require >1 h.

Sixty millimeter cell culture dish *lids* have a depth and diameter convenient for microinjection. We typically coat lids in batches of 10. The dishes are coated with 1% protamine sulfate solution for 2 min (we have compared the performance of lids coated for 1, 2, or 3 min and find no difference). Arrange the lids in a row and pour enough 1% protamine sulfate solution into the first lid to completely cover the surface with gentle swirling (3–4 ml/lid). Begin timing with a watch or laboratory timer when the protamine sulfate solution is added to the first lid. Proceed down the row, adding protamine sulfate solution to each lid. It will take about 30 s to cover all 10 lids. When the protamine sulfate solution has been in contact with the first lid for 2 min, pour the excess solution back into the original 50 ml conical tube for subsequent use and place the lid in a 1-liter beaker filled with deionized water. Place this beaker in a sink under running deionized water to agitate and rinse the lids. Proceed down the row of lids in the same order that the protamine sulfate was added. It will take about 30 s to transfer each lid to the beaker of water, so that each lid will have been in contact with the protamine sulfate solution for approximately 2 min. After the last lid has been placed in the beaker, continue to rinse all the lids for an additional 5 to 10 min. Then remove the lids from the beaker of water and place them coated surface down on clean benchcoat to air dry.

When the dishes are dry, use a black lab marker to draw a straight line on the outside (i.e., the uncoated side) of each lid, about 2.2 cm from the edge (Fig. 3). This line will be used as a guide when rowing the eggs and is also useful for locating the row of eggs in the microscope during injection. Invert the lid and place a small piece of double-sided Scotch tape directly to the left of the line on the inside (coated side) of the lid. Use a tiny dot of stopcock grease to attach a small glass coverslip fragment directly below the tape. The tape and coverslip should meet at approximately the midline of the plate (Fig. 3). The coverslip fragment is used as a hard surface for breaking open microinjection pipettes. The tape is used to remove cellular debris from the tip of the injection pipette. After attaching the tape and coverslip fragment, reassemble the lid with a 60 mm dish bottom for easy storage and to keep the lid clean. Store assembled dishes in a clean, dry plastic storage box at room temperature. Injection dishes typically retain their adhesivity for up to 1 month.

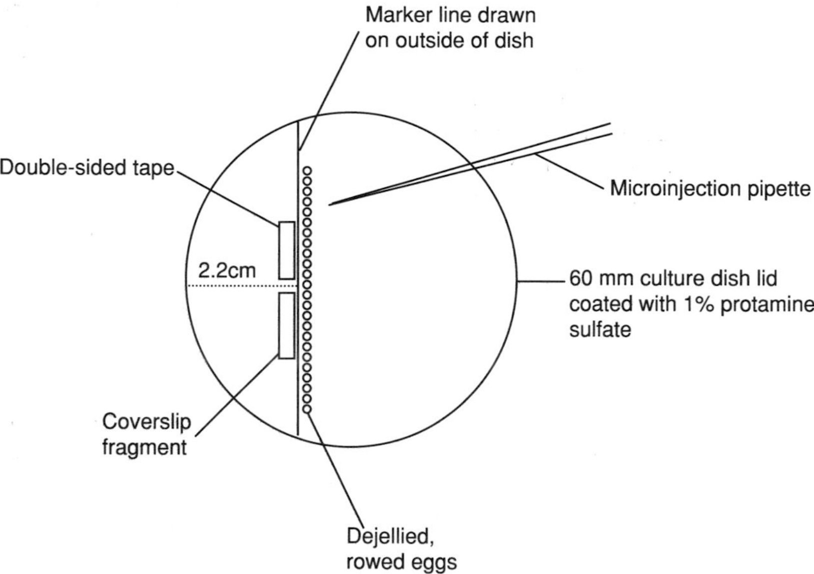

Fig. 3 Configuration of an injection dish. The inner surface of a 60 mm cell culture dish *lid* is coated with 1% protamine sulfate. A line is drawn with a marker on the outer (uncoated) surface of the lid, about 2.2 cm from the edge. A small strip of double-sided tape and a fragment of a glass coverslip are attached to the inner (coated) surface of the lid, as shown. Dejellied eggs are rowed in single file on the coated surface of the lid, using the marker line as a guide. The glass microinjection pipette approaches the eggs from the right side at a slight downward angle (15–20°).

C. Preparation of Microinjection Pipettes

Injection pipettes are pulled from Kwik-Fil Borosilicate Glass Capillaries, 1.0 mm × 6″. These capillary tubes have an internal filament and injection pipettes pulled from them can therefore be "backfilled" by capillary action. For microinjection of RNase-free injection solutions (i.e., mRNAs), bake the capillaries overnight at 400 °C and handle them only with gloves. Following the manufacturer's instructions, experiment with a horizontal pipette puller to determine the settings appropriate to pull pipettes that taper through a length of 65 μm from a width of 10 μm at the shoulder to 10 to 20 nm at the tip. Finding a setting that works well can be tedious but is important for successful injections. Up to three injection pipettes can be pulled from each 6″ capillary tube without making the pipettes too short. If pipettes have pulled tips at both ends, one tip should be broken off using watchmaker's forceps. We generally find that injection pipettes load best soon after they are pulled and recommend pulling fresh pipettes on the day they are to be loaded with injection solutions.

To protect the pipettes and inspect their quality, attach them to a standard glass microscope slide by pressing the blunt end of the pipettes against a piece of double-sided tape that has been placed at one end of the slide. Up to 16 injection pipettes can be conveniently attached to a single microscope slide. Place the slide in a 100 mm plastic petri plate if the pipettes are to be transported or stored. Examine the quality of the pipettes by placing the microscope slide on a compound microscope (100× total magnification). Discard any pipettes with tips that are broken or poorly shaped.

Prior to loading the pipettes, transfer them to a humid chamber that consists of a 100 mm plastic petri dish with a moistened, tightly rolled Kimwipe pressed against one side of the dish (Fig. 4). Place a small strip of modeling clay near the other side of the dish. Press the blunt ends of several (up to 10) injection pipettes into the clay, angling them such that their tips point slightly downward toward the dish bottom. Take care not to allow the tips of the pipettes to come in contact with the dish or with the moist Kimwipe. Cover the humid chamber while preparing the injection solutions.

D. Preparation of Injection Solutions

1. Morpholino Antisense Oligonucleotides

Morpholino antisense oligonucleotides (Gene Tools, LLC, Philomath, OR) are reconstituted as 5 mM stock solutions in sterile distilled water following the manufacturer's instructions. Four millimolar injection solutions are prepared by combining 20 μl 5 mM morpholino stock, 5 μl 100% sterile glycerol, and 0.5 μl of a 5% fluorescent dextran solution (5 mg dextran + 100 μl sterile distilled water). If it is desirable to dilute the injection solution further (100 μM–2 mM), do so by mixing with an appropriate volume of 20% sterile glycerol/0.1% fluorescent dextran in sterile-filtered distilled water. Morpholino antisense oligo stock solutions

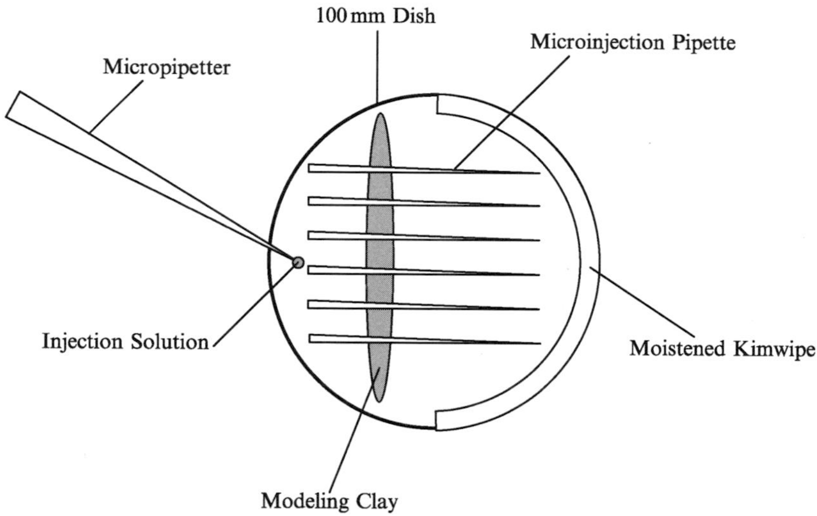

Fig. 4 Loading microinjection pipettes. A humid chamber is prepared by placing a tightly wrapped, moistened Kimwipe along one side of a 100 mm plastic dish. A strip of modeling clay is placed near the opposite side of the dish. Several (1–10) microinjection pipettes are held in position by pressing them into the clay, as shown. A small volume of injection solution (~1.5 μl) is drawn into the tip of a micropipetter (e.g., a Gilson P2 micropipetter). By pressing slightly on the plunger, a droplet of the injection solution (~0.3 μl) protrudes from the pipetter tip. When the droplet touches the back of a microinjection pipette, it is pulled quickly into the pipette by capillary action and is gradually drawn to the tip.

and injection solutions can be stored at −20 °C for many months. It is best to bring them to room temperature prior to use, and some researchers recommend warming them to 60 °C to ensure that the oligonucleotide is fully dissolved. It has been recommended that morpholino injection solutions be buffered with 10 mM Tris, pH 8.0 (Angerer and Angerer, Chapter 28).

Centrifuge the injection solution (10 min, 15,000*g*) to remove any particulate matter that might clog the injection pipettes or hinder capillary action. To load the injection pipettes, draw approximately 1.5 μl of injection solution into a micropipetter (e.g., a P2 Gilson pipetter). Gently depress the plunger until a droplet of injection solution (~0.3 μl) bulges slightly from the pipette tip (Fig. 4). Touch the protruding injection solution to the *back* of a microinjection pipette. Capillary action will pull the solution away from the pipetter tip and into the microinjection pipette. Proceed down the row of injection pipettes in this manner. One and a half microliters of injection solution should be enough to fill 4 to 5 injection pipettes. The loaded injection solution will gradually be drawn from the back of the injection pipettes to the tip by capillary action. With the naked eye, it is usually possible to see fluid accumulate at the pipette tip within 2 to 3 min, although we allow at least 5 to 10 min for loading to occur. Occasionally a pipette will fail to load properly and must be discarded, but this is rare.

Morpholino antisense oligonucleotides are highly stable (Angerer and Angerer, Chapter 28). Loaded injection pipettes can be stored at 4 or −20 °C and used repeatedly for several weeks if they are kept from drying out. We typically store loaded pipettes in the same humid chamber used to load them, after sealing the dish with Parafilm. It is important to inspect the pipettes before their next use to ensure that condensate has not accumulated inside during storage. This risk is decreased if the pipettes are kept at −20 °C. Warm the injection pipettes to room temperature before use.

2. mRNAs

mRNA injection solutions are prepared and loaded in much the same way as morpholinos. To prevent contamination by RNases, bake the capillary tubes overnight at 400 °C and always wear gloves when handling the capillaries or microinjection pipettes. We typically do not add fluorescent dextran to mRNA solutions because it appears that the level of mRNA expression is sometimes reduced even when RNase-free water is used to prepare the stock solution of dextran. (Note that we have not systematically analyzed this effect and there may be many fluorescent dyes that can be mixed with mRNAs and used as markers.) Stock mRNA solutions are therefore mixed only with 100% sterile glycerol and RNase-free water to yield a final glycerol concentration of 20% and a final mRNA concentration of 1 to 10 mg/ml. Injection pipettes are loaded as has been described, using RNase-free tips. Injection pipettes are generally used on the day they are loaded but can be stored at −20 °C for at least a few days.

3. Plasmid DNA

Methods for preparing plasmid DNA for microinjection are given in Chapter 25 of this volume (Arnone *et al.*).

E. Collecting and Preparing Eggs for Injection

In this section, we briefly describe methods that are useful for collecting gametes of *Lytechinus variegatus* and *Strongylocentrotus purpuratus* and preparing them for microinjection.

1. Gamete Collection

General methods for collecting sea urchin gametes and fertilizing eggs are described in Chapter 3 (Foltz *et al.*) and Chapter 33 (Epel *et al.*). We typically induce spawning by intracoelomic injection of 0.5 M KCl. A useful alternative is to collect gametes by the electric shock method, which allows one to repeatedly collect small numbers of gametes from the same adult sea urchin (Foltz *et al.*, Chapter 3). Microinjection experiments typically require only small numbers of eggs, and therefore this method conserves animals. Electric shock works well for *S. purpuratus* but has not been used successfully with *L. variegatus*.

Collect the sperm dry and store them on ice or at 4 °C until use. Concentrated, dry sperm can be used for up to one week. Eggs are collected in seawater of the appropriate temperature. *S. purpuratus* is a cold-water species and the eggs and embryos must be kept at 4 to 15 °C. *L. variegatus* is a warm-water species and its eggs and embryos should be maintained at 18 to 25 °C. It is not necessary to rinse the eggs at this point, as they will be rinsed extensively when they are dejellied (Section III.E.2). Unfertilized *S. purpuratus* eggs can be stored at 4 °C as a monolayer in a clean glass culture dish with sterile-filtered seawater. Depending on the quality of eggs obtained, eggs stored in this manner will fertilize well up to 48 h post-collection. We have not had success storing *L. variegatus* eggs in this way, however, and use the eggs within a few hours after spawning. Some batches of *L. variegatus* eggs are viable up to 6 h after spawning, while others decline more rapidly. Epel *et al.* (Chapter 33) describe methods for long-term (>7 day) storage of unfertilized sea urchin eggs in the presence of antibiotics.

Assess the quality of the eggs and sperm collected by fertilizing a test bowl. It is generally not advisable to perform injection on batches of eggs that yield fertilization rates of <95%, or on eggs that have a ragged periphery or a vesiculated cytoplasm. If the fertilization envelope of many of the eggs elevates only partially, discard the eggs. The quality of the sperm can be independently assessed by examining a few drops of diluted sperm at ~200X total magnification on a compound microscope. If the sperm are actively motile, they are usually of high quality. If fewer than 50% of the sperm are swimming vigorously 2 to 3 min after dilution, however, discard them and collect fresh sperm from a different male. Note that sperm that have been stored at 4 °C may require several additional minutes to begin to swim after they are mixed with seawater.

Once the quality of the gametes has been confirmed, prepare the eggs for injection by removing the external jelly layer. It is best to move directly from egg dejellying to injection, so assemble all necessary components before proceeding.

2. Dejellying Eggs

To dejelly *S. purpuratus* eggs, transfer about 200 μl of unwashed eggs to a clean, "embryological grade," 15 ml glass centrifuge tube. Pellet the eggs by spinning them for 30 s at 1/2 speed in a clinical centrifuge. Carefully pour off the supernatant or aspirate it using a clean Pasteur pipette attached to a vacuum flask. Add about 10 ml of fresh, cold seawater and, if necessary, resuspend the eggs by gentle pipetting. Repeat this washing procedure 2 to 3 times. Resuspend the eggs in acid seawater (seawater brought to pH 5.0 with 0.5 M citric acid). Quickly spin the eggs and resuspend them again in fresh acid seawater. Allow the eggs to incubate on ice in the acid seawater for 10 min. Collect the eggs by centrifugation and wash 2 to 3 times with normal seawater. Note that after the jelly layer is removed some eggs may adhere to the walls of the centrifuge tube. If excessive clumping occurs, decrease the duration of the acid seawater treatment.

After the final wash, transfer the dejellied *S. purpuratus* eggs to a 60 mm petri dish that has been coated with 2% agarose to prevent the eggs from adhering to the plastic. Keep the dish on ice. (Note: to prepare agarose-coated dishes, dissolve 1 g of agarose in 50 ml seawater by microwaving the solution in a clean, detergent-free Erlenmeyer flask. After the agarose has dissolved, pour the hot solution into 60 mm culture dishes and allow it to solidify. Rinse the dish several times with seawater before use.) Agarose-coated dishes can be reused. At the end of the injection session, rinse the dish several times with distilled water, wrap it with Parafilm to prevent it from drying out, and store it at room temperature. Immediately before the next use, rinse the dish several times with seawater.

L. variegatus eggs can be dejellied by mechanical means alone and do not require treatment with acid seawater. Simply wash the eggs by centrifugation using regular seawater, as described above. We routinely wash the eggs 8 times to ensure that they have been effectively dejellied. One can monitor the dejellying process by the appearance of the pellet of eggs following each successive centrifugation. Initially, the eggs stratify into two distinct layers: a darkly colored pellet of tightly packed, dejellied eggs at the bottom of the tube and an overlying, lightly colored zone of loosely packed eggs that retain their jelly coat. The lightly colored zone decreases in size after each wash and eventually disappears, leaving only densely packed, dejellied eggs. After the final spin, resuspend the eggs in a small volume of seawater and transfer them to an embryological grade glass dish. *L. variegatus* eggs treated in this way do not adhere to glass and thus it is not necessary to place the eggs on an agarose-coated dish. Store the dejellied eggs at room temperature and proceed to rowing.

3. Rowing Dejellied Eggs

"Rowing" involves using a mouth pipette to arrange dejellied eggs in neat files on the surfaces of protamine sulfate-coated injection dishes. Rowing greatly facilitates rapid and successful microinjection and is essential if one wishes to inject most or all of the eggs on a dish. The secrets to successful rowing are: (1) experience with a mouth pipette so that the flow of seawater and eggs from the tip can be precisely regulated and (2) a mouth pipette of correct size and shape. The most common problem is to use a mouth pipette with too large a tip opening. With a large tip, even slight positive pressure expels a large volume of seawater and eggs, scattering the eggs over a large region of the dish rather than depositing them in a neat row.

The preparation of mouth pipettes is described in detail by Sweet *et al.* (Chapter 11) and Wessel *et al.* (Chapter 5). We use plastic or glass mouthpieces attached to flexible Tygon tubing (ID = 3/16″, OD = 5/16″). Mouth pipettes are pulled by hand from 9″ Pasteur pipettes that have been heated in a small gas flame. The pipettes are pulled at a slight angle such that the tip is bent at about 120° relative to the thicker shaft. Ideally, pull a slender mouth pipette that has a tip opening just slightly larger than an egg. The walls of the pipette should be parallel over a considerable distance near the tip.

Working under a dissecting microscope and with a dish of dejellied eggs on the stage, draw several hundred eggs into the pipette. It is important to draw up eggs from a region of the dish where they are densely packed (i.e., where they form at least a solid monolayer) so that they will be highly concentrated within the mouth pipette. If necessary, swirl the dish to concentrate the dejellied eggs near the center. While holding the eggs in the mouth pipette with suction, remove the dish of dejellied eggs from the stage and replace it with a protamine sulfate-coated injection dish that has been filled with sterile-filtered seawater of the appropriate temperature. Position the dish so that the black marker line is aligned horizontally. Carefully lower the tip of the mouth pipette into the seawater and maneuver it alongside the marker line (Fig. 5). Hold the pipette such that the tip lies parallel to the black marker line, and place the tip in gentle contact with the surface of the dish. Starting at one end, slowly expel eggs in a single row adjacent to the marker line by exerting slight positive pressure while sliding the dish across the stage of the microscope in a smooth motion. Note that the dish is moved rather than the mouth pipette. If the mouth pipette has the correct diameter and the eggs are sufficiently concentrated within it, they will be extruded in a compact stream and form a neat row, one cell wide. The eggs will adhere to the surface of the dish immediately upon contact. To ensure firm attachment, however, do not disturb the dish for at least 1 min after rowing. Note that while a narrow-bore mouth pipette is essential for successful rowing, it is also possible to have too narrow an

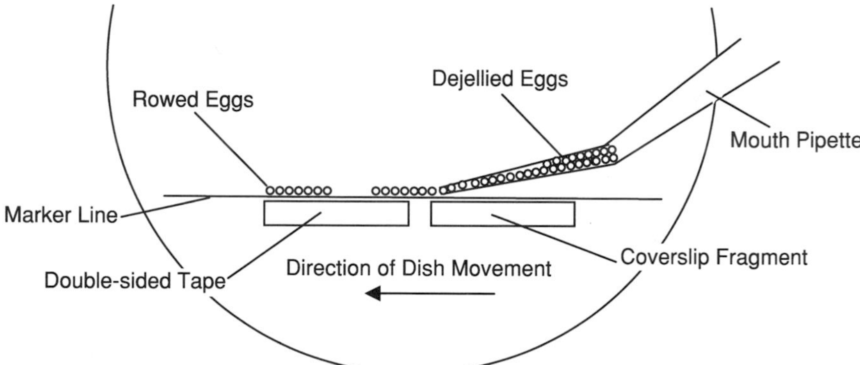

Fig. 5 Rowing eggs. Working under a dissecting microscope, a concentrated suspension of dejellied unfertilized eggs is drawn into the tip of an angled mouth pipette. The opening of the mouth pipette should be slightly larger than the diameter of an egg. The tip of the pipette is placed parallel to the marker line and gentle pressure is used to expel a file of eggs. As the eggs are expelled, the dish is slid along the stage of the microscope in a smooth motion (arrow) while the mouth pipette remains stationary. For injecting eggs of *S. purpuratus*, it is useful to row 2 to 3 short files of eggs that are separated slightly from each other, as shown here (see text).

opening. If the eggs appear misshapen or torn when extruded from the mouth pipette, a larger opening is needed. It may be possible to use the same pipette by breaking it further up the shank.

Although the most common approach is to row a single, long file of 150 to 200 eggs in a dish, eggs can be rowed in different configurations, depending on the experiment. A shorter row of fewer eggs may be useful if it is important to inject every egg on the dish. Alternatively, two or three long, parallel rows can be generated if one wishes to maximize the number of injected eggs/per dish. As will be discussed, the eggs of *S. purpuratus* (but not those of *L. variegatus*) become difficult to inject soon after fertilization, and it is often useful to arrange these eggs in two or three separate, short rows per dish (Fig. 5). Each short row can be fertilized independently by adding a few drops of diluted sperm to the seawater directly above the eggs. This approach allows one to inject a greater number of eggs on a single dish and conserves protamine sulfate-coated dishes.

A well-shaped mouth pipette is precious and may be reused many times. After an experiment, the pipette should be rinsed by drawing distilled water in and out of the tip several times. If this step is omitted, salt crystals will form when the pipette dries and may be impossible to dissolve completely. The pipette should be stored tip-up in a safe location.

F. Injecting Eggs

Next, we describe a general protocol for injecting fertilized eggs, followed by a description of some special considerations when injecting using epifluorescent or brightfield optics. We also discuss differences in methods used to inject *L. variegatus* and *S. purpuratus* eggs.

1. General Injection Procedure

Attach a filled injection pipette to an appropriate pipette holder and mount the holder on the micromanipulator. The injection pipette should point downward at an angle of 15 to 20°. Place a dish of rowed eggs on the stage of the inverted microscope, orienting it such that the row of eggs will be aligned vertically when viewed through the eyepieces. Using brightfield illumination, adjust the stage to bring the row of eggs to the middle of the field of view and focus on the row of attached eggs. Slowly lower the injection pipette using the micromanipulator, adjusting its position as necessary to gradually bring the tip of the pipette into the same plane of focus as the eggs. When positioning the micropipette, it is generally advisable to close the condenser aperture of the microscope almost completely in order to maximize the contrast and depth of field in the transmitted light image. This makes it easier to locate the micropipette if it is far above the plane of focus (it will appear as a dark shadow). When the tip of the micropipette comes into focus, confirm that the tip has not been broken inadvertently. Set the picospritzer to the upper end of the usable range (~300 msec) and depress the foot pedal. If solution is expelled from the micropipette, the tip is open, and one can then adjust the pulse of the picospritzer to deliver an appropriate injection volume (see following text).

If no injection solution is emitted from the micropipette (and this is usually the case), the tip must be broken open. Carefully raise the injection pipette slightly above the row of eggs and move it toward the glass coverslip fragment. Focus on the front edge of the coverslip fragment and bring the tip of the micropipette into the same plane of focus using the z-axis movement of the micromanipulator. Using the XY stage movement of the microscope, carefully bring the tip of the micropipette into contact with the front edge of the coverslip fragment. Only the slightest touch is necessary to fracture the tip! Depress the foot pedal and confirm that injection solution is delivered from the pipette tip. If not, repeat the process. Once the tip is open, adjust the injection volume to the desired amount. As will be discussed, pipettes with large openings can puncture eggs but are subject to backflow problems.

When the injection pipette is open, raise it slightly above the substrate and move the XY stage controller such that the top of the row of eggs is in the center of the field of view. Focus on the approximate equator of the eggs and bring the tip of the micropipette into the same plane of focus. Prepare a dilute sperm solution and fertilize the eggs by placing one or more drops of dilute sperm above the row. Higher concentrations of sperm are usually needed to fertilize eggs on protamine sulfate-coated dishes than eggs in suspension, because many sperm adhere to the charged surface of the dish. Abnormal cleavage of uninjected eggs, however, may indicate that too much sperm was added and that the embryos are polyspermic. When the fertilization envelopes elevate, the tip of the injection pipette should be readjusted to the midline of the eggs as the raising of the fertilization envelope lifts the eggs slightly. Note that eggs can be fertilized in the presence of para-aminobenzoic acid to prevent hardening of the fertilization envelope (McClay, Chapter 14), but we have not found this to be necessary for the injection pipette to penetrate the envelope.

Puncture the egg at the top of the row by using the stage movement of the microscope to drive the egg into the micropipette (Fig. 6). A sharp movement will assist in puncturing the membrane. It is also useful to deliver a pulse of injection solution at the same time the egg is forced against the pipette, as the flow of solution from the tip aids in penetrating the cell. When the pipette tip contacts the egg surface, the fertilization membrane and egg membrane should deform almost simultaneously. If the fertilization envelope is pushed inward but the plasma membrane of the egg is not deformed, the tip of the injection pipette is positioned too far above or below the center of the egg. The height of the pipette tip must then be adjusted with the z-axis movement of the micromanipulator. It is sometimes useful to raise the tip of the micropipette slightly above the equator of the egg so that the tip pierces the upper hemisphere of the egg, but injections at the equator generally work well.

Note that our injection method involves using the microscope stage controller to move eggs laterally against a stationary microinjection pipette which is positioned

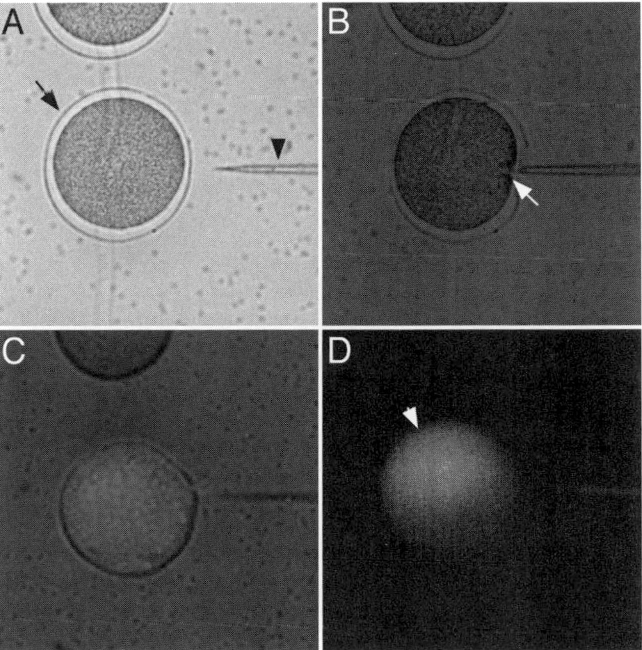

Fig. 6 Microinjection of rhodamine dextran. A fertilized egg is attached to a protamine sulfate-coated injection dish. (A) Just prior to injection (transmitted light only). The egg has been fertilized and the fertilization envelope (arrow) has elevated. Numerous sperm are attached to the surface of the dish. The injection pipette is indicated by an arrowhead. (B) Penetration of the egg (combined transmitted light/epifluorescence). The tip of the injection pipette has pierced the fertilization envelope and is indenting the plasma membrane of the egg (arrow), immediately before puncturing it. (C, D) Delivery of injection solution (C—combined transmitted light/epifluorescence, D—epifluorescence only). A pulse of pressure has been delivered with the picospritzer and fluorescent injection solution has been expelled into the cytoplasm of the egg (arrowhead, D). (See Color Insert.)

at a slight angle. The shear of the pipette tip against the egg membrane may assist in penetrating the membrane. Other investigators use a different approach, and inject by moving the injection pipette with the micromanipulator while keeping the egg stationary. To do this, one can use a remote, hydraulic single-axis micromanipulator (e.g., Narishige Model MO-22) or other style of micromanipulator that allows one to produce sharp movements of the injection pipette along its long axis. For rapid, repetitive injections, it is important that the experimenter is required to make only small hand movements.

Depress the foot pedal to expel injection solution into the cytoplasm. If the picospritzer adjusted to a pressure of 50 psi, a pulse duration of 60 to 300 msec is usually sufficient to deliver an amount equal to 1/100 to 1/5 of the volume of the egg cytoplasm. The volume of solution injected is dependent not only on the pressure (which is held constant at 50 psi) and pulse duration, but also on the size of the pipette opening, which varies from pipette to pipette. Moreover, the effective opening of a particular injection pipette can change during an experiment if the tip becomes partially clogged. If too much solution is injected into the eggs they will cleave abnormally or not at all and in extreme cases they will lyse immediately after injection. With experience one learns to judge reasonable injection volumes. We typically inject a range of volumes in each experiment by delivering one, two, four, or eight pulses into different eggs with the same injection pipette, while keeping the pulse duration constant. It is simpler to vary the injection volume by varying the number of pulses than by constantly readjusting the pulse duration.

When injecting fluorescent solutions, it is usually possible to determine retrospectively which embryos received a given injection volume based on the relative intensity of the fluorescence. When injecting nonfluorescent solutions, however, this is not possible and it is often advisable to inject eggs in separate dishes with small, medium, or large volumes. Whatever the approach, if the eggs injected with the largest volumes fail to cleave normally one can be confident that the upper end of the usable range has been reached, at least for that particular injection solution. As a general rule, keep injection solutions concentrated to minimize the volume of solution that must be injected into the eggs to see the desired effect. Also, note that some injection solutions (i.e., certain morpholinos) are more toxic than others, irrespective of the volumes injected.

After successfully injecting the first egg, use the stage controller of the microscope to withdraw the pipette and move to the next egg in the row. With experience, it is possible to inject ~15 eggs/minute. When injections are proceeding rapidly the operator primarily uses the stage movement of the microscope, with occasional adjustments to the pipette height using the z-axis controller of the micromanipulator. If the tip of the injection pipette becomes clogged during an experiment, it is usually possible to reopen the tip by breaking it against the glass coverslip fragment. If the injection pipette becomes covered with sperm or egg cytoplasm, it can be cleaned by pushing the tip gently into the small strip of double-sided tape (see Section III.G.6).

When injections have been completed, remove the protamine sulfate-coated dish from the microscope stage and place it in a humid chamber to allow the embryos to develop. If development is to continue for several days, it is advisable to change the seawater to eliminate the excess sperm. Store the humid chamber at a temperature appropriate for the particular species. To harvest free-swimming embryos after they have hatched from the fertilization envelope, simply pour the seawater from the dish into a centrifuge tube and pellet the embryos. To harvest embryos before hatching, they must be detached from the protamine sulfate-coated dish by pipetting a strong stream of seawater across the surface of the dish or by using a glass needle to strip them from the fertilization envelope (a microinjection pipette works well for this). The latter operation must be carried out under a microscope and can be performed freehand or with a micromanipulator.

2. Special Considerations

a. Injecting Using Epifluorescence

When injecting fluorescent solutions, such as a mixture of a morpholino and a fluorescent dextran, a combination of brightfield and epifluorescence optics is used (Fig. 6). Begin by bringing the injection pipette into focus with the row of eggs, as has been described. Dim the transmitted light source so that only the fluorescent image is visible. The fluorescent injection solution should be apparent inside the pipette. Depress the foot pedal to expel injection solution. A substantial plume of fluorescent injection solution should be emitted from the pipette tip. If the pipette is fluorescent but no fluorescent solution is emitted, the tip is not open and must be broken against the coverglass fragment, as has been described. If the injection pipette is visible but not fluorescent, it may have too large a tip opening, which can result in backflow of seawater into the pipette (see Section III.G.5). Depress the foot pedal repeatedly and observe whether fluorescent injection solution moves down to the pipette tip. If it does, but retreats rapidly into the pipette after each pulse is delivered, the diameter of the tip is too large and the pipette should be discarded. Increase the intensity of the transmitted light source slightly so that the eggs can barely be visualized. Add one drop of dilute sperm solution above the row of eggs to be injected. Wait a few seconds for the fertilization envelope to elevate and begin to inject. Successful injection can be confirmed by observing diffusion of the fluorescent injection solution throughout the egg cytoplasm (Fig. 6).

A valuable approach to learning microinjection is to practice by injecting a solution of 20% glycerol and 0.1% fluorescent dextran. Inject the eggs as has been described, and then assess subsequent development of the fluorescent, injected embryos. Cleavage of uninjected and injected embryos can be compared by observing the protamine sulfate-coated injection dish at frequent intervals. After hatching, embryos can be collected by centrifugation and examined. If the great majority of the dextran-labeled embryos develop on schedule with control embryos and appear phenotypically normal, one can proceed to experimental injection solutions.

b. Injecting Using Transmitted Light

We typically do not mix mRNAs with fluorescent dextrans (see preceding text). The experimenter therefore relies on visualizing the mixing of the glycerol in the injection solution with seawater or the cytoplasm of the egg. Without a fluorescent marker for distinguishing injected eggs from uninjected ones, it may be important to inject all or most of the eggs on a dish. This may require rowing fewer eggs per dish. In addition, it is generally advisable to raise uninjected, sibling embryos in a separate dish for comparison with the injected embryos. Distinguishing injected from uninjected embryos is less of an issue when injecting GFP-tagged constructs, as the GFP fluorescence can usually be used to identify eggs that have been injected.

Injection of mRNAs is carried out using brightfield illumination. It is generally advisable to close the condenser aperture almost completely in order to maximize contrast and depth of field. Focus on the eggs and bring the pipette tip into the plane of focus. Depress the foot pedal several times and watch the tip of the pipette for signs of glycerol emission. The glycerol in the injection solution requires some time to mix with seawater and appears as an oily swirl when first discharged from the pipette tip. If no injection solution is expelled even after several pulses, move the pipette to the coverglass fragment and crack the tip open slightly. It is difficult to discern the glycerol solution from seawater in the pipette, so it is important that the pipette opening be kept small to reduce the likelihood that seawater will be drawn back into the pipette, thereby diluting the injection solution. If one finds that after waiting several seconds without expelling solution from the pipette, several pressure pulses are required to bring the glycerol solution to the pipette opening (i.e., to see the glycerol–seawater mixing effect at the tip), a new pipette with a smaller opening may be required. Sometimes this effect becomes less pronounced during the injection process, as the injection pipette tip becomes partially obstructed with cellular debris and effectively smaller. When injecting mRNAs it is advisable to continuously tap the foot pedal to minimize problems with backflow (see Section III.G.5).

When the pipette opening is suitable, bring the tip into contact with the midline of an egg. Depress the foot pedal twice in quick succession to displace any seawater and immediately move the egg sharply against the pipette tip using the XY stage controller of the microscope. Expel the injection solution into the cytoplasm of the egg. One can be confident that the egg was injected if the glycerol diffuses throughout the cytoplasm. Use the size of the bolus of glycerol solution in the egg cytoplasm as a rough indication of the volume injected.

c. Species Differences

There are important differences in the methods used to microinject *S. purpuratus* and *L. variegatus* eggs. *L. variegatus* eggs are very easy to inject. They are large, develop at room temperature (18–25 °C), and can be injected at essentially any time after fertilization. In contrast, *S. purpuratus* eggs must be kept constantly cool and can be injected only during a short period of time following fertilization. The agarose-coated dish containing dejellied eggs should be kept on ice and the seawater used to fill the protamine sulfate-coated dishes should be precooled on ice. After a dish of eggs has been injected it should immediately be placed in a 15° incubator. If these steps are taken, it is not necessary to use a cooled microscope stage for injections. After fertilizing *S. purpuratus* eggs one must work very quickly because the cortex of the egg changes within minutes, making the cell very difficult to penetrate. Some batches of eggs are injectable for 10 to 20 min while others become impossible to work with after only 3 to 5 min. For that reason, *S. purpuratus* eggs must be fertilized and injected one dish at a time. As has been described (see also Fig. 5), if eggs are rowed in several groups per plate and a very dilute sperm solution is used to fertilize each region independently of the others, it is possible to inject a greater number of eggs on each dish. When fertilizing each short row of eggs we typically place a single drop of dilute sperm near the top of the row. As the sperm disperse in the medium they fertilize the eggs asynchronously, beginning with the eggs at the top of the row. The operator can then inject the eggs from the top of the row to the bottom, following the wave of fertilization. This makes it possible to inject a greater number of eggs in each individual row.

G. Troubleshooting

1. The Injected Egg is Surrounded by a Halo of Fluorescence that Appears Larger than the Egg

Sometimes the injection pipette penetrates the fertilization envelope but not the plasma membrane. The injection solution is then delivered into the space between the envelope and membrane. When injecting fluorescent solutions, this creates a halo of fluorescence that surrounds the egg. The effect may be caused either by difficulty in puncturing the eggs (see following text) or by positioning the tip of the micropipette too far above or below the equator of the egg. To ensure that the tip of the micropipette is positioned near the equator of the egg, follow the procedure described in Section III.F.1.

2. The Tip of the Injection Pipette Pushes the Egg Across the Plate or Causes the Egg to Roll

This indicates that the eggs are not adhering firmly to the surface of the injection dish. It is almost impossible to inject under these conditions, and if too many eggs are nonadherent, one must discard the plate. The problem can lie with the eggs, the dishes, or both. Occasionally the lack of stickiness results from eggs that have not been fully dejellied and additional washing can eliminate the problem. If the eggs have been dejellied as has been described, however, this is rarely the cause of the problem. Ensure that adequate time has been allowed for the rowed eggs to adhere to the protamine sulfate substrate before disturbing the dish (1–2 min). Another possibility is that the dejellied eggs are getting old. A good batch of washed eggs will retain their stickiness for several hours, but this stickiness decreases over time. In practice, plan to inject eggs as soon after dejellying as possible. If none of these steps makes a difference, then the problem probably lies with the injection dishes. If the injection dishes are old (>1 month) they may have lost some of their adhesivity, and the problem can often be solved by coating new plates. Occasionally the 1% protamine sulfate solution must be replaced. If another member of the laboratory has a batch of injection dishes that have been working well, test the adherence of the "problematic" eggs to these dishes to help pinpoint the cause of the problem.

3. The Injection Pipette Does Not Puncture the Egg

Assuming that the microinjection pipette is positioned properly (see preceding text), one possible problem is that the tip of the injection pipette is too large or blunt. In practice, surprisingly large tips (with openings of several microns) can puncture eggs if other conditions are optimal. Switching to a new injection pipette is a simple option, however. When working with *S. purpuratus*, the impenetrability of the membrane is probably an indication that too much time has elapsed following fertilization. This is certainly the case if a particular row of fertilized eggs is at first easy to inject with a given micropipette but becomes more difficult to inject over time. Give yourself more time to inject the row by rowing fewer eggs on each plate or by rowing the eggs in several separate sections and using a very dilute sperm solution to fertilize each group independently. Try depressing the foot petal repeatedly as the injection pipette approaches and then contacts the surface of the egg. The added pressure exerted by the injection solution often aids in penetrating the plasma membrane.

4. Cytoplasm Oozes Out During Injection

Cytoplasm oozing from an injected egg indicates that the tip of the microinjection pipette is too large, too much solution is being injected, or the pipette is inserted too deeply into the cell.

5. Fluorescent Injection Solution is Not Apparent in the Tip of the Injection Pipette or Can be Seen to Withdraw into the Pipette After Each Pressure Pulse

When using the Picospritzer II, if the opening of the injection pipette tip is too large, seawater will be drawn up into the pipette after each pulse of pressurized air is delivered. It is critical that this potential problem be recognized because when extensive backflow occurs, each egg is actually being injected primarily with seawater rather than the injection solution. There are two options when using the Picospritzer II. First, no appreciable backflow occurs when the tip opening of the injection pipette is small. Switching to a new pipette with a smaller opening will solve the problem. Second, when slight backflow is occurring, constantly tap the foot pedal (2–3 times/s) to keep the concentrated injection solution at the pipette tip. We do this routinely whenever injecting, unless the tip of the pipette is inside the cytoplasm of an egg. After penetrating the egg, continual tapping is not necessary because the viscous egg cytoplasm is not drawn up into the pipette after a pulse is delivered. Some companies (e.g., Harvard Apparatus, Holliston, MA, USA) manufacture picospritzers that provide a "holding pressure" intended to eliminate this backflow problem. These instruments are more expensive and we have not compared them directly with the Picospritzer II in this regard, although they would be worth considering. Finally, some investigators inject using a constant flow of injection solution. This eliminates backflow but also prevents one from carefully controlling the amount of solution injected.

6. The Tip of Injection Pipette is Coated with Eggs or Cellular Debris

To dislodge eggs or cellular debris from the injection pipette, carefully lower the tip onto the double-sided tape and draw it along the surface, leaving the debris on the tape. The tip can also be cleaned effectively by pushing it into the front edge of the tape and then withdrawing it (note: avoid hitting the plastic layer in the middle of the double-sided tape that separates the two softer layers of adhesive). It may be necessary to repeat these operations several times. Alternatively, lifting the tip of the injection pipette through the air/water interface will dislodge any large debris.

7. The Injection Pipette Clogs

First, try the quick fix of gently flexing the pipette tip against the bottom of the plate while depressing the foot petal. Try this in combination with changing the time setting on the picospritzer to a longer pulse (~400 ms). If this fails, try to reopen the tip by breaking it against the glass coverslip fragment. If this also fails, discard the clogged pipette and try a fresh one.

8. The Fertilization Envelope Elevates Before Any Sperm is Added

Artificial activation suggests that the eggs are of suboptimal quality. If the eggs are activated in a culture bowl without sperm present, discard them. Artificial activation following rowing may occur if the injection dishes are too heavily coated with protamine sulfate (e.g., if they were not rinsed with water or if the eggs were damaged during rowing).

9. After Adding Sperm, Many of the Eggs on the Injection Dish are Not Fertilized

A low fertilization rate is often an indication of poor gamete quality and it may be necessary to obtain fresh eggs and/or sperm. If a test bowl of washed eggs fertilizes well, however (>95% fertilization), there can be several other explanations. If the sperm are too dilute or weakly motile, most will adhere to the surface of the dish rather than the eggs. Try adding more sperm or making a fresh dilution of sperm. Occasionally, eggs may not fertilize well if they sit for too long after rowing. Eggs can also be damaged by mouth pipetting, particularly if the pipette has an unusually small opening. If the eggs appear misshapen or ragged after rowing (but not prior to rowing), this may be the problem. Note, however, that poor batches of eggs are sometimes especially susceptible to damage during rowing. If a test bowl of washed eggs does not fertilize well then the problem lies with the quality of the gametes, provided the quality of the seawater is not an issue. If the sperm appear nonmotile, try making a fresh dilution of the same sample of concentrated sperm. If these are also nonmotile it will be necessary to obtain fresh sperm from a different male. If the sperm are highly motile, the problem is likely to lie with the eggs. It may be necessary to obtain fresh eggs from a different female.

References

Colin, A. M. (1986). Rapid repetitive microinjection. *In* "Echinoderm Gametes and Embryos" (T. Schroeder, ed.), *Methods Cell Biol.* **27,** 395–406.

Conner, S. D., and Wessel, G. M. (2000). A rab3 homolog in sea urchin functions in cell division. *FASEB J.* **14,** 1559–1566.

Chui, K. K., Rogers, G. C., Kashina, A. M., Wedaman, K. P., Sharp, D. J., Nguyen, D. T., Wilt, F., and Scholey, J. M. (2000). Roles of two homotetrameric kinesins in sea urchin embryonic cell division. *J. Biol. Chem.* **275,** 38005–38011.

Hiramoto, Y. (1956). *Exp. Cell Res.* **11,** 630–636.

Hiramoto, Y. (1962). Microinjection of live spermatozoa into sea urchin eggs. *Exp. Cell Res.* **27,** 416–426.

Kiehart, D. P. (1981). Studies on the *in vivo* sensitivity of spindle microtubules to calcium ions and evidence for a vesicular calcium-sequestering system. *J. Cell Biol.* **88,** 604–617.

Kishimoto, T. (1986). Microinjection and cytoplasmic transfer in starfish oocytes. *In* "Echinoderm Gametes and Embryos" (T. Schroeder, ed.), *Methods Cell Biol.* **27,** 379–394.

Kumano, M., Carroll, D. J., Denu, J. M., and Foltz, K. R. (2001). Calcium-mediated inactivation of the MAP kinase pathway in sea urchin eggs at fertilization. *Dev. Biol.* **236,** 244–257.

Kurita, M., Kondoh, H., Mitsunaga-Nakatsubo, K., Shimotori, T., Sakamoto, N., Yamamoto, T., Shimada, H., Takata, K., and Akasaka, K. (2003). Utilization of a particle gun DNA introduction system for the analysis of cis-regulatory elements controlling the spatial expression pattern of the arylsulfatase gene (HpArs) in sea urchin embryos. *Dev. Genes Evol.* **213,** 44–49.

Rogers, G. C., Chui, K. K., Lee, E. W., Wedaman, K. P., Sharp, D. J., Holland, G., Morris, R. L., and Scholey, J. M. (2000). A kinesin-related protein, KRP(180), positions prometaphase spindle poles during early sea urchin embryonic cell division. *J. Cell Biol.* **150,** 499–512.

Ruffins, S. W., and Ettensohn, C. A. (1996). A fate map of the vegetal plate of the sea urchin (*Lytechinus variegatus*) mesenchyme blastula. *Development* **122,** 253–263.

Steinhardt, R., Zucker, R., and Schatten, G. (1977). Intracellular calcium release at fertilization in the sea urchin egg. *Dev. Biol.* **58,** 185–196.

CHAPTER 14

Methods for Embryo Dissociation and Analysis of Cell Adhesion

David R. McClay
Department of Biology
Duke University
Durham, North Carolina 27708

I. Introduction

This chapter provides methods for studying cell adhesion in the sea urchin embryo. Cell adhesion is an important component of any multicellular organism and dynamic changes in adhesion are crucial to morphogenesis. Thus, it is of great value to have assays for assessment of adhesive mechanisms and for experimental analysis of adhesive changes that occur during development. This chapter gives assays and formulary for several sea urchin species commonly in use around the world. The sea urchin embryo has been used for many cell adhesion and cell transplantation studies over the years and a number of overlapping approaches

METHODS IN CELL BIOLOGY, VOL. 74

0091-679X/04 $35.00

and reagents have been used for both. These include simple to more quantitative adhesion assays, *in vitro* culture of micromeres to produce spicules in culture, adherence of cells during cell transplantation, and recombination experiments using whole or fragmented embryos. Of importance, adhesion assays began by examining adhesive phenomena long before analysis of the detailed mechanisms was possible. In contrast, current research questions examine the causal molecular bases of those phenomena. In order to ask ever more complex questions, the assays for testing become more quantitative and, in some cases, more complicated. This chapter reflects those advances.

Before performing the adhesion assays, it is necessary to obtain cell suspensions from the embryos, so the adhesion of the cells can then be measured in some way. This immediately introduces an artifact into the whole adhesion-measuring process. It is important to understand from the outset that single cells in suspension cannot be assumed to behave in the same way as they behave in the embryo. In fact, it is very unlikely they will behave in the same way. This means that every adhesion assay is performed under a cloud of uncertainty. It might then be asked, why do adhesion assays at all, if this is true? The simple answer to this question, or at least the answer provided by one who has studied cell adhesion for many years, is that even though many artifacts are present, assays that measure cell-to-cell or cell-to-substrate interactions do reveal some properties about how cells adhere to substrates and to one another in a multicellular organism. Even against the artifactual background, the adhesion assay becomes more powerful when one compares two populations of cells, one of which expresses a modified, adhesion molecule, or in which some other directed molecular perturbation exists. Since this capacity for single molecule perturbation is now possible in the sea urchin, adhesion assays once again become valuable tools for understanding mechanisms of cell associations and of morphogenetic change.

This chapter is divided into three sections. First, methods are presented for eliminating extraembryonic and extracellular matices. Second, a series of non-quantitative and semiquantitative adhesion assays are described, and finally, a more quantitative adhesion assay is described. This latter assay is adapted to a microscale for analysis of control cells vs cells from embryos perturbed by injection of an RNA or morpholinos. The quantitative assay allows the investigator to assess the adhesion of cells from 100 to 200 embryos, a number too small for most assays, but adequate for the assay presented.

II. Gaining Access to the Egg and Embryo

The egg is contained within a jelly layer and a vitelline envelope. After fertilization, the vitelline envelope is elevated, crosslinked, and becomes the fertilization envelope (FE). If one wants to examine cell adhesion prior to hatching, it is necessary to remove one or more of these layers. In addition, the embryo is surrounded by the hyaline layer, composed of the protein hyalin and a number of other matrix proteins. Fortunately, access to the egg is relatively simple, and none of the methods described here affect normal development. Prior to fertilization, the jelly coat and vitelline envelope may need to be removed for an experiment, while after fertilization, the only envelopes of concern are the FE and the hyaline layer.

A. Removal of the Jelly Layer

Two methods of choice for this process are mechanical shearing and pH shift. Passing jelly-coated eggs through 102 μm Nitex mesh several times easily removes the jelly coat of *Lytechinus variegatus*. To score for removal of the jelly coat, simply allow the eggs to settle in a concave glass dish. If the eggs touch without any space in between, the jelly has been removed. An alternative way to remove jelly is to add citric acid in SW (pH 4.6) or acid SW (SW plus 0.05N HCl to pH 5) dropwise to a beaker of eggs while measuring the pH. When the pH drops to about 5 to 5.2, immediately neutralize the eggs with NaOH (0.1N) or with 1 M glycine to pH 8. Then, let the eggs settle, decant, and replace with artificial seawater (ASW (see formulary in Table I)). For removal of the jelly layer from *Strongylocentrotus purpuratus*, most labs use the acid wash approach but the shearing method also works, using 60 μm Nitex mesh. Either method also works for *Paracentrotus lividus*, but because the jelly is so sticky in this species, the acid wash is preferable.

B. Penetration of the Vitelline Layer

Two methods are available for elimination of the vitelline envelope. These methods simply punch holes in the vitelline envelope and do not entirely remove it.

1. Trypsin Method

Incubate eggs with a tiny amount of crude trypsin (the cheapest grade from Sigma), added as a powder to the top of the beaker containing the eggs (use about 0.5 mg per 250 ml of ASW). Stir the ASW to circulate the eggs and dissolve the trypsin. Incubate for the time it takes for the eggs to settle to the bottom of the beaker. Decant the trypsin-containing SW and wash 3× in ASW before fertilizing the eggs. This approach eliminates the species-specific sperm–egg interaction mechanism so the eggs can be fertilized efficiently with sperm from a different species if one wishes to produce interspecific hybrids (McClay and Hausman, 1975). A complication with this approach is that the fertilization envelope is absent, so assessment of fertilization cannot be made conveniently until the two-cell stage.

Table I
Formulae for Dissociation and Culture Reagents (All recipes are calculated per liter)

Artificial Seawater (ASW)
28.3 g NaCl
0.77 g KCl
5.41 g $MgCl_2 \cdot 6H_2O$
3.42 g $MgSO_4$ or 7.13 g $MgSO_4 \cdot 7H_2O$
0.2 g $NaHCO_3$
1.56 g $CaCl_2$ dihydrate (add last)
pH to 8.2, salinity should be in the range of 34–36 ppt
Calcium-Free Seawater (CF)
26.5 g NaCl
0.7 g KCl
5.81 g $MgSO_4$ or 11.9 g $MgSO_4 \cdot 7H_2O$
0.5 g $NaHCO_3$
pH to 8.2, salinity should be in the range of 34–36 ppt
Calcium/Magnesium Free Seawater (CMF)
31 g NaCl
0.8 g KCl
0.2 g $NaHCO_3$
1.6 g Na_2SO_4
pH to 8.2, salinity should be in the range of 34–36 ppt
Hyaline Extraction Medium (HEM)
17.5 g NaCl
0.75 g KCl
1.25 g $MgSO_4$ or 2.5 g $MgSO_4 \cdot 7H_2O$
22.5 g Glycine
1.21 g Tris
0.76 g EGTA
pH to 8.2, salinity should be in the range of 34–36 ppt

2. DTT (Epel *et al.*, 1970)

Treat eggs with 10 mM dithiothreitol in ASW (dilute eggs 1:1 into 20 mM DTT, pH 9.4). Treat 5 to 7 min. Wash 2 to 3× with ASW. The eggs can then be fertilized. As with trypsin-treated eggs, the DTT treatment eliminates the fertilization envelope.

C. Removal of the Fertilization Envelope (FE)

The following protocol is the standard one for removal of the FE prior to experimenting with embryos between fertilization and hatching. Because the FE is elevated but is not crosslinked covalently, it is a better method than vitelline envelope removal because one can assess the success of fertilization before removing the membrane. Removal of the FE is a simple procedure and can be done in bulk. The FE normally is hardened within a minute or two after fertilization by covalently crosslinking a number of proteins. To block FE hardening, 2 to

10 μM PABA (paraminobenzoic acid) in ASW is used. Published methods also describe use of aminotriazole, but because it is far more carcinogenic than PABA, we prefer the latter reagent. Both reagents work in the same way to block protein crosslinking. Transfer unfertilized eggs to the PABA-ASW and fertilize them in this solution. The PABA can either be made fresh or can be made as a 1000X stock solution (2.0 M) and stored in the refrigerator for a year or more. In PABA-ASW, the fertilization envelope elevates with the same efficiency and with the same dynamics as in control fertilized eggs, yet the envelope is soft and can be removed at any time. After about 2 min., the zygotes can be washed out of the PABA-ASW and FE hardening is prevented permanently. For most applications, however, simply leave the zygotes in the PABA-ASW for several hours prior to an experiment and the embryos will suffer no adverse effects. To remove the soft FE, pass the embryos through a Nitex mesh filter. The following mesh sizes are optimal: for *Lytechinus variegatus*, 102 μm mesh; for *Strongylocentrotus purpuratus*, 73 μm mesh; for *Arbacia punculata*, 73 μm; for *Paracentrotus lividus*, 60 μm mesh. The size of mesh is not correlated with the size of the egg as much as it is correlated with the distance the FE is elevated above the egg's surface. A larger mesh can be used if there is a high elevation and one should push the mesh to a smaller size if the elevation distance is small. Of course, there is a lower limit to the size of the mesh because the egg must be able to pass.

D. Removal or Penetration of the Hyaline Layer and Embryo Dissociation

The sea urchin is a good system for adhesion experiments because the embryo is rather easily dissociated into a cell suspension in the absence of enzymatic treatments that digest adhesion molecules. The most significant impediment to embryo dissociation is the hyaline layer, which is an extracellular matrix surrounding the embryo. Fortunately, the hyaline layer is not covalently cross-linked and can be removed almost entirely to gain access to the embryo. The hyaline layer is difficult to remove from some species and very sticky in others. The following protocol and several modifications work for all the species used by our lab. Hyaline Extraction Medium (McClay and Marchase, 1979; HEM, Table I) quickly softens the membrane so that washing the embryos in HEM and then CF (Table I) eliminates most of the hyaline layer. On a macroscale, place embryos in a centrifuge tube, pellet in a hand centrifuge, replace the medium with HEM, pellet the embryos again, and replace the HEM with CF. By trial, one can determine the length of time necessary to remove the hyaline layer simply by placing embryos in a concave dish and determining whether the embryo surfaces touch or remain separated by the clear hyaline layer. On a microscale, transfer embryos from well to well in glass plates to remove the hyaline layer. For embryo recombination experiments, a brief wash in HEM (1–5 min) softens the membrane so the experiment can be performed; then if the embryo is returned to ASW, the hyaline membrane becomes tough again in the presence of the calcium in the ASW. Embryos cultured in HEM for up to 30 min and then returned to ASW develop into normal plutei. However, if one dissociates embryos in HEM, the cells do not survive well. For this reason, after softening or removing the hyaline membrane, the embryos are transferred to CF or CMF for dissociation.

The following gives more specific approaches we use for several species, for treatment of embryos in preparation for both transplantation experiments and cell adhesion experiments. Of importance, embryos vary from batch to batch, and as the embryo grows older, it becomes progressively more difficult to dissociate.

Lytechinus variegatus and *Arbacia punctulata:* A forgiving hyaline layer surrounds the embryos of these species. Soften the hyaline layer with a 1 to 2 min incubation in CF (HEM is not necessary until later stages; see following text). Embryos incubated in CF for hours develop normally as long as they are not jostled (which will easily separate the cells from one another). At later stages, the hyaline layer progressively toughens and cell adhesions strengthen. It becomes necessary to wash the embryos in HEM, then in CF. Batches of embryos, age of embryos, and the different tissues in the embryos vary in their responsiveness to this treatment. Therefore, to optimize embryo dissociation, alternate washes in HEM and CF, using a hand centrifuge to pellet the embryos with resuspension in the alternating solution. Each time the embryos are returned to the CF, remove a small sample to an Eppendorf tube, pipette up and down with a Pasteur pipette several times, and examine the suspension by microscopy. This will establish the number of alternate washes necessary to dissociate the embryos at any stage of interest. As an important control, if embryos are returned to ASW without dissociation at the end of this alternating solution treatment, they should develop normally. When the embryos in the small test sample dissociate easily by pipetting, repeat the protocol with its particular modifications that day on the embryos to be used for adhesion experiments. As a caution, early embryos dissociate quite easily, but the large cells are very delicate, especially the macromeres of 16-cell stage embryos. After hatching, cells adhere more tightly, so that one often obtains persistent clumps in the cell suspension. To remove the clumps, filter the cells through 15 μm Nitex. This step may need to be repeated several times to obtain a single cell suspension. Keep the cell suspension in CF until the beginning of the adhesion experiment, at which time return the cells to ASW. If embryos are to be dissociated at the 16-cell stage for isolation of micromeres, macromeres, or mesomeres, most protocols (Fink and McClay, 1985; Hynes and Gross, 1970; Okazaki, 1975) transfer the embryos at the 2- to 4-cell stage into CF (after removing the FEs), and culture the embryos to the 16-cell stage in CF. The embryos dissociate quite easily by gentle pipetting.

Strongylocentrotus purpuratus: This species has a much tougher hyaline layer than *Lytechinus*. Therefore, to dissociate embryos, start with a 5 min incubation in HEM plus BSA (2 mg/ml) and at 15 °C. The BSA is necessary to eliminate the stickiness of the softened hyaline layer in this species. After about 5 min (again, this varies with batch and age of embryos and an empirical test should be performed first), transfer the embryos to CF. The cells of this species are quite hardy and can withstand considerable punishment in preparation for cell adhesion experiments. In practice, however, do not leave blastomeres in CF for more than 15 min and test whole embryos for their tolerance to length of incubation in HEM. Keep the cell suspension on ice until the experiment is started.

Paracentrotus lividus (also *Anthocidaris crassispina, Pseudocentrotus depressus, and Hemicentrotus pulcherrimus*): A protocol originally from Giudice (1965) was used to dissociate *Paracentrotus*, and later the same protocol was used for Japanese sea urchins (Amemiya, 1971). Wash embryos in CF, then with 0.44 M sucrose containing 1 mM EDTA and 0.05 M tris-citrate buffer, pH 8.0. Pass the embryos through a pipette several times gently to dissociate. Again, as the embryos grow older, the stringency of the treatment must be increased. Alternatively, *Paracentrotus lividus*, at least, is dissociated easily after a brief rinse in HEM and transfer to CF for dissociation.

III. Cell Adhesion Assays: Nonquantitative and Semiquantitative

Many assays measure cell adhesion, though none is perfect. This is because cell adhesion is a complicated series of steps, and no single assay provides adequate information on that series of events. The selection of technology therefore depends on the particular question being asked. The following protocols provide a range of choices, from simple to complex, for addressing several questions related to adhesion. The advantages and shortcomings of each method are given in the description.

A. Cell Aggregation Assays—Nonquantitative

The original assays for cell adhesion used these approaches (Giudice, 1965; Spiegel and Spiegel, 1975). The phenomenon of cell aggregation, originally demonstrated by Wilson in reconstitution experiments with sponge cells (Wilson, 1907), allows one to ask about the inherent capacity of embryonic cells first to associate with one another and then to reorganize an embryo to a greater or lesser extent. With increasing knowledge of molecules involved in patterning, this approach again becomes useful for analysis of patterning decisions in the embryo. This assay is not very useful for molecular analysis of adhesion molecules because it measures an endpoint far beyond the initial adhesion. Obviously, if that initial adhesion is blocked or manipulated, the endpoint will be different, but there are many other cellular events that also may be modified in the experimenter's hands, often making the aggregation assay an unsatisfactory way to assess a single desired molecular event. Thus, while the aggregation assay is the simplest assay to employ, interpretations of data should be considered with care.

If the experimenter's interest is in whether the cells of the embryo maintain the capacity to sort into different cell types, simple procedures are readily available. An easy way to make cell aggregates is to pellet a cell suspension in ASW and then gently break up the pellet into clusters of cells of approximately embryo size with a Pasteur pipette. This is admittedly a crude approach but if one simply wants to start with aggregates to study the capacity of cells to sort out, this approach is useful. To employ this method, dissociate the embryos as has been described. Ensure that the cell suspension is entirely composed of single cells. This warning is not given lightly. If clusters of cells remain, those clusters, in an aggregation assay, tend to share in a community effect (Standley *et al.*, 2001) and give the investigator a false positive by reconstituting more pattern than single cells in the suspension are capable of providing. Therefore, after making sure the starting cell suspension is of single cells, centrifuge the cells in SW or ASW at the lowest speed necessary to pellet them (about $100\times g$ for 3 min is sufficient, or at a setting of 3 on a clinical centrifuge). Gently pipet the pellet and, by eye, break it up to give roughly embryo-size aggregates. Culture these aggregates in fresh Millipore-filtered SW or in sterile-filtered ASW. Over the first hour, simply rotate the aggregates on a shaker or a rocker at low speed. Let the aggregates settle several times during that first hour, and replace about three-quarters of the ASW. Then replace the ASW less frequently over the remaining incubation period. After a day for warm-water species, or longer for coldwater species, examine the embryoids for their capacity to sort. This examination can be done using any of a number of cell-specific markers available from different laboratories or, in some cases, commercially. For example, in an early reconstitution series of experiments, we used a simple commercially available cytochemical alkaline phosphatase reaction (Sigma) to show reconstitution of the midgut (McClay *et al.*, 1977). Modifications of this simple approach can be used to measure the capacity of perturbed cells to engage in cell sorting. This approach was applied effectively to vertebrate cells to show that different members of the cadherin family have a differential capacity to help a cell sort to a position in an aggregate (outside, middle, or inside) (Duguay *et al.*, 2003; Steinberg and Takeichi, 1994).

A second way to form aggregates is by rotation. Place the cell suspension in ASW in a 25 ml Erlenmeyer flask. Cells at 1×10^6 cells/ml will aggregate if they are rotated at about 50 to 70 rpm. By adjusting the speed of rotation, one can obtain larger or smaller aggregates of fairly uniform diameter. Not all the cells aggregate, so if the aggregates are to be cultured, it is best to remove the single cells remaining after several hours of aggregation. Allow the aggregates to settle to the bottom of the flask and carefully replace the medium with fresh ASW. This second assay adds one feature to the adhesion phenomenon not seen in the first aggregation approach. Cells tend to associate with one another more selectively in the rotation since a fluid shear is present to challenge the adhesion environment. This approach, though quite old, is an application that can be useful for examining perturbed cells (have they lost the ability to adhere relative to controls?) or for examining qualitatively any of the many components involved in an adhesion sequence (adhesion molecules, bifunctional binding proteins in the cytoplasm such as β-catenin, cytoskeletal proteins, energetics). In each case, perturbation of the molecule will alter the ability of the cells to aggregate. This aggregation approach is qualitative and not very useful for quantitative avidity determinations of cell adhesion. The assay is also difficult to quantify with any precision. This handicap means that perturbations must be significant to provide a convincing difference between experimental and control cells.

B. Semiquantitative Approaches to Measurement of Cell Adhesion

Quantitative methods for analysis of cell adhesion offer the investigator a greater capacity to learn details of adhesion function. Again, however, a given assay offers the investigator only a very limited window into the adhesion sequence. But, if the assay is designed properly, the investigator can choose which functional window of many possible to explore. In the following treatment, several assays will be described, some in more detail than others because the assays become increasingly complex as the desire for detailed information becomes greater. This section is far from exhaustive. Cell adhesion experts seem to be people who enjoy inventing assays, so assay design varies from deployment of a two-cylinder viscometer or a Boydon chamber (Elvin *et al.*, 1983; Evans and Proctor, 1978) to recent adaptations of optical tweezers to examine adhesion and cytoskeletal events on a nanoscale that is quantitatively measurable (Fallman *et al.*, 2004; Lambert *et al.*, 2002; Schmidt *et al.*, 1993; Simpson *et al.*, 2002). None of these approaches has been adapted for research on sea urchin embryonic cells, though all are possibilities, and a method like the optical tweezers should be adapted for use with sea urchin embryonic cells.

C. Simple Assays

The earliest approach at quantification of adhesion was to measure disappearance of single cells from suspensions of cells aggregating by rotation (Oppenheimer *et al.*, 1973). Data from this approach provides information on the progression of an adhesion of a cell population but offers little insight into adhesion itself unless combined with selected perturbations. It has a weakness as a measure since cell disappearance also occurs when cells lyse. Nevertheless, if cell lysis is modest, this approach can be used for many applications. Several methods are available for measuring disappearance of single cells but the two simplest methods are to count the number of cells at several timepoints with a hemocytometer, or to measure the OD of the cell suspension at a given wavelength (the cells scatter light in the cuvette, which is the readout). For different species, this wavelength may differ so an initial experiment to optimize scatter sensitivity will establish the wavelength of choice. To measure scatter, transfer a sample of the culture medium containing your cell suspension to a cuvette and take an OD reading at 540 nm. If there are no cells in the cuvette, there should be no light scatter. The more cells in the cuvette, the more light scatter, and therefore the more "apparent" absorbance (actually, the measure is largely the light scatter only). As the aggregates grow, there will be fewer particles in suspension and therefore less light scatter.

Alternatively, if a particle counter is available, it will also work to count the cells in suspension. Follow the directions from the manufacturer to use the particle counter. The result of each of these approaches is the same; measure of the loss of single cells. Another simple semiquantitative method is measurement of aggregate size. Place a cell suspension (about 1×10^6 cells/ml) in 3 ml in a 10 to 25 ml Erlenmeyer flask and rotate the flask at about 50 to 70 rpm. Image the aggregates at different timepoints by removing a small sample from the aggregation flask (use a pipet with a bore large enough to prevent shear as the aggregates enter the pipette. If a micropipettor is used, enlarge the bore of the yellow plastic tip by cutting off about 3 mm of the tip with a sharp blade). Image the aggregates, and then measure their diameter to obtain a simple measure of the progress of an aggregation.

These simple methods have limited usage over and above asking the question: can the cells aggregate? They provide some idea of rate but supply no details about mechanism, unless a particular perturbation is compared to controls and something about the molecule being perturbed is known from independent studies.

IV. Quantitative Centrifugation Assay

A method that allows the investigator to determine the avidity of an adhesion between cells or between cells and a matrix component uses centrifugal force to quantify the avidity of that cell interaction. The assay also is easily adapted to follow the strengthening of an adhesion due to cytoskeletal participation. In the assay, test cells are spun onto the bottom of flat microtiter wells containing either a matrix component or a freshly deposited monolayer of cells. The assay plate is then flipped and centrifuged to supply a force tending to remove the attached test cells. The measure of adhesion is the ability of the test cells to resist the centrifugal force and remain adherent to the substrate or to the cell monolayer. The method described in the following text is miniaturized compared to the original published protocol (Hertzler and McClay, 1999; McClay *et al.*, 1981). The reason for miniaturizing the assay is for utility in molecular perturbation studies. The experimenter often has relatively few cells to work with because the perturbation requires microinjection of a construct into eggs, so that several hundred embryos are the maximum amount of material available. The assay described in the following text provides specific information that allows one to measure adhesive forces of a small number of cells, or the assay can be used simply to assess rather qualitatively the capacity of perturbed cells to adhere relative to their control counterparts. A discussion of the variations to the assay at the end provides examples of experimental applications of this technology.

Materials needed: 384-well microtiter well plates (Nunc#142761). These are black-walled plates with glass optical bottom. In addition, one needs 3 M clear packing tape, a darkroom roller, and a good fluorescent microscope with imaging capability. For repetitive quantitative applications, it greatly speeds the cell counting to acquire software to analyze the images of bound cells. Several products are available that will automate the cell counting, e.g., Metamorph software (Universal Imaging Corp), Open Lab (Improvision), Compix, Inc., or equivalent software for image analysis. For fluorescent tags, use tetramethyrhodamine isothiocyanate (Sigma T3163) and CellTracker Green (5-chloromethylfluorescein diacetate) (Molecular Probes C-7025). Other dyes may be used as well, but these are the two that have proven to be most versatile in our lab. Both label embryos intensely at concentrations that have no affect on development. In addition, a refrigerated centrifuge with microtiter plate carriers is necessary (Beckman TJ6, Eppendorf 5810, Jouan 4.22 centrifuge, or a comparable centrifuge). It is important that the centrifuge be well balanced and that minimal vibrations occur in the centrifuge during either acceleration or deceleration.

Cell preparation: Dissociate embryos as described. To the cell suspension(s), add a small amount of either red (TRITC) or green (CellTracker green) dyes. Use the dyes sparingly as both will stain cells intensely at very low concentrations. Our method is to dip the dry yellow tip of a pipettor into the dry stock dye. The few grains of dye that attach to the pipet tip are dissolved into 50 μl DMSO. A microliter or less of the DMSO-dye solution is added to a 1 ml solution of cell suspension. Alternatively, if there is only a small volume of cell suspension, place the suspension in a round-bottom glass spotplate well and, with the tip of a P2 pipettor containing dye, watch under the dissecting microscope as a tiny amount of DMSO dye is released from the tip. Against a white background, the red dye should be added until it turns the CF just slightly pink. The CellTracker DMSO-dye initially is colorless so gauge the dye addition by the amount of DMSO introduced. Though CellTracker is colorless, the cells will quickly become green and it is easy to overstain with this dye. After incubating cells in the dyes for up to 15 min on ice, pellet the cells by centrifugation (about $100\times g$ for 3 min) and resuspend in CF. A check of the cell suspensions under fluorescence should reveal intensely red and green cells.

Substrate preparation: The 384-well plate bottoms are flat and tissue culture-treated, which provides an excellent surface for attachment of an extracellular matrix molecule or a monolayer of cells that serves as the target for the fluorescent cell suspensions. Note that because the cell monolayer is unstained, it is easy to distinguish these cells from the fluorescent test cells. The first difficulty encountered is that the individual wells of the 384-well plates are small, and it is easy to confuse the wells. With a Magic Marker, delineate the wells to be used for the assay, and cover the remaining wells with the clear packing tape. The 384-well plate can be used for many assays if the sectors are carefully labeled and the tape removed from a sector at a time.

For a cell-matrix attachment assay, first, determine the minimum concentration of substrate molecule that will provide coverage of the well bottom. This is done empirically by measuring attachment of cells to increasing dilutions of substrate. Choose the lowest concentration that gives attachment, as defined by the ability of the cells to resist removal from that substate when the plate is flipped and centrifuged at a force that tends to remove the cells from a blank well (see following text). For fibronectin, useful concentrations range from 1 to 25 ng/μl with 10 μl of solution applied to the well. The reason for this determination is that cells will be spun onto the matrix, and then pulled off the matrix again as a measure of the adhesive force of the interaction. The detachment occurs between the weakest-adhering components at the bottom of the plate. If the cell–matrix attachment is stronger than the matrix–matrix attachment, then in the adhesion assay, excess matrix deposition will not provide a measure of cell–matrix adhesion but instead will measure matrix–matrix interactions. For matrix molecules that are strong adhesive substrates, this dilution is easy to establish. Poorer adhesive substrates are difficult to effectively titer so use a concentration that supports adhesion of positive control cells, such as tissue culture cells. Add the matrix molecules to wells in dH_2O and incubate on the well bottoms for 30 min to overnight at 4 °C. Remove the water by aspiration and wash once with BSA (1 mg/ml) that has been freshly heat-inactivated (65 °C for 5 min). BSA sometimes has inherent adhesion properties itself but these are eliminated by the heat inactivation. The BSA binds to unbound charged sites on the substrate so when cells are added the BSA is nonadhesive and the only potentially adhesive component remaining is the matrix molecule being tested. Several kinds of controls should be used. Wells containing no substrate and no BSA should be very adhesive for cells. Wells treated with BSA only should not be adhesive. BSA is the best blocking agent we are aware of, though we have not searched exhaustively. In many experiments with fibronectin, laminin, collagen, and other matrix molecules, the BSA-treatment does not appear to interfere with the subsequent cell–matrix interaction.

For quantitative cell–cell adhesion studies, cells of an applied monolayer provide the target for adhesion on the well bottoms. It is important that the cells stick to the well bottoms very tightly and not be removed during the assay. To establish a good monolayer of cells, it may be necessary to experimentally modify the substrate in advance of centrifuging a suspension of cells onto the well bottom. Poly-l-lysine (1 mg/ml), protamine sulphate (0.025%), or even paraformaldehyde (4%), in each case, followed by a wash with distilled H_2O, can facilitate attachment of the cell monolayer. An important control for monolayer formation is to carefully examine the wells following a deposition of the monolayer by centrifugation at $30\times g$ for 5 min, and a removal force of $50\times g$ for 5 min. If the monolayer is intact after the removal force, the monolayer is ready for the assay. If, however, some of the cells come off, exposing areas of substrate, try other modifications. Again, as with the cell–matrix assay, deposit no more than a monolayer of cells; otherwise, the assay may well measure the adhesion of the monolayer cells toward each other rather than the adhesion of the test cells toward the monolayer cells. Also, use a freshly deposited monolayer rather than a monolayer that has been cultured for any period of time, because the adhesive surface of the cells shifts from the exposed surface if the cells are cultured. Carefully wash the monolayers with ASW containing heated-inactivated BSA (1 mg/ml) to block all exposed potential adhesive sites on the substrate.

Adhesion assay: Figure 1 shows a diagram of the assay. Add fluorescent cells to the wells at a dilution of less than 10^5 cells per ml (you want to avoid clumping of cells on the substrate either from incomplete dissociation or from piling on top of one another on the bottom of the well). Each well of the 384-well plate holds 80 μl. Fill the wells to a total of 85 μl to provide a positive meniscus on the top of the well. Add the fluorescent test cells in 10 μl of CF, and fill the balance of the well with ASW, all at 4 °C. Calcium is added as the wells are filled, but it is important to keep the cells in a nonadherent solution until the wells are filled, so the slight reduction in calcium due to addition of cells in 10 μl CF is a small compromise, as it results in only about a 12% reduction in calcium in the well. To keep track of the wells, we use sectors of 9 wells (3 × 3) surrounded by empty wells for each assay. The empty wells are necessary to receive the overflow of ASW when the wells are sealed with the clear packing tape. After filling the wells to be tested, roll the packing tape over the wells and press it with the roller to firmly attach the tape to the tops of the wells. This will squeegee away the few μl of excess medium in each well, but it is necessary to avoid air bubbles, which disrupt the adhesion being measured.

Place the microtiter plates on a plate carrier in a refrigerated tabletop centrifuge at 4 °C and spin the fluorescent cells onto the substrate (30×g for 5 min). A control plate that contains cells spun on only provides the on–only control for quantitation.

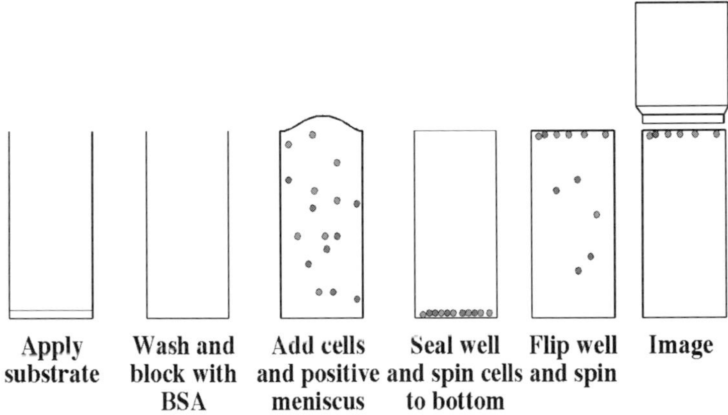
Apply substrate
Wash and block with BSA
Add cells and positive meniscus
Seal well and spin cells to bottom
Flip well and spin
Image

Fig. 1 Quantitative adhesion assay. This assay modification allows one to measure strength of adhesion of small numbers of cells. The chambers represent the steps in the treatment of one well in a 384-well plate. First, the substrate is added and then background blocked with BSA. Then cells are added. In the experiment depicted, cells of two colors are added to the same well, which allows a direct comparison between these two cell types. A positive meniscus is left at the top so when the well is sealed, no air bubbles remain trapped in the well. The cells are spun to the bottom of the well. The plate is flipped and centrifuged (a dislodgement force) at a speed selected to challenge the adhesive avidity of the cells. In the diagram, the green cells are more adherent than the red cells. To see that difference, the plates are transferred to a microscope stage and the well bottoms imaged by red and then green fluorescence. The adherent cells then are counted from the images. (See Color Insert.)

In the pilot experiments, examine the on–only microtiter plates with an inverted fluorescent microscope to determine the well-bottom coverage of the fluorescent cells. The cells should be distributed evenly on the substrate or monolayer and fluorescent cells should not be so concentrated that they touch other fluorescent cells. Invert the experimental plate and return it to the centrifuge. Minimally, an adhesion should survive incubation in the inverted position at $1\times g$ for 5 min. If not, the cells are nonadhesive to the substrate or to other cells. This is easily determined by allowing the cells to sit in the inverted plate for 5 min, then examining the well bottoms, this time on an upright fluorescent microscope. If the fluorescent cells are at the same density as the spin-on only samples, then 100% of the cells adhere at $1\times g$. This process is repeated at progressively higher centrifugal forces until the cells are completely removed from the substrate with centrifugation. If, by the time the maximal centrifugal force recommended by the centrifuge manufacturer is reached for the plate carriers, the cells are still not dislodged, their adhesiveness exceeds the capacity of the assay to measure adhesion by this method. In early studies of this sort, using an ultracentrifuge, strongly attached fibroblasts were not removed from a fibronectin substrate at $40,000\times g$, but the nuclei were spun out of the cells at this speed (Rich, 1981). Thus, the forces measured in the present assay include only those involved in the initial cell interactions, and up to a modestly strengthened adhesion.

To quantify the adhesion of the fluorescent cells use the spin-on only wells as 100% binding controls. Use BSA-treated blank wells to verify that 0% of the cells or a very low number bind to nonadherent substrates. Image each of these wells on the fluorescent microscope with a 5X or, maximally, a 10X objective to cover as much surface area as possible of the well bottom. There are two ways to count the cells. Crudely, one can simply manually count the cells on the image, or one can use imaging software to automatically quantify the number of cells. This is done in the software (see previous text for products with these capabilities) by asking the program to identify all spots above a selected pixel intensity. Then, to distinguish cells from debris, select minimal diameters of the spots, and finally, if spots are above a certain size (for example, if two cells encompass a single recognized spot), most software programs include directions for including that spot as multiples. While this phase of the assay design can be time-consuming at first, a macro can be established such that the computer will automatically count the cells. Then, from an Excel spreadsheet of the data, the computer will also plot curves of cell adhesion vs centrifugal force. For simple applications, this analysis need not be pursued, but if one is counting cells in hundreds of images for quantitative purposes, use of the software for this purpose greatly speeds the analysis.

A. Assay Variations

1. Red–green Challenge

Often, the experimenter wishes to compare two cell populations, one of which is a control and the other experimental. The difficulty with this comparison in many assays is that experimental error greatly confounds the comparison, making differences on the order of 50% or more necessary for significance. Error is introduced at every phase of any adhesion assay and is unavoidable in even the best hands. The goal is to minimize that error. The easiest way to do this is to include the two test cell populations, one labeled red and the other green, in the same well. Then, perform the assay, image the red cells and the green cells on the same substrate, and count. To minimize the error, add the red and green cell suspensions together at 4 °C prior to addition to the microtiter wells. That way, most pipetting errors are eliminated because the 100% binding well control, the 0% binding well control, and all experimental wells receive the same ratio of red vs green cells. There may be a small pipetting error in adding cells to the different wells, but the red-vs-green ratio will not be affected by that error. This assay is very useful when comparing perturbed cells vs control cells. In most cases, this comparison is very simple, and the experiment asks which cells adhere best. As an initial control, compare the adhesion of red control cells vs green control cells. Establish the ratio of red vs green on the 100% binding well. Now, spin the cells off the plate at a force such that around 50% remain bound to the substrate. Examine the well bottoms again. The red-vs-green ratio should be the same as in the 100% binding wells. Next, make the experimental cells one of the colors and the control cells the other and repeat the test. If the centrifuged red-vs-green ratio is different from the 100% binding ratio, then the perturbation altered the adhesiveness of the experimental cells relative to the control cells.

2. Initial Binding Test

If the entire adhesion assay is performed at 4 °C, it tends to measure the molecule-to-molecule affinity between cell surface-to-substrate or cell-to-cell. Four degrees essentially removes ATP-dependent steps from the adhesion so that only the avidity between the two surfaces holds the cells in contact with the substrate. These forces are quite modest on a molecule-to-molecule level (Dembo and Bell, 1987), and it is not particularly sensitive to number of molecules per cell area or per substrate area (many theoretical and experimental treatments suggest that the cells peel away from the substrate so that, with the time involved in the centrifugation run, the adhesion is no better than the force of a single bond (Dembo and Bell, 1987; Hammer and Lauffenburger, 1987; Kuo and Lauffenburger, 1993; Saterbak and Lauffenburger, 1996). This means, practically, that when comparing experimental vs control cell populations, very low centrifugal forces are necessary. A simple analogy is to imagine peeling Scotch tape away from the substrate. The entire surface of the tape adheres, but in peeling the tape, the only molecules that matter at any given time are the molecules at the immediate site of the peeling. In the same way, cells peel away from their substrate if the centrifugal force exceeds the strength of the molecule-to-molecule bond between cells, and with time under that force, the cell eventually peels away from the substrate.

3. Strengthened Attachments

Many cellular events strengthen an adhesion with time in an ATP-dependent fashion. The number of adhesion molecules per unit area increases by crosslinking attachment at the sites of close cell–cell or cell–substrate contact. The cell adhesion molecules attach to the cytoskeleton. Most importantly, multiple cell adhesion molecules, when attached to the cytoskeleton, cooperatively increase the adhesive strength of the attachment enormously. Each of these events may occur with time of contact between cells. To gain a sense of these properties, or to manipulate them, bring the cells into contact with the substrate, or with the cell monolayer at 4 °C, and spin off immediately to determine the initial avidity. Compare this to wells in which cells were spun on at 4 °C and spun off only after an incubation for some period at a higher temperature. Use the temperature at which embryos normally are grown. With time at that temperature, the attached cells should increase their avidity to the substrate, sometimes so dramatically that they are not removed even at very high centrifugal forces. Again, examination of red vs green cells provides an internally controlled comparison between experimental and controlled. This modification of the adhesion assay will effectively compare cells expressing a modified cytoskeletal protein relative to the control cells, and a difference will be seen if that cytoskeletal protein interferes with the normal progression of adhesive strengthening. In this way, multiple components of the adhesive complex can be examined.

4. Other Considerations

A lot of effort is required to adapt the centrifuge assay to a laboratory, so it is important to know how to optimize the assay for use. The centrifugation assay has the capacity to measure and distinguish relatively small perturbation differences between experimental and control cells, where the simpler aggregation methods do not. The most important aspect of the assay is to establish the 100% binding vs 0% binding differences. Conditions must be worked out to determine what allows control cells to bind at 100% to a particular substrate or to a particular cell. Under those same conditions, if substrate or cell monolayer is absent, then 0% of the cells should bind. In reality, it is difficult to achieve these conditions absolutely. In some assays, a background of 5 to 10% cell binding to the nonadhesive substrate is acceptable, especially if in the 100% binding test, 100% test cells remain bound to the substrate under the same conditions. If these conditions are established first, the assay comparison between control red vs green cells is very simple. If this cannot be achieved, because the BSA doesn't provide a good background coating, try other molecules, such as reconstituted dried milk, glycine, or other potential nonadhesive molecule. Of importance, that background coating must not cover the adhesive substrate to be tested. To be confident the blocking agent is not interfering with adhesion to the substrate molecule, adhere the cells to the substrate molecule in the absence of the blocking agent, and then after treatment with the blocking agent. That adhesion should not differ by much. Compare that difference to a blank well to which cells were added vs a blank well previously treated with a blocking agent. If conditions are perfect, the blocking agent should eliminate adhesion to the blank well but not interfere with adhesion to the substrate molecule of choice. Again, in reality, 100% blocking may not be achieved, and a small amount of interference with the substrate molecule may occur. Nevertheless, both the red and the green cells will encounter the same adhesive environment, so a comparison is valid.

For simple assays, it is unnecessary to go to the lengths described here for counting cells by software or for computationally building avidity-vs-force curves. These are necessary, however, if one wants to obtain highly quantitative information about an adhesion. It should be pointed out, however, that even if all the trouble is taken to make the assay as highly quantitative as possible, the window of adhesion seen is very limited. This is why single molecule perturbations vs control cells offers added power. There, the adhesion is assessed under conditions where potentially there is only one variable in the adhesion reaction.

The assay works well for sea urchin cells. A complicating factor occurs with these cells during development, however. Early in cleavage, the cells are very large. If those cells are used in the assay, only a very small percentage of the cell surface comes into contact with the substrate, and there is an enormous volume of cell tending to pull the cell away from the substrate when centrifugal forces are applied. As the cells become smaller and smaller, the actual adhesive force remains roughly the same but the volume of the cell is much smaller, and therefore the detachment force at a given centrifugal force is less than that for larger blastomeres (the detachment force is provided by the mass difference between the cells and the medium in which they are contained; McClay *et al.*, 1981). For that reason a realistic red-vs-green comparison should test cells of the same developmental age.

The adhesion assay, like any assay, has its limitations. If, for example, the cells do not adhere when the plate is simply inverted at $1 \times g$, then the cells, if adhesive at all, are too weakly adhesive to be measured by this means. Alternatively, if the cells stick so avidly that even high centrifugal forces cannot remove them, then the adhesion exceeds the capacity of the assay to measure. In practice, the assay has been used successfully with a wide variety of cell types under many different conditions. Even with mammalian fibroblasts which adhere quite avidly at 37 °C, that adhesive strength requires time to increase to the point where it exceeds the capacity of the assay to measure it. In that case, if one does the assay within the times at which strengthening is still measurable, it is possible to learn about the strengthening process.

Another limitation of the assay is the inherent variation both in the test cells and in the substrate or monolayer of cells. For example, the monolayer is not a flat plane but is a cobblestone-like surface. It is easy to imagine that the test cells adhere both to the rounded tops of cells and also in the crevices. The apparent avidity of those two interactions may be different when the assay is performed.

With this variable describing but one of many that will exist for each assay, the property being measured is the average experienced by the population of cells in the presence of multiple variables. Of value, one can consider the red–green cell comparison to vary only according to the perturbation experienced by one of those two populations. This red–green comparison then provides the greatest power in the assay because it limits the inherent experimental variation.

Anyone who has looked closely at early cleavage stages knows that during the cell cycle, the avidity of adhesiveness of cells changes dramatically. As development progresses, both the hyaline layer and a newly formed basement membrane become adhesive surfaces. The cells then switch from primary reliance on the hyaline layer to a greater adhesive reliance on the basement membrane. As the germ layers are specified, cells acquire differential adhesive properties. These and many other properties of the embryo involve adhesion but many of these may not have assayed appropriately by any one of the assays described herein. Thus, adhesion assays provide but one tool for the experimenter to examine how the dynamics of cell associations change during development. Nevertheless, for a restricted examination of adhesion molecules to their receptors in a biological context, for assembly of adhesion supporting cytoskeletons, and especially for the comparison between control vs perturbed, these assays have good utility, especially when the molecules in question can be selectively perturbed.

References

Amemiya, S. (1971). Relationship between cilia formation and cell association in sea urchin embryos. *Exp. Cell Res.* **64,** 227–230.

Dembo, M., and Bell, G. I. (1987). The thermodynamics of cell adhesion. *Current Topics in Membranes and Transport* **29,** 71–89.

Duguay, D., Foty, R. A., and Steinberg, M. S. (2003). Cadherin-mediated cell adhesion and tissue segregation: Qualitative and quantitative determinants. *Dev. Biol.* **253,** 309–323.

Elvin, P., Drake, B. L., and Evans, C. W. (1983). Adaptation of the Boyden chamber to flow conditions: Rheological effects on chemotaxis. *J. Immunol. Meth.* **64,** 295–301.

Epel, D., Weaver, A. M., and Mazia, D. (1970). Methods for removal of the vitelline membrane of sea urchin eggs. I. Use of dithiothreitol (Cleland Reagent). *Exp. Cell Res.* **61,** 64–68.

Evans, C. W., and Proctor, J. (1978). A collision analysis of lymphoid cell aggregation. *J. Cell Sci.* **33,** 17–36.

Fallman, E., Schedin, S., Jass, J., Andersson, M., Uhlin, B. E., and Axner, O. (2004). Optical tweezers-based force measurement system for quantitating binding interactions: System design and application for the study of bacterial adhesion. *Biosens. Bioelectron.* **19,** 1429–1437.

Fink, R. D., and McClay, D. R. (1985). Three cell recognition changes accompany the ingression of sea urchin primary mesenchyme cells. *Dev. Biol.* **107,** 66–74.

Giudice, G. (1965). The mechanism of aggregation of embryonic sea urchin cells: A biochemical approach. *Devel. Biol.* **12,** 233–247.

Hammer, D. A., and Lauffenburger, D. A. (1987). A dynamical model for receptor-mediated cell adhesion to surfaces. *Biophys. J.* **52,** 475–487.

Hertzler, P. L., and McClay, D. R. (1999). alphaSU2, an epithelial integrin that binds laminin in the sea urchin embryo. *Dev. Biol.* **207,** 1–13.

Hynes, R. O., and Gross, P. R. (1970). A method for separating cells from early sea urchin embryos. *Dev. Biol.* **21,** 383–402.

Kuo, S. C., and Lauffenburger, D. A. (1993). Relationship between receptor/ligand binding affinity and adhesion strength. *Biophys. J.* **65,** 2191–2200.

Lambert, M., Choquet, D., and Mege, R. M. (2002). Dynamics of ligand-induced, Rac1-dependent anchoring of cadherins to the actin cytoskeleton. *J. Cell Biol.* **157,** 469–479.

McClay, D. R., Chambers, A. F., and Warren, R. G. (1977). Specificity of cell–cell interactions in sea urchin embryos. Appearance of new cell-surface determinants at gastrulation. *Dev. Biol.* **56,** 343–355.

McClay, D. R., and Hausman, R. E. (1975). Specificity of cell adhesion: Differences between normal and hybrid sea urchin cells. *Dev. Biol.* **47,** 454–460.

McClay, D. R., and Marchase, R. B. (1979). Separation of ectoderm and endoderm from sea urchin pluteus larvae and demonstration of germ layer specific antigens. *Dev. Biol.* **71,** 289–296.

McClay, D. R., Wessel, G. M., and Marchase, R. B. (1981). Intercellular recognition: Quantitation of initial binding events. *Proc. Natl. Acad. Sci. USA* **78,** 4975–4979.

Okazaki, K. (1975). Spicule formation by isolated micromeres of the sea urchin embryo. *Amer. Zool.* **15,** 567–581.

Oppenheimer, S. B., Potter, R. L., and Barber, M. L. (1973). Alteration of sea urchin embryo cell surface properties by mycostatin, a sterol binding antibiotic. *Dev. Biol.* **33,** 218–223.

Rich, A. (1981). PhD Thesis. University of North Carolina. Chapel Hill, North Carolina.

Saterbak, A., and Lauffenburger, D. A. (1996). Adhesion mediated by bonds in series. *Biotechnol. Prog.* **12,** 682–699.

Schmidt, C. E., Horwitz, A. F., Lauffenburger, D. A., and Sheetz, M. P. (1993). Integrin-cytoskeletal interactions in migrating fibroblasts are dynamic, asymmetric, and regulated. *J. Cell Biol.* **123,** 977–991.

Simpson, K. H., Bowden, M. G., Hook, M., and Anvari, B. (2002). Measurement of adhesive forces between S. epidermis and fibronectin-coated surfaces using optical tweezers. *Lasers Surg. Med.* **31,** 45–52.

Spiegel, M., and Spiegel, E. (1975). The reaggregation of dissociated embryonic sea urchin cells. *Amer. Zool.* **15,** 583–606.

Standley, H. J., Zorn, A. M., and Gurdon, J. B. (2001). eFGF and its mode of action in the community effect during Xenopus myogenesis. *Development* **128,** 1347–1357.

Steinberg, M. S., and Takeichi, M. (1994). Experimental specification of cell sorting, tissue spreading, and specific spatial patterning by quantitative differences in cadherin expression. *Proc. Natl. Acad. Sci. USA* **91,** 206–209.

Wilson, H. V. (1907). On some phenomena of coalescence and regeneration in sponges. *J. Exp. Zool.* **5,** 245–258.

PART III

Cell Biological Approaches

CHAPTER 15

Analysis of Sea Urchin Embryo Gene Expression by Immunocytochemistry

Judith M. Venuti,* Carmen Pepicelli,† and Vera Lynn Flowers*

*Department of Cell Biology and Anatomy
Louisiana State University Health Sciences Center
New Orleans, Louisiana 70112

†Curis, Inc.
Cambridge, Massachusetts 02138

METHODS IN CELL BIOLOGY, VOL. 74

0091-679X/04 $35.00

I. Introduction

In 1941, Albert Coons initiated a major revolution in immunology and cell biology when he developed a technique for labeling specific antibodies with fluorescent dyes (Coons *et al.*, 1941). The technique of immunocytochemistry permitted the detection of virtually any antigenic protein in cells and tissues (Coons and Kaplan, 1950) and has now become an essential tool in most fields of biomedical research.

Since immunocytochemistry involves specific antigen–antibody interactions, it has advantages over traditionally used histological and enzyme staining techniques, which can identify only a limited number of proteins, enzymes, or cell structures. Immunocytochemistry enables demonstration of not only the presence, but also the intracellular localization, of an antigen. In addition, it can be used to simultaneously detect multiple antigens, allowing comparison of their relative distributions.

Immunocytochemistry has been widely used to identify embryonic territories and specific cell lineages during normal sea urchin development (Wessel and McClay, 1985). It is also an extremely useful tool for assaying the consequences of microinjections (for examples, see Angerer *et al.*, 2000; Gross *et al.*, 2003; Sherwood and McClay, 1999; Vonica *et al.*, 2000) or micromanipulations (for examples, see Logan *et al.*, 1999; McClay *et al.*, 2000; Oliveri *et al.*, 2003; Wikramanayake *et al.*, 1998). With the increased information gained from modern molecular techniques that alter gene expression in the embryo by microinjection of antisense morpholinos (see Chapter 28) or *in vitro* transcribed RNAs (see Chapter 27), indirect immunofluorescent labeling of whole embryos provides a rapid means of assaying perturbations that alter territorial specification and embryonic patterning. Since microinjection/micromanipulation techniques are usually applied to relatively few embryos at a time, this chapter will focus on general immunocytochemical techniques and methods for handling small numbers of embryos. To prevent embryo loss during processing, the researcher should use a minimal number of steps. Whole mount immunofluorescence uses the fewest number of steps, compared to other staining procedures. When combined with confocal microscopy, whole mount immunocytochemistry eliminates the need for time-consuming embedding and sectioning steps, which also can destroy antigens and reduce antibody accessibility. The advantage of whole mount preparations is that they provide three-dimensional information about the localization of antigens without the need for reconstruction from sections. We provide a general scheme for indirect immunofluorescence localization in whole embryos. It is recommended that familiarity with the general protocol be gained by staining unperturbed embryos by a batch procedure (Section II) prior to assaying small numbers of experimentally altered embryos (Section III).

The identification and localization of intracellular antigens in sea urchin eggs and embryos, particularly those involved in cytoskeletal changes during cytokinesis and fertilization, has a long history. An excellent reference for general cytological and immunocytochemical techniques that focuses on intracellular localization of antigens in sea urchin embryos can be found in Harris (1986). Many of the procedures outlined there are still applicable and their utility is improved by the use of confocal microscopy for imaging. Chapter 16 of this volume also outlines immunocytochemical methods for detection of cytoskeletal components, while immunocytochemical procedures optimized for use in other species can be found in Chapters 6 and 9. For those interested in nonfluorescent immunolocalization techniques in sea urchin eggs and embryos, we recommend Wright and Scholey (1993). Extremely useful information on the generation, characterization, and handling of antibodies can be found in Harlow and Lane (1988, 1998).

II. Immunocytochemistry on Whole Embryos

A. Collecting Embryos and Treatments Prior to Fertilization

1. Large Numbers of Embryos

When embryos are abundant, they can be processed in 1.5 ml microcentrifuge tubes and gently centrifuged between washes and incubations. We find that embryos often stick to plastic tubes, especially if one is not careful to spin them down or one does not include detergents in the wash buffers. Small glass tubes of various sizes also work well and have the added advantage that the embryo pellet (and any embryos that stick to the walls) can be seen clearly through the glass. Gentle centrifugation can be accomplished in a hand centrifuge, a clinical centrifuge, or in a variable speed microcentrifuge at its lowest setting ($\sim 1000 \times g$) for 1 to 2 min. Alternatively, embryos can be allowed to settle by gravity, as they sink readily once fixed. If centrifugation is to be avoided, swimming embryos can be settled by adding a few drops of fixative to the culture. Once embryos have settled to the bottom of the tube, the seawater to which fixative was added should be replaced with the 100% fixative solution. Swimming embryos can also be settled by the addition of a few drops of 10% sodium azide in Millipore-filtered seawater (MFSW). Gently invert the tube, and remove the sodium azide/MFSW once the embryos have settled. Work rapidly when using sodium azide to settle embryos because sodium azide kills embryos and morphological deterioration will occur if the embryos are not fixed immediately.

Even when large numbers of embryos are available, researchers may still choose to use the microwell plate staining method described later (with 100–200 embryos in the well). Although this is more laborious than processing in tubes, we find it generally gives more uniform labeling of all embryos (less variability in staining from embryo to embryo).

2. Limited Numbers of Embryos

If embryos to be processed for immunostaining are derived from microinjection or micromanipulation experiments, their numbers are likely to be limited and processing for immunostaining should be performed under a dissecting microscope. Visual monitoring of the embryos through all stages will assure limited loss and damage. Fixed embryos tend to stick to pipettes and dishes and are easily lost, regardless of the precautions taken; therefore, it is preferable to change solutions rather than move embryos through the different immunostaining reagents. When working with small numbers of embryos, it is also better to process a few that can be rapidly counted (~20) rather than a larger number (~100). By using an easily manageable number of embryos, loss of embryos can be assessed at different steps and a lost embryo found before proceeding to the next step. Recommendations for processing limited numbers of embryos are detailed in Section III. These procedures require a good dissecting microscope, small diameter pulled pipettes with proper bends, and an aspiration device (either a mouth pipettor or a handheld syringe), as illustrated in Fig. 2.

Eggs and embryos that retain their fertilization envelope (FE) may be impermeable to antibodies. Procedures to remove vitelline envelopes and FEs before fixing embryos are outlined in Chapter 20. Batch removal of the FE is generally not necessary for embryos that have been microinjected or micromanipulated since embryos can be manually dissected from FEs when they are collected individually by micropipette. It is generally easier to remove the FE than to remove the vitelline envelope (VE) from unfertilized eggs. To remove the VE from unfertilized eggs, one must use harsh treatments that can potentially damage the egg plasma membrane. With any treatment that removes the VE, no FE will be elevated, so fertilization success cannot be assessed until after first cleavage. It should also be remembered that eggs lacking fertilization envelopes or "naked" eggs are susceptible to polyspermy and agglutination. Polyspermy can be avoided by using small amounts of diluted sperm. Agglutination can be avoided by rapid dilution of the eggs after fertilization or by culture in CaFSW for 30 min following removal of membranes (Epel *et al.*, 1970).

B. General Indirect Immunofluorescence Steps

Standard indirect immunofluorescent staining proceeds through the following steps: fixation, permeabilization, post-fixation washes, blocking, primary antibody incubation, post-primary incubation washes, secondary antibody incubation, post-secondary incubation washes, and mounting (Table I). If enzyme-conjugated secondary antibody is employed, then an additional chromogenic development step and its termination is required before mounting. Since the latter requires additional steps and is generally not required for most commonly used sea urchin antibodies (Table II), this chapter focuses on indirect immunofluorescent procedures.

1. Fixation

Selection of a suitable immunocytochemical method depends on the type of specimen under investigation and the degree of sensitivity required. Tissue preparation is the cornerstone of immunohistochemistry. To ensure the preservation of tissue architecture and cell morphology, prompt and adequate fixation is essential.

Table I
General Immunostaining Procedure

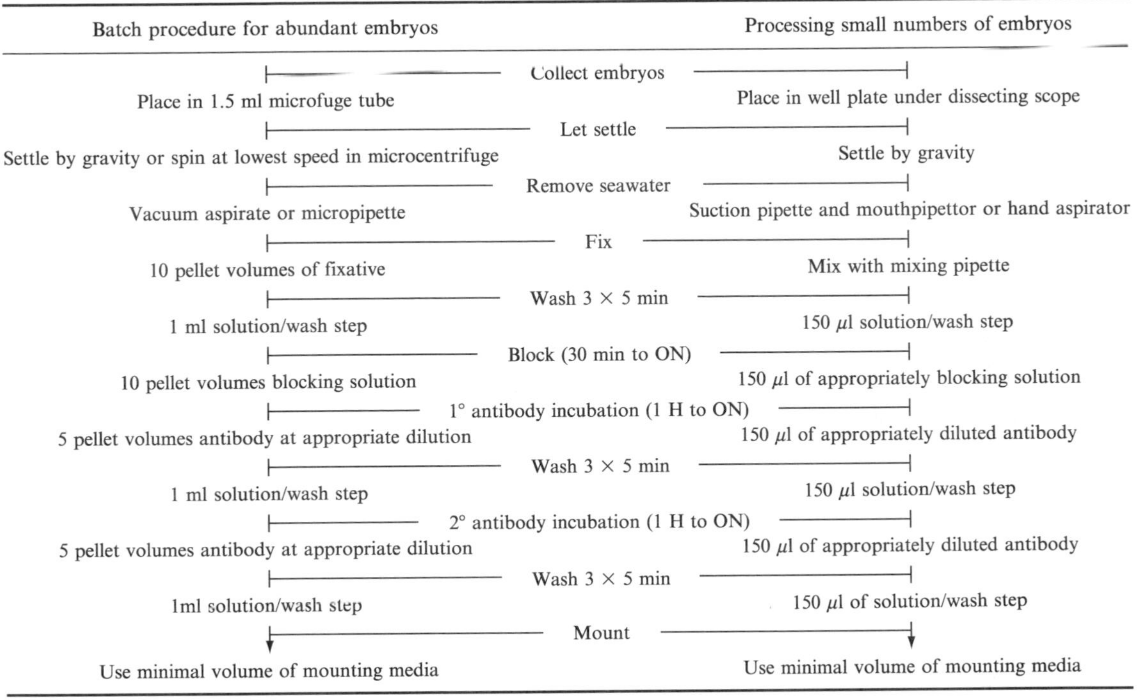

Batch procedure for abundant embryos		Processing small numbers of embryos
	Collect embryos	
Place in 1.5 ml microfuge tube		Place in well plate under dissecting scope
	Let settle	
Settle by gravity or spin at lowest speed in microcentrifuge		Settle by gravity
	Remove seawater	
Vacuum aspirate or micropipette		Suction pipette and mouthpipettor or hand aspirator
	Fix	
10 pellet volumes of fixative		Mix with mixing pipette
	Wash 3 × 5 min	
1 ml solution/wash step		150 μl solution/wash step
	Block (30 min to ON)	
10 pellet volumes blocking solution		150 μl of appropriately blocking solution
	1° antibody incubation (1 H to ON)	
5 pellet volumes antibody at appropriate dilution		150 μl of appropriately diluted antibody
	Wash 3 × 5 min	
1 ml solution/wash step		150 μl solution/wash step
	2° antibody incubation (1 H to ON)	
5 pellet volumes antibody at appropriate dilution		150 μl of appropriately diluted antibody
	Wash 3 × 5 min	
1ml solution/wash step		150 μl of solution/wash step
	Mount	
Use minimal volume of mounting media		Use minimal volume of mounting media

However, inappropriate or prolonged fixation may significantly diminish antibody binding ability. Therefore, fixation procedures for indirect immunocytochemistry are chosen to balance the preservation of soluble antigens and the maintenance of recognizable structure. Fixation is the most important step in the entire immunocytochemical procedure. There is no one universal fixative that is ideal for the demonstration of all antigens. The type of fixative used and the duration of fixation varies for different antigens. Some antigens can be easily destroyed by overfixation while others can be easily solubilized. If a "tried and true" protocol is not available for the antigen of interest, the best fixation must be determined empirically. Several different fixation protocols have been used successfully to localize a variety of antigens in sea urchin embryos. In our experience, the simplest fixation that frequently preserves morphology and provides antigen access is treatment in −20 °C MeOH for 20 min. MeOH fixes and permeabilizes simultaneously, whereas fixation with aldehydes requires inclusion of nonionic detergents such as Triton X-100 (0.1–0.5%) or Tween 20 (0.1–0.5%) during fixation or subsequent steps, to permeabilize cells. Post-fixation in ice-cold acetone or MeOH can also be used in combination with an initial aldehyde fixation to permeabilize.

Table II
Sea Urchin Antibodies

Antibody name	Antigen recognized	Monoclonal or polyclonal	Species specificty	Stages detected	Best fixation	Whole mount or sections	Citation	Email address of contributor	Notes
295	ciliated band	Mono	Lv, Sp, others?					dmcclay@duke.edu	
α–catenin	α–catenin	Poly	Lv, others?					dmcclay@duke.edu	
β–catenin	β–catenin	Poly	Lv, others?					dmcclay@duke.edu	
β-glucanase	β–glucanase	Poly	Lv, others?					dmcclay@duke.edu	
brachyury	brachyury	Poly	Lv, others?					dmcclay@duke.edu	(very low supply)
cadherin	cadherin	Poly	Lv, others?					dmcclay@duke.edu	
echinonectin	echinonectin	Poly	Lv, others?					dmcclay@duke.edu	
Ecto V	Oral Ectoderm and foregut	Mono	Lv, Sp, others?	late gastrula on	M		Coffman and McClay, 1990	dmcclay@duke.edu	
Endo 1	midgut and hindgut	Mono	Lv, Sp, others?	gastrula-pluteus	M		Wessel and McClay, 1985	dmcclay@duke.edu	
groucho	groucho	Poly	Lv, others?					dmcclay@duke.edu	available when paper published fringe - available when paper published
gsc	goosecoid	Poly	Lv, others?					dmcclay@duke.edu	new supply in progress
hedgehog	hedgehog	Poly	Lv, others?					dmcclay@duke.edu	works on blots but doesn't stain Hyalin
Hyalin	Hyalin	Poly	Lv, others?					dmcclay@duke.edu	
Hyalin 183	cell adhesion site of hyalin and outer ECM	Poly	Lv, others?				Adelson and Humphreys, 1988	dmcclay@duke.edu	
IbIO	several collagens and basal laminar vesicles in eggs	Mono	Lv, others?				Wessel and McClay, 1987	dmcclay@duke.edu	
Ig8 or JHID10	PMC-specific	Mono	Lv, others?					dmcclay@duke.edu	
LvSU2 integrin	LvSU2 integrin	Poly	Lv, others?					dmcclay@duke.edu	not a splendid reagent but it stains L1 (adhesion molecule)
moesin	moesin	Poly	Lv					dmcclay@duke.edu	
Notch	Notch	Poly	Lv					dmcclay@duke.edu	
Tbx2/3	Tbx2/3	Poly	Lv					dmcclay@duke.edu	new supply in progress
2.8h10	Ciliary band	Mono	Lv, others?	gastrula-pluteus	P	WM	unpublished	ettenson@andrew.cmu.edu	
3g6 (ECM1)	Basal lamina (ECM1 proteins)	Mono	Many	late blastula-pluteus	F	Both	Ingersoll and Ettensohn, 1994	ettenson@andrew.cmu.edu	
4g7	Basal lamina	Mono	Many	late blastula-pluteus	F	Both	unpublished	ettenson@andrew.cmu.edu	
6a9	PMC-specific (MSP130 proteins	Mono	Many	late blastula-pluteus	M/P/F	Both	Ettenshon and McClay, 1988	ettenson@andrew.cmu.edu	
6e10	PMC-specific (MSP130 proteins)	Mono	Lv, others?	late blastula-pluteus	M/P/F	WM	Hodor *et al.*, 2000	ettenson@andrew.cmu.edu	

Alx1	Large micromere-PMC lineage, Alx1 protein	Poly	Lv, Sp, Lp	early blastula-prism	P	WM	Ettensohn *et al.*, submitted	ettenson@andrew.cmu.edu	
ECM3	Ectodermal basal lamina	Mono	Lv, others?	early blastula-pluteus	P	WM	Hodor *et al.*, 2000	ettenson@andrew.cmu.edu	
SMC1	Non-skeletogenic. Mesoderm	Mono	Lv	blastula-pluteus	P	WM	Sweet *et al.*, 1999	ettenson@andrew.cmu.edu	
SMC2	Non-skeletogenic. Mesoderm	Mono	Lv	late blastula-pluteus	P	WM	Sweet *et al.*, 1999	ettenson@andrew.cmu.edu	
Ach	Acetylcholine	Poly	Pl	zygote-gastrula-pluteus		S		falugi@unige.it	Dako
AChE	Acetylcholinesterase		Pl	zygote-gastrula-pluteus		S		falugi@unige.it	Genesys
ChAT	Choline-acetyltransferase	Poly	Pl	zygote-gastrula-pluteus		S		falugi@unige.it	Genesys
m1	muscarinic receptor type 1	Mono	Pl	zygote-gastrula-pluteus		S		falugi@unige.it	Sigma
m2	muscarinic receptor type 2	Poly	Pl	zygote-gastrula-pluteus		S		falugi@unige.it	Sigma
Nos brain	Nitric oxyde synthase		Pl	Egg				falugi@unige.it	Sigma
OTX1	human OTX1 protein		Pl	zygote-gastrula-pluteus		S		falugi@unige.it	Prof. G. Corte
OTX2	human OTX2 protein		Pl	zygote-gastrula-pluteus		S		falugi@unige.it	Prof. G. Corte
1g8	msp130	Mono	All urchins tested (>15 species, representing all major groups)	mesenchyme blastula - juvenile	insensitive	Both	Lowe and Wray 1997; Lowe *et al.*, 2002	gwray@duke.edu	Source: David McClay
4D9	engrailed		As	juvenile	2% P	Both	Wray and McClay, 1989	gwray@duke.edu	Not in urchins (due to an amino acid substitution within epitope); Source Nipam Patel
6a3		Mono	All urchins tested (8 species, including most major groups)	gastrula-late larva (juvenile not examined)	insensitive	Both	Wray and McClay, 1989	gwray@duke.edu	Source: David McClay
distal-less	distal-less	Poly	Urchins: Sp, Lv, Sd; Seastars: Po, Et, Di. Sea cucumbers: Sc, Pc	late larva, juvenile (all)	2% P, 10–20 min	WM	Lowe and Wray, 1997	gwray@duke.edu	Source: Grace Panganiban, University of Wisconsin, Madison
anti-Epith-1mAb	Epithelial antigen ectoderm and endoderm	Mono	Tn	fertilized eggs and older	E	WM	Kanoh *et al.*, 2001	hKatow@mail.cc.tohoku.ac.jp	also works for immunoblotting and immunoprecipitation
anti-Epith-2mAb	Epithelial antigen ectoderm and endoderm	Mono	Hp, Cj, Si	fertilized eggs and older	E	WM	Kanoh *et al.*, 2001	hKatow@mail.cc.tohoku.ac.jp	also works for immunoblotting and immunoprecipitation

(*continues*)

Table II *(continued)*

Antibody name	Antigen recognized	Monoclonal or polyclonal	Species specificty	Stages detected	Best fixation	Whole mount or sections	Citation	Email address of contributor	Notes
anti-FR-1R pAb	RGDS peptide binding protein	Poly	Cj	unfertilized eggs and older	E	late blastula-pluteus	Katow and Sofuku, 2001	hKatow@mail.cc.tohoku.ac.jp	immunoblotting and immunohistochemistry
VIIIE6 · 2	pamlin	Mono		morlula and older	E	WM	Katow, 1995	hKatow@mail.cc.tohoku.ac.jp	
Aryllufatase	aboral ectoderm-specific	Poly	Sp	gastrula-pluteus				lang@mail.rochester.edu	Rabbit Polyclonal, requires affinity purification
FGFR	muscle	Poly	Sp, Lp					lang@mail.rochester.edu	requires affinity purification
Gsc	goosecoid	Poly	Sp	20–72hr	4% PF	WM	NA	lang@mail.rochester.edu	Guinea pig polyclonal needs to be affinity purified for immunostaining.
MHC	myosin heavy chain	Poly	Sp	late gastrula on	M/P/F	Both	Wessel *et al.*, 1990	lang@mail.rochester.edu	less effective with paraformaldehyde fixation
SoxB1	SoxB1	Poly	Sp, Lp, Lv, He	0–72hr	4% PF, NOT M	WM	Kenny *et al.*, 1999	lang@mail.rochester.edu	unpurified; use at 1:1000
SP1	pigment cells	Mono	Sp				Gibson and Burke, 1985	lang@mail.rochester.edu	Developmental hybridoma bank:
Spec-1	aboral ectoderm-specific	Poly	Sp		P		Chuang *et al.*, 1996	lang@mail.rochester.edu	works poorly in Lp
BEVIB12b8	toposome	Mono	Pl	egg to pluteus	M, B	WM	Noll *et al.*, 1985; Matranga *et al.*, 1986; Cervello and Matranga, 1989	matranga@ibim.cnr.it	
BEVIID4c3	toposome	Mono	Pl	egg to pluteus	M, B	WM	Noll *et al.*, 1985; Matranga *et al.*, 1986; Cervello and Matranga, 1989	matranga@ibim.cnr.it	
NE IC2c8	Pl-nectin	Mono	Pl	egg to pluteus	M, B	WM	Matranga *et al.*, 1995; Zito *et al.*, 1998, 2000	matranga@ibim.cnr.it	
NEVIIIB5g4	Pl-nectin	Mono	Pl	egg to pluteus	M, B	WM	Matranga *et al.*, 1995; Zito *et al.*, 1998, 2000	matranga@ibim.cnr.it	
echinonectin	echinonectin	Poly/Mono	Lv	egg-pluteus	B	S	Alliegro *et al.*, 1988, 1990	mallie@lsuhsc.edu	
hyl	hyalin	Poly	Sp, Lv	Oocytes-eggs-embryos-larvae	F, P, B	both	Wessel *et al.*, 1998	rhet@brown.edu	
SFE 1	Soft fertilization envelope protein 1	Poly	Sp, Lv	Oocytes-eggs-embryos	F, P	both	Laidlaw and Wessel, 1994	rhet@brown.edu	
SFE 9	Soft fertilization envelope protein 9	Poly	Sp, Lv	Oocytes-eggs-embryos	F, P	both	Laidlaw and Wessel, 1994, Wessel, 1995	rhet@brown.edu	
Pln	proteoliaisin	Poly/Mono	Sp, Lv	Oocytes-eggs-embryos	F, P	both	Somers *et al.*, 1989; Wessel *et al.*, 2001	rhet@brown.edu	

OP	ovoperoxidase	Poly	Sp, Lv	Oocytes-eggs-embryos	F, P	both	LaFleur *et al.*, 1998	rhet@brown.edu	
CGSP1	Cortical granule serine protease	Poly	Sp, Lv	Oocytes-eggs-embryos	F, P	both	Haley and Wessel, 1999	rhet@brown.edu	
YP30	Yolk platelet protein of 30 kDa	Poly	Sp, Lv	Oocytes-eggs-embryos-larvae	F, P	both	Wessel *et al.*, 2001	rhet@brown.edu	
MYP	Major yolk protein	Poly	Sp, Lv	Oocytes-eggs-embryos-larvae, adult gut	F, P	both	Brooks and Wessel, 2002	rhet@brown.edu	
Gi	G*i* alpha subunit of the heterotrimeric G protein	Poly	Sp, Lv	Oocytes-eggs-embryos-larvae	F, P, B	both	Voronina and Wessel, 2004	rhet@brown.edu	
Gq	G*i* alpha subunit of the heterotrimeric G protein	Poly	Sp, Lv	Oocytes-eggs-embryos-larvae	F, P, B	both	Voronina and Wessel, 2004	rhet@brown.edu	
G12	G*i* alpha subunit of the heterotrimeric G protein	Poly	Sp, Lv	Oocytes-eggs-embryos-larvae	F, P, B	both	Voronina and Wessel, 2004	rhet@brown.edu	
tagmin	synaptotagmin	Poly	Sp, Lv	Oocytes-eggs-embryos-larvae	F, P	both	Conner *et al.*, 1997	rhet@brown.edu	
sec	n-sec 1	Poly	Sp, Lv	Oocytes-eggs-embryos-larvae	F, P	both	Leguia and Wessel, 2004	rhet@brown.edu	
VAMP	Vesicle associated membranous protein (synaptobrevin)	Poly	Sp, Lv	Oocytes-eggs-embryos-larvae	F, P	both	Conner *et al.*, 1997	rhet@brown.edu	
taxin	syntaxin	Poly	Sp, Lv	Oocytes-eggs-embryos-larvae	F, P	both	Conner *et al.*, 1997	rhet@brown.edu	
Rab 3	Rab 3	Poly	Sp, Lv	Oocytes-eggs-embryos-larvae	F, P	both	Conner and Wessel, 1998	rhet@brown.edu	
dyn	dynamin	Poly	Sp, Lv	Oocytes-eggs-embryos-larvae	F, P	both	Brooks and Wessel, 2004	rhet@brown.edu	
Cyclin B	Cyclin B	Poly	Lv	Most prevalent in oocytes and cleaving embryos	F, P	both	Voronina *et al.*, 2003	rhet@brown.edu	
suPKc	Protein kinase C	Poly	Sp, Lp	egg to pluteus				sshen@iastate.edu	western and immunostaining
suPLC-beta4	phospholipase C-beta4		Lp					ssshen@iastate.edu	westerns
anti-SM30	SM30 of PMC's and spicules	Poly	Sp		M		Urry *et al.*, 2000	wilt@socrates.Berkeley.edu	absorb with *E. coli* and egg acetone powder
anti-SM30	SM30 of PMC's and spicules	Poly	Lp		M		Urry *et al.*, 2000	wilt@socrates.Berkeley.edu	absorb with *E. coli* and egg acetone powder
anti-SM50	SM 50 of PMC's and spicules	Poly	Sp		P, M		Urry *et al.*, 2000	wilt@socrates.Berkeley.edu	absorb with *E. coli* and egg acetone powder
anti-SM50	SM 50 of PMC's and spicules	Poly	Lp		P, M		Urry *et al.*, 2000	wilt@socrates.Berkeley.edu	absorb with *E. coli* and egg acetone powder

As, Amphipholis squamata; B, Bouin's fixative; Cj, Clypeaster japonicus; Di, Dermasterias imbricata; E, ethanol fixative; Et, Evasterias troscheli; F, formaldehyde fixative; Hp, Hemicentrotus pulcherrimus; Lp, Lytechinus pictus; Lv, Lytechinus variegatus; M, methanol fixative; Mono, monoclonal; P, paraformaldehyde fixative (4% unless otherwise noted); Po, Pisaster ochraceus; Poly, polyclonal; Ps, Psolus chitonoides; S, sections; Sc, Stichopus californicus; Sd, Strongylocentrotus droebachiensis; Si, Strongylocentrotus intermedius; Sp, Strongylocentrotus purpuratus; Th, Temnopleulus hardiwicki; WM, whole mount

Permeabilization is essential if intracellular antigens are to be detected, or if antigens are localized deep within the embryo. If a surface antigen is to be detected, however, permeabilization should be avoided.

a. MeOH Fixation

MeOH fixation acts by simultaneous denaturation and precipitation of proteins. Since MeOH also permeabilizes cell membranes, fewer steps are required prior to immunostaining. Fixation by precipitation does not preserve the morphology of specimens as well as aldehyde fixation methods do and specimens fixed in MeOH can shrink significantly. However, MeOH fixation does not covalently modify the antigen's epitopes like aldehyde fixatives do, thereby preserving the antigen binding sites. Another advantage of MeOH fixation is that it does not introduce autofluorescence into a specimen as is sometimes the case with aldehyde fixation. One disadvantage of using MeOH is that the addition of MeOH to seawater causes a precipitate to form. Complete removal of all seawater before the addition of MeOH minimizes this effect. Other cooled organic solvents, such as acetone, can also be used to fix and permeabilize embryos. Some protocols use cold MeOH or acetone after an initial aldehyde fixation for permeabilization.

b. Formaldehyde Fixation

Formaldehyde fixes cellular structures by cross-linking proteins. It has a single aldehyde-containing carbon and does not cross-link as effectively as glutaraldehyde. Because formaldehyde has a low molecular weight, it can penetrate cells and tissues rapidly. Formaldehyde is available commercially as "formalin," which is typically provided as either a 10 or a 37% stock, and is generally used at 1 to 4% working concentrations. If left standing for long periods, formaldehyde readily oxidizes to form formic acid and polymerizes to form paraformaldehyde (Humason, 1972). Thus, commercially available formaldehyde solutions which have been opened for some time are not as effective as freshly prepared formaldehyde. In addition, commercially prepared formaldehyde solutions frequently have additives to help stabilize the formaldehyde. A common additive is MeOH, which may be unsuitable for certain applications (see Section IV.C). It is therefore preferable to prepare fresh formaldehyde from paraformaldehyde immediately before use (see Section II.B.3.d). Unlike glutaraldehyde (see following text), formaldehyde fixation does not generate significant autofluorescence.

c. Glutaraldehyde Fixation

While glutaraldehyde is extraordinarily effective at preserving cellular structure, it has several disadvantages over other fixatives for immunocytochemistry. Glutaraldehyde is slow to penetrate due to its large molecular size and it cross-links as it penetrates, further reducing its ability to penetrate the tissue. The efficient cross-linking of proteins by glutaraldehyde can also mask epitopes and reduce antigenicity. Since glutaraldehyde possesses two functional aldehyde groups per molecule, unreacted aldehyde groups can cause autofluorescence. For this reason, glutaraldehyde fixation is not recommended for immunofluorescent applications. Although subsequent treatment of samples with sodium borohydride (Clancy and Cauller, 1998) or glycine can quench autofluorescence originating from unreacted aldehyde groups, the introduction of these additional steps increases the risk of embryo loss.

d. Choice of Buffers

Another consideration when designing a fixation protocol is the choice of buffer in which the fixative will be dissolved. Ideally, it should be isotonic with the cell so as not to perturb cellular structure, and it should maintain the cell in as close to its physiological state as possible. The most frequently used buffer is phosphate-buffered saline (PBS), as it is inexpensive and easy to prepare. When using aldehydes as fixatives, amine-containing buffers such as Tris should never be used, as they will react with the fixative. In certain published protocols, seawater is frequently used but it has no actual buffering capacity.

2. Post-Fixation Washes and Permeabilization

a. Wash Buffers

Once fixation is complete, removal of the fixative is required before proceeding to subsequent steps. While some experimenters prefer to use MFSW during fixation and washing steps, we have found that phosphate buffered saline (PBS) can be substituted for most applications. However, it is best to maintain the same buffer and pH from start to finish. It is also recommended that once embryos have been washed from fixative, they be processed immediately for immunofluorescence. Although the addition of sodium azide to storage buffers will help prevent bacterial growth, prolonged post-fixation storage in aqueous buffers is not recommended. Since aldehyde fixation is reversible, fixed and washed embryos can deteriorate within a week if stored in aqueous solutions. In contrast, for some applications, embryos can be stored for several months without loss of morphology or antigenicity in $-20\,^{\circ}\mathrm{C}$ MeOH (for example, myosin heavy chain remains detectable in circumesophageal muscles of the late gastrula and later stages after prolonged storage in $-20\,^{\circ}\mathrm{C}$ MeOH). Optimal storage conditions should be determined empirically for each antigen.

All buffers should be filtered before use since particulates and debris will readily stick to the surfaces of embryos. This can be accomplished either in bulk immediately after preparation or during addition to the samples by delivering the buffers through a syringe capped with a $45\,\mu M$ syringe filter. Microcentrifugation of all protein-containing solutions is also recommended to remove protein aggregates.

b. Permeabilization

Following aldehyde fixation, for most applications, the cell membranes are permeabilized to allow antibody access to intracellular components or internal regions of the embryo. This can be accomplished with organic solvents, such as

MeOH or acetone, or with detergents. Permeabilization should be avoided when detecting extracellular membranous epitopes, as has been described under "MeOH fixation."

The most commonly used detergents for permeabilization are Triton X-100 and Tween-20. These detergents extract membrane lipids without interfering with protein interactions. Because of the viscosity of most detergents, a 10% stock solution is generally prepared and used to prepare working solutions. Detergents are used at 0.1 to 1% working concentrations in wash buffers and antibody diluents.

3. Recommended Fixation Protocols

a. Methanol Fixation I (after Harris et al., 1980)

This protocol is particularly suited for preservation of microtubules and other cytoskeletal elements. Ethylene glycol-bis(β-aminoethyl ether)-N,N, N′,N′-tetraacetic acid (EGTA) is included to chelate any calcium that might be released from the cell and cause depolymerization of cytoskeletal components (Harris, 1986).

1. Collect embryos by gentle centrifugation and wash once with MFSW.
2. Resuspend pelleted embryos in at least 10 pellet volumes of 90% methanol, 10 mM EGTA in MFSW at −20 °C.
3. Fix 20 min at −20 °C.
4. For immediate use, wash several times with MFSW or PBS. Embryos can be stored only briefly (a few days to a week), as long as 5 mM sodium azide is added to the MFSW or PBS to prevent bacterial growth and embryo deterioration.

b. Methanol Fixation II

Alternatively, −20 °C, 100% MeOH (without EGTA) can be used, as has been described, and is an effective fixative for most immunofluorescence applications, However, it is necessary to remove as much seawater as possible from the embryos before adding the MeOH since a precipitate will form. This precipitate may obscure embryos but will disappear after several washes with PBS.

c. Formaldehyde Fixation

Aldehyde fixation crosslinks proteins but does not permeabilize the cell membrane. Detergent permeabilzation should be employed either simultaneously or subsequent to aldehyde fixation by including detergent in the post-fixation wash buffers. Alternatively, post-fixation with an organic solvent, such as ice-cold acetone or MeOH, can be employed prior to the wash steps. In addition, different antigens may require more or less incubation time in the fixative.

Formaldehyde Fixation

1. Collect embryos by gentle centrifugation and wash once with MFSW.
2. Suspend embryos in at least 10 pellet volumes of 3.7% formaldehyde (10% formalin) in PBS and incubate 15 min on ice.

3. Wash 3 times with PBS.
4. Collect by gentle centrifugation, decant supernatant, and add at least 10 pellet volumes of −20 °C acetone or MeOH to permeabilize. Incubate 15 min at −20 °C.
5. Add an equal volume of PBS, mix thoroughly, and incubate at −20 °C for 15 min.
6. Collect by gentle centrifugation and wash with PBS 3 times for 5 min.
7. Proceed with staining or store short-term in 5 mM sodium azide in PBS at 4 °C.

d. Paraformaldehyde Fixation

Freshly prepared fixative is sometimes required for best preservation of antigenicity. The length of fixation should be determined empirically for each antigen of interest. However, prolonged fixation is likely to decrease accessibility of antigens.

1. Freshly prepare 4% paraformaldehyde as follows:
   a. Dissolve 4 g of paraformaldehyde in 80 ml of deionized water by heating on a hot plate in a fume hood with constant stirring. (Dissolution is aided by adding 1 to 2 drops of 10N NaOH.)
   b. Continue to stir until completely dissolved.
   c. Cool to room temperature and add 10 ml of 10× PBS.
   d. Bring to 100 ml with deionized water and store at 4°C.
2. Fix embryos for 30 min to overnight (ON) in fixative. If incubating ON, incubate at 4 °C.
3. Wash 3 times for 5 min each in PBST (PBS with 0.1% Triton X-100 or Tween-20 added).
4. Proceed with staining or store short-term in 5 mM sodium azide in PBST at 4 °C.

4. Washes

Washes are required post-fixation, as has been described, and also for post-incubation with antibodies. The most commonly used wash solution and diluent for antibodies and blocking agents is PBS. Some procedures use MFSW, but PBS can generally be employed instead of MFSW in most applications. Addition of nonionic detergents such as Triton X-100 or Tween-20 (at 0.1 to 0.5%) is generally recommended to reduce background, aid permeabilization, and also help minimize embryos sticking to the walls of tubes or dishes.

5. Blocking

A blocking step is usually required before initial antibody incubations to prevent antibody binding to nonspecific sites. Nonspecific background staining is usually due to the primary or secondary antibodies binding to the specimen at sites unrelated to the target antigen (Harlow and Lane, 1998). Monoclonal antibodies are frequently applied as tissue culture supernatants and contain sufficient proteins to obviate a blocking step. However, a blocking step is generally recommended and is required for polyclonal antibodies. Ideally, the blocking agent should be a nonspecific protein that is not recognized by either the primary or secondary antibodies. A 3 to 5% solution of normal serum from the same species as the secondary antibody is the best choice. For example, commercially available secondary antibodies are frequently raised in goats, so normal goat serum would be the best choice as a blocking agent. Bovine serum albumen (BSA, 3%) or nonfat dry milk (3%) can be used as alternatives.

Incubation in blocking solution should be for a minimum of 30 min but can be extended to ON. For this and any of the steps to follow, shorter incubations can proceed at RT while 4 °C is recommended for incubations that extend ON. Addition of sodium azide to a final concentration of 0.02% is also recommended for longer incubations unless horseradish peroxidase (HRP) is to be used as the detection method (sodium azide will inhibit HRP activity). Detergents are included in blocking steps (if the samples have been permeabilized) to compete for hydrophobic interactions with antibodies.

6. Antibody Incubations

a. Primary Antibody Incubation

Once the blocking step is complete, incubate the embryos in the primary antibody at the appropriate dilution. There is no need to wash specimens after the blocking step. The appropriate dilution of the primary antibody may be based on published protocols (where available) or may have to be determined empirically. Dilutions should be made in filtered PBS or PBST and a brief centrifugation in a microcentrifuge is recommended to remove protein aggregates. The primary antibody incubation can proceed from 30 min to ON.

All antibodies that are described in the published literature generally will be available upon request. However, polyclonal antibodies may be in limited supply, since only a finite amount is usually generated. Monoclonal antibodies are not usually limited but may take some time for the supplier to generate. If large amounts are needed, it may be preferable to request the hybridoma line to generate supernatants inhouse.

Monoclonal Antibodies to Sea Urchin Antigens. Several different lineage-specific monoclonal antibodies are currently available from various sea urchin labs and the Developmental Studies Hybridoma Bank (//www.uiowa.edu/~dshbwww/). Many of these antibodies are useful for distinguishing among different tissues after gastrulation and are therefore particularly useful for determining whether alterations in embryonic patterning have occurred as the result of experimental perturbation. Most of these antibodies give consistent staining and are recommended for those trying immunostaining for the first time. Although the antibodies listed in

Table II have been shown to work reliably, their inclusion in Table II does not guarantee their availability.

Polyclonal Antibodies to Sea Urchin Antigens. A wide variety of polyclonal antibodies have been raised to sea urchin antigens and many are listed in Table II. Unfortunately, several of these polyclonal antibodies are limited in their availability, since the animals used to generate them are no longer in existence. As has been stated, while every effort has been made to compile a comprehensive list, inclusion in Table II does not guarantee their availability. Since a number of these antibodies were generated to fusion proteins produced in bacteria, the plasmids that encode them may be requested and used to generate additional antisera.

b. Post-Incubation Washes

To remove excess primary antibody, embryos should be washed several times before incubating in secondary antibody. Generally, 3 washes of 5 min each in PBS or PBST is sufficient. If not washed out, unbound primary antibody can bind to free secondary antibody, reducing its effective concentration and causing excessive background staining. Antibodies are sensitive to changes in the composition of the washing solution; therefore, it is recommended that the buffer used to dilute them and the washing buffer be the same.

c. Secondary Antibodies

The most common choices for detection methods include: direct immunofluorescence by epifluorescence microscopy or confocal microscopy; indirect immunofluorescence by standard epifluorescence microscopy or by confocal microscopy; and enzyme-linked detection methods using conventional light microscopy. Direct immunofluorescence requires direct cross-linking of a fluorochrome to the primary antibody and is rarely used unless a large quantity of primary antibody is available. Indirect immunofluorescence requires a second antibody which is conjugated to a fluorochrome and directed against the species used to produce the primary antibody. Enzyme-linked detection methods require a secondary antibody that is conjugated to an enzyme and the development of a colored reaction product after addition of enzyme substrate. The latter requires several additional steps over fluorochrome-based detection methods and is recommended only when fluorescent-based detection methods are not viable (i.e., specimen autofluorescence) or if amplification of the signal is required. An excellent resource for application of enzyme-linked detection methods to the analysis of sea urchin antigens can be found in Wright and Scholey (1993).

Fluorescence Detection Methods. Indirect immunofluorescence is the most common detection method used for whole mount staining of sea urchin embryos. Primary antibodies can be either monoclonal (raised in mice) or polyclonal (usually, raised in rabbits). More recently, other species have been used to raise polyclonal antibodies as well, including guinea pig, hamster, chicken, and sheep. For detection of monoclonal antibodies, a variety of commercially available anti-mouse secondary antibodies conjugated to a wide range of fluorochromes are available. Similarly, a wide choice of fluorochromes coupled to other species-specific secondary antibodies can be obtained. The choice of fluorochrome will be determined by the method of microscopic detection and specific application. For localization of a single antigen by conventional epifluorescence microscopy, either fluorescein isothiocyanate (FITC) or tetramethylrhodamine isothiocyanate (TRITC) conjugated to the secondary antibody is usually sufficient. However, if performing double or triple labeling to detect multiple antigens simultaneously (Section IV.A and B) or combining antigen localization with nuclear or cytoskeletal staining (Section IV.C and D), one must take into consideration the excitation and emission spectra of the combined fluorochromes. Another consideration is whether the specimens will be visualized by conventional epifluorescence or confocal microscopy, since the intense light of a confocal microscope laser may cause rapid quenching of a fluorochrome, particularly if the signal is weak. A variety of fluorochromes have been developed that are less susceptible to photobleaching than TRITC and FITC and which are more suitable for confocal microscopic applications. These include the Alexa Fluor dyes (Panchuk-Voloshina *et al.*, 1999), and Cy3 and Cy5 (Mujumdar *et al.*, 1993). While the embryos of some sea urchin species are weakly autofluorescent, judicious choice of dichroic filter sets can help alleviate this problem.

Enzyme-Linked Detection Methods. Immunofluorescence may not be ideal for all applications. If specimen autofluorescence is a problem or if the signal is particularly weak, enzymes conjugated to secondary antibodies can be employed for antigen detection. These enzyme-linked detection methods have the disadvantage of requiring additional incubation and washing steps beyond those used in immunofluorescence procedures. On the other hand, they do not require an epifluorescent or confocal microscope, and do provide permanent records of the experiment. Although fluorescent techniques provide high resolution and allow for detection of multiple labels, they will fade over time. Enzyme-linked detection methods are highly sensitive, allowing controlled development of a colored reaction product at the site of antibody binding. Another advantage of enzyme-detection methods for immunocytochemistry is the possibility of combining them with histological counterstains.

Enzyme-linked detection methods are rarely direct, where the enzyme is coupled to the primary antibody. More commonly, they are indirect and enzyme is conjugated to the secondary antibody. However, enzyme-linked detection methods can be bridged and the signal amplified. For example, biotin conjugated secondary antibody is sandwiched between the primary antibody and avidin conjugated to an enzyme or fluorochrome. The most common enzyme detection systems for immunocytochemistry use either horseradish peroxidase (HRP) or alkaline phosphatase (AP) conjugates. A number of kits with all necessary reagents, buffers, and chromogens are commercially available.

i. Horseradish Peroxidase. Perhaps the most common enzyme used for enzyme-linked detection is HRP. HRP conjugates are sometimes avoided because diaminobenzidine (DAB), the most commonly used HRP substrate, is a known carcinogen. The advantage of HRP over other enzyme-linked detection methods is that the chromogen is insoluble in both aqueous and alcoholic solutions and adding metal salts, such as $NiCl_2$ at 0.03% final concentration, enhances its sensitivity. HRP labeling is compatible with a wide array of histological stains.

ii. Alkaline Phosphatase. The second most common enzyme-linked detection method for immunocytochemistry utilizes AP. The most frequently used substrate for AP development is a combination of bromochloroindolyl phosphate/nitroblue tetrazolium (BCIP/NBT). These chromogens generate a purple-blue precipitate at the site of enzyme binding. The sensitivity of AP staining can be controlled by the length of incubation and is usually terminated by replacing the substrate solution with 20 mM EDTA in PBS.

iii. Avidin/Biotin Techniques. In addition to conjugation with enzymes and fluorochromes, antibodies can be conjugated to biotin and then detected by the addition of avidin or streptavidin coupled to enzymes or fluorochromes. Avidin has high affinity for biotin and can bind as many as 4 biotin molecules, thereby increasing the sensitivity of antigen detection. Biotin conjugation allows a single antibody to be detected by a broad range of methods. In the simplest procedures, biotinylated antibodies are added to the specimen and detected with labeled avidin or streptavidin. In a complex permutation of the procedure, the "ABC method" (Vector Laboratories, Burlingame, CA), a primary antibody is used to detect the antigen within the cells or tissues, then biotinylated secondary antibody is used to detect the primary antibody. Subsequently, the secondary antibody is detected by enzyme or fluorochrome conjugated to avidin, considerably increasing sensitivity (Hsu *et al.*, 1981). Due to the likelihood of nonspecific background, this technique should be avoided when using primary antibodies that give high backgrounds.

7. Mounting Specimens

a. Choice of Mounting Media

Mounting media are chosen to be compatible with the detection method used. Generally, aqueous mounting media are used for immunofluorescence applications since many nonaqueous mounting media, such as Permount, are fluorescent. The latter is suitable for preparations involving enzyme-linked detection methods but requires the extra steps of dehydration before it can be applied.

A simple mounting medium suitable for immunofluorescence is glycerol:PBS (1:1) followed by sealing of the coverslip with nail polish. Clear nail polish is recommended because colored nail polish can fluoresce and leach into the sample. Also suitable for immunofluorescent mounting are semisolid media using polyvinyl alcohols (Harlow and Lane, 1988). These include Elvanol (Dupont, Wilmington, DE), Mowiol 4-88 (Calbiochem, San Diego, CA) and Gelvatol (Air Products, Allentown, PA). The polyvinyl alcohols dry as a gel and do not require nail polish sealing of coverslips to the slides.

Another factor to consider when choosing mounting media is photobleaching, which reduces fluorescent signal and sensitivity. Factors that contribute to photobleaching include high light intensity and prolonged sample exposure. Photobleaching can be avoided by including anti-fade reagents in the final mounting media. These anti-fade agents usually increase the photostability of fluorochromes by scavenging free radicals released by excitation of the fluorochromes and damaging them. Several antioxidants can be added to mounting media such as PBS:glycerol or the aqueous semi-solid media described previously. They include p-phenylene diamine (1 mg/ml), n-propylgallate (2–5%), and 1,4–diazbicyclo [2,2,2] octane (DABCO; 100 mg/ml). Anti-fade mounting media are also available commercially. Among these are Vectashield (Vector Laboratories, Bellingame, CA), and ProLong, or Slowfade (Molecular Probes, Eugene, OR). Another means of avoiding photobleaching is to reduce the intensity of light exciting the specimen, either by reducing light intensity at the source or by using neutral density filters.

To prepare semisolid mounting media from Gelvatol, Mowiol, or Elvanol:

1. Add 2.4 g of powder to 6 g of glycerol and mix.
2. Add 6 ml of dH_2O and mix; let sit several hours.
3. Add 12 ml of 0.2 M Tris (pH 8.5) and heat to 50 °C for 10 min with occasional mixing.
4. Once dissolved, clarify by centrifugation at 500 *g* for 15 min.
5. Anti-fade agents can be added at this stage at the appropriate concentrations listed previously.

These media harden into a gelatinous form that does not require sealing the coverslip with nail polish. However, they can become progressively opaque over time, so it is recommended that specimens prepared with these media be observed shortly after mounting.

8. Controls

Specificity of staining should always be determined by inclusion of appropriate controls. If using known antibodies to determine the effects of an experimental perturbation, the pattern of antigen expression is compared between experimentals and unperturbed control embryos. Additional controls are necessary if developing or testing a new antibody. For example, when testing a new antibody raised in rabbits, preimmune serum from that rabbit should be tested on parallel specimens. If preimmune serum is not available, nonimmune serum from another rabbit can be used. If developing monoclonal antibodies, the appropriate control for monoclonal supernatants is supernatant from the parental hybridoma cell line. A control against a known positive antigen should also be included for comparison, as should a control applying the secondary antibody alone. Immunoblotting can be used not only to demonstrate the specificity of the antibody for the antigen of interest, but also to determine the timing and levels of expression at different stages of development. If purified antigen is available, preabsorbtion of the antibody with antigen should eliminate staining while preabsorption with other related or unrelated antigens should not eliminate staining. These types of controls are necessary and ideal for the characterization and evaluation of new antibodies. Specific protocols for these procedures can be found in Harlow and Lane (1988, 1998).

III. Immunostaining Small Numbers of Embryos

Small numbers of embryos cannot be processed in microcentrifuge tubes as has been described because the embryos tend to stick to the tubes and are easily lost during processing. When collecting small numbers of prehatched embryos that have been microinjected or micromanipulated, the FE usually can be manually removed when collecting the embryos by aspiration.

Before processing small numbers of embryos that have either been injected or micromanipulated, it is advisable to practice staining on large numbers of control embryos, as described previously (Section II) to ascertain the best (1) fixative, (2) fixation duration, and (3) permeabilization procedure.

A. Processing Small Numbers of Embryos for Immunostaining

1. Well Plate Technique

To process small numbers of embryos, the same steps described previously for batch processing are performed in flexible 96-well "U-Bottom" plates (Falcon # 353911). These flexible multiwell plates offer advantages over other devices: they allow the processing of one to several hundred embryos; they minimize the loss of embryos, as the specimens tend to settle in a pile in the center (rather than at the edges, as is the case in flat-bottom wells); they reduce the reagent and buffer volumes used; and all steps can be followed under a dissecting microscope. Flexible well plates from other manufacturers do not have the same curvature at the bottom, and are sometimes molded rather than extruded in one piece; this makes the removal of solutions more difficult, and can cause embryos to dry out and flatten.

Only a portion of the 96-well plate is used. The plates are cut into smaller sections (4 wells by 4 wells) and set in a 5 cm petri dish (Fig. 1A) filled with just enough distilled water to hold it in place and make the embryos on the sides of the wells visible (Fig. 1B). Every step of the well plate procedure is continuously monitored under a dissecting microscope, with the light coming through from below. All solutions are changed by aspirating the supernatant with a specially prepared, bent Pasteur pipette (Fig. 2A), using a mouthpiece and mouth pipettor (described in the following text and in Fig. 2C), or using a handheld suction pipette (Fig. 2D).

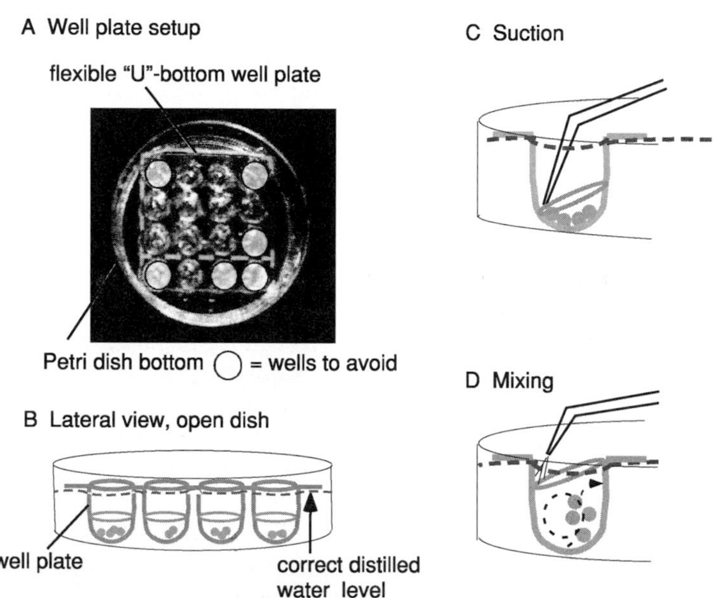

Fig. 1 Well plate setup. (A) A piece of flexible 96-well plate is cut 4 wells by 4 wells and put into the bottom of a 5-cm petri dish. It is preferable to cut this square as large as possible and clip the corners in order to fit it into the dish. For right-handed individuals, the wells indicated are to be avoided, as they tend to get flooded more easily than wells in other positions. (B) To keep the well plate in position and for optimal viewing of the embryos in the wells, the petri dish is filled with distilled water until the surface tension and capillary forces just begin to fill the spaces between the wells. Avoid overfilling the plate as this may result in spillage. (C) For liquid removal, wait until embryos have settled at the bottom of the well, then position the tip of the suction pipette (Fig. 2A) at the edge of the well opposite to the hand holding the pipette (i.e., the left edge for right-handers). Begin aspirating the bulk of the liquid (approximately 120 μl of a full well), then gently tap the entire plate against the bench top, while holding the well plate in position with two or three fingers; then, move the entire petri dish in a circular motion a few times so as to concentrate the embryos in the center of the bottom again. Afterwards, remove most of the remaining liquid. At no point are the embryos to dry out completely, so a small amount of liquid must remain in the well. Carefully monitor the entire process under a dissecting scope, always focusing on the tip of the micropipette. After removal of the solution, fill the well with 150 μl or 4 drops of liquid. (D) For mixing, position the tip of the mixing pipette (Fig. 2B) a few millimeters above the liquid level at the opposite edge of the well and blow a constant stream of air, until all embryos are suspended. Monitor this process by focusing on the bottom of the well. After mixing, focus through the entire depth of the well to detect embryos stuck to the sides of the well. They can be dislodged by gently blowing a stream of liquid at them with a suction pipette.

2. Pipetting

Mouth pipetting is used to collect embryos and change solutions in the wells. The mouth pipettor consists of a mouthpiece connected to a piece of flexible latex tubing at one end and to a Pasteur pipette (that has been pulled as will be described) at the other (Fig. 2C). Mouthpieces that are supplied with calibrated

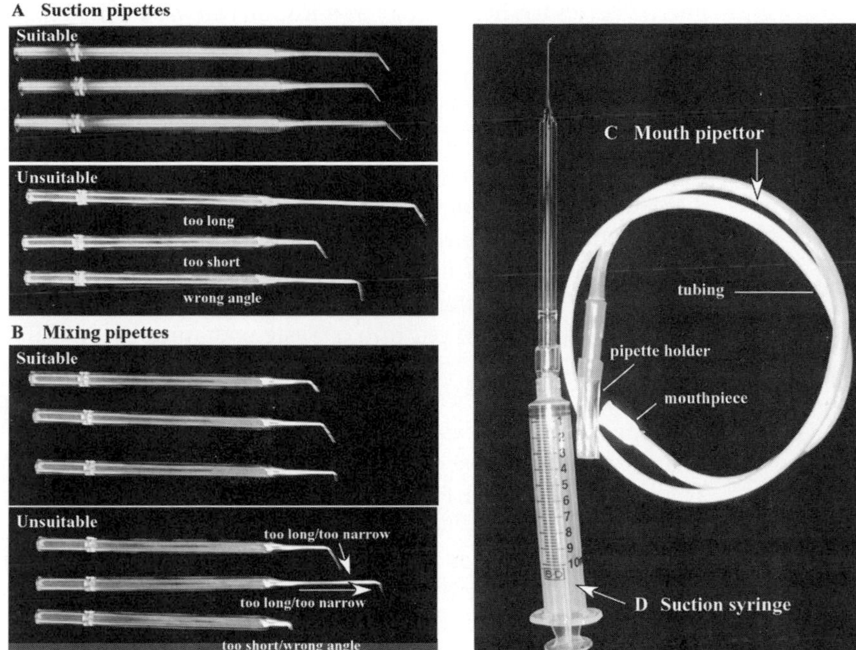

Fig. 2 Micropipettes and suction devices. (A) Suitable suction pipettes are pulled by briefly melting the thin part of a long Pasteur pipette approximately 1.5 cm from the area where it begins to narrow, taking it out of the flame and quickly pulling downward with the right hand (if right-handed), while keeping the left hand steady and the rest of the pipette horizontal. This will generate the correct angle and an evenly tapered tip at the same time. Avoid pulling the tip while the glass is still positioned in the microflame. Good suction pipettes are wide enough to easily accommodate embryos (several embryo diameters wide) and narrow enough to allow for good control when aspirating off solutions. The ideal diameter will vary with an individual's preference. After pulling out the tip of a suction pipette, the thin end is broken off between 0.75 and 1.5 cm from the bend. It is advisable to flame-polish the tip by quickly moving it through a microflame, to smooth the edge and avoid shearing embryos. Avoid using tips too narrow, too long, or too short to easily remove solutions, or tips with angles close to 90°. (B) Suitable mixing pipettes are pulled as described in the previous paragraph, with the exception that their tips are wider in diameter and shorter than those of the suction pipettes. Good mixing pipettes are wide enough to easily generate enough airstream against the liquid surface to suspend all embryos in the solution, and narrow enough to easily control airflow and avoid splashing. They do not need to be flame-polished, as they are not used to manipulate embryos. (C) A mouth pipettor consists of a plastic mouthpiece such as those used for hematology applications attached to a 60 to 80 cm long piece of latex tubing and connected at the other end to a suitable rubber or plastic adapter (e.g., a piece of a blue Gilson pipette tip and a wider piece of latex tubing) into which the wide end of a Pasteur pipette can be inserted. (D) A suction syringe can be used alternatively to remove solutions from wells and to transfer embryos. This is constructed from a 10 ml syringe to which a piece of tubing is added, which has a wide enough internal diameter to fit snugly over both the end of the syringe and the Pasteur pipette.

micropipettes for hematology applications are best suited for this purpose. If these are not available, a 1 ml pipette tip works adequately. The optimal length of the latex tubing should be between 50 and 80 cm long, depending on the height of the experimenter, to prevent condensing breath and saliva from getting into the wells, but not so long as to make handling unmanageable. An adapter at one end (e.g., a cutoff 1 ml pipette tip) is used to attach the latex tubing to a piece of wider diameter tubing, which will hold the Pasteur pipettes (Fig. 2C). The latex tubing of the mouth pipettor can be wrapped around the neck once, resting on the shoulder on the same side that holds the suction or mixing pipette. The mouth pipettor is designed for moving very small volumes of solutions. The fluid removed from one well should be discharged into a waste container before removing solution from another well, or the experimenter runs the risk of aspirating fluids into their mouth. Similarly, the experimenter should watch for movement of oral fluid into the mouth pipettor, as it will eventually be introduced into the sample wells. Enzymes in the saliva can digest embryos. A good rule of thumb is to never aspirate more than a few hundred microliters at a time when using the mouth pipettor.

Alternatively, a 5 or 10 ml plastic syringe can be used to apply suction to the pipettes when changing solutions and collecting embryos (Fig. 2D). The syringe is attached to the Pasteur pipette by a piece of tubing approximately 1″ long and wide enough to fit over both the Pasteur pipette and the end of the syringe. The syringe is held in one hand and suction is applied by pulling on the syringe plunger with the thumb and index finger of the same hand. Practice is needed before an individual is proficient at regulating the amount of suction necessary to change the small volumes of solutions involved. An advantage of the handheld syringe over mouth pipetting is that the risk of saliva passing along the tubing of a mouth pipettor and contaminating the wells is eliminated. As has been stated, monitoring of the process under a dissecting microscope is critical for success and practicing on control embryos is recommended before processing experimental specimens.

Two forms of pipettes are used for this procedure—straight suction pipettes to pick up embryos from petri dishes and bent pipettes to mix and remove solutions. Pipettes are pulled in a Bunsen microflame or butane microtorch to be slightly wider than the diameter of an embryo ($\sim$200 to 300 μm inner diameter, $\sim$ the tip of a staple). Mixing pipettes should be somewhat wider. To pull straight pipettes, melt the narrow part of the pipette (Fig. 3), take it out of the flame, and then pull in a straight line. Bent pipettes can be made similarly by melting a piece of the narrow part of the pipette, taking it out of the flame, then pulling down the thin part so as to introduce an angle and pulling a narrow tip at the same time. The angle should be between 110 and 140°. If it is less, drops may catch in the curved part and the tip will not be visible under the microscope; if it is wider, it cannot reach the bottom of the well. It is important to always keep these pipettes clean, to avoid touching the tip with one's fingers, and to rinse the tips with distilled water a few times between solutions. Bubbles are to be avoided as well. When transferring

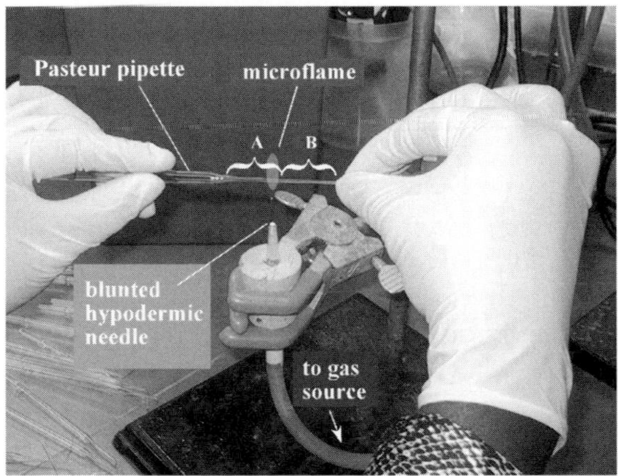

Fig. 3 Microflame Bunsen burner used to pull micropipettes from long Pasteur pipettes. The microburner consists of a piece of Tygon tubing, attached to a laboratory gas source on one side and a wide gauge needle at the other. The length of the tubing is chosen to allow the burner to be placed close to the edge of the bench, so that one's arms can be positioned approximately one foot from the torso while pulling pipettes. The piece of tubing is threaded through the center hole of a cork or rubber stopper and is connected to a blunted 16 or 18 gauge syringe needle either directly with Parafilm, or by using a piece of a blue Gilson pipette tip of suitable diameter as an adapter. The syringe needle should be blunted with a sanding stone or sanding paper. The cork or rubber stopper is held securely in place by adjustable metal clamps attached to a heavy metal base, and its height adjusted so that the experimenter's arms are at chest level while pulling micropipettes. "A" denotes the area to be heated when pulling mixing pipettes, while "B" denotes the area to be heated when pulling suction pipettes.

embryos, be sure to keep them in the narrow part of the pipette, as pulling them into the wider portion of the Pasteur pipette will likely result in their loss.

3. Collecting Embryos

a. Transfer embryos to be stained to the 4 × 4 section of flexible well plate with a straight pipette pulled in a microburner (Fig. 3).

b. After the embryos have settled, most of the supernatant in the wells can be removed with a suction pipette (Fig. 2A).

c. To settle the embryos in the very center and bottom of the well, the plate is moved in circles and tapped gently against the bench a few times.

d. Do not leave embryos in the wells uncovered; add solutions immediately to prevent embryos from drying out.

Following these steps will minimize the loss of material because the embryos tend to move away from the tip of the pipette when the last 30 to 50 μl of seawater is removed (Fig. 1C). However, at no point are the embryos allowed to dry out.

Viscous solutions (blocking solution, for example) are more difficult to remove, and suction pipettes of different diameters may be required (Fig. 2A).

Whenever solutions are added to the wells, they need to be mixed by blowing against the surface of the solution with a mixing pipette of a wider diameter than that used to collect the embryos (Fig. 1C). Mixing should be done preferably by using a laboratory air line. This requires some practice, since the airflow has to be controlled carefully to prevent the splashing of solutions and embryos out of the wells. Be sure to mix the entire liquid column in the well so that the embryos come off the bottom during mixing, checking under the microscope to determine whether the solutions are mixing properly by focusing at the bottom of the well. After mixing, allow embryos to settle at $1\times$ gravity.

The most widely used wash buffer, PBST, is added to the wells with a syringe connected to a $25\,\mu l$ syringe filter. The filter helps minimize dirt and salt precipitates in the wells. Antibody and blocking solutions should also be diluted to working concentrations with filtered PBS or PBST, vortexed, and spun at full speed in a microcentrifuge to remove any additional particulates that could adhere to the embryos. Debris and particulates adhering to embryos will fluoresce and can obscure specific labeling as well as affect the aesthetics of the embryos and image quality.

4. Fixation

Embryos grown to the appropriate stage in MFSW are transferred using a straight pipette to the well plate in a petri dish on ice, as has been described. If they are swimming, three drops of the appropriate fixative can be put into the seawater to settle them. The embryos will become immobilized within 1 to 2 min, sink to the bottom, and can then be collected quickly. All seawater is removed from the wells by using a bent pipette and a mouth pipettor or suction syringe. Fixative is added ($\sim 150\,\mu l$ or 4 drops per well) and the solution mixed.

When fixing embryos in MeOH, the addition of MeOH to the MFSW will result in a white precipitate that can obscure the specimens in the bottom of the wells. Subsequent washes with PBS or PBST will dissolve the precipitate. However, during these washes, embryos can become trapped at the surface of the wash solutions due to surface tension. Careful monitoring under the dissecting microscope is essential to prevent the loss of embryos. It is recommended that the removed washes be placed in neighboring wells and checked for stray embryos. The use of aldehyde fixatives does not cause similar precipitates but since the subsequent permeabilization step usually requires addition of MeOH or non-ionic detergents, precipitates or bubbles can be created during mixing. Bubbles are generated when air is introduced through the pipette beneath the surface of the solution. They obscure the embryos at the bottom of the wells, and embryos can also become trapped by the surface tension created by bubbles at the surface. Bubbles therefore should be avoided when adding and mixing solutions.

5. Washing

It is essential that all fixative be removed before adding blocking solution to the wells. A minimum of three washes is recommended. Remove all fixative by aspiration as has been described for removing seawater and replace the fixative with ~150 μl of wash buffer and mix. The addition of non-ionic detergents, either Tween-20 or Triton X-100 at 0.1 to 0.5% is recommended.

6. Immunostaining

a. Blocking

Blocking solution is added to the wells so that the embryos are completely submerged.

b. Primary antibody incubation

Since the well plate procedure uses minimal volumes of solutions, it is important that most of the blocking solution be removed from the wells to avoid diluting the antibody.

c. Washes

A minimum of three washes for 5 min each is recommended.

d. Secondary antibody

Secondary antibody is added at the appropriate dilution. If nuclear staining is required, add the nuclear stain (see Section IV) into the secondary antibody solution to give the appropriate final concentration. When co-localizing multiple antigens, remember to consider the species used for both primary and secondary antibodies.

e. Washes

A minimum of three washes at 5 min each is recommended.

f. Mounting

After removal of the final wash solution, add sufficient PBS ~100 μl to allow transfer of the embryos to a glass slide. Use PBS rather than PBST to prevent bubbles from forming. Remove the embryos from the well and place on the center of the slide using a mouth pipette or suction syringe and pipette. Avoid collecting more solution than will fit in the straight piece of the micropipette, as the embryos will stick to the curved region of the pipette. Typically, more embryos are lost during mounting than at any other step in the procedure. Carefully aspirate off as much PBS as possible and place mounting media in a ring around the embryos. When the coverslip is added, this will push the embryos toward the center of the coverslip, rather than to the edges. Put clay feet (1 mm thick) at each corner of the coverslip, place this on top of the embryos, and attach the feet by carefully applying pressure at the four corners. If necessary, fill in the space under the cover glass with more mounting media applied with a straight, wide pipette. The clay feet will prevent the embryos from being squashed by the weight of the coverslip and will allow them to be repositioned by gently moving the coverslip, thus rolling the embryos underneath. The coverslips can be sealed with nail polish to prevent evaporation and contamination. Sealed coverslips allow the use of oil immersion lenses and can be cleaned with EtOH or Windex without causing damage to the embryos. If properly stored (in the dark and in the cold), immunofluorescent staining can be preserved for many months on well-sealed slides.

For high-magnification work with oil immersion lenses, an alternative mounting approach is recommended. With standard coverslips (1–1/2 thickness) and clay feet, one cannot observe anything but the upper surface of the specimen with oil immersion lenses. In addition, the lens will push on the coverslip, flex it, and compress or roll the embryos. For high-magnification work, embryos should be mounted using the thinnest possible spacer. Scotch #665 double-sided tape, which is ~100 microns thick (just thinner than a Lytechinus embryo) can be used with a #1 or #0 coverslip.

B. Processing Larvae for Immunocytochemistry

Since older sea urchin larvae are large and fragile, special handling is required to maintain their integrity during fixation and processing for immunostaining. Because they become sticky after fixation and are easily damaged, it is advisable to collect, process, and view them in the same dish.

Preparation of glass bottom dishes

1. Cut a hole (~0.5 inches in diameter) in the bottom of a 35 mm tissue culture dish using a sheet metal hand punch (Roper Whitney Model XX Hand punch, Roper Whitney of Rockford, Inc., Rockford, IL).
2. Invert the dish and attach a coverslip with Crazy Glue™ or nail polish to the outside bottom covering the hole. This creates a well the thickness of the plastic dish.
3. Transfer larvae to the well and process through all stages of immunostaining and view on an inverted microscope directly in the dish.

To assure that the thickness of the larvae is maintained, it is important that solutions never be completely removed from the well, or the larvae will flatten under their own weight. These special glass-bottom dishes are also commercially available (MatTeK Corporation, Ashland, MA). The large size and thickness of sea urchin larvae makes them ideally suited for confocal microscopic analysis of fluorescence immunostaining. Phalloidin staining (Section IV.C) alone or coupled with antibody staining can be used to view the basic shape and developmental stage of the larvae (Fig. 4).

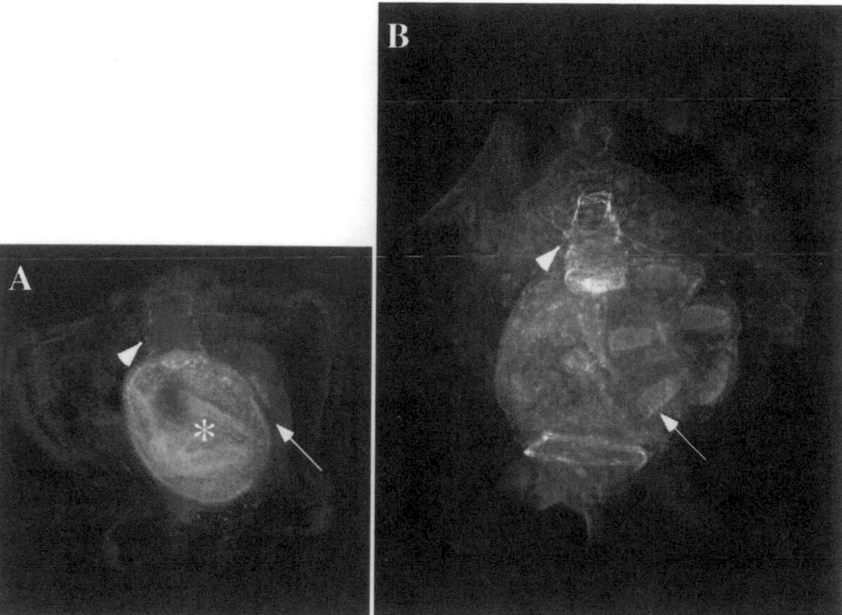

Fig. 4 Phalloidin staining of larval stage embryos. Larvae were collected and labeled with phalloidin, as detailed in the text. An early rudiment (arrow) can be seen on one side of the 3-week-old larva in (A) that was also immunostained with Endo1 monoclonal antibody (*). A fully developed rudiment (arrow) containing well-defined tube feet is detected in the 6-week-old larva (B) by phalloidin staining. Arrowhead points to circumesophageal muscles. (See Color Insert.)

C. Variations

Terasaki dishes (microwell minitrays with lids; Nalge, Nunc #438733, Naperville, IL) can also be used to process small numbers of embryos for immunostaining. Terasaki dishes contain multiple wells; each well has sloping sides that hold ~10 μl of solution and can therefore save on antibody and other solutions. These are ideal for processing only a few embryos. Small numbers of embryos (1–20 embryos) can also be processed in glass depression slides. The wells are larger than those of the 96-well plate and so the embryos are easier to track individually.

IV. Co-localization

Immunofluorescence can be combined with nuclear stains (Section IV.D), fluorescent lineage tracers, or fluorescent phalloidin (Section IV.C) to label microfilamentous actin. Be sure to use appropriate fluorochrome-conjugated secondary antibodies to allow distinction between the different fluorophores used. For example, propidium iodide is not compatible with TRITC or Texas Red-conjugated antibodies, and DAPI or Hoechst staining should not be used with AMCA-conjugated antibodies.

Be certain that the appropriate dichroic filter sets are available on the conventional epifluorescence microscope you will use for detection. Also, some confocal microscopes are limited in their ability to detect UV, so be sure that the one you will use has the appropriate blue diode laser (see Chapter 16).

A. Double Antibody Labeling

1. Simultaneous Incubation

Combinations of antibodies generated in different species can be used to detect multiple antigens in embryos simultaneously. Fix and block as has been described. Incubate the embryos in the two primary antibodies (each at the appropriate dilution), wash as indicated in the standard protocol, then incubate in the secondary antibodies (at the appropriate dilution for each), wash again and mount. Be sure to choose secondary antibodies that cross-react with only the primary antibodies and not with each other. For example, a monoclonal antibody raised in mice and a polyclonal antibody raised in rabbits can be combined and detected by a combination of goat anti-mouse FITC and goat anti-rabbit TRITC (Fig. 5A and B). Some commercial suppliers sell secondary antibodies that have been preabsorbed against the Igs of other species and are particularly useful for these kinds of studies. Alternatively, samples can be labeled by indirect immunofluorescence and counterstained with fluorescent phalloidin or a DNA counterstain (see Section D). Controls for double-labeling should include labeling with each antibody individually.

2. Consecutive Incubation

It is also possible to detect two antigens recognized by different monoclonal antibodies sequentially by inserting an additional blocking step in the protocol. The embryos are fixed, washed, blocked, incubated in the first primary antibody, washed, incubated in the first secondary antibody, washed again, then blocked again before the process is repeated using a different monoclonal antibody and a second anti-mouse antibody conjugated to a fluorochrome different from that coupled to the first secondary antibody. This technique is not recommended if the objective is to ascertain co-localization because the chance of cross-reactivity is high. However, it is useful if the objective is to demonstrate the pattern of expression of markers of different lineages.

Isotype-specific secondary antibodies can also be used to distinguish IgM (μ chain) from IgG (γ chain) monoclonal antibodies and avoid potential antibody "cross-talk." This technique can be used for antibodies regardless of species or subtype specificity, and uses commercially available reagents (Wessel and McClay, 1986).

B. Triple Antibody Staining

The preceding procedure can also be modified with the addition of a polyclonal antibody that is used simultaneously with either the first or second monoclonal antibody incubation. While this procedure allows visualization of three

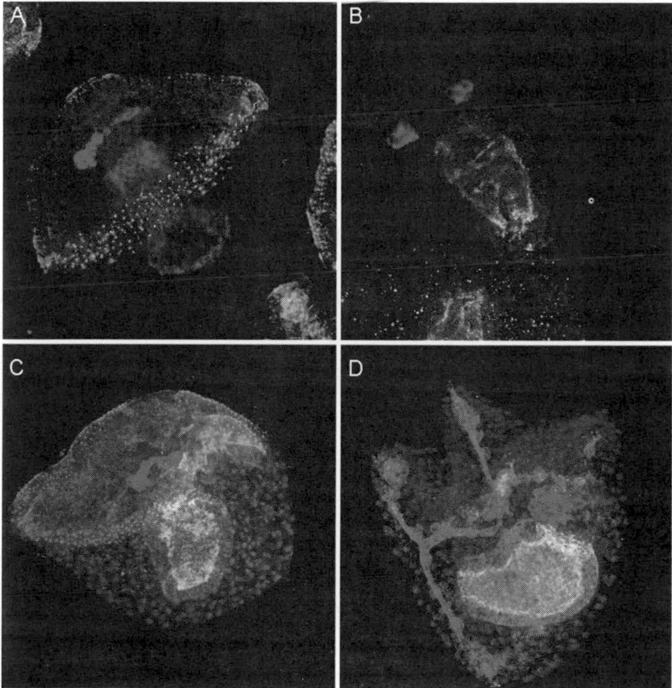

Fig. 5 Double and triple immunofluorescent labeling of embryos. (A and B) For double labeling, embryos were simultaneously incubated with (A) MHC (polyclonal) and CBA (monoclonal) or (B) MHC (polyclonal) and Endo1 (monoclonal) primary antibodies, followed by simultaneous incubation in goat anti-mouse FITC and goat anti-rabbit TRITC secondary antibodies. The muscles cells appear red, and the ciliary band (A) or gut (B) are green. (C and D) For triple staining, embryos were incubated in MHC (polyclonal) and Endo1 (monoclonal) simultaneously, followed by simultaneous incubation in goat anti-mouse Cascade Blue and goat anti-rabbit TRITC secondary antibodies. After a second blocking step in 5% NGS/PBST, embryos were incubated with CBA (C) or 6A9 (D) monoclonal antibodies, followed by a final incubation in FITC-conjugated goat anti-mouse secondary antibody. Muscle cells appear red, the guts blue, and the ciliary band (C) and primary mesenchyme cells (D) are green. All embryos were fixed in MeOH. Images were obtained on a BioRad Radiance 2000 Confocal Microscope. (See Color Insert.)

antigens simultaneously, it works best when one of the monoclonal's targets is extremely abundant (Fig. 5C and D).

C. Phalloidin Labeling

Filamentous actin can be labeled using fluorescently conjugated phallotoxins, bicyclic peptides isolated from the poisonous mushroom *Amanita* (Barak *et al.*, 1980). The most commonly used member of this family, phalloidin, may be purchased conjugated to a wide variety of fluorescent dyes. Phalloidin selectively binds and stabilizes polymerized, filamentous actin without binding monomeric actin, and its nonspecific staining is negligible. These properties make phalloidin more attractive for fluorescence microscopy than actin-specific antibodies. Furthermore, phalloidin binds to actin from different species (including plant, animal, and fungal cells) and does not discriminate between actin isoforms. Fluorescently labeled phalloidin can be used to visualize actin microfilaments particularly well in late sea urchin gastrulae and larvae (Burke and Alvarez, 1988). This will allow visualization of muscle at the prism stage and the developing adult rudiment in the late larva.

Phalloidin is usually purchased as a lyophilized powder (Sigma, St. Louis, MO; Molecular Probes, Portland, OR) and is reconstituted in MeOH or DMSO. Phalloidin is prepared as a 1000× stock of 3 μM and stored in small aliquots below 0 °C. For staining, this stock should be diluted into PBS or may be included with fluorescent secondary antibodies for double-labeling experiments. Phalloidin is cell-impermeable, so specimens must be fixed and detergent-permeabilized. Embryos to be stained should be fixed with either aldehydes or cold acetone but not MeOH, as MeOH fixation destroys the phalloidin-binding site on actin. Phalloidin-stained specimens need to be imaged within a few days because phalloidin will dissociate into the mounting medium over time and produce a background autofluorescence that may obscure fine detail.

D. Nuclear Staining

Nuclear stains are useful for providing an image of the entire embryo in conjunction with immunostaining, particularly when only a few cells express the antigen of interest. Nuclear stains are also an important adjunct when confirming nuclear localization of a particular antigen, such as a transcription factor (Kenny *et al.*, 1999). There are numerous DNA-specific fluorescent stains available; the choice depends on the fluorochromes conjugated to the secondary antibody as well as the detection method used. The most common fluorescent DNA stains used in combination with immunofluorescent staining of embryos are 4′, 6′-diamidino-2phenylindole, dihydochloride (DAPI), bisbenzimide (Hoechst 33345 and 33258), and propidium iodide (PI). These are intercalating agents that fluoresce only when bound to nucleic acids.

1. DAPI and Hoechst Labeling

These most commonly used nuclear counterstains are applied simply by including dye with the secondary antibody or in a subsequent wash step. The absorption and emission of these fluorophores do not overlap with those for most conventional fluorophores such as FITC or TRITC, and the necessary filters for epifluorescence microscopes are relatively inexpensive. However, for confocal microscopy, DAPI and Hoechst use requires an additional laser source. Since both dyes emit in the blue wavelength, they can be combined with both FITC and TRITC conjugated secondary antibodies and are particularly useful for triple labeling. Also, both DAPI and Hoechst are membrane-permeant and can be used to label DNA in living cells.

DAPI or Hoechst as Nuclear Counterstains for Immunofluorescence

1. Fix, permeabilize, and process embryos as described in Section II or III for indirect immunofluorescence.
2. During the secondary antibody incubation, add DAPI or Hoechst dye diluted 1:1000 from a 10 mg/ml stock (prepared in water and stored at 4 °C) along with the secondary antibody.
3. Perform postsecondary antibody incubation washes for 3 × 5 min in PBST.
4. Mount and coverslip.

2. Propidium Iodide

PI is a robust, photostable DNA intercalating agent that stains nuclei red. However, PI binds RNA as well as DNA; therefore, an RNA digestion step may be required to reduce cytoplasmic staining. Since PI is generally detected with the same filter sets as Texas Red or TRITC (Table II), PI should be avoided as a counter stain with secondary antibodies conjugated to these fluorochromes. Unlike DAPI and Hoechst dyes, PI is not membrane-penetrant and cannot be used to label living cells.

Propidium Iodide as a Nuclear Counter Stain for Immunofluorescence

1. Fix and permeabilize embryos.
2. Treat with 1 μg/ml of RNAse A in PBST at 37 °C for 30 min (prepare RNAse from a 10 mg/ml stock, boiled 15 min to destroy DNAses).
3. Wash 3 × 5 min in PBST.
4. Process embryos as described in Section II or III for indirect immunofluorescence.
5. During the secondary antibody incubation, add PI diluted 1:1000 from a 10 mg/ml stock (prepared in water and stored at 4 °C) along with the secondary antibody.
6. Perform postsecondary antibody incubation washes for 3 × 5 min in PBST.
7. Mount and coverslip.

3. Alternative Approaches

Mounting media containing either DAPI or PI as nuclear counterstains can be obtained commercially (Vectashield with DAPI or PI, Vector Laboratories, Burlingame, CA; Slowfade with DAPI, Molecular Probes, Inc., Eugene, OR). While these have the advantage of combining the counterstaining and mounting steps, not all situations require nuclear staining so they are appropriate only when nuclear counterstaining is routinely performed.

V. Troubleshooting

Common problems encountered during processing small numbers of embryos for immunocytostaining are highlighted in Table III. The following section describes more generally applicable problems that may be encountered whenever performing immunostaining.

A. Loss of Embryos

1. Loss of embryos is a serious problem when working with small numbers of embryos but occurs whenever embryos are processed for immunostaining. To minimize specimen loss, always check for embryos that may have stuck to the sides of the wells or pipettes.
2. When transferring embryos, avoid aspirating the embryos too far into the micropipettes; keep them in a small volume of liquid near the tip.
3. When working with flexible "U" bottom well plates, have enough water in the petri dish so that the entire depth of the well is visible.
4. Don't mix solutions so vigorously that specimens will be transported out of the well.

Table III
Hints for Successful Handling of Small Numbers of Embryos

1. Do not remove liquid with a Gilson or other pipettman without checking under the microscope.
2. Never leave the pipette you use to aspirate supernatant standing in a rack; always leave it on a clay pad, with the tip tilted downward. Solutions will run along the inside of the pipette and will pick up condensing water (from your breath) that might contain enzymes and solutions from other steps and may be accidentally transferred to the wells.
3. Always use fresh and clean pipettes. Used pipettes that have not been carefully rinsed can have attached salt crystals and debris that will contaminate the wells.
4. Wipe the surface of the well plate between the wells periodically, when the wells don't contain much solution. This removes dry crystals that may have been swept up during mixing and can be transferred into wells.
5. The more viscous the solution, the less volume is used.
6. Keep your mixing pipette dry.
7. Do not try to put more than 200 μl of any solution in one well.
8. Avoid bubbles in the pipettes. Either the embryos will stick to the inside of the glass or you will contaminate the solution.
9. When using PBS or PBST, put the buffer in a 20 to 40 ml syringe filter through a 22 or 45 μm syringe filter; 5 drops per well.
10. Never let the wells dry out completely.
11. When mixing solutions in the wells, always control the airflow. If you don't apply any air pressure at all, solution or embryos will be sucked into the pipette because of capillary forces. Use a laboratory air line where possible.

B. No Staining

1. Fixation may cause loss of antigenicity. The appropriate fixative, fixative concentration, and time of fixation should be determined empirically.

2. An important aspect of this is that some antibodies may recognize epitopes that are buried in the native protein. This may be the case if the antibody was raised against denatured protein or against a synthetic peptide that is in a region buried in the native protein. Such antibodies may recognize the denatured protein on immunoblots but not the fixed protein by immunostaining.

3. The permeabilization procedure may solubilize the antigen of interest. Apply a different permeabilization procedure. Use aldehyde-based fixation followed by MeOH, acetone, or detergent permeabilization instead of MeOH fixation alone, or vice versa.

4. The antibody (primary and/or secondary) may be too dilute; try a different concentration.

5. You may have chosen the wrong species-specific secondary antibody.

6. The antibody may be inactive and you should try a different antibody.

C. Overstaining and High Background

1. The most common cause of overstaining is too high an antibody concentration; you should test a range of dilutions.

2. Reduce the antibody incubation time to help lower background staining.

3. The choice of protein or protein concentration used in the blocking step may be inappropriate or too dilute.

4. Include the blocking protein in all incubations and washes except the last few washes to avoid high background staining.

5. Nonspecific background in the well plate procedure can be caused by inadequate mixing.

6. If the wrong species-specific antibody is used in double labeling experiments, unexpected background staining may be observed.

7. Endogenous enzymes in embryos may react with the chromogenic substrates when using enzyme-linked detection methods. To block endogenous enzyme activity:

   a. Block endogenous peroxidase activity by incubating embryos in 4 parts MeOH to 1 part H_2O_2 (3%) for 20 min before staining.
   b. Block endogenous alkaline phosphatase activity by adding 0.1 μM levamisole to the enzyme substrate solution.

8. Similarly, unrelated proteins may bind your antibody nonspecifically. Preabsorption of the primary antibody with an acetone powder prepared from an embryonic stage where the antigen is not expressed (see Harlow and Lane, 1988; Harris, 1986) may help reduce nonspecific staining. Affinity purification of the

primary antibody may help by eliminating antibodies in the serum that recognize unrelated proteins in the specimen. Neither of these methods will solve the problem of a shared epitope; i.e., the specific epitope recognized by your antibody on the protein of interest may also be present on other proteins.

D. Flat or Collapsed Embryos

1. Embryos collapse and become flat when allowed to dry out, so do not let specimens become dry.

2. Embryos collapse if there are dramatic changes in buffer concentration and/or viscosity. If you cannot remove and replace solutions quickly, you can either treat one well at a time or leave some solution in the well but increase the number of washes (so that the total volume of solution through which the embryos are passed is still the same).

E. Broken Embryos

1. Broken embryos are caused by sticky precipitates at the bottom of the wells or faulty or dirty pipettes that tear the embryos.

2. Embryos may not have been fixed adequately; increase fixative concentration or fixation duration.

F. Precipitates and "Sticky Embryos"

1. Precipitates may occur during transfer from fixative into PBS or PBST, and continued washing will generally remove the precipitates.

2. Protein aggregates may be present in the blocking or antibody solutions; filter solutions and centrifuge antibody and other protein solutions briefly in a microcentrifuge.

Acknowledgments

We thank Dr. Peter Cserjesi and Ms. Mary Anne Alliegro, Department of Cell Biology and Anatomy, LSUHSC, New Orleans, LA, for critically reading the manuscript and also thank all our colleagues who generously provided information on antibodies useful for immunostaining sea urchin embryos. J.M.V. is supported by grants from the N.S.F (IBN-0196065) and the Muscular Dystrophy Association.

References

Adelson, D. L., and Humphreys, T. (1988). Sea urchin morphogenesis and cell-hyalin adhesion are perturbed by a monoclonal antibody specific for hyalin. *Development* **104,** 391–402.

Alliegro, M. C., Burdsal, C. A., and McClay, D. R. (1990). *In vitro* biological activities of echinonectin. *Biochem.* **29,** 2135–2141.

Alliegro, M. C., Ettensohn, C. A., Burdsal, C. A., Erickson, H. P., and McClay, D. R. (1988). Echinonectin: A new embryonic substrate adhesion protein. *J. Cell Biol.* **107,** 2319–2327.

Angerer, L. M., Oleksyn, D. W., Logan, C. Y., McClay, D. R., Dale, L., and Angerer, R. C. (2000). A BMP pathway regulates cell fate allocation along the sea urchin animal–vegetal embryonic axis. *Development* **127,** 1105–1114.

Barak, L. S., Yocum, R. R., Nothnagel, E. A., and Webb, W. W. (1980). Fluorescence staining of the actin cytoskeleton in living cells with 7-nitrobenz-2-oxa-1,3-diazole-phallacidin. *Proc. Natl. Acad. Sci. USA* **77,** 980–984.

Brooks, J. M., and Wessel, G. M. (2002). The major yolk protein in sea urchins is a transferrin-like, iron binding protein. *Dev. Biol.* **245,** 1–12.

Brooks, J. M., and Wessel, G. M. (2004). The major yolk protein of sea urchins is endocytosed by a dynamin-dependent mechanism. *Biol. Reprod.* **71,** 705–713.

Burke, R. D., and Alvarez, C. M. (1988). Development of the esophageal muscles in embryos of the sea urchin Strongylocentrotus purpuratus. *Cell Tissue Res.* **252,** 411–417.

Cervello, M., and Matranga, V. (1989). Evidence of a precursor–product relationship between vitellogenin and toposome, a glycoprotein complex mediating cell adhesion. *Cell Diff. Dev.* **26,** 67–76.

Chuang, C. K., Wikramanayake, A. H., Mao, C. A., Li, X., and Klein, W. H. (1996). Transient appearance of Strongylocentrotus purpuratus Otx in micromere nuclei: Cytoplasmic retention of SpOtx possibly mediated through an alpha–actinin interaction. *Dev. Gen.* **19,** 231–237.

Clancy, B., and Cauller, L. J. (1998). Reduction of background autofluorescence in brain sections following immersion in sodium borohydride. *J. Neurosci. Methods* **83,** 97–102.

Coffman, J. A., and McClay, D. R. (1990). A hyaline layer protein that becomes localized to the oral ectoderm and foregut of sea urchin embryos. *Dev. Biol.* **140,** 93–104.

Coons, A. H., Creech, H. J., and Jones, R. N. (1941). Immunological properties of an antibody containing a fluorescent group. *Proc. Soc. Exp. Biol. Med.* **47,** 200–205.

Coons, A. H., and Kaplan, N. H. (1950). Localization of antigens in tissue cells. *J. Exp. Med.* **91,** 1–13.

Conner, S., Leaf, D., and Wessel, G. (1997). Members of the SNARE hypothesis are associated with cortical granule exocytosis in the sea urchin egg. *Mol. Reprod. Dev.* **48,** 106–118.

Conner, S., and Wessel, G. M. (1998). rab3 mediates cortical granule exocytosis in the sea urchin egg. *Dev. Biol.* **203,** 334–344.

Epel, D., Weaver, A. M., and Mazia, D. (1970). Methods for removal of the vitelline membrane of sea urchin eggs. II. Use of dithiothreitol (Cleland Reagent). *Exp. Cell Res.* **61,** 69–70.

Ettensohn, C. A., and McClay, D. R. (1988). Cell lineage conversion in the sea urchin embryo. *Dev. Biol.* **125,** 396–409.

Gibson, A. W., and Burke, R. D. (1985). The origin of pigment cells in embryos of the sea urchin. *Strongylocentrotus purpuratus. Dev. Biol.* **107,** 414–419.

Gross, J. M., Peterson, R. E., Wu, S. Y., and McClay, D. R. (2003). LvTbx2/3: A T-box family transcription factor involved in formation of the oral/aboral axis of the sea urchin embryo. *Development* **130,** 1989–1999.

Haley, S. A., and Wessel, G. M. (1999). The cortical granule serine protease CGSP1 of the sea urchin, Strongylocentrotus purpuratus, is autocatalytic and contains a low-density lipoprotein receptor-like domain. *Dev. Biol.* **211,** 1–10.

Harlow, E., and Lane, D. (1988). "Antibodies: A Laboratory Manual," p. 726. Cold Spring Harbor Laboratory.

Harlow, E., and Lane, D. (1998). "Using Antibodies: A Laboratory Manual," p. 495. Cold Spring Harbor Laboratory.

Harris, P., Osborn, M., and Weber, K. (1980). Distribution of tubulin-containing structures in the egg of the sea urchin Strongylocentrotus purpuratus from fertilization through first cleavage. *J. Cell Biol.* **84,** 668–679.

Harris, P. (1986). Cytology and immunocytochemistry in echinoderm gametes and embryos. *Meth. Cell Biol.* **27,** 243–262.

Hodor, P. G., Illies, M. R., Broadley, S., and Ettensohn, C. A. (2000). Cell-substrate interactions during sea urchin gastrulation: Migrating primary mesenchyme cells interact with and align extracellular matrix fibers that contain ECM3, a molecule with NG2-like and multiple calcium-binding domains. *Dev. Biol.* **222,** 181–194.

Hsu, S. M., Raine, L., and Fanger, H. (1981). The use of antiavidin antibody and avidinbiotin-peroxidase complex in immunoperoxidase techniques. *Am. J. Clin. Pathol.* **75,** 816–821.

Humason, G. L. (1972). "Animal Tissue Techniques," p. 641. W. H. Freeman and Company, San Francisco.

Ingersoll, E. P., and Ettensohn, C. A. (1994). An N-linked carbohydrate-containing extracellular matrix determinant plays a key role in sea urchin gastrulation. *Dev. Biol.* **163,** 351–366.

Kanoh, K., Aizu, G., and Katow, H. (2001). Disappearance of an epithelial cell surface-specific glycoprotein (Epith-1) associated with epithelial–mesenchymal conversion in sea urchin embryogenesis. *Dev. Growth Diff.* **43,** 83–95.

Katow, H. (1995). Pamlin, a primary mesenchyme cell adhesion protein, in the basal lamina of the sea urchin embryo. *Exp. Cell Res.* **218,** 469–478.

Katow, H., and Sofuku, S. (2001). An RGDS peptide-binding receptor, FR-1R, localizes to the basal side of the ectoderm and to primary mesenchyme cells in sand dollar embryos. *Dev. Growth Diff.* **43,** 601–610.

Kenny, A. P., Kozlowski, D., Oleksyn, D. W., Angerer, L. M., and Angerer, R. C. (1999). SpSoxB1, a maternally encoded transcription factor asymmetrically distributed among early sea urchin blastomeres. *Development* **126,** 5473–5483.

LaFleur, G. J., Jr, Horiuchi, Y., and Wessel, G. M. (1998). Sea urchin ovoperoxidase: oocyte-specific member of a heme-dependent peroxidase superfamily that functions in the block to polyspermy. *Mech. Dev.* **70,** 77–89.

Laidlaw, M., and Wessel, G. M. (1994). Cortical granule biogenesis is active throughout oogenesis in sea urchins. *Development* **120,** 1325–1333.

Leguia, M., and Wessel, G. M. (2004). Selective expression of a secl/muncl8 member in sea urchin eggs and embryos. *Mech. Dev.* in press.

Logan, C. Y., Miller, J. R., Ferkowicz, M. J., and McClay, D. R. (1999). Nuclear beta-catenin is required to specify vegetal cell fates in the sea urchin embryo. *Development* **126,** 345–357.

Lowe, C. J., and Wray, G. A. (1997). Radical alterations in the roles of homeobox genes during echinoderm evolution. *Nature* **389,** 718–721.

Lowe, C. J., Issel-Tarverm, L., and Wray, G. A. (2002). Gene expression and larval evolution: Changing roles of distal-less and orthodenticle in echinoderm larvae. *Evol. Dev.* **4,** 111–123.

Matranga, V., Kuwasaki, B., and Noll, H. (1986). Functional characterization of toposomes from sea urchin blastula embryos by a morphogenetic cell aggregation assay. *EMBO J.* **5,** 3125–3132.

Matranga, V., Yokota, Y., Zito, F., Tesoro, V., and Nakano, E. (1995). Biochemical and immunological relationships among fibronectin-like proteins from different sea urchin species. *Roux's Arch. Dev. Biol.* **204,** 413–418.

McClay, D. R., Peterson, R. E., Range, R. C., Winter-Vann, A. M., and Ferkowicz, M. J. (2000). Micromere induction signal is activated by beta-catenin and acts through notch to initiate specification of secondary mesenchyme cells in the sea urchin embryo. *Development* **127,** 5113–5122.

Mujumdar, R. B., Ernst, L. A., Mujumdar, S. R., Lewis, C. J., and Waggoner, A. S. (1993). Cyanine dye labeling reagents: Sulfoindocyanine succinimidyl esters. *Bioconjug. Chem.* **4,** 105–111.

Noll, H., Matranga, V., Cervello, M., Humphreys, T., Kuwasaki, B., and Adelson, D. (1985). Characterization of toposome from sea urchin blastula cells: A cell organelle mediating adhesion and expressing positional information. *Proc. Natl. Acad. Sci. USA* **82,** 8062–8066.

Oliveri, P., Davidson, E. H., and McClay, D. R. (2003). Activation of pmarl controls specification of micromeres in the sea urchin embryo. *Dev. Biol.* **258,** 32–43.

Panchuk-Voloshina, N., Haugland, R. P., Bishop-Stewart, J., Bhalgat, M. K., Millard, P. J., Mao, F., Leung, W. Y., and Haugland, R. P. (1999). Alexa dyes, a series of new fluorescent dyes that yield exceptionally bright, photostable conjugates. *J. Histochem. Cytochem.* **47,** 1179–1188.

Sherwood, D. R., and McClay, D. R. (1999). LvNotch signaling mediates secondary mesenchyme specification in the sea urchin embryo. *Development* **126,** 1703–1713.

Somers, C. E., Battaglia, D. E., and Shapiro, B. M. (1989). Localization and developmental fate of ovoperoxidase and proteoliaisin, two proteins involved in fertilization envelope assembly. *Dev. Biol.* **131,** 226–235.

Sweet, H. C., Hodor, P. G., and Ettensohn, C. A. (1999). The role of micromere signaling in Notch activation and mesoderm specification during sea urchin embryogenesis. *Development* **126,** 5255–5265.

Urry, L. A., Hamilton, P. C., Killian, C. E., and Wilt, F. H. (2000). Expression of spicule matrix proteins in the sea urchin embryo during normal and experimentally altered spiculogenesis. *Dev. Biol.* **225,** 201–213.

Vonica, A., Weng, W., Gumbiner, B. M., and Venuti, J. M. (2000). TCF is the nuclear effector of the beta-catenin signal that patterns the sea urchin animal–vegetal axis. *Dev. Biol.* **217,** 230–243.

Voronina, E., Marzluff, W. F., and Wessel, G. M. (2003). Cyclin B synthesis is required for sea urchin oocyte maturation. *Dev. Biol.* **256,** 258–275.

Voronina, E., and Wessel, G. M. (2004). Regulatory contribution of heterotrimeric G-proteins to oocyte maturation in the sea urchin. *Mech. Dev.* **121,** 247–259.

Wessel, G. M., and McClay, D. R. (1985). Sequential expression of germ-layer specific molecules in the sea urchin embryo. *Dev. Biol.* **111,** 451–463.

Wessel, G. M., and McClay, D. R. (1986). Two embryonic, tissue-specific molecules identified by a double-label immunofluorescence technique for monoclonal antibodies. *J. Histochem. Cytochem.* **34,** 703–706.

Wessel, G. M., and McClay, D. R. (1987). Gastrulation in the sea urchin embryo requires the deposition of crosslinked collagen within the extracellular matrix. *Dev. Biol.* **121,** 149–165.

Wessel, G. M., Zhang, W., and Klein, W. H. (1990). Myosin heavy chain accumulates in dissimilar cell types of the macromere lineage in the sea urchin embryo. *Dev. Biol.* **140,** 447–454.

Wessel, G. M. (1995). A protein of the sea urchin cortical granules is targeted to the fertilization envelope and contains an LDL-receptor-like motif. *Dev. Biol.* **167,** 388–397.

Wessel, G. M., Berg, L., Adelson, D. L., Cannon, G., and McClay, D. R. (1998). A molecular analysis of hyalin-a substrate for cell adhesion in the hyaline layer of the sea urchin embryo. *Dev. Biol.* **193,** 115–126.

Wessel, G. M., Brooks, J. M., Green, E., Haley, S., Voronina, E., Wong, J., Zaydfudim, V., and Conner, S. (2001). The biology of cortical granules. *Int. Rev. Cytol.* **209,** 117–206.

Wikramanayake, A. H., Huang, L., and Klein, W. H. (1998). β-Catenin is essential for patterning the maternally specified animal–vegetal axis in the sea urchin embryo. *Proc. Natl. Acad. Sci. USA* **95,** 9343–9348.

Wray, G. A., and McClay, D. R. (1989). Molecular heterochronies and heterotropies in early echinoid development. *Evolution.* **43,** 803–813.

Wright, B. D., and Scholey, J. M. (1993). Nonfluorescent immunolocalization of antigens in mitotic sea urchin blastomeres. *Meth. Cell. Biol.* **37,** 223–240.

Zito, F., Tesoro, V., McClay, D. R., Nakano, E., and Matranga, V. (1998). Ectoderm cell–ECM interaction is essential for sea urchin embryo skeletogenesis. *Dev. Biol.* **196,** 184–192.

Zito, F., Nakano, E., Sciarrino, S., and Matranga, V. (2000). Regulative specification of ectoderm in skeleton disrupted sea urchin embryos treated with monoclonal antibody to Pl-nectin. *Growth Diff.* **42,** 499–506.

CHAPTER 16

Light Microscopy of Echinoderm Embryos

Laila Strickland,* George von Dassow,† Jan Ellenberg,‡ Victoria Foe,† Peter Lenart,‡ and David Burgess*

*Department of Biology
Boston College
Chestnut Hill, Massachusetts 02167

†Friday Harbor Laboratories
University of Washington
Seattle, Washington 98195

‡European Molecular Biological Lab
Heidelberg, Germany

0091-679X/04 $35.00

I. Introduction (D. Burgess and L. Strickland)

Echinoderm embryos have had a long and remarkable history under the microscope. Developmental and cell biologists have marveled at these cells for more than one hundred years, and have been able to answer many fundamental questions about complex cellular processes simply by watching them happen. Indeed, the year was 1876 when Oscar Hertwig first observed sperm and egg pronuclear fusion after mixing the gametes of a Mediterranean sea urchin (Hertwig, 1877). At an average 100 micrometers in diameter, the egg of the sea urchin has provided a powerful system for visualization of events covering all stages of early development. The size and durability of this cell make it ideal for microinjection and micromanipulation. At later stages of development, individual blastomeres can be manually isolated and transplanted between embryos. Oocyte maturation, sperm–egg recognition, fertilization, mitotic division, and gastrulation are just a few of the fields of study that have been advanced by significant work in echinoderm embryos.

Building on this illustrative foundation, the study of echinoderm embryos is certain to remain important. With ever improving optical equipment and the expanding innovation of methods for observing structures and processes inside the cell, as well as the advent of ever more sophisticated biochemical and molecular tools, we are able to ask increasingly complex questions of these cells.

Compiled in this chapter are notes on commonly used procedures for visualizing both fixed and live echinoderm embryos.

II. Formaldehyde Fixation of Cleavage Stage Sea Urchin Embryos

There are several important considerations to make when choosing a method of fixation. In general, one must decide if the experiment calls for a large number of embryos, such as would be required for immunofluorescence at multiple timepoints during development. In this case, a large-scale "free fix" would be appropriate, in which several milliliters of embryos are harvested by hand centrifugation and fixed while mixing in a large (50 ml) Falcon tube. For other types of experiments, it may be advantageous to fix a small number of cells that have settled onto a coverslip coated with protamine sulfate or poly-lysine. For example, individual cells that have been microinjected or otherwise micromanipulated and then fixed in this way can be used for immunofluorescence or various staining methods. By following a few general guidelines, a fixation protocol that is optimized for a specific application can be developed.

Described in this section are several procedures for fixing early sea urchin embryos with formaldehyde, as well as a discussion of how these procedures may be scaled appropriately to accommodate a variety of experiments. Fixation with cold methanol is also described as an alternative to the aldehyde-based methods. Finally, different fixatives, detergents, and buffers are described, with specific considerations about their use.

Note on gamete preparation: In preparation for all of the protocols presented, we collect eggs and sperm from sea urchins by intercoelemic injection of 0.5 M KCl into the aboral surface of the urchin. Prior to fertilization, the jelly coats of the eggs are dissolved in Calcium Free Sea Water (CaFSW), and then resuspended seawater (SW). Sperm is diluted 1000-fold with SW and activated by vigorous aeration. Within 3 min of fertilization, the eggs are pelleted by hand centrifuge, resuspended in CaFSW, and poured through Nitex mesh of an appropriate diameter to strip the fertilization envelopes. To prevent the demembranated eggs from sticking together in clumps, we raise them to the desired stage in CaFSW with gentle stirring. There are numerous methods for obtaining and preparing gametes, described elsewhere in this volume.

A. Protocols for Fixation

1. Formaldehyde Fixation of Early Sea Urchin Embryos (Fig. 1.A)

a. Large-Scale Free Fix

1. At the desired stage of development, gently pellet 1 to 3 ml of cells in a 50 ml Falcon tube by hand centrifugation.

2. Aspirate the supernatant, leaving enough as to avoid an air–water interface at the pellet. Passage of eggs through an air–water interface can cause the cells to lyse.

3. Wash the cells once by resuspending in 50 ml of isolation buffer (refer to Section II.A.1.e for recipe). This wash equilibrates the cells to the solution in which they will be fixed. It contains all of the components of the fixation buffer, except for the fixative and detergent. The 1 M glycerol used in isolation buffer stabilizes some of the components of the cell, and can give a pre-clearing effect of the yolk and pigment. Isolation buffer should be applied to the cells at the same temperature as that at which the cells have been cultured.

4. Pellet the cells by hand centrifugation; aspirate the supernatant as before.

5. Resuspend in 50 ml of fixation buffer (refer to Section II.A.1.e for recipe). This buffer contains 3.7% formaldehyde as a fixative and 0.1% NP-40. The detergent permeabilizes the cells, which can help with clearing of yolk and pigment that may interfere with fluorescence microscopy. Permeabilization is also required to allow for full penetration of many stains and antibodies to be used for visualization.

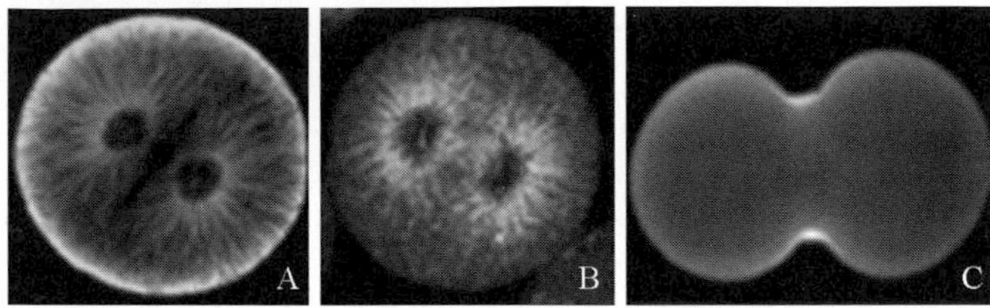

Fig. 1 Common techniques for fluorescent labeling of fixed cells. (A) Tubulin immunofluorescence in an early anaphase *S. purperatus* egg. The cells were fixed in a large batch by the free fix method in section II.A.1.a. After fixation, and washing by hand centrifuging 3X through PBS, a 1 ml aliquot of cells was taken and brought to 3% with BSA. Primary antibody (E7; Developmental Studies Hybridoma Bank) was added at a dilution of 1:50, and the cells were incubated on a rotator for 2 h. The sample was again washed 3X in PBS, and resuspended in PBS with 3% BSA. Secondary antibody (Fluorescein-conjugated anti-mouse; Chemicon) was added at a dilution of 1:200. All post-fix incubations and washes were performed at room temperature. The cells were incubated on a rotator for 1 h, washed 3X, and imaged on a Leica confocal microscope using an argon/krypton laser for illumination. (Image: Laila Strickland) (B) Tubulin immunofluorescence in a metaphase *S. purperatus* egg. The cell was fixed on a glass coverslip by the method in Section II.A.1.d.i. The antibodies and dilutions used were the same as in (A). Incubations were performed at room temperature in a small tissue culture dish on a gentle rocker. Images were acquired as in (A). The punctate appearance of the microtubules and higher level of background fluorescence is characteristic of immunofluorescence in cells that have been fixed in this way. (Image: Laila Strickland) (C) Filipin-staining of sterol-rich membrane domains (lipid rafts) at the cleavage furrow of an *S. drobachiensis* egg. Lipid rafts are insoluble in 0.1% ice-cold Triton-X 100 and can be visualized after fixation and solubilization, by incubating with filipin for 30 min. (Image: Michelle Ng). (See Color Insert.)

6. Incubate the cells in fixation buffer at room temperature, rotating end-over-end, or otherwise mixing for at least 1 h, and not exceeding 3 h.
7. After the fixation, wash the cells three times by pelleting and resuspending in Phosphate Buffered Saline (PBS), using fresh PBS for each wash. At this stage, the cells may be stored at 4 °C for up to 5 days.

b. Scaling Down for Smaller Samples

The formaldehyde fixation protocol given in the previous section can be easily scaled down to fix a small batch of embryos. As a rule of thumb, the volume of isolation buffer and fixation buffer must exceed 10 to 20 times that of the cells being fixed. For example, a cell pellet of 0.5 ml should be isolated and fixed in volumes of at least 10 ml of isolation buffer and fixation buffer, respectively. Typically, we use 15 ml Falcon tubes. Likewise, a small cell pellet of 50 to 100 ul can be isolated and fixed in a 1.7 ml microfuge tube; however, loss of sample is an important consideration when fixing such a small number of cells. Because fixation requires spinning down the cells multiple times throughout the protocol, a certain percentage of the cells will inevitably be lost at each step.

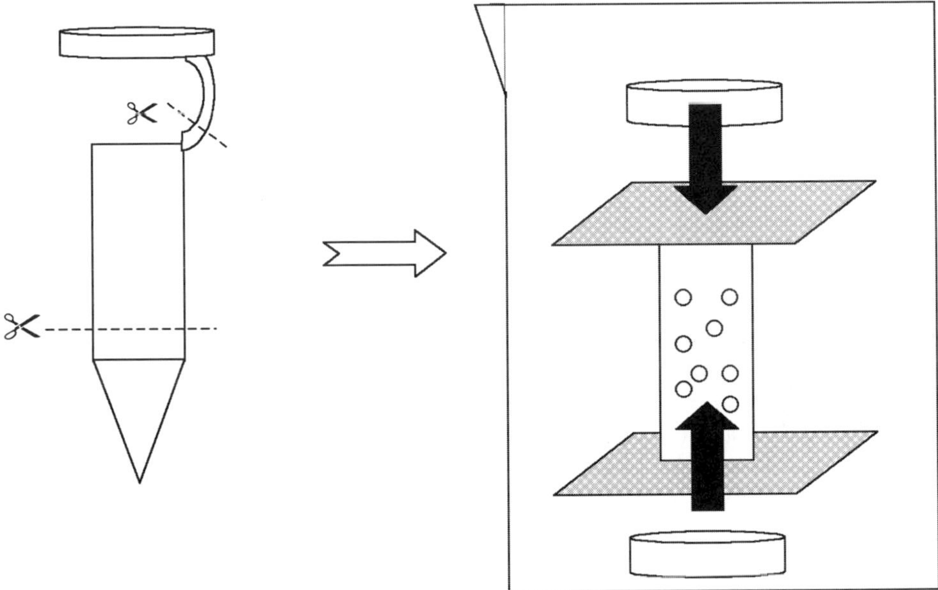

Fig. 2 Construction of a microfuge chamber for small scale fixation. The tops are cut off of two microfuge chambers, and the bottom of one tube is removed with a razor blade. The two microfuge tops are used to cap either end of the cylinder, fastened over two squares of Nitex mesh. The Nitex should have a square size smaller than the eggs so that they will be retained in the chamber. Because air–water interfaces are detrimental to eggs, it is important to keep one end of the chamber submerged during all volume exchanges of buffer and fixative during the procedure.

c. Micro-Scale Free Fix

Described here is a protocol for fixation of a very small number of cells with minimal sample loss. In this case, rather than pelleting the cells at each step, the volume exchanges required for fixation are carried out in a beaker, while the cells are maintained in a small chamber constructed of a microfuge tube and Nitex mesh. The chamber can be constructed from either 1.7 ml or 0.7 ml microfuge tubes (Fig. 2).

Construction of a microfuge/Nitex chamber:

1. Cut the caps off of two microfuge tubes and bore holes through the center of each. The caps will provide the top and bottom pieces of the chamber. The holes will be screened off with Nitex, through which fluid can flow, but the cells cannot.
2. Use a razor blade to cut the bottom half off of one of the microfuge tubes; the cylindrical top half will be the fixation chamber.
3. Cut out two squares of Nitex, large enough to cover the two ends of the chamber. The mesh size of the Nitex should be small enough such that the eggs will not pass through. For example, 80 micron Nitex would work well for *L. pictus* eggs, which have a diameter of approximately 150 microns.

4. Snap one of the caps in place over a Nitex square at one end of the chamber.

5. Submerge the capped end of the chamber into a small beaker of seawater or CaFSW.

6. Transfer the cells to be fixed into the chamber, close the top end with the remaining square of Nitex, and snap the microfuge cap into place.

7. Submerge the chamber fully into the beaker. Volume changes of isolation buffer, fixation buffer, and PBS can be carried out in the beaker containing the chamber, though care must be taken to always leave one end of the chamber submerged to avoid passage of the eggs through an air–water interface.

d. Coverslip Fixation

Allowing a small number of cells to settle onto a glass coverslip and then manually performing volume exchanges by gently pipetting isolation buffer, fixation buffer, and PBS is another way to preserve small samples, and even to monitor specific cells throughout the procedure. A key advantage of this fixation technique is that it can be used to fix individual cells that have been either microinjected or micromanipulated. The size and durability of the echinoderm egg lends the system to spectacular microscopy, and also renders it amenable to direct physical manipulation. Indeed, by using glass tools fashioned on a microforge, Rappaport was able to show that microtubules of the mitotic apparatus are required for induction of a cleavage furrow at the cell cortex (Rappaport, 1996).

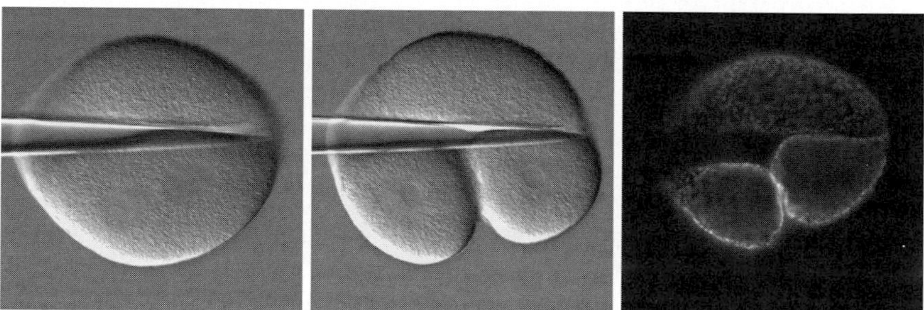

Fig. 3 Coverslip fixation, micromanipulation, hyaline immunofluorescence. At cytokinesis of the first mitotic division, hyaline is secreted at the cleavage furrow (Shuster and Burgess; 2002). Cells were fixed in CaFSW with 5% formaldehyde for 20 min, washed 3X with PBS, and incubated at room temperature with a monoclonal primary antibody directed against hyaline at a 1:50 dilution. This was followed by three washes with PBS and room temperature incubation for 1 h with 1:200 rhodamine-conjugated anti-mouse IgG (Chemicon). Images were acquired on a Nikon TE 200 microscope. An equatorial image of the cell is shown by DIC (left), and the corresponding hyaline immunofluorescence is also shown (center). A cortical section at a plane of focus near the cell surface also shows hyaline immunofluorescence concentrating at the cleavage furrow (right). Cells were fixed on a protamine sulfate-coated coverslip in CaFSW in 5% formaldehyde for 20 min. In the cell shown, an elongated glass needle was brought down onto the cells and used to manipulated the mitotic apparatus of a metaphase *L. pictus* egg toward the cell cortex. The cell is shown just after MA manipulation (left), and again as the cortex furrows on the side of the displaced MA (center). At this stage, the cell was fixed on the coverslip, and treated for hyaline immunofluorescence, as in (A).

The potential to fix individual cells that have been manipulated in such a way and then using immunofluorescence to visualize various cellular structures is certainly an attractive application. Furthermore, fixing cells onto a coverslip has been a fruitful way of visualizing surface proteins, such as the secretion of hyaline at the site of cleavage furrow ingression (Fig 3; Shuster and Burgess, 2002).

Coverslip Fixation for Intracellular Visualization (Fig 1B). 1. For best adhesion to the coverslip, the jelly coats around the cells need to be thoroughly dissolved, and the fertilization envelopes must be removed. The coverslip should be treated with fresh protamine sulfate (10 mg/ml) (Sigma, catalog no. P-4505). Spread a drop of the protamine sulfate solution over the coverslip and allow it to sit undisturbed for at least 10 min. Alternatively, poly-L-lysine can also be used to create an adherent coverslip (Sigma, catalog no. P-8920). Pipette the poly-L-lysine over the glass and set on end, against a paper towel or Kimwipe to air dry. We find protamine sulfate and poly-L-lysine to be equally effective for creating an adherent surface on cover glass. We use protamine sulfate most frequently because it is significantly less expensive.

2. Rinse the coverslip with distilled water, and wick off the excess with a paper towel. Note: Air-dried, poly-L-lysine coated slides do not need to be rinsed.

3. Secure the treated coverslip to a chamber slide and fill with CaFSW. Refer to Section V.B and Figs. 4 and 5 for a description and diagram of our chamber slides.

4. It is convenient to make an identifying mark on the bottom surface of the coverslip with a Sharpie to use as a reference point to locate specific cells later on.

5. Use a transfer pipette to evenly spread a drop of embryos over the surface of the coverslip. The embryos should be well spaced on the coverslip. Allow them to settle onto the glass, then fill the chamber with CaFSW.

6. When ready for fixation, wash the cells once by drawing off the full volume of CaFSW with a transfer pipette and replace it quickly with isolation buffer.

7. Draw off the Isolation Buffer and replace the full volume with an initial fixation buffer that has been prepared without detergent (refer to Section II.A.1.e). An important consideration for fixing embryos onto coverslips is that the permeabilizing detergent included in the fix can adversely affect the adhesion of the cells to the glass. As the plasma membrane is dissolved by the detergent, so is the contact between the cell and the protamine sulfate or poly-L-lysine coating. We have found that allowing the cells on the coverslip to incubate in detergent-free fixation buffer for 1 h and then replacing the full volume with fixation buffer with 0.5% TritonX-100 will allow adequate preservation of the cells prior to permeabilization. An incubation of 20 min with the permeabilizing fixation buffer will clear the cells enough for immunofluorescence and minimize the loss of contact with the coverslip.

8. Observe the cells periodically during the permeabilizing incubation. Clearing of the cells can be visualized with bright field optics, and the adherence of the cells to the coverslip can be assessed by gently jiggling the slide. The cells will clear better if they are well spaced on the coverslip.

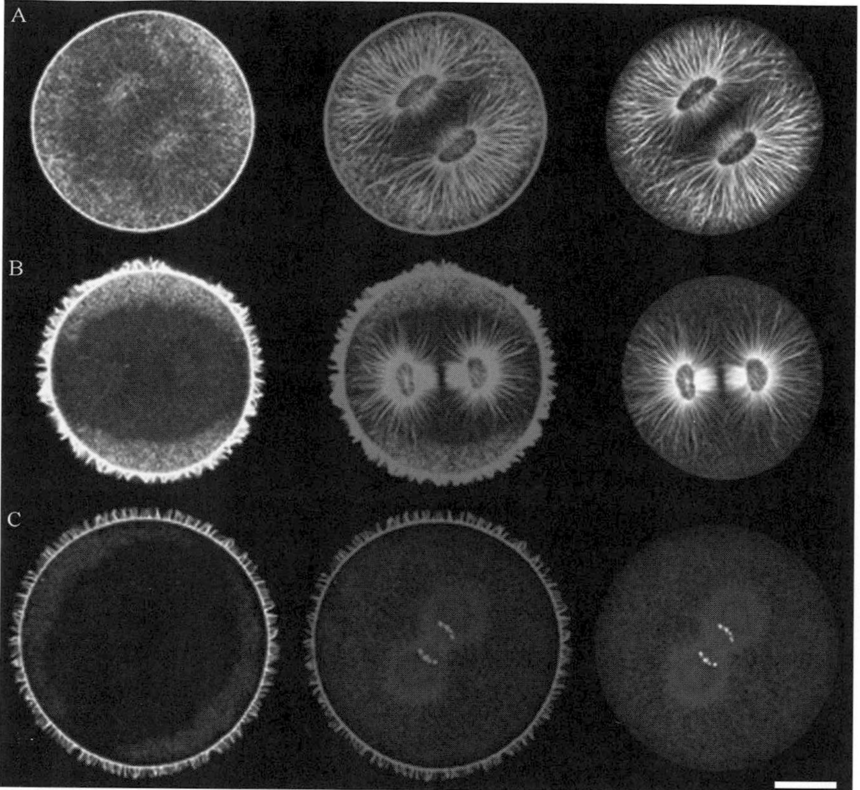

Fig. 4 Simultaneous preservation and staining of actin filaments and microtubules. (A) Top row: A sand dollar embryo stained for F-actin with phalloidin (left) and tubulin by immunofluorescence (right). In the overlay (center), actin staining is shown in red and tubulin staining in green. Fixation and staining were performed by the protocols in Section IV. (B) One blastomere from a two-cell sea urchin embryo, fixed and stained as in (A). (C) A sea urchin embryo stained for F-actin with phalloidin (left) and for DNA with propidium iodide (right). In the overlay (center), actin staining is shown in red and nucleic acid staining is shown in green. (See Color Insert.)

9. After the cells have been fixed and cleared, draw off the fixation buffer and use forceps to press the coverslip out of the chamber and transfer to a small dish of PBS. We use 35 mm tissue culture dishes from Corning.
10. Wash the coverslip three times in PBS for 5 min each wash on a gentle rocker.

Coverslip Fixation for Visualization of Surface Proteins (Fig. 3). Proteins that are secreted by the cell, such as hyaline, or proteins that are present at the cell surface can be fixed by the same protocol, but they require no permeabilization, since the stain or antibody used to detect the feature of interest does not need to penetrate the cell. In fact, if intracellular preservation is not of concern, we have found 3.7% formaldehyde in seawater (SW) or artificial seawater (ASW) to be an

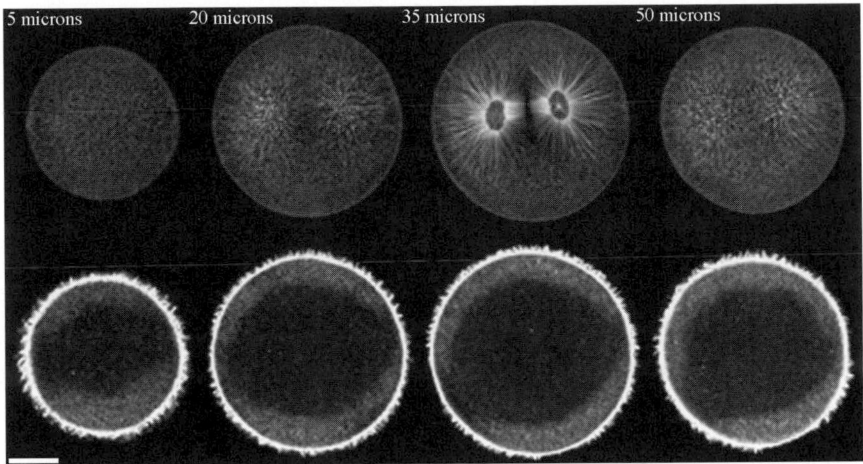

Fig. 5 Murray Clear and deep-cell staining. The cells were fixed and stained by the methods described in Section IV. They were stained for tubulin by immunofluorescence (top row) or for F-actin with phalloidin (bottom row). Following staining, the embryos were treated with Murray Clear to enhance the clarity of confocal sections deep within the cell. Focal planes at increasing depth are shown.

adequate fixative. It is worth mentioning that endo- and exocytotic events will require the presence of calcium, and so culture in CaFSW will not be useful for visualizing many surface events. To avoid the clumping together of cells that occurs in SW as a result of the secretion of various glycoproteins after fertilization, we distribute a small number of cells across a protamine sulfate-treated coverslip in a chamber slide filled with SW. If the cells are dispersed enough that they are not contacting each other, they will not clump up.

e. Recipes for Formaldehyde Fixation

Isolation Buffer
80 mM PIPES, pH 7.2
5 mM EGTA
5 mM $MgCl_2$
1 M Glycerol

Fixation Buffer*
80 mM PIPES, pH 7.2
5 mM EGTA
5 mM $MgCl_2$
1 M Glycerol
3.7% Formaldehyde
0.1% NP-40

*For coverslip fixation, prepare an initial aliquot of fixation buffer that lacks NP-40, and an aliquot of permeabilizing fixation buffer in which 0.5% TritonX-100 is substituted for 0.1% NP-40. (See text.)

Phosphate Buffered Saline (PBS)
137 mM NaCl
2.7 mM KCl
1.5 mM KH_2PO_4
8 mM Na_2HPO_4 or $Na_2HPO_4{*}7H_2O$

2. Preservation of the Actin Cytoskeleton

In many cases, permeabilization of fixed mitotic cells is disruptive to the actin cytoskeleton at the cortex. Certain fixation protocols have been developed that are specifically optimized for preserving the morphology of the actin cytoskeleton. The following procedure is ideal for actin maintenance, particularly in the cortex, because it calls for fixation of the cells prior to permeabilization with detergent. Treating the cells with the cross-linking agent prior to permeabilizing them with detergent may be important for preserving many cortical and cell-surface components. The following protocol from Begg (1996) provides a useful method for fixation of the actin cytoskeleton.

a. Protocol for Actin Fixation

1. At the desired timepoint, pellet eggs by hand centrifugation.
2. Resuspend the cell pellet in at least 20X volume of Millonig's Phosphate-Buffered Fixative (See Sec II.A.2.b for recipe.)
3. Incubate the cells in the fixative while mixing at room temperature for 30 min.
4. Pellet the cells by hand centrifuge and aspirate off the fixative.
5. Resuspend the cells in Wash Buffer containing 0.1% Triton X-100 (Sec. II.A.2.b) and mix for 10 min at room temperature to permeabilize the cells.
6. Pellet by hand centrifugation, and resuspend in Wash Buffer; mix for 10 min at room temperature.
7. Repeat the wash an additional two times.

At this stage, the actin cytoskeleton can be visualized by immunofluorescence (see Sections III.A and IV.B.1 or with phalloidin/phallacidin (see Section IV.B.2).

b. Recipes for Actin Fixation

Millonig's Phosphate Buffered Fixative
3% formaldehyde
0.2 M $NaH_2PO_4{*}H_2O$
0.136 M NaCl
pH 7.0

Wash Buffer*
50 mM HEPES
50 mM PIPES
0.6 M mannitol
3 mM $MgCl_2$
pH 7.0

*Prepare Wash Buffer with 0.1% Triton X-100 at the permeabilization step of the protocol for actin preservation (step 5, Section II.A.2.a).

3. Cold Methanol Fixation as an Alternative to Aldehyde Fixation

Fixation in cold methanol (−20 °C) is an alternative that may allow the investigator to avoid some of the complications arising from aldehyde fixation, although potentially at the expense of structural preservation. Moreover, some stains and antibodies are ineffective in cells that have been fixed in methanol. In some cases, more extensive washing of the fixed cells can overcome this difficulty. The catalogue information provided with most dyes and antibodies will often contain information about the particular reagent's compatibility with methanol fixation. Protocol for fixation in cold methanol adapted from Henson (1995).

1. Pellet the cells to be fixed by hand centrifuge.
2. Resuspend in a volume 10 times greater than that of the pellet in −20 °C methanol plus 50 mM EGTA.
3. Incubate the cells at −20 °C for 1 h.
4. After fixation, wash the cells three times with PBS by pelleting and resuspending in fresh PBS.

B. Considerations for Optimal Fixation

1. Alternative Fixatives

Aldehyde fixatives have been used extensively to preserve echinoderm embryos. The fixative should be prepared fresh on the day of use.

Formaldehyde or formalin, and its solid form counterpart paraformaldehyde, are commonly used at concentrations ranging from 3 to 5%. Formaldehyde is often commercially available in a 37% solution (Sigma catalog no. F-1635) and so it is convenient to use a 10X dilution of this solution. Formaldehyde solutions usually contain a certain percentage of methanol to prevent polymerization and precipitation. The methanol content can interfere with staining by some dyes and antibodies, and so it is sometimes desirable to prepare a paraformaldehyde fixative. Paraformaldeyde is polymerized formaldehyde, and contains no methanol (Sigma catalog no. P-6148). It comes in a powdered form and is usually brought to the desired concentration, often 4% (w/v), in Phosphate Buffered Saline (PBS). Getting paraformaldehyde into solution requires the dropwise addition of 0.1 M

NaOH while stirring, with moderate heat. The solution of fixative should be well chilled on ice before it is applied to cells.

Gluteraldehyde is also a frequently employed aldehyde fixative, usually at concentrations of 0.5 to 2% (v/v). In some case, gluteraldehyde can be combined with formaldehyde or paraformaldehyde in the same fix (Wessel, 1989, and Section IV.A). Formaldehyde creates a single-carbon cross-link between cellular components, making it a somewhat "looser" fix than gluteraldehyde, which can cross-link over the span of its five carbons. Presumably, the greater length of the gluteraldehyde molecule makes it a somewhat more thorough fixative than formaldehyde because of its five-carbon cross-linking range.

In staining and immunofluorescence applications, antibodies may have a more difficult time penetrating deeply into cells that have been fixed with gluteraldehyde alone because of the higher density of the cross-linked cellular matrix that results from this treatment. Because of this, cells fixed in gluteraldehyde will generally require longer incubation in various stains or antibodies.

Furthermore, the gluteraldehyde molecule has two aldehyde groups, as opposed to the single aldehyde of formaldehyde. This can result in an increased proportion of reactive aldehydes that are available to nonspecifically cross-link with antibodies, and lead to unwanted background fluorescence. Treatment with a reducing agent such as 0.1% (w/v) sodium borohydride is sufficient to eliminate most of the background fluorescence that is artifactual to gluteraldehyde fixation (see Section II.A.2.b).

2. Alternative Detergents

TritonX-100, Nonidet P40, saponin, and glycerol are all commonly used to permeabilize echinoderm embryos, rendering them more easily penetrated by antibodies and stains, and also more transparent for microscopy applications. A detergent may be employed in the isolation buffer and/or the fixation buffer at concentrations ranging from 0.1 to 1% (v/v). Glycerol is frequently used at 1 M to give a moderate clearing and permeabilization effect. We achieve the best results with TritonX-100 and NP-40. Saponin is a somewhat more gentle detergent, and glycerol is very mild.

Ice-cold TritonX-100 has been useful in permeabilization of the plasma membrane without solubilizing sterol-rich microdomains (Belton, 2001). In this procedure, ice-cold TritonX-100 is applied to formaldehyde-fixed cells. Sterol-specific stains such as filipin can then be used to detect the microdomains that have been retained (Fig. 1C).

III. Staining and Imaging Fixed Embryos

There are numerous options for fluorescent labeling of structures and proteins inside the cell. The echinoderm embryo is ideal for cellular localization studies using these techniques, due to its size, durability, and relative clarity.

Immunofluorescence is a widely used technique, as there are many commercially available antibodies that will cross-react beautifully with endogenous echinoderm proteins. The Developmental Studies Hybridoma Bank is an outstanding resource for many valuable antibodies (http://www.uiowa.edu/~dshbwww/). Furthermore, there are many small molecule stains that will bind to specific cellular structures. It is important to consider that not all stains are compatible with all fixation protocols; for example phalloidin staining of F-actin will be ineffective in methanol-fixed cells, and treatment of embryos with Murray Clear will disrupt the fluorescence of some conjugates.

Provided in this section are general protocols for immunofluorescence of formaldehyde-fixed cells, and for staining nucleic acids with various small-molecule dyes.

A. Protocol for Immunofluorescence

Starting with embryos that have been fixed and permeabilized:

1. Wash three times with PBS by pelleting in a hand centrifuge and resuspending in fresh PBS. Alternatively, PBS with 0.1% TritonX-100 (PBT) may improve penetration of the antibodies during staining. This is particularly true if the cells have been fixed in a recipe that contains gluteraldehyde and/or dextrose, such as that in Section IV.A. Washes should be at an excess of 10X the volume of the cell pellet.
2. After the final wash, resuspend the cells in a volume of PBS or PBT that is 5X that of the pellet. This will concentrate the cells to maximize the sample to be stained.
3. Transfer 1 ml of the suspension to a 1.7 ml microfuge tube and spin down by hand centrifuge. This should yield a cell pellet of approximately 100 μl. If there are significantly more cells, then the sample should be diluted.
4. Resuspend the cells in PBS or PBT with 3% BSA. Incubating the embryos for 1 h in PBS + 3% BSA before adding the primary antibody will sufficiently block most nonspecific binding.
5. Add the appropriate dilution of primary antibody to the cells in PBS (or PBT) +3% BSA. Generally, we use a 10-fold greater concentration of primary for immunofluorescence than would be required for a Western Blot. For example, we use a monoclonal tubulin antibody raised in mouse (E7 from DSHB) at 1:500 on Western Blots and 1:50 for immunofluorescence (Figs. 1A and B).
6. Incubate the cells in primary antibody for 1 to 3 h with constant, gentle mixing. Longer incubation (3–4 h) will be required if fixed by the method in Section II.A.2.a.
7. Wash the cells three times with PBS, as previously described.

8. After the third wash, resuspend the cells in PBS or PBT + 3% BSA with the appropriate dilution of secondary antibody. We use secondaries with a variety of fluorescent conjugates, including fluorescein, rhodamine, and a number of different Alexafluors.

9. Incubate the cells in secondary antibody for 1 h with mixing, then wash three times with PBS.

10. Mount the cells onto a slide for imaging.

B. Staining Nucleic Acids: DAPI, Hoechst, and SYTO Dyes

Hoechst is a widely used marker of DNA in live and fixed cells. Like DAPI and propidium iodide, Hoechst is an intercalating agent that inserts into the DNA double helix between the stacked base pairs. Unlike DAPI, Hoechst is cell-permeant, making it a valuable stain for live cells. DAPI staining requires that the cells be treated with a detergent in order to penetrate the membrane, and is therefore only useful in fixed embryos. Hoechst and DAPI are both excitable by ultraviolet light, and easily viewed under a mercury light source. Due to their UV excitability, however, Hoechst and DAPI are not visible on most scanning laser confocal systems.

Molecular Probes also offers a variety of both cell-permeant and -impermeant SYTO dyes that bind to nucleic acids and are excitable by argon and krypton laser light sources; however, we have found these stains to bind to RNA with high affinity, and therefore to yield a great deal of background fluorescence. Treatment of fixed cells with RNase may alleviate this effect if one does not wish to stain RNA.

It is important to note that because these dyes are intercalating agents, they can interfere with DNA replication and gene expression. They will therefore have adverse effects on development if they are used in a culture that is grown beyond activation of the embryonic genome at the mid-blastula transition. For studies of later development, the nucleic acid stain should be added later, or the treatment should be done after the cells are fixed. Protocol for staining cells with DAPI, Hoechst, or SYTO dye:

1. Starting with fixed cells in PBS, or live cells in CaFSW if staining with Hoechst, or a cell-permeant SYTO dye, pellet embryos by hand centrifuge.

2. Resuspend fixed cells in PBS, or live cells in CaFSW, with the appropriate concentration or dilution of dye:

    DAPI (Molecular Probes catalog no. D-1306): 300 nM
    Hoechst (Molecular Probes catalog no. H3570): 1:10,000
    SYTO 16 (Molecular Probes catalog no. S-7578): 1:5000

3. Incubate the cells in the dye while rotating or mixing for 30 min. This amount of time is sufficient for all three of the above dyes.

4. Wash the cells three times by pelleting by hand centrifuge and reuspending in fresh PBS (fixed) or CaFSW (live).

IV. Simultaneous Fixation and Visualization of the Actin and Microtubule Cytoskeletons (G. von Dassow and V. Foe)

A. Fixation Procedure for Simultaneous Preservation of Actin Filaments and Microtubules in Early Urchin Embryos

1. General Explanation of Fixation

We have developed a fixative cocktail that resembles Gard's fixative and similar formulas, which we use extensively for simultaneously preserving actin filaments and microtubules in urchin, mollusc, and jellyfish embryos. The key points appear to be (1) the inclusion of low-percentage glutaraldehyde, (2) balancing rapid permeabilization with the prevention of change in cell shape through additives, and (3) elimination of free calcium ions with a high concentration of EGTA. Actin filaments in the deep cytoplasm are much more sensitive to these parameters than are microtubules. Our recipe is adjusted to work on these particular cytoskeletal structures in first through fourth cleavage, and may not be not such a good fixative once the embryo has internal fluid cavities.

2. Protocol and Fixative Recipe

1. Make up fixative on the day it is to be used. We usually use it within a couple hours of mixing, but there seems to be little change over the course of a working day.
2. Chill fixative to the same temperature as are the embryos to be cultured.
3. Embryos should be devitellinized one way or another, but it seems not to matter whether they retain their hyaline layer or not—in any case, the hyaline layer dissolves rapidly in this fixative.
4. Concentrate embryos by swirling in a dish or settling in a tube, and add, e.g., 150 μl of embryos to 1350 μl fix, and mix gently but quickly.
5. Fix 30 to 60 min on a nutator or a rotator that can invert the tubes. Check periodically to see if embryos are clumping together, and if so, pipette them up and down, or shake the tube briskly, to disperse them. If clumping is a chronic problem, try a different shape of tube or a different agitator.
6. Settle embryos in the fixative. This can take a while and, for example, purple urchins fixed in the presence of 1 M dextrose may take so long to settle completely that one loses control of the time of fixation. Therefore, once they have settled part of the way, replace as much of the fix as possible with PBS containing 0.1% Triton X-100 (PBT). Then they will settle faster.
7. Pass embryos through several changes of PBT and, if possible, let them sit overnight in PBT. Embryos will tend to clump if not stirred up at each change. Once the embryos are in PBT, if you examine them on a microscope slide, you will

find they have swelled significantly and cleared substantially. Swelling is no longer a concern; after fixation is complete, the embryos swell and shrink isometrically, like a sponge.

Before storage, samples should be changed completely to PBS without detergent; otherwise, the detergent will slowly degrade the flimsier populations of actin filaments. Also, it is preferable to reduce the samples with sodium borohydride and block them with BSA and normal goat serum (see following text). This will prevent to some degree clumping of the cells during storage.

Recipe for fixative:
100 mM HEPES pH 7.0 titrated with KOH
50 mM EGTA titrated to pH 7.0 with KOH
10 mM $MgSO_4$
400 mM – 1 M Dextrose
2% Formaldehyde from EM-grade 20%
0.2% Glutaraldehyde from 50% unbuffered stock
0.2% Acrolein from 2% stock in Cacodylate (optional; available from EMS)
0.2% Triton-X100

We make up this recipe leaving out 10% of the volume of distilled water, then add 1/10 volume of embryos in seawater. HEPES and EGTA stocks should be made from the free acid rather than the sodium salts, first of all, because if they are made from the salt, then a substantial increase in both tonicity and ionic strength will result and, second, because we have found that rapid permeabilization of urchin embryos in the presence of a high concentration of sodium ions leads to distortion and often complete rupture of the cortex.

Note: Acrolein is a monaldehyde that supposedly penetrates tissue very rapidly. Unfortunately, it is also volatile and rather nasty to work with. We include it because it seems to improve the overall morphology of the cells and increase their resistance to processing damage. But it is not essential to preservation and labeling with any probe we've used.

Because it contains dextrose, the fixative doesn't mix readily, so one needs to do something (like quickly add excess fix to embryos waiting in a tube) to avoid distortion. The dextrose concentration is adjusted for each species of embryo and to match salinity. Local salinity can vary over the course of a single year from 2.5 to 3.3%, and is an important parameter in fixation. Furthermore, different species seem to have a different effective internal tonicity; for example, purple urchins can require 1 M dextrose in this formula to prevent swelling, whereas green urchin embryos require usually 400 to 600 mM. We periodically do test fixes with different dextrose concentrations, and measure egg diameter 5 to 10 min after the start of fixation, then choose the concentration that causes the least change from live embryos. We tolerate an apparently unavoidable 3 to 5% change in cell diameter; at this level, the swelling appears to be isometric, but (especially in purple urchins) swelling to greater than 10% increase in diameter usually reflects a sudden inflation of the cortex, separating it from the deep cytoplasm.

3. Drawbacks, Modifications, and Alternative Recipes

One major drawback of this recipe is that the dextrose prevents extraction of soluble cytoplasmic proteins, leaving embryos dense and difficult for antibodies to diffuse in an out of. For urchins, this is not a serious drawback; soaking them for longer during the staining steps will usually still yield adequate penetration.

Another feature of the recipe given is that the high concentration of TritonX-100 will eliminate many membrane proteins and soluble cytoplasmic constituents. If staining for the cytoskeleton, then this is a virtue, but otherwise it may make this recipe unsuitable. For example, we have found that, while we can stain for Rho (using Sigma's rabbit anti-rho A/B/C), we lose beta-catenin. It is therefore advisable, when trying a new antibody, to try a simple non-permeabilizing fixative for comparison. We find that 4% formaldehyde in 80% seawater (i.e., one part 20% EM-grade formaldehyde plus four parts filtered seawater) causes little morphological change. Microvilli are mostly lost, microtubules will look a little like they've had an unsuccessful perm, and cytoplasmic actin filaments will be completely absent, but (for instance) beta-catenin will still be present. Furthermore, if one reduces the detergent concentration 10-fold in our recipe, beta-catenin labeling is retained, and the microtubule morphology is better preserved than with 4% formaldehyde in seawater. Again, however, cytoplasmic actin filaments will not usually be preserved without rapid permeabilization.

Glutaraldehyde and formaldehyde can cause problems of their own. Many antibodies simply do not work on glutaraldehyde-fixed cells. If cytoplasmic actin filaments and nice straight microtubules are not your primary concern, then try reducing the glutaraldehyde concentration or leaving it out entirely. Perhaps surprisingly, formaldehyde is not essential; leaving out both formaldehyde and acrolein, so that 0.2 to 0.4% glutaraldehyde is the only reactive agent present, yields reasonably good microtubule and actin filament preservation as long as the fixation time is roughly doubled.

4. Sodium Borohydride Treatment to Reduce Autofluorescence

We find it necessary to add a low percentage of glutaraldehyde to the fixative in order to accurately preserve cortical morphology (microvilli, filopodia, blebs) or dynamic intracellular structures like microtubules and F-actin. However, in many cells, glutaraldehyde renders the cytoplasm autofluorescent. Why some cells do this and others don't remains obscure to us. Whatever the cause, treating the tissue after fixation with sodium borohydride will reduce and often eliminate glutaraldehyde-induced autofluorescence. This is not a cure-all, because some kinds of yolky autofluorescence remain after borohydride treatment.

Borohydride is a reducing agent and seems to have a very short half-life in aqueous solution, so one must dissolve it *immediately* before use. We use 0.1%

sodium borohydride in PBS. The solution bubbles, so DO NOT CAP YOUR SAMPLES during borohydride treatment or they will explode. More concentrated solutions will bubble more vigorously, and too much bubbling will damage embryos by trapping them and preventing settling. Embryos should NOT be treated with borohydride in the presence of detergent; instead of bubbles you will get foam, and the foam traps the embryos (as even small bubbles can do), and the embryos dry out, get deformed, etc.

PBS seems to be a good buffer for borohydride treatment. Among other commonly used buffers, Tris-buffered saline seems acceptable but PEM (PIPES/EGTA/$MgSO_4$) foams even without detergent.

Protocol for treating gluteraldehyde-fixed embryos with sodium borohydride:

1. Wash embryos in detergent-free PBS, at least once.
2. Add a dash of dry borohydride to PBS, approx. 0.1%, no more than 5 min before use.
3. Replace PBS with borohydride solution.
4. Let sit for at least 30 min uncapped; since we work with small embryos that easily get trapped by small bubbles, we usually let samples sit for an hour or two to debubble.
5. If your embryos settle nicely, you may want to replace the borohydride with freshly dissolved stuff after about 10 to 20 min.
6. Replace most of the borohydride solution with PBS; watch for embryos stuck to the side of the tube or resting on top of a bubble on the wall. This wash will bubble a bit, too.
7. Replace PBS with PBT and proceed to blocking and staining.

B. Staining and Imaging Techniques

1. Antibody Staining (Fig. 4A–B center and right panels; Fig. 5, top row) Protocol:

1. Starting with samples in PBT, replace PBT with 5% normal goat serum (NGS; Jackson Labs; heat-inactivate for 1 h at 55° and freeze aliquots) in PBT containing 0.1% BSA (PBT+BSA). Leave embryos at room temperature, agitating them if possible, for at least half an hour. With glutaraldehyde-fixed cells, we block for several hours.
2. If we plan to store samples for any length of time, at this step we replace the blocking solution with two changes in PBS, and keep them in the refrigerator. Blocked embryos are not as likely to clump together during storage.
3. Otherwise, replace the blocking solution with primary antibodies diluted to appropriate concentration in PBT+BSA. For example, we get excellent results with the rat anti-tubulin YL1/2 supplied by ImmunologicalsDirect.com diluted 1:1000. The addition of BSA helps prevent nonspecific background staining, which is exacerbated in glutaraldehyde-fixed cells.

4. Incubate primary antibodies with gentle agitation if possible (nutator or rotator), for 3 to 6 h at room temperature if embryos were fixed with formaldehyde alone. However, longer incubation at room temperature is required for embryos fixed with even 0.2% glutaraldehyde. We stain purple urchin 1-cell embryos for 18 to 24 h, but the much larger green urchin 1-cell embryos require as much as 2 days for full labeling of the deepest microtubules.

5. Wash off the primary antibodies with 3 changes in PBT over the course of an hour for samples fixed in formaldehyde alone or longer for samples fixed with glutaraldehyde.

6. Replace PBT with secondary antibodies diluted in PBT+BSA. We recommend the Alexa-conjugated goat-anti-mouse, -rat, or -rabbit secondaries available from Molecular Probes, and we use them at 1:500. We also find that Molecular Probes' isotype-specific secondaries work well, and that when using two primaries from different mammal species, the "highly cross-adsorbed" products ameliorate cross-reactivity, despite reducing the intensity somewhat. Incubation times same as in step 4.

7. Wash off secondary antibodies as in step 5; if counterstaining with phalloidin or propidium iodide, one of the wash steps can include the counterstain.

8. Replace PBT with two washes in PBS before mounting.

2. Phalloidin Staining (Fig. 4A–C, Left and Center Panels; Fig. 5, Bottom Row)

Phalloidin is a fungal toxin that binds to filamentous but not monomeric actin. Even if you aren't interested in actin per se, phalloidin is a tremendously useful counterstain because of the intense F-actin accumulation at most cell boundaries. Phalloidin is available conjugated to an astonishing variety of fluorescent dyes. Nevertheless, it can be an extremely frustrating reagent to use; if what you're after is wispy, hard-to-fix filaments deep in murky eggs, the sensitivity of phalloidin staining to alcohol treatment raises several related problems.

Phalloidin conjugates are usually supplied as a dry smudge in the bottom of a tube. Molecular Probes is the major supplier of fluorescent phalloidins. They recommend dissolving phalloidin conjugates in methanol at 1 Unit/5 μl of methanol, which is then kept in the freezer. The unit definition is based on tissue culture cells; our substitute is that one unit is about enough to fully stain 100 fly embryos or 500 urchin embryos. Instead of methanol, we dissolve phalloidins in DMSO at 1 U/μl and leave it frozen, thawing it each time before use. Phalloidin conjugates stored this way and regularly thawed and refrozen remain effective for at least several months. Methanol stocks are probably stable for years.

Before staining with phalloidin, the methanol stock must be dried, because ANY trace of methanol at ANY step before or after phalloidin staining will a good deal of the most interesting staining. The same goes for ethanol. (There is one exception: formalin, which is a 37% solution of formaldehyde, is a perfectly good fixative, despite the fact that it contains methanol.) One can use a SpeedVac to dry down the phalloidin. Alternatively, place however much you want to use into an open dish or an unsealed tube, and leave it in a drawer for a while to protect it from light while the methanol evaporates. Check under the microscope to make sure it's all dry! (The reason we switch to a DMSO stock is that we don't have to remember to start the phalloidin drying in advance.)

Protocol for phalloidin staining of fixed embryos, starting with embryos in PBS + 0.1% Triton X-100:

1. Dissolve dry phalloidin or DMSO stock at 3 to 10 U/ml in PBT. Pipette the solution a bit to make sure all the crystals dissolve.
2. Replace PBT with phalloidin staining solution, and incubate embryos for 30 min to 2 h. Longer probably doesn't help and could hurt.
3. Wash embryos once in PBT, and 3× in PBS, not PBT; in general, we do not let these post-phalloidin washes take more than 20 min apiece. Detergents will make the phalloidin slowly come off the sample.
4. Mount immediately, whether in Murray Clear or aqueous medium. The phalloidin signal progressively diminishes and becomes fuzzy if stained samples are stored for any length of time. If using Murray Clear, one *must* use an isopropanol series instead of methanol or an ethanol series.

An important note: A bewildering array of phalloidin conjugates are available. One can get practically any color from Molecular Probes, using some really wonderful fluorophores. Our favorite is the BODIPY-FL phallacidin, which is a fluorescein-like dye. It is very bright, very stable compared to fluorescein or rhodamine, and it is the least sensitive of any we've tried to our isopropanol/Murray Clear procedure. If you are primarily interested in visualizing cell outlines using the confocal, or if you choose to mount embryos in an aqueous medium, the Alexa 488 and 568 dyes are quite bright. However, they seem not to be quite compatible with isopropanol/Murray Clear mounting (see following text), which is our method of choice. Although samples often stain very nicely with Alexa-conjugated phalloidin, the stain disappears rapidly as one examines the embryo. This is unlikely to be due to photobleaching, as it takes place at illumination levels at which Alexa-conjugated antibodies are rock-solid, and the effect seems to be somewhat dependent on the embryo—we don't notice this problem as much with fly embryos, and it is most severe in mollusc embryos—and the fixation (shorter is better). Our conclusion is that if you plan to use Murray Clear, you are best off using BODPIY-FL phallacidin.

Note that there are two "phalloidins": phalloidin and phallacidin. With regard to Murray Clear, it is conceivable that it is the phallacidin that is important, not the choice of fluorophore. The fluorophore gets coupled to the opposite side of the molecule in Molecular Probes' phallacidin derivatives. Unfortunately, Molecular Probes makes only four phallacidin derivatives: BODIPY-FL, BODIPY-TR-X (Texas Red-like), NBD, and coumarin. Coumarin is UV-excited, and NBD photobleaches very rapidly. We have tried the BODIPY-TR-X phallacidin, but without success.

3. Staining Nucleic Acids with Propidium Iodide (Fig. 4C, Center and Right Panels)

Many of the small molecular weight dyes that bind to DNA are either unsuitable for standard confocal microscopes (unless you are lucky enough to have a UV laser) or are impossible to use in Murray Clear. Propidium iodide (PI) is an exception. It is cheap, bright, relatively stable, fast-acting, and it works just fine in Murray Clear, whether you use methanol or isopropanol.

Propidium iodide stains both RNA and DNA, which can be very useful. However, if you want to see just the DNA, you need to treat your sample with DNase-free RNase before staining. We use a DNase-free RNase supplied by Sigma in 50% glycerol (R-4642). Otherwise, one may use RNase A, boiled for 15 min to denature DNase, dissolved as a stock solution of 100 mg/ml in PBT, aliquoted and frozen.

PI can be dissolved in water to make a stock at 2 mg/ml. This seems to become less and less potent over the course of a few months, so should be made fresh occasionally.

Protocol for staining nucleic acids with PI:

1. Starting with embryos in PBS+0.1% TritonX-100 (PBT), add RNase solution (~1:100 Sigma RNase in glycerol or ~0.5 ug/ml boiled RNase A) and soak for 30 min to 2 h at 30 to 37°. For small embryos fixed with mild glutaraldehyde, 1 h seems more than adequate. Omit this step if you want to visualize RNA.
2. Rinse in PBT by pelleting the cells and resuspending in fresh PBT.
3. Apply PI in PBT at 2 to 10 ug/ml for RNA labeling, or 15 to 20 ug/ml if samples have been RNase-treated, and soak for 30 min to 2 h. We often combine PI staining with phalloidin.
4. Wash 3× in PBS, 10 to 20 min. each. Embryos will be quite pink, but some of the stain will come out as they are washed.
5. Mount immediately; otherwise, the stain will dissipate.

One difficulty with PI is that it has a large Stokes shift, and absorbs about as well at 488 and 568 nm. This can cause bleedthrough and crosstalk and precludes one from using a long-pass filter set with blue-excited dyes like BODIPY-FL phallicidin. If you are blessed with an UV laser or multi-photon confocal, or if you intend to use epifluorescence only, DAPI and Hoechst will be preferable. Unlike PI, DAPI and Hoechst stain only the chromatin, and thus don't require the RNase step as described. Both dyes can be dissolved as stock solutions at 0.1 to 1 mM in distilled water, and used at a 1:1000 dilution.

4. Murray Clear (Fig. 5)

While the embryos of sea urchins, starfish, and many other invertebrates are often very close to clear already, and can be made to nearly disappear with such simple media as glycerol, to get the best deep sectioning out of the confocal microscope, it nonetheless behooves one to mount embryos in some medium that nearly matches the refractive index of glass. Some aqueous mounting media, such as Fluoromount G, come close, but we vastly prefer Murray Clear, a 2:1 mix of benzyl benzoate and benzyl alcohol. Using Murray Clear, we achieve nearly as good signal and resolution on the far side of the embryo as on the near side. Although it might seem like overkill for urchin embryos, Murray Clear will also render yolky fly and frog embryos clear enough to read the newspaper through (unlike glycerol, Fluoromount-G, VectaShield, or anything else we've tried).

The major disadvantages of Murray Clear are that (1) it is immiscible with water, and therefore requires a dehydration series during which embryos tend to shrink or which strips off certain small-molecule stains; (2) although it is compatible with most lab plastics, it dissolves plastic tissue cultureware and several useful sorts of tape; and (3) since it never hardens, and prevents nail polish from hardening, truly permanent preps are impossible. Despite these drawbacks, the improvement in optics is worth the trouble.

Murray Clear will absorb a small amount of water and should therefore be kept in tightly capped containers. We prefer short-form screw-cap Coplin jars, and we adhere embryos to slides coated with poly-L-lysine so that we can change them rapidly by dipping. With embryos that settle readily, one may perform everything up to the final mounting in microfuge tubes, which are not affected by Murray Clear.

A methanol transition from PBS to Murray Clear works beautifully for antibody-labeled embryos, but cannot be used with phalloidin, which will be stripped by even small amounts of methanol. Ethanol also removes phalloidin. Many people therefore believe Murray Clear cannot be used with phalloidin. However, we found that a rapid isopropanol series *is* compatible with phalloidin, although it does not lead to quite as effective clearing as does methanol. Isopropanol can slowly cause the phalloidin to come off, but with short steps, the staining is indistinguishable from preps with aqueous mounting media.

Protocol for treating embryos with Murray Clear:

1. Start with embryos in PBS and poly-lysine slides to which have been affixed shims of appropriate thickness (single-sided Scotch tape, either standard for embryos smaller than 100 um, or a thicker metal-backed tape for large specimens). Pipet embryos onto the slide in 100 to 200 ul PBS, and allow to settle until stuck.
2. For antibody-labeled preps:
   Drain as much PBS as possible and dip slides in 100% methanol; leave them for 1 to 5 min. Dip through two more 1 to 5 min changes in 100% methanol, the last one kept as close to anhydrous as possible.
3. For phalloidin-labeled preps:
   Drain PBS and immediately dip slides in 70% isopropanol in distilled water; leave them for 30 to 60 sec. Thick embryos may require longer steps, thus sacrificing some labeling.

4. Change through 85%, then 95%, and then 2× 100% isopropanol, again allowing only 30 to 60 sec per step.

5. Wipe/daub off as much methanol or isopropanol as possible without touching the embryos before transferring to the first jar of Murray Clear.

6. Dip through a total of 3 changes of Murray Clear, 1 to 5 min each. There is no harm in letting the samples sit in these steps for a while.

7. Drain slides only lightly so as to leave enough to fill the chamber between the shims; gently lay a coverslip on top, being careful not to shear the embryos, which will be quite brittle and will crush rather than squish at this stage.

8. Using filter paper, wick off excess Murray Clear from around the edges of the coverslip, being careful not to shear it.

9. For convenience but shorter-lifetime preps, seal with clear nail polish (after extensive testing we recommend Wet'n'Wild or Cover Girl). Be sure to use a generous amount and check for gaps! Don't use colored polish; the Murray Clear soaks up the pigment. Nail polish won't truly harden and remains a goo that can be removed with forceps and patience, i.e., if the chamber needs refilling. But even stored in the freezer, some Murray Clear will seep through it and evaporate.

10. For longer-lasting preps, seal the coverslip down with Devcon 5-minute epoxy, which can be spread with a toothpick. Again, watch for gaps. The major problem here is to smooth out the epoxy so that it doesn't prevent you from using short working distance objectives. If these preps are sealed carefully, they can last months in the freezer.

Murray Clear recipe:
2 parts benzyl benzoate
1 part benzyl alcohol
Store tightly sealed to prevent uptake of water.

V. Observation of Live Embryos (D. Burgess and L. Strickland)

The embryos of echinoderms are beautifully amenable to live observation, accompanied by time-lapse video microscopy. Different aspects of the cells can be emphasized using various imaging techniques, such as Differential Interference Contrast (DIC). Shinya Inoue and others have produced amazing images of the mitotic apparatus using polarized light and birefringence (Inoue and Oldenbourg, 1998). In addition to the appropriate microscope equipment and settings, some other basic items may be required, such as a method for keeping the cells at an ideal temperature during viewing, chamber slides, and microinjectors or micromanipulators. Discussed next are methods of temperature control and construction of basic chamber slides.

A. Temperature Control

Different echinoderm species have different optimal temperatures for culture. While some warm-water species, such as *L. pictus* or *L. variegatus*, can grow easily at or near room temperature, others require a method of cooling. In some cases, the temperature of the microscopy room can be kept suitably low, but for species such as *S. drobachiensis* that require temperatures below 10 °C, this is not a feasible option, and some method of cooling must be employed to keep the cells at their natural temperature.

1. Thermal Stage Cooling

One way of maintaining the temperature of the cells is to control the temperature of the microscope stage itself. We use a Piezoelectric cooling stage that is cooled by circulation to a fluid heat sink from Brook Enterprises. Our chamber slides are made from stainless steel, and thus are able to conduct the temperature of the thermal stage around the medium that contains the cells. The bottom face of the chamber slide is milled to be perfectly flat for full contact against the thermal stage and maximally accurate control of temperature.

2. Circulating Chamber Slide

Another possibility is to use a chamber slide that is equipped with a mechanism for fluid exchange, as shown schematically in Fig. 6. This way, a constant input of fresh, temperature-controlled culture medium can be maintained.

Alternatively, the full volume of the chamber slide can be manually exchanged just by pipetting off the CaFSW or SW and replacing it with the same volume, cooled to the ideal temperature for the embryos. Such manual circulation of the slide will not keep the slide at a very constant temperature, but we find that exchange every 10 to 15 min for 12 to 17 °C embryos such as *L. pictus*, is sufficient to keep them alive and healthy through several cell divisions.

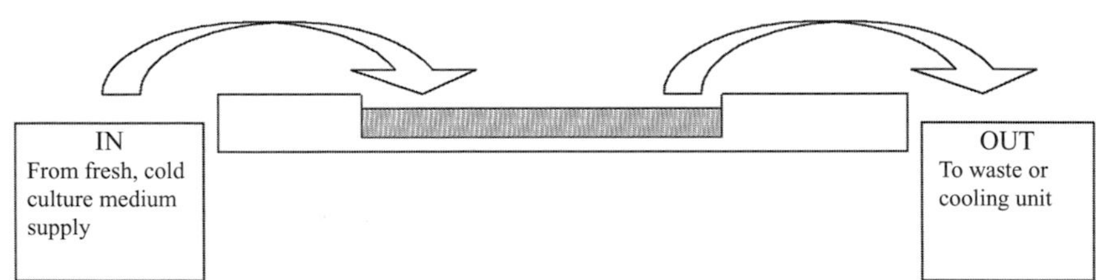

Fig. 6 Circulating chamber slide. A schematic diagram of a chamber slide is shown, with circulation of medium into and out of the chamber. Circulation of the chamber volume in this way can be used to expose the cells to various treatments, as well as to regulate temperature. Circulation can be performed manually by pipetting off the slide volume and replacing with new media, or by a slow water pump.

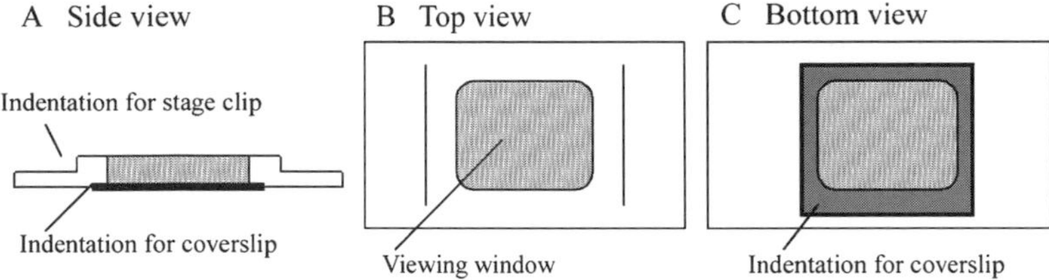

Fig. 7 Basic chamber slide. We use chamber slides that are carved from stainless steel to fit our stage and stage clips. The bottom surface of the slide is milled to be completely smooth. This maximizes the surface contact between the slide and the temperature-controlled plate installed on the microscope stage. The stainless steel helps to conduct the cooling around the entire chamber and ensure overall temperature stability. A side view of the slide (left) shows the indentations for stage clips, and the indentation that is carved around the viewing window on the bottom surface of the chamber. This indentation is carved to be the exact thickness of a glass coverslip, so that when the coverslip is sealed in place, it is flush against the bottom surface of the slide. This allows for observation of live cells under high-power oil objectives that have short working distance. Coverslips are sealed into the viewing window using melted VALAP (Vaseline:lanolin:paraffin 1:1:1). Also diagrammed are a top view (center) and bottom view (right) of the same chamber slide design.

B. Chamber Slides

While live embryos can easily be observed as wet mounts on a standard slide or in a perfusion chamber, the coverslip prevents direct access to the cells. Wet mounts, and perfusion chambers therefore, are not useful if one wishes to micromanipulate or microinject the cells. Chamber slides are used on an inverted microscope and allow full access to the cells. A variety of chamber slides have been described for live observation of echinoderm embryos (Kiehart, 1982). Ours are carved from stainless steel at a machine shop, according to our measurements and specifications. The basic features are diagrammed in Fig. 7.

Indentations for the stage clips are carved into the top of the chamber. Also, an indentation with the same dimensions as a coverslip is carved into the bottom of the chamber, around the viewing window. We use 22X40-1 coverslips from Fisher Scientific (catalog no. 12-548-5C). The coverslip is then sealed into place with heated VALAP (Vaseline:lanolin:parafin, 1:1:1). Sealing the coverslip underneath the viewing window in this way allows for observation on inverted microscopes that have a short working distance (Fig. 7).

C. Intracellular Labeling

1. Introducing Fluorescent Markers Into Oocytes and Eggs From Sea Urchin and Starfish for Live-cell Imaging (J. Ellenberg and P. Lenart)

Sea urchin eggs can be injected in chamber slides, as has been described. We find this method more appropriate if relatively large number of cells need to be injected. For imaging single cells and embryos (both sea urchin eggs and starfish

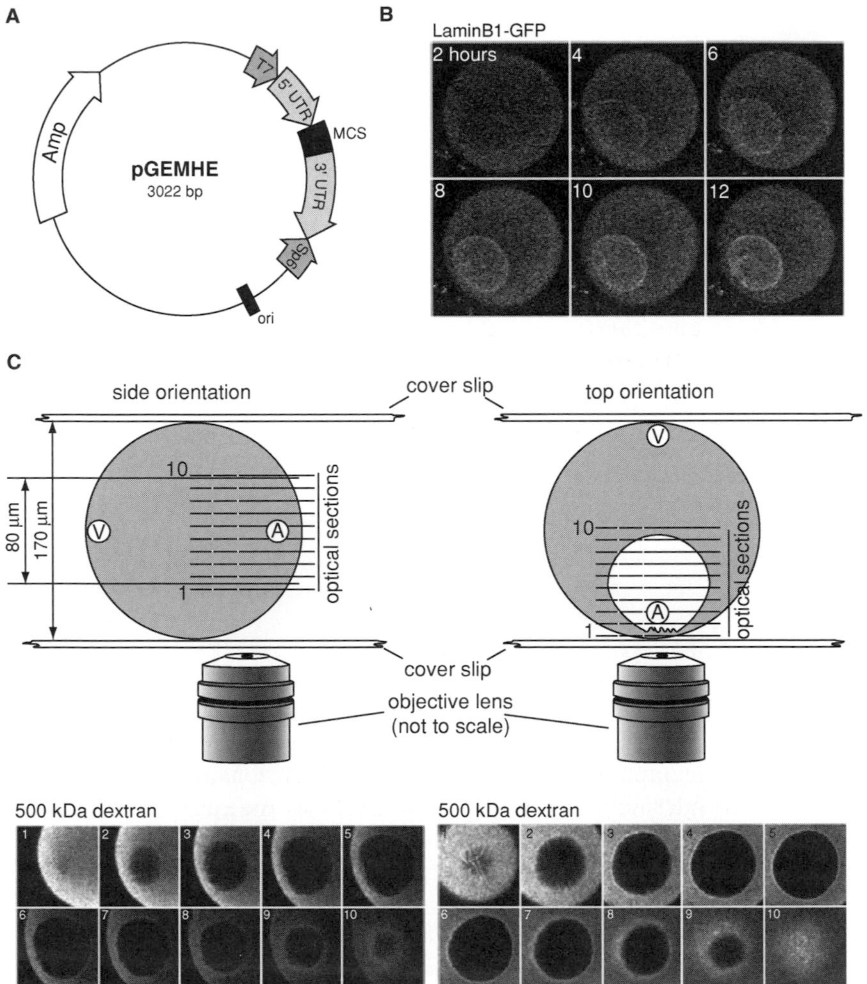

Fig. 8 Imaging starfish oocytes. (A) Map of pGEMHE, the plasmid used for *in vitro* transcription. (B) Expression time-course of lamin B1-GFP. Time is given in hours after injection. (C) Two possible orientations for imaging the animal hemisphere of oocytes. Objective lens is not drawn to scale. A, animal pole; V, vegetal pole. Bottom panels: z-series of oocytes injected with a 500 kDa fluorescently labeled dextran in the two orientations.

oocytes), we use the "coverslip shelf" injection chambers (Fig. 8C) and mercury-filled needles, which allow injection of precisely defined amounts of the sample. This method is described in detail in Chapter 10 of this book and elsewhere (Kiehart, 1982; Terasaki and Jaffe, 1993. See also online injection manual at http://terasaki.uchc.edu/panda/injection/). In this way, mRNAs, fluorescent

RNAs, recombinant proteins, antibodies, dextrans, membrane dyes, and fluorescent beads (up to 0.5 μm diameter) can be introduced into the cells (e.g., Lénárt *et al.*, 2003; Terasaki, 1994; Terasaki *et al.*, 2001). In a single injection, not more than 5% of the oocyte volume can be injected, but injection can be repeated several times at 5 to 10 min intervals if necessary (in starfish oocytes, urchin eggs are more sensitive and can typically be injected only once or, at most, twice). In immature starfish oocytes, which have large germinal vesicles, markers can also be injected into the cell nucleus. However, the nuclear envelope does not seal, which leads to the collapse of the nucleus after several minutes to several hours, depending on the size of the perforation (Galbraith and Terasaki, 2003; Lénárt *et al.*, 2003).

Due to the large cell volume and the time required for diffusion from the injection site, typically, all markers require at least 30 min to distribute homogenously in the entire cell. After injection of mRNAs, oocytes have to be incubated for even longer times; fluorescence from FP (Fluorescent Protein)-fusions is first detected after ~4 hours (Fig. 8B). This is more problematic in sea urchin eggs that do not express exogenous mRNAs very efficiently before fertilization. If mature eggs are injected, FP fluorescence reaches levels appropriate for imaging, typically, around the 3rd embryonic division. To visualize earlier stages, immature oocytes have to be isolated, which then mature spontaneously in a couple of hours, leaving sufficient time for protein expression (see Section IV.C.3.b). Alternatively, recombinant proteins that are labeled with chemical dyes have to be injected (Section IV.C.3.a).

We typically leave starfish oocytes overnight for sufficient expression and induce maturation the next day (see Section VI). Protein levels typically reach steady-state after ~1 day (Terasaki *et al.*, 1996). Since oocytes are resting cells, some protein complexes are extremely stable (e.g., the nuclear pore complex or nucleosomes); thus, the newly synthesized GFP-tagged proteins may not immediately incorporate into their target structure. In this case, oocytes have to be incubated for an additional 1(–2) days to achieve correct localization. By regularly changing the seawater and removing dead cells from the injection chamber, oocytes can be incubated for up to 2 weeks, but maturation becomes unreliable after 3 to 4 days.

2. Visualizing the Microtubule Cytoskeleton in Live Embryos (D. Burgess and L. Strickland)

Microtubule dynamics have been visualized in live tissue culture cells (Waterman-Storer and Salmon, 1998), the embryos of *D. melanogaster* (Kellog, 1988), and *C. elegans* (Dechant and Glotzer, 2003), as well as egg extracts from *X. laevis* (Belmont *et al.*, 1990). Some of these methods may be applicable to the echinoderm egg. Tubulin can be rendered fluorescent either by expression as a GFP fusion or by conjugation to a fluorochrome such as rhodamine.

a. Fluorescently Conjugated Tubulin

Tubulin that has been covalently labeled with a fluorochrome such as rhodamine can either be purchased from Molecular Probes, or made in the lab, and then microinjected into cells. Alternatively, plasmids encoding EGFP-tubulin fusion proteins (Clontech catalog no. 6117-1) can be bacterially expressed and then purified for microinjection. An important consideration in either case, however, is that the high density of microtubules in mitotic cells can obscure the imaging of specific microtubule populations or individual microtubules. One possibility might be to microinject less labeled tubulin into the egg than has been recommended for other systems, which would disperse the signal somewhat.

Molecular Probes provides 10 ul of a 10 mg/ml solution of tetramethylrhodamine-labeled tubulin (Molecular Probes catalog no. T-7640), and it is also available with several fluorescent conjugates. The protein is in a solution of PEM buffer (PIPES, EGTA, and $MgCl_2$) with 10% glycerol. Microinjection of a 1:4 dilution of this solution at a volume that is 5 to 10% that of the cell being injected has been sufficient for observation of microtubule dynamics in the systems already mentioned. The use of a higher dilution of fluorescent tubulin and confocal microscopy may allow for the visualization of microtubule dynamics in the embryos of echinoderms.

A detailed protocol for preparing fluorochrome-labeled tubulin has been published by Hyman (1991). This procedure has been used extensively by Mitchison and many others and is very similar to the preparation sold by Molecular Probes. A detailed version of the protocol is available on the Mitchison group's website, (http://mitchison.med.harvard.edu/protocols/label.html).

b. Microinjection of GFP-Tubulin mRNA

The GFP-tubulin construct mentioned in the previous section can also be cloned into a plasmid that contains a promoter sequence, allowing for *in vitro* transcription of the recombinant gene. The mRNA encoding GFP-tubulin can then be microinjected into echinoderm oocytes or eggs. Preparation of mRNA for microinjection into starfish oocytes is described by Lenart and Ellenberg in Section VI of this chapter and in Lénárt *et al.* (2003). For imaging GFP-tubulin in early cleavage-stage embryos, the issue is complicated by the fact that GFP itself can require up to 4 h to fold into a conformation in which it is capable of fluorescence. The first two cell cycles are optimal for imaging the dynamic activities of microtubules, and so if GFP is to be useful, it will need to be fluorescent during the earliest hours of an embryo's life.

One solution may be to inject the mRNA encoding the fusion into immature oocytes, which will require a number of hours to mature, such that the GFP fusion will have time to be expressed and become fluorescent prior to initiation of the first cell cycle. Alternatively, mature eggs can be injected with recombinant mRNA encoding the GFP fusion, and given approximately 4 h for translation prior to fertilization.

Obtaining sea urchin oocytes for mRNA injection:

1. mRNA encoding the EGFP-tubulin fusion is generated as described in Section VI.

2. Acquire oocytes by using sharp forceps to dissect the ovaries from a female urchin.

3. Dissociate the ovaries and release the cells by mincing with forceps in CaFSW.

4. Remove large pieces of ovary with a transfer pipette. Collect the cells and place them in new beaker of CaFSW.

5. Apply the cells to a protamine sulfate or poly-L-lysine coated chamber slide.

6. Identify fully grown oocytes by the presence of the large germinal vesicle.

7. Microinject 1 to 5% of oocytes' volume with an RNA solution of 1 to 5 ug/ul, and culture to maturation for 10 to 16 h at the temperature appropriate for that species, until GFP fluorescence is detectable. Alternatively, microinject mature eggs and fertilize them after 4 h, or when fluorescence is detectable. If mature eggs are to be injected, it is necessary to perform the injection in CaFSW to prevent an influx of calcium ions when the needle is introduced. Calcium ions will result in artificial activation of the egg, and the egg will not be able to be fertilized later on.

8. Add a 1:2000 dilution of sperm, aerated in seawater, and image by fluorescence or confocal microscopy through the early cell cycles.

VI. 4-D Imaging of Fluorescent Markers in Live Starfish Oocytes (J. Ellenberg and P. Lenart)

Starfish oocytes provide an excellent model system to study meiotic maturation in live cells. Oocytes isolated from female animals are arrested in G2/prophase of meiosis I (MI). Upon simple addition of the maturation hormone 1-methyladenine to seawater, the cells reliably resume meiosis with highly reproducible timing: after 20 min, the nuclear envelope breaks down, followed after 60 min by the extrusion of the first (MI) and shortly thereafter, the second polar body (MII) (times apply for *Asterina miniata*; timing varies across different species) (Kishimoto, 1999). After completing meiosis, the female pronucleus forms and fuses with the sperm nucleus, whereas unfertilized oocytes die by apoptosis (Yuce and Sadler, 2001).

Besides the precise timing of maturation, a further advantage of starfish oocytes is that they are sufficiently small and transparent to follow cellular processes *in vivo* by high-resolution confocal microscopy (e.g., Terasaki, 1994). On the other hand, their large diameter of $\sim$160 μm still allows quantitative injection of defined amounts of diverse fluorescent markers and easy manipulation of the cells (see Chapter 10 in this volume; Kiehart, 1982). Moreover, although they are transcriptionally inactive, exogenous proteins, such as GFP-fusions, can be expressed in oocytes from recombinant mRNAs (Lénárt *et al.*, 2003; Shilling *et al.*, 1990; Terasaki *et al.*, 2001).

Considering these advantages and the increasing power of live cell imaging techniques, starfish oocytes constitute an ideal system to study oocyte maturation in intact cells. In *Xenopus* oocytes, the workhorse system for biochemical studies, imaging of intact cells is nearly impossible due to their low transparency. Compared to sea urchin and mouse oocytes, starfish oocytes stand out because of the reliability of maturation and ease of handling of the cells.

Here we show examples from our studies on nuclear envelope dynamics during maturation of starfish oocytes using quantitative confocal 3-D time-lapse microscopy (4-D imaging) of GFP-fusion proteins and other fluorescent markers. The same techniques can also be used in other fields of research, including microtubule and chromosome dynamics as well as cell cycle regulation.

A. Plasmids for *In Vitro* Transcription of mRNAs Encoding GFP Fusion Proteins

Despite the advantages already listed, a major drawback of the starfish system is the limited number of genes isolated, although arrayed oocyte cDNA libraries are available (Kalinowski *et al.*, 2003). To overcome this difficulty, we have successfully expressed heterologous mammalian proteins in starfish oocytes (Lénárt *et al.*, 2003). Furthermore, in frame of the sea urchin genome project, cDNA libraries were set up and more and more sequence information is available (see at http://sugp.caltech.edu). We expect that many proteins of this echinoderm species will be functional in starfish; however, the evolutionary distance is quite large between these two groups. In any case, the sequence information will be useful to identify and clone starfish homologues.

To generate mRNA encoding for heterologous GFP fusion proteins for injection into starfish oocytes, we have found it most efficient to clone the cDNA of the gene of interest initially into the appropriate mammalian expression vector containing the desired fluorescent protein (FP) tag (e.g., plasmids available from Clontech Inc., Palo Alto, CA) and then test the fusions for proper subcellular localization and, if possible, other functions in mammalian cells by transient transfection. Then, the coding sequence of the fusion protein is transferred to a suitable vector for *in vitro* transcription. For this purpose, we generally use pGEMHE (Fig. 8A), which contains a multiple cloning site, a T7 promoter, and 3′ and 5′ UTRs from the *Xenopus* beta-globin gene to enhance expression (Liman *et al.*, 1992). However, in some cases, we could also obtain sufficient expression of proteins from constructs without such UTRs. Others and we also found that changing the Kozak sequence (the few nucleotides preceding the start codon) has a large effect on expression levels in some cases. However, to our knowledge, echinoderm Kozak sequences were not yet tested systematically.

Capped mRNAs from linearized DNA templates are then synthesized by using the mMessage mMachine *in vitro* transcription kit (Ambion, Austin, TX). So far we could detect the fluorescence of all FPs tested in starfish oocytes, including ECFP, EGFP, EYFP, and HcRed.

B. Image Acquisition

Starfish oocytes are polarized cells with a large 80 μm diameter nucleus, the germinal vesicle, close to the animal pole of the cell, while the vegetal hemisphere contains mostly cytoplasm. Therefore, prior to injection of fluorescent markers, one has to bear in mind in which orientation the oocyte is going to be imaged afterwards on the confocal microscope. If one is interested in processes on the animal half of the cell, two possible orientations are depicted on Fig. 9: the "side view" is more appropriate when a single optical section is acquired over time and gives an overview of the cell along the animal–vegetal axis. The "top view" is better suited for acquiring 3-D stacks and to observe cellular structures close to the animal pole, such as centrosomes and the meiotic spindle. After loading oocytes in the injection chamber, one should choose cells with the right orientation for injection or, alternatively, rotate them after injection with the microneedle. The injection chambers are compatible with imaging on both upright and inverted microscopes. However, a setup to image oocytes during the injection is easier to implement on upright microscopes. This can be especially useful to image the kinetics of equilibration of fluorescent markers in different cellular compartments (Lénárt *et al.*, 2003) or to follow processes induced by micromanipulation, such as cell wounding (Terasaki *et al.*, 1997).

The use of a confocal microscope is unavoidable for most applications when imaging oocytes. Wide-field deconvolution microscopy cannot be used in most cases, because deconvolution algorithms do not work well at such low magnifications and extended z dimensions, especially when fluorescence intensity is inhomogenous in z due to the thickness of the specimen (Swedlow *et al.*, 2002). We commonly use high NA 20× air or 40× water immersion objective lenses, because these allow focusing through the whole depth of the oocyte. If fine subcellular details, such as single microtubules, are of interest, the use of a high NA 60/63× water immersion objective can be advantageous. Typically, the critical parameter for fluorescence imaging of oocytes is not the resolution but the high transmission of the objective lens. Of the ones tested, we found best the Nikon SuperFluar 20× 0.7 NA air and the Zeiss 40× 1.2 NA C-Apochromat water immersion objective. In Table I, we give examples for dye combinations and appropriate filter-sets we have used successfully for multicolor imaging. Especially for triple labeling experiments, it is a very useful option if different laser lines can be switched rapidly line-by-line by an AOTF to avoid cross-excitation between fluorophores (this is available on the Zeiss LSM 510, Leica TCS, and BioRad Radiance 2100 microscopes). On systems without AOTFs, the fluorescence intensities for the different fluorophores have to be balanced carefully to minimize cross-talk. For 3-D time-lapse imaging applications, it is also essential that the microscope is equipped with a rapid and reproducible z-stepping device, such as a piezo-stepper or z-scanning

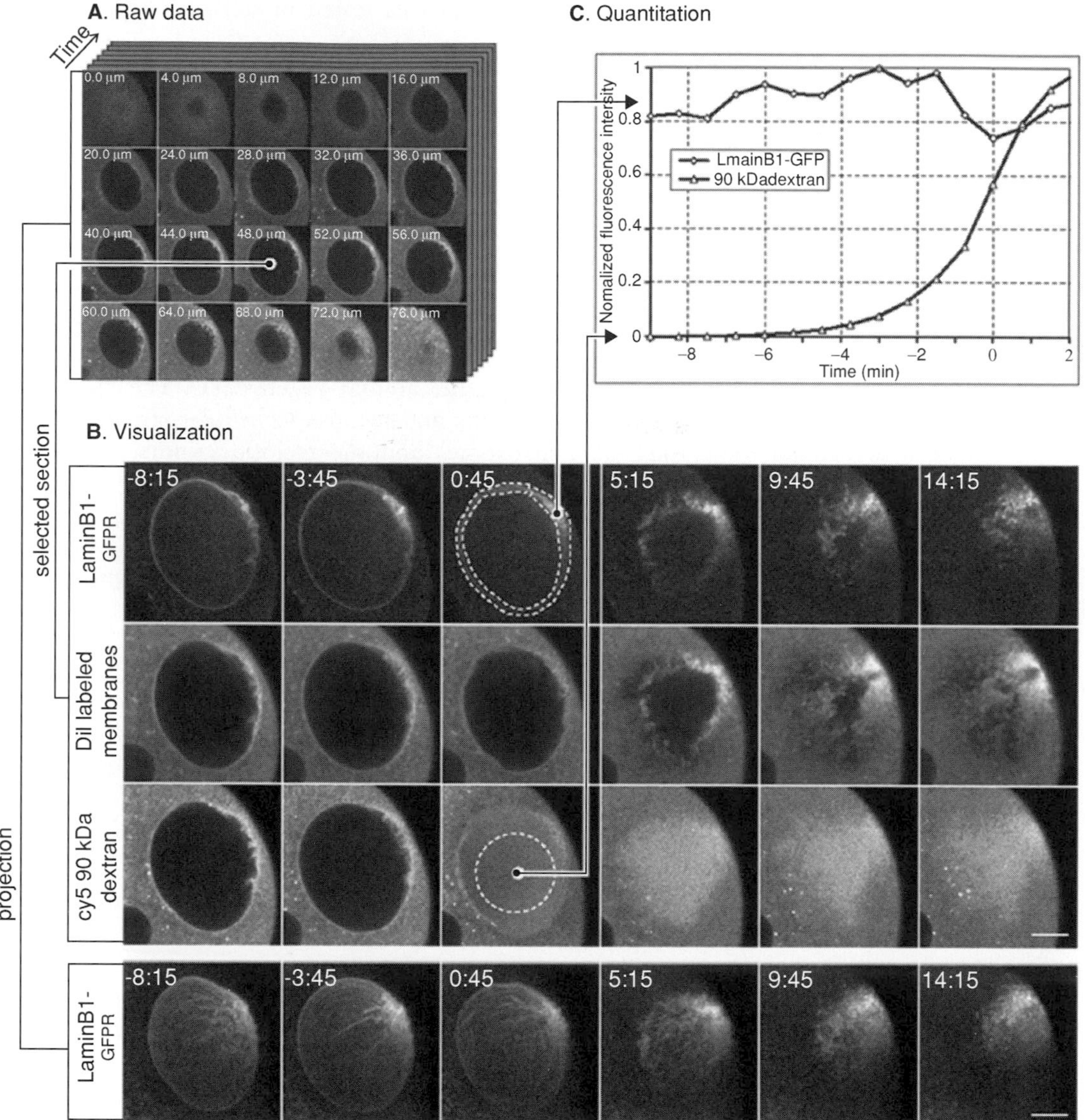

Fig. 9 Dynamics of the nuclear lamina during maturation of starfish oocytes. (A) Raw data. A starfish oocyte expressing laminB1-GFP was injected with Cy5 labeled 90 kDa dextran and $DiIC_{16}$ and imaged during maturation. For each timepoint, 20 optical sections were taken; stacks were acquired every 45 s. (B) Two possible ways of visualizing the 4-D dataset: selected optical section for representative timepoints are shown for all three channels separately (upper panels); maximum intensity projection of the laminB1-GFP channel (lower panel). (C) Chart showing quantitation of the image series. Fluorescence intensities were measured in the nucleoplasm and on the nuclear rim for the dextran and laminB1, respectively. Mean intensities were then normalized from minimum to maximum values and plotted over time. Time is mm:ss. Bars 25 μm.

Table I
Filter Combination for Triple Labeling Experiments

	Channel 1	CFP	GFP, Aloxa 488	GFP, Aloxa 488
	Channel 2	YFP	TMRTRITC, Alexa 546, DiIC16	HcRed, Alexa 594
Fluorophores	Channel 3	HcRed, Cy5	Cy5	Cy5
Lasocs	Scan 1	413, 633	488, 633	488, 633
	Scan 2	514	543	543
Dichroic mirrors	Main	413/514/633	488/543/633	488/543/633
	1st	490	560	560
	2rd	635	635	635
Filters	Channel 1	430–470	500<	500<
	Channel 2	525–575	560–615	560–615
	Channel 3	650<	650<	650<
Notes			XXX	XXX

Note. All filter data is in nanometers (nm).

stage. For a typical time-series, we acquire frames every 5 to 15 s when a single optical section is imaged or a stack of 15 to 20 z-sections every 40 to 60 s.

Bleaching experiments can be carried out in oocytes similar to those in other cells (Klonis *et al.*, 2002); however, their different cellular architecture has to be taken into account. For example, bleaching experiments aiming to deplete most of the cellular fluorescence (such as fluorescence loss in photobleaching, FLIP, or inverse fluorescence recovery after photobleaching, iFRAP) might take very long, because of the large cytoplasmic volume.

Moreover, it is not possible to image and bleach oocytes homogenously in z direction with standard confocal optics because of their thickness, and higher laser powers are necessary to bleach regions deep within the cells due to the light scattering by the specimen. To overcome these limitations, the use of photoactivatable GFP can provide a solution (Patterson and Lippincott-Schwartz, 2002) because a much smaller region has to be illuminated at high intensity. Photoactivation is also possible by two-photon excitation, which can provide a well-defined activation volume in the z-direction. However, to our knowledge, these techniqueshave not been tested in oocytes yet. Because of the inhomogeneities in z, FRAP and photoactivation experiments in oocytes have to be analyzed in three-dimensions, which complicates quantitation and modeling of kinetic processes.

C. Image Processing

Acquisition of multicolor time-lapse sequences generates large amounts of data. A typical 2-D time-lapse sequence of $512 \times 512 \times 100 \times 3 \times 12$ bit (x, y, t, channels, pixel depth) has a size of ~120 Mbytes; a comparable 3-D series

$512 \times 512 \times 20 \times 100 \times 3 \times 12$ bit (x, y, z, t, channels, pixel depth) is about 2 Gbyte. To process such datasets, computers with large hard disk space, fast data transfer, and sufficient memory are necessary. The first step of processing is typically to visualize the time-series in an intuitive way, e.g., by converting raw sequences to movies. If this is not an in-built function of the microscope operating software, an additional image analysis software application is required. Although many other packages are available commercially, we found the open-source software, ImageJ (http://rsb.info.nih.gov/ij/) especially useful when working with time series. The software runs on all operating systems, most microscope suppliers' image formats can be directly imported, all operations can be executed on time series (cropping, filtering, combining channels, aligning slices, etc.), and then exported in Microsoft avi or Apple QuickTime mov format.

When acquiring 3-D stacks over time, it is easiest to convert z-stacks to single images for each timepoint to generate a movie. This can be achieved simply by selecting a single optical section, which shows the structure of interest (Fig. 9B), projecting the z-slices to a single plane (Fig. 9B), or by more advanced volume or surface rendering methods (e.g., isosurface rendering; Fig. 10, Gerlich *et al.*, 2001).

After visualization to assess the overall quality of the experiment, images can be further analyzed to extract quantitative information. Especially in oocytes, as a consequence of quantitative injection, fluorescence intensities of introduced fluorescent molecules (e.g., recombinant proteins or dextrans) can be directly converted to number of molecules (Lénárt *et al.*, 2003). For FP fusions expressed from injected mRNAs, intensities can be calibrated by comparing to oocytes injected with known amounts of purified recombinant FPs (Fink *et al.*, 1998; Terasaki *et al.*, 1996). As a consequence of the large thickness of oocytes, fluorescence intensities significantly differ at different focal positions because the cytoplasm scatters more and more light with the increasing distance from the objective lens. Therefore, only fluorescence intensities from the same focal plane should be compared or 3-D stacks have to be normalized before analysis. After this correction and subtraction of background, fluorescence intensities are proportional to the concentration of the fluorophore, if the images are not saturated.

D. Protocol for Imaging the Dynamics of Nuclear Lamina during Nuclear Envelope Breakdown (Fig. 9)

1. To produce LaminB1-GFP mRNA, a DNA fragment encoding for human laminB1-GFP (Daigle *et al.*, 2001) was transferred to pGEMHE. After linearization downstream of the 3′ UTR, this construct was used as template in an *in vitro* transcription reaction (T7 mMessage mMachine Kit, Ambion). The reaction mixture was extracted by phenol/chloroform and the mRNA was dissolved in 10 μl water (RNase-free, non-DEPC treated, Ambion) resulting in ~5 μg/μl final concentration.

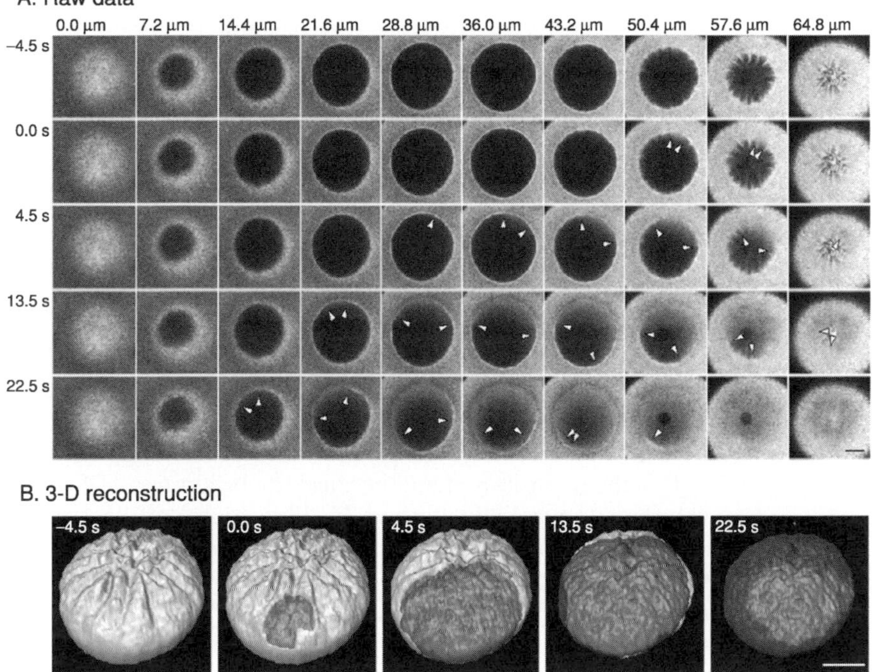

Fig. 10 Starfish oocytes injection: Oocyte injected with Alexa 488 500 kDa dextran imaged in 3-D over time during maturation. (A) 3-D stacks for selected timepoints. The original data set contained 20 zslices; only every second is shown. Arrowheads mark the boundary between permeabilized and intact NE. (B) Isosurface visualization of the 4-D dataset Intact (light gray) and permeabilized (dark gray) areas of the NE were segmented and reconstructed in 3-D. Bars 25 μm.

2. Different amounts of the mRNA (30–150 pl, corresponding to 1–5% of the oocyte volume) were injected. All injections and imaging were done at 20 °C. Oocytes were then incubated overnight at 16 °C in the injection chamber placed in a petri dish with a wet Kimwipe to prevent evaporation of seawater from the chamber. Injection of too much mRNA can be toxic to the oocytes, irrespective of the encoded protein; in this case, the mRNA has to be further diluted. Although lamins have extremely low turnover once incorporated into the lamina (Daigle *et al.*, 2001), newly synthesized lamin molecules assemble into the lamina rapidly, presumably by lateral association (Fig. 8B).

3. The next day, to label the endoplasmic reticulum and nuclear membrane, oocytes were also injected with $DiIC_{16}$ (cat. no. D-384, Molecular Probes; ~1% of oocyte volume) dissolved in household vegetable oil to saturation (Terasaki, 1994). Furthermore, as a marker for nuclear envelope permeabilization, oocytes were injected with Cy5 labeled 90 kDa dextran to ~1% of oocyte volume (from the 0.2 μM stock; a variety of different sizes of fluorescent and amino dextrans are

available from Molecular Probes. The latter can easily be labeled with amine-reactive fluorescent dyes, such as succinimidyl esters of Alexa dyes (Molecular Probes) or Cy-dyes (Amersham)). Both markers diffuse only slowly in the cell; thus, cells have to be incubated for at least 1 h until they distribute homogenously.

4. Imaging was started after addition of the maturation hormone to the chamber (1-methyladenine, 10 μM final concentration in seawater; Sigma). We use an inverted Zeiss LSM510 confocal microscope with a 40x C-Apochromat NA 1.2 water immersion objective (Carl Zeiss, Göttingen, Germany). We decided to image in the "side" orientation to best visualize the collapse of the nuclear envelope from the vegetal toward the animal pole of the cell. In the first scan, GFP and Cy5 were excited simultaneously with the 488 and 633 nm lasers. In the second scan, $DiIC_{16}$ was excited by the 543 nm laser. These two scan modes were switched in each line of a frame. The filter settings are shown in Table I. Because $DiIC_{16}$ is very bright and the excitation and emission spectra are relatively broad, a 500 to 530 nm filter was used to image GFP to avoid bleed-through of $DiIC_{16}$ fluorescence.

For the quantitation and because the brightness of LaminB1-GFP and $DiIC_{16}$ changes dramatically during the time-course of the experiment, images were acquired in 12-bit/pixel depth. Since we wanted to see when gaps can be detected on the lamina and on the nuclear membranes, we imaged at high xy resolution (512×512 pixels). To still achieve reasonable time resolution, we sacrificed the z resolution and acquired only 20 z sections every 4 μm to sample the entire 80 μm nucleus (Fig. 8C). Stacks were acquired every 45 s during the 30 min experiment. The resulting dataset contained about 1 Gbyte.

5. Images were exported as a series of 16-bit TIFF images for each channel and imported into ImageJ as a single stack. This stack (xyz over time) was split up into 20 substacks of single optical sections over time by using the "Hypervolume_Shuffler" plug-in. After selecting the appropriate z-slice, representative timepoints were chosen. After brightness and contrast adjustment, images were converted to 8-bit and assembled to a montage by using ImageJ's "montage" function (Fig. 9B). The maximum intensity projection was done with the "grouped Zprojector" plug-in (Fig. 9B). Fluorescence intensities in the GFP and dextran channel were quantified on the selected optical section throughout the whole time series by measuring mean pixel intensities in regions outlining the nuclear rim and in the nucleoplasm, respectively (Fig. 9). Values were then pasted into Excel and normalized from minimum to maximum intensity and plotted over time (Fig. 9C).

E. 4-D Imaging of Dextran Entry during NEBD (Fig. 10)

1. We observed that during NEBD, large dextrans (e.g., 500 kDa) first enter the nucleus at a single location. The permeabilization of the nuclear envelope then spreads from this initial site, engulfing the whole nucleus in less than a minute (Lénárt *et al.*, 2003). To resolve the details of this rapid and spatially complex process, we optimized the imaging setup for fast acquisition. To avoid the delay

caused by alternating multiple scan modes (see Table I), we only injected a single dextran and did not image other colors. We imaged the oocyte in the "top" orientation (Fig. 8C), first, to observe the spreading of the dextran wave perpendicular to the animal–vegetal axis, and second, because it brings the nucleus closer to the objective lens, resulting in brighter signal. The oocytes were injected with Alexa 488 500 kDa dextran (from a 0.28 μM stock dissolved in water; Molecular Probes) to ~1% of their volume and incubated for ~1 h to allow the dextran to equilibrate in the cytoplasm. Then, the maturation hormone was added, but imaging was started just a few minutes before start of the dextran entry (15 min after the hormone addition) to avoid photodamage due to the continuous image acquisition. Injections and imaging were done at 20 °C.

2. Imaging was done on the same system described in Protocol I. However, in this case, only the 488 nm laser line was used to excite the fluorophore and the filter settings were also chosen for maximum light collection efficiency: a single wavelength 488 nm dichroic mirror and a >500 nm longpass filter. We scanned with the maximum speed in xy without averaging and reduced resolution to 256 × 256 pixels. To acquire the stacks rapidly, we made use of a fast z-scanning stage (HRZ200, Zeiss). In this manner, we could acquire a stack of 20 z-slices in 4.5 s. Stacks were taken continuously during the 10 to 15 min of the experiment, producing a dataset of ~100 MBytes (Fig. 10).

3. Next, we wanted to analyze the dataset to determine the starting point of the entry wave and the speed and direction of spreading of the permeabilization. A simple way to initially evaluate the data is to browse through the complete 4-D dataset. Most microscope operating software includes such browsing function, where the dataset can be displayed as a gallery or, alternatively, ImageJ can be used (HyperVolume_Browser plug-in) (Fig. 10A). Distances can be measured on single optical sections; however, to measure in 3-D, software with 3-D browsing and measurement capabilities is required. We used Amira 2.3 (TGS Inc., San Diego, CA); other alternatives are Imaris (Bitplane AG, Zürich, Switzerland) or Volocity (Improvision).

To quantitate the dextran wave, we first manually marked the boundary between the permeabilized and intact parts of the NE on single optical sections (see arrows on Fig. 10A). Then, we split the dataset to two subsets, one outside, one inside the boundaries. These two sets were visualized by isosurface rendering, again using Amira (Fig. 10B). The initial point of entry as well as the spreading speed could then easily be measured on the reconstructed dataset. We consistently found that the permeabilization started on the animal hemisphere and spread with an average speed of 1 to 2 μm/s.

References

Begg, D. A., *et al.* (1996). Stimulation of cortical actin polymerization in the sea urchin egg cortex by NH_4Cl, procaine and urethane: Elevation of cytoplasmic pH is not the common mechanism of action. *Cell Mot. Cytoskel.* **35,** 210–224.

Belmont, L. D., *et al.* (1990). Real-time visualization of cell cycle-dependent changes in microtubule dynamics in cytoplasmic extracts. *Cell* **62,** 579–589.

Belton, R. J., *et al.* (2001). Isolation and characterization of the sea urchin egg lipid rafts and their possible function during fertilization. *Mol. Reprod. Dev.* **59,** 294–305.

Daigle, N., *et al.* (2001). Nuclear pore complexes form immobile networks and have a very low turnover in live mammalian cells. *J. Cell Biol.* **154,** 71–84.

Dechant, R., and Glotzer, M. (2003). Centrosome separation and central spindle assembly act in redundant pathways that regulate microtubule density and trigger cleavage furrow formation. *Dev. Cell* **4**(3), 333–344.

Fink, C., *et al.* (1998). Intracellular fluorescent probe concentrations by confocal microscopy. *Biophys. J.* **75,** 1648–1658.

Galbraith, A. J., and Terasaki, M. (2003). Controlled damage in thick specimens by multiphoton excitation. *Mol. Biol. Cell* 10.1091/mbc.E02-03-0163.

Gerlich, D., *et al.* (2001). Four-dimensional imaging and quantitative reconstruction to analyze complex spatiotemporal processes in live cells. *Nat. Cell Biol.* **3,** 852–855.

Henson, J. H., *et al.* (1995). Immunolocalization of the heterotrimeric related protein KRP(85/95) in the mitotic apparatus of sea urchin embryos. *Dev. Biol.* **171,** 182–194.

Hertwig, O. (1877). Beitrage zur Kenntniss der Bildung, Befruchtung, und Theilung des theirischen Eies. *Morphol. Jahr.* **1,** 347–452.

Hyman, A. (1991). Preparation of modified tubulins. *Meth. Enzymol* **196,** 478–485.

Inuoe, S., and Oldenbourg, R (1998). Microtubule dynamics in mitotic spindle displayed by polarized light microscopy. *Mol. Biol. Cell.* **9,** 1603–1607.

Kalinowski, R. R., *et al.* (2003). A receptor linked to a Gi-family G-protein functions in initiating oocyte maturation in starfish but not frogs. *Dev. Biol.* **253,** 139–149.

Kellog, D. R., *et al.* (1988). Behavior of microtubules and actin filaments in living *Drosophila* embryos. *Development* **130,** 675–686.

Kiehart, D. P. (1982). Microinjection of echinoderm eggs: Apparatus and procedures. *Methods Cell Biol.* **25**(Pt. B), 13–31.

Kishimoto, T. (1999). Activation of MPF at meiosis reinitiation in starfish oocytes. *Dev. Biol.* **214,** 1–8.

Klonis, N., *et al.* (2002). Fluorescence photobleaching analysis for the study of cellular dynamics. *Eur. Biophys. J.* **31,** 36–51.

Lénárt, P., *et al.* (2003). Nuclear envelope breakdown in starfish oocytes proceeds by partial NPC disassembly followed by a rapidly spreading fenestration of nuclear membranes. *J. Cell Biol.* **160,** 71055–71068.

Liman, E. R. (1992). Subunit stoichiometry of a mammalian K+ channel determined by construction of multimeric cDNAs. *Neuron.* **9,** 861–871.

Patterson, H. G., and Lippincott-Schwartz, J. (2002). Photoactivatablle GFP for selective photolabeling of proteins and cells. *Science* **297,** 1873–1877.

Rappaport, R. (1996). Cytokinesis in Animal Cells. Cambridge University Press, Cambridge, UK.

Shilling, F., Mandel, G., and Jaffe, L. A. (1990). Activation by serotonin of starfish eggs expressing the rat serotonin 1c receptor. *Cell Regul.* **1,** 465–469.

Shuster, C. B., and Burgess, D. R. (2002). Targeted new membrane addition in the cleavage furrow is a late, separate event in cytokinesis. *Proc. Natl. Acad. Sci.* **99**(6), 3633–3638.

Swedlow, J. R., *et al.* (2002). Measuring tubulin content in Toxoplasma gondii: A comparison of laser-scanning confocal and wide-field fluorescence microscopy. *Proc. Natl. Acad. Sci. USA* **99,** 2014–2019.

Terasaki, M., and Jaffe, L. A. (1993). Imaging of the endoplasmic reticulum in living marine eggs. *In* "Cell Biological Applications of Confocal Microscopy," (B. Matsumoto, ed.), Vol. 38, pp. 211–220. Academic Press, Orlando, FL.

Terasaki, M. (1994). Redistribution of cytoplasmic components during germinal vesicle breakdown in starfish oocytes. *J. Cell Sci.* **107**(Pt. 7), 1797–1805.

Terasaki, M. (1996). Actin filament translocations in sea urchin eggs. *Cell Motil. and the Cytoskeleton* **34,** 148–156.

Terasaki, M., *et al.* (1996). Structural change of the endoplasmic reticulum during fertilization: Evidence for loss of membrane continuity using the green fluorescent protein. *Dev. Biol.* **179,** 320–328.

Terasaki, M., *et al.* (1997). Large plasma membrane disruptions are rapidly resealed by Ca2+-dependent vesicle–vesicle fusion events. *J. Cell Biol.* **139,** 63–74.

Terasaki, M., *et al.* (2001). A new model for nuclear envelope breakdown. *Mol. Biol Cell* **12,** 503–510.

Waterman-Storer, C. M., and Salmon, E. D. (1998). Endoplasmic reticulum membrane tubules are distributed by microtubules in living cells using three distinct mechanisms. *Curr. Biol.* **8,** 798–806.

Wessel, G. M. (1989). Cortical granule-specific components are present within oocytes and accessory cells during sea urchin oogenesis. *J. Histochem. Cytochem.* **37,** 1409–1420.

Yuce, O., and Sadler, K. C. (2001). Postmeiotic unfertilized starfish eggs die by apoptosis. *Dev. Biol.* **237,** 29–44.

CHAPTER 17

TEM and SEM Methods

Bruce J. Crawford* and Robert D. Burke†

*Department of Anatomy and Cell Biology
University of British Columbia
Vancouver, British Columbia V6T 1Z4, Canada

†Departments of Biology and Biochemistry/Microbiology
University of Victoria
Victoria, British Columbia V8W 3P6, Canada

METHODS IN CELL BIOLOGY, VOL. 74

0091-679X/04 $35.00

I. Introduction

The importance of accurate and complete morphological data should not be underestimated. Almost all modern functional studies work upon a foundation of knowing what cell types are involved and how they are spatially organized. The need for this type of data is especially true for invertebrate deuterostomes, where there are numerous stages of development and tissue types for which our knowledge of the cells and their organization is incomplete.

Transmission and Scanning electron microscopy (TEM and SEM) remain useful principally because of the range of magnification and the resolution that can be achieved. These advantages become most useful when coupled with immunological methods or methods that permit identification of specific components. However, because we must fix and process tissues, the images we create are imperfect reflections of what exists in life. The interpretation of images is directly dependent upon how the specimens were prepared.

Here, we describe several methods that have proven useful in preparing embryos and larvae of several species of invertebrate deuterostomes. The procedures represent a useful starting point for people who want to answer a specific morphological question, and have been prepared assuming only moderate experience. There are several additional sources of information that include detailed procedures for TEM and SEM which should be consulted (Hayat, 1978, 1981). Most of the illustrations are of asteroids embryos and larvae (Figs. 1–5); however, some of the preparations were developed and used successfully with sea urchins (Burke *et al.*, 1991, 1998; Tamboline and Burke, 1989) and ascidians (Figs. 6 and 7) (Cavey and Cloney, 1972; Cloney and Florey, 1968; Dunlap, 1966).

II. Basic Fixation and Preparation

A. Seawater Fixation

The simplest method and one that gives acceptable results for LM and TEM is primary fixation in glutaraldehyde in seawater, followed by post-fixation in OsO_4 in a sodium bicarbonate buffer (Crawford and Chia, 1978). The concentrations of glutaraldehyde used in primary fixatives may vary; however, it has been our experience that higher concentration, such as 2.5% glutaraldehyde in seawater, tends to cause shrinkage in echinoderm embryos. Although a certain amount of trial and error may be necessary when working with the embryos and larvae of other organisms, for echinoderm embryos, the following formulation of 0.5 to 1% glutaraldehyde in 80% seawater lowers the osmolarity to more closely match that of seawater and gives acceptable results (Fig. 1a,b).

The salinity of open ocean seawater is roughly 34% with an osmolarity of roughly 1000 milliosmoles (mOms). The composition and salinity, hence, the osmolarity of seawater from coastal locations, varies from location to location and from month to month at the same location, depending on the amount of fresh water runoff. If you

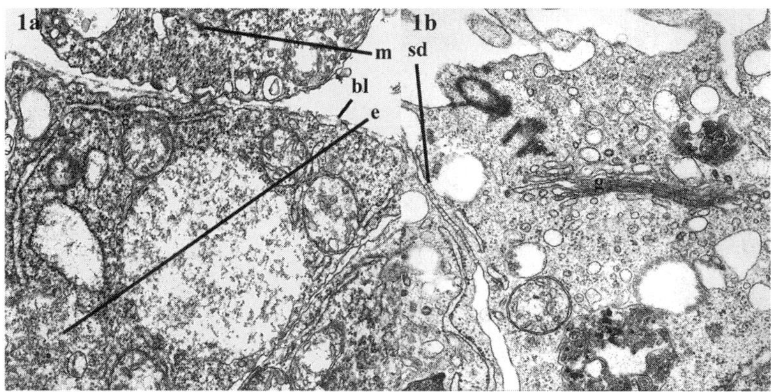

Fig. 1 (a) A TEM through the esophageal region of a 5-1/2-day-old larva of the starfish *Pisaster ochraceus* fixed in 1% glutaraldehyde in 80% seawater showing an esophageal cell (e) and a developing smooth muscle cell (m) ×21,940. (b) A TEM through the apex of an endodermal cell from a 7-day-old bipinnaria larva fixed in 0.5% glutaraldehyde in 80% seawater, ×26,730. (sd; septate desmosome).

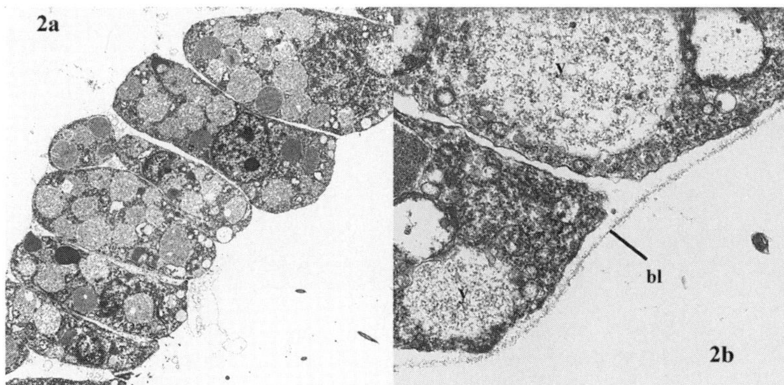

Fig. 2 (a) A low power view of ectodermal cells from a mid gastrula stage embryo of the starfish *Pisaster ochraceus* fixed in phosphate-buffered glutaraldehyde and post-fixed in osmium bicarbonate; ×2400. (b) A higher power view of material fixed in phosphate-buffered glutaraldehyde showing the base of two ectoderm cells and the basal lamina (bl) ×16,200.

are using artificial seawater, this should not pose a problem. If natural seawater is used, it is difficult to determine the osmolarity of the seawater in which your specimens are currently living without directly measuring it. It has been our experience that the best fixation occurs when the osmolarity of the fixative is slightly hypotonic to the seawater in which the animals have been raised (i.e., roughly 960 mOms). We therefore suggest that the seawater that is used for the fixative should be from the same source as that in which the animals are being raised. This will ensure that

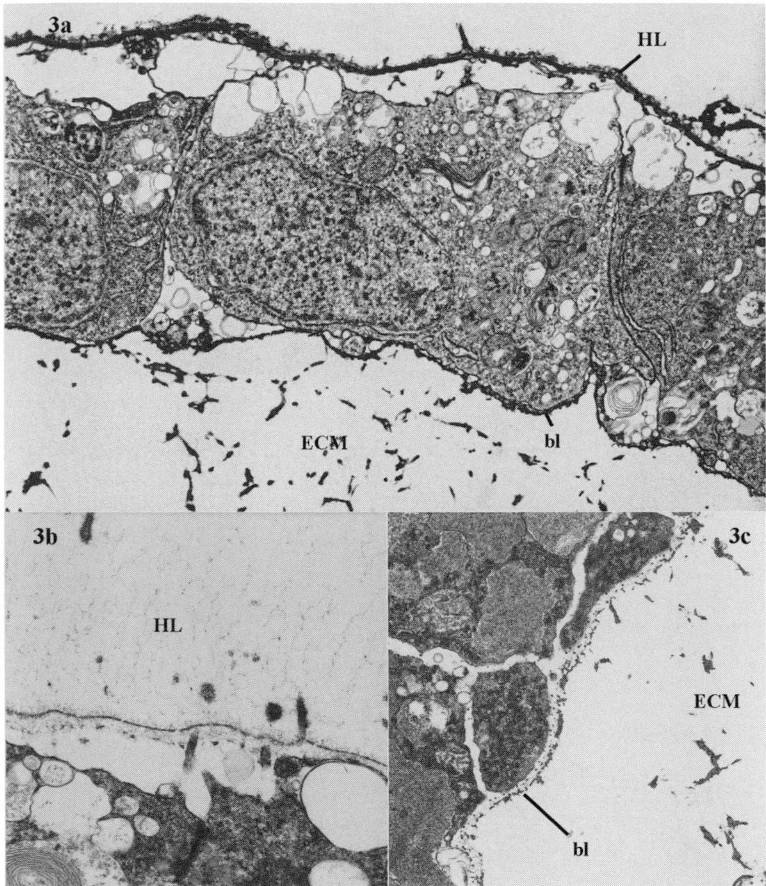

Fig. 3 (a) Ectodermal cells from a 5-1/2-day-old *P. ochraceus* larva fixed in 1% glutaraldehyde in 80% seawater saturated with alcian blue. The embryo had been opened in the fixative to allow the stain to gain access to the blastocoel. Note the excellent preservation of the hyaline layer (HL) and the ECM in the blastocoel (ECM), ×11,700. (b and c) View of the apex (b) and the base (c) of an ectodermal cell from a 5-day-old *P. ochraceus* larva fixed in cacodylate-buffered glutaraldehyde with ruthenium red. Note the preservation of the HL and the ECM; (bl; basal lamina) ×19,500; ×13,410.

the fixative will contain all of the ions and trace elements that are present in the animal's environment and that it will have the same osmolarity as that in which the animal has been living. Dilution of this seawater to roughly 80% with distilled water compensates for the addition of 1% glutaraldehyde (110 mOms). This means that the osmolarity of the fixative will be slightly hypotonic to that of the seawater in which the animal lives.

If the echinoderm embryos are only being prepared for light microscopy or if they are to be kept and photographed later for whole mounts, the primary

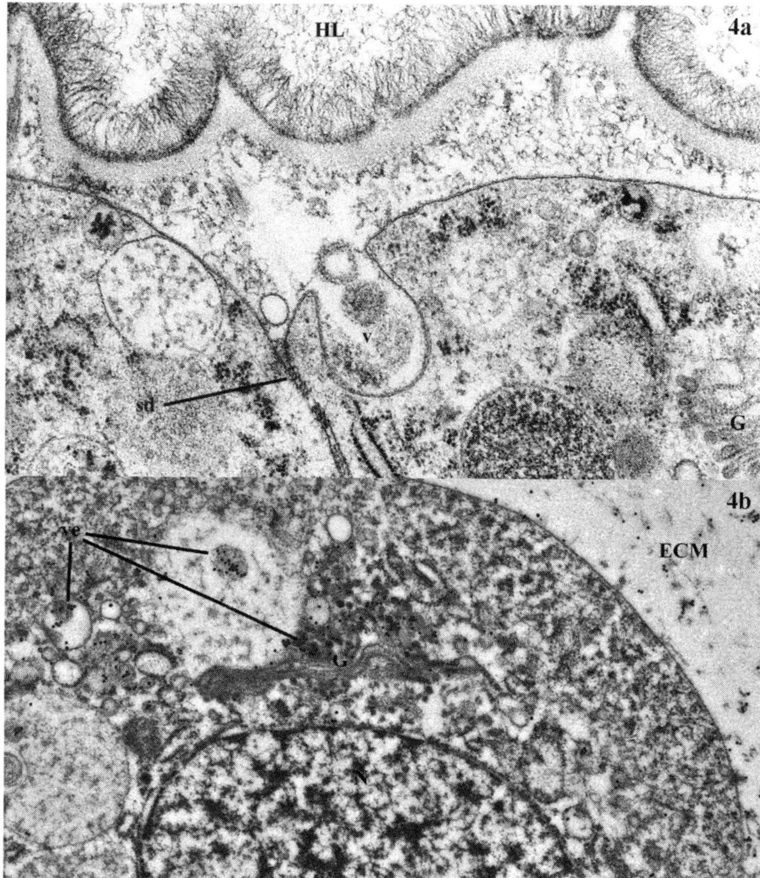

Fig. 4 (a) The apical region of an ectoderm cell of the starfish *P. ocraceus* that has been fixed by freeze substitution in ethanol, post-fixed in OsO_4 in acetone and embedded in Epon. Note the excellent preservation of the hyaline layer (HL) and a vacuole (v) that is either discharging or phagocytosing material of the HL. A Golgi (g) body, septate desmosome (sd) ribosomes and numerous granules and vacuoles are also well preserved. ×33,375 (b) A mesenchyme cell located in the blastocoel of a 5-1/2-day-old bipinnaria larva that has been freeze substituted in ethanol saturated with alcian blue, embedded in LR white and stained with colloidal gold using a primary mouse monoclonal antibody against one component of the blastocoel ECM. The secondary antibody was prepared by conjugating colloidal gold to a rabbit anti-mouse IgG, as described in the text. Note that the tissue is reasonably well fixed. Label is seen over the ECM and Golgi associated vesicles (ve). ×19,500. (Fig. 4b reprinted from Crawford *et al.*, 1997, with permission).

fixative gives excellent preservation. If the preserved whole embryos are to be photographed in color, it is best to transfer them to seawater or, if long-term storage is desired, to buffer with seawater containing 0.1% sodium azide, as the embryos will yellow with time if kept in the primary fixative.

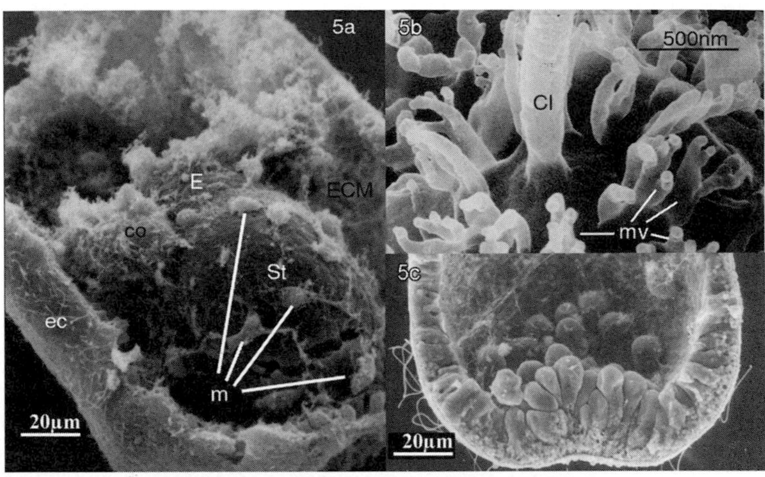

Fig. 5 (a) An SEM of a 5-day-old starfish embryo fixed according to the methods described in this chapter that has been dissected to show the ectoderm (ec), the coeloms (co), the esophagus (E), stomach (st) ECM, and mesenchyme cells (m), ×550. (b) A high magnification SEM of a specimen fixed according to the protocols described in this chapter showing the base of a cilium (CI) and staghorn microvilli (mv) located on the surface of an ectodermal cell from a 5-day-old starfish embryol; ×30,000. (c) An SEM image of an embryo of *Strongylocentrotus purpuratus* fixed using the SEM methods described here. The embryo was fractured after freeze point drying to reveal cellular and internal details; ×488.

1. Material

a. Primary Fixative: 1% glutaraldehyde in 80% seawater

0.22 µm filtered seawater	78 ml
Distilled water	18 ml
25% TEM grade glutaraldehyde	4 ml

Adjust pH to 8.0–8.2 with 0.1 M NaOH.

Filter through Whatman #1 filter paper.

Seawater is a poor buffer. When adjusting the pH, only 0.1 M NaOH should be used and it should be added slowly so as not to overshoot the desired pH. If the pH accidentally goes above 10, an insoluble precipitate will be formed. If this happens, the fixative should be discarded.

Addition of 10 mM Tris (base) to the seawater improves its buffering capacity without changing the pH.

b. Washing Buffer: 2.5% $NaHCO_3$

$NaHCO_3$	1.25 g
Distilled water	50 ml

Adjust pH to 7.2–7.4 with HCl (about 5 drops of 10 N in 50 ml, then adjust with 1 N as necessary).

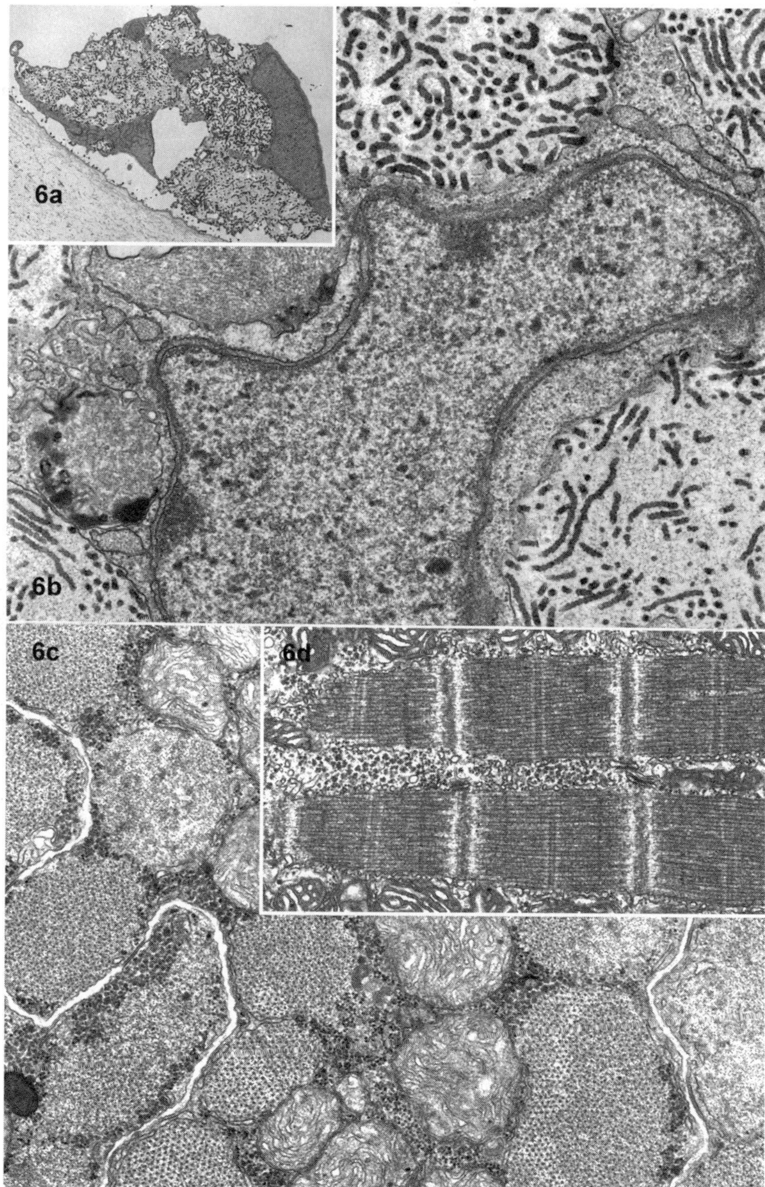

Fig. 6 (a) Section of the nucleus and the perinuclear cytoplasm of a test cell in the perivitelline space of an egg of the ascidian *Distaplia occidentalis*. The nucleus is irregular in shape, and small patches of condensed chromatin flank the inner membrane of the nuclear envelope. The cytoplasm contains massive vacuoles with electron-dense ornaments in various stages of maturation; ×37,870. (b) Section of a test cell releasing ornaments onto the external surface of the differentiating tunic; ×4,190. Fixation is phosphate-buffered glutaraldehyde and bicarbonate-buffered osmium tetroxide. (c) Transverse section of embryonic muscle cells of *Distaplia occidentalis*. Myofibrils occupy the cortical sarcoplasm,

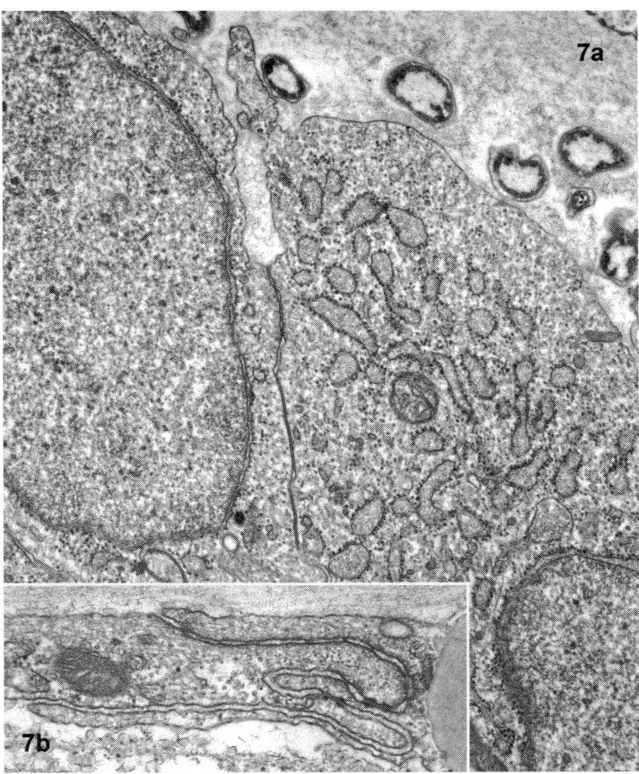

Fig. 7 (a) Vertical section of the larval epidermis in the trunk of the ascidian *Distaplia occidentalis* epidermis. The epidermis in the trunk is a simple cuboidal epithelium invested by a cellulose tunic. Epidermal cells, exhibiting oblong nuclei and distended cisternae of granular endoplasmic reticulum, are joined laterally by occluding and communicating junctions. Spiriform bacteria inhabit the innermost lamina of the tunic. Fixation is phosphate-buffered glutaraldehyde and bicarbonate-buffered osmium tetroxide; ×29,110 (b) Vertical section of the epidermis in the tail. The epidermis is a simple squamous epithelium, and the caudal tunic is amplified to form a pair of larval fins. Epidermal cells, exhibiting flattened nuclei (not shown), are interdigitated laterally and joined by junctional complexes similar to those between truncal cells. Fixation is phosphate-buffered glutaraldehyde and bicarbonate-buffered osmium tetroxide; ×48,460. Images contributed by Dr. Cavey, University of Calgary, Canada.

---

situating close to the sarcolemma. The medullary sarcoplasm contains numerous mitochondria with elaborate cristae. Multigranular glycogen rosettes are apparent throughout the cell. Fixation is cacodylate-buffered glutaraldehyde with ruthenium red and bicarbonate-buffered osmium tetroxide; ×24,380. (d) Longitudinal section of a larval cell of *Diplosoma macdonaldi*. Myofibrils reside in the medullary sarcoplasm. Robust mitochondria with elaborate cristae infiltrate the contractile apparatus, and they also accumulate in the cortical sarcoplasm, segregating the myofibrils from the sarcolemma. Fixation is phosphate-buffered glutaraldehyde and bicarbonate-buffered osmium tetroxide; ×14,630. Images contributed by Dr. M. J. Cavey, University of Calgary, Canada.

Some authors suggest that the osmolarity of the washing solution and the buffer should be the same as the osmolarity of the primary fixative (Hayat, 1981). The following wash is quite hypotonic to the primary fixative and therefore results in a large decrease in osmolarity. We have found that although the osmolarity of the primary fixative is very important, that of the wash and the secondary fixative(s) seem to be much less so. If it is thought that osmolarity might be a problem in the washing step, a washing buffer consisting of equal parts of 0.2 M Millonig's phosphate buffer, pH 7.6, and 0.3 M NaCl may be substituted for the 2.5% NaHCO3 buffer. This provides a buffer of roughly equivalent osmolarity to seawater (~980 moms).

c. Secondary Fixative: 2% OsO_4 in 1.25% $NaHCO_3$ (Wood and Luft, 1965)

OsO_4 4% in distilled water	0.5 ml
2.5% NaHCO3 buffer pH 7.2–7.4	0.5 ml

This fixative should be used for 1 h at room temperature and is used to stabilize lipids and increase contrast.

d. Phosphate-buffered osmium tetroxide (Dunlap, 1966)

If osmolarity is thought to be a problem this formulation will provide a fixative that is roughly isotonic to seawater.

4% OsO_4 in distilled water	0.5 ml
0.4 M Millonig's phosphate buffer (pH 7.6)	0.5 ml

OsO_4 has a low vapor pressure and exposure to even small amounts of vapor can do serious damage to corneas, nasal, oral, and respiratory mucosa and skin. In addition, breathing even small amounts of it can cause severe asthma. It is often used as a 2 to 4% solution in either distilled water or acetone. Handling and particularly dissolving OsO_4 is both difficult and dangerous and must always be done in a fume cabinet! In addition to being dangerous, OsO_4 is very expensive and therefore should be used in small amounts. Glassware used to contain OsO_4 solutions should be carefully cleaned and rinsed extensively. Pure OsO_4 is usually sold in sealed glass ampules containing 0.5 or 1.0 g, although larger sizes are available (EM Sciences, Fort Washington, PA; Marivac, Halifax, NS). Prepared stock solutions of OsO_4 stored under nitrogen are available and eliminate the need for handling (EM Sciences, Fort Washington, PA).

2. Method

a. Concentrate the animals into a small amount of seawater, either by gentle centrifugation or by placing them in a conical centrifuge tube in ice for 10 to 20 min.

When changing solutions, never remove all of the liquid from the vial before adding the next solution and never let the specimens remain on the glass or get caught in the surface tension. If they do so, they will pick up air and will float on the surface. If this happens, they are usually either lost during further processing or they collapse during embedding and do not embed properly.

b. Add the embryos/larvae to a clean glass vial, containing 10 ml of primary fixative in 1 to 5 drops of seawater. After the specimens have settled, remove as much fixative as possible without letting them dry and replace with fresh fixative. Allow the specimens to fix for 1 h at room temperature (RT).

c. Remove the primary fixative and wash in two changes of the buffer that will be used with the secondary fixative.

d. After the second rinse, remove buffer and replace with the secondary fixative.

e. Post fix for a further 1 h at room temperature.

f. Rinse in 2 changes of distilled water.

g. Dehydrate in 5-min changes in the following alcohol series as follows: 2 changes of 30%; 50%; 70%; 90%; 2 changes of 95%; 2 changes of 100%; 2 changes of propylene oxide.

B. Phosphate-Buffered Glutaraldehyde, Osmium Tetroxide (Cloney and Florey, 1968; Dunlap, 1966)

This primary fixative gives good results with ascidian and echinoderm embryos. It uses Millonig's phosphate-buffered glutaraldehyde and adjusts the osmolarity of the fixative to that of seawater using sodium chloride (Figs. 2, 6, and 7).

1. Material

a. Primary fixative:

The fixative is best prepared from stock solutions:

25% glutaraldehyde	5 ml
0.34 M sodium chloride	20 ml
0.4 M Millonig's phosphate buffer	25 ml

The final concentrations are:

2.5% (0.25 M) glutaraldehyde

0.2 M Millonig's phosphate buffer pH 7.6

0.14 M Sodium chloride

Mix and check that pH is approximately 7.6.

Millonig's Phosphate Buffer (0.4 M).

Sodium phosphate (monobasic)	11.8 g
Sodium hydroxide	2.85 g
Distilled water	200 ml

The pH should be 7.6 and, after dilution 1:1, the osmolarity should be 420 mOms.

b. Secondary fixative

Either of these fixatives can be used. The first has a low osmolarity relative to the primary fixative and gives less background. As has been noted, the second has an osmolarity that is similar to that of the primary fixative and to that of seawater.

2% OsO_4 in 1.25% $NaHCO_3$ (see preceding for preparation)

2% OsO_4 in 0.2 M Millonig's phosphate buffer

4% osmium tetroxide	0.5 ml (1 part)
0.4 M Millonig's phosphate buffer (pH 7.6)	0.5 ml (1 part)

2. Method

a. Concentrate specimens and remove most of the seawater, without letting specimens dry.
b. Add primary fixative and fix for 1 h at room temperature.
c. Rinse for 10 to 15 min in the buffer to be used for the secondary fixative.
d. Post-fixation for 1/2 to 1 h with osmium tetroxide is necessary to stabilize lipids and improve contrast.
e. Rinse with distilled water twice for 10 min and dehydrate with a graded series of ethanol (30%, 50%, 70%, 90%, Absolute), as has been described.

III. Embedding Media

Once the animals have been fixed and dehydrated, they must be embedded in a material that will support them during sectioning and viewing in the transmission electron microscope. The use of two of these media will be discussed here: Epon for general use and L. R. White for use with specimens that have been prepared for immunocytochemistry. There are numerous embedding media available, many designed to meet special requirements and beyond the scope of this chapter. For a review of this subject, see Hayat (1989, 2000).

A. Embedding in Epon 812 (Luft, 1961)

Epon 812 (EmBed 812, EM Sciences) is an epoxy resin. Epoxy resins cause minimal shrinkage during embedding, are strongly cross-linked, and form a strong, highly inert plastic that is remarkably stable in the electron beam (Hayat, 1989, 2000). They provide excellent tissue support and, although the contrast between the unstained tissue and the plastic is low, staining with heavy metals easily enhances it. Epoxy resins are excellent for detailed morphological work but they are hydrophobic and strongly cross-linked. This means that they

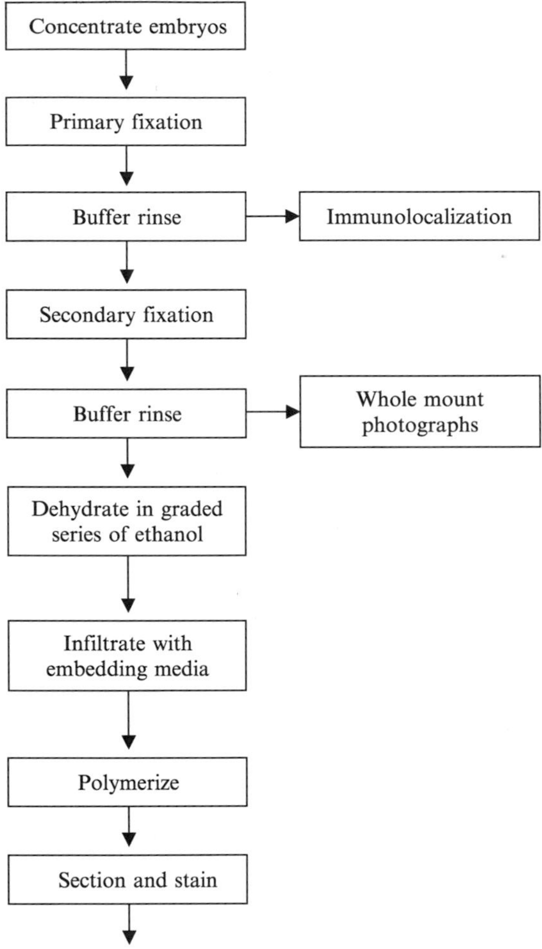

TEM Flow Chart

have poor penetration by aqueous solutions, such as those used in enzyme cytochemistry and immunohisto/immunocytochemical work. We do not recommend them for this use.

When embedding small organisms, you can simply use commercial embedding molds. A better method is to flat embed in an aluminium weighing dish or other flat receptacle that is not dissolved by the embedding resin and which can be removed after polymerization. Once the resin has set, the block is removed from the dish, the specimens are marked out, cut from the block using a hand coping saw, and trimmed with a razor blade. The embryo/larvae can then be mounted on an aluminium stub or a blank block from an embedding mold that fits the microtome chuck using epoxy glue. By mounting the trimmed embryos on stubs rather than simply placing them in an embedding mold, one can determine the orientation of the embryo or larvae with respect to the microtome knife and has accurate control of the plane of section. In high humidity environments where curing epoxy is difficult, attaching a specimen securely to a stub may not be possible. Mounting a block cut from the flat embedded disc directly in a viselike specimen holder may be necessary.

1. Material (EM Sciences, Ft. Wn., PA)

   a. Epon 812
   b. DDSA Dodecenyl Succininc Anhydride
   c. NMA Nadic Methyl Anhydride
   d. DMP-30 2,4,6 Tridimethyl Amino Methane Phenol
   e. propylene oxide

   To prepare the Epon resin stock solutions:

Part A:	
Epon 812	62 ml
DDSA	100 ml
Part B:	
Epon 812	100 ml
NMA	89 ml

Parts A and B are mixed individually by stirring without creating lots of bubbles. They can be stored at 4 °C for several months. Before use, they should be brought to room temperature, preventing condensation from contaminating them.

Mix 7 parts A with 3 parts B and add 1.6 ml of DMP-30 for each final volume of 100 ml. After all components are thoroughly mixed and the DMP-30 added, the mixture is drawn into 10 ml syringes, wrapped, and can be stored at −20 °C for several weeks. Ensure the mixture is brought to room temperature without contamination by condensation. In high humidity environments, a 6:4 ratio of parts A and B may be necessary to obtain blocks that are suitable for sectioning.

2. Method

a. Mix Epon with propylene oxide (1:1) and add it to vial containing specimens and only a scant amount of propylene oxide. Propylene oxide is extremely volatile and specimens will dry within a few minutes if left in an open vial.

b. Infiltrate for 4 to 8 h. It is not necessary to rotate or agitate specimens; this frequently results in mechanical damage. A toothpick or equivalent can be used to move the delicate, sticky specimens.

c. Mix Epon/propylene oxide (75%:25%) and replace; infiltrate a further 4 to 8 h.

d. Finally, place specimens in 100% Epon in an embedded mold and polymerize for 24 to 48 h at 60 °C.

B. Embedding in LR White

LR White (Polysciences, Warrington, PA; London Resin Co., Basingstoke, UK) is an acrylic resin. It can be polymerized either in the cold or using heat but polymerization requires anerobic conditions. LR White resin has a low viscosity so that it infiltrates the tissue rapidly and it exhibits minimal nonspecific staining. The polymerized medium is not as strongly cross-linked as epoxy resins and therefore supplies less support and has a somewhat lower contrast. The great advantage of this material is that the sections are hydrophilic, allowing good penetration of aqueous solutions (Hayat, 1989). We therefore strongly recommend its use for immunocytochemical studies.

1. Material

L.R. White resin

2. Method

Dehydrate with alcohol as acetone will react with the resin

a. After 2 changes in absolute ethanol, remove ethanol and replace with LR White resin.

b. After 60 min, replace the resin with fresh resin.

c. Leave at room temperature for at least another 60 min, up to overnight. Specimens will become translucent and sink when fully infiltrated.

d. Put specimens in an embedding mold with fresh resin.

e. Polymerization requires an oxygenfree atmosphere at 50 $\pm 2^{\circ}$ for 20 to 24 h. A vacuum oven, purged several times with nitrogen, works well to exclude oxygen from the curing process. If an oxygen free oven is not available, a simple device made from 2 syringes (Volker *et al.*, 1985) is described on page 111 of Hayat (1989).

IV. Preservation of Extracellular Matrix

Fixation with glutaraldehyde containing Ruthenium Red or Alcian Blue (Fig. 3) preserves extracellular matrix, including the hyaline layer in echinoderms, and gives excellent fixation of ascidian tissues. Addition of the anionic dyes to fixatives preserves elements of the extracellular matrix (ECM) not preserved by routine fixation in glutaraldehyde and osmium tetroxide. In addition, fixatives containing these dyes also give improved cellular preservation of marine deuterostome larvae. They do, however, cause some collapse and shrinkage of the ECM when compared with material prepared by freeze substitution (see Following Text).

A. Cacodylate-Buffered Glutaraldehyde Fixative with Ruthenium Red (Luft, 1961, modified by Cavey and Cloney, 1972)

This fixative was originally designed to preserve the extracellular matrix in vertebrates. The version presented here is one developed by M. Cavey (Cavey and Cloney, 1972) for preservation of marine invertebrates, particularly ascidians. The osmolarity has been adjusted to that of seawater by the addition of sucrose. Marine embryos and larvae should be fixed for 3 to 6 h at room temperature in the primary fixative and in the secondary fixative for 1 to 2 h on ice or 45 min at room temperature.

1. Material

a. Primary Fixative:

2% glutaraldehyde
0.2 M cacodylate buffer
0.275 M sucrose
0.5% ruthenium red

To prepare fixative add;

8% glutaraldehyde	1 part
0.8 M cacodylate buffer	1 part
0.55 M sucrose containing 0.1% unpurified ruthenium red	2 parts

check pH and adjust to 7.4, if necessary

b. Secondary Fixatives:

The recommended secondary fixative is:

2% osmium tetroxide in
0.625% sodium bicarbonate

2. Method

a. Concentrate specimens and add primary fixative; fix for 3 to 6 h at room temperature (longer fixations are preferred).
b. Remove primary fixative and rinse with buffer of the secondary fixative.
c. Resuspend specimens in secondary fixative.
d. Rinse with distilled water and dehydrate with a graded series of ethanols.

B. Glutaraldehyde with Alcian Blue (Abed and Crawford, 1986a,b; Crawford and Abed, 1983, 1986)

A simpler formula that also gives good preservation of the ECM consists of 1% glutaraldehyde in 80% seawater saturated with Alcian blue 8GX (Marivac, Halifax, NS) (Crawford, 1989; Crawford and Abed, 1986). This is a modification of a vertebrate fixative developed by Behnke and Zelander (1970). It gives good preservation of the cells, the hyaline layer, and the ECM of the blastocoel of echinoderm embryos.

1. Material

Prepare 1% glutaraldehyde in 80% seawater as has been described and add 1% Alcian Blue 8GX and stir for 1 to 4 h. Filter through Whatman #3 filter paper.

2. Method

a. Concentrate specimens as previously described and fix for 1 h at room temperature.

b. The embryos should settle to the bottom of the vial but they will be light colored and the fixative will be dense blue so that the embryos can not be readily seen. To prevent losing the embryos, remove about 2/3 of the fixative, being careful not to stir the fix. Add buffer to the vial and wait 5 to 10 min for the embryos to settle; then, repeat the procedure until the buffer solution is clear.

c. Post-fix the specimens for 1 h at room temperature, as directed with the secondary fixative of choice. 2% OsO4 in $NaHC0_3$ buffer made as previously described works well.

V. Fixation for SEM

Cleanliness is extremely important in preparing specimens for SEM. It is necessary to make sure that all solutions used in SEM preparation are clean so as to minimize the deposition of contaminating particles on the outer surfaces. It is strongly recommended that all solutions used to fix and stain specimens for SEM be filtered (0.22 μm) before use. Solutions containing solvents such as amyl acetate cannot be filtered, since they will dissolve commonly used membrane filters (Fig. 5).

Preservation of specimens for SEM will also require the same high quality fixation used for TEM (Crawford, 1990; Crawford and Campbell, 1993). Secondary fixation with OsO_4 and the use of TEM heavy metal stains is also recommended, particularly when older SEMs are all that are available. Although specimens can be preserved using only a primary fixative, it is often desirable to view specimens at high magnification. Operation of the instrument at high magnifications requires the beam to be focused to a very small spot. In order to achieve a beam with a small spot size that produces a usable signal, high accelerating voltages must be used. This causes the beam to penetrate the specimen, which reduces the resolution. Although the accelerating voltages used in modern SEMs are lower than those in older machines, beam penetration is still a problem at high magnifications. Impregnation of the specimen with heavy metals such as osmium and uranium decrease beam penetration of the specimen and increase resolution.

1. Material

a. Primary Fixative:
   1% glutaraldehyde in 80% seawater (see preceding)

b. 1.25% $NaHCO_3$ buffer (see preceding)

c. Secondary Fixative:
   2%OsO_4 in 1.25% $NaHCO_3$

2. Method

a. Concentrate embryos and add primary fixative. Fix for 1 h at room temperature.

b. Rinse twice with 1.25% $NaHCO_3$ for a total of 15 min.

c. Post-fix for 1 h at room temperature.

d. Wash 2 times with distilled water.

e. Incubate for 1 h in 0.22 μm filtered, saturated aqueous uranyl acetate solution.

f. Wash 2 times with distilled water.

g. Dehydrate in ethanol as for TEM specimens.

SEM specimens must be dried. Cellular surface structures are extremely delicate and will be destroyed by surface tension forces if the specimens are air-dried. Two methods to dry small delicate specimens which eliminate surface tension effects are freeze-drying and critical point drying. Freeze-drying is a relatively slow process and is often associated with shrinkage of small specimens. Critical point drying is usually the method of choice for small specimens such as marine invertebrate larvae and works very well. A thorough discussion of critical point drying is beyond the scope of this chapter. For an excellent summary, see Hayat (1978).

h. Critical point drying using CO_2 as a transition fluid is probably the most commonly used method. Specimens can be dried either from ethanol or amyl acetate as an intermediate fluid. If the amyl acetate technique is used, the specimens should be passed through a graded series of ethanol amyl acetate mixtures to pure amyl acetate.

i. Put specimens in Teflon microporous specimen capsules [EM Sciences, Ft. Washington. PA. (cat.# 70187–88)] (Atwood *et al.*, 1975). This will protect them during critical point drying and will serve to retain the intermediate fluid during transfer to the critical point drying apparatus and immersion in CO_2. The simplest way to do this is to place the open Teflon microcapsules in a glass petri dish containing enough intermediate fluid to go roughly halfway up the sides of the capsules. The specimens can then be carefully pipetted into the capsules and the capsules closed with their lids. The capsules can be quickly transferred to the critical point drying apparatus; the apparatus is then closed and flooded with liquid CO_2.

j. When amyl acetate is used as a transition fluid, extensive washing (2–3 h) with liquid CO_2 is required to remove the amyl acetate before critical point drying. This is costly in CO_2. If ethanol is to be used as the intermediate fluid, the specimens must be transferred rapidly to the apparatus and must be quickly immersed in liquid CO_2. Speed is necessary so as to prevent air-drying. As has been noted, the use of the Teflon microporous capsules helps to retain some ethanol and prevent air-drying during transfer. Only minimal washing with CO_2 is required (circa 5 min) if the ethanol CO_2 method is used.

k. Once the specimens have been dried, they can be transferred with a fine camel--hair brush to an SEM stub coated with double-sided tape. The specimens can either be coated whole or dissected with fine glass or steel needles prior to coating. In order to achieve good grounding of the specimens, the edge of the tape should be overlapped with silver paint (Marivac, Halifax, NS). The thickness of the coating will depend on the type of machine used and the magnification that is desired. The coating should be as thin as possible, since thicker coatings obscure fine details.

VI. Quick Freezing and Freeze-Substitution

In most light and electron microscopic studies, chemical fixatives containing formaldehyde, glutaraldehyde, and/or osmium tetroxide are used to prepare the materials (Hayat, 1981). Ultrastructural studies often use several different fixatives as the morphology often varies somewhat, depending upon the technique employed. These chemicals are thought to act by causing molecular cross-linking, thus stabilizing molecules that may otherwise be extracted or relocated during dehydration and infiltration of the tissues. While these techniques give excellent morphological preservation, they tend to alter or destroy the immunological characteristics of many compounds, denature proteins, and rearrange cellular and extracellular components. In the last 25 years, more and more studies depend on the use of immunological techniques (Hayat, 1981). In order to use immunological techniques effectively, preservation techniques must be developed that minimize or eliminate the alterations or artifacts caused by chemical fixation (Fig. 4).

Rapidly freezing cells, tissues, or small embryos is one method that could significantly reduce such artifacts. To achieve this, the tissues and/or embryos are frozen so rapidly that all cellular processes are stopped instantaneously. In order to achieve this, very small pieces of tissue were slammed on to copper blocks cooled with liquid helium (Robards and Sleytr, 1985) by plunging them into supercooled liquids such as liquid ethane or propane (Campbell *et al.*, 1991) or preserving them by a high-pressure freezing technique. This latter technique is excellent but it involves the use of equipment that is expensive to buy and to use. This then preserves the cells and tissues in a lifelike state without using aldehydes and heavy metals. Nonaqueous solvents such as ethanol are then substituted for the water ice and these, in turn, are used as a vehicle to infiltrate the tissue with a fixative, if desired, and a resin.

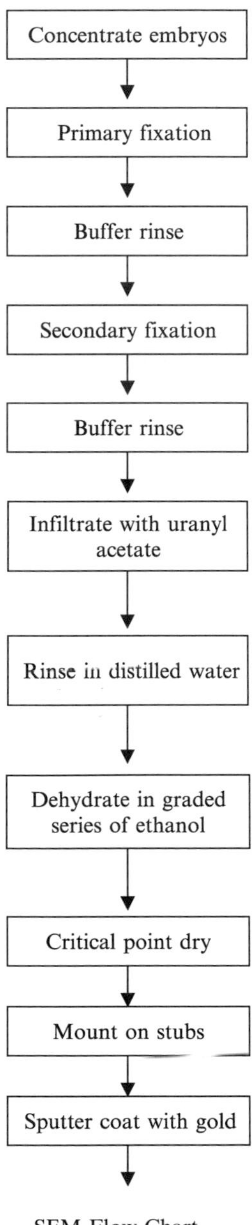

SEM Flow Chart

Experiments with starfish and sea urchin eggs/embryos/larvae (Campbell *et al.*, 1991; Crawford *et al.*, 2000; Pang *et al.*, 2002, 2003; Reimer and Crawford, 1995) have shown that rapidly plunging the specimens into liquid propane gives good preservation of the eggs/embryos/larvae, particularly those closest to the cryogen.

The cryogen is contained in a metal cup that sits in a Dewar flask containing liquid nitrogen in a fume cabinet. Adjusting the level of the liquid nitrogen in the cup allows the operator to maintain the liquid propane at a temperature very close to its freezing point (−192 °C). (Liquid propane is potentially explosive; it will also cause severe frostbite if it is allowed to come in contact with the skin. Unless you are familiar with handling liquefied gasses, you should read the reference cited before attempting this technique.)

Cryoprotectants decrease damage due to ice crystal formation during rapid freezing. This is achieved either by removing water from the cells (nonpenetrating cryoprotectant) and increasing the intracellular solute concentration or by penetrating the cells and interfering with nucleation of the ice crystals (penetrating cryoprotectant). Tests with different cryoprotectants on echinoderm embryos (Campbell *et al.*, 1991) demonstrated that better freezing can be achieved if the embryos are incubated in a cryoprotectant prior to freezing. Use of these cryoprotectants raises the possibility that the cells of the specimens are damaged by the cryoprotectant. Tests with 2,3 butenediol on starfish larvae demonstrated that although they stopped swimming when placed in the 15% solution, they resumed swimming when the cryoprotectant was removed by washing, suggesting that the damage was minimal.

The water ice is dissolved from the frozen specimens using 100% ethanol or 100% ethanol containing Alcian Blue. The dye will help to preserve the ECM and it appears to help preserve cellular morphology as well. The cells are fixed as the ice is removed. Once the ice has been completely substituted with ethanol, the specimens are gradually brought up to room temperature, where they may undergo further fixation or may be embedded for immunohistochemical or immunocytochemical studies without further fixation.

A. Freezing

1. Material (see Campbell *et al.*, 1991 for details)

Liquid propane. This can be bought at the local hardware store. To get the propane as a liquid, the valve of a standard propane torch can be modified by removing the nozzle and turning the torch upside down

Liquid Nitrogen

2 Dewar flasks, one fitted with a metal cup and an eddy-current (spark free) stirring motor

Forceps (cross-closing ones are best)

Copper TEM freeze fracture or 50 mesh EM grids

Liquid nitrogen freezer vials

Liquid nitrogen freezer canes

Snap cap vials

Ethanol, Alcian Blue, 2,3 butanediol
Micropipet capable of delivering 1 μl

2. Method

a. Set up the freezing equipment as described in Campbell *et al.* (1991). Fill the Dewar flask with liquid nitrogen and the metal cup with liquid propane and start the spark free stirring motor. You should also set up a second Dewar flask filled with liquid nitrogen and containing a cane with short freezer vials clipped in position. The tops on the vials should be loose to allow them to fill with liquid nitrogen. Alternatively, you should have some small open snap cap glass vials filled with 100% ethanol or 100% ethanol saturated with Alcian Blue at −85 °C available in a small −85 °C freezer or a cold box. The ethanol should never be allowed to warm up above −85 °C.

b. Collect embryos/larvae of the desired stage into a conical glass centrifuge tube and concentrate them, either by allowing them to sit at 0 °C until they have fallen to the bottom of tube or by mild centrifugation (250 ×*g* for 1–3 min).

c. Cryoprotect the specimens, if desired, by placing them in ice-cold seawater containing 15% 2,3 butenediol or 15% propylene glycol for 10 to 15 min.

d. Grip a 50 mesh copper EM grid in fine-tipped cross-closing forceps. Pick up 1 μl of loosely packed embryos using a plastic tipped micro pipet and place them on the copper grid. Test to make sure that the opening in the pipet tips are wide enough so as not to damage the embryos. If they are too narrow, cut the tip off with a scalpel blade until the opening is large enough to pass the specimens through without damage.

e. Under a dissection microscope, remove as much water as possible by touching the tip of a triangle of Whatman #3 filter paper to the water surface. Try to get the specimens spread out in a monolayer on the surface of the grid with as little water on them as possible. You will certainly lose egg/embryos/larvae when you do this, so be prepared.

f. Take a toothpick in your left hand (if you are right-handed) and the cross-closing forceps containing your specimen in your right and plunge the grid into the cryogen; move it around in the cryogen for 30 sec, then rapidly transfer it to the liquid nitrogen. Cool the toothpick in the liquid nitrogen and use it to push the grid into one of the vials containing the 100% ethanol for immediate ethanol substitution or into one of the freezer vials for storage.

g. Repeat the process as required. If you have sufficient material, several grids can be placed in each vial.

h. If the grids are to be stored for future use, close the vials and place the cane in a liquid nitrogen refrigerator. Specimens have been successfully stored in this manner for several years.

B. Substitution

1. Material

a. Ethanol, acetone, 2% OsO4 in acetone Epon
b. JB-4 or LR White (Polysciences)
c. Snap cap vials (Fisher)
d. Freezer (−85 °C)

2. Method

a. Although methanol has often been used for substitution, ethanol has been chosen here because of its lower freezing point and superior results with echinoderm embryos (for discussion, see Campbell *et al.*, 1991).

b. If they are not already in the 100% ethanol Alcian Blue solution at −85 °C, open the freezer vial and transfer the grids to a glass vial containing fresh 100% ethanol/Alcian Blue at −85 °C. You may have to push the grids out of the freezer vial with a toothpick. Try to make sure that the grids do not warm up during the transfer, since if the grids warm above −80 °C, ice-crystal formation will begin and will cause cell damage.

c. Cover the vials and allow the specimens to undergo substitution for roughly 5 days at −85 °C. (The time required for substitution will vary with the material. We have found that most small specimens are substituted by 5 days.)

d. The grids are then washed 2 to 4 times in anhydrous 100% ethanol at −85 °C. The specimens are washed from the grids using a pipette, the grids are removed, and the vials transferred to a freezer at −20 °C for 2 h.

e. They are then transferred to a refrigerator for a further 2 h and finally to the bench top.

If the specimens are to be used for immunocytochemistry, they should be embedded as will be described.

If the specimens are to be used for morphological studies, they can be osmicated and embedded in Epon as follows:

f. Wash the specimens with 100% acetone 3× over a 1 h period.

g. Immerse them in 2% OsO_4 in acetone for 2 h at room temperature.

h. Wash with 100% acetone and transfer to a 1:1 acetone Epon mixture for 12 h.

i. Transfer the specimens to pure Epon and cure the Epon at 60 °C for 48 h.

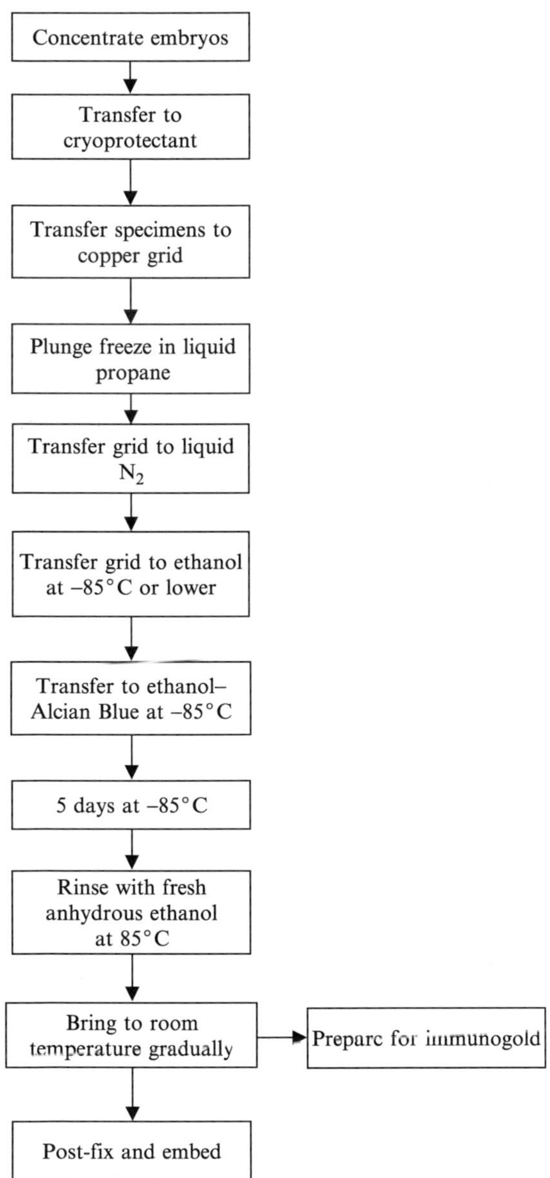

Cryopreservation Flow Chart

VII. Fixation for Immunohistochemistry and Immunocytochemistry

While conventional fixation techniques give excellent morphological preservation, they tend to alter or destroy the immunological characteristics of many compounds, denature proteins, and rearrange cellular and extracellular components. Thus, they are not usually suitable for studies involving immunolocalization of antigens at the TEM level. Since 1980, more and more studies depend on the use of immunological techniques (Hayat, 1981). Effective preservation techniques must therefore be developed that minimize or eliminate the alterations or artifacts caused by chemical fixation. Rapidly freezing cells, tissues, or small embryos either by high-pressure liquid freezing or by the method previously described, is one method that could significantly reduce such artifacts. Fixation with ethanol causes much less denaturation of cellular components, however, preservation at the ultrastructural level is poor. Studies on starfish embryos (Campbell *et al.*, 1991) suggest that while 100% ethanol is a poor fixative at room temperature, it appears to give acceptable preservation when used as a fixative at $-85\,^{\circ}$C, particularly if it contains Alcian Blue.

Epon is not the resin of choice for immunological studies. Epon forms a three-dimensional lattice during polymerization. While this gives excellent tissue support and contrast in the TEM, the high degree of cross-linking will probably exclude most antibodies. Etching the Epon surface is possible but the solutions used could themselves denature the antigens. Resins with lower levels of cross-linking, such as LR White, Lowacryl, HM20, or Ultracryl for TEM, or JB4 for light microscopic studies, provide good support while allowing antigens access to the tissues. Luckily, all of these resins are soluble in 100% ethanol so the tissues prepared by the quick-freeze/freeze substitution method already described are easily embedded in them.

Some echinoderm embryos collapse if embedded directly into 100% resin from alcohol. If this occurs, the specimens should be embedded by gradually increasing the strength of the resin over 1 or even 2 days.

VIII. Immunogold Methods

This method had been used successfully with echinoderm embryos and larvae prepared with fixatives, or with freeze substitution, embedded in LR White resin, and sections supported on nickel grids. Before embedding specimens, prepare embryos for whole mount immunofluorescence to ensure that the fixations have preserved antigens and antibodies are able to bind them (Fig. 4).

1. Material

a. TBS—20 mM Tris, 150 mM NaCl, pH 7.4–7.6
b. TBS/BSA—TBS with 1% BSA, pH 7.4–7.6
c. TBS/BSA/Tween—TBS/BSA with 0.2% Tween20, pH 7.4–7.6
d. TBS/BSA/Tween/LS— TBS/BSA/Tween with 5% Normal Lamb Serum, pH 7.4–7.6

2. Methods

a. Block grids 15 to 30 min at room temperature, section side down on 50 μl drops of TBS/BSA/Tween/LS on a piece of Parafilm.

b. Transfer grids to drops of TBS/BSA/Tween/LS containing your primary antibody on a piece of Parafilm. Start with a dilution that has given satisfactory results with indirect immunofluorescence, usually, 1:250 or 1:750. These should be in a glass petri dish lined with a piece of damp filter paper in order to prevent the grids from drying out (the Parafilm sits on top of the wet filter paper). These can be incubated at 4 °C overnight or for a few hours at RT on a slow rotating shaker platform.

c. Wash the grids by placing them section side down on drops of TBS/BSA/Tween/LS and allowing them to shake for approximately 5 min at RT. Repeat 3×. Washing can also be done by floating the grids section side down on the surface tension of the wash buffer in a 96-well plate.

d. If you are using a gold-conjugated secondary Ab (see Section IX or EM Sciences, Ft Wn., PA), then you may wish to continue to use TBS/BSA/Tween/LS for your secondary incubation and wash buffers. However, if you are using a gold-conjugated protein-A secondary (EM Sciences), then you may use TBS/BSA/Tween for the secondary incubation and wash buffers (for available sizes for conjugated gold particles, see EM Sciences catalog). Dilute your secondary in the appropriate buffer, start with a 1:50 dilution. Place grids section side down on drops of secondary on a piece of Parafilm in the glass petri dish, as for primary incubation. Incubate at RT on a slow rotating shaker for 1 to 2 h.

e. Wash grids 3×, using the same procedure as for primary washes in TBS/BSA.

f. Wash grids 1 or 2× in TBS.

g. Remove grids and place section side down onto drops of 2% glutaraldehyde in TBS; fix for 10 to 15 min at RT.

h. Wash grids 1× in TBS, 1× in dH_2O.

i. Touch the edge of the grid to a piece of filter paper to remove excess fluid and allow the grid to dry.

j. Prior to viewing, you may wish to stain using 2% uranyl acetate to improve contrast, but prolonged staining may obscure gold particles.

IX. Ultrastructural Immunoperoxidase

This method has proven useful as a quick and reliable procedure for ultrastructural localization of ECM components (Burke *et al.*, 1998; Tamboline and Burke, 1989). The combination of whole mount immunocytochemistry with ultrastructural methods has a broad utility. An advantage of the method is that the immunological reactions can be assessed and confirmed before the ultrastructural analysis. The electron-dense reaction product is readily identified in thin sections, but the lack of contrast in tissues may hamper precise interpretations of localization.

1. Material

a. Primary Fixative (as outlined in Basic Fixation, already described)

b. PBS, PBS containing 5% serum, horseradish peroxidase conjugated secondary antibody

c. DAB substrate solution (prepared immediately before use):

Diaminobenzidine	0.005 g
PBS	10 ml
Bovine Serum Albumin	0.125 g
Hydrogen peroxide (30%)	10 μl

d. Secondary Fixative (2% OsO_4 in 0.2 M Millonig's phosphate buffer, as previously described)

2. Method

a. Fix with Primary Fixative, as previously described. Either 1% glutaraldehyde in 80% seawater or 2.5% glutaraldehyde in Millonig's phosphate buffer gives good preservation. If epitopes are not preserved with glutaraldehyde, use 4% formaldehyde in 80% seawater or Millonig's phosphate.

b. Rinse 3× with PBS, for 10 min to remove excess fixative.

c. Block for 30 min with PBS containing 5% serum from the species in which the secondary antibody was prepared.

d. Incubate with the primary antibody diluted in PBS. The time can range from 1 h to overnight. The exact dilution will have to be determined empirically, as each serum or antibody will differ. Good, high titre antibodies should be diluted about 1:200.

e. Rinse specimens 3× with PBS, for 10 min.

f. Incubate specimens in horseradish peroxidase conjugated secondary antibody solution diluted in PBS for a minimum of 1 h.

g. Rinse specimens 3× with PBS, for 10 min.

h. Resuspend the specimens in the DAB substrate. The duration of this step will vary. Normally, specimens should be reacted with substrate until a deep, black reaction product has formed.

i. Rinse specimens twice with PBS for 10 min.

j. Post-fix with osmium containing fixative, rinse in distilled water, and dehydrate with a graded series of ethanol. Embed in Epon as described. Cut thin sections, mount on grids, and examine with TEM.

X. Preparation of Colloidal Gold Reagents (Horisberger, 1981; Slot and Geuze, 1985) as modified by Campbell (1990) and Reimer (1994) (Table I)

Colloidal gold reagents for immmunocytochemistry are often expensive and do not store well. The reagents are actually not difficult to prepare. Also, by preparing them yourself, you can make use of a range of particle sizes for double labeling experiments and you can conjugate the colloidal gold particles to primary antibodies or other proteins if desired.

1. Material

To make 50 mls:

a. Solution I: 0.005 g $HAuCl_4$ (Chloroauric acid) in 39.5 ml H_2O

b. Solution II (Reducing Mixture):

2 ml of a 1% Sodium Citrate solution in H_2O
H_2O to make 10 ml.

2. Method

a. Bring Solutions I and II to 60 °C separately in a 60 °C water bath.

b. Add Solution II to Solution I quickly while stirring vigorously, maintaining the temperature at 60 °C.

c. Continue stirring at 60 °C until the solution turns a wine red. Once this has occurred, heat the solution to boiling, cool, and store at 4 °C.

d. The formula previously given should yield a particle size of 25 to 30 nm. If smaller sizes are desired, the following solutions of 25 mM potassium carbonate (K_2CO_3) and/or 1% tannic acid should be added to Solution II prior to diluting it to 10 ml. The time required for the reaction will increase as the size of the gold particles increases.

Table I
Determination of Size of Colloidal Gold Particles

Particle size	Volume 25 mM K_2Co_3	Volume 1% tannic acid
4 nm	1 ml	1 ml
6 nm	0 ml	0.25 ml
10–15 nm	0 ml	0.062 ml
20–25 nm	0 ml	0.015 ml
25–30 nm	0 ml	0 ml

Once the colloidal gold has been produced, it must be titrated against the protein you wish to conjugate it to so as to determine the optimal amount of the protein required to stabilize the gold colloid.

e. Dilute the protein to be conjugated to 1.0 mg/ml (minimum volume should be 100–125 μl).

f. Add 100 μl of ddH20 to each tube.

g. Add 100 μl of the protein solution prepared in #1 to tube #1 and serially dilute the protein solution to tube #9 (see Table II). Tube #10 should be left protein-free.

h. At room temperature, adjust the pH of 3 to 5 ml of the colloidal gold solution to pH 9.0 with 0.2 M K_2CO_3.

i. Add 500 μl of the colloidal gold solution to each tube and stir well. Allow to stand for 15 min.

j. Add 100 μl of 10% NaCl to assess resistance to flocculation.

k. Tubes containing the higher concentrations of protein should remain red; those toward the end of the line should turn blue.

l. Select the highest numbered red well (i.e., the last well to remain red). This is the concentration of protein required to stabilize the colloidal gold.

m. Calculate the amount of protein required to stabilize each milliliter of colloidal gold.

In order to ensure good stabilization, you should use double this concentration.

In the example in this table, the endpoint equals 0.0625 mg/ml $\times$ 0.1 ml $=$ 0.00625 mg (6.25 μg) of protein is required to stabilize 500 μl of colloidal gold, or 12.5 μg is required to stabilize 1 ml. To ensure good stabilization of the colloid, 25 μg of protein should be used to stabilize each 1 ml of colloidal gold.

If tannic acid is used in the colloidal gold prep, the color change may be slow and/or incomplete. To remove excess tannic acid, H_2O_2 can be added to a final concentration of 0.1 to 0.2% prior to inducing flocculation with the NaCl solution.

Conjugation of Protein to Colloidal Gold:

Table II
Sample Calculations

Tube #	Serial Dilution	Color	[Protein] mg/ml
1	1/2	red	0.5
2	1/4	red	0.25
3	1/8	red	0.125
4	1/16	red	0.0625
5	1/32	blue	0.0313
6	1/64	blue	0.0156
7	1/128	blue	0.0078
8	1/256	blue	0.0039
9	1/512	blue	0.0020
10		blue	0.00

1. Material

a. Protein stock solution
b. 2 mM sodium borate [$Na_2B_4O_7$(10H20)]-HCl buffer, pH 9.0
c. 0.2 M K_2CO_3
d. 10% bovine serum albumin
e. 10 mM Tris with 50 mM NaCl and 0.1% BSA (pH 8.2)
f. 2% Sodium azide (100× stock)

2. Method

a. Dilute the protein stock solution to give a final concentration of double that determined in the titration procedure already described in 500 μl dH_20. (In this case, if we wished to make 10 ml of colloidal gold solution using the protein, 250 μl of protein in 500 μl of H_2O would be required.)

b. Dialize the protein solution against 2 mM sodium borate [$Na_2B_4O_7$(10H20)]-HCl buffer, pH 9.0, with one change over 12 to 18 h.

c. Adjust the pH of 10 ml of colloidal gold solution to pH 9.0 with 0.2 M K_2CO_3. Do this with care. If pH 9.0 is exceeded, do not attempt to bring the pH down with acid; discard the colloidal gold solution and start again.

d. With the gold solution stirring, rapidly add a bolus of 500 μl of the dialyzed protein solution. Continue to stir for a further 30 min.

e. Once the protein has been bound to the surface of the gold colloid, 10% bovine serum albumin (BSA) is often added to ensure that all sites on the surface of the colloid are coated with protein. This prevents binding of other nonspecific proteins. We have found that this may result in a non-resuspendable pellet following centrifugation (see following text). This step can be omitted without creating nonspecific binding. If this step is deemed desirable, 0.5 ml of 10% BSA in H_2O, pH 9.0, can be added (final concentration 0.5% BSA). The solution should then be stirred for a further 5 min.

f. Centrifuge the solution for 35 min at 4 °C at 15,000 ×*g*.

g. Resuspend the pellet in 10 mM Tris with 50 mM NaCl and 0.1% BSA (pH 8.2) and dialyze slowly, starting from 10 mM Tris with 50 mM NaCl and 0.1% BSA (pH 8.2) and gradually adding NaCl over 24 h to bring the dialysate up to a final concentration of 10 mM Tris with 150 mM NaCl and 0.1% BSA (pH 8.2).

h. Add sodium azide to a final concentration of 0.02%.

Acknowledgments

We are grateful to Dr. M. J. Cavey, Department of Cell Biology and Anatomy, University of Calgary, for generously contributing unpublished TEM images of ascidian larvae. A Canadian Institute of Health Operating Grant to RDB and Discovery Grants from the Natural Sciences and Engineering Research Council of Canada to RDB and BJC supported this work. Dr. Yoko Nakajima, Keio University, provided helpful comments on the manuscript.

References

Abed, M., and Crawford, B. J. (1986a). Ultrastructural aspects of mouth formation in the starfish, *Pisaster ochraceus*. *J. Morphol.* **188,** 239–250.

Abed, M., and Crawford, B. J. (1986b). Changes in structure and location of extracellular matrix in the blastocoel during development of the starfish, *Pisaster ochraceus*. *Can. J. of Zool.* **64,** 1436–1443.

Atwood, D. G., Crawford, B. J., and Braybrook, G. D. (1975). A technique for processing mucous coated marine invertebrate spermatozoa for scanning electron microscopy. *J. Microsc.* **103,** 259–264.

Behnke, O., and Zelander, T. (1970). Preservation of intracellular substances by the cationic dye alcian blue in preparative proceedures for electron microscopy. *J. Ultrastruct. Res.* **31,** 424–438.

Burke, R. D., Myers, R. L., Sexton, T. L., and Jackson, C. (1991). Cell movements during the initial phase of gastrulation in the sea urchin embryo. *Dev. Biol.* **146,** 542–557.

Burke, R. D., Lail, M., and Nakajima, Y. (1998). The apical lamina and its role in cell adhesion in sea urchin embryos. *Cell Adhes. Commun.* **5,** 97–108.

Campbell, S. S. (1990). "Cryofixation of Asteroid Extracellular Matrix." M.Sc. Thesis, University of British Columbia, Vancouver, BC, Canada.

Campbell, S. S., Crawford, B. J., and Reimer, C. L. (1991). A simple ethanol-based freeze-substitution technique for marine invertebrate embryos that allows retention of antigenicity. *J. Microsc.* **164,** 197–215.

Cavey, M. J., and Cloney, R. A. (1972). Fine structure and differentiation of Ascidian muscle. I. Differentiated caudal musculature of Distaplia occidentalis tadpoles. *J. Morphol.* **138,** 349–373.

Cloney, R. A., and Florey, E. (1968). Ultrastructure of cephalopod chromatophore organs. *Z. Zellforsch. Mikrosk. Anat.* **89,** 250–280.

Crawford, B. J. (1989). Ultrastructure of the basal lamina and its relationship to extracellular matrix of embryos of the starfish *Pisaster ochraceus* as revealed by anionic dyes. *J. Morphol.* **199,** 349–361.

Crawford, B. J. (1990). Changes in the arrangement of the extracellular matrix, larval shape, and mesenchyme cell migration during asteroid larval development. *J. Morphol.* **206,** 147–161.

Crawford, B. J., and Abed, M. (1983). The role of the basal lamina in mouth formation in the embryo of the starfish *Pisaster ochraceus*. *J. Morphol.* **170,** 235–246.

Crawford, B. J., and Abed, M. (1986). Ultrastructural aspects of the surface coatings of eggs and larvae of the starfish, *Pisaster ochraceus*, revealed by alcian blue. *J. Morphol.* **187,** 23–37.

Crawford, B. J., and Campbell, S. S. (1993). The microvilli and hyaline layer of embryonic asteroid epithelial collar cells: A sensory structure to determine the location of the sensory cilia? *Anat. Rec.* **236,** 697–709.

Crawford, B. J., and Chia, F. S. (1978). Coelomic pouch formation in the starfish *Pisaster ochraceus* (Echinodermata: Asteroidia). *J. Morphol.* **157,** 99–120.

Crawford, B. J., Campbell, S. S., and Reimer, C. L. (1977). Ultrastructure and synthesis of the extracellular matrix of *Pisaster ochraceus* embroys preserved bt freeze substitution. *J. Morphol.* **232,** 133–153.

Crawford, B. J., Reimer, C. L., and Pang, T. (2000). Localization and partial characterization of a molecule found in the plasma membrane of starfish and sea urchin embryos using a novel monoclonal antibody. *Biochem. Cell Biol.* **78,** 1–10.

Dunlap, H. L. (1966). "Oogenesis in Ctenophora." Ph.D. Thesis. University of Washington, Seattle, WA.

Hayat, M. A. (1978). "Introduction of Biological Scanning Electron Microscopy," p. 323. University Park Press, Baltimore, MD.

Hayat, M. A. (1981). "Fixation for Electron Microscopy," p. 501. Academic Press, New York, NY.

Hayat, M. A. (1989). "Principles and Techniques of Electron Microscopy: Biological Applications," 3rd ed., p. 469. CRC Press, Boca Raton, FL.

Hayat, M. A. (2000). "Principles and Techniques of Electron Microscopy. Biological Applications," p. 543. Cambridge University Press, New York, NY.

Horisberger, M. (1981). Colloidal gold: A cytochemical marker for light and fluorescent microscopy and for transmission and scanning electron microscopy. *Scan Electron Microsc.* 9–31.

Luft, J. H. (1961). Improvement of epoxy resin embedding methods. *J. Biochem. Biophys. Cytol.* **9,** 409–414.

Pang, T., Crawford, B. J., and Campbell, S. S. (2002). Ultrastructural aspects of the development of the hyaline layer and extracellular matrix lining the intestinal tract in embryos and larvae of the starfish, *Pisaster ochraceus*, preserved by freeze substitution. *J. Morphol.* **251,** 169–181.

Pang, T., Crawford, B. J., and Mahgsoodi, B. (2003). Synthesis and secretion of molecules exhibiting the HL-1 epitope during development of the hyaline layer of the asteroid, *Pisaster ochraceus*. *J. Morphol.* **255,** 58–68.

Reimer, C. L. (1994). "Characterization of Starfish Yolk and Cortical Granule Proteins, and of a Novel Extracellular Matrix Proteoglycan Implicated in Digestive Tract Morphogenesis." Ph.D. Thesis. University of British Columbia, Vancouver, BC, Canada.

Reimer, C. L., and Crawford, B. J. (1995). Identification and partial characterization of yolk and cortical granule proteins in eggs and embryos of the starfish, *Pisaster ochraceus*. *Dev. Biol.* **167,** 439–457.

Robards, A. W., and Sleytr, U. B. (1985). *In* "Low Temperature Methods in Biological Electron Microscopy," (A. Glauert, ed.), p. 551. Elsevier, NY.

Slot, J. W., and Geuze, H. J. (1985). A new method of preparing gold probes for multiple labelling cytochemistry. *Eur. J. Cell. Biol.* **38,** 87–93.

Tamboline, C. R., and Burke, R. D. (1989). Ontogeny and characterization of mesenchyme antigens of the sea urchin embryo. *Dev. Biol.* **136,** 75–86.

Volker, W., Frick, B., and Robenek, H. (1985). A simple device for low temperature polymerization of Lowicryl K4M resin. *J. Microsc.* **138,** 91.

Wood, R. L., and Luft, J. H. (1985). The influence of the buffer system on fixation with osmium tetraxide. *J. Ultrastruct. Res.* **12,** 22–45.

CHAPTER 18

Calcium Imaging

Michael Whitaker

School of Cell & Molecular Biosciences
Faculty of Medical Sciences
University of Newcastle upon Tyne
Framlington Place, NE2 4HH
United Kingdom

I. Introduction

Calcium ions signal developmental transitions. The purpose of calcium imaging in embryos is to identify calcium signals during development as a means to understanding how they are generated and what they do. I will first provide a

METHODS IN CELL BIOLOGY, VOL. 74

0091-679X/04 $35.00

précis of work in the field to provide a justification and context for calcium imaging in embryos and to give a sense of what the method has so far delivered. I will then provide broad details of the methods used to obtain these data. I will end with a brief section on likely developments in methods over the next few years.

II. Calcium Signals in Embryos

A. Aequorin

Perhaps the most striking developmental transition is that of fertilization, during which the cell cycle resumes and the embryo sets off on its developmental journey. Calcium signals define fertilization; without them, development does not begin (Stricker, 1999; Whitaker, 1993; Whitaker and Steinhardt, 1982). The first measurement of the fertilization calcium signal was an imaging experiment in large (1 mm) diameter fish eggs (Berridge *et al.*, 2000; Gilkey *et al.*, 1978). Calcium-dependent luminescence from the jellyfish protein aequorin was measured using an image intensifier. The calcium signal took the form of a wave propagating from the point of sperm entry. At that time, aequorin was the only show in town for detection of intracellular calcium. Only later did intracellular calcium electrodes became available (Busa and Nuccitelli, 1985). Aequorin was also used to demonstrate fertilization calcium waves in sea urchin and starfish (Eisen and Reynolds, 1984; Eisen *et al.*, 1984; Swann and Whitaker, 1986), ascidian (Speksnijder, 1992; Speksnijder *et al.*, 1989, 1990a,b), annelid (Eckberg and Miller, 1995), and mammalian (Miyazaki, 1988) eggs or oocytes. These later experiments showed that fertilization calcium signals could also take the form of repetitive oscillations. It is a hallmark of calcium signals that they vary in both space and time (Berridge *et al.*, 2000), requiring the use of imaging methods to capture their spatiotemporal complexity.

B. Designer Dyes

Aequorin is a light-emitting protein isolated from a jellyfish. It is sensitive to calcium at the concentrations found in the cytoplasm because calcium is the intracellular signal that the jellyfish uses to control the amount of light it emits. These concentrations are of the order of 0.1 to 1 μM. The rational design of fluorescence indicators based on calcium chelators with affinity constants of this order (Grynkiewicz *et al.*, 1985; Tsien, 1986) rapidly outshone aequorin. These indicators had a number of advantages: they were easier to detect, emitting many more photons than aequorin at ambient calcium concentrations, at levels of excitation that were not phototoxic to living cells; they were small, chemically stable molecules relative to aequorin, a medium-sized protein; they were true indicators in a sense that aequorin was not: the light-emitting prosthetic group of aequorin (coelenterazine) is consumed as aequorin binds to calcium and emits light. This last property is perhaps the most serious drawback in the use of native aequorin, as exposure of aequorin to even relatively low levels of calcium as it is extracted, purified, stored, and then introduced into cells consumes the protein and renders it inactive, so requiring unusual care in its handling.

The designer indicators, fura2, indo1, calcium green, and their colleagues, swept the imaging field, offering for the first time a robust and sensitive method for calcium measurement. They were sufficiently convenient to lend themselves to routine use to study the causes and consequences of calcium signals at fertilization, during meiosis and in the early embryonic cell cycle (Brownlee and Dale, 1990; Carroll and Swann, 1992; Crossley *et al.*, 1988; Deguchi and Osanai, 1994; Fissore *et al.*, 1992; Gillot and Whitaker, 1994; Gould *et al.*, 2001; Hafner *et al.*, 1988; Jones *et al.*, 1998; Lawrence *et al.*, 1997; McDougall *et al.*, 1995, 2000; Mohri and Hamaguchi, 1991; Muto *et al.*, 1996; Poenie *et al.*, 1986; Stricker, 1995, 1996; Stricker *et al.*, 1994; Sun *et al.*, 1992; Vincent *et al.*, 1992). At first, calcium imaging of the designer dyes used what has come to be known as wide-field fluorescence imaging, that is, image capture using sensitive video or charged coupled device (CCD) cameras to capture images from epifluorescence microscopes (Brooker *et al.*, 1990; Cheek *et al.*, 1993; Hafner *et al.*, 1988). The invention of the laser scanning confocal microscope made it the method of choice for calcium imaging in embryos, the reason being that early embryos are composed of large cells. Though not perhaps intuitively obvious, the thickness of a cell limits resolution in the plane of focus as well as compromising focal depth resolution (Silver *et al.*, 1992), a defect that confocal microscopy cures (Takahashi *et al.*, 1999).

C. Ratio Imaging

One of the first designer dyes, fura2, was a paragon in that it was designed as a ratiometric dye (Grynkiewicz *et al.*, 1985); it underwent a shift in excitation wavelength as it bound calcium. A ratio of the emission intensity at two sequentially presented excitation wavelengths provided a signal that was monotonically related to the calcium concentration but independent of the concentration of the dye itself. This is essential for calcium imaging experiments where it is paramount to distinguish true differences in calcium distribution from difference in dye distribution that may arise from variations in dye concentration in different parts of the cell. Neglecting to differentiate dye concentration distributions from calcium signals by ratiometric imaging led, for example, to the erroneous idea that concentrations of calcium in the nucleus at fertilization were greater than those in the cytoplasm (Gillot and Whitaker, 1994; Stricker *et al.*, 1992). Fura2 does not lend itself readily to confocal calcium imaging for two reasons: its excitation spectrum lies in the UV, while UV laser confocal imaging is compromised by the embryo's sensitivity to UV irradiation; it requires dual wavelength excitation in the UV, which is very difficult to achieve with the laser lines available for laser confocal imaging. It is nonetheless possible to carry out ratiometric confocal calcium imaging using a single dye: a related ratio dye, Indo1, has the useful property of shifting emission wavelength as calcium binds, making it more suitable for use in laser scanning confocal microscopes (Fontanilla and Nuccitelli, 1998; Jones *et al.*, 1998). Its disadvantage is that it requires UV excitation and undergoes substantial photobleaching, its signal fading more rapidly on excitation than that of the other designer dyes.

In the absence of a suitable ratiometric calcium dye for confocal calcium imaging, the pragmatic approach has been to use a calcium-sensitive dye in combination with a second fluorescent dye with distinct spectral characteristics that is insensitive to calcium. To ensure that both dyes share as closely as possible the same spatial distribution within cells, both are conjugated to dextrans. This approach has been used to image calcium signals at fertilization (Gillot and Whitaker, 1994), during mitosis (Groigno and Whitaker, 1998; Wilding *et al.*, 1996), cleavage (Muto *et al.*, 1996), and in early development (Fig. 1). Its disadvantages are the need to introduce the calcium indicators by microinjection (though this is not too difficult in embryos, whose cells are large) and the fact that even large molecules, such as dextrans, are taken up out of the cytoplasm by unknown mechanisms, so ceasing to report cytoplasmic calcium concentrations (Wilding *et al.*, 1996), instead misleadingly reporting calcium concentrations inside organelles.

D. Aequorin Makes a Comeback

One major advantage of aequorin as a calcium indicator is its highly nonlinear photon emission as calcium increases (Caswell, 1979). Small spatial differences in calcium distribution in embryos produce larger differences in the spatial pattern of photon emission. Transgenic expression of recombinant aequorin overcomes the handling problems and eliminates the need for microinjection. The recombinant protein, of course, lacks the prosthetic coelenterazine, but this small molecule can be isolated separately and is cell permeant. The best available images of calcium signals later in development have been obtained using aequorin (Creton *et al.*, 1998; Jaffe, 1995; Leung *et al.*, 1998; Webb *et al.*, 1997). Nonetheless, aequorin still has its disadvantages. Its photon emission is very low, requiring very sensitive detection systems and rigorous attention to blackout in an experimental system; its spatial resolution is limited: there is no equivalent of confocal imaging for luminescence, so its spatial resolution in thick specimens, such as embryos, is as poor as wide-field fluorescence microscopy.

E. FRET-Based Calcium Indicators

There is the promise of recombinant calcium indicators that can be expressed as transgenes so removing the need for microinjection, that do not require the addition of cofactors such as coelenterazine and that can be imaged in confocal microscopes, so offering the highest spatial resolution light microscopy can achieve. These indicators are the equivalent of fura2 or indo1—designer dyes, but protein based.

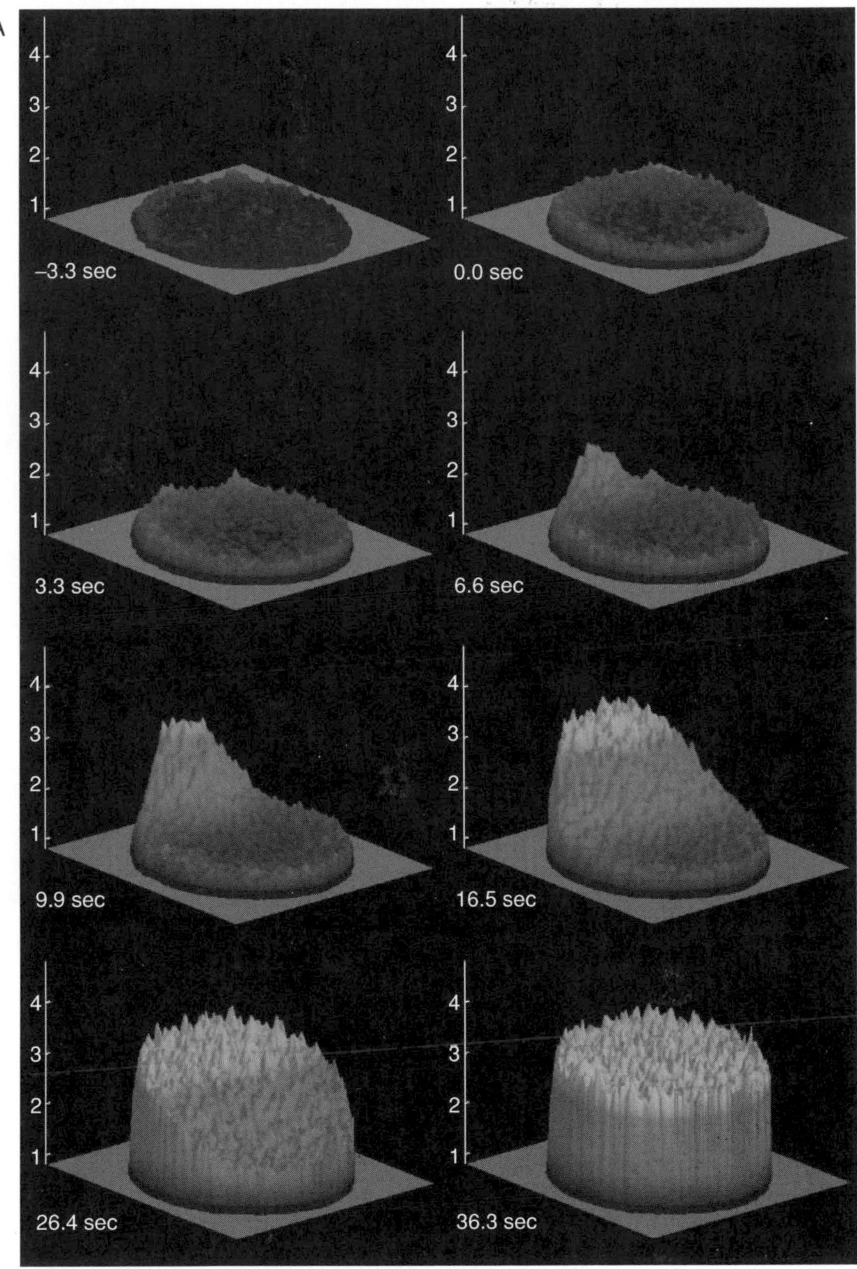

Fig. 1 (*Continued*)

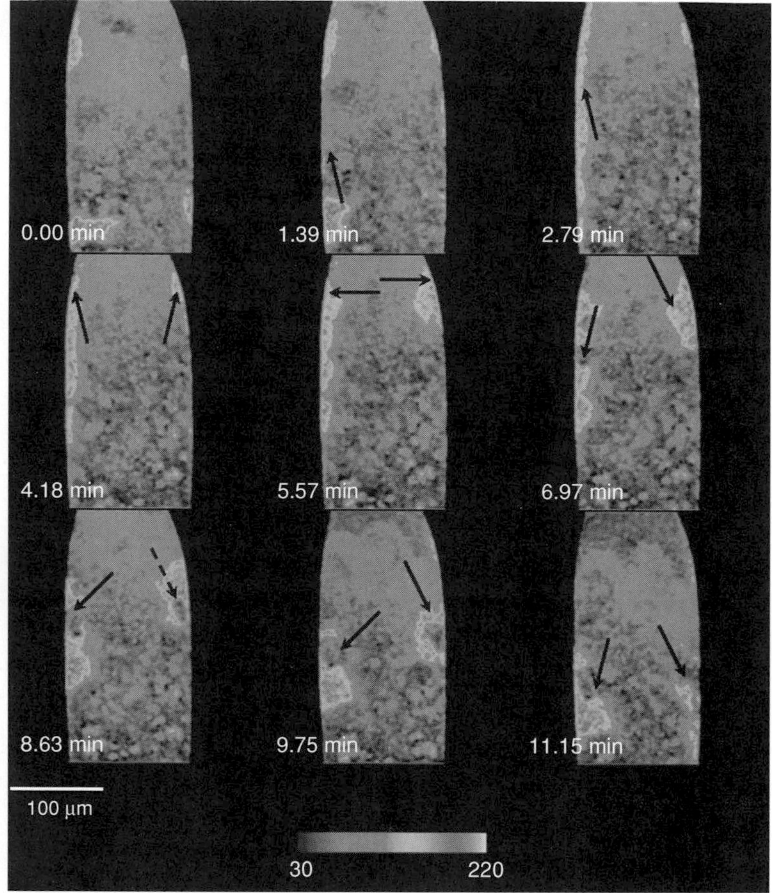

Fig. 1 Calcium imaging using dextran-linked fluorescent calcium indicators. (A) The fertilization calcium wave in a sea urchin egg (L. pictus) imaged using calcium green dextran introduced by pressure microinjection (10 μM in the egg). Pseudoratio images are shown. Images were captured using a confocal microscope every 3.3 s. The time of fertilization ($t = 0$) is known from the rapid (ms) calcium influx that occurs when sperm and egg fuse; this can be seen as a small peripheral calcium increase. The pseudoratio series was obtained by dividing each image pixel by pixel by a resting image obtained at $t = -6.6$. The topographical representation of the data shows the ratio on the z-axis in a single equatorial confocal section. The first indication of the fertilization calcium wave initiation can be seen in the image at 6.6 s. Thereafter, the wave spreads across the egg. Martin Wilding's data.

EXPERIMENTAL METHODS

Unfertilized eggs were attached to a coverslip using 100 μg/ml poly-L-lysine and pressure-injected with a buffer containing 100 μM EGTA and 10 mM calcium green dextran (10,000 mwt). A 0.1% injection was made to give a final dye concentration of 10 μM. The 488 nm laser line of an Ar-Kr laser was used to excite the dye. A 520 nm dichroic filter of 20 nm bandwidth was used for the emitted light.

(B) Traveling calcium waves in a gastrulating *Drosophila* embryo (Bownes stage 9). In circumstances in which cells or their contents move during an experiment, as is the case here, pseudoratio imaging is

The optical phenomenon that the designers exploited is FRET (Forster resonance energy transfer[1]). This phenomenon can be observed between two fluorophores with overlapping emission and excitation spectra if they are in close proximity. Exciting the bluer partner can result in nonradiative transfer of excitation energy to the redder partner through dipole–dipole coupling: FRET (Truong *et al.*, 2001). The important point is that the extent of the coupling depends very heavily on the distance between the two fluorophores in space (it follows a sixth power relationship). Tsien and Miyawaki designed a recombinant protein based on green fluorescent protein (GFP) recombinant technology (Miyawaki *et al.*, 1997). It is a coincidence, or perhaps synchronicity, that the green fluorescent protein was isolated from the very jellyfish that gave us aequorin, the first calcium indicator used in imaging. The indicator is a chimeric protein that uses calmodulin as a calcium sensor. Linked to the N-terminal of calmodulin is a GFP variant, CFP, with a blue excitation spectrum and green emission.

misleading or impossible. An alternative, as here, is to pair calcium green dextran with rhodamine dextran, a dye indifferent to calcium. The assumption is that the dyes attached to the large dextran molecule will co-distribute in the cell, a supposition largely borne out by experiment. The successive images show calcium waves traveling intercellularly in the embryo. They first progress toward the anterior, then reverse and travel posteriorly. Their function is unknown. Jaime-Ann Tweedie's data. (See Color Insert.)

EXPERIMENTAL METHODS

Drawn borosilicate glass micropipettes (Clarke Electromedical, GC150F-10) were loaded with injection solution and advanced towards immobilized *Drosophila* embryos using an Eppendorf microinjection system. All fluorescent probes for microinjection were dissolved in injection solution (Ashburner, 1989), except Xestospongin C which was dissolved in DMSO for microinjection. The embryos were injected using gas pressure (pneumatic picopump, WPI, Herts. UK). Cytoplasmic concentrations were calibrated by first measuring the size of droplet injected into the oil before injection into the embryo. Embryos are approximately 470 by 160 μm but can vary in length and diameter considerably. The approximate volume of an embryo is 6.5 nl, calculated by considering the volume of an ellipsoid of the above dimensions. The volume of liquid injected into the embryo was estimated by measuring the diameter of a droplet injected under mineral oil. This was 28 μm, giving an injected volume of 12pl (i.e. approximately 1/500th embryo volume).

A Leica inverted confocal microscope (model DMIRBE) was used. The light source was an argon-krypton laser with two excitation beams available at 488 nm and 568 nm. Calcium measurements were performed using two fluorescent dyes, one calcium sensitive: Calcium green dextran (10 kDa: CaGr) and one calcium-insensitive: tetramethyl rhodamine dextran (10 kDa: TMR). CaGr was excited at 488 nm and TMR at 568 nm with a dichroic mirror at 580 nm. Emission filters were a FITC bandpass (530 +/− 30 nm) and a 590 nm (longpass). Ratio images were performed for each image pair. All image processing was on a Silicon graphics computer using IDL software (Research Systems International (UK) Ltd., Berks.) and the images displayed in pseudo-color using a rainbow look-up-table. All experiments were carried out at 18 °C.

[1]FRET is also known as Fluorescence Resonance Energy Trasfer, but this is a misnomer, as the energy transfer itself is not carried by photons, but by nonradiative dipole–dipole coupling.

Linked to the C-terminal is a short peptide sequence from myosin light chain kinase (MLCK) to which is, in turn, attached YFP (green excitation, yellow emission). When calcium binds to calmodulin, its affinity for the MLCK target peptide increases, folding the chimera and bringing CFP and YFP together, and so increasing FRET (Miyawaki *et al.*, 1997). A number of variants of the prototype have been made (Miyawaki, 2003; Miyawaki *et al.*, 1999; Nagai *et al.*, 2001; Shimozono *et al.*, 2002; Truong *et al.*, 2001); in some cases, they have been used successfully when expressed in living cells by transfection (Ikeda *et al.*, 2003; Miyawaki, 2003; Robert *et al.*, 2001). It should be possible to generate transgenic embryos in which these proteins are expressed. A start has been made in *Drosophila* (Fig. 2).

III. Getting Calcium Sensors into Embryos

A. Permeation

Calcium chelator-based calcium indicators contain four carboxyl groups, making them highly negatively charged at physiological pH. This means that they do not readily cross the cell membrane. Tsien developed a method for masking the charges on the carboxyl groups by esterification with acetoxymethanol (Tsien, 1984). The AM esters are uncharged, and so will cross the plasma membrane. Once inside the cell, the ester bond is cleaved by cytoplasmic esterases, regenerating the carboxyl groups essential for high affinity calcium binding. There are disadvantages to this approach that need to be borne in mind: the esters find their way into cellular organelles, such as the endoplasmic reticulum (ER) and lysosomes, where they are cleaved and then report the calcium concentration in these compartments, not in the cytoplasm. This fact can occasionally be used to advantage to measure, say, ER calcium concentrations (Terasaki and Sardet, 1991) but, in general, leads to an ambiguous and misleading indicator signal, since calcium concentrations in these organelles is high relative to cytoplasm and makes a major contribution to the overall indicator signal.

Moreover, early embryos have low surface-to-volume ratios relative to somatic cells: the cytoplasmic concentration of AM esters rises relatively slowly in large cells as the rate of rise is directly proportional to the surface-to-volume ratio. Despite this disadvantage, adequate loading is possible in mammalian oocytes and embryos (Kline and Kline, 1992; Swann, 1990), but not in sea urchin, where there is the added problem of the low solubility of the AM esters in seawater, presumably because of its high ionic strength. It is possible to load sea urchin embryos using the AM method by lowering the ionic strength of seawater with osmotic substituents such as sucrose (Whitaker, unpublished), but the long-term viability of embryos treated thus is poor.

Because of the artifacts inherent in the AM loading method and the difficulties with loading, this approach is not recommended.

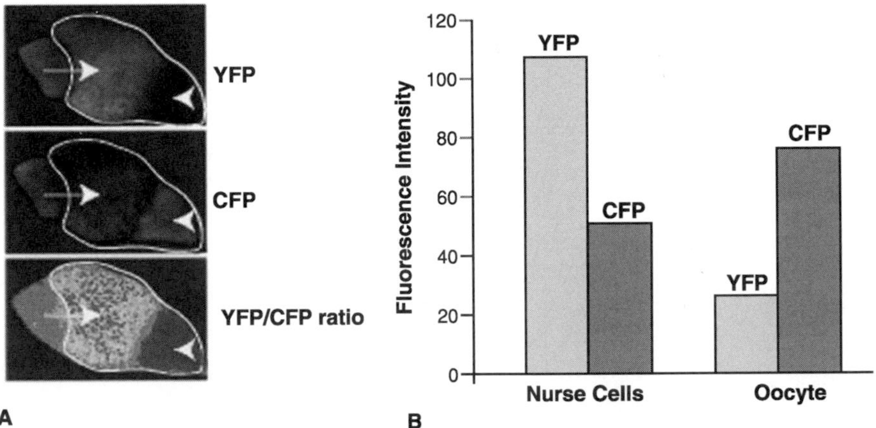

Fig. 2 Expression of the transgenic chameleon indicator in *Drosophila* ovary. Transgenic flies expressing the chameleon indicator were constructed by expressing the chameleon transgene under the control of promoter sequences that permit germ line expression when crossed with a suitable Gal4 enhancer trap line. A single ovariole is shown (A). The nurse cells (arrows) show a strong YFP signal, while the oocyte shows a predominant CFP signal. Relative intensities are shown in B. The YFP/CFP ratio is 2.13 in the nurse cells and 0.34 in the ovary, indication that calcium concentrations are substantially higher in the nurse cells than in the oocyte. Jun-Yong Huang's data. (See Color Insert.)

EXPERIMENTAL METHODS

Virgin female transgenic flies expressing the Venus cameleon indicator under the control of UASp promoter sequence that permits germ line expression were crossed with males of a Gal4 VP16 enhancer trap line to establish a transgenic line that carried both UASp-Venus Cameleon and Gal4 VP16 constructs. The ovaries from 3–4 days old well fed female offspring from this established line were dissected in *Drosophila* Ringer's (Ashburner, 1989). 2–3 pairs of ovaries are quickly transferred to a coverslip (BDH cover glass 22 × 64 mm, thickness No.1) on a glue stripe. The glue was made by dissolving Scotch tape in heptane. The individual ovarioles were carefully spreaded using a soft paintbrush and the remaining *Drosophila* Ringer's was quickly drained away using a piece of tissue. The ovaries then were covered immediately with Voltalef 10S oil to prevent desication of the ovaries.

A Leica TCS SP2 inverted confocal system was used. CFP fluorescence intensity was detected by using a UV laser light source with excitation at 405 nm and emission window range from 465–506 nm. YFP fluorescence intensity was detected by using an argon-krypton laser with excitation at 514 nm and emission window range from 528–600 nm. Ratio images were made for each image pair. All the image processing was performed using MetaMorph software from Universal Imaging Corp. All experiments were carried out at 20 °C.

B. Microinjection

1. Delivery

The classical method for introducing nonpermeant low molecular weight organic molecules and proteins into cells is by microinjection. One advantage of oocytes and embryos is their relatively large size. The target is thus larger and the damage smaller relative to somatic cells. To minimize the latter, the micropipette tip diameter should be as small as possible, taking into account its cargo. Lower molecular weight molecules permit a smaller tip diameter (perhaps 0.5 μm) than do proteins, which tend to aggregate and block the micropipette. One way to minimize blockage of protein solutions is to centrifuge protein solutions in the microfuge at around 10 000*g* immediately before microinjection.

There are two approaches to microinjection. The first is a method developed by Kiehart (Kiehart, 1982), in which an oil-filled micropipette (Wesson soya bean oil is favored) is dipped into the solution to be microinjected under oil and the solution drawn into the pipette tip as a bolus using relatively low hydraulic pressures applied by a micrometer-controlled hydraulic cylinder. The procedure is observed using a dissecting or compound microscope: once a certain volume has been drawn into the tip (judged by the linear displacement of the oil in the pipette shank), the tip is moved back into the oil and a small quantity of oil drawn up to seal the bolus of injection solution inside the pipette. Once the pipette is presented to the oocyte or embryo to be microinjected, this cap of oil protects the injection solution from contamination by the extracellular medium. The pipette is then driven through the plasma membrane and, once inside, the micrometer is advanced to deliver the injection solution to the cytoplasm. The hallmark characteristic of this method is an oil droplet in the cytoplasm, since the oil sealing the micropipette must be expelled into the cytoplasm in order to deliver the pipette's cargo.

The advantages of this method are precision and the complete protection of the cargo from contamination by extracellular fluid. Relative to the pulse pressure injection method, the procedure is time-consuming.

Pulse pressure injection applies large hydraulic pressures (2000 kPa, 20 bar) for very short times (10–100 ms) to force defined volumes from a micropipette tip. Pipettes are backfilled by applying a small volume of cargo solution into the pipette shank using a Hamilton syringe or a Gilson yellow tip drawn out finely by melting and stretching. For backfilling to deliver the solution to the very tip of the pipette, so-called filament tubing must be used to draw pipettes: this class of glass capillary has a thin glass filament fused to its lumen, which draws the injectate into the tip through interfacial surface tension; without the filament, backfilling will not deliver solution to the tip. Pressure pulses are applied to the pipette using gas pressure delivered through hydraulic valves controlled electronically (Eppendorf and World Precision Instruments are brand names). Contamination of the cargo through contact with the extracellular solution is minimized by applying pressure pulses continuously to the pipette at 1 to 5 s intervals. Once the pipette is introduced into the egg or embryo, the volume leaving the tip can be estimated by measuring in the microscope the volume of cytoplasm displaced by the injectate during a single pulse, using an eyepiece graticule to measure the diameter of the spherical displacement and the formula for spherical volume ($4\pi d^3/24$, where d is the diameter of the displacement).

The advantage of pulse pressure microinjection is speed, particularly when combined with automatic or semiautomatic cell penetration using digitally controlled micropipette positioners sold by Eppendorf and others. The disadvantage is a relative lack of precision when compared to the low-pressure method.

Solutions can also be introduced into eggs and embryos using so-called patch pipettes, which are relatively blunt, polished pipettes that form a tight seal with plasma membranes (Dale *et al.*, 1996). Once the seal is made, the plasma membrane under the tip of the pipette can be ruptured using negative pipette pressure, thus connecting the pipette lumen with the cytoplasm. The advantage of the approach is that injection can be combined with recording and manipulation of membrane potential. The disadvantages are that the method is not quantitative, except at equilibrium, when cell contents have diffused into the pipette and pipette contents into the cell; and that most embryos have extracellular coats that preclude the formation of an adequate seal between pipette and plasma membrane.

2. Immobilizing Embryos

It is futile to chase an egg or embryo across a petri dish or slide with a microinjection pipette. The cell must be immobilized to be penetrated. Echinoderm eggs can be immobilized by coating the slide or petri dish with a polycation such as poly-L-lysine (Whitaker and Irvine, 1984) before adding the eggs to the substrate. Embryo coats can make adhesion more difficult. Egg jelly coat can best be removed by passing the eggs through a nylon (Nitex) mesh. Echinoderm embryos elevate a fertilization envelope above the plasma membrane after fertilization, separated from the plasma membrane by the colloid osmotic pressure of the perivitelline space. Fertilized eggs can be immobilized by shrinking down the perivitelline space by adding dextran to the bathing solution, provided that the hardening of the fertilization envelope itself is prevented by adding a peroxide scavenger such as p-amino-benzoic acid (Wilding *et al.*, 1996). Alternatively, the entire fertilization envelope can be stripped off by passing PABA-treated eggs through a fine nylon mesh (Becchetti and Whitaker, 1997).

Ascidian embryo coats are poorly adhesive and do not stick to polycation surfaces. The preferred method is to create a wedge-shaped chamber into which the embryo is introduced, with the micropipette presented to the embryo so as to force it further into the wedge (McDougall and Sardet, 1995).

Mouse, human, and other mammalian embryos are held in a blunt, polished pipette as they are injected with a sharp microinjection pipette introduced diametrically opposite the holding pipette (Marangos *et al.*, 2003).

3. Penetrating Embryos

The egg or embryo plasma membrane and its associated extracellular coat and intracellular cytoskeleton are surprisingly elastic: it is quite easy to advance a micropipette deep within an egg without gaining entry to the cytoplasm. The micropipette must be advanced very rapidly to overcome this elasticity and penetrate the plasma membrane. Piezo advancers (for example, those manufactured by Marzhauser Wetzlar) are available to achieve this. In mouse oocytes, penetration of the plasma membrane can be achieved using a short oscillating electrical pulse (Marangos *et al.*, 2003).

IV. Measuring Emitted Light

A. Optics

All fluorescence and luminescence imaging is photon-limited. It is therefore important to collect as many emitted photons as possible from the specimen. High numerical aperture objectives should be used and it is sometimes preferable to choose an achromat over better-corrected objectives, as it is likely to have better transmission characteristics. Achromats themselves are corrected for aberration at three defined wavelengths. In regions of the visible spectrum outside and even between these wavelengths, correction for chromatic aberration may be poor. For critical multiple wavelength imaging, the chromatic correction data for the objective should be obtained from the manufacturer.

B. Luminescence from Aequorin

Aequorin-loaded eggs and embryos typically emit only tens of photons per second at resting calcium concentrations. When imaging, the average irradiation of each pixel on the detector is thus small fractions of a photon per second. The appropriate detectors at these light levels are image-intensified charge coupled device (CCD) cameras or imaging photon detectors (IPD). There is little to choose in sensitivity between these two approaches. CCD cameras accumulate equivalent photons in an array of charge wells on the detector. Periodically, the charge in each well is read out and transformed into a digital data array (image). Reading out the charge accumulated on the detector introduces readout noise into the image. Frequent readout of photon-sparse images thus degrades the signal-to-noise ratio relative to infrequent readout. IPDs are intensifiers whose readout is the time of detection of a photon and its coordinates (pixel) in the detector. From this information, an image is built up cumulatively using appropriate software. One very large advantage of this approach is that the raw data are a stream and can be subsequently reprocessed to display images accumulated over any desired time period within the experimental sequence. Suppose that a relatively brief calcium signal occurs within a developing embryo, but that its precise time of occurrence cannot be predicted with an accuracy comparable to its duration. Using a CCD-based approach, the sampling interval would need to be comparable to the likely duration of the calcium transient, with the attendant disadvantages of degraded signal-to-noise ratio due to readout noise and the need to acquire many images, only one or two of which will contain the calcium signal. With an IPD, the calcium signal can be detected in the datastream by plotting photon counts with time. Images can then be synthesized around, before, or after the signal using appropriate accumulation times. The relative disadvantage of the IPD is its maximum photon detection rate, which is around 100 000 s^{-1}. This is nonetheless greater than the rate at which most aequorin-loaded specimens emit photons, but makes the IPD ill-suited for fluorescence applications, where rates of photon emission are orders of magnitude greater. Interestingly, since the photon detection rate is the saturation limit for the detector, the dynamic range pixel by pixel is very large. If all the 100 000 photons were emitted from an area of the specimen corresponding to a single pixel in the image, they would be faithfully attributed to that pixel, giving it a dynamic range of 10^5. This is an advantage that plays to one of the strengths of aequorin mentioned in Section II.D: its nonlinear response to calcium means that relatively small changes in the spatial distribution of calcium lead to very marked disparities in signal from one pixel to the next, so pixel dynamic range is important.

C. Fluorescence from Designer Dyes and Proteins

1. Wide-Field Microscopy

A straightforward adaptation of a compound epifluorescence microscope made by mounting a CCD camera on an output port is sufficient for wide-field imaging. The sensitivity of the CCD camera required will depend upon the specimen and dye loading. The most sensitive cameras are cooled; these are recommended. A further increase in sensitivity can be obtained by using a back-illuminated CCD. If a versatile setup is needed or if multiple wavelength imaging is being undertaken, then digitally driven filter wheels can be added and operated under the control of the acquisition software. The illumination source may be either a mercury vapor or a xenon lamp. The mercury lamp has the advantage of well-defined emission lines that may fall at an excitation frequency of interest, but a disadvantage is that the mercury arc is inherently unstable and variations in intensity are common, with some lamps more variable than others. Xenon lamps provide a broadband, stable excitation source.

A disadvantage of wide-field microscopy is the loss of resolution in all dimensions due to the unavoidable collection of out-of-focus emitted light from the specimen (see Chapters 16 and 19). Resolution in thick specimens, such as eggs and embryos, is poor. Thinner specimens a few microns thick are very adequately imaged, however.

2. Laser Scanning Confocal Microscopy

The principle of confocal microscopy is to overcome the collection of out-of-focus light by use of a confocal aperture to block it. The simplest confocal microscope can collect light from a single point in the specimen; an image can be formed by moving the specimen on a scanning stage relative to the illuminated point (Gorelik *et al.*, 2002; Shevchuk *et al.*, 2001). This is slow, but is useful in combination with other slow scanning techniques, such as scanning ion conductance microscopy (Gorelik *et al.*, 2002; Korchev *et al.*, 1997). Rapid acquisition in confocal mode can be obtained by sequential raster scanning of the specimen using laser light focused to point illuminations. Only lasers provide a bright enough light source for this method to be readily realized.

Laser scanning confocal microscopes (CLSM) are now a mature technology. Competition between manufacturers is strong, driving rapid changes in specification. A recent comparison of different CLSM in our lab found no marked differences in sensitivity, but it is worth pointing out some key differences in implementation and their consequences.

a. Scanning Speed

The classical implementation of raster scanning is to use a galvanometrically controlled mirror. The physical properties of the mirror (size and weight) set a limit to its maximum frequency of oscillation. A carefully designed resonance system has been used to deliver much higher scan speeds, but the design is not versatile enough to deliver slow scan speeds as well, making it very application-specific. Acousto-optically modulated deflection has also been implemented. There have been problems of reliability in the past, but these problems may have been solved. Scanning speed can be traded off against image resolution. Fastest acquisition is achieved in line scan mode, where the specimen is scanned in one axis only (thus offering only single pixel resolution in the other axis (Callamaras and Parker, 1999; Cheng *et al.*, 1993; Gillot and Whitaker, 1994)). The smaller the total number of pixels in the image, the faster a specimen may be scanned. Scanning speed must also necessarily be traded against sensitivity: in a given specimen, emitted photons are acquired at a fixed rate and if scanning speed is increased, the number of photons per image decreases. In many applications, scan speed is determined not by the speed of the CLSM, but by the brightness of the specimen.

b. Separation of Excitation and Emission Wavelength

Most straightforward epifluorescence applications use a dichroic mirror held at 45°, whose property is to reflect light below a specified wavelength and transmit light at longer wavelengths. This is straightforward and convenient for single excitation wavelengths. Problems arise when multiple excitation wavelengths are required, as custom dichroic mirrors are needed with multiple bands of reflection and transparency with wavelength. Finding a suitable mirror to suit the absorption (excitation) and emission spectra of multiple dyes can be difficult or impossible. Two solutions to this difficulty have emerged. One uses a semi-silvered mirror in place of the dichroic, providing broad spectrum reflection and transmission, with the mirror designed to transmit a greater proportion of the light than it reflects. Disadvantages are the loss of a large proportion of the exciting light and a smaller proportion of the emitted light. The former is not a serious problem, given the increasing availability of high power lasers, while the latter is not optimal, but a price worth paying for the versatility. The second implementation uses acousto-optic modulators in tandem to divert the light path of the exciting beam away from the detector. This is a sophisticated solution that allows excitation wavelengths to be closely apposed to the lower wavelength boundary of the emission detector bandwidth, but it may suffer from excitation intensity variation because of its exquisite sensitivity to laser polarization.

A useful way of handling multiple excitation wavelengths is to use Pockel cells, acousto- or electro-optic modulators to attenuate (indeed, switch on and off) individual laser beams. Sequential line or frame scans can be made with different laser lines, providing images at each excitation wavelength.

c. Analysis of Emission Spectra

Again, the simplest implementation uses interference filters as dichroic mirrors and bandpass and long pass filters to segregate the emitted light to separate detectors. The disadvantage is having a supply of filters sufficient to suit any dye spectrum. One simple but imaginative way around part of this difficulty is to use long- and shortpass filters in aligned filter wheels: the bandwidth seen by the detector is determined by the appropriate choice of long- and shortpass filter pair. This particular solution does not dispense with the need for dichroic mirrors to separate detection channels, but it does permit the sequential acquisition of emission spectra. An alternative implementation that also permits the acquisition of emission spectra uses dispersion by a prism and slits variable in position and width set in front of the detectors, allowing continuous variation in the wavelengths arriving at the detectors and dispensing with dichroic mirrors. This appears to have no substantial disadvantages, though it should be noted that the true dye excitation spectrum is not obtained, but a convolution of the spectrum with the spectral sensitivity curve of the detector. A third implementation uses a dedicated array of detectors onto which all the emitted light is dispersed. The detectors are calibrated for equivalent sensitivity, offering the advantage of a close approximation to the true emission spectrum, but by definition reducing detector sensitivity at the detectors' most sensitive spectral ranges. This latter implementation is combined with conventional filters and dichroics to allow separation of emission channels, with the disadvantages already mentioned.

3. Multiphoton Imaging

At very high light intensities, photon density is so high that a molecular absorption event can take place in which two or more photons are captured simultaneously. The photons involved in a two-photon event will, to a first approximation, have half the energy of a single-photon absorption (Takahashi *et al.*, 1999; Tauer, 2002). UV-absorbing fluorophores can thus be excited by high intensity infrared light. The main advantage of this method is that it removes the need for a confocal aperture. Only in the object plane of the microscope are intensities high enough for multiphoton excitation; moreover, multiphoton excitation follows a steep power law, so excitation is limited to a very narrow depth of focus (Centonze and White, 1998). As the object is excited only at the focal plane, all light emitted at a particular pixel as the laser scans the specimen comes from a defined focal volume. It need not therefore be brought into focus at the detector, allowing even scattered light to contribute to the image.

One main advantage of multiphoton imaging is reduced photodamage and photobleaching, as only molecules in the focal plane are subject to high intensity excitation; in general, infrared light is not strongly absorbed by eggs and embryos. Another advantage is that the longer wavelength illumination is not only absorbed less than bluer light but also scattered less, allowing much deeper (up to 0.5 mm) penetration into specimens. Disadvantages include the need for high-intensity pulsed lasers that are expensive and require great thought to safety and the fact that not all embryos are transparent to infrared light (which leads to rapid damage). It is not possible to undertake multiple wavelength multiphoton excitation very readily, as the lasers are so costly relative to lower power visible light lasers, but this disadvantage is offset by the fact that multiphoton absorption spectra are usually much broader than their single photon counterparts.

Multiphoton imaging of calcium so far have been reported in mammalian oocytes using indo1 (Jones *et al.*, 1998) and calcium waves have been imaged in zebrafish (Gilland *et al.*, 2003).

4. Nipkow Disc Confocal Microscopy

Confocal microscopy was invented before lasers were used to implement CLSM. A spinning disc with many carefully spaced holes provides a multiplicity of confocal apertures that scan the specimen. The holes are spaced to ensure that out-of-focus light due to excitation through one hole does not enter an adjacent hole. The big disadvantage of this method was that only a small fraction of the spinning disc comprised holes, so that 98% or more of the illuminating light was wasted, reflected back from the surface of the disc. Light sources were not bright enough to permit detection with high sensitivity, though stereomicroscopy is possible (Boyde, 1985; Boyde *et al.*, 1990).

A modern implementation of the Nipkow disc uses silicon wafer technology to etch lenses in a second disc apposed to and spinning with the Nipkow disc (Tanaami *et al.*, 2002). A much greater fraction of the illuminating light is transmitted through the disc which, coupled with the use of an expanded laser beam as an illumination source and CCD camera detectors, provides images at a sensitivity comparable to CLSM at single excitation and emission wavelengths (Egner *et al.*, 2002). One disadvantage is that the hole spacing in the Nipkow disc cannot cope well with thick specimens such as eggs and embryos. Out-of-focus light in these thick specimens does stray to adjacent holes. Another relative disadvantage is the need for filter wheels and dichroic mirrors for multiple wavelength measurements.

V. Fluorescence Lifetime Imaging

The aim of quantitative calcium imaging is to obtain a signal that is proportional to intracellular calcium concentration but independent of other parameters, for example, dye concentration (see Section II.C). As we have seen, this can be achieved by using dyes that are inherently ratiometric, by choosing dye partners with similar distributions within the cell, and by using FRET. In all three cases, the ratio is constructed from two measurements of fluorescence intensity. The latter two are vulnerable to artifacts caused by the differential photobleaching of one dye relative to another. Emission intensity is, perhaps confusingly, an extensive dye property: the more molecules, the more signal. Fluorescent molecules also have a characteristic intensive property, the fluorescence lifetime. Lifetime measurements are independent of the number of dye molecules, except that the signal-to-noise ratio of the lifetime measurement will improve as the number of dye molecules increases. The fluorescence lifetime of a FRET donor molecule is sensitive to the distance from the donor to the acceptor. Donor lifetime is thus a way of quantitating FRET. The method has been used to measure protein–protein interactions inside cells (Bastiaens and Squire, 1999; Elangovan *et al.*, 2002) but has not yet been used to measure calcium, though, in theory, any calcium dye that alters its intensity in response to calcium can be imaged using a change in fluorescence lifetime. The advantage of the method is that it produces a signal independent of dye concentration, even for non-ratio dyes. The disadvantages are the need for visible or mutiphoton pulsed lasers and the acquisition speed, which is currently of the order of tens of seconds.

VI. Calibration

The change in fluorescence of a calcium-sensitive dye as calcium changes is predicted by the proportion of dye molecules to which calcium is bound at any instant, relative to the proportion of unbound dye molecules. The proportion of each form is predicted by the equation for chemical equilibrium for the binding of calcium to the dye, governed and characterized by the dissociation constant for each calcium binding site. Most dyes have a single site. The calibration parameters of dyes are thus well defined. The difficulty is in calibrating them in a specific experimental situation.

For ratio dyes, the value of the ratio measured at a given calcium concentration for a given dye is apparatus-specific and depends on the transmission properties of objectives as well as on the filter sets used. Each system must thus be calibrated and each calibration checked reasonably frequently. The simplest method uses small droplets of dye solution of known calcium concentration held under oil on a microscope slide or coverslip (Crossley *et al.*, 1991). The ratio measured for each calcium concentration will provide a calibration curve. In fact, knowing the ratio value at zero calcium concentration (achieved by using a millimolar concentration of a calcium chelator, such as EGTA) and at saturating calcium concentration (achieved by using millimolar calcium) is sufficient to calibrate the signal if the dissociation constant is known (Grynkiewicz *et al.*, 1985; Poenie *et al.*, 1985). It turns out, however, that the dissociation constant inside the cell is not always identical to that measured *in vitro*, requiring a more sophisticated method described by Poenie and his co-workers (Poenie *et al.*, 1985).

Single wavelength dyes are very difficult to calibrate satisfactorily, as the signal measured is a mixture of both dye concentration and calcium concentration. For these dyes, each experiment must be calibrated separately by permeabilizing the egg or embryo to calcium (often, using a calcium ionophore) and adding high- and low-calcium solutions (Tsien *et al.*, 1982). This is cumbersome. An alternative approach is to use a ratio dye to characterize the calcium changes and to use the resting calcium concentration and peak of the calcium response as givens with which to calibrate the single wavelength signal (Gillot and Whitaker, 1993).

In some cases, knowledge of the absolute value of cytoplasmic calcium may not be necessary. It is then possible to use a ratio value as a proxy for the cytoplasmic calcium concentration. The ratio signal is immediately available from a ratio dye. Using single wavelength dyes, it is possible to construct a ratio that eliminates the dye concentration by using the fold increase over resting signal (a so-called pseudoratio (Groigno and Whitaker, 1998)). Using this method, a comparison between experiments assumes that the resting calcium concentration is relatively invariant from cell to cell. The shape of a calcium response represented as a ratio understimates the calcium increase if calcium concentrations rise very much above the value of the dye dissociation constant and overestimate the calcium increase if calcium concentration does not rise to values comparable to the dye dissociation constant.

VII. Manipulating Intracellular Calcium

When investigating the role of an embryonic calcium signal, it is essential to determine its causal significance. This can be achieved by blocking the signal to demonstrate that it is necessary and artificially inducing an appropriately sized calcium signal to show that the calcium signal is sufficient for a cellular response.

A. Chelators

1. Calcium Chelators

Calcium signals can be blocked by using calcium chelators, tetracarboxylic acids that bind calcium with high affinity. EGTA was the first calcium chelator to be used in embryos to show that the fertilization calcium transient was an essential signal for egg activation (Zucker and Steinhardt, 1978a). EGTA was more effective than EDTA, because the latter has a high affinity for magnesium, making its apparent affinity for calcium very dependent on the magnesium concentration (Martell and

Sillen, 1964; Zucker and Steinhardt, 1978a). EGTA also has its defects: its affinity for calcium is very dependent on pH in the physiological range (Kim and Padilla, 1978; Martell and Sillen, 1964). With this in mind, Tsien developed BAPTA, a tetracarboxylic acid calcium chelator that is much less sensitive to pH in the physiological range (Lew *et al.*, 1982). BAPTA then became the backbone of the fluorescent calcium dyes that I have already discussed (Tsien, 1986). BAPTA is also available as a permeant acetoxymethyl ester (Lew *et al.*, 1982).

Microinjection of both EGTA and BAPTA has been widely used to demonstrate that calcium signals control events at fertilization and in early embryogenesis (Aizawa *et al.*, 1996; Chang and Lu, 2000; Groigno and Whitaker, 1998; Hamaguchi and Hiramoto, 1981; Kawahara and Yokosawa, 1994; Kline, 1988; Mohri and Hamaguchi, 1991; Steinhardt and Alderton, 1988; Stith *et al.*, 1994; Swann *et al.*, 1992; Tosti and Dale, 1994; Twigg *et al.*, 1988; Webb *et al.*, 1997; Zucker and Steinhardt, 1978b). It has also proved possible to use the permeant BAPTA-AM proto-chelator in mammalian and ascidian embryos. (He *et al.*, 1997; Kawahara and Yokosawa, 1994; Kline and Kline, 1992; Lawrence *et al.*, 1998; McDougall *et al.*, 1995; Raz *et al.*, 1998; Winston *et al.*, 1995). Nonetheless, the difficulties that are encountered with AM loading that I discussed earlier in the context of calcium indicator dyes are compounded using BAPTA-AM because millimolar levels of chelator are necessary to block calcium signals in embryos (Groigno and Whitaker, 1998; Wilding *et al.*, 1996), in contrast to the tens of micromolar levels needed to obtain an adequate fluorescence indicator dye signal.

All the tetracarboxylate chelators bind trace metal ions such as Mn^{2+}, Co^{2+}, and Zn^{2+} (Arslan *et al.*, 1985). A third of all proteins require metal ions for their proper function (Rosenzweig, 2002), so it is often important to exclude the possibility that the effects of calcium chelators are due not to chelation of calcium itself but to chelation of trace metal ions. A simple control for this is to use TPEN, a permeant trace metal ion chelator that does not affect calcium itself (Arslan *et al.*, 1985). It has also been found that BAPTA inhibits protein synthesis (Lawrence *et al.*, 1998), though it is not clear whether this is due to trace metal ion chelation or to another chemistry.

2. $InsP_3$ Chelators

Embryonic calcium signals are often controlled by the $InsP_3$ signaling pathway (Whitaker, 2004). Another way of blocking calcium signals is thus to interfere with $InsP_3$ production. The classical inhibitor of this pathway is lithium, which interferes with regeneration of $PtdInsP_2$, leading to its depletion (Becchetti and Whitaker, 1997). However, lithium has also been reported to inhibit glycogen synthase kinase, an important enzyme in the pathway that specifies dorso–ventral axis formation (Klein and Melton, 1996); for embryos, therefore, something more specific is necessary. An antibody to the $InsP_3$ receptor has proved effective (Kume *et al.*, 1997; Nuccitelli *et al.*, 1993). It is also possible to use $InsP_3$ binding sites themselves to chelate $InsP_3$ once it is produced. We have used p130, a protein of the PLCδ family that lacks phospholipase activity but that binds $InsP_3$ tightly (Takeuchi *et al.*, 2000), to block calcium signals and cell cycle progression in early *Drosophila* embryos (unpublished). The $InsP_3$ binding domain of the $InsP_3$ receptor achieves the same effect. The recombinant $InsP_3$ binding domain is also available with a single point mutation that markedly reduces its affinity for $InsP_3$, providing a very useful control (Walker *et al.*, 2002).

B. Photolabile "Caged" Compounds

Showing that a calcium signal is sufficient to bring about a particular cellular event can be achieved by microinjection of calcium-buffered solutions or $InsP_3$ (Brind *et al.*, 2000; Hamaguchi and Hiramoto, 1981; Kline, 1988; Miyazaki, 1988; Steinhardt and Alderton, 1988; Swann and Whitaker, 1986; Twigg *et al.*, 1988). In some circumstances, however, this is technically very demanding; for example, microinjecting embryos at metaphase of mitosis offers a time window of only a few minutes once the cell has entered mitosis and there is a high probability that the mitotic spindle will be disrupted by the microinjection itself. A good way to sidestep these problems is to use photolabile compounds.

Photolabile "caged" compounds are synthesized by attaching a photolabile leaving group. This group masks the biological function of the molecule until it is irradiated with ultraviolet light, at which point it very rapidly falls off, unmasking (or uncaging) the active biological molecule. The first caged compound to be used in biology was caged ATP (Goldman *et al.*, 1982; McCray *et al.*, 1980). Using the same chemistry, nitr-5, DM-nitrophen, and NP-EGTA, calcium chelators that lose their high affinity for calcium when irradiated with UV light [i.e., caged calcium (Ellis-Davies *et al.*, 1996; Ellis-Davies Graham, 2003; Kao *et al.*, 1989; Kaplan and Ellis-Davies, 1988; Tsien and Zucker, 1986)] were developed. Caged $InsP_3$ is also available (Groigno and Whitaker, 1998; Parker and Miledi, 1989).

These caged compounds can be microinjected into embryos long before they are activated at the appropriate time by UV light from a UV flashlamp or laser (Ellis-Davies *et al.*, 1996). Uncaging with a multiphoton laser leads to very well-defined release of calcium from a small cytoplasmic volume (Brown *et al.*, 1999). The original caged calcium compounds developed for UV flashlamps and single-photon laser excitation perform poorly with mutiphoton excitation (Brown *et al.*, 1999). Two new compounds, azid-1 (Brown *et al.*, 1999) and DMPE-4 (Soeller *et al.*, 2003) with good multiphoton absorption cross-sections have been developed.

VIII. Conclusions and Perspectives

I do not foresee any major advances in technology. At present, there are several novel technologies (multiphoton imaging, FRET, lifetime imaging) whose utility in calcium imaging in eggs and embryos is yet to be proven and developed. What

I believe is necessary and feasible is advances in imaging in multiple modalities, in which calcium measurements are combined with other approaches to image cell proteins and link calcium signals with changes in protein distribution and activation state. Examples that point the way are the simultaneous imaging of calcium signals and microtubules imaged using fluorescent tubulin (Groigno and Whitaker, 1998) and imaging of the activation state of the calcium target calmodulin (Torok *et al.*, 1998). Simultaneous imaging of calcium signals and protein function (Groigno and Whitaker, 1998; Torok *et al.*, 1998; Whitaker, 2000), using multiple methods such as intensity, FRET, and lifetime imaging, may help us understand how calcium signals control cell function and developmental fate.

Acknowledgments

I thank Michael Aitchison for help in preparing the figures. Work in the lab is supported by the Wellcome Trust and the BBSRC.

References

Aizawa, H., Kawahara, H., Tanaka, K., and Yokosawa, H. (1996). Activation of the proteasome during Xenopus egg activation implies a link between proteasome activation and intracellular calcium release. *Biochem. Biophys. Res. Commun.* **218,** 224–228.

Arslan, P., Di Virgilio, F., Beltrame, M., Tsien, R. Y., and Pozzan, T. (1985). Cytosolic Ca2+ homeostasis in Ehrlich and Yoshida carcinomas. A new, membrane-permeant chelator of heavy metals reveals that these ascites tumor cell lines have normal cytosolic free Ca2+. *J. Biol. Chem.* **260,** 2719–2727.

Ashburner, M. J. (1989). "Drosophila. A Laboratory Handbook." Cold Spring Harbor Laboratory Press.

Bastiaens, P. I., and Squire, A. (1999). Fluorescence lifetime imaging microscopy: Spatial resolution of biochemical processes in the cell. *Trends Cell Biol.* **9,** 48–52.

Becchetti, A., and Whitaker, M. (1997). Lithium blocks cell cycle transitions in the first cell cycles of sea urchin embryos, an effect rescued by myo-inositol. *Development. (Cambridge, England)* **124,** 1099–1107.

Berridge, M. J., Lipp, P., and Bootman, M. D. (2000). The versatility and universality of calcium signalling. *Nat. Rev. Mol. Cell Biol.* **1,** 11–21.

Boyde, A. (1985). Stereoscopic images in confocal (tandem scanning) microscopy. *Science* **230,** 1270–1272.

Boyde, A., Jones, S. J., Taylor, M. L., Wolfe, L. A., and Watson, T. F. (1990). Fluorescence in the tandem scanning microscope. *J. Microsc.* **157,** 39–49.

Brind, S., Swann, K., and Carroll, J. (2000). Inositol 1,4,5-trisphosphate receptors are downregulated in mouse oocytes in response to sperm or adenophostin A but not to increases in intracellular Ca^{2+} or egg activation. *Devel. Biol.* **223,** 251–265.

Brooker, G., Seki, T., Croll, D., and Wahlestedt, C. (1990). Calcium wave evoked by activation of endogenous or exogenously expressed receptors in *Xenopus* oocytes. *Proc. Natl. Acad. Sci. USA* **87,** 2813–2817.

Brown, E. B., Shear, J. B., Adams, S. R., Tsien, R. Y., and Webb, W. W. (1999). Photolysis of caged calcium in femtoliter volumes using two-photon excitation. *Biophys. J.* **76,** 489–499.

Brownlee, C., and Dale, B. (1990). Temporal and spatial correlation of fertilization current, calcium waves, and cytoplasmic contraction in eggs of *Ciona intestinalis. Proc. R. Soc. London. Ser. B Biol. Sci.* **239,** 321–328.

Busa, W. B., and Nuccitelli, R. (1985). An elevated free cytosolic Ca^{2+} wave follows fertilization in eggs of the frog, *Xenopus laevis. J. Cell Biol.* **100,** 1325–1329.

Callamaras, N., and Parker, I. (1999). Construction of line-scan confocal microscope for physiological recording. *Methods Enzymol.* **307,** 152–169.

Carroll, J., and Swann, K. (1992). Spontaneous cytosolic calcium oscillations driven by inositol trisphosphate occur during *in vitro* maturation of mouse oocytes. *J. Biol. Chem.* **267,** 11196–11201.

Caswell, A. H. (1979). Methods of measuring intracellular calcium. *Intl. Rev. Cytology* **56,** 145–181.

Centonze, V. E., and White, J. G. (1998). Multiphoton excitation provides optical sections from deeper within scattering specimcus than confocal imaging. *Biophys. J.* **75,** 2015–2024.

Chang, D. C., and Lu, P. (2000). Multiple types of calcium signals are associated with cell division in zebrafish embryo. *Microsc. Res. Tech.* **49,** 111–122.

Cheek, T. R., McGuinness, O. M., Vincent, C., Moreton, R. B., Berridge, M. J., and Johnson, M. H. (1993). Fertilisation and thimerosal stimulate similar calcium spiking patterns in mouse oocytes but by separate mechanisms. *Development (Cambridge, England)* **119,** 179–189.

Cheng, H., Lederer, W. J., and Cannell, M. B. (1993). Calcium sparks: Elementary events underlying excitation–contraction coupling in heart muscle. *Science* **262,** 740–744.

Creton, R., Speksnijder, J. E., and Jaffe, L. F. (1998). Patterns of free calcium in zebrafish embryos. *J. Cell Sci.* **111,** 1613–1622.

Crossley, I., Swann, K., Chambers, E., and Whitaker, M. (1988). Activation of sea urchin eggs by inositol phosphates is independent of external calcium. *Biochem. J.* **252,** 257–262.

Crossley, I., Whalley, T., and Whitaker, M. (1991). Guanosine 5′-thiotriphosphate may stimulate phosphoinositide messenger production in sea urchin eggs by a different route than the fertilizing sperm. *Cell Reg.* **2,** 121–133.

Dale, B., Fortunato, A., Monfrecola, V., and Tosti, E. (1996). A soluble sperm factor gates Ca^{2+}-activated K+ channels in human oocytes. *J. Assisted Reprod. Genet.* **13,** 573–577.

Deguchi, R., and Osanai, K. (1994). Repetitive intracellular Ca^{2+} increases at fertilization and the role of Ca^{2+} in meiosis reinitiation from the first metaphase in oocytes of marine bivalves. *Devel. Biol.* **163,** 162–174.

Eckberg, W. R., and Miller, A. L. (1995). Propagated and nonpropagated calcium transients during egg activation in the annelid, *Chaetopterus*. *Devel. Biol.* **172,** 654–664.

Egner, A., Andresen, V., and Hell, S. W. (2002). Comparison of the axial resolution of practical Nipkow-disk confocal fluorescence microscopy with that of multifocal multiphoton microscopy: Theory and experiment. *J. Microsc.* **206,** 24–32.

Eisen, A., Kiehart, D. P., Wieland, S. J., and Reynolds, G. T. (1984). Temporal sequence and spatial distribution of early events of fertilization in single sea urchin eggs. *J. Cell Biol.* **99,** 1647–1654.

Eisen, A., and Reynolds, G. T. (1984). Calcium transients during early development in single starfish (Asterias forbesi) oocytes. *J. Cell Biol.* **99,** 1878–1882.

Elangovan, M., Day, R. N., and Periasamy, A. (2002). Nanosecond fluorescence resonance energy transfer–fluorescence lifetime imaging microscopy to localize the protein interactions in a single living cell. *J. Microsc.* **205,** 3–14.

Ellis-Davies, G. C., Kaplan, J. H., and Barsotti, R. J. (1996). Laser photolysis of caged calcium: Rates of calcium release by nitrophenyl-EGTA and DM-nitrophen. *Biophys. J.* **70,** 1006–1016.

Ellis-Davies, G. C. R. (2003). Development and application of caged calcium. *Methods Enzymol.* **360,** 226–258.

Fissore, R. A., Dobrinsky, J. R., Balise, J. J., Duby, R. T., and Robl, J. M. (1992). Patterns of intracellular Ca^{2+} concentrations in fertilized bovine eggs. *Biol. Reprod.* **47,** 960–969.

Fontanilla, R. A., and Nuccitelli, R. (1998). Characterization of the sperm-induced calcium wave in *Xenopus* eggs using confocal microscopy. *Biophy. J.* **75,** 2079–2087.

Gilkey, J. C., Jaffe, L. F., Ridgway, E. B., and Reynolds, G. T. (1978). A free calcium wave traverses the activating egg of the medaka, *Oryzias latipes*. *J. Cell Biol.* **76,** 448–466.

Gilland, E., Baker, R., and Denk, W. (2003). Long duration three-dimensional imaging of calcium waves in zebrafish using multiphoton fluorescence microscopy. *Biol. Bull.* **205,** 176–177.

Gillot, I., and Whitaker, M. (1994). Calcium signals in and around the nucleus in sea urchin eggs. *Cell Calcium* **16,** 269–278.

Gillot, I., and Whitaker, M. J. (1993). Imaging calcium waves in eggs and embryos. *J. Exp. Biol.* **184,** 231–239.

Goldman, Y. E., Hibberd, M. G., McCray, J. A., and Trentham, D. R. (1982). Relaxation of muscle fibres by photolysis of caged ATP. *Nature* **300,** 701–705.

Gorelik, J., Shevchuk, A., Ramalho, M., Elliott, M., Lei, C., Higgins, C. F., Lab, M. J., Klenerman, D., Krauzewicz, N., Korchev, Y., *et al.* (2002). Scanning surface confocal microscopy for simultaneous topographical and fluorescence imaging: Application to single virus-like particle entry into a cell. *Proc. Natl. Acad. Sci. USA* **99,** 16018–16023.

Gould, M. C., Stephano, J. L., Ortiz-Barron, B. J., and Perez-Quezada, I. (2001). Maturation and fertilization in *Lottia gigantea* oocytes: Intracellular pH, Ca^{2+}, and electrophysiology. *J. Exp. Zool.* **290,** 411–420.

Groigno, L., and Whitaker, M. (1998). An anaphase calcium signal controls chromosome disjunction in early sea urchin embryos. *Cell* **92,** 193–204.

Grynkiewicz, G., Poenie, M., and Tsien, R. Y. (1985). A new generation of $Ca2^{+}$ indicators with greatly improved fluorescence properties. *J. Biol. Chem.* **260,** 3440–3450.

Hafner, M., Petzelt, C., Nobiling, R., Pawley, J. B., Kramp, D., and Schatten, G. (1988). Wave of free calcium at fertilization in the sea urchin egg visualized with fura-2. *Cell Motil. Cytoskel.* **9,** 271–277.

Hamaguchi, Y., and Hiramoto, Y. (1981). Activation of sea urchin eggs by microinjection of calcium buffers. *Exp. Cell Res.* **134,** 171–179.

He, C. L., Damiani, P., Parys, J. B., and Fissore, R. A. (1997). Calcium, calcium release receptors, and meiotic resumption in bovine oocytes. *Biol. Reprod.* **57,** 1245–1255.

Ikeda, M., Sugiyama, T., Wallace Christopher, S., Gompf Heinrich, S., Yoshioka, T., Miyawaki, A., and Allen Charles, N. (2003). Circadian dynamics of cytosolic and nuclear Ca2+ in single suprachiasmatic nucleus neurons. *Neuron* **38,** 253–263.

Jaffe, L. F. (1995). Calcium waves and development. *Ciba Foundation Symposium* **188,** 4–12.

Jones, K. T., Soeller, C., and Cannell, M. B. (1998). The passage of Ca^{2+} and fluorescent markers between the sperm and egg after fusion in the mouse. *Development (Cambridge, England)* **125,** 4627–4635.

Kao, J. P., Harootunian, A. T., and Tsien, R. Y. (1989). Photochemically generated cytosolic calcium pulses and their detection by fluo-3. *J. Biol. Chem.* **264,** 8179–8184.

Kaplan, J. H., and Ellis-Davies, G. C. (1988). Photolabile chelators for the rapid photorelease of divalent cations. *Proc. Natl. Acad. Sci. USA* **85,** 6571–6575.

Kawahara, H., and Yokosawa, H. (1994). Intracellular calcium mobilization regulates the activity of 26 S proteasome during the metaphase–anaphase transition in the ascidian meiotic cell cycle. *Devel. Biol.* **166,** 623–633.

Kiehart, D. P. (1982). Microinjection of echinoderm eggs: Apparatus and procedures. *Methods Cell Biol.* **25,** 13–31.

Kim, Y. S., and Padilla, G. M. (1978). Determination of free Ca ion concentration with an ion-selective electrode in the presence of chelating agents in comparison with calculated values. *Anal. Biochem.* **89,** 521–528.

Klein, P. S., and Melton, D. A. (1996). A molecular mechanism for the effect of lithium on development. *Proc. Natl. Acad. Sci. USA* **93,** 8455–8459.

Kline, D. (1988). Calcium-dependent events at fertilization of the frog egg: Injection of a calcium buffer blocks ion channel opening, exocytosis, and formation of pronuclei. *Devel. Biol.* **126,** 346–361.

Kline, D., and Kline, J. T. (1992). Repetitive calcium transients and the role of calcium in exocytosis and cell cycle activation in the mouse egg. *Devel. Biol.* **149,** 80–89.

Korchev, Y. E., Bashford, C. L., Milovanovic, M., Vodyanoy, I., and Lab, M. J. (1997). Scanning ion conductance microscopy of living cells. *Biophys. J.* **73,** 653–658.

Kume, S., Muto, A., Inoue, T., Suga, K., Okano, H., and Mikoshiba, K. (1997). Role of inositol 1,4,5-trisphosphate in ventral signalling in Xenopus embryos. *Science* **278,** 1940–1943.

Lawrence, Y., Ozil, J. P., and Swann, K. (1998). The effects of a Ca^{2+} chelator and heavy-metal-ion chelators upon Ca^{2+} oscillations and activation at fertilization in mouse eggs suggest a role for repetitive Ca^{2+} increases. *Biochem. J.* **335,** 335–342.

Lawrence, Y., Whitaker, M., and Swann, K. (1997). Sperm–egg fusion is the prelude to the initial Ca^{2+} increase at fertilization in the mouse. *Development (Cambridge, England)* **124,** 233–241.

Leung, C. F., Webb, S. E., and Miller, A. L. (1998). Calcium transients accompany ooplasmic segregation in zebrafish embryos. *Devel. Growth Diff.* **40,** 313–326.

Lew, V. L., Tsien, R. Y., Miner, C., and Bookchin, R. M. (1982). Physiological [Ca2+]i level and pump-leak turnover in intact red cells measured using an incorporated Ca chelator. *Nature* **298,** 478–481.

Marangos, P., FitzHarris, G., and Carroll, J. (2003). Ca^{2+} oscillations at fertilization in mammals are regulated by the formation of pronuclei. *Development* **130,** 1461–1472.

Martell, A., and Sillen, L. G. (1964). "Stability Constants. Special Publication No. 17." The Chemical Society, London, UK.

McCray, J. A., Herbette, L., Kihara, T., and Trentham, D. R. (1980). A new approach to time-resolved studies of ATP-requiring biological systems: Laser flash photolysis of caged ATP. *Proc. Natl. Acad. Sci. USA* **77,** 7237–7241.

McDougall, A., and Sardet, C. (1995). Function and characteristics of repetitive calcium waves associated with meiosis. *Curr. Biol.* **5,** 318–328.

McDougall, A., Sardet, C., and Lambert, C. C. (1995). Different calcium-dependent pathways control fertilization-triggered glycoside release and the cortical contraction in ascidian eggs. *Zygote (Cambridge, England)* **3,** 251–258.

McDougall, A., Shearer, J., and Whitaker, M. (2000). The initiation and propagation of the fertilization wave in sea urchin eggs. *Biology of the Cell/Under the Auspices of the European Cell Biology Organization* **92,** 205–214.

Miyawaki, A. (2003). Visualization of the spatial and temporal dynamics of intracellular signaling. *Develop. Cell* **4,** 295–305.

Miyawaki, A., Griesbeck, O., Heim, R., and Tsien, R. Y. (1999). Dynamic and quantitative Ca2+ measurements using improved cameleons. *Proc. Natl. Acad. Sci. USA* **96,** 2135–2140.

Miyawaki, A., Llopis, J., Heim, R., McCaffery, J. M., Adams, J. A., Ikura, M., and Tsien, R. Y. (1997). Fluorescent indicators for Ca2+ based on green fluorescent proteins and calmodulin. *Nature* **388,** 882–887.

Miyazaki, S. (1988). Inositol 1,4,5-trisphosphate-induced calcium release and guanine nucleotide-binding protein-mediated periodic calcium rises in golden hamster eggs. *J. Cell Biol.* **106,** 345–353.

Mohri, T., and Hamaguchi, Y. (1991). Propagation of transient Ca^{2+} increase in sea urchin eggs upon fertilization and its regulation by microinjecting EGTA solution. *Cell Struct. Func.* **16,** 157–165.

Muto, A., Kume, S., Inoue, T., Okano, H., and Mikoshiba, K. (1996). Calcium waves along the cleavage furrows in cleavage-stage *Xenopus* embryos and its inhibition by heparin. *J. Cell Biol.* **135,** 181–190.

Nagai, T., Sawano, A., Park, E. S., and Miyawaki, A. (2001). Circularly permuted green fluorescent proteins engineered to sense Ca2+. *Proc. Natl. Acad. Sci. USA* **98,** 3197–3202.

Nuccitelli, R., Yim, D. L., and Smart, T. (1993). The sperm-induced Ca^{2+} wave following fertilization of the Xenopus egg requires the production of Ins(1, 4, 5)P_3. *Devel. Biol.* **158,** 200–212.

Parker, I., and Miledi, R. (1989). Nonlinearity and facilitation in phosphoinositide signaling studied by the use of caged inositol trisphosphate in Xenopus oocytes. *J. Neuroscience—Off. J. Soc. Neurosci.* **9,** 4068–4077.

Poenie, M., Alderton, J., Steinhardt, R., and Tsien, R. (1986). Calcium rises abruptly and briefly throughout the cell at the onset of anaphase. *Science* **233,** 886–889.

Poenie, M., Alderton, J., Tsien, R. Y., and Steinhardt, R. A. (1985). Changes of free calcium levels with stages of the cell division cycle. *Nature* **315,** 147–149.

Raz, T., Ben-Yosef, D., and Shalgi, R. (1998). Segregation of the pathways leading to cortical reaction and cell cycle activation in the rat egg. *Biol. Reprod.* **58,** 94–102.

Robert, V., Gurlini, P., Tosello, V., Nagai, T., Miyawaki, A., Di Lisa, F., and Pozzan, T. (2001). Beat-to-beat oscillations of mitochondrial [Ca2+] in cardiac cells. *EMBO J.* **20,** 4998–5007.

Rosenzweig, D. C. (2002). Metallochaperones: Bind and deliver. *Chem. Biol.* **9,** 673–677.

Shevchuk, A. I., Gorelik, J., Harding, S. E., Lab, M. J., Klenerman, D., and Korchev, Y. E. (2001). Simultaneous measurement of Ca2+ and cellular dynamics: Combined scanning ion conductance and optical microscopy to study contracting cardiac myocytes. *Biophys. J.* **81,** 1759–1764.

Shimozono, S., Fukano, T., Nagai, T., Kirino, Y., Mizuno, H., and Miyawaki, A. (2002). Confocal imaging of subcellular Ca2+ concentrations using a dual-excitation ratiometric indicator based on green fluorescent protein. *Science's STKE Electronic Resource—Signal Transduction Knowledge Environment* p14.

Silver, R. A., Whitaker, M., and Bolsover, S. R. (1992). Intracellular ion imaging using fluorescent dyes: Artifacts and limits to resolution. *Pflugers Archiv—European J. Physiol.* **420,** 595–602.

Soeller, C., Jacobs, M. D., Donaldson, P. J., Cannell, M. B., Jones, K. T., and Ellis-Davies, G. C. R. (2003). Application of two-photon flash photolysis to reveal intercellular communication and intracellular Ca2+ movements. *J. Biomed. Optics* **8,** 418–427.

Speksnijder, J. E. (1992). The repetitive calcium waves in the fertilized ascidian egg are initiated near the vegetal pole by a cortical pacemaker. *Devel. Biol.* **153,** 259–271.

Speksnijder, J. E., Corson, D. W., Sardet, C., and Jaffe, L. F. (1989). Free calcium pulses following fertilization in the ascidian egg. *Devel. Biol.* **135,** 182–190.

Speksnijder, J. E., Sardet, C., and Jaffe, L. F. (1990a). The activation wave of calcium in the ascidian egg and its role in ooplasmic segregation. *J. Cell Biol.* **110,** 1589–1598.

Speksnijder, J. E., Sardet, C., and Jaffe, L. F. (1990b). Periodic calcium waves cross ascidian eggs after fertilization. *Devel. Biol.* **142,** 246–249.

Steinhardt, R. A., and Alderton, J. (1988). Intracellular free calcium rise triggers nuclear envelope breakdown in the sea urchin embryo. *Nature* **332,** 364–366.

Stith, B. J., Espinoza, R., Roberts, D., and Smart, T. (1994). Sperm increase inositol 1,4,5-trisphosphate mass in *Xenopus laevis* eggs preinjected with calcium buffers or heparin. *Devel. Biol.* **165,** 206–215.

Stricker, S. A. (1995). Time-lapse confocal imaging of calcium dynamics in starfish embryos. *Devel. Biol.* **170,** 496–518.

Stricker, S. A. (1996). Repetitive calcium waves induced by fertilization in the nemertean worm *Cerebratulus lacteus. Devel. Biol.* **176,** 243–263.

Stricker, S. A. (1999). Comparative biology of calcium signaling during fertilization and egg activation in animals. *Devel. Biol.* **211,** 157–176.

Stricker, S. A., Centonze, V. E., and Melendez, R. F. (1994). Calcium dynamics during starfish oocyte maturation and fertilization. *Devel. Biol.* **166,** 34–58.

Stricker, S. A., Centonze, V. E., Paddock, S. W., and Schatten, G. (1992). Confocal microscopy of fertilization-induced calcium dynamics in sea urchin eggs. *Devel. Biol.* **149,** 370–380.

Sun, F. Z., Hoyland, J., Huang, X., Mason, W., and Moor, R. M. (1992). A comparison of intracellular changes in porcine eggs after fertilization and electroactivation. *Development (Cambridge, England)* **115,** 947–956.

Swann, K. (1990). A cytosolic sperm factor stimulates repetitive calcium increases and mimics fertilization in hamster eggs. *Development (Cambridge, England)* **110,** 1295–1302.

Swann, K., McCulloh, D. H., McDougall, A., Chambers, E. L., and Whitaker, M. (1992). Sperm-induced currents at fertilization in sea urchin eggs injected with EGTA and neomycin. *Devel. Biol.* **151,** 552–563.

Swann, K., and Whitaker, M. (1986). The part played by inositol trisphosphate and calcium in the propagation of the fertilization wave in sea urchin eggs. *J Cell Biol.* **103,** 2333–2342.

Takahashi, A., Camacho, P., Lechleiter, J. D., and Herman, B. (1999). Measurement of intracellular calcium. *Physiol. Rev.* **79,** 1089–1125.

Takeuchi, H., Oike, M., Paterson, H. F., Allen, V., Kanematsu, T., Ito, Y., Erneux, C., Katan, M., and Hirata, M. (2000). Inhibition of Ca2+ signalling by p130, a phosolipase-C-related catalytically inactive protein: Critical role of the p130 pleckstrin homology domain. *Biochem. J.* **349,** 357–368.

Tanaami, T., Otsuki, S., Tomosada, N., Kosugi, Y., Shimizu, M., and Ishida, H. (2002). High-speed 1-frame/ms scanning confocal microscope with a microlens and Nipkow disks. *Appl. Opt.* **41,** 4704–4708.

Tauer, U. (2002). Advantages and risks of multiphoton microscopy in physiology. *Exp. Physiol.* **87,** 709–714.

Terasaki, M., and Sardet, C. (1991). Demonstration of calcium uptake and release by sea urchin egg cortical endoplasmic reticulum. *J. Cell Biol.* **115,** 1031–1037.

Torok, K., Wilding, M., Groigno, L., Patel, R., and Whitaker, M. (1998). Imaging the spatial dynamics of calmodulin activation during mitosis. *Curr. Biol.* **8,** 692–699.

Tosti, E., and Dale, B. (1994). Regulation of the fertilization current in ascidian oocytes by intracellular second messengers. *Mol. Reprod. Dev.* **37,** 473–476.

Truong, K., Sawano, A., Mizuno, H., Hama, H., Tong, K. I., Mal, T. K., Miyawaki, A., and Ikura, M. (2001). FRET-based *in vivo* Ca2+ imaging by a new calmodulin-GFP fusion molecule. *Nat. Struct. Biol.* **8,** 1069–1073.

Tsien, R. Y. (1984). Measuring and manipulating cytosolic Ca2+ with trapped indicators. *Kroc Foundation Series* **17,** 147–155.

Tsien, R. Y. (1986). New tetracarboxylate chelators for fluorescence measurement and photochemical manipulation of cytosolic free calcium concentrations. *Soc. Gen. Physiol. Series* **40,** 327–345.

Tsien, R. Y., Pozzan, T., and Rink, T. J. (1982). Calcium homeostasis in intact lymphocytes: Cytoplasmic free calcium monitored with a new, intracellularly trapped fluorescent indicator. *J. Cell Biol.* **94,** 325–334.

Tsien, R. Y., and Zucker, R. S. (1986). Control of cytoplasmic calcium with photolabile tetracarboxylate 2-nitrobenzhydrol chelators. *Biophys. J.* **50,** 843–853.

Twigg, J., Patel, R., and Whitaker, M. (1988). Translational control of InsP3-induced chromatin condensation during the early cell cycles of sea urchin embryos. *Nature* **332,** 366–369.

Vincent, C., Cheek, T. R., and Johnson, M. H. (1992). Cell cycle progression of parthenogenetically activated mouse oocytes to interphase is dependent on the level of internal calcium. *J. Cell Sci.* **103,** 389–396.

Walker, D. S., Gower, N. J. D., Ly, S., Bradley, G. L., and Baylis, H. A. (2002). Regulated disruption of Inositol 1,4,5-Trisphosphate signalling in Caenorhabditis elegans reveals new functions in feeding and embryogenesis. *Mol. Biol. Cell* **13,** 1329–1337.

Webb, S. E., Lee, K. W., Karplus, E., and Miller, A. L. (1997). Localized calcium transients accompany furrow positioning, propagation, and deepening during the early cleavage period of zebrafish embryos. *Devel. Biol.* **192,** 78–92.

Whitaker, M. (2000). Fluorescent tags of protein function in living cells. *BioEssays—News and Reviews in Mol. Cell. Devel. Biol.* **22,** 180–187.

Whitaker, M., and Swann, K. (1993). Lighting the fuse at fertilization. *Development* **117,** 1–12.

Whitaker, M. J. (2004). Calcium signalling in eggs and embryos. *Physiol. Rev.* in press.

Whitaker, M. J., and Steinhardt, R. A. (1982). Ionic regulation of egg activation. *Quart. Rev. Biophys.* **15,** 593–666.

Wilding, M., Wright, E. M., Patel, R., Ellis-Davies, G., and Whitaker, M. (1996). Local perinuclear calcium signals associated with mitosis-ENTRY in early sea urchin embryos. *J. Cell Biol.* **135,** 191–200.

Winston, N. J., McGuinness, O., Johnson, M. H., and Maro, B. (1995). The exit of mouse oocytes from meiotic M-phase requires an intact spindle during intracellular calcium release. *J. Cell Sci.* **108,** 143–151.

Zucker, R. S., and Steinhardt, R. A. (1978a). Prevention of the cortical reaction in fertilized sea urchin eggs by injection of calcium-chelating ligands. *Biochim. Biophys. Acta* **541,** 459–466.

Zucker, R. S., and Steinhardt, R. A. (1978b). Prevention of the cortical reaction in fertilized sea urchin eggs by injection of calcium-chelating ligands. *Biochim. Biophys. Acta* **541,** 459–466.

CHAPTER 19

Labeling of Cell Membranes and Compartments for Live Cell Fluorescence Microscopy

Mark Terasaki and Laurinda A. Jaffe

Department of Cell Biology
University of Connecticut Health Center
Farmington, Connecticut 06032

I. Introduction

During the last 10 years, the development of new fluorescence and optics technologies has resulted in revolutionary advances in imaging living cells. This chapter describes methods for fluorescently labeling the surface and organellar membranes as well as the spaces they enclose. We describe in detail methods with which we have had direct experience and provide information for other procedures. Fluorescence imaging of cytoskeletal elements is discussed in Chapter 16.

METHODS IN CELL BIOLOGY, VOL. 74

0091-679X/04 $35.00

It would be very convenient if one could pull off the shelf a particular fluorescent dye or GFP chimera to label each type of membrane compartment specifically. However, at this time, many useful methods are actually not specific. For example, a method may label several types of compartments or may only "work" for a period of time. In fact, it is sometimes easier to think that one is labeling a "process" than an entity. Therefore, using a method for labeling membrane compartments involves not just a protocol but also some understanding of membrane interrelationships and dynamics in order to interpret the labeling patterns well. Table I is a summary of fluorescent labels and their uses.

To begin with, the plasma membrane provides the boundary between the extracellular space and the intracellular space. The intracellular space contains additional membranes which form the boundaries for the organelles. The organelles are relatively small except for the endoplasmic reticulum (ER), which has some similarities to the plasma membrane, in that it is an extensive, continuous membrane. The intracellular space between the plasma membrane and organelles is one continuous space. A special part of the ER, the nuclear envelope, separates one part of this space and forms the boundary of the nucleus while the rest of the intracellular space is the cytosol.

The membranes of the cell undergo complex interactions in which they fuse to combine compartments, bud to create new ones, or move to different locations within cells. For instance, most membrane components are made in the ER, and then pass through the Golgi apparatus to become part of the other organelles, and there is often enormous turnover of plasma membrane through endocytosis and exocytosis. How fusion and budding are regulated, how organelles are able to establish and maintain a functional "identity," how membranes can recover

Table I
Some representative dyes for labeling membrane compartments

Dye	Excit. max	Emiss. max	Stock solution	Solvent	Application
Fluorescein dextran	488 nm	515 nm	10 mg/ml	seawater/IB	extracellular/intracellular
Rhodamine dextran	540	580	10 mg/ml	seawater/IB	extracellular/intracellular
calcein	494	517	0.5 mM	seawater/IB	extracellular/intracellular
FM 1-43	475	595	2 mM	ethanol	plasma membrane
R18	556	578	10 mM	ethanol	plasma membrane
DiI	549	565	2.5 mg/ml	ethanol	plasma membrane
			saturated	Wesson oil	ER
mitotracker	490	516	0.2–1.0 μM	seawater	mitochondria
nile red	552	636	50 μg/ml	DMSO	yolk

Other useful dyes are described in the text. IB = intracellular buffer (100 mM potassium glutamate, 10 mM Hepes, pH 7).

from catastrophic wounds to reprogram themselves (e.g., Terasaki *et al.*, 1997), and what regulates organelle movements are subjects of much current research.

II. Labeling of Extracellular Space/Endocytosis/Exocytosis

The extracellular space is easily labeled by including an impermeant fluorescent marker in the seawater. Fluorescent dextrans or calcein work well for this purpose. Dextrans are water-soluble polysaccharides that can be coupled to various fluorophores and are available in sizes ranging from 3 kDa to >1000 kDa. They are available as lysine-fixable versions which can be fixed in place by formaldehyde. Calcein (Molecular Probes #C-481) is fluorescein modified to have two negative charge groups. This is the smallest fluorescent space marker (MW = 623 Da) with which we have extensive experience. Alexa fluor 350 hydrazide (MW = 349 Da; Molecular Probes #A-10439) has been useful in preliminary experiments.

Extracellular labeling can be used to visualize exocytosis of cortical granules during sea urchin egg fertilization (Terasaki, 1995). The cortical granules are docked at the plasma membrane of unfertilized eggs and are triggered to undergo fusion by the calcium wave at fertilization. Fusion of the large (~1 μm diameter) cortical granules leaves a craterlike depression in the egg surface that can be imaged due to the movement of the extracellular marker in the seawater into the newly created space (Fig. 1A). In some of the original experiments, fluorescently labeled ovalbumin was used as an extracellular marker, but it is probably better to use lower molecular weight dextrans, which diffuse into the depression faster. As

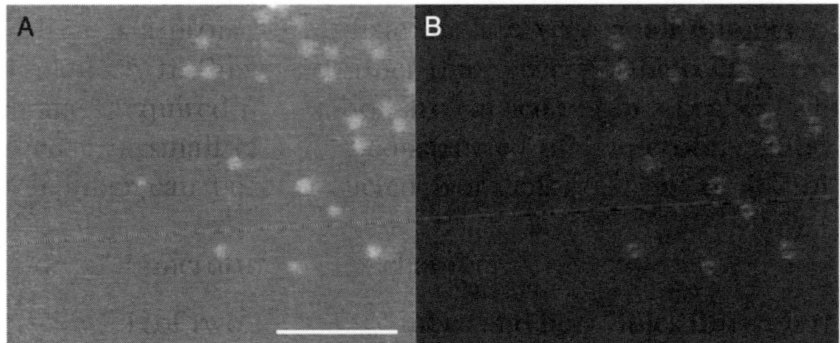

Fig. 1 Simultaneous visualization of cortical granule exocytosis by (A) an extracellular space marker and (B) a plasma membrane marker. A sea urchin egg was fertilized in the presence of 0.2 mg/ml Texas Red conjugated ovalbumin (which behaves similarly to fluorescent dextrans) and 2 μM FM 1–43, both in the seawater. A 63×, 1.4 NA objective lens of a confocal microscope was focused on the surface of the egg adjacent to the coverslip; the fluorescence of the ovalbumin and FM 1–43 was imaged simultaneously through use of a double labeling filter set. As described in the text, cortical granule exocytosis results in a craterlike depression in the cell surface. Labeling of this depression by an extracellular marker results in a disk pattern, while labeling by a plasma membrane marker results in a ring pattern. Bar = 10 μm. Modified from Terasaki, 1995.

an example of a labeling protocol, make a stock solution of 10 kDa tetramethylrhodamine dextran at a concentration of 10 mg/ml in seawater (most fluorescent dextrans can be made at this concentration). Store the stock solution in a microfuge tube in the refrigerator; stock solutions appear to be stable for at least several years. As a labeling solution, make a 3:100 dilution to 0.3 mg/ml (~30 μM dye) in seawater. If one is using an injection chamber (see Chapter 10) for imaging, make 1 ml of labeling solution and replace the reservoir of the chamber with the solution. Allow ~5 to 10 min for the solution to equilibrate between the coverslips of the chamber.

During endocytosis, a portion of the plasma membrane buds inward and the lumen of the new intracellular compartment contains fluid that was formerly part of the extracellular space. Uptake of fluorescent markers in the seawater is an excellent marker for endocytosis. However, there is an important precaution. The smallest endocytic compartments may not contain enough fluorescence to be detectable as an image of a vesicle in light microscopic images, even with good optical conditions.

A prolonged period of endocytotic reuptake of membrane follows the massive exocytosis at fertilization. Extracellular fluorescent markers were used to document the appearance of large endosomes following fertilization, and information about the time course of the endocytosis was obtained by doing pulse chase experiments with dextrans conjugated with different fluorophores (Whalley *et al.*, 1995). Vogel and coworkers have subsequently investigated mechanisms regulating this compensatory endocytosis that occurs after cortical granule exocytosis (e.g., Smith *et al.*, 2000). A quantitative fluid phase assay uses 100 μM 3 kDa tetramethylrhodamine dextran as a fluorescent marker in the seawater (Vogel *et al.*, 1999). Procedures are similar to that already described for imaging cortical granule exotcytosis.

Fluorescent extracellular markers were used to assay for membrane rupture in studies of wound healing of the plasma membrane (Terasaki *et al.*, 1997). While observing an egg by confocal microscopy, a large plasma membrane disruption was made in the presence of seawater containing 100 μg/ml fluorescein-conjugated stachyose (a tetrasaccharide; we now prefer calcein because of its lower cost and smaller molecular size). Very little fluorescent marker entered the egg, which is evidence that the disruption was repaired within a few seconds.

III. Labeling of Plasma Membrane/Exocytosis/Endosomes

The plasma membrane should be the easiest cell membrane to label because it is the most accessible, but the process is not always completely straightforward. Fluorescent dyes that label the plasma membrane intercalate into the membrane bilayer, and therefore have significant hydrophobicity. Hydrophobic molecules can often pass through membranes and label interior membranes, and are often not soluble in seawater, so that it is more difficult to deliver the dye to the plasma membrane. Effective plasma membrane dyes have electric charges or hydrophilic portions which result in properties that can help circumvent these problems.

Another kind of problem is that labeling of the plasma membrane bilayer will end up labeling intracellular compartments due to continuing endocytosis. This is desired for some applications, but there are some applications where this is inconvenient or even detrimental. Lastly, microscopic images of the plasma membrane sometimes do not correspond to one's intuition—when seen in *en face* views, the microvilli and similar structures, along with the undulations into and out of the image plane, can make it difficult to interpret confocal images of the plasma membrane.

Currently, the most commonly used dyes for labeling the plasma membrane are probably those of the FM 1-43 family. The pioneering use of these dyes was in investigations of membrane turnover at synapses (Betz and Bewick, 1992). These dyes are soluble in both water and the membrane, but are much less fluorescent in water than in the membrane. They also have the important property that they do not cross membranes. To use them, an aqueous solution of FM 1-43 is put onto cells; after a short while, a dynamic equilibrium is established in which a fraction of the dye has partitioned into and stained the plasma membrane. The dyes can be washed out of the seawater, resulting in loss of plasma membrane staining but retention of dye within any endosomes that have formed while the plasma membrane was stained.

FM 1–43 (Molecular Probes #T-3163) has been used in many studies as an endocytic marker (e.g., Frejtag *et al.*, 2003; Wessel *et al.*, 2002; Whalley *et al.*, 1995). It was also used to visualize exocytosis after fertilization (Terasaki, 1995). As has been mentioned, exocytosis of a cortical granule results in the formation of a crater-shaped depression in the cell surface. This depression appears as a ring in FM 1–43 labeled eggs for the following reason. The depression lies completely within the optical section of a confocal microscope, which is several microns in thickness. By considering the geometry of the depression, there is more membrane mass (and FM 1–43 molecules) in the wall of the depression (perpendicular to the image plane) than in the floor of the depression (parallel to the image plane); this results in the ring staining (Fig. 1B).

To use FM 1–43 for either endocytosis or cortical granule exocytosis, make a stock solution of 2 mM FM 1–43 in ethanol and store it at −20 °C. Make a 1:1000 dilution of this stock solution in seawater (i.e., 2 μM final concentration) to label the eggs. If using an injection chamber, replace the seawater in the chamber with the FM 1–43 solution and allow 10 min for the labeling to come to equilibrium before adding sperm. Even under conditions of optimal mixing, it takes on the order of 1 min for aqueous FM 1–43 to come to equilibrium with the plasma membrane; it is important to take this into account in brief, quantitative experiments but it is not an issue in long-term experiments. FM 1–43 fluorescence is high with fluorescein optics, with some spillover into the rhodamine channel, so FM 1–43 is not an ideal marker for double labeling experiments. There are a number of dyes, such as FM 4–64 (Molecular Probes, T-3166) or RH 414 (Molecular Probes T-1111), which have similar staining properties but different fluorescent spectra and could therefore be useful.

Long-chain dicarbocyanine dyes such as DiI ($DiIC_{18}(3)$) are insoluble in water but will incorporate into and diffuse in membrane bilayers; these have been widely used in neurobiology to stain the outlines of neurons (Honig and Hume, 1986). They do not cross membranes and, in contrast to FM 1–43, they remain associated once they have become incorporated into a membrane bilayer. The difficulty in using these dyes is delivery. A DiI solution in organic solvent or dried crystals is usually deposited in the region of neurons. This does not seem to work with unfertilized sea urchin eggs because the dye does not get past the vitelline envelope. DiI dissolved in soybean oil can be used to touch the surface of fertilized embryos to label the plasma membrane (D. Burgess, personal communication), though this does not seem to work in unfertilized eggs, perhaps due to the vitelline envelope.

A dye similar to DiI, R18 (octadecyl rhodamine Molecular Probes #O-246), has been used to label the sea urchin egg plasma membrane (Smith *et al.*, 2000). R18 is rhodamine with a single 18-carbon alkyl chain attached to it. A stock solution of 10 mM in ethanol was diluted to a concentration of 20 μM in seawater. R18 is insoluble in seawater and forms many small micelles or aggregates that are visible by fluorescence microscopy. Eggs were exposed to this suspension for 5 min, then washed to get rid of unincorporated micelles.

R18 was also used for labeling the starfish sperm plasma membrane and observing acrosome formation (Terasaki, 1998) (Fig. 2); this method appeared to work better than with FM 1–43 and offers the possibility of observing sperm–egg fusion. To use this method, make a stock solution of 10 mM R18 in ethanol; store at −20 °C. Make a 1 ml 1:1000 sperm suspension in seawater in a 1.5 ml Eppendorf tube, then add 1 μl of R18 stock, invert, and incubate for 20 min. During the incubation time, the R18 micelles/aggregates apparently collide with and label the sperm. Unlike starfish sperm, sea urchin sperm do not remain active for longer than 5 min after suspension in seawater, and this is not long enough to get good labeling with R18.

There are a large number of related dyes available from Molecular Probes or Avanti Polar Lipids which may have useful properties for plasma membrane labeling. Bodipy sphingomyelin (Molecular Probes #D-7711) is an example. It is soluble in water and labels the plasma membrane, but in contrast to FM 1–43, does not come out of the membrane when the staining solution is replaced with dye-free seawater. To label eggs, make a stock solution of 1 mM in ethanol, then use 0.1 μM in seawater for 1 min to stain eggs, followed by several seawater washes.

Another approach has been to chemically couple plasma membrane proteins to fluorescent markers. This has been done by removing the vitelline envelope, then incubating eggs with Alexa 488 maleimide (Molecular Probes; Smith *et al.*, 2000). Also, it is possible to use Oregon green conjugated concanavalin A, a lectin that

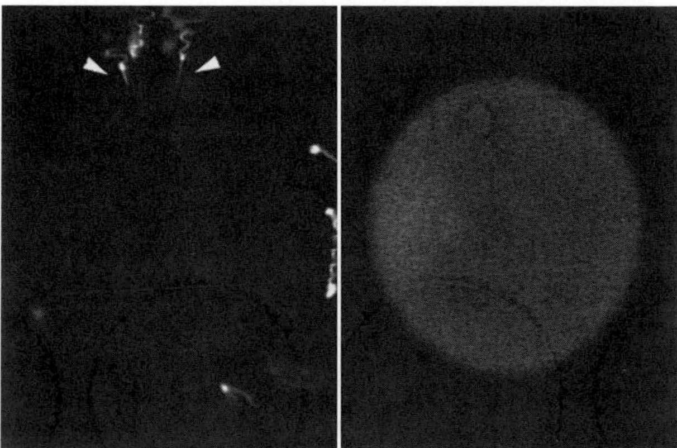

Fig. 2 Visualization of the sperm acrosomal process by labeling of the plasma membrane. Starfish sperm were mixed with R18 (octadecyl rhodamine) and, after an incubation period, were added to eggs previously injected with calcium green dextran and imaged by confocal microscopy (20×, 0.5 NA objective lens). The jelly layer surrounding the egg induces the sperm acrosome reaction, resulting in extension of the acrosomal process, which in starfish is particularly long (~10 μm). In the left panel, the acrosomal processes of two sperm have extended to contact the egg surface (arrowheads); the right panel shows the outline of the calcium green injected egg.

binds to specific carbohydrates of glycoproteins on the cell surface (Smith *et al.*, 2000).

In some cases, it would be useful to label just the plasma membrane without labeling of endocytic compartments. This would be most feasible using a membrane protein which is not normally endocytosed. A GFP chimera of casein kinase 1-gamma is localized to the plasma membrane and apparently does not become endocytosed in *Drosophila* embryos (http://biodev.obs-vlfr.fr/gavdos/); this chimera is untested in sea urchins.

IV. Labeling of the Cytosol

The "cytosol" is the intracellular space excluding the nucleus and the organelles, while the "cytoplasm" is the intracellular space excluding the nucleus but including the organelles. The cytosol is easy to label because aqueous fluorescence markers injected into eggs diffuse throughout the cytosol. As with extracellular space labeling, fluorescent dextrans and calcein work well.

As an example, to use 10 kDa or 70 kDa fluorescein dextran (Molecular Probes #D-1821), make a stock solution of 10 mg/ml in 100 mM potassium glutamate (or potassium chloride), 10 mM Hepes, pH 7. Make a 2% volume injection (200 μg/ml final concentration); this will result in bright staining. Calcein (Molecular Probes #C-481) is convenient for some applications because it diffuses rapidly due to its small size (623 Da). To use this, make a stock solution of 0.5 mg/ml in sea water and make a 2% volume injection (i.e., 10 μg/ml final concentration) to obtain bright staining.

Confocal microscope images of fluorescent markers in the cytosol often do not show a uniform distribution. When the cytoplasm is observed with high numerical aperture water or oil immersion lenses (numerical aperture > 1.0), the numerous 1 to 2 μm diameter yolk platelets, which take up about half of the volume of the cytoplasm, are seen as dark negative images (Fig. 3A; see also next Section). Small fluorescent markers ($<\sim$25 kDa; Lénárt *et al.*, 2003) injected into the cytosol diffuse passively through the nuclear pores of the nuclear envelope and label the interior of the nucleus, which becomes brighter than the cytosol in confocal microscope images. However, the local concentration within the nuclear space and the cytosol are probably the same; the difference in brightness is due to the cytoplasmic space occupied by the yolk platelets (Terasaki, 1994). Larger markers do not pass through the pores so the nucleus appears dark (Fig. 3B). During mitosis, the yolk platelets are excluded from the mitotic pole regions, so the mitotic pole regions are brighter than the surrounding regions in confocal microscope images (Terasaki, 2000) (Fig. 3B).

Imaging of the cytosol can be particularly important when using fluorescent ion indicators. With many indicators, binding of an ion causes the amplitude of the excitation and emission curves to change without any change in the shapes of those curves. Calcium green, for instance, has fluorescein-like fluorescence that increases several fold when it binds to calcium; it is coupled to dextran to prevent it from crossing membranes (Molecular Probes #C-3713). When injected into eggs, calcium green dextran diffuses throughout the cytosol; thus, the cytosolic fluorescence increases when cytosolic Ca^{2+} increases. However, the cytosolic fluorescence also depends on cell thickness or the relative abundance of organelles (which occupy space that would otherwise be cytosol). To correct for this, the calcium dependent fluorescence can be "normalized" or "ratioed" to the amount of cytosol present. This has been accomplished by co-injecting calcium green dextran and tetramethylrhodamine dextran and dividing the calcium green signal by the rhodamine signal at each pixel (e.g., Stricker, 1995; Wilding *et al.*, 1996). This co-injection method is not necessary for other ion indicators, such as fura-2 or indo-1; with these dyes, the shape as well as the amplitude of the excitation or emission curves change, so it is possible to normalize the fluorescence by measuring fluorescence at two different wavelengths.

V. Yolk Platelets/Reserve Granules

Echinoderm eggs have abundant numbers of large organelles of about 1 to 2 μm diameter that are commonly called "yolk platelets" or "yolk granules." Approximately half the volume of the egg is taken up by these organelles. It has been assumed that these organelles serve as a food supply for the growing embryo, which is the function of similar-appearing organelles in embryos of

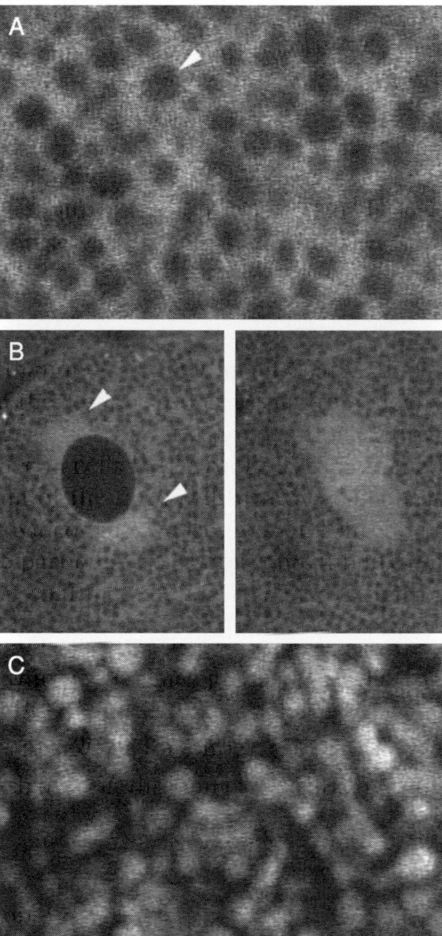

Fig. 3 (A) Labeling of the sea urchin egg cytosol by injection of fluorescein conjugated 70 kD dextran. The fluorescence was observed by confocal microscopy at high zoom with a 63× 1.4 NA objective lens. The yolk platelets are abundant organelles and because of their relatively large size (1–2 μm diameter), the space each yolk platelet occupies is clearly seen as a dark oval region in the cytosolic fluorescence image (one of the yolk platelets is indicated by the arrowhead). (B) Labeling of the cytosol before and during mitosis. An egg injected with fluorescent 70 kD dextran was fertilized and observed after a few cleavage cycles using imaging conditions as in the previous panel but at a lower zoom. The large dextran does not cross the nuclear pores, so that the interior of the nucleus appears dark (left panel); this image was taken a few minutes before nuclear envelope breakdown, and the two spindle pole regions have become brighter due to exclusion of yolk platelets from these regions (arrowheads). After the nucleus has broken down (right panel), the fluorescent 70 kD dextran enters the former nuclear region. This region becomes brighter than the peripheral cytoplasm due to the absence of yolk platelets. (C) Labeling of yolk platelets by nile blue observed by confocal microscopy with a 63×, 1.4 NA objective lens.

other phyla. However, there is evidence against a nutrient role in sea urchin embryos (Scott *et al.*, 1990) and more recently, evidence for a role in calcium regulation (Churchill *et al.*, 2002), so that these organelles are also called "reserve granules" (Chestkov *et al.*, 1998).

The yolk platelets/reserve granules can be fluorescently stained by nile blue (Danilchik and Gerhart, 1987; Sigma N5632) or nile red (Molecular Probes N-1142) (Fig. 3C). To label by this method, make a stock solution in DMSO (nile blue at 2 mg/ml; nile red at 50 μg/ml; store at room temperature protected from light). Add 1 μl of stock solution to 1 ml of seawater in a glass test tube and swirl. Add sea urchin eggs in ~50 μl and swirl again. The staining should be complete by ~10 to 20 min. The dye can be left in the seawater, and the fluorescence of either dye is observed with rhodamine filters. The time required for uniform staining is ~10 min. To stain eggs that are already loaded in a microinjection chamber, simply change the large reservoir to seawater with a 1:1000 dilution of nile blue/nile red stock solution. The time for complete staining is somewhat longer (~20 min), due to restricted access. These dyes do not label yolk platelets in starfish oocytes, where they only stain sparse large vesicles in the cortex (these do not undergo exocytosis at fertilization).

VI. Mitochondria

In many cell types, mitochondria are easily labeled by water-soluble, lipophilic, positively charged fluorescent molecules which cross the plasma membrane and become concentrated in mitochondria due to the large negative membrane potential of the mitochondria. The first of these dyes to be described was rhodamine 123 (Johnson *et al.*, 1980), but many other dyes have these properties, including short-chain dicarbocyanine dyes and the "mitotracker" series of dyes.

Mitochondria have been labeled in ascidian eggs using $DiOC_2(3)$ (Molecular Probes, #D-14730) or Mitotracker Green FM (Molecular Probes, #M-7514) (e.g., Roegiers *et al.*, 1999). Eggs are incubated for 15 to 20 min (0.5 μg/ml $DiOC_2(3)$ or 1 μM mitotracker) in seawater buffered with TAPS, pH 8.3. Both of these dyes have fluoresceinlike fluorescence, but mitotracker is also available with different excitation/emission spectra. Mitotracker Green FM may be useful for staining mitochondria in sea urchin embryos using similar conditions (200 nM for 10 min at 15 °C in the dark; Coffman and Davidson, 2001).

VII. Other Organelles

It would be convenient to have fluorescent markers for some other organelles found in echinoderm eggs. We are unaware of any method to specifically label the cortical granules in live eggs. They can be visualized in negative image by injecting cytosolic markers (e.g., Terasaki, 1995). There are large "acidic vesicles" in the cytoplasm of eggs. These are pigmented in *Arbacia punctulata* eggs and can be distinguished by transmitted light microscopy. The dye nile blue, which labels yolk platelets in sea urchin eggs, does not label yolk platelets in starfish oocytes. Instead, it labels large vesicles in the cortex which are not exocytosed at fertilization. We are unaware of methods to label lysosomes except perhaps by a specific GFP chimera targeted to the lysosome.

VIII. Endoplasmic Reticulum

The endoplasmic reticulum (ER) is a membrane compartment concerned with synthesis of membrane proteins, secretory proteins, and membrane lipids, as well as with regulation of intracellular calcium. As has been mentioned, it is distributed throughout the cell and includes the nuclear envelope.

Short-chain dicarbocyanine dyes, such as $DiOC_6(3)$, are effective for staining ER in cultured fibroblasts (Terasaki *et al.*, 1984). These dyes stain many organelles nonspecifically and the ER pattern is possible to identify in the thin-spread regions of fibroblasts, but it is not possible to do this in spherical echinoderm eggs. Instead, the ER in echinoderm eggs has been successfully stained by two other methods: injecting an oil drop saturated with a long-chain dicarbocyanine dye, or expressing a GFP chimera.

DiI Method The name "DiI" refers to either $DiIC_{16}(3)$ or $DiIC_{18}(3)$ (D-384, D382; Molecular Probes). Both are dicarbocyanine dyes that consist of a fluorescent portion and two long alkyl chains, and both can be used to label the ER. DiI incorporates into the membrane bilayer through intercalation of the long hydrocarbon chains (Axelrod, 1979). Once it is incorporated in the bilayer, the dye is able to diffuse freely within the bilayer, but does not transfer out of the membrane. DiI was first used to trace the plasma membrane of neurons by Honig and Hume (1986). Since then, it and several other related dyes have been used extensively for this purpose (see Haugland, 2002, for references).

For labeling the ER, DiI is dissolved in Wesson soybean oil and microinjected into eggs. The oil droplet contacts and stains many different membranes at the site of contact. However, of the organelles in the cell, the only organelle that has extensive continuity is the ER. The dye spreads in the continuous membranes of the ER, so that in regions away from the oil drop, only the ER is stained.

One limitation of this method is that bright punctate labeling accumulates with time and represents dye leaving the ER network via membrane traffic with the Golgi apparatus; this first becomes noticeable ~1 hr after injection. Another possible disadvantage is that the oil drop usually must be relatively large to get bright staining. In our experiments, the oil drop was ~10% of the size of the egg, and this sometimes was a barrier for imaging or caused the mitotic apparatus to become displaced. If necessary, the oil drop can be removed from the egg by suction with a micropipet, after the DiI has diffused into the ER.

Using DiI in sea urchin eggs, we observed a transient change in the ER structure at fertilization, in which large cisternae became more finely divided (Terasaki and Jaffe, 1991). We also fixed eggs in glutaraldehyde and then injected DiI-saturated oil drops. The fixation eliminates spread of the dye by membrane traffic. We observed that the dye spread throughout the fixed eggs, providing stronger evidence that the internal cisternae, cortical network, and nuclear envelope are all part of one membrane (Jaffe and Terasaki, 1993). We also observed that the dye spread throughout eggs fixed 10 min after fertilization, but that it did not spread significantly from the oil drop in eggs fixed 1 min after fertilization. This provided evidence that the ER is fragmented at the time of calcium release during fertilization, and that the ER subsequently regains its continuity. This was corroborated by later experiments with GFP-KDEL, which will be described. The DiI method has also been used in ascidians (Speksnijder *et al.*, 1993) and starfish (Jaffe and Terasaki, 1994) to label the ER and observe its dynamics during maturation, fertilization, and early development.

To prepare a saturated solution of DiI in oil, crystals of DiI (several milligrams) were sprinkled into the bottom of a 1.5 ml microfuge tube. Approximately 300 μl of Wesson oil (purchased from a grocery store) was then added and the tube was inverted a few times. DiI dissolves relatively slowly, so it should be left for a few hours or overnight, by which time the the oil should be a bright red solution, with some crystals remaining in the bottom of the tube. To remove the crystals and transfer the saturated oil to a new tube, centrifuge at 10,000*g* for a few minutes. The DiI in oil is then stored at room temperature protected from light. This is good for at least several months. Because Wesson oil eventually becomes rancid, it is probably better to use a bottle that has been purchased within the last 6 months to 1 year.

In the original experiments, we used Wesson soybean oil. At the time, this was unavailable in Europe, so an attempt was made with corn oil; DiI dissolved better in the corn oil, but when this was injected, most of it remained in the oil drop and did not spread into the ER (A. Speksnijder, unpublished observations). We then tested different oils but most appeared to work as well as soybean oil in sea urchin eggs (e.g., olive, canola, walnut).

DiI fluorescence is best observed using rhodamine fluorescence filters. For some double labeling experiments, it would be useful to observe the ER with fluorescein fluorescence filters. This can be done by using the long-chain dicarbocyanine "DiO" (DiOC18(3); Molecular Probes #D275). This dye does not dissolve directly in oil, as does DiI, so the procedure is to dissolve it first in DMSO at 10 mM, then to make a 3% dilution of this solution into Wesson oil (Feng *et al.*, 1994).

GFP Method— The green fluorescent protein (GFP) from the jellyfish *Aequorea* has revolutionized cell biology and is described in detail in other chapters. The construct "GFP-KDEL" improved the specificity of labeling of the ER over the DiI method (Terasaki *et al.*, 1996). GFP-KDEL, as with most membrane proteins or secretory proteins, cannot be injected as a recombinant protein and must be synthesized by the cell's own machinery. Currently, the best way to accomplish this in echinoderms is to inject mRNA encoding for the protein. We provide a detailed discussion of what was involved in designing and expressing GFP-KDEL as a guide for designing and expressing other GFP chimeras of membrane proteins or secreted proteins.

The GFP-KDEL labeling method was based on how proteins are thought to be targeted to, and retained in, the ER. As a protein is being synthesized in the cytoplasm, a "signal sequence" protrudes from the ribosome and is recognized by the signal recognition particle (SRP). The SRP stops translation until the SRP/ribosome complex binds to a receptor on the ER, where translation resumes and the nascent protein is directed into the ER. GFP is a soluble protein of the cytosol, and was targeted to the ER as a soluble protein of the ER lumen. We used an N terminal signal sequence from ECast/PDI, a lumenal ER protein of sea urchin eggs (Lucero *et al.*, 1994). It was necessary to make a second alteration in order for GFP to be retained in the ER. It is thought that the default fate for soluble proteins in the ER is to be processed by the Golgi and secreted out of the cell. Many resident ER lumenal proteins have a characteristic four amino acid C terminal retention sequence KDEL (lysine glutamate aspartate leucine), and artificial proteins with this C terminal sequence are retained in the ER (Munro and Pelham, 1987). To ensure that our GFP construct remained in the ER, we designed it with a C terminal KDEL retention sequence.

Injection of mRNA is a routine procedure in *Xenopus* oocytes (Wormington, 1991), and many of the methods and reagents that have been worked out for *Xenopus* can be applied to echinoderm eggs. We chose the vector pSP64-RI (Tang *et al.*, 1995), variants of which are widely used for *Xenopus* oocytes. This vector has a promoter for the RNA polymerase SP6 for making mRNA *in vitro* (Krieg and Melton, 1984). We were initially interested to observe the ER during fertilization. Expression of proteins for fertilization experiments can be easily accomplished with starfish oocytes by injecting mRNA into immature oocytes, incubating overnight to allow for expression, maturing the oocyte, then fertilizing (Shilling *et al.*, 1990). This is not as easy in sea urchins, due to low rates of protein synthesis in unfertilized sea urchin eggs, but incubation for >6 h has resulted in enough GFP chimera expression for some fertilization experiments (J. Ellenberg, personal communication).

The GFP-KDEL construct was made by John Hammer (NHLBI, NIH) starting with DNA coding for the signal sequence of ECast/PDI (provided by H. Lucero and B. Kaminer) and DNA coding for the S65T mutant of GFP (provided by R. Tsien). This GFP mutant is brighter and folds faster after its synthesis (Heim *et al.*, 1995), so it is clearly desirable to use this mutant for GFP chimeras (the EGFP variant (Clontech) is preferable for mammalian cell studies because it folds better at 37 °C than wild-type or S65T GFP but has no advantage for echinoderms).

Oligonucleotide sequences were designed so that PCR reactions would add two desired sequences as well as provide restriction sites that would allow the PCR products to anneal to each other, and also to integrate into the multiple cloning site of the vector. The two sequences added by this procedure were the C terminal KDEL sequence and the "Kozak sequence" from sea urchin ECast/PDI. The Kozak sequence is the first few nucleotides preceding the start codon (ATG) and the first nucleotide after the start codon (Kozak, 1999); these nucleotides are thought to be involved in ribosome association and have been found to have a very significant effect on expression of exogenous proteins. If a construct from another species does not express well, we, in agreement with others, have found that changing the Kozak sequence (the pre-ATG nucleotides seem to be sufficient) to one of the recipient species will frequently improve the expression level.

The plasmid containing GFP-KDEL was grown in bacteria and handled by standard techniques to isolate and cut DNA for making mRNA. We have used the mRNA synthesis kit from Ambion (mMessage mMachine; Austin, TX), which makes 10 to 20 μg mRNA from 1 μg DNA. The mRNA is resuspended in nuclease-free water supplied in the kit, and then stored in 2 μl aliquots at $-70\,^{\circ}$C. RNA is easily degraded by RNases in the environment, so microfuge tubes and pipetman tips from boxes designated "for RNA use only" were used, and gloves were used to handle the tubes. The mRNA aliquots can be frozen and thawed several times.

Experimental results from Terasaki *et al.* (1996) are summarized here. We found that the oocytes tolerated an injection of ~10 to 20 μg/ml final concentration of GFP-KDEL mRNA; higher concentrations led to cell death within a few hours. We monitored the appearance of GFP fluorescence and found a lag of 1 to 2 h, then a linear increase phase, and a plateau by about 24 h. The lag is probably due to the time needed for recovery from injection, synthesis, and folding. The fluorescence of GFP-KDEL expressing eggs was compared with that of eggs injected with a known amount of fluorescein dextran. By using published values for the quantum yield and extinction coefficient of fluorescein and GFP, the plateau concentration of GFP in the cell was estimated as 0.7 μM; the concentration in the ER lumen must be higher because the ER occupies a small fraction of the total cell volume. When imaged at high magnification, the GFP-KDEL fluorescence pattern was the same as seen previously with DiI (Jaffe and Terasaki, 1994), but there was no apparent transfer to other compartments with time, such as seen with DiI.

For studies of fertilization or development, it is necessary for the oocytes to undergo "maturation." Starfish oocytes are arrested at prophase of meiosis I ("GV stage"); we injected the mRNA at this stage. Starfish oocytes are induced to mature by the hormone 1-methyladenine and become competent to undergo normal fertilization at ~30 min (at the time of first metaphase). The eggs remain at first metaphase for another ~30 to 60 min (depending on the temperature). Fertilization should be carried out during this period prior to first polar body formation.

During starfish maturation and fertilization, changes in GFP-KDEL labeled ER organization were identical to those seen earlier with DiI (Jaffe and Terasaki, 1994). In immature oocytes, randomly oriented cisternae were present, while in mature oocytes, circular profiles were common, which are likely to correspond to incomplete spherical shells around yolk platelets. After fertilization, there was a transient disruption of this pattern, lasting about 10 to 15 min. As with DiI, the resolving power of the confocal microscope was not quite sufficient to determine whether the ER had become converted to tubules or had become vesiculated. We used FRAP (fluorescence redistribution after photobleaching) to address this issue.

In FRAP experiments, the fluorescence in a small region of the cell is photobleached by exposure to high-intensity excitation light. The redistribution of fluorescence from the unbleached regions is then monitored. The way in which the redistribution occurs can give very useful information, for instance, on the kinetics that underlie steady state distributions. GFP is very well-suited for photobleaching experiments; this is probably because the fluorophore of GFP is buried within the protein (Ormo *et al.*, 1996; Yang *et al.*, 1996), so that photodynamic damage involves the GFP protein rather than neighboring proteins.

When a small region of GFP-KDEL was photobleached in unfertilized eggs, the fluorescence in that region recovered within about 1 min. A similar recovery from photobleaching was seen in eggs 20 min after fertilization. However, when eggs were photobleached 1 min after fertilization (when the structural change could already be detected), the fluorescence took approximately 10 min to recover. These results are consistent with a continuous network of ER that normally provides pathways for diffusion throughout the cell, but which is transiently disrupted at the time of fertilization. The disruption is very probably related to the release of Ca^{2+} from the ER which occurs at fertilization.

In later experiments, mRNA for GFP-KDEL was injected into sea urchin eggs. Due to the increase in protein synthesis that occurs after fertilization, GFP fluorescence developed so that it became possible to observe labeling after about the 4th or 5th cleavage (Terasaki, 2000). It was found that the ER did not vesiculate during mitosis, and that it accumulated around the spindle poles (Fig. 4A).

IX. Golgi Apparatus

Proteins and lipids synthesized in the ER must pass through the Golgi apparatus before they become part of other organelles. By electron microscopy, there are many stacks of Golgi apparatus scattered throughout the cytoplasm of echinoderm eggs; later in embryonic development, when the cells take on epithelial characteristics, there appears to be only one large Golgi complex (Gibbins *et al.*, 1969).

Bodipy ceramide (Molecular Probes #D-3521) is a fluorescent marker for the Golgi apparatus that can be applied from the outside of cultured mammalian cells (Pagano *et al.*, 1991); however, it does not appear to work well in echinoderm eggs. The most effective way to label the Golgi apparatus has been with GFP chimeras (Terasaki, 2000). The chimeras were originally used in cultured mammalian cells and were targeted to the Golgi apparatus using a galactosyl transferase fragment or an ELP mutant ($KDELR_m$) which is retained in the Golgi (Cole *et al.*, 1996). mRNA was injected into unfertilized sea urchin eggs, which were then fertilized. Staining was bright enough for imaging by around the 4th cleavage. In time-lapse sequences, the scattered Golgi stacks underwent large changes during

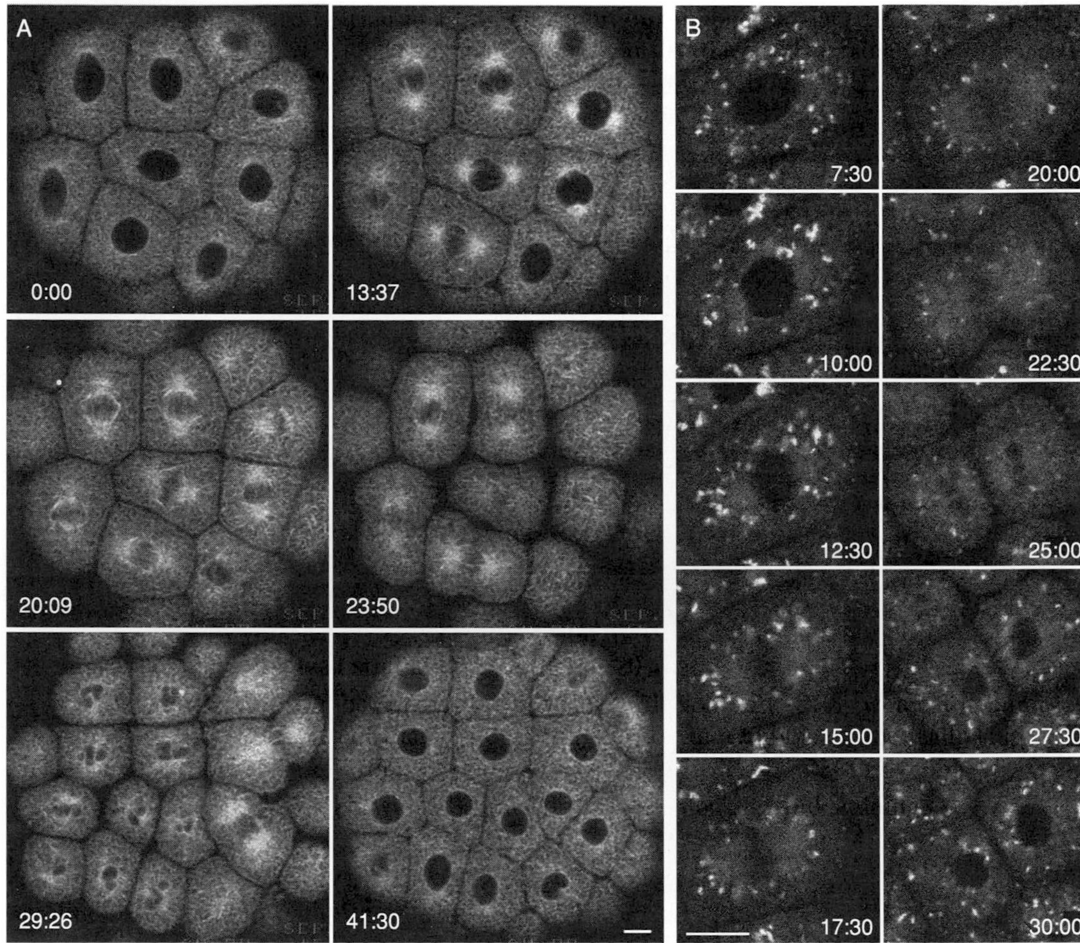

Fig. 4 (A) Labeling of the ER by GFP-KDEL. Sea urchin eggs were injected with mRNA coding for GFP-KDEL, then fertilized. Images were obtained by confocal microscopy at low zoom with a 63×, 1.4 NA objective lens as the blastomeres went through the sixth cleavage; the timing for images is indicated (min:sec). The ER undergoes a reversible accumulation at the spindle poles during mitosis; the breakdown and reformation of the nucleus can also be seen. Bar = 10 μm. (B) Labeling of the Golgi apparatus by GFP-KDELR$_m$. Eggs were injected with mRNA for GFP-KDELR$_m$, then fertilized. Fluorescence was observed as in the previous panel while the embryos underwent seventh cleavage; the timing of the images is indicated. During interphase, the Golgi is present as numerous, separated stacks. The number and size of these stacks decrease drastically during mitosis; the stacks then return as the blastomeres exit from mitosis. Bar = 10 μm. Modified from Terasaki, 2000.

mitosis, where most of the staining appeared to relocate to the ER, and the Golgi stacks re-formed after the end of mitosis (Fig. 4B). Time-lapse observations also showed that the scattered Golgi coalesced into one Golgi after the 9th cleavage, which is one cleavage cycle before the embryo secretes a hatching enzyme.

X. Nucleus

The nucleus is bounded by the nuclear envelope, a part of the endoplasmic reticulum that has closed in on itself. The membranes of the nuclear envelope form a permeability barrier between the nucleus and the cytosol. Passage across the nuclear envelope occurs through the nuclear pores; small molecules of less than ~25 kD can diffuse passively through the pores (Lénárt *et al.*, 2003), while larger molecules must be actively transported. When large fluorescent dextrans are injected into the cytoplasm, the nucleus remains dark (e.g., Fig. 3B). Smaller fluorescent dextrans enter the nucleus and reach a level that is approximately twice as bright as the surrounding cytoplasm. This is due to the absence of organelles in the nucleus and to the presence of the large abundant yolk platelets in egg cytoplasm (Terasaki, 1994).

The nuclear envelope is easily visible by transmitted light microscopy, but it has been useful to label its components (nuclear pores, lamina) by GFP chimeras to investigate how the nuclear envelope breaks down during meiosis or mitosis (Lénárt *et al.*, 2003; Terasaki *et al.*, 2001).

During meiosis and mitosis, the cellular DNA becomes condensed into the chromosomes so that the DNA can be efficiently partitioned between two daughter cells. The most convenient way to label chromosomes remains incubation in seawater with Hoechst 33258, 33342, or DAPI (H-33258, H33342, D-9564 from Sigma Chemical Co., St. Louis, or H-1398, H-1399, D-1306 from Molecular Probes). See Chapter 16 for details. The main disadvantage of the Hoechst dyes or DAPI is the requirement for ultraviolet excitation, which is not available on many confocal microscopes. Other methods for labeling chromosomes require microinjection. Fluorescently labeled chromosomal proteins, such as rhodamine histone (Minden *et al.*, 1989), or GFP chimeras can potentially be used.

YOYO-1 (Molecular Probes #Y-3601) binds directly to nucleic acids and can be used to label chromosomes, but there is a large background staining of RNA in the cytoplasm. YOYO comes as a 1 mM solution in DMSO. Make a 1:10 dilution (to 100 μM) in an injection buffer consisting of 100 mM potassium glutamate, 10 mM Hepes, pH 7. Make a 3% volume injection, resulting in a final concentration in the egg of 3 μM. There are a number of dyes related to YOYO-1 that are available from Molecular Probes and which may also be useful.

Another possibility is Oregon Green dUTP (Molecular Probes #C-7630), a fluorescent nucleotide that is incorporated into newly synthesized DNA (Carroll *et al.*, 1999). This produces bright staining but only after one round of cellular DNA synthesis (Fig. 5). Oregon Green dUTP comes as a 1 mM solution in 10 mM Tris, 1 mM EDTA, pH 7.6. Make a 100 μM stock solution in the injection buffer described previously and then make a 1% volume injection to a final concentration in the egg of 1 μM. Observe the fluorescence with fluorescein optics.

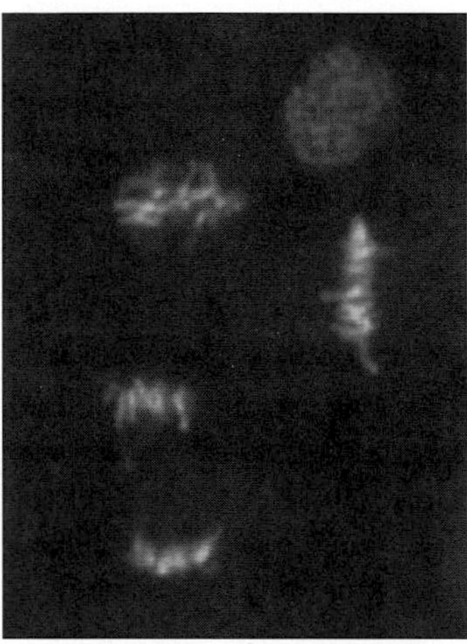

Fig. 5 Labeling of chromosomes by a fluorescent nucleotide. A sea urchin egg was injected with Oregon green dUTP, then fertilized. The fluorescent nucleotide becomes incorporated into the DNA during the S period of the cell cycle. The embryo was imaged after several cleavage cycles by confocal microscopy with a 63×, 1.4 NA lens. In this image, a nucleus with condensing chromosomes is seen at the top, a cell in early metaphase is at left top, a cell in metaphase is seen at right middle, and an anaphase pair is seen at the left bottom.

XI. Microscopy Considerations

The way in which the egg or embryo is kept on the microscope stage is critical for the success of a live cell experiment. The observation chamber must be conducive for normal development as well as for microscopy. We most frequently use the same chamber used for microinjection (see Chapter 10), though there are certainly other ways. The eggs are held securely between two coverslips and can be observed by oil or water immersion objective lenses. The eggs can be injected, fertilized, and their development observed in the same chamber. There does not seem to be much difference in using an upright versus inverted microscope, though in cases where it is desired to image while injecting, it is much easier to use an upright microscope configuration.

Another very significant problem in live cell imaging is phototoxicity. From its excited state, the fluorescent molecule sometimes does not decay to make light but instead reacts with molecular oxygen to form triplet state oxygen, which can then react with and damage neighboring molecules (e.g., Tsien and Waggoner, 1994). In live cell imaging experiments, the challenge is to obtain useful data without perturbing the system by phototoxicity. It is important to use the microscope efficiently so as to use minimal amounts of light in obtaining data (e.g., Terasaki and Dailey, 1994). It is also important to be able to evaluate whether phototoxicity is affecting the process or structure that one is studying. With regard to this problem, there is an advantage to working with echinoderm eggs or embryos. It is possible to fertilize and watch normal development over relatively short periods, which provides a straightforward and easy way to evaluate whether damage has occurred. As an example, if one were imaging events in fertilization, one way to test for phototoxicity is to let the embryo develop and see if the first cleavage occurs on time and in a normal fashion.

A related issue is that the labeling procedure itself can affect normal processes. GFP chimeras of endogenous proteins are of necessity overexpressed and could cause toxicity. Fluorescent dyes can be used at higher concentrations to provide brighter signals but may be toxic at the high levels. These should also be easier to evaluate in echinoderm embryos by observing whether the labeling procedures are compatible with normal development.

XII. Future Directions

There will most certainly be continuing improvements in techniques for live cell imaging. The most likely advances will involve fluorescent proteins, from the development of useful GFP chimeras to new types of fluorescent proteins. It also seems likely that there will be an increasing use of light microscopy to make quantitative physiological measurements, such as with sensors for kinases. With multiphoton microscopy, imaging of tissues and older embryos should become more feasible, and fluorescent labels for components of the extracellular matrix, and specialized structures and tissues such as the skeleton or nervous system should become available. Optical methods for sperm should improve as well. Lastly, mundane details such as keeping embryos alive or immobilized on the microscope, can enable the newest, most exciting technique to be used well.

Acknowledgments

We thank several colleagues for telling us about fluorescent probes: Steve Vogel for octadecyl rhodamine, Kristien Zaal for bodipy sphingomyelin, and John Newport for Oregon green dUTP.

References

Axelrod, D. (1979). Carbocyanine dye orientation in red cell membrane studied by microscopic fluorescence polarization. *Biophys. J.* **2,** 557–574.

Betz, W. J., and Bewick, G. S. (1992). Optical analysis of synaptic vesicle recycling at the frog neuromuscular junction. *Science* **255,** 200–203.

Carroll, D. J., Albay, D. T., Terasaki, M., Jaffe, L. A., and Foltz, K. R. (1999). Identification of PLCγ-dependent and independent events during fertilization of sea urchin eggs. *Dev. Biol.* **206,** 232–247.

Chestkov, V. V., Radko, S. P., Cho, M. S., Chrambach, A., and Vogel, S. S. (1998). Reconstitution of calcium-triggered membrane fusion using "reserve" granules *J. Biol. Chem.* **273,** 2445–2451.

Churchill, G. C., Okada, Y., Thomas, J. M., Genazzani, A. A., Patel, S., and Galione, A. (2002). NAADP mobilizes Ca(2+) from reserve granules, lysosome-related organelles, in sea urchin eggs. *Cell* **111,** 703–708.

Coffman, J. A., and Davidson, E. H. (2001). Oral–aboral axis specification in the sea urchin embryo. I. Axis entrainment by respiratory asymmetry. *Dev. Biol.* **230,** 18–28.

Cole, N. B., Smith, C. L., Sciaky, N., Terasaki, M., Edidin, M., and Lippincott-Schwartz, J. (1996). Diffusional mobility of Golgi proteins in membranes of living cells. *Science* **273,** 797–801.

Danilchik, M. V., and Gerhart, J. C. (1987). Differentiation of the animal–vegetal axis in *Xenopus laevis* oocytes. I. Polarized intracellular translocation of platelets establishes the yolk gradient. *Dev. Biol.* **122,** 101–112.

Feng, J. J., Carson, J. H., Morgan, F., Walz, B., and Fein, A. (1994). Three-dimensional organization of endoplasmic reticulum in the ventral photoreceptors of *Limulus. J. Comp. Neurol.* **341,** 172–183.

Frejtag, W., Burnette, J., Kang, B., Smith, R. M., and Vogel, S. S. (2003). An increase in surface area is not required for cell division in early sea urchin development. *Dev. Biol.* **259,** 62–70.

Gibbins, J. R., Tilney, L. G., and Porter, K. R. (1969). Microtubules in the formation and development of the primary mesenchyme in *Arbacia punctulata. J. Cell Biol.* **41,** 201–226.

Haugland, R. P. (2002). *In* "Handbook of Fluorescent Probes and Research Chemicals." p. 966. Molecular Probes Inc., Eugene, OR.

Heim, R., Cubitt, A. B., and Tsien, R. Y. (1995). Improved green fluorescence. *Nature* **373,** 663–664.

Honig, M. G., and Hume, R. I. (1986). Fluorescent carbocyanine dyes allow living neurons of identified origin to be studied in long-term cultures. *J. Cell Biol.* **103,** 171–187.

Jaffe, L. A., and Terasaki, M. (1993). Structural changes of the endoplasmic reticulum of sea urchin eggs during fertilization. *Dev. Biol.* **156,** 556–573.

Jaffe, L. A., and Terasaki, M. (1994). Structural changes in the endoplasmic reticulum of starfish oocytes during meiotic maturation and fertilization. *Dev. Biol.* **164,** 579–587.

Johnson, L. V., Walsh, M. L., and Chen, L. B. (1980). Localization of mitochondria in living cells with rhodamine 123. *Proc. Natl. Acad. Sci. (USA)* **77,** 990–994.

Kozak, M. (1999). Initiation of translation in prokaryotes and eukaryotes. *Gene* **234,** 187–208.

Krieg, P. A., and Melton, D. A. (1984). Functional messenger RNAs are produced by SP6 *in vitro* transcription of cloned cDNAs. *Nucleic Acids Res.* **12,** 7057–7070.

Lénárt, P., Rabut, G., Daigle, N., Hand, A. R., Terasaki, M., and Ellenberg, J. (2003). Nuclear envelope breakdown in starfish oocytes proceeds by partial NPC disassembly followed by a rapidly spreading fenestration of nuclear membranes. *J. Cell Biol.* **160,** 1055–1068.

Lucero, H. A., Lebeche, D., and Kaminer, B. (1994). ERcalcistorin/protein disulfide isomerase (PDI). Sequence determination and expression of a cDNA clone encoding a calcium storage protein with PDI activity from endoplasmic reticulum of the sea urchin egg. *J. Biol. Chem.* **269,** 23112–23119.

Minden, J. S., Agard, D. A., Sedat, J. W., and Alberts, B. M. (1989). Direct cell lineage analysis in *Drosophila melanogaster* by time-lapse, three-dimensional optical microscopy of living embryos. *J. Cell Biol.* **109,** 505–516.

Munro, S., and Pelham, H. R. (1987). A C-terminal signal prevents secretion of luminal ER proteins. *Cell* **48,** 899–907.

Ormo, M., Cubitt, A. B., Kallio, K., Gross, L. A., Tsien, R. Y., and Remington, S. J. (1996). Crystal structure of the *Aequorea victoria* green fluorescent protein. *Science* **273,** 1392–1395.

Pagano, R. E., Martin, O. C., Kang, H. C., and Haugland, R. P. (1991). A novel fluorescent ceramide analogue for studying membrane traffic in animal cells: Accumulation at the Golgi apparatus results in altered spectral properties of the sphingolipid precursor. *J. Cell Biol.* **113,** 1267–1279.

Roegiers, F., Djediat, C., Dumollard, R., Rouviere, C., and Sardet, C. (1999). Phases of cytoplasmic and cortical reorganizations of the ascidian zygote between fertilization and first division. *Development* **126,** 3101–3117.

Scott, L. B., Leahy, P. S., Decker, G. L., and Lennarz, W. J. (1990). Loss of yolk platelets and yolk glycoproteins during larval development of the sea urchin embryo. *Dev. Biol.* **137,** 368–377.

Shilling, F., Mandel, G., and Jaffe, L. A. (1990). Activation by serotonin of starfish eggs expressing the rat serotonin 1c receptor. *Cell Regulation (*now called *Mol. Biol. Cell)* **1,** 465–469.

Smith, R. M., Baibakov, B., Ikebuchi, Y., White, B. H., Lambert, N. A., Kaczmarek, L. K., and Vogel, S. S. (2000). Exocytotic insertion of calcium channels constrains compensatory endocytosis to sites of exocytosis. *J. Cell Biol.* **148,** 755–767.

Speksnijder, J. E., Terasaki, M., Hage, W. J., Jaffe, L. F., and Sardet, C. (1993). Polarity and reorganization of the endoplasmic reticulum during fertilization and ooplasmic segregation in the ascidian egg. *J. Cell Biol.* **120,** 1337–1346.

Stricker, S. A. (1995). Time-lapse confocal imaging of calcium dynamics in starfish embryos. *Dev. Biol.* **170,** 496–518.

Tang, T. L., Freeman, R. M., O'Reilly, A. M., Neel, B. G., and Sokol, S. Y. (1995). The SH2-containing protein-tyrosine phosphatase SH-PTP2 is required upstream of MAP kinase for early *Xenopus* development. *Cell* **80,** 473–483.

Terasaki, M. (1994). Redistribution of cytoplasmic components during germinal vesicle breakdown in starfish oocytes. *J. Cell Sci.* **107,** 1797–1805.

Terasaki, M. (1995). Visualization of exocytosis during sea urchin egg fertilization using confocal microscopy. *J. Cell Sci.* **108,** 2293–2300.

Terasaki, M. (1998). Imaging of echinoderm fertilization. *Mol. Biol. Cell* **9,** 1609–1612.

Terasaki, M. (2000). Dynamics of the ER and Golgi apparatus during early sea urchin development. *Mol. Biol. Cell* **11,** 897–914.

Terasaki, M., and Dailey, M. E. (1994). Confocal microscopy of living cells. *In* "Handbook of Biological Confocal Microscopy" (J. Pawley, ed.), pp. 327–346, Plenum Press, New York.

Terasaki, M., and Jaffe, L. A. (1991). Organization of the sea urchin egg endoplasmic reticulum and its reorganization at fertilization. *J. Cell Biol.* **114,** 929–940.

Terasaki, M., Song, J., Wong, J. R., Weiss, M. J., and Chen, L. B. (1984). Localization of endoplasmic reticulum in living and glutaraldehyde fixed cells with fluorescent dyes. *Cell* **38,** 101–108.

Terasaki, M., Jaffe, L. A., Hunnicutt, G. R., and Hammer, J. A., III (1996). Structural change of the endoplasmic reticulum during fertilization: Evidence for loss of membrane continuity using the green fluorescent protein. *Dev. Biol.* **179,** 320–328.

Terasaki, M., Miyake, K., and McNeil, P. L. (1997). Large plasma membrane disruptions are rapidly re-sealed by Ca^{2+}-dependent vesicle–vesicle fusion events. *J. Cell Biol.* **139,** 63–74.

Terasaki, M., Campagnola, P., Rolls, M. M., Stein, P., Ellenberg, J., Hinkle, B., and Slepchenko, B. (2001). A new model for nuclear envelope breakdown. *Mol. Biol. Cell* **12,** 503–510.

Tsien, R. Y., and Waggoner, A. (1994). Fluorophores for confocal microscopy: Photophysics and photochemistry. *In* "Handbook of Biological Confocal Microscopy" (J. Pawley, ed.), pp. 267–279, Plenum Press, New York.

Vogel, S. S., Smith, R. M., Baibakov, B., Ikebuchi, Y., and Lambert, N. A. (1999). Calcium influx is required for endocytotic membrane retrieval. *Proc. Natl. Acad. Sci. (USA)* **96,** 5019–5024.

Wessel, G. M., Conner, S. D., and Berg, L. (2002). Cortical granule translocation is microfilament mediated and linked to meiotic maturation in the sea urchin oocyte. *Development* **129,** 4315–4325.

Whalley, T., Terasaki, M., Cho, M. S., and Vogel, S. S. (1995). Direct membrane retrieval into large vesicles after exocytosis in sea urchin eggs. *J. Cell Biol.* **131,** 1183–1192.

Wilding, M., Wright, E. M., Patel, R., Ellis-Davies, G., and Whitaker, M. (1996). Local perinuclear calcium signals associated with mitosis-entry in early sea urchin embryos. *J. Cell Biol.* **135,** 191–199.

Wormington, M. (1991). Preparation of synthetic mRNAs and analyses of translational efficiency in microinjected *Xenopus* oocytes. *Meth. Cell Biol.* **36,** 167–183.

Yang, F., Moss, L. G., and Phillips, G. N. (1996). The molecular structure of green fluorescent protein. *Nature Biotech.* **14,** 1246–1251.

CHAPTER 20

Isolation of Organelles and Components from Sea Urchin Eggs and Embryos

Gary M. Wessel* and Victor D. Vacquier[†]

*Department of Molecular and Cell Biology and Biochemistry
Brown University
Providence, Rhode Island 02912

[†]Center for Marine Biotechnology and Biomedicine
Scripps Institute of Oceanography
University of California San Diego
La Jolla, California 92093

0091-679X/04 $35.00

I. Overview

Although sea urchin eggs and embryos are only about 100 microns in diameter and about 50 ng in total mass, the numbers obtainable from each adult reach many millions. The eggs are each at an identical stage when shed, and embryos are remarkably synchronous during development. Thus, one can isolate significant amounts of organelles and components from a uniform population of cells for specific biochemical and molecular analysis. Because of the ability to isolate various embryonic cell-types during development, and the conservation of many of the specialized egg and embryo organelles, this embryo has served as a rich source for experimentation, and development of techniques for organelle isolation.

This chapter documents many of the prevalent techniques in organelle and component isolation from sea urchin eggs and embryos. Isolation and manipulation of sperm is dealt with in a separate chapter (Chapter 21). As might be expected for this classic system, the techniques have evolved over the years and so, where possible, we give the current, general isolation scheme followed by reference to significant variations on this theme. We emphasize here the protocols that have been developed since the first volume of this Methods book on echinoderms (Schroeder, 1986).

II. Egg Jelly Molecules Affecting Sperm

A. Introduction

Sea urchin eggs of various species are from 80 to 120 μm in diameter and are surrounded by a jelly coat that, when hydrated, can be roughly 0.5 egg diameters in thickness. The egg jelly coat has two functions: to protect the egg surface and to induce the sperm acrosome reaction. Quick-Freeze, deep-etch rotary shadowing electron microscopy has been applied to the egg jelly layer and to isolated egg jelly molecules. The egg jelly layer *in situ* is a dense meshwork of three-dimensional fibers. Isolated jelly macromolecules resemble fibrous pearls on a necklace (Bonnell and Chandler, 1990; Bonnell *et al.*, 1994).

Three types of molecules have been isolated from egg jelly that have profound effects on sperm physiology: (1) sperm-activating peptides (Darszon *et al.*, 2001; Garbers, 1989; Kaupp *et al.*, 2003; Suzuki, 1995); (2) the fucose sulfate polymer (FSP; Hirohashi and Vacquier, 2002a; SeGall and Lennarz, 1979; Vacquier and Moy, 1997; Vilela-Silva *et al.*, 2002); and (3) a sialoglycan that is a form of polysialic acid (Hirohashi and Vacquier, 2002b; Kitazume *et al.*, 1994). Considerable research has also been accomplished on the jelly layer of starfish eggs, the molecules of which also have profound effects on sperm (Hoshi *et al.*, 2000).

1. Egg Jelly Peptides

The sequences of approximately 100 egg jelly peptides from various echinoid species have been determined (Suzuki, 1995). The egg jelly peptide speract from sea urchins of the genera *Strongylocentrotus, Hemicentrotus*, and *Lytechinus* has the sequence GFDLNGGGVG. Resact, the egg jelly peptide from sea urchins of the genus *Arbacia*, has the sequence $\text{CVTGAPGCVGGGRL}_{\text{NH2}}$. Nowadays, these peptides can be synthesized commercially at reasonable prices. Approximately 150 to 200 million years separate the genera *Strongylocentrotus* and *Arbacia* (Gonzales and Lessios, 1999), both speract and resact bind to sperm receptors to up-regulate ion channels that activate sperm respiration and motility (reviewed in Darszon *et al.*, 2001). Speract can also act synergistically with FSP in the induction of the sperm acrosome reaction (Hirohashi and Vacquier, 2002c; Yamaguchi *et al.*, 1989). Speract binds to a 77 kDa receptor protein in the sperm tail plasma membrane that, in turn, up-regulates guanylyl cyclase (Garbers, 1989). However, resact binds directly to the *Arbacia* sperm tail guanylyl cyclase to up-regulate the enzyme. The concentration of cGMP increases and acts directly on cation channels that regulate chemo-attraction toward the egg (Garbers, 1989; Ward *et al.*, 1985). A single *Arbacia* sperm cell can change its swimming behavior upon binding one molecule of resact (Kaupp *et al.*, 2003). The *Arbacia* guanylyl cyclase was the first receptor guanylyl cyclase to be cloned (Tamura *et al.*, 2001).

2. The Sialoglycan (SG)

The SG can be isolated from egg jelly by DEAE chromatography after pronase digestion, or β-elimination of total egg jelly. If pronase digestion is used, at least one amino acid will remain on the SG chain. This residue can be utilized for the coupling of the SG to a solid support, such as beaded Affigel (BioRad). In *Hemicentrotus pulcherrimus*, the polysialic acid chains are *O*-glycosidically attached to a 180 kDa glycoprotein, whereas in *S. purpuratus* the size of the intact molecule is 250 kDa. The structure of the polysialic acid was determined to be (5-O-glycolyl-Neu5Gcα2)$_n$, which is usually abbreviated "Neu5Gc." It has a degree of polymerization of 4 to 40 residues in *H. pulcherrimus* and 25 residues in *S. purpuratus* (Kitazume *et al.*, 1994). Purified SG greatly potentiates the FSP-induced acrosome reaction, but by itself is totally noninductive. SG elevates intracellular pH without increasing intracellular calcium (Hirohashi and Vacquier, 2002b). SG must bind a yet unidentified receptor on the sea urchin sperm plasma membrane. The potentiation by SG of the FSP-induced acrosome reaction is the first physiological response to be found for a polysialic acid (Hirohashi and Vacquier, 2002b).

3. The Fucose Sulfate Polymer (FSP)

Purified FSP of egg jelly can by itself induce the sperm acrosome reaction. FSP is known to bind REJ1 (receptor for egg jelly-1), a glycoprotein on the sperm plasma membrane localized to the acrosome and the flagellum (Vacquier and Moy, 1997). FSP opens two pharmacologically different sperm calcium channels that are required for acrosome reaction induction (reviewed in Darszon *et al.*, 2001). REJ1 contains the ~1000 residue "REJ module" found in members of the polycystin-1 family of orphan receptors that are mutated in autosomal dominant polycystic kidney disease (Moy *et al.*, 1996).

FSPs from different sea urchin species are species-specific inducers of the sperm acrosome reaction. Most FSPs are (3-α-L-fucopyranosyl-1)$_n$ homopolymers that differ from each other in the position of sulfation at the *2-O-* and *4-O-*positions of the fucosyl unit. Females of *S. purpuratus* have one of two isotypes of FSP. One isotype is a trisaccharide repeat in which the first fucosyl residue is sulfated on the *2-O-* and *4-O-*positions and the second and third fucosyl residues are sulfated on only the *4-O-*position. The second isotype of FSP of this species is a homopolymer of-1→3-linked fucosyl residues sulfated on the *2-O-* and *4-O-*positions (Alves *et al.*, 1998). The species *S. franciscanus* has the simplest FSP, which consists of α-L-1→3-fucopyranosyl residues sulfated at only the *2-O-*position (Vilela-Silva *et al.*, 1999). Females of *S. droebachiensis* also synthesize two different, female specific FSPs, but in this species, the glycosidic bond linking fucosyl residues is -1→4-, which is highly unusual considering the evolutionary closeness of this species to *S. purpuratus* (Vilela-Silva *et al.*, 2002). FSP is the only known pure polysaccharide that induces a signal transduction event in an animal sperm. The surprising finding is that the specific molecular recognition between sperm receptors and FSP is based on the pattern of FSP's sulfation and the glycosidic linkage.

B. Purification of FSP and SG from Sea Urchin Egg Jelly: Preparing Crude Egg Jelly

1. Female sea urchins are injected with 0.5 M KCl and spawned into Millipore-filtered seawater (MFSW) and the eggs with hydrated jelly coats are allowed to settle. The supernatant is then aspirated away and the settled egg mass is resuspended in an equal volume of MFSW.

2. A combination pH electrode is placed in the beaker and the egg suspension stirred rapidly by hand with a spatula while 0.1 N HCl is added dropwise. When

the pH reaches 5.0, the stirring continues for 2 min before the dropwise addition of 0.1 N NaOH to carefully bring the pH back to between 7 and 8.

3. The egg suspension is poured into 50 ml conical plastic centrifuge tubes that are centrifuged in a hand-driven centrifuge at ~900 rpm for 2 min.

4. The supernatant is carefully removed with a 10 ml pipette and centrifuged at 30,000×g for 30 min (4 °C). The supernatant is called "crude egg jelly" and is aliquoted and frozen at −20 °C. It retains its acrosome reaction-inducing ability for years. The macromolecules of crude egg jelly do not stain with Coomassie blue. However, silver-stained gels of crude egg jelly show over 10 bands, varying from 350 to 20 kDa (Vacquier and Moy, 1997). Future work should be directed toward the purification of these macromolecules and the study of their effects on sperm physiology.

C. β-Elimination

1. Crude egg jelly is precipitated by the addition of an equal volume of 95% ethanol. The precipitate is collected by a 1000×g centrifugation (5 min) and dissolved in distilled water. β-elimination is performed by addition of 0.05 N NaOH containing 1 M $NaBH_4$ (final concentrations) and incubation at 45 °C for 24 h.

2. An equal volume of 50 mM sodium acetate, pH 5.0, is then added and the final pH brought to 5.0 by the dropwise addition of 2 M acetic acid. Silver staining of SDS-PAGE shows that all protein is degraded by this treatment (Hirohashi and Vacquier, 2002a).

D. Pronase Digestion

Crude egg jelly is dissolved in 50 mM Tris, 10 mM sodium azide, pH 7.5, and pronase or proteinase-K added to 0.05 mg/ml. Digestion is at 37 °C for 6 to 18 h. The egg jelly polysachairde chains are then recovered by precipitation after the addition of two volumes of 95% ethanol.

E. DEAE Cellulose Chromatography

1. SG and FSP are then separated and purified from each other by DEAE cellulose chromatography. We prefer to use Whatman DE-52 Cellulose.

2. Either β-eliminated or pronase digested egg jelly polysaccharides are dissolved in 50 mM sodium acetate, pH 5.0, and loaded onto DEAE-cellulose equilibrated in this buffer. After washing with 10 column volumes of this buffer, a linear gradient of 0 to 3 M NaCl in 50 mM acetate, pH 5.0, is applied. For a 50 ml column of DEAE, a 200 ml gradient, collected as forty 5 ml fractions, will work well.

3. Fractions are tested for sulfated glycans in the metachromatic assay (Farndale *et al.*, 1986), for neutral sugars by the phenol sulfuric acid assay (Dubois *et al.*, 1956), and sialic acid by the thiobarbituric acid assay (Aminoff, 1961). Two

peaks of eluted material are obtained. The fractions eluting before 1.0 M NaCl contain the SG, and the fractions eluting between 2 and 3 M NaCl contain FSP.

4. Peak fractions of SG and FSP are pooled and precipitated by addition of an equal volume of 95% ethanol and chilling for more than 30 min. The precipitates are collected by low-speed centrifugation, dissolved in distilled water, dialyzed against distilled water, and stored at −20 °C. As little as 1 μg per ml of FSP will induce the acrosome reaction in approximately 50% of *S. purpuratus* sperm. The relative molecular mass of SG determined by gel filtration chromatography is approximately 100,000, and of FSP, approximately 1 million Daltons (Hirohashi and Vacquier, 2002a,b). The SG has a molar ratio of fucose to sialic acid of approximately 1:6 (Hirohashi and Vacquier, 2002b), whereas the FSP is essentially 100% fucose sulfate (reviewed in Vilela-Silva *et al.*, 2002).

III. Isolation of the Vitelline Layer from Sea Urchin Eggs

A. Introduction

Transmission electron microscopy of the vitelline layer (VL) of unfertilized eggs shows this glycocalyx to be a lacy, fibrous extracellular matrix intimately bonded to the plasma membrane (Chandler and Heuser, 1980; Kidd, 1978; Kidd and Mazia, 1980). The VL cannot be separated from the unfertilized egg without destroying the egg. The VL appears to be attached to the plasma membrane by regularly spaced posts that embed into the plasma membrane. During normal fertilization, the exocytosing cortical granule protease (Haley and Wessel, 1999) cleaves the bonds between the VL and the plasma membrane, thus allowing the VL to elevate and act as a template for the formation of the fertilization envelopes (see Weidman and Kay, 1986, for a comprehensive review). Eight monoclonal antibodies, that reacted species-specifically with the outer surface of the VL, inhibited sperm binding in proportion to the amount of antibody bound (Gache *et al.*, 1983). Other work showed that the VL contains over 20 macromolecules of between 30 and 370 kDa (Correa and Carroll, 1997; Niman *et al.*, 1984). A proteomics approach should now be applied to identify all the proteins comprising the VL.

B. Isolation of Vitelline Layers by Homogenization

1. The method described here is taken from Glabe and Vacquier (1977). Eggs of *S. purpuratus* are obtained by 0.5 M KCl injection and egg jelly coats dissolved by a 2 min exposure to pH 5.0 seawater, as previously described. All procedures are at 4 °C.

2. The settled eggs are washed three times in excess volumes of fresh MFSW. At this step, the eggs may be treated, or not treated, for 5 min with 0.2 mM *N*-bromosuccinimide in seawater (NBS, pH 7.5) and then washed with excess

MFSW. The mild oxidation provided by the NBS results in VL that isolate as intact envelopes which still retain their sperm-binding capacity and ultrastructural appearance (Glabe and Vacquier, 1977). If NBS is excluded, the VLs fragment during homogenization but their ultrastructure is still preserved.

3. After NBS treatment, the egg pellet is resuspended in 10 volumes of 20 mM EDTA, 50 mM sodium acetate (pH 6.0), 0.4% Triton X-100, 10 mM benzamidine, 0.1 mg/ml soybean trypsin inhibitor (SBTI), and 1 mM PMSF (phenylmethylsulfonylfluoride, made as a 1 M stock in acetone).

4. The egg suspension is immediately poured into a loose-fitting glass ball Thomas tissue grinder and several cycles of homogenization done with periodic microscopic observation; 100% egg lysis and VL liberation can be obtained.

5. The homogenate is centrifuged $800 \times g$ for 10 min, the supernatant discarded, and the VL pellet washed in fresh homogenization medium. Egg nuclei are found at the bottom of the tube. The VLs are stable to high and low salt buffers at neutral pH.

6. Forty ml of dejellied *S. purpuratus* eggs yields about 1 ml of $800 \times g$ packed VLs, which are about 50 to 60 mg protein (Glabe and Vacquier, 1977). SDS-PAGE analysis shows many protein components from 25 to 250 kDa. VLs are 90 to 95% protein and 4% carbohydrate by weight. Because Triton X-100 is used in their isolation, they are devoid of lipid. Electron microscopy shows VLs have an average thickness of 30 nm and that the spacing of the microvillar casts of the egg surface is retained. When returned to seawater and mixed with sperm, the sperm are seen to bind only to the outer surface of the isolated VL (Glabe and Vacquier, 1977).

C. Isolation of Thin Vitelline Layers

1. This method is based on papers by Acevedo-Duncan and Carroll (1986) and Correa and Carroll (1997). All previous work on this subject was reviewed in detail by Weidman and Kay (1986).

2. The egg jelly coats are removed by treatment for 2 min to pH 5.0 seawater, the pH adjusted down with 0.1 N HCl. After readjustment to pH 7 with 0.1 N NaOH, the eggs are settled twice through fresh MFSW.

3. The eggs are fertilized in MFSW with a heavy concentration of sperm. At 20 s after insemination, the eggs are diluted 10-fold into divalent cation-free medium (575 mM NaCl, 10 mM KCl, 10 mM EGTA, 5 mM ethylene glycol, 10 mM Tris pH 8.0, 5 mM benzamidine, and 1 mM 3-amino –1,2,3-triazole, and 50 μg/ml soybean trypsin inhibitor).

4. As an alternative to sperm fertilization, the eggs can be activated in ionophore A23187. Ionophore is made up as a 2 mg/ml stock solution in dimethylsulfoxide. The divalent cation-free medium is made 1% in the ionophore stock immediately before addition to the eggs. The final A23187 concentration is 38 μM (Glabe and Vacquier, 1978). Each 1 ml of dejellied eggs is resuspended in 10 ml of

ionophore-divalent cation-free medium. The cortical granules exocytose in unison throughout the entire egg population, causing the VL to elevate at once over the entire egg surface.

5. In either sperm or ionophore activation, the eggs with thin, elevated VLs are washed several times by settling in fresh divalent cation-free medium. The thin VLs are then stripped from the eggs by passing the egg suspension through a Nytex nylon screen of about 100 μm pore size. An alternative method is to gently homogenize the eggs with a loose-fitting glass ball homogenizer. With either method, the suspension is checked microscopically to monitor release of the thin VLs.

6. Egg debris is then sedimented by gentle hand centrifugation, visually checking for eggs still in the supernatant. The egg-free supernatant is finally centrifuged at 100 to 300$\times g$ for 10 min to sediment the thin VLs. The supernatant is discarded and the pellet of thin VLs is resuspended in fresh divalent cation-free medium, stirred with a spatula, and then resedimented. This washing procedure is repeated until the VLs are free of cytoplasm.

7. These thin VLs contain a 350 kDa glycoprotein that has the properties expected of a sperm receptor (Correa and Carroll, 1997).

IV. Isolation of the Cell Surface Complex and the Plasma Membrane–Vitelline Layer (PMVL) Complex from Sea Urchin Eggs

A. Introduction

The first method to isolate the cell surface complex (CSC) in large quantity was developed in the laboratory of W. J. Lennarz (Detering *et al.*, 1977). The method underwent several revisions and was presented in detail by Kinsey (1986). Since then, several workers (in particular, N. Hirohashi) have modified the procedure, which works especially well for eggs of *S. purpuratus*. The CSC consists of the cortical granules (CG) (Fig. 1), the plasma membrane (PM), and the vitelline layer (VL). After isolation of the CSC, the CG are dissociated from the PMVL (Kinsey, 1986). The isolation methods (Kinsey, 1986) have been slightly altered to prepare the CSC for identification of a 350 kDa glycoprotein with the characteristics expected of a sperm receptor (Hirohashi and Lennarz, 2001; Ohlendieck *et al.*, 1993).

B. Isolation of the Cell Surface Complex

1. All procedures are at 4 °C. Eggs are obtained by spawning females into MFSW. Eggs are dejellied by exposure to pH 5.0 for 2 min and then washed 3 times by resuspension, settling, and aspiration of MFSW.

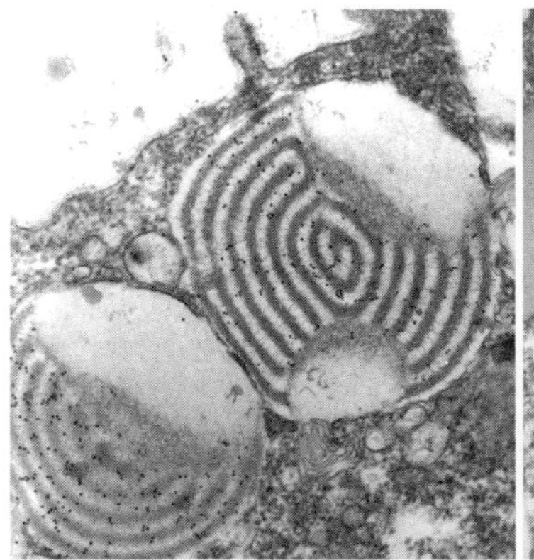

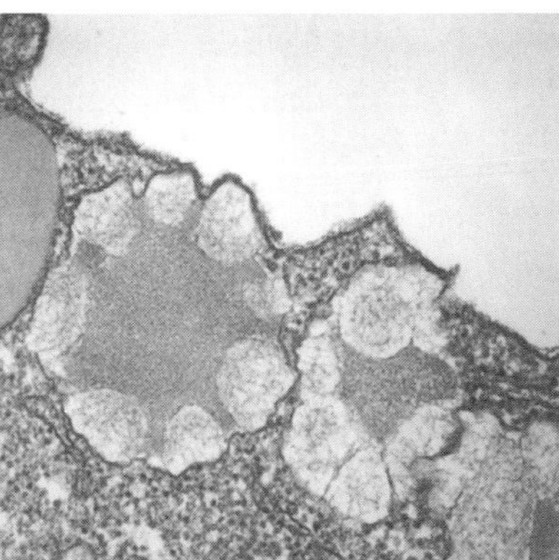

Fig. 1 Cortical granules from two species of sea urchins. Each is attached to the plasma membrane of the egg, and contains a distinct substructural morphology. Each granule from *Strongylocentrotus purpuratus* is 1.5 microns in diameter, and from *Arbacia punctulata* is approximately 1.2 microns. The granules shown from *S. purpuratus* are labeled with antibodies to a content protein, and show that it is present only in the lamellar region, and not in the lucent homogeneous region.

2. The eggs are washed twice in calcium-magnesium free artificial seawater (CMFSW; 500 mM NaCl, 10 mM KCl, 2.5 mM $NaHCO_3$, 25 mM EGTA, adjusted to pH 8.0 with 1 N NaOH). A 10% vol/vol suspension of the eggs in CMFSW containing a protease inhibitor cocktail is then prepared (CMFSW-PIC).

3. The protease inhibitor cocktail (PIC) can be of two formulations. For example, the Sigma PI (catalog number P-8340) can be diluted 1:500 in CMFSW in addition to final concentrations of 1 mg/ml soybean trypsin inhibitor and 10 mM benzamidine. Alternatively, a stock of leupeptin 1 mg/ml, antipain 2 mg/ml, benzamidine 10 mg/ml, pepstatin 1 mg/ml can be prepared in 0.5 M EDTA pH 7.5 and added to CMFSW at a 1:500 dilution in addition to 1 mg/ml soybean trypsin inhibitor.

4. The egg suspension in CMFSW-PIC is transferred to a prechilled glass/Teflon homogenizer and the eggs homogenized slowly, checking continuously with the microscope. It is important not to overhomogenize the CSC, resulting in their fragmentation into small pieces. When >90% of the eggs are lysed, add an additional 10 volumes of CMFSW-PIC, and stir by hand for 2 min, and then centrifuge $800 \times g$ for 1 min.

5. Discard the supernatant. Carefully remove the upper pellet of white CSC without picking up the bottom yellow pellet of broken eggs, and transfer the CSC to a clean tube.

6. Resuspend CSC in fresh CMFSW-PIC, stir, and repeat the 1 min centrifugation at $800 \times g$. After the final wash, the pellet should be fairly pure CSC.

C. Separation of the Plasma Membrane–Vitelline Layer Complex from the Cortical Granules

1. Remove as much CMFSW-PIC from the CSC pellet as is possible. Add 2 ml ice-cold 1 M sucrose-PIC, allowing the solution to flow gently down the side of the tube. Allow this to wash over the top of the CSC pellet without disturbing the pellet. Pipette off the sucrose solution.

2. Resuspend the pellet of CSC in 10 vol 1 M sucrose-PIC (the PIC being diluted 1:200 in the 1 M sucrose). Swirl the tube by hand in ice. The cortical granules will detach in 15 to 60 min. Check with the phase contrast microscope every 10 min and gently swirl the tube when checking.

3. When the detachment of CG is complete, centrifuge at $1000 \times g$ for 20 min. The supernatant containing the cortical granules is removed to a clean tube.

4. Resuspend the final PMVL pellet in CMFSW-PIC and centrifuge $1000 \times g$ for 1 min. Repeat this wash until the PMVL pellet is clean.

5. Resuspend the final PMVL pellet in CMFSW-PIC. Aliquot and freeze at −80 °C. Electron microscopy shows the membrane preparation is quite pure (Kinsey, 1986).

V. The Cytolytic Isolation of the Egg Cortex

A. Introduction

The cortex of the egg consists of the plasma membrane (PM) and the first 5 μm of cytoplasm that is associated with the PM. The cortex is rich in actin and changes in its mechanical properties during fertilization and the cell cycle (reviewed in Vacquier, 1981). Morphogenetic information in the form of specific mRNAs could be differentially localized to the cortex. A simple method was discovered where the unfertilized egg, or the zygote, spontaneously contracts so hard as to lyse the cell and roll the cortex off as an intact structure that is rich in actin (Vacquier and Moy, 1980). Isolation of mRNA from these cortices could be done to characterize what messages are stored in this region.

B. Cytolytic Isolation of the Egg Cortex

1. *Lytechinus pictus* eggs are spawned by injection of adult sea urchins with 0.5 M KCl. The egg jelly coats are dissolved by a 2 min exposure to pH 5.0 seawater, as has been described.

2. Five ml of dejellied eggs is suspended in 100 ml seawater containing 2 mg of pancreatic trypsin. The eggs are stirred gently for 10 min and allowed to settle and the supernatant aspirated away.

3. The eggs are resuspended and then allowed to settle in 50 ml of seawater containing 0.01 M dithiothreitol (DTT), pH 9.1. The eggs are washed in seawater pH 8.0 and allowed to settle. The trypsin/DTT treatment completely removes the egg VL, stripping the surface down to the plasma membrane.

4. For unfertilized eggs, the isolation medium (IM) is 280 mM NaCl, 350 mM glycine, 20 mM EGTA, 2 mM $MgCl_2$, 42 mM Tris-OH, 5 mM sodium azide, 5 mM benzamidine HCl, 1 mM ATP, pH 7.0. The osmotic pressure of IM is 980 mOsm. Each 1 ml of eggs is suspended in 100 ml of IM at 21 °C. Microscopic observation shows that the egg cortex contacts, the egg ruptures, and the cortex peels off as a single cap-shaped structure.

5. When about 90% of the eggs have lysed, the lysate is poured into 50 ml conical tubes, which are hand-centrifuged 2 min at 216 rpm (radius 17 cm). The pellet of unlysed eggs is removed with a long Pasteur pipette, the contents of the tube mixed gently by inversion and the 2-min centrifugation repeated until no whole eggs remain in the pellet.

6. The intact cortices are then sedimented by a $100\times g$ centrifugation for 3 min and washed in fresh IM. The cortices contain the plasma membrane and the cortical granules.

7. To isolate cortices from zygotes, the trypsin/DTT treated eggs are inseminated with a heavy dose of sperm in seawater and then poured into a large petri dish and cultured without agitation. The zygotes are washed by aspiration and addition of fresh seawater (21 °C) until the supernatant is clear.

8. At 40 min post-insemination, the zygotes are poured into a 50 ml tube and allowed to settle and the supernatant removed.

9. Each 0.5 ml of zygote pellet is resuspended in 50 ml IM and allowed to stand 30 min at 21 °C. The lysed cortices are then collected and washed as previously directed for the unfertilized eggs. Actin is the major component of the zygote cortices (Vacquier and Moy, 1980).

VI. Isolation of Cortical Granules

A. Introduction

Almost all holoblastically cleaving eggs have cortical granules (CG) bound to the inner surface of the egg plasma membrane. In *S. purpuratus*, each egg has about 18,000 cortical granules, with an average diameter of 1 μm. The CG migrate to the plasma membrane at a late stage in oogenesis (reviewed by Wessel *et al.*, 2001). During normal fertilization, the calcium transient in eggs triggers the CG to fuse with the egg cell membrane. Exocytosis of the CG begins at the point of sperm fusion and radiates around the egg as a circular wave, taking about 30 s to complete. The CG membrane surface area of roughly 57,000 square μm is incorporated into the egg cell membrane, also of about 41,000 square μm. Thus, during 30 s, the egg's surface area is roughly doubled (Schroeder, 1979). The extra surface area is extended as elongated surface microvilli and then most of the membrane is retrieved in a massive wave of endocytosis within the first 10 min after completion of exocytosis (Walley *et al.*, 1995). The cortical granule reaction is one of the most massive, synchronous examples of exocytosis known. Sea urchin egg cortical granule exocytosis has been studied as a model system for exocytosis and much has been learned.

B. Preparation of Cortical Granule Lawns

1. Eggs of *S. purpuratus* are dejellied and washed extensively in fresh MFSW. Plastic culture dishes, coverslips, or microscope slides are exposed for 5 min to a 1 mg/ml solution of protamine sulfate, or poly-L-lysine, and the surface washed in tap water and air dried. Tissue culture dishes of 5 cm diameter are perfect for this procedure. A 10% suspension of eggs is then applied to the surface. The negatively charged egg VL surfaces bond tightly to the positively charged substrate and the eggs flatten out.

2. The dish is then flooded with CaFSW containing 10 mM EGTA, pH 7.5. However, any solution of approximately 1000 milliosmoles and pH 7 to 8 will work. For example, for low ionic strength, one could use 1 M glycerol with 10 mM EGTA, 10 mM Hepes pH 7.5. After swirling for 2 min, the liquid is flicked out of the dish and the eggs irrigated with a jet of iso-osmotic medium containing 10 mM EGTA, 10 mM Hepes, pH 7.5. The cytoplasm of the eggs is sheared away with the strong jet of calcium chelating medium, leaving the cortices of the eggs firmly bound to the solid support. The negatively charged VL is electrostatically bound to the solid support, the VL is bound to the plasma membrane, and on the inside of the PM is the layer of tightly bound cortical granules (Vacquier, 1975; Fig. 2).

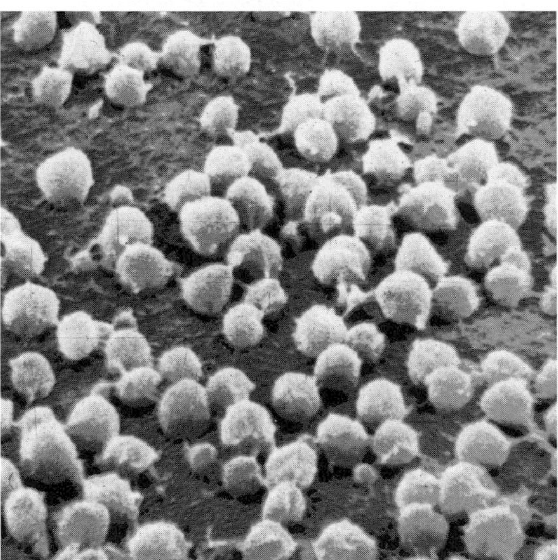

Fig. 2 Isolation of cortical granule lawns. The cortical granules remain attached to the plasma membrane and can further be isolated by shear, or activated by exocytosis by introduction of calcium.

VII. Isolation of Yolk Platelets

Echinoderms are isolecithal animals, meaning that a relatively small amount of yolk (compared to amphibians, avians, or dipterans) is evenly distributed throughout the cells. The major yolk components are packed into distinct membrane-bound organelles termed yolk platelets or granules. The yolk platelets of sea urchins contain several major molecules made either by the oocyte (YP30, protease) or the adult gut (Major Yolk Protein, or MYP) and reaching the oocyte by selective transport and endocytosis. It should be noted that the overall mass of the yolk platelets does not change during early development (e.g., Harrington and Easton, 1982), calling into question the function of the contents. Several procedures are available for the isolation of yolk platelets. Here, we present a general method for obtaining highly enriched preparations of yolk platelets based on Yokota and Kato (1988) with modifications (Brooks and Wessel, 2002). Additional protocols are available that are variations of the presented protocol, or that take advantage of density gradient media for obtaining purified yolk platelets (e.g., Armant *et al.*, 1986; Harrington and Easton, 1982; Schuel *et al.*, 1975; Unuma *et al.*, 1998).

1. Gametes are obtained by intercoelomic injection of 0.5 M KCl, and the eggs dejellied and washed twice in artificial seawater.
2. To avoid contamination of the yolk preparation by the very abundant cortical granule proteins, the yolk platelet preparation is isolated following fertilization and removal of the fertilization envelope [see Fertilization Envelope section of this chapter for detail]. The preferred approach is to treat eggs with 10 mM dithiothreitol (DTT) to remove the vitelline layer before activation (Epel *et al.*, 1970) so the cortical granule proteins are freely washed away.
3. Eggs are activated by the addition of the calcium ionophore A23187 (10 micrograms/ml) to minimize sperm contamination.
4. Following activation, the cells are kept on ice and washed (by settling) twice with calcium-free seawater and then twice with KCl solution (0.55 M KCl; 1 mM EDTA, pH 7.0).
5. Resuspend the cells in 5 volumes of KCl solution to which a general protease inhibitor cocktail is added and immediately homogenize in a Dounce tissue grinder by hand several times until complete lysis is achieved.
6. Centrifuge the egg homogenate for 4 min at 400×*g* at 4 °C, collect the supernatant, and re-centrifuge for 10 min at 2400×*g*.
7. Resuspend the pelleted precipitate in KCl solution and re-centrifuge at 2400×*g*. The final precipitate is the yolk platelet preparation.
8. This preparation is greatly enriched for yolk platelets, as assessed by electron microscopy, with the majority of contamination by mitochondria. For MYP isolation, this preparation is resolved on a SDS-PAGE gel and the 180 kDa MYP band is excised.

9. For biochemical analysis, we recommend aliquoting the preparation and avoid freeze–thawing, as this results in the loss of a major yolk platelet protein, YP30.

The MYP precursor protein can also be isolated from the coelomic fluid of adults (both male and female; see Shyu *et al.*, 1986 and Unuma *et al.*, 1998) and is described in Harrington and Easton (1982). Briefly:

1. Adults (males or females) are opened by cutting the peristomial membrane around Aristotle's lantern. The perivisceral coelomic fluid is poured out and collected in a beaker on ice to allow the coelomocytes to clot.
2. This clot is removed by decanting the supernatant, and the supernatant is centrifuged for 10 min at 10,000×*g* at 4 °C.
3. The supernatant is passed through a Corning disposable sterile syringe filter (0.2 micron; Corning, NY) to remove any residual cellular material, dialyzed at 4 °C overnight against several changes of deionized water, and vacuum dried.
4. The relative abundance of MYP is variable in the coelomic fluid but can reach over 75% in well-fed animals.

Recently, another constituent of the coelomic fluid of adult sea urchins was identified as an olfactomedin-containing protein, amassin. This protein functions in intercellular adhesion of the coelomic fluid cell population, the coelomocytes, to form the large clots that form in native coelomic fluid (Hillier and Vacquier, 2003).

VIII. Isolation of Mitochondria from Eggs and Embryos

Mitochondrial isolation has been used for many studies on the biochemistry of transcriptional regulation, DNA replication, and metabolism. Procedures are prevalent for mitochondrial isolation, and presented here are two of the most general schemes. The first scheme utilizes a sucrose gradient for excellent purification of mitochondria for studies in the biochemistry of DNA replication and transcriptional control (Cantatore *et al.*, 1974; Roberti *et al.*, 1997). The resultant mitochondria are, however, uncoupled from oxidative phosphorylation. A preparation amenable for the studies of oxidative phosphorylation mechanisms will follow, based on the protocol of Selak and Scandella (1987).

Protocol for sucrose-gradient purification of mitochondria (Cantatore *et al.*, 1974; Roberti *et al.*, 1997):

1. All steps are performed at 4 °C.
2. One ml of packed eggs or embryos is resuspended in 10 mls of 0.25 M sucrose in TEK buffer (100 mM Tris pH 7.6, 1 mM EDTA, 240 mM KCl) and homogenized using a Dounce type glass homogenizer.
3. The homogenate is centrifuged at 600*g* for 10 min and the supernatant recentrifuged at 2400*g* for 10 min in order to remove yolk.

4. The supernatant is again centrifuged at 7500*g* for 10 min and the pellet containing mitochondria is resuspended in 0.25 M sucrose in TEK buffer (original volume) and centrifuged at 15,000*g* for 10 min. The enriched mitochondria, at this point, exhibit oxidative phosphorylation but replacement of sucrose with mannitol (see next protocol by Selak and Scandella, 1987) enhances the respirational capabilities of mitochondria.

5. For sucrose-gradient purification of mitochondria, the pellet is resuspended in 0.25 M sucrose/TEK buffer and layered onto a step gradient of 1.0 and 1.5 M sucrose in TEK buffer, and spun in a swinging bucket rotor at 20,000*g* for 3 h.

6. The mitochondria sediment on top of the 1.5 M sucrose cushion, whereas contaminating yolk settle on top of the 1.0 M sucrose step.

7. The mitochondria are removed by pipetting, diluted into 0.25 M sucrose/TEK, and repelleted at 15,000*g* for 10 min.

8. The mitochondria resulting from this protocol are highly purified and have excellent transcription and translational capabilities when an energy source is provided.

Protocol for the isolation of mitochondria for studies of respiration (Selak and Scandella, 1987):

1. All steps in the protocol are at 4 °C.

2. Washed and dejellied eggs (see preceding text) are resuspended to 10% in isolation medium (300 mM mannitol, 4 mM $MgCl_2$, 4 mM MOPS pH 7.2, 10 mM EGTA, and soybean trypsin inhibitor at 0.5 mg/ml).

3. The cells are homogenized with a Teflon pestle until complete lysis occurs.

4. The lysate is centrifuged twice at 500*g* for 5 min and then at 12,000*g* for 10 min.

5. The pellet is resuspended using a loose-fitting Dounce homogenizer in isolation medium at the original volume and recentrifuged at 12,000*g* for 10 min.

6. The resultant pelleted mitochondria show some contamination by yolk and are held on ice until use. These mitochondria show excellent respirational response to the addition of multiple rounds of ADP and inhibition by addition of oligomycin (blocking the mitochondria F1 ATPase) and the uncouplers CICCP and FCCP.

IX. Isolation of Plasma Membranes and Lipid Rafts from Eggs and Zygotes

Understanding fertilization and early development is greatly enhanced by an ability to isolate and analyze the cell surface molecules. Many protocols are available for plasma membrane isolation, and here we present a general isolation scheme, as well as the current approach to separating lipid rafts from the membranes of eggs and zygotes. Lipid rafts are microdomains of plasma membranes with a highly ordered lipid phase that partitions and concentrates select proteins involved in cell interactions and signal transduction. Such raft structures are found in many eukaryotic cells, including sperm (see also Chapter 21 by Vacquier and Hirohashi for sperm raft isolation).

Membrane raft isolation procedure (Belton *et al.*, 2001):

1. Eggs or zygotes are pelleted at 250*g*. The seawater is aspirated and the cells resuspended in ice-cold buffer 1 (50 mM Hepes pH 7.4, 250 mM KCl, 5 mM EGTA, 1 mM EDTA, 10 mM NaF, 1 mM sodium vanadate, 1 mM PMSF, and 10 micromolar each of aprotinin, leupeptin, and benzamidine.
2. The cells are then homogenized on ice with a Teflon pestle until 100% lysis is obtained.
3. The homogenate is centrifuged at 150,000*g* in a swinging bucket rotor for 30 min at 2 °C and the supernatant is designated as the cytosolic fraction.
4. The pellet (total particulate fraction) is resuspended in five volumes of buffer 2 (50 mM Hepes pH 7.4, 150 mM NaCl, 5 mM EGTA, 1 mM EDTA, 2% Triton X-100, 10 mM NaF, 1 mM sodium vanadate, 1 mM PMSF and 10 micromolar each of aprotinin, leupeptin, and benzamidine), and the pellet is lightly homogenized until it appears completely dissolved. It is then incubated for 15 min more on ice.
5. The dissolved pellet is then centrifuged at 150,000*g* for 30 min at 2 °C to separate the Triton X-100 insoluble and soluble fractions. For analysis of the insoluble fraction at this point, use SDS sample buffer containing 10 mM NaF and 1 mM sodium vanadate to inhibit phosphatase activity.
6. The Triton X-100-insoluble fraction is resuspended in an equal volume of 80% sucrose in buffer 2 to give a final sucrose concentration of 40%.
7. Three to 4 mls of this fraction is layered into an ultracentrifuge tube designed for sucrose gradients, and overlayed with 6 mls of 30% sucrose buffer 2, then with 6 mls of 5% sucrose in buffer 2. The samples are centrifuged at 250,000*g* for 16 h at 2 °C and 1 ml fractions are collected from the top of the gradient.
8. Prior to SDS analysis, aliquots of each fraction are dialyzed against cold buffer 2 without Triton X-100.
9. To disrupt the lipid rafts by removal of cholesterol, treat with the cholesterol-binding agent, methyl-β-cyclodextrin (MβCD) dissolved in buffer 2 with protease and phosphatase inhibitors, but without Triton X-100, at 16 °C for 1 h. The levels of MβCD used for extraction have been examined (0–50 mM) and are documented (Belton *et al.*, 2001).
10. The MβCD soluble and insoluble fractions are separated by centrifugation at 21,000*g* for 1 h at 4 °C. Several phosphotyrosine-containing proteins are known to be present in the rafts, including the protein kinase Src, and are released (MβCD-soluble) following cholesterol treatment.

Several other plasma membrane isolation procedures for eggs and embryos are available. Readers are especially encouraged to explore the detailed protocols of Kinsey (1986) and modifications given in Giusti *et al.* (1997) as well as the isolation procedure and analysis of membranes from embryos by Cestelli *et al.* (1975).

X. Isolation of Microsomes Containing the Endoplasmic Reticulum

The endoplasmic reticulum of eggs contains calcium stores required for egg activation. At fertilization, signal transduction events leading to the formation of inositol tris-phosphate stimulates calcium release from the endoplasmic reticulum. Procedures have been developed to enrich for the endoplasmic reticulum in functional form i.e., uptake and release of calcium. We document here the procedure for isolation of the microsomes, which, by enzymatic and microscopic analysis, are enriched in the endoplasmic reticulum (Oberdorf *et al.*, 1986).

Procedure (All steps are performed at 4 °C.):

1. Eggs are washed three times in calcium-free seawater and resuspended in 10 volumes of buffer B (0.5 M KCl, 10 mM $MgCl_2$, 10 mM 2-[N-morpholino]propane sulfonic acid, 10 mM EGTA, 1 mM dithiothreitol, 1 mM phenylmethylsulfonyl fluoride, and 5 mM benzamidine, pH 7.0).
2. The eggs are then homogenized with a Teflon homogenizer and sequentially pelleted in a fixed-angle rotor as follows: 200*g* for 3 min (egg cortex); 3000*g* for 20 min (yolk); 15,000*g* for 20 min (mitochondria).
3. The 15,000*g* supernatant is spun at 100,000*g* for 30 min to pellet the microsomes. The microsomes pack loosely over the more dense, aggregated pellet.
4. The microsomes are carefully separated from the lower pellet and layered on a 0.3 M sucrose, 0.7 M glycine pH 7.0 cushion and the pellets from a 100,000*g* spin for 30 min were stored at −80 °C.

This protocol is modified from an earlier procedure (Inoue and Yoshioka, 1982).

XI. Nuclear Isolation Procedures

Several procedures are available for isolating nuclei from echinoderm oocytes, eggs and, embryos, and the choice of procedures depends on the application. Two major applications of isolated nuclei in sea urchins are (1) nuclear run-on (or run-off) experiments to measure as closely as possible the *in vivo* transcriptional activity of specific genes in select cell-types, and (2) biochemical analysis of transcription factors from nuclear extracts. Both applications and procedures are well documented, and the reader will be referred to other sources for detail.

Transcription of RNA in isolated nuclei is used to mimic the transcriptional activity in intact cells. Chromatin is maintained in its native state following nuclear isolation and the newly synthesized transcripts reflect the activity of that nucleus in the cell. By isolation, though, one is able to readily label the newly synthesized transcripts and measure their abundance in relative units. In one reaction, an investigator can measure the transcriptional activity of the cell for any one of the thousands of genes that may be transcribed in the cell at that time. We found Marzluff and Huang (1984) the most useful and comprehensive source for the entire procedure of nuclear isolation and *in vitro* transcription. Much of the development of the procedures by these investigators was using sea urchin embryos and the reference given specifically addresses sea urchin nuclear isolation.

Procedures for isolating nuclei for biochemical extracts of trans-factors is given in Chapter 26 by Coffman and Yuh in this volume, and isolation of pronuclei from both eggs and sperm are given in Poccia and Green (1986). Extensive detail is given in the isolation, extraction, and application of nuclei in those sources and will not be repeated here.

Isolation procedures are also available for germinal vesicles from oocytes of starfish (Chiba *et al.*, 1995). The procedure given below (Nemoto *et al.*, 1992) relies on first disrupting the actin–cytoskeletal matrix, which enables the germinal vesicles to be released intact by the cells.

Procedure:

1. Starfish oocytes denuded of their extracellular layers (see Chapter 3 by Foltz *et al.* for procedures) were resuspended in calcium-free seawater containing 10 micrograms/ml cytochalasin B (stock 2 mg/ml in DMSO).

2. Approximately 5 mls of the treated oocytes are collected by sedimentation and gently layered on a discontinuous sucrose gradient in a 50 ml tube containing different sucrose (1 M) mixtures with calcium-free seawater (CaFSW). From top to bottom, the gradient steps are: (1) 3 parts sucrose:7 parts CaFSW (7 mls); (2) 4 parts sucrose:6 parts CaFSW (11 mls); (3) 5 parts sucrose; 5 parts CaFSW (7 mls); and (4) 1.2 M sucrose (5 mls).

3. The preparations are then centrifuged in a swinging bucket rotor (5500*g* for 20 min at 4 °C). The oocytes sediment to the top of step 4 whereas the germinal vesicles sediment to the top of step 2. The time and speed of centrifugation will vary for different species—the values given are for the Japanese starfish *Asterias amurensis* and need to be adjusted for use in other species.

4. The germinal vesicles are gently removed by pipetting and are washed three times by centrifugation (1000*g* for 3 min) in cold washing medium (0.2 M sucrose, 0.3 M KCl, 5 mM $MgCl_2$, and 10 mM Tris pH 7.5).

5. The final pellet is kept in an ice bath or stored at −80 °C until use.

XII. Removal and Isolation of the Fertilization Envelope

The fertilization envelope forms within 60 s of insemination, and becomes stabilized by chemical cross-linking within a few minutes. The envelope forms largely by the secreted contents of the cortical granules mixing with the vitelline layer scaffolding as it lifts off the plasma membrane surface by hydration forces. This envelope protects the embryo from mechanical and biochemical insults until hatching but may also thwart investigators wishing to access the blastomeres for experimental manipulation. Several protocols have been developed to remove the envelope following fertilization or to prevent it from forming in the first place (Fig. 3). Other protocols have been established to isolate the envelope, or various parts of the envelope and its constituents (see preceding text for vitelline layer; see Weidman and Kay (1986) for detailed description of biochemical approaches for isolating cortical granule content proteins, and fertilization envelope intermediates; Fig. 4). Three approaches will be documented here to either (1) remove the envelope at some point following fertilization; (2) isolate the envelope; or (3) prevent the envelope from forming.

A. Removal of the Envelope during Embryogenesis

For access to blastomeres early in development, the fertilization envelope may have to be removed. This can be performed shortly after fertilization, though if prolonged culture of the embryos is needed before experimentation, it is often advisable to keep the envelope on the embryo until just prior to experimentation. The envelope reduces the stickiness of the exposed hyalin layer and resultant clumping of the embryos.

The stabilization of the fertilization envelope is due to chemical cross-linking of tyrosine residues by ovoperoxidase. If one blocks the activity of ovoperoxidase, the envelope still forms normally, it just does not stabilize and is thus easy to remove at any point following fertilization.

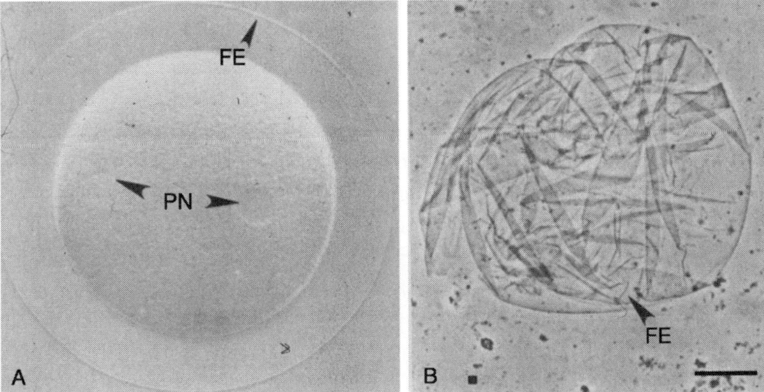

Fig. 3 Fertilization envelope isolation. The envelopes are stripped off the early embryo by first inhibiting the cross-linking reaction of ovoperoxidase, and then pouring the embryos through a nylon mesh to shear the envelopes off the cell. FE, fertilization envelope; PN, male (left) and female (right) pronuclei. Bar = 25 microns.

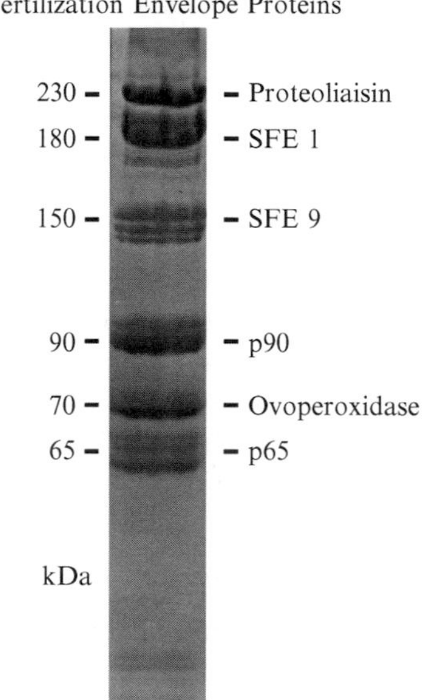

Fig. 4 The isolated fertilization envelope from *Strongylocentrotus purpuratus* contains a small population of abundant proteins as seen by SDS-PAGE.

To inhibit activity of ovoperoxidase, two popular reagents are used (Weidman and Kay, 1986): (a) 2 mM 3-amino-1,2,3-triazole in seawater (from a 1 M stock in deionized water), and (b) 10 mM para-aminobenzoic acid in seawater (make fresh before using, keep out of direct light, and pH the seawater to pH 8.0 following dissolution of the pABA). Eggs should be washed once in the reagent and then fertilized and cultured in the inhibitor. Embryos develop normally to larvae in either reagent condition.

When envelopes need to be removed, pass the embryos through Nitex nylon mesh. Sizes of Nitex used depend on the embryo. The strategy is to choose the size through which the embryo will pass, but which the envelope would need to deform. This deformation is usually sufficient to strip the envelope off the embryo and both it and the embryo will then pass through the Nitex. If too large a mesh is selected, the embryos pass through unaltered, whereas if too small a mesh is used, the embryos collect on the top of the Nitex and the envelopes are not removed. Representative Nitex sizes are usually as follows: for *S. purpuratus*, 64 microns, for *L. variegatus*, 80 microns.

A convenient apparatus for the Nitex mesh is a plastic beaker that has had its bottom cut off. The size of the beaker depends on the volume of embryos to be processed. Usually, a 100 to 400 ml beaker is sufficient for a wide range of applications. Place the Nitex flat over the top of the beaker, and affix a rubber band around the rim to hold the Nitex snugly in place. Pour the treated embryos through the Nitex and collect with another beaker. Do not swirl the embryos because this will increase shear and lysis. Monitor envelope removal on the microscope and repeat once or twice if necessary.

B. Isolation of Fertilization Envelopes

Following removal of the envelopes from embryos (see #1 in preceding text), both fractions can be collected by differential centrifugation. The embryos can be isolated by low-speed spins (500×*g* or with a hand centrifuge for 2 min). Repeat until effectively all the embryos are removed from the supernatant. To collect the fertilization envelopes, spin at high speed on a clinical centrifuge (4 – 5000×*g* for 5 min). The pelleted envelopes should appear white on top of any residual embryos or debris, and can usually be removed by gentle resuspension with a Pasteur pipette without disturbing the contaminating debris.

C. Prevent Fertilization Envelopes from Forming

An effective way to prevent envelopes from forming is to disrupt the vitelline layer needed to scaffold the cortical granule contents. A simple and effective method to disrupt the vitelline layer is by treatment with pH 8.0 seawater containing 10 mM dithiothreitol (DTT; Epel *et al.*, 1970). Simply resuspend the embryos in the DTT for 10 min and then wash the eggs several times in seawater. Removal of the egg jelly is first required for effective DTT treatment (see preceding text for acid treatment of eggs to remove jelly).

XIII. Isolation of Cilia from Embryos

Deciliation is accomplished by immersing embryos briefly in hypertonic medium, as used originally in Auclair and Siegel (1966; see also Stephens, 1986, for background, and more recent applications by Casano *et al.*, 1998). Sea urchin embryos are best deciliated by 2X seawater (1.0 M NaCl).

1. To make 2X seawater, add an additional 0.5 M NaCl (29.2 gm/l) to normal seawater.

2. To deciliate embryos, first pellet the embryos gently (1000×*g* for 2 min), remove the supernatant, and then add 10 volumes of 2X seawater at the normal growth temperature of the embryo. Gently resuspend embryos quickly and completely, and keep embryos resuspended during the deciliation process. It is imperative that the embryos be handled gently to minimize lysis. Even if only a small percent of the embryos lyse during the procedure, abundant cytoplasmic components, e.g., yolk, will contaminate the cilia preparation. Monitor deciliation continuously with a phase microscope, leaving embryos in hypertonic medium no longer than is necessary, usually about 2 min.

3. Sediment the deciliated embryos by low-speed centrifugation (1000×*g* for 2 min) and remove the supernatant carefully with a pipette or by careful decanting. Repeat with increased centrifugation if residual embryos remain.

4. For examination of regenerating cilia in the embryos, resuspend the embryos in original growth conditions. Embryos will regenerate their cilia rapidly and will develop normally, even after multiple deciliations (see Stephens, 1986).

5. For recovery of the cilia, pellet them at 10,000×*g* for 10 min at 4 °C.

Notes:

1. Prior to deciliation, it is sometimes helpful to first wash the embryos through nylon mesh, collecting the embryos while washing through any contaminants. This wash will minimize any microbial contamination or residual lysed embryos in the cilia preparation.

2. All cells of the swimming blastula generate and then, following retraction at mitosis, regenerate their cilia. Cilia growth ceases in ingressing mesechyme cells, e.g., primary mesenchyme cells. Cilia are also lost from invaginating endoderm although the foregut retains significant cilia function.

XIV. Isolation of Extracellular Matrices from Embryos and Larvae

The embryo has many extracellular matrix molecules divided among its two extracellular environments. The inside of the embryo contains a blastocoel, composed of a mixture of ECM molecules, through which cells of the developing embryo will migrate and extend cellular processes. Surrounding the blastocoelar matrix, and immediately underlying the epithelium of the embryonic cells, is the basal lamina (sometimes erroneously referred to as a basement membrane or basal membrane, though the structure has no lipid bilayer, and the terms are easily confused with the basal membrane domain of the plasma membrane of an epithelial cell). The basal lamina is a concentrated extracellular matrix, is readily apparent in the electron microscope, and has distinct molecular composition.

The outside of the embryo (apical aspect of the epithelial cells) is surrounded by an extracellular matrix distinct from the blastocoel environment. The apical, or extraembryonic matrix, consists of an apical lamina immediately adjacent to the epithelial cells, surrounded by a hyaline layer on the outermost surface. Each of the extracellular matrix domains can be isolated and such procedures will be documented here.

A. Isolation of Hyalin

Hyalin is the major protein of the hyaline layer, the outermost layer of the embryo. Hyalin is made from two sources: maternal—the developing oocytes package the protein into cortical granules, and release the hyalin at fertilization;

and zygotic—new hyalin mRNA is synthesized in most cells during gastrulation, and is secreted to form a new hyalin extracellular matrix pool (Wessel *et al.*, 1998). It appears that the mRNA used for translation in embryos is derived from the same gene as the oocyte version, and both maternal and zygotic forms of the mRNA and protein are indistinguishable. Hyalin appears to be a calcium-binding protein and removal of calcium solubilizes hyalin, while addition of calcium to solubilized hyalin re-precipitates, or gelates, the hyalin (Kane, 1973). Hyalin is most readily isolated one hour following fertilization, though the same procedure can be used to isolate hyalin from any stage of development.

1. If hyalin is to be removed prior to hatching, the fertilization envelope must first be removed (see section on fertilization envelope isolation). This is accomplished by fertilizing eggs in reagents that block the activity of ovoperoxidase-mediate crosslinking of the fertilization envelope, e.g., 10 mM para-aminobenzoic acid (note that the seawater must be re-pHed following addition of pABA or 2 mM 3-aminotriazole (Showman and Foeder, 1979; Weidman and Kay, 1986). The envelope will still form; it will just not be stabilized. One hour following fertilization in either of the above reagents, pass the zygotes through Nitex mesh (see fertilization envelope isolation procedure for details of sizes). Remove embryos by settling or gentle centrifugation; use caution as they are now very sticky with their exposed hyalin layers.

2. Resuspend the embryos in 10 volumes of hyalin extraction medium (see Appendix) at the temperature of normal embryo culture. Gently swirl embryos and keep in suspension for 10 min (caution: do not lyse embryos, because the hyalin preparation will become contaminated with abundant cytoplasmic components, e.g., yolk).

3. Remove embryos by gentle centrifugation (1000×*g* for 2 min) and collect the supernatant. (Note: if the embryos are to be recultured, first wash them once in seawater and then replace them into culture conditions more sparse than normal—no more than 0.2%—to minimize the clumping introduced by fertilization envelope removal.) Repeat centrifugation as necessary to remove all embryos and cells. Recentrifuge the supernatant at higher force (10,000×*g* for 10 min) and again collect the supernatant containing the solubilized hyalin and place on ice.

4. Add calcium (from a 1 M stock of $CaCl_2$) gradually with gentle stirring at 4 °C to a final concentration of 20 mM and continue stirring for 30 min.

5. Centrifuge the precipitated hyalin for 10 min at 10,000×*g* for 10 min, and decant the supernatant. Hyalin appears as a translucent, gelatinous pellet.

6. If significant purity of hyalin is needed, or if the hyalin preparation is discolored by contamination, the process of calcium-free solubilization and calcium-mediated precipitation is employed. To re-solubilize the precipitated hyalin, add calcium- and magnesium-free seawater to which 2 mM EDTA (CMFE) has been added. Use the original volume of CMFE used to solubilize hyalin. Gently swirl the hyalin resuspension for 30 min, and recentrifuge at 10,000×*g* for 10 min.

7. Repeat the calcium-mediated precipitation process as above. Continual calcium-precipitation/EDTA-solubilization does help purify the hyalin but at a significant expense of yield. Sometimes it is beneficial to dialyze the solubilized hyalin against CMFE for several hours to aid in solubilization and to remove additional contaminants. It is best to start the purification process with as little lysis of embryos as possible.

B. The Apical Lamina

The apical lamina (AL) is a fibrous extracellular layer intimately associated with the clear hyaline layer (HL) that forms on the sea urchin egg plasma membrane after cortical granule exocytosis. The HL proteins can be solubilized from the egg surface by treatment with 1.1 M glycine, leaving the fibrous AL bound to the egg surface (Hall and Vacquier, 1982). The AL is composed of three major glycoproteins of 175, 145, and 110 kDa. Uronic acid, sialic acid, and collagen are not found in the AL, but sulfate is present. Unlike hyalin, the soluble AL glycoproteins do not precipitate in the presence of calcium. The AL does not change in composition between the zygote and the hatched blastula stage. The AL is extremely sticky and difficult to work with as an intact layer. It is hypothesized that the AL plays a role in blastomere adhesion (Hall and Vacquier, 1982). The cloning of the AL proteins showed them to contain EGF repeats and they were named "fibropillins." At the blastula stage, fibropillins become organized into fibers on the surface of the embryo (Bisgrove and Raff, 1993; Bisgrove *et al.*, 1991). Phylogenetic studies indicate that the fibropillins of the AL have been conserved for at least 250 million years (Bisgrove *et al.*, 1995). Antibodies to one fibropillin immunoprecipitate a complex of all three AL fibropillins. The AL fibropillins appear to be synthesized throughout cleavage to the hatching blastula stage (Burke *et al.*, 1998). Apextrin, another AL component, has been identified in the AL of the direct developing sea urchin *Heliocidaris erythrogramma* (Haag *et al.*, 1999). The AL isolation method described in the following text is taken from Hall and Vacquier (1982).

1. Hyaline layer-apical lamina (HL-AL) complexes are isolated from eggs which have been treated with 10 mM dithiothreitol (DTT), pH 9.1, for 10 min. This prevents the vitelline envelope from elevating and forming the fertilization envelope (Epel *et al.*, 1970). The DTT-treated eggs are fertilized and cultured at 15 °C under normal conditions.

2. The embryos are concentrated to 10% vol/vol, packed in ice, and 20% Triton X-100 added to a final concentration of 1%.

3. The detergent-extracted embryos are then pressurized in a nitrogen cavitation bomb to 1200 psi. (Yeda Press, a Parr Bomb, or a French Press will accomplish the disruption.) The exact pressure has to be worked out empirically for each species and each apparatus.

4. The embryos are passed through the small orifice of the pressure release valve and are lysed and collected in a beaker. The HL-AL complexes are

sedimented by centrifugation at $180 \times g$ for 1 min. The pellet is then resuspended in 50 vol ice-cold filtered seawater with 1% Triton X-100. The pellet is resuspended by vigorous pipetting. The HL-AL complexes are then sedimented again. This wash is repeated a total of three times.

5. To isolate the AL fibropillins, HL-AL complexes are suspended in 30 vol of ice-cold 1.1 M glycine, 2 mM EGTA adjusted to pH 8.0. The material is allowed to stand for 30 min on ice with gentle mixing. The glycine extraction solubilizes the hyalin protein, while the fibropillin containing fibrous AL remain insoluble. The insoluble AL can be washed twice in the glycine buffer to rid them of contaminating proteins. The AL are greatly enriched in the 175, 245, and 110 kDa fibropillin proteins.

6. The hyalin protein can be purified by preparing a $10{,}000 \times g$ supernatant of the glycine extracted complexes. The supernatant is dialyzed against 20 mM $CaCl_2$, 20 mM Tris-HCl, pH 8.0. The calcium–proteinate precipitate is then recovered by low-speed centrifugation.

7. Living embryos can also be extracted in the glycine buffer. The result is that the hyalin protein goes into solution, leaving the fibrous AL on the surface of the embryo, where it is clearly visible by scanning EM.

C. The Basal Lamina/Blastocoel Matrix

Two general methods are used to isolate the blastocoelar extracellular matrix. One is used to obtain an intact blastocoel with mesenchyme cells intact. These preparations are amenable to cellular manipulation and observation *in situ*. The second approach is a biochemical preparation. Both will be documented here.

For intact blastocoel preparations, the method of Harkey and Whiteley (1980) is used. This preparation is used to isolate intact ECM with mesenchyme cells in their native environment, but in it the cells and ECM can be readily manipulated. This procedure was also used for mesenchyme cell isolation but now other, larger scale and easier protocols are available (see Chapter 14). The basis of this protocol is rapid dissociation of embryos with minimal mechanical shearing. The basal lamina bags are very delicate, and normal procedures to strip epithelial cells from embryos are sufficiently harsh as to fragment the bags. The best dissociation occurs at early gastrula stages, following primary mesenchyme cell ingression, and about 1/4 invagination of the archenteron. The procedures were developed for a variety of embryo types—*Strongylocentrotus purpuratus, S. droebachiensis*, and the sand dollar, *Dendraster excentricus*—and minor adjustment of times and centrifugation forces may be necessary for other species.

1. Procedure

Pellet embryos by hand centrifugation or settling and gently wash 3 times in 15 volumes of CMFSW, twice in 5 to 15 volumes of DM, and then resuspend in BIM.

These steps are all carried out on ice. Monitor the dissociation process microscopically, especially the second DM wash, as one wants to get loosening of the epithelial cells in DM so that when they are washed into BIM, complete dissociation occurs, with occasional gentle mixing to maximize dissociation.

Bag isolation procedure A: This procedure exploits a difference in sedimentation of bags and dissociated single cells.

1. Bags are partially purified by 4 cycles of sedimentation at $250 \times g$ for 2 min, followed by resuspension in 20 volumes of BIM.
2. The resuspension is layered as 10 ml aliquots over several sucrose step gradients, each consisting of 10 ml steps of 30% sucrose stock–70% BIM, 50% sucrose stock–50% BIM, and a sucrose stock cushion.
3. The preparation is centrifuged 3 min at 650 g in a swinging bucket rotor. The vast majority of bags with their constituent cells inside are found on the 50% step.

Bag isolation procedure B: This approach utilizes the difference in densities between bags and dissociated epithelial cells. Their densities were estimated from a linear Percoll gradient: dissociated cells from early gastrulae band at 22.64% Percoll (1.062 g/ml), while bags banded loosely with a peak at 16.73% Percoll (1.054 g/ml). A range of densities is seen for bags that are empty (lighter) versus bags with many adherent cells.

1. A freshly dissociated cell suspension is settled on ice and resuspended with BIM and Percoll stock making the suspension 30 to 40% cells and 19% Percoll by volume.
2. The preparation is centrifuged for 20 min at 650 g in a swinging bucket rotor, with a BIM overlay and a cushion of sucrose stock. To reduce mixing during deceleration, use of small diameter centrifuge tubes is preferable.
3. Bags accumulate at the BIM interface. These are collected by pipette, resuspended in BIM, and further purified on a sucrose gradient, as in Procedure A.

2. Blastocoel Bag Isolation Reagents

CMFSW (calcium- and magnesium-free seawater): see Appendix, containing 100 micromolar EDTA, pH 8.0

DM (dissociation medium): 1 M glycine, 100 micromolar EDTA, pH 8.0

BIM: (bag isolation medium): 40% CMFSW, 40% 1 M dextrose, 20% distilled water, pH 8.0.

Percoll stock: 84% Percoll (Pharmacia Fine Chemicals), 16% 5× CMFSW

Sucrose stock: 1 M sucrose, 100 micromolar EDTA, 1 mM Tris, pH 8.0

An alternative method has been use for large-scale biochemical preparations of extracellular matrix (Wessel *et al.*, 1984). These preparations have been used for biochemical analysis, as a substrate for cell attachment, and as a source to generate antibodies.

1. Embryos are pelleted gently, and then resuspended in ice-cold extraction buffer (0.1% Triton X-100, 10 mM NaHCO3, protease inhibitors (especially 0.01% phenylmethylsulfonylfluoride), and 15 micrograms per ml of DNase. The combination of low ionic strength and detergent removes cellular materials and enriches for the inherently insoluble extracellular matrix. Extract for 15 min while keeping the embryos in suspension on a rotator or rocking table at 4 °C.

2. The matrix remaining can either be pelleted at 15,000×*g* for 30 min at 4 °C or collected by running the extraction through 80 micron Nitex mesh.

3. Repeat extractions 2 to 3 times until cellular debris is removed.

4. Enhanced preparations can be obtained by first removing the hyalin layer (see preceding text) and the epithelium (see chapter 14 by McClay for cell dissociation) prior to the extraction.

Notes:

1. The older the embryos, the more stable the extracellular matrix. Usually, embryos prior to gastrulation are difficult to isolate by this procedure, but following gastrulation, the extracellular matrix becomes more robust and is stable to extraction. A trade-off though is increased contamination by mesenchyme cells within the blastocoel.

2. If the extracellular matrix preparation is to be used for live cell work, care must be taken to remove the detergent in the extraction buffer. If the preparation is to be used as a substrate plated on plastic dishes, it can be washed repeatedly with seawater. Other applications may require extensive dialysis against seawater.

3. These preparations have been successfully used to study invasion of metastatic cells from mammals (Livant *et al.*, 1995). To prepare the extracellular matrix for incubation with live cells, wash the preparation several times by pelleting in sterile seawater, followed by washing in culture media.

XV. Isolation of Sea Urchin Larval Skeletons

The skeleton of larvae is a prominent structure and is intensely birefringent. Significant enrichments of the spicule can be made by first lysing the larvae in hypotonic buffer, and then repeatedly washing the lysate in detergent-containing buffers. The major contaminant in such preparations is the basal lamina/blastocoel extracellular matrix, which is difficult to solubilize. Urea helps in removal of the contaminants (see Benson *et al.*, 1986), but a more effective way is to briefly wash the enriched skeletons with alkaline sodium hypochlorite. The method is rapid, effective, and results in skeletal preparations relatively clean of contaminating proteins. The protocol for skeleton isolation will be given from the detailed analysis and utilization of these preparations in Benson *et al.* (1986). These preparations were instrumental in the identification of the spicule matrix proteins, e.g., SM50 and SM 30.

1. Collect prism stage embryos or larvae and wash once in calcium- and magnesium-free seawater and twice in ice-cold 1.5 M glucose. Pellet the larvae.
2. Resuspend the larvae in 10 mM Tris pH 7.4 containing broad spectrum protease inhibitors and lightly resuspend in a loose-fitting Dounce homogenizer. Allow the lysate to sit on ice for 10 min and centrifuge for 3 min at 800*g* in plastic tubes. Throughout the procedure, use plastic vessels wherever possible because low yields have been seen when preparations are made in glass.
3. Discard the supernatant and repeat the homogenization procedure.
4. Resuspend the pellet in 10 volumes of 2% Triton X-100, 4% sodium deoxycholate, 20 mM Tris pH 7.4, and homogenize in a loose-fitting Dounce homogenizer.
5. Centrifuge (800×*g* for 3 min) the suspension and repeat extraction of the pellet 4 to 5 times until most of the cellular debris is removed and the pellet is light brown to off-white.
6. Resuspend again in the same buffer and, while lightly homogenizing again, add 5 original embryo volumes of cold 5% sodium hypochlorite, 10 mM Tris pH 8.0. Continue homogenization for 30 to 60 s. As the preparation turns clear white, quickly centrifuge at 800*g* for 1 min and wash by resuspending the skeletons in cold 2% hypochlorite 10 mM Tris pH 8.0.
7. Centrifuge again as previously directed and wash the skeletons 2 to 3 times with ice-cold distilled water.
8. The isolated skeletons can be demineralized by resuspending the water-washed pellet in 50 mM EDTA pH 8.0 or in 0.1 N acetic acid. Some residue usually remains after demineralization and is usually attributed to dust and salt crystals, though some insoluble skeleton debris may be present as well. Remove the debris by centrifugation at 1000*g* for 2 min.
9. The soluble skeleton matrix is then dialyzed against several changes of cold distilled water, lyophilized, and stored frozen. The lyophilized powder resuspends well in aqueous solutions.

Skeletons may also be analyzed and isolated from cultures in which micromeres have been allowed to develop. See chapter 14 for further information.

XVI. Resources for the Isolation of Additional Organelles

A. Mitotic Spindles

Because of the great synchrony in development of early sea urchin embryos, embryos will progress through the cell cycle en masse, and protocols have been developed to isolate mitotic spindles at different stages of mitosis. In some protocols, these spindles are capable of progressing through the cell cycle—exhibiting pole–pole elongation similar to that observed *in vivo*. Readers are referred especially to the following references: Palazzo *et al.*, 1991; Rebhun and Palazzo, 1988; reviews on the subject by Kuriyama, 1986; Silver, 1986.

References

Acevedo-Duncan, M., and Carroll, E. J. (1986). Immunological evidence that a 305 kilodalton vitelline envelope polypeptide isolated from sea urchin eggs is a sperm receptor. *Gamete Res.* **15,** 337–359.

Alves, A. P., Mulloy, B., Moy, G. W., Vacquier, V. D., and Mourão, P. A. S. (1998). Females of the sea urchin *Strongylocentrotus purpuratus* differ in the structures of their egg jelly sulfated fucans. *Glycobiology* **8,** 939–946.

Aminoff, D. (1961). Methods for quantitative estimation of *N*-acetylneuraminic acid and their application to hydrolysates of sialomucoids. *Biochem. J.* **81,** 384–392.

Armant, D. R., Carson, D. D., Decker, G. L., Welply, J. K., and Lennarz, W. J. (1986). Characterization of yolk platelets isolated from developing embryos of *Arbacia punctulata*. *Dev. Biol.* **113,** 342–355.

Auclair, W., and Siegel, B. W. (1966). Cilia regeneration in the sea urchin embryo: Evidence for a pool of ciliary proteins. *Science* **154,** 913–915.

Belton, R. J., Adams, N. L., and Foltz, K. R. (2001). Isolation and characterization of sea urchin egg lipid rafts and thier possible function during fertilization. *Mol. Reprod. Dev.* **59,** 294–305.

Benson, S. C., Jones, E. M., Benson, N. C., and Wilt, F. (1986). Morphology of the organic matrix of the spicule of sea urchin larvae. *Exp. Cell Res.* **148,** 249–253.

Benson, S. C., Benson, N. C., and Wilt, F. (1986). The organic matrix of the skeletal spicule of sea urchin embryos. *J. Cell Biol.* **102,** 1878–1886.

Bisgrove, B. W., Andrews, M. E., and Raff, R. A. (1991). Fibropillins, products of an EGF repeat containing gene, form a unique extracellular matrix structure that surrounds the sea urchin embryo. *Dev. Biol.* **146,** 89–99.

Bisgrove, B. W., and Raff, R. A. (1993). The SpEGFIII gene encodes a member of the fibropillins: EGF repeat-containing proteins that form the apical lamina of the sea urchin embryo. *Dev. Biol.* **157,** 526–538.

Bisgrove, B. W., Andrews, M. E., and Raff, R. A. (1995). Evolution of the fibropillin gene family and patterns of fibropillin gene expression in sea urchin phylogeny. *J. Mol. Evol.* **41,** 34–45.

Bonnell, B. S., and Chandler, D. E. (1990). Visualization of the *Lytechinus pictus* egg jelly coat in platinum replicas. *J. Struct. Biol.* **105,** 123–132.

Bonnell, B. S., Keller, S. H., Vacquier, V. D., and Chandler, D. E. (1994). The sea urchin egg jelly layer consists of globular glycoproteins bound to a fibrous fucan superstructure. *Dev. Biol.* **162,** 313–324.

Brooks, J. M., and Wessel, G. M. (2002). The major yolk protein in sea urchins is a transferrin-like, iron binding protein. *Dev. Biol.* **245,** 1–12.

Burke, R. D., Lail, M., and Nakajima, Y. (1998). The apical lamina and its role in cell adhesion in sea urchin embryos. *Cell Adhes. Commun.* **5,** 97–108.

Cantatore, P., Nicotra, A., Loria, P., and Saccone, C. (1974). RNA synthesis in isolated mitochondria from sea urchin embryos. *Cell Diff.* **3,** 45–53.

Casano, C., Roccheri, M. C., Onorato, K., Cascino, D., and Gianguzzi, F. (1998). Deciliation: A stressful event for *Paracentrotus lividus* embryos. *Biochem. Biophys. Res. Comm.* **248,** 628–634.

Cestelli, A. M., Albeggiani, G., Allotta, S., and Vittorelli, M. L. (1975). Isolation of the plasma membrane from sea urchin embryos. *Cell Diff.* **4,** 305–311.

Chandler, D. E., and Heuser, J. E. (1980). The vitelline layer of the sea urchin egg and its modification during fertilization. *J. Cell Biol.* **84,** 618–632.

Chiba, K., Nakano, T., and Hoshi, M. (1995). Induction of germinal vesicle breakdown in a cell-free preparation from starfish oocytes. *Dev. Biol.* **205,** 217–223.

Correa, L. M., and Carroll, E. J. (1997). Characterization of the vitelline envelope of the sea urchin *Strongylocentrotus purpuratus. Develop. Growth & Differ.* **39,** 69–85.

Darszon, A., Beltran, C., Felix, R., Nishigaki, T., and Trevino, C. L. (2001). Ion transport in sperm signaling. *Dev. Biol.* **240,** 1–14.

Detering, N. K., Decker, G. L., Schmell, E. D., and Lennarz, W. J. (1977). Isolation and characterization of plasma membrane-associated cortical granules from sea urchin eggs. *J. Cell Biol.* **75,** 899–914.

Dubois, M., Gilies, J. A., Hamilton, J. K., Robers, P. A., and Smith, F. (1956). Colorimetric method for determination of sugars and related substances. *Anal. Chem.* **28,** 350–356.

Epel, D., Weaver, A. M., and Mazia, D. (1970). Methods for removal of the vitelline membrane of sea urchin eggs. 1. Use of dithiothreitol (Cleland's reagent). *Exp. Cell Res.* **61,** 64–68.

Farndale, R. W., Buttle, D. J., and Barrett, A. J. (1986). Improved quantitation and discrimination of sulfated glycosaminoglycans by use of dimethylmethylene blue. *Biochim. Biophys. Acta* **883,** 173–177.

Gache, C., Niman, H. L., and Vacquier, V. D. (1983). Monoclonal antibodies to the sea urchin egg vitelline layer inhibit fertilization by blocking sperm adhesion. *Exp. Cell Res.* **147,** 75–84.

Garbers, D. L. (1989). Molecular basis of fertilization. *Annu. Rev. Biochem.* **58,** 719–742.

Giusti, A. F., Hoang, K. M., and Foltz, K. R. (1997). Surface location of the sea urchin egg receptor for sperm. *Dev. Biol.* **184,** 10–24.

Glabe, C. G., and Vacquier, V. D. (1977). Isolation and characterization of the vitelline layer of sea urchin eggs. *J. Cell Biol.* **75,** 410–421.

Glabe, C. G., and Vacquier, V. D. (1978). Egg surface glycoprotein receptor for sea urchin sperm binding. *Proc. Natl. Acad. Sci. USA* **75,** 881–885.

Gonzales, P., and Lessios, H. A. (1999). Evolution of sea urchin retroviral-like (SURL) elements: Evidence from 40 echinoid species. *Mol. Biol. Evol.* **16,** 938–952.

Haag, E. S., Sly, B. J., Andrews, M. E., and Raff, R. A. (1999). Apextrin, a novel extracellular protein associated with larval ectoderm evolution in *Heliocidaris erythrogramma. Dev. Biol.* **211,** 77–87.

Haley, S. A., and Wessel, G. M. (1999). The cortical granule serine protease CGSP1 of the sea urchin, *Strongylocentrotus purpuratus,* is autocatalytic and contains a low density lipoprotein receptor-like domain. *Dev. Biol.* **211,** 1–10.

Hall, H. G., and Vacquier, V. D. (1982). The apical lamina of the sea urchin embryo: Major glycoproteins associated with the hyaline layer. *Dev. Biol.* **89,** 168–178.

Harkey, M. A., and Whitely, A. H. (1980). Isolation, culture, and differentiation of echinoid primary mesenchyme cells. *Wilhelm Roux's Arch.* **189,** 111–122.

Harrington, F. E., and Easton, D. P. (1982). A putative precursor to the major yolk protein of the sea urchin. *Dev. Biol.* **95,** 505–508.

Hillier, B. J., and Vacquier, V. D. (2003). Amassin, an olfactomedin protein, mediates the massive intercellular adhesion of sea urchin coelomocytes. *J. Cell Biol.* **160,** 597–604.

Hirohashi, N., and Lennarz, W. J. (2001). Role of a vitelline layer-associated 350 kDa glycoprotein in controlling species-specific gamete interaction in the sea urchin. *Develop. Growth Differ.* **43,** 247–255.

Hirohashi, N., and Vacquier, V. D. (2002a). High molecular mass egg fucose sulfate polymer is required for opening both calcium channels involved in triggering the sea urchin sperm acrosome reaction. *J. Biol. Chem.* **277,** 1182–1189.

Hirohashi, N., and Vacquier, V. D. (2002b). Egg sialoglycans increase intracellular pH and potentiate the acrosome reaction of sea urchin sperm. *J. Biol. Chem.* **277,** 8041–8047.

Hirohashi, N., and Vacquier, V. D. (2002c). Egg fucose sulfate polymer, sialoglycan, and speract all trigger the sea urchin sperm acrosome reaction. *Biochem. Biophys. Res. Commun.* **296,** 833–839.

Hoshi, M., Nishigaki, T., Kawamura, M., Ikeda, M., Gunaratne, J., Ueno, S., Ogiso, M., Moriyama, H., and Matsumoto, M. (2000). Acrosome reaction in starfish: Signal molecules in the jelly coat and their receptors. *Zygote* **8**(Suppl.), S26–S27.

Inoue, H., and Yoshioka, T. (1982). Comparison of calcium uptake characteristics of microsomal fractions isolated from unfertilized and fertilized sea urchin eggs. *Exp. Cell Res.* **140,** 283–288.

Kane, R. E. (1973). Hyaline release during normal sea urchin development and its replacement after removal at fertilization. *Exp. Cell Res.* **81,** 301–311.

Kaupp, U. B., Solzin, J., Hildebrand, E., Brown, J. E., Helbig, A., Hagen, V., Beyermann, M., Pampaloni, F., and Weyland, I. (2003). The signal flow and motor response controlling chemotaxis of sea urchin sperm. *Nat. Cell. Biol.* **5,** 109–117.

Kidd, P. (1978). The jelly and vitelline coats of the sea urchin egg: New ultrastructural features. *J. Ultrastruc. Res.* **64,** 204–215.

Kidd, P., and Mazia, D. (1980). The ultrastructure of surface layers isolated from fertilized and chemically stimulated sea urchin eggs. *J. Ultrastruc. Res.* **70,** 58–69.

Kinsey, W. (1986). Purification and properties of the egg plasma membrane. *Meth. Cell Biol.* **27,** 139–152.

Kitazume, S., Kitajima, K., Inoue, S., Troy, II, F. A., Cho, J.-W., Lennarz, W. J., and Inoue, Y. (1994). Identification of polysialic acid-containing glycoprotein in the jelly coat of sea urchin eggs. *J. Biol. Chem.* **269,** 22712–22718.

Kuriyama, R. (1986). Isolation of sea urchin spindles and cytasters. *Meth. Enzymol.* **134,** 190–199.

Livant, D. L., Linn, S. L., Markwart, S., and Shuster, J. (1995). Invasion of selectively permeable sea urchin embryo basement membranes by metastatic tumor cells, but not their normal counterparts. *Cancer Res.* **55,** 5085–5093.

Marzluff, W. F., and Huang, R. C. (1984). Transcription of RNA in isolated nuclei. *In* "Transcription and Translation: A Practical Approach" (B. D. Hames and S. J. Huggins, eds.), pp. 89–130. IRL Press, Washington DC.

Moy, G. W., Mendoza, L. M., Schulz, J. R., Swanson, W. J., Glabe, C. G., and Vacquier, V. D. (1996). The sea urchin sperm receptor for egg jelly is a modular protein with extensive homology to the human polycystic kidney disease protein, PKD1. *J. Cell Biol.* **133,** 809–817.

Nemoto, S-I., Yamamoto, K., and Hashimoto, N. (1992). A nuclear extract, prepared from mass-isolated germinal vesicles, retains a factor able to sustain a cytoplasmic cycle of starfish oocytes. *Dev. Biol.* **151,** 485–490.

Niman, H. L., Hough-Evans, B. R., Vacquier, V. D., Britten, R. J., Lerner, R. A., and Davidson, E. H. (1984). Proteins of the sea urchin egg vitelline layer. *Dev. Biol.* **102,** 390–401.

Oberdorf, J. A., Head, J. F., and Kaminer, B. (1986). Calcium uptake and release by isolated cortices and microsomes from the unfertilized egg of the sea urchin *Strongylocentrotus droebachiensis*. *J. Cell Biol.* **102,** 2205–2210.

Ohlendieck, K., Dhume, S. T., Partin, J. S., and Lennarz, W. J. (1993). The sea urchin egg receptor for sperm: Isolation and characterization of the intact, biologically active receptor. *J. Cell Biol.* **122,** 887–895.

Palazzo, R. E., Lutz, D. A., and Rebhun, L. I. (1991). Reactivation of isolated mototic apparatus: Metaphase versus anaphase spindles. *Cell Motil. Cytoskeleton* **18,** 304–318.

Poccia, D. L., and Green, G. R. (1986). Nuclei and chromosomal proteins. *Meth. Cell Biol.* **27,** 153–174.

Rebhun, L. I., and Palazzo, R. E. (1988). *In vitro* reactivation of anaphase B in isolated spindles of the sea urchin egg. *Cell Motil. Cytoskeleton* **10,** 197–209.

Roberti, M., Musicco, C., Loguercio Polosa, P., Gadaleta, M. N., Quagliariello, E., and Cantatore, P. (1997). Purification and characterization of a mitochondria, single-stranded-DNA-binding protein from *Paracentrotus lividus* eggs. *Eur. J. Biochem.* **247,** 52–58.

Schroeder, T. E. (1979). Surface area change at fertilization: Resorption of the mosaic membrane. *Dev. Biol.* **70,** 306–326.

Schroeder, T. E. (1986). Methods in Cell Biology: The Echinoderm, Vol. 27. Academic Press, San Diego.

Schuel, H., Wilson, W. L., Wilson, J. R., and Bressler, R. S. (1975). Heterogeneous distribution of lysosomal hydrolases in yolk platelets isolated from unfertilized sea urchin eggs by zonal centrifugation. *Dev. Biol.* **46,** 404–412.

SeGall, G. K., and Lennarz, W. J. (1979). Chemical characterization of the component of the jelly coat from sea urchin eggs responsible for induction of the acrosome reaction. *Dev. Biol.* **71,** 33–48.

Selak, M. A., and Sandella, C. J. (1987). Respirational capacity of mitochondria isolated from unfertilized and fertilized sea urchin eggs. *Exp. Cell Res.* **169,** 369–378.

Shuy, A. B., Raff, R. A., and Blumenthal, T. (1986). Expression of the vitellogenin gene in female and male sea urchins. *PNAS* **83,** 3865–3869.

Silver, R. B. (1986). Isolation of native, membrane-containing mitotic apparatus from sea urchin embryos. *Meth. Enzymol.* **134,** 200–217.

Stephens, R. E. (1986). Isolation of embryonic cilia and sperm flagella. *Meth. Cell Biol.* **27,** 217–227.

Suzuki, N. (1995). Structure, function and biosynthesis of sperm-activating peptides and fucose sulfate glycoconjugate in the extracellular coat of sea urchin eggs. *Zool. Sci.* **12,** 12–27.

Tamura, N., Chrisman, T. D., and Garbers, D. L. (2001). The regulation and physiological roles of the guanylyl cyclase receptors. *Endocr. J.* **48,** 611–634.

Unuma, T., Suzuki, T., Kurokawa, T., Yamamoto, T., and Akiyama, T. (1998). A protein identical to the yolk protein is stored in the testis in the male red sea urchin, *Pseudocentrotus depressus. Biol. Bull.* **194,** 92–97.

Vacquier, V. D. (1975). The isolation of intact cortical granules from sea urchin eggs: Calcium ions trigger granule discharge. *Dev. Biol.* **43,** 62–74.

Vacquier, V. D. (1981). Review: Dynamic changes of the egg cortex. *Dev. Biol.* **84,** 1–26.

Vacquier, V. D., and Moy, G. W. (1980). The cytolytic isolation of the cortex of the sea urchin egg. *Dev. Biol.* **77,** 178–190.

Vacquier, V. D., and Moy, G. W. (1997). The fucose sulfate polymer of egg jelly binds to sperm REJ1 and is the inducer of the sea urchin sperm acrosome reaction. *Dev. Biol.* **192,** 125–135.

Vilela-Silva, A. C., Alves, A. P., Valente, A. P., Vacquier, V. D., and Mourão, P. A. S. (1999). Structure of the sulfated alpha-L-fucan from the egg jelly coat of the sea urchin *Strongylocentrotus franciscanus*: Patterns of preferential *2-O-* and *4-O-*sulfation determine sperm cell recognition. *Glycobiology* **9,** 927–933.

Vilela-Silva, A. C., Castro, M. O., Valente, A. P., Biermann, C. H., and Mourão, P. A. S. (2002). Sulfated fucans from the egg jellies of the closely related sea urchins *Stronyglocentrotus droebachiensis* and *S. pallidus* ensure species-specific fertilization. *J. Biol. Chem.* **277,** 379–387.

Walley, T., Terasaki, M., Cho, M.-S., and Vogel, S. S. (1995). Direct membrane retrieval into large vesicles after exocytosis in sea urchin eggs. *J. Cell Biol.* **131,** 1183–1192.

Ward, G. E., Brokaw, C. J., Garbers, D. L., and Vacquier, V. D. (1985). Chemotaxis of *Arbacia punctulata* spermatozoa to resact, a peptide from the egg jelly layer. *J. Cell Biol.* **101,** 2324–2329.

Weidman, P. J., and Kay, E. S. (1986). Egg and embryonic extracellular coats: Isolation and purification. *Meth. Cell Biol.* **27,** 113–137.

Wessel, G. M., Marchase, R. M., and McClay, D. R. (1984). Ontogeny of the basal lamina in the sea urchin embryo. *Dev. Biol.* **103,** 235–245.

Wessel, G. M., Brooks, J. M., Haley, S., Voronina, E., Wong, J., Zaydfudim, V., and Conner, S. (2001). The biology of cortical granules. *Int. Rev. Cytol.* **209,** 117–206.

Wessel, G. M., Berg, L., Adelson, D. L., Cannon, G., and McClay, D. R. (1998). A molecular analysis of hyalin-A substrate for cell adhesion in the hyaline layer of the sea urchin embryo. *Dev. Biol.* **193,** 115–126.

Yamaguchi, M., Krita, M., and Suzuki, N. (1989). Induction of the acrosome reaction of *Hemicentrotus pulcherrimus* spermatozoa by the egg jelly molecules, fucose-rich glycoconjugate and sperm-activating peptide I. *Develop. Growth & Differ.* **31,** 233–239.

Yokota, Y., and Kato, K. H. (1988). Degradation of yolk proteins in sea urchin eggs and embryos. *Cell Differ.* **23,** 191–200.

CHAPTER 21

Sea Urchin Spermatozoa

Victor D. Vacquier* and Noritaka Hirohashi†

*Center for Marine Biotechnology and Biomedicine
Scripps Institution of Oceanography
University of California, San Diego
La Jolla, California 92093

†Department of Biology
Ochanomizu University
2-2-1 Otsuka, Tokyo, 112-8610 Japan

METHODS IN CELL BIOLOGY, VOL. 74

0091-679X/04 $35.00

I. Introduction

This chapter updates a previous chapter on sea urchin spermatozoa (Vacquier 1986a). These spermatozoa can be obtained in vast numbers with little expenditure of time or money. The cells in the population are uniform in morphology (Fig. 1) and in their physiological response to egg molecules that activate their ion channels. Sea urchin sperm are specialized for five functions: oxidative phosphorylation to produce ATP, flagellar motility, the acrosome reaction, binding to the egg, and fusion with the egg. Fusion with the egg accomplishes three things: it restores the diploid genome, it biochemically activates the egg, setting it on a mitogenic pathway, and it gives to the egg the centrosome that will nucleate the first and all subsequent mitotic spindles. Currently, there is much interest in these cells in terms of the ionic regulation of motility and induction of the acrosome reaction (reviewed in Darszon *et al.*, 2001). Also, these sperm contain metabolites such as nicotinic acid adenine dinucleotide phosphate (NAADP) that might be the sperm molecule involved in the metabolic activation of the egg (Churchill *et al.*, 2003).

Sea urchin sperm have been used extensively in research on ciliary/flagellar motility by microtubule sliding, axonemal bending (Brokaw, 2002) and axonemal wave form generation (Woolley and Vernon, 2001). Sea urchin sperm axonemes

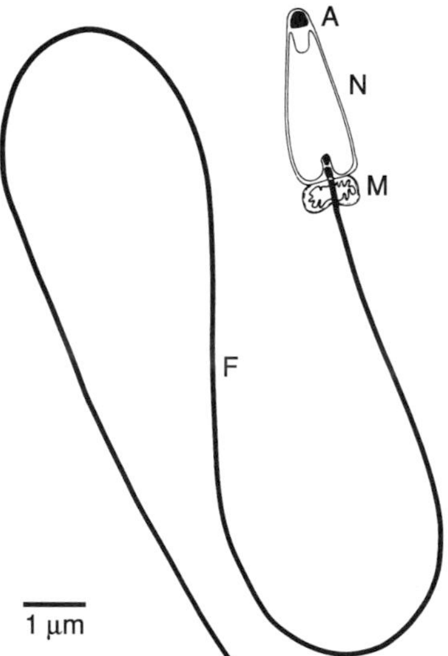

Fig. 1 Drawing of an *S. purpuratus* spermatozoon. The head consists of the phase-contrast dense acrosome vesicle (A) containing the sperm-to-egg adhesive protein bindin, the nucleus (N) with an anterior fossa containing profilamentous actin, and a posterior fossa containing a pair of centrioles, and a midpiece which is a single large mitochondrion (M). The flagellum (F) is $50 \times 0.1\,\mu m$ and consists of an axonemal complex covered by a plasma membrane. The entire cell is covered by a continuous plasma membrane (taken from Vacquier, 1986a).

show surprising specializations, such as a differential distribution of glutamylated tubulin (Huitorel *et al.*, 2002) and different tubulin variants comprising the A and B tubules of the outer doublet microtubules (Multigner *et al.*, 1996). Kinesin II is found in association with the axoneme, kindling speculation that it is a microtubule motor during sperm differentiation (Henson *et al.*, 1997).

Sea urchin sperm are immotile and their respiration is repressed in the concentrated semen coming out of the gonopores after males are injected with 0.5 M KCl. Upon dilution into seawater, respiration and motility explosively activate. The internal pH (pHi) increases from about 6.8 to 7.6, which activates the dynein ATPase to hydrolyze ATP, producing the ADP needed to run the Krebs cycle in the single, giant mitochondrion. The energy reservoir for oxidative phosphorylation is granules of phosphatidylcholine that are stored within the mitochondrion at the posterior end of the sperm head (Mita and Ueta, 1988). ATP utilization by the dynein ATPase produces protons that are removed from the cell by a coupled H^+/Na^+ exchange. The Na^+ is then removed by the Na^+/K^+ ATPase and the K^+ level is controlled by an outwardly directed, passive, K^+ leak though the sperm membrane (Gatti and Christen, 1985).

Sea urchin sperm are excellent model cells for studying the link between mitochondrial respiration and flagellar motility, and the link between the regulation of intracellular pH and Ca^{2+} and the regulation of ion channels that drive the acrosome reaction (Darszon *et al.*, 2001). For example, these sperm have a unique 145 kDa creatine kinase that evolved by a gene triplication (Wothe *et al.*, 1990). The enzyme is plasma membrane-associated. It is used to shuttle the terminal high energy phosphate of ATP from the sperm mitochondrion down the 50 μm length of the flagellum through the cyclic rephosphorylation of ADP at the expense of creatine phosphate (Tombes *et al.*, 1987; van Dorsten *et al.*, 1997). Both myristoylated and nonmyristoylated forms of this unique enzyme exist together in the sperm flagellum (Quest *et al.*, 1997).

Three molecules derived from sea urchin egg jelly—the peptide speract, the fucose sulfate polymer (FSP), and a sialoglycan (SG)—are known to regulate ion channels in the sperm membrane. There are probably other egg jelly molecules that regulate sperm behavior awaiting discovery. About 100 different sperm-activating peptides have been sequenced from various sea urchin species (Suzuki, 1995). Speract, from species of the genera *Strongylocentrotus, Hemicentrotus*, and *Lytechinus*, is a 10 amino acid peptide (GFDLNGGGVG) that binds to a 77 kDa flagellar membrane receptor that controls guanylyl cyclase (Darszon *et al.*, 2001; Garbers, 1989). Speract binding results in the hyperpolarization of the sperm membrane by a cyclic nucleotide-regulated K^+ channel (Galindo *et al.*, 2000) and an increase in Na^+ permeability involving Na^+/H^+ exchange and possibly channels (Rodríguez and Darszon, 2003). Speract also causes an increase in the frequency of the pulsatory calcium transients in sperm (Wood *et al.*, 2003). Resact, a 14 amino acid peptide ($CVTGAPGCVGGGRL_{NH2}$) from the egg jelly of *Arbacia* sea urchins, binds directly to sperm flagellar guanylyl cyclase to up regulate the production of cGMP and cause the chemotaxis of sperm toward eggs

(Garbers, 1989). *Arbacia* sperm respond to 1 molecule of bound resact and their behavioral response saturates at 60 to 80 molecules per sperm (Kaupp *et al.*, 2003). Speract can also be a positive cofactor in acting synergistically with FSP in the triggering of the sperm acrosome reaction (Hirohashi and Vacquier, 2002a).

The FSP is the major component of egg jelly. It is a pure polysaccharide with no associated amino acids. Depending on the species, the fucosyl residues are connected by either α-L-1,3- or α-L-1,4-glycosidic linkages. The sulfation of the fucosyl residue is on the *2-O* and/or *4-O* position. The polymers are either tri- or tetrasaccharide repeats, or homopolymers. In two species, *S. purpuratus* and *S. droebachiensis*, individual females make one of two different FSPs. Females of all other species synthesize only one form of the polymer (Vilela-Silva *et al.*, 2002). FSPs are species-specific inducers of the sperm acrosome reaction (Fig. 2). It is unusual for a pure polysaccharide to induce a signal transduction event in a eukaryotic cell. Structurally, the species-specificity involves both the glycosidic bond and the pattern of sulfation (Hirohashi *et al.*, 2002; Vilela-Silva *et al.*, 2002). Intact FSP, of about 1 million Daltons, opens two distinct Ca^{2+} channels in sperm, both of which are needed to induce the acrosome reaction (Darszon *et al.*, 2001). If FSP is degraded to approximately 60 kDa, it only opens the second Ca^{2+} channel (Hirohashi and Vacquier, 2002b). The second channel is a store-operated channel (Gonzáles-Martinez *et al.*, 2001) whose activity results in the exocytosis of the sperm acrosomal vesicle without altering intracellular pH (Hirohashi and Vacquier, 2003). Affinity chromatography indicates that the sperm membrane protein REJ1 (receptor for egg jelly-1) is at least one of the sperm membrane proteins binding FSP (Moy *et al.*, 1996; Vacquier and Moy, 1997).

Sialoglycan (SG) is a polysialic acid with an apparent size of about 100 kDa by gel filtration chromatography. It is produced by the β-elimination of egg jelly in a way that destroys all protein while leaving the carbohydrate chains intact. Following neutralization, the SG and the FSP are purified by DEAE chromatography. The first fraction to elute is SG. It can be further purified by electrodialysis. SG greatly potentiates the FSP-induced acrosome reaction, but in the absence of FSP, it has no inductive activity of its own (Hirohashi and Vacquier, 2002c). SG acts by increasing intracellular pH without increasing intracellular Ca^{2+}. The biological activity of SG is destroyed by neuraminidase treatment and by mild periodate treatment that specifically degrades sialic acid. The molar ratio of sialic acid to fucose in SG is approximately 6:1. This is the first demonstration of physiological signaling mediated by a polysialic acid. SG must interact with an unknown sperm membrane receptor (Hirohashi and Vacquier, 2002c).

Several important membrane proteins have been cloned from sea urchin sperm. The first was the resact receptor guanylyl cyclase. This provided the foundation for the cloning of receptor guanylyl cyclases, which are key regulators of mammalian physiology (reviewed in Garbers, 1989; Tamura *et al.*, 2001). Guanylyl cyclases appear to be key regulators of most sperm thus far studied (Revelli *et al.*, 2002). The hyperpolarization activated, cAMP-regulated, pacemaker

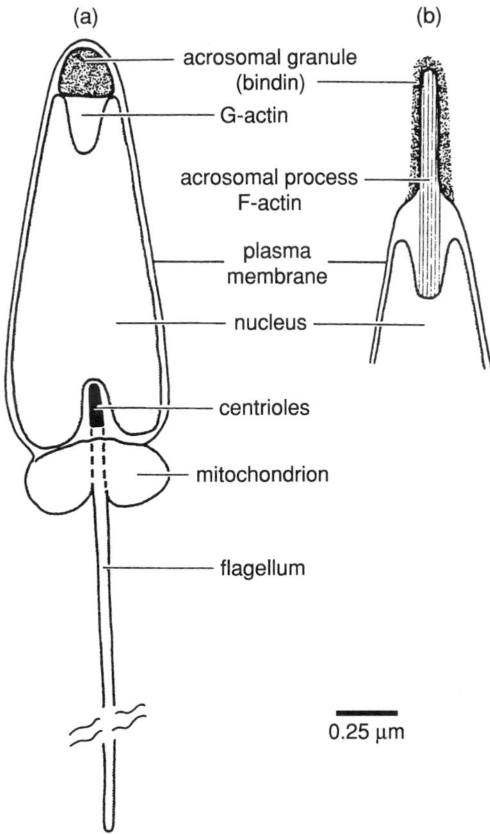

Fig. 2 A sea urchin spermatozoon before (a) and after (b) the acrosome reaction (AR). The AR consists of the exocytosis of the acrosomal vesicle and polymerization of actin to form the acrosomal process, which is coated with bindin. Bindin attaches the sperm to the egg. The membrane at the tip of the acrosomal process fuses with the egg cell membrane (taken from Vacquier, 1986b).

channel, I_h, from *S. purpuratus* sperm, is heavily localized on the sperm flagellar membrane and has similarity to other pacemaker channels of mammalian heart and neural tissue (Gauss *et al.*, 1998). Its relative simplicity has provided a model for understanding the pore topology (Roncaglia *et al.*, 2002) and voltage gating of this type of channel (Männikkö *et al.*, 2002; Rothberg *et al.*, 2002). Receptor for egg jelly-1 (REJ1) and REJ3 are interesting proteins involved in triggering the AR. Monoclonal antibodies to REJ1 will trigger the acrosome reaction (Moy *et al.*, 1996) and compete for FSP binding. REJ1 has 1 transmembrane spanning segment, whereas REJ3 has 11 (Mengerink *et al.*, 2002). REJ1 is localized to the membrane over the acrosomal vesicle and the flagellum, whereas REJ3 is localized only over the acrosomal vesicle. Both proteins have significant homology to the human polycystin-1 protein family, the proteins mutated in autosomal dominant polycystic kidney disease (Igarashi and Somlo, 2002). Finally, a K^+ dependent Na^+/Ca^{2+} exchanger has been cloned from *S. purpuratus* sperm and is the first such exchanger known in nonneural cells. It could represent the mechanism by which the sperm keep intracellular Ca^{2+} in the submicromolar range when seawater is 10 mM in Ca^{2+} (Su and Vacquier, 2002). Sea urchin sperm also possess an unusual adenylyl cyclase (Bookbinder *et al.*, 1990) that is regulated by membrane potential (Beltrán *et al.*, 1996).

Upon fusion with the egg, the sea urchin sperm undergoes a dramatic transformation into the sperm pronucleus, which is destined to fuse with the egg pronucleus in the center of the egg. The nuclear envelope is shed and within 5 min, a new nuclear envelope, derived from egg cytoplasmic vesicles, has been acquired. Within 15 min, the sperm-specific histones have been exchanged for egg cytoplasmic histones and DNA topoisomerase activity associates with the remodeled sperm chromatin (Poccia, 1989; Poccia and Green, 1986). An egg cytoplasm extract has been developed that supports the *in vitro* transformation of detergent-permeabilized sea urchin sperm heads into male pronuclei, making possible the biochemical dissection of the transformation process (Collas, 2000). For example, two separate protein kinase activities are needed for the decondensation of the tightly condensed sperm chromatin (Stephens *et al.*, 2002).

II. Obtaining Sperm and Removing Coelomocytes

These methods have been worked out for *Strongylocentrotus purpuratus* sperm and may have to be altered for other species. As much as 5 ml of undiluted semen can be collected from a single male. It is possible to work with 100 ml of fresh semen. In this species, 1 μl of semen contains 4×10^7 spermatozoa, equaling 100 μg of sperm protein. Each sperm contains 0.85 pg of DNA. No other animal model provides such high numbers of spermatozoa, and so much plasma membrane, at such a low cost in time, labor, and money. The plasma membrane surface area of the sperm head is 6 μm^2 and 25 μm^2 for the flagellum (Figs. 1, 2; Cross, 1983).

A. Collecting Concentrated Semen, "Dry Sperm"

Adult sea urchins are injected intracoelomically with 0.5 M KCl (23 °C). The tip of the needle is placed at a sharp angle just under the peristomial membrane around the jaws to prevent puncturing the gonad. The creamy white semen comes out of the gonopores and is collected in 1.5 ml microcentrifuge tubes, which are capped and stored in ice. This is termed "dry" sperm and contains 4×10^{10} spermatozoa per ml. These numbers are surprisingly constant for *S. purpuratus*. The dry sperm can be stored packed in ice for 5 days without noticeable changes in the plasma membrane proteins. It is important for such long-term viability to store the sperm packed in ice.

B. Removing Coelomocytes

This procedure is done if the sperm are to be fractionated into component parts. Coelomocytes are sea urchin immune cells that have access to all tissues of the animal. Their job is to phagocytose all foreign (nonself) material. It is critical to remove these cells from the sperm because they are rich in protease activity that will degrade all proteins isolated from sperm. They have a red pigment that makes them easy to spot as a red pellet after low-speed centrifugation.

1. Resuspend dry sperm in 20 volumes filtered seawater (4 °C) in conical 50 ml plastic centrifuge tubes. Stir the cell suspension by hand with a spatula.
2. Centrifuge at 500× *g* for 8 min.
3. Draw off the sperm suspension with a 10-ml pipette and discard the small red pellet at the tube bottom.
4. Repeat steps 2 and 3 twice. This should remove all the coelomocytes.
5. Pipette the sperm suspension into round-bottom 50 ml tubes.
6. Sediment the sperm by centrifugation at 1500× *g* for 20 min in a swinging bucket rotor. Resuspend the cell pellet in the appropriate buffer. Use a flat, plastic spatula to break up the pellet and get the cells back into suspension.
7. The sperm cells can be washed by resuspension in fresh medium and resedimentation at 1500× *g* for 20 min. These cells are termed "washed sperm." Some species, such as *Lytechinus pictus*, have such fragile sperm that a high percentage will lyse and demembranate if sedimented twice. One way to decrease the lysis of such sperm is to add 1 mg/ml ovalbumin to all media.

III. Isolating Sperm Heads, Flagella, and Chromatin

The sperm head consists of the acrosomal vesicle, the profilamentous actin that will form the acrosomal process, the nucleus, the centrosome with centrioles, and the mitochondrion. The flagellum consists of a plasma membrane covering an axoneme (Fig. 1). The head and flagellum can be efficiently separated from each other. To prevent the acrosome reaction from occurring, we prefer to suspend the washed sperm in seawater buffered at pH 6.0 with 10 mM MES (morpholinoethanesulfonic acid). All procedures are performed at 4 °C.

A. Method

1. Suspend 1 to 4 ml of washed sperm (Section II.B) in filtered seawater/MES, pH 6.0, and pour the suspension into a tissue homogenizer that uses a tight-fitting Teflon pestle or a glass ball (preferred).
2. Perform 10 cycles of slow homogenization and then observe the homogenate with a 100× phase contrast objective. Make a gentle squash of the slide to remove

extra liquid so that you can observe individual cells in the same focal plane. Estimate the percentage of cells in which the flagellum has detached from the head. Most heads should completely lack a flagellum.

3. Continue to perform 10 cycles of homogenization until over 95% of the sperm are "tail-less". This usually requires 30 to 40 cycles.

4. Centrifuge the homogenate at 1000× *g* for 15 min in a swinging bucket rotor. The heads will form a pellet, on top of which will be a white fluffy layer of flagella. Remove all but 5 ml of the supernatant with a pipette and swirl the tube very gently to resuspend the sedimented flagella. Remove these last 5 ml and combine them with the supernatant.

5. To purify the isolated heads, continue to resuspend them in fresh pH 6.0 seawater and resediment them at 1500× *g* for 15 min. Save all the resultant supernatants that contain broken flagella. After about three sedimentations, the heads are usually free of almost all flagellar fragments.

6. To purify flagella, sediment the residual heads by several centrifugations at 1000× *g* for 15 min. Remove the thin white pellet of broken flagella over the pellet of heads and recentrifuge the supernatant. Check for head contamination by microscopy under low power so that a large field is viewed.

7. SDS-PAGE of isolated sperm heads and flagella should show minimal core histone contamination of the flagella and minimal contamination of the sperm heads by tubulin and the doublet bands of sperm tail creatine kinase (140 kDa) and guanylyl cyclase (160 kDa) as shown for *Arbacia punctulata* sperm (Fig. 3).

8. A relatively pure preparation of sperm head chromatin can be obtained by exposing the isolated sperm heads to seawater containing 1% Triton X-100 and buffered with 10 mM MES and pH 5.7. After stirring for 15 min at 4 °C, the chromatin can be sedimented by centrifugation at 10,000× *g* for 20 min. The extracted sperm heads form a gel at the tube bottom.

IV. The Isolated Sperm Flagellum as a Sealed Compartment

The echinoderm sperm flagellum consists of an axonemal complex of the typical "9 plus 2" pattern of microtubule doublets that is covered by a plasma membrane. There are no dense fibers as in mammalian sperm and there are no internal membrane systems. The flagella have a large surface area-to-volume ratio and when they break off at the base of the sperm head, the membrane around the broken end reseals (Lee, 1984, 1995). Lee showed that isolated sperm flagella were excellent for studying ion flux activities of this special membrane compartment (Lee, 1984). In Na^+ free media, the flagella maintain a pH of 6.7 when the external pH is 7.9. When Na^+ is added, the internal pH of the flagella goes up by 0.6 units. The voltage dependence of the Na^+/H^+ exchanger is unchanged in isolated flagella (Lee, 1985).

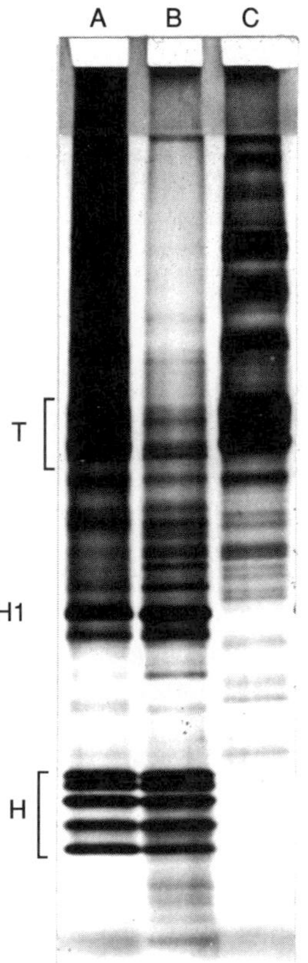

Fig. 3 Separation of sperm heads and flagella of *Arbacia punctulata* spermatozoa as demonstrated by SDS-PAGE and silver staining of 7.5 μg protein per lane. Whole sperm (A), isolated heads (B), and isolated flagella (C). Histone H1 and the core histones (H) are not found in the flagella. The tubulins (T) are heavily represented in the flagella, but not in the heads (taken from Vacquier, 1986a).

V. Extracting Intact Sperm or Isolated Flagella and Heads with Detergents

We routinely use 1% Triton X-100, 1% Zwittergent 3–10 (Sigma catalog no. D-4266), or 1% CHAPS (Calbiochem catalog no. 220201) to extract sperm to obtain membrane proteins for immunoprecipitations and lectin affinity chromatography. The buffer can be 150 to 500 mM NaCl buffered at pH 7.5 with 10 mM HEPES.

A. Method

1. Protease inhibitor cocktail (Sigma catalog no. P-8340) diluted 1:100 from the purchased stock, plus a final concentration of 10 mM benzamidine (Sigma catalog

no. B-6506) is made with the detergent concentration at 2X the desired final concentration at 4 °C).

2. Each 1 ml of washed sperm, isolated heads, or isolated flagella, is suspended in 5 ml of 150 to 500 mM NaCl/HEPES pH 7.5, and then an equal volume of detergent solution is added. The suspension is gently stirred or agitated for 30 min at 4 °C and then centrifuged at different speeds, depending on what is desired.

3. Centrifugation at 10,000× *g* for 30 min will sediment all insoluble cell debris. Centrifugation at 40,000× *g* for 60 min will sediment cell debris and large molecular aggregates. Centrifugation at 170,000× *g* for 60 min will sediment several of the membrane proteins, such as the guanylyl cyclase and the 150 kDa creatine kinase (Podell *et al.*, 1984).

VI. Wheat Germ Agglutinin (WGA) Affinity Chromatography of Sperm Membrane Proteins

The sperm plasma membrane proteins REJ1 (Moy *et al.*, 1996), REJ3 (Mengerink *et al.*, 2002), and ATP binding cassette transporter (Mengerink and Vacquier, 2002) all bind to the lectin WGA that is purchased immobilized on agarose beads (E-Y Laboratories). WGA treatment of living sperm also blocks the egg jelly-induced acrosome reaction (Podell and Vacquier, 1984).

A. Method

1. The 10,000× *g* 1% nonionic detergent supernatant of sperm is incubated with WGA-agarose for 1 to 18 h in a sealed tube on a slowly rotating wheel at 4 °C.

2. The WGA agarose are poured into a column and washed with 30 column volumes of the background buffer (the detergent plus 150–500 mM NaCl, 10 mM HEPES, pH 7.5, and 10 mM sodium azide). If 5 ml of WGA agarose is used, then 1 ml of elution buffer is applied and each ml collected in a 1.5 ml centrifuge tube; 20 fractions are collected. The elution buffer is the background buffer plus 100 mM N-acetyl-D-glucosamine.

3. Fifty μl of each fraction is mixed with 50 μl of 2X concentrated Laemmli sample buffer and applied to a 7.5% SDS-PAGE gel, which is silver-stained. The gel shows that REJ1, REJ3, the ATP binding cassette protein, and at least two unknown proteins are enriched in the eluate (see Fig. 1 in Mengerink and Vacquier, 2002).

VII. Isolating Sperm Plasma Membranes

There is much interest in the sea urchin sperm plasma membrane because it is possible to obtain enough of the membrane to attempt to isolate ion channels, ion exchangers, and receptors. Methods for isolating the plasma membrane of intact sperm, isolated sperm heads, and isolated flagella have been previously described (Darszon *et al.*, 1994; Lee, 1995; Vacquier, 1986a, 1987). The resultant membranes can be used for recording single ion channel activity (Darszon *et al.*, 2001) and for the isolation of proteins which will facilitate amino acid sequencing and cloning of receptor proteins (Mengerink and Vacquier, 2002; Mengerink *et al.*, 2002; Moy *et al.*, 1996).

A. The pH 9 Method to Isolate Sperm Membrane Vesicles (SMV)

1. Sperm are spawned, the coelomocytes are removed, and the cells are washed and resuspended in seawater (without added buffer) as has been described (Section II.B). All procedures are performed at 0 to 4 °C. Isolated sperm heads and isolated flagella can also be used as the starting material to isolate either head or flagellar plasma membrane (Section III.A).

2. Seawater containing 20 mM benzamidine, 40 mM Tris-OH, and 1:200 Sigma protease inhibitor cocktail is adjusted to pH 9.15.

3. While constantly stirring the sperm suspension by hand, an equal volume of the pH 9.15 medium is added. The suspension is allowed to stand without agitation for 5 to 18 h.

4. Following this incubation, the suspension is vigorously stirred for 2 min with a magnetic stir bar.

5. The suspension is centrifuged 30 min at 6000× *g* in a swinging bucket rotor. The supernatant is carefully removed to clean tubes and the 6000× *g* centrifugation is repeated.

6. The resulting supernatant is centrifuged 60 min at 40,000× *g*. The pellet contains the SMV that have a highly reproducible protein composition. The SMV are in the form of tightly sealed, right-side-out, unilamellar vesicles. The protein composition of these pelleted SMV is identical to the composition of SMV banded in sucrose gradients. The membranes are enriched 6-fold in adenylyl cyclase activity and 4- to 8-fold in specific radioactivity when they are isolated from surface radioiodinated sperm. Mitochondrial contamination appears minimal as judged from the 23-fold decrease in cytochrome oxidase in SMV as compared to whole sperm (Podell *et al.*, 1984; Vacquier, 1986a).

B. Nitrogen Cavitated Membrane Vesicles (CMV)

1. Sperm are washed into filtered seawater as has been described (Section II.B.) and then pelleted again by centrifugation at 1500× *g* for 20 min. The pellet is resuspended in at least 10 volumes of cavitation buffer (480 mM NaCl, 10 mM $MgCl_2$, 10 mM KCl, 1:200 Sigma protease inhibitor cocktail, and 10 mM MES (morpholino-ethanesulfonic acid, pH 6.0). Isolated sperm heads and flagella can also be used as starting material (Section III.A).

2. The suspension is placed in a Parr 4635 high pressure bomb (Parr Instruments Company, Moline, IL) and pressurized with nitrogen gas to 400 psi. After 10 min, the valve is released slowly and the cavitate collected in a beaker packed in ice.

3. The cavitate is centrifuged at $8000\times g$ for 20 min and the resulting supernatant centrifuged again at $11{,}000\times g$ for 20 min. This supernatant is then centrifuged at $75{,}000\times g$ for 1 h to sediment the CMV.

4. The CMV have been used as the starting material for the isolation of the sperm flagellar guanylyl cyclase (Ward *et al.*, 1985) and to study the phosphorylation of this enzyme (Ward *et al.*, 1986). CMV are tightly sealed vesicles, half of which appear to be right-side-out due to the fact that 50% of the total CMV protein binds to wheat germ agglutinin (WGA) agarose (Vacquier, unpublished).

5. To separate inside-out from right-side-out CMV, the CMV pellet is resuspended in cavitation buffer and pipetted or vortexed to break up all clumps. The tube is centrifuged at $8000\times g$ for 20 min. The supernatant CMV are applied to WGA agarose for 1 to 18 h and then the agarose beads are washed with 20 volumes of cavitation buffer. The CMV which do not bind the WGA beads are the inside-out vesicles. The right-side-out vesicles can be eluted from the washed WGA agarose beads by exposure to cavitation buffer containing 100 mM N-acetyl-D-glucosamine.

C. Isolation of Flagellar Membranes by Hypotonic Lysis

This method was originally worked out by Gray and Drummond (1976), modified by Cross (1983), and used extensively by Darszon and collaborators (Darszon *et al.*, 1994; García-Soto *et al.*, 1988).

1. Sperm are first separated into flagellar and head fractions as has been described (Section III.A).

2. The flagella are pelleted by centrifugation at $10{,}000\times g$ for 20 min. The supernatant is completely removed and the flagella suspended in a hyposmotic buffer (10 mM Tris, pH 8.0, 10 mM KCl, 10 mM NaCl, and 0.1 mM dithiothreitol).

3. The flagella are gently homogenized and the homogenate layered on a 40% sucrose cushion (40% sucrose, 10 mM Tris, pH 8.0, 1 mM $CaCl_2$, 0.1 mM dithiothreitol) and centrifuged $17{,}000\times g$ for 75 min. The fraction above the sucrose interface constitutes the "top" membranes, and the band at the sucrose interface is referred to as the "middle" membranes.

4. The middle membranes are collected and diluted with 4 volumes of hyposmotic lysis buffer. Both fractions of membranes can be sedimented by centrifugation at $100{,}000\times g$ for 1 h.

5. Electron microscopy shows the top membranes to be composed of vesicles and sheets. Microtubules are associated with the middle membranes. SDS-PAGE

shows that the top membranes have 9 major bands and the middle membranes have these 9 bands plus tubulin. Top membranes isolated from ^{125}I-surface labeled sperm show a 12-fold increase in specific radioactivity compared to intact cells (Cross *et al.*, 1983). Iodination patterns of whole cells are similar to those of isolated flagella, indicating that most of the proteins in the head membrane must be represented in the flagellum. However, we do know that at least one protein, REJ3, is exclusively localized in the head over the acrosomal vesicle (Mengerink *et al.*, 2002).

D. Isolation of Sperm Head Plasma Membranes by Affinity to Cationic Beads

The plasma membrane from isolated sperm heads can be isolated by both the pH 9 and the nitrogen cavitation methods already presented. A third method takes advantage of the net negative charge on the surface of the sperm head (Darszon *et al.*, 1994; Garcia-Soto *et al.*, 1988).

1. Isolated sperm heads are washed by resuspension in solution A (486 mM NaCl, 10 mM KCl, 2.4 mM $NaHCO_3$, pH 8.0) and sedimentation at 3000× *g* for 10 min.
2. The heads isolated from 1 ml of semen are resuspended in 2 ml solution A and are incubated with 500 mg Affi-gel 731 (BioRad) with gentle agitation for 15 min.
3. 15 ml of solution A containing 40 U/ml of heparin is added to block free cationic groups on the resin for 15 min.
4. The Affi-gel beads are then washed several times by centrifugation at 120× *g* for 5 min with removal of the supernatant and resuspension in fresh solution A.
5. The beads are resuspended in 10 volumes of lysis buffer (10 mM Tris, 1 mM $MgCl_2$, 0.1 mM EGTA, 10 μM leupeptin, 1 μM pepstatin A, 0.1 μM phenylmethylsulfonylfluoride, 0.2 μg/ml DNase/ml, pH 8.0). The sperm heads lyse and the plasma membrane remains bound to the beads.
6. The beads are washed twice in excess lysis buffer and then resuspended in solution A. The beads are sonicated 3 times for 2 min each time in a bath sonicator. This releases the membrane vesicles from the beads.
7. The tube is centrifuged 7500× *g* for 10 min and the supernatant containing the membrane vesicles is removed.
8. The vesicles can be concentrated by centrifugation at 100,000× *g* for 60 min. The head vesicles have been extensively characterized (Darszon *et al.*, 1994).

VIII. Isolation of Acrosome Reaction Vesicle Membranes (ARV)

During the sperm acrosome reaction, the sperm plasma membrane and the acrosomal vesicle membrane fuse in many places, creating membrane vesicles that are a composite of both membranes and that are shed from the sperm. These small vesicles, termed acrosome reaction vesicles (ARV), contain signaling proteins and ion channels involved in the acrosome reaction. Thus, ARV will be important for a deeper understanding of the acrosome reaction. Shedding of ARV from the acrosome reacting sperm can be a protein selective event, as shown by the fact that all the detectable syntaxin, VAMP (Schulz *et al.*, 1997), and SNAP-25 (Schulz *et al.*, 1998) are shed from sperm in the ARV. ARV have been incorporated into planar lipid bilayers and the activity of single ion channels detected (Schulz, 1999). To make ARV, one must begin with isolated sperm heads. If whole sperm are used, flagellar membrane vesicles will contaminate the final ARV pellet.

A. Method

1. Isolate sperm heads as previously described (Section III.A).
2. Suspend sperm head pellet in 20 volumes of filtered seawater.
3. While stirring rapidly with a spatula, add 1 volume of stock ionophore A23187 and 1 volume of stock ionophore nigericin to 98 volumes of sperm suspension and incubate for 5 min. (Stock A23187 is 2 mg nigericin dissolved in 1 ml dimethylsulfoxide; stock nigericin is 2 mg nigericin dissolved in 1 ml dimethylsulfoxide.) The ionophores induce the acrosome reaction.
4. Remove reacted sperm heads by centrifuging at 10,000× *g* for 30 min.
5. Collect the supernatant and sediment the ARV by ultracentrifugation at 200,000× *g* for 1 h.

IX. Isolating Lipid Rafts from Sperm

When eukaryotic cells are extracted with 1% Triton X-100, a fraction of the cell membrane with a unique group of proteins remains associated as a large particulate structure that can be banded in a sucrose density gradient. This band of organized membrane proteins has been termed "lipid rafts." The current idea is that the rafts represent macromolecular organizations of proteins that are physiologically relevant to signal transduction. Rafts have been isolated from sea urchin sperm and partially characterized (Ohta *et al.*, 1999, 2000).

A. Method

1. A 150 to 300 μl aliquot of undiluted semen is suspended in 1 ml of 10 mM Tris-HCl (pH 7.5), 1% Triton X-100, 150 mM NaCl, 5 mM EDTA, and 75 units of aprotinin. The tube is incubated for 20 min on ice.
2. The tube contents are homogenized by 10 strokes in a tight-fitting Dounce homogenizer. The cellular debris is sedimented at 1300× *g* for 5 min.
3. The supernatant is mixed with an equal volume of 85% w/v sucrose in 10 mM Tris-HCl, pH 7.5, 150 mM NaCl, and 5 mM EDTA (TNE).

4. The mixture is placed in an ultracentrifuge tube and layered with 6 ml of 30% sucrose in TNE and with 3.5 ml 5% sucrose in TNE. The tube is centrifuged at 200,000× *g* for 18 h in a Beckman SW41 rotor. Eleven 1 ml fractions are collected from the tube and characterized. The low density-detergent insoluble membranes (LD-DIM) form a tight band in the center of the tube.

5. Roughly 50% of the sperm glycosphingolipids are present in the LD-DIM fraction (Ohta *et al.*, 1999).

6. Western immunoblots with antibodies show that 7 sperm membrane proteins isolate with the raft fraction: REJ1, the 63 kDa GPI-anchored protein (Mendoza *et al.*, 1993), the 80 kDa speract receptor (Garbers, 1989), the 190 kDa adenylyl cyclase (Bookbinder *et al.*, 1990), a 48 kDa $G_{s\alpha}$ subunit, the 133 kDa guanylyl cyclase (Garbers, 1989), and a 47 kDa protein kinase A (Ohta *et al.*, 2000).

X. Scoring the Acrosome Reaction by Phase Contrast Microscopy

The acrosome reaction (AR) of sea urchin sperm consists of the exocytosis of the acrosomal vesicle, which releases the sperm-to-egg attachment protein, bindin, and the polymerization of acrosomal actin to form the fingerlike, 1 μm long acrosomal process that is coated with bindin (Fig. 2). The phase contrast method works well *for S. purpuratus, S. franciscanus, A. punctulata*, and *L. pictus*. However, because of the smaller size of the acrosomal vesicle, short length of the acrosomal process, and shape of the sperm head, it does not work for *S. pallidus* and *S. droebachiensis* sperm. Test molecules can be used to see if they induce or inhibit the AR. Total egg jelly is always used as a positive control and seawater only as a negative control. It is possible to obtain 100% AR with fresh sperm.

For those species in which it is impossible to discern the AR by phase contrast microscopy, fluorescent phalloidin can be used to discern the polymerization of actin forming the acrosomal process (Biermann and Marks, personal communication). This method is adapted from one used to stain filamentous actin in the fertilization tubules of the single-cell alga *Chlamydomonas* (Wilson *et al.*, 1997). Sperm are fixed in 3% paraformaldehyde in seawater and then washed into 150 mM NaCl, 10 mM Tris, pH 7.5. Drops of sperm suspension are placed on coverslips, which have been coated with polylysine, protamine sulfate, or alcian blue. Sperm bind to the positively charged surface. They are then permeabilized by treatment with acetone or non-ionic detergent and subsequently treated with fluorescent phalloidin. Also, a rabbit antibody to bindin, the major protein of the acrosomal vesicle, can be used to assay for vesicle exocytosis in the absence of the polymerization of acrosomal actin (Hirohashi and Vacquier, 2003). Antibindin is available from Vacquier.

A. Phase Contrast Method

1. Sperm are spawned and stored as undiluted semen in microcentrifuge tubes stored in ice as undiluted semen. For critical experiments in which precious molecules are to be tested, the semen is not used after 12 h storage. For *S. purpuratus* and *S. franciscanus*, all media and tubes are preincubated 20 min in a 15 to 18°C room.

2. The molecule to be tested as an AR inducer or inhibitor is titered down in microcentrifuge tubes as 50% dilutions in 50 μl volumes of seawater buffered with 10 mM HEPES, pH 7.9. (All volumes can be adjusted.)

3. Ten μl of fresh semen is mixed with 2 ml of HEPES-buffered seawater. Twenty seconds later, 50 μl of this 1:200 suspension of dry sperm is added to each tube, using a wide mouth micropipette tip diameter of 1.2 mm. The sperm are mixed with the test solution by 5 up-and-down fillings of the micropipette. The pipette tips are discarded after addition of each ingredient.

4. After 5 min incubation, 100 μl of 10% glutaraldehyde in seawater is mixed with the sperm suspension with 5 rapid up-and-down fillings of the micropipette.

5. The fixed sperm are sedimented by 500× *g* centrifugation for 5 min in a microcentrifuge and the glutaraldehyde is carefully removed. The fixed sperm are resuspended in 50 μl of 2 M glycine, pH 8.0.

6. Thirty μl of fixed sperm suspension is placed on a clean microscope slide and a clean coverslip is added. The slide is turned on edge on a tissue to drain off excess fluid. The slide is then flipped over onto a folded tissue and a thumb squash is performed to flatten the preparation.

7. The slide is observed using the oil immersion phase contrast optics at approximately 1200× magnification. Sperm can be scored rapidly for being, or not being, acrosome reacted (Fig. 2).

B. Fluorescent Phalloidin

1. Dry sperm are diluted 1:10 into 10 mM HEPES buffered seawater, pH 7.9. The temperature varies from 9 to 23 °C, depending on the species.

2. Within 2 min, 10 μl of diluted sperm is mixed with 100 μl of test solution. Egg jelly is used as a positive control and seawater as a negative control.

3. After 5 min, the cells are fixed in 200 μl of ice-cold 3% paraformaldehyde in seawater. Pipette tips are changed for each addition.

4. After 30 min of fixation, the cells are gently sedimented at 500× *g* for 5 min and the supernatant discarded. The fixed cells are resuspended in 1 ml of seawater and resedimented by 500× *g* centrifugation for 5 min.

5. Coverslips are prepared by being exposed for 5 min to either 1 mg/ml protamine sulfate, polylysine, or a 0.1% solution of alcian blue. The coverslips are washed in water and air dried. The positive charge on the coverslip will bind the negatively charged sperm surface.

6. Drops of fixed sperm suspension are placed on the coverslips and allowed time to adhere. The length of time has to be worked out empirically. Each coverslip is then placed in a 5 cm diameter tissue culture dish.

7. The coverslips are then exposed for 10 min to 0.05% Triton X-100 to permeabilize the plasma membrane.

8. The coverslips are then exposed to excess 100 mM glycine, 1 mg/ml BSA, 0.02% sodium azide, 150 mM NaCl, 10 mM HEPES, pH 7.5. One unit of Alexafluor 488-conjugated phalloidin (Molecular Probes) is added to the cells for 60 min.

9. The coverslips are then washed twice in 1 ml 150 mM NaCl, 10 mM HEPES, pH 7.5. A drop of 50% glycerol is put on the coverslip, which is flipped over on a microscope slide. The cells are viewed with the 100X oil immersion, phase contrast objective. The fluorescein fliters are used for viewing the phalloidin bound to actin filaments.

C. Antibindin

1. Sperm are treated with agents that may induce the acrosome reaction and then fixed 10 min in 3% paraformaldehyde seawater.

2. The sperm are sedimented in 12 × 75 mm tubes at 500× *g* for 5 min for all washings. The supernatant is discarded and the sperm are resuspended for a minimum of 30 min in 1 ml 150 mM NaCl, buffered with 10 mM HEPES, pH 7.5.

3. The cells are washed 3X by sedimentation and resuspension in 2 to 3 ml fresh buffer. The cells are exposed for at least 60 min to a 20 μg/ml solution of rhodamine-conjugated rabbit antibindin. The cells are then washed twice in 1 ml 150 mM NaCl, 10 mM HEPES, pH 7.5. The final pellet is resuspended in 50% glycerol and the cells are placed on a microscope slide. The antibindin is available from Vacquier.

4. If unconjugated antibindin is used, an Alexafluor-conjugated goat anti-rabbit IgG can be used as a secondary detecting antibody.

XI. Sea Urchin Sperm Bindin

The acrosome vesicle of sea urchin sperm contains bindin as its major protein. Bindin from *S. purpuratus* is the 24 kDa, species-specific, adhesive protein that is the bonding material (glue) between the acrosomal process of the sperm and the vitelline layer of the egg (Fig. 2). All references to bindin from 1977 to 1995 can be found in a review article (Vacquier *et al.*, 1995). In addition to its role in sperm–egg adhesion, bindin may also be the protein that mediates the fusion of sperm and egg. An 18 amino acid peptide in the central region of the highly conserved central domain of ~60 residues appears to have the same quantitative membrane fusagenic potential as the entire bindin protein of 246 residues (Binder *et al.*, 2000; Glaser *et al.*, 1999). Continued study of the biochemistry of bindin will require the determination of its crystallographic structure. Methods to isolate bindin protein depend on the species used. References for these methods are found in Vacquier *et al.* (1995).

The differentiation of bindin within a species into two or more sequence variants could be the basis for the evolution of species-specific fertilization and, hence, barriers to prezygotic reproduction. Thus, bindin's evolution might be involved in the formation of new sea urchin species. Bindin consists of three domains: a highly conserved central region of ~60 amino acids, an N-terminal region that shows the signal of positive selection, and a C-terminal region with short repeats of 7 or 9 residues (Biermann, 1998; Debenham *et al.*, 2000; Metz and Palumbi, 1996; Minor *et al.*, 1991; Zigler and Lessios, 2003a,b). Within the genus *Echinometra*, bindin varies both within species and among species. Strong reproductive barriers have developed between sympatric species of *Echinometra* that involve failure of acrosome-reacted sperm to bind to the egg surface (Metz *et al.*, 1994). Even within the species *Echinometra mathaei*, the isotype of bindin carried by individual sperm influences the rate of fertilization of eggs (Palumbi, 1999). The central domain of bindin has existed for at least 250 million years of echinoid evolution (Zigler and Lessios, 2003a,b).

Hopefully, further studies of bindin variation, both within and among species, will shed more light on the perplexing problem of speciation in the sea. The sequences of primers used to amplify bindin sequences from various sea urchin species are available in the references already cited.

Acknowledgments

We thank C. Beltrán, C. H. Biermann, A. Darszon, J. A. Marks, G. W. Moy, and A. T. Neill for help with the manuscript. Much of the work presented was supported by NIH Grant HD12986.

References

Beltrán, C., Zapata, O., and Darszon, A. (1996). Membrane potential regulates sea urchin sperm adenylylcyclase. *Biochemistry* **35,** 7591–7598.

Biermann, C. H. (1998). The molecular evolution of bindin in six species of sea urchins (*Echinoidea: Strongylocentrotidae*). *Mol. Biol. Evol.* **15,** 1761–1771.

Binder, H., Arnold, K., Ulrich, A. S., and Zschornig, O. (2000). The effects of Zn^{2+} on the secondary structure of a histidine-rich fusagenic peptide and its interaction with lipid membranes. *Biochim. Biophys. Acta* **1468,** 345–358.

Bookbinder, L. H., Moy, G. W., and Vacquier, V. D. (1990). Identification of sea urchin sperm adenylate cyclase. *J. Cell Biol.* **111,** 1859–1866.

Brokaw, C. J. (2002). Computer simulation of flagellar movement VIII: Coordination of dynein by local curvature control can generate helical bending waves. *Cell Motil. Cytoskel.* **53,** 103–124.

Churchill, G. C., O'Neill, J. S., Masgrau, R., Patel, S., Thomas, J. M., Genazzani, A. A., and Galione, A. (2003). Sperm deliver a new second messenger: NAADP. *Curr. Biol.* **13,** 125–128.

Collas, P. (2000). Formation of the sea urchin male pronucleus in cell-free extracts. *Mol. Reprod. Dev.* **56,** 265–270.

Cross, N. L. (1983). Isolation and electrophoretic characterization of the plasma membrane of sea urchin sperm. *J. Cell Sci.* **59,** 13–25.

Darszon, A., Beltrán, C., Felix, R., Nishigaki, T., and Treviño, C. L. (2001). Review: Ion transport in sperm signaling. *Dev. Biol.* **240,** 1–14.

Darszon, A., Labarca, P., Beltrán, C., García-Soto, J., and Liévano, A. (1994). Sea urchin sperm: An ion channel reconstitution study case. *Methods* **6,** 37–50.

Debenham, P., Brzezinski, M. A., and Foltz, K. R. (2000). Evaluation of sequence variation and selection in the bindin locus of the red sea urchin, *Strongylocentrotus franciscanus. J. Mol. Evol.* **51,** 481–490.

Galindo, B. E., Beltrán, C., Cragoe, E. J., Jr., and Darszon, A. (2000). Participation of a K+ channel modulated directly by cGMP in the speract induced signaling cascade of *Strongylocentrotus purpuratus* sea urchin sperm. *Dev. Biol.* **221,** 285–294.

García-Soto, J., Mourelle, M., Vargas, I., De La Torre, L., Ramírez, E., López-Colomé, A. M., and Darszon, A. (1988). Sea urchin sperm head plasma membranes: Characteristics and egg jelly induced Ca^{2+} and Na^{+} uptake. *Biochim. Biophys. Acta* **944,** 1–12.

Garbers, D. L. (1989). Molecular basis for fertilization. *Ann. Rev. Biochem.* **58,** 719–742.

Gatti, J.-L., and Christen, R. (1985). Regulation of internal pH of sea urchin sperm. *J. Biol. Chem.* **260,** 7599–7602.

Gauss, R., Seifert, R., and Kaupp, U. B. (1998). Molecular identification of a hyperpolarization activated channel in sea urchin sperm. *Nature* **393,** 583–587.

Glaser, R. W., Grüne, M., Wandelt, C., and Ulrich, A. S. (1999). Structural analysis of a fusogenic peptide sequence from the sea urchin fertilization protein bindin. *Biochemistry* **58,** 2300–2309.

Gonzáles-Martinez, M. T., Galindo, B. E., De La Torre, L., Zapata, O., Rodriguez, E., Florman, H. M., and Darszon, A. (2001). A sustained increase in intracellular Ca^{2+} is required for the acrosome reaction in sea urchin sperm. *Dev. Biol.* **236,** 220–229.

Gray, J. P., and Drummond, G. I. (1976). Guanylate cyclase from sea urchin sperm: Subcellular localization. *Arch. Biochem. Biophys.* **172,** 31–38.

Henson, J. H., Cole, D. G., Roesener, C. D., Capuano, S., Mendola, R. J., and Scholey, J. M. (1997). The heterotrimeric motor protein kinesin-II localizes to the midpiece and flagellum of sea urchin and sand dollar sperm. *Cell Motil. Cytoskel.* **38,** 29–37.

Hirohashi, N., and Vacquier, V. D. (2002a). Egg fucose sulfate polymer, sialoglycan, and speract all trigger the sea urchin sperm acrosome reaction. *Biochem. Biophys. Res. Comm.* **296,** 833–839.

Hirohashi, N., Vilela-Silva, A.-C. E. S., Mourão, P. A. S., and Vacquier, V. D. (2002). Structural requirements for species-specific induction of the sperm acrosome reaction by sea urchin egg sulfated fucan. *Biochem. Biophys. Res. Comm.* **298,** 403–407.

Hirohashi, N., and Vacquier, V. D. (2002b). High molecular mass egg fucose sulfate polymer is required for opening both Ca^{2+} channels involved in triggering the sea urchin sperm acrosome reaction. *J. Biol. Chem.* **277,** 1182–1189.

Hirohashi, N., and Vacquier, V. D. (2002c). Egg sialoglycans increase intracellular pH and potentiate the acrosome reaction of sea urchin sperm. *J. Biol. Chem.* **277,** 8041–8047.

Hirohashi, N., and Vacquier, V. D. (2003). Store-operated calcium channels trigger exocytosis of the sea urchin sperm acrosomal vesicle. *Biochem. Biophys. Res. Commun.* **304,** 285–292.

Huitorel, P., White, D., Fouquet, J. P., Kann, M. L., Cosson, J., and Gagnon, C. (2002). Differential distribution of glutamylated tubulin isoforms along the sea urchin sperm axoneme. *Mol. Reprod. Dev.* **62,** 139–148.

Igarashi, P., and Somlo, S. (2002). Genetics and pathogenesis of polycystic kidney disease. *J. Am. Soc. Nephrol.* **13,** 2384–2398.

Kaupp, U. B., Solzin, J., Hildebrand, E., Brown, J. E., Helbig, A., Hagen, V., Beyermann, M., Pampaloni, F., and Weyand, I. (2003). The signal flow and motor response controlling chemotaxis of sea urchin sperm. *Nat. Cell Biol.* **5,** 109–117.

Lee, H.-C. (1984). Sodium and proton transport in flagella isolated from sea urchin spermatozoa. *J. Biol. Chem.* **259,** 4957–4963.

Lee, H.-C. (1985). The voltage-sensitive Na^+/H^+ exchange in sea urchin spermatozoa flagellar membrane vesicles studied with an entrapped pH probe. *J. Biol. Chem.* **260,** 10794–10799.

Lee, H.-C. (1995). Isolation of flagella and their membranes from sea urchin spermatozoa. *Meth. Cell. Biol.* **47,** 43–46.

Männikkö, R., Elinder, F., and Larsson, H. P. (2002). Voltage-sensing mechanism is conserved among ion channels gated by opposite voltages. *Nature* **419,** 837–841.

Mendoza, L. M., Nishioka, D., and Vacquier, V. D. (1993). A GPI-anchored sea urchin sperm membrane protein containing EGF domains is related to human uromodulin. *J. Cell Biol.* **121,** 1291–1297.

Mengerink, K. J., Moy, G. W., and Vacquier, V. D. (2002). suREJ3, a polycystin-1 protein, is cleaved at the GPS domain and localizes to the acrosomal region of sea urchin sperm. *J. Biol. Chem.* **277,** 943–948.

Mengerink, K. J., and Vacquier, V. D. (2002). An ATP-binding cassette transporter is a major glycoprotein of sea urchin sperm membranes. *J. Biol. Chem.* **277,** 40729–40734.

Metz, E. C., and Palumbi, S. R. (1996). Positive selection and sequence rearrangements generate extensive polymorphism in the gamete recognition protein bindin. *Mol. Biol. Evol.* **13,** 397–406.

Metz, E. C., Kane, R. E., Yanagimachi, H., and Palumbi, S. R. (1994). Fertilization between closely related sea urchins is blocked by incompatibilities during sperm–egg attachment and early stages of fusion. *Biol. Bull.* **187,** 23–34.

Minor, J. E., Fromson, D. R., Britten, R. J., and Davidson, E. H. (1991). Comparison of the bindin proteins of *Strongylocentrotus franciscanus, S. purpuratus, Lytechinus variegatus*: Sequences involved in the species specificity of fertilization. *Mol. Biol. Evol.* **8,** 781–795.

Mita, M., and Ueta, N. (1988). Energy metabolism of sea urchin spermatozoa, with phosphatidylcholine as the preferred substrate. *Biochim. Biophys. Acta* **959,** 361–369.

Moy, G. W., Mendoza, L. M., Schulz, J. R., Swanson, W. J., Glabe, C. G., and Vacquier, V. D. (1996). The sea urchin sperm receptor for egg jelly is a modular protein with extensive homology to the human polycystic kidney disease protein, PKD1. *J. Cell Biol.* **133,** 809–817.

Multigner, L., Pignot-Paintrand, I., Saoudi, Y., Job, D., Plessmann, U., Rudiger, M., and Weber, K. (1996). The A and B tubules of the outer doublets of sea urchin sperm axonemes are composed of different tubulin variants. *Biochemistry* **35,** 10862–10871.

Ohta, K., Sato, C., Matsuda, T., Toriyama, M., Lennarz, W. J., and Kitajima, K. (1999). Isolation and characterization of low density detergent-insoluble membrane (LD-DIM) fraction from sea urchin sperm. *Biochem. Biophys. Res. Commun.* **258,** 616–623.

Ohta, K., Sato, C., Matsuda, T., Toriyama, M., Vacquier, V. D., Lennarz, W. J., and Kitajima, K. (2000). Co-localization of receptor and transducer proteins in the glycosphingolipid-enriched, low density, detergent-insoluble membrane fraction of sea urchin sperm. *Glycoconjugate J.* **17,** 205–214.

Palumbi, S. R. (1999). All males are not created equal: Fertility differences depend on gamete recognition polymorphisms in sea urchins. *Proc. Natl. Acad. Sci. USA* **96,** 12632–12637.

Poccia, D. L. (1989). Reactivation and remodeling of the sperm nucleus after fertilization. *In* "The Molecular Biology of Fertilization" (H. Schatten and G. Schatten, eds.), pp. 115–131. Academic Press, San Diego.

Poccia, D. L., and Green, G. R. (1986). Nuclei and chromosomal proteins. *Meth. Cell Biol.* **27,** 153–174.

Podell, S. B., and Vacquier, V. D. (1984). Wheat germ agglutinin blocks the acrosome reaction in *Strongylocentrotus purpuratus* sperm by binding a 210,000-mol-wt membrane protein. *J. Cell Biol.* **99,** 1598–1604.

Podell, S. B., Moy, G. W., and Vacquier, V. D. (1984). Isolation and characterization of a plasma membrane fraction from sea urchin sperm exhibiting species specific recognition of the egg surface. *Biochim. Biophys. Acta* **778,** 25–37.

Quest, A. F., Harvey, D. J., and McIlhinney, R. A. (1997). Myristoylated and nonmyristoylated pools of sea urchin sperm flagellar creatine kinase exist side-by-side: Myristoylation is necessary for efficient lipid association. *Biochemistry* **36,** 6993–7002.

Revelli, A., Ghigo, D., Moffa, F., Massobrio, M., and Tur-Kaspa, I. (2002). Guanylate cyclase activity and sperm function. *Endoc. Rev.* **23,** 484–494.

Rodríguez, E., and Darszon, A. (2003). Intracellular sodium changes during the speract response and the acrosome reaction in sea urchin sperm. *J. Physiol.* **546,** 80–100.

Roncaglia, P., Mistrik, P., and Torre, V. (2002). Pore topology of the hyperpolarization-activated cyclic nucleotide-gated channel from sea urchin sperm. *Biophys. J.* **83,** 1953–1964.

Rothberg, B. S., Shin, K. S., Phale, P. S., and Yellen, G. (2002). Voltage-controlled gating at the intracellular entrance to a hyperpolarization-activated cation channel. *J. Gen. Physiol.* **119,** 83–91.

Schulz, J. R. (1999). Sea urchin sperm acrosomal exocytosis: Identification of acrosome reaction vesicle associated proteins, p. 71. Ph.D. Thesis. University of California, San Diego.

Schulz, J. R., Wessel, G. M., and Vacquier, V. D. (1997). The exocytotic regulatory proteins syntaxin and VAMP are shed from sea urchin sperm during the acrosome reaction. *Dev. Biol.* **191,** 80–87.

Schulz, J. R., Sasaki, J. D., and Vacquier, V. D. (1998). Increased association of synaptosome-associated protein of 25 kDa with syntaxin and vesicle-associated membrane protein following acrosomal exocytosis of sea urchin sperm. *J. Biol. Chem.* **273,** 24355–24359.

Stephens, S., Beyer, B., Balthazar-Stablein, U., Duncan, R., Kostacos, M., Lukoma, M., Green, G. R., and Poccia, D. L. (2002). Two kinase activities are sufficient for sea urchin sperm chromatin decondensation *in vitro*. *Mol. Reprod. Dev.* **62,** 496–503.

Su, Y.-H., and Vacquier, V. D. (2002). A flagellar K^+-dependent Na^+/Ca^{2+} exchanger keeps Ca^{2+} low in sea urchin spermatozoa. *Proc. Natl. Acad. Sci. USA* **99,** 6743–6748.

Suzuki, N. (1995). Structure, function, and biosynthesis of sperm-activating peptides and fucose sulfate glycoconjugate in the extracellular coat of sea urchin eggs. *Zool. Sci.* **12,** 12–27.

Tamura, N., Chrisman, T. D., and Garbers, D. L. (2001). The regulation and physiological roles of the guanylyl cyclase receptors. *Endocr. J.* **48,** 611–634.

Tombes, R. M., Brokaw, C. J., and Shapiro, B. M. (1987). Creatine kinase-dependent energy transport in sea urchin spermatozoa. *Biophys. J.* **52,** 75–86.

Vacquier, V. D. (1986a). Handling, labeling, and fractionating sea urchin spermatozoa. *Meth. Cell Biol.* **27,** 15–40.

Vacquier, V. D. (1986b). Activation of sea urchin spermatozoa during fertilization. *Trends Biochem. Sci.* **11,** 77–81.

Vacquier, V. D. (1987). Plasma membranes isolated from sea urchin spermatozoa. *In* "New Horizons in Sperm Cell Research" (H. Morhi, ed.), pp. 217–233. Japan Science Society Press, Tokyo.

Vacquier, V. D., Swanson, W. J., and Hellberg, M. E. (1995). What have we learned about sea urchin sperm bindin? *Develop. Growth Differ.* **37,** 1–10.

Vacquier, V. D., and Moy, G. W. (1997). The fucose sulfate polymer of egg jelly binds to sperm REJ and is the inducer of the sea urchin sperm acrosome reaction. *Dev. Biol.* **192,** 125–135.

van Dorsten, F. A., Wyss, M., Wallimann, T., and Nicolay, K. (1997). Activation of sea urchin sperm motility is accompanied by an increase in the creatine kinase exchange flux. *Biochem. J.* **325,** 411–416.

Vilela-Silva, A.-C. E. S., Castro, M. O., Valente, A.-P., Biermann, C. H., and Mourão, P. A. S. (2002). Sulfated fucans from the egg jellies of the closely related sea urchins *Strongylocentrotus droebachiensis* and *S. pallidus* ensure species-specific fertilization. *J. Biol. Chem.* **277,** 379–387.

Ward, G. E., Garbers, D. L., and Vacquier, V. D. (1985). Effects of egg extracts on sperm guanylate cyclase. *Science* **227,** 768–770.

Ward, G. E., Moy, G. W., and Vacquier, V. D. (1986). Phosphorylation of membrane-bound guanylate cyclase of sea urchin spermatozoa. *J. Cell Biol.* **103,** 95–101.

Wilson, N. F., Foglesong, M. J., and Snell, W. J. (1997). The *Chlamydomonas* mating type plus fertilization tubule, a prototypic cell fusion organelle: Isolation, characterization, and *in vitro* adhesion to mating type minus gametes. *J. Cell Biol.* **137,** 1537–1553.

Wood, C. D., Darszon, A., and Whitaker, M. (2003). Speract induces calcium oscillations in the sperm tail. *J. Cell Biol.* **161,** 89–101.

Woolley, D. M., and Vernon, G. G. (2001). A study of helical and planar waves on sea urchin sperm flagella, with a theory of how they are generated. *J. Exp. Biol.* **204,** 1333–1345.

Wothe, D. D., Charbonneau, H., and Shapiro, B. M. (1990). The phosphocreatine shuttle of sea urchin sperm: Flagellar creatine kinase resulted from a gene triplication. *Proc. Natl. Acad. Sci. USA* **87,** 5203–5207.

Zigler, K. S., and Lessions, H. A. (2003a). Evolution of bindin in the pantropical sea urchin *Tripneustes*: Comparisons to bindin of other genera. *Mol. Biol. Evol.* **20,** 220–231.

Zigler, K. S., and Lessios, H. A. (2003b). 250 million years of bindin evolution. *Biol. Bull.* **205,** 8–15.

CHAPTER 22

Measuring Ion Fluxes in Sperm

Alberto Darszon,* Christopher D. Wood,* Carmen Beltrán,* Daniel Sánchez,† Esmeralda Rodríguez,* Julia Gorelik,† Yuri E. Korchev,† and Takuya Nishigaki*

*Department of Developmental Genetics and Molecular Physiology
Institute of Biotechnology
Universidad Nacional Autónoma de México
Cuernavaca, Morelos, México

†Division of Medicine
Imperial College London
MRC Clinical Sciences Centre
London W12 0NN, United Kingdom

0091-679X/04 $35.00

I. Overview*

Ion permeability changes are deeply involved in how sperm sense environmental cues and signals from the outer envelope of the egg to achieve fertilization. Combining *in vivo* measurements of intracellular ions and membrane potential in sperm populations and single cells with electrophysiological approaches, such as the smart patch-clamp, and reconstitution strategies in planar bilayers is revealing how sperm ion channels participate in sperm motility and the acrosome reaction. Improvements in the spatial and temporal resolution of these complementary strategies are contributing to unraveling the mechanism of gamete signaling and fertilization.

II. Introduction

Cell communication is central to the behavior and preservation of organisms. Fertilization, a vital process in the generation of a new individual, depends on sperm–egg communication. Spermatozoa are specialized to deliver their genetic material to the egg; in many cases, they only possess a nucleus, a centriole, mitochondria, a flagellum, and an acrosomal vesicle. These cells lack the machinery for protein or nucleic acid synthesis. Ion channels and transporters in the sperm plasma membrane participate crucially in fertilization (reviewed in Darszon *et al.*, 2001, 1999).

Spermatozoa are tiny cells (head diameter 2–6 μm), making their electrophysiological characterization difficult. Understanding how sperm ion fluxes participate in fertilization requires complementary experimental strategies such as planar bilayer ion channel reconstitution; *in vivo* measurements of membrane potential (Em), intracellular Ca^{2+} ($[Ca^{2+}]$i), intracellular pH (pHi), intracellular Na^+ ($[Na^+]$i); and patch clamp techniques.

---

*Abbreviations: ADP, Adenosine 5′-diphosphate; AR, Acrosome reaction; ASW, Artificial seawater; ASWL, ASW containing 1mM Ca^{2+}, pH 7.0; ATP, Adenosine 5′-triphosphate; BCECF AM, 2′,7′-bis-(2-carboxyethyl)-5-(and-6)-carboxyfluorescein, acetoxymethyl ester; BLMs, Black lipid bilayers; $[Ca^{2+}]$i, Intracellular Ca^{2+}; cAMP, Cyclic adenosine monophosphate; CCCP, Carbonyl cyanide m-chlorophenyl hydrazone; CCD, Charge-coupled device; cGMP, Cyclic guanosine monophosphate; DASW, 1/10 diluted ASW plus 20mM $MgSO_4$, pH 6.8; DHPs, Dihydropyridines; DMSO, Dimethylsulfoxide; DPPC, diphytanoyl PC; $diSC_3(5)$, 3,3′-dipropylthiadicarbocyanine iodide; DMCF, Dimethylcarboxyfluorescein; diSBAC2(3), bis-(1,3-diethylthiobarbituric acid)trimethine oxonol; EDTA, Ethylenediaminetetraacetic acid; EGTA, Ethylene glycol-bis(2-aminoethylether)-N-N′N′-tetraacetic acid; E_K, K^+ potential; Em, Membrane potential; E_{Na}, Na^+ potential; E_R, Resting Em; FSP, Fucose sulfate polymer; F, Faraday's constant; FCCP, Carbonyl cyanide 4-(trifluoromethoxy)phenylhydrazone; GC, Guanylate cyclase; HEPES, N-(2-hydroxyethyl)piperazine-N′-(ethanesulfonic acid); I, Specific ion; [I]e, External concentrations of the ion; [I]i, Internal concentrations of the ion; IP3, Inositol 1,4,5 trisphosphate; $[K^+]$i, Intracellular K^+; $[Na^+]$i, Intracellular Na^+; PC, Phosphatidylcholine; PE, Phosphatidylethanolamine; PHi, Intracellular Ph; PS, Phosphatidylserine; R, Gas Constant; SBFI AM, Fluorescent probe used to measure $[Na^+]$i; SOCs, Store operated Ca^{2+} channels; T, Absolute temperature; TEA^+, Tetraethylammonium chloride; VDCCs, Voltage-dependent Ca^{2+} channels; Z, Valence of ion.

Sea urchins have been a preferred model in the study of fertilization. They are external fertilizers; thus, mimicking the physiological conditions requires only seawater. Mature males can spawn about 10^{10} cells, thus providing large amounts of biological material. Spermatozoa respond to environmental changes and egg-ligands rapidly, synchronously, and in a compulsory order to fuse with the egg and create a zygote. Diffusion of chemotactic factors from the outer layer of the egg attract spermatozoa toward their target (Eisenbach, 1999; Kaupp *et al.*, 2003). Contact with the egg jelly triggers the acrosome reaction (AR), a set of significant morpho-physiological changes in spermatozoa necessary for fertilization. Ion transport systems are, in one way or another, at the heart of these responses. These fundamental events occur in many marine organisms, in which interesting studies of ion transport have been conducted (Hoshi *et al.*, 1994; Morisawa, 1994; Vacquier, 1998). However, many of the strategies used to study ion transport in sperm have been developed in sea urchins, an ideal system for studying signal transduction. This chapter therefore describes ion transport protocols for sea urchin sperm, although they are applicable for other marine species.

III. Sperm Physiology is Deeply Influenced by Ion Channels and Transporters

A. Sea Urchin Spermatozoa Are Immotile

Sea urchin spermatozoa are immotile in the gonads since the high CO_2 tension in semen keeps pHi acidic (~7.2) with respect to seawater. Dynein, the ATPase that drives the flagellum, is inactive below pH 7.3. Spawning quickly activates sperm respiration and motility as CO_2 tension decreases, protons are released, and pHi increases to ~7.4. A Na^+/H^+ exchange system of unknown molecular identity is involved in this process. At this pHi, dynein hydrolyzes ATP to ADP, activating mitochondrial respiration 50-fold. A phosphocreatine shuttle allows mitochondrial energy to reach the flagellum (reviewed in Darszon *et al.*, 1999).

B. Sperm Responses to Egg Peptides

After spawning, spermatozoa are attracted to the egg by a concentration gradient of peptides diffusing from the egg's outer investment, the jelly (Cook *et al.*, 1994; Ward *et al.*, 1985). This coat is formed mainly by a high molecular weight fucose sulfate polymer (FSP), containing also protein and small peptides (Mengerink and Vacquier, 2001). Among sea urchins, chemotaxis has been demonstrated only in *Arbacia punctulata*. Spermatozoa are attracted by nanomolar concentrations of resact, a 14 amino acid peptide isolated from the homologous jelly. This peptide activates respiration at picomolar concentrations. Resact binds to guanylate cyclase (GC), a 160 kD plasma membrane protein, activating it transiently and increasing cGMP, cAMP, and pHi in a Na^+-dependent manner. This peptide has been shown to induce at least two types of $[Ca^{2+}]$i increases, one fast (msecs) and one slower. The fast transient increase appears to correlate with swimming trajectory changes involved in chemotaxis (Kaupp *et al.*, 2003). Speract, an analogous decapeptide from *Strongylocentrotus purpuratus* egg jelly, binds to a 77 kD protein that modulates GC. This decapeptide activates respiration and lipid oxidation, elevates cGMP, cAMP, $[Na^+]$i, pHi, $[Ca^{2+}]$i, and causes a K^+-dependent hyperpolarization in homologous sperm, sperm flagella, and flagellar plasma membrane vesicles, probably mediated by opening of K^+ channels (reviewed by Darszon *et al.*, 2001). In addition to regulating respiration and sperm motility, these egg peptides may cooperate with other factors in the egg jelly to promote fertilization (Yamaguchi *et al.*, 1987).

C. Sea Urchin Sperm Acrosome Reaction

Exposure of spermatozoa to FSP from the egg jelly layer triggers the AR. This exocytotic reaction requires external Ca^{2+} ($[Ca^{2+}]$e) for the fusion of the acrosomal vesicle to the plasma membrane and for extension of the acrosomal tubule, which is surrounded by the membrane destined to fuse with the egg. Within seconds, FSP induces Na^+ and Ca^{2+} influx and H^+ and K^+ efflux in sperm, leading to interrelated changes in Em, $[Ca^{2+}]$i, pHi, and $[Na^+]$i (Rodriguez and Darszon, 2003). In addition, FSP produces increases in the concentration of cAMP, protein kinase activity, turnover of inositol 1,4,5 trisphosphate (IP3), and stimulation of phospholipase D (reviewed in Darszon *et al.*, 2001).

FSP transiently hyperpolarizes and then depolarizes *L. pictus* spermatozoa. The hyperpolarization is K^+-dependent and probably mediated by K^+ channels. The opening of this channel could remove inactivation from voltage-dependent Ca^{2+} channels (VDCCs) and somehow link the FSP-induced change in Em with the pHi increase (Gonzalez-Martinez and Darszon, 1987). The Na^+/H^+ exchange and the increase in pHi associated to the AR depend on $[Ca^{2+}]$e (Rodriguez and Darszon, 2003).

Antagonists of K^+ channels (TEA^+) and Ca^{2+} channels (verapamil, Ni^{2+}, and dihydropyridines (DHPs)) inhibit the AR, indicating their mandatory participation in this process (reviewed in Darszon *et al.*, 1999). Two different, interrelated Ca^{2+} channels participate in the sea urchin sperm AR. FSP binding to its receptor (REJ; Mengerink *et al.*, 2002) transiently opens a Ca^{2+} selective channel that is blocked by verapamil and DHPs. A second channel, insensitive to the latter blockers, activates 5 s later and leads to the AR. This second channel inactivates very slowly (many seconds), is permeable to Mn^{2+}, and is pHi-dependent (Guerrero *et al.*, 1998). Blocking the first channel inhibits the second and the AR. The second channel appears to belong to the family of store operated Ca^{2+} channels (SOCs) (Gonzalez-Martinez *et al.*, 2001) and is important in the AR of many species (O'Toole *et al.*, 2000). It is likely that the internal store is the acrosome since sperm do not have endoplasmic reticulum. *S. purpuratus* sperm produce IP3 (Domino and Garbers, 1988) during the AR and possess IP3 receptors (Zapata *et al.*, 1997).

IV. Strategies to Study Sperm Ion Channels and Transporters

The small size and complex morphology of spermatozoa has hampered their electrophysiological characterization. Thus, the study of sperm ion transport has required combining *in vivo* measurements using fluorescent dyes for Em, $[Ca^{2+}]i$, pHi, and $[Na^{+}]i$, (Fig. 1) with patch clamp recordings in swollen and normal sperm (Fig. 4) and planar bilayer techniques (Fig. 5).

A. General Procedures

1. Collection of Gametes (Details in Vacquier's Chapter 21)

1. Place a live sea urchin upside down on ice.
2. Perform an intracoelomic injection of 0.5 to 1.5 ml of 0.5 M KCl with a hypodermic syringe.
3. Leave the sea urchin to spawn and observe the gametes emerge from the gonopores. If white, they are sperm and it is a male, these are called dry sperm; if yellow to orange, they are eggs and it is a female.
4. Collect the dry sperm with a Pasteur pipette and transfer them to a 1.5 ml Eppendorf tube on ice until use. Take care never to cause bubbles by allowing too much air inside the pipette.
5. It is preferable to use sperm as fresh as possible and to collect a new batch of sperm for each day's experiments. If insufficient resources are available, dry sperm left overnight at 4 °C may be used the following day in species like *S. purpuratus* (but not *L. pictus*). There is often a noticeable degradation in the quality of such samples.
6. The eggs are recovered into a beaker containing artificial seawater (ASW) (mM: 486 NaCl, 26 $MgCl_2$, 27 $MgSO_4$, 10 $CaCl_2$, 10 KCl, 2.5 $NaHCO_3$, 0.1 EDTA, 10 HEPES, pH 8.0). The egg jelly and FSP are obtained as described in (Garbers *et al.*, 1983). The percentage of AR is determined by phase contrast microscopy (Vacquier, 1986).

B. Labeling of Sperm

Sperm for use in ion flux measurement must be loaded with the relevant fluorescent indicator. Unfortunately, the small size of sperm prevents the use of invasive bulk loading techniques, such as microinjection. On the other hand, these cells do not carry out protein synthesis so they can not be transfected to express the new generation of fluorescent proteins (Zhang *et al.*, 2002). Therefore, many of the dyes used in sperm must be membrane permeant. Such dyes fall into two broad categories: lipophilic dyes, such as $diSC_3(5)$ and acridine orange, which can pass through lipid bilayers and distribute according to the parameter they are sensitive

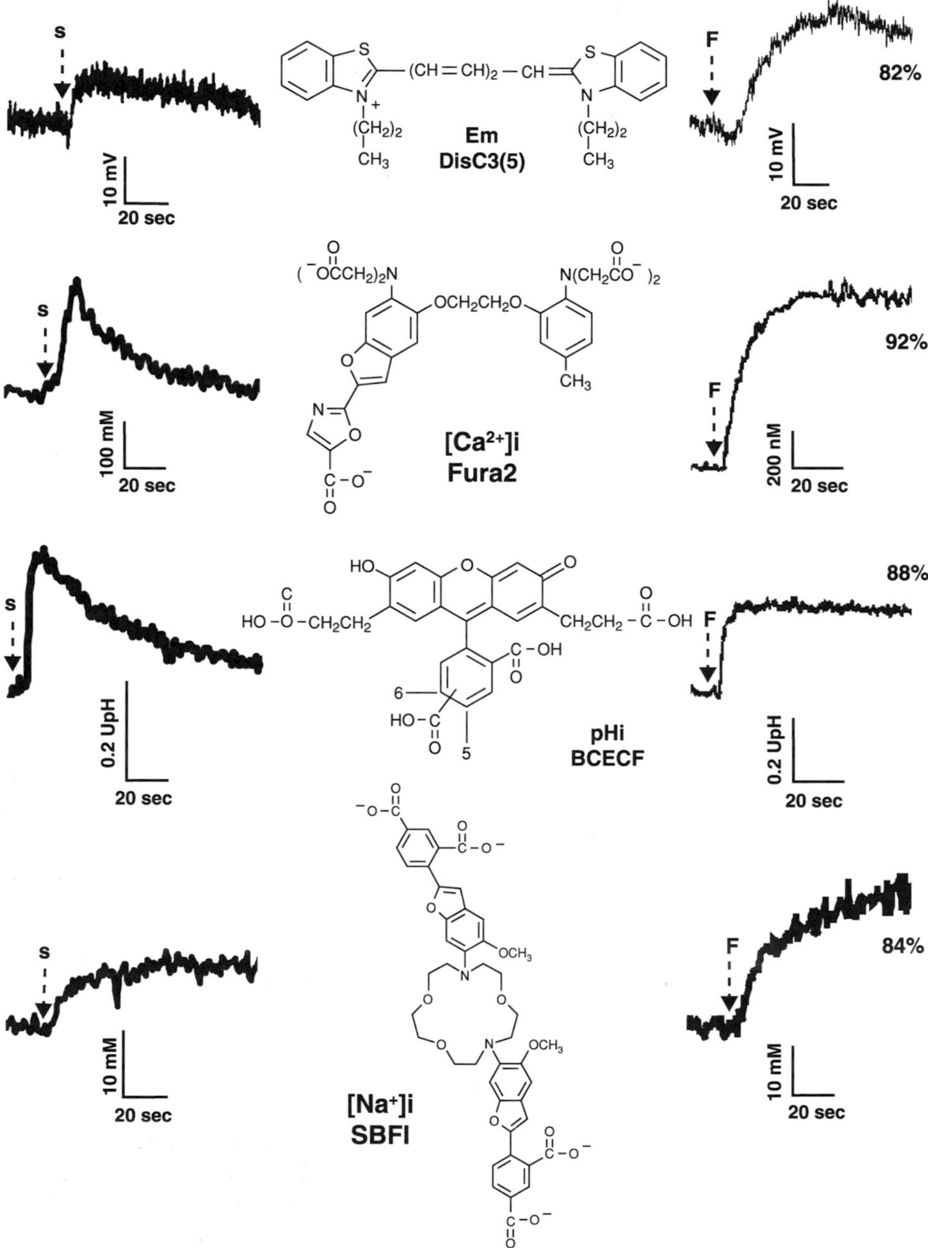

Fig. 1 Changes of membrane potential (Em), intracellular Ca^{2+} concentration ($[Ca^{2+}]$i), intracellular pH (pHi), and intracellular Na^{+} concentration ($[Na^{+}]$i) induced by speract (s) and FSP (F) in *S. purpuratus* sea urchin sperm. Dotted arrows indicate the speract (100 nM) and FSP addition to the sea urchin sperm population. Upward and downward deflections indicate increases or decreases

to, and esterified derivatives of membrane impermeant dyes. These esterified dye derivatives generally can, in principle, enter all cellular compartments, where upon they become substrates for cellular esterase activity. This releases their active, fluorescent form in a membrane-impermeant state within the cell. Use of such dyes is dependent on there being sufficient esterase activity to release enough active dye within the cell. It is necessary to test several dyes for a specific ion to see which one is loaded better and displays adequate responses. For instance, *S. purpuratus* sperm load very well with fura-2 for $[Ca^{2+}]$i and BCECF for pHi (Guerrero and Darszon, 1989b) but *L. pictus* sperm do not; they must be loaded with quin2 or fluo-3/4 for $[Ca^{2+}]$i and dimethylcarboxyfluorescein (DMCF) for pHi (Gonzalez-Martinez *et al.*, 1992).

Regardless of the dye, it is important to consider the phenomenon of compartmentalization, whereby active dye accumulates unevenly in different cellular compartments. The specific factors underlying this phenomenon vary among cell types, but in general, compartmentalization increases with the duration of labeling, and incubation times should be kept to the minimum required to achieve an adequate fluorescence signal (see Rodriguez and Darszon, 2003).

1. Suggested Protocol for Ion-Specific Dye Loading in Sea Urchin Sperm

1. Dilute sperm 1:10 in modified ASW containing 1 mM Ca^{2+}, pH 7.0 (ASWL). Use of this modified ASW recipe prevents spontaneous AR in sea urchin sperm.
2. Stock solutions (1 mM) of the dyes (i.e., SBFI AM, fura-2 AM, fluo-3 AM, DiS-C-(5), and BCECF AM) are made in DMSO. Add dye to a final concentration of 0.3 to 25 μM to the incubation medium. It is common to include the detergent Pluronic F-127 (Molecular Probes) in the incubation mixture (~0.05–0.1% w/v). It has been claimed that this detergent increases the solubility of esterified dye precursors and helps prevent compartmentalization of dyes within the cell. It is advisable to carefully resuspend the dye in ASW, plus or minus Pluronic F-127, and then add sperm.
3. Incubate at 12 to 16 °C for upwards of 1 h. The incubation time will vary from dye to dye; it is suggested to find the optimum balance between the time elapsed, extent of labeling, and degree of compartmentalization for each dye used.
4. Dilute sperm 100-fold in modified ASW (1 mM Ca^{2+}, pH7.0).
5. Centrifuge at 1000*g* for 8 min.

---

in Em, $[Ca^{2+}]$i, pHi, and $[Na^{+}]$i. In records with FSP, percent numbers at the end of each trace correspond to AR determined by phase contrast microscopy. The fluorescent probes used to measure Em (diSC$_3$(5)), $[Ca^{2+}]$i fura-2, pHi (BCECF), and $[Na^{+}]$i (SBFI) are shown in the middle of the figure. Sperm were loaded with fura-2 AM (20 μM), BCECF-AM (20 μM), SBFI-AM (25 μM) for 3 h at 15 °C in 1 mM Ca^{2+} artificial seawater, pH 7.0. For Em determinations, sperm were pre-equilibrated with diSC$_3$(5) (500 nM) during 2 to 3 min (see text for general loading and measuring conditions).

6. Resuspend pellet in approximately 10× volume of ASWL and store on ice, wrapped in foil, prior to use. The time elapsed after washing the external dye may influence the response since intracellular dye may still be hydrolyzed; controls should be performed to determine the extent of this effect.

2. Considerations/Modifications to Loading Conditions

1. Care should be taken at all stages to avoid excessive exposure of sperm to light during and after the labeling process. Consider performing the procedure in a darkroom and/or keeping the sample(s) wrapped in foil as much as possible. This is especially important when including caged compounds in any experimental design.
2. Sperm are delicate. Always use pipette tips with wide apertures, which are available commercially or can be made simply by snipping the first few millimeters off the end of conventional pipette tips with scissors. Take care to be gentle while resuspending the pellet after centrifugation.
3. Sperm from some species, such as *L. pictus*, are especially susceptible to damage during labeling. In this case, consider omitting steps 1. 4–6 and using the sperm directly from the labeling mixture. This approach is particularly suitable for experiments with single cells that require adhesion of the labeled sperm to a glass coverslip, whereafter the sperm may be gently washed *in situ* once the imaging chamber has been assembled. This technique has the disadvantage that intracellular dye concentrations may alter during the course of successive experiments as more active, de-esterified dye accumulates inside the cells. Pluronic F-127 has also been reported to interfere with adhesion of sperm to glass coverslips (Kaupp *et al.*, 2003).

3. Estimating Concentration of Intracellular Dye

The following method is a modification from that of Levi *et al.*, 1994:

1. Divide the sperm suspension into two identical samples; one of them is loaded with the dye as described earlier and the other treated equally but without dye (i.e., same amount of Pluronic-F 127 and DMSO).
2. After incubating the cells for an appropriate time, wash to remove external dye, as indicated in Section IV.B.1.
3. Resuspend each pellet in its original volume and add a 10 μl aliquot to a tube containing the volume of media used in the spectrofluorometer cuvette.
4. Lyse cells by adding 5% Triton X-100. Centrifuge each sample briefly to eliminate cellular debris.
5. Record the fluorescence of the dye-loaded and control cell supernatants in a spectrofluorometer. The difference between the spectra of both samples gives the dye-dependent fluorescence. This value is then compared to a standard curve of the free acid of the dye to estimate its concentration in the supernatant. The

number of cells and the cytoplasmic volume (~15 fl/sperm; Schackmann *et al.*, 1981) are considered to obtain the dye concentration in the cell.

4. Estimating Dye Distribution within Cells

1. Digitonin (3 μM) and Triton X-100 (2%) are used to release dye from the cytoplasm and organelles, respectively (Levi *et al.*, 1994). Digitonin, a non-ionic detergent, has been used to permeabilize the sea urchin sperm plasma membrane (Castellano *et al.*, 1995).

2. To assure that fluorescence signals caused by digitonin are not due to ion concentration changes to which the dye is sensitive, measurements are carried out at the dye's isosbestic point with sperm suspended in intracellular media (IM; mM: 750 mannitol, 175 KCl, 30 NaCl, 5 $MgCl_2$, and 50 HEPES, pH 7.2). The isosbestic point is where the dye's fluorescence spectra with and without ion crosses. A change in fluorescence is observed upon digitonin addition. Many of the dyes exhibit altered spectral properties in the cytoplasm of intact cells compared with those they display *in vitro*.

3. Comparison of emission spectra of loaded sperm in intracellular media before and after digitonin addition will reveal the dye-specific change, which can be used to estimate the percentage of dye present in the cytoplasm.

4. The subsequent fluorescence decrease caused by the addition of Triton X-100 is due to the release of the dye localized in intracellular organelles (Rodriguez and Darszon, 2003).

C. Measuring [Ca^{2+}]i, pHi, [Na^+]i, and Membrane Potential in Populations of Sperm (Fig. 1)

Ion-sensitive dye fluorescence measurements in sperm suspensions are performed in a spectrofluorometer (e.g., an SLM 8000) with a temperature-controlled cell holder equipped with a magnetic stirrer, at 14 to 16 °C. In each case, 5 to 30 μl of the suspension of dye-loaded sperm, or in the case of membrane potential measurements, unlabeled sperm (see Section IV.C.4) is added to ~800 μl of ASW in a round glass cuvette, stirred constantly, and left to equilibrate for 2 min prior to commencing fluorescence measurements. Sperm autofluorescence is usually less than 10% of the total fluorescence and may be considered insignificant in most experiments.

1. Measuring [Na^+]i

1. SBFI fluorescence is monitored at 500 nm using dual excitation at 340 and 380 nm.

2. The relationship between [Na^+]i and SBFI fluorescence is determined by adding gramicidin D (20 μM) or palytoxin (20 nM) to equilibrate [Na^+]e and [Na^+]i

in a medium where $[Na^+]e$ is varied from 0 to 75 mM (Levi *et al.*, 1994; Negulescu *et al.*, 1990).

3. Two calibration solutions are prepared and mixed to obtain the desired $[Na^+]e$: solution 1 (mM) 135 NaCl, 364 Na^+ methanesulphonate, 10 $CaCl_2$, 26 $MgCl_2$, 30 $MgSO_4$, 10 HEPES, and 0.1 EDTA at pH 7.2; solution 2 (mM) 135 KCl, 364 K^+ methanesulphonate, 10 $CaCl_2$, 26 $MgCl_2$, 30 $MgSO_4$, 10 HEPES, and 0.1 EDTA at pH 7.2.

4. The Cl^- concentration in the calibration solutions is chosen to be approximately equal to the $[Cl^-]i$ in sperm to minimize Cl^- fluxes and the consequent changes in cell volume that may occur when the transmembrane monovalent cation gradients are collapsed by gramicidin or palytoxin. The relationship between the fluorescence ratio and $[Na^+]i$ is plotted and used as a calibration curve (Rodriguez and Darszon, 2003).

2. Measuring $[Ca^{2+}]i$

1. Fluo-3 fluorescence can be measured at 525 nm, following excitation at 505 nm.
2. Fura-2 fluorescence may be measured at 500 nm following excitation at 340 nm, although it is recommended to exploit the ratiometric properties of fura-2 (340/380 nm excitation) to record Ca^{2+} fluxes independently of dye concentration.
3. The calibration of fura-2 fluorescence signal is detailed in Guerrero and Darszon (1989a) and Grynkiewicz *et al.* (1985). The isosbestic wavelength for fura-2 inside the sperm was found to be 357 nm.

3. Measuring pHi

1. BCECF fluorescence may be measured at 540 nm following excitation at 500 nm.

2. As with fura-2, it is recommended to exploit the ratiometric properties of BCECF (500/440 nm excitation) to record pHi changes independently of dye concentration.

3. BCECF fluorescence is calibrated as indicated in (Hallam and Tashjian, 1987). Sperm are suspended in ASW with 210 mM of external $[K^+]$ (Na^+ is replaced by an equimolar $[K^+]$) at different pH values (6.4–7.4), and the fluorescence value recorded after the addition of 15 μM nigericin or Triton X-100 ($\sim$0.1% of final volume). In either case, a linear curve of fluorescence vs pH is obtained but they are not strictly parallel. Having both curves, it is possible to derive a correction to transform the detergent values to those obtained with nigericin. At the end of each experiment, the BCECF fluorescence signal is calibrated using two

different external pHs in the presence of Triton X-100 (~0.1% of final volume) and the values corrected as has been described (Guerrero and Darszon, 1989b).

4. Measuring Membrane Potential

The movement of an ion through an open channel across a cell membrane is mainly determined by (a) the difference in the concentration of the ion on both sides of the membrane and (b) the electric potential across the membrane, Em. The value of Em at which these two forces are equal but of opposite direction is called the equilibrium potential and no net ion flux will occur under this condition. Thus, each ion (I) will have its own equilibrium potential (E_I) according to its concentration gradient and valence. This potential can be calculated using the Nernst equation: $E_I = RT/ZF \times \ln([I]e/[I]i)$, where I is the specific ion; R is the gas constant; T is the absolute temperature; Z is the valence of ion I; F is Faraday's constant; and [I]e and [I]i are the external and internal concentrations of the ion I. If the cell's permeability is dominated by K^+ channels, its resting Em (E_R) will be close to E_K, typically ~ -80 mV. However, cells are permeable to other ions that have a positive E_I, such as Na^+, and this will displace their E_R to a value between E_K (−80 mV) and E_{Na} (+40 mV), which will depend on the permeability ratio to the two ions. It is easy to see that the cell will be able to change its Em by opening or closing certain ion selective channels (Hille, 1992). Because some channels and ion transporters are voltage-dependent, these Em changes can modulate second messengers like Ca^{2+} and change pHi in milliseconds.

Several fluorescent dyes have been used to measure Em in sperm. Successful examples in sea urchin and mammalian sperm are diSBAC2(3) and $diSC_3(5)$ (Espinosa and Darszon, 1995; Galindo *et al.*, 2000; Garcia-Soto *et al.*, 1987; Lee and Garbers, 1986). These fluorescent dyes are charged and distribute across cellular membranes in response to electrochemical gradients. They have been reported to bind to intracellular proteins and to undergo emission changes and a slight shift in their spectrum (Plasek and Hrouda, 1991). diSBAC2(3) is an anionic oxonol and $diSC_3(5)$ is a cationic carbocyanine dye; thus, they distribute in opposite directions across the cell membrane at an equivalent Em. Cationic dyes will also partition into internal organelles having a negative Em, such as mitochondria. Because of this, it is necessary to eliminate the mitochondrial Em with uncouplers, for instance, FCCP or CCCP (~1 μM). These compounds make the inner mitochondrial membrane permeable to H^+. It is worth keeping in mind that since the uncoupler will change the H^+ gradient and permeability of the plasma membrane also, it can have some influence on Em and pHi. Usually, these complications are dismissed, though sometimes they should be considered (Guzman-Grenfell *et al.*, 2000). Neither dye appears to be toxic on sperm function at concentrations <1 μM. Both the dye and cell concentrations must be varied to find optimal conditions for the specific sperm species to be studied. Once optimized, $diSC_3(5)$ provides reproducible estimates of Em (reviewed in Darszon *et al.*,

1999) (Fig. 1). Because this dye appears to contain residual Zn^{2+}, when used with *L. pictus* sperm that are very sensitive to this cation (Clapper *et al.*, 1985), the dye solution or ASW should contain enough EDTA.

Protocol:

1. Diluted spermatozoa (1:10 in ASWL, 5–20 μl) are added to a round cuvette containing ASW (0.7–2 ml) at 14 to 16 °C under constant stirring, then $diSC_3(5)$ (stock 1 mM in DMSO) added to a final concentration of 0.5 μM.
2. The fluorescence (620 nm excitation – 670 nm emission) is recorded continuously in the spectrofluorometer as the dye is taken by the cells due to its positive charge. Thereafter, uncoupler (1 μM final) is added and Em changes induced by the egg components recorded once the fluorescence reaches equilibrium (~3 min).
3. The fluorescence signal is calibrated using valinomycin (1 μM), a K^+ selective ionophore. Sequential additions of KCl are used to determine the relationship between the K^+ equilibrium potential (E_K), calculated according to the Nernst equation, and the fluorescence. It is necessary to know or assume from the literature the $[K^+]i$ (180 mM for *S. purpuratus* sperm; Babcock *et al.*, 1992). The $[K^+]$ added at which Em equals the fluorescence value before the addition of valinomycin gives an estimate of the resting membrane potential (E_R; see Galindo *et al.*, 2000, for an example). However, because sperm have voltage-dependent ion channels, estimations of E_R can be difficult, depending on the cell type and ionic conditions. The fluorescence emission of dyes depends on many variables; therefore, internal calibrations must be performed with all compounds and media utilized.

D. Time-Resolved Measurements of Sperm Responses (pHi and $[Ca^{2+}]i$) to Egg Ligands

Since egg ligands immediately induce sperm responses, it is important to measure the rapid kinetics of their responses to understand the underlying signaling mechanisms. In order to perform a successful time-resolved measurement, it is essential that sperm be exposed to egg ligands as rapidly as possible. There are two different methods to achieve this: (1) *rapid mixing* (stopped-flow fluorometry) and (2) *photolysis of caged (photoactivatable) ligands* (Fig. 2).

1. Rapid Mixing

Stopped-flow fluorometry (or spectrophotometry) is commonly used to study kinetics of chemical reactions in the millisecond time range (Gibson and Milnes, 1964). This strategy has been successfully applied to *S. purpuratus* (Nishigaki *et al.*, 2001) and *A. punctulata* (Kaupp *et al.*, 2003) sea urchin sperm. Figure 2A, shows a stopped-flow system composed of two mixing syringes, one stop syringe, and a small volume flow cell. A photomultiplier with high temporal resolution (submillisecond) measures fluorescence (or absorbance) in the flow cell. Mixing syringes

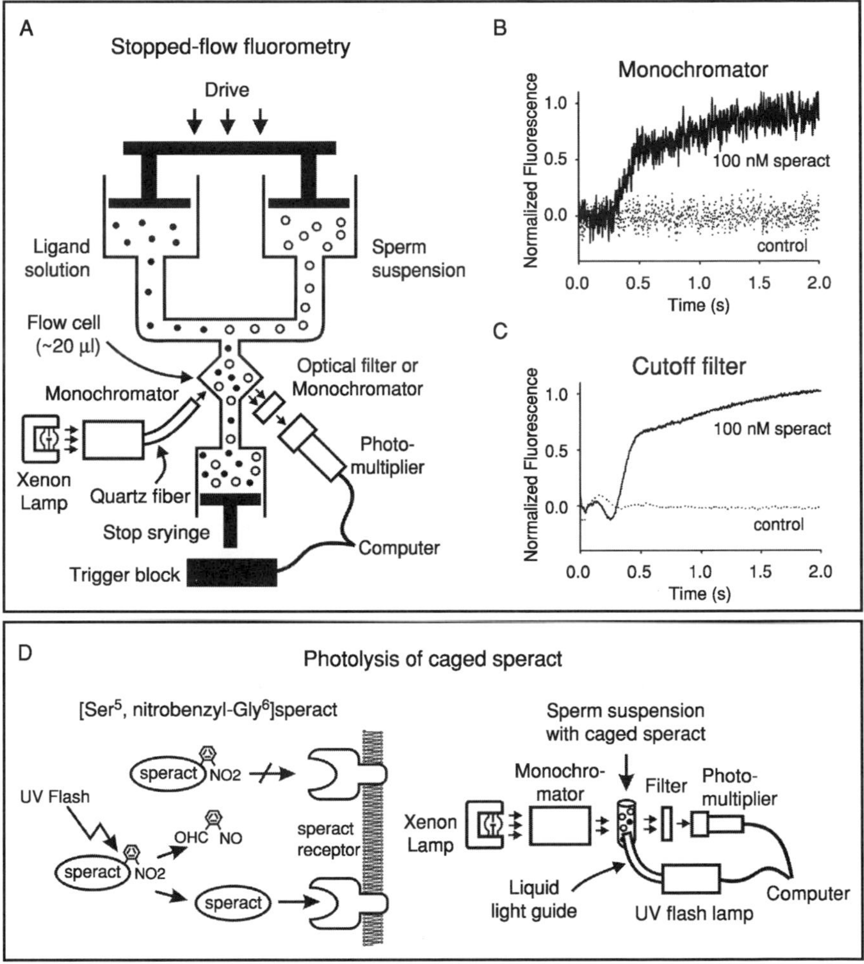

Fig. 2 Schematic representation of "time-resolved sperm [Ca^{2+}]i changes induced by the egg ligand speract." (A) A stopped-flow spectrophotometer, SX 17 MV (Applied Photophysics), equipped with a 150 W xenon arc lamp. Excitation (490 nm) and emission wavelengths were selected with monochromators (ex. 525 nm) or band-pass filter (>515 nm) and connected to the flow cell (20 μl) through quartz fibers. Fluo-3-loaded sperm (75 μl), prepared as indicated in the dye-loading section, and a speract (100 nM, final) solution (75 μl), were placed in separate syringes and rapidly mixed by a pressure-driven piston (nitrogen gas at 26 psi). Most sperm passed through the machine were motile under this condition, although head and tail separation occurred in a small population (<10%). Excited at 490 nm, the control (dotted line) without speract and the speract (continuous line) traces are shown either using the monochromator (em: 520 nm, (B)) or band-pass filter (em: >515 nm, (C)). The traces represent the average of 40 and 20 measurements, respectively. (D) Photolysis of caged speract. A speract analog whose backbone amide (between Ser^5 and Gly^6) is modified by a 2-nitrobenzyl group has a vastly reduced affinity for its receptor. UV irradiation of this speract analog (caged speract) photocleaves the caging group and converts it into an active speract analog ([Ser^6]speract). Sperm pHi and [Ca^{2+}]i measurements using caged speract were performed with an Aminco SLM 8000 (SLM

are driven by nitrogen gas pressure (Applied Photophysics Limited, UK), mechanical force (Bio-Logic Science Instruments, France), or manually, depending on the system. A fluorescence measurement starts when the stop syringe gets to the trigger block. Though the sperm suspension (loaded with fluorescent indicators) and ligand solution are mixed just before the flow cell, there is a dead time of a few milliseconds (depending on the flow rate) at the beginning of the measurement; the faster the flow rate, the more damage to the biological material. Therefore, it is important to determine the optimal flow rate so as to keep the sample functional but with the shortest dead time possible, depending on the biological material. If fluorescence intensity noticeably changes in a control experiment without egg ligands, sperm may be damaged and leaking the dye.

Fluorescence measurements using a small volume of sample with high temporal resolution (short integration time for photon counting) tend to be noisy since few photons reach the photomultiplier (Nishigaki *et al.*, 2001). Using a cutoff filter instead of a monochromator to select emission light improves the fluorescence signal to noise ratio (Fig. 2B and C), as the emission spectra of fluorescent indicators are relatively broad and the monochromator cuts off a significant amount of the indicator-emitted light. This rule is also applicable for time-resolved fluorescence measurements using caged ligands, as described later, but not for indicators whose emission peak shifts upon a change in the measuring parameter, such as indo-1 or SNARF-1. Also, indicators having higher quantum yield are recommended for time-resolved measurement for the same reason.

Since sperm suspensions cause a scattering effect on fluorescence measurements, a fluorescence value at the beginning (~50 ms) tends to be perturbed even after the signal-to-noise ratio is improved by using a cutoff filter. Therefore, it is always recommended to repeat measurements as many times as possible to obtain a reliable average fluorescence measurement.

2. Photolysis of Caged (Photoactivatable) Ligands

One disadvantage of stopped-flow fluorometry is that cells are subjected to pressure during accelerated mixing. Furthermore, stopped-flow techniques cannot be adapted to a single cell imaging approach on a microscope. Caged ligands may offer another option to perform time-resolved measurements and compensate for the disadvantages of stopped-flow fluorometry. Caged compounds are derivatives of biologically active molecules whose activities are masked (or sufficiently reduced) by the introduction of photocleavable groups. Their biological activity is

---

Instruments). For rapid kinetic measurements, samples were excited at 490 nm and emission was selected by a cutoff filter (>515 nm, Applied Photophysics). The emission signals were acquired every 5 ms (200 Hz). Photolysis of the caged speract was achieved using a UV flash lamp system (JML-C2, Rapp Opto Electronic, Hamburg) with a UV band-pass filter (270–400 nm) connected to the spectrofluorometer through a liquid light guide (2 mm diameter). Pyrex glass tubes (6 × 50 mm) were used as optical cells with 200 μl of sperm suspension and attached to the end of the liquid light guide.

unmasked by irradiation with UV light. There are several commercially available caged compounds for small bioactive molecules, such as ATP, amino acids, Ca^{2+} (Ca^{2+} chelator), and cyclic nucleotides (Molecular Probes, OR; BIOLOG Life Science Institute, Germany).

Although not commercially available, there are a few successful examples of caged egg ligands, namely, caged speract (Tatsu *et al.*, 2002) and caged resact (Kaupp *et al.*, 2003). The caged peptides are modified by a nitrobenzyl group on a backbone amide or by 4,5-dimethoxy-2-nitrobenzyl on a side chain (cysteine), respectively, and both are approximately 1000 times less potent than the unmodified peptide. In the presence of the caged peptides, changes in sperm pHi and $[Ca^{2+}]i$ are elicited by irradiation of UV light (Tatsu *et al.*, 2002).

To carry out photolysis experiments with a high temporal resolution, it is necessary to use a UV laser or UV flash lamp (Fig. 2D) as a light source (Rapp, 1998). A continuously operating mercury arc lamp, combined with a mechanical shutter, may be a suitable alternative if a high temporal resolution is not required.

E. Measuring Ion Fluxes in Individual Sperm Cells—A Microscopic Approach

Much valuable information has been collected over the years on ion fluxes in sea urchin sperm populations subjected to a variety of conditions and treatments (e.g., Babcock *et al.*, 1992; Cook and Babcock, 1993a,b; Cook *et al.*, 1994; Schackmann and Chock, 1986). Sperm are, however, highly polarized cells, and there is evidence that components of signal transduction pathways that may control vital aspects of sperm function, such as motility, chemotaxis, and the AR, are unevenly distributed in the sperm cell, particularly between the head and flagella (Cardullo *et al.*, 1994; Su and Vacquier, 2002; Toowicharanont and Shapiro, 1988). Researchers performing measurements of ion fluxes in populations of sperm have generally presumed that their characteristics are directly relevant to the biological function under consideration. However, a recent study has shown that, for the case of intracellular Ca^{2+} measurements, between 85 and 95% of the fluorescent signal from a sea urchin sperm originates in the head, and that the properties of agonist-stimulated Ca^{2+} fluxes vary between the head and flagellum of the same cell (Wood *et al.*, 2003) (Fig. 3). This potential for spatially distributed signaling mechanisms exposes some of the limitations of experiments performed on populations of sperm. Increasingly, researchers are turning to the techniques of microscopy and fast fluorescence imaging to obtain dynamic, spatially resolved information on ion fluxes in a variety of cell model systems (Zhang *et al.*, 2002). This section explains techniques used in the study of ion fluxes in single sea urchin sperm.

1. Single Cell $[Ca^{2+}]i$ Measurements

Though fluorescent dyes are available to measure $[Ca^{2+}]i$, pHi, Em, $[Na^{+}]i$, and other second messengers and activities, the only ion fluxes measured successfully so far in single marine sperm are $[Ca^{2+}]i$ and pHi (Sase *et al.*, 1995; Wood *et al.*,

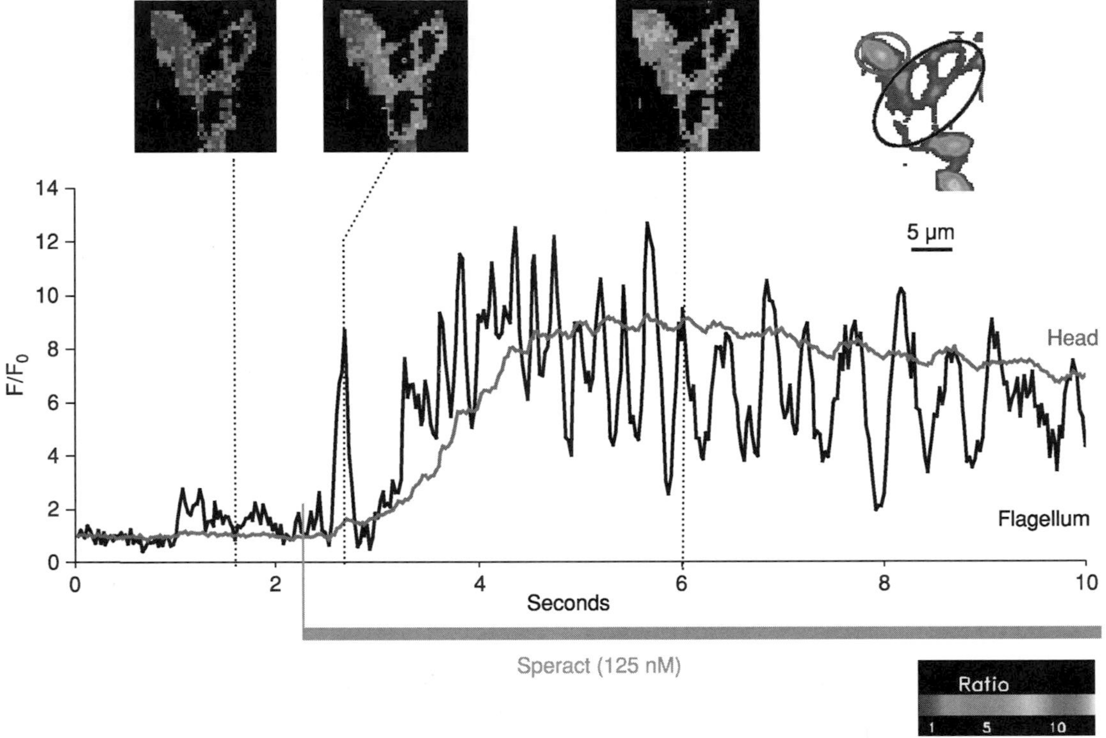

Fig. 3 Complex, spatially discrete patterns of $[Ca^{2+}]$i increase as measured by the single-cell imaging techniques outlined in Section IV.E. Graph shows average per-pixel ratio increase (against value at time 0) of fluo-4 fluorescence in head and flagellum of a single *S. purpuratus* sperm following application of 125 nM speract. Regions of interest are as shown in the image at top right. Images above graph are ratio images against the frame immediately preceding speract addition. Images collected at 40 frames/s with a frame exposure time of 25 ms. (See Color Insert.)

2003). The following sections are limited to procedures to measure $[Ca^{2+}]$i in single sea urchin sperm; however, as improvements are continually being made in fluorescent dye characteristics and the sensitivity and speed of fluorescence measuring equipment, it should be borne in mind that these methods are readily adaptable for measuring all such ion fluxes.

A detailed discussion of the relative merits of different dyes for measuring Ca^{2+} is beyond the scope of this section, and readers are referred to Takahashi *et al.* (1999) and Whitaker (1999). Dyes for measuring intracellular Ca^{2+} may be broadly divided into two classes:

a. Ratiometric Dyes

Fluorescence measurements are made at two different wavelengths (either two different excitation wavelengths with one emission wavelength or one excitation wavelength with two emission wavelengths). The ratio of these two measurements is then used to calculate the concentration of intracellular ions. Examples of Ca^{2+} reporting dyes in this class include fura-2 and indo-1 (Molecular Probes). One advantage of using these ratiometric dyes is that the values they give for Ca^{2+} concentration are independent of both cell morphology and dye concentration. The main disadvantage of ratiometric dyes is the need to collect two images for each Ca^{2+} measurement, together with the delay introduced by any need to switch filters, which degrades the ability to accurately record rapidly changing ion fluxes. Thus, unless two CCD cameras are used simultaneously (Sase *et al.*, 1995), if the recorded events are very fast, as Ca^{2+} fluxes in sea urchin sperm appear to be, then differences between images at different wavelengths in each pair of images may introduce errors into the $[Ca^{2+}]i$ calculation.

To calibrate the recorded $[Ca^{2+}]i$ using ratiometric dyes, measurements must also be taken of the intracellular dye signal in its Ca^{2+}-saturated and Ca^{2+}-unbound states. The former is relatively simple to measure by addition of a calcium ionophore to the cells, as $[Ca^{2+}]$ in seawater (approximately 10 mM) is sufficient to saturate most dyes. The latter has been performed by addition of Ca^{2+} ionophore to cells in Ca^{2+} free ASW in the presence of high concentrations of EGTA. Care must be taken in both cases to allow time for the dye to reach its saturated and unbound states, respectively—particularly in the latter case, which may require several minutes. It is also worth noting that the Kd for all dyes is dependent on a number of environmental variables. Fura-2, the most widely used ratiometric dye, is reported to be sensitive to the viscosity of the medium surrounding the dye, which varies among cells, and among different cellular compartments (Busa, 1992; Poenie, 1990). Care must be taken when performing calculations of $[Ca^{2+}]$ to use an appropriate Kd for the cell system and compartment being studied.

b. Nonratiometric Dyes

Examples of Ca^{2+}-reporting dyes in this class include calcium green, fluo-3, and fluo-4 (Molecular Probes). They are known as single wavelength dyes, in that measurements are restricted to one excitation and one emission wavelength. The advantage of this approach is that successive images may be collected with a minimum delay between measurements—the speed of acquisition is effectively limited by the sensitivity and frame rate of the recording equipment. One particular advantage of the dyes already mentioned that makes them suitable for experiments in single cells is their relatively high quantum efficiency and large increase in fluorescence upon Ca^{2+} binding. Furthermore, their blue light excitation makes them compatible with the use of caged compounds, usually excited in the near UV. The main disadvantage of this class of dyes is that quantitative measurement of $[Ca^{2+}]$ is much more difficult, as they are susceptible to errors due to differences in individual cell morphology and dye concentrations. As such, these dyes are largely restricted to measurements of relative $[Ca^{2+}]$ changes within and among cells. Despite these limitations, these dyes have proven the most useful in uncovering the spatial and temporal characteristics of Ca^{2+} fluxes in single sea urchin sperm (Fig. 3).

2. Imaging Single Sea Urchin Sperm

The imaging system should be designed to capture as much light as possible from the sperm. High magnification, high numerical aperture objective lenses perform best. Charge-coupled device (CCD) cameras are now available with sufficient sensitivity to record the fluorescence from individual sperm, and have become the method of choice for recording ion fluxes in a wide variety of cell systems. Their sensitivity, combined with relatively high frame rates, have allowed measurements of Ca^{2+} changes in single sea urchin sperm to be recorded at high spatiotemporal resolutions, uncovering complex patterns of Ca^{2+} flux that differ between head and flagella regions (Fig. 3).

As a first approach, measurements of ion fluxes in single sperm are made in immobilized cells on a suitable substrate. As these measurements are usually made on microscope systems, glass coverslips are commonly used. Sea urchin sperm will adhere readily to untreated glass coverslips. However, we have observed more consistent speract responses in sperm adhered to poly-L-lysine treated coverslips, although other substances can be used, such as Cell-Tak (Becton Dickinson Biosciences) (Arnoult *et al.*, 1996b). The imaging chamber is preferably temperature regulated to prevent overheating of the sperm. Chamber incubators are available commercially (e.g., Harvard Apparatus), although alternatives, such as air-conditioning of the room in which experiments are performed, are equally feasible.

Protocol:

1. Place a small drop (5–10 μl) of a 50 μg/ml aqueous solution of poly-L-lysine (Sigma) onto the center of a coverslip using a glass Pasteur pipette.
2. Spread the drop over the whole surface of the coverslip using the side of the pipette, employing rapid side-to-side movements.
3. Wipe off the excess liquid gently using a tissue and leave to air-dry face up until all residual liquid has evaporated.
4. Place the prepared coverslips into imaging chambers. Such chambers are available commercially (e.g., Harvard Apparatus) or may be constructed from a small ring of suitable material attached to the coverslip using silicon grease.
5. Dilute labeled sperm 1:40 in ASW (or whichever medium required for the particular experimental conditions, e.g., ASW modified by the augmentation or omission of particular components).
6. Transfer a small quantity (20 μl) to the coated coverslip in the chamber, and leave for up to a minute.
7. Gently wash chamber with fresh media to remove unbound sperm.
8. In our system, the chamber is transferred to a Nikon Diaphot 300 (40× fluor, numeral aperture 1.4 objective) and fluorescence changes are recorded on a Coolsnap FX camera (Photometrics), used in continuous (stream) acquisition mode and utilizing Metamorph (Universal Imaging).

3. Future Perspectives

The next step essential for our understanding of sperm motility is to measure ion fluxes in swimming sperm with spatial and temporal resolution. Such imaging is technically challenging, as it requires a camera that combines high frame rates of acquisition (>50 frames/second) with high sensitivity to record sufficient fluorescence signal over short integration times. Upon encountering a glass surface, motile sea urchin sperm swim in circles (a phenomenon known as thigmotaxis); thus, they remain in the microscope's field of view, often for many seconds, permitting extended observation of their motile behavior. Therefore, sea urchins, and ascidians, which possess similar sperm motility characteristics, are attractive models with which to conduct this research. Adding motility modulators to swimming sperm presents a problem, as the turbulence of addition disturbs their trajectories. Recent reports have begun exploiting UV flash photolysis of caged substrates (Kaupp *et al.*, 2003; Tatsu *et al.*, 2002) and other strategies to overcome this problem (Yoshida *et al.*, 2002). Though no ion fluxes have been measured in motile single sperm yet, soon we will be hearing about these exciting experiments. As new fluorescent dyes are developed with higher quantum yields and increased sensitivity, and the speed and sensitivity of CCD technologies improve while their price decreases, it will be possible to measure other parameters such as pH, membrane potential, and Na^+ fluxes, uncovering more and more information about this most fundamental process of cell regulation.

F. Detecting Ion Channels Directly on Sea Urchin Spermatozoa

As mentioned in the introduction, spermatozoa are very small and morphologically complex cells. The head is conical and they have a long and thin flagellum. These characteristics make conventional patch clamping impractical. To overcome some of the problems involved in recording ion channel activity in spermatozoa, several strategies have been developed and applied through time: (1) Recording from immotile sperm or isolated sperm heads in suspension using suction; (2) Patch clamping suspended swollen spermatozoa also by suction; (3) The smart patch clamp—a new patch clamp system which allows precise control of small-diameter patch pipettes, improving the success rate of gigaseal formation in the sperm head (Gorelik *et al.*, 2002). So far, no reproducible whole-cell recordings have been obtained, regardless of the approach. Success has been most frequent in the on-cell configuration and only few inside-out patches have been documented (reviewed in Darszon *et al.*, 1999, 2001). The characterization of single ion channels in patch-clamp experiments and in planar bilayers follows similar steps, in some respects; the reader can obtain an in depth view of this field in the book by Sakmann and Neher (1983). A brief discussion of the steps followed to characterize single channels is presented in the section on planar bilayers. As has been indicated, some of the procedures are valid for patch-clamp studies.

1. Equipment

Sperm can be visualized under phase contrast microscopy, using an inverted microscope (Nikon Corp., Tokyo, Japan) and patch clamped in the cell-attached or inside-out configuration using an amplifier (Axon, Warner, or Dagan Instruments). Single channel currents are digitized at 5 to 10 kHz, filtered on line at 2 kHz, and stored on a PC computer using pClamp software (Axon Instruments). The pulse protocols are also generated by the analog/digital converter (Axon Instruments) under the control of pClamp. It is recommended to perform the experiments at 20 °C to maintain sperm as healthy as possible. The Axon guide for Electrophysiology and Biophysics Laboratory Techniques (//www.axon.com/mr_Axon_Guide.html) is helpful in setting up the proper acquisition and analysis environment for ion channel work.

2. Patch Clamping Spermatozoa or Sperm Heads

1. Patch clamp electrodes for recording on sea urchin sperm or sperm heads are made by the conventional "two pull" method (Hamill *et al.*, 1981), with either soft (VWR) or hard (Pyrex) glass capillaries. Very small tips have to be used; pipettes filled with ASW should have resistances between 5 and 20MΩ. Narishige or Sutter pullers have been employed. Pipette heat polishing has not improved the success of gigaseal formation in our hands (Fig. 4A).
2. The composition of the pipette and bath solutions will be determined by the type of ion channel to be studied. However, all solutions have to be passed through a 0.22 μm filter to remove particles, which can interfere with gigaseal formation. The patch electrodes are filled by briefly immersing the tip in the desired solution and then backfilling it with a hypodermic needle (Guerrero *et al.*, 1987; Sanchez *et al.*, 2001).
3. Fresh dry spermatozoa are diluted 50-fold in ASWL and washed by centrifugation (1000*g*, 10 min). The washed sperm are diluted 2500-fold in ASW.
4. A recording chamber with a glass bottom (e.g., a coverslip) is used for better visualization under an inverted microscope.
5. Spermatozoa are highly motile under these conditions, making it impossible to position the patch electrode near them. Increasing KCl to 50 mM in seawater stops motility in time (minutes). Thereafter, the air–water interface is penetrated applying positive pressure to the electrode, and once in the vicinity of a cell, negative pressure is applied, thus bringing the sperm into contact with the microelectrode tip and, with good luck, allowing gigaseal formation. The success of gigaseal formation using whole sperm is very low (3%), at least in part due to flagella getting in the electrode.
6. Immotile functional heads, suitable for patch-clamping, can be obtained by gently detaching flagella from spermatozoa (see section on membrane isolation in Vacquier's chapter). The isolated sperm heads are diluted 2500-fold with ASW

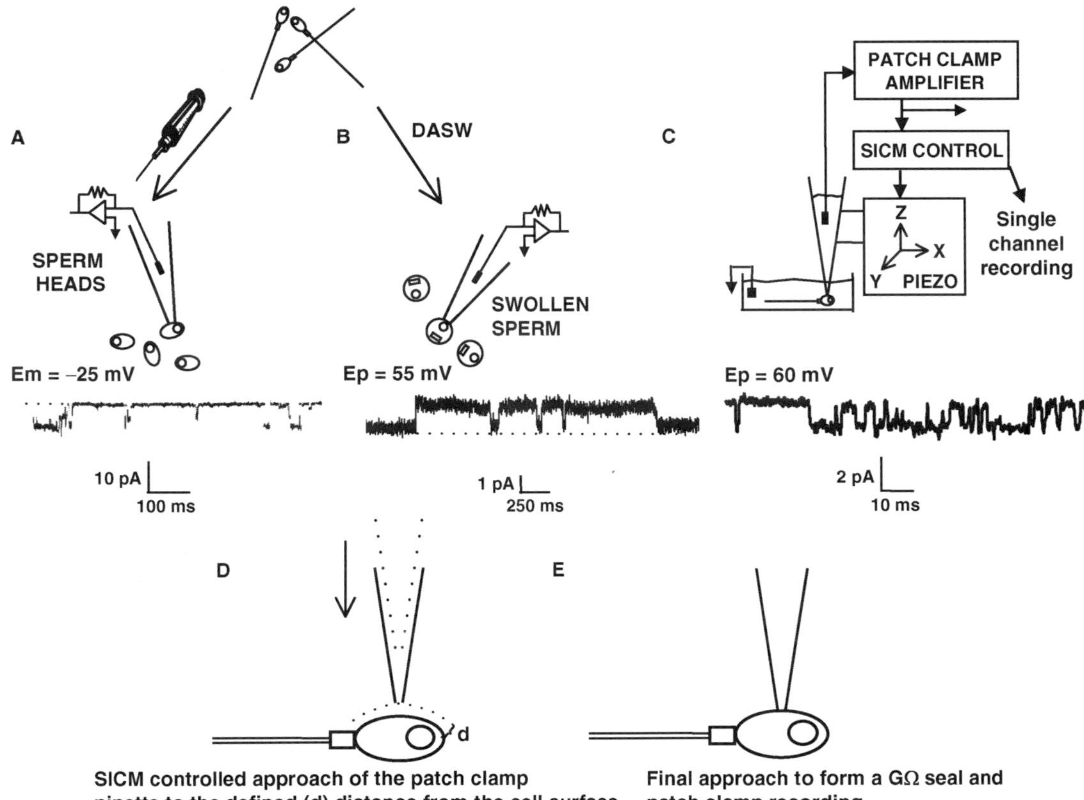

Fig. 4 Electrophysiological strategies to study ion channels in sperm. (A) Flagella can be detached from sperm by passing them through a 21-gauge needle, leaving the immotile heads suitable for patch clamping. (B) Intact sperm can be swollen if resuspended in 10 times diluted ASW (DASW). An example of a single channel detected with each technique is illustrated (A and B, respectively). Both recordings were obtained in the cell-attached configuration. The membrane potential (Em) or the pipette potential (Ep) at which the recording was made is indicated. (C) Schematic diagram of the scanning patch-clamp setup. The micropipette is mounted on a three-axis piezo actuator controlled by a computer. A patch-clamp amplifier measures the ion current that flows through the pipette; the current is used for the feedback control to keep a constant distance between the micropipette and the cell during approaching and scanning. Upon completion of the scanning procedure, computer control is used to position the micropipette at a place of interest based on the topographic image acquired, and finally the same patch-clamp amplifier is used for electrophysiological recording. (D) Cell-attached patch-clamp recordings from the head of intact sea urchin sperm. (E) Scanning patch-clamp principle of operation. A micropipette approaches the sperm surface and reaches a defined separation distance (d), whereupon the distance is kept constant by SICM feedback control. The micropipette is lowered to form the gigaohm (GΩ) seal for patch-clamp recording from the selected structure.

and added to the recording chamber. Gigaseals are formed, applying suction when the patch pipette is close to a swollen sperm head. Removal of the flagellum increases the success by a factor of roughly 2.5 (Guerrero *et al.*, 1987).

7. An additional complication is that single-channel on-cell recordings are attenuated and distorted due to the cell's resistance and capacitance product (RC) network (Guerrero *et al.*, 1987). "Silent patches" with no single channel activity are often observed. This difficulty is sometimes overcome by breaking the cell membrane facing the bath by adding EGTA (16 mM) to the medium. Single channel cationic conductances of 65 and 170 pS have been recorded using this procedure in ASW (Fig. 4A).

3. Swollen Sperm Cells

To increase the success rate of gigaseal formation in sea urchin spermatozoa, these cells can be osmotically swollen without rupturing them (Babcock *et al.*, 1992). In brief, *S. purpuratus* or *L. pictus* dry sperm are diluted 10 times in pH 7.0 ASW, and then diluted 100-fold in 1/10 diluted ASW plus 20 mM $MgSO_4$, pH 6.8 (DASW), which has in mM: 49 NaCl, 2.5 $MgCl_2$, 23 $MgSO_4$, 1 KCl, 1 $CaCl_2$, 0.3 $NaHCO_3$, 0.01 EDTA, 1-10 HEPES. In swollen sperm, the axoneme is internalized and the cells become spherical (diameter $\sim 4\,\mu m$), improving gigaseal formation from 6% in sperm heads to ~35 to 50%. This strategy has allowed the detection of K^+ channels activated by picomolar concentrations of speract (Babcock *et al.*, 1992) and to cAMP-regulated cation channels (Sanchez *et al.*, 2001) (Fig. 4B).

4. The Smart Patch

The smart patch clamp method combines the Scanning Ion Conductance Microscopy (SICM) and patch-clamp recording through a single glass nanopipette probe. The patch clamp electrode can gather topographic information of the cell surface that can be used to facilitate single-channel recordings from small cells and submicron cellular structures that are inaccessible by conventional methods (Gorelik *et al.*, 2002). The instrument is based on an inverted optical microscope (Nikon Corp., Tokyo, Japan), with a mechanism for coordinating optical and scanned images through a video camera. For living cells, displacements in excess of 30 to 50 μm in the vertical direction are required. We have, therefore, used a three-axis piezo translation stage (Tritor, Piezosystem Jena, Germany) with a 100 μm travel distance in the *x, y*, and *z* directions. To control the voltage and measure the current through the pipette, we use an Axopatch 200B patch-clamp amplifier (Axon Instruments, Foster City, CA).

The smart patch method uses a patch clamp microelectrode positioned perpendicularly to the sample. The pipette may first be used to scan the cell surface, controlled by a computer three-axis translation stage with a feedback system. The position of the tip relative to the sample surface strongly influences the ion current through the pipette, which provides the feedback signal to control the vertical position of the tip (Korchev *et al.*, 1997, 2000). As soon as the pipette reaches a determined distance from the surface, the SICM feedback control maintains a constant tip–sample separation. This procedure makes the approach straightforward and safe, because the patch pipette is prevented from touching the cell membrane until it is desired to do so to form a seal. It is important to note that in the case of sperm, scanning the cell surface is not necessary for seal formation. It is possible to center the pipette tip on the sperm head optically and then perform the approximation of the pipette under the feedback control to a precise position on the cell. Thereafter, the feedback control is switched off, the pipette is lowered, suction is applied, and seal formation achieved (Fig. 4C and D).

The sperm suspension is pretreated with 0.01 mg/ml trypsin (Sigma, Poole, UK) solution during 10 min on ice and plated for 1 to 5 min on round coverslip chambers treated with a 0.01% solution of poly-L-lysine, (Sigma, Poole, UK) in ASW. The precise controlled approach of the patch pipette to the cell at a perpendicular angle increases the probability of achieving high resistance seals on the sperm cell body to 45% (Gorelik *et al.*, 2002), compared with the 6% obtained in sperm heads (Guerrero *et al.*, 1987) and 35% in swollen sperm (Sanchez *et al.*, 2001). In addition, because the cells are well adhered to the bottom of the chamber, the probability to obtain an inside-out patch increases enormously. It is possible to switch to this configuration in practically all the attempts where the cell-attach high resistance seals are achieved. A voltage-dependent multistate Ca^{2+} channel with a high main-state conductance (Gorelik *et al.*, 2002), which is similar to the multistate conductance channel seen in planar bilayer recordings (Lievano *et al.*, 1990), was documented using this approach.

5. Future Perspectives

Until now, it has been impossible to carry out sperm whole cell recordings. The smart patch technique opens the possibility of achieving this goal, either directly or using the perforated patch technique. A success in this approach would allow a quantum leap in our understanding of how sperm channels participate in the most important sperm functions that lead to fertilization: motility and the acrosome reaction.

G. Incorporation of Sperm Ionic Channels into Planar Lipid Bilayers

Ion channel incorporation into planar bilayers is an ideal strategy to detect and characterize these proteins from tiny cells difficult to patch-clamp, where plasma membranes and their purified components can be isolated readily (reviewed in Darszon *et al.*, 1999). In fact, it has even been possible to transfer ion channels directly from sperm to planar bilayers (Beltran *et al.*, 1994) (Fig. 5C). The limits of the bilayer assay are related to its capacity to detect single molecular activity.

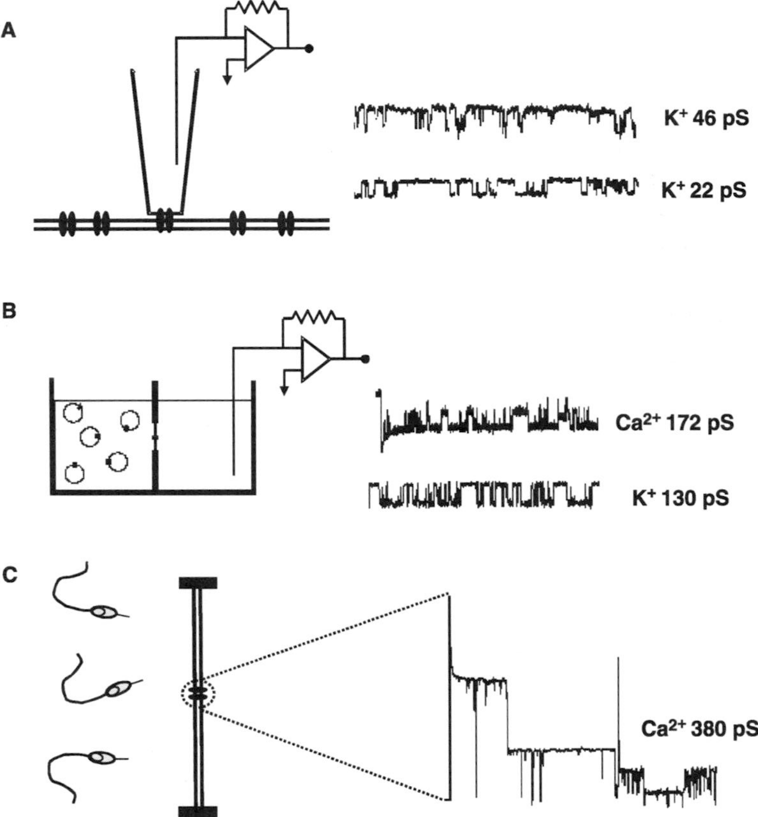

Fig. 5 Study of sperm ion channels in planar bilayers. Examples of single channel activity detected with each technique are illustrated, indicating main ion transported and single-channel conductance in pS. (A) Bilayers formed at tip of a patch-clamp pipette. (B) Black lipid membranes (BLMs) with fused sperm plasma membranes. (C) Direct ion channel transfer from sperm to BLMs.

Problems of contaminant membranes or isolated proteins not derived from the plasmalemma can only be discarded using complementary experimental strategies, including ion measurements, in single sperm or in sperm populations which reveal the pharmacology and modulation of the key ion channels involved in sperm physiology. To avoid channel artifacts, only lipids of the highest purity should be used (such as Avanti Polar Lipids Inc., Birmingham, AL).

1. Isolation of Sea Urchin Sperm Head and Flagella Plasma Membranes

Flagellar and head plasma membranes have been prepared from several sea urchin sperm species by different techniques (reviewed in Darszon *et al.*, 1994). The flagellar and head sperm plasma membranes preparations we use for reconstitution into planar bilayers and liposomes are described in Cross (1983) Darszon *et al.* (1984) and Garcia-Soto *et al.* (1988). We prefer these preparations to that involving a long incubation (>4 h) at pH 9.0 (Chapter 21 of this volume), since these conditions favor AR (Collins and Epel, 1977) and possibly proteolysis of some ion channels.

The head preparation contains homogeneous vesicular membranes as revealed by electron microscopy. They contain less than 5% of the mitochondrial marker cytochrome oxidase and 10% of the total DNA/mg protein. ^{125}I surface labeling indicates a 2.5- to 3-fold enrichment in specific activity in the head membranes with respect to whole sperm. When the flagellar or head membranes are reassembled into liposomes, they increase their permeability to Ca^{2+} and Na^{+} in response to egg jelly in a species-specific manner (Darszon *et al.*, 1984; Garcia-Soto *et al.*, 1988). As in whole sperm, Ca^{2+} uptake is inhibited by the Ca^{2+} channel blocker nisoldipine. A close analog of this compound, [^{3}H]nitrendipine binds with high affinity in a saturable and reversible manner, showing a Kd and B_{Max} of 31 nM and 5.3 pMol/mg protein, respectively. The phosphatidylcholine (PC) content of head membranes is higher, while that of phosphatidylserine (PS) and phosphatidylethanolamine (PE) is lower than that in flagellar membranes. The ouabain-sensitive Na^{+}/K^{+}-ATPase and Ca^{2+}-ATPase activities are 2.5-fold lower and 1.7- to 2.3-fold higher, respectively, than that in the flagellar membranes. Adenylyl cyclase activity is ~10-fold higher in flagella than in sperm heads (Mourelle *et al.*, 1984).

2. Bilayers at the Tip of a Patch Electrode

A patch clamp setup (see Fig. 4C) can also be used to form planar lipid bilayers at the tip of the patch electrode (Coronado and Latorre, 1983; Suarez-Isla *et al.*, 1983; Wilmsen *et al.*, 1983) (Fig. 5A). This approach produces small-diameter bilayers having less intrinsic noise than other planar bilayer strategies. This method derives from the Montal-Müeller approach to lipid bilayer assembly from the apposition of two monolayers; it allows asymmetric membrane formation (reviewed in Montal *et al.*, 1981).

1. Lipid vesicle suspensions spontaneously generate lipid monolayers at the air–water interface (Schindler and Rosenbusch, 1978). The monolayer composition is determined by the composition of the lipid vesicles. If the liposome suspension is mixed with sperm plasma membranes, the lipid monolayer generated will have sperm membrane proteins, including ionic channels (Suarez-Isla *et al.*, 1983). Multilamellar lipid vesicles are formed by mechanical dispersion of PC (1 mg/ml) or mixtures of PC:PS in the desired buffer containing $CaCl_2$ (0.1–5 mM). Membrane protein (1–20 μg) is added and mixed thoroughly. The membrane-containing suspension is then deposited in a small plastic cup and left undisturbed for 5 min to allow for monolayer self assembly.
2. Glass electrodes are made by the two pull technique (Hamill *et al.*, 1981) (1–5 MΩ in 0.1 M KCl, HEPES 5 mM, pH 7.6). Pyrex glass pipettes (melting point

capillaries) allow stable bilayer formation (10–200 GΩ). Strict cleanliness must be exerted at every step since this technique is extremely sensitive to surface-active contaminants. As in the patch-clamp technique, all solutions are passed through 0.22 μm millipore filters before use.

3. Passing the patch-pipette tip twice through the generated lipid monolayer yields a lipid bilayer that, depending on the sperm membranes used, will sometimes have ion channels incorporated. Several K^+ channels were detected using flagellar membrane fractions (Lievano *et al.*, 1985) (Fig. 5A).

3. Black Lipid Bilayers

The most successful ion channel reconstitution approach so far involves vesicle fusion to black lipid bilayers (BLMs) (Miller and Racker, 1976; Montal *et al.*, 1981) (Fig. 5B). BLMs are spontaneously assembled when phospholipids dispersed in a nonvolatile solvent, usually decane, are deposited into a small hole on a hydrophobic partition (see Miller, 1986).

1. Bilayer chambers can be built in the machine shop following already published guidelines (Darszon, 1986; Miller, 1986) or purchased (Warner Instruments). BLMs are generated in a septum that separates two aqueous compartments, called from hereon *cis* and *trans*. The *cis* chamber is connected to a voltage generator, the *trans* to a virtual ground (I-V transducer (amplifier) input). Accordingly, a positive-applied voltage means *cis* positive. A positive-applied voltage, under symmetrical ionic conditions, produces an upward current due to cation flux from *cis* to *trans*. Anion flux from *cis* to *trans* yields a negative, downward current, at positive-applied voltages.

2. BLMs are formed in the hole of the plastic or Teflon partition separating the two aqueous compartments, as previously described (Lievano *et al.*, 1990). Initially, it is advisable to use a lipid mixture containing a significant fraction of acidic lipid to form the planar bilayer (brain PS/PE 1:1, 20 mg/ml, total concentration; DPPC/PE/PS 70:15:15, 20 mg/ml, or diphytanoyl PC, 20 mg/ml in n-decane). The bathing solutions will depend on the ion channel to be studied; for instance, they might consist of a 0.1 M KCl, 5 mM HEPES, pH 7.0, buffer for K^+ or Cl^- channels.

3. Under constant magnetic stirring in *cis*, add plasma membrane fragments or a purified channel reconstituted in liposomes (10–50 μl aliquots, 0.1–2 mg/ml protein stock), loaded with the 0.4 to 0.6 M sucrose, HEPES 10 mM, pH 7.0. Thereafter, elevate $[Ca^{2+}]$ in *cis* to 1 to 5 mM. Fusion to the BLM will be promoted by Ca^{2+} and the swelling of the sucrose-loaded vesicles upon contact with the 0.1 M KCl solution. Fusion can be further enhanced by raising the osmolarity in *cis* by adding aliquots of a concentrated solute. We use 3 M KCl but other solutes, like urea and glycerol, have also been employed (Cohen, 1986).

Single channel current fluctuations appear when fusion occurs. When this happens, the *cis* bilayer chamber is perfused with a buffer of choice to remove excess vesicles and lower [Ca^{2+}], thus preventing more fusion events.

4. Characterizing Single Ion Channel Activity in Planar Bilayers

1. Regardless of the type of planar bilayer system used, after channel insertion, its anion or cation selectivity must be determined. Insertion of a cation-selective channel will result in positive, upward current transitions at zero-applied voltage, when the electrolyte concentration is higher in the *cis* chamber. Its E_R (the voltage at which the net current flowing across the open channel becomes zero) will be negative. In contrast, insertion of an anion channel will give downward current transitions at zero-applied voltage and a positive E_R (Hille, 1992). E_R values will equal those predicted by the Nernst equation for perfectly cation or anion selective channels.

2. To determine if a cation-selective channel present in the bilayer is K^+-selective, perfuse the *cis* chamber with 5 to 10 volumes of a 0.1 M NaCl buffer. Since the *trans* chamber contains 0.1 M KCl, opening of a K^+-selective channel will result in downward current transitions at zero-applied voltage and a positive E_R. If this is the case, building the current-voltage relation for the channel allows an accurate determination of E_R and an estimate of the PK^+/PNa^+ permeability ratio (e.g., Labarca and Latorre, 1992).

3. Assuming a K^+-selective channel is present in the bilayer, it may be interesting to determine its pharmacological characteristics. Useful sets of agents known to block K^+ channels are commercially available (e.g., Alomone). TEA^+, a common K^+ channel blocker, might block from the *cis* or *trans* side, or from both, at micromolar to millimolar concentrations (Coetzee *et al.*, 1999). Blocking will be evident either because of a reduction in the size of the single channel current, accompanied by an increase in channel open time, or by the occurrence of noisy "flickery" open–close transitions (Hille, 1992).

4. To study Ca^{2+}-selective channels, the *trans* bilayer chamber may contain $BaCl_2$ (25–50 mM) and the *cis* solution made with 0.1 M KCl (Liévano *et al.*, 1990). Ba^{2+} is a good divalent since it permeates Ca^{2+}-selective pores and it is a blocker of K^+-selective channels. There are also commercially available kits of blocking agents that will help to define the type of Ca^{2+}-selective channel present in the bilayer.

5. Assuming a K^+ or a Ca^{2+}-selective channel is reproducibly inserted in the planar bilayer, it will be necessary to determine its open-state conductance properties; that is, how many conductance substates are present in a given channel (Fox *et al.*, 1987; Labarca and Latorre, 1992).

6. Define to what extent the channel-open probability is affected by the membrane electric field. This involves determining the fraction of time that a channel

spends in the open conformation(s) as a function of the applied voltage, over a given voltage range.

Consider, for simplicity, a voltage-dependent channel with only one open conductance state; the fraction of time the channel spends open as a function of voltage can be expressed as: $fo = (1 + e^{zF(V-V_{0.5}/RT})^{-1}$. Here, z represents the equivalent charge that moves across the membrane when a channel changes from the open to the closed state (Hille, 1992; Moczydlowski *et al.*, 1986), V is the applied voltage, $V_{0.5}$ is the voltage at which fo is 0.5 and F, R, and T have their usual meaning. Of course, the channel may exhibit a weak or no voltage-dependence. On the other hand, the channel may have several open conductance substates. Then it is necessary to determine the conductance of each substate and the effect of the applied voltage on the fraction of time the channel spends in each substate. An example of this is the Ca^{2+}-selective channel derived from *S. purpuratus* sperm plasma membranes (Lievano *et al.*, 1990). It is then possible to study the mechanisms underlying the substates (Hille, 1992; Miller, 1986; Sakmann and Neher, 1983).

H. Role of Ion Channels in Sperm Physiology

Ultimately, the fundamental objective of studying sperm ion channels is to relate and understand their role in sperm physiology. As mentioned earlier, ion channel blockers are powerful tools to establish this correlation. K^+ and Ca^{2+} channel blockers are known to interfere with sperm physiology (Darszon *et al.*, 1999). If the concentration-dependence of a channel blocker on a specific sperm function matches that obtained in single-channel current patch-clamp or planar lipid bilayer studies, it would strongly suggest that the ion channel in question is involved in the physiological function. For a detailed discussion of ion channel blockade, see Hille (1992) and Moczydlowski *et al.* (1986).

In addition to voltage, ion channels can be regulated by multiple factors, such as ligands, second messengers (Ca^{2+}, cyclic nucleotides, IP3, G proteins, pH, phosphorylation state, etc.). It is crucial to understand how the sperm ion channels involved in motility and AR are regulated. To this end, the single channel activity of the ion channel in question with regard to conductance, selectivity, kinetics, pharmacology, voltage, and pH dependence must be compared with that displayed by the channel after being subjected to the modulators mentioned (Labarca *et al.*, 1996).

Determining the molecular identity of a specific ion channel is a difficult quest. Because sperm do not synthesize proteins, the strategies of molecular biology are not easily applied until it becomes possible to perform experiments on spermatogenic cells, as has been done in mammalian systems (Arnoult *et al.*, 1996a; Lievano *et al.*, 1996). Another approach to relate molecular identity and function is to isolate membrane protein complexes suspected of participating in motility or the AR and studying them functionally in planar bilayers.

References

Arnoult, C., Cardullo, R. A., Lemos, J. R., and Florman, H. M. (1996a). Activation of mouse sperm T-type Ca2+ channels by adhesion to the egg zona pellucida. *Proc. Natl. Acad. Sci. USA* **93,** 13004–13009.

Arnoult, C., Zeng, Y., and Florman, H. M. (1996b). ZP3-dependent activation of sperm cation channels regulates acrosomal secretion during mammalian fertilization. *J. Cell Biol.* **134,** 637–645.

Babcock, D. F., Bosma, M. M., Battaglia, D. E., and Darszon, A. (1992). Early persistent activation of sperm K+ channels by the egg peptide speract. *Proc. Natl. Acad. Sci. USA* **89,** 6001–6005.

Beltran, C., Darszon, A., Labarca, P., and Lievano, A. (1994). A high-conductance voltage-dependent multistate Ca2+ channel found in sea urchin and mouse spermatozoa. *FEBS Lett.* **338,** 23–26.

Busa, W. B. (1992). Spectral characterization of the effect of viscosity on Fura-2 fluorescence: Excitation wavelength optimization abolishes the viscosity artifact. *Cell Calcium* **13,** 313–319.

Cardullo, R. A., Herrick, S. B., Peterson, M. J., and Dangott, L. J. (1994). Speract receptors are localized on sea urchin sperm flagella using a fluorescent peptide analog. *Dev. Biol.* **162,** 600–607.

Castellano, L. E., Lopez-Godinez, J., Aldana, G., Barrios-Rodiles, M., Obregon, A., Garcia de De la Torre, L., Darszon, A., and Garcia-Soto, J. (1995). The acrosome reaction in digitonin-permeabilized sea urchin sperm in the absence of the natural inducer. *Eur. J. Cell Biol.* **67,** 23–31.

Clapper, D. L., Davis, J. A., Lamothe, P. J., Patton, C., and Epel, D. (1985). Involvement of zinc in the regulation of pHi, motility, and acrosome reactions in sea urchin sperm. *J. Cell Biol.* **100,** 1817–1824.

Coetzee, W. A., Amarillo, Y., Chiu, J., Chow, A., Lau, D., McCormack, T., Moreno, H., Nadal, M. S., Ozaita, A., Pountney, D., Saganich, M., Vega-Saenz de Miera, E., and Rudy, B. (1999). Molecular diversity of K+ channels. *Ann. NY Acad. Sci.* **868,** 233–285.

Cohen, F. S. (1986). "Ion Channel Reconstitution." Plenum, New York.

Collins, F., and Epel, D. (1977). The role of calcium ions in the acrosome reaction of sea urchin sperm: Regulation of exocytosis. *Exp. Cell Res.* **106,** 211–222.

Cook, S. P., and Babcock, D. F. (1993a). Activation of Ca2+ permeability by cAMP is coordinated through the pHi increase induced by speract. *J. Biol. Chem.* **268,** 22408–22413.

Cook, S. P., and Babcock, D. F. (1993b). Selective modulation by cGMP of the K+ channel activated by speract. *J. Biol. Chem.* **268,** 22402–22407.

Cook, S. P., Brokaw, C. J., Muller, C. H., and Babcock, D. F. (1994). Sperm chemotaxis: Egg peptides control cytosolic calcium to regulate flagellar responses. *Dev. Biol.* **165,** 10–19.

Coronado, R., and Latorre, R. (1983). Phospholipid bilayers made from monolayers on patch-clamp pipettes. *Biophys. J.* **43,** 231–236.

Cross, N. L. (1983). Isolation and electrophoretic characterization of the plasma membrane of sea-urchin sperm. *J. Cell Sci.* **59,** 13–25.

Darszon, A. (1986). Planar bilayers: A powerful tool to study membrane proteins involved in ion transport. *Methods Enzymol.* **127,** 486–502.

Darszon, A., Beltran, C., Felix, R., Nishigaki, T., and Trevino, C. L. (2001). Ion transport in sperm signaling. *Dev. Biol.* **240,** 1–14.

Darszon, A., Gould, M., De La Torre, L., and Vargas, I. (1984). Response of isolated sperm plasma membranes from sea urchin to egg jelly. *Eur. J. Biochem.* **144,** 515–522.

Darszon, A., Labarca, P., Beltran, C., Garcia-Soto, J., and Lievano, A. (1994). Sea urchin sperm: An ion channel reconstitution study case. *Methods: A Companion to Methods in Enzymology* **6,** 37–50.

Darszon, A., Labarca, P., Nishigaki, T., and Espinosa, F. (1999). Ion channels in sperm physiology. *Physiol. Rev.* **79,** 481–510.

Domino, S. E., and Garbers, D. L. (1988). The fucose-sulfate glycoconjugate that induces an acrosome reaction in spermatozoa stimulates inositol 1,4,5-trisphosphate accumulation. *J. Biol. Chem.* **263,** 690–695.

Eisenbach, M. (1999). Sperm chemotaxis. *Rev. Reprod.* **4,** 56–66.

Espinosa, F., and Darszon, A. (1995). Mouse sperm membrane potential: Changes induced by Ca2+. *FEBS Lett.* **372,** 119–125.

Fox, A. P., Nowycky, M. C., and Tsien, R. W. (1987). Single-channel recordings of three types of calcium channels in chick sensory neurones. *J. Physiol.* **394,** 173–200.

Galindo, B. E., Nishigaki, T., Rodriguez, E., Sanchez, D., Beltran, C., and Darszon, A. (2000). Speract-receptor interaction and the modulation of ion transport in Strongylocentrotus purpuratus sea urchin sperm. *Zygote* **8**(Suppl 1), S20–S21.

Garbers, D. L., Kopf, G. S., Tubb, D. J., and Olson, G. (1983). Elevation of sperm adenosine 3′:5′-monophosphate concentrations by a fucose-sulfate-rich complex associated with eggs: I. Structural characterization. *Biol. Reprod.* **29,** 1211–1220.

Garcia-Soto, J., Gonzalez-Martinez, M., de De la Torre, L., and Darszon, A. (1987). Internal pH can regulate Ca2+ uptake and the acrosome reaction in sea urchin sperm. *Dev. Biol.* **120,** 112–120.

Garcia-Soto, J., Mourelle, M., Vargas, I., de De la Torre, L., Ramirez, E., Lopez-Colome, A. M., and Darszon, A. (1988). Sea urchin sperm head plasma membranes: Characteristics and egg jelly induced Ca2+ and Na+ uptake. *Biochim. Biophys. Acta.* **944,** 1–12.

Gibson, Q. H., and Milnes, L. (1964). Apparatus for rapid and sensitive spectrophotometry. *Biochem. J.* **91,** 161–171.

Gonzalez-Martinez, M., and Darszon, A. (1987). A fast transient hyperpolarization occurs during the sea urchin sperm acrosome reaction induced by egg jelly. *FEBS Lett.* **218,** 247–250.

Gonzalez-Martinez, M. T., Galindo, B. E., de De La Torre, L., Zapata, O., Rodriguez, E., Florman, H. M., and Darszon, A. (2001). A sustained increase in intracellular Ca(2+) is required for the acrosome reaction in sea urchin sperm. *Dev. Biol.* **236,** 220–229.

Gonzalez-Martinez, M. T., Guerrero, A., Morales, E., de De La Torre, L., and Darszon, A. (1992). A depolarization can trigger Ca2+ uptake and the acrosome reaction when preceded by a hyperpolarization in L. pictus sea urchin sperm. *Dev. Biol.* **150,** 193–202.

Gorelik, J., Gu, Y., Spohr, H. A., Shevchuk, A. I., Lab, M. J., Harding, S. E., Edwards, C. R., Whitaker, M., Moss, G. W., Benton, D. C., Sanchez, D., Darszon, A., Vodyanoy, I., Klenerman, D., and Korchev, Y. E. (2002). Ion channels in small cells and subcellular structures can be studied with a smart patch-clamp system. *Biophys. J.* **83,** 3296–3303.

Grynkiewicz, G., Poenie, M., and Tsien, R. Y. (1985). A new generation of Ca2+ indicators with greatly improved fluorescence properties. *J. Biol. Chem.* **260,** 3440–3450.

Guerrero, A., and Darszon, A. (1989a). Egg jelly triggers a calcium influx which inactivates and is inhibited by calmodulin antagonists in the sea urchin sperm. *Biochim. Biophys. Acta.* **980,** 109–116.

Guerrero, A., and Darszon, A. (1989b). Evidence for the activation of two different Ca2+ channels during the egg jelly-induced acrosome reaction of sea urchin sperm. *J. Biol. Chem.* **264,** 19593–19599.

Guerrero, A., Garcia, L., Zapata, O., Rodriguez, E., and Darszon, A. (1998). Acrosome reaction inactivation in sea urchin sperm. *Biochim. Biophys. Acta.* **1401,** 329–338.

Guerrero, A., Sanchez, J. A., and Darszon, A. (1987). Single-channel activity in sea urchin sperm revealed by the patch-clamp technique. *FEBS Lett.* **220,** 295–298.

Guzman-Grenfell, A. M., Bonilla-Hernandez, M. A., and Gonzalez-Martinez, M. T. (2000). Glucose induces a Na(+),K(+)-ATPase-dependent transient hyperpolarization in human sperm. I. Induction of changes in plasma membrane potential by the proton ionophore CCCP. *Biochim. Biophys. Acta.* **1464,** 188–198.

Hallam, T. J., and Tashjian, A. H., Jr. (1987). Thyrotropin-releasing hormone activates Na+/H+ exchange in rat pituitary cells. *Biochem. J.* **242,** 411–416.

Hamill, O. P., Marty, A., Neher, E., Sakmann, B., and Sigworth, F. J. (1981). Improved patch-clamp techniques for high-resolution current recording from cells and cell-free membrane patches. *Pflugers Arch.* **391,** 85–100.

Hille, B. (1992). "Ionic Channels of Excitable Membranes." Sinauer Associates Inc., Sunderland, MA.

Hoshi, M., Nishigaki, T., Ushiyama, A., Okinaga, T., Chiba, K., and Matsumoto, M. (1994). Egg-jelly signal molecules for triggering the acrosome reaction in starfish spermatozoa. *Int. J. Dev. Biol.* **38,** 167–174.

Kaupp, U. B., Solzin, J., Hildebrand, E., Brown, J. E., Helbig, A., Hagen, V., Beyermann, M., Pampaloni, F., and Weyand, I. (2003). The signal flow and motor response controling chemotaxis of sea urchin sperm. *Nat. Cell Biol.* **5,** 109–117.

Korchev, Y. E., Bashford, C. L., Milovanovic, M., Vodyanoy, I., and Lab, M. J. (1997). Scanning ion conductance microscopy of living cells. *Biophys. J.* **73,** 653–658.

Korchev, Y. E., Negulyaev, Y. A., Edwards, C. R., Vodyanoy, I., and Lab, M. J. (2000). Functional localization of single active ion channels on the surface of a living cell. *Nat. Cell Biol.* **2,** 616–619.

Labarca, P., and Latorre, R. (1992). Insertion of ion channels into planar lipid bilayers by vesicle fusion. *Methods Enzymol.* **207,** 447–463.

Labarca, P., Santi, C., Zapata, O., Morales, E., Beltran, C., Lievano, A., and Darszon, A. (1996). A cAMP regulated K+-selective channel from the sea urchin sperm plasma membrane. *Dev. Biol.* **174,** 271–280.

Lee, H. C., and Garbers, D. L. (1986). Modulation of the voltage-sensitive Na+/H+ exchange in sea urchin spermatozoa through membrane potential changes induced by the egg peptide speract. *J. Biol. Chem.* **261,** 16026–16032.

Levi, A. J., Lee, C. O., and Brooksby, P. (1994). Properties of the fluorescent sodium indicator "SBFI" in rat and rabbit cardiac myocytes *J. Cardiovasc. Electrophysiol.* **5,** 241–257.

Lievano, A., Sanchez, J. A., and Darszon, A. (1985). Single-channel activity of bilayers derived from sea urchin sperm plasma membranes at the tip of a patch-clamp electrode. *Dev. Biol.* **112,** 253–257.

Lievano, A., Santi, C. M., Serrano, C. J., Trevino, C. L., Bellve, A. R., Hernandez-Cruz, A., and Darszon, A. (1996). T-type Ca2+ channels and alpha1E expression in spermatogenic cells, and their possible relevance to the sperm acrosome reaction. *FEBS Lett.* **388,** 150–154.

Lievano, A., Vega-Saenz de Miera, E. C., and Darszon, A. (1990). Ca2+ channels from the sea urchin sperm plasma membrane. *J. Gen. Physiol.* **95,** 273–296.

Mengerink, K. J., Moy, G. W., and Vacquier, V. D. (2002). suREJ3, a polycystin-1 protein, is cleaved at the GPS domain and localizes to the acrosomal region of sea urchin sperm. *J. Biol. Chem.* **277,** 943–948.

Mengerink, K. J., and Vacquier, V. D. (2001). Glycobiology of sperm–egg interactions in deuterostomes. *Glycobiology* **11,** 37R–43R.

Miller, C. (1986). "Ion Channel Reconstitution." Plenum, New York.

Miller, C., and Racker, E. (1976). Ca++-induced fusion of fragmented sarcoplasmic reticulum with artificial planar bilayers. *J. Membr. Biol.* **30,** 283–300.

Moczydlowski, E., Uehara, A., Guo, X., and Heiny, J. (1986). Isochannels and blocking modes of voltage-dependent sodium channels. *Ann. NY Acad. Sci.* **479,** 269–292.

Montal, M., Darszon, A., and Schindler, H. (1981). Functional reassembly of membrane proteins in planar lipid bilayers. *Q. Rev. Biophys.* **14,** 1–79.

Morisawa, M. (1994). Cell signaling mechanisms for sperm motility. *Zoolog. Sci.* **11,** 647–662.

Mourelle, M., Vargas, I., and Darszon, A. (1984). Adenylate cyclase activity of membrane fractions isolated from sea urchin sperm. *Gamete Research* **9,** 87–97.

Negulescu, P. A., Harootunian, A., Tsien, R. Y., and Machen, T. E. (1990). Fluorescence measurements of cytosolic free Na concentration, influx, and efflux in gastric cells. *Cell Regul.* **1,** 259–268.

Nishigaki, T., Zamudio, F. Z., Possani, L. D., and Darszon, A. (2001). Time-resolved sperm responses to an egg peptide measured by stopped-flow fluorometry. *Biochem. Biophys. Res. Commun.* **284,** 531–535.

O'Toole, C. M., Arnoult, C., Darszon, A., Steinhardt, R. A., and Florman, H. M. (2000). Ca(2+) entry through store-operated channels in mouse sperm is initiated by egg ZP3 and drives the acrosome reaction. *Mol. Biol. Cell* **11,** 1571–1584.

Plasek, J., and Hrouda, V. (1991). Assessment of membrane potential changes using the carbocyanine dye, diS-C3-(5): Synchronous excitation spectroscopy studies. *Eur. Biophys. J.* **19,** 183–188.

Poenie, M. (1990). Alteration of intracellular Fura-2 fluorescence by viscosity: A simple correction. *Cell Calcium* **11,** 85–91.

Rapp, G. (1998). Flash lamp-based irradiation of caged compounds. *Methods Enzymol.* **291,** 202–222.

Rodriguez, E., and Darszon, A. (2003). Intracellular sodium changes during the speract response and the acrosome reaction in sea urchin sperm. *J Physiol.* **546,** 89–100.

Sakmann, B., and Neher, E. (1983). "Single-Channel Recording." Plenum Press, New York.

Sanchez, D., Labarca, P., and Darszon, A. (2001). Sea urchin sperm cation-selective channels directly modulated by cAMP. *FEBS Lett.* **503,** 111–115.

Sase, I., Okinaga, T., Hoshi, M., Feigenson, G. W., and Kinosita, K., Jr. (1995). Regulatory mechanisms of the acrosome reaction revealed by multiview microscopy of single starfish sperm. *J. Cell Biol.* **131,** 963–973.

Schackmann, R. W., and Chock, P. B. (1986). Alteration of intracellular [Ca2+] in sea urchin sperm by the egg peptide speract. Evidence that increased intracellular Ca2+ is coupled to Na+ entry and increased intracellular pH. *J. Biol. Chem.* **261,** 8719–8728.

Schackmann, R. W., Christen, R., and Shapiro, B. M. (1981). Membrane potential depolarization and increased intracellular pH accompany the acrosome reaction of sea urchin sperm. *Proc. Natl. Acad. Sci. USA* **78,** 6066–6070.

Schindler, H., and Rosenbusch, J. P. (1978). Matrix protein from Escherichia coli outer membranes forms voltage-controlled channels in lipid bilayers. *Proc. Natl. Acad. Sci. USA* **75,** 3751–3755.

Su, Y. H., and Vacquier, V. D. (2002). A flagellar K(+)-dependent Na(+)/Ca(2+) exchanger keeps Ca(2+) low in sea urchin spermatozoa. *Proc. Natl. Acad. Sci. USA* **99,** 6743–6748.

Suarez-Isla, B. A., Wan, K., Lindstrom, J., and Montal, M. (1983). Single-channel recordings from purified acetylcholine receptors reconstituted in bilayers formed at the tip of patch pipets. *Biochemistry* **22,** 2319–2323.

Takahashi, A., Camacho, P., Lechleiter, J. D., and Herman, B. (1999). Measurement of intracellular calcium. *Physiol. Rev.* **79,** 1089–1125.

Tatsu, Y., Nishigaki, T., Darszon, A., and Yumoto, N. (2002). A caged sperm-activating peptide that has a photocleavable protecting group on the backbone amide. *FEBS Lett.* **525,** 20–24.

Toowicharanont, P., and Shapiro, B. M. (1988). Regional differentiation of the sea urchin sperm plasma membrane. *J. Biol. Chem.* **263,** 6877–6883.

Vacquier, V. D. (1986). Handling, labeling, and fractionating sea urchin spermatozoa. *Methods Cell Biol.* **27,** 15–40.

Vacquier, V. D. (1998). Evolution of gamete recognition proteins. *Science* **281,** 1995–1998.

Ward, G. E., Brokaw, C. J., Garbers, D. L., and Vacquier, V. D. (1985). Chemotaxis of Arbacia punctulata spermatozoa to resact, a peptide from the egg jelly layer. *J. Cell Biol.* **101,** 2324–2329.

Whitaker, M. (1999). Ways of looking at calcium. *Microsc. Res. Tech.* **46,** 342–347.

Wilmsen, V., Methfessel, C., Hanke, N., and Boheim, G. (1983). Channel current fluctuation studies with solvent-free lipid bilayers using Neher-Shakmann pipettes. *In* "Physical Chemistry of Transmembrane Ion Motions" (G. Spach, ed.), pp. 479–485. Elsevier, Amsterdam, Netherlands.

Wood, C. D., Darszon, A., and Whitaker, M. (2003). Speract induces calcium oscillations in the sperm tail. *J. Cell Biol.* **161,** 89–101.

Yamaguchi, M., Niwa, T., Kurita, M., and Suzuki, N. (1987). The participation of speract in the acrosome reaction of Hemicentrotus pulcherrimus. *Dev. Growth Differ.* **30,** 159–167.

Yoshida, M., Murata, M., Inaba, K., and Morisawa, M. (2002). A chemoattractant for ascidian spermatozoa is a sulfated steroid. *Proc. Natl. Acad. Sci. USA* **99,** 14831–14836.

Zapata, O., Ralston, J., Beltran, C., Parys, J. B., Chen, J. L., Longo, F. J., and Darszon, A. (1997). Inositol triphosphate receptors in sea urchin sperm. *Zygote* **5,** 355–364.

Zhang, J., Campbell, R. E., Ting, A. Y., and Tsien, R. Y. (2002). Creating new fluorescent probes for cell biology. *Nat. Rev. Mol. Cell Biol.* **3,** 906–918.

PART IV

Molecular Biological Approaches

CHAPTER 23

Isolating DNA, RNA, Polysomes, and Protein

Bruce P. Brandhorst

Department of Molecular Biology and Biochemistry
Simon Fraser University
Burnaby, British Columbia V5A 1S6, Canada

METHODS IN CELL BIOLOGY, VOL. 74

0091-679X/04 $35.00

I. Overview

Embryos of sea urchins and other marine deuterostomes are convenient material for biochemical fractionation on large or small scales and for analysis of gene expression. Methods are described for the purification of total RNA and cytoplasmic RNA, with a consideration of ways to minimize ribonuclease-mediated degradation of RNA and contamination by DNA. A rapid method for preparing total polyadenylated RNA from small samples is described. Methods for the preparation and analysis of polysomes are outlined. A simplified classic method for the preparation of high molecular weight DNA from sperm is described, as well as preparation of mitochondrial DNA from eggs and embryos. Methods are described for preparing extracts of proteins for electrophoresis in one and two dimensions on polyacrylamide gels, with some guidance concerning the electrophoretic analysis of proteins.

II. Introduction

An attractive feature of embryos of sea urchins and other marine invertebrates is that they are suitable for biochemical analyses using a wide variety of fractionation and extraction methods. In most instances, no unusual adaptations of methods are required and new methods that arise for other cells are usually readily applicable. I outline methods for preparing RNA, DNA, polysomes, and proteins from the embryos or sperm of *Strongylocentrotus purpuratus*, which are widely applicable to the embryos of many other marine species. Analysis of gene expression focuses on the synthesis, presence, and distribution of specific RNA transcripts or proteins, facilitated by the methods described in this chapter and others in this book. Genomic DNA is required for gel blotting, PCR analyses, and preparation of recombinant libraries. The methods described are not the only ones available, and they may not be the most appropriate for all applications. But they work well for sea urchins and should be effective when applied to a broad range of marine embryos.

III. Purification of Total RNA

A. General Considerations about RNA Purification and Handling

Undegraded RNA is required for cell free translation and gel blot hybridizations (Northerns), methods that remain useful for the analysis of gene expression. Assays of gene expression based on RNAase protection and reverse transcriptase-polymerase chain reactions (RT-PCR) are much more sensitive, and can be quantitative. Preparation of complementary DNA from RNA for hybridization to DNA arrays, sometimes following subtractive hybridization, allows large-scale analyses of gene expression. Many investigations involve very small amounts of starting material, such as one or a few embryos injected with transgenes or morpholino antisense oligonucleotides. It is thus important to obtain good yields of undegraded RNA. While there are instances in which it is important to analyze cytoplasmic or polysomal RNA, most investigations involve extraction of total RNA, sometimes followed by selection of polyadenylated RNA by chromatography on oligo(dT). For very sensitive methods such as RT-PCR, it is critical that there be no significant DNA contamination and minimal contamination with inhibitors of the enzymatic reactions.

The almost universal method for purifying total RNA from marine embryos is based on the single-step method of Chomczynski and Sacchi (1987), in which cells are homogenized in guanidinium isothiocyanate and extracted with acidified phenol and chloroform. This method was modified by Chomczynski (1993) to allow for rapid preparation of RNA, DNA, and protein using a patented TRI Reagent that is commercially available from several sources. The Tri-reagent method described in the following text includes optional steps to reduce contamination by polysaccharides and proteoglycans that may inhibit solubilization of RNA and RT-PCR (Chomczynski and Mackey, 1995a), though this is not normally a problem for extracts of echinoid embryos. The Tri-reagent method has emerged as the almost universal method of choice among marine embryologists for purification of total RNA because of its speed, simplicity, utility for high throughput analyses, low level of DNA contamination, and the availability of commercial reagents. When there is significant and undesirable DNA contamination, several options are available for eliminating this problem. The original single-step method of Chomczynski and Sacchi (1987) is also effective for isolating total RNA and none of the components are proprietary; it is described in detail by Sambrook and Russell (2001).

Sea urchin embryos have low levels of nucleases and proteases, but these increase dramatically when feeding begins. The chaotropes of the Tri-reagent promote rapid denaturation of proteins, protecting the extracts from degradation. Other methods for minimizing the effects of nucleases and proteases are mentioned in the following text. A very rapid method for preparing total polyadenylated RNA from small samples is described in Section III.E.

B. Tri-Reagent Method for Purification of Total RNA

1. Collect embryos by light centrifugation and vigorously resuspend by shaking or vortexing in 10 to 20 vol of Tri-reagent in a glass or polypropylene centrifuge tube at room temperature. If this does not solubilize the sample, homogenize by grinding with a Teflon pestle or with a Dounce homogenizer (Kimble/Kontes Cat. Co. 885300-0002), keeping samples on ice; mechanical homogenizers such as a Polytron™ (Brinkman) can be used, but are not usually required except for large pieces of tissue. Pulverize frozen samples in liquid nitrogen before adding Tri-reagent to frozen powder; to avoid nuclease degradation of RNA, it is critical not to let frozen samples thaw before being dispersed in Tri-reagent. Let samples stand

for 5 min at room temperature to allow for complete dissociation of nucleoprotein complexes. When collecting eggs or embryos by centrifugation prior to homogenization, use a hand centrifuge or clinical centrifuge set to an intermediate speed to avoid rupturing embryos and attendant release of nucleases.

2. Centrifuge the lysate at 12,000*g* (10,000 rpm in a Sorvall SS-34 rotor) at 4 °C for 10 min; transfer supernatant to another tube. This optional step removes debris, large genomic DNA, polysaccharides, proteoglycans, and insoluble extracellular components.

3. Add 0.2 ml chloroform/ml starting volume of Tri-reagent. Shake vigorously for 1 min or vortex for 15 s. Let stand for 10 min at room temperature. If the extract is viscous, vortex more or sonicate to shear the high molecular weight DNA. The chloroform can be replaced with 1-bromo-3-chloropropane (BCP), a less toxic and less volatile compound that may reduce DNA contamination (Chomczynski and Mackey, 1995b).

4. Centrifuge at 12,000*g* for 15 min at 4 °C to separate phases.

5. Transfer the aqueous (upper) phase into microcentrifuge tubes, being careful not to transfer any DNA trapped at the interface.

6. To each 1 ml of initial Tri-reagent used for extraction, add 0.25 ml isopropanol and 0.25 ml RNA precipitation solution. This step removes polysaccharides and proteoglycans. If these are not a problem, RNA can be precipitated with 0.5 ml isopropanol. Mix vigorously with the vortex, then incubate for 10 min at room temperature. Prior to the addition of isopropanol, 10 to 50 μg/ml glycogen carrier can be added (see Section III.C.1).

7. Centrifuge at highest speed in a microfuge for 10 min at 4 °C. The pellet may be loose, so keep the supernatant in case recentrifugation is required.

8. To the pellet, add 1 ml 75% ethanol for each ml of Tri-reagent used and vortex; repeat centrifugation for 5 min. Repeat this ethanol wash (optional). Remove any residual ethanol from the pellet using a yellow plastic pipette tip. Pellets may be loose.

9. Allow the ethanol to evaporate for 10 min at room temperature. If the pellet is completely dried, it may be difficult to dissolve.

10. Dissolve pellets in DEPC-H_2O containing 1 mM EDTA (pH 7.5) water or TE buffer (pH 7.6) by passing repeatedly in and out of a plastic pipette tip; add a volume consistent with the desired final concentration, typically, about 1 mg/ml. Inclusion of 0.5% sodium dodecyl sulfate (SDS; make sure pH is 7.0–7.5) and heating at 65 °C will facilitate solubilization of the RNA, though this is usually not necessary. Note that precipitated RNA will be deposited on the walls of the tube as well as at the bottom; use the beads of fluid at the tip of the pipette to wash the insides of the tube to dissolve this material.

11. Dilute an aliquot and measure A_{260} and A_{280}. The ratio of A_{260}/A_{280} for pure RNA should be about 2.0; a lower ratio suggests contamination with protein or phenol. The concentration of RNA (μg/ml) is $A_{260} \times 44$. Alternatively, assays

based on binding of a fluorescent dye can used. For example, the RiboGreen™ RNA quantitation reagent (Molecular Probes, Product Code R11491) is convenient and sensitive, though it does not distinguish RNA and DNA.

C. Notes and Suggestions

1. Small Samples

For small samples, the volume of Tri-reagent should be decreased to the limit of reagent volumes that can be conveniently handled. RNA can be prepared from single embryos or fragments. If small quantities of RNA are being processed, add 10 to 50 μg/ml glycogen from a 2 mg/ml stock solution in water (stored at $-20\,^{\circ}$C) to the extract prior to precipitation with isopropanol (see Section III.B.6). This carrier may increase the yield of dilute RNA, and does not seem to interfere with quantification of RNA or enzymatic reactions performed on RNA, such as reverse transcription. Transfer RNA is sometimes used instead as a carrier, but it introduces an RNA contaminant. Nucleic acids precipitated with ethanol can be nearly fully recovered without a carrier by centrifugation at 100,000*g* for an hour at 4 °C (Sambrook and Russell, 2001). For RT-PCR, protocols have been developed that do not involve organic extraction of proteins from lysates. For instance, polyadenylated RNA can be bound to oligo(dT) on a solid matrix that is then removed from lysates prior to reverse transcription; the magnetic DynaBeads™-oligo(dT) (Dynal Biotech, Oslo) have been used effectively for this purpose. RT-PCR amplifications can be performed directly in lysates of small samples with no RNA purification; for example, Chiang and Melton (2003) have used RT-PCR of lysates of single pancreas cells to prepare probes for hybridization to microarrays.

2. Expected Yields

The eggs and embryos of *S. purpuratus* each contain about 3.3 ng of total RNA. One ml of packed eggs represents about 2×10^6 eggs containing about 6.6 mg RNA.

3. DNA Cleanup

Tests for DNA contamination in RT-PCR assays include leaving out the reverse transcription step, hydrolysis of the RNA prior to RT-PCR, and designing primers such that intron sequences are amplified from contaminating DNA. While the Tri-reagent method can produce RNA from urchin embryos without co-purifying detectable DNA, any contaminating DNA must be removed. One simple approach is to reprecipitate the RNA by addition of an equal volume of 4 M LiCl (see Section IV.B.4), but some DNA may precipitate and Li ions may interfere with cell free translation and reverse transcriptase. Of particular convenience is the DNA-*free*™ kit (Ambion Cat. No. 1906). The DNA of the sample is digested with DNase I (free of RNase), which is then removed with a resin that is collected by centrifugation. An alternative is to use the RNase-free DNase that includes RNase inhibitors from Qiagen (Cat. No. 79254) followed by use of the RNeasy™ Mini kit (Qiagen Cat. No. 74104); this involves the use of a spin column to remove DNase and digested DNA. The DNase digestion and removal can be performed on the aqueous phase after step 5, prior to the precipitation of the RNA. Preparation of polyadenylated RNA by oligo(dT) chromatography eliminates DNA.

4. RNA Storage

The dissolved RNA can be stored at −80 °C; avoid repeated freeze–thaw cycles by storing aliquots. Any SDS should be extracted with chloroform, followed by ethanol precipitation. RNA can also be stored at −20 °C as an ethanol precipitate. Use of 1.5 ml siliconized RNase-free microfuge tubes (Ambion Cat. No. 12450) is convenient for storage. RNA can be dissolved at up to 4 mg/ml and stored at −20 °C in deionized formamide (Chomczynski, 1992). At higher concentrations, this RNA in formamide can be used directly for gel electrophoresis, RT-PCR, and RNase protection, or it can be precipitated with 4 vol ethanol.

5. Storage at Intermediate Steps

According to GIBCO/Invitrogen, after homogenization in Tri-reagent, the sample can be stored at room temperature for a few hours or at −70 °C for at least a month prior to processing. The sample can be stored overnight at 4 °C after addition of isopropanol. The ethanol-precipitated RNA pellet can be stored in 75% ethanol for a week at 4 °C or a year at −20 °C prior to resuspension.

6. Reduction of RNase Contamination

All work involving RNA requires great care to eliminate RNAase contamination, particularly in handling the purified RNA. Clean gloves and decontaminated glassware must be used in a dustfree environment. Water and salt solutions (but not solutions of amine buffers, such as Tris) should be treated with the potent alkylating agent diethylpyrocarbonate (DEPC) to eliminate RNase activity. Buffers and aqueous solutions should be filtered through 0.2 μm membranes and used aseptically. Glassware, spatulas, and stir bars should be baked for 4 h at 300° and should be segregated for RNA use, as should pipetting tools and electrophoresis equipment. In my experience, disposable plastic pipette tips and centrifuge tubes certified to be RNase-free do not require decontamination as long as they are not touched where contact may be made with RNA solutions. Glassware and plasticware can be decontaminated by filling with 0.1% DEPC in water for 1 h at 37 °C or overnight at room temperature. Rinse several times with DEPC-H_2O (see Section III.D.3) and autoclave on the liquid cycle for 15 min; store in a dustfree environment and handle with gloves or decontaminated forceps. Work surfaces and labware can be decontaminated using RNaseZap™ (Ambion Cat. No. 9780.9782). Alternatively, electrophoresis apparati can be cleaned with detergent solutions, rinsed with water, rinsed with ethanol, and dried. They are then filled with 3% H_2O_2 for at least 10 min at room temperature, and rinsed thoroughly with DEPC-H_2O (Sambrook and Russell, 2001).

D. Reagents and Solutions

1. Tri-reagent: TRI Reagent® (Molecular Research Center Cat. No. TR 118, or Sigma Cat. No. T9424) or TRIzol Reagent® (Invitrogen Life Technologies, Cat. No. 15596-018; we routinely use this product). Examples of apparently similar proprietary reagents include (Sambrook and Russell, 2001): Tri-Pure® (Boehringer-Mannheim), RNA Stat-60® (Tel-Test), Isogen® (Nippon Gene, Toyama, Japan). RNAzol® A and B (Tel-Test) are older formulations that may result in more DNA contamination. Tri-reagent contains phenol and can cause severe burns. It and other phenol solutions should be handled using gloves and eye protection, preferably in a fume hood to avoid breathing vapors. It should be stored as recommended by the supplier, usually in the refrigerator protected from light for up to a year. Precipitated components should redissolve at room temperature. Molecular Research Center has an informative website about using its patented TRI-reagent: http://www.mrcgene.com
2. DEPC (diethylpyrocarbonate); e.g., Sigma Cat. No. D5758. Store DEPC refrigerated in small aliquots protected from moisture and open only at room temperature to avoid condensation; traces of water promote conversion to diethylcarbonate, effectively inactivating DEPC (Sambrook and Russell, 2001).
3. DEPC-H_2O: Treat water with 0.1% DEPC at room temperature for at least 4 h or 37° for 1 h, then autoclave for at least 15 min at 15 psi on liquid cycle.
4. TE buffer: 100 mM Tris-Cl, pH 7.4 or 8.0.
5. RNA precipitation solution: 1.2 M NaCl, 0.8 M disodium citrate.
6. Other reagents needed: reagent-grade chloroform, isopropanol, and ethanol.

E. Rapid Preparation of Total Polyadenylated RNA from Small Samples

We often use the QuickPrep™ Micro mRNA Purification Kit (Amersham Biosciences, product code 27-9255-01) to quickly prepare total polyadenylated RNA for cDNA synthesis, RT-PCR, and Northern blot analysis. Samples are homogenized in a solution of guanidinium thiocyanate and N-lauroyl sarcosine, diluted with an aqueous buffer, cleared by centrifugation, and applied to a spin column of oligo(dT)-cellulose; after washing the column to remove unbound protein, DNA, and carbohydrates, polyadenylated RNA bound to the column is eluted in a warmed low-salt buffer. It takes only about 15 min to complete the purification of RNA ready for precipitation, and the method can be applied to single embryos or cells. We follow the manufacturer's instructions. In a standard extraction, about 50 μl packed embryos are suspended in 400 μl Extraction Buffer. Vigorous vortexing or passage through a Pasteur pipette is often sufficient to create a uniform homogenate, but mechanical homogenization may be required for some samples; probably twice as much sample volume could be handled effectively. We normally wash three times and elute twice from the oligo(dT) spin column, resulting in a volume of 400 μl in a screw-capped 2 ml microcentrifuge tube. After removing 16 μl for quantitation of RNA, 10 μl 10 mg/ml glycogen carrier in DEPC-treated water and 38.4 μl 2.5 M potassium acetate (pH 5.0) is added. After mixing, 1 ml cold 95% ethanol is added. After precipitation at −20 °C for 30 min, RNA is collected at 4 °C for 5 min in a microcentrifuge at top speed.

IV. Purification of Cytoplasmic and Polysomal RNA

A. General Considerations

For analysis of polyribosomes and preparation of extracts of cytoplasmic RNA, embryos are lysed with a nonionic detergent. Nuclei are collected by centrifugation, and the supernatant can be placed on a sucrose gradient for centrifugation or processed for RNA extraction. Some mRNA bound to the cytoskeleton is not solubilized by the detergent and may fractionate with the nuclear fraction. The cytoskeleton can usually be dispersed by light homogenization (3 strokes with a Dounce homogenizer with a large clearance (type A) pestle (Kimble/Kontes Cat. No. 885301-0002) or by repeated aspiration through a Pasteur pipette. Sea urchin embryos contain a nuclease that is dependent on Ca^{++} ions for activity. Thus, removal of Ca^{++} ions is important for preserving the integrity of RNA while preparing the cytoplasmic extracts. Urea promotes dissociation of ribonucleoprotein complexes and solubilization of proteins, while LiCl selectively precipitates RNA, but not tRNA and DNA.

B. Preparation of Cytoplasmic Extracts and RNA Purification

1. Collect embryos by light centrifugation (see Section III.B.1), and resuspend embryos briefly in cold artificial seawater lacking Ca^{++} ions. Collect by light centrifugation and resuspend embryos in 10 to 20 vol lysis buffer on ice.

2. Transfer the suspension of embryos to a centrifuge tube. Collect by centrifugation (this may require higher relative force, such as top speed in the clinical centrifuge, because embryos may begin to dissociate) and resuspend embryos in 10 to 20 vol lysis buffer containing 0.5% Triton X-100 on ice; optional inclusion of a RNase inhibitor, such as 30 μg/ml placental inhibitor (Sigma cat. no. R7253), may be helpful if problems with nuclease activity arise. Aspirate vigorously or vortex briefly.

3. Centrifuge at 12,000*g* for 15 min at 4° to collect nuclei, mitochondria, and debris. Remove the supernatant. It can be layered on a sucrose gradient for polysome analysis or used to prepare cytoplasmic RNA. It is likely that some mitochondria will be ruptured and mitochondrial RNA will be included in the cytoplasmic fraction.

4. For extraction of cytoplasmic RNA, add an equal volume of Li-urea solution. Mix and store at −20 °C for at least 3 h (overnight is fine).

5. Collect the RNA by centrifugation at 15,000*g* for 15 min at 4 °C.

6. At this point, there is a choice. Dissolve the pellet in Tri-reagent (same volume as in step 2) and follow the protocol for preparation of total RNA, as described in Section III.B. Alternatively, dissolve the pellet in the same volume of RNA extraction buffer and proceed to step 7.

7. Add 1 volume phenol:chloroform at room temperature. Shake vigorously and separate phases by centrifugation at 12,000*g* for 5 min. Repeat the extraction of the organic (lower) phase with RNA extraction buffer if there is significant flocculent material at the interface; this optional step is often not necessary. Extract the aqueous supernatant with 1 vol chloroform and repeat the centrifugation.

8. Collect the aqueous supernatant and add 0.1 vol 2M sodium acetate, pH 5.0, and 2.7 vol cold ethanol and incubate at −20° for at least an hour, preferably overnight.

9. Collect RNA precipitate by centrifugation at 14,000*g* for 10 min at 0 °C. Suspend RNA in DEPC-H_2O or TE buffer, pH 7.4. Store frozen at −80 °C.

C. Preparation and Analysis of Polysomes

There are many protocols for analysis of polysomes profiles. One that works well is to layer 1 ml of cytoplasmic extract (corresponding to about 150,000 *S. purpuratus* embryos) on a linear gradient of 10 to 55% sucrose (W/V) in lysis buffer (Bédard and Brandhorst, 1986; Gong and Brandhorst, 1988). Centrifuge for 130 min at 32,000 rpm at 4 °C in a Beckman SW41 rotor. The polysomes are monitored at A_{260} using a flow cell in a photometer. RNA in polysome fractions can be extracted by conventional hot phenol:chloroform methods after dilution with an equal volume of lysis buffer containing 0.5% SDS. The sucrose should be diluted to less than 12.5%; at sucrose concentrations above about 10%, the aqueous phase will partition below phenol:chloroform. Alternatively, diluted polysomal fractions can be precipitated by addition of 2.5 vol cold ethanol; after centrifugation at 15,000*g* for 15 min at 4 °C, the pellets are dissolved and extracted using the Tri-reagent protocol or conventional phenol:chloroform extraction (Sambrook and Russell, 2001). Nearly all of the ribosomal subunits of sea urchin eggs and embryos are either in polysomes or in 80S monosomes; very few 40S and 60S subunits are present.

D. Notes and Suggestions

1. Contaminating Nonpolysomal RNA

RNA in large ribonucleoprotein complexes can be in the fractions of the sucrose gradient corresponding to polysomes. This RNA can be selectively isolated by bringing a parallel sample of cytoplasmic extract to 0.2M EDTA for 10 min prior to centrifugation; this dissociates polysomes, releasing ribosomes and mRNA that will now fractionate toward the top of the gradient. RNA that is extracted from the fractions of the gradient corresponding to polysomes after this release is not engaged in protein synthesis on polysomes and can be purified for comparison with polysomal RNA. Thus, mRNA is RNA encoding protein (and usually polyadenylated) in the polysomal fraction that is displaced toward the top of the sucrose gradient by this treatment.

2. Preparation of Nuclei

If nuclei having minimal contamination with cytoplasmic components are desired (e.g., for use in a nuclear run-on transcription assay or analysis of nuclear proteins or RNA), the lysate should be layered over 2 ml of 1 M sucrose in 5 mM $MgCl_2$, 100 mM Tris/HCl, pH 7.4; centrifuge in a swinging bucket rotor at 3000*g* for 5 min to collect nuclei.

E. Reagents and Solutions

1. Lysis buffer: 400 mM NH_4Cl, 12 mM $MgCl_2$, 25 mM EGTA (ethylene glycol-bis(2-aminoethylether)-N,N,N′,N′-tetracetic acid, tetrasodium salt), 50 mM Pipes (piperazine-N,N′-bis(2-ethanesulfonic acid), pH 6.5 in DEPC-H_2O; filter through a 0.22 μm membrane and store in the refrigerator. It is best made up from 10–20× stock solutions.

2. LiCl-Urea solution: 4 M LiCl, 8M urea, 0.5 mM EDTA, 20 mM Tris, pH 7.5. Filter at 0.22 μm, and store in refrigerator.

3. RNA extraction buffer: 100 mM sodium acetate, pH 5.0, 25 mM EGTA, 0.5% SDS; filter.

4. Phenol: Use liquefied phenol, stored at −20 °C. Do not use if it is colored. Phenol is toxic and caustic; work in a fume hood with gloves. Pre-saturate the phenol with RNA extraction buffer by adding an equal volume and stirring with a magnetic stirrer for 15 min. After the phases have separated, aspirate off the aqueous (upper phase), and repeat the equilibration until the pH of the phenol is about 5.0, using pH paper to test pH. Add 0.2% hydroxylquinoline and 0.1 vol lysis buffer containing 0.2% β-mercaptoethanol. Store at −20 °C for up to a month.

5. Phenol:chloroform: 49% buffer-saturated phenol, 49% chloroform, 1.9% isoamyl alcohol. Store as for equilibrated phenol.

V. Purification of Genomic DNA

A. General Considerations

High molecular weight genomic DNA is conveniently prepared from sperm, where it is highly concentrated, using modifications of the classic method of Blin and Stafford, 1976. Care is required to avoid shearing DNA to be used for cloned DNA libraries having large inserts and for pulsed field analysis. Depending on how much shearing occurs, the DNA isolation step of the Tri-reagent protocol produces DNA in the 20 to 70 kb range, a size range that may be suitable for PCR and Southern blots. Since it can provide a quantitative recovery of DNA, the Tri-reagent method can be used to normalize RNA extracts to cell numbers (see the supplier's instructions or the Molecular Research Center website for protocols and information). It is likely that many of the commercial kits available for rapid preparation of small quantities of DNA will work well for embryos and tissues of marine deuterostomes. It is helpful to begin with a preparation of nuclei, such as those produced in Section IV.B.3, to reduce contaminating RNA and protein. DNA can be extracted from these nuclei digested with Proteinase K, as described here for sperm. DNA extracted from sperm will include mitochondrial DNA. Supercoiled mitochondrial DNA can be separated from nuclear DNA by equilibrium centrifugation on a density gradient of CsCl, but nicked mitochondrial DNA will contaminate the nuclear DNA unless it has a distinct base composition; nicking is effectively unavoidable.

B. DNA Purification Protocol

1. Place 20 μl dry sperm into a 1.5 ml microfuge tube. Aliquots of dry sperm can be quick-frozen in liquid nitrogen and stored at $-80\,^{\circ}$C with little change in quality. To scale up, increase the number of tubes processed.

2. Add 0.75 ml Sperm Digestion Buffer. Incubate overnight at $50\,^{\circ}$C with gentle rocking.

3. Cool the samples to room temperature. The solution should be clear and viscous. Centrifuge at top speed for 10 min to remove any debris or clumps of undigested sperm. Remove the supernatant to a fresh tube. For transfer of DNA, use a plastic pipette tip from which 5 to 7 mm of the tip has been removed; this reduces shearing.

4. Add 0.5 ml 1:1 phenol:chloroform (saturated with TE, pH 8.0). Mix gently for 15 min at room temperature on an orbital shaker with end-over-end action, e.g., Clay Adams Nutator™, Model 1105. Alternatively, mix by rotating the tube at 20 rpm on a wheel. An emulsion should form; if not, incubate longer or mix slightly more vigorously. To reduce shearing of DNA, do not vortex or shake vigorously. Centrifuge at top speed for 10 min. Transfer the supernatant to another tube, taking care to avoid the interface. Repeat the phenol:chloroform extraction of the aqueous phase.

5. Extract the aqueous phase with chloroform containing 4% isoamyl alcohol. This optional step removes residual phenol that may interfere with enzymes.

6. Add 0.75 ml isopropanol to the aqueous extract. Mix by inverting the tube several times. Allow DNA to precipitate for 5 to 15 min at room temperature and centrifuge at top speed for 5 min at room temperature. Wash the pellet twice with 70% ethanol at room temperature by resuspension and centrifugation; it is important to remove any residual phenol that can interfere with DNA enzymes. Remove all residual ethanol with a micropipetter and allow the pellet to dry for 15 to 20 min at room temperature; a completely dry pellet may be difficult to re-dissolve.

7. Dissolve the DNA pellet in 100 μl TE, pH 8.0, overnight at 4 °C, with gentle rotation. Some DNA will be on the sides of the tube.

8. Quantify the DNA by spectrophotometry. The A_{260}/A_{280} ratio should be about 1.8. If less, repeat the extraction to remove residual protein or phenol. The concentration of DNA (μg/ml) is $A_{260} \times 50$. For viscous solutions of DNA, cut off the end of the yellow pipette tip to transfer DNA and vortex vigorously prior to reading. Store the DNA in TE at 4 °C. If stored frozen, do not repeat freezing after thawing to minimize DNA breakage. To test quality, analyze 1 μg aliquots of uncut and restriction digested DNA by electrophoresis on 0.7% agarose. The uncut DNA should run as a tight band near the origin, while the restricted DNA should run as smear. The yield of DNA should be 0.5 to 1.0 mg.

C. Notes and Suggestions

1. RNA Contamination

Sperm cells contain very little RNA, so normally no efforts are made to eliminate RNA. For situations in which RNA contamination poses a problem, the sample can be dispersed in sperm digestion solution containing 20 μg/ml DNase-free pancreatic RNase. Incubate the suspension for 1 h at 37 °C. Then add the Proteinase K to 0.2 mg/ml and continue from Section V.B.2.

2. Nucleases

The concentration of EDTA is high enough to inhibit hydrolysis of DNA by nucleases and heavy metals. Such nucleases are not usually a problem for urchin sperm, and the concentration of EDTA may be reduced to 25 mM.

3. Phenol pH

The phenol should be presaturated with 0.5 M Tris-HCl, pH 8.0, prepared as in Section IV.E. It is important that the pH during the phenol extraction be about 8 so that DNA will partition into the aqueous phase. For preferential extraction of RNA, the pH of the phenol and solutions should be slightly acidic.

4. Minimization of Shearing

About 5 mm should be cut off from the tips of pipettes used for transferring DNA solutions using a toenail clipper, scissors, or razor blade. The tips should then be sterilized by autoclaving or dipping into 70% ethanol, followed by drying.

5. Mitochondrial DNA and Polysaccharide Contamination

Their high ratio of mitochondria to nuclei makes eggs an excellent source for preparations of mitochondrial DNA; differential fractionation of mitochondria or mitochondrial DNA (e.g., by equilibrium density centrifugation) is unnecessary for most purposes, such as PCR amplification, prior to sequencing. The preparation of mitochondrial DNA from crinoid eggs has been described and should be generally applicable. The protocol is similar to that described in Section V.B, but is followed by an extraction of the aqueous fraction with CTAB (hexadecyltrimethylammonium bromide). One-sixth vol of 5 M NaCl is added, followed by one-eighth volume of CTAB/NaCl (10% w/v CTAB in 0.7 M NaCl); the sample is incubated at 65 °C for 30 min with occasional mixing, and then extracted with 0.5 vol 24:1 chloroform/isoamyl alcohol and centrifuged. DNA is precipitated by the addition of 2.5 vol 95% ethanol. The CTAB extraction reduces polysaccharide contamination of the DNA, improving restriction digestion and PCR (Scouras and Smith, 2001).

6. Dissolving DNA Pellets

If the pellet does not dissolve well, incubate the open tube for 3 to 5 min at 45 °C to evaporate the ethanol (Sambrook and Russell, 2001). DNA may be difficult to dissolve in solutions containing $MgCl_2$ or more than 0.1 M NaCl. Alternatively, dissolving the DNA in 8 mM NaOH may be faster. The pH is then adjusted with HEPES: to 1 ml, add 101 or 159 μl 0.1 M HEPES to obtain pH 8.0 or 7.5, respectively; add 23 μl 1.0 M HEPES to obtain pH 7.0.

7. Fluorometric Determination of DNA Concentration

DNA can be measured by spectrophotometry in the range of 5 to 90 μg/ml. Lower concentrations (as low as 10 nanogram quantities) can be measured by comparison of binding of Hoechst 33258 dye to DNA in an unknown sample to a standard curve (Sambrook and Russell, 2001). Dilute 50 μl 0.2 mg/ml Hoechst stock (stored at room temperature in a vessel wrapped with foil to protect from light) into 100 ml 2 M NaCl, 50 mM sodium phosphate. Add 3 ml of this dye solution to each DNA sample. Read the fluorescence, using an excitation wavelength of 365 nm and emission wavelength of 458 nm. The sensitivity of DNA fluorometry can be adjusting by altering the dye concentration. This method applies to DNA of at least 1 kb. Alternatively, the PicoGreen™ dsDNA quantitation reagent (Molecular Probes Cat. No. P7581) offers convenience and sensitivity over a broad range of DNA concentrations.

D. Reagents and Solutions

1. Sperm Digestion Solution: 100 mM NaCl, 100 mM EDTA, 10 mM Tris-HCl, pH 8.0, 1% SDS. Add Proteinase K just before use to 0.2 mg/ml from a stock solution of 20 mg/ml in TE.
2. Proteinase K. Use a genomic DNA grade Proteinase K that is certified to be free of DNase and RNase. Store aliquots of stock at −20 °C; they can be refrozen a few times.

VI. Preparation of Protein Samples for One- and Two-Dimensional Gel Electrophoresis

A. General Considerations

Many protocols for preparing proteins for electrophoretic analysis have been described, some serving special purposes. It is often useful to compare the utility of different methods. I describe methods that work well on sea urchin embryos for preparing samples for one-dimensional discontinuous SDS polyacrylamide gel electrophoresis (SDS-PAGE) (Laemmli, 1970), and for conventional two-dimensional gel electrophoresis (2D-PAGE) with isoelectrofocusing (IEF) for the first dimension and SDS-PAGE for the second dimension (Brandhorst, 1976; O'Farrell, 1975; Tufaro and Brandhorst, 1979). Samples to be compared should be consistently prepared: they should consist of similar volume, solutes (including SDS, if used), protein concentration, and the same lot of ampholytes, where applicable. Proteins are ampholytes and will alter the pH gradient of IEF gels in a concentration-dependent manner. For radioactive samples, it may be necessary to adjust the specific activity of samples by the addition of nonradioactive extract to achieve uniformity. While the IEF gels used have very low conductivity, isofocusing is somewhat tolerant of salt ions introduced in sample preparation, but the pH gradient formed will be influenced by the ions in the sample. It is thus important that samples to be compared have similar salt compositions.

For 2D-PAGE, it is important that the urea concentration be 9.5 M (essentially saturated) to maintain protein solubility; otherwise, streaking in the isoelectric dimension may occur because proteins tend to precipitate as they approach their isoelectric point. It is also crucial that samples in urea never be heated (certainly not above 37 °C) to prevent charge modifications: proteins are carbamylated by the cyanate generated by heating the urea. Samples for 1-D SDS-PAGE are usually boiled to promote complete solubilization and reduction of proteins, including many membrane proteins (Ames and Nikaido, 1976). Some highly hydrophobic proteins may precipitate at high temperatures but not at more modest temperatures such as 65 °C. Suspension of feeding echinoid larvae in 95 °C SDS solubilization buffer followed by immediate immersion in boiling water effectively inactivates their potent protease activities. These samples can be modified for 2-D PAGE as will be described, though some charge modification may occur during preparation, altering the isoelectric point. Cocktails of protease inhibitors added to solubilization solutions may be tried as an alternative method for suppressing protease activities, but in my experience are not entirely effective.

B. Sample Preparation

1. Protein solubilization for 1-D SDS-PAGE. Samples consisting of pellets of embryos or cells are suspended in at least 10 vol of SDS solubilization buffer and immersed in a boiling water bath for 3 to 5 min; homogenization is not normally required. If proteases are known to be a problem, the sample should be resuspended in SDS solubilization buffer at 95 °C and then immediately immersed in boiling water for 3 min. For samples already in solution, dilute with an equal volume of 2X SDS solubilization buffer prior to heating. Centrifuge the sample at top speed in a microcentrifuge for 5 min to remove any debris. Store samples frozen; repeat centrifugation just prior to loading samples on gels.

2. Protein solubilization for 2-D PAGE. Suspend a sample in 20 to 100 vol of IEF lysis solution. Dispersal of the sample by repeated passage through a yellow pipette tip is usually sufficient, but a small homogenization pestle can be used (the shaft of a 1 ml disposal syringe from which the rubber tip is removed is convenient for this purpose when using microfuge tubes). Centrifuge at top speed in a microcentrifuge for 5 min to remove debris. Samples can be flash-frozen and stored frozen at −80 °C for at least several months. These samples can be used for 1-D SDS-PAGE by addition of an equal volume of 2X SDS solubilization buffer followed by boiling.

3. To convert a sample prepared for SDS-PAGE as in Section VI.B.1 for use for IEF in the first dimension of 2-D-PAGE, add 10 μl lysate cooled to room temperature to 9 mg urea and then add 80 μl IEF lysis solution; mix. In sufficient excess, the non-ionic detergent will form micelles with the SDS, removing it from proteins (Ames and Nikaido, 1976). These micelles will focus near pH 5 in the isoelectric dimension, resulting in local distortion of the pH gradient (and convenient marker of the bottom of the IEF gel).

4. Preparation of large samples or those containing high concentrations of nucleic acids: Viscous nucleic acids can interfere with solubilization of some proteins and the uniformity of IEF. Garrels (1983) has described a widely applicable sample preparation protocol that alleviates these problems and provides consistent results; these operations should be done quickly. Suspend the sample in micrococcal nuclease buffer on ice (about 0.5 ml per estimated mg of protein), and rupture the cells by several passages through a 26 to 28 gauge syringe needle; this shears and digests DNA, reducing the viscosity. Quickly add 0.12 vol 3% SDS, 10% β-mercaptoethanol, and 0.12 vol cold nuclease digestion buffer to the sample and mix. The viscosity of the sample should disappear within 10 to 15 s and the sample should then be quick-frozen and lyophilized. Immediately upon releasing

the vacuum, resuspend the residue in IEF lysis solution (1.2× the volume of the original lysate, such that the final concentration of SDS is 0.3% and there is sufficient excess of NP-40). Freeze and store the sample at −80 °C; centrifuge the sample at top speed on a microcentifuge for 5 min at room temperature prior to loading on an IEF gel.

C. Notes and Considerations

1. Preparing Samples Containing Nucleic Acids

If samples are viscous (not common for extracts of embryos having low nuclear content), the DNA can be sheared by passage through a 26 gauge syringe needle prior to loading. Interactions of some proteins with nucleic acids even in high concentrations of urea can sometimes be a problem, causing streaking on the IEF gel. This can be alleviated by the inclusion of 0.2 to 0.3% SDS in the IEF lysis solution for the isofocusing (first) dimension of 2-D gels, as described. The inclusion of SDS also results in the solubilization of some otherwise resistant proteins that bind nucleic acids (Garrels, 1983). SDS will influence the pH gradient of the IEF gel, so samples to be compared should contain consistent amounts. Samples can be predigested with DNase and RNase as described in Section VI.B.4, but this may introduce or allow the activity of endogenous proteases (this has not been a notable problem for sea urchin embryos prior to pluteus stage). Use nucleases certified to be free of protease activities and consider using protease inhibitors. For preparation of proteins from purified nuclei (prepared as in Section IV.B.3), a nuclease digestion step is essential. Alternatively, nucleic acids can be removed by ultracentrifugation at 100,000 to 200,000*g* for 1 to 2 h. This reduces the dark background staining sometimes observed on silver-stained gels (Dunbar, 1987), but some proteins are likely to be selectively removed as well.

2. Sample Size

For optimal resolution, 2D-PAGE samples on standard size gels (0.25–3 mm tubes; 14 cm × 16 cm slabs, 0.75 mm thick) should not exceed 10 to 20 μg. The volume of the sample has little influence, but should be consistent for samples being compared because it can influence the pH gradient. An *L. pictus* egg contains 54 ng total protein, while *S. purpuratus* eggs have about half that amount. For preparation of extracts from large samples, it is advisable to pulverize the sample in liquid nitrogen prior to solubilization.

3. Measuring Protein Content of Samples

The reductants, detergents, chaotropes, and ampholytes in the solutions used to prepare samples for 1-D or 2-D electrophoresis interfere with many commonly used assays for protein content. To circumvent this problem for the Coomassie

Blue dye binding assay (Bradford, 1976), dilute 10 μl samples containing 1 to 50 μg protein with 10 μl 0.1 N HCl and 80 μl water (Ramagli and Rodriguez, 1985). Add 3.5 ml of the Coomassie Blue reagent (Pierce, Cat. No. 23200) to each tube and vortex. After 5 min, read the optical absorbance at 595 nm against the solution lacking protein. For preparing a standard curve using bovine serum albumin or other convenient purified protein, it is important to prepare the protein standards and the blank in the same solution as the sample.

Another convenient and sensitive dye assay was described by Esen (1978). Spot 5 μl samples onto the center of 1 cm squares drawn with pencil on Whatman #1 filter paper. Dry the samples and stain the paper in 0.2% Coomassie Brilliant Blue R250™ in 10% acetic acid and 50% methanol for 20 min in a flat dish; interfering agents are dissolved, while protein is precipitated onto the filter. Destain with several changes of water for 5 min each. Cut out the squares and immerse in 1.5 ml 1% SDS at 37 °C to elute stain from the spot. Read the absorbance of the eluate at 595 nm, and compare to a standard curve. Several manufacturers sell proprietary kits for precipitation of proteins prior to resolubilization for protein determination or electrophoresis. For removal of salts and many other contaminants, most proteins can be quantitatively precipitated from cold 80% acetone, collected by centrifugation, quickly dried and resuspended in an appropriate solubilization solution prior to electrophoresis or quantification; resuspension of the pellet is sometimes difficult.

Sensitive fluorescence assays are available that are insensitive to some interfering agents or allow them to be diluted so that they do not interfere. For example, the NanoOrange™ (N6666) and CBQCA™ (C6667) protein quantitation assays from Molecular Probes are very sensitive and convenient to use with a spectrofluorimeter.

4. Radiolabeled Proteins

For analysis of newly synthesized proteins, incubate 5000 to 10,000 embryos in 1 ml of seawater and a tracer such as ^{35}S-methionine (50–125 μCi/ml); lyse in 0.5 to 1 ml IEF lysis solution, and load samples of 10 to 25 μl. Incorporation into protein can be determined by precipitation with hot 10% trichroloacetic acid (TCA) with 100 μg bovine serum albumin added as carrier and collection on a Whatman GF/A glass fiber or Millipore HA filter. The filters are washed with 10% TCA and then with ethanol and dried prior to scintillation counting.

5. Staining Gels

Coomassie Brilliant Blue R-250 staining can generally detect 0.2 to 0.5 μg protein in a spot or sharp band. There is a limited range of quantitative binding of the dye, depending on the protein, up to about 10 to 15 μg (Hames, 1981). On 2-D gels, some proteins may be displaced or obscured by more abundant proteins of similar mobility. Single spots containing more than 0.1 μg protein will be larger, reducing local resolution; this is of particular concern for fluorography of radioactive protein samples in which the expanded area of highly radioactive spots can be considerable, obscuring adjacent spots. In practice, loading samples having 5 to 20 μg total protein on a standard sized 2-D gel (0.25–3 mm tubes; 14 cm × 16 cm slabs, 0.75 mm thick) usually provides excellent resolution of extracts of embryo. Many silver-staining methods and kits are available that increase sensitivity by 10- to 100-fold in comparison with Coomassie. SYPRO^{TM} Ruby (Molecular Probes) interacts with lysine, arginine, and histidine and provides simple staining with detection limits similar to silver staining and a linear dynamic range of about 1000, compared to about 10 for conventional silver staining. Another sensitive fluorescent stain is Deep Purple^{TM} (Amersham Biosciences RPN6305). These fluorescent dyes are compatible with Edman sequencing and mass spectrometry, and can be detected with laser scanning imagers, such as the Typhoon, or CCD cameras using UV or visible light boxes. Different protein extracts can be covalently labeled with spectrally different fluorescent dyes and then compared in 2-D on the same PAGE gel; Amersham Biosciences refers to this as fluorescence difference gel electrophoresis. Sensitive staining methods specific for protein modifications are available. For instance, while SYPRO^{TM} Ruby detects all proteins, Pro-Q Diamond^{TM} (Molecular Probes) is specific for phosphates and Pro-Q EmeraldTM (Molecular Probes) is specific for glycoproteins. The GelCode^{TM} Phosphoprotein Staining Kit (Pierce Cat. No. 24550) stains phosphoproteins green, while the GelCode Blue Reagent (Pierce 24590) stains proteins blue in the gel.

6. Preparative 2D-PAGE

Much more protein can be loaded for preparative gels, but with possible localized overloading artifacts (see Section VI.B.4). Consistent, well-resolved patterns can be obtained when consistent conditions are used. For instance, extracts of echinoid embryos containing 150 μg protein prepared as described in Section VI.B.4 are well resolved on standard-sized gels (14×16 cm, 0.75 mm thick; loads should be adjusted in proportion to the volume of the gel). This allows for detection of more proteins by staining, and for extraction or electro-elution of protein spots from gels. Proteins can be extracted from gels dried onto Whatman 3 MM filter paper by resuspending in 3 ml 1% SDS. After rehydration of the gel, the paper is removed with forceps and the gel crushed with a glass homogenizer. After incubation overnight at 37 °C, the acrylamide particles are collected at 2000 rpm for 5 min; the supernatant is saved and gel pellet is re-extracted twice for several hours. The pooled supernatants are cooled to 4 °C for 1 h to precipitate the SDS, and centrifuged at 10,000 rpm for 20 min at 4 °C in a JA-20 rotor. The supernatant is quick-frozen and lyophilized; the residue is then dissolved in 1 ml water. This procedure recovers more than 90% of the protein. Dried, stained gels can be stored in a vacuum dessicator indefinitely prior to recovery of proteins for raising antisera and possibly other uses. Using larger or thicker gels allows more sample protein to be loaded while maintaining good resolution.

7. Tri-Reagent

Proteins can be prepared using the Tri-reagent according to the supplier's instructions. Such proteins are denatured, but may be useful for PAGE and immunoblotting.

8. Cell-Free Translation Products

Products of translation of messenger RNA in rabbit reticulocyte lysates should be lyophilized and then resuspended in IEF lysis solution or SDS solubilization buffer for electrophoretic analysis (Bédard and Brandhorst, 1986). The heme will focus on the IEF gel.

9. Optimal Electrophoresis Conditions

Conditions should be adjusted empirically for the sample being analyzed. For 2D-PAGE, we closely follow the protocol originally described by O'Farrell (1975), who provided a good guide to troubleshooting. Another guide to troubleshooting is in the book by Dunbar (1987). We normally pre-run IEF gels for 400 V-h; then load the samples and run at 800 V for 12 h (9600 V-h) at about 22 °C. For the second dimension, and for SDS-PAGE, an exponential gradient of 10 to 16% polyacrylamide (with 2.66% of the acrylamide as bisacrylamide) provides excellent resolution of proteins from echinoid embryos; alternatively, a 14% polyacrylamide gel works well, though a narrower size range of proteins will be well resolved.

10. Nuclear Proteins

Harrington *et al.* (1992) have described a protocol for large-scale preparation of nuclear proteins from sea urchin embryos and chromatographic fractionation on cationic exchange columns, prior to analysis by 2D-PAGE. These nuclear proteins having basic domains include DNA binding proteins, and some are known transcription factors that have been purified by sequential DNA affinity chromatography (Coffman *et al.*, 1992).

11. SDS Variation

Migration of some proteins on SDS-PAGE can differ dramatically when using SDS from different sources, presumably depending on contaminants. For Edman sequencing of proteins eluted from gels, even electrophoresis grade SDS should be purified as described by Hunkapiller *et al.* (1983).

D. Reagents and Solutions

1. Reagents. Use electrophoresis-quality reagents including acrylamide, urea, and SDS. Ampholytes are of variable quality even from the same company, so the same lot should be used for a series of comparative investigations. It can be advantageous to blend ampholytes from different sources to suppress flat zones on the pH gradient. Water quality is crucial; it should be deionized or double-distilled water with low conductivity. The 50 mM NaOH anode solution for IEF should be extensively degassed and stored under vacuum until used. Filter all stock solutions used for gels through a 0.45 μm membrane. Degas solutions under vacuum and refilter through Whatman #1 paper just prior to adding TEMED and ammonium persulfate just prior to pouring gels. The 10% stock solution of ammonium persulfate catalyst should be prepared fresh daily to avoid distorted separation of proteins on the SDS slab gel. The correct pH of the upper and lower SDS gel buffers is critical for good resolutions and must be adjusted at room temperature using an electrode certified for Tris buffers.

2. SDS solubilization buffer: 2% SDS (Biorad Cat. No. 161-0302 works well), 10% glycerol, 0.0625M Trizma base (e.g., Sigma Cat. No. T-6791), pH 6.8. Filter through a 0.22 μm membrane and store at room temperature. Add β-mercaptoethanol to 5% (v/v) or 100 mM dithiothreitol just prior to use.

3. IEF lysis solution: 9.5 M urea, 2% (w/v) NP-40 (Sigma Cat. No. N-3516); use a 10% w/v stock), 2% ampholytes (usually 1.6% 5–7 range, and 0.4% 3–10 range), 0.2 to 0.3% SDS (optional), 5% β-mercaptoethanol. Store frozen at $-80\,^{\circ}$C in small aliquots until just prior to use. The zwitterionic detergent CHAPS (e.g., Sigma Cat. No. C9426) is a good substitute for NP-40, possibly superior for some applications (Perdew *et al.*, 1983); it should also substitute for NP-40 in the IEF gel.

4. Micrococcal nuclease lysis buffer: 50 μg/ml staphylococcal nuclease (e.g., Worthington code NFCP) in 2 mM $CaCl_2$, 20 mM Tris, pH 8.8.

5. Nuclease digestion buffer: 1 mg/ml DNase I (e.g., Worthington code DPRF), 500 μg/ml RNase A (e.g., Worthington code RPDF), 50 mM $MgCl_2$, 0.5 M Tris, pH 7.0. Store aliquots at $-80\,^{\circ}$C.

Acknowledgments

I thank the members of the marine embryo community who provided helpful input, especially Andy Ransick, Sandeep Dayal, Billie Swalla, and Andrea Scouras. I am pleased to acknowledge the help over the years of Frank Tufaro, André Bédard, Zhiyuan Gong, Linda Hougan, Hung Fang, Corina Dumitrescu, and Sharon Hourihane. My research program is funded by NSERC.

References

Ames, G. F., and Nikaido, K. (1976). Two dimensional gel electrophoresis of membrane proteins. *Biochemistry* **15,** 616–623.

Bédard, P-A., and Brandhorst, B. P. (1986). Translational activation of maternal mRNA encoding the heat-shock protein HSP90 during sea urchin embryogenesis. *Devel. Biol.* **117,** 286–293.

Blin, N., and Stafford, D. W. (1976). A general method for isolation of high molecular weight DNA from eukaryotes. *Nucleic Acids Res.* **3,** 2303–2308.

Brandhorst, B. P. (1976). Two dimensional gel patterns of protein synthesis before and after fertilization of sea urchin eggs. *Devel. Biol.* **52,** 310–317.

Bradford, M. (1976). A rapid and sensitive method for the quantitation of microgram quantities of protein utilizing the principles of protein-dye binding. *Anal. Biochem.* **72,** 248–254.

Bravo, R. (1984). Two dimensional gel electrophoresis: A guide for the beginner. *In* "Two Dimensional Gel Electrophoresis of Proteins. Methods and Applications" (J. E. Celis and R. Bravo, eds.), pp. 3–36. Academic Press, Orlando, FL.

Chiang, M-K., and Melton, D. A. (2003). Single-cell transcript analysis of pancreas development. *Devel. Cell* **4,** 383–393.

Chomczynski, P. (1992). Solubilization in formamide protects RNA from degradation. *Nucleic Acids Res.* **20,** 3791–3792.

Chomczynski, P. (1993). A reagent for the single-step simultaneous isolation of RNA, DNA, and proteins from cell and tissue samples. *BioTechniques* **15,** 532–534.

Chomczynski, P., and Mackey, K. (1995a). Modification of the TRI Reagent procedure for isolation of RNA from polysaccharide- and proteoglycan-rich sources. *BioTechniques* **19,** 942–945.

Chomczynski, P., and Mackey, K. (1995b). Substitution of chloroform by bromo-chloropropane in the single-step method of RNA isolation. *Anal. Biochem.* **225,** 163–164.

Chomczynski, P., and Saccchi, N. (1987). Single-step method of RNA isolation by acid guanidinium thiocynate-phenol-chloroform extraction. *Anal. Biochem.* **162,** 156–159.

Coffman, J. A., Moore, J. G., Calzone, F. J., Britten, R. J., Hood, L. E., and Davidson, E. H. (1992). Automated sequential affinity chromatography of sea urchin embryo DNA binding proteins. *Mol. Mar. Biol. Biotechnol.* **1,** 136–146.

Dunbar, B. S. (1987). "Two-dimensional Electrophoresis and Immunological Techniques." Plenum Press, New York and London.

Esen, A. (1978). A simple method for quantitative, semiquantitative, and qualitative assay of protein. *Anal. Biochem.* **89,** 264–273.

Garrels, J. I. (1983). Quantitative two-dimensional gel electrophoresis of proteins. *Methods Enzymol.* **100,** 411–423.

Gong, Z., and Brandhorst, B. P. (1988). Stabilization of tubulin mRNA by inhibition of protein synthesis in sea urchin embryos. *Mol. Cell. Biol.* **8,** 3518–3525.

Hames, B. D. (1981). An introduction of PAGE. *In* "Gel Electrophoresis of Proteins: A Practical Approach" (B. D. Hames and D. Rickwood, eds.), pp. 1–91. IRL Press, Oxford, UK.

Harrington, M. G., Coffman, J. A., Calzone, F. J., Hood, L. E., Britten, R. J., and Davidson, E. H. (1992). Complexity of sea urchin embryo nuclear proteins that contain basic domains. *Proc. Natl. Acad. Sci. USA* **89,** 6252–6256.

Hunkapiller, M. W., Lujan, E., Ostrander, F., and Hood, L. E. (1983). Isolation of microgram quantities of proteins from polyacrylamide gels for amino acid sequence analysis. *Methods Enzymol.* **91,** 227–236.

Laemmli, U. K. (1970). Cleavage of structural proteins during the assembly of the head of bacteriophage T4. *Nature* **277,** 680–685.

O'Farrell, P. H. (1975). High resolution two-dimensional electrophoresis of proteins. *J. Biol. Chem.* **250,** 4007–4021.

Perdew, G. H., Schaup, H. W., and Selivonchick, D. P. (1983). The use of a zwitterionic detergent for two-dimensional gel electrophoresis of trout liver microsomes. *Anal. Biochem.* **135,** 453–455.

Ramagli, L. S., and Rodriguez, L. M. (1985). Quantitation of microgram amounts of protein in two-dimensional polyacrylamide gel electrophoresis. *Electrophoresis* **6,** 559–563.

Sambrook, J., and Russell, D. W. (2001). "Molecular Cloning. A Laboratory Manual," 3rd ed., Cold Spring Harbor Laboratory Press, Cold Spring Harbor, MI.

Scouras, A., and Smith, M. J. (2001). A novel mitochondrial gene order in the crinoid echinoderm *Florometra serratissima. Mol. Biol. Evol.* **18,** 61–73.

Tufaro, F., and Brandhorst, B. P. (1979). Similarity of proteins synthesized by isolated blastomeres of early sea urchin embryos. *Devel. Biol.* **72,** 390–397.

CHAPTER 24

Detection of mRNA by *In Situ* Hybridization and RT-PCR

Andrew Ransick
Division of Biology
California Institute of Technology
Pasadena, California 91125

I. Introduction

Understanding the molecular bases of the developmental processes of sea urchin embryogenesis will undoubtedly accelerate upon completion of the genome project. Important tools for this impending era of discovery include accurate and sensitive methods to monitor gene expression—spatially, temporally, and quantitatively. This chapter provides an overview of several approaches currently being used that rely on messenger RNA (mRNA) from a relatively small amount of starting material as a means to obtain detailed information on the expression of a single gene or an entire network of genes in the embryo.

METHODS IN CELL BIOLOGY, VOL. 74

0091-679X/04 $35.00

II. *In Situ* Hybridization

A. Whole Mount *In Situ* Hybridization (WMISH)

Lepage *et al.* (1992a,b) and Harkey *et al.* (1992) were the first to apply nonradioactive digoxygenin (DIG)-based whole-mount *in situ* hybridization (WMISH) methodology to sea urchin embryos. DIG-WMISH is now widely used, with over 100 papers published using the basic method. Not surprisingly, numerous protocols exist in that literature detailing modifications to fixation, buffering and processing procedures. Suffice it to say that in terms of reagents, solutions, and incubation times, there are clearly multiple approaches that will generate satisfactory results. On the other hand, there are a number of steps and handling measures that are required if one wants to produce an aesthetically pleasing preparation with well-preserved morphological details. Additionally, experimental situations often produce very limited numbers of specimens for analysis, which can demand successful processing of individual embryos. Consequently, this chapter emphasizes handling the embryos at various stages in the protocol and identifying pitfalls. The core method presented in detail here is modified from those cited, hereafter to be referred to as the "Lepage method" or the "Harkey method." However, as presented here, the protocol is optimized for preserving specimen morphology and handling small numbers of specimens. Noteworthy alternative steps or solution formulations are listed separately in the appropriate context. The procedures described here are mainly drawn from the author's own experience working with *S. purpuratus* embryos (Ransick and Davidson, 1995, 1998; Ransick *et al.*, 1993), but should be applicable to other sea urchin or echinoderm embryos with reasonably similar characteristics (e.g., Hinman *et al.*, 2003).

WMISH is a relatively easy procedure, resulting in specimens with long-term stability and great utility for deciphering gene expression patterns. However, it can be difficult (particularly at some embryonic stages) to consistently generate well-preserved specimens with strong WMISH signals but low background. Early cleavage stages tend to be more prone to higher background staining, while the natural prismatic shape of the embryos between gastrulation and formation of the pluteus is difficult to maintain. Additionally, despite one's best efforts, application of the same protocol sometimes produces variable or unsatisfactory results. Consequently, different researchers mention supplemental steps that have been taken to reduce noise while maximizing signal and morphological preservation. These "tips" or alternatives are indicated in the appropriate context. An exception to this style relates to specifics of the Arenas-Mena method, which are given together in Section II.A.9.

1. Culture and Collection of Embryos

If cleavage stage (i.e., prehatching) embryos will be used for WMISH, it is necessary that the fertilization membrane (FM) be removed in order to obtain satisfactory results.

a. Standard methods for inhibiting hardening of the FM (e.g., presence of 0.3 mM p-aminobenzoic acid at the time of fertilization) should be employed in these situations.
b. One hour after fertilization is an ideal timepoint for removing the FM by passing the embryos through Nitex screening (60 μm mesh for *S. purpuratus*) or by aspirating the embryos in and out of a drawn-out Pasteur pipette.
c. Embryos should be washed in fresh filtered seawater (FSW) to remove all debris. Although gentle centrifugation (lowest setting) is certainly an effective means of concentrating embryos, an alternative for washing live material in excess volume is to use wide, shallow glass dishes (sometimes referred to as "coasters" or "Syracuse dishes"). These dishes easily hold 15 ml volumes but have less than 1 cm of depth, so all specimens rapidly settle to the bottom. The specimens are very efficiently concentrated by gently moving the dish in a circular pattern along a horizontal surface.
d. *S. purpuratus* embryos are cultured at 15 to 16 °C in temperature-controlled rooms or incubators, and chilled water circulated through bored anodized aluminum plates provides a temperature-regulated surface on the bench.

2. Materials and Specimen Handling

All of the steps of the WMISH protocol can be carried out on batches containing from 1 to 200 embryos using flexible plastic round-bottom plates (Falcon #3911). This arrangement makes it very convenient to observe the specimens on a dissecting microscope at any point in the procedure. Solutions can be added quickly to all the wells of a plate using a pipettor (e.g., Gilson P200). Whenever possible traditional mouth-pipetting techniques should be avoided, since some of the solutions used for WMISH contain toxic components. Glass pipettes prepared in a standard fashion are very useful for removing solutions via a low-strength suction line.

a. Glass pipettes are prepared from long shank (9 inch) borosilicate glass Pasteur pipettes pulled over a microflame to produce a one-inch shank at a 45° angle that tapers to ~100 μm diameter opening. Fire-polish all rough edges.
b. Falcon plates are cut into sets of 16 wells and placed in 60 mm plastic petri dishes. Ten to 12 milliliters of water in the petri dish improves both the optics and temperature control. The individual wells in these dishes will hold a maximum volume of 200 μl, although between 50 and 150 μl is routinely the volume employed.
c. Solutions in the wells are mixed thoroughly by directing a stream of air (from an airline) into the well at an oblique angle to create a stirring vortex. As consecutive solutions required in the protocol are often of significantly different consistency, the ability to routinely achieve complete mixing in the wells is critical.

d. Following the mixing process visually under the microscope provides an assurance of having completely transitioned the specimens into the next solution. This approach also provides an opportunity to selectively remove micro-debris that inevitably shows up in the wells.

e. Removal of solutions is a more painstaking process and several factors that can affect the success of this routine step should be noted.
   1. It is important that the specimens have completely settled into a compact grouping on the bottom of the well so that the maximum amount of solution can be removed while not accidentally removing the specimens.
   2. Settling of the specimens into a tight grouping can be encouraged by tapping the dish against the deck and by a relatively gentle stream of air that slightly vortexes the fluid.
   3. A minimum of 5 to 10 μl of solution must be left in the wells to adequately protect the fragile specimens from compression effects at air–fluid interfaces.

f. To reduce the stickiness of specimens to each other and to well walls, all solutions (post-fixation) contain 0.1% Tween-20. Deciliation of embryos also tends to reduce adherence to well walls, to other specimens, or to micro-debris.
   Deciliation procedure by hypertonic shock with 2× seawater:
   1. Add 12 μl of 4.45 M NaCl per 100 μl FSW.
   2. Mix and allow embryos to settle.
   3. Rinse twice with normal FSW (embryos should recover natural shape).
   4. Fix embryos promptly.

Additional Notes: If a large number of specimens (>200) are to be processed, a silicon-coated 0.7 ml snap-cap polypropylene tube (Phenix #MN-650GSL) can be used to carry the specimens through the WMISH protocol. Mixing solutions is accomplished by inverting such tubes, while mild centrifugation is used to collect the specimens. The in-tube approach obviously requires relatively larger volumes of solutions, but it does eliminate the more time-consuming and demanding aspects of exchanging solutions from wells. Less satisfactory is the increased tendency for the in-tube processed specimens to collect micro-debris and to be damaged because of the repeated pelleting and resuspension required by this approach.

3. Fixation of Embryos

Good fixation is critical to preserving embryo morphology and retaining the mRNA in its endogenous location. Weak fixation results in deterioration of embryo integrity during the rigors of workup and leads to weak signals due to degradation or extraction of mRNAs. The recommended fixative uses freshly prepared buffered-glutaraldehyde solutions with added salt to match the osmolarity of seawater. By scaling up these fixation and washing steps, one can prepare a large supply of fixed embryos that may be stored in a refrigerator and provide specimens for WMISH for literally years.

a. Good results are routinely obtained with ice-cold 2.5% glutaraldehyde in 0.2 M sodium-phosphate (Na-phos) buffer, pH 7.4, with 0.14 M NaCl added (after Harkey *et al.*, 1992; Cloney and Florey, 1968). The fixative is freshly prepared from high-grade 25% stock solutions and fixation should proceed on ice for at least 2 h. Fixation can continue for longer durations, although it tends to result in some background staining in early cleavage stages. [An alternative fixation is 1 h on ice using 1% glutaraldehyde in 50 mM potassium phosphate, pH 7.4, plus either 0.43 M NaCl (after Angerer *et al.*, 1987) or 0.55 M NaCl (after Lepage method)].

b. The fixative is washed out with three exchanges of cold 0.2 M Na-phos buffer containing 0.3 M NaCl and 0.1% Tween-20.

c. If proceeding immediately toward hybridization, the specimens are transferred into fresh wells containing TRIS-buffered saline with 0.1% Tween-20 (TBST; or PBST by substituting Na-phos for TRIS buffering), [TBS recipe: 0.15 M NaCl; 0.2 M Tris buffer, pH 7.5; PBS recipe: 0.2 M Na-phos buffer, pH 7.4, replaces TRIS]. There are no apparent disadvantages to using TBST throughout the protocol instead of PBST. In fact, use of TBST is recommended since any phosphates will form a precipitate with magnesium ions in the post-hybridization histochemical staining reactions (see Section II.A.7.d).

d. To store the fixed embryos long-term, skip the previous step and proceed from washing out the fix to dehydration of the specimens up to 70% ethanol (EtOH) in the following steps: 0.5 M NaCl wash; 10%; 20%; 30%; 50%; 70%; 70% EtOH. (Lepage method rehydrates using Methanol/PBST to facilitate an extraction in 1% CHAPS/PBST for 10 min prior to proceeding with prehybridization.)

e. Another optional step at this point is to dehydrate specimens completely to 100% EtOH and proceed to extract with toluene for 5 min, before rehydrating and storing in 70% EtOH. This toluene extraction step is designed to reduce background staining (after Fang and Brandhorst, 1996 and Tautz and Pfeifle, 1989). However, it can also be particularly useful when later-stage embryos are processed, as toluene-extracted embryos are significantly firmer and undergo less deformation during the later hybridization steps.

4. Proteinase K and Post-Fixation

The efficiency of protein cross-linking during glutaraldehyde fixation makes it necessary to digest specimens briefly with dilute proteinase K (PK) in order for the probes to have access to the mRNAs. However, the concentration of PK and duration of this digestion should be carefully controlled, as specimen integrity and signal quality can be rapidly degraded if PK is used too aggressively. Importantly, the specimens show no visible ill effects from moderate PK treatment after adequate glutaraldehyde fixation. However, it is a clear indication that fixation was inadequate and/or the PK treatment too aggressive when specimens begin to shear or shred in subsequent washing steps.

a. Satisfactory results are routinely obtained by incubating specimens in 5 μg/ml PK in TBST for 5 to 10 min. However, be aware that early cleavage stage embryos are often more susceptible to PK, while pluteus stages may require stronger PK treatment for optimal results.
b. Follow immediately with two washes with TBST containing 25 mM glycine and a wash with TBST alone. (Lepage method uses washes with 0.1 M glycine after PK).
c. Post-fix for 30 min in 4% paraformaldehyde at RT.
d. Thoroughly wash out the post-fix with three rounds of TBST.

Additional Notes: Stock solutions of 2 mg/ml PK and 4% paraformaldehyde (and any other perishable reagents) are routinely maintained as small volume, single-use frozen aliquots in nondefrosting −20 °C freezers. Another variation intended to reduce background staining involves a 1 h incubation in 6% hydrogen peroxide in PBST prior to PK treatment, followed by adding 0.2% glutaraldehyde to the post-fixation (Wikramanayake, personal communication).

5. Probe Preparation

Single-stranded RNA probes, optimally 400 to 600 nucleotides long and incorporating digoxygenin-11-UTP (DIG), are routinely synthesized utilizing commercially available RNA polymerases (Sp6, T3, or T7) with clean Bluescript (Stratagene) or pSport (Invitrogen) plasmid DNA template linearized appropriately to produce run-off antisense-transcripts. All details provided here pertain to DIG-containing probes, although UTP derivatives with fluorescein, biotin, and dinitrophenol side groups are also available for multicolor WMISH (Hauptmann, 2001; Hauptmann and Gerster, 1996; Long and Rebagliati (2002).

a. It is convenient to use a 10X DIG RNA Labeling Mix (Roche #1-277-073) (10 mM ATP, CTP, and GTP; 6.5 mM UTP and 3.5 mM DIG-UTP) with the supplied 10× buffer (400 mM Tris-HCl, pH 8; 60 mM $MgCl_2$; 100 mM dithiothreitol; 20 mM spermidine). Follow the supplier's instructions for building reactions and make sure the 10× transcription buffer is completely resuspended.

b. Run transcription reaction for 2 h, then selectively precipitate the RNA product by adding $\frac{1}{10}$ volume 7.5 M LiCl; 3 volumes 100% EtOH; 5 μg glycogen (Roche #901–393). Hold at −20 °C for 2 h, pellet, wash, and resuspend in nuclease-free water (see Ambion Technical Bulletin #160). As these reactions typically yield a 10:1 mass ratio of product to template and LiCl selectively precipitates RNA, it is unnecessary to digest the template DNA.

c. Aliquots of DIG-RNA probes should be stored at −70 °C as small-volume, concentrated stocks (50 ng/μl), then thawed just before addition to hybridization buffer to produce a 1 ng/μl working stock. The working stock typically can be stored at −20 °C for at least a month without any deterioration of the probe.

Additional Notes on WMISH Probes: WMISH with DIG-RNA probes is surprisingly sensitive to probe concentration, and it is routinely found that desirable labeling is obtained at final probe concentrations ranging from 0.005 ng/μl to 0.05 ng/μl. It is painfully obvious when one is outside of the effective probe range, as rapid overstaining artifacts appear, or no signal at all develops even with prolonged staining. It is, therefore, a good practice when titering a new probe preparation, or exploring an unknown gene expression pattern, to rely on the same 1 ng/μl working stock in consecutive WMISH attempts, because this intermediate dilution step facilitates identification of the most advantageous final probe concentration.

It is worth noting that very impressive results are now being obtained with multicolor fluorescent WHISH probes using fruit fly embryos. Through the use of multiple fluors, tyramide signal amplification, and direct fluorescent probe labeling, over six separate gene expression patterns have been visualized in single specimens. Currently, there are few papers in the literature employing this methodology (e.g., Wilkie and Davis, 1998), although its usefulness suggests wider application can be expected. For the present, the best summary of the current state fluorescent WMISH methods is from D. Kosman at http://www.biology.ucsd.edu/~davek/.

6. Hybridization and Washing

Hybridization buffers are complex mixtures of reagents that each, in some manner, affects the rate and specificity of the molecular interactions. The objective is to create a milieu in which the salts, buffers, blocking agents, crowding agents, and chaeotropic agents interact to not only allow, but favor, the hybridization of the probe specifically with its target. Once achieved, the specific hybrids of cellular mRNAs with complementary (antisense) probe RNAs are highly stable and will easily persist through the series of high stringency post-hybridization washes that dissociate nonspecific interactions.

a. The routinely used hybridization buffer (HB), which is preferably freshly prepared from stocks, contains 50% formamide; 5% PEG 8000 mw (from 25% stock); 0.6 M NaCl; 20 mM Tris-base, pH 7.5; 5 mM EDTA; 2× Denhardts; 0.1% Tween-20, and 500 μg/ml yeast tRNA. This HB formula is only slightly modified from Harkey's original formula, and even though developed for use with single-stranded DNA probes, it works well with DIG-RNA probes. An alternative hybridization buffer developed specifically for RNA probes differs in replacing the PEG and NaCl components with 5× SSC and 10% dextran sulfate, as well as eliminating EDTA and adjusting down to 1× Denhardts and up to 0.3% Tween-20 (after Lepage method).

b. The specimens are transitioned from TBST into HB in three steps (35%HB, 65%HB, 100%HB), and then are incubated for 1 h at 60 to 65 °C as a prehybridization step.

c. Probe is added to HB in the appropriate dilution from a 1 ng/μl working stock. For example, if four wells of specimens are to be hybridized with the same probe, a 200 μl solution with final probe concentration of 0.05 ng/μl is made by adding 10 μl of probe working stock to 190 μl HB in a snap-cap 0.7 ml tube.

d. That tube is heated to at least 70 °C for several min, transferred to ice, then 50 μl of HB with probe is added to each of the specimen-containing wells.

e. The individual wells are then sealed by overlaying a piece Microseal Film (MJ Research # MSA-5001; designed to seal wells during thermal cycling applications) trimmed to fit without excess hangover, and then pressing down firmly over the top of each well.

f. The entire petri dish, with wells containing specimens, is then transferred to a hybridization oven set at 50 to 55 °C and hybridization allowed to proceed overnight (12 to 16 h).

g. HB with probe is removed and replaced with 50% HB:TBST at room temp as a first low stringency wash, and is followed by two washes in 100% TBST at 60 to 65 °C for 10 min each. Use at least 100 μl volume in each of these wash steps and pay particular attention to thorough mixing of solutions during each exchange.

h. Transfer the specimens to fresh Falcon plate wells, then proceed with three high stringency washes using 1× SSC plus 0.1% Tween-20 (SSC+T) at 65 °C for 15 to 20 min each.

i. The Lepage method opts to hybridize at 46 °C, followed by a digestion step using 20 μg/ml RNase A and 10 U/ml RNase T1 at 37 °C for 15 min, followed by 15 to 30 min washes in 2× SSC+T and 0.1 SSC+T at temperatures ranging from RT to 50 °C (see Lepage *et al.*, 1992a, for details). The RNase steps specifically target single-stranded RNA for degradation, and thus can significantly improve the final signal-to-noise ratio in cases where a significant amount of probe is bound nonspecifically or is trapped in the specimen. However, unless probe-derived background staining is a persistent problem, routine application of RNases is unnecessary. It can actually diminish the strength of specific hybridization signals via fragmentation of target mRNAs or probe molecules, followed by their elution in subsequent high-stringency washes.

7. Immunohistochemical Detection

Specimens hybridized and washed as has been described are ready for immunohistochemical detection of the DIG-RNA probe.

a. Excellent results are routinely obtained by blocking in 5% sheep serum (Sigma S3772) in TBST for 30 min at RT.

b. Follow with a 1 h/RT incubation in sheep anti-DIG antibodies conjugated to bacterial alkaline phosphatase (Fab fragments; Roche #1-093-274) prepared as a 1:500 dilution in 5% sheep serum/TBST.

c. Follow with several TBST washes to remove excess and unbound antibody.

d. IMPORTANT: If one is using PBST instead of TBST, at this point it is critical that the specimens are washed twice in TBST to remove all traces of phosphate, which otherwise precipitates in the $MgCl_2$-containing staining reaction.

e. Wash the specimens once in alkaline phosphatase staining buffer (APB) [APB: 100 mM Tris-base, pH 9.5; 100 mM NaCl; 50 mM $MgCl_2$; 1 mM levamisole; 0.1% Tween-20]. Levamisole is a specific inhibitor of eukaryotic AP and will therefore block residual endogenous AP activity in the specimens.

f. Initiate staining by replacing with APB containing the histochemical substrates BCIP (5-bromo-4-chloro-3-indolyl phosphate) and NBT (nitro blue tetrazolium). Ready to use NBT/BCIP tablets and buffers are commercially available (Roche #1-697-471) and do produce good results. However, it is more economical to obtain the substrates in powder form and prepare separate 100 μl (or larger) stock solutions that are stable for many months when stored at $-20\,^{\circ}$C. BCIP (Sigma #B-8503) stock is 5 mg/100 μl of 100% dimethylformamide and NBT (Sigma #N-6876) stock is 7.5 mg/100 μl of 70% dimethylformamide. The final staining mixture, which should be protected from light, is prepared immediately before use and consists of 1 ml APB plus 4.5 μl NBT stock solution and 3.5 μl BCIP stock solution.

g. Staining is halted by removing the staining mix and washing twice with 1 mM EDTA/TBST.

h. Stained specimens can be transitioned to 50% glycerol (e.g., 15%, 30%, 50%), containing 5 mM Na-Azide as a preservative and stored long-term by maintaining the wells sealed with Parafilm or Microseal. Alternatively, stained specimens can be dehydrated to 100% EtOH (which tends to have a bluing effect on the stain), then cleared and semipermanently mounted in terpineol (Sigma #T3407). Specimens stored in 50% glycerol remain pliable and tend to clear more over time. Specimens stored in terpineol are well cleared, but are more brittle and tend to fragment under even slight coverslip pressure.

Additional Notes on Staining: Following the protocol outlined here, using a good DIG-probe and properly processed specimens, one can typically expect some specific staining to appear within 30 to 60 min. It should be kept in mind that since there isn't a 1:1 relationship between mRNA and AP signal, but rather an amplification of signal over time, WMISH is not routinely intended as a quantitative procedure. The general rule of thumb should be to continue the staining reaction only as long as is necessary to see a distinct specific signal. On the other hand, WMISH can be an effective technique to assay for potential differences in expression of an mRNA, provided all specimens see the BCIP/NBT reagents for the same interval.

Progress of the staining can be monitored periodically under a good quality dissecting microscope using low light. Resist the temptation to linger at observing the specimens during staining, as the histochemical substrates are sensitive to light and this practice will increase background staining. It is not uncommon to stop the staining reaction after 1 h, or less, as a result of seeing adequate discrete staining. However, if necessary, the staining reaction can be allowed to continue for 3 or even 6 h, provided that background staining remains low. The absence of any specific staining, or a rapid appearance of general staining, doesn't necessarily mean the target mRNA is, respectively, absent or ubiquitous. Even an experienced individual will occasionally make a critical mistake in a protocol containing this many steps and reagents. Slightly modifying the probe concentration or taking extra care with the washing steps can produce striking improvements on repeat attempts.

8. Use of Positive Controls

Regardless of one's experience level with WMISH, inclusion of a positive control is always a wise choice, and it is a must for those with little experience using this protocol. Process some specimens using a probe to a target mRNA known to be localized and reasonably abundant (e.g., Wnt 8, Endo16, SM50, or Spec1). There is considerable value in obtaining well-stained positive control specimens that were fixed and processed in parallel from the same batch of embryos. In particular, this control reveals the general quality of the mRNA preservation in that preparation by way of the strength of the specific hybridization signal, while showing the overall adeptness with which the protocol has been executed by the level of background staining. Traditional "sense strand controls," which are "negative controls" and are expected to produce no hybridization signal whatsoever, are only useful in identifying staining ascribable to nonspecific sources. When investigating an unknown expression pattern, only the positive controls provide confidence that many potential pitfalls capable of wiping out signals have been eliminated. One should be able to routinely obtain WMISH specimens with low background and high specific hybridization signals using such a positive control before employing WMISH to explore unknown expression patterns or assay gene expression in experimental material.

9. Arenas-Mena Method

Arenas-Mena *et al.* (2000) introduced an alternative protocol for WMISH that was developed in connection with describing HOX cluster expression profiles in later-stage sea urchin larva (up to several weeks old). This protocol is being applied to earlier developmental stages (Hinman *et al.*, 2003; Minokawa *et al.*, 2004) and is producing dramatic results with respect to increasing the strength of specific hybridization signals while maintaining low background staining. References to this protocol have been omitted up to this point in the chapter due to numerous significant differences that might have been confusing in the previous context. Some noteworthy steps in Arenas' protocol that are unique include:

a. Primary fixation with 4% paraformaldehyde in MOPS buffer.
b. Hybridization for 1 week in HB containing 70% formamide.
c. Using DIG-RNA probe at final concentration of 0.1 ng/μl.
d. Extended times for Ab reaction (3 h) and Ab washout (3 h).
e. Allowing BCIP/NBT staining to proceed up to overnight.

Although by no means a fast procedure, this protocol should be considered an attractive alternative in cases where the mRNA under investigation is expressed at relatively low levels, and especially when the other protocols are yielding unsatisfactory results. The conditions of the Arenas' WMISH method appear to increase sensitivity, enabling detection and amplification of hybridization signals from low abundance mRNAs while maintaining strikingly low background signals. Keep in mind, however, the cautions regarding the nonquantitative nature of WMISH become particularly applicable when using extended staining durations.

B. *In Situ* Hybridization on Sections Using Isotope-Labeled Probes (This Section Contributed by L. Angerer)

In Situ hybridization methods using sectioned tissue and radioactively labeled RNA probes were developed in the 1980s in the Angerer lab (Cox *et al.*, 1984a,b). These workers found that RNA probes provided much greater sensitivity than DNA probes because they are single-stranded, they form hybrids with higher stability, and unhybridized probes can be effectively removed by RNase digestion. Whole-mount *in situ* hybridization has largely replaced radioactive methods with sectioned tissue because of its relative speed, the long-term stability of the probes, and the ease of three-dimensional interpretation. However, there are two instances where the sectioning/radioactive probe approach is advantageous. First, it provides a more quantitative assessment of relative mRNA target levels. For example, relative signals of a given mRNA between stages that were quantitated by measuring grain counts per area were in excellent agreement with measurements made by RNase protection. Consistent with this, *in situ* signals for a large number of different mRNAs as a function of developmental stage are always consistent with other hybridization assays, such as RNA blotting or RNase protection. Second, it offers a good way to compare distributions of different mRNAs in the same embryo by probing adjacent thin (1 μm) sections (e.g., Angerer *et al.*, 1989; Howard *et al.*, 2001; Wei *et al.*, 1999). The number of different mRNAs that can be monitored by this method is considerable since the fixed sea urchin embryo is about 80 to 200 μm, depending on stage.

The success of radioactive *in situ* depends, in part, on the choice of isotope. Initially, probes were labeled with ^{3}H, but because the emissions are weak, the use of these probes is confined to abundant target mRNAs. One advantage of ^{3}H-labeled probes is that they offer good resolution because the path length of the emitted beta particle is short. To detect less abundant mRNAs, the sensitivity of the technique was increased significantly by using other isotopes, such as ^{125}I (Angerer and Angerer, 1991), ^{35}S (reviewed in Angerer *et al.*, 1987; Angerer and Angerer, 1991), and ^{33}P UTP (e.g., Angerer *et al.*, 2001). Of these, ^{125}I offers the best combination of sensitivity and resolution, but it has not been used because of limited availability of the isotope and the complicated precautions in handling that are required to avoid damaging exposure. ^{35}S-UTP was the isotope of choice until ^{33}P-UTP became available. The latter offers an important advance in sensitivity because noise levels are significantly lower. ^{35}S and ^{33}P have similar emission characteristics, although the half-lives of ^{33}P and ^{35}S are 25 and 87 days, respectively. Controlling background using ^{35}S requires extensive washes, as has previously been discussed in detail (Angerer and Angerer, 1989, 1991, 1992).

The reader is referred to Angerer *et al.* (1987) and Angerer and Angerer (1991, 1992) for detailed accounts of tissue preparation, sectioning, probe preparation, hybridization conditions, post-hybridization washes, autoradiography with liquid emulsion, staining, and microscopy. Here, emphasis is on several modifications of the protocol that were developed for the use of ^{33}P-labeled probes.

a. ^{33}P-probe specific activity: The specific activity of these probes is adjusted to between 0.5 and 5×10^8 dpm/μg, depending on the abundance of the target RNA. RNAs present at 5 to 10 copies/cell can be detected with the highest specific activity probes in a week or two of exposure time.
b. The hybridization buffer is 50% formamide, 0.3 M NaCl, 20 mM Tris, pH 8.0, 5 mM EDTA, 0.02% polyvinylpyrollidone, 0.02% BSA, 0.02% polyvinylsulfate, 10% dextran sulfate; the hybridization temperature is 50 °C for probes ~300 bases long with normal GC content (Tm − 25 °C).
c. Post-hybridization washes:
    1. 4× SSC, 3 × 1 min in 50 ml (Coplin jars) to remove most of the probe. These washes are discarded in radioactive waste; this is followed by additional washes in high salt (4× SSC) 2 × 15 min in 250 ml.
    2. 1 × 10 min in 50% formamide, 0.3 M NaCl, 20 mM Tris, pH 8.0, 5 mM EDTA at 60 to 65 °C, depending on the GC composition of the probe (Tm − 5 to 10 °C). This high stringency wash is important to eliminate non-specific hybridization.
    3. RNase treatment, as described previously. This treatment is also important in order to achieve low backgrounds.
    4. RNase removal washes: 0.5 M NaCl, 20 mM Tris, pH 8.0, 5 mM EDTA, 4 × 250 ml, 10 min each.
    5. 0.1× SSC, 50 °C, 10 min.
    6. Dehydration as described previously.

Additional Comment on Controls: Although it is common to use the sense strand as a control for background, the level of noise is highly variable among different control probes. Therefore, the best control for background is a tissue that lacks the target RNA.

III. Quantitative PCR

A long-standing constraint for experimentalists attempting to quantitatively assay levels of gene expression in developing embryos has been the sensitivity limits of detection methods, which generally require sample sizes that are relatively large and, consequently, make the final per-embryo estimate an average with a substantial N value. Although such methods can be quite accurate, they are not accessible when the numbers of specimens are limited. Many of the manipulations that are currently employed by experimentalists, such as microinjections of mRNAs, morpholino antisense oligonucleotides (MASO), or engrailed-fusion constructs, often produce dramatic effects on gene expression leading to equally dramatic phenotypes. However, without an effort of herculean proportions, these experiments typically yield sample sizes of several hundred specimens or fewer. One can immediately see the attractiveness of being able to reduce the requisite sample size as much as possible, as long as this doesn't sacrifice accuracy in interpretation.

Virtually all such constraints are overcome by combining the extremely small amounts of starting template DNA required in polymerase chain reaction (PCR) technology with the sensitivity of quantitating DNA in real time using fluorescence detection (Higuchi *et al.*, 1992, 1993; Livak *et al.*, 1995). As with WMISH, alternative methods and apparatus that are equally satisfactory exist for real time quantitative-PCR (RTQ-PCR). Applied Biosystems (ABI) offers several grades of their Sequence Detection Systems (5700, 7000, 7700, and 7900HT), and the Lightcycler (Roche) and SmartCycler (Cepheid) represent alternative RTQ-PCR apparatuses. A comprehensive discussion of RTQ-PCR methodology will not be included here, although an excellent primer entitled "DNA/RNA Real Time Quantitative PCR" can be found on the ABI website at http://docs.appliedbiosystems.com/pebiodocs/00777904.pdf. Rather, the emphasis here will be to detail the method that has been developed in our group, which has proven to be a versatile and sensitive assay of gene (mRNA) expression levels in small, experimentally derived samples. The method can be broken down into 3 phases: RNA isolation, cDNA synthesis, and RTQ-PCR.

A. RNA Isolation

In preparation for cDNA synthesis, total RNA is isolated from 100 to 1000 embryos, although samples with as few as 20 embryos have been successfully processed for RTQ-PCR. Several excellent primers on critical factors for isolating intact RNA exist in the literature (MacDonald *et al.*, 1987). Mehra (1996) provides an overview of RNA isolation methods, including preparation of reagents for the single-step acid phenol/guanidinium thiocyanate procedure (Chomczynski and Sacchi, 1987) used here.

a. Embryos in FSW are transferred to a plastic snap-cap tube and pelleted in a microfuge.
b. FSW is removed and replaced with 100 to 200 μl of RNAzol-B (Leedo Med. Lab. #CS-105).
c. Rapidly aspirate the sample several times using a P200 pipetman, striving for complete dissolution of the specimens. These preps can be processed immediately or quick-frozen in liquid nitrogen and stored at $-70\,^{\circ}$C.
d. Induce a phase separation that extracts the RNA by adding $\frac{1}{10}$ volume of chloroform, then vortex 15 s to vigorously mix. Place the tube on ice for 5 min, then centrifuge for 15 min at 12,000*g* at 4 °C.
e. Carefully transfer the upper 80% of the aqueous phase (containing RNA) to a new tube, taking care not to remove the lower 20% or disturb the interphase.
f. Precipitate the RNA by adding 5 μg glycogen and an equal volume of isopropanol, then placing the tube at $-20\,^{\circ}$C for 30 min.
g. Centrifuge at 12,000*g* for 15 min at 4 °C; then wash the pellet ($\times$2) with 75% EtOH and allow the pellet to air dry. As a pellet may not be visible in many instances, orientation of the tubes in the centrifuge should be noted.
h. Resuspend RNA in 17 μl nuclease-free water; if necessary, dissolve the pellet at 37 °C for 10 min.

To ensure that no genomic DNA is present during cDNA synthesis, a DNase I digestion step can be carried out with the aid of the DNA-Free Kit (Ambion #1906).

i. To the resuspended RNA, add 2 μl ($\frac{1}{10}$ volume) of 10$\times$ DNase buffer [100 mM Tris, pH 7.5; 25 mM $MgCl_2$, 1 mM $CaCl_2$] and 1 μl DNase I (2 units), then incubate at 37 °C for 20 min.
j. Next, add 5 μl of well-suspended DNase Inactivation Reagent, mix, and let sit 2 min at RT before pelleting the resin; then transfer the RNA containing supernatant to a fresh tube. If proceeding directly to cDNA synthesis, transferring the RNA sample to a thin-wall 0.2 ml PCR-tube facilitates use of a programmable thermalcycler for the various temperature-dependent steps.
k. (optional) To ensure an even more pure RNA sample, another round of RNA precipitation can be performed at this point by adding $\frac{1}{10}$ volume 5 M ammonium acetate and 2.5 volumes 100% EtOH, holding at $-20\,^{\circ}$C for 30 min, and recovering the RNA by repeating steps (g) and (h).

Additional Comments: An alternative method of RNA isolation is via silica membrane columns, such as the GenElute Total RNA Kit (Sigma #RTN-70) or RNAqueous-4PCR Kit (Ambion #1914). Per the GenElute kit instructions, samples are lysed and homogenized in guanidine thiocyanate and 2-mercaptoethanol to release RNA and inactivate RNases. Lysates are spun through a filtration column to remove cellular debris and shear DNA. The filtrate is mixed with an ethanol solution, then applied to a high-capacity silica column to bind total RNA, followed by washing and elution. Up to 150 μg of total RNA can be recovered per prep in 100 μl of water.

Several coprecipitants are commercially available to aid recovery and visualization of the RNA pellets, including linear acrylamide (Ambion #9520), glycogen (Ambion #9510), and glycogen covalently bound to a blue dye (Ambion #9515). Linear acrylamide may be the best for RTQ-PCR, as this synthetic product is sure to have no contaminating nucleic acids or residual proteases.

B. cDNA Synthesis

For the majority of labs using molecular cloning technologies, the synthesis of complementary DNA (cDNA) is a routine procedure. Not surprisingly, there are many reagents and kits available from suppliers that will adequately accomplish this. In particular, reagents and specific kits are available which have been optimized for the purpose, using small starting amounts of RNA to generate first-strand cDNA. Reverse transcription can be primed using random primers, oligo-dT primers, or gene-specific primers. Using random primers provides a broad sampling of the mRNA set present in your sample and will allow a large number of different PCR primer sets. Consequently, random primed first-strand cDNA is a highly suitable template for RTQ-PCR.

First strand cDNA synthesis kits that have been successfully employed by our group are TaqMan Reverse Transcription Reagents (ABI #N808-0234) and Superscript First-Strand Synthesis System for RT-PCR (Invitrogen #11904-018). Per the TaqMan Kit instructions,

a. Build a 100 μl RT reaction (can be scaled down to 50 μl) in a thin-walled 0.2 ml PCR tube with the RNA sample in nuclease-free water (from isolation step) and the following final concentrations of reagents: 20 mM Tris-HCl, pH 8.4, 50 mM KCl, 5.5 mM $MgCl_2$, 500 μM each dNTPS, 2.5 μM random hexamer.

b. As step 1 in a programmable thermalcycler, the reaction constituents are heated to at least 70 °C for 5 min to denature the RNA.

c. Step 2 is a hold at 25 °C for 10 min that allows the random hexamers to anneal.

d. Halfway through that step, add to the reaction 0.4 U/μl RNase Inhibitor and 1.25 U/μl Multiscribe Reverse Transcriptase; mix contents and spin the tubes briefly.

e. Step 3 is a 48 °C hold for 1 to 3 h for cDNA synthesis.

f. Step 4 is a 5 min hold at 95 °C to kill the reaction.

Additional Comments: As a guideline, satisfactory RTQ-PCR results are routinely obtained with three embryo equivalents of cDNA per reaction. Thus, starting with a sample of 100 embryos should provide enough cDNA template for about 30 PCR reactions. If more cDNA is produced than is required immediately, store it at −20 °C in a nondefrosting freezer RT. Keep in mind that single-stranded cDNA can fragment if subjected to repeated freezing/thawing, so aliquot before freezing.

C. Real Time Quantitative PCR (RTQ-PCR)

Several options are available in RTQ-PCR machines and quantitation methodologies. As experimentally oriented developmental biologists, our choices in turning to RTQ-PCR were guided by wanting to obtain accurate relative quantitation of mRNA expression for as many genes as possible in direct comparisons between differentially manipulated experimental samples using relatively small quantities of starting material.

A combination that suits that goal very satisfactorily is the PE Applied Biosystem GeneAmp 5700, used in combination with the SYBR Green I double-stranded DNA binding dye (Rast *et al.*, 2000). There is an alternative methodology of fluorogenic probes that capitalizes on the inherent 5′ nuclease activity of DNA polymerase to provide real time quantitation of specific amplification products (Holland *et al.*, 1991; Lee *et al.*, 1993). However, that method requires the generation of a sequence-specific labeled probe for every template to be assayed. Since the synthesis of such probes is costly and we sought to monitor a large number of target genes in our preparations, it was not practical to use fluorogenic probe methodology on a routine basis. By contrast, the binding characteristics of SYBR Green provide that fluorescent signal will be generated from any double-stranded product in the PCR reaction, and so this dye can be applied to quantitate the amplification of any template. This feature of SYBR Green made it relatively inexpensive and provided great versatility. However, an obvious potential drawback to this approach is that the dye doesn't distinguish specific amplicons from primer–dimer artifacts. Fortunately, the GeneAmp 5700 provides a means to detect the presence of nonspecific products through running a dissociation curve protocol on the PCR products at the end of the cycling. Reactions containing only template-specific products will have a highly characteristic and relatively sharp melting temperature (T_m) profile, whereas the T_m profile of reactions that contain significant nonspecific products will be broad and will indicate multiple T_m peaks. Consequently, quantitation by SYBR Green is quite reliable when the dissociation curve profile shows amplification of a single (or very dominant) molecular species.

There are several precautions routinely implemented to ensure reliable quantitation with SYBR Green. First, as even minor primer artifacts can amplify to detectable levels after 40 cycles, and most templates reach the optimal levels for quantitation in the 15 to 25 cycle range, the GeneAmp 5700 is programmed to terminate after 30 to 35 cycles. Second, the dissociation curve of every new batch of primers synthesized is checked to ensure they are not prone to generating amplification artifacts. Only primer sets that generate clean dissociation curves, without significant artifacts, are placed among the lab stocks for use in data generation. Third, primers are designed to generate products in the narrow range of 125 to 175 nucleotides, in keeping with the manufacturer's literature suggesting that short amplicons are the most efficiently amplified, produce the most consistent results, and are least likely to exhaust components in the reaction. In following this guideline, approximately the same number of SYBR Green molecules is bound per product molecule, thereby providing uniform signal generation across all of the different products that are assayed.

2× SYBR Green PCR Master Mix (PE Applied Biosystems #4309155) is a convenient reagent for use in the GeneAmp 5700. Aside from containing the DNA binding dye, dNTPs, and optimized buffer components, this mix contains components specifically tailored to RTQ-PCR using the GeneAmp 5700. AmplitaqGold DNA Polymerase is chemically modified to be inactive at room temperature, but become active after heating, providing an automated hot-start capability. A second dye that does not participate in the PCR reaction is included as a passive reference. The GeneAmp5700 reads the level of this dye and automatically uses it to normalize for well-to-well fluorescence fluctuations. Finally, dideoxyUTP replaces dideoxyTTP in the nucleotide mix, making all products generated specifically susceptible to uracil-N-glycosylase (UNG). This provides the optional capability to treat PCR reagents with AmpErase UNG (PE Biosystems #N808–0096) to eliminate carryover contamination. As residual UNG itself can subsequently cause product degradation, and the wells containing amplified products are not routinely opened anyway, to date it has been unnecessary to employ this option. A typical run in the GeneAmp 5700 utilizing a 96-well plate format is prepared as follows:

a. Prepare each well to contain three-embryo equivalents of cDNA (usually, 2 μl of 100 μl cDNA reaction), 5 to 7.5 picomoles of each primer (2–3 μl, premixed at 2.5 pm/μl of both forward and reverse primer), 12.5 μl of 2x PCR Master Mix. Add enough water to bring the total volume to 25 μl.

b. Seal the wells of the entire plate using an adhesive optical transparent cover (PE Biosystems #4313663). The contents of the plate are mixed by inverting it several times.

c. The contents are concentrated into the well bottoms by centrifuging the plate at 1000*g* for 1 min in a 96-well-plate centrifuge (SIGMA 4–15C; Sigma Laborzentrifugen Gmbh, available through Qiagen as #81010 and #81031).

d. The plate is then transferred to the GeneAmp 5700, which has been programmed for the desired parameters.

Additional Comments: Once the raw output of the RTQ-PCR is obtained, the first task is to normalize all samples against an internal standard. It is optimal to pick a gene that is expressed at relatively constant levels throughout embryonic development. It is also important to demonstrate that the expression of this gene is unaffected by any experimental perturbations that might have been carried out on the embryos. The level of ubiquitin message has routinely been used for normalizing samples with satisfactory results (Ransick *et al.*, 2002). Ubiquitin message level is relatively abundant, nearly constant through 24 h into development, and increases by less than a factor of two through 48 h into development (Nemer *et al.*, 1991). Alternatives that have been used for normalization include ISP-1 (Calzone *et al.*, 1988) and 18S ribosomal mRNA levels, although many other mRNAs could serve as the internal standard with which to normalize cDNA samples—as long they fit the criteria already described.

After normalization, the RTQ-PCR results can be used to derive either absolute or relative quantitation of the expression level(s) of a particular gene or set of genes. In order to estimate expression in terms of molecules per embryo, it is necessary to separately generate a standard curve for each template amplicon against which the RTQ-PCR results from the unknown sample can be plotted using standard methods. Keep in mind, however, that the efficiency of extraction and conversion of mRNA to cDNA is typically only in the 50% range, and thus expression values obtained by this method will typically be lower than actual per-molecule levels in the cells.

A much more routine use of the RTQ-PCR output is the relative quantitation of gene expression levels in two differentially treated samples. Typically, for example, in a single 96-well RTQ-PCR run, the level of expression of a dozen different marker genes can be assayed in both the cDNAs generated from embryos injected with a MASO directed against a specific target and in the control cDNAs generated from embryos injected with a nonspecific MASO. The difference in cycle number, if any, between the two samples with respect to when amplification of the template amplicon reaches a set threshold provides a relative quantitation of the difference in gene expression levels under the different conditions. This method has proven to be extremely sensitive in providing valuable information pertaining to gene regulatory networks. We currently are choosing from among 75 different gene-specific primer sets in designing our RTQ-PCR assays—with the number of genes that can be assayed steadily increasing. See Chapter 32 in this volume by Oliveri and Davidson for more detailed specific examples that apply the RTQ-PCR method described here in a gene regulatory network analysis.

Acknowledgments

I thank L. Angerer for contributing the section on isotope-labeled probes. Also, I thank T. Minokawa and A. Wikramanayake for providing details from their unpublished work. This work was supported by NIH Grant HD-37105 and by a Grant from the Office of Science (BER), U.S. Department of Energy, No. DE-FG02-03ER63584.

References

Angerer, L. M., Cox, K. H., and Angerer, R. C. (1987). Demonstration of tissue-specific gene expression by *in situ* hybridization. *In* "Methods in Enzymology, Guide to Molecular Cloning Techniques" (S. Berger and A. Kimmel, eds.), Vol. 152, pp. 649–661.

Angerer, L. M., and Angerer, R. C. (1989). *In situ* hybridization with ^{35}S-labeled RNA probes. *Biotech. Update* **4,** 1–6.

Angerer, L. M., Dolecki, G. J., Gagnon, M. L., Lum, R., Yang, Q., Humphreys, T., and Angerer, R. C. (1989). Progressively restricted expression of a homeo box gene within aboral ectoderm of developing sea urchin embryos. *Genes Dev.* **3,** 370–383.

Angerer, L. M., and Angerer, R. C. (1991). Localization of mRNAs by *in situ* hybridization. *"Methods in Cell Biology"* **35,** 37–71.

Angerer, L. M., and Angerer, R. C. (1992). *In situ* hybridization to cellular RNA with radiolabeled RNA probes. *In* "*In Situ* Hybridization: A Practical Approach" (D. G. Wilkinson, ed.), pp. 15–32. IRL Press at Oxford University Press, New York.

Angerer, L. M., Oleksyn, D. W., Levine, A., Li, X., Klein, W. H., and Angerer, R. C. (2001). Sea urchin goosecoid function links fate specification along the animal–vegetal and oral–aboral embryonic axes. *Development* **128,** 4393–4404.

Arenas-Mena, C., Cameron, A. R., and Davidson, E. H. (2000). Spatial expression of *Hox* cluster genes in the ontogeny of a sea urchin. *Development* **127,** 4631–4643.

Calzone, F. J., Lee, J. J., Le, N., Britten, R. J., and Davidson, E. H. (1988). A long, nontranslatable poly(A) RNA stored in the egg of the sea urchin *Strongylocentrotus purpuratus. Genes Dev.* **2,** 305–318.

Chomczynski, P., and Sacchi, N. (1987). Single-step method of RNA isolation by acid guanidinium thiocynate-phenol-chloroform extraction. *Anal. Biochem.* **162,** 156–159.

Cloney, R. A., and Florey, E. (1968). Ultrastructure of cephalopod chromatophore organs. *Z Zellforsch Mikrosk Anat.* **89,** 250–280.

Cox, K. H., Angerer, L. M., Lee, J. J., Davidson, E. H., and Angerer, R. C. (1984a). Cell lineage-specific programs of expression of multiple actin genes during sea urchin embryogencsis. *J. Mol. Biol.* **188,** 159–172.

Cox, K. H., DeLeon, D. V., Angerer, L. M., and Angerer, R. C. (1984b). Detection of mRNAs by *in situ* hybridization with asymmetric RNA probes. *Dev. Biol.* **101,** 485–502.

Fang, H., and Brandhorst, B. P. (1996). Expression of the actin gene family in embryos of the sea urchin *Lytechinus pictus. Dev. Biol.* **173,** 306–317.

Harkey, M. A., Whiteley, H. R., and Whiteley, A. H. (1992). Differential expression of the msp130 gene among skeletal lineage cells in the sea urchin embryo: A three-dimensional *in situ* hybridization analysis. *Mech. Dev.* **37,** 173–184.

Hauptmann, G., and Gerster, T. (1996). Multicolor whole-mount *in situ* hybridization to *Drosophila* embryos. *Dev. Genes Evol.* **206,** 292–295.

Hauptmann, G. (2001). One-, two-, and three-color whole-mount *in situ* hybridization to *Drosophila* embryos. *Methods* **23**(4), 359–372.

Higuchi, R., Dollinger, G., Walsh, P. S., and Griffith, R. (1992). Simultaneous amplification and detection of specific DNA sequences. *Biotechnology* **10,** 413–417.

Higuchi, R., Fockler, C., Dollinger, G., and Watson, R. (1993). Kinetic PCR analysis: Real-time monitoring of DNA amplification reactions. *Biotechnology* **11,** 1026–1030.

Hinman, V., Nguycn, A., and Davidson, E. H. (2003). Expression and function of starfish Otx ortholog, AmOtx: A conserved role for Otx proteins in endoderm development that predates divergence of the eleutheroza. *Mech. Dev.* **120,** 1165–1176.

Holland, P. M., Abramson, R. D., Watson, R., and Gelfand, D. H. (1991). Detection of specific polymerase chain reaction product by utilizing the 5′—3′ exonuclease activity of *Thermus aquaticus* DNA polymerase. *Proc. Natl. Acad. Sci. USA* **88,** 7276–7280.

Howard, E. W., Newman, L. A., Oleksyn, D. W., Angerer, L. M., and Angerer, R. C. (2001). SpKrl: A direct target of b-catenin regulation required for endoderm differentiation in sea urchin embryos. *Development* **128,** 365–375.

Lee, L. G., Connell, C. R., and Bloch, W. (1993). Allelic discrimination by nick-translation PCR with fluorogenic probes. *Nucleic Acids Res.* **21,** 3761–3766.

Lepage, T., Sardet, C., and Gache, C. (1992a). Spatial expression of the hatching enzyme gene in the sea urchin embryo. *Dev. Biol.* **150,** 23–32.

Lepage, T., Ghiglione, C., and Gache, C. (1992b). Spatial and temporal expression pattern during sea urchin embryogenesis of a gene coding for a protease homologous to the human protein BMP-1 and to the product of the Drosophila dorsal–ventral patterning gene tolloid. *Development* **114,** 147–163.

Livak, K. J., Flood, S. J. A., Marmaro, J., Giusti, W., and Deetz, K. (1995). Oligonucleotides with fluorescent dyes at opposite ends provide a quenched probe system useful for detecting PCR product and nucleic acid hybridization. *PCR Methods and Applications* **4,** 357–362.

Long, S., and Rebagliati, M. (2002). Sensitive two-color whole-mount *in situ* hybridizations using digoxygenin- and dinitrophenol-labeled RNA probes. *BioTechniques* **32,** 494–500.

MacDonald, R. J., Swift, G. H., Przybyla, A. E., and Chirgwin, J. M. (1987). Isolation of RNA using guanidinium. *In* "Methods in Enzymology, Guide to Molecular Cloning Techniques" (S. Berger and A. Kimmel, eds.), Vol. 152, pp. 219–227.

Mehra, M. (1996). RNA isolation from cells and tissues. *In* "A Laboratory Guide to RNA: Isolation, Analysis and Synthesis" (P. Krieg, ed.), pp. 1–20. Wiley-Liss.

Minokawa, T., Rast, J. P., Arenas-Mena, C., Franco, C. B., and Davidson, E. H. (2004). Expression patterns of four different regulatory genes that function during sea urchin development. *Gene Expr. Patterns* **4,** 449–456.

Nemer, M., Rondinelli, E., Infante, D., and Infante, A. A. (1991). Polyubiquitin RNA characteristics and conditional induction in sea urchin embroys. *Dev. Biol.* **145,** 255–265.

Ransick, A., Ernst, S., Britten, R. J., and Davidson, E. H. (1993). Whole-mount *in situ* hybridization shows Endo 16 to be a marker for the vegetal plate territory in sea urchin embryos. *Mech. Dev.* **42,** 117–124.

Ransick, A., and Davidson, E. H. (1995). Micromeres are required for normal vegetal plate specification in sea urchin embryos. *Development* **121,** 3215–3222.

Ransick, A., and Davidson, E. H. (1998). Late specification of Veg1 lineages to endodermal fate in the sea urchin embryo. *Dev. Biol.* **195,** 38–48.

Ransick, A., Rast, J. P., Minokawa, T., Calestani, C., and Davidson, E. H. (2002). New early zygotic regulators expressed in endomesoderm of sea urchin embryos discovered by differential array hybridization. *Dev. Biol.* **246,** 132–147.

Rast, J. P., Arnone, G., Calestani, C., Liri, C. B., Ransick, A., and Davidson, E. H. (2000). Recovery of developmentally defined gene sets from high-density cDNA macroarrays. *Dev. Biol.* **228,** 270–286.

Tautz, D., and Pfeifle, C. (1989). A non-radioactive *in situ* hybridization method for the localization of specific RNAs in *Drosophila* embryos reveals translational control of the segmentation gene hunchback. *Chromosoma* **98,** 81–85.

Wei, Z., Angerer, L. M., and Angerer, R. C. (1999). Spatially regulated SpEts4 transcription factor activity along the sea urchin embryo animal–vegetal axis. *Development* **126,** 1729–1737.

Wilkie, G. S., and Davis, I. (1998). Visualizing mRNA by *in situ* hybridization using high resolution and sensitive tyramide signal amplification. Elsevier Trends Journals,Technical Tips Online: T01458.

CHAPTER 25

Using Reporter Genes to Study *cis*-Regulatory Elements

Maria I. Arnone,* Ivan J. Dmochowski,† and Christian Gache‡

*Stazione Zoologica Anton Dohrn
Villa Comunale
80121 Napoli, Italy

†Department of Chemistry
University of Pennsylvania
Philadelphia, Pennsylvania 19104

‡CNRS-Université Pierre et Marie Curie (Paris VI)
Observatoire Océanologique
06230 Villefranche-sur-Mer, France

0091-679X/04 $35.00

Overview

Chapter 25 highlights important practical aspects of studying *cis*-regulatory gene elements and describes four reporter gene systems: chloramphenicol acetyltransferase (CAT), green fluorescent protein (GFP), luciferase (*luc*), and β-galactosidase (*lacZ*). These assays are currently of widespread use in studies of gene regulation in invertebrate deuterostome embryogenesis. This chapter provides protocols and analysis of each method to guide the researcher in choosing an assay with appropriate spatial and temporal resolution, as well as quantitation.

I. Introduction

Reporter genes have been largely used in cell culture and embryos to study regulatory elements that are important for gene expression during development. These reporter genes are often of prokaryotic origin and encode products not normally associated with eukaryotic cells and tissues. Some of these gene products can only be detected in cell and tissue homogenates or *in situ* at the single cell level, whereas some of them offer both possibilities. For two decades, sea urchin embryos have served as a model system to study gene regulation during development. Many experimental advantages render this embryo one of the most suitable models for functional analysis of *cis*-regulatory elements. The methods for spatial localization, and temporal and quantitative measurements of expression of reporter genes in the sea urchin embryo will be the subject of this chapter.

To perform *cis*-regulatory analysis, the DNA fragments of interest are inserted into an expression vector to drive a reporter gene. After injection into eggs, the regulatory activity of these constructs is assessed by measuring the quantitative output of the reporter gene and its spatial pattern of expression at different stages of development. Depending on the specific questions addressed, some reporters are more suitable for analyzing spatial expression, and some allow more quantitative measurements. Particular care must be taken in the choice of the reporter gene when studying temporal regulation, such as turning gene expression on and off. In these cases, reporters giving rise to unstable products are preferred. A special requirement is when reporter gene expression is used as a marker for clonal incorporation of exogenous DNA, e.g., for selecting embryos on the basis of the domain of transgene integration. In these cases, it is advantageous to visualize reporter gene expression nondestructively in living embryos to allow further analysis. In the following sections, the advantages and disadvantages of each reporter will be discussed.

A. Gene Transfer in the Sea Urchin Embryo

The methodology of gene transfer in the sea urchin by microinjection of plasmid DNA into egg cytoplasm was developed almost 20 years ago. The fate of the exogenous DNA has been well characterized in *Strongylocentrotus purpuratus* (for a review, see Arnone *et al.*, 1997). The injected DNA is concatenated in the egg cytoplasm within minutes of injection and the concatenates are stably incorporated, usually in a single nucleus, at 2nd, 3rd, or 4th cleavage, leading invariably to a mosaic pattern of incorporation (Flytzanis *et al.*, 1985; Franks *et al.*, 1990; Hough-Evans *et al.*, 1988; Livant *et al.*, 1991; McMahon *et al.*, 1984, 1985). The incorporation site is entirely random with respect to cell lineage and cell fate (Hough-Evans *et al.*, 1987, 1988). The extent of mosaicism can be reduced by multiple injections (Livant *et al.*, 1991), but this is an inconvenient procedure in applications where large numbers of experimental embryos are required, as is the case when performing *cis*-regulatory analysis. Only linear fragments of DNA are correctly incorporated and then amplified during embryo development (McMahon *et al.*, 1985). Expression vectors must therefore be linearized prior to microinjection. It has also been demonstrated that the presence of carrier DNA in the injection solution increases the efficiency of expression of the transgene (possibly due to transcriptional interference in the absence of carrier; Franks *et al.*, 1990). Carrier DNA can be prepared by enzymatic digestion of DNA purified from sperm, size selected to have an average length of 5 to 10 kb (as in the case of *HindIII*-digested *S. purpuratus* DNA). Optimal conditions of embryo survival, incorporation, and expression of transgenes occur when 1500 to 2000 molecules of DNA are injected together with a 5-fold mass excess of carrier DNA in a 2 pl volume of 0.12 M KCl, with or without 20% glycerol. Following is the protocol for a typical injection solution calculated for a 5000-bp plasmid.

Injection solution
For 20 μl final solution mix:

2 μl of 50 ng/μl linearized plasmid DNA
1 μl of 500 ng/μl carrier DNA
2.4 μl of 1 M KCl
14.6 μl of H_2O (or 4.6 μl H_2O plus 10 μl 40% glycerol)

Injection solutions are preferentially prepared the day they are used. Caution should be taken to avoid formation of precipitates during microinjection. To this goal, all solutions except DNA must be Millipore-filtered (0.22 μm) before use and injection solutions must be spun for at least 15 min in a microcentrifuge at maximum speed immediately before filling the needle. Embryos are extremely sensitive to traces of phenol in the injection solution. It is recommended to always check by UV spectrophotometry the quality of DNA solutions purified by standard phenol extraction followed by ethanol precipitation. To avoid phenol contamination, a double chloroform extraction prior to precipitation is therefore advisable. A full description of the techniques for microinjection into sea urchin egg cytoplasm is presented in Chapter 10 of this volume.

B. Specific Examples

Following is the description of three general approaches that can be used to perform *cis*-regulatory analysis:

1. Study of the *cis*-Regulative System of a Given Gene

The most widely used approach is to fuse a large (2–10 kb long) upstream DNA fragment of the gene of interest, including transcription and translation start sites, in frame with the chosen reporter gene. Once the basic construct is generated, and the spatial and temporal accuracy of expression of the reporter gene with respect to the endogenous one is assessed, mutated forms can be obtained by nested deletions and/or site-specific mutagenesis. The expression constructs containing these variants of the natural regulatory DNA sequence are then compared to the basic construct. This method has so far been exploited in a number of detailed functional *cis*-regulatory studies on developmentally regulated genes of the sea urchin embryo (Arnone *et al.*, 1998; Gan and Klein, 1993; Ghiglione *et al.*, 1997; Kirchhamer and Davidson, 1996; Kozlowski *et al.*, 1991; Makabe *et al.*, 1995; Mao *et al.*, 1994; Martin *et al.*, 2001; Palla *et al.*, 1999; Wei *et al.*, 1995; Xu *et al.*, 1996; Yuh and Davidson, 1996; Yuh *et al.*, 1996).

2. Study of Patchy Regulative Elements of DNA and Enhancer/Promoter-Trap Approach

A different approach can be taken when studying *cis*-regulatory elements of DNA that are dispersed within a large expanse of genomic DNA. *Cis*-regulatory systems can be very complex and extended: control sequences may occur anywhere within tens of kilobases of the transcription start site(s) of the gene, downstream, or within its introns. In such cases, a more systematic, high throughput approach to study *cis*-regulatory elements is required. A possibility is to test and scan for spatial expression of either mapped genomic DNA fragments or sufficient randomly generated fragments to provide coverage of the region flanking the gene of interest, cloned into an appropriate expression vector. In order to narrow the search, orthologous regions of genomic DNA belonging to two species from the same phylum can be computationally compared to detect all patches of sequence that are conserved and thus may contain *cis*-regulatory elements (Yuh *et al.*, 2002). On the other hand, a completely random approach can be used in searches of *cis*-regulatory elements within the genome using the enhancer/promoter-trap methodology. In all these cases, the DNA elements of interest must be cloned in an expression vector equipped with a promiscuous basal promoter, which can service any *cis*-regulatory module and work in any domain of the embryo, fused to a reporter gene. Two expression vectors with these features have been characterized thus far. Both of them carry the *Endo 16* basal promoter (Yuh and Davidson, 1996), but differ in the reporter gene used: these are the *Ep-GFPII* and *CE-CAT* expression constructs, whose maps are presented in Fig. 1. These reporter constructs are described by C.H. Yuh and colleagues for the study of the regulatory system of the *otx* gene (Yuh *et al.*, 2002), and by R. A. Cameron and colleagues for a random search of *cis*-regulatory elements using an enhancer/promoter-trap approach (Cameron *et al.*, 2004).

3. Use of Reporter Constructs as Markers for Clonal Incorporation

As observed in many experimental animals, DNA incorporation in transgenic sea urchin embryos occurs in a mosaic fashion (see Section I.A). This is detrimental for *cis*-regulatory analysis because it increases the number of transgenic embryos needed to delimit precisely the domain in which the transgene construct is expressed. For this and several other applications, it is useful to identify the clone of embryonic cells in which a trangene has been incorporated. It is then possible to localize the transgenic domain with respect to the embryonic structures and the normal domain of expression. This facilitates assessment of the spatial fidelity and allows sorting of embryos for further experiments. A second reporter gene can help identify the clone, since it has been demonstrated that co-injected transgene and marker constructs are incorporated into the same clone of cells (Arnone *et al.*, 1997). The expression marker should be equipped with a strong, broadly active *cis*-regulatory element that drives a reporter gene, preferably detectable in living embryos. Marker expression constructs that carry GFP as the reporter gene have been characterized in the sea urchin embryo. These are *H2b-GFP* (Arnone *et al.*, 1997) and *HE-GFP* (Bogarad *et al.*, 1998), which contain a *cis*-regulatory element derived from the *SpH2b* early histone gene (Zhao *et al.*, 1990) and the *SpHE* hatching enzyme gene (Wei *et al.*, 1995), respectively. A map of *HE-GFP* is reported in Fig. 1. GFP permits nondestructive visualization in a living embryo, thus allowing presorting of the embryos bearing the transgene in the domain(s) of interest prior to further experiments (Arnone *et al.*, 1997; Bogarad *et al.*, 1998; Rast *et al.*, 2002). An example of clonal incorporation tested at gastrula stage using the *HE-GFP* expression vector is reported in panel F of Fig. 2. *SpHE* drives expression in all domains of the embryo except skeletogenic mesenchyme.

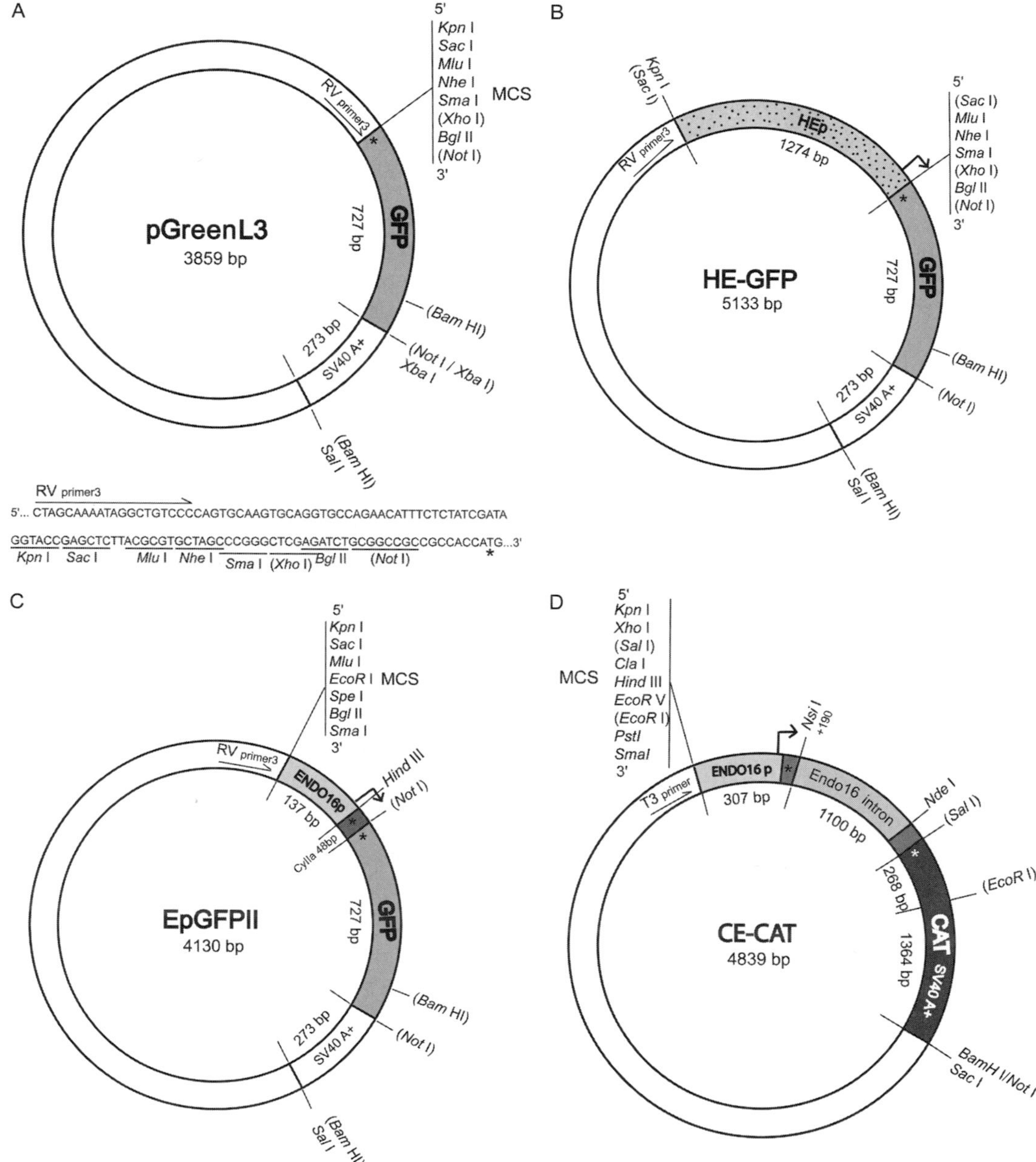

Fig. 1 Maps of some expression vectors of use in sea urchin. (A) *pGreenL3* (Arnone *et al.*, 1997) is a derivative of Green Lantern (Gibco) and pCAT3 (Promega). The sequence upstream of the GFP start codon (asterisk) is reported in the insert. (B) *HE-GFP* (Bogarad *et al.*, 1998) is a derivative of *pGreenL3*, which contains a regulatory element, from −1255 to +19, of the *SpHE* gene (Wei *et al.*, 1995). (C) *EpGFPII* (Cameron *et al.*, 2004) is a derivative of *EpGFP* (Arnone *et al.*, 1997) which contains the basal promoter of *Endo16*, from −117 to +20 (Yuh and Davidson, 1996) and the Kozak

II. Using Chloramphenicol Acetyltransferase (CAT) Reporter Gene

A. General Explanation

The chloramphenicol acetyltransferase (CAT) gene is one of the most widely used expression reporter genes in eukaryotes. It encodes a bacterial enzyme that transfers acetyl groups from acetyl-CoA to chloramphenicol (acetyl-CoA: chloramphenicol 3-*O*-acetyltransferase, EC 2.3.1.28). There are several commercially available expression vectors carrying the CAT reporter gene. They typically bear a multiple cloning site upstream of CAT coding sequence for insertion of *cis*-regulatory elements; a canonical Kozak sequence at the 5′ end of the CAT gene for efficient translation initiation; a SV40 polyadenylation signal sequence at the 3′ end of CAT coding sequence for efficient transcription termination. A commonly used vector is pCAT-3 from Promega, which carries the SV40 late poly(A) signal sequence at the usual position and a synthetic poly(A) and transcriptional pause site, placed upstream of the multiple cloning site to terminate spurious transcription. CAT reporter expression can be detected in sea urchin embryos, at the mRNA or protein level, using different methods as described in the following sections.

1. CAT mRNA Detection by Whole-Mount *In Situ* Hybridization (WMISH)

Localization of CAT mRNA expression in transgenic sea urchin embryos is conveniently detected by whole-mount *in situ* hybridization (WMISH) using a digoxigenin labeled CAT antisense riboprobe transcribed from a 268 bp long *Sal* I–*EcoR* I fragment of CAT coding sequence (see map of CE-CAT in Fig. 1) that has been cloned in pBS-SK for transcription (A. Ransick, unpublished). The principles, protocols, and tips for the WMISH procedure are described in Chapter 24. By this technique, single cells can be easily detected and counted in the clone of expressing cells (Fig. 2, panel A). This method is effective in showing expression driven by exogenous *cis*-regulatory elements in every territory of the sea urchin embryo (Arnone *et al.*, 1998; Coffman *et al.*, 1997; Kirchhamer *et al.*, 1996; Makabe *et al.*, 1995; Wang *et al.*, 1995; Yuh and Davidson, 1996; Yuh *et al.*, 1994, 2001b, 2002).

---

sequence plus first 14 codons of the *CyIIa* gene (Arnone *et al.*, 1998) in frame with GFP start codon. Plasmid sequences are from *pGreenL3*. The complete sequence of this vector is available at http:// sugp.caltech. edu/resources/methods/epgfp.py. (D) *CE-CAT* (Yuh *et al.*, 2002) contains a fragment of the *Endo 16* gene from −117 to +1283, which includes a 1100 bp long intron that can be excised using the *Nsi* I and *Nde* I sites. Plasmid sequence is from pMOB (Strathmann *et al.*, 1991). Bent arrows and asterisks indicate transcription and translation start sites, respectively. Restriction sites in parenthesis are not unique in the vector sequence. Fragment size is not proportional to fragment length (bp). MCS, Multiple Cloning Site.

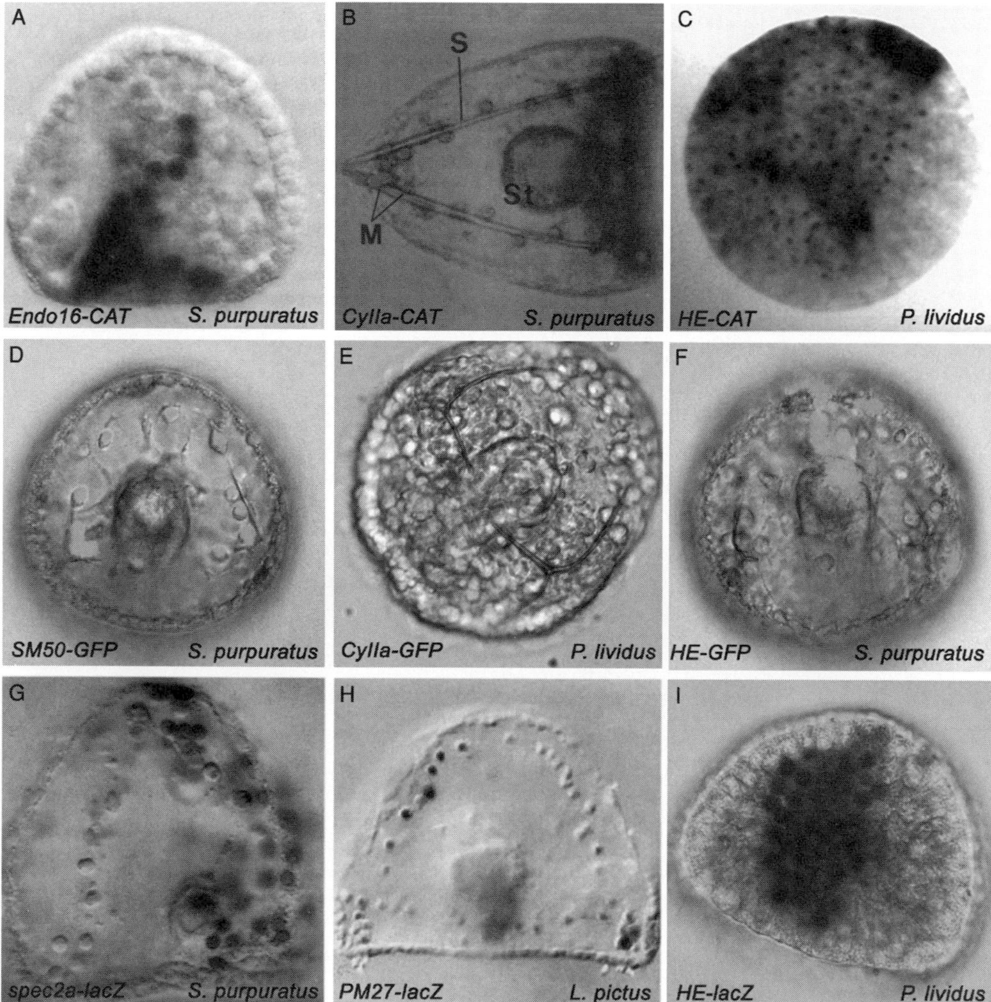

Fig. 2 Visualization of expression of various reporter genes in transgenic sea urchin embryos. The construct and the sea urchin species injected are indicated at the bottom left and right corner of each panel, respectively. (A–C) Expression of the CAT reporter gene detected with different methods: (A) Whole mount *in situ* hybridization to detect CAT mRNA in the gut of a *S. purpuratus* gastrula driven by the *Endo16* regulatory region [adapted with permission from Yuh *et al.* (1994)]. (B) *CyIIa-CAT* expression detected in skeletogenic cells and stomach of a *S. purpuratus* pluteus using a rabbit polyclonal anti-CAT primary antibody and (FITC)-labeled goat anti-rabbit secondary antibody. Composite false color video images of bright field (blue) and fluorescent (red) exposures are shown superimposed [adapted with permission from Zeller *et al.* (1992)]. (C) *P. lividus* blastula showing ectodermal expression of *HE-CAT* visualized by immunolabeling (large spots; CAT protein diffuses in the cytoplasm of the expressing cell) using a sheep anti-CAT polyclonal Ig coupled to digoxigenin and a sheep alkaline-phosphatase-conjugated anti-digoxigenin Ig. The domain of expression of the endogenous HE gene in this embryo has been simultaneously visualized by immunolabeling of the gene product (small single spot/cell; the hatching enzyme concentrates in a subcellular compartment

2. CAT Protein Localization by Immunostaining

An alternative to CAT mRNA *in situ* hybridization is CAT protein localization by immunostaining. There are several commercially available CAT antibodies, some conjugated, which allow multiple labeling for simultaneous detection of exogenous and endogenous expression (example shown in panel C of Fig. 2, from Ghiglione *et al.*, 1997). The use of fluorescently labeled secondary antibodies permits detailed visualization of exogenous reporter gene expression in the intact three-dimensional structure by confocal microscopy, as shown in panel B of Fig. 2 (Zeller *et al.*, 1992). For the immunostaining procedure, see Chapter 15 of this volume.

3. Enzymatic Assay of CAT Activity

The quantitative output of gene expression driven by a *cis*-regulatory element of DNA fused to the CAT reporter gene can be biochemically measured taking advantage of the enzymatic properties of its gene product. A CAT unit is defined as the amount of enzyme that converts 1.0 nmol of chloramphenicol and acetyl-CoA to chloramphenicol 3-acetate and CoA per min at pH 7.8, 25 °C. The CAT activity assay is described for transgenic sea urchin embryos in Section II.B.

B. Detailed Method for Enzymatic Assay of CAT Activity

The method described here summarizes previous studies employing quantitative measurements of CAT activity in transgenic sea urchin embryos (McMahon *et al.*, 1984; Yuh and Davidson, 1996).

Materials

Tris-HCl, pH 8.0 (0.25 M)

D-threo-[dichloroacetyl-1-^{14}C] chloramphenicol (Amersham, 2.15 GBq/mmol)

Acetyl coenzyme A (Sigma) (4 mM)

---

between the nucleus and the apical surface of the cell) using a rabbit anti-HE primary antibody and a goat alkaline-phosphatase-conjugated anti-(rabbit IgG) Ig [adapted with permission from Ghiglione *et al.* (1997)]. (D–F) GFP expression in different domains of gastrula stage embryos. GFP fluorescence is shown in green in these digital false color images superimposed to bright-field (gray) exposures obtained with DIC. The images in (D) and (F) are *S. purpuratus* specimen courtesy of P. Oliveri and show expression of the *SM 50* gene in skeletogenic cells and of the *SpHE* gene in ectoderm and gut cells, respectively. The *P. lividus* shown in (E) expresses GFP in skeletogenic cells driven by a regulatory element of the *CyIIa* gene of *S. purpuratus* (M. I. Arnone, unpublished). (G–I) Expression of the *lacZ* gene driven by the promoters of the indicated genes: (G) *Spec2a* in aboral ectoderm cells of *S. purpuratus* embryo at the prism stage [adapted with permission from Gan *et al.* (1990)]. (H) *PM27* in PMCs of *L. pictus* at the prism stage [adapted with permission from Klueg *et al.* (1997)]. (I) *HE* in ectoderm cells of *P. lividus* embryo at the gastrula stage (C. Ghiglione and C. Gache, unpublished). (See Color Insert.)

CAT enzyme standard (Pharmacia)
Ethyl acetate, chloroform, methanol
TLC plate (20×20 silica gel on polyester; Sigma)

1. Embryo Collection and Lysis

a. Collect samples of 50 to 150 embryos at each desired stage of development and spin for 5 min at 1500 rpm in Eppendorf tubes. Quickly remove seawater from the pellet to avoid detaching embryos from the bottom of the tube.
b. Resuspend embryo pellets in 100 μl of 0.25 M Tris-HCl (pH 8.0).
c. Freeze–thaw samples at least three times by cycling from $-70\,^{\circ}$C for 1 min to $37\,^{\circ}$C for 2 min.
d. Incubate samples at $65\,^{\circ}$C for 10 min to destroy elements that inhibit CAT enzyme activity.

2. CAT Enzymatic Assay

a. To each tube containing 100 μl of embryo lysate add 1 μl (0.5 μCi) of D-threo-[dichloroacetyl-1-^{14}C] chloramphenicol (2.15 GBq/mmol), 20 μl of 4 mM acetyl coenzyme A, and 29 μl of 0.25 M Tris-HCl (pH 8.0).
b. Incubate sample reactions at $37\,^{\circ}$C for 12 h.
c. Add 1 ml of ethyl acetate to each tube and extract the organic phase by vortexing for 30 s, followed by centrifugation for 5 min.
d. Transfer the organic phase to new Eppendorf tubes and dry them by speed vacuum.
e. Resuspend the dried samples in 20 μl of ethyl acetate and quickly spot them on a TLC plate for ascending chromatography.
f. Perform chromatography in a closed chamber saturated with 95% chloroform and 5% methanol until the solvent runs 3/4 of the TLC plate.
g. Air dry TLC plates and expose to autoradiographic film overnight.
h. After developing the film, using it as a reference to locate radioactive spots, cut equal areas of the TLC that contain the acetylated products and the nonacetylated substrate and count them separately in a scintillation counter.

3. Standard Curve and Units Conversion

a. To construct a standard curve, prepare 5 Eppendorf tubes containing 0.5, 1, 2, 4, and 8×10^{-3} units of bacterial CAT enzyme standard in total volumes of 100 μl Tris buffer (0.25 M, pH 8.0).

b. Process reactions of standards together with samples as described in Section II.B.2.
c. Calculate percentage of conversion (%Con) for both standard and sample reactions using the formula:

$$\%\text{Con} = 100 \cdot \text{cpm}_{(\text{acetylated products})}/\text{cpm}_{(\text{non acetylated substrate})}$$

d. Construct standard curve of $\%\text{Con}_{\text{standard}}$ as a function of CAT enzyme units (U_{standard}). Determine units of CAT for each sample (U_{sample}) from $\%\text{Con}_{\text{sample}}$ values by reference to the standard curve using linear least squares analysis.
e. Convert units of CAT enzyme activity to CAT enzyme molecules per embryo using the following formula:

$$\text{CAT molecules/embryo} = U_{\text{sample}} \cdot 2.6 \times 10^{11} \cdot N_E^{-1}$$

Where N_E is the number of embryos used per reaction and 2.6×10^{11} is a conversion factor calculated from the equation, 2.6×10^{7} molecules of CAT enzyme $= 10^{-4}$ units of activity, estimated by McMahon *et al.* (1984).

C. Advantages and Disadvantages of the CAT Reporter System

Most advantageously, the CAT reporter system offers, at the same time, precise spatial and accurate quantitative measurements of gene expression. The CAT enzymatic assay is sensitive and can detect fewer than 3×10^{6} CAT molecules. This means that 100 injected embryos are sufficient to detect as few as 1000 CAT enzyme molecules per cell expressed by an average clone of 20 to 30 cells per embryo (as calculated from Fig. 2B in Yuh and Davidson, 1996). This allows sophisticated quantitative measurements and comparison of the kinctics of gene expression profiles during development among different regulatory modules (Arnone *et al.*, 1998; Kirchhamer and Davidson, 1996; Kirchhamer *et al.*, 1996; Yuh and Davidson, 1996; Yuh *et al.*, 1996, 1998, 2001a).

The CAT gene reporter system also affords precise temporal measurements at the mRNA level. CAT mRNA is, in fact, relatively unstable in sea urchin embryos (Flytzanis *et al.*, 1987). This allows the shutoff of gene expression to be measured more precisely than using GFP, as tested for the late regulatory module of the *Endo 16* gene (Arnone *et al.*, 1997). It also provides a sensitive indication of cellular expression in a defined time frame (Yuh *et al.*, 2002). Similar to other methods of detecting expression by measuring the mRNA transcribed from a reporter gene, there is little delay in detecting the onset of expression. CAT expression driven by a sea urchin enhancer element has been detected by WMISH as early as 32 cells stage (T. Minokawa and E. H. Davidson, personal communication). The limitations of this reporter system are linked to the WMISH technique (laborious, time-consuming, difficult to apply to three-dimensional structures for visualization of complex spatial patterns of expression).

III. Using Green Fluorescent Protein (GFP) Reporter Gene

A. General Explanation

Since 1997, the Green Fluorescent Protein (GFP) from *Aqueora victoria* jellyfish has been efficiently used as reporter gene in sea urchin embryos and larvae (Arnone *et al.*, 1997, 1998; Bogarad *et al.*, 1998; Consales and Arnone, 2002; Dmochowski *et al.*, 2002; Haruguchi *et al.*, 2002; Kurita *et al.*, 2003; Martin *et al.*, 2001; Rast *et al.*, 2002; Yuh *et al.*, 2001b). The GFP gene encodes a naturally fluorescent protein requiring no substrates for visualization. A map of the plasmid *pGreenL*, created for use in sea urchin embryos (Arnone *et al.*, 1997) and formerly designated as *pGL3-Basic* (Bogarad *et al.*, 1998) is reported in Fig. 1. It bears the Green Lantern (Gibco) version of GFP in which serine at position 65 has been mutated into a threonine (S65T GFP) to enhance fluorescence peaking at 510 nm with a single excitation peak at 490 nm. Some examples of GFP expression in sea urchin embryos and larvae are reported in Fig. 2 (D–F).

B. Methods of GFP Visualization

1. Embryo Preparation and Mounting

Living embryos can be directly mounted in Millipore-filtered seawater (MFSW) under a coverslip for observation by fluorescence microscopy. To prevent embryos from being flattened, spacers are placed between the coverslip and slide prior to mounting. Spacers (0.2–0.3 mm thick) can be prepared using modeling clay (Plasticine) or Blue-Tac (four spots positioned at the four corners of the coverslip) or double-sided tape (two stripes positioned at two parallel sides of the coverslip). Total 35 μl of MFSW are sufficient to fill the space offered by a 18 $\times$ 18 mm coverslip and such spacers. The poor oxygenation offered by these mounting conditions and/or a gentle pressure at the corners of the coverslip are usually enough to immobilize the embryos for proper imaging. When necessary, a drop of methanol can be used as anesthetic to prevent movement of blastulae and gastrulae (plutei are easily immobilized by simple pressure, given their elongated shape). It is not advisable to use high doses of fixatives or to seal the coverslip with nail polish, because most of the chemicals contained in them diminish GFP fluorescence. Slides of mounted embryos must be kept in a moisture chamber at 4 to 10 °C to reduce evaporation and prevent embryos from drying out. Mounted embryos can be stored only up to 12 h in these conditions.

2. Microscopy and Imaging

Depending on the final application, GFP-expressing cells can be visualized under epifluorescence or confocal microscopy, or by using a high magnification stereomicroscope equipped with fluorescence illumination. In the last case, or when using upright microscopes equipped with water immersion objectives, embryos can be directly observed on a petri dish containing a large volume of seawater. GFP fluorescence is typically resistant to photobleaching. Long exposure of gastrula and pluteus-stage embryos to blue light excites pigment cells, and the fluorescence is particularly visible when using GFP filter sets. To determine the exact cells that express the fluorescent protein, the same embryos or embryo sections are imaged under bright-field and fluorescent illumination. The corresponding bright-field and fluorescent images can be overlaid using imaging software.

C. Advantages and Disadvantages of the GFP Reporter System

The GFP reporter system allows nondestructive detection of spatial expression in living embryos and larvae. This is extremely useful in applications that require identification and sorting of embryos by clonal incorporation, as described in Section I.B.3. Another advantage of GFP is that it freely diffuses within cytoplasmic cables, thus revealing connection between cells. This is particularly useful in sea urchin embryos for detection of expression in mesenchyme cells (examples are shown in Fig. 2, D and E). Some diffusion in the skeletogenic syncytium is seen with CAT protein (Zeller *et al.*, 1992), as shown in the pluteus depicted in panel B of Fig. 2, but not with CAT mRNA detected by WMISH. The GFP reporter system is inappropriate for most temporal measurements. The post-translational oxidation of the protein leading to chromophore formation requires time, which interferes with measurements of the onset of transcription. This time delay will vary among organisms and species, as it is dependent on temperature and the concentration of oxygen, among other factors. Because GFP is an extremely stable protein in sea urchin embryos (Arnone *et al.*, 1997), the fluorescence measured at any given time is the sum of all prior expression. This represents a drawback when measuring shutoff of gene expression, but it is useful for detecting the spatial domain of gene expression driven by weak promoters.

IV. Quantitative Imaging of GFP in Living Embryos

A. General Explanation

Advances in biological imaging will continue to facilitate the study of embryogenesis in nonvertebrate deuterostomes. This section focuses, in particular, on recently developed methods for measuring GFP in living sea urchin embryos using quantitative confocal laser scanning microscopy (QCLSM) (Dmochowski *et al.*, 2002). Some of these techniques have also been applied to studies of starfish, zebrafish, and mouse embryos. The protocols described herein form a basis for extending these methods to other nonvertebrate deuterostomes.

GFP imaging allows observations of the same embryo over time, in contrast to CAT, luciferase, and *lacZ* assays and *in situ* hybridizations that rely on fixed specimens and pooled tissue homogenates. In fact, vital imaging of fluorescent reporters has expanded from GFP to encompass many new technologies. Other markers for gene expression include novel fluorescent reporters based on β-lactamase (Raz *et al.*, 1998; Zlokarnik *et al.*, 1998) and GFP color variants (e.g., CFP, YFP, RFP) (Cotlet *et al.*, 2001; Garcia-Parajo *et al.*, 2001; Miyawaki *et al.*, 1997). Fluorescent proteins with shortened lifetimes ($\tau_{1/2} < 1$ h *in vivo*) provide greater temporal resolution (Dantuma *et al.*, 2000), and tissue-specific and inducible promoters allow better spatial resolution. Most recently, GFP mutants have been described whose spectral properties can be modified with a laser *in vivo* to increase the intensity or tune the wavelength of the fluorescence (Ando *et al.*, 2002; Lippincott-Schwartz and Patterson, 2003; Patterson and Lippincott-Schwartz, 2002). Thus, photoactivation can fluorescently label specific cells or regions of interest within an embryo. Other important advances include new hardware and software (Dickinson *et al.*, 2001) for multispectral analysis. These technologies distinguish among numerous fluorescent reporters with overlapping spectral features. Within a few years, it will be possible to monitor the activity of dozens of genes *in vivo* simultaneously, in real-time. But, for spatial and temporal analyses of gene expression patterns, either within the same embryo or across multiple embryos, challenges in *fluorescence quantitation* still remain. Work by Dmochowski and associates addresses these challenges in the sea urchin embryo (Dmochowski *et al.*, 2002) and quantifies *cis*-regulatory activity in endomesoderm specifying genes using GFP as a reporter. This work expands the utility of GFP for developmental biology applications, but also highlights some of the remaining challenges in quantitative microscopy using fluorescent reporters.

A fundamental problem in biological optical imaging is the interaction of light with living samples (Stephens and Allan, 2003). Quantifying GFP within cultured or living cells presents slightly different challenges than with embryos, and can often be accomplished using a fluorescent plate reader (Chen *et al.*, 2003; Daelemans *et al.*, 2001; Hughes *et al.*, 2002; Niswender *et al.*, 1995). In thick biological specimens, refractive index changes within the sample can cause much more severe aberrations (Ferrer-Martinez and Gomez-Foix, 2002; Kam *et al.*, 2001). Such tissues differentially absorb and scatter light in CLSM experiments, perturbing the delivery of photons from a laser to the sample and the collection of resulting fluorescence at the detector (Fig. 3A). Even in typically small, nonvertebrate deuterostome embryos (tens to a few hundred microns in diameter), depth and geometry effects can diminish the observed fluorescence intensity as much as 50-fold (Dmochowski *et al.*, 2002). For many studies of nonvertebrate deuterostomes, it will be important to evaluate and compensate for these losses.

Some quantitative imaging experiments with indicator dyes have employed a ratiometric analysis of fluorescence intensities at two different wavelengths (Terasaki, 2000). This approach can be extended to the use of Texas Red (TR)-dextran for the quantitation of GFP in three-dimensional structures, such as sea urchin embryos. Injected TR-dextran diffuses homogeneously throughout the embryo, and its red fluorescence is spectrally resolvable from GFP. By taking a ratio of the TR intensity at each depth relative to its maximum value, a "depth correction factor" is obtained for each fluorescent voxel. Thus, TR-dextran serves

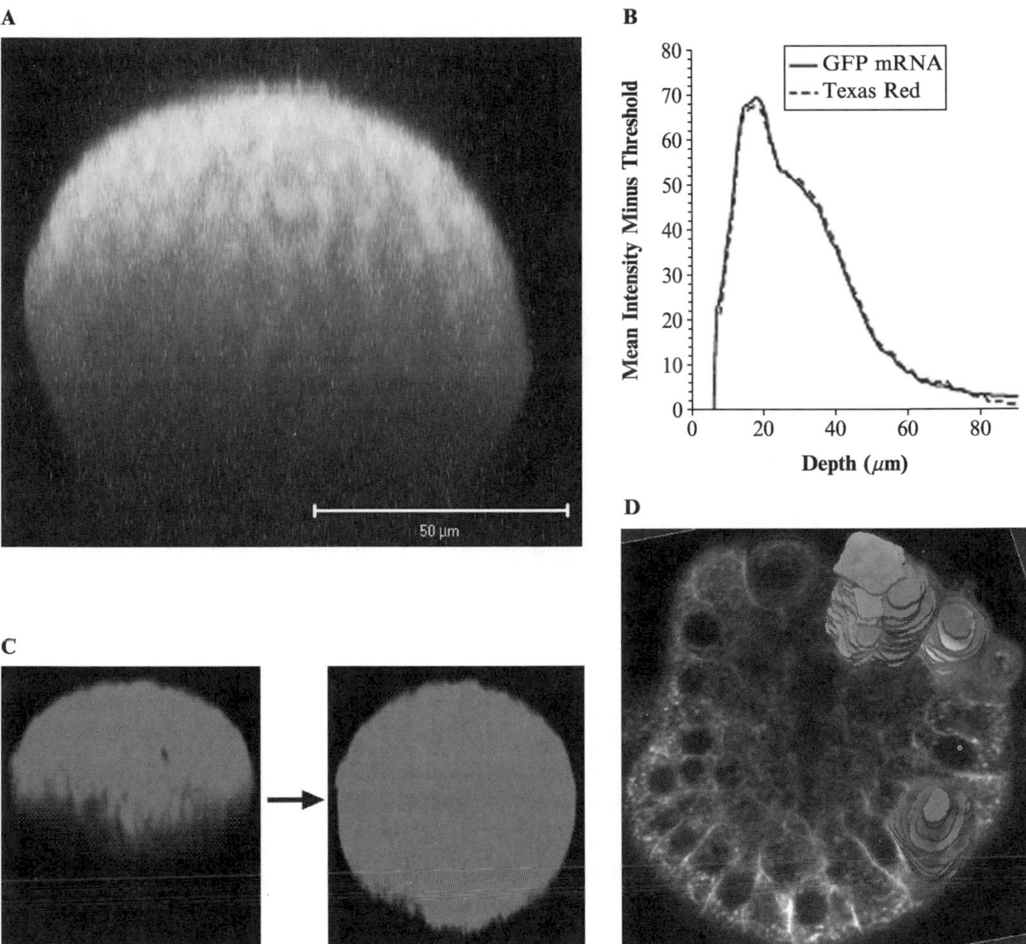

Fig. 3 (A) Maximum intensity projection of confocal laser scanning microscopy (CLSM) fluorescence intensity (TR, red; GFP, green) in a 18 hpf *S. purpuratus* embryo. The embryo is coinjected with TR and *Krox-GFP* mRNA, and uniformly labeled. Image reconstruction shows marked loss of signal with increasing depth (imaging performed from the top.) (B) Mean intensity vs depth for both GFP and TR channels. The mean intensities are virtually superimposable, deviating only slightly at the brightest and dimmest regions of the embryo. This indicates that freely diffusing TR-dextran can serve as an internal fluorescent standard for quantitative measurements of GFP. (C) CLSM 3-dimensional image reconstructions of a *Krox-GFP* (mRNA) injected sea urchin embryo. The image on the left has not been depth corrected. After image processing, the GFP fluorescence (green) is restored to its proper intensity in the bottom half of the embryo (right image). Quantifying fluorescence within cell boundaries in a sea urchin blastula. Boundaries in a single plane (white) are indicated by the membrane dye FM 4-64 (Molecular Probes). The colorful stacks mark the entire boundary of each 3-dimensional cell, and are regions in which GFP has been quantified on a cell-by-cell basis (M. Souren and I. J. Dmochowski, unpublished). (See Color Insert.)

as an internal fluorescent standard and allows the GFP fluorescence to be determined, even in regions of the embryo where signal-to-noise is very low. By lysing the embryos, their fluorescent contents can be measured with a fluorometer. In this way, the corrected *in vivo* fluorescence intensity can be correlated with actual numbers of GFP molecules.

B. Detailed Method

1. Embryo Preparation and Injection

Petri dishes are prepared with a "window" sealed by a glass coverslip (170 μm thick, 25 mm diameter). The coverslip is coated with a poly-L-lysine solution (molecular mass, 150–300 kDa; 50 μg/ml in 10 mM Tris, pH 8.0; 10 min exposure) (Dmochowski *et al.*, 2002). Gametes from *Strongylocentrotus purpuratus* are harvested, and the eggs rowed onto plates and fertilized just before injection (Kirchhamer and Davidson, 1996). Embryos are microinjected with solutions containing a freely diffusing internal fluorescent standard, TR-labeled dextran (neutral; molecular weight, 10,000; Molecular Probes). As an example, the TR is coinjected with the well-characterized hatching enzyme (HE) *cis*-regulatory system (Wei *et al.*, 1995) fused to a "bright" GFP (S65T) reporter gene. For purposes of correlating *in vivo* fluorescence with absolute numbers of GFP molecules, it is also necessary to inject mRNA coding for GFP. (Injected mRNA is translated into GFP in every cell, thereby maximizing the fluorescent signal from the embryo and facilitating subsequent fluorometry). The start codon from the *Krox* gene contains a good Kozak sequence, and can be put in frame with the coding sequence for GFP. Both *HE-GFP* (DNA) and *Krox-GFP* (mRNA) injection solutions should have a total volume of 10 μl, and should contain 0.65 mM TR and 0.12 M KCl (Dmochowski *et al.*, 2002). After injection, the embryos are typically cultured at 15 °C overnight.

2. Collection of *In Vivo* Fluorescence Data

a. *Image Acquisition by CLSM*

To maintain *S. purpuratus* embryos at an optimal culture temperature (10–15 °C) during imaging, the universal stage adaptor can be fitted with an aluminum Peltier cooling device (Fig. 4). The embryos develop normally on this cold stage and can be imaged repeatedly over several hours with no adverse effects. Because *S. purpuratus* embryos typically hatch at 18 hpf, it is easiest to conduct the imaging on 14 to 18 hpf embryos. At this stage, the embryos are still firmly attached by their fertilization membrane to the poly-lysine coated coverslip in the imaging chamber. To image later developmental stages, the embryos can be deciliated through brief exposure to 2× seawater and transferred to the imaging chamber. When kept cold (~10 °C), these embryos can be imaged for 30 min before the cilia regenerate.

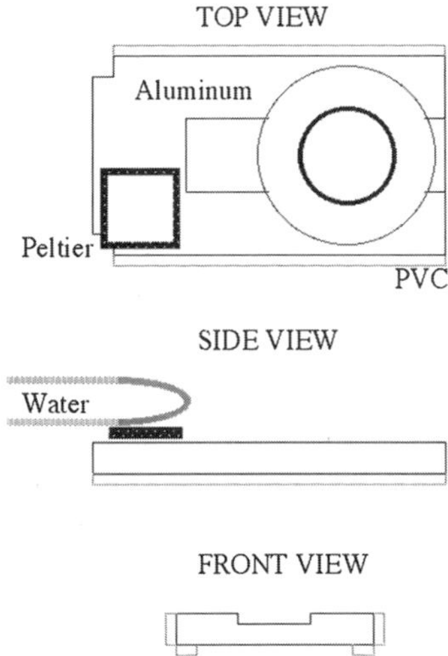

Fig. 4 Home-built aluminum cold stage. The imaging chamber sits in the center hole (top view) and the seawater is kept at 10 to 15 °C by a Peltier device. Circulating water serves as a heat sink (side view), and plastic strips (front view) isolate the cold plate from the microscope stage.

Confocal data sets can be collected by a variety of techniques, including using the Axiovert 100 M inverted microscope configured for CLSM (LSM 5 PASCAL, Zeiss). The embryos (~80 μm diameter) are imaged well using a 40×, 1.2-numerical aperture water immersion lens (Zeiss) with coverslip correction. The pinhole should be maintained at ~100 μm (1 Airy unit), with identical detector gain (typically, 525 V, red channel; 650 V, green channel; 215 V, transmitted light) and offset settings (−0.1, red; −0.1, green; −0.05, transmitted light) during each experiment. Laser power (argon ion, 20% power, 488 nm; helium–neon, 80% power, 543 nm) should be monitored to ensure reproducibility. Using appropriate filters (for example, green, 505–530 nm bandpass filter; red, 605 nm long-pass filter, Chroma Technology, Brattleboro, VT), no crosstalk is observed between channels. *S. purpuratus* embryos can be imaged at intermediate scan speeds (typically, 1.76 msec per voxel, less than 1 sec per image) in 1-μm-thick z sections, for a total of 91 slices.

b. Embryo Lysis and Fluorometry

An embryo lysis procedure removes all of the GFP from within the embryos and preserves the fluorescence intensity of the GFP in solution.

1. Image embryos by CLSM.
2. Transfer 20 embryos in minimal volume (~10 μl) to an Eppendorf tube on ice.

3. Add an equal volume of 2× lysis buffer: 3 mg/ml BSA, 100 mM Tris-HCl (pH 8.0), 300 mM NaCl, 20% glycerol, 1 mM EGTA, 0.2% Triton X-100, and 0.2% Nonidet P-40.
4. Store embryos at −20 °C overnight.
5. Microcentrifuge tubes at 16,000×g at 4 °C to remove cell debris.
6. Transfer 20 μl solution to a 1-cm pathlength, small-volume fluorescence cuvette (WG-16/10F-Q-10-1, Wilmad, Buena, NY).
7. Measure fluorescence using maximum gain (950 V), slit settings (10-nm excitation, 20-nm emission), while exciting at 475 nm and using a 500-nm long-pass filter on a F-4500 spectrophotometer (Hitachi, Tokyo). Integrate fluorescence from 500 to 700 nm using commercial software.
8. Standardize measurements using a fluorescein stock solution (0.1 M NaOH, quantum yield 0.95) and purified S65T GFP standards (quantum yield 0.65).
9. Confirm that the lysis conditions do not adversely affect the fluorescence of purified S65T GFP, and confirm that the embryos are completely lysed.

3. Image Analysis

a. Ascertaining That a Ratiometric Analysis is Suitable for Quantitation

An important first step in the depth correction procedure is determining whether TR-dextran is a valid internal fluorescent standard by which to correct the depth-dependent losses in the GFP channel. For this purpose, it is necessary to show that plots of the average intensity vs depth in each channel can be superimposed (Fig. 3B). De-noised images can be obtained using MATLAB, as will be described, and the mean intensity (excluding noise and stray light) computed for each optical section. These plotted TR "depth profiles" are remarkably reproducible for embryos of a given organism (sea urchin, mouse, starfish, zebrafish) at a given stage of development.

b. Algorithms for Correcting Depth Dependent Losses in Fluorescence

The MATLAB code employs image processing functions for removing noise from the raw confocal data sets. The Strel function removes scattered light of defined size that is noncontiguous with the rest of the embryo. Otsu's Method finds the threshold for the TR signal that maximizes the difference between the histograms of the signal and noise distributions, and provides dynamic thresholding on a slice-by-slice basis. This is important, due to large changes in signal-to-noise throughout the embryo. More details of the image processing are available (Dmochowski *et al.*, 2002).

Correction of the GFP fluorescence involves filtering with a low threshold and multiplication by a correction factor:

$$\frac{\text{Mean TR intensity at the brightest slice}}{\text{Mean TR intensity at any other slice}}$$

This correction procedure produces much more uniformly labeled TR and *Krox-GFP* (mRNA) embryos, and particularly improves the calculation of fluorescence intensity in the deepest third of the embryo, where signal/noise is very low (Fig. 3C).

c. *Correlating In Vivo Fluorescence with Numbers of GFP Molecules*

This exercise is performed in several steps:

1. Use CLSM to image embryos coinjected with TR and *Krox-GFP* (mRNA).
2. Lyse the embryos and quantify the GFP by fluorometry, as described.
3. Use image processing software to correct the observed *in vivo* fluorescence.
4. Divide the integrated corrected *in vivo* fluorescence by the number of molecules per embryo (computed by fluorometry).
5. Apply this factor, number of GFP molecules per corrected fluorescence unit, to *HE-GFP* (or other DNA–injected) embryos. This value is typically 30 GFP molecules per intensity unit.

d. *Quantifying Fluorescence within Cells or Tissues*

The capabilities of the image analysis software are being extended (M. Souren, I. J. Dmochowski, unpublished results) to permit fluorescence quantitation in defined regions of an embryo (Fig. 3D). Cellular boundaries can be distinguished by putting a diffusible membrane-specific fluorescent dye (FM 4-64, Molecular Probes, Eugene, OR) in solution with the embryos. Although the dye is only resident in the membranes for ~30 min, it can be reintroduced multiple times. Like TR, FM 4-64 is spectrally distinguishable from GFP. FM 4-64 also serves as an internal fluorescent standard, although the loading levels and homogeneity achieved by diffusion are lower than those achieved with microinjected TR.

C. Advantages and Disadvantages of Quantitative GFP Analysis

New methods to quantify actual numbers of GFP molecules or other fluorescent reporters within living embryos represent a significant opportunity for developmental studies of nonvertebrate deuterostomes. This work will particularly benefit studies of transgenic embryos, where the gene of interest is expressed in a mosaic fashion. In these cases, as it is with *S. purpuratus*, the orientation of the embryo profoundly affects on the apparent "brightness" of the fluorescent reporter. Quantitative CLSM methods reduce the effects of sample orientation and variation by correcting the GFP intensity in each sample relative to an internal fluorescent standard. In addition, CLSM provides spatial information that is unavailable by many other techniques. Because the TR is contained within each embryo, small variations in injection volume (typically, a factor of two) are normalized. However, greater than 10-fold variations in TR and GFP labeling cause significant problems in the image analysis, since under the current implementation, the same microscope settings are used for every embryo. Future improvements should include a calibration feature and, through a feedback loop, increase the PMT voltage (detector sensitivity) as signal intensity decreases.

Ongoing improvements in image processing algorithms for CLSM and differential interference contrast microscopy (DICM) hold considerable promise for better optical imaging in the near future. New computational approaches can map the refractive index of a sample using DICM. This allows proper assignment of photons to their origin and opens possibilities for studying weakly labeled live specimens (Kam *et al.*, 2001).

However, there are still many problems with quantitative GFP imaging. The software is not widely accessible, and processing times make it impractical for many applications. In addition, the methods remain untested for most organisms, and the algorithms and protocols will need to be adapted for each case. But, it is encouraging that, despite their larger size, it is possible to image mouse, starfish, and zebrafish embryos (diameters of ~100, 200, and 500 microns, respectively) and perform quantitative measurements by the techniques described. The fact that embryos from many invertebrate deuterostomes are more transparent than *S. purpuratus* should facilitate quantitative fluorescence imaging of these species.

V. Using Luciferase (*luc*) Reporter Gene

A. General Explanation

The machinery responsible for the light emission of the firefly *Photinus pyralis* has been diverted into a powerful reporter system that has been used in a variety of organisms. In the presence of Mg^{2+} and O_2, the firefly luciferase catalyzes the ATP-dependent oxidative decarboxylation of luciferin with emission of light at 562 nm:

$$\text{Luciferin} + \text{ATP} + O_2 \rightarrow \text{oxyluciferin} + \text{AMP} + \text{PPi} + CO_2 + \text{light}$$

The reaction has a quantum yield of 0.88, the highest known for a chemiluminescence reaction. The light emission is very transient, with a peak intensity at about 0.5 s followed by a rapid decay, but the addition of coenzyme A increases the turnover of a reaction intermediate and allows constant light emission over several minutes. The photons emitted are detected with a luminometer and the enzyme activity can be measured over a broad dynamic range and with high sensitivity. Less than 10^4 luciferase molecules can be detected. Luciferase is not present or detectable in most cell types so there is no background activity.

The luciferase reporter system began recently to be applied in sea urchin embryos. It has been used almost exclusively in the Japanese species *H. pulcherrimus* to study the promoter elements responsible for the level of activity and temporal regulation of *HpArs* (Kiyama *et al.*, 1998, 2000; Koike *et al.*, 1998; Morokuma *et al.*, 1997) and *HpOtx* (Kiyama *et al.*, 1998; Kobayashi *et al.*, 2002).

B. Detailed Method

1. Plasmids

The pGL3 series of vectors (Promega) incorporate several features that ensure a highly efficient production of luciferase. These vectors contain downstream of the *luc* coding sequence a SV40 late poly (A) signal for improved RNA processing and a poly (A) and transcription pause site positioned upstream of the MCS to terminate spurious transcription. The translation initiation site conforms to the Kozak consensus. The coding sequence of the wild-type *luc* has been modified. The C-terminal has been changed to eliminate targeting to the peroxisome and the codon usage has been improved to increase expression in animal cells. Promoters to be tested should be inserted into the *pGL3-Basic* vector that is devoid of promoter and enhancer sequences. This vector should not be confused with the *pGL3-Basic* vector designed by Arnone *et al.* (1997) to express GFP, hereafter renamed as *pGreenL3* (see Section III.A and Fig. 1).

In order to minimize experimental variability, it is advisable to normalize the activity of the construct tested to the activity of another reporter system working as an internal control. Any reporter gene like *CAT* or *lacZ* could be used. However, this requires performing two separate assays, which may be time-consuming, or using two very different but compatible methods, which may not be possible. These difficulties have been overcome by the introduction of another luciferase gene isolated from the coelenterate *Renilla reniformis*. The *Renilla* luciferase uses coelenterazine as the substrate and its activity can be measured under conditions compatible with those of the luciferase and with a similar dynamic range. The *Renilla* luciferase is placed under the control of different strong and promiscuous promoters in vectors of the pRL series. The pGL3 vector and the pRL vector can be coinjected in a ratio of 10 to 1 or greater. The Dual-Luciferase Reporter Assay System from Promega allows sequential measurements of the two luciferases in a single sample.

2. Simple Luciferase Assay

a. Reagents

Lysis buffer:

Tris-phosphate, pH 7.8 (25 mM)
EDTA (2 mM)
$MgSO_4$ (1 mM)
Dithiothreitol (2 mM)
Triton X-100 (1%)
(10% glycerol can be included)

Assay buffer:

Tris-phosphate, pH 7.8 (25 mM)
$MgSO_4$ (20 mM)
EDTA (1 mM)
Dithiothreitol (5 mM)
Coenzyme A (500 μM)
ATP (1 mM)
D-luciferin (500 μM)

b. Measurements

The reagents must be pre-warmed at room temperature.

1. Collect about 100 embryos at the desired stage. Centrifuge at 12,000 rpm for 2 min at 4 °C. Remove most of the supernatant; repeat the centrifugation and remove the rest of the supernatant.
2. To the pellet, add 35 μl of lysis buffer. Hold at room temperature for 30 min with occasional gentle mixing. Avoid multiple freeze/thaw cycles that lead to loss of luciferase activity.
3. Centrifuge for 5 min at 12,000 rpm to pellet the cell debris.
4. Transfer the lysate supernatant to a new tube. Freeze at −80 °C to store.
5. Mix 20 μl of the lysate supernatant with 50 μl of the assay buffer. Read immediately in a luminometer after a 2 s delay over a 10 s window.

3. Notes

The number of embryos to use in a single assay depends on the promoter strength, usually ranging between 50 and 300. The settings of the luminometer depend on the instrument and on the light intensity. It is advisable to construct a standard curve to determine the linear range of detection. When using an automated luminometer, the assay buffer will be injected into the tube. For manual measurements, it can be convenient to add the luciferin separately to trigger the reaction. The composition of the buffers and the volumes used are indicative. Conditions have to be adjusted, depending on the number of embryos available, their stage and, possibly, their species. Very convenient assays are commercially available as kits in which the composition of the buffers has been optimized. The CCLR (Cell Culture Lysis Reagent freshly diluted from 5× stock) and the Luciferase Assay Reagent from Promega work well. It is recommended to follow the instructions from the manufacturers. This is particularly important when expressing two reporter genes simultaneously. For example, Triton X-100 partially inhibits CAT and intensifies coelenterazine autoluminescence. The Dual-Luciferase Reporter Assay System from Promega has been successfully operated with *H. pulcherrimus*, but was found not to be reliable with *S. purpuratus*. In the latter case, the lysis buffer compatible with the *Renilla* luciferase measurements does not provide consistent lysis, and the background in the *Renilla* assay is very high. Both inconveniences are stage-specific. Therefore, it is advisable to normalize luciferase assays by measuring the amount of luciferase DNA present in the lysate. A 10 μl aliquot of the lysis medium (step 2) is extracted by standard methods, blotted with a slot apparatus, and hybridized with a luciferase probe (Wei *et al.*, 1995; L. Angerer, unpublished).

C. Advantages and Disadvantages of the Luciferase Reporter System

The luciferase system is extremely sensitive and quantitative. The assay allows detection of fewer than 10^4 molecules of luciferase, i.e., it is about 100-fold more sensitive than the radioactive CAT assay. It is linear with respect to enzyme concentration over 6 to 7 orders of magnitude. In addition, utilization of the *Renilla* luciferase as a control increases accuracy. This allows the study of both weak and strong promoters, and the assay is simple and fast. Even when the firefly and the *Renilla* luciferase are assayed successively, measurements take less than one minute. The method, however, requires preferably a luminometer or a scintillation counter.

The firefly luciferase is stable in the sea urchin embryo, and therefore the system is inappropriate to follow rapid temporal changes. However, this inconvenience should be overcome since an unstable form of the luciferase protein has been designed (Leclerc *et al.*, 2000). Although in the sea urchin embryo the luciferase gene has been only used for quantitative global measurements of cell lysates, the luciferase activity can be recorded by imaging systems providing a very sensitive method to analyze spatial expression of reporters in living cells (Greer and Szalay, 2002).

VI. Using *lacZ* Reporter Gene

A. General Explanation

The *lacZ* gene from *E. coli* is an excellent reporter gene that has been used in many cell types. The β-galactosidase coded for by the *lacZ* gene is one of the most useful enzyme tools for cell and molecular biology. Background activity is low and often undetectable in eukaryotic cells under the conditions of the assay. β-galactosidase can hydrolyze a variety of synthetic substrates that consist of galactose linked through a β-D-glycosidic bond to a moiety that can be tailored for specific purposes. Several substrates release upon hydrolysis soluble products that allow sensitive and quantitative photometric measurements on cell lysates. In order to localize gene activity within a cell, a tissue, or an embryo, the moiety released from galactose must undergo a change of its spectral characteristics and must precipitate at the site of β-galactosidase activity. The most commonly used substrate for monitoring *in situ* localized gene activity is X-gal (5-bromo-4-chloro-3-indolyl β-D-galactopyranoside). Hydrolysis by β-galactosidase produces galactose and 5-bromo-4-chloro-indoxyl. The colorless indoxyl product non-enzymatically dimerizes and is oxidized to form a blue halogenated derivative of indigo (5,5′-dibromo-4,4′-dichloro-indigo) that is insoluble and forms a blue precipitate. The latter part of the reaction is facilitated by electron transfer to electron acceptors like the Fe^{3+} ions present in the reaction medium. The deep blue precipitate can be easily detected under the microscope with very little or no background.

The *lacZ* gene was first used as reporter to localize gene activity in sea urchin embryos by Klein and colleagues in the early 1990s (Gan *et al.*, 1990). It has been used since in *S. purpuratus, L. Pictus, L. variegatus, P. lividus, H. pulcherrimus*, and for several genes including *Spec1, Spec2a*, and *Spec2c* (Gan and Klein, 1993; Gan *et al.*, 1990; Mao *et al.*, 1994), *SpHE* (Wei *et al.*, 1995, 1997, 1999), *SpAN* (Kozlowski *et al.*, 1996), *LpS1* (Xiang *et al.*, 1991), *HE* (Ghiglione *et al.*, 1997), *HpArs* (Kurita *et al.*, 2003), *Sp-msp130, Ht-msp130, He-msp130*, and *PM27* (Klueg *et al.*, 1997). Examples of these results are shown in Fig. 2G–I.

B. Detailed Method

1. Plasmids

Two plasmids have been used to express *lacZ* in the sea urchin embryo: plasmid pNASSβ from Clontech, in one case, and the pNL plasmid in all others. The pNASSβ plasmid has been discontinued by Clontech but could be replaced by the newer pβgal-Basic. In this plasmid, intron and polyadenylation signal from SV40 located downstream of the *lacZ* gene ensure correct processing of the transcript and another polyadenylation signal upstream of the MCS decreases spurious transcription. The pNL plasmid was designed by Gan *et al.* (1990). pNL derives from Bluescript (Stratagene) by insertion of the *lacZ* gene followed by the splice sites and polyadenylation signal from SV40 T-antigen. The *lacZ* gene is modified by addition at the 5′ end of the ORF of a short sequence coding for the nuclear localization signal of the SV40 T-antigen.

2. Enzymatic Assay of β-Galactosidase Using X-gal

a. Reagents

X-gal stock solution:

40 mg/ml in DMSO or DMF
(aliquots stored at −20 °C protected from light)

Staining solution:

$MgCl_2$ (1 mM)
$K_3Fe(CN)_6$ (5 mM)
$K_4Fe(CN)_6$ (5 mM)
NaCl (150 mM)
$NaPO_4$, pH7.2 (10 mM)
(stored at 4 °C, protected from light)

Including detergents like 0.02% NP-40, 0.01–0.1% SDS, or 0.3% Triton X-100 may increase staining. The pH must be in the neutral range to avoid background due to endogenous lysosomal β-galactosidase active at low pH.

b. Fixation for X-gal Staining

1. Collect embryos at the desired stage. If the embryos have to be collected before hatching, 2 mM ATA or 2 mM PABA must be added before fertilization to prevent hardening of the fertilization envelopes. To collect the embryos, create streams of seawater with a pipette so as to tear the weakened envelopes stuck to the dish until embryos are liberated.
2. Fix embryos following one of the following procedures:
   a. Ca^{2+}-free artificial seawater containing 1% glutaraldehyde for 6 min at room temperature.
   b. 2% paraformaldhehyde in seawater for 6 min (seawater can be replaced by PBS).

Although protocols used for vertebrate cells call for long fixation times, it has been reported that, in the sea urchin embryo, overfixation inhibits activity of β-galactosidase. Therefore, fixation time can be critical and should not be increased without careful control.

c. X-gal Staining

1. Pre-warm the staining solution at room temperature to prevent precipitation of X-gal. Dilute X-gal from the stock solution at a final concentration of 2 mg/ml in the staining solution. If precipitation still occurs, the concentration can be reduced to 1.5 mg/ml.
2. Wash embryos with 3 vol of staining solution without X-gal.
3. Incubate embryos in the staining solution containing X-gal, in the dark, at room temperature. Depending on the promoter strength, incubation times may vary from half an hour to 2 days. For long incubation times, it is recommended to add antibiotics (50 μg/ml streptomycin, 50 μg/ml penicillin).

3. Simultaneous Assay of β-Galactosidase Using X-gal and Antigen Immunolocalization

Revealing on the same embryo the activity of β–galactosidase with X-gal and the presence of an antigen using antibodies may be difficult. The fixation procedure and the conditions for the X-gal reaction and immunolabeling must be compatible, and should be tested for each antigen. The intensity of the blue precipitate should be controlled to avoid masking the fluorescence. A successful procedure was described by Wei *et al.* (1995) to identify PMCs with the monoclonal antibody 6e10 in embryos expressing a *lacZ* reporter. This procedure could be easily adapted for any antibody. The embryos fixed following this procedure are fragile and must be handled with care.

1. Fix embryos with 2% paraformaldehyde as described in Section VI.B.2.b, 2a, and reveal β–galactosidase activity with X-gal, as has been described.
2. Refix in 2% paraformaldehyde in seawater for 1 h.
3. Wash embryos 3 times with seawater.
4. Permeabilize by addition of an equal volume of 0.5% Triton X-100 in seawater. Incubate 20 min.
5. Transfer to seawater containing 0.25% Triton X-100 and the antibody at the adequate dilution (to be tested). Incubate overnight at room temperature.
6. Wash embryos with seawater, 3 times for 10 min each.
7. Mount on slides and observe under a microscope.

4. Immunodetection of the β-Galactosidase

If X-gal staining is to be avoided (for example, if the fixation procedure is not compatible with other labelings or if the blue precipitate interferes with optical measurements), β–galactosidase can be detected with commercial antibodies. The following protocol was routinely applied to *P. lividus* using alkaline phosphatase conjugated antibodies (Ghiglione *et al.*, 1997). Fluorescent antibodies can be used in the same way except for the staining steps.

1. Collect embryos at the desired stage by low-speed centrifugation. Wash the embryos once with MFSW and resuspend the pellet in 10 vol of MFSW.
2. Add an equal volume of a freshly made twofold concentrated solution of fixative. Final concentration is: 4% paraformaldhyde, 0.1% glutaraldehyde, 10 mM EPPS pH 8.0 in MFSW. Allow fixation for 15 min at room temperature under gentle agitation. Wash embryos twice in MFSW.
3. Block by incubating embryos for 5 min in 1 M ethanolamine pH 8.0.
4. Wash embryos with MFSW and quickly transfer to methanol at −20 °C. Store at −20 °C.

Steps 5 to 12 are carried out at room temperature on a rotating wheel.

5. Rehydrate embryos in 50% MeOH, 50% TBST. Rinse several times in TBST.
6. Block for 20 min in 5% nonfat dry milk.
7. Incubate for 1 h with a rabbit anti β–galactosidase antibody diluted according to instructions of the manufacturer.
8. Wash embryos 6 times for 5 min each in TBST.
9. Incubate embryos for 1 h in TBST with an alkaline-phosphatase-conjugated goat anti-rabbit IG antibody diluted 1/7500. The antibody has been previously adsorbed against fixed embryos.
10. Wash as in step 8.
11. Rinse briefly in buffer AP (100 mM Tris-HCl pH 9.5, 100 mM NaCl, 5 mM $MgCl_2$).

12. Develop staining in the same buffer containing 330 μg/ml NBT and 165 μg/ml BCIP.

C. Advantages and Disadvantages of the *lacZ* Reporter System

The *lacZ* reporter system is powerful, versatile, and well developed. It can be used in a variety of experimental situations. As with any other reporter system, it is possible to monitor directly the products of the *lacZ* gene. The mRNA could be detected by WMISH, but this has never been done in sea urchin embryos. Neither the sensitivity of the method nor the stability of the messenger has been determined. Many commercial antibodies allow detection of the protein by whole-mount immunolocalization. It is thus possible to choose among a large variety of enzymatic or fluorescent labels and to use additional amplification steps if required. However, the most interesting detection methods rely on the enzymatic activity of the protein, which provides an important amplification and so a greater sensitivity. Several substrates allow quantitative measurements of global level of expression in cell lysates. Although they have not been used yet with the sea urchin embryo, they could be useful tools. Spectrophotometric, fluorescent, and luminescent assays have been developed and are often commercially available as kits. These measurements in solution have a large linear range of detection and are very sensitive. In cell lysates, luminescent assays allow detection of about 10^4 molecules. The most useful and frequently used substrate is X-gal, which is an invaluable tool to localize gene activity in embryos from many species including sea urchin. The procedure is simple, with a short fixation step and a single staining step that allow detection in whole-mount embryos. As it results from an enzymatic reaction, staining appears progressively. It can be monitored under any microscope and interrupted when the desired intensity is reached. The sensitivity of the method in sea urchin embryos has not been assessed. It is probably very sensitive since *lacZ* expression was detected in embryos that had received about 5 copies of plasmid by the gun particle method (Kurita *et al.*, 2003). The pNL plasmid facilitates measurements by targeting the β-galactosidase to the nucleus. Cells expressing this *lacZ* transgene have their nucleus strongly colored in blue, making them more easily identified and counted (Fig. 2I). Furthermore, the concentration of dye in the nucleus makes more discernible simultaneous labeling in the cytoplasm or in other compartments of the cell. This feature is also favorable when using whole-mount immunolabeling (Ghiglione *et al.*, 1997).

As explained in Section III.C, when reporter genes are integrated in a fraction of the PMCs, their protein products diffuse easily within the syncytium and label every cell, including the thin connecting cables. This should be observed also with β-galactosidase, although its large size may decrease its ability to diffuse from cell to cell. However, β-galactosidase targeted to the nucleus, like the one produced by the pNL plasmid, should barely label the connecting cables and should not label all the PMCs, but should more closely reflect the mosaic pattern of the transgene incorporation (Fig. 2H). It is to be noted that microinjection often artifactually leads to the appearance within the blastocoel of abnormal cells that express the *lacZ* reporter. These cells can be differentiated from the normal PMCs by using a specific antibody like 6e10 (Wei *et al.*, 1995).

The *lacZ* system has a number of disadvantages. The methods currently used do not allow observations on live embryos, which is one the major strengths of the GFP system. In addition, the β-galactosidase is stable in sea urchin embryos, and thus it is not adequate to study temporal variations of activity and particularly shutting off of gene expression. It is also inappropriate to study dynamic patterns of expression, as the spatial pattern observed at any time is the sum of previous phases of expression. However, in some cases, the stability of the protein may be an advantage. If there is a single phase of expression, and provided the lineage relationship is well established, it is possible to detect expression of the transgene after the expression of the endogenous gene, at a stage where the pattern may be more easily analyzed using morphological cues or molecular markers. The *lacZ* system remains extremely useful to determine spatial patterns when high sensitivity and accurate localization are required.

VII. Summary, Prospects, Concluding Remarks

This chapter summarizes four powerful assays for analyzing gene expression in *cis*-regulatory studies. The enzymatic assays (CAT, luciferase, *lacZ*) are currently limited by their application to embryo homogenates or fixed samples, but offer more robust analysis of gene activity than GFP. Assays based on CAT enzymatic activity or on CAT mRNA detection by WMISH are laborious but are well established for accurately quantifying gene expression and to determine spatial patterns at defined timepoints during development. *LacZ* assays are the current standard for spatially visualizing gene products in whole-mount fixed embryos. They are very sensitive but they provide limited temporal or quantitative information due to the perdurance of β-galactosidase and the subtleties of the staining technique. Recently developed luciferase assays promise to be even more sensitive and accurate than the CAT and *lacZ* assays, and applicable to living cells and embryos. But, they have not yet been well established in invertebrate deuterostome research. GFP allows visualization of gene expression within living embryos. But because this is not an enzymatic assay, sensitivity can be a problem, particularly for weak promoters. Furthermore, imaging live embryos and quantifying gene expression in space and time (due to scattering of light by tissue, the perdurance of GFP, and other experimental details) is currently fraught with challenges. Ongoing improvements in imaging technology and the advent of multiple fluorescent proteins, as well as fluorescent and luminescent assays for vital imaging, will dramatically facilitate studies of gene expression in the coming decade.

Acknowledgments

We are grateful to Dr. C.-H. Yuh (California Institute of Technology) and Dr. L. M. Angerer (University of Rochester) for methodological information on CAT and luciferase reporter systems, respectively. We also thank Dr. T. Minokawa and Dr. E. H. Davidson (California Institute of Technology) for providing unpublished data. The helpful discussions on the GFP reporter system and the gift of unpublished pictures by Dr. P. Oliveri (California Institute of Technology) and Mr. M. Souren (University of Heidelberg) are gratefully acknowledged.

References

Ando, R., Hama, H., Yamamoto-Hino, M., Mizuno, H., and Miyawaki, A. (2002). An optical marker based on the UV-induced green-to-red photoconversion of a fluorescent protein. *Proc. Natl. Acad. Sci.* **99,** 12651–12656.

Arnone, M. I., Bogarad, L. D., Collazo, A., Kirchhamer, C. V., Cameron, R. A., Rast, J. P., Gregorians, A., and Davidson, E. H. (1997). Green fluorescent protein in the sea urchin: New experimental approaches to transcriptional regulatory analysis in embryos and larvae. *Development* **124,** 4649–4659.

Arnone, M. I., Martin, E. L., and Davidson, E. H. (1998). *Cis*-regulation downstream of cell type specification: A single compact element controls the complex expression of the *CyIIa* gene in sea urchin embryos. *Development* **125,** 1381–1395.

Bogarad, L. D., Arnone, M. I., Chang, C., and Davidson, E. H. (1998). Interference with gene regulation in living sea urchin embryos: Transcription factor knock out (TKO), a genetically controlled vector for blockade of specific transcription factors. *Proc. Natl. Acad. Sci. USA* **95,** 14827–14832.

Cameron, R. A., Oliveri, P., Wyllie, J., and Davidson, E. H. (2004). *cis*-Regulatory activity of randomly chosen fragments from the sea urchin. *Gene Expr. Patterns* **4,** 205–213.

Chen, N., Hsiang, T., and Goodwin, P. H. (2003). Use of green fluorescent protein to quantify growth of Colletotrichum during infection of tobacco. *J. Microbiol. Meth.* **53,** 113–122.

Coffman, J. A., Kirchhamer, C. V., Harrington, M. G., and Davidson, E. H. (1997). SpMyb functions as an intramodular repressor to regulate spatial expression of *CyIIIa* in sea urchin embryos. *Development* **124,** 4717–4727.

Consales, C., and Arnone, M. I. (2002). Functional characterization of Ets-binding sites in the sea urchin embryo: Three base pair conversions redirect expression from mesoderm to ectoderm and endoderm. *Gene* **287,** 75–81.

Cotlet, M., Hofkens, J., Habuchi, S., Dirix, G., Guyse, M. V., Michiels, J., Vanderleyden, J., and Schryver, F. C. D. (2001). Identification of different emitting species in the red fluorescent protein DsRed by means of ensemble and single-molecule spectroscopy. *Proc. Natl. Acad. Sci.* **98,** 14398–14403.

Daelemans, D., Clercq, E. D., and Vandamme, A.-M. (2001). A quantitative GFP-based bioassay for the detection of HIV-1 Tat transactivation inhibitors. *J. Viological Meth.* **96,** 183–188.

Dantuma, N. P., Lindsten, K., Glas, R., Jellne, M., and Masucci, M. G. (2000). Short-lived green fluorescent proteins for quantifying ubiquitin/proteasome-dependent proteolysis in living cells. *Nature Biotechnology* **18,** 538–543.

Dickinson, M. E., Bearman, G., Tille, S., Lansford, R., and Fraser, S. E. (2001). Multi-spectral imaging and linear unmixing add a whole new dimension to laser scanning fluorescence microscopy. *Biotechniques* **31,** 1272–1278.

Dmochowski, I. J., Dmochowski, J. E., Oliveri, P., Davidson, E. H., and Fraser, S. E. (2002). Quantitative imaging of *cis*-regulatory reporters in living embryos. *Proc. Natl. Acad. Sci. USA* **99,** 12895–12900.

Ferrer-Martinez, A., and Gomez-Foix, A. M. (2002). Laser scanning cytometry to quantify gene transfer efficiency and transcriptional activity of EGFP constructs. *Biotechniques* **32,** 62–66.

Flytzanis, C. N., Britten, R. J., and Davidson, E. H. (1987). Ontogenic activation of a fusion gene introduced into sea urchin eggs. *Proc. Natl. Acad. Sci. USA* **84,** 151–155.

Flytzanis, C. N., McMahon, A. P., Hough-Evans, B. R., Katula, K. S., Britten, R. J., and Davidson, E. H. (1985). Persistence and integration of cloned DNA in postembryonic sea urchins. *Dev. Biol.* **108,** 431–442.

Franks, R. R., Anderson, R., Moore, J. G., Hough-Evans, B. R., Britten, R. J., and Davidson, E. H. (1990). Competitive titration in living sea urchin embryos of regulatory factors required for expression of the *CyIIIa* actin gene. *Development* **110,** 31–40.

Gan, L., and Klein, W. H. (1993). A positive *cis*-regulatory element with a bicoid target site lies within the sea urchin *Spec2a* enhancer. *Dev. Biol.* **157,** 119–132.

Gan, L., Wessel, G. M., and Klein, W. H. (1990). Regulatory elements from the related *spec* genes of Strongylocentrotus purpuratus yield different spatial patterns with a *lacZ* reporter gene. *Dev. Biol.* **142,** 346–359.

Garcia-Parajo, M. F., Koopman, M., van Dijk, E. M., Subramaniam, V., and van Hulst, N. F. (2001). The nature of fluorescence emission in the red fluorescent protein DsRed, revealed by single-molecule detection. *Proc. Natl. Acad. Sci.* **98,** 14392–14397.

Ghiglione, C., Emily-Fenouil, F., Lhomond, G., and Gache, C. (1997). Organization of the proximal promoter of the hatching-enzyme gene, the earliest zygotic gene expressed in the sea urchin embryo. *Eur. J. Biochem.* **250,** 502–513.

Greer, L. F., 3rd, and Szalay, A. A. (2002). Imaging of light emission from the expression of luciferases in living cells and organisms: A review. *Luminescence* **17,** 43–74.

Haruguchi, Y., Horii, K., Suzuki, G., Suyemitsu, T., Ishihara, K., and Yamasu, K. (2002). Genomic organization of the gene that encodes the precursor to EGF-related peptides, exogastrula-inducing peptides, of the sea urchin *Anthocidaris crassispina. Biochim. Biophys. Acta* **1574,** 311–320.

Hough-Evans, B. R., Britten, R. J., and Davidson, E. H. (1988). Mosaic incorporation and regulated expression of an exogenous gene in the sea urchin embryo. *Dev. Biol.* **129,** 198–208.

Hough-Evans, B. R., Franks, R. R., Cameron, R. A., Britten, R. J., and Davidson, E. H. (1987). Correct cell-type-specific expression of a fusion gene injected into sea urchin eggs. *Dev. Biol.* **121,** 576–579.

Hughes, E. H., Hong, S.-B., Shanks, J. V., San, K.-Y., and Gibson, S. I. (2002). Characterization of an inducible promoter system in *Catharanthus roseus* hairy roots. *Biotechnol. Prog.* **18,** 1183–1186.

Kam, Z., Hanser, B., Gustafsson, M. G. L., Agard, D. A., and Sedat, J. W. (2001). Computational adaptive optics for live three-dimensional biological imaging. *Proc. Natl. Acad. Sci. USA* **98,** 3790–3795.

Kirchhamer, C. V., and Davidson, E. H. (1996). Spatial and temporal information processing in the sea urchin embryo: Modular and intramodular organization of the *CyIIIa* gene *cis*-regulatory system. *Development* **122,** 333–348.

Kirchhamer, C. V., Yuh, C. H., and Davidson, E. H. (1996). Modular *cis*-regulatory organization of developmentally expressed genes: Two genes transcribed territorially in the sea urchin embryo, and additional examples. *Proc. Natl. Acad. Sci. USA* **93,** 9322–9328.

Kiyama, T., Akasaka, K., Takata, K., Mitsunaga-Nakatsubo, K., Sakamoto, N., and Shimada, H. (1998). Structure and function of a sea urchin orthodenticle-related gene (*HpOtx*). *Dev. Biol.* **193,** 139–145.

Kiyama, T., Sasai, K., Takata, K., Mitsunaga-Nakatsubo, K., Shimada, H., and Akasaka, K. (2000). CAAT sites are required for the activation of the *H. pulcherrimus Ars* gene by Otx. *Dev. Genes Evol.* **210,** 583–590.

Klueg, K. M., Harkey, M. A., and Raff, R. A. (1997). Mechanisms of evolutionary changes in timing, spatial expression, and mRNA processing in the *msp130* gene in a direct-developing sea urchin, *Heliocidaris erythrogramma. Dev. Biol.* **182,** 121–133.

Kobayashi, A., Akasaka, K., Kawaichi, M., and Kokubo, T. (2002). Functional interaction between TATA and upstream CACGTG elements regulates the temporally specific expression of Otx mRNAs during early embryogenesis of the sea urchin, *Hemicentrotus pulcherrimus. Nucleic Acids Res.* **30,** 3034–3044.

Koike, H., Akasaka, K., Mitsunaga-Nakatsubo, K., and Shimada, H. (1998). Proximal *cis*-regulatory elements of sea urchin arylsulfatase gene. *Dev. Growth Differ.* **40,** 537–544.

Kozlowski, D. J., Gagnon, M. L., Marchant, J. K., Reynolds, S. D., Angerer, L. M., and Angerer, R. C. (1996). Characterization of a *SpAN* promoter sufficient to mediate correct spatial regulation along the animal–vegetal axis of the sea urchin embryo. *Dev. Biol.* **176,** 95–107.

Kozlowski, M. T., Gan, L., Venuti, J. M., Sawadogo, M., and Klein, W. H. (1991). Sea urchin USF: A helix–loop–helix protein active in embryonic ectoderm cells. *Dev. Biol.* **148,** 625–630.

Kurita, M., Kondoh, H., Mitsunaga-Nakatsubo, K., Shimotori, T., Sakamoto, N., Yamamoto, T., Shimada, H., Takata, K., and Akasaka, K. (2003). Utilization of a particle gun DNA introduction system for the analysis of *cis*-regulatory elements controlling the spatial expression pattern of the arylsulfatase gene (*HpArs*) in sea urchin embryos. *Dev. Genes Evol.* **213,** 44–49.

Leclerc, G. M., Boockfor, F. R., Faught, W. J., and Frawley, L. S. (2000). Development of a destabilized firefly luciferase enzyme for measurement of gene expression. *Biotechniques* **29,** 590–598.

Lippincott-Schwartz, J., and Patterson, G. H. (2003). Development and use of fluorescent protein markers in living cells. *Science* **300,** 87–91.

Livant, D. L., Hough-Evans, B. R., Moore, J. G., Britten, R. J., and Davidson, E. H. (1991). Differential stability of expression of similarly specified endogenous and exogenous genes in the sea urchin embryo. *Development* **113,** 385–398.

Makabe, K. W., Kirchhamer, C. V., Britten, R. J., and Davidson, E. H. (1995). *Cis*-regulatory control of the *SM50* gene, an early marker of skeletogenic lineage specification in the sea urchin embryo. *Development* **121,** 1957–1970.

Mao, C. A., Gan, L., and Klein, W. H. (1994). Multiple Otx binding sites required for expression of the *Strongylocentrotus purpuratus Spec2a* gene. *Dev. Biol.* **165,** 229–242.

Martin, E. L., Consales, C., Davidson, E. H., and Arnone, M. I. (2001). Evidence for a mesodermal embryonic regulator of the sea urchin *CyIIa* gene. *Dev. Biol.* **236,** 46–63.

McMahon, A. P., Flytzanis, C. N., Hough-Evans, B. R., Katula, K. S., Britten, R. J., and Davidson, E. H. (1985). Introduction of cloned DNA into sea urchin egg cytoplasm: Replication and persistence during embryogenesis. *Dev. Biol.* **108,** 420–430.

McMahon, A. P., Novak, T. J., Britten, R. J., and Davidson, E. H. (1984). Inducible expression of a cloned heat shock fusion gene in sea urchin embryos. *Proc. Natl. Acad. Sci. USA* **81,** 7490–7494.

Miyawaki, A., Llopis, J., Helm, R., McCaffrey, J. M., Adams, J. A., Ikura, M., and Tsien, R. Y. (1997). Fluorescent indicators for Ca2+ based on green fluorescent proteins and calmodulin. *Nature* **388,** 882–887.

Morokuma, J., Akasaka, K., Mitsunaga-Nakatsubo, K., and Shimada, H. (1997). A *cis*-regulatory element within the 5′ flanking region of arylsulfatase gene of sea urchin, *Hemicentrotus pulcherrimus*. *Dev. Growth Differ.* **39,** 469–476.

Niswender, K. D., Blackman, S. M., Rohde, L., Magnuson, M. A., and Piston, D. W. (1995). Quantitative imaging of green fluorescent protein in cultured cells—Comparison of microscopic techniques, use in fusion proteins, and detection limits. *J. Microsc.-Oxford* **180,** 109–116.

Palla, F., Melfi, R., Di Gaetano, L., Bonura, C., Anello, L., Alessandro, C., and Spinelli, G. (1999). Regulation of the sea urchin early H2A histone gene expression depends on the modulator element and on sequences located near the 3′ end. *J. Biol. Chem.* **380,** 159–165.

Patterson, G. H., and Lippincott-Schwartz, J. (2002). A photoactivatable GFP for selective photolabeling of proteins and cells. *Science* **297,** 1873–1877.

Rast, J. P., Cameron, R. A., Poustka, A. J., and Davidson, E. H. (2002). *Brachyury* target genes in the early sea urchin embryo isolated by differential macroarray screening. *Dev. Biol.* **246,** 191–208.

Raz, E., Zlokarnik, G., Tsien, R. Y., and Driever, W. (1998). Beta-lactamase as a marker for gene expression in live zebrafish embryos. *Dev. Biol.* **203,** 290–294.

Stephens, D. J., and Allan, V. J. (2003). Light microscopy techniques for live cell imaging. *Science* **300,** 82–86.

Strathmann, M., Hamilton, B. A., Mayeda, C. A., Simon, M. I., Meyerowitz, E. M., and Palazzolo, M. J. (1991). Transposon-facilitated DNA sequencing. *Proc. Natl. Acad. Sci. USA* **88,** 1247–1250.

Terasaki, M. (2000). Dynamics of the endoplasmic reticulum and Golgi apparatus during early sea urchin development. *Mol. Biol. Cell* **11,** 897–914.

Wang, D. G., Kirchhamer, C. V., Britten, R. J., and Davidson, E. H. (1995). SpZ12-1, a negative regulator required for spatial control of the territory-specific *CyIIIa* gene in the sea urchin embryo. *Development* **121,** 1111–1122.

Wei, Z., Angerer, L. M., and Angerer, R. C. (1997). Multiple positive *cis* elements regulate the asymmetric expression of the *SpHE* gene along the sea urchin embryo animal–vegetal axis. *Dev. Biol.* **187,** 71–78.

Wei, Z., Angerer, L. M., and Angerer, R. C. (1999). Spatially regulated SpEts4 transcription factor activity along the sea urchin embryo animal–vegetal axis. *Development* **126,** 1729–1737.

Wei, Z., Angerer, L. M., Gagnon, M. L., and Angerer, R. C. (1995). Characterization of the *SpHE* promoter that is spatially regulated along the animal–vegetal axis of the sea urchin embryo. *Dev. Biol.* **171,** 195–211.

Xiang, M. Q., Ge, T., Tomlinson, C. R., and Klein, W. H. (1991). Structure and promoter activity of the *LpS1* genes of *Lytechinus pictus*. Duplicated exons account for LpS1 proteins with eight calcium binding domains. *J. Biol. Chem.* **266,** 10524–10533.

Xu, N., Niemeyer, C. C., Gonzalez-Rimbau, M., Bogosian, E. A., and Flytzanis, C. N. (1996). Distal *cis*-acting elements restrict expression of the *CyIIIb* actin gene in the aboral ectoderm of the sea urchin embryo. *Mech. Dev.* **60,** 151–162.

Yuh, C. H., Bolouri, H., and Davidson, E. H. (1998). Genomic *cis*-regulatory logic: Experimental and computational analysis of a sea urchin gene. *Science* **279,** 1896–1902.

Yuh, C. H., Bolouri, H., and Davidson, E. H. (2001a). *Cis*-regulatory logic in the *Endo16* gene: Switching from a specification to a differentiation mode of control. *Development* **128,** 617–629.

Yuh, C. H., Brown, C. T., Livi, C. B., Rowen, L., Clarke, P. J., and Davidson, E. H. (2002). Patchy interspecific sequence similarities efficiently identify positive *cis*-regulatory elements in the sea urchin. *Dev. Biol.* **246,** 148–161.

Yuh, C. H., and Davidson, E. H. (1996). Modular *cis*-regulatory organization of *Endo16*, a gut-specific gene of the sea urchin embryo. *Development* **122,** 1069–1082.

Yuh, C. H., Li, X., Davidson, E. H., and Klein, W. H. (2001b). Correct expression of *spec2a* in the sea urchin embryo requires both Otx and other *cis*-regulatory elements. *Dev. Biol.* **232,** 424–438.

Yuh, C. H., Moore, J. G., and Davidson, E. H. (1996). Quantitative functional interrelations within the *cis*-regulatory system of the *S. purpuratus Endo16* gene. *Development* **122,** 4045–4056.

Yuh, C. H., Ransick, A., Martinez, P., Britten, R. J., and Davidson, E. H. (1994). Complexity and organization of DNA-protein interactions in the 5′-regulatory region of an endoderm-specific marker gene in the sea urchin embryo. *Mech. Dev.* **47,** 165–186.

Zeller, R. W., Cameron, R. A., Franks, R. R., Britten, R. J., and Davidson, E. H. (1992). Territorial expression of three different trans-genes in early sea urchin embryos detected by a whole-mount fluorescence procedure. *Dev. Biol.* **151,** 382–390.

Zhao, A. Z., Colin, A. M., Bell, J., Baker, M., Char, B. R., and Maxson, R. (1990). Activation of a late H2B histone gene in blastula-stage sea urchin embryos by an unusual enhancer element located 3′ of the gene. *Mol. Cell Biol.* **10,** 6730–6741.

Zlokarnik, G., Negulescu, P. A., Knapp, T. E., Mere, L., Burres, N., Feng, L., Whitney, M., Roemer, K., and Tsien, R. Y. (1998). Quantitation of transcription and clonal selection of single living cells with β-lactamase as reporter. *Science* **279,** 84–88.

CHAPTER 26

Identification of Sequence-Specific DNA Binding Proteins

James A. Coffman* and Chiou-Hwa Yuh[†]

*Stowers Institute for Medical Research
Kansas City, Missouri 64110

[†]Division of Biology
California Institute of Technology
Pasadena, California 91125

0091-679X/04 $35.00

Overview

By providing large quantities of synchronously developing embryos, sea urchins lend themselves readily to a biochemical analysis of developmental gene regulation. This chapter describes methods for the identification of nuclear proteins that bind specific DNA sequences within the *cis*-regulatory domains of genes. Protocols for preparation of nuclear extract, quantitative electrophoretic mobility shift analysis, oligonucleotide affinity chromatography, and protein preparation for amino acid sequence analysis are presented, along with examples of the successful application of these methods to the *cis*-regulatory analysis of the *CyIIIa* and *Endo16* genes. It is expected that with the completion of the *Strongylocentrotus purpuratus* genome project and the availability of highly sensitive methods of protein sequencing by mass spectrometry, the biochemical approach described here for the identification of transcription factors will become even more efficient and adaptable to rapid high-throughput technologies.

I. Introduction

Over the last decade, studies of the sea urchin embryo have provided a paradigm for the molecular analysis of transcriptional regulatory systems that control development. This is, in large part, due to two remarkable features of the sea urchin: on the one hand, its embryos are extremely amenable to gene transfer by microinjection and, on the other, it readily provides the large quantities of raw material required for protein biochemistry. This makes it possible to functionally dissect the *cis*-regulatory DNA sequences that control transcription during development, and in parallel, to identify the *trans*-acting factors that interact with each regulatory sequence. This chapter presents methodology for the latter, from the preparation of nuclear extracts, through the identification by electrophoretic mobility shift analysis of specific sites of protein-DNA interaction, to the purification and identification of the proteins that bind those sites. Methodology for large-scale culture, freezing, and storage of embryos was described previously (Coffman and Leahy, 2000).

The approaches described here have the following rationale. First, since transcription factors tend to be very low-abundance proteins, all of the methods are greatly facilitated by the preparation of nuclear extracts. This procedure enriches for transcription factors at the expense of other proteins, thus increasing the signal-to-noise ratio for all subsequent analyses. In fact, we have found that nuclear extracts provide sufficient enrichment to allow direct purification of transcriptional regulatory proteins by one or two rounds of affinity chromatography, without preliminary purification by other chromatographic methods. Second, one of the properties of transcriptional regulatory proteins is that they bind DNA sequence-specifically. Thus, their identification in nuclear extract is greatly facilitated by the use of quantitative electrophoretic mobility shift analysis (EMSA), in which binding of proteins to labeled probe sequence is performed in the presence of a large excess of nonspecific competitor DNA. This selects proteins that specifically bind the probe sequence and provides an activity assay for monitoring their purification. Third, affinity purification of transcription factors on oligonucleotide columns bearing their target DNA sequences typically enriches the protein of interest to an extent that allows either its unambiguous identification by SDS polyacrylamide gel electrophoresis (SDS-PAGE), using one of a number of available methods, or its final purification by reverse-phase high performance liquid chromatography (HPLC). This, in turn, facilitates the fourth and final step in the procedure, which is the production of peptides for amino acid sequencing by protolytic digestion of the purified protein followed by HPLC.

The *cis*-regulatory analysis of the *CyIIIa* and *Endo16* genes provides examples of the successful application of the methods described in this chapter. The binding sites of nine different transcriptional regulatory proteins that interact with the 2.3 kb *CyIIIa cis*-regulatory domain were identified by EMSA (Calzone *et al.*, 1988; Theze *et al.*, 1990), and six of the proteins that bind these sites were purified by affinity chromatography and identified by amino acid sequence analysis (Calzone *et al.*, 1991; Char *et al.*, 1993; Coffman *et al.*, 1992, 1996, 1997; Zeller *et al.*, 1995). Similarly, all of the target sequences within the 2.3 kb *cis*-regulatory domain of the *Endo16* gene were mapped by EMSA (Yuh *et al.*, 1994), and these sequences have been used to purify and identify the proteins that interact with them (C.H.Y., unpublished data). Amino acid sequences obtained from purified transcription factors have been used directly to design probes for cDNA cloning (Calzone *et al.*, 1991; Coffman *et al.*, 1996; Zeller *et al.*, 1995) or, alternatively, for the validation of clones that were obtained by other methods (Char *et al.*, 1993; Coffman *et al.*, 1997).

II. Preparation of Nuclear Extracts

The preparation of nuclear extracts is the first step in the biochemical identification of sequence specific DNA binding proteins. The method described here results in approximately 20-fold enrichment for most DNA binding proteins, and produces extracts of sufficient quality for electrophoretic mobility shift analysis and for affinity chromatography. It consists essentially of the following steps: (1) isolation of nuclei, (2) lysis of nuclei, and (3) recovery and concentration of the soluble nuclear proteins.

A. Nuclear Extract Procedure

1. Buffers (4 °C):

a. Buffer A

	Stock solution	Per liter
10 mM Tris-HCl (pH 7.4)	1 M	10 ml
1 mM EDTA	0.5 M	2 ml
1 mM EGTA	0.5 M	2 ml
1 mM DTT	1 M	1 ml
1 mM Spermidine Tri-HCl		255 mg
0.36 M Sucrose		123 g

b. Buffer B

	Per liter
Buffer A	1000 ml
0.1% NP-40	1 ml

c. Buffer C

	Stock solution	Per liter
20 mM Hepes–KOH (pH 7.9)	1 M	20 ml
40 mM KCl	2 M	20 ml
0.1 mM EDTA	0.5 M	0.2 ml
1 mM DTT	1 M	1 ml
20% glycerol	100%	200 ml
0.1% NP-40	100%	1 ml
0.1 mM PMSF	100 mM	1 ml

d. Buffer D

	Stock solution	Per liter
10 mM Hepes-KOH (pH7.9)	1 M	10 ml
1 mM EDTA	0.5 M	2 ml
1 mM EGTA	0.5 M	2 ml
1 mM DTT	1 M	1 ml
1 mM spermidine Tri-HCl		255 mg
10% glycerol	100%	100 ml

2. For extracts from prehatching embryos, eggs need to be fertilized in 2 mM para-aminobenzoic acid (PABA, in seawater) and the embryos demembranated by passage through 60 μm nylon mesh at the time of collection. Pellet embryos by centrifugation at 1000*g* for 5 min at 4 °C. Wash pellet once with 10 to 20 volumes of ice-cold 1 M glucose, then repellet by centrifugation at 2000*g* for 5 min at 4 °C.

3. Resuspend embryo pellet in 10 to 20 volumes of Buffer A. Freeze in liquid nitrogen and store at −70 °C until ready to process for nuclear extract.

4. Thaw frozen embryos in a beaker or bucket set in a lukewarm water bath, with vigorous stirring to lyse the embryos (for large-scale preps, begin with a Teflon or hard rubber mallet, followed by a stiff PVC or glass stir rod until the embryos become slushy, then continue stirring until the embryos melt into liquid).

5. Pellet the nuclei by centrifugation at 3500*g* for 10 min at 4 °C. Resuspend pellet in 20 to 30 volumes of ice-cold Buffer A (shake vigorously to resuspend), and repeat centrifugation. Repeat Buffer A washes 2 to 3 times.

6. Resuspend nuclear pellet from final Buffer A wash in 20 to 30 volumes of ice-cold Buffer B. Repeat this wash step 2 to 3 times, until the nuclear pellet is colorless. A homgenizer can be used to resuspend the nuclei in Buffer B, in which case it is important that the homogenizer be put on ice and kept cold before being used.

7. Resuspend nuclear pellet from final Buffer B wash in an accurately measured volume of Buffer D. The volume of Buffer D should be about 2 to 3× the nuclear pellet volume; for example, if the nuclear pellet is about 50 ml, use 100 to 150 ml Buffer D to resuspend the nuclei. The volume of the nuclear suspension is then accurately determined in a graduated cylinder; if needed, more Buffer D is added to ensure that the volume is 2 to 3× the original nuclear pellet volume. For example, if the measured volume of the nuclear suspension is 170 ml after adding 100 ml of Buffer D, add another 40 ml of Buffer D (total Buffer D added is then 2× nuclear pellet volume).

8. Aliquot the nuclear suspension to Oak Ridge screw cap polycarbonate centrifuge tubes. The tubes should not be filled more than $\sim\frac{1}{3}$ full. Lyse the nuclei by the addition of 0.1 volume of 4 M ammonium sulfate, pH 7.9, while vortexing, followed by vigorous shaking. The nuclear suspension should become highly viscous after release of the chromatin.

9. Incubate the lysed nuclei on ice for 1 h with occasional vigorous shaking.

10. Spin the lysed nuclei in a fixed-angle rotor at 35,000 rpm for 2 h to remove the chromatin.

11. Recover and combine the supernatants from the ultracentrifuge spin, and accurately determine the volume. Add 0.3 g ammonium sulfate per ml of supernatant, invert several times until the ammonium sulfate is completely dissolved, and incubate at 4 °C overnight in order to precipitate proteins.

12. Recover the precipitated nuclear proteins by centrifugation at 16,000*g* for 30 min at 4 °C. Resuspend the pellet in 1/2 the original nuclear pellet volume of ice-cold Buffer C and let dissolve.

13. Dialyze the resuspended nuclear proteins against 50 to 100 volumes Buffer C overnight at 4 °C. Following dialysis, there is generally some precipitate, which should be removed by centrifugation at 16,000*g* for 30 min at 4 °C. The supernatant is now ready for use in gel mobility shift assays or affinity chromatography, and should be stored in aliquots at −70 °C until ready for use.

III. Identification of Transcription Factor Target Sites by Footprint and Electrophoretic Mobility Shift Analysis (EMSA)

There are two general approaches to identifying the DNA target sequences of transcription factors. One approach uses various methods of "footprint" analysis that rely on fragmentation of radioactively labeled probe DNA, most often by DNAase I digestion, in the presence of sufficient protein to occupy all of the binding sites in a given preparation of probe. The fragmented DNA is then run out on a DNA sequencing gel, along side probe that has been similarly fragmented in the absence of protein, and Maxam-Gilbert type G and G + A ladders of the probe DNA. Autoradiography of the gel and comparison of the lanes will then reveal a footprint, which consists of the absence or diminished intensity of a series of DNAase I fragments from the probe that had been complexed with protein, as compared to the naked probe. Reference to the G and G + A ladders shows the sequence location of the protected region. While footprint analysis has been an effective and reliable method of locating transcription factor target sites, it suffers from being very labor-intensive, requiring extensive optimization of conditions for every probe. For this reason, it is not particularly adaptable to rapid, high-throughput approaches to *cis*-regulatory analysis, and will not be discussed further here. For the interested reader, various protocols for footprint analysis that work well with sea urchin embryo nuclear extract have been described (DiLiberto *et al.*, 1989; Ganster *et al.*, 1992; Katula *et al.*, 1998).

The second approach to identifying transcription factor binding sites uses quantitative electrophoretic mobility shift analysis (EMSA; Fig. 1A). Unlike in footprint analysis, the DNA–protein complexes in EMSA are formed in the presence of excess probe. The complexes and free probe are then resolved by native electrophoresis in a polyacrylamide gel. The sequence specificity of a given interaction is ensured by including a large molar excess (with respect to the probe) of nonspecific DNA in the binding reaction, and it can also be checked by performing competition with unlabeled homologous versus heterologous target site DNA. Like footprint analysis, EMSA requires the use of a labeled DNA probe with high specific activity. There are a number of different options available for the probe construction, all of which use standard molecular biological procedures that will not be described in detail here. Suffice it to say that EMSA probes consist of a double-stranded DNA sequence, which is constructed either from complementary synthetic oligonucleotides as small as a 20 bases or from restriction fragments of cloned DNA as large as 0.5 kb. Complementary synthetic oligonucleotides need to be annealed, typically by mixing in equal molar proportions, heating to 95 °C, and allowing to cool to room temperature over a period of several hours. Probes are traditionally labeled at their end(s) with ^{32}P: either by end-filling with α-^{32}P deoxynucleotides that are complementary to bases in a 5′ overhang, using the klenow fragment of DNA polymerase, or by 5′ end-labeling with γ-^{32}P ATP using T4 polynucleotide kinase. The probe DNA is then purified

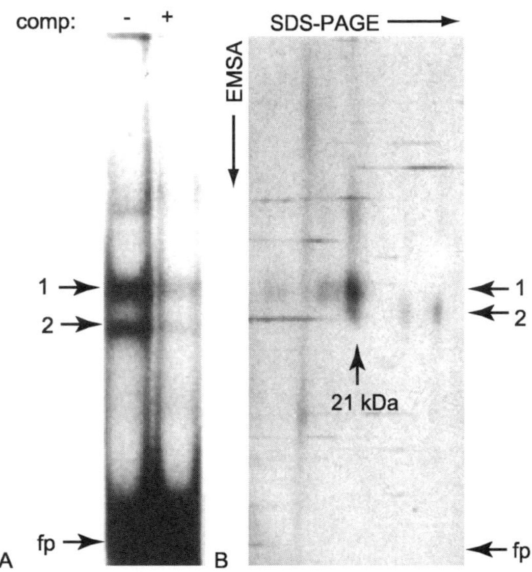

Fig. 1 Identification of the transcriptional regulatory protein SpRunt by EMSA and 2-D EMSA, using the SpRunt target site from the *CyIIIa* cis-regulatory domain as probe (Coffman *et al.*, 1992, 1996). (A) EMSA of SpRunt in crude nuclear extract, using a radioactively labeled probe. The free probe (fp) and protein-bound probe bands are indicated. SpRunt protein from blastula stage nuclear extract migrates in EMSA as two different protein-DNA complexes (1 and 2). The binding reaction was performed with (+) or without (−) a 100-fold molar excess of unlabeled SpRunt target site DNA as competitor, as indicated. The reaction conditions for this EMSA were as described in the text, using 5 μg poly dIdC as nonspecific competitor DNA and 1 μg nuclear extract per 10 μl of binding reaction. (B) 2-D EMSA of SpRunt protein eluted from an oligonucleotide affinity column (from Coffman *et al.*, 1992), with the first dimension (EMSA) oriented to allow direct comparison to the gel in (A). Both of the DNA–protein complexes formed in the first dimension (1 and 2, as in (A)) release a protein that migrates as a 21 kDa band in the second dimension, which by amino acid sequence analysis was subsequently shown to be SpRunt (Coffman *et al.*, 1996). Note that both complexes also release additional proteins in the second dimension, suggesting that SpRunt normally binds DNA in a heteromeric complex with one or more partner proteins, which accounts for the multiplicity of EMSA bands.

by polyacrylamide gel electrophoresis or over a G-50 Sephadex spin column. More recently, it has become possible to perform EMSA using nonradioactive probes, either by end filling with a fluorescently labeled deoxynucleotide (e.g., Cy5-dATP), or by custom synthesis of fluorescently end-labeled DNA oligonucleotides (a service provided by most commercial suppliers of oligonucleotides). This has several major advantages over the traditional radioactive method (probes are not hazardous and have a much longer life, gels do not require drying and can be visualized immediately after electrophoresis, etc.), and offers comparable sensitivity.

Since many proteins bind DNA nonspecifically, the key to EMSA is that the binding reaction be performed in the presence of a large excess of nonspecific competitor DNA. Typically, synthetic double-stranded polymers of poly dAdT or poly dIdC are used. These can be purchased as dried DNA from a number of different vendors. They are prepared by dissolving in water or TE, heating to 95 °C for 5 min to denature, and slow cooling over several hours to anneal. The DNA is then purified by phenol/chloroform extraction, precipitated in ethanol, resuspended in water, and its concentration measured by optical density. It can then be stored at −20 °C until ready for use.

Mapping specific sites of protein–DNA interaction within extended *cis*-regulatory sequences typically begins by subcloning restriction fragments of ≤0.5 kb that cover the entire sequence (Fig. 2). Inserts of each subclone are isolated, labeled at both ends, and digested with a restriction enzyme that recognizes an internal site, yielding two asymmetrically labeled fragments that are purified by native polyacrylamide gel electrophoresis. These asymmetrically labeled probes are then reacted with nuclear extract and subjected to EMSA. In the event that a specific complex is formed with a given probe, nested deletions are prepared from the unlabeled end of that probe (e.g., by restriction digests using 4-hitter enzymes), which allows one to further map where sequence-specific sites of protein–DNA interaction occur (Theze *et al.*, 1990; Yuh *et al.*, 1994). Once binding sites are mapped in this way to a resolution of ≤100 bp fragments, synthetic oligonucleotides of 25 to 30 base pairs can be constructed and used to continue the mapping procedure at finer-scale resolution. Finally, unlabeled synthetic oligonucleotides are used in pairwise competition experiments to determine whether a given protein interacts with a single or multiple sites within a regulatory sequence (Yuh *et al.*, 1994).

A. EMSA Procedure

1. Buffers:
   a. 5× Binding Buffer

		Per ml
50 mM Hepes-KOH, pH 7.9	1 M	50 μl
2.5 mM DTT	1 M	2.5 μl
25 mM $MgCl_2$	0.5 M	5 μl

   b. Loading Buffer

	Stock solution	Per 100 ml
20 mM Hepes-KOH (pH 7.9)	1 M	2 ml
20% glycerol	100%	20 ml
0.1% NP-40	100%	0.1 ml

2. Prepare a labeled DNA probe by your favorite method.

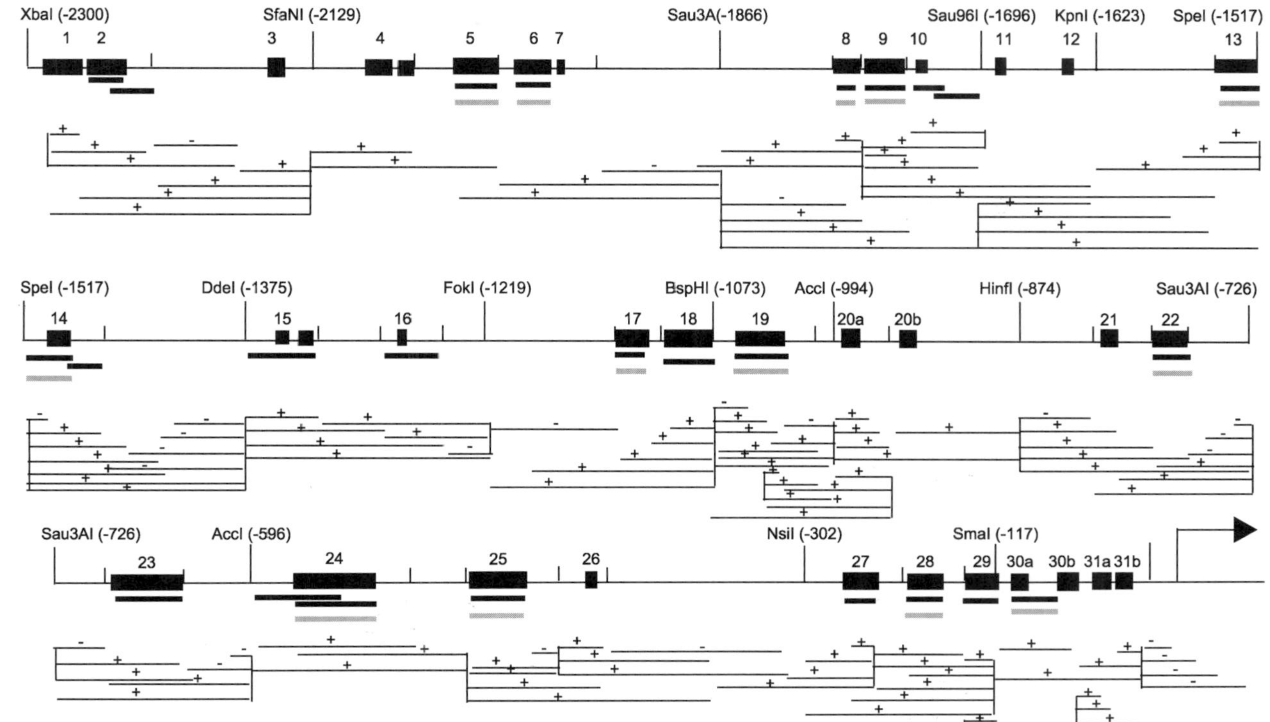

Fig. 2 EMSA mapping of the protein binding sites in the *Endo16 cis*-regulatory domain (redrawn from Fig. 3 of Yuh *et al.*, 1994). The *Endo16* cis-regulatory domain is depicted as a horizontal line, with numbered boxes indicating specific sites of protein–DNA interaction identified by EMSA. Vertical lines demarcate restriction sites; enzymes used to generate subclones are indicated. Horizontal lines below the regulatory region represent DNA fragments that were asymmetrically labeled at the end indicated by a vertical line, subjected to nested deletions from the unlabeled end using restriction enzymes, and used as probe in EMSA experiments, as described in the text. A plus (+) sign indicates that a specific complex was formed when the probe DNA was reacted with blastula stage nuclear extract in the presence of excess nonspecific competitor, while a minus (−) sign indicates that no specific complex was detected. The black bars underneath the protein binding sites represent oligonucleotides that were used to map the interactions in finer-scale resolution. The grey bars represent oligonucleotides that were used for DNA affinity chromatography, as described in the text. The transcription start site is indicated by a bent arrow.

3. Assemble the binding reaction on ice as follows (10 μl total volume): in a microfuge tube, mix 2 μl 5× binding buffer, 0.5 μl 2M KCl (the amount of KCl may need to be optimized for each protein), 1 μl nonspecific competitor DNA (1–5 μg poly dIdC, poly dAdT, or poly dI/dC; the optimal competitor and concentration need to be determined empirically for each protein), 1 μl probe (50,000 c.p.m. in about 0.2 ng DNA for ^{32}P labeled probes), 1 μl nuclear extract (typically containing ~10 μg protein; the optimal amount needs to be determined empirically for each complex), and 4.5 μl loading buffer. Spin down in a microfuge and incubate on ice for 10 min.

4. Load the reactions on a native polyacrylamide/TBE gel, and run at 200 to 300 V. The loading buffer in the reactions does not contain any dye; however, a small amount of loading buffer containing bromphenol blue and xylene cyanol can be run in an empty lane to monitor progress of the gel. The optimal polyacrylamide percentage in the gel and TBE concentration will depend on the size of the probe and the nature of the complex, and should be determined empirically for each complex of interest. Either 5% (for large probes and crude nuclear extract) or 10% (for small probes and affinity purified proteins) polyacrylamide gels and 1× TBE work for most proteins. After running, dry the gel and either perform autoradiography or use a phosphorimager to visualize the complex (for radioactive probes), or directly visualize the gel using a phosphorimager (for fluorescent probes).

IV. Affinity Purification of Sequence-Specific DNA Binding Proteins

Sequence-specific DNA binding proteins can be purified directly by affinity chromatography on columns bearing the target DNA sequence of the protein of interest. While this does not typically result in purification to homogeneity, it provides enough enrichment to allow for final purification by preparative SDS-PAGE or by reverse-phase HPLC on a C4 column. The DNA that is linked to the column matrix consists of concatenated double-stranded oligonucleotides. The concatenation of binding sites facilitates affinity chromatography by providing multiple copies of binding site per oligonucleotide (potentially enhancing cooperative interactions), and also by creating a spacer between the column matrix and most of the binding sites. In order to construct concatemers, the double-stranded oligonucleotides need to be designed such that each end contains a 5′ overhang. The simplest way to do this is to place the complement of the 5′-terminal 4 to 6 nucleotides of one oligonucleotide at the 5′ end of the complementary oligonucleotide (e.g., if one oligo has the sequence 5′-ACCTAATTCGCGT-3′, its partner would be 5′-AGGTACGCGAATT-3′). Following annealing and phosphorylation, the oligonucleotides are ligated into concatemers that are then purified by size-exclusion chromatography on Sephadex G-200.

High-quality nuclear extracts are key to the success of the procedure described here. Another important factor is that the amount of DNA on the affinity column be matched to the amount of sequence-specific DNA binding protein contained in the volume of nuclear extract to be chromatographed—an excess of DNA binding sites will tend to cause high background by favoring nonspecific binding. Finally, it is sometimes useful to do two rounds of affinity chromatography, performing the second round in the presence of an excess of nonspecific DNA. This will greatly reduce the number of nonspecific proteins in the final eluate.

A. Construction of Affinity Columns

1. Buffers:
   a. G-200 Buffer:

	Stock solution	Per liter
50 mM Tris-HCl (pH 7.4)	1 M	50 ml
0.3 M NaCl	5 M	60 ml
1 mM EDTA	0.5 M	2 ml
0.1% SDS	10%	10 ml

   b. Column Storage Buffer:

	Stock solution	Per liter
10 mM Tris-HCl (pH 7.6)	1 M	10 ml
0.3 M NaCl	5 M	60 ml
1 mM EDTA	0.5 M	2 ml
0.02% NaN_3	1%	5 ml

2. Preparation of double-stranded oligonucleotide concatemers:
   a. Mix complementary oligonucleotides (50–500 μg each) in an equal molar ratio (in H_2O, final volume 200–500 μl).
   b. Add 1/10 volume of 10× T4 polynucleotide kinase buffer containing ATP (NEB), an appropriate quantity of T4 polynucleotide kinase (NEB), and incubate at 37 °C for 1 to 2 h.
   c. Heat to 90 °C to denature, then slow cool to room temperature to anneal.
   d. Add 1/10 volume of 10× ligation buffer (NEB), H_2O as needed to adjust the volume, an appropriate quantity of T4 DNA ligase (NEB), and incubate overnight at 15 °C.
   e. Run an aliquot of the ligation out on a 2% agarose gel to check the efficacy of the reaction. The mass average of the DNA should be ~10 mers or greater (i.e., ≥200 bp for a monomeric oligo of 20 bp). If it is not, extract the mix once with phenol/chloroform, precipitate the DNA with ethanol, and repeat the kinase and ligase steps (a, b, d, and e),

omitting the denaturation step (c). Usually, two rounds of kinase/ligation are sufficient to produce concatemers of sufficient length.

f. Purify the final concatemers by phenol/chloroform extraction and ethanol precipitation.

g. Dissolve final pellet in 90 μl H_2O and add 10 μl DNA gel loading buffer.

h. Run DNA through a Sephadex G-200 column that has been pre-equilibrated in G-200 Buffer. A useful column size is 28 cm (length) $\times$ 1 cm (diameter). Collect column eluate in 0.5 ml fractions, monitoring the absorbance at 260 nm. DNA will usually begin eluting between fractions 5 and 10.

i. Run an aliquot ($\sim$10 μl) of each peak fraction out on a 2% agarose gel to assess the efficacy of the size separation.

j. Pool fractions that contain DNA of size $\geq$5 mers, extract with phenol/chloroform, and precipitate with ethanol.

k. Dissolve the DNA pellet in 100 μl H_2O and take an OD_{260} reading to measure the DNA concentration.

3. Coupling of double-stranded oligonucleotide concatemers to CNBr-activated Sepharose 4 Fast Flow (Pharmacia):

a. Swell the required amount of preactivated Sepharose in water. Typically, affinity columns use between 0.1 and 1 ml resin, depending on the scale of the purification.

b. Wash Sepharose with 10 to 15 bed volumes of cold 1 mM HCl as follows: add 1 bed volume of cold 1 mM HCl, stir for 5 min, and then remove the liquid; repeat 10 to 15 times.

c. Wash Sepharose 3$\times$ with 0.1 M potassium phosphate buffer, pH 8 (prepare buffer by 10$\times$ dilution of 1 M stock prepared by mixing 1 M stocks of mono-basic and di-basic potassium phosphate to achieve pH 8).

d. Add the DNA concatemer to the activated Sepharose (250 μg DNA/ml resin as a 50% slurry in 0.1 M potassium phosphate buffer, pH 8); incubate overnight at room temperature on a rotating wheel.

e. Spin down the Sepharose in a microfuge and remove the supernatant. Take an OD_{260} reading of undiluted supernatant to ensure that coupling has occurred—the value should be <0.01.

f. Deactivate the Sepharose by adding 1 ml of 0.1 M Ethanolamine-HCl, pH 8, and incubate for 4 h at room temperature.

g. Add Sepharose to column, and wash with 15 bed volumes of H_2O, 15 bed volumes of 1 M potassium phosphate, pH 8, 15 bed volumes of 0.1 M potassium phosphate, and finally 15 bed volumes of column storage buffer. After adding an excess of storage buffer to the column, cap the column and store at 4 °C until ready for use in chromatography.

B. Affinity Chromatography

1. Buffers:
   a. 10× Column Buffer

	Stock solution	Per liter
200 mM Hepes-KOH (pH7.9)	1 M	200 ml
1 mM EDTA	0.5 M	2 ml
10 mM DTT	1 M	10 ml
1% NP-40	100%	10 ml
1 mM PMSF	100 mM	10 ml

   b. Column Washing and Elution Buffers (Buffer 0.1, 0.2, 0.3, 0.4, 0.5, 0.6, 0.7, 0.8, 0.9, and 1.0)

	Stock solution	Per liter
Column Buffer	10×	100 ml
0.1–1.0 M KCl*	2 M	50–500 ml
20% glycerol	100%	200 ml

*KCl concentration in step gradient used for elution (i.e., 0.1, 0.2, 0.3 ··· 1.0 M).

2. Affinity chromatography (all steps at 4 °C):
   a. Condition affinity column by washing with 10 bed volumes of buffer 0.1, 10 bed volumes of buffer 1.0, and 10 more bed volumes of buffer 0.1.
   b. Bring a measured amount of nuclear extract (containing a quantity of DNA binding protein that is matched to the size of the affinity column) to 0.1 M KCl by adding an appropriate volume of 2 M KCl, and spin out any precipitate.
   c. Load the nuclear extract onto the column, letting it flow through the bed at a rate of approximately 1 ml/10 min.
   d. Wash the column with 30 bed volumes of buffer 0.1, with a flow rate of 1 ml/min (the flow rate for all subsequent steps).
   e. Elute the column with a step gradient of 3 bed volumes each of buffers 0.2, 0.3, 0.4, 0.5, 0.6, 0.7, 0.8, 0.9, and 1.0. Collect each elution step in a single fraction, analyze 1 to 2 μl each by EMSA and 10 to 20 μl each by SDS-PAGE, and store the remainder at −70 °C.
   f. If further purification is necessary (as determined by examination of the SDS-PAGE results), pool the fractions containing the peak DNA binding activity, and dialyze against buffer 0.1 overnight.
   g. Add nonspecific competitor DNA (same as used in EMSA, final concentration of 20 μg/ml) to the dialyzed sample, and repeat steps a–c. It should now be possible to identify the DNA binding protein of interest by SDS-PAGE or reverse phase HPLC.

V. Identification of Affinity-Purified DNA-Binding Proteins in SDS Gels

Sometimes, two rounds of affinity chromatography purify a DNA binding protein to near homogeneity, in which case a final round of purification can be achieved by reverse phase HPLC on a C4 column (see following text). More often than not, however, the DNA binding protein of interest is in a fairly complex mixture, even after two rounds of affinity chromatography. Nonetheless, it is usually pure enough to be detected as a well-separated band on an SDS gel. We have successfully used four different methods for identifying affinity purified transcription factors as bands in SDS gels: (1) "Southwestern" blot analysis, in which the SDS gel is transferred to nitrocellulose and the membrane probed with a ^{32}P-labeled DNA fragment containing the target site of the transcription factor of interest; (2) two-dimensional (2-D) EMSA, consisting of a preparative EMSA in a tube gel followed by SDS-PAGE; (3) pull-down assays using biotinylated oligonucleotides and magnetic avidin beads; and (4) renaturation analysis, in which candidate bands are removed from a gel, and their protein contents eluted, renatured, and subjected to EMSA. Each of these methods has its advantages and disadvantages, and it is best, if possible, to use at least two different methods before concluding that a given band is the band of interest. The biggest advantage of Southwestern blot analysis is its relative simplicity; the disadvantage is that nonspecific DNA binding proteins can often give false-positive signals. This can be avoided to some extent by the use of end-labeled monomeric probes (as opposed to nick-translated concatemers, which was the traditional method). Candidate bands identified by this method should in any case be verified by renaturation analysis. The advantage of 2-D EMSA is that it uses a scaled-up version of the same activity assay that was originally used to identify the protein of interest, allowing one to directly identify each component of an EMSA complex as a band on the SDS gel, including non-DNA binding "piggyback" proteins (Fig. 1B). The disadvantages of this procedure are that, since it relies on silver staining, it doesn't work for very complex mixtures (as some proteins will enter the first dimension without the help of DNA), and it requires the use of tube gel electrophoresis. The latter problem can be avoided by using a variation of the method, in which the DNA complex in a traditional (but preparative scale) EMSA gel is visualized by autoradiography, excised, and then subjected to SDS-PAGE. Micro-scale affinity chromatography ("pull-down") using biotinylated oligonucleotides and magnetic avidin beads in conjunction with silver-stained SDS-PAGE is a reliable method for identifying DNA-binding proteins, and is somewhat simpler than 2-D EMSA. The most reliable method for identification of a DNA binding protein as a band in an SDS gel is to subject the contents of the band to renaturation analysis. This method is particularly attractive when only a few bands are present in a column fraction that contains the activity of interest—in this case, each band can be excised, eluted, renatured, and subject to EMSA. Renaturation analysis becomes impractical as a primary method for identifying DNA-binding proteins in complex mixtures, in which case it is best used as a secondary procedure for verifying candidates identified by one of the other methods.

A. Southwestern Blot Analysis

1. Buffers:

a. Denaturation buffer

	Stock solution	Per liter
20 mM Hepes-KOH (pH7.9)	1 M	20 ml
0.1 mM EDTA	0.5 M	0.2 ml
1 mM DTT	1 M	1 ml
6 M Guanidium-HCl		573.18 gm
1 mM $MgCl_2$	1 M	1 ml

b. 10× Buffer D

	Stock solution	Per liter
0.2 M Hepes-KOH (PH7.9)	1 M	200 ml
1 mM EDTA	0.5 M	2 ml
10 mM DTT	1 M	10 ml
10 mM $MgCl_2$	1 M	10 ml
1% NP-40	100%	10 ml
1.5 M KCl	2 M	750 ml

c. 0.3 M KCl in Buffer D

	Stock solution	Per liter
Buffer D	10×	100 ml
0.3 M KCl	2 M	75 ml
20%Glycerol	100%	200 ml

d. 0.6 M KCl in Buffer D

	Stock solution	Per liter
Buffer D	10×	100 ml
0.6 M KCl	2 M	225 ml
20% Glycerol	100%	200 ml

e. 1.0 M KCl in Buffer D

	Stock solution	Per liter
Buffer D	10×	100 ml
1 M KCl	2 M	500 ml
20% Glycerol	100%	200 ml

2. Perform SDS-PAGE on affinity column fractions containing the sequence-specific DNA binding activity of interest (see protocol in following text), and transfer to nitrocellulose filters using your favorite transfer method.
3. Incubate the filters for 10 min with gentle shaking in 100 ml denaturation buffer.
4. Remove 50 ml denaturation buffer and add 50 ml of 1× buffer D, shaking for 5 min.
5. Repeat step 4 four times.
6. Incubate the filters in 200 ml of buffer D + 5% nonfat dry milk for 10 min.
7. Remove buffer and repeat with fresh buffer D + 5% nonfat dry milk for 10 min.
8. Wash the filters twice with buffer D + 0.25% nonfat dry milk, 5 min each.
9. Incubate the filters with buffer D + 0.25% nonfat dry milk + 25 ng/ml of probe (either multimeric or monomeric binding site labeled with ^{32}P; about 3-6×10^6 cpm/ml) + 10 μg/ml of sonicated calf thymus DNA in 10 ml buffer. Incubate overnight at 4 °C with shaking.
10. Wash and visualize:
    a. Wash with 200 ml 1× buffer D four times, at 5 min/wash.
    b. Use autoradiography or a phosphorimager to visualize bands.
    c. Wash filter with 200 ml buffer D containing 0.3 M KCl four times, at 5 min/wash.
    d. Use autoradiography or a phosphorimager to visualize bands.
    e. Wash with 200 ml buffer D containing 0.6 M KCl four times, at 5 min/wash.
11. Use autoradiography or a phosphorimager to visualize bands. To regenerate filter, wash with 200 ml buffer D containing 1 M KCl four times, at 5 min/wash.

B. Two-Dimensional (2-D) EMSA

1. Buffers: same as for one-dimensional EMSA and SDS-PAGE.
2. Prepare 10% polyacrylamide tube gels (1× TBE buffer) of a size that is compatible with the second dimension SDS gel.
3. Perform preparative-scale EMSA (typically, 20–100× the concentration of probe and protein used in an analytical EMSA, in a final volume of 40 μl) in the tube gel. For this EMSA, the probe DNA does not need to be labeled, since the final detection is by silver-staining. Also, much less nonspecific competitor DNA can be used (typically, 1 μg/reaction).

4. Extrude the tube gel into a Pyrex test tube (with a screw cap) containing 1 ml 2× SDS gel sample buffer, and equilibrate with agitation for 30 min.
5. Heat the tube gel to 95 °C for 5 min to denature protein contents.
6. Lay the tube gel on top of a SDS gel (8–12% polyacrylamide) with a 3% stacking gel, and run at 25 mA until the bromphenol blue dye front is 1 cm from the bottom.
7. Stain the gel by your favorite silver-staining method. Typically, most nonspecific proteins will not enter the first dimension and will be seen as a column of bands on the side of the gel corresponding to the top of the tube gel, whereas the band of interest will appear as a spot or band within a column of the SDS gel that corresponds to the expected location of the DNA–protein complex in the first dimension (see Fig. 1B). Oligonucleotide probe DNA will usually run with the dye front.

C. Magnetic Bead Purification Using Biotinylated Oligonucleotides

1. Dynabeads M-280 streptavidin and Dynabeads KilobaseBINDER kit can be obtained from Dynal company. The latter contains both washing solution and binding solutions, and is suited for binding of large DNA fragments. For oligonucleotides, use M-280 Dynabeads.
2. Biotinylated oligonucleotides can be purchased from Qiagen. Only one strand of a double-stranded oligonucleotide should be biotinylated. Anneal complementary oligonucleotides as has been described.
3. Wash 100 μl resuspended Beads with 200 μl of binding buffer, and capture the beads with magnet. Then, add 200 μl of binding buffer containing 100 picomoles of biotinylated double-stranded oligonucleotide, and rotate at room temperature for 3 h.
4. Wash twice with washing buffer, once with water, and then add 40 μl nuclear extract along with 40 μl of polydIdC (5 μg/μl), 80 μl of 5× binding buffer and 220 μl of H_2O. Rotate at 4 °C for 3 h to overnight.
5. Transfer unbound fraction to another tube, wash once with 500 μl 1× binding buffer, 4 °C for 30 min.
6. Wash once with 500 μl 1× binding buffer with polydI-dC (40 μl of 5 μg/μl), 4 °C for 30 min.
7. Wash once with 1× binding buffer for 30 min.
8. Elute with 50 μl 1× binding buffer containing 1 M KCl, 4 °C, 30 min.
9. Elute with 50 μl 1× binding buffer containing 1 M KCl.
10. Analyze eluates by SDS-PAGE/silver staining (along with affinity column fractions for comparison).

D. Renaturation Analysis

1. Buffers:
   a. Elution buffer:

	Stock solution	Per 100 ml
50 mM Tris-HCl (pH 7.6)	1 M	1 ml
0.1 mM EDTA	0.5 M	0.2 ml
5 mM DTT	1 M	0.5 ml
150 mM NaCl	5 M	3 ml
0.1% SDS	10%	1 ml

   b. Denaturation buffer: Same as for Southwestern blot Analysis (see Preceding text).
   c. Renaturation buffer:

	Stock solution	Per 100 ml
10 mM Hepes-KOH (pH 7.9)	1 M	1 ml
100 mM KCl	2 M	5 ml
1 mM DTT	1 M	0.1 ml
10% glycerol	100%	10 ml
0.1% NP-40	100%	0.1 ml

2. Perform SDS-PAGE on an affinity column fraction containing the activity of interest. It may be necessary to first concentrate the contents by acetone precipitation (or TCA precipitation), as will be described.
3. Visualize the bands using the BioRad Zinc stain kit: Place the SDS-PAGE gel in a container with diluted Imidazole, Solution A, rotating for 10 min.
4. Transfer the gel to diluted zinc sulfate, Solution B. Completely immerse the gel for even staining. Allow 30 to 60 s for the gel to develop.
5. To stop the reaction, transfer the gel to a container filled with H_2O, and rinse for 3 min, replacing with fresh H_2O.
6. Cut the band out and transfer to a 1.5 ml Eppendorf tube.
7. Destain with 1× Tris/Glycine buffer three times, 1 ml each.
8. Crush the gel slice in 0.5 ml elution buffer, and incubate for 5 to 6 h at room temperature. (Note: alternatively, electroelution can be used here; for this, we prefer the Centrilutor apparatus from Amicon).
9. Spin at 12,000 rpm for 5 min, remove supernatant, wash pellet twice with 0.2 ml H_2O, and combine washes with original supernatant (final volume = 0.9 ml). Add 5 ml ice-cold acetone, mix, and allow to precipitate overnight at −20 °C.
10. Spin down acetone precipitate at 10,000*g* for 20 min, then wash protein pellet once in ice-cold acetone.
11. Dissolve pellet in 100 μl denaturation buffer.

12. Layer denatured protein onto a 0.5 ml Biogel P6 (BioRad) column that has been preequilibrated in renaturation buffer and allow to enter the bed.
13. Layer 1 ml of renaturation buffer onto the column bed, and allow to flow through by gravity while collecting 100 μl fractions. Typically, the renatured protein elutes beginning in the second fraction.
14. Analyze the fractions by EMSA to determine whether or not the SDS gel slice contains the DNA binding activity of interest.

VI. Preparation of Affinity-Purified Transcription Factors for Protein Sequencing

The most technically unforgiving step in the biochemical identification of transcription factors is the preparation of the purified protein for amino acid sequence analysis. The reason is the large loss of material that occurs during the procedure, which is particularly vexing when dealing with low-abundance transcription factors. In general, direct amino acid sequencing of purified proteins by Edman degradation does not work because the N-terminus is usually blocked. Therefore, most proteins must be fragmented into peptides by proteolytic digestion, and the peptides separated by HPLC for amino acid sequencing. Until recently, the lack of a complete genome sequence in sea urchins has made it impractical to use mass spectrometric approaches to obtaining amino acid sequences, necessitating purification of enough peptide to allow sequencing by Edman degradation. This has been facilitated somewhat by high-sensitivity protein micro-sequencers, such as the Procise system from ABI. However, with the imminent completion of the *Strongylocentrotus purpuratus* genome project, mass spectrometric methods of protein sequencing will become practical. The protocols to be described are compatible with protein sequencing by either Edman degradation or mass spectrometry. (For mass spectrometric sequencing, it may be useful to substitute trypsin for LysC in the digestion steps.)

We have successfully used two different approaches for final preparation of transcription factors for protein sequence analysis. The first method involves preparative SDS-PAGE, followed by excision of the band of interest and *in situ* digestion of the band's contents with *Achromobacter* lysylendopeptidase C (lysC) or trypsin. This has been the method of choice for complex mixtures of proteins. For more highly purified proteins that are not too large, final purification can be achieved by HPLC using a C4 column, followed by digestion of fractions that contain the protein/activity of interest. With this method, losses are minimized compared to purification by SDS-PAGE. Also, with the HPLC method, an SDS gel band does not necessarily have to be identified prior to final purification, as EMSA can be applied to the HPLC column fractions to locate the activity of interest, which will often be purified to near-homogeneity. Both the SDS-PAGE and HPLC methods of protein preparation are described here.

A. Protein Preparation by SDS-PAGE

1. Concentrate sample by TCA precipitation (15% final concentration), and precipitate overnight at −20 °C.
2. Centrifuge at 14K rpm for 30 min.
3. Remove supernatant, and add 1 ml of ethanol.
4. Centrifuge at 14K rpm for 20 min.
5. Remove supernatant, speed-vac dry.
6. Resuspend in 2X sample buffer containing β-mercaptoethanol, either 20 μl or 40 μl.
7. Incubate at 50 °C for 1 to 2 h.
8. Add 1 μl of 4-vinylpyridine per 20 μl total volume to block sulfhydryls.
9. Incubate in dark for 30 to 60 min.
10. Heat at 65 °C for 3 min.
11. Immediately load on the SDS gel (aged for more than 3 days), and run at 25 mA.
12. Stain the gel with Coommassie Blue, Sypro Orange, or Sypro Ruby protein stain.
13. Cut the band out and transfer to a 1.5 ml Eppendorf tube.
14. Destain using the ProteoProfile trypsin in-gel digest kit (Sigma), 2×, 1 ml each.
15. Wash the gel pieces with 250 μl 200 mM NH_4HCO_3/50% CH_3CN, 30 min at room temperature.
16. Remove liquid phase, and speed-vac.
17. Prepare a 0.033 mg/ml solution of lysC (WAKO; 25 μl of 1 mg/ml stock solution with 725 μl of 200 mM NH_4HCO_3), or use trypsin from ProteoProfile trypsin in-gel digest kit (Sigma).
18. Add enough lysC or trypsin solution to just cover the gel pieces.
19. Incubate overnight at 37 °C.
20. Extract peptides by adding 150 ul of 0.1% TFA, 60% CH_3CN and shaking on a rocker table at room temperature for 60 min.
21. Remove supernatant containing peptides and repeat extraction with 150 ul 0.1% TFA, 60% CH_3CN.
22. Speed-vac the combined washes to a volume of ~100 μl, and proceed to HPLC purification of the peptides.

B. Protein Preparation by HPLC

1. Dialyze pooled affinity column fractions containing the activity of interest against affinity column buffer 0.1 *without glycerol.*

2. Concentrate to $\leq$500 μl in a Centricon 10 (Amicon).
3. Add 1/10 volume of 10% TFA, and proceed with reverse phase HPLC (on a C4 column), as will be described. As for HPLC in general, it may be necessary to perform multiple runs to optimize the gradient profile.
4. Perform EMSA and SDS-PAGE analysis on fractions to locate the activity of interest and assess its purity.
5. Pool the peak fractions and concentrate in a speed vac to ~100 μl.
6. If the activity of interest is present as a single band (or, at most, 2 or 3 bands), proceed to step 7; if not, bring up in HPLC buffer A, and repeat the HPLC (if necessary, with an adjusted gradient profile).
7. Add lysC to a final concentration of 0.033 mg/ml, incubate overnight at 37 °C, and proceed to HPLC purification of the peptides.

C. HPLC

(Note: A flow splitter can be used to divert a portion of the HPLC eluate to a mass spectrometer equipped for ion spray, thus allowing the simultaneous collection of mass spectrometric data for each peak eluted from the HPLC.)

1. Equilibrate 0.2 mm i.d. reverse phase column (C4 for proteins, C18 for peptides) with buffer A (0.1% TFA, 2% CH_3CN in HPLC-grade water).
2. Load sample (in 300 ul buffer A) to the sample loop (500 ul volume).
3. Start the gradient after 35 min of equilibration and transfer the sample from the sample loop to the column.
4. Elute column with a linear gradient from 0% buffer B (0.09% TFA and 70% CH_3CN in HPLC-grade water) at 35 min to 100% buffer B at 95 minutes. Collect 200 μl fractions for further analysis (e.g., SDS-PAGE, activity assays, or amino acid sequencing).
5. Monitor HPLC using UV with wavelength set at 214 and 280.

VII. Prospects

With the completion of the *Strongylocentrotus purpuratus* genome project and the availability of high-sensitivity mass spectrometric methods of protein identification (Wolters *et al.*, 2001), the biochemical approach described here will become even more efficient and applicable to smaller quantities of protein. Moreover, most of the methods can be adapted to high-throughput approaches; for example, simultaneous affinity chromatography of several different DNA binding proteins can be performed in tandem on an automated platform (Coffman *et al.*, 1992). In addition to providing the protein biochemical component of hard core *cis*-regulatory analysis, this facilitates the rapid biochemical validation of regulatory interactions that are predicted computationally and/or by network perturbation analyses (Brown *et al.*, 2002; Davidson *et al.*, 2002). Even with the high sensitivity afforded by mass spectrometry, however, biochemical analyses inherently require relatively large amounts of starting material. Because of this, the sea urchin embryo will remain the model system of choice for biochemical approaches to the study of developmental gene regulation.

References

Brown, C. T., Rust, A. G., Clarke, P. J., Pan, Z., Schilstra, M. J., De Buysscher, T., Griffin, G., Wold, B. J., Cameron, R. A., Davidson, E. H., and Bolouri, H. (2002). New computational approaches for analysis of cis-regulatory networks. *Dev. Biol.* **246,** 86–102.

Calzone, F. J., Hoog, C., Teplow, D. B., Cutting, A. E., Zeller, R. W., Britten, R. J., and Davidson, E. H. (1991). Gene regulatory factors of the sea urchin embryo. I. Purification by affinity chromatography and cloning of P3A2, a novel DNA-binding protein. *Development* **112,** 335–350.

Calzone, F. J., Theze, N., Thiebaud, P., Hill, R. L., Britten, R. J., and Davidson, E. H. (1988). Developmental appearance of factors that bind specifically to cis-regulatory sequences of a gene expressed in the sea urchin embryo. *Genes Dev.* **2,** 1074–1088.

Char, B. R., Bell, J. R., Dovala, J., Coffman, J. A., Harrington, M. G., Becerra, J. C., Davidson, E. H., Calzone, F. J., and Maxson, R. (1993). SpOct, a gene encoding the major octamer-binding protein in sea urchin embryos: Expression profile, evolutionary relationships, and DNA binding of expressed protein. *Dev. Biol.* **158,** 350–363.

Coffman, J. A., Kirchhamer, C. V., Harrington, M. G., and Davidson, E. H. (1996). SpRunt-1, a new member of the runt domain family of transcription factors, is a positive regulator of the aboral ectoderm-specific CyIIIA gene in sea urchin embryos. *Dev. Biol.* **174,** 43–54.

Coffman, J. A., Kirchhamer, C. V., Harrington, M. G., and Davidson, E. H. (1997). SpMyb functions as an intramodular repressor to regulate spatial expression of CyIIIa in sea urchin embryos. *Development* **124,** 4717–4727.

Coffman, J. A., and Leahy, P. S. (2000). Large-scale culture and preparation of sea urchin embryos for isolation of transcriptional regulatory proteins. *Methods Mol. Biol.* **135,** 17–23.

Coffman, J. A., Moore, J. G., Calzone, F. J., Britten, R. J., Hood, L. E., and Davidson, E. H. (1992). Automated sequential affinity chromatography of sea urchin embryo DNA binding proteins. *Mol. Mar. Biol. Biotechnol.* **1,** 136–146.

Davidson, E. H., Rast, J. P., Oliveri, P., Ransick, A., Calestani, C., Yuh, C. H., Minokawa, T., Amore, G., Hinman, V., Arenas-Mena, C., Otim, O., Brown, C. T., Livi, C. B., Lee, P. Y., Revilla, R., Schilstra, M. J., Clarke, P. J., Rust, A. G., Pan, Z., Arnone, M. I., Rowen, L., Cameron, R. A., McClay, D. R., Hood, L., and Bolouri, H. (2002). A provisional regulatory gene network for specification of endomesoderm in the sea urchin embryo. *Dev. Biol.* **246,** 162–190.

DiLiberto, M., Lai, Z. C., Fei, H., and Childs, G. (1989). Developmental control of promoter-specific factors responsible for the embryonic activation and inactivation of the sea urchin early histone H3 gene. *Genes Dev.* **3,** 973–985.

Ganster, R., Paul, H., and Katula, K. S. (1992). Analysis of the DNA binding proteins interacting with specific upstream sequences of the S. purpuratus CyI actin gene. *Mol. Reprod. Dev.* **33,** 392–406.

Katula, K. S., Dukes, R. L., Paul, H., and Franks, R. R. (1998). Modifications in protein binding to upstream sequences of the sea urchin cytoplasmic actin gene CyIIa in comparison to its linked neighbors, CyI and CyIIb. *Gene* **213,** 195–203.

Theze, N., Calzone, F. J., Thiebaud, P., Hill, R. L., Britten, R. J., and Davidson, E. H. (1990). Sequences of the CyIIIa actin gene regulatory domain bound specifically by sea urchin embryo nuclear proteins. *Mol. Reprod. Dev.* **25,** 110–122.

Wolters, D. A., Washburn, M. P., and Yates, J. R., 3rd. (2001). An automated multidimensional protein identification technology for shotgun proteomics. *Anal. Chem.* **73,** 5683–5690.

Yuh, C. H., Ransick, A., Martinez, P., Britten, R. J., and Davidson, E. H. (1994). Complexity and organization of DNA-protein interactions in the 5′-regulatory region of an endoderm-specific marker gene in the sea urchin embryo. *Mech. Dev.* **47,** 165–186.

Zeller, R. W., Coffman, J. A., Harrington, M. G., Britten, R. J., and Davidson, E. H. (1995). SpGCFI, a sea urchin embryo DNA-binding protein, exists as five nested variants encoded by a single mRNA. *Dev. Biol.* **169,** 713–727.

CHAPTER 27

Expression of Exogenous mRNAs to Study Gene Function in the Sea Urchin Embryo

Thierry Lepage and Christian Gache

Laboratory of Developmental Biology
CNRS-Université Pierre et Marie Curie (Paris VI)
Observatoire Océanologique
06230 Villefranche-sur-Mer, France

Overview

Expression of exogenous mRNAs in the embryo is a basic method to study gene function during development of invertebrate deuterostomes. Chapter 27 shows how the method can be used to selectively perturb gene expression and how it can be combined with blastomere manipulation or differential screening. It highlights examples of successful experiments, describes the different steps involved, and details protocols for the *in vitro* transcription of the mRNA.

0091-679X/04 $35.00

I. Introduction

A. Background

Analysis of gene function in the embryos of basal deuterostomes is required to unravel the specific features of their development and to understand the evolution of animal body plans. Unfortunately, those animals which are models for developmental studies cannot be easily submitted to methods from classical genetics and thus we can not benefit from the powerful gain-and-loss-of-function analyses based on this approach. However, for some years now, methodologies have been developed to circumvent this limitation. These methods are based on the misexpression of genes through expression of exogeneous mRNAs. The microinjection technique first designed for transferring plasmid DNA encoding reporter genes into eggs can be used to introduce a variety of molecules that can potentially interfere with gene function in the embryo: DNA constructs coding for active molecules, antisense oligonucleotides including morpholino-oligonucleotides, double-stranded RNAs, mRNAs, and proteins including antibodies. The utilization of antibodies and morpholino-oligonucleotides is reviewed by Angerer and Angerer in Chapter 28. Here, we focus on the approaches based on expression of exogenous mRNAs in the sea urchin embryo.

B. Purpose of Approaches

Expression of exogenous mRNA is one of the most effective methods to study gene function in marine invertebrate embryos. In the sea urchin embryo, microinjection of synthetic mRNA can be carried out at an early stage. Translation is strongly activated at fertilization, and mRNAs injected into eggs or blastomeres are rapidly translated into their protein products. Since mRNAs are synthesized *in vitro* from DNA templates that can be easily manipulated, virtually any native or modified (recombinant) protein can be expressed in the developing embryo. The developmental defects obtained are analyzed at the morphological and at the molecular levels using Q-PCR, *in situ* hybridization, or immunolocalization. The method is simple, powerful, and versatile. It gives insights about gene function and regulation through several approaches.

1. Gain/Loss of Function

The major aim of the method is to mimic the effects of a gain of function/loss of function at the gene level by misexpressing a protein, or by increasing or decreasing its activity. Gain of function can be obtained by expression or misexpression of wild-type proteins, proteins whose activity has been increased or proteins that have been modified to be constitutively active. Loss of function can be obtained by taking advantage of the dominant-negative effects produced by modified molecules that specifically compete with the endogenous gene product. For most of the proteins that are key regulators of development, like kinases, receptors, transcription factors, secreted ligands, and other signaling molecules, it is possible to design variants whose activity is increased, totally abolished, or deregulated. For example, kinases can be inactivated by point mutation at the ATP binding site or by deletion of a large part of their kinase domain; proteins that are regulated by phosphorylation can be switched to constitutively active or constitutively inactive forms by emulating phosphorylation through mutation into E or D of the amino acid residues that receive the phosphate group; some receptors truncated from part of their extracellular domain become constitutively active; transcription factors can be deleted from their regulatory domain while their DNA binding domain is left intact; transcriptional activators can have their activity increased by addition of the VP16 domain or can be turned into repressors by replacing their activation domain with the repressor domain from engrailed, and conversely, repressors can be changed into activators or their repressor activity can be enhanced. In all cases, useful controls are provided by injection of appropriate mRNA constructs: mRNAs in which a frame-shift is introduced near the beginning of the ORF; constructs in which all the coding sequence of the protein tested, except the initiation codon and a few downstream codons, are replaced by the coding sequence of one of the GFPs.

In addition, mRNA injection can complement experiments with morpholino-oligonucleotides. Constructs harboring the same coding sequence but a 5′ UTR different from that of the endogenous mRNA can potentially rescue the effects of morpholino-oligonucleotides and thus are useful to assess their specificity.

2. Gain/Loss of Function. Spatial Targeting and Temporal Restriction

In the basic version of the method, mRNA microinjection into the egg leads to uniform expression of a gene product in the embryo. For a spatially restricted gene, this corresponds to a massive misexpression. In many cases, it would be also appropriate to express the mRNA only in a defined subset of cells, which may or may not correspond to the normal expression territory of the gene under study. To this aim, mRNA can be microinjected into selected blastomeres. However, consistent injection into single blastomeres is difficult beyond the 16-cell stage. The combination of microinjection into eggs with blastomere manipulation offers a powerful alternative. Blastomeres from microinjected embryos can be transplanted to normal embryos either by replacing their normal counterpart or by grafting at ectopic sites. Therefore, single cells or groups of cells expressing an exogenous mRNA can be placed in a native surrounding, at normal or ectopic sites, which allows assessing the properties endowed by the injected mRNA. For sea urchin embryos this method is limited to early cleavage stages, when micromeres, macromeres, mesomeres, and their first progeny can be identified and manipulated.

Since microinjection of mRNAs is most conveniently carried out in eggs or early blastomeres, the exogenous protein product is present almost continuously from the very early stages and throughout development if the mRNA or the protein is stable. For genes that are temporally regulated, such an anachronic expression may lead to artifacts. Unfortunately, no tool allowing experimental control of translation is available. However, nuclear proteins are a special case. Their activity can be controlled through fusion to the hormone-inducible glucocorticoid receptor. The fusion protein is retained in the cytoplasm until the addition of dexamethasone at the desired time, thereby releasing the hsp90 block and allowing entry into the nucleus.

3. Search for Downstream Genes

Interfering with the expression of a particular gene by mRNA injection is a tool to identify downstream targets. Known genes whose expression pattern and function make them likely candidates as target genes can be individually tested for their response to an upstream perturbation. The variation of their expression level is measured by Q-PCR and the alteration of their expression territory visualized by *in situ* hybridization. Combining RNA expression with new methods for differential screening gives a more powerful and unbiased approach that allows discovering new genes. The synthesis of the probes required by this method involves amplification steps by PCR and thus can be obtained from a relatively small number of embryos. Two complex probes are derived from two populations of embryos in which the gene under study has been differentially altered. Subtractive hybridization between the two probes produces enrichment in transcripts that are differentially expressed. Hybridization of arrayed libraries using the unsubtracted and the subtracted probes permits detection of transcripts from genes that are likely to be direct or indirect targets of the gene studied.

4. Studying Structure–Function Relationships *In Vivo*

Various features of mRNAs and proteins can be studied *in vivo* following their expression in embryos. The sequence of 3'UTRs can be manipulated to understand their role in RNA localization and stability. The function of isolated protein domains can be investigated and GFP chimeras allow monitoring the stability and subcellular localization of proteins.

C. Specific Examples

Expression of exogenous mRNAs has been used successfully to study the function of several genes in embryos from most sea urchin species currently used, including *S. purpuratus, L. variegatus, P. lividus*, and *H. pulcherrimus*. Only a few illustrative examples will be cited here within their conceptual framework. Some results are presented in Fig. 1. The method was also used to study other genes including *Bep* (Montana *et al.*, 1998), *ets1* (Kurokawa *et al.*, 1999), *otx* (Li *et al.*, 1999), *brachyury* (Gross and McClay, 2001), *Krl* (Howard *et al.*, 2001), *otp* (Cavalieri *et al.*, 2003), *coquillette-Tbx2/3* (Croce *et al.*, 2003; Gross *et al.*, 2003), *Alx1* (Ettensohn *et al.*,

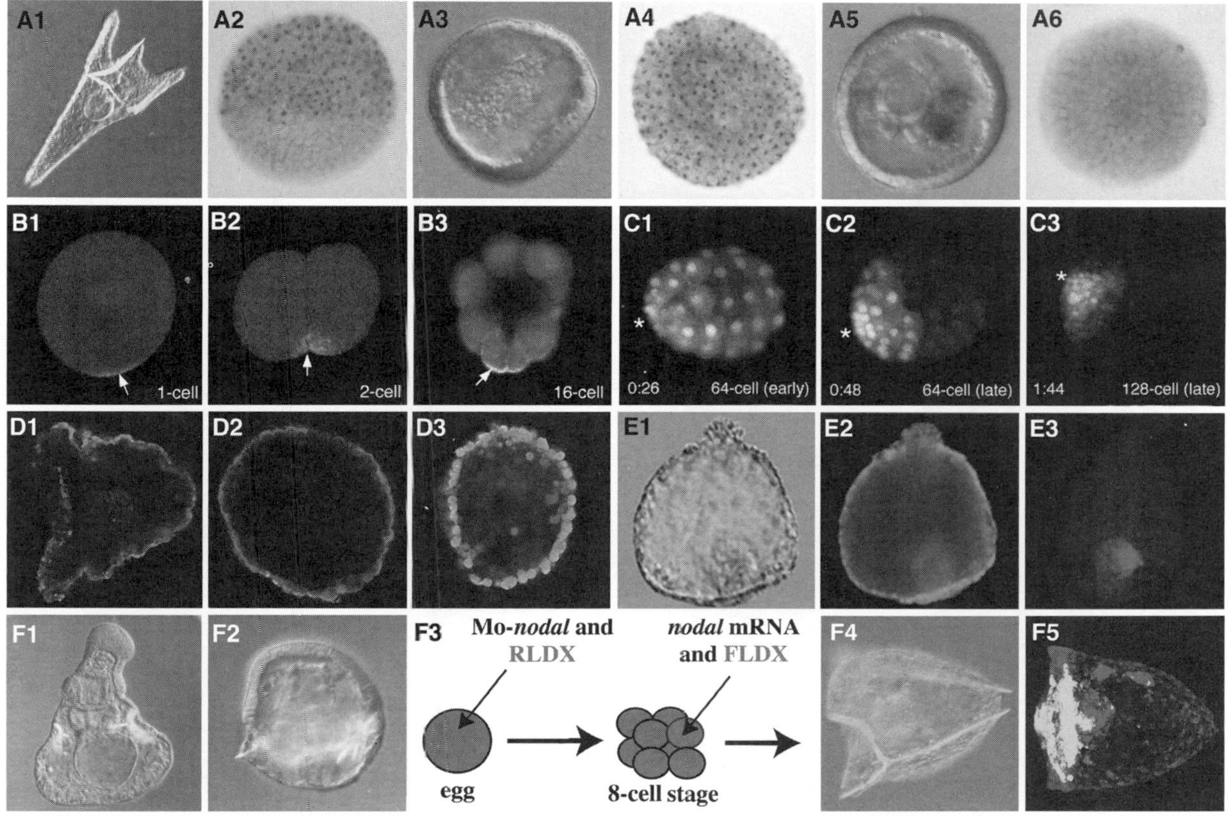

Fig. 1 (*continued*).

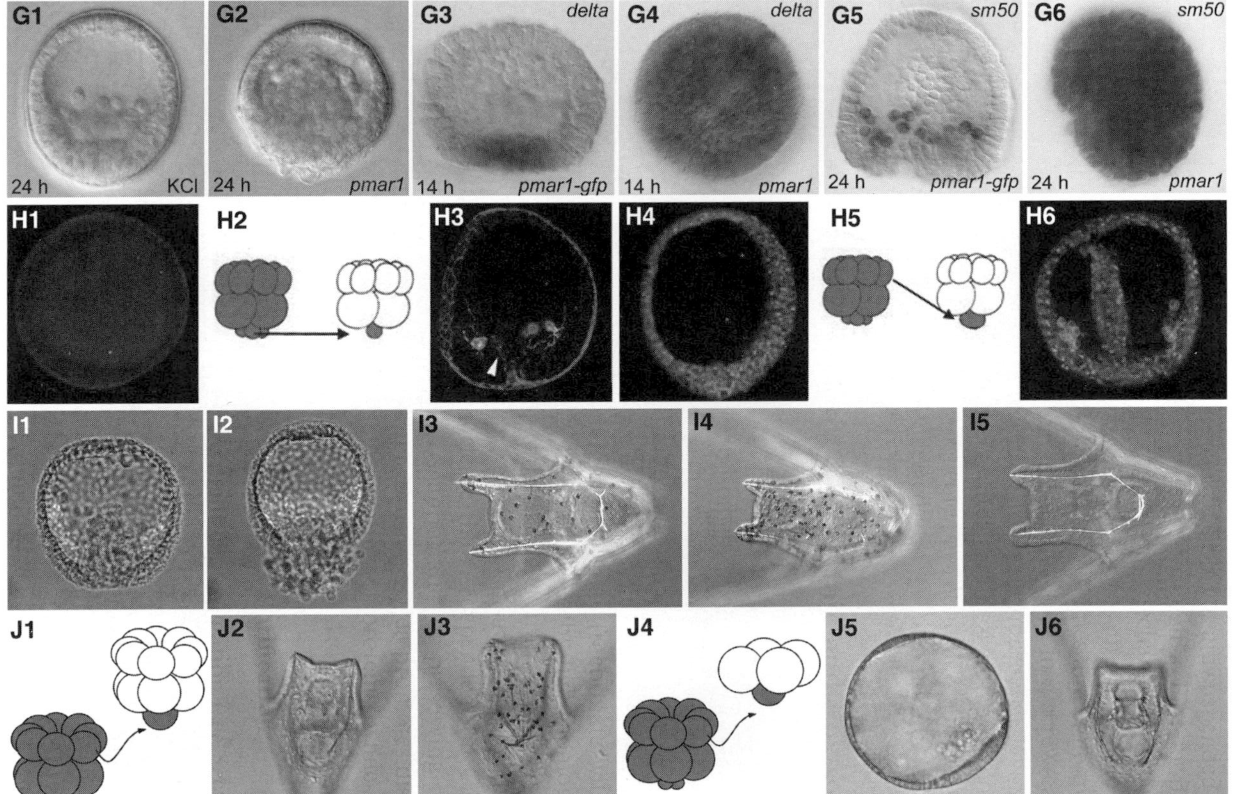

Fig. 1 Sea urchin embryos expressing synthetic mRNAs encoding wild-type or mutated forms for (A) GSK3β; (B) Dishevelled-GFP; (C) β-catenin-GFP; (D) Goosecoid; (E) BMP2/4; (F) Nodal and antivin; (G, H) Pmar1; (I) Notch, and (J) Delta. (A) GSK3β. DIC images of 48 h embryos and HE immunolabeling at the blastula stage. A1, 2, control embryos; A3, 4, embryos overexpressing wt-GSK3β; A5, 6, embryos expressing dn-GSK3β. Increasing or decreasing the activity of GSK3β shifts the presumptive ectoderm–endoderm border and the border of the *HE* expression domain toward the vegetal pole or the animal pole, respectively (from Emily-Fenouil *et al.*, 1998). (B) Dishevelled-GFP. A mRNA construct coding for Dishevelled-GFP is injected into the egg. The Dishevelled-GFP fusion protein localizes to the vegetal cortex of the egg (B1). This cortical area is bisected by the first cleavage plane (B2) and is inherited later by the vegetal-most blastomeres (B3) (from Weitzel *et al.*, 2004). (C) β-catenin-GFP. Following injection of β-catenin-GFP mRNA into the egg, fluorescence is monitored by 4-D confocal imaging. Frames from a time-lapse sequence at the indicated time (h:min) and stages. Initially, the GFP-tagged β-catenin is present in all blastomeres (C1) but over one or two rounds of cell division, it becomes restricted

2003), *SoxB1* (Kenny *et al.*, 2003), and *ERK* (Fernandez-Serra *et al.*, 2004; Rottinger *et al.*, 2004).

1. The Wnt Pathway and the Animal–Vegetal Axis

The role of the maternal components of the Wnt pathway in patterning the animal–vegetal axis was studied using a set of synthetic mRNAs coding for wild-type or modified proteins from sea urchin or Xenopus (Emily-Fenouil *et al.*, 1998; Huang *et al.*, 2000; Logan and McClay, 1999; Vonica *et al.*, 2000; Weitzel *et al.*, 2004; Wikramanayake *et al.*, 1998). The DIX domain of Dishevelled is required for signaling by the canonical Wnt pathway and behaves as a dominant negative when expressed alone. A dominant-negative form of GSK3-β is made by mutating of a critical lysine residue at the active center. β-catenin mutated on residues that are phosphorylated by GSK3-β is not addressed for degradation, which increases

---

to the nuclei of the vegetal-most blastomeres (C2, C3) (from Weitzel *et al.*, 2004). (D) Goosecoid. Three-day-old embryos stained with antibodies against Ecto V(red) and Spec1 (green). (D1) control embryo expressing EctoV on the oral side and Spec1 on the aboral side of the ectoderm. (D2) embryo injected with *gsc* mRNA express EctoV uniformly in the ectoderm. (D3) embryo injected with *gsc-VP16* mRNA express Spec1 in all ectodermal cells (from Angerer *et al.*, 2001). (E) BMP2/4. Three day-old embryos overexpressing BMP2/4. (E1) DIC. (E2) Immunolabeling of Spec 1 (green, aboral ectoderm). (E3) Immunolabeling of Ecto V (red; oral ectoderm and foregut). BMP2/4 promotes aboral fates. (from Angerer *et al.*, 2000). (F) Nodal. (F1) embryo radialized by overexpression of *nodal.* (F2) Embryo radialized by injection of *antivin* mRNA (Microinjection of a morpholino oligonucleotide directed against *nodal* transcripts (MO-*nodal*) produces an identical phenotype). (F3) After injection of MO-*nodal* and RLDX into the egg, embryos are rescued by injection of *nodal* mRNA in one blastomere at the 8-cell stage together with FLDX as a lineage marker. Rescued embryo at the pluteus stage: (F4) DIC. (F5) Confocal image showing that the progeny of *nodal* injected cells contributes to the oral field. (from Duboc *et al.*, 2004). (G) Pmar1. Embryos injected with the constructs indicated in the lower right corner. (G1, G2) DIC. (G3–G6) WMISH with probes for the genes indicated in the upper right corner. Misexpression of *p-mar1* convert any cell into mesenchyme cell. (H) Pmar1. Micromereless embryos do not form spicules and do not gastrulate. Micromeres taken from embryos depleted of β-catenin by injection of *cadherin* mRNA or mesomeres from normal embryos cannot rescue micromeless embryos when grafted at their vegetal pole (H1, H4). When micromeres from embryos coinjected with *cadherin* and *pmar1* mRNAs or mesomeres from embryos injected with pmar mRNA are grafted at the vegetal pole of micromereless embryos, (H2, H5), PMC formation and gastrulation are rescued (H3, H6) (from Oliveri *et al.*, 2002, 2003). (I) Notch. (I1) normal embryo at the gastrula stage; (I2) gastrula stage embryo expressing an activated form of Notch. The PMCs (red label) are in a characteristic ring-pattern. The cells that extrude from the vegetal pole are SMCs. Pluteus stage embryos (I3) noninjected, (I4) overexpressing Notch, (I5) expressing dn-Notch. Embryos over-expressing Notch have a large number of pigment cells, while in the absence of Notch function, the embryo is albino (from Sherwood and McClay, 1999). (J) Delta. (J1) transplantation of mesomeres from embryos expressing truncated or normal forms of Delta at the vegetal pole of micromereless embryos. (J2) transplantation of mesomeres expressing a truncated form of Delta results in embryos that have few pigmented cells. (J3) Embryos grafted with mesomeres expressing the full-length Delta develop supernumerary pigmented cells. (J4) Transplantation of mesomeres expressing truncated and normal forms of Delta at the vegetal side of animal halves. (J5) Isolated animal caps develop into "dauer blastulae," even when grafted with mesomeres expressing a truncated form of Delta. (J6) Mesomeres expressing delta induce animal caps to develop into plutei (from Sweet *et al.*, 2002).

its stability and thus its activity level. Conversely, overexpression of C-cadherin fragment, which binds β-catenin, depletes the β-catenin pool. Controls can be performed using an inactive β-catenin deleted with the armadillo repeats 5–13 and the C-terminal domain. Deletion of the N-terminal β-catenin binding domain of TCF produces a dominant negative-form, while replacing the same domain by the activation domain of VP16 yields a constitutive activator. In addition, a GR-TCF construct allows releasing an active TCF form at the desired time. The DIX domain of Dishevelled, wild-type GSK3-β, C-cadherin, and dn-TCF animalize embryos. In contrast, dnGSK3-β, stable β-catenin, and activated TCF promote vegetal fates and thus vegetalize the embryo (Fig. 1, A1–6). Using the GR-TCF chimera, released at various times by addition of dexamethasone, it has been shown that TCF is required before the 60-cell stage, confirming that the maternal components of the wnt pathway are required at very early stages.

Dishevelled is the most upstream of the maternal components of the wnt pathway that have been identified. A dsh-GFP fusion protein expressed from injected mRNA localizes to the vegetal cortex of the egg (Fig. 1, B1–3). During early cleavage, this cortical area is inherited by blastomeres of the vegetal area where activation of the Wnt pathway signaling leads to the cellular autonomous nuclear accumulation of β-catenin. Expression of a series of mutated Dishevelled mRNAs indicates that the N-terminal half of the protein is required for its targeting to the vegetal cortex.

The β-catenin level, which is critical for signaling, is directly related to its stability. Microinjection into the egg of a mRNA coding for a GFP-tagged Xenopus βcatenin leads to the production of a fusion protein whose level can be monitored in live embryos. When fluorescence is first detected, the protein is uniformly distributed. However, during early cleavage stages, the fluorescence decreases rapidly in the animal blastomeres and the βcatenin-GFP becomes restricted to the nuclei of the vegetal-most blastomeres (Fig. 1, C1–3), following a pattern similar to that of the endogenous βcatenin. This restriction is suppressed if an mRNA coding for a dominant-negative form of GSK3β is coinjected with the β-catenin-GFP mRNA or if β-catenin is stabilized by mutation of the phosphorylation sites for GSK3β. Measurements of the rate of fluorescence decay in the mesomere, macromere, and micromere territories demonstrate differential stability of β-catenin along the AV axis.

2. The Nodal/BMP/Goosecoid Pathway and the Oral–Aboral Axis

Several molecules that are implicated in the specification of the oral–aboral axis have been identified recently—goosecoid (gsc), BMP2/4, antivin, and nodal—that are all exclusively expressed on the oral side of the embryo (Angerer *et al.*, 2000, 2001; Duboc *et al.*, 2004).

Goosecoid is a well-known transcriptional repressor and injection of *gsc* mRNAs alters the oral–aboral (OA) polarity. Misexpression of wild-type gsc drives ectoderm toward oral fates while expression of a gsc-VP16 chimera promotes aboral fates (Fig. 1, D1–3). Embryos expressing BMP2/4 lack morphological polarity in the ectoderm along the OA axis. They are composed of mostly squamous epithelium and express uniformly an aboral marker (Fig. 1, E1–3). Thus, ectopic expression of BMP2/4 promotes aboral fates whereas it is normally expressed in oral ectoderm. Misexpression of *nodal* by mRNA injection into the egg provokes a strong radialization (Fig. 1, F1). Markers of the oral ectoderm are expressed radially in extended territories and, in contrast, the expression of aboral markers is abolished. Expression of antivin, an antagonist of nodal, produces phenotypes identical to those obtained by injection of a morpholino-oligonucleotide directed against *nodal* transcripts (MO-*nodal*) (Fig. 1, F2).

Microinjection of *nodal* mRNA can potentially rescue the effect of MO-*nodal*, provided that the complementary sequence is not included in the synthetic mRNA. Microinjection of MO-*nodal* into the egg, followed by microinjection of *nodal* mRNA into a single blastomere at the 8-cell stage, rescues, to a large extent, the OA polarity in a large fraction of the embryos. Furthermore, labeling of the progeny of the cells injected with the *nodal* mRNA shows that the *nodal* expressing cells contribute exclusively to the oral ectoderm in nearly all rescued embryos (Fig. 1, F3–5). This suggests that *nodal* contributes to the orientation of the OA axis.

3. Micromere Specification

Pmar1, a transcription factor zygotically expressed in the micromere lineage, plays a central role in the specification of the micromeres (Oliveri *et al.*, 2002, 2003). This was established through expression of exogenous mRNAs, one RNA coding for the wild-type protein and two constructs in which the repressor domain of engrailed was fused to the N-terminal of pmar1 (en-pmar1-tot) or to the homeodomain of pmar1 (en-pmar1-hd). Expression of these mRNAs results in similar phenotypes, suggesting that pmar1 is a transcriptional repressor. Embryos expressing ectopically any of these constructs develop apparently normally to the hatching blastula stage. While in non-injected embryos the 16 PMC precursors located at the vegetal pole ingress into the blastocoel, in embryos expressing a *pmar1* construct, a large number of cells ingress (Fig. 1, G1, 2). The area of ectopic ingression is dose-dependent. It extends from the vegetal pole to the presumptive endoderm territory and often to a large part of the animal blastomere progeny. When high doses of the *engrailed-pmar1* fusion mRNAs are microinjected, almost every cell in the embryo is converted into migrating cells. Q-PCR measurements showed that expression of endoderm markers was decreased while that of genes involved in key functions carried out by the micromeres were strongly increased, including the ligand Delta, skeletogenic transcription factors (*ets1, tbr, dri*), and terminal differentiation genes of the primary mesenchyme lineage (*sm50, sm30, msp130, cyclophylin*). *In situ* hybridization shows that genes whose expression level is increased are expressed in almost every cell of the embryo (Fig. 1, G3–6). Ectopic expression of pmar1 converts any cell into a mesenchyme-like cell by derepressing regulatory genes that control specification, signaling, and differentiation of the micromere lineage.

Micromere specification depends on the nuclear accumulation of β-catenin. Micromeres derived from embryos in which β-catenin has been depleted by expression of an mRNA encoding cadherin are not properly specified. When grafted at the vegetal pole of micromereless embryos, they fail to induce archenteron formation and they do not form PMCs.

However, micromeres originating from embryos coinjected with *cadherin* and *pmar1* mRNAs behave as normal micromeres when swapped with micromeres from normal hosts (Fig. 1, H1–H3). Furthermore, mesomeres taken from embryos microinjected with *pmar1* RNA (with or without co-injection of *cadherin* mRNA) and grafted at the vegetal pole of micromereless embryos induce endoderm, trigger archenteron formation, and differentiate into skeletogenic mesenchyme (Fig. 1, H4–6). Pmar1 thus can rescue β-catenin depletion in micromeres and endows any blastomere with micromere properties.

4. The Delta/Notch Pathway

mRNA expression was instrumental to demonstrate that the Notch signaling pathway also contributes to the positioning of the ectoderm–endoderm boundary and to cell specification (McClay *et al.*, 2000; Sherwood and McClay, 1999, 2001; Sweet *et al.*, 2002). Two forms were derived from the wild-type Notch receptor: the intracellular domain alone is constitutively active while deletion of this intracellular domain, leaving only the extracellular and the transmembrane domains, gives a dominant-negative form. As controls, activated Notch and dn-Notch were silenced by deletions in critical domains, a short deletion in ankyrin repeat 5 for activated Notch and the deletion of EGF-like repeats 7–29 for dnNotch. The ectoderm–endoderm border is shifted toward the animal pole by overexpression of activated Notch and toward the vegetal pole by dn-Notch. By injecting each construct in single animal or vegetal blastomeres at the 8-cell stage, and labeling adjacent blastomeres at the 16-cell stage, it was shown that the Notch pathway acts cell autonomously in the animal region but through a non-autonomous pathway in the vegetal half. Notch signaling has a strong impact on secondary mesenchyme cells (SMC) specification. Embryos overexpressing activated Notch have an increased number of SMC, which can lead to mesenchyme extrusion, and a greater number of pigment cells than in normal embryos. When Notch signaling is blocked with dnNotch, the SMC are almost absent and the embryos are albino (Fig. 1, I1–5). SMC specification is triggered by signaling from the micromeres. Combining mRNA expression and micromere transplantation gave insights into this process. In embryos that express dnNotch, normal micromeres fail to specify SMC, while in embryos expressing activated Notch, SMC are specified even in the absence of micromeres. In addition, micromeres expressing dnNotch can signal to specify SMC. Therefore, Notch signaling is required for SMC specification by micromeres but not for the generation of the inductive signal in micromeres.

One of the ligands of Notch, Delta, is expressed successively in the micromere and in the macromere progeny. Exogenous Delta mRNA endows cells in which it is expressed with unique properties. Single mesomeres taken from embryos microinjected with an inactive truncated form or with the full-length Delta were transplanted at the vegetal pole of non-injected embryos from which micromeres had been removed. Hosts receiving a mesomere expressing the inactive Delta form develop few pigmented cells, while numerous pigmented cells were present in embryos grafted with a mesomere-expressing Delta (Fig. 1, J1–3). Similar transplantations were carried out using animal caps isolated from a non-injected embryo. Animal halves receiving a mesomere devoid of Delta activity developed as "dauer" blastulas. Those recombined with a mesomere-expressing Delta were able to gastrulate and to develop into pluteus (Fig. 1, J4–6). Lineage tracing indicates that descendants of the transplanted cells and of host cells both contribute to mesoderm and endoderm.

5. Search for Downstream Target Genes by Differential Array Hybridization

The early asymmetrical nuclear accumulation of β-catenin is an essential step for specification of the mesendoderm and alteration of the β-catenin level leads to changes in cell fate. Differential screening of macroarrays using complex probes provides a tool to identify known genes or to discover new genes that are direct or indirect targets of β-catenin (Ransick *et al.*, 2002; Rast *et al.*, 2000). Two cDNA pools, the selectate and the driver, are prepared from two populations of embryos in which the β-catenin level has been increased or decreased. The selectate was obtained from lithium-treated embryos and the driver from 15,000 embryos that had been microinjected with *cadherin* mRNA. Complex probes derived from the selectate before and after subtractive hybridization were used to screen a macroarrayed library constructed from 12 h embryos, in the early phase of mesendoderm specification. Genes that are strongly upregulated by lithium treatment and down-regulated in embryos injected with cadherin mRNA include homologs to transcription factors *even skipped* (*eve*), *glial cells missing* (*gcm*), *FoxC* (Forkhead/winged helix class), *Krox1*, and *CA 150*. All these genes are expressed in endomesoderm and are likely regulators of cell fates. The same approach was used to isolate genes specifically expressed in the SMCs (Calestani *et al.*, 2003). The SMC presumptive territory is part of the endomesodermal territory that is extended by lithium treatment and its specification requires Notch signaling. The selectate was prepared from lithium-treated embryos and the driver from about 6000 embryos injected with a mRNA coding for a dominant-negative form of Notch. An arrayed hatched blastula library was hybridized sequentially with probes derived from the selectate, the subtracted selectate, and the driver. Six genes specifically expressed in SMCs were identified: glial cells missing (*gcm*), which encode a transcription factor, the polyketide synthase gene cluster (*pks-gc*), members of the family of flavin-containing monooxygenease (*fmo*), and the sulfotransferase gene (*sult*).

II. Methods for Expressing Synthetic mRNA in the Sea Urchin Embryo

A. General Explanation

Synthetic mRNA are prepared by *in vitro* transcription with bacteriophage RNA polymerases, such as the SP6, T3, or T7 RNA polymerase. The basis of the method is to insert the protein coding sequence of the cDNA of interest into a transcription vector containing a bacteriophage promoter. The template for *in vitro* transcription is obtained by linearization of the plasmid downstream of the cloned sequence using restriction enzymes.

In eukaryotic cells, translation initiation involves both ends of eukaryotic messenger RNAs as well as multiple sequential protein–RNA and protein–protein interactions. As a consequence, there are several essential structural requirements for the 3′ and 5′ ends of synthetic mRNAs in order for them to be stable and efficiently translated when injected into embryos. First, the synthetic transcript must either have a poly (A) tail at its 3′ end or contain sequences that will allow polyadenylation in the injected cells. A series of vectors have been designed to achieve these goals (Table I). The second important requirement for efficient translation of the synthetic mRNA is the presence at its 5′ end a 7-methyl guanosine linked by a triphosphate bridge to the mRNA that is normally found in eukaryotic mRNA. This modified nucleotide (called the Cap) is necessary for efficient translation of the mRNA because it binds the translation initiation factor eIF-4E, which then recruits other components of the translation initiation complex. Synthesis of capped mRNA is made possible by adding in the transcription reaction mixture a molar excess of a Cap analog, which, because of its special structure, can be incorporated only once at the 5′ end in place of the guanosine, which is the first base incorporated by the SP6, T3, or T7 RNA polymerases in a template-independent fashion. Finally, the sequence context of the translation initiation codon is critical for efficient translation of the synthetic mRNA and should not depart too much from an optimal sequence for translation initiation identified by studies from Kozak (1987).

1. Preparation of the DNA Template

The first step in the preparation of the DNA template is to insert the desired protein coding sequence into the transcription vector in the correct orientation. If possible, the presumed initiation codon should be the first AUG contained in the 5′ end of the transcript. In practice, we usually amplify by PCR the open reading frame of the desired cDNA using specific oligonucleotide primers and a high fidelity DNA polymerase. We systematically exclude the 5′ or 3′ UTR sequences that are often very long in sea urchin transcripts because their presence can potentially interfere with synthesis of full-length transcripts. We typically incorporate restriction sites into the oligonucleotides to allow efficient directional

Table I
Vectors for *In Vitro* Transcription of mRNA for Injection in Sea Urchin Embryos

Transcription vector	Promoter	Cloning sites	3′ UTR	5′ UTR	Linearization sites (5′ overhang)	References
pSP64T	SP6	BglII	Xenopus β-globin	Xenopus β-globin	SalI, XbaI, SmaI, EcoRI	Krieg and Melton, 1984
pSP64 Poly(A)	SP6	HindIII, PstI, SalI, AccI, HincII, XbaI, BamHI, AvaI, SmaI, SacI			EcoRI	Promega
pCMVTnT	SP6, T7	XhoI, EcoRI, MluI, KpnI, XbaI, SalI, AccI, SmaI, BstZI, NotI	Xenopus β-globin	SV40 late poly(A)	BamHI	Promega
pSP64TS	SP6	BglII, EcoRV, SpeI,	Xenopus β-globin	Xenopus β-globin	SalI, XbaI, SmaI, EcoRI	(derivative of pSP64T, Johnson and Krieg, unpublished data)
pT3TS	T3	BglII, EcoRV, SpeI	Xenopus β-globin	Xenopus β-globin	SalI, XbaI, SmaI, EcoRI	(derivative of pSP64T, Ekker, unpublished data)
pBluescript RN3	T3	BglII, EcoRI, PstI, SmaI, BamHI	Xenopus β-globin	Xenopus β-globin	SfiI	Lemaire *et al.*, 1995
pCS2	SP6	BamHI, ClaI, BstB1, ecoRI, StuI, XhoI, XbaI		SV40 late poly(A)	NotI, NsiI, Asp718	Turner (web site) Turner and Weintraub, 1994

cloning into the transcription vector, paying attention to add a few additional nucleotides 5′ to the restriction site to allow efficient cutting by the restriction enzymes. Unless the sequence surrounding the initiator AUG is in a good context, we include the motif ACCATG into the 5′ oligonuleotide sequence, which allows efficient translation of the mRNA. Finally, the 3′ oligonucleotide is designed to contain the end of the protein coding sequence and the translation termination codon.

2. Vectors

Special vectors are required for production of translatable synthetic mRNA and special attention must be given to the choice of the transcription vector. As seen also by our colleagues, we have found that transcripts generated from Bluescript vectors (Stratagene) are poorly translated in sea urchin embryos. A number of vectors have been designed and successfully used for production of synthetic mRNA and overexpression studies in various species including Xenopus, zebrafish, sea urchin, and ascidians. They are listed in Table I. A common feature of these vectors is the presence of a bacteriophage promoter and of a polylinker sequence for insertion of the sequence to be transcribed and for linearization of the DNA template. In addition, some of these vectors contain 5′ and 3′ UTR sequences from efficiently translated mRNA. For example, the pSP64T and pBluescript RN3 vectors have the 5′ and untranslated sequences of the Xenopus globin message and a poly (A) tail of 30 bases followed by a stretch of poly(C). The pCS2 series of vectors is also useful for overexpresion in sea urchin embryos. It contains an SP6 promoter followed by a polylinker and the SV40 late polyadenlyation site. mRNAs derived from these vectors have been reported to generate more protein than the corresponding mRNA from other vectors (Turner and Rupp, unpublished data). These effects may reflect polyadenylation of the injected mRNA in the cytoplasm. A number of derivatives of pCS2 have been constructed that allow fusions to epitope tags and other marker proteins, as well as to nuclear localization signals. Vectors of the CS2+ series are highly recommended, although pRNP3 seems to work properly.

3. *In Vitro* Transcription and Purification of the Synthetic mRNA

In Vitro transcription reactions are very robust and allow the preparation of large quantities (10–50 μg) of any single mRNA of average size. Detailed explanations of the protocol have been described elsewhere (Wormington, 1991; mMESSAGE mMACHINE Kit manual, Ambion Inc, Austin, TX). Following *in vitro* transcription, the DNA template is removed by treatment with RNAse-free DNase. For overexpression studies, we found that it is important to purify the mRNA extensively following *in vitro* transcription. This is usually obtained by extraction with phenol/chloroform and chromatography through Sephadex G50 columns.

4. Microinjection

The procedures and equipment used for microinjecting mRNA are largely the same as for injecting DNA. Synthetic mRNA is injected into eggs or embryos immobilized onto protamine-coated petri plates. While injecting plasmid DNA invariably results in mosaic expression of the transgene, injection of mRNA results in translation of the mRNA throughout the embryo. Injections are made into unfertilized or fertilized eggs if widespread expression of the protein is the desired goal. In contrast, if one wishes to mis-express the gene of interest in only part of the embryo, the injection can be targeted to a single blastomere. In practice, single cell injections are readily feasible up to the 8-cell-stage. After this stage, they become much more difficult due to the small size of the blastomeres. In order to identify the injected cells, it is very convenient to include a fluorescent dextran into the injection mixture (see following text for details). This ensures that the embryos were properly injected and that a similar volume of injection solution was delivered to the embryos. Alternatively, mRNA encoding a reporter such as β-galactosidase or EGFP or its fast maturing variant YFP can be co-injected along with the mRNA. However, we found that mRNA encoding the reporter can sometimes interfere with translation of the mRNA studied. After injection, the embryos are allowed to develop in a humid chamber kept in an incubator at the appropriate temperature.

5. Analyzing the Phenotypes

If the mRNA is tested for developmental function, the embryos should be observed at regular intervals and their morphology compared to control embryos of the same age. It should be kept in mind that injection usually results in a slightly delayed development and characteristic events, such as PMC ingression or gastrulation, may occur with some delay compared to uninjected embryos. Development of control uninjected sea urchin embryos is remarkably consistent and even phenotypes occurring at low frequencies (10–20%) which result from overexpression of the mRNA of interest can be readily identified.

If the expression of molecular markers is going to be analyzed, embryos can be fixed at the desired developmental stages along with control embryos and processed for *in situ* hybridization. In order to obtain semi-quantitative data, it is important to process control and experimental embryos in the same way and to develop the color reaction for the same time. Alternatively, perturbation of gene expression caused by the injected mRNA can be studied by Real Time Quantitative PCR (QPCR), which is particularly well suited for the analysis of a limited number of embryos. In this case, RNA is extracted from groups of injected and control embryos and used to synthesize cDNA that will serve to quantitate expression of particular genes by QPCR.

6. Controls and Commonly Encountered Injection Artifacts

In the course of mRNA overexpression experiments, it is particularly important to compare control and injected embryo at equivalent stages and to assess the specificity of the observed phenotypes. The trivial controls are to verify that the phenotypes do not reflect toxicity due to injection of nucleic acids at too high concentrations or that they are not caused by the stress of the microinjection. The cells should have divided at the same time as control embryos. The presence of abnormally large cells is usually the sign of interference with the cell cycle caused by nonspecific interference with translation of endogenous mRNAs. A commonly used control experiment consists of injecting in parallel a RNA devoid of biological activity. Control RNAs frequently used are GFP or β-galactosidase mRNAs. When injected in parallel at similar or higher concentrations, these RNAs should not perturb development. These control RNAs should be prepared in parallel, if possible, to the tested RNAs to ensure that differences between control and experimental RNA do not result from the presence of toxic contaminants.

The most common nonspecific phenotype observed in microinjection experiments with sea urchin embryos is toxicity leading to death, an effect also frequently observed with antisense oligonucleotides. Toxicity is usually associated with a reduced motility and the presence of supernumerary mesenchymal cells in the blastocoele. Also, an abnormal color of the embryos, which might appear brownish, can be a sign of toxicity and can reflect cell death.

A very important control for specificity is to test whether the effect observed is dose-dependent. For each RNA tested, the optimal concentration should be determined empirically. When using an RNA for the first time, we recommend testing different concentrations covering 2 orders of magnitude (e.g., from 10 μg/ml to 1 mg/ml).

More elaborated controls can be transcripts encoding a protein with altered function. For example, a frame shift can be introduced to interrupt the protein coding sequence, resulting in a truncated protein, which may have no activity or a different activity. In the case of homeobox, point mutations in the homeodomain have been used to demonstrate the specificity of the phenotypes caused by injection of mRNA encoding the wild-type protein (ex homeobox K > E). When using mRNA encoding dominant negative versions, a very stringent test for specificity is to demonstrate that normal development can be rescued by injection of mRNA encoding a wild-type protein. Another test for specificity would be to antagonize the effect by the injection of a second mRNA, or to rescue by injection of mRNA encoding a downstream component.

Finally, it should be realized that injection of high levels of some mRNAs might well give a very consistent but meaningless phenotype. This is why, in any overexpression experiment, the observed phenotypes should be interpreted cautiously. This is particularly true when the developmental function of genes that are expressed after gastrulation is tested by overexpression of their mRNAs in the oocyte. It is indeed possible that, in this case, the effects observed might result from the abnormal timing of mis-expression and interference with similar pathways acting earlier in development.

B. Detailed Method for Preparation of Synthetic mRNA for Microinjection

Materials

Appropriate restriction enzymes:

TE pH 7,6 (Tris-HCl, pH 7.6 10 mM; EDTA 0.1 mM)

In vitro transcription kit: (mMessage mMachine, Ambion cat.# 1340, 1344, 1348)

or:

RNase inhibitor (RNAsin®, Promega, cat.# N2111; SUPERase • In™, Ambion, cat.# 2694; Protector RNase Inhibitor, Roche, cat.# 3 335 399)

RNA polymerases: T3 (Ambion, cat.# 2060, Promega, cat.# P2083), T7 (Ambion, cat.# 2082; Bioloabs, cat.# M0251; Promega, cat.# P2075), SP6 (Ambion, cat.# 2071; Biolabs, cat.# M0207; Promega, cat.# P1081)

Yeast inorganic pyrophosphatase

RNase free DNase I

Sephadex G-50 (50% slurry in 0.3 M sodium acetate, pH 5.5, 0.1% SDS)

To prepare Sephadex, rehydrate the resin with nuclease-free water or TE and wash it with several volumes of 0.3 M sodium acetate, pH 5.5. Add DEPC to 0.1% and autoclave. After autoclaving. SDS is added to 0.1% and the resin can be stored at room temperature.

10× SP6 RNA polymerase buffer:	800 mM Hepes-KOH pH 7.5 160 mM $MgCl_2$ 20 mM spermidine 400 mM DTT
10× SP6 enzyme mix	50% glycerol 1000 u/ml RNAse inhibitor 1800 u/ml SP6 RNA polymerase 50 u/ml yeast inorganic pyrophosphatase
10× T3/T7 RNA polymerase	buffer 800 mM Hepes-KOH pH 7.5 120 mM $MgCl_2$ 20 mM spermidine 400 mM DTT
10× T3/T7 enzyme mix	50% Glycerol 1000 u/ml RNAse inhibitor 1800 u/ml T3 or T7 RNA polymerase 50 u/ml yeast inorganic pyrophosphatase
2× Ribonucleoside triphosphate for SP6	ATP 10 mM CTP 10 mM UTP 10 mM GTP 2 mM M^7 Gppp(5′)G 8 mM

2× Ribonucleoside triphosphate for T3/T7	ATP 15 mM
	CTP 15 mM
	UTP 15 mM
	GTP 3 mM
	M^7Gppp(5′)G 12 mM

Phenol:Chloroform
Nuclease-free water
Gel loading buffer 2× (95% formamide, 0.025% bromophenol blue, 20 mM EDTA, 0.025% SDS)
Dextran conjugated with fluorophores 25 mg/ml in RNase-free water (Molecular Probes)
TRIzol Reagent (Invitrogen Life Technologies)

Note: When working with mRNA, meticulous care should be given to maintaining an RNase-free environment. Solutions should be treated with diethylpyrocarbonate (DEPC) when possible and autoclaved. Sets of autoclaved tubes and pipet tips and gel tanks should be dedicated to RNA work. Gloves should be worn at all times.

Protocol

a. Subclone the protein coding sequence of interest into a transcription vector (Table I).
b. Prepare highly pure plasmid DNA by your favorite method.
c. Digest 10 μg of plasmid DNA with the suitable restriction enzyme that cuts downstream of the termination codon, 3′ UTR, and poly(A) or polyadenylation sequence. Analyze 100 ng of digested DNA by agarose gel electrophoresis to check the efficiency of cleavage. *Note that if RNase was used during minipreparations of plasmid DNA, traces of RNase should be removed by proteinase K digestion. In this case, add 0.05 volume of SDS 10% and proteinase K to a final concentration of 100 μg/ml and continue incubation for 30 min at 50 °C.*
d. Purify the DNA by phenol/chloroform extraction and precipitation with ethanol.
e. Redissolve the DNA in RNase-free TE (pH 7.6) at 0.5 μg/μl.
f. Prepare a 20 μl *in vitro* transcription reaction at room temperature.

water	6 μl
2× ribonucleoside mix	10 μl
linearized template DNA	2 μl
enzyme mix	2 μl

g. Incubate for 2 h at 37 °C.
h. Add one unit of RNase-free DNase I and incubate for 15 min at 37 °C.

Note: at this stage an aliquot of the reaction can be run on a standard non-denaturing agarose gel to check the efficiency of synthesis. In this case, take 0.5 μl of the reaction and mix it with 10 to 20 μl of 1× gel loading buffer. Heat the sample 1 min at 90 °C and load on a mini gel containing ethidium bromide or Syber Green.

i. Add 130 μl of RNAse-free H_2O and purify the RNA by extraction with phenol:chloroform. Remove most of the unincorporated nucleotides on a 5-ml Sephadex G-50 spin column.
j. Prepare the column as follows: pipet 5 ml of a 50% slurry of Sephadex G50 into a 5 ml spin column. Place the column into a 15 ml plastic centrifuge tube and spin 4 min at 1500 rpm in a swinging-bucket rotor. Wash the column once with 150 μl of sodium acetate/SDS solution and spin as described previously. Place the end of the column into a clean 0.5 ml microcentrifuge tube. Load the RNA sample at the center of the resin bed and spin as above. Collect the eluate (usually 80% of the loaded volume) into a new microcentrifuge and store on ice.
k. Purify the RNA by extraction with phenol:chloroform.
l. Add 2.5 volumes of 100% ethanol to the eluate to precipitate the RNA.
m. Resuspend RNA in nuclease-free water, aliquot, and keep at −80 °C until use.
n. To prepare the injection solution, make dilutions of the RNA and the fluorescent dextran (5 mg/ml final concentration) with RNase-free water.

Before any injection experiment it is important to centrifuge the injection solution 5 min at 13,000 g at 4 °C to pellet any insoluble material. The sample to be injected should be kept on ice during the injection.

III. Summary, Prospects, Concluding Remarks

Expression of exogenous mRNAs has become part of the standard approach to studying gene function during development of the sea urchin. The method is simple and reliable, protocols for the preparation of synthetic mRNAs are well described, and the technique to transfer them into eggs is efficient. The protein encoded by these mRNAs can be designed to address a variety of biological questions and their DNA matrices are easily constructed by standard molecular biology techniques. The method aims to simulate gain or loss of gene function, and the phenotypes obtained are characterized using an increasing number of molecular markers.

With the completion of the *S. purpuratus* genome project, the complete set of genes from the sea urchin will become available. Expression of mRNA will be an invaluable tool to study the function of newly identified genes and their protein products and to determine their positions within the networks of gene and protein interactions that control development.

Acknowledgments

We thank Patrick Chang for carefully reading the manuscript and L. and R. Angerer, E. Davidson, C. Ettensohn, and D. McClay for kindly providing pictures.

References

Angerer, L. M., Oleksyn, D. W., Levine, A. M., Li, X., Klein, W. H., and Angerer, R. C. (2001). Sea urchin goosecoid function links fate specification along the animal–vegetal and oral–aboral embryonic axes. *Development* **128,** 4393–4404.

Angerer, L. M., Oleksyn, D. W., Logan, C. Y., McClay, D. R., Dale, L., and Angerer, R. C. (2000). A BMP pathway regulates cell fate allocation along the sea urchin animal–vegetal embryonic axis. *Development* **127,** 1105–1114.

Calestani, C., Rast, J. P., and Davidson, E. H. (2003). Isolation of pigment cell specific genes in the sea urchin embryo by differential macroarray screening. *Development* **130,** 4587–4596.

Cavalieri, V., Spinelli, G., and Di Bernardo, M. (2003). Impairing Otp homeodomain function in oral ectoderm cells affects skeletogenesis in sea urchin embryos. *Dev. Biol.* **262,** 107–118.

Croce, J., Lhomond, G., and Gache, C. (2003). Coquillette, a sea urchin T-box gene of the Tbx2 subfamily, is expressed asymmetrically along the oral–aboral axis of the embryo and is involved in skeletogenesis. *Mech. Dev.* **120,** 561–572.

Duboc, V., Rottinger, E., Besnardeau, L., and Lepage, T. (2004). Nodal and BMP2/4 signaling organizes the oral–aboral axis of the sea urchin embryo. *Dev. Cell* **6,** 397–410.

Emily-Fenouil, F., Ghiglione, C., Lhomond, G., Lepage, T., and Gache, C. (1998). GSK3β/shaggy mediates patterning along the animal–vegetal axis of the sea urchin embryo. *Development* **125,** 2489–2498.

Ettensohn, C. A., Illies, M. R., Oliveri, P., and De Jong, D. L. (2003). Alx1, a member of the Cart1/Alx3/Alx4 subfamily of paired-class homeodomain proteins, is an essential component of the gene network controlling skeletogenic fate specification in the sea urchin embryo. *Development* **130,** 2917–2928.

Fernandez-Serra, M., Consales, C., Livigni, A., and Arnone, M. I. (2004). Role of the ERK-mediated signaling pathway in mesenchyme formation and differentiation in the sea urchin embryo. *Dev. Biol.* **268,** 384–402.

Gross, J. M., and McClay, D. R. (2001). The role of Brachyury (T) during gastrulation movements in the sea urchin Lytechinus variegatus. *Dev. Biol.* **239,** 132–147.

Gross, J. M., Peterson, R. E., Wu, S. Y., and McClay, D. R. (2003). LvTbx2/3: A T-box family transcription factor involved in formation of the oral/aboral axis of the sea urchin embryo. *Development* **130,** 1989–1999.

Howard, E. W., Newman, L. A., Oleksyn, D. W., Angerer, R. C., and Angerer, L. M. (2001). SpKrl: A direct target of beta-catenin regulation required for endoderm differentiation in sea urchin embryos. *Development* **128,** 365–375.

Huang, L., Li, X., El-Hodiri, H. M., Dayal, S., Wikramanayake, A. H., and Klein, W. H. (2000). Involvement of Tcf/Lef in establishing cell types along the animal–vegetal axis of sea urchins. *Dev. Genes Evol.* **210,** 73–81.

Kenny, A. P., Oleksyn, D. W., Newman, L. A., Angerer, R. C., and Angerer, L. M. (2003). Tight regulation of SpSoxB factors is required for patterning and morphogenesis in sea urchin embryos. *Dev. Biol.* **261,** 412–425.

Kozak, M. (1987). An analysis of 5′-noncoding sequences from 699 vertebrate messenger RNAs. *Nucleic Acids Res.* **15,** 8125–8148.

Krieg, P. A., and Melton, D. A. (1984). Functional messenger RNAs are produced by SP6 *in vitro* transcription of cloned cDNAs. *Nucleic Acids Res.* **12,** 7057–7070.

Kurokawa, D., Kitajima, T., Mitsunaga-Nakatsubo, K., Amemiya, S., Shimada, H., and Akasaka, K. (1999). HpEts, an ets-related transcription factor implicated in primary mesenchyme cell differentiation in the sea urchin embryo. *Mech. Dev.* **80,** 41–52.

Lemaire, P., Garrett, N., and Gurdon, J. B. (1995). Expression cloning of Siamois, a Xenopus homeobox gene expressed in dorsal-vegetal cells of blastulae and able to induce a complete secondary axis. *Cell* **81,** 85–94.

Li, X., Wikramanayake, A. H., and Klein, W. H. (1999). Requirement of SpOtx in cell fate decisions in the sea urchin embryo and possible role as a mediator of beta-catenin signaling. *Dev. Biol.* **212,** 425–439.

Logan, C. Y., Miller, J. R., Ferkowicz, M. J., and McClay, D. R. (1999). Nuclear beta-catenin is required to specify vegetal cell fates in the sea urchin embryo. *Development* **126,** 345–357.

McClay, D. R., Peterson, R. E., Range, R. C., Winter-Vann, A. M., and Ferkowicz, M. J. (2000). A micromere induction signal is activated by beta-catenin and acts through notch to initiate specification of secondary mesenchyme cells in the sea urchin embryo. *Development* **127,** 5113–5122.

Montana, G., Sbisa, E., Romancino, D. P., Bonura, A., and Di Carlo, M. (1998). Folding and binding activity of the 3′UTRs of Paracentrotus lividus bep messengers. *FEBS Lett.* **425,** 157–160.

Oliveri, P., Carrick, D. M., and Davidson, E. H. (2002). A regulatory gene network that directs micromere specification in the sea urchin embryo. *Dev. Biol.* **246,** 209–228.

Oliveri, P., Davidson, E. H., and McClay, D. R. (2003). Activation of pmar1 controls specification of micromeres in the sea urchin embryo. *Dev. Biol.* **258,** 32–43.

Ransick, A., Rast, J. P., Minokawa, T., Calestani, C., and Davidson, E. H. (2002). New early zygotic regulators expressed in endomesoderm of sea urchin embryos discovered by differential array hybridization. *Dev. Biol.* **246,** 132–147.

Rast, J. P., Amore, G., Calestani, C., Livi, C. B., Ransick, A., and Davidson, E. H. (2000). Recovery of developmentally defined gene sets from high-density cDNA macroarrays. *Dev. Biol.* **228,** 270–286.

Rottinger, E., Besnardeau, L., and Lepage, T. (2004). A Raf/MEK/ERK signaling pathway is required for development of the sea urchin embryo micromere lineage through phosphorylation of the transcription factor Ets. *Development* **131,** 1075–1087.

Sherwood, D. R., and McClay, D. R. (1999). LvNotch signaling mediates secondary mesenchyme specification in the sea urchin embryo. *Development* **126,** 1703–1713.

Sherwood, D. R., and McClay, D. R. (2001). LvNotch signaling plays a dual role in regulating the position of the ectoderm–endoderm boundary in the sea urchin embryo. *Development* **128,** 2221–2232.

Sweet, H. C., Gehring, M., and Ettensohn, C. A. (2002). LvDelta is a mesoderm-inducing signal in the sea urchin embryo and can endow blastomeres with organizer-like properties. *Development* **129,** 1945–1955.

Turner, D. L. http://sitemaker.umich.edu/dlturner.vectors.

Turner, D. L., and Weintraub, H. (1994). Expression of achaete-scute homolog 3 in Xenopus embryos converts ectodermal cells to a neural fate. *Genes Dev.* **8,** 1434–1447.

Vonica, A., Weng, W., Gumbiner, B. M., and Venuti, J. M. (2000). TCF is the nuclear effector of the beta-catenin signal that patterns the sea urchin animal–vegetal axis. *Dev. Biol.* **217,** 230–243.

Weitzel, H. E., Illies, M. R., Byrum, C. A., Xu, R., Wikramanayake, A. H., and Ettensohn, C. A. (2004). Differential stability of beta-catenin along the animal–vegetal axis of the sea urchin embryo mediated by dishevelled. *Development* **131,** 2947–2956.

Wikramanayake, A. H., Huang, L., and Klein, W. H. (1998). beta-Catenin is essential for patterning the maternally specified animal–vegetal axis in the sea urchin embryo. *Proc. Natl. Acad. Sci. USA* **95,** 9343–9348.

Wormington, M. (1991). Preparation of synthetic mRNAs and analyses of translational efficiency in microinjected Xenopus oocytes. *Methods Cell Biol.* **36,** 167–183.

CHAPTER 28

Disruption of Gene Function Using Antisense Morpholinos

Lynne M. Angerer and Robert C. Angerer

Department of Biology
University of Rochester
Rochester, New York 14627

I. Introduction

The most important approaches for determining the function of genes during development are either mis/overexpression or loss-of-function assays. Although the former can provide some useful information, in general, they are difficult to interpret without knowing the consequences of interfering with gene activity. Depending on the system, interference may be accomplished either by mutation or through some biochemical or pharmacological perturbation that reduces either the production or the activity of the encoded protein. Since forward genetics is feasible in only a few systems, the latter strategy is required in most embryos. Even in "genetic" systems, the use of interference techniques is gaining in popularity

0091-679X/04 $35.00

because the availability of genome sequences allows one to develop loss-of-function tests easily and quickly.

These new techniques are relatively inexpensive in time and cost compared to mutational analyses. The two most useful approaches are RNA interference (RNAi) and morpholino antisense oligonucleotide inhibition. RNAi has been shown to affect gene expression at the levels of chromatin structure, mRNA stability, and mRNA translation (reviewed by Hannon, 2002), while morpholinos have been shown to inhibit the initiation of translation with high specificity (Hannon, 2002; Heasman, 2002) and, alternatively, to alter splicing (Draper *et al.*, 2001; Schmajuk *et al.*, 1999). In lower deuterostome embryos, principally, the sea urchins (*Strongylocentrotus, Hemicentrotus, Lytechinus*, and *Paracentrotus* species) and the tunicates (*Ciona* and *Halocynthia*), the use of morpholinos has revolutionized the analysis of gene function during development (Angerer *et al.*, 2001; Davidson *et al.*, 2002b; Fuchikami *et al.*, 2002; Howard *et al.*, 2001; Moore *et al.*, 2002; Satou *et al.*, 2001a,b; Sweet *et al.*, 2002; Wada and Saiga, 2002). In addition to demonstrating the developmental roles of individual genes, this approach has allowed investigations of the epistatic relationships among gene regulatory factors and signaling pathways, which has led to the elucidation of gene regulatory networks that control cell fate specification and differentiation (Davidson *et al.*, 2002a, 2003). The availability of whole genome sequence in the purple sea urchin and in *Ciona* greatly simplifies identifying translation start or splice sites, thereby increasing the number of different genes whose functions are easily testable.

The mechanism of morpholino-mediated antisense interference is fundamentally different from that of other antisense approaches. It does not require degradation of target mRNAs by cellular RNase H activity, but instead physically blocks translation or access to the splicing apparatus. In the following text, we review briefly the properties of morpholinos and why these offer significant advantages over other antisense and loss-of-function approaches. In subsequent sections, we present current methods for using morpholinos and for analyzing their effects on the development of basal deuterostome embryos. Finally, we review the kinds of information that can be gained from loss-of-function morpholino approaches and consider other possible applications of this technology. The reader is referred to a recent review on this subject (Heasman, 2002) and the Gene Tools web site (www.gene-tools.com), which includes an extensive bibliography of work employing morpholinos.

II. Morpholino-Mediated Loss-of-Function: Advantages and Limitations

A. Advantageous Properties of Morpholinos

(Detailed discussion can be found at gene-tools.com.)

1. Stability and Lack of Toxicity

Morpholinos are ribooligonucleotides, usually 25 nucleotides long, in which the ribose moiety has been substituted with a morpholine ring. Uracil bases are replaced by thymine to increase thermal stability. These modified oligonucleotides are resistant to degradation by cellular nucleases and remain highly stable after years in storage. Concentrations that elicit specific and distinct phenotypes and cause reductions of target proteins to undetectable levels are well below those that elicit toxic effects (<10 μM final concentration in the injected sea urchin egg). The high stability and lack of toxicity of morpholinos are significant advantages over other traditional antisense oligonucleotides.

2. Specificity

A major concern with methods employing conventional antisense is that these oligonucleotides may adventitiously hybridize to other mRNAs in addition to the desired target. Oligonucleotides as short as 10-mers can form transient hybrids *in vivo* that allow RNase H digestion (Woolf *et al.*, 1992). This problem cannot be overcome and is, in fact, exacerbated by increasing oligonucleotide length. In contrast, hybrids formed with morpholino oligonucleotides are not RNaseH-sensitive and function by blocking translation initiation or splicing. These blocking reactions require morpholinos 25 nucleotides in length that are well-matched, have fewer than 4 mismatches per 25-mer, and that bind stably to the target RNA. For translational interference, this is what is required to block initiation of movement of mRNA through the ribosome. Standard oligonucleotides targeted to the same site are not as effective, because the resulting hybrids are not sufficiently stable.

Morpholinos are effective at blocking translation by forming hybrids in a restricted region of the mRNA between ~ -80 and $+20$ of the translational start codon ($+1$). Once translation has initiated and proceeded about 20 to 30 nucleotides into the protein coding region, the morpholino is unable to block translation elongation. This property is the major reason why morpholinos are so specific compared to other antisense oligonucleotides, because most (>95%) of the sequences in an mRNA population are eliminated as possible nonspecific targets.

3. Hybrid Stability

Hybrids formed with morpholinos are more stable than those formed with other oligonucleotides. This property facilitates translational interference and it allows hybrid formation even when there is secondary structure in the RNA, which is often the case in 5′ untranslated leader sequences.

Several recent studies have shown that morpholinos targeted to an exon/intron border at the 5′ donor site can interfere with splicing, thereby leading to production of altered gene products (Draper *et al.*, 2001; P. Marcos, gene-tools.com; Schmajuk *et al.*, 1999). An advantage of this variation of the technique is that the extent of knockdown can by monitored at the RNA level. It also offers the possibility of interfering with specific alternatively spliced transcripts. However, the extent of interference may not be as great as with translation-blocking morpholinos and, in some cases, the splicing defect may not create a null phenotype, but instead might lead to production of a protein with altered properties. Thus far, the only embryo system in which this approach has been used is the zebrafish (Draper *et al.*, 2001).

4. Independence from Host Components

In many conventional antisense techniques, interference depends on mRNA degradation mediated by RNase H. Because some embryos, such as those of *Xenopus*, lack this enzyme at early stages, morpholinos are required. RNAi also requires the Dicer enzyme and other proteins whose activities have not yet been demonstrated in sea urchin or tunicate embryos.

5. Perdurance through Development

In our experience, morpholinos are effective through at least the 3 days that are required to complete sea urchin embryogenesis. For example, SoxB1 mRNA that is abundant throughout development is not translated to a detectable level in embryos derived from morpholino-injected eggs (Fig. 1, right). The majority of genes transcribed during embryogenesis are activated by the mesenchyme blastula stage in sea urchin embryos. Several that are not transcribed until a day after fertilization have been shown to be effectively knocked down by morpholino injection into the egg (Fuchikami *et al.*, 2002; Rast *et al.*, 2002).

B. Limitations

1. Spatial and Temporal Resolution

Interference is limited to where and when morpholinos can be introduced into the embryo. In general, they are microinjected which, in practice, is limited to the egg or early blastomeres. An alternative is to construct chimeric embryos via

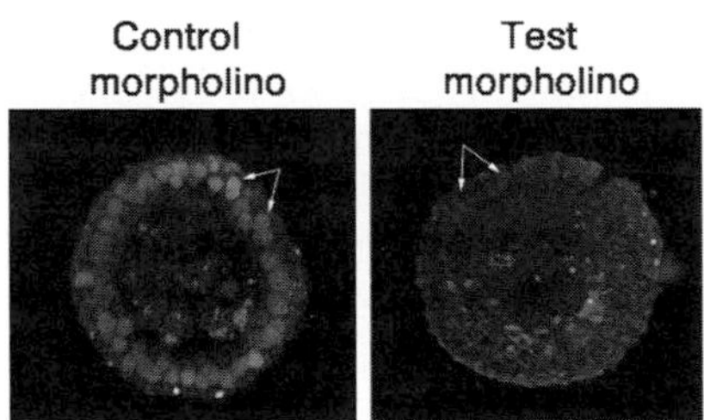

Fig. 1 A morpholino knockdown of SpSoxB1 to undetectable levels in sea urchin blastulae. Embryos were stained with a polyclonal antibody to SpSoxB1 (green) and a monoclonal antibody, 6e10 (red), that specifically recognizes primary mesenchyme cells. (See Color Insert.)

blastomere recombination (Fuchikami *et al.*, 2002; Sweet *et al.*, 2002), but this approach requires special equipment and expertise, limits the number of embryos analyzed, and is applicable to a restricted subset of early blastomeres.

2. Early Maternal Functions

Morpholinos introduced into the egg interfere with synthesis of protein in the zygote, but do not block the function of maternal proteins that persist for variable lengths of time. In these cases, immunological detection of the target protein during development is necessary to evaluate the timing of morpholino effects. Alternative approaches for interfering with maternal protein function are to mis/overexpress proteins with dominant negative function or to introduce neutralizing antibodies.

3. Cost

Each morpholino currently costs $450, which is relatively inexpensive in comparison to the time and manpower needed to generate null phenotypes via standard forward genetic methods. On the other hand, the cost is sufficient to prohibit genomewide gene function analyses, such as those that have been done in *C. elegans* using RNAi.

C. Comparison to Other Loss-of-Function Approaches

1. Dominant Negative Variant Peptides

This method has the advantage of interfering at the protein level and, in principle, could interfere with maternal protein function, depending on how rapidly the embryo translates sufficient blocking protein. However, because this method is based on competition between normal and mutated forms, it requires injection of high mRNA concentrations, leading to correspondingly high protein concentrations that could result in abnormal protein–protein interactions. Overexpression may also have toxic effects. Loss-of-function via dominant negative interference is also limited by the stability of the injected mRNA and the protein that it encodes. If the dominant negative protein is expressed from injected mRNA, as is the standard approach, then there is no control on spatial expression. While this limitation can be overcome by synthesizing dominant negative proteins under the control of a tissue-specific promoter, that requires marking the expressing cells because incorporation of exogenous DNA is mosaic. Lastly, the effect of a dominant negative may not specifically interfere with the function of only one protein; it could affect the activity of members of the same family of proteins or that of other interacting proteins in intersecting pathways. In contrast, a morpholino specifically reduces the amount of one component.

2. TKO

One report has described the isolation of a gene encoding a single-chain antibody that binds to a specific sea urchin transcription factor with high affinity (Bogarad *et al.*, 1998). Microinjection of mRNA encoding this antibody into the egg blocked the function of that factor, as shown by *in vivo* promoter/reporter transgene assays. While this is an interesting and potentially powerful approach, it is relatively difficult and is not guaranteed to succeed. Each knockout requires the cloning of an effective, neutralizing antibody and this process is labor-intensive. In addition, its success depends on the stability of the injected mRNA and the protein it encodes, and it requires demonstrating the specificity of the antibody, which can be limited by assay sensitivity. By comparison, morpholino-mediated knockouts are considerably easier to execute and interpret.

3. Injection of Neutralizing Antibodies

This approach is also relatively time-consuming and depends on the production of a good antibody that not only recognizes the target protein but also neutralizes its function. Because it interferes at the protein level, it has the potential for blocking both maternal and zygotic functions. However, it is limited by the perdurance of the antibody and the possibility of epitope masking.

4. RNAi

Although this approach has achieved spectacular success in other systems, notably *C. elegans* and *D. melanogaster* (reviewed in Hannon, 2002), no reports using this approach in basal deuterostome embryos have appeared. A potential and significant advantage over morpholinos is that spatially and temporally regulated interference might be achieved by expressing specific double-stranded (hairpin) RNAs under the control of tissue and/or stage-specific promoters.

III. Morpholino Methods

A. Design

Gene Tools and Funikoshi (Japan) are the only companies that currently supply morpholinos. Sequences can be submitted and the company will select an optimal target sequence within −80 to +20 of the translation initiation site. The two main criteria for selection are that the morpholino sequence should have no self-complementarity longer than 4 consecutive base pairs and that sequences containing three (or more) consecutive G residues should be avoided. In addition, if experiments involving rescue of the morpholino phenotype by co-injection of synthetic mRNA (see following text) are planned, then it is useful if the morpholino complement is upstream of the initiator AUG.

B. Preparation of Injection Solutions

Stock solutions are made by diluting the morpholino to a final concentration of 8 mM oligomer with glass-distilled, diethylpyrocarbonate-treated (RNase-free) water. When they are stored at $-20\,^{\circ}$C, we have detected no change in potency over a 2-year interval. The concentration of morpholino in the final injection solution varies between 100 and 800 μM, depending on the particular morpholino. The optimal concentrations are determined empirically, aiming for the lowest one that gives consistent phenotypes. Injection solutions should be buffered with 10 mM Tris, pH 8.0, because some batches of morpholinos contain impurities that can reduce the pH. Techniques for microinjecting embryos are described by McMahon *et al.*, 1985 (sea urchins); Imai *et al.*, 2000 (tunicates); L. Jaffe, this volume.

C. Controls

1. Test Whether the Morpholino Recognizes the mRNA of Interest

a. The most direct and conclusive approach is to use a specific antibody that monitors the level of the target protein. Fig. 1 illustrates the morpholino-mediated knockdown of the transcription factor, SoxB1, that is stained green. This protein normally accumulates in the nuclei of most cells of the *S. purpuratus* embryo. It is absent from those of the vegetal-most cells, the primary mesenchyme, which have entered the blastocoel and are stained red.

b. When an antibody is not available,

1. Inject chimeric mRNAs that contain 5′ sequence that either is (Fig. 2E,F; test) or is not (Fig. 2A–D; controls) complementary to the morpholino linked upstream of an open reading frame that encodes an easily detectable marker (e.g., GFP; Howard *et al.*, 2001; Rast *et al.*, 2002). Co-inject morpholino with each of these mRNAs and monitor the difference in marker expression between control and test. Compare the signals shown in panels B and D with that in F, which is not detectable. For this test, in

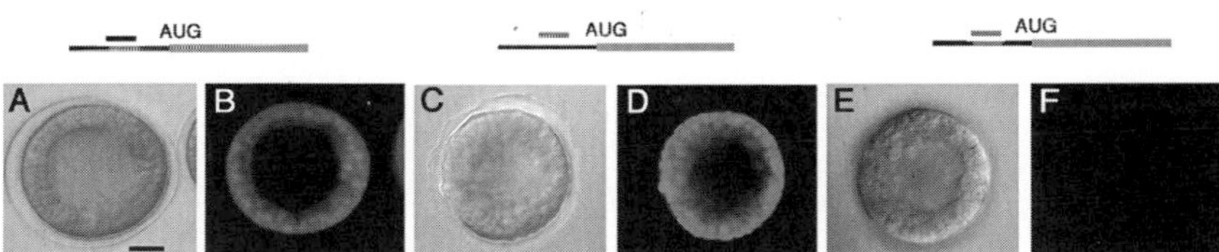

Fig. 2 Demonstration that a morpholino can efficiently block translation *in vivo*. Early blastulae derived from eggs injected with (A, B) GFP mRNA containing the morpholino target sequence (red) and a control morpholino (black bar); (C, D) GFP mRNA lacking the morpholino target sequence and the test morpholino (yellow bar); or (E, F) GFP mRNA containing the morpholino target sequence and the test morpholino (yellow bars). (Figure from Howard *et al.*, 2001). (See Color Insert.)

sea urchin embryos, we injected a very large number (500,000 copies) of synthetic GFP mRNA transcripts (2 pl of 1 μg RNA/μl injection solution). Since a signal above background can be detected when 100-fold fewer copies of GFP mRNA are injected (our unpublished observations), the morpholino effectively blocked greater than 99% of the translation of GFP mRNA. No adverse effects on development were observed at the concentrations of morpholino required to achieve this level of inhibition. Since very few endogeneous mRNAs are more abundant than 500,000 copies/egg, it follows that the translation of most endogenous messages can also be effectively blocked at concentrations of morpholino that are otherwise inert.

2. Test whether the morpholino blocks translation in an *in vitro* coupled transcription/translation system (Moore *et al.*, 2002). This *in vitro* certification is not as stringent as the *in vivo* tests previously mentioned. However, it does have the advantage of being rapid because no additional cloning operations are required.
3. For morpholinos targeted at splice junctions, the appearance of altered mRNA and the disappearance of normal mRNA should be assayed. For example, see Draper *et al.* (2001), in which the morpholino generated several aberrantly spliced Fgf 8 pre-mRNAs, which were detectable by RT-PCR.

2. Test Whether the Morpholino is Specific for the Desired mRNA

A significant problem with standard antisense approaches is that the oligonucleotides can form transient short hybrids, 10 to 12 nucleotides in length, which could serve as targets for RNAse H digestion (Woolf *et al.*, 1992). This can lead to the targeting of unrelated or unknown mRNAs, a problem which increases in proportion to the complexity of the RNA population. As has been discussed, the potential for such nonspecific interactions is greatly reduced with morpholino technology. A second concern is that the phenotypes observed may result from some general toxic effect of the morpholino itself, rather than loss of a specific protein. Two approaches to control for this artifact are:

a. Rescue the phenotype by injecting, along with the morpholino, a series of concentrations of mRNA encoding the target protein but lacking the sequence complementary to the morpholino. For example, a morpholino against mRNA, encoding a protein expressed in most cells of the sea urchin embryo, prevents gastrulation and correct patterning of the ectoderm (Fig. 3, left). An embryo co-injected with morpholino and rescuing mRNA is shown on the right of Fig. 3. More than 95% of the rescued embryos show the same normal development: They contain a fully differentiated tripartite gut (arrows) and coelomic pouches (arrowheads). Similar tests have been carried out in tunicates (Imai *et al.*, 2002; Wada and Saiga, 2002). For example, knockdown of Cs-ZicL, a zinc finger protein,

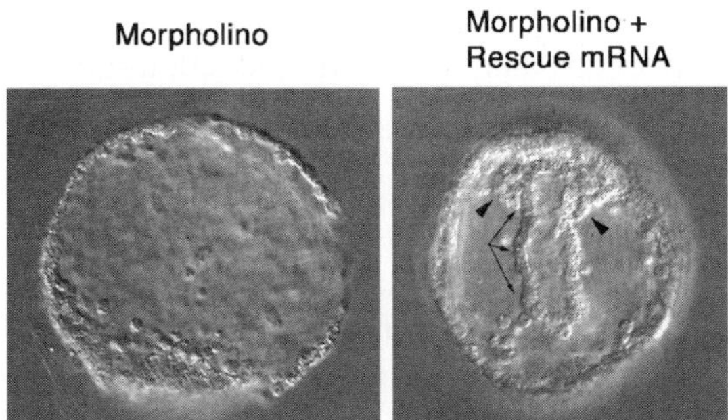

Fig. 3 Demonstration that a morpholino knockdown is specific. Three-day-old embryos derived from eggs injected with either (left panel) the same morpholino against SpSoxB1 as in Fig. 1, which prevents gastrulation and correct ectoderm patterning, or (right panel) the morpholino coinjected with synthetic SpSoxB1 mRNA lacking the morpholino target sequence. Arrows and arrowheads indicate rescue of a tripartite gut and coelomic pouches, respectively.

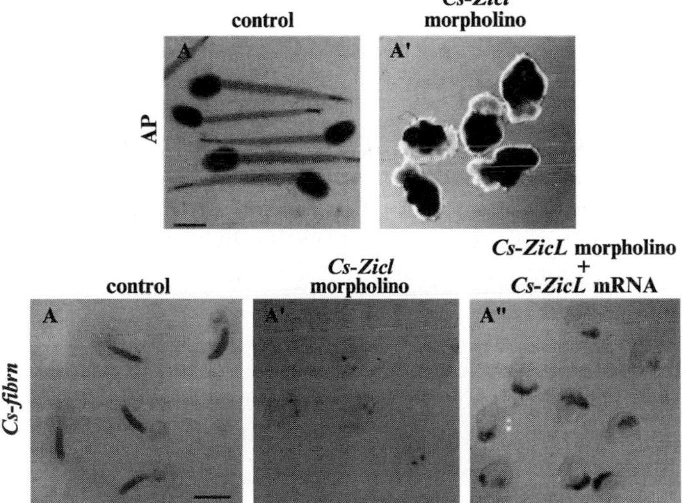

Fig. 4 Demonstration of a specific morpholino-mediated knockdown in *Ciona*. Top panels demonstrate a dramatic loss-of-tail phenotype upon introduction of a Cs-ZicL morpholino. Embryos are stained for alkaline phosphatase (AP), which is expressed in the endoderm. Bottom panels show the reduction in number of cells expressing Cs-fibrn, a fibrinogen-like protein, in the presence of the morpholino (*cf.* A and A′) and a significant recovery of activity when CsZicL mRNA is co-injected (*cf.* A′ and A″). (Figures taken from Imai *et al.*, 2002.) (See Color Insert.)

perturbs notochord development as shown by morphology (Fig. 4, top panels, cf A and A′) and by sharp reduction in the number of cells expressing Cs-fibrn (fibrinogen-like), a notochord marker (Fig. 4, bottom panels A and A′). Co-injection of mRNA increases the number of fibribinogen-like-positive cells (Fig. 4, bottom panels, A″).

Of course, in most cases, reversing the phenotype will not lead to complete rescue of normal development but instead is expected to produce a phenotype like that which results from mis/overexpression due to ectopic, elevated, and precocious prodution of the protein. Therefore, the phenotypes of embryos injected with only the rescuing RNA must also be defined. The type of assay, i.e., morphological or molecular, that should be used to determine whether rescue was achieved will depend on how restricted the sites of expression are in the normal embryo and whether ectopic misexpression phenotypes occur. In some cases, observation of reciprocal phenotypes for morpholino-mediated loss-of-function and mRNA misexpression gain-of-function provides support for morpholino specificity. For example, knocking down the transcriptional repressor goosecoid in sea urchin embryos causes ectoderm to differentiate as aboral ectoderm, whereas misexpressing goosecoid prevents differentiation of this cell type (Angerer *et al.*, 2001).

If severe phenotypes or embryo lethality occur as a result of morpholino injection, the rescue experiment is important to distinguish between possible nonspecific toxic effects of the morpholino and an essential function of the protein.

b. Use two morpholinos targeted to different sequences within the translation initiation region and test whether the phenotypes are the same. (e.g., see Howard *et al.*, 2001).

c. Although control morpholinos at the same concentration as test morpholinos have been employed in a number of different studies, they provide a much less rigorous control for specificity than do the two methods just described. They do not rule out either nonspecific hybridization of the test morpholino or batch-specific morpholino toxic effects. Control morpholinos that have been used include scrambled sequence of the test morpholino, sequence corresponding to the sense strand, or antisense sequence targeted greater than 30 nucleotides downstream of the initiator AUG. Gene Tools offers a "control" morpholino that we have found does not interfere in any detectable way with normal morphological development of *S. purpuratus* embryos. In ascidian embryos, a morpholino against lacZ has been shown not to interfere as well (Satou *et al.*, 2001b).

D. Interpretation of Phenotypes

Several factors must be considered when interpreting phenotypes generated by morpholinos that block either translation or splicing. First, the amount, activity, and persistence of maternal protein may prevent detection of early phenotypes.

Even in cases where no maternal RNA is detectable, maternal protein may reduce the knockdown effect (Edelmann *et al.*, 1998). In most cases, no antibody will be available to evaluate the amount of maternal protein. Consequently, it may be helpful to check whether morpholino-generated phenotypes are similar to those caused by interfering with the function of the protein of interest, e.g., dominant negatives. However, as has been discussed, the phenotypes produced by these two approaches will not always be the same. A second consideration is that morpholino-generated phenotypes will reflect predominantly the earliest functions of a protein, because morpholinos are introduced at the beginning of development.

1. Morphology Throughout Development

This is, of course, the first criterion and it provides the first indications of what cell types and tissues are sensitive to loss of the protein of interest. It also provides a guide to whether the expression of specific genes should or should not be affected.

2. Expression of Other Genes

a. Immunohistochemistry. The reader is referred to Chapter 15 for a list of antibodies specific for different cell types and tissues.

b. mRNA measurements. Because the number of embryos that can be microinjected practically is relatively small (hundreds to thousands), sensitive methods, such as RT-PCR and RNase protection, are required. The former is more sensitive, requiring only about 5 to 10 embryo equivalents per assay, whereas the latter requires several hundred. For discussion of assays of RNA concentrations from limited numbers of embryos, see Chapter 24.

IV. Concluding Remarks

Morpholino-mediated loss-of-function is a major advance for analyzing developmental mechanisms in systems not amenable to classic forward genetic analysis. It provides rapid analysis of the developmental functions of genes and, in some cases, assays can be constructed to test the role of individual proteins in biochemical reactions or cellular processes. The combination of morpholino knockdowns with molecular assays of expression of other genes is establishing the genetic pathways that underlie fate specification and tissue differentiation (Davidson *et al.*, 2002a, 2003; http://www.its.caltech.edu/~mirsky/endomes.htm). Discoveries of new genes and new pathways are feasible using sensitive molecular screens that select for mRNA sequences that are affected by introduction of a particular morpholino. The newly completed genome sequences of the sea urchin and the sea squirt will allow rapid identification of morpholino target sequences for many genes, greatly accelerating the pace of studies on the functions of many developmentally important proteins.

References

Angerer, L. M., Oleksyn, D. W., Levine, A. M., Li, X., Klein, W. H., and Angerer, R. C. (2001). Sea urchin goosecoid function links fate specification along the animal–vegetal and oral–aboral embryonic axes. *Development* **128,** 4393–4404.

Bogarad, L. D., Arnone, M. I., Chieh, C., and Davidson, E. H. (1998). TKO, a general approach to functionally knock out transcription factors *in vivo*, using a novel genetic expression vector. *Proc. Natl. Acad. Sci. USA* **95,** 14827–14832.

Davidson, E., McClay, D., and Hood, L. (2003). Regulatory gene networks and the properties of the developmental process. *Proc. Natl. Acad. Sci. USA* **100,** 1475–1480.

Davidson, E. H., Rast, J. P., Oliveri, P., Ransick, A., Calestani, C., Yuh, C. H., Minokawa, T., Amore, G., Hinman, V., Arenas-Mena, C., Otim, O., Brown, C. T., Livi, C. B., Lee, P. Y., Revilla, R., Rust, A. G., Pan, Z., Schilstra, M. J., Clarke, P. J., Arnone, M. I., Rowen, L., Cameron, R. A., McClay, D. R., Hood, L., and Bolouri, H. (2002a). A genomic regulatory network for development. *Science* **295,** 1669–1678.

Davidson, E. H., Rast, J. P., Oliveri, P., Ransick, A., Calestani, C., Yuh, C. H., Minokawa, T., Amore, G., Hinman, V., Arenas-Mena, C., Otim, O., Brown, C. T., Livi, C. B., Lee, P. Y., Revilla, R., Schilstra, M. J., Clarke, P. J., Rust, A. G., Pan, Z., Arnone, M. I., Rowen, L., Cameron, R. A., McClay, D. R., Hood, L., and Bolouri, H. (2002b). A provisional regulatory gene network for specification of endomesoderm in the sea urchin embryo. *Dev. Biol.* **246,** 162–190.

Draper, B., Morcos, P., and Kimmel, C. (2001). Inhibition of zebrafish fgf8 pre-mRNA splicing with morpholino oligos: A quantifiable method for gene knockdown. *Genesis* **30,** 154–156.

Edelmann, L., Zheng, L., Wang, Z. F., Marzluff, W., Wessel, G. M., and Childs, G. (1998). The TATA binding protein in the sea urchin embryo is maternally derived. *Dev. Biol.* **204,** 293–304.

Fuchikami, T., Mitsunaga-Nakatsubo, K., Amemiya, S., Hosomi, T., Watanabe, T., Kurokawa, D., Kataoka, M., Harada, Y., Satoh, N., Kusunoki, S., Takata, K., Shimotori, T., Yamamoto, T., Sakamoto, N., Shimada, H., and Akasaka, K. (2002). T-brain homologue (HpTb) is involved in the archenteron induction signals of micromere descendant cells in the sea urchin embryo. *Development* **129,** 5205–5216.

Hannon, G. (2002). RNA interference. *Nature* **418,** 244–251.

Heasman, J. (2002). Morpholino oligos: Making sense of antisense? *Dev. Biol.* **243,** 209–214.

Howard, E. W., Newman, L. A., Oleksyn, D. W., Angerer, R. C., and Angerer, L. M. (2001). *SpKrl*: A direct target of β-catenin regulation required for endoderm differentiation in sea urchin embryos. *Development* **128,** 365–375.

Imai, K., Satou, Y., and Satoh, N. (2002). Multiple functions of a zic-like gene in the differentiation of notochord, central nervous system, and muscle in *Ciona savignyi* embryos. *Development* **129,** 2723–2732.

Imai, K., Takada, N., Satoh, N., and Satou, Y. (2000). β-catenin mediates the specification of endoderm cells in ascidian embryos. *Development* **127,** 3009–3020.

McMahon, A. P., Flytzanis, C. N., Hough-Evans, B. R., Katula, K. S., Britten, R. J., and Davidson, E. H. (1985). Introduction of cloned DNA into sea urchin egg cytoplasm: Replication and persistence during embryogenesis. *Dev. Biol.* **108,** 420–430.

Moore, J. C., Sumerel, J. L., Schnackenberg, B. J., Nichols, J. A., Wikramanayake, A., Wessel, G. M., and Marzluff, W. F. (2002). Cyclin D and cdk4 are required for normal development beyond the blastula stage in sea urchin embryos. *Mol. Cell Biol.* **22,** 4863–4875.

Rast, J. P., Cameron, R. A., Poustka, A. J., and Davidson, E. H. (2002). Brachyury target genes in the early sea urchin embryo isolated by differential macroarray screening. *Dev. Biol.* **246,** 191–208.

Satou, Y., Imai, K., and Satoh, N. (2001). Action of morpholinos in *Ciona* embryos. *Genesis* **30,** 103–106.

Satou, Y., Imai, K., and Satoh, N. (2001). Early embryonic expression of a LIM-homeobox gene Cs-lhx3 is downstream of beta-catenin and responsible for the endoderm differentiation in *Ciona savignyi* embryos. *Development* **128,** 3559–3570.

Schmajuk, G., Sierakowska, H., and Kole, R. (1999). Antisense oligonucleotides with different backbones. *J. Biol. Chem.* **274,** 21783–21789.

Sweet, H. C., Gehring, M., and Ettensohn, C. A. (2002). LvDelta is a mesoderm-inducing signal in the sea urchin embryo and can endow blastomeres with organizer-like properties. *Development* **129,** 1945–1955.

Wada, S., and Saiga, H. (2002). HrzicN, a new Zic family gene of ascidians, plays essential roles in the neural tube and notochord development. *Development* **129,** 5597–5608.

Woolf, T., Melton, D., and Jennings, C. (1992). Specificity of antisense oligonucleotides *in vivo*. *Proc. Natl. Acad. Sci. USA* **89,** 7305–7309.

CHAPTER 29

Generation and Use of Transgenic Ascidian Embryos

Robert W. Zeller

Department of Biology
San Diego State University
San Diego, California 92182

Overview

Since the late 1990s, there has been renewed interest in using ascidians (phylum Chordata, subphylum Urochordata, class Ascidiacea) to explore fundamental questions regarding the origins of chordates and the evolution of vertebrates from an invertebrate chordate ancestor. The ascidian embryo has many advantages for experimental developmental biology: (1) rapid embryogenesis resulting in a tadpole larva of ~2500 cells just 18 h after fertilization; (2) a well-defined cell lineage; (3) a small, compact genome of ~160 MB with relatively simple gene regulatory

METHODS IN CELL BIOLOGY, VOL. 74

0091-679X/04 $35.00

networks; (4) nearly complete genome sequences from two closely related species with an extensive EST sequence collection; and (5) a wide assortment of microsurgical and molecular and cellular biological experimental techniques. The development of a simple and rapid electroporation technique to generate transgenic embryos has made the ascidian one of the best-suited animal models for gene regulatory analysis. This chapter describes the development and use of electroporation in ascidian embryos and presents a detailed electroporation protocol. Electroporation has been successfully used on a variety of ascidian species to generate transgenic embryos and this protocol can likely be adapted for use on other marine invertebrate embryos.

I. Ascidian Embryos as Model Chordate Embryos

Since the 1990s, there has been a renewed interest in using the ascidian (phylum Chordata, subphylum Urochordata, class Ascidiacea) as a model for chordate development. Chordates are broadly categorized into three subgroups: urochordates, cephalochordates, and vertebrates (Gee, 1996). The relationships among these subgroups, as well as the origin of chordates from a deuterostome ancestor, have fostered renewed experimental interest. Cephalochordates and vertebrates are believed to be sister groups, with the ascidians occupying a more basal position during the evolution of chordates (Gee, 1996). Ascidians are an early offshoot of the last common chordate ancestor and thus represent an interface between invertebrates and vertebrates. Ascidians are a well-suited invertebrate chordate system in which experimental embryological techniques may be coupled with modern molecular and cell biological approaches to analyze developmental mechanisms. These properties make ascidians an ideal animal model in which to study two fundamental questions: (1) How did chordates evolve from an ancestral deuterostome? and (2) How did vertebrates evolve from an ancestral chordate?

Ascidians are marine invertebrates; no freshwater species are known. Adults are sessile filter feeders, usually self-sterile hermaphrodites, which reproduce via a motile, tadpole larva (reviewed by Satoh, 1994). Alexander Kowalevsky first described the chordate affinity of ascidians in the late nineteenth century (Kowalevsky, 1866, 1871). In his studies, Kowalevsky correctly described the ascidian tadpole notochord and dorsal, hollow nerve chord as definitive chordate characteristics. In *Ciona intestinalis*, the most commonly studied ascidian for developmental biology, the notochord is flanked by three rows of muscle cells and occupies a central position within the tail. The hollow nerve cord runs dorsally to the notochord and the endodermal strand runs ventrally along the tail (see Katz, 1983). Embryonic development in ascidians is rapid and is complete in about one day in *C. intestinalis*. The tadpole larva is used for dispersal and, after swimming for several hours to a few days, will attach to a substrate and metamorphose into the adult. During metamorphosis, the notochord and dorsal hollow nerve cord are lost. The adult ascidian uses a pharyngeal basket for filter feeding and this is where the pharyngeal (gill) slits are located. The endostyle, thought to be equivalent to the vertebrate thyroid gland, secretes mucous to assist with feeding (Satoh, 1994).

In 2002, the genome sequence of *C. intestinalis* was published (Dehal *et al.*, 2002). The *C. intestinalis* genome is approximately 160 million base pairs (Simmen *et al.*, 1998) and is only about 5% the size of the human genome. The current genome assembly is believed to contain about 95% of the protein-coding genes and it is estimated that there are about 15,800 total genes (Dehal *et al.*, 2002). In addition to the genome sequence, nearly 500,000 EST sequences are available from a variety of embryonic stages and adult tissues (Fujiwara *et al.*, 2002; Inaba *et al.*, 2002; Kusakabe *et al.*, 2002; Nishikata *et al.*, 2001; Ogasawara *et al.*, 2002; Satou *et al.*, 2001, 2002a,b; Takamura *et al.*, 2001). Comparisons of the *Ciona* genome with the genomes of flies, nematodes, pufferfish, and mammals have provided clues to the origins of many different genes and gene families (Dehal *et al.*, 2002). From this analysis, genes could be classified as (1) common to all bilaterians, (2) common to invertebrate bilaterians, (3) present only in deuterostomes, and (4) ascidian-specific (Dehal *et al.*, 2002). In addition to the *C. intestinalis* genome, the genome of the related species *C. savignyi* has been sequenced by the Whitehead Center for Genome Research (http://www-genome.wi.mit. edu/annotation/ciona/) but still requires assembly and annotation.

In addition to the vast amount of sequence data available, the ascidian embryo offers a number of additional features that are useful for developmental biology research. The *C. intestinalis* tadpole completes embryogenesis about 18 h post-fertilization (PF) at 18 °C. The tadpole is formed after fewer than 12 cell divisions and is composed of about 2500 cells with a limited number of tissue types. For example, the *C. intestinalis* notochord is composed of exactly 40 cells (Nishida, 1987) and the simple nervous system contains only ~330 cells (Nicol and Meinertzhagen, 1988, 1991). Like the nematode *Caenorhabditis elegans*, the *C. intestinalis* cell lineage has been well documented. The original cell-lineage description was published in 1905 by Edwin Conklin (Conklin, 1905) and additional minor corrections were contributed by several researchers (Nishida, 1987; Nishida and Satoh, 1983, 1985; Ortolani, 1955, 1957, 1962, 1971). Gastrulation begins at the 110-cell stage, about 5 h PF and the lineage of nearly every cell is known (reviewed by Satoh, 1994). The embryo is amenable to a wide variety of microsurgical techniques including blastomere isolation and partial embryo analysis (Meedel and Whittaker, 1984; Meedel *et al.*, 2002), cleavage-arrest experimentation (Crowther and Whittaker, 1983, 1986; Crowther *et al.*, 1990; Whittaker, 1973, 1983; Whittaker and Meedel, 1989), and cytoplasmic transfer experimentation (Whittaker, 1980, 1982). Several laboratories have begun to perform genetic screens in both *C. intestinalis* (Sordino *et al.*, 2000, 2001) and *C. savignyi* (Moody *et al.*, 1999; Nakatani *et al.*, 1999). Because ascidian embryos are generally transparent, small, and develop rapidly, the entire period of embryogenesis may be monitored by high-resolution microscopy (Miyamoto and Crowther, 1985; Munro and Odell, 2002a,b). A number of histochemical techniques are available to identify specific tissues in the embryo, such as endoderm, muscle, and the pigmented cells within the CNS (see Whittaker, 1973). In the late 1990s, a number of laboratories have identified *Ciona* orthologues of developmentally important genes such as Brachyury (Corbo *et al.*, 1997b) and FoxA2 and Snail (Corbo *et al.*, 1997a). From these studies and others, it has become apparent that ascidians and vertebrates share many aspects of tissue specification at the genetic level.

II. Transgenic Ascidian Embryos

Ascidian embryos are probably best suited as a system in which to study gene regulation. The small genome size and the lack of extensive gene duplication events suggest that developmentally important gene regulatory networks are likely to be greatly simplified compared to their vertebrate counterparts. The rapid embryonic development of ascidian embryos allows experimental results to be obtained in less than one day when using *C. intestinalis* embryos. Perhaps the most important technological advance in modern ascidian embryological research was the development of a simple and extremely efficient electroporation technique for creating transiently transfected embryos (Corbo *et al.*, 1997b). Electroporation of ascidian embryos is easy and rapid, generating thousands of transgenic embryos in about 30 min. This procedure revolutionized the study of gene regulation in ascidians and has led to tremendous advances in our understanding of ascidian gene regulation and gene regulatory networks.

Early attempts to create transgenic ascidians relied on microinjection of reporter gene constructs into fertilized eggs (Corbo *et al.*, 1997b). This experimental procedure worked well on *Halocynthia roretzi* which has an egg diameter of ~260 μm (Hikosaka *et al.*, 1994). Application of this technique to *C. intestinalis* proved more troublesome (egg size ~160 μm diameter), despite the fact that *C. savignyi* (egg size ~160 μm diameter; Hikosaka *et al.*, 1993) and sea urchin species such as *Strongylocentrotus purpuratus* (egg size 80 μm in diameter; Flytzanis *et al.*, 1987) could be easily microinjected. Although microinjection in *C. intestinalis* was possible, it was not efficient. After fertilization, ascidian eggs undergo two phases of cytoplasmic streaming, called ooplasmic segregation. The first phase results in the accumulation of peripheral cytoplasmic components at the vegetal pole of the egg. One of these components is the myoplasm, the "yellow crescent" material observed by Conklin in his lineage experiments (Conklin, 1905). The second phase of ooplasmic segregation occurs shortly thereafter and results in the myoplasm and additional cytoplasmic components segregating to the future posterior embryonic pole (reviewed by Satoh, 1994). Microinjection into *C. intestinalis* zygotes was only possible in the few minutes between these two phases because injections during cytoplasmic streaming resulted in egg lysis. There is no overt pigmentation of egg cytoplasm; it was therefore necessary to time the injection "window of opportunity" from the time of fertilization and any alterations in room temperature caused havoc with the timing. With the need to fertilize and time small batches of eggs for microinjection, only about 10 to 20 zygotes could be injected per fertilization. In addition to these limitations, the microinjection setup required extensive equipment resources including a glass micro-needle puller, microscope, micromanipulator apparatus, and specialized injection chambers developed by Dan Kiehart (Kiehart, 1982).

To try and overcome this problem, efforts were made to develop an alternative procedure for generating transgenic embryos. During the early 1990s, several research groups reported on the use of electroporation as a means of introducing DNA constructs into a variety of aquatic organisms including abalone (Powers *et al.*, 1995), zebrafish (Buono and Linser, 1992; Inoue, 1992; Powers *et al.*, 1992), and nonaquatic organisms such as *Drosophila* (Kamdar *et al.*, 1992). Electroporation is a process in which an electrical pulse is used to create transient pores in cell membranes, allowing exogenous material to enter the cell (Andreason and Evans, 1988; Shigekawa and Dower, 1988; Weaver, 1993, see Section III). At about this time, Nishida was performing a series of ascidian blastomere and egg fragment fusion experiments. The blastomeres or egg fragments were placed in a 0.77 M D-mannitol solution (iso-osmotic with seawater) while an electrical current was applied to fuse the fragments (Nishida, 1992). Because ascidian embryos survived this treatment, the 0.77 M D-mannitol solution was used as an iso-osmotic medium for electroporations.

The identification of suitable electroporation parameters is often a trial-and-error process. In bacteria, electroporation conditions are typically at high voltage with a small capacitance setting, while in eukaryotic cells the voltage is usually much lower with higher capacitance (Shigekawa and Dower, 1988). Experiments with *C. intestinalis* focused on using low voltage, high capacitance, exponentially decaying pulses. Initially, electroporator parameters were varied until a setting was obtained that resulted in about 50% embryonic mortality. An important observation was that the follicle cells and chorion surrounding the fertilized eggs had to be removed for embryonic survival. A chemical dechorionation method (Mita-Miyazawa *et al.*, 1985) was thus used to isolate the large quantities of dechorionated zygotes required for the electroporation procedure. Various amounts of supercoiled plasmid DNA, a 3.5 kb *cis*-regulatory domain from the *C. intestinalis* Brachyury gene fused to a green fluorescent protein (GFP) reporter (Corbo *et al.*, 1997b) was added to the D-mannitol solution and the electroporations repeated. Embryos were scored for GFP expression and suitable electroporation conditions were obtained after just a few days of experimentation. Once GFP expression was observed, electroporation parameters were adjusted further to try to improve the efficiency of transfection and decrease the mortality of zygotes. The final settings (125 VDC/cm and 960 μF capacitance, 20 millisecond time-constant) resulted in efficiencies approaching 75% transgenic embryo creation with a 75% survival rate.

Exogenous DNA in ascidians is expressed mosaically as is the situation in sea urchin embryos (Franks *et al.*, 1988; Hough-Evans *et al.*, 1988; Livant *et al.*, 1991; Zeller *et al.*, 2004, in press) . The degree of mosaic expression can be controlled, to a degree, by the amount of DNA used during the electroporation procedure.

Using about 25 μg of DNA per electroporation results in embryos that are highly mosaic. The level of mosaicism decreases as the amount of DNA used per electroporation approaches 100 μg. Above this amount of DNA, embryonic survival decreases. In addition to varying the amount of DNA used, the level of mosaic expression can be reduced by decreasing the amount of time between fertilization and electroporation—the fastest time practically achievable for *C. intestinalis* is about 12 to 13 min (Zeller, unpublished observations). Although square wave electroporator devices have been used successfully to generate transgenic *C. intestinalis* embryos (Zeller, Embryology Course, Woods Hole, unpublished observations), the protocols described later in this chapter (Section V) will focus on electroporation devices that output an exponentially decaying pulse. A 2004 report details the construction and characterization of a simple-to-build electroporator for use with ascidian embryos (Zeller *et al.*, 2004). Using this device, high efficiencies of transformation, approaching 100%, and high levels of embryo survival, near 90%, may be obtained. In addition, optimized parameters for use with this electroporation device are reported.

Electroporation has been used to express a wide variety of DNA constructs in ascidian embryos (reviewed by Di Gregorio and Levine, 2002). These constructs are either reporter genes used to assay *cis*-regulatory regions of various genes or constructs designed to ectopically express or overexpress protein coding regions in ascidian embryos. Two main reporter genes have been utilized: *lacZ*, derived from vectors used to generate transgenic *Caenorabditis elegans* embryos (Fire *et al.*, 1990) and GFP, which has been shown to function as a reporter gene in many experimental systems (Chalfie *et al.*, 1994; Gerdes and Kaether, 1996; Haseloff, 1999; Prasher, 1995; Tsien, 1998; Welsh and Kay, 1997). Reporter genes have proven invaluable for identifying *cis*-regulatory domains of genes expressed specifically in muscle (Hikosaka *et al.*, 1994; Vandenberghe *et al.*, 2001), epidermis (Ishida and Satoh, 1999; Ueki and Satoh, 1995), and notochord (Corbo *et al.*, 1997b). Additionally, genes such as Snail and FoxA2 are expressed in multiple tissue types in *Ciona* embryos (Corbo *et al.*, 1997a; Di Gregorio *et al.*, 2001; Erives *et al.*, 1998). Many of these studies have identified minimal enhancer elements required for tissue-specific expression by systematically truncating *cis*-regulatory DNA. Because large numbers of transgenic ascidian embryos can be generated, efforts have also been made to identify tissue-specific enhancers by electroporating random genomic DNA fragments fused upstream of a naive basal promoter and *lacZ* reporter (Harafuji *et al.*, 2002).

Using electroporation, it is also possible to ectopically express or overexpress protein coding genes in ascidian embryos to create "mutant" embryos. The best example of this approach was the use of electroporation to identify downstream targets of the notochord-specific Brachyury transcription factor in *C. intestinalis* embryos (Takahashi *et al.*, 1999). In this study, the *C. intestinalis* Brachyury gene coding sequence was fused in-frame with the promoter for the FoxA2 gene to create a transgene that would express Brachyury in every cell type that normally expressed FoxA2. In normal *Ciona* embryos, Brachyury (CiBra) is only expressed in the ascidian notochord (Corbo *et al.*, 1997b), while FoxA2 is expressed in a variety of tissues including the notochord and endoderm (Corbo *et al.*, 1997a). Embryos transgenic for the FoxA2-CiBra construct contained many cells (mostly endoderm cells) that ectopically expressed Brachyury. Using subtractive hybridization techniques, the researchers identified nearly 40 genes predominately expressed in the notochord (Hotta *et al.*, 1999, 2000; Takahashi *et al.*, 1999). The cells ectopically expressing Brachyury also expressed one or more of these additional notochord genes, consistent with at least a partial transformation of blastomeres from an endoderm to notochord fate. Previous experiments using *Xenopus* animal caps clearly demonstrated that notochord fate is specified by the combined actions of both Brachyury and FoxA2 (O'Reilly *et al.*, 1995) and the *Ciona* results are consistent with these observations (see also Shimauchi *et al.*, 2001). Electroporation experiments have been used to study dominant-negative transcription factors such as Supressor of Hairless (Corbo *et al.*, 1998), to study cellular morphogenesis (Di Gregorio *et al.*, 2002), and to test the function of hemichordate and echinoderm Brachyury orthologues in ascidian embryos (Satoh *et al.*, 2000).

III. Fundamentals of Electroporation

Electroporation is a technique that is used to introduce exogenous materials into cells (Andreason and Evans, 1988; Shigekawa and Dower, 1988). In the context of making transgenic ascidian embryos, the exogenous material is a DNA transgene. Electroporation has been successful on a wide variety of cell types, both eukaryotic and prokaryotic. Several experiments have also used electroporation for introducing DNA into chicken embryos *in ovo* (reviewed by Itasaki *et al.*, 1999) and for introducing therapeutic DNAs (reviewed by Bigey *et al.*, 2002; Li and Benninger, 2002). During the electroporation process, cells are exposed to an electric field, thus creating a voltage potential across the cell membrane. This electric field is applied in one or more "pulses," depending on the particular application and cell type. If the voltage potential exceeds a threshold level, the cell membrane will break down, forming localized pores that can take up exogenous materials such as DNA (Shigekawa and Dower, 1988). Experiments using single cells and fluorescent microscopy have demonstrated that while pores form all around a cell, DNA only interacts at the side of the cell facing the negative electrode (Golzio *et al.*, 2002). DNA-membrane aggregates are formed in this region so long as the electric field pulse is applied above a threshold value. These aggregates eventually disperse into the cytoplasm and the DNA then enters the cell's nucleus (Golzio *et al.*, 2002).

The success of the electroporation technique depends on a number of parameters (Shigekawa and Dower, 1988). The strength of the electric field (E) is related to the potential across the electrodes by

$$E = V/d$$

where d is the distance between the electrodes and V is the voltage applied (Shigekawa and Dower, 1988). To generate an exponentially decaying pulse, a capacitor is discharged across a resistance (the sample) and the voltage declines over time by

$$V_t = V_0 e^{-t/\tau}$$

$$\text{where } \tau = RC$$

R is the resistance in ohms of the sample, C is the capacitance measured in farads, and τ is the time constant in seconds (Shigekawa and Dower, 1988). Pore formation depends on both the field strength (E) and the size of the cell. Since the electrodes in the electroporation cuvette are a fixed distance apart, the field strength follows the capacitor discharge (Shigekawa and Dower, 1988):

$$E_t = E_0 e^{-t/\tau}$$

For ascidian embryos, the critical parameters are (1) the size of the capacitor (960 μF), (2) the voltage applied (125 VDC/cm), and (3) the electrode gap of the electroporation cuvette (0.4 cm). Using these settings on a typical electroporation device, such as the Biorad Gene Pulser I, time constants (τ) of about 20 milliseconds are obtained using the protocol described in Section V.

IV. Transgene Construction

The typical reporter genes used in ascidian embryos are generally based around a standardized vector (Corbo *et al.*, 1997b) that contains a short polylinker, a reporter gene coding region such as *lacZ* or GFP, and the SV40 polyadenylation sequence derived from the vector pPD1.27 (Fire *et al.*, 1990). To create novel transgenes, the *C. intestinalis* genome sequence may be used as a reference to design PCR primers to amplify the desired *cis*-regulatory domain. Typically, the reporter protein is fused in-frame about 4 to 5 amino acids downstream of the endogenous ATG initiation codon, although fusions to other regions, such as the 5′ UTR, are also possible. Using genomic DNA as a template, perform a PCR reaction to amplify the genomic region of interest and subclone the *cis*-regulatory fragment into the electroporation vector. The transgenes are first verified by restriction mapping or PCR and then should be tested in embryos before large-scale DNA preparations are made. For rapidly testing the constructs, sufficiently clean quantities of plasmid DNA may be obtained by combining four Qiagen minipreps of the construct, precipitating, and resuspending the plasmid in 10 μl of water. The entire DNA preparation is then used for electroporations. Large-scale plasmid preps (preferably cesium chloride banded) may then be prepared once the test electroporations have identified functional transgenes. Similar approaches may be used to construct transgenes to ectopically express proteins by exchanging GFP for the protein-coding region. Enhancers, introns, and other regions of genomic DNA may also be incorporated into transgenes. In these instances, it may be necessary to fuse a *cis*-regulatory element with a basal promoter to create a functional transgene. Several basal promoters have been tested in *C. intestinalis*, including a *Halocynthia* muscle actin promoter (Corbo *et al.*, 1997b) and the *C. intestinalis* FoxA2 basal promoter (Harafuji *et al.*, 2002). The exact sequences and additional details regarding the use of these basal promoters may be found in the respective references.

V. Detailed Protocols for Electroporation of Ascidian Zygotes

The electroporation procedure outlined here is described for *Ciona intestinalis*. This procedure has also been successfully used with *C. savignyi* (Nakatani *et al.*, 1999; Zeller, unpublished observations), with *Boltenia villosa* and *Phallusia* sp. (Di Gregorio and Levine, 2002), and with *Ascidiella aspersa* and *Styela plicata* (Zeller, unpublished observations). This procedure does not work well for *Halocynthia roretzi* (Di Gregorio and Levine, 2002). Electroporation in other species will likely require modifications to this protocol. The key steps include the dechorionation of eggs and the optimization of the electroporation parameters. The procedure is not difficult, but will take some time to master. The basic steps arc (1) fertilization (Section V.A), (2) dechorionation (Section V.B), and (3) electroporation (Section V.C). Be sure to obtain all necessary reagents and supplies before starting (Section VI). Read through the protocol carefully and prepare all solutions, dishes, filtered seawater, and other items as indicated. All solutions (thioglycolic acid and DNA dilutions) and seawater should be equilibrated to 18 °C for *Ciona*. Use the appropriate incubation temperature for other ascidian species. Time is of the essence during this procedure. Fertilization and washes should be completed about 3 to 4 min post-fertilization (PF). Dechorionation should be completed by 12 to 15 min PF. Ideally, electroporations should be completed by 20 min PF.

A. Fertilization

1. Preparation for Fertilization

Prepare the following before beginning the procedure:

a. Filtered Seawater (FSW)

All seawater should be filtered with a pleated filter prior to use to remove large particulate matter. Fertilizations and washes are performed using FSW; all further steps require sub-micron-filtered seawater (SMFSW). Approximately 1 liter of FSW is required per fertilization/electroporation procedure.

b. Sub-Micron-Filtered Seawater (SMFSW)

Use a disposable 0.45 um bottletop filter and filter FSW into clean 1 liter bottles. Store SMFSW in the dark. Disposable filters may be reused several times (4–5) until the flow rate drops to an unreasonable level. Approximately 150 ml of SMFSW is required for 1 round of dechorionation and 10 electroporations.

c. Coated Dishes

A number of dishes are required for electroporation. Disposable plastic dishes are routinely used (35 and 60 mm sizes, Falcon numbers 351008 and 351008, tend to work best for agar coating). Dechorionated eggs and embryos will adhere to all noncoated surfaces. Agar-coated dishes may be prepared several hours in advance by melting a 1% agar solution in FSW in the microwave. Dishes are arrayed on a benchtop and the agar poured form one dish to the next bucket-brigade style. Only a very thin coat of agar is required and may be obtained by rapidly pouring the agar from one dish to the next.

Alternatively, gelatin-coated dishes may be prepared ahead of time in large quantities (Henry and Martindale, 2001). A 5X solution of gelatin-coating solution is prepared by adding 250 mg of Knox gelatin and 250 μl of 37% formaldehyde to 50 ml of distilled water. Dilute 5X stock solution to 1X concentration with distilled water immediately before use. Pour 1X solution from one dish to the next bucket-brigade style. Allow gelatin coating to completely dry (several hours to overnight). Rinse out dishes three times with distilled water and allow to dry. Kept dry in sealed plastic sleeves at room temperature, gelatin-coated dishes are good for several months. One 35 mm and four 60 mm coated dishes are required to dechorionate fertilized eggs. At least one 60 mm coated dish is required per electroporation.

d. Thioglycolic Acid

Immediately prior to fertilization, prepare 10 ml of 1% thioglycolic acid (100 mg/10 ml) in SMFSW. Incubate at 18 °C.

2. Procedure for Fertilization

Ascidians are self-sterile hermaphrodites. Adults should be maintained under constant lighting conditions for one to several days to induce gamete production. Sufficient numbers of gametes are obtained from two gravid individuals for performing several electroporations. Adjust the number of eggs for more electroporations; however, ten electroporations per fertilization is about the maximum number that can be easily handled. Fertilizations are performed in FSW (see also Zeller, 1998). Add approximately 150 ml of FSW into a four-inch finger bowl. Surgically remove the eggs from two individuals and add to the seawater in the finger bowls. Add one small drop of sperm from each individual, stir and mix well until the sperm is dissolved. Allow 2 min for fertilization. While fertilization commences, add 25 ul 10 M NaOH to 10 ml of 1% thioglycolate solution to bring pH close to 9.5. Excess sperm is washed away from fertilized eggs by passing through a plastic container covered with 80 um nylon mesh (i.e., Nytex). Be sure to wash eggs through 3 to 4 exchanges of FSW. Zygotes are now ready for dechorionation.

B. Dechorionation

1. Preparation for Dechorionation

Prepare the following before beginning the procedure:

a. Protease Stock Solution

Protease (Sigma type XIV) should be dissolved in SMFSW to a final concentration of 2.5 mg of protease per 100 μl of SMFSW. Aliquot protease into 100 μl aliquots and snap freeze in liquid nitrogen. Store at −80 °C until use.

2. Procedure for Dechorionation

Pour enough ~pH 9.5 thioglycolate solution into a 35 mm coated dish to barely cover the bottom. With a 9-inch Pasteur pipet, transfer zygotes into coated dish containing thioglycolate solution. The number of eggs transferred should not make more than three or four layers of settled eggs. Using a Pasteur pipet, gently swirl eggs to mix with thioglycolate. Allow eggs to settle, remove most of the thioglycolate solution, and add 2 to 3 ml fresh thioglycolate solution. Swirl eggs and repeat wash. Reserve last 3 to 4 ml of thioglycolate for dechorionation. After all washes are completed, add reserved thioglycolate and one tube of protease (2.5 mg, 0.05% final concentration). Mix well and allow eggs to settle, and then mix again. Continue this process until eggs are dechorionated (about 3–6 min after protease addition). After protease is added, the chorion and follicle cells will begin to turn yellowish in color. As dechorionation proceeds, the solution will become quite cloudy. Dechorionated eggs will be reddish-brown in color and can be distinguished from the yellow-colored eggs that remain within their chorions. It may be necessary to transfer small aliquots of eggs to a coated dish with clean SMFSW to monitor dechorionation. The process is considered complete when about 85 to 90% of the eggs are dechorionated. The dechorionated zygotes must now be washed four times by serially transferring into four 60 mm coated dishes filled with SMFSW. Four washes are usually sufficient; however, if larger numbers of zygotes are dechorionated, then add additional wash steps here. Pasteur pipets should be changed at every wash to reduce solution carryover. The washes dilute the dechorionation solution and lower the pH, inactivating the protease. This step of the procedure is the most important and critical. Dechorionation takes practice, but can be easily mastered. Note: The 1% thioglycolate solution should be relatively transparent when dissolved. If the solution is cloudy, the thioglycolate is old and the dechorionation will be difficult. New thioglycolic acid should thus be obtained.

C. Electroporation

1. Preparation for Electroporation

a. Equipment

Obtain a suitable electroporator and 0.4 cm gap electroporation cuvettes. The original protocol was performed on a Biorad Gene Pulser 1 with capacitance extender unit (Corbo *et al.*, 1997b). This machine outputs exponentially decaying pulses (while square wave pulses will work, extensive modifications of the protocol may be necessary to obtain optimal results). Any equivalent machine should be suitable as long as it is capable of outputting 125 VDC/cm at a setting of about 1000 μF. The 0.4 cm electroporation cuvettes hold about 800 μl of solution.

b. DNA Samples for Electroporation

In a 1.5 ml microfuge tube, add 500 μl 0.77 M D-mannitol and plasmid DNA (up to ~100 μg). DNA should be concentrated and no more than 50 μl of DNA should be added to the mannitol solution. The use of volume-calibrated microfuge tubes allows for the easy measurement of fertilized eggs in seawater later in the procedure. The D-mannitol serves to reduce the ion concentration of seawater and has been used in ascidian egg fusion studies (Nishida, 1992). Incubate DNA solutions at 18 °C until electroporation.

2. Procedure for Electroporation

For each tube containing the ~500 μl DNA sample, add ~300 μl dechorionated zygotes in SMFSW with a Pasteur pipet. Hold the pipet vertically to minimize egg lysis. Gently pipet several times to mix, add entire ~800 μl volume to 0.4 cm cuvette, place in holder, and electroporate sample. Remove eggs with the Pasteur pipet and gently transfer to 60 mm coated dish filled with SMFSW. Allow zygotes to recover for about 5 min, then gently mix dish to spread out eggs. Eggs should not be crowded in dish and should be distributed to additional coated dishes if necessary. Monitor embryos for DNA construct expression as appropriate (i.e., GFP or *lacZ* staining). If desired, antibiotics such as penicillin–streptomycin (10 IU/ml and 10 μg/ml, respectively) may be added to electroporated embryos.

D. *lacZ* Staining

1. Preparation for Staining

a. Staining Buffer

The *lacZ* staining buffer consists of 5 mM potassium ferricyanide, 5 mM potassium ferrocyanide, 1 mM $MgCl_2$, and 0.01% SDS in SMFSW. The staining buffer should be made immediately before use from stock solutions (i.e., 100 mM ferricyanide, 100 mM ferrocyanide, 10% SDS, and 1 M $MgCl_2$).

2. Procedure for *lacZ* Staining

At the desired embryonic stage, lightly fix electroporated embryos by adding 100 μl 37% formaldehyde to the dish containing the embryos. Fix for 10 min at room temperature and then transfer embryos, in a minimal amount of seawater, into staining dish (i.e., a gelatin-coated well in a 24-well tissue culture dish) containing 500 μl of staining buffer and 5 μl of 3% X-gal. Add more staining buffer and X-gal if necessary. Allow color development to proceed until desired (monitor under stereo microscope), and then transfer embryos to gelatin-coated dish of SMFSW with 1% formaldehyde for post-fixation. Fix for 30 min, wash well, and mount on slide for observations.

VI. List of Required Chemicals and Supplies

A. Required Chemicals

1. Dish Preparation

a. Knox gelatin
b. 37% formaldehyde
c. Seawater
d. Agarose

2. Dechorionation

a. Thioglycolic acid, Sodium salt (e.g., Sigma T0632)
b. Protease, Type XIV (e.g., Sigma P5147)
c. D-mannitol, 0.77 M (e.g., Sigma M4125)
d. NaOH, 10 M (e.g., Sigma S0899)
e. Plasmid DNA containing construct of interest, preferably cesium chloride banded. At least 25 μg is required for a single electroporation.

3. Reagents for *lacZ* Staining

a. Potassium ferrocyanide trihydrate, 100 mM (e.g., Sigma P2569)
b. Potassium ferricyanide, 100 mM (e.g., Sigma P3667)
c. SDS, 10% solution (e.g., Sigma L5750)
d. X-gal in DMF, 3% solution (e.g., Sigma B9146 and D4254)

B. Required Supplies

1. Dechorionation

a. 35 mm and 60 mm plastic dishes (e.g., Falcon 351008 and 351007)
b. 15 ml screw cap centrifuge tubes (e.g., Corning 430766)
c. 0.45 μm bottletop filters (e.g., Corning 430514) and bottles

2. Electroporation
   a. 0.4 cm gap electroporation cuvettes
   b. 1.5 ml graduated microfuge tubes

C. Required Equipment

1. Electroporator, exponential decay output with high capacitance (e.g., Biorad gene pulser)
2. 18 °C incubator for rearing embryos
3. Aquarium system with constant lighting for maintaining adults
4. Finger bowls for fertilizations
5. Nytex-lined baskets for washing eggs after fertilization.

References

Andreason, G. L., and Evans, G. A. (1988). Introduction and expression of DNA molecules in eukaryotic cells by electroporation. *Biotechniques* **6**(7), 650–660.

Bigey, P., Bureau, M. F., and Scherman, D. (2002). *In vivo* plasmid DNA electrotransfer. *Curr. Opin. Biotechnol.* **13**(5), 443–447.

Buono, R. J., and Linser, P. J. (1992). Transient expression of RSVCAT in transgenic zebrafish made by electroporation. *Mol. Mar. Biol. Biotechnol.* **1**(4–5), 271–275.

Chalfie, M., Tu, Y., Euskirchen, G., Ward, W. W., and Prasher, D. C. (1994). Green fluorescent protein as a marker for gene expression. *Science* **263**(5148), 802–805.

Conklin, E. G. (1905). The organization and cell lineage of the ascidian egg. *J. Acad. Nat. Sci. (Philadelphia)* **13,** 1–119.

Corbo, J. C., Erives, A., Di Gregorio, A., Chang, A., and Levine, M. (1997a). Dorsoventral patterning of the vertebrate neural tube is conserved in a protochordate. *Development* **124**(12), 2335–2344.

Corbo, J. C., Fujiwara, S., Levine, M., and Di Gregorio, A. (1998). Suppressor of hairless activates brachyury expression in the Ciona embryo. *Dev. Biol.* **203**(2), 358–368.

Corbo, J. C., Levine, M., and Zeller, R. W. (1997b). Characterization of a notochord-specific enhancer from the Brachyury promoter region of the ascidian, Ciona intestinalis. *Development* **124**(3), 589–602.

Crowther, R. J., Meedel, T. H., and Whittaker, J. R. (1990). Differentiation of tropomyosin-containing myofibrils in cleavage-arrested ascidian zygotes expressing acetylcholinesterase. *Development* **109**(4), 953–959.

Crowther, R. J., and Whittaker, J. R. (1983). Developmental autonomy of muscle fine structure in muscle lineage cells of ascidian embryos. *Dev. Biol.* **96**(1), 1–10.

Crowther, R. J., and Whittaker, J. R. (1986). Differentiation without cleavage: Multiple cytospecific ultrastructural expressions in individual one-celled ascidian embryos. *Dev. Biol.* **117**(1), 114–126.

Dehal, P., Satou, Y., Campbell, R. K., Chapman, J., Degnan, B., De Tomaso, A., Davidson, B., Di Gregorio, A., Gelpke, M., Goodstein, D. M., Harafuji, N., Hastings, K. E., Ho, I., Hotta, K., Huang, W., Kawashima, T., Lemaire, P., Martinez, D., Meinertzhagen, I. A., Necula, S., Nonaka, M., Putnam, N., Rash, S., Saiga, H., Satake, M., Terry, A., Yamada, L., Wang, H. G., Awazu, S., Azumi, K., Boore, J., Branno, M., Chin-Bow, S., DeSantis, R., Doyle, S., Francino, P., Keys, D. N., Haga, S., Hayashi, H., Hino, K., Imai, K. S., Inaba, K., Kano, S., Kobayashi, K., Kobayashi, M., Lee, B. I., Makabe, K. W., Manohar, C., Matassi, G., Medina, M., Mochizuki, Y., Mount, S., Morishita, T., Miura, S., Nakayama, A., Nishizaka, S., Nomoto, H., Ohta, F., Oishi, K., Rigoutsos, I., Sano, M., Sasaki, A., Sasakura, Y., Shoguchi, E., Shin-i, T., Spagnuolo, A., Stainier, D., Suzuki, M. M., Tassy, O.,

Takatori, N., Tokuoka, M., Yagi, K., Yoshizaki, F., Wada, S., Zhang, C., Hyatt, P. D., Larimer, F., Detter, C., Doggett, N., Glavina, T., Hawkins, T., Richardson, P., Lucas, S., Kohara, Y., Levine, M., Satoh, N., and Rokhsar, D. S. (2002). The draft genome of Ciona intestinalis: Insights into chordate and vertebrate origins. *Science* **298**(5601), 2157–2167.

Di Gregorio, A., Corbo, J. C., and Levine, M. (2001). The regulation of forkhead/HNF-3beta expression in the Ciona embryo. *Dev. Biol.* **229**(1), 31–43.

Di Gregorio, A., Harland, R. M., Levine, M., and Casey, E. S. (2002). Tail morphogenesis in the ascidian, Ciona intestinalis, requires cooperation between notochord and muscle. *Dev. Biol.* **244**(2), 385–395.

Di Gregorio, A. D., and Levine, M. (2002). Analyzing gene regulation in ascidian embryos: New tools for new perspectives. *Differentiation* **70**(4–5), 132–139.

Erives, A., Corbo, J. C., and Levine, M. (1998). Lineage-specific regulation of the Ciona snail gene in the embryonic mesoderm and neuroectoderm. *Dev. Biol.* **194**(2), 213–225.

Fire, A., Harrison, S. W., and Dixon, D. (1990). A modular set of lacZ fusion vectors for studying gene expression in Caenorhabditis elegans. *Gene* **93**(2), 189–198.

Flytzanis, C. N., Britten, R. J., and Davidson, E. H. (1987). Ontogenic activation of a fusion gene introduced into sea urchin eggs. *Proc. Natl. Acad. Sci. USA* **84**(1), 151–155.

Franks, R. R., Hough-Evans, B. R., Britten, R. J., and Davidson, E. H. (1988). Direct introduction of cloned DNA into the sea urchin zygote nucleus, and fate of injected DNA. *Development* **102**(2), 287–299.

Fujiwara, S., Maeda, Y., Shin, I. T., Kohara, Y., Takatori, N., Satou, Y., and Satoh, N. (2002). Gene expression profiles in Ciona intestinalis cleavage-stage embryos. *Mech. Dev.* **112**(1–2), 115–127.

Gee, H. (1996). "Before the Backbone." Chapman and Hall, London, UK.

Gerdes, H. H., and Kaether, C. (1996). Green fluorescent protein: Applications in cell biology. *FEBS Lett.* **389**(1), 44–47.

Golzio, M., Teissie, J., and Rols, M. P. (2002). Direct visualization at the single-cell level of electrically mediated gene delivery. *Proc. Natl. Acad. Sci. USA* **99**(3), 1292–1297.

Harafuji, N., Keys, D. N., and Levine, M. (2002). Genome-wide identification of tissue-specific enhancers in the Ciona tadpole. *Proc. Natl. Acad. Sci. USA* **99**(10), 6802–6805.

Haseloff, J. (1999). GFP variants for multispectral imaging of living cells. *Methods Cell Biol.* **58,** 139–151.

Henry, J. Q., and Martindale, M. Q. (2001). Multiple inductive signals are involved in the development of the ctenophore Mnemiopsis leidyi. *Dev. Biol.* **238**(1), 40–46.

Hikosaka, A., Kusakabe, T., and Satoh, N. (1994). Short upstream sequences associated with the muscle-specific expression of an actin gene in ascidian embryos. *Dev. Biol.* **166**(2), 763–769.

Hikosaka, A., Satoh, N., and Makabe, K. W. (1993). Regulated spatial expression of fusion gene constructs with the 5′ upstream region of Halocynthia roretzi muscle actin gene in Ciona savignyi. *Roux's Arch. Dev. Biol.* **203,** 104–112.

Hotta, K., Takahashi, H., Asakura, T., Saitoh, B., Takatori, N., Satou, Y., and Satoh, N. (2000). Characterization of Brachyury-downstream notochord genes in the Ciona intestinalis embryo. *Dev. Biol.* **224**(1), 69–80.

Hotta, K., Takahashi, H., Erives, A., Levine, M., and Satoh, N. (1999). Temporal expression patterns of 39 Brachyury-downstream genes associated with notochord formation in the Ciona intestinalis embryo. *Dev. Growth and Diff.* **41**(6), 657–664.

Hough-Evans, B. R., Britten, R. J., and Davidson, E. H. (1988). Mosaic incorporation and regulated expression of an exogenous gene in the sea urchin embryo. *Dev. Biol.* **129**(1), 198–208.

Inaba, K., Padma, P., Satouh, Y., Shin, I. T., Kohara, Y., Satoh, N., and Satou, Y. (2002). EST analysis of gene expression in testis of the ascidian Ciona intestinalis. *Mol. Reprod. Dev.* **62**(4), 431–445.

Inoue, K. (1992). Expression of reporter genes introduced by microinjection and electroporation in fish embryos and fry. *Mol. Mar. Biol. Biotechnol.* **1**(4–5), 266–270.

Ishida, K., and Satoh, N. (1999). Genomic organization and the 5′ upstream sequences associated with the specific spatio-temporal expression of HrEpiC, an epidermis-specific gene of the ascidian Halocynthia roretzi. *Cell Mol. Biol.* **45**(5), 523–536.

Itasaki, N., Bel-Vialar, S., and Krumlauf, R. (1999). "Shocking" developments in chick embryology: Electroporation and *in ovo* gene expression. *Nat. Cell Biol.* **1**(8), E203–E207.

Kamdar, P., Von Allmen, G., and Finnerty, V. (1992). Transient expression of DNA in Drosophila via electroporation. *Nucleic Acids Res.* **20**(13), 3526.

Katz, M. (1983). Comparative anatomy of the tunicate tadpole, Ciona intestinalis. *Biol. Bull.* **164,** 1–27.

Kiehart, D. P. (1982). Microinjection of echinoderm eggs: Apparatus and procedures. *Meth. Cel. Biol.* **25,** 13–31.

Kowalevsky, A. (1866). Entwickelungsgeschichte der einfachen Ascidien. *Mem. Acad. Imp. Sci. St. Petersbourg* **10 Series 7,** 1–19.

Kowalevsky, A. (1871). Weitere Studien uber die Entwicklung der einfachen Ascidien. *Arch. Mikr. Anat.* **5,** 459–463.

Kusakabe, T., Yoshida, R., Kawakami, I., Kusakabe, R., Mochizuki, Y., Yamada, L., Shin-i, T., Kohara, Y., Satoh, N., Tsuda, M., and Satou, Y. (2002). Gene expression profiles in tadpole larvae of Ciona intestinalis. *Dev. Biol.* **242**(2), 188–203.

Li, S., and Benninger, M. (2002). Applications of muscle electroporation gene therapy. *Curr. Gene. Ther.* **2**(1), 101–105.

Livant, D. L., Hough-Evans, B. R., Moore, J. G., Britten, R. J., and Davidson, E. H. (1991). Differential stability of expression of similarly specified endogenous and exogenous genes in the sea urchin embryo. *Development* **113**(2), 385–398.

Meedel, T. H., Lee, J. J., and Whittaker, J. R. (2002). Muscle development and lineage-specific expression of CiMDF, the MyoD-family gene of Ciona intestinalis. *Dev. Biol.* **241**(2), 238–246.

Meedel, T. H., and Whittaker, J. R. (1984). Lineage segregation and developmental autonomy in expression of functional muscle acetylcholinesterase mRNA in the ascidian embryo. *Dev. Biol.* **105**(2), 479–487.

Mita-Miyazawa, I., Ikegami, S., and Satoh, N. (1985). Histospecific acetylcholinesterase development in the presumptive muscle cells isolated from 16-cell-stage ascidian embryos with respect to the number of DNA replications. *J. Embry. Exp. Morph.* **87,** 1–12.

Miyamoto, D. M., and Crowther, R. J. (1985). Formation of the notochord in living ascidian embryos. *J. Embry. Exp. Morph.* **86,** 1–17.

Moody, R., Davis, S. W., Cubas, F., and Smith, W. C. (1999). Isolation of developmental mutants of the ascidian Ciona savignyi. *Mol. Gen. Genet.* **262**(1), 199–206.

Munro, E. M., and Odell, G. (2002a). Morphogenetic pattern formation during ascidian notochord formation is regulative and highly robust. *Development* **129**(1), 1–12.

Munro, E. M., and Odell, G. M. (2002b). Polarized basolateral cell motility underlies invagination and convergent extension of the ascidian notochord. *Development* **129**(1), 13–24.

Nakatani, Y., Moody, R., and Smith, W. C. (1999). Mutations affecting tail and notochord development in the ascidian Ciona savignyi. *Development* **126**(15), 3293–3301.

Nicol, D., and Meinertzhagen, I. A. (1988). Development of the central nervous system of the larva of the ascidian, Ciona intestinalis L. I. The early lineages of the neural plate. *Dev. Biol.* **130**(2), 721–736.

Nicol, D., and Meinertzhagen, I. A. (1991). Cell counts and maps in the larval central nervous system of the ascidian Ciona intestinalis L. *J. Comp. Neur.* **309**(4), 415–429.

Nishida, H. (1987). Cell lineage analysis in ascidian embryos by intracellular injection of a tracer enzyme. III. Up to the tissue restricted stage. *Dev. Biol.* **121**(2), 526–541.

Nishida, H. (1992). Regionality of egg cytoplasm that promotes muscle differentiation in embryo of the ascidian, Halocynthia roretzi. *Development* **116**(3), 521–529.

Nishida, H., and Satoh, N. (1983). Cell lineage analysis in ascidian embryos by intracellular injection of a tracer enzyme. I. Up to the eight-cell stage. *Dev. Biol.* **99**(2), 382–394.

Nishida, H., and Satoh, N. (1985). Cell lineage analysis in ascidian embryos by intracellular injection of a tracer enzyme. II. The 16- and 32-cell stages. *Dev. Biol.* **110**(2), 440–454.

Nishikata, T., Yamada, L., Mochizuki, Y., Satou, Y., Shin-i, T., Kohara, Y., and Satoh, N. (2001). Profiles of maternally expressed genes in fertilized eggs of Ciona intestinalis. *Dev. Biol.* **238**(2), 315–331.

Ogasawara, M., Sasaki, A., Metoki, H., Shin-i, T., Kohara, Y., Satoh, N., and Satou, Y. (2002). Gene expression profiles in young adult Ciona intestinalis. *Dev. Genes Evol.* **212**(4), 173–185.

O'Reilly, M. A., Smith, J. C., and Cunliffe, V. (1995). Patterning of the mesoderm in Xenopus: Dose-dependent and synergistic effects of Brachyury and Pintallavis. *Development* **121**(5), 1351–1359.

Ortolani, G. (1955). The presumptive territory of the mesoderm in the ascidian germ. *Experientia* **11,** 445–446.

Ortolani, G. (1957). Il territorio precoce della corda nelle Ascidie. *Acta Embryol. Morphol. Exp.* **1,** 33–36.

Ortolani, G. (1962). Territorio presuntivo del systema nervoso nelle larve di ascidie. *Acta Embryol. Morphol. Exp.* **5,** 189–198.

Ortolani, G. (1971). Sul cell-lineage delle ascidie. *Boll. Zool.* **38,** 85–88.

Powers, D. A., Hereford, L., Cole, T., Chen, T. T., Lin, C. M., Kight, K., Creech, K., and Dunham, R. (1992). Electroporation: A method for transferring genes into the gametes of zebrafish (Brachydanio rerio), channel catfish (Ictalurus punctatus), and common carp (Cyprinus carpio). *Mol. Mar. Biol. Biotechnol.* **1**(4–5), 301–308.

Powers, D. A., Kirby, V. L., Cole, T., and Hereford, L. (1995). Electroporation as an effective means of introducing DNA into abalone (Haliotis rufescens) embryos. *Mol. Mar. Biol. Biotechnol.* **4**(4), 369–375.

Prasher, D. C. (1995). Using GFP to see the light. *Trends Genet.* **11**(8), 320–323.

Satoh, G., Harada, Y., and Satoh, N. (2000). The expression of nonchordate deuterostome Brachyury genes in the ascidian Ciona embryo can promote the differentiation of extra notochord cells. *Mechanisms of Development* **96**(2), 155–163.

Satoh, N. (1994). "Developmental Biology of Ascidians." Cambridge University Press, Cambridge, UK.

Satou, Y., Takatori, N., Fujiwara, S., Nishikata, T., Saiga, H., Kusakabe, T., Shin-i, Y., Kohara, Y., and Satoh, N. (2002a). Ciona intestinalis cDNA projects: Expressed sequence tag analyses and gene expression profiles during embryogenesis. *Gene* **287**(1–2), 83–96.

Satou, Y., Takatori, N., Yamada, L., Mochizuki, Y., Hamaguchi, M., Ishikawa, H., Chiba, S., Imai, K., Kano, S., Murakami, A., Nakayama, A., Nishino, A., Sasakura, Y., Satoh, G., Shimotori, T., Shin, I. T., Shoguchi, E., Suzuki, M. M., Takada, N., Utsumi, N., Yoshida, N., Saiga, H., Kohara, Y., and Satoh, N. (2001). Gene expression profiles in Ciona intestinalis tailbud embryos. *Development* **128**(15), 2893–2904.

Satou, Y., Yamada, L., Mochizuki, Y., Takatori, N., Kawashima, T., Sasaki, A., Hamaguchi, M., Awazu, S., Yagi, K., Sasakura, Y., Nakayama, A., Ishikawa, K., Inaba, K., and Satoh, N. (2002b). A cDNA resource from the basal chordate Ciona intestinalis. *Genesis* **33**(4), 153–154.

Shigekawa, K., and Dower, W. J. (1988). Electroporation of eukaryotes and prokaryotes: A general approach to the introduction of macromolecules into cells. *Biotechniques* **6**(8), 742–751.

Shimauchi, Y., Chiba, S., and Satoh, N. (2001). Synergistic action of HNF-3 and Brachyury in the notochord differentiation of ascidian embryos. *Int. J. Dev. Biol.* **45**(4), 643–652.

Simmen, M. W., Leitgeb, S., Clark, V. H., Jones, S. J., and Bird, A. (1998). Gene number in an invertebrate chordate, Ciona intestinalis. *Proc. Natl. Acad. Sci. USA* **95**(8), 4437–4440.

Sordino, P., Belluzzi, R., De Santis, R., and Smith, W. C. (2001). Developmental genetics in primitive chordates. *Philos. Trans. R. Soc. Lond. B. Biol. Sci.* **356**(1414), 1573–1582.

Sordino, P., Heisenberg, C.-P., Cirino, P., Toscano, A., Giuliano, P., Marino, R., Pinto, M. R., and DeSantis, R. (2000). A mutational approach to the study of development of the protochordate Ciona intestinalis (Tunicata, Chordata). *Sarsia* **85,** 173–176.

Takahashi, H., Hotta, K., Erives, A., Di Gregorio, A., Zeller, R. W., Levine, M., and Satoh, N. (1999). Brachyury downstream notochord differentiation in the ascidian embryo. *Genes and Development* **13**(12), 1519–1523.

Takamura, K., Oka, N., Akagi, A., Okamoto, K., Okada, T., Fukuoka, T., Hogaki, A., Naito, D., Oobayashi, Y., and Satoh, N. (2001). EST analysis of genes that are expressed in the neural complex of Ciona intestinalis adults. *Zoolog. Sci.* **18**(9), 1231–1236.

Tsien, R. Y. (1998). The green fluorescent protein. *Annu. Rev. Biochem.* **67,** 509–544.

Ueki, T., and Satoh, N. (1995). Sequence motifs shared by the 5′ flanking-regions of two epidermis-specific genes in the ascidian embryo. *Dev. Growth & Diff.* **37,** 597–604.

Vandenberghe, A. E., Meedel, T. H., and Hastings, K. E. (2001). mRNA 5′-leader trans-splicing in the chordates. *Genes Dev.* **15**(3), 294–303.

Weaver, J. C. (1993). Electroporation: A general phenomenon for manipulating cells and tissues. *J. Cell Biochem.* **51**(4), 426–435.

Welsh, S., and Kay, S. A. (1997). Reporter gene expression for monitoring gene transfer. *Curr. Opin. Biotechnol.* **8**(5), 617–622.

Whittaker, J. R. (1973). Segregation during ascidian embryogenesis of egg cytoplasmic information for tissue-specific enzyme development. *Proc. Natl. Acad. Sci. USA* **70**(7), 2096–2100.

Whittaker, J. R. (1980). Acetylcholinesterase development in extra cells caused by changing the distribution of myoplasm in ascidian embryos. *J. Embryol. Exp. Morphol.* **55,** 343–354.

Whittaker, J. R. (1982). Muscle lineage cytoplasm can change the developmental expression in epidermal lineage cells of ascidian embryos. *Dev. Biol.* **93**(2), 463–470.

Whittaker, J. R. (1983). Quantitative regulation of acetylcholinesterase development in the muscle lineage cells of cleavage-arrested ascidian embryos. *J. Embryol. Exp. Morphol.* **76,** 235–250.

Whittaker, J. R., and Meedel, T. H. (1989). Two histospecific enzyme expressions in the same cleavage-arrested one-celled ascidian embryos. *J. Exp. Zool.* **250**(2), 168–175.

Zeller, R. W. (1998). Culture of marine invertebrate embryos. *In* "Cells: A Laboratory Manual" (D. L. Spector, R. D. Goldman, and L. A. Leinwand, eds.), pp. 1–23. Cold Spring Harbor Laboratory Press, Plainview, NY.

Zeller, R. W., Virata, M. J., and Cone, A. C. (2004). Predictable mosaic transgene expression in ascidian embryos produced with a simple electroporation device. *Dev. Biol.*, in press.

PART V

Genomics

CHAPTER 30

Genomic Resources for the Study of Sea Urchin Development

R. Andrew Cameron, Jonathan P. Rast, and C. Titus Brown

Division of Biology and the Center for Computational Regulatory Genomics
Beckman Institute
California Institute of Technology
Pasadena, California 91125

METHODS IN CELL BIOLOGY, VOL. 74

0091-679X/04 $35.00

Overview

As the whole genome sequence assemblies of *Strongylocentrotus purpuratus* emerge, the computational tools for manipulating this large body of information become ever more crucial. Arrayed libraries and characterized clones are now easily identified and obtained. Software that predicts genome features such as genes, proteins, cis-regulatory regions, and even micro-RNAs resides on the desktop computer of every investigator studying sea urchin cell and developmental biology. Beginning with the establishment of the Sea Urchin Genome Project in 1998 and the later inception of the Center for Computational Regulatory Genomics, there continues a concerted effort to gather, invent, and explain these tools to the community of experimentalists. Here, we describe the status of these resources at the present time.

I. Introduction

Genomic resources in support of cell and developmental studies of sea urchin gametes and embryos are just now emerging. Indeed, a whole genome sequencing project for the purple sea urchin *Strongylocentrotus purpuratus* is underway at the time of this writing and is expected to be completed in 2004. The availability of genomic sequences from bacterial artificial chromosome (BAC) inserts, assembled gene regions, and sets of cDNA sequences extends the scope of investigations that can now be undertaken with sea urchins. As cDNA studies give way to defined whole genome gene sets, the use of macroarray libraries will be superseded by microarray diagnostics. Until that happens, the macroarray resources of the Sea Urchin Genome Project will remain quite useful.

The genome of the sea urchin *Strongylocentrotus purpuratus* is fairly well characterized for a nonchordate deuterostome. There are 42 chromosomes (Gerhart, 1983) and the mass of the diploid genome is 1.8 pg (Hinegardner, 1974). This corresponds to about 800 megabases of sequence. Like all other deuterostomes that have been studied, the sea urchin displays a pattern of short period interspersion of single copy sequences such that about 25% of the genome is represented in many thousands of repetitive element families (reviewed in Davidson, 1986) and simple sequence repeats (Cameron *et al.*, 1999). Thus, the genome contains about 2700 families of repeats of >70 copies per genome (Cameron *et al.*, 2000) if the degree of sequence similarity in computational match is set to be equivalent to the standard experimental hybridization condition at which much earlier work on *S. purpuratus* repetitive DNA sequences has been done (~25 °C below Tm, i.e., a criterion of 60 °C, 0.12 per M Na+). With the accumulation of genomic sequence from several small genome-wide sequencing projects, a first rough estimate of total gene number in the range 22,000 ± 5000 emerges.

Besides their obvious advantage for the study of cell and developmental biology, sea urchins are an apt comparison for the chordates. They are nonchordate deuterostomes and the phylum to which they belong, Echinodermata, together with hemichordates, forms the sister group to the chordates. Whole genome sequences are currently available for several chordates, including human, mouse, rat, and for two species in the ascidian genus *Ciona*. Sequence comparisons between the genomes of these organisms are proving to be a very useful approach to predicting genome sequence features computationally. In view of the explanations for deuterostome genome evolution which invoke sequence duplication on scales ranging from single genes to entire genomes, the sea urchin genome will become a standard for comparisons with the chordates due, in part, to its relative simplicity. Expanded gene families in vertebrates are represented by only a single member in the genome of *S. purpuratus*. For example, sea urchins and the cephalochordate, amphioxus, have one cluster of *hox* genes, whereas mammals have four (Bailey *et al.*, 1997; Do and Lonai, 1988; Scott, 1992). Invertebrate deuterostomes have few *pax* genes, whereas mammals have many (Czerny *et al.*, 1997). In the regulation of exocytosis, there appears to be one gene in the sea urchin for *syntaxin*, whereas in humans there are eight (Conner and Wessel, 1999). There is one gene in urchins for *synaptotagmin*, whereas in humans there are at least ten (Conner *et al.*, 1997). Furthermore, this absence of duplicated genomes and genome regions makes the selection of orthologous genes rather simple compared to vertebrates.

Due to the ease of gene transfer and a broad base of experimental embryology at both the cell and molecular levels, the sea urchin system has proven to be an ideal platform to elucidate the gene regulatory networks (GRN) that control the developmental process. *cis*-Regulatory elements that are hard-wired in the genome are the primary building blocks of GRNs and the sites through which regulatory genes that make up networks interact. If these transcriptional control regions average 2 to 3 Kbases in length, they could constitute as much as 6 to 8% of the genome. Operationally, a *cis*-regulatory module is a fragment of genomic DNA that, when fused to a reporter gene and introduced into a zygote, will recapitulate some aspect of the expression of the gene from which the fragment was isolated. The array of sites in the DNA fragment to which transcription factors bind contain the heritable genomic sequence code necessary and sufficient for the function of that module (reviewed by Arnone and Davidson, 1997). *cis*-Regulatory modules are analogous to computational machines that integrate at the gene a variety of input information, presented in the form of activated transcription factors (see Davidson, 2001, for review). It is clear from all examples examined in sufficient detail that developmental *cis*-regulatory systems function as information processing elements, and that what they do depends on their target site sequences: but in order to discover what they do, they have first to be found in the vast expanse of DNA surrounding each gene and then isolated. Attempts to do so rely on a multidisciplinary approach that combines genomic BAC library construction, selection, and sequencing of specific clones. As a set of rules for the rapid and efficient discovery of *cis*-regulatory elements emerges, it is clear that comparative sequence analysis at the genomic level is a key component of this process.

II. Sea Urchin Arrayed Library Resources

Arrayed cDNA and genomic libraries have become standard tools for investigating animal genomes. Bacterial arrays offer several advantages over traditional pooled libraries. Since they are maintained as large collections of monocultures and are thus immune to the skewing effects of differential clone growth, they can better represent the prevalence distributions of their source DNA or mRNA. Because their organization is fixed, arrays can form the foundation for a database with information from individual clones. Notably, this information can be accumulated from an entire research community. The most important advantage for genome-scale projects is that arrayed libraries significantly reduce screening time, while enabling quantitative and multiplex screening strategies that would be impossible with traditional libraries.

With the aid of the sea urchin research community, we have constructed a collection of bacterial clone arrays. These include cDNA libraries from different embryonic stages and adult tissues, and Bacterial Artificial Chromosome (BAC) libraries from a range of echinoid species and an asteroid and hemichordate (see Table I). The following sections are meant to summarize the types and characteristics of available clone arrays and to present detailed protocols for some of the more commonly used procedures. Complex protocols such as those for library construction, filter spotting, and differential screening are beyond the scope of this article, but are briefly described and referenced.

A. Characteristics of Bacterial Clone Arrays

All of the libraries described here, whether genomic or cDNA, have much in common in terms of their construction and use. They are maintained in nearly identical bacterial strains, arrayed by the same procedure and spotted onto filters in the same high-density pattern. The arrays are derived from unamplified transformants that are selected on antibiotic containing LB-agar trays and robotically picked into 384-well microtiter plates. For storage, growth media is supplemented with 7% glycerin to allow freezing at −80 °C. Libraries are arrayed from 92,160 to 110,592 clones in either 240 or 288 384-well plates, respectively. For each filter, 18,432 clones from 48 384-well plates are robotically spotted onto 22 cm × 22 cm Hybond N^+ nylon membranes (Clark *et al.*, 1999). Clones are spotted in duplicate in a pattern that allows the well-plate position to be determined. The X–Y position in any of 6 fields of 24 × 16 subarrays of 16 spots gives the X–Y location in the plate. The number of the plate in the group being spotted emerges from the hybridizing spot pair's position and angle within the 16-spot subarray. Figure 1 illustrates the logic of this spotting pattern. Colonies are grown to a diameter of approximately 1 mm and then processed in an alkaline lysis procedure, leaving a portion of the plasmid DNA from each clone covalently bound to the filter in the position of the original colony (modified from Nizetic *et al.*, 1991).

Table I
The Complement of Arrayed Libraries Prepared as Part of the Sea Urchin Genome Project[a]

Source	Made	Arrayed	Spotted
7 h cDNA	X	X	X
9.5 h cDNA	X	X	X
15 h cDNA	X	X	X
20 h cDNA	X	X	X
30 h cDNA	X	X	X
40 h cDNA	X	X	X
72 h cDNA	X	X	X
Larval cDNA	X	X	X
Lantern cDNA	X	X	
PMC	X	X	X
Gut cDNA	X	X	
Mesentery	X	X	
Egg cDNA	X	X	X
Ovary cDNA	X	X	X
Testes cDNA	X	X	X
Sp BAC genomic	X	X	X
Sp small BAC	X	X	X
Sp BAC FR2	X	X	X
Lv BAC	X	X	X
Et small BAC	X	X	X
Lv small BAC	X	X	X
Lv large BAC	X	X	X
Pl small BAC	X	X	X
Am small BAC	X	X	X
Am large BAC	X	X	X

[a]Small BAC libraries have an average insert size of 50 Kb, while large ones are 120 Kb or larger. Am, *Asterinea miniata*; Pl, *Paracentrotus lividus*; Lv, *Lytechinus variegatus*; Et, *Eucidaris tribuloides*; Sp, *Strongylocentrotus purpuratus*. The cDNA libraries are all made from Sp RNA.

All libraries are maintained in the DH10B *E. coli* strain or a phage-resistant derivative of this strain, DH10B T1^{R} (Invitrogen Life Technologies). This strain is suited to library construction and is useful for its high electroporation efficiency. Clones from the libraries are grown in LB media supplemented with either 100 μg/ml ampicillin for cDNA libraries or 20 μg/ml chloramphenicol for BAC libraries. The vectors used for library construction will be described.

B. cDNA Libraries

1. Library Construction

All of the cDNA libraries described here are made in the pSPORT1 vector (GibcoBRL/Invitrogen; GenBank accession number U12390). This vector carries a β-lactamase gene to confer ampicillin resistance, a high copy number colE1

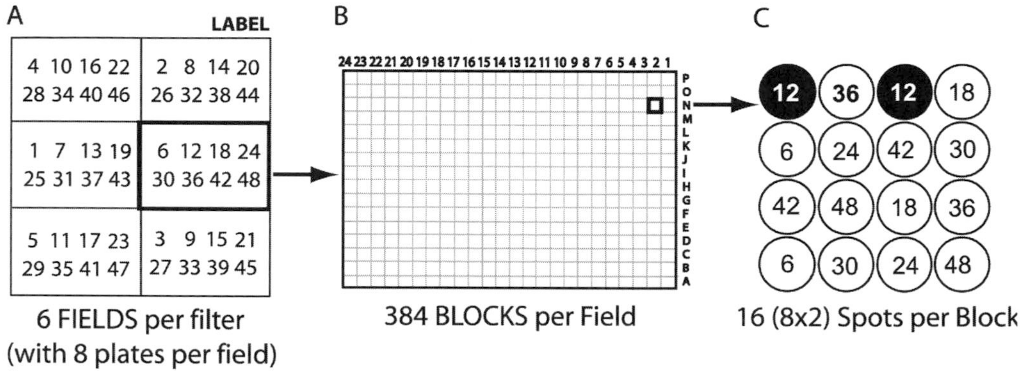

Fig. 1 The organization of the arrayed library spot pattern. Each library usually consists of five or six *filters* (A). Forty-eight libraries plates are spotted on 22 × 22 cm nylon membranes in six *fields*, each corresponding to the pattern of a 384-well plate (designated 1–6 in the order that they are spotted). Eight such plates are represented by offset spotting in each field. These are shown as numbers in each field for the first 48 plates. (B) Each field is made up of 384 *blocks* arranged in a matrix that is 24 blocks wide (numbered 1–24) and 16 blocks high (lettered A–P), corresponding to the arrangement of wells in the original 364-well plate. (C) The block contains 16 spots coming from identical well positions in each of eight plates in the field, each spotted in duplicate (here, numbered for field 6). The duplicate spots serve a dual purpose. They control for artifactual spots, and their relative angle is used to distinguish among the eight field plates. The 48 filter plates are spotted in ascending order beginning from the first in the middle left field, following the order shown in A, and repeating eight times. The filters of the library are designated by letters that correspond to sets of 48 plates (usually, A–E or A–F). Filters are labeled in the upper right corner over field 2. The letters correspond to plate numbers as follows: A, 1–48; B, 49–96; C, 97–144; D, 145–192; E, 193–240; F, 241–288. The positions of spots from plate 12, well N-2 (12N02) are represented by the bold lines and dark spots progressing through the three diagrams.

replication origin, and sites for *in vitro* transcription using T7 (sense) and Sp6 (antisense) RNA polymerase. pSPORT1 also carries a polycloning site surrounding the *Sal*1/*Not*1 cloned insert. The vector is 4084 bp in length.

Insert cDNA is directionally cloned and random-primed. First-strand cDNA is synthesized from a primer containing a *Not*1 restriction site and a random 8-mer sequence at its 3′ end (primer sequence: 5′-AAAGGAAG-GAAAAAA**GCGGCCGC**TACANNNNNNNNT-3′). (Only the *Not*1 [bold] and 3′ region are incorporated upon cloning [underlined region].) The 5′ region is removed in the cloning process and is present to protect the *Not*1 site from exonucleolytic digestion. The final T is present in early libraries. Others contain no defined nucleotides beyond the random 8-mer. Second-strand cDNA is synthesized by standard RNaseH mediated nick translation (Gubler and Hoffman, 1983). The cDNA is then ligated with a *Sal*1 adapted (cloned sequence: 5′-GTCGACCCACGCGTCCG-3′) and digested with *Not*1. After size selection, the cDNA is ligated into *Sal*1/*Not*1 cut pSPORT1 vector and electroporated into DH10B cells.

2. cDNA Library Characteristics

Typical cDNA inserts for the pSPORT1 cDNA libraries average in size from 1.5 to 2 kb. The random priming and size selection process generally biases representation to the 5′ end of the transcript. Probe excess measurements show that there is about 250 pg of insert target DNA available for hybridization in each spot on the arrayed filters (Rast *et al.*, 2000).

3. Notes for Arrayed cDNA Library Use

The sea urchin arrayed cDNA libraries are of sufficient size to contain representatives of most transcripts originating from genes that are active at the embryonic stage or in the tissue from which they are constructed. This includes relatively rare message species, such as those encoding transcription factors. The strong hybridization signal obtained from arrayed filters ensures that, in a typical screening assay, clones will be detected when present and, because of this screening, may be more comprehensive than is typical of lambda libraries. On the other hand, these libraries are not as large as traditional lambda libraries, which usually derive from >500,000 primary recombinants, so very rare transcripts may not be represented. The random primed inserts have the advantage of containing a higher-level coding sequence than is found in poly-T primed cDNA, but long transcripts are often truncated compared to lambda libraries.

One artifact of the cDNA construction process is worth noting. cDNA from transcripts that contain rare endogenous *Not*1 recognition sites can be cloned in opposite orientation and in skewed prevalence, relative to inserts from typical transcripts. In such cDNAs, the TACA sequence between the *Not*1 site and random portion of the primer will be absent.

These caveats aside, the arrayed libraries have significant advantages over traditional library formats, especially when the project scale demands highly efficient or multiplexed procedures. For the vast majority of genes expressed in the embryo, library coverage and clone size is more than adequate to ensure representation.

C. BAC Libraries

1. BAC Library Construction

Large insert, bacterial artificial chromosome (BAC) libraries have been constructed and arrayed from a phylogenetic range of nonchordate deuterostome species. These were chosen to be useful both as research models and for sequence comparisons directed at uncovering regulatory sequence. A brief description of library construction will be presented. Detailed BAC library construction procedures can be found elsewhere (Amemiya *et al.*, 1996; Osoegawa *et al.*, 1998; Strong *et al.*, 1997). High molecular weight DNA is prepared from agarose-embedded sperm cells. The DNA is cleaved to the appropriate size (usually, about 150 kb) by partial restriction enzyme digestion with either *Eco*R1 or *Mbo*1. DNA fragments are size selected in a process of repeated separation by pulsed-field gel electrophoresis. Insert DNA is then ligated into *Eco*R1- or *Bam*H1-cut vector. The library is constructed in the pBACe3.6 vector (Frengen *et al.*, 1999; GenBank Accession No. U80929). pBACe3.6 contains a number of useful features for large insert cloning including rare cutting restriction enzyme sites flanking the insert and in the vector; T7 and Sp6 polymerase binding sites for making BAC-end riboprobes; and LoxP and Tn7 recognition sites for BAC retrofitting. Ligations are electroporated into DH10B or DH10BT1^{R} cells and plated onto LB-agar-chloramphenicol (20 μg/ml) plates supplemented with 2% sucrose. The sucrose is used to negatively select for clones containing small or no inserts (activation of the SacB gene in the pBACe3.6 vector will kill cells in the presence of sucrose [Osoegawa *et al.*, 2000]). After the initial selection, clones are grown in LB-Agar 20 μg/ml chloramphenicol without sucrose.

2. Growing BAC Clones

BAC clones are grown like other plasmid-containing bacteria. Typically, single colonies are isolated after streaking on 20 μg/ml chloramphenicol plates (LB with 1.5% agar). Small 2 to 5 ml liquid cultures are grown for 14 to 16 h at 37 °C, shaking at 250 rpm. Larger cultures are started by dilution of smaller fresh cultures. Large preparations of BAC DNA can be isolated on CsCl gradients from one to three liter cultures. Smaller preps can quickly be prepared by alkaline lysis.

3. BAC Minipreps

For insert size determination, clones are prepped as described in Amemiya *et al.* (1996). Once the bacteria have been lysed, BAC DNA solutions are handled gently to avoid shearing; they are mixed by inversion, never by vortexing. Two milliliter microcentrifuge tubes are convenient for this procedure.

1. 2 ml cultures are grown from fresh isolated colonies by shaking at 250 rpm at 37 °C for 16 h.
2. Cultures are transferred to 2 ml microcentrifuge tubes and centrifuged at 3500×*g* for 2 min to pellet the cells.
3. Pellets are thoroughly resuspended in 300 μl of cold P1 solution.
4. Then, 300 μl of P2 solution is added and the tubes are mixed by gentle inversion. The suspension is given time to clear (no more than 5 min).
5. 300 μl of cold P3 Solution is added. The tubes are thoroughly but gently mixed, and held on ice for 5 min.
6. The tubes are then centrifuged at 13,000×*g* for 10 min and 800 μl of the clear supernatant is added to 800 μl of room-temperature isopropyl alcohol.
7. The tubes are then centrifuged for 10 min at 13,000×*g* and the supernatant is completely removed. A small pellet will be visible.

8. The pellet is carefully washed with 500 μl ice-cold 70% ethanol. After briefly air drying, the BAC DNA is resuspended in 30 μl TE (10 mM Tris, 1 mM EDTA, pH 8.0).

Solutions for BAC Miniprep

P1 Solution: 15 mM Tris, pH 8.0; 10 mM EDTA; 100 μg/ml RNaseA. Stored at 4 °C.

P2 Solution: 0.2 M NaOH, 1% SDS; freshly prepared at room temperature.

P3 Solution: 3M Potassium Acetate, pH 5.5. Stored at 4 °C.

4. Insert Size Determination

For insert size determination, 5 μl of this BAC DNA solution is digested with *Not*1 in a 20 μl reaction. Most often, the insert will be released as one or two fragments. The products of this digestion are analyzed by pulsed-field electrophoresis (on a Biorad CHEF system, the parameters are 1–15 s switch time, 14 °C, 120° switch angle, 16 hrs, and 6 V/cm). The more numerous bands of digestions with frequent cutting enzymes are separated on standard agarose gels. As with all large DNA preparations, BAC DNA should be stored at 4 °C, not frozen.

5. Notes for BAC Library Use

BAC libraries and clones are handled similarly to any plasmid system, though there are certain peculiarities which are worth mentioning. A very small number of clones in the libraries (fewer than 1%) contain vector molecules that retain a pUC19 stuffer that is normally removed during vector preparation. Because this DNA fragment includes a high copy number origin of replication, clones that contain it produce a relatively high mass of plasmid. Unlike the normal BAC vector, this insert contains sequence common to many vectors (such as the β-lactamase gene). Therefore, these clones are readily detectable by even small amounts of contaminating vector in probes. The distinguishing feature of these clones is that, upon electrophoresis after *Not*1 digestion, they produce a strong band that is smaller than the pBACe3.6 vector, usually with no apparent insert band. The best way to avoid these clones is to gel purify probe fragment DNA.

Finally, it is important to mention the direct use of BAC DNA in sea urchin transgenics. A byproduct of the *S. purpuratus* genome project will be a set of sequenced BACs that can be immediately accessed for a particular gene of interest. Recombinant methods based on an inducible λ phage recombinase system can be used to efficiently and accurately modify BAC DNA (Court, 2002; Lee *et al.*, 2001; Yu *et al.*, 2000). Large-scale modifications, such as inserting reporters and deleting regulatory modules, and more directed modifications, such as changing single nucleotides (Swaminathan *et al.*, 2001), are readily made. These BAC constructs can then be microinjected and their activity observed during embryogenesis.

D. Probes for Arrayed Library Screening

^{32}P radiolabeled probes for arrayed library screening are made by standard methods, including oligolabeling of DNA fragments, endlabeling of oligonucleotides, and *in vitro* transcription of radiolabeled RNA. Oligoprimed DNA probes are used most often. A few precautions may help avoid some common problems, especially when screening BAC libraries. Probes are usually between 200 and 500 bp in length. Screening problems usually involve clone misidentification as a result of cross-hybridization to repetitive sequence or to aberrant vector sequence. Repetitive sequence can be present at high levels relative to single-copy gene sequence, so very small amounts of probe contamination can lead to intense hybridization. In our hands, the best probes are generated by PCR amplification of an isolated clone DNA source (usually, plasmid cDNA), followed by gel purification of the probe fragment. The amount of initial template should be low (100–500 pg) and the PCR cycle number should be minimized to reduce amplification of nontarget sequence. Use of complex cDNA or genomic DNA as template, without an intermediate cloning step for isolation, often results in small amounts of contaminating repetitive DNA, even in gel purified probes. Agarose gel purification and elution by standard methods can minimize the problem of contaminating vector sequence.

E. Screening Arrayed Libraries

Arrayed BAC and cDNA libraries are screened similarly, although the much lower mass of target sequence in BAC library spots typically requires a longer exposure time after hybridization. It is advisable to subject new filters to the stripping procedure (described in the following text) prior to hybridization in order to remove bacterial debris and to ensure complete denaturation of target DNA.

Screening Protocol: Sets of 5 or 6 high-density filters are typically exposed to probe in 35×250 mm hybridization oven bottles. Filters are sandwiched with 22×22 cm Nitex mesh separators to enhance circulation of probe solution. Conditions for hybridization are essentially identical to other nylon membrane procedures. Probe DNA is usually added to approximately 2×10^5 cpm/ml. The ^{32}P radiolabeled probe can be in the form of random-primed DNA, *in vitro* transcribed RNA, or end-labeled oligonucleotides. For oligonucleotide probes, hybridization and wash stringencies should be adjusted according to the melting temperature of the probe, but the same basic solution constituents can be used for all probe types.

1. Standard Filter Hybridization Procedure

1. If filters are new, expose them to the stripping procedure (see following text) once prior to hybridization to ensure complete spot DNA denaturation and to completely remove any residual bacterial debris.

2. To aid in identification of hybridizing clones, it is helpful to mark the array corners. This can be done on dry filters using a ballpoint pen to make an indentation directly on the four corner bacterial spots. These are easily seen under a dissection microscope under oblique lighting.

3. Arrange the dry filters, each on top of a Nitex separator, into a stack. The dry filter-separator stack is rolled up and placed in the hybridization bottle.

4. Hybridization solution (100–150 ml) is added to wet the filters. The filter bottle is rotated in the hybridization oven at 65 °C for 15 min to completely wet the filters and to remove air bubbles.

5. A volume of hybridization solution is removed to leave 15 to 20 ml per filter.

6. Denatured probe is added to a final concentration of approximately 2×10^5 cpm/μl.

7. Hybridize approximately 16 h at 65 °C (a lower temperature and shorter time may be called for with oligonucleotide probes).

8. Wash bottles three times quickly (1 min each) with room temperature 2 × Wash (see formula in following text) to reduce unbound probe levels.

9. Remove filters from bottles into a 27 × 27 cm (approximately) square plastic container and wash 2 × 20 min at 65 °C in 2 × Wash in a shaking water bath. It is advisable to remove the Nitex separators during this wash phase to avoid abrasion of the nylon filters.

10. Continue washing 2 × 20 min, 65 °C in 1 × Wash.

11. For higher stringency, wash another 20 min in 0.1 × Wash.

12. Sandwich filters in plastic wrap, being very careful not to allow drying. This can best be done by lifting the filters from the final wash solution for approximately 30 s to allow excess wash solution to run off, then laying the filter carefully on top of one layer of plastic wrap, and using the entire plastic wrap roll to directly apply a top layer without bubbles. Any bubbles that do appear can be forced out by wiping with a tissue. Finally, the edges are trimmed to about 4 cm around the filter and folded over, leaving a 1 cm clear edge around the filter.

13. Place on film for 4 h (strong cDNA signal) to 48 h (BAC filters). Intensifying screens are helpful when screening BAC libraries, but are often not necessary for screening cDNA libraries.

Solutions for Filter Hybridization

Hybridization Solution: 5 × SSPE, 5% SDS, 0.01% sodium pyrophosphate (1 × SSPE = 0.15 M NaCl, 0.01 M NaH_2PO_4, 1 mM EDTA; pH 7.4)

Wash Solution: SSPE, 0.01% SDS, 0.01% sodium pyrophosphate. The wash SSPE concentration varies with the stringency required. Usually, the washes are made in series from 2 × SSPE (2 × Wash) to 1 × SSPE (1 × Wash), then 0.1 × SSPE (0.1 × Wash) if high stringency is required.

2. Procedure for Stripping Filters of Probe

High density array filters can usually be hybridized up to 10 times before signal intensity is seriously impaired. Filter use can be maximized by screening with multiple probes simultaneously, then sorting positive clones in a separate secondary screen. Once filters have been hybridized and exposed to film, they can be stripped of probe as follows:

1. Unwrap wet filters and immediately submerge in 1 L Stripping Solution. Incubate in a shaking water bath at 45 °C for 30 min.
2. Replace Stripping Solution with 2 L of Neutralization Solution. Incubate in a shaking water bath at 65 °C for 30 min.
3. Replace solution with 1 L fresh Neutralization Solution supplemented with 20 mM EDTA.
4. Remove filters and check with survey meter to assess stripping efficacy. Filters may also be exposed to film to confirm loss of signal. Once they are known to be free of probe they can be air dried and stored in plastic wrap.

Solutions for Filter Stripping

Stripping Solution: 0.2M NaOH, freshly prepared

Neutralization Solution: 0.2 M Tris, pH 7.5; 0.1% SDS; 0.1 × SSC (SSC = 0.15 M NaCl, 0.015 M Sodium Citrate).

F. Subtractive Probe Screening on Macroarray Filters

The *S. purpuratus* cDNA library filters have been invaluable as tools for gene discovery (Calestani *et al.*, 2003; Ransick *et al.*, 2002; Rast *et al.*, 2002). Unlike microarrays, relatively comprehensive macroarray libraries can be constructed relatively cheaply and with incomplete knowledge of the transcript populations. This makes them particularly valuable in evolutionary studies and in studies of emerging genomic model systems. However, their size and relatively high hybridization volume present a challenge in terms of detection sensitivity. The detection limit for macroarray screens with complex cDNA probes corresponds to a messages prevalence of about 30 copies per average embryo cell (Rast *et al.*, 2000). For comparison, transcription factors that are active in the embryo are often present at about 2 to 5 copies per average embryo cell. In fact, the majority of the ~8500 gene species expressed in the early sea urchin embryo are present at <10 copies per average embryo cell (Davidson, 1986). Transcripts from these genes must be enriched if they are to be detectable in macroarray screens. We have developed a strategy that accomplishes this goal. By this subtractive probe method, macroarrays filters can be used to obtain sets of candidate. These genes can then be further verified by more precise, yet labor intensive, means (usually, quantitative PCR). Here, we briefly discuss the overall strategy. Detailed protocols are available in published papers (Rast *et al.*, 2000, 2002) and on the Sea Urchin Genome website (http://sugp.caltech.edu/resources/methods/subtrmethods.py).

The procedure is illustrated in Fig. 2. The strategy amplifies two message sequence populations: (1) a population in which differential transcripts are expected to be present (the *selectate*) and (2) a population in which these genes are absent (the *driver*). A variety of techniques are used to maintain critical prevalence relationships throughout the procedure. cDNA is random primed in order to (1) overlap in sequence composition as much as is possible with the library cDNA, (2) break message sequences into independent fragments, and (3) allow selection

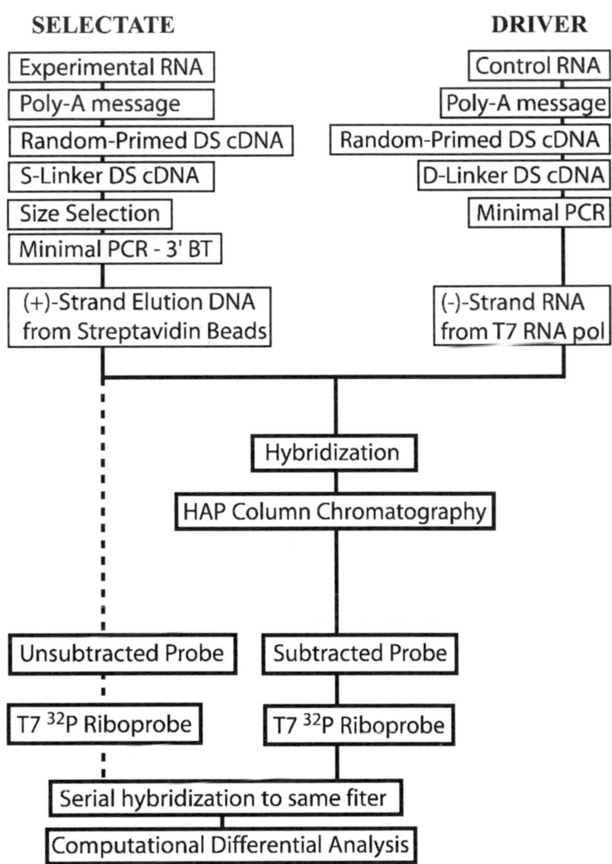

Fig. 2 A flow diagram summarizing the progression through a subtractive probe screen. The diagram begins with experimental and control RNA at the top and proceeds to computational analysis of the filters at the bottom. *Selectate* is the sequence population that is expected to contain message that is absent in the *Driver*. The sequence that is unique to the selectate will be enriched upon subtraction. Selectate synthesis is at the left of the diagram and Driver synthesis at the right. The heavy lines trace selectate and driver through hybridization, HAP column chromatography, and subtractive probe synthesis. The dashed line follows the synthesis of the unsubtracted probe. Hybridizations of each of these probes are compared to find clones representing differential message.

of small fragments that can be efficiently amplified by PCR. Special care is taken to minimize PCR cycle number, especially for the driver population in which it is critical to maintain a genuine prevalence distribution. Amplification steps are made with T7 polymerase when possible (see Phillips and Eberwine, 1996). The comparison is restricted to hybridization of a pre-subtracted and subtracted probe; thus, limited skewing in the selectate, during the amplification or hybridization process, is largely irrelevant. Unique selectate sequence is isolated after hydroxylapatite column chromatography. Finally, comparisons are always made serially on the same library filters, so that filter variation is not a factor.

Using this subtractive probe strategy, we typically achieve 25- to 45-fold enrichment of sequence that is unique to the selectate. This is sufficient to detect low prevalence transcripts. Empirically, we know that the system works for two reasons: (1) we are able to recover low prevalence controls that are planted in the selectate but not in the driver; and (2) we are able to recover known transcription factors in screens where differential transcription has been partially characterized *a priori* (e.g., Ransick *et al.*, 2002). The method excludes 96 to 98% of common sequence from the subtracted probe, but because differential species usually make up just a small percentage of expressed genes, false positives are still recovered. These are ultimately excluded in secondary screens and can be minimized by control screens, such as selectate minus selectate or reverse screens (driver minus selectate).

III. Computational Tools

A. BioArray

BioArray is a program that identifies clone spots on autoradiographs of macroarray filters and quantifies the signal from each spot according to user-specified parameters; it can also superimpose images and highlight differences, both qualitatively and quantitatively (Brown *et al.*, 2002). In its first version, it has been used extensively for differential screens (Rast *et al.*, 2000, 2002). This version works only with filters spotted in the pattern already described, which includes all filters created at the Caltech Genomics Technology Facility.

BioArray was developed by Zhengjun Pan and Hamid Bolouri at the University of Hertfordshire, in collaboration with the Davidson Lab at Caltech.

1. How BioArray Works

BioArray uses the known spot patterning relations to align images on a local level and interpolates between local alignments to position the image globally. The intensities at individual pairs of spots are evaluated based on a circular mask, and the intensity of each spot can be determined as the mean, median, sum, or 80th and 90th quantile of pixel intensities within the spot. The local and global intensities can be measured in a variety of ways as well, and the readout can be limited to those spots with intensities a specific level above that of the background. This combination of options allows the readout to be tailored to lab-specific hybridization and wash protocols.

Even in automatic mode (with no human intervention), BioArray offers about 50% greater accuracy than VisualGrid and ArrayVision as measured by root mean square error between the intensity of spot pairs (Brown *et al.*, 2002). Greater accuracy can be achieved with manual adjustments to local grid alignment and spot positioning.

2. Acquiring BioArray

BioArray v1.1 is the version described in Brown *et al.* (2002). This version is available (upon request; see *Online Resources*) for Windows 98/2000/XP only. Genetix (Genetix USA Inc., 7913 SW Nimbus, Beaverton, OR 97008) has acquired the rights to future releases and is continuing development; please contact them for more information.

3. Using BioArray

An extensive user manual comes with BioArray 1.1. BioArray can read the two most common phosphorimager formats, TIFF and GEL, and has a standardized Microsoft-style interface. BioArray can output spot intensities directly into Excel.

B. Cartwheel and FamilyRelations

Cartwheel and FamilyRelations are tools to support the analysis, annotation, and comparison of BAC-sized genomic sequences (50–150 kb). They have been used extensively by the Davidson Lab in the creation of the Endomesoderm Gene Regulatory Network (Davidson *et al.*, 2002a,b) and serve as the core of our automatic annotation system.

Cartwheel is a Web-based system with a user-friendly interface that allows users to upload sequences and analyze them with a variety of tools, including BLAST and several gene finders. The computation is done on compute nodes attached to the Cartwheel server, and the results are stored on the Web site where they can be retrieved and viewed with the FamilyRelations software. FamilyRelations is a graphical user interface written in Java Programming language that lets users explore the results of analyses done with Cartwheel. FamilyRelations can display analyses of a single sequence as well as pairwise comparisons, and allows interactive viewing of the analyses as well as sequence extraction and motif search.

The Center for Computational Regulatory Genomics at Caltech runs a publicly accessible Cartwheel server; see the Cartwheel Web site at http://cartwheel.caltech.edu/ for more information. FamilyRelations is available through that site as well, or at http://family.caltech.edu/. There is an online tutorial to help users get started at http://family.caltech.edu/tutorial/.

1. Single-Sequence Analyses and SUGAR

FamilyRelations can be used to visualize the results of one or more Cartwheel analysis programs and align them against the sequence analyzed; this is often useful when confronting a previously uncharacterized region of the genome, where one has no information about known genes, repeat elements, or potential cis-regulatory sites. The SUGAR analyses ("Sea Urchin Genome Annotation Resource") posted on the Sea Urchin Genome Project (SUGP) Web site make use of this type of analysis to produce a graph of known genes, ESTs, cDNAs, and repeats, as well as computationally predicted genes, against the BAC sequences stored in the SUGP. This type of analysis can also be used via the Cartwheel Web interface to analyze other sequences from essentially any metazoan genome.

Before creating an analysis, you must log in and create a folder in which to put the sequence and subsequent analyses; instructions for doing this are available at the Web site. Once an analysis folder has been created, the sequence to analyze can be uploaded under the "Sequences" section of the folder. Cartwheel can receive sequences either via cut-and-paste into a Web form or via an upload of a FASTA-format file; we strongly recommend that sequences over 15 kb be uploaded in FASTA format, because there are often limitations on how much text can be submitted via a Web form. Most sites (including the SUGP and NCBI Entrez) offer a FASTA download format that produces files suitable for uploading to Cartwheel. Note also that Cartwheel can autodetect sequence type (the default) or force the sequence to DNA or protein, replacing incorrect bases with N (DNA) or X (protein).

Once the sequence is present and listed in the sequence section of an analysis folder, the next step is to create a single-sequence analysis group. To do this, select "create analysis group···" and then use the pull-down menu to pick either "Single sequence analysis" or "SUGAR analysis," depending on whether you wish to establish your own set of analyses or use the pre-programmed set for SUGP BAC sequences. The next menu asks for a name and the sequence to use, and you will then be deposited into a screen that allows you to add analyses or (in the case of SUGAR analyses) allows you to set whether or not to use an extant sequence for the cDNA or comparative BLAST analysis.

There are many different kinds of analyses available for single-sequence analysis groups, and new analysis are continually being added. Currently, one can choose from among the following: a BLAST against another sequence; a BLAST against a sequence database; execution of three distinct gene finding programs, hmmgene, geneid, and genscan; a search for motifs in several formats, including strictly literal matches and IUPAC motif matches (e.g., including "R" and "W" in the motif to match); and a simple hand-annotation mechanism that allows for the addition of comments and features for given sets of coordinates. All of these are available under the "add analysis···" menu accessible from the single-sequence analysis group.

Visualization of the analysis results is done primarily with FamilyRelations, although direct export to PDF is possible with some additional libraries. To load the results of your analyses into FamilyRelations, start FamilyRelations and navigate through the tree of labs and folders on Cartwheel to find your analysis folder, select the analysis you just created, and load it into FamilyRelations. You can now browse the results of the analysis and export selected regions of the sequence.

a. Pairwise Comparisons and seqcomp

Cartwheel and FamilyRelations also enable comparisons of two genomic sequences for the purpose of locating conserved regions that are potential regulatory regions (Fig. 3); this approach has been used with great success to locate regulatory elements for over 10 distinct genes in the Endomesoderm Network (Davidson *et al.*, 2002a,b).

To establish a pairwise analysis, first upload the two sequences you wish to compare (see "Single-sequence analyses and SUGAR" in the preceding text for more information) and create a pairwise analysis group. At this point, there are two types of analyses you can create: "BLAST" will use NCBI BLAST to compare the two sequences, and "seqcomp" will use a fixed-width window of 10 to 100 bases to exhaustively compare the two sequences. You can also go and set up analyses to display against the top or bottom sequences, such as matches to known cDNA sequences or known protein-binding sequences.

We suggest using the fixed-width window analysis, seqcomp, to look for conserved regions for several reasons. seqcomp does a global exhaustive analysis with only two parameters, the windowsize and the threshold over which similarities should be plotted. Unlike BLAST, it does no gapping and does not preferentially select regions of extensive similarity over regions with minimal similarity; instead, it displays both kinds of regions according to the given threshold, thus showing small patches in addition to the large patches that BLAST will display. This results in a more complete analysis of the region, which we have found to be useful in determining all the potential regulatory regions to analyze. In addition, seqcomp parameters are easy to understand and are directly linked to the display you see; generally, very little parameter tweaking is needed to obtain a satisfactory analysis.

There are several issues to keep in mind when setting up a pairwise analysis:

- The current version of FamilyRelations has trouble loading seqcomp comparisons between sequences that are individually larger than 100 kb because of memory limitations in the programming language used;
- FamilyRelations will likewise have difficulty with seqcomp comparisons when a window size smaller than 20 bp is used;
- Partly because of these limitations, we suggest starting out with a 50 bp seqcomp analysis with a hard threshold set at 50%; this will show broadly conserved patches that can then be further investigated with a 20 bp window. For 20 bp windows, thresholds lower than 70% (14/20 bases matching) are statistically irrelevant, and we suggest using a hard threshold of 80 or 90%.

These issues are being addressed in a rewrite of FamilyRelations ("FRII") that should be available in 2004. Together, FamilyRelations and Cartwheel allow the characterization and analysis of genomic sequence in a variety of ways.

IV. Sea Urchin Genome Project Web Site

A. Purpose

The Web site in support of the Sea Urchin Genome Project (SUGP) was originally conceived as a focus for the exchange of information related to research activities at the level of the genome, viz. high-throughput sequencing efforts such as EST projects; broad molecular analyses pertaining to general features of the genome, for example, the distribution of repeat families; and studies of the functional aspects above the level of single transcription units like the elucidation of gene regulatory networks. This site is designed to avoid any duplication with existing publicly available sites and databases. For example, the SUGP will not attempt to curate complete sets of sea urchin sequences outside of those generated through the efforts of the genome sequencing consortium since the NCBI Genbank databases hold published sequences for sea urchin genes and genomic sequence. Classes of information that are specifically useful to investigators who work with sea urchins are presented on the Web site.

B. Description

The Web server machine itself is an Intel Xeon dual-processor computer running Linux, hosted at the Center for Computational Regulatory Genomics, Beckman Institute, California Institute of Technology. The Web server is linked to a Beowulf cluster of 35 machines divided into two sets by CPU speed, 2.0 GHz and 2.8 GHz, respectively. The cluster machines are used to conduct sequence searches by distributed processing methods and accessed over 100 Mb network communications through a local high-speed switch. The software used to run the site including the search capabilities is based in open-source software packages. The HTTP server is Apache Web Server (http://httpd.apache.org/) backed by a PostgresSQL database (http://www.postgresql.org/). Much of the data manipulation and page posting is managed with Python (http://www.python.org/) and Webware (http://webware.sourceforge.net/).

In preparation for the posting of a purple sea urchin genome assembly and to communicate ongoing developments, several institutions have mounted useful Web sites. The National Center for Biotechnological Information has a Sea Urchin Genome Resources page with links to a BLAST page that only searches sea urchin sequences (http://www.ncbi.nlm.nih.gov/genome/guide/sea_urchin/). The Baylor College of Medicine Hunman Genome Sequencing Center has a Sea Urchin Genome Project page (http://hgsc.bcm.tmc.edu/projects/seaurchin/) with several links to useful search pages, including a "BACFisher" query page where it is possible to build an assembly around a provided sequence.

C. Web Site Data

The data contained on the Web site is divided into two classes: a database of searchable DNA sequences unique to the SUGP and a collection of static documents including descriptions and protocols pertaining to the research materials of the SUGP specifically and work with sea urchin eggs and embryos in general. These include protocols for working with the macroarray research materials provided through the Genomics Facility, Beckman Institute, California Institute of Technology.

The sequence data comes from several sources and was accumulated by different approaches. Thus, the quality differs, depending on the source. There are, as of May 2003, a total of 89,051 sequences in the sequence database, for a total of about 60 mb of sequence. The classes are summarized in Table II and their location within Genbank is indicated. Each of these sequences is linked to spot on one of the macroarray filters in The Genomics Technology Facility (see following text). The genome sequencing project traces are maintained in a separate location because they are not linked to filter clones. Many of the sequences are single-pass sequences, namely, the cDNAs, the ESTs, and the BAC-end or STC sequences. There are 95 BAC sequences in the database, of which 38 are from *S. purpuratus*. These sequences were produced at the Joint Genome Institute, Department of Energy, and The Institute for Systems Biology, Seattle, WA, and they exist as ordered and oriented contigs from the BAC insert. In order to provide ordering and orientation, they were sequenced by shotgun methods to a level of 3 to 6×

Table II
Sequence Data Collection: The Distribution of the Different Classes of Sequences Contained in the SUGP

Number of sequences	Source	Database at Genbank
43	cDNA collection	dbEST
95	BACs sequenced for endomesoderm network project	HTG division
770	Contigs from PMC EST sequence collection	dbEST
1144	Repeat collection	Not in Genbank as such
1558	Filter data	Not applicable; in other categories
9436	PMC EST collection	DbEST
76,020	STC/BAC end collection	DbSTS or GSS

coverage. Similarly, the repeats are consensus sequences from aligned members of each repeat family. In both of these cases, the error rate is reduced.

The most unique, information-rich component of the Sea Urchin Genome Project Web site is the database of macroarray filter information. Here, all of the sequences already mentioned and the gene annotation information collected in the process of screening these library filters, for whatever purpose, is stored. This includes sequence collections from complex probes screens, such as those used for the identification of genes in the endomesoderm specification pathway (Calestani *et al.*, 2003; Ransick *et al.*, 2002; Rast *et al.*, 2002); the results of homology screening strategies (Giusti *et al.*, 2003) and random EST projects (Lee *et al.*, 1999; Poustka *et al.*, 1999, 2003; Zhu *et al.*, 2001). Because the data are coupled to a filter location that contains an individual clone from the library, the clone is immediately recoverable. As more clones are characterized in a library, that library becomes more valuable. Eventually, the several well-characterized libraries can be used to confirm *ab initio* gene predictions and gene catalogs for the sea urchin.

D. How to Use The Web Site

The information on the Web site can be divided into two groups: static text information and sequence information which is dynamic in the sense that it is linked to various other data such as hit lists and filter-well locations. The textpages are self-explanatory. They provide information on protocols and other-kinds of stable information. The information linked to the sequences can change and various links will be evident, depending on how you access the sequences.

We will consider the textual information first. To facilitate use of the resources at the center, a number of recipes and protocols are listed. They are specifically chosen to support the use of our macroarray resources. Protocols for making both cDNA libraries and genomic libraries are present. Also, the methods we use for screening the libraries and preparing complex probes are in evidence.

How one proceeds to use the sequence data is limited by both what is available and how it is made available. The most frequent use is for sequences or clones to aid in gene discovery projects focused on cell and developmental biology of sea urchin gametes and embryos. If one chooses to use the data from the Web site locally, it may be downloaded in various forms from the download page at http://sugp.caltech.edu/ftp_page/. To explore the sequence collections on the Web site, peruse http://sugp.caltech.edu/resources/. Searches begin at the search page http://sugp.caltech.edu/databases/. We will consider some methods of searching that have been employed at the Web site. It is important to note that almost all of the sea urchin sequences in the Web site database are included in Genbank. Only sequences that we have collected as part of laboratory studies or those sent to us by others who have accumulated such sequences are unique to the Sea Urchin Genome Project Web site. Primarily, these are cDNA sequences that were not useful to the goal at hand and were therefore not examined further.

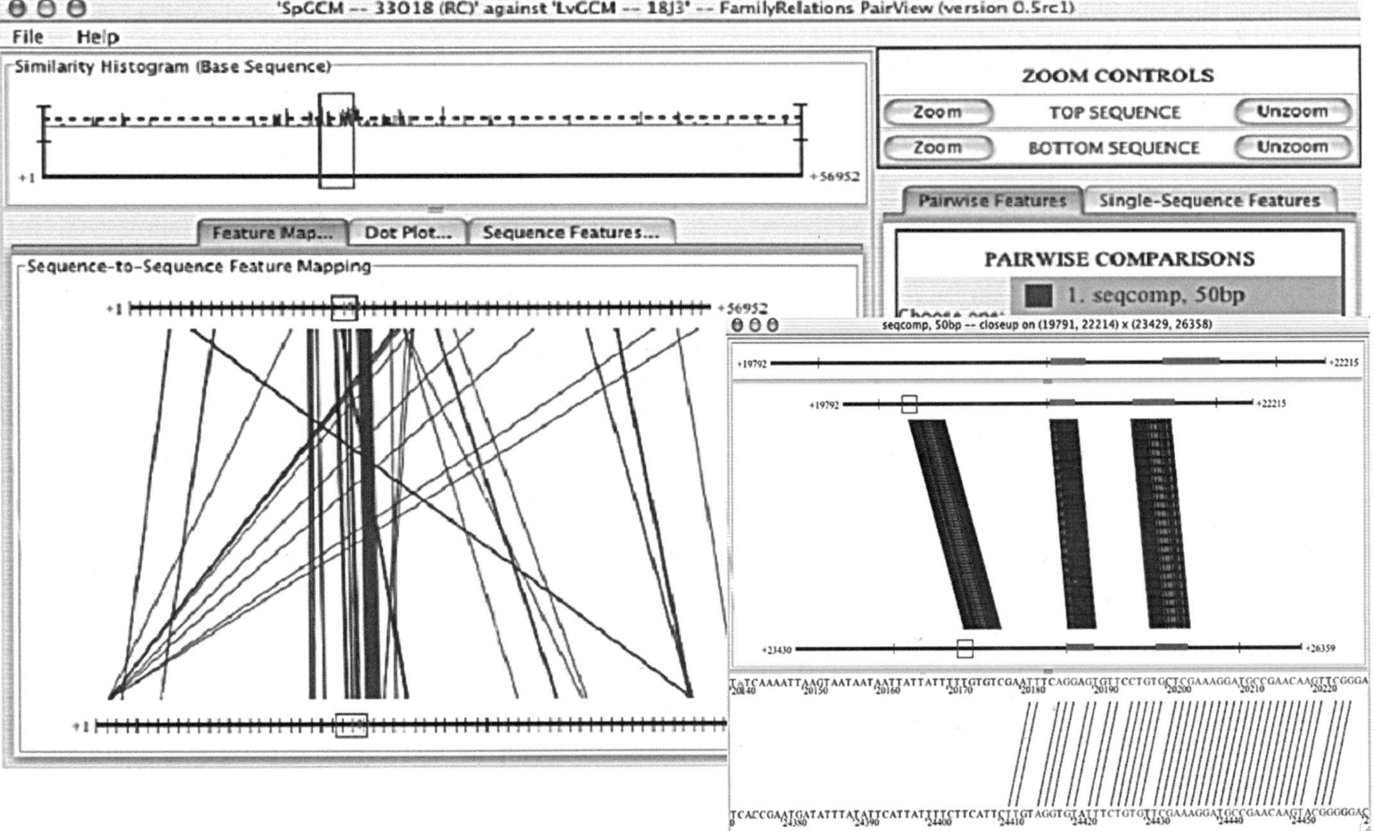

Fig. 3 A Family Relations pairwise analysis of two BACs containing *gcm*, one sequence from *S. purpuratus* (top sequence) and the other from *L. variegatus*. The large panel shows an overall view of the pairwise analysis, with lines drawn between the two BACs where similarities exist above a threshold of 80% between any two 50 bp windows. The inset panel shows a zoomed view where matches between nucleotides are graphed. In both panels, the exons are plotted in red. (See Color Insert.)

1. Start with a sequence

One can search the various subsets of purple sea urchin sequences on the Blast search page. We employ a very simple Blast Web page. The sequences must be pasted into the query window as nucleic acid or protein designations without extraneous characters. Several Blast programs are available and subsets of the sequence database can be selected. Due to its simplicity and the small database, the searches are very fast. A number of matches and alignments are displayed in the standard Blast format. Furthermore, a link is provided to the matched sequence so that the sequence and ancillary information about it can be copied.

Example 1: In order to look for a clone in the collection that will extend the known sequence used as a query, one searches with the known sequence, chooses a high scoring match, and follows the link to the sequence. The sequences thus derived can be used in an alignment to produce a longer stretch of sequence for further investigation.

Example 2: A conserved protein sequence can be searched against the EST part of the database to find a DNA sequence for which sequence-based reagents such as PCR primers can be designed.

2. Start with a filter well spot

If you have identified a clone on a library filter, there might be prior information on that clone's identity and/or sequence. Using the coordinate system from the library filter, you can recall and view the information on a particular clone.

Example: A screen of the 40 h cDNA library using a degenerate probe for E-hand transcription factor identifies 10 spots on the filters. When the filter spot information is collected, there are sequences for two of the spots. These sequences contain a part of the probe sequence. A contig from the assembly of these fragments produces a match to a Genbank sequence at a high value.

3. Start with a gene name

Given the name of a candidate gene known to be involved in a developmental process in other organisms, a search of the gene list can reveal a clone and sequence found by a match to Genbank "nr" database. This sequence can be used to develop PCR primers or the clone itself used to generate a probe for further investigation of this gene product.

Example 1: A search of the hit list reveals a match to the vertebrate lunatic fringe signaling molecule. There is no sequence in the record, only the annotation. Using the plate-well location information, a search of Genbank dbESt provides the sequence.

Example 2: A search of the gene hit list for snail shows a match to 2 cDNA filter clones and a BAC-end sequence. The two cDNA sequences are to different parts of the cDNA. The clones are requested and sequenced in their entirety. The resultant assembled sequence provides an entire open reading frame with stretches of 5′ and 3′ untranslated regions. The BAC genomic clone is requested and sequenced, revealing the intron structure of the gene.

4. Prospective Probe Analysis

Genomic regions and even 3′ untranslated regions (3′UTR) of cDNAs which are otherwise suitable for hybridization probes or included PCR products may contain a repeat that would match many clones in a genomic library or multiple 3′UTRs of cDNAs that contain a coding region different from the one of interest. As has been discussed, the sea urchin genome is made up of approximately 25% repeated sequence. Thus, there is a very good chance of inadvertently including a repeat in a probe sequence. Until sufficient sequence is assembled from the genome to produce a reliable repeat masking library, the best approach is to use existing databases of repeat sequences as Blast subjects to identify repeats in a prospective probe sequence. There are three such sequence collections that can be used: a repeat library (Cameron *et al.*, 2000); the BAC-end sequences; and the sea urchin trace collection. While the last two are not repeat collections per se, a large number of hits in these databases that are all to the same or overlapping sequence regions would be indicative of a repeat. The rationale is as follows: The trace collection is currently at about 3.5× genome coverage and expanding. The BAC-end sequences are 76,000 in number and cover about 5% of the genome. A single copy sequence would be represented only a few times in these genomic sequences. Blast results giving many hits to identical or overlapping sequences are most likely repeats. If only a few sequences are matched, then the probe sequence in question is likely to be a single copy sequence and suitable for screening. Of course, any match to the repeat library is unsuitable.

Acknowledgments

The original research described in this review was supported by grants from NIH (HD-37105, RR-06591, and GM-61005); Office of Science (BER), Department of Energy DE-FG02-03ER63584; NASA's Fundamental Space Biology program (NAG2-1368); the NSF (IBN-9604454), and the Stowers Institute of Medical Research, and the Caltech Beckman Institute.

References

Amemiya, C. T., Ota, T., and Litman, G. W. (1996). Construction of P1 artificial chromosome (PAC) libraries from lower vertebrates. *In* "Nonmammalian Genomic Analysis: A Practical Guide" (B. Birren and E. Lai, eds.), pp. 223–256. Academic Press, San Diego, CA.

Arnone, M., and Davidson, E. H. (1997). The hardwiring of development: Organization and function of genomic regulatory systems. *Development* **124,** 1851–1864.

Bailey, W. J., Kim, J., Wagner, G. P., and Ruddle, F. H. (1997). Phylogenetic reconstruction of certebrate Hox cluster duplications. *Mol. Biol. Evol.* **14,** 843–853.

Brown, C. T., Rust, A. G., Clarke, P. J. C., Pan, Z., Schilstra, M. J., De Buysscher, T., Griffin, G., Wold, B. J., Cameron, R. A., Davidson, E. H., and Bolouri, H. (2002). New computational approaches for analysis of *cis*-regulatory networks. *Dev. Biol.* **246,** 86–102.

Calestani, C., Rast, J. P., and Davidson, E. H. (2003). Isolation of pigment cell specific genes in the sea urchin embryo by differential macroarray screening. *Development* **130,** 4587–4596.

Cameron, R. A., Leahy, P. S., Britten, R. J., and Davidson, E. H. (1999). Microsatellite loci in wild-type and inbred *Strongylocentrotus purpuratus. Dev. Biol.* **208,** 255–264.

Cameron, R. A., Mahairas, G., Rast, J. P., Martinez, P., Biondi, T. R., Swartzell, S., Wallace, J. C., Poustka, A. J., Livingston, B. T., Wray, G. A., Ettensohn, C. A., Lehrach, H., Britten, R. J., Davidson, E. H., and Hood, L. (2000). A sea urchin genome project: Sequence scan, virtual map, and additional resources. *Proc. Natl. Acad. Sci. USA* **97,** 9514–9518.

Clark, M. D., Panopoulou, G. D., Cahill, D. J., Bussow, K., and Lehrach, H. (1999). Construction and analysis of arrayed cDNA libraries. *Methods Enzymol.* **303,** 205–233.

Conner, S., Leaf, D., and Wessel, G. (1997). Members of the SNARE hypothesis are associated with cortical granule exocytosis in the sea urchin egg. *Mol. Reprod. Dev.* **48,** 106–118.

Conner, S. D., and Wessel, G. M. (1999). Syntaxin is required for cell division. *Mol. Biol. Cell* **10**(8), 2735–2743.

Court, D. L., Sawitzke, J. A., and Thomason, L. C. (2002). Genetic engineering using homologous recombination. *Annu. Rev. Genet.* **36,** 361–388.

Czerny, T., Bouchard, M., Kozmik, Z., and Busslinger, M. (1997). The characterization of novel Pax genes of the sea urchin and Drosophila reveals an ancient evolutionary origin of the Pax2/5/8 subfamily. *Mech. Dev.* **67,** 179–192.

Davidson, E. H. (1986). "Gene Activity in Early Development." Academic Press, Orlando, FL.

Davidson, E. H. (2001). Genomic Regulatory Systems. *Development and Evolution.* Academic Press, San Diego, California.

Davidson, E. H., Rast, J. P., Oliveri, P., Ransick, A., Calestani, C., Yuh, C.-H., Minokawa, T., Amore, G., Hinman, V., Arenas-Mena, C., Otim, O., Brown, C. T., Livi, C. B., Lee, P. Y., Revilla, R., Schilstra, M. J., Clarke, P. J. C., Rust, A. G., Pan, Z. J., Arnone, M. I., Rowen, L., Cameron, R. A., McClay, D. R., Hood, L., and Bolouri, H. (2002a). A genomic regulatory network for development. *Science* **295,** 1669–1678.

Davidson, E. H., Rast, J. P., Oliveri, P., Ransick, A., Calestani, C., Yuh, C.-H., Minokawa, T., Amore, G., Hinman, V., Arenas-Mena, C., Otim, O., Brown, C. T., Livi, C. B., Lee, P. Y., Revilla, R., Schilstra, M. J., Clarke, P. J. C., Rust, A. G., Pan, Z. J., Arnone, M. I., Rowen, L., Cameron, R. A., McClay, D. R., Hood, L., and Bolouri, H. (2002b). A provisional regulatory gene network for specification of endomesoderm in the sea urchin embryo. *Dev. Biol.* **246,** 162–190.

Do, M. S., and Lonai, P. (1988). Gene organization of murine homeobox-containing gene clusters. *Genomics* **3,** 195–200.

Frengen, E., Weichenhan, D., Zhao, B., Osoegawa, K., van Geel, M., and de Jong, P. J. (1999). A modular, positive selection bacterial artificial chromosome vector with multiple cloning sites. *Genomics* **58,** 250–253.

Gerhart, S. G. (1983). Sea urchin cytogenetics. Thesis, Dept. Medical Sciences, University of Calgary, Canada.

Giusti, A. F., O'Neill, F. J., Yamasu, K., Foltz, K. R., and Jaffe, L. A. (2003). Function of a sea urchin egg Src family kinase in initiating Ca2_ release at fertilization. *Dev. Biol.* **256,** 367–378.

Gubler, U., and Hoffman, B. J. (1983). A simple and very efficient method for generating cDNA libraries. *Gene* **25,** 263–269.

Hinegardner, R. (1974). Cellular DNA content of the Echinodermata. *Comp. Biochem. Physiol.* **49B,** 219–226.

Lee, E. C., Yu, D., Martinez de Velasco, J., Tessarollo, L., Swing, D. A., Court, D. L., Jenkins, N. A., and Copeland, N. G. (2001). A highly efficient Escherichia coli-based chromosome engineering system adapted for recombinogenic targeting and subcloning of BAC DNA. *Genomics* **73,** 56–65.

Lee, Y.-H., Huang, M., Cameron, R. A., Graham, G., Davidson, E. H., Hood, L., and Britten, R. J. (1999). EST analysis of gene expression in early cleavage-stage sea urchin embryos. *Development* **126,** 3857–3867.

Nizetic, D., Drmanac, R., and Lehrach, H. (1991). An improved bacterial colony lysis procedure enables direct DNA hybridization using short (10, 11 bases) oligonucleotides to cosmids. *Nucleic Acids Res.* **19,** 182.

Osoegawa, K., Woon, P. Y., Zhao, B., Frengen, E., Tateno, M., Catanese, J. J., and de Jong, P. J. (1998). An improved approach for construction of bacterial artificial chromosome libraries. *Genomics* **52,** 1–8.

Osoegawa, K., Tateno, M., Woon, P. Y., Frengen, E., Mammoser, A. G., Catanese, J. J., Hayashizaki, Y., and de Jong, P. J. (2000). Bacterial Artificial Chromosome libraries for mouse sequencing and functional analysis. *Genome Res.* **10,** 116–128.

Phillips, J., and Eberwine, J. H. (1996). Antisense RNA amplification: A linear amplification method for analyzing the mRNA population from single living cells. *Methods: Companion Methods Enzymol.* **10,** 283–288.

Poustka, A. J., Herwig, R., Krause, A., Hennig, S., Meier-Ewert, S., and Lehrach, H. (1999). Toward the gene catalogue of sea urchin development: The construction and analysis of an unfertilized egg cDNA library highly normalized by oligonucleotide fingerprinting. *Genomics* **59,** 122–133.

Poustka, A. J., Groth, D., Hennig, S., Thamm, S., Cameron, R. A., Beck, A., Reinhardt, R., Herwig, R., Panopoulou, G., and Lehrach, H. (2003). Generation, annotation, evolutionary analysis, and database integration of 20,000 unique sea urchin EST clusters. *Genome Research* **13,** 2736–2746.

Ransick, A., Rast, J. P., Minokawa, T., Calestani, C., and Davidson, E. H. (2002). New early zygotic regulators expressed in endomesoderm of sea urchin embryos discovered by differential array hybridization. *Dev. Biol.* **246,** 132–147.

Rast, J. P., Amore, G., Calestani, C., Livi, C. B., Ransick, A., and Davidson, E. H. (2000). Recovery of developmentally defined gene sets from high-density cDNA macroarrays. *Dev. Biol.* **228,** 270–286.

Rast, J. P., Cameron, R. A., Poustka, A. J., and Davidson, E. H. (2002). *Brachyury* target genes in the early sea urchin embryo isolated by differential macroarray screening. *Dev. Biol.* **246,** 191–208.

Scott, M. P. (1992). Vertebrate homeobox gene nomenclature. *Cell* **71,** 551–553.

Strong, S. J., Ohta, Y., Litman, G. W., and Amemiya, C. T. (1997). Marked improvement of PAC and BAC cloning is achieved using electroelution of pulsed-field gel-separated partial digests of genomic DNA. *Nucleic Acids Res.* **25,** 3959–3961.

Swaminathan, S., Ellis, H. M., Waters, L. S., Yu, D., Lee, E. C., Court, D. L., and Sharan, S. K. (2001). Rapid engineering of bacterial artificial chromosomes using oligonucleotides. *Genesis* **29,** 14–21.

Yu, D., Ellis, H. M., Lee, E. C., Jenkins, N. A., Copeland, N. G., and Court, D. L. (2000). An efficient recombination system for chromosome engineering in *Escherichia coli*. *Proc. Natl. Acad. Sci. USA* **97,** 5978–5983.

Zhu, X., Mahairas, G., Cameron, R. A., Davidson, E. H., and Ettensohn, C. A. (2001). Large-scale analysis of mRNAs expressed by primary mesenchyme cells of the sea urchin embryo. *Development* **128,** 2615–2627.

CHAPTER 31

Genomic Resources for Ascidians: Sequence/Expression Databases and Genome Projects

Nori Satoh

Department of Zoology
Graduate School of Science
Kyoto University, Kyoto 606-8502, Japan

Overview

Ascidians provide an appealingly simple experimental system to investigate the molecular mechanisms that underlie cell-fate specification, morphogenesis, and metamorphosis during development. The fertilized egg develops quickly into a

0091-679X/04 $35.00

tadpole-type larva. The ascidian tadpole is composed of only about 2600 cells that constitute a small number of organs including epidermis, central nervous system, endoderm and mesenchyme in the trunk, and notochord and muscle in the tail. This configuration of the ascidian tadpole represents the basic chordate body plan. The draft genome sequence of the most studied ascidian, *Ciona intestinalis*, has been determined. Its ~159 Mbp genome contains 15,852 protein-coding genes, many of which are single copy representatives of vertebrate gene families engaged in a variety of signaling and regulatory processes. Thus, the streamlined nature of the ascidian genome should have an enormous impact on unraveling complex developmental processes in vertebrates. In addition, recent *Ciona* cDNA projects provide a great quantity of information concerning the transcriptome of this ascidian. Large-scale EST analyses identify ~18,000 independent transcripts that are expressed in embryos, larvae, and various organs of the adult. All of the cDNA clones have been re-arrayed to release as *Ciona* gene resources. Microarrays with cDNA chips as well as oligonucleotide chips have been prepared from the resources. The sequences of three-fifths of the cDNAs have been characterized with respect to open reading frames. Furthermore, *in situ* hybridization analyses reveal the spatial expression profiles of ~5000 developmental genes. All together, recent advances in the cDNA projects and genome projects of *Ciona intestinalis* highlight this ascidian as an attractive experimental system in developmental biology and genome sciences.

I. Introduction

Ascidians, or sea squirts, are sessile marine invertebrate chordates found throughout the world seas (see also Chapters 6, 7, and 29). The adults are simple filter feeders encased in a fibrous tunic supporting incurrent and outcurrent siphons (Fig. 1a) and were classified previously as mollusks. However, their so-called tadpole larva possesses a notochord, which places them at a unique evolutionary position between nonchordate deuterostomes and chordates. These features have attracted developmental and evolutionary biologists since the turn of twentieth century (reviewed by Corbo *et al.*, 2001; Jeffery, 2001; Nishida, 2002a,b; Satoh, 1994, 2001, 2003; Satoh and Jeffery, 1995).

In particular, ascidians provide an appealingly simple experimental system for investigating the molecular mechanisms that underlie cell-fate specification, morphogenesis, and metamorphosis during development. First, as shown in Fig. 1, the fertilized egg develops rather quickly, through bilaterally symmetrical cleavage, gastrulation, neurulation, and tailbud embryo formation, into a tadpole larva. The ascidian tadpole is composed of only ~2600 cells, which constitute a small number of organs including epidermis, central nervous system (CNS), endoderm and mesenchyme in the trunk, and notochord and muscle in the tail (Fig. 1l). This configuration of the tadpole with notochord and dorsal nervous system represents the basic chordate body plan. Second, extensive information on the cell lineage

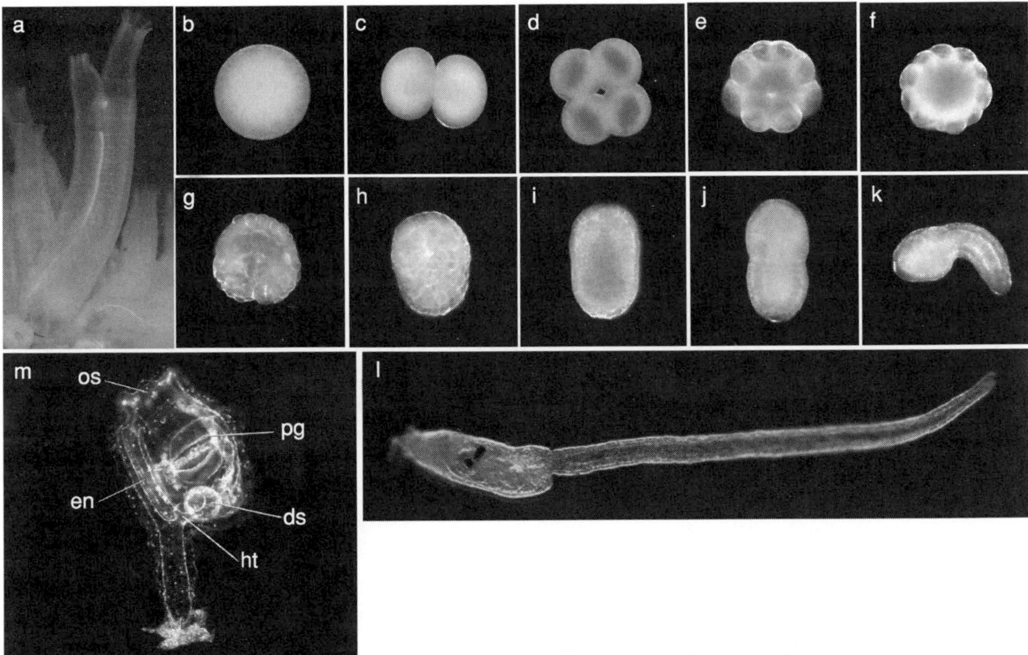

Fig. 1 The sea squirt *Ciona intestinalis*. (a) Adults are filter feeders with incurrent and outcurrent siphons. The white duct is the sperm duct, while the orange duct paralleling it is the egg duct. (b–l) Embryogenesis. (b) Fertilized egg, (c) 2-cell embryo, (d) 4-cell embryo, (e) 16-cell embryo, (f) 32-cell embryo, (g) gastrula (about 150 cells), (h) neurula, (i–k) tailbud embryos, and (l) tadpole larva. Embryos were dechorionated to show their outer morphology clearly. (m) A juvenile a few days after metamorphosis, showing the internal structure. ds, digestive system; en, endostyle; ht, heart; os, neuronal complex; and pg, pharyngeal gill. (See Color Insert.)

of most major organs of the larva (Conklin, 1905; Nishida, 1987; Nishida and Satoh, 1985) as well as adult organs and tissues (Hirano and Nishida, 1997, 2000) is available. Third, the early embryonic cells with definite developmental fates are large (especially in *Halocynthia roretzi*) and easy to manipulate (Nishida, 2002a,b). Fourth, the embryonic cells also permit the detailed visualization of differential gene expression during development (Satou *et al.*, 1995; Yasuo and Satoh, 1993). *In situ* hybridization analysis shows that zygotic expression of developmental genes is first detectable in the nuclei of certain blastomeres, permitting us to distinguish which cells, as well as when, begin the expression of the gene. Fifth, novel functions of developmental genes can be determined by misexpressing or overexpressing a variety of regulatory genes that encode transcription factors or signaling molecules or by the functional suppression of genes with antisense oligonucleotides (Imai *et al.*, 2002a,b,c; Satou *et al.*, 2001b). Sixth, transgenic DNAs can be introduced into fertilized eggs using simple electroporation methods that permit the simultaneous transformation of hundreds, even thousands, of synchronously developing embryos, providing a special advantage in exploring *cis*-regulatory systems governing gene expression (see Chapter 29). Seventh, embryogenesis of the most studied ascidians *Ciona intestinalis* and *Ciona savignyi* is rapid and the entire life cycle takes less than 3 months, facilitating mutagenesis and mutant screening (see Chapter 7).

In addition, the ascidian tadpole has occupied center stage in recent evolutionary developmental biology (Corbo *et al.*, 2001; Satoh, 2003; Satoh and Jeffery, 1995) because it represents the closest living form of the ancestral chordates. Disclosure of molecular mechanisms underlying development of this larva would facilitate our understanding of the ancestor of humans and other chordates.

For all these reasons, several ascidian cDNA projects and genome projects have been carried out over the last several years. The reading of the draft genome of *Ciona intestinalis* was finished by the end of 2002 (Dehal *et al.*, 2002). In addition, cDNA projects have accumulated a large quantity of information concerning the ascidian transcriptome. In this chapter, I discuss the genomic resources for ascidians, especially for *Ciona intestinalis*.

II. The cDNA Projects: cDNA Sequence and Expression Databases

The cDNA projects of ascidians include analyses of Expressed Sequence Tags (ESTs), cDNA sequence information, and spatial expression profiles of developmental genes. Besides small-scale analyses from various ascidian species, a large-scale analysis was first aimed at maternal transcripts of *Halocynthia roretzi* (Kawashima *et al.*, 2000; Makabe *et al.*, 2001). This project, called "MAGEST," includes ~40,000 ESTs and spatial expression profiles of ~1200 maternally expressed genes (Makabe *et al.*, 2001; Makabe, personal communication). All its current information is available at the project's Web site (http://www.genome.ad.jp/magest/).

Larger scale and more comprehensive cDNA projects have been carried out in *Ciona intestinalis*, which is an ascidian species used by researchers throughout the world. I discuss here only *Ciona intestinalis* cDNA projects which have been carried out within the last 2 to 3 years. All information concerning the *Ciona intestinalis* cDNA projects is accessible at our website (http://ghost.zool.kyoto-u.ac.jp/indexr1.html).

A. Expressed Sequence Tags (ESTs)

1. EST Data Set

Two large-scale EST analyses have been conducted to determine both 3′ and 5′ end sequences of transcripts expressed in *Ciona intestinalis*, and a third project is now ongoing. The first project examined transcripts expressed at five different developmental stages: fertilized eggs (Nishikata *et al.*, 2001), cleaving embryos (Fujiwara *et al.*, 2002), tailbud embryos (Satou *et al.*, 2001a), tadpole larvae (Kusakabe *et al.*, 2002), and whole young adults (Ogasawara *et al.*, 2002), and also examined adult testis (Inaba *et al.*, 2002). A total of 167,602 ESTs, 81,637 from the 3′ ends and 85,965 from the 5′ ends, have been read. The sequences of the 3′ ends were compared using the program FASTA (Pearson and Lipman, 1988) to identify overlapping of cDNA clones, and this analysis categorized the 167,602 ESTs into 13,464 independent cDNA clusters (Satou *et al.*, 2002a,b).

The second series expanded the first by adding more data for the same five developmental stages, and by adding new data of transcripts expressed in gastrulae/neurulae and those expressed in the gonad (ovary), endostyle (the primordium of vertebrate thyroid gland), neural complex, heart, and blood (coelomic) cells of the adult (Satou *et al.*, unpublished). As summarized in Table I, the present data set exceeds 480,000 ESTs (this number ranks the third highest position in DDBJ/Genbank/EMBL EST database entry, following humans and mice, as of Feb. 21, 2003). The ESTs are categorized into 17,834 independent cDNA clusters. As discussed later, the *Ciona intestinalis* genome is estimated to contain ~16,000 protein-coding genes (Dehal *et al.*, 2002). Therefore, this number of independent cDNA clusters is 1.12 times larger than the number of *Ciona* protein-coding genes.

The third series is now ongoing to determine transcripts expressed in the digestive system, body-wall muscle, and pharyngeal gill of the adult. Altogether, the ultimate data set of *Ciona intestinalis* ESTs may cover transcripts expressed at almost all embryonic stages and in most adult organs and/or tissues.

Table I
ESTs of Transcripts Expressed in *Ciona intestinalis*

Library	Lib ID	5′ EST	3′ EST	Total
Embryogenesis				
Egg	EG	29,810	29,444	59,254
Cleaving embryo	CL	31,156	26,796	57,952
Gastrula/Neurula	GN	23,066	23,475	46,541
Tailbud embryo	TB	30,282	31,209	61,491
Larva	LV	24,282	24,680	48,962
Young adult	AD	28,547	29,138	57,685
Adult organs/tissues				
Gonad	GD	16,048	16,239	32,287
Testis	TS	4,655	4,717	9,372
Endostyle	ES	2,241	2,497	4,738
Neural complex	NC	10,116	10,029	20,145
Heart	HT	12,904	12,414	25,318
Blood cells	BD	28,412	28,596	57,008
Total		241,519	239,234	480,753

2. Application of the EST Database

The information in the extensive EST databases is useful for studying mechanisms involved in the control of the developmental gene expression, especially the genome-wide global control of gene expression. For example, because the cDNA libraries used for EST analyses were not amplified or normalized, the appearance of cDNA clones (or EST counts) occurs in proportion of their abundance at the corresponding stage or in the certain organ. Taking advantage of the enormous amount of EST data, an intensive survey of dynamic changes in *Ciona* developmental gene expression has been carried out (Satou *et al.*, 2004). Results demonstrate that (a) the fertilized eggs store a great variety of maternal transcripts and, as development proceeds, a progressively smaller repertoire of genes is expressed, (b) a significant portion of genes involved in embryogenesis are downregulated during metamorphosis, at which point the adult appears to utilize a different set of genes to form its body, and (c) at least 25% of the genes involved in development are used multiple times.

The expression patterns of a considerable number of *Ciona* developmental genes have been fully characterized by Northern blot analysis and/or whole-mount *in situ* hybridization (e.g., Corbo *et al.*, 1997; Fujiwara *et al.*, 1998; Imai *et al.*, 2002a,b,c). Detailed comparison between the expression profiles of certain developmental genes obtained by these methods and changes in EST counts of the same genes shows fairly good agreement (Satou *et al.*, 2003b), suggesting that analysis of EST counts can, at least partially, substitute for Northern blot analysis. For example, if ESTs corresponding to a given gene are identified only from samples of fertilized eggs but not from embryos at other developmental stages, this gene is highly likely to be expressed only maternally. Similarly, if ESTs corresponding to a gene are identified only in samples of heart muscle, these genes may be specifically expressed there.

B. Gene (cDNA) Resources

As has been mentioned, *Ciona intestinalis* has a treasure box of ESTs. The 13,464 independent cDNA clones obtained from the first series were re-arrayed in 36 384-well plates. This re-arrayed set is now released from Nori Satoh's lab as "*Ciona intestinalis* Gene Collection Release 1 (CiGCR1)" for free academic-research usage (Satou *et al.*, 2002b). An additional 4370 cDNA clones identified from the second series are ready to release as "the supplement of CiGCR1." Therefore, a total of 17,834 independent cDNA clones are now available as a *Ciona* gene resource. This cDNA resource covers ~85% of the transcripts expressed in *Ciona intestinalis*.

C. Microarrays: cDNA Chips and Oligonucleotide Chips

Microarray technologies will be useful in investigating basic gene networks involved in the formation of basal chordate body plan as well as chordate-specific biological processes. Analysis of gene networks in this simple system may also facilitate our understanding of more complex gene networks in vertebrates. Thus, to date, three types of chips have been and are now being prepared in our laboratory. The first one, or "*Ciona* cDNA chip version 1," is a cDNA chip comprising 13,464 cDNAs from the CiGCRI, spotted on a set of four glass slides. Preliminary examination with probes from fertilized eggs and from larvae showed that this chip can be used to distinguish genes specifically expressed at either developmental stage (Azumi *et al.*, 2003). The second *Ciona* cDNA chip version 2 is now being prepared to cover 17,834 cDNAs including the supplement cDNA clones, spotted on a single glass slide. In addition, a third oligonucleotide chip including 17,834 cDNAs has been made commercially (Agilent Tech).

The microarrays may be used for many kinds of investigations. Gene networks now being targeted in our lab and other labs include those for downstream of T-box genes, β-catenin target genes (see next section), *Ci-macho 1* target genes, and genes involved in apoptosis, immune response, senescence, circadian rhythms, or environmental pollution. All of the chips may be used by academic researchers (please contact Nori Satoh).

D. Analyses of Gene Networks

Although this is one of the main subjects discussed in Chapter 29, here I wish to discuss this subject briefly in relation to future application of *Ciona* cDNA information and microarray techniques.

The electroporation method as well as other gene introduction techniques works efficiently in *Ciona* embryos and can be used to identify downstream target genes of transcription factors or signaling molecules (see Chapter 29). For example, a T-box gene, *Brachyury*, is expressed exclusively in notochord cells in ascidian embryos (Corbo *et al.*, 1997; Yasuo and Satoh, 1993) and its function is essential for notochord cell differentiation (Satou *et al.*, 2001c; Yasuo and Satoh, 1998). Experimental embryos in which *Ci-Bra* was overexpressed by electroporation were used to screen for *Ci-Bra* target genes, which would be expressed in differentiating notochord cells (Takahashi *et al.*, 1999). The *Ci-Bra* target notochord-specific genes thus obtained include genes encoding Prickle, netrin, ERM (ezrin-radixin-moesin), tropomyosin-like protein (Di Gregorio and Levine, 1999), cdc45, and several proteins without sequence similarity to known proteins (Hotta *et al.*, 2000). It is expected that the use of *Ciona* microarrays may facilitate an isolation of more *Ci-Bra* target genes. Once the candidate cDNAs are identified, the present *Ciona intestinalis* transcriptome data set will provide information concerning their sequences and the expression profiles of the corresponding genes.

Another example is β-catenin target downstream genes. In early *Ciona* embryos, β-catenin initiates the specification of endoderm cells (Imai *et al.*, 2000). β-catenin overexpressing embryos and β-catenin down-regulating embryos were used to identify β-catenin target genes in early *Ciona savignyi* embryos. Thus, *FoxA, FoxD, otx, ttf1, FGF9/16/20*, and *Zic-like* have been identified as candidate genes (Imai, 2003; Imai *et al.*, 2002a,b,c; Satou *et al.*, 2001b). Of special interest is *FoxD*, which is transiently expressed in endoderm cells and functions in notochord specification but not endoderm differentiation (Imai *et al.*, 2002c). Similar to the case of *Ci-Bra* target gene isolation, the microarrays and cDNA information would be used in further characterization of β-catenin target downstream genes.

Future studies with microarrays and cDNA information may disclose targets of various *Ciona* genes with developmentally important functions.

E. cDNA Sequence Information

Sequences of the 5′ and 3′ ends of cDNA clones examined by EST analyses have been used to obtain protein information by searching cDNA databases and protein databases using BLAST algorithms (BLASTN and BLASTX). Using the information, the proteins have been categorized into four major classes, basically according to the classification of Lee *et al.* (1999): Class A, proteins associated with functions that many kinds of cells use; Class B, proteins associated with cell–cell communication; Class C, proteins that function as transcription factors and other gene regulatory proteins; and Class D (I and II), DI in which sequences are matches to ESTs (mostly from mice and humans) or reported proteins without known function, and DII with no significant sequence similarities to other known proteins (for details, see, for example, Satou *et al.*, 2001a). In addition, Class A proteins are categorized further into nine subclasses, Class B into three subclasses, and Class C into three subclasses (for details, see Lee *et al.*, 1999; Satou *et al.*, 2001a). For example, of 1003 nonredundant cDNAs for genes that are expressed in fertilized eggs, 362 were genes of Class A, 65 were those of Class B, 25 were Class C, and 551 were Class D (Nishikata *et al.*, 2001). And, of 1213 genes expressed in the tailbud embryos, 390 were genes of Class A, 85 were those of Class B, 27 were Class C, and 711 were Class D (Satou *et al.*, 2001a).

Projects to determine full-length sequences of cDNAs are also progressing. To date, both strands have been determined on 1000 cDNAs to reveal their ORFs. Single-strand sequencing has covered over 4500 cDNAs. In addition, multiple coverage by ESTs of comparatively short cDNAs frequently gave complete nucleotide information of the cDNAs. Nearly full-length sequences of more than 4500 cDNAs have been obtained by the EST assembly. Thus, in total, ~10,000 cDNAs (about 56% of 17,834) have their own and nearly complete identity (Satou *et al.*, 2002b).

F. Functional Analyses of Developmental Genes with Morpholino Oligos

It has been shown that morpholino oligos can be efficiently applied to various systems including *Xenopus*, zebrafish, and sea urchins (Heasman, 2002; see also Chapter 28). Morpholinos are highly effective at disrupting the function of genes in the ascidian embryo (Satou *et al.*, 2001c), targeting the genes encoding transcription factors (Imai *et al.*, 2002b,c; Satou *et al.*, 2001b) and signaling molecules (Imai *et al.*, 2002a). The specificity of the effect of morpholinos is assessed by rescuing the effect by injection of *in vitro* synthesized mRNA which is designed to escape the morpholino-target sequences (Satou *et al.*, 2001c). A large-scale analysis of targeting 200 *Ciona* genes including genes encoding zinc-finger proteins, neural-specific genes, and genes encoding proteins with vertebrate counterparts with unknown function suggests that nearly one-fifth of the genes examined show detectable morphological effects when their function is suppressed with specific morpholinos (Yamada *et al.*, 2003). This high efficiency, gene knock-out strategy with morpholinos in *Ciona* embryos will be useful in the future identification of novel genes with developmental function.

G. Spatial Expression of Developmental Genes

One of the long-term goals of research on *Ciona intestinalis* is to obtain a complete set of information on the temporal and spatial expression of every developmental gene and to discover molecular mechanisms of the complex network of gene expression. As mentioned previously, the EST database has contributed enormously to determining the temporal expression profiles of *Ciona* developmental genes. Therefore, what we need now is to collect information on the spatial expression profiles of as many *Ciona* developmental genes as possible. This line of research has already been conducted at the five developmental stages including fertilized eggs (Nishikata *et al.*, 2001), cleavage-stage embryos (Fujiwara *et al.*, 2002), tailbud embryos (Satou *et al.*, 2001a), tadpole larvae (Kusakabe *et al.*, 2002), and young adults (Ogasawara *et al.*, 2002). At each stage, the spatial expression of a randomly selected set of 1000 or so developmental genes has been determined (Fig. 2).

Results from the tailbud embryo and larva, for example, show that each of the larval organs is characterized by the expression of specific sets of developmental genes (Kusakabe *et al.*, 2002; Satou *et al.*, 2001a). For example, most of the genes that are specifically expressed in muscle encode proteins that are categorized into subclass AIV of cytoskeleton proteins including muscle structural proteins. The larval CNS is characterized by the expression of genes categorized into three major groups: subclass AIV of cytoskeleton and membrane proteins, subclass BII of intracellular signal transduction pathway molecules including kinases and signal intermediates, and subclass BII of extracellular matrix proteins and cell adhesion. This feature is reflected by the function of the CNS.

H. "Ghost Database" for *Ciona intestinalis* cDNAs and Its Application

The complete cDNA and expression databases are accessible through our website, http://ghost.ascidian.zool.kyoto-u.ac.jp. Using this database, one can obtain *Ciona intestinalis* cDNA information from *Ciona* gene ID, BLAST search, keyword search, gene expression patterns, or sequence. All together, recent

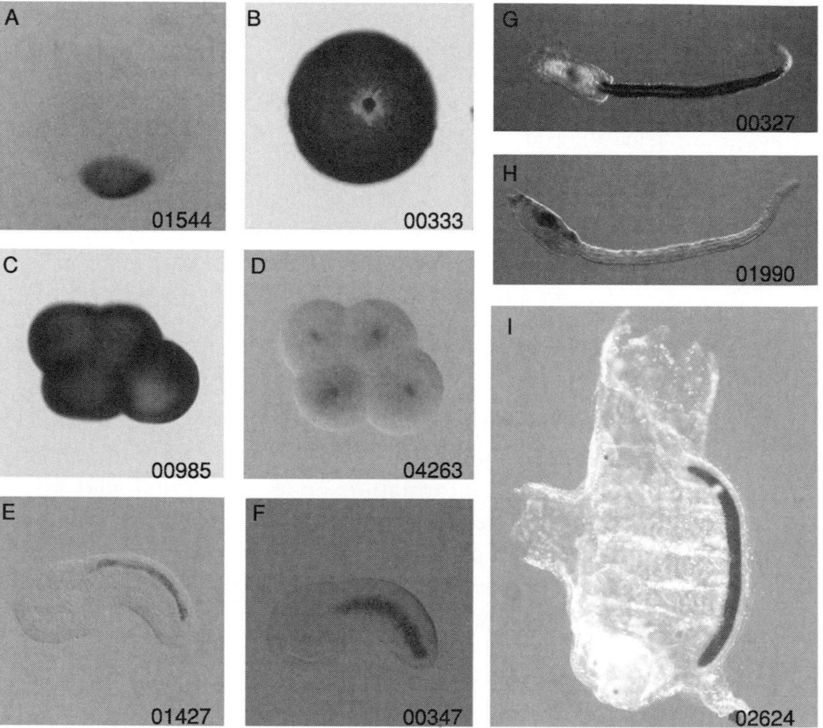

Fig. 2 Whole-mount *in situ* hybridization showing the expression profiles of genes in *Ciona intestinalis*. Gene IDs are shown at the right bottom corner of each photograph. (A, B) Two examples of the distribution of maternal transcripts in fertilized eggs. (C, D) Examples of the distribution of (C) a maternal transcript and (D) a zygotic transcript in 8-cell embryos. (E) Expression of a nerve cord-specific gene and (F) notochord-specific gene in tailbud embryos. (G) Expression of a muscle-specific gene and (H) neuron-specific gene in larvae. (I) Expression of an endostyle-specific gene in a young adult. (See Color Insert.)

advances in the *Ciona intestinalis* cDNA project provide us with an opportunity to investigate developmental genes of the basal chordates. If one is interested in *Ciona* homologues of a certain vertebrate gene with an important function in a biological process, one may access the Ghost to identify the desired cDNA by keyword search or sequence similarity search. In some cases, one may obtain information on the entire sequence of the cDNA, temporal expression profile by EST counts, and spatial expression pattern of the gene. It is also possible to obtain the cDNA from the CiGCR1 and related genome information by accessing the JGI website. It should therefore be possible to immediately begin to analyze the function of the *Ciona* gene or the gene network associated with the gene.

III. The Genome Project

A. Reading of the *Ciona intestinalis* Draft Genome is Finished

The draft genome of *Ciona intestinalis* has been read by the whole-genome shotgun method with eight-fold coverage (Dehal *et al.*, 2002). Together with large quantities of transcriptomic information, BAC end sequencing, and cosmid library sequencing, the genome sequences have been assembled into "contigs" and then into "scaffolds" (for details, see Dehal *et al.*, 2002). The present *Ciona intestinalis* assembly spans 116.7 Mbp, of nonrepetitive sequence in 2501 scaffolds longer than 3 kbp. Sixty Mbp, or about half the assembly, is reconstructed in only 177 scaffolds longer than 190 kbp, and more than 85% of the assembled sequence (total of 104.1 Mbp) is found in 905 scaffolds longer than 20 kbp. At present, available mapping data do not yet allow the placement of the assembled sequences on *Ciona*'s 14 chromosomes. FISH is now being applied to overcome this problem.

Altogether, the ~159 Mbp-long *Ciona intestinalis* genome comprises ~117 Mbp of nonrepetitive and euchromatic sequence, 16~18 Mbp of high-copy number tandem repeats (rRNAs, clusters of tRNAs, and so forth), and 15~17 Mbp of additional low-copy transposable elements and repeats. The genome is notably AT-rich (65%) compared with the human genome. A high level of allelic polymorphism is found in the sequences from the single individual, with 1.2% of the nucleotides differing between alleles—nearly 15-fold more than in humans, and 3-fold more than in *Fugu*. The *Ciona intestinalis* genome is estimated to contain a total of 15,852 protein-coding genes (Dehal *et al.*, 2002). Table II compares characteristic features of the *Ciona* genome with those of six animals including *Drosophila melanogaster* and humans. It is evident that *Ciona* genes are organized in the genome more compactly than those of the protostomes (except *C. elegans*) and vertebrates, indicating an advantage of the *Ciona* system for future studies of *cis*-regulatory networks, as discussed in Chapter 29.

Table II
Comparison of the Genomic and cDNA Information Among Seven Animals of Which Genomic Sequences have been Read

	Size of euchromatic sequence	Number of protein-coding genes	Size of genome/gene
Human	2,916 Mbp	31,778	92 Kbp
Mouse	2,493	22,011	113
Fugu rubripes	333	31,059	11
Ciona intestinalis	117	15,852	7
Drosophila melanogaster	122	13,601	9
Anopheles gambiae	278	13,683	20
Caenorhabditis elegans	97	19,099	5

Analysis of the *Ciona intestinalis* genome clearly demonstrates that ascidians contain the basic ancestral chordate complement of genes. Namely, it contains single copy genes for the present-day complement of gene families engaged in a variety of signaling and regulatory processes seen in vertebrate development, including FGFs (Satou *et al.*, 2002c), Smads (Yagi *et al.*, 2003), and T-box genes (Takatori *et al.*, unpublished). Therefore, the streamlined nature of the *Ciona* genome should have an enormous impact on unraveling complex developmental processes in vertebrates. The ascidian genome has also acquired a number of lineage-specific innovations (Dehal *et al.*, 2002). The *Ciona* genome appears to have lost several Hox genes (Hox7, 8, 9, and 11). In contrast, it includes a group of genes engaged in cellulose metabolism that are related to those in bacteria and fungi (Nakashima *et al.*, 2004).

In addition, comparison of the *Ciona* genes with invertebrate and vertebrate genes suggests evolutionary innovations that might have occurred during the evolution of ancestral chordates or ancestral vertebrates. For example, genes encoding nuclear receptors, genes for thyroid hormone receptors, retinoic acid receptors, and peroxisome proliferator-activated receptor are likely to be chordate-specific innovations, while genes encoding estrogen receptors and steroid hormone receptors are vertebrate-specific innovations. Further analysis of the *Ciona* genome is now underway (Kawashima *et al.*, 2003; Satou *et al.*, 2003a; Yagi *et al.*, 2003). All of the *Ciona intestinalis* genomic information is accessible at http://jgi.doe.gov/ciona.

The *Ciona intestinalis* genome projects have been carried out in collaboration with the Joint Genome Institute (Department of Energy, USA: Daniel Rokhsar), Univ. of California, Berkeley (Michael Levine), Kyoto University (Nori Satoh and Yutaka Satou), and National Institute of Genetics, Japan (Yuji Kohara). To avoid the possibility that high polymorphisms caused by mixed samples would interfere with an appropriate assembly of the genome, the published genome sequences were produced from a single individual collected at Half Moon Bay, California. In parallel with sequencing of the California individual, the draft genome of a mixture of three Japanese individuals has been sequenced by the whole-genome shotgun method with 6-fold coverage (Kohara *et al.*, unpublished). Thus, when the Japanese *Ciona* sequences are assembled by late-2004, we may compare the entire *Ciona intestinalis* genome sequences from two different populations. This comparison may answer several basic problems of population genetics as well as of genome sciences.

B. Reading of the *Ciona savignyi* Draft Genome

In addition to the *Ciona intestinalis* draft genome, the *Ciona savignyi* genome project consortium (Arend Sidow, Stanford Univ.) has conducted the reading of *Ciona savignyi* draft genome sequencing by the whole-genome shotgun method to cover the genome 11-fold, in collaboration with the Whitehead Institute, MIT. The genome sequence of *Ciona savignyi* is expected to have higher polymorphism than *Ciona intestinalis* (Shungo Kano, personal communication). This seems to make it difficult to assemble the shotgun sequences. At present, many unassembled sequences have been made public through NCBI Megablast site (http://www.ncbi.nlm.nih.gov/blast/tracemb.html). Nevertheless, the high level of *Ciona intestinalis* genome sequence assembly may facilitate the assembly of *Ciona savignyi* sequences. Release of a draft genome sequence of *Ciona savignyi* is expected in the very near future. If so, we may compare the entire genome sequences of two taxonomically closely related *Ciona* species, and those of *Ciona intestinalis* from two different populations. These sequence data will provide us many opportunities for investigating various questions of genome sciences.

IV. Conclusion

Since the turn of the twentieth century, ascidians have provided an attractive experimental system in developmental and evolutionary biology. For the last two decades, a more precise description of embryonic cell lineages and introduction of various molecular biological techniques have made ascidians a simple experimental system for the investigation of molecular mechanisms that underlie cell-fate specification during development. As discussed here, the reading of the draft genome of the most studied ascidian, *Ciona intestinalis*, and related cDNA projects have provided a great quantity of developmental gene information in this animal. The cDNA resource may cover almost all of the 16,000 gene transcripts estimated from the genome project, nearly 60% of them with cDNA sequence information. A systematic *in situ* hybridization analysis may disclose in the very near future the spatial expression profiles of all of the developmental genes. Altogether, *Ciona* has become one of the best-characterized animals with respect to its genome (Satoh *et al.*, 2003).

Ciona intestinalis has an appealingly compact genome that contains a basic set of genes for the basal chordate, with much less redundancy than is found in vertebrates. This situation not only promotes analysis of specific gene expression regulation and gene function but also future *Ciona* studies should have an enormous impact on unraveling complex developmental processes in vertebrates. In addition, all of this information—gene-expression profiles and the gene-disruption data—will be compiled along with large-scale screens for *cis*-regulatory networks underlying the development of basic chordate features such as the neural tube and notochord, and thus facilitate an understanding of our ancestors and their evolution.

Acknowledgments

I thank all colleagues who have contributed to studies included in this chapter, and Charles A. Ettensohn for critical reading of the manuscript. The research of my laboratory is supported by Grants-in-Aid from MEXT, Japan, CREST project of JST, Human Frontier Science Program, and 21COE (A14).

References

Azumi, K., Takahashi, H., Miki, Y., Fujie, M., Ishikawa, H., Kitayama, A., Satou, Y., Ueno, N., and Satoh, N. (2003). Construction of a cDNA microarray derived from the ascidian *Ciona intestinalis*. *Zool. Sci.* **20,** 1223–1229.

Conklin, E. G. (1905). The organization and cell lineage of the acidian egg. *J. Acad. Nat. Sci. (Philadelphia)* **13,** 1–119.

Corbo, J. C., Levine, M., and Zeller, R. W. (1997). Characterization of a notochord-specific enhancer from the *Brachyury* promoter region of the ascidian, *Ciona intestinalis*. *Development* **124,** 589–602.

Corbo, J. C., Di Gregorio, A., and Levine, M. (2001). The ascidian as a model organism in developmental and evolutionary biology. *Cell* **106,** 535–538.

Dehal, P., Satou, Y., Campbell, R. K., *et al.* (2002). The draft genome of *Ciona intestinalis*: Insights into chordate and vertebrate origins. *Science* **298,** 2157–2167.

Di Gregorio, A., and Levine, M. (1999). Regulation of *Ci-tropomyosin-like*, a Brachyury target gene in the ascidian, *Ciona intestinalis*. *Development* **126,** 5599–5609.

Fujiwara, S., Corbo, J. C., and Levine, M. (1998). The Snail repressor establishes a muscle/notochord boundary in the *Ciona* embryo. *Development* **125,** 2511–2520.

Fujiwara, S., Maeda, Y., Shin-i, T., Kohara, Y., Takatori, N., Satou, Y., and Satoh, N. (2002). Gene expression profiles in *Ciona intestinalis* cleavage-stage embryos. *Mech. Dev.* **112,** 115–127.

Heasman, J. (2002). Morpholino oligos: Making sense of antisense? *Dev. Biol.* **243,** 209–214.

Hirano, T., and Nishida, H. (1997). Developmental fates of larval tissues after metamorphosis in ascidian *Halocynthia roretzi*. I. Origin of mesodemal tissues of the juvenile. *Dev. Biol.* **192,** 199–210.

Hirano, T., and Nishida, H. (2000). Developmental fates of larval tissues after metamorphosis in the ascidian, *Halocynthia roretzi*. II. Origin of endodermal tissues of the juvenile. *Dev. Genes Evol.* **210,** 55–63.

Hotta, K., Takahashi, H., Asakura, T., Saitoh, B., Takatori, N., Satou, Y., and Satoh, N. (2000). Characterization of *Brachyury*-downstream notochord genes in the *Ciona intestinalis* embryo. *Dev. Biol.* **224,** 69–80.

Imai, K. S. (2003). Isolation and characterization of β-catenin downstream genes in early embryos of the ascidian *Ciona savignyi*. *Differentiation* **71,** 346–360.

Imai, K., Takada, N., Satoh, N., and Satou, Y. (2000). β-catenin mediates the specification of endoderm cells in ascidian embryos. *Development* **127,** 3009–3020.

Imai, K. S., Satoh, N., and Satou, Y. (2002a). Early embryonic expression of *FGF4/6/9* gene and its role in the induction of mesenchyme and notochord in *Ciona savignyi* embryos. *Development* **129,** 1729–1738.

Imai, K. S., Satou, Y., and Satoh, N. (2002b). Multiple functions of a Zic-like gene in the differentiation of notochord, central nervous system, and muscle in *Ciona savignyi* embryos. *Development* **129,** 2723–2732.

Imai, K. S., Satoh, N., and Satou, Y. (2002c). An essential role of a *FoxD* gene in notochord induction in *Ciona* embryos. *Development* **129,** 3441–3453.

Inaba, K., Padma, P., Satouh, Y., Shin-i, T., Kohara, Y., Satoh, N., and Satou, Y. (2002). EST analysis of gene expression in testis of the ascidian *Ciona intestinalis*. *Mol. Reproduction Dev.* **62,** 431–445.

Jeffery, W. R. (2001). Determinants of cell and positional fate in ascidian embryos. *Int. Rev. Cytol.* **203,** 3–62.

Kawashima, T., Kawashima, S., Kanehisa, M., Nishida, H., and Makabe, K. W. (2000). MAGEST: MAboya Gene Expression patterns and Sequence Tags. *Nucl. Acids Res.* **28,** 133–135.

Kawashima, T., Tokuoka, M., Awazu, S., Satoh, N., and Satou, Y. (2003). A genomewide survey of developmentally relevant genes in *Ciona intestinalis*: VIII. Genes for PI3K signaling and cell cycle. *Dev. Genes Evol.* **213,** 284–290.

Kusakabe, T., Yoshida, R., Kawakami, I., Kusakabe, R., Mochizuki, Y., Yamada, L., Shin-i, T., Kohara, Y., Satoh, N., Tsuda, M., and Satou, Y. (2002). Gene expression profiles in tadpole larvae of *Ciona intestinalis*. *Dev. Biol.* **242,** 188–203.

Lee, Y.-H., Huang, G. M., Cameron, R. A., Graham, G., Davidson, E. H., Hood, L., and Britten, R. J. (1999). EST analysis of gene expression in early cleavage-stage sea urchin embryos. *Development* **126,** 3857–3867.

Makabe, K. W., Kawashima, T., Kawashima, S., *et al.* (2001). Large-scale cDNA analysis of the maternal genetic information in the egg of *Halocynthia roretzi* for a gene expression catalog of ascidian development. *Development* **128,** 2555–2567.

Nakashima, K., Yamada, L., Satou, Y., Azuma, J.-I., and Satoh, N. (2004). The evolutionary origin of animal cellulose synthase. *Dev. Genes Evol.* **214**, 81–88.

Nishida, H. (1987). Cell lineage analysis in ascidian embryos by intracellular injection of a tracer enzyme. III. Up to the tissue restricted stage. *Dev. Biol.* **121,** 526–541.

Nishida, H. (2002a). Patterning the marginal zone of early ascidian embryos: Localized maternal mRNA and inductive interactions. *BioEssays* **24,** 613–624.

Nishida, H. (2002b). Specification of developmental fates in ascidian embryos: Molecular approach to maternal determinants and signaling molecules. *Int. Rev. Cytol.* **217,** 227–276.

Nishida, H., and Satoh, N. (1985). Cell lineage analysis in ascidian embryos by intracellular injection of a tracer enzyme. II. The 16- and 32-cell stages. *Dev. Biol.* **110,** 440–454.

Nishikata, T., Yamada, L., Mochizuki, Y., Satou, Y., Shin-i, T., Kohara, Y., and Satoh, N. (2001). Profiles of maternally expressed genes in fertilized eggs of *Ciona intestinalis. Dev. Biol.* **238,** 315–331.

Ogasawara, M., Sasaki, A., Metoki, H., Shin-i, T., Kohara, Y., Satoh, N., and Satou, Y. (2002). Gene expression profiles in young adult *Ciona intestinalis. Dev. Genes Evol.* **212,** 173–185.

Pearson, W. R., and Lipman, D. J. (1988). Improved tools for biological sequence comparison. *Proc. Natl. Acad. Sci. USA* **85,** 2444–2448.

Satoh, N. (1994). "Developmental Biology of Ascidians." Cambridge University Press, New York.

Satoh, N. (2001). Ascidian embryos as a model system to analyze expression and function of developmental genes. *Differentiation* **68,** 1–12.

Satoh, N. (2003). The ascidian tadpole larva: Comparative molecular development and genomics. *Nature Rev. Genet.* **4,** 285–295.

Satoh, N., and Jeffery, W. R. (1995). Chasing tails in ascidians: Developmental insights into the origin and evolution of chordates. *Trends Genet.* **11,** 354–359.

Satoh, N., Satou, Y., Davidson, B., and Levine, M. (2003). *Ciona intestinalis*: An emerging model for whole-genome analyses. *Trends Genet.* **19,** 376–381.

Satou, Y., Kusakabe, T., Araki, I., and Satoh, N. (1995). Timing of initiation of muscle-specific gene expression in the ascidian embryo precedes that of developmental fate restriction in lineage cells. *Dev. Growth Differ.* **37,** 319–327.

Satou, Y., Takatori, N., Yamada, L., Mochizuki, Y., Hamaguchi, M., Ishikawa, H., Chiba, S., Imai, K., Kano, S., Murakami, S. D., Nakayama, A., Nishino, A., Sasakura, Y., Satoh, G., Shimotori, T., Shin-i, T., Shoguchi, E., Suzuki, M. M., Takada, N., Utsumi, N., Yoshida, N., Saiga, H., Kohara, Y., and Satoh, N. (2001a). Gene expression profiles in *Ciona intestinalis* tailbud embryos. *Development* **128,** 2893–2904.

Satou, Y., Imai, K. S., and Satoh, N. (2001b). Early embryonic expression of a LIM-homeobox gene *Cs-lhx3* is downstream of β-catenin and responsible for the endoderm differentiation in *Ciona savignyi* embryos. *Development* **128,** 3559–3570.

Satou, Y., Imai, K. S., and Satoh, N. (2001c). Action of morpholinos in *Ciona* embryos. *Genesis* **30,** 103–106.

Satou, Y., Takatori, N., Fujiwara, S., Nishikata, T., Saiga, H., Kusakabe, T., Shin-i, T., Kohara, Y., and Satoh, N. (2002a). *Ciona intestinalis* cDNA projects: Expressed sequence tag analyses and gene expression profiles during embryogenesis. *Gene* **287,** 83–96.

Satou, Y., Yamada, L., Mochizuki, Y., Takatori, N., Kawashima, T., Sasaki, A., Hamaguchi, M., Awazu, S., Yagi, K., Sasakura, Y., Nakayama, A., Ishikawa, H., Inaba, K., and Satoh, N. (2002b). A cDNA resource from the basal chordate *Ciona intestinalis. Genesis* **33,** 153–154.

Satou, Y., Imai, K. S., and Satoh, N. (2002c). *Fgf* genes in the basal chordate *Ciona intestinalis. Dev. Genes Evol.* **212,** 432–438.

Satou, Y., Imai, K. S., Levine, M., Kohara, Y., Rokhsar, D., and Satoh, N. (2003a). A genomewide survey of developmentally relevant genes in *Ciona intestinalis*: I. Genes for bHLH transcription factors. *Dev. Genes Evol.* **213,** 213–221.

Satou, Y., Kawashima, T., Kohara, Y., and Satoh, N. (2003b). Large scale EST analysis in *Ciona intestinalis*: Its applications as Northern blot analyses. *Dev. Genes Evol.* **213,** 314–318.

Satou, Y., Murakami, S. D., Kawashima, T., and Satoh, N. (2004). Dynamic changes in developmental gene expression in a basal chordate *Ciona intestinalis*. *Dev. Biol.* Submitted.

Takahashi, H., Hotta, K., Erives, A., Di Gregorio, A., Zeller, R. W., Levine, M., and Satoh, N. (1999). *Brachyury* downstream notochord differentiation in the ascidian embryo. *Genes Dev.* **13,** 1519–1523.

Yagi, K., Satou, Y., Mazet, F., Shimeld, S. M., Degnan, B., Rokhsar, D., Levine, M., Kohara, Y., and Satoh, N. (2003). A genomewide survey of developmentally relevant genes in *Ciona intestinalis*: III. Genes for Fox, ETS, nuclear receptors, and NFκB. *Dev. Genes Evol.* **213,** 235–244.

Yamada, L., Shoguchi, E., Wada, S., Kobayashi, K., Mochizuki, Y., Satou, Y., and Satoh, N. (2003). Morpholino-based gene knockdown screen of novel genes with developmental function in *Ciona intestinalis*. *Development* **130,** 6485–6495.

Yasuo, H., and Satoh, N. (1993). Function of vertebrate *T* gene. *Nature* **364,** 582–583.

Yasuo, H., and Satoh, N. (1998). Conservation of the developmental role of *Brachyury* in notochord formation in a urochordate, the ascidian *Halocynthia roretzi*. *Dev. Biol.* **200,** 158–170.

CHAPTER 32

Gene Regulatory Network Analysis in Sea Urchin Embryos

Paola Oliveri and Eric H. Davidson

Division of Biology
California Institute of Technology
Pasadena, California 91125

I. Introduction

Gene regulatory networks (GRNs) provide logic maps of the control functions that direct development, and they relate these maps directly to the genomic regulatory sequence. A GRN, if not too incomplete, indicates what inputs into each regulatory gene are responsible for its spatial and temporal patterns of expression. It thereby explains the progression of regulatory states that underlies spatial specification of domain fate in development, and thereafter the institution of regional cell differentiation. Additionally, it shows how intercellular signaling and developmentally important maternal anisotropies are linked into the gene regulatory apparatus of the embryonic cells, that is, why signaling interactions and localizations in the egg affect the developmental process. As will be discussed, a properly constructed GRN has an immense predictive, as well as explanatory,

METHODS IN CELL BIOLOGY, VOL. 74

0091-679X/04 $35.00

potentiality. The sea urchin embryo has turned out to be a material of choice for the system-level, high-throughput kinds of research required for GRN analysis. A large-scale provisional GRN for embryonic endomesoderm specification has been worked out for sea urchin (*Strongylocentrotus purpuratus*) embryos (Davidson *et al.*, 2002a,b, 2003; for current version, see http://www.its.caltech.edu/~mirsky/endomes.htm). In this chapter, we consider special methodologies developed for GRN analysis, and for the use of GRNs, that are not dealt with elsewhere in this volume.

To work out the architecture of a GRN, we need to determine how each regulatory gene specifically involved in a developmental process is linked into the network. This requires many forms of experimental information. A list of the kinds of information that goes into the process of solving a GRN must include the following:

- Identification of the genes involved: these may be recovered by differential screening to isolate those genes expressed specifically in the relevant developmental context, particularly genes encoding transcription factors and signaling molecules (Calestani *et al.*, 2003; Ransick *et al.*, 2002; Rast *et al.*, 2000, 2002; see Chapter 30 of Cameron *et al.* in this volume).
- Exact evidence on when and where the gene is expressed: such evidence requires detailed, whole-mount *in situ* hybridization (WMISH) observations as well as quantitative measurement of the time course of transcript accumulation (see Chapter 24 of Ransick in this volume).
- Evidence on *cis*-regulatory inputs: this type of evidence is typically obtained by direct gene transfer experiments. Sea urchin eggs and embryos provide easy and accessible material for high-throughput *cis*-regulatory analysis (e.g., Kirchhamer *et al.*, 1996; Yuh *et al.*, 2001a,b, 2002; see Chapter 25 of Arnone *et al.*, this volume).
- The results of a complete as possible perturbation analysis: the activity of each individual gene or of different sets of genes is perturbed, and the consequences on transcript level are assayed for all other relevant genes in the GRN. Methods for perturbing gene experiments in sea urchin embryos are discussed in Chapter 28 of Angerer and Angerer in this volume.
- Computational analyses: to work with GRNs requires application of many computational methods (c.f. Brown *et al.*, 2002). These include methods for resolution of differential screens that include many thousands of clones, methods for detection of *cis*-regulatory elements by interspecific sequence comparison (Brown *et al.*, 2002; Yuh *et al.*, 2002; see Chapter 30 of Cameron *et al.*, this volume), and methods for annotating genomic sequence data. Ultimately, the GRN must be represented, simulated, and its functions analyzed in a computational model, but as this is written, that goal has not yet been accomplished for any real life-scale developmental GRN.

In this chapter, we focus on two specific aspects of GRN analysis. Our main objective is to outline the design and execution of measurements of the effects of gene expression perturbations using multiplexed quantitative PCR (QPCR). This method permits the consequences of given gene perturbations to be determined at different developmental times for dozens of other genes. Second, we discuss how a GRN can be used, by generating and testing its implicit predictions.

II. Perturbation Analysis

A. General Explanation

In this section, we describe the general strategies used in our lab to perturb the expression and/or function of a gene, and to measure the consequent effects on any other possible target genes. We compare the effects caused by different perturbation approaches and describe the tools used for accurate quantification of gene expression. We also provide some general tools for data processing and for the calculation of the relative abundance and the absolute number of transcripts per embryo.

For the study of gene regulatory networks, analysis of a single gene and its effects on a few other downstream genes is never sufficient. A quantitative high-throughput approach is instead required. The methodologies and techniques adopted must be sensitive enough to quantify variations in transcript abundance for genes usually expressed at relatively low levels. For sea urchin embryos under normal conditions, a typical, spatially restricted transcription factor is expressed in the range of a mere hundred to several thousand transcripts per embryo (Davidson, 1986; Lee *et al.*, 1999; Poustka *et al.*, 1999). The quantification can be either relative or absolute; the results must be accurate, consistent, and reproducible; ideally, the whole procedure must not be too time-consuming. A suitable technique for this task is real time QPCR. This method has the sensitivity to detect two-fold differences between samples by means of automated measurement of the kinetics of amplification in the early phases of the PCR reaction (for technical details of instrumental performance in these respects, see the website www.appliedbiosystems.com). As described by Ransick in Chapter 24, in the specific case of the endomesoderm sea urchin embryo, one to five embryos per amplification reaction are sufficient for an accurate estimate of relative abundance of any specific transcript.

The alternative approach to measurement of relative change in RNA levels is the use of DNA microarray hybridization. Microarrays consist of short synthetic oligos or cDNA fragments spotted on a glass slide or membrane (www.affymetrix.com,www.rii.com). Microarrays are used as templates for hybridization with complex, fluorescently labeled cDNA probes derived from the samples between which a comparison is desired. The main advantage of microarray technology is its capability of analyzing several thousand genes at the same time. On the other hand, there are also severe disadvantages. One problem is that the amount of starting material necessary to synthesize a standard complex microarray probe is on the order of several micrograms of total RNA (Shulze and Downward, 2001), though, of course, an appropriate amplification technology can be applied to solve this problem (e.g., see Rast *et al.*, 2000). However, one microgram of total RNA corresponds to 300 to 400 embryos, so where multiple samples are required, as for replicates or timepoints, and the embryos are hand-manipulated (i.e., injected), these quantitative RNA requirements are prohibitive. There is a second, and even more important difficulty for the particular purposes of GRN analysis: since transcription factors are usually encoded by rare transcripts, accurate analysis of their expression profiles demands an extremely high sensitivity that in total animal embryo cDNAs is difficult to achieve with conventional microarray technology. For these reasons, QPCR is the technique of choice and will be focused on here.

B. Detailed Methods

1. Perturbation of Gene Expression

The key strategy for GRN analysis centers around the specific alteration of expression of a single regulatory gene and the measurement of the consequences on all potential target genes. Table I lists the different approaches used for perturbing gene function that have been used in studying the sea urchin endomesoderm GRN. Here, we briefly describe their mode of action and the different controls used.

The most successful approach to knockdown specific gene expression employs the use of antisense synthetic oligonucleotides and, more specifically, of

Table I
Perturbation Agents Used for the Sea Urchin Gene Regulatory Network Analysis

Perturbation	Mode of Action	Effects	References
Morpholino-antisense oligonucleotide (MASO)	Complementary to the 5′UTR or ATG, blocks translation	No targeted protein is produced	Angerer *et al.*, 2001 and Chapter 28
mRNA injection (MOE)	Synthetic capped transcripts coding for TF	Global ectopic expression of a spatially restricted TF	For review, see Chapter 27
Fusion with engrailed repressor domain (En)	Engrailed repressor domain fused to a specific DNA binding domain transforms the TF in obligate repressor	Repression of genes normally regulated by the TF	Li *et al.*, 1999, and Chapter 27
Clonal expression	Expression of a TF coding sequence under an exogenous promoter	Ectopic expression of a TF only in a restricted clone of cells	Rast *et al.*, 2002; Oliveri *et al.*, 2003
Cadherin	Blocks β-catenin nuclearization	Disrupts endomesoderm specification	Logan *et al.*, 1999
Dn-Notch	Blocks notch signaling	Disrupts secondary mesenchyme specification	Sherwood and McClay, 1999

morpholino-substituted antisense oligos (MASO; Gene Tools, Inc., Philomath, OR). The oligonucleotides are synthesized to be complementary to the sequence around the translation start site of the target mRNA, and their mode of action is to inhibit the translation of the mRNA (Summerton and Weller, 1997; for review, see Heasman, 2002). The result is absence or severe decrease in the amount of the protein encoded by the targeted mRNA. In our experience, MASOs targeted against mRNAs encoding transcription factors almost always produce a specific phenotype. For use in sea urchin embryos, MASOs are injected in the one cell zygote. The useful concentrations vary, depending on the mRNA target. The range of effective concentrations extends in our experience all the way from 10 to 300 μM MASO in the injection solution, and preliminary experiments are in order to determine the best operating concentration in each new case. Controls for specificity are obtained by injecting either a standard control MASO (available from Gene Tools, Philomath, OR), the sequence of which is 5′-CCTCTTACCT-CAGTTACAATTTATA-3′; an inverted antisense sequence; a sense sequence; or an antisense sequence that includes at least 4 mismatches. The efficacy of the MASO can be assessed for control purposes by co-injection of the MASO with a synthetic mRNA coding for GFP that has been fused with the natural mRNA sequence around the translation start site (roughly, 300–500 bp) of the targeted mRNA, or by synthesizing a second MASO against a different sequence of the same RNA (Davidson *et al.*, 2002a; Howard *et al.*, 2001). The two different MASOs should lead to identical effects. For further experimental details, see Chapter 28 of this volume, and the Gene Tools web site (www.gene-tools.com). It is to be stressed that a limit of the MASO approach is that it cannot affect proteins already present maternally. Of course, this same feature means that MASOs can be used to discriminate between maternal and zygotic phases of gene expression.

The second and widely accepted perturbation agent of gene expression is mRNA overexpression (MOE). In this approach, mRNA encoding a transcription factor or signaling molecule is injected into the egg. Since the injected RNA diffuses to every part of the egg prior to first cleavage, the result is to create global ectopic expression or mRNA overexpression so that the protein is produced in every cell of the embryo. The phenotypes thus generated are often drastic. But a very serious class of artifact is likely to arise if excess transcription factor is introduced into the embryo, in that additional lower affinity target sites may be bound by the factor if it is present at concentrations more than a few fold higher than normal. To avoid this, the amount of injected mRNA must be quantified carefully and must be set relatively close to the normal levels of the endogenous mRNA per cell in which it is present (for review and technical tips on mRNA injection, see Chapter 27 of this volume).

Perturbation of gene expression can also be achieved when the injected mRNA encodes a negatively acting version of a transcription factor, such as can be obtained by fusing the DNA binding domain to the repressor domain of the Engrailed protein (En) (Li *et al.*, 1999; Oliveri *et al.*, 2002). This form of perturbation reagent transforms any transcription factor into an obligate repressor, causing the shutdown of all genes controlled by *cis*-regulatory elements to which the factor normally binds (whether it normally acts as an activator or a repressor). In the case that the DNA binding domain is well characterized, and specific mutations that reduce the affinity of the protein for the target DNA by several orders of magnitude are known, an optimal control is the introduction of an mRNA encoding the En fusion to the mutated version of the protein (Kenny *et al.*, 2003; Li *et al.*, 1999).

Of course, mRNAs encoding many different kinds of protein aside from transcription factors can be used in specific ways to perturb the process of development. Of particular importance in the analysis of sea urchin embryo GRNs has been the injection of *in vitro* transcribed RNA encoding the transmembrane and intracellular domains of Cadherin (Logan *et al.*, 1999). The global excess of Cadherin protein that ensues completely abolishes β-catenin nuclearization, an essential process necessary for the initial specification of the endomesoderm (Logan *et al.*, 1999). mRNAs encoding parts of signaling ligands or receptors have also been used, for example, to cause ectopic *wnt8* expression (A. Wikramanayake, *et al.*, 2004), or to block Delta-Notch signaling (Sherwood and McClay, 1999). mRNA encoding a Notch protein that lacks the intracellular domain traps the Delta ligand and thus interferes with Notch signaling. The injection of mRNA encoding such a protein into sea urchin eggs impairs secondary mesenchyme specification (Sherwood and McClay, 1999).

A different approach to altering gene expression experimentally is the use of clonally expressed constructs. Here, the sequence encoding a regulatory molecule is expressed ectopically under the control of an exogenous *cis*-regulatory element (Cavalieri *et al.*, 2003; Oliveri *et al.*, 2002; Rast *et al.*, 2002). The DNA construct is injected in the zygote along with a linearized control plasmid containing a marker gene (e.g., GFP) under the control of the same or an analogous *cis*-regulatory element. The two constructs co-integrate in a blastomere during very early cleavage, and the GFP expression serves as an easily detectable marker for the descendant clone of cells that ectopically expresses the regulatory molecule. This mode of producing ectopic expression has several important advantages compared to global MOE: First, only a minority of the cells of the embryo is subjected to the unnatural expression of the regulatory protein, resulting in a less developmentally compromised embryo and, hence, a much cleaner interpretation of the effects. This becomes a critically important factor in postgastrular stages, when developmental failures caused by earlier insults to the normal process tend to snowball and spread secondarily to additional spatial domains. Second, ectopic clonal expression can be driven specifically in any desired set of cells. Third, the timing of the clonal expression is similarly controlled according to the *cis*-regulatory element used, rather than starting automatically from the beginning of development. Fourth, the levels of mRNA produced by integrated expression constructs are in the same range of magnitude as for the endogenous gene.

2. Quantification of Transcript Levels

Figure 1 summarizes the major steps required for quantification of a given mRNA molecule starting from living embryos. The method depicted involves six steps, as follows.

1. The starting point is culturing and handling the embryos.
2. The samples are collected at different developmental stages or from embryos perturbed in different ways, as just described.
3. Total RNA is extracted from each sample.
4. A first strand cDNA is synthesized using random primers.
5. The amount of transcript of the desired species is measured by QPCR.
6. The values derived from the QPCR measurements are mathematically processed and translated into differences in relative abundance of transcripts among different samples or into absolute number of transcripts for each given timepoint in development.

In Chapter 24, Ransick gives detailed protocols for culturing and collecting of embryos, RNA isolation, cDNA synthesis, and QPCR (steps 1 to 5). He also highlights technical problems and provides practical information for the correct design of QPCR experiments. Here, we focus on mathematical reduction of the data (step 6).

The results of the PCR reactions are given in amplification plots (Fig. 1) that display cycle number versus the fluorescence produced by an indicator dye (i.e., SYBR Green), which binds to the double-stranded DNA product. The first step in analyzing raw data from PCR reactions is to set an arbitrary threshold. The number of cycles required to attain the threshold in a PCR reaction (C_T) is directly dependent on the input concentration of the target transcript (I_0), so long as the threshold is set within the exponential phase of the reaction. In practice, the threshold is usually set somewhere in the range between 0.2 and 0.6, where completion of the reaction is considered as 1.0. The threshold should be set so that the standard deviation around the C_T of replicates is minimal, that is, the plots of replicates overlie to the greatest degree (see Fig. 1). In the instruments used for QPCR measurements in our lab, the GeneAmp 5700 and 7900 HT Sequence Detection System (Applied Biosystems, Foster City, CA), the C_T values for the 96 or 384 reactions accommodated per plate are reported as a table of numbers. As summarized in Fig. 2, the C_T values from the identical replicates are averaged. For precise relative quantification comparing embryos treated in different ways, each averaged C_T value resulting from the amplification of a specific set of primers with a particular cDNA ($C_{T\ \text{sample}}$), must be normalized to an internal control amplified in the same cDNA sample ($C_{T\ \text{internal control}}$). This normalization is necessary in order to correct for minor differences in cDNA input. The internal control must be a transcript constantly present in the embryo mRNA, and not subject to variation under different perturbation conditions, so that the C_T value of the internal control depends directly on the amount of cDNA in each reaction. The

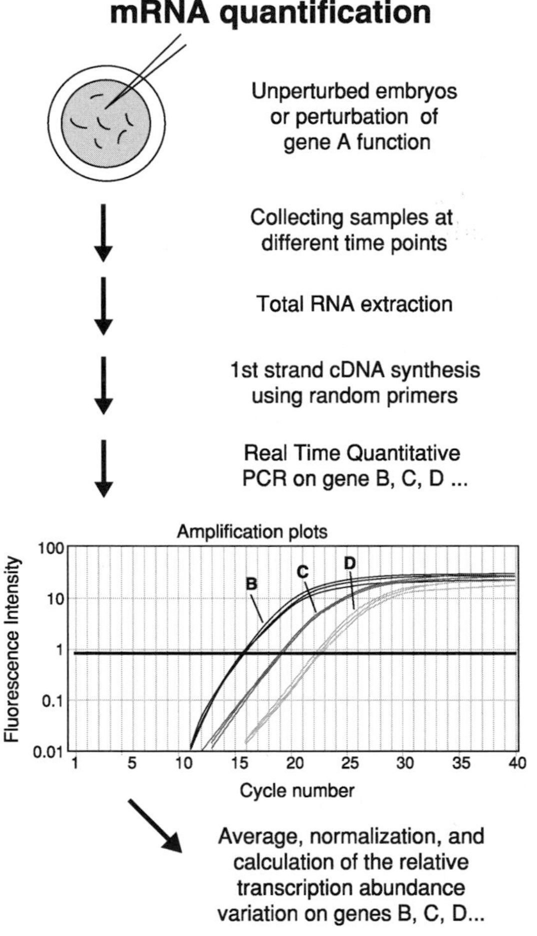

Fig. 1 Quantitative PCR (QPCR) analysis. Diagram showing the main steps involved for a quantitative analysis of gene expression. The embryo cultures are the living material used for RNA extraction, followed by cDNA synthesis and QPCR. The result is given in amplification plots where the values of fluorescence intensity are reported at each PCR cycle for each amplification reaction. Here, an example of an amplification plot for thee different genes (B, C, and D) is reported. In each experiment, the reactions for each gene are performed in triplicate and, from the amplification plot, it can be seen that the triplicates are almost perfectly coincident. The reproducibility of the amplification plot for each reaction shows that variations among experiments are external to the QPCR system. The numerical data obtained by QPCR technology are mathematically processed, and absolute or relative quantification of the number of transcripts is obtained, as described in the text.

internal controls used in our lab are ubiquitin mRNA and 18S rRNA. Ubiquitin expression is maintained at a constant level for the first 72 h of sea urchin development (Nemer *et al.*, 1991; A. Ransick, P. Oliveri, and E. Davidson, unpublished data). When using rRNA as a standard, the cDNA template is diluted

Mathematical formulas
for relative quantification of gene expression

Average of C_T values for replicates	$(C_{T\,replica1}+C_{T\,replica2}\ldots)/n$
Normalization of cDNA samples by reference to an internal control	$C_{T\,Sample} - C_{T\,Internal\,Control} = C_{T\,exp}$
Calculation of the amount of PCR product at threshold	$P = I_0 \cdot 1.94^{C_T}$
Quantification relative to a known transcript needed for calibration ("Control")	$P_{Exp}/P_{Control} = (I_0 \cdot 1.94^{C_T})_{Exp}/(I_0 \cdot 1.94^{C_T})_{Control}$
Folds of difference	$P_{Exp}/P_{Control} = 1.94^{\Delta C_T}$
For convenience	$\Delta C_T = C_{T\,Exp} - C_{T\,Control}$

C_T = number of cycles at threshold
n = number of replicas P = PCR product at cycle x
I_0 = initial transcripts
1.94 = efficiency of amplification at exponential phase

Fig. 2 Mathematical tools for the analysis of QPCR raw data. A brief guide for the processing of data coming from real time PCR reactions is provided.

500 times before the PCR reaction. The normalized averaged C_T value ($C_{T\,exp}$) is now ready to be compared to a similar value derived from the nonperturbed control sample. The comparison is expressed as the difference in cycles (ΔC_T) between the sample from the perturbed embryos and the control sample for each specific gene analyzed. This value (ΔC_T) can be converted to fold difference between experimental and control samples, since the amount of an amplified molecule in the exponential phase of the PCR reaction is directly dependent on the starting number of molecules of that species present in the cDNA, and since the primers will amplify the target sequence with the same efficiency (1.94-fold per cycle) in experimental and control cDNAs. A good control must show no difference in abundance of any transcript analyzed relative to normal embryos (called uninjected in Fig. 3). For this reason, each QPCR experiment must also be controlled for the batch of embryos used. Today, computational tools, particularly useful for high-throughput approaches, are available for easy mathematical and statistical analysis, storage, and graphical presentation of the data (Muller *et al.*, 2002).

The data analysis so far considered provides accurate relative quantification for the effects of a given perturbation on particular target transcript levels. Relative

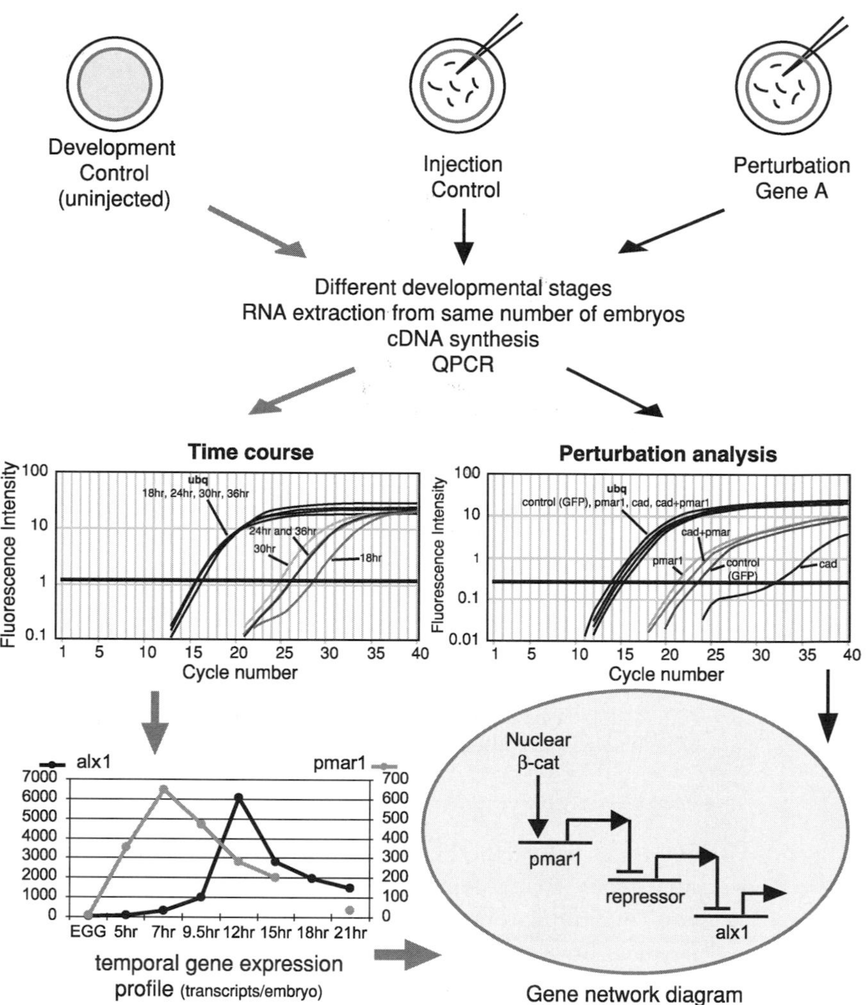

Fig. 3 QPCR measurements for GRN construction. Summary of the important steps required for comparison of different samples differently treated (black arrows) and for absolute quantification of the number of transcripts during normal development (gray arrows). Both strategies are necessary for building the network diagram and for understanding the relationships among regulatory genes. The transcript abundances are measured as described in Fig. 1, and the numerical data are processed as described in Fig. 2. This diagram shows two examples of amplification plots. The amplification plot on the left is an example of a time-course analysis for the gene *gsc* (P. Oliveri and E. Davidson, unpublished data; Angerer *et al.*, 2001) using *ubiquitin* (ubq) as internal control. The QPCR data can easily be translated into absolute number of transcripts by reference to an internal standard and plotted in a graph for easy visualization. The graph at the lower left shows the temporal expression profiles of the spatially restricted micromere genes *pmar1* (Oliveri *et al.*, 2002) and *alx1* (Ettensohn *et al.*, 2003). The graph shows the number of transcripts per embryo of both genes during the first 21 h of embryonic development. The amplification plot on the right shows the behavior of the *alx1* gene in embryos perturbed in different ways. The internal control used for normalizing each cDNA input in

abundance can be converted to absolute number of transcripts per embryo by using an internal standard. This is required, for example, in determining the quantitative temporal expression profile of a gene. An example of an absolute kinetic gene expression measurement is shown in Fig. 3. The internal standard most used for this purpose in our lab is the mRNA of the gene SpZ12.1, the abundance of which was quantified by single strand probe excess RNA titration (Lee and Costlow, 1987; Wang *et al.*, 1995). One useful aspect of a thorough knowledge of the temporal gene expression profile is that it affects the choice of timepoints to be analyzed in perturbation experiments. Thus, when studying the effect of a given regulatory factor on its target genes, it is important to quantify the effects of disrupting its function at, or shortly after, its normal climax of expression, that is, often well before the disruption causes any complex phenotypic changes to arise. The peak of expression is the most likely point when the factor is normally active, and phenotypes generally become evident only after much later secondary effects are superimposed.

C. Possible Problems

One of the main concerns with real time QPCR is the nature of the molecule amplified during the reaction. Everyone familiar with any kind of PCR knows that certain artifacts occur easily, for example, the accumulation of primer-dimers or amplification of nonspecific bands. It is extremely important in QPCR measurements that only the desired product be amplified. The C_T value obtained is a function of the fluorescence of dye interacting with all double-stranded DNA present in the reaction, and so the quantification is immediately impaired by the presence of adventitious or nonspecific double-stranded DNA in the reaction tube. The purity of the amplified fragment can usually be assessed by applying the dissociation protocol included in the GeneAmp 5700 and 7900HT Sequence Detection Systems (Applied Biosystems, Foster City, CA). The dissociation curve expresses the change in fluorescence as a function of temperature, that is, it tracks the release of the dye as the double-stranded DNA reaction product melts. The dissociation is plotted in derivative form, and the desired result is a single peak centered on the melting temperature (T_m) of a unique amplified DNA fragment. Dissociation curves can be used as a quality control for the QPCR reaction because different amplified fragments have different T_ms depending on their length

---

the PCR reaction is again *ubiquitin* mRNA (ubq). The different amplifications are performed in samples injected with *cadherin* mRNA (cad), *pmar1* mRNA (pmar1), *cadherin* and *pmar1* (cad + pmar1) mRNAs, and GFP mRNA as an injection control. The data resulting from the temporal profile and the perturbation analysis on *alx1*, together with other data from similar experiments, can be combined in a network diagram, as shown in the bottom right corner of this figure. This diagram shows how, though a double negative sequential input, pmar1 activates the *alx1* gene only in the micromere lineage; and it also indicates why introduction of *pmar1* in micromeres rescues the negative effect on *alx1* expression caused by inhibition of β-catenin nuclearization (via injection of *cadherin* mRNA).

and the nature of their sequence. Thus, artifacts are usually flagged by a dissociation curve that displays multiple peaks, representing multiple species of double-stranded DNA in the same reaction. Such a result can be confirmed by gel electrophoresis and, if need be, sequencing. If it occurs, new primers are in order, as those being employed are reacting with more than one sequence present in the mixture.

Another feature to be aware of is that amplification of very rare templates (Fewer than 100 molecules per reaction) drastically lowers the quantitative precision of the system, due mainly to decrease in the efficiency of amplification after about 30 cycles. This systematic change in amplification efficiency can be neglected when calculating relative abundance variation between two different samples deriving from differently treated embryos, especially if both contain the same low numbers of copies. But it can be drastically important in calculating absolute numbers of transcripts per embryo relative to a known internal standard. Ways to solve this problem are to use more starting material or to use a different amplification biochemistry, such as a different thermostable polymerase preparation (M. I. Arnone, personal communication).

When quantitative data are required in a comparison of different samples from differently treated embryos, it is very important that the C_Ts of the internal controls for each of the cDNA samples match at least within two cycles. Otherwise, the normalization for cDNA content in these exponential reactions cannot be expected to yield accurate values; it is absolutely necessary to compare cDNA preparations of about the same overall concentration per reaction tube. The most important parameter affecting cDNA content per sample is variation in RNA recovery. We generally use the same number of embryos for each sample (i.e., 100–500 embryos per sample), and the RNA extraction must be as efficient as possible. Ransick describes several methods for RNA extraction from small numbers of sea urchin embryos in Chapter 24 of this volume. This problem does not exist for large-scale RNA extractions, since here the total RNA used for cDNA synthesis can be precisely quantified by spectrophotometry so that equal amounts can be used for each PCR reaction.

QPCR is inherently quite accurate and reproducible. In our experience, the variation in cycle number at threshold is practically nil, and generally triplicate samples differ by only a few tenths of a cycle (Fig. 1). The only significant source of the experimental variation observed among experiments, if the experiments are executed according to the precepts described, is the inherent biological variation among individual embryo batches. Based on our experience in this area, we recommend imposition of an arbitrary threshold of significance in fold change between an experimental and a control sample, and only above this are differences considered real: we consider that a difference of 1.6 cycles in C_T, corresponding to ~3-fold change in transcript level, is likely to be significant, but that anything less is likely to fall within the expected experimental variation. Of course, this conservative standard practice could, in certain cases, mask potentially important and reproducible biological changes that for good reasons happen to affect transcript level less than 3-fold. Though great statistical depth can resolve the reality of small differences, in such situations, some additional independent evidence is clearly indicated (i.e., WMISH) if the treatment affects spatial expression.

III. Assembling and Testing the GRN Logic Map

A. Putting the Network Together

The mass of information required to solve a GRN can only be acquired over a period of time and, as in any research program, this is best done intelligently rather than blindly. That is, the experiments producing this information must be informed by the current state of GRN assembly, at each point as the work goes along. The alternative, trying to set up a formally complete set of measurements *a priori* and then trying to deconvolve the results later is not the best way to proceed, for it throws away that which is the essence of science: experimentally testing propositions, here, particular proposed GRN linkages.

The key problem in assembling a GRN is application of diverse constraints. No one source of evidence provides the constraints available from other sources of evidence. For example, a perturbation analysis might reveal that expression of Gene B requires expression of Gene A. But if WMISH reveals that Gene B is not expressed in the same cells as is Gene A, then the relationship between them must depend on an intercellular signaling pathway. On the other hand, the two genes could be expressed in the same cells, with temporal relations that are plausibly explained if Gene A encodes a direct transcriptional activator of Gene B. In this case, the testable prediction is that the *cis*-regulatory element responsible for the relevant phase of activity of Gene B must include necessary target sites for the transcription factor encoded by Gene A. The constraint is applied by performing *cis*-regulatory analysis on the relevant module of Gene B. If this shows the proposition false, then the relation between Genes A and B is indirect. So expression data, perturbation data, and *cis*-regulatory data all provide unique constraints for one another, that when applied together uniquely resolve GRN architecture. The obverse of this coin is that it is futile to attempt to erect a GRN by entirely computational means using only one form of data, e.g., perturbation data or gene expression arrays.

A developmental GRN is not an edifice composed solely of relationships devolving from new experimental data, seated upon a smooth, blank surface. Rather, it must be built up from a deep foundation of biological knowledge. The metaphor is further useful: The developmental biology of well-known systems lays out the floor plan of the GRN edifice, so to speak. It tells us where each structure of the embryo comes from, where the intercellular signaling functions are required, and what they determine, and usually these days it provides a starting list of genes involved in the development of each spatial domain. The biology of the system, in other words, provides the most general and important constraint,

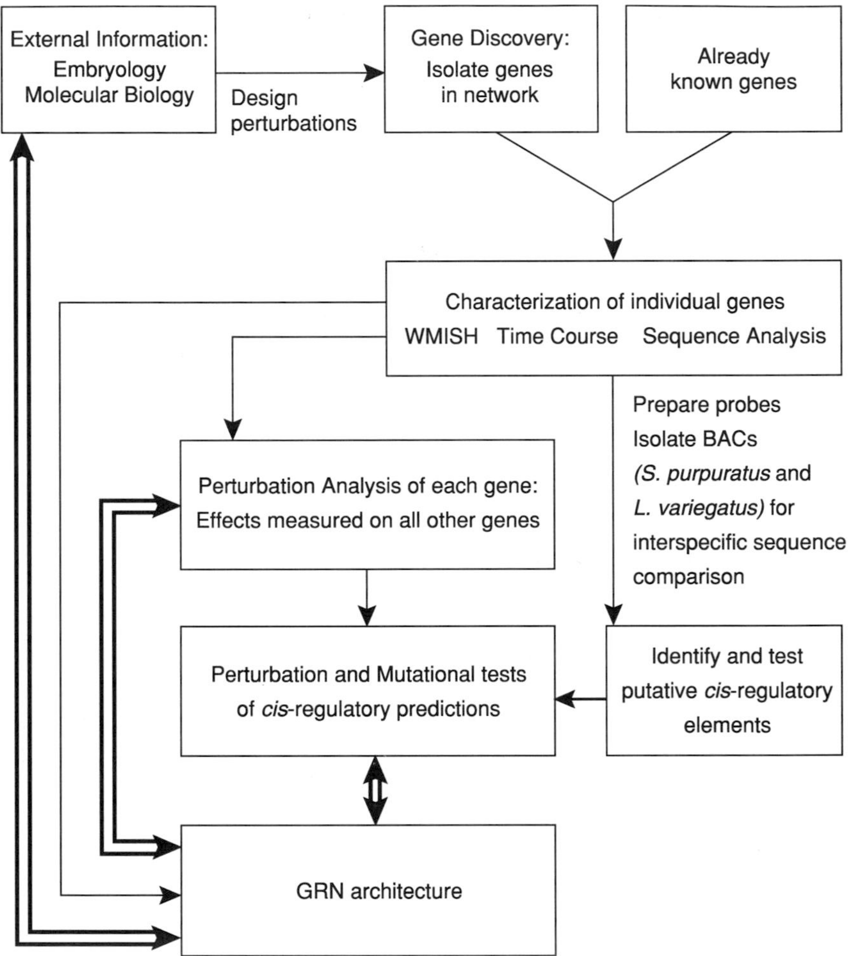

Fig. 4 Use of various forms of evidence in the solution of the sea urchin endomesoderm GRN and their continuing and recursive interplay (modified from Brown *et al.*, 2002).

for the GRN must fit onto the biology, must explain, and must predict the phenomena that constitute a biological description.

In Fig. 4 is a diagram (simplified from Brown *et al.*, 2002) that indicates the recursive process by which evidential constraints are applied to the process of GRN assembly. The ultimate strength of the analysis is that by resolving contradictions and eliminating alternative relationships, it will converge on the solution. Each node of a GRN consists of a regulatory gene, the inputs to which derive from other elements in the GRN, and the outputs of which go to other genes within the GRN. Because the inputs and outputs at each node are always multiple, the position of the node in the architecture will eventually be found to be determined by multiple items of evidence.

Following are some working tips based on our experience in assembling the endomesoderm GRN:

- Perturbations by overexpression of mRNA (MOE): A class of experiment that is particularly difficult to interpret is MOE producing a global excess of a positively acting transcription factor. As discussed in the previous section, it is mandatory in this kind of experiment to ensure that the number of exogenous mRNA molecules per cell is well less than an order of magnitude different from what is normally present at the relevant stage. The difficulties in interpreting MOE experiments are compounded by secondary effects, if phenotypic or QPCR effects are observed at a late stage, after gastrulation, for instance.
- MOE for study of repression: A profitable use of an MOE experiment is the comparison of the effects of injecting mRNA encoding a transcription factor and mRNA encoding an Engrailed domain fusion constructed using the DNA binding domain of that factor (e.g., Li *et al.*, 1999; Oliveri *et al.*, 2002). If the factor normally acts as a repressor, the two experiments will yield the same result; otherwise, they will yield opposite results.
- Interpretation of Engrailed domain fusion perturbations: When these reagents interact directly with a target *cis*-regulatory element, they often produce a very dramatic shutdown of transcript level ($\geq$97% knockdown of expression, i.e., $\Delta C_T \geq 5$; many examples are to be seen in the QPCR data on the previously cited Web site). A sure sign that an engrailed domain fusion effect is, on the other hand, indirect, is if the level of expression of the putative target gene is not affected at earlier times, and a decrease in expression is only observed later, for example, after 24 h. Once an Engrailed domain fusion sits down on a target site, it will silence transcription, whether that site is normally serviced by the factor at that time or not. Nor does a direct effect of an Engrailed fusion indicate the sign of the normal input (i.e., whether it acts as an activator or repressor), only that there is a specific target site for that DNA-binding domain.
- Clonal expression as a tool for GRN analysis: In sea urchin embryos, exogenous DNA injected into the egg is stably incorporated in a clonal manner. As has been discussed, this affords the opportunity to express, or prevent the expression, of a given gene in a clone of cells marked by GFP activity (Arnone *et al.*, 1997; Rast *et al.*, 2002). The downstream effects of the regulatory state imposed within the clone can now be observed by assaying expression of other genes using WMISH (Oliveri *et al.*, 2003). This method of ectopic expression avoids the gross, nonphysiological levels of transcription factor expression often seen in conventional MOE experiments. It also avoids the secondary, if not tertiary and quaternary, effects of the global disorganization of development.

• Regulatory gene functions: It is the rule, not the exception, that given transcription factors play multiple roles in development. Some regulatory genes are indeed expressed only in a given lineage (an example in the sea urchin embryo is *gcm*, a gene activated in response to Notch signaling in pigment cell precursors, that continues to be used in these cells; Calestani *et al.*, 2003; Ransick *et al.*, 2002). Some genes are expressed transiently for only a few hours in only one place (e.g., *pmar1*; Oliveri *et al.*, 2002). But more usually, a regulatory gene will be expressed at multiple times, and often in multiple territories, under the control of multiple *cis*-regulatory elements (e.g., see Amore *et al.*, 2003; Davidson *et al.*, 2002a,b; Minokawa *et al.*, 2004; Yuh *et al.*, 2002). For such genes, it is particularly important to interpret QPCR perturbation results in respect to the particular phase of expression relevant to the timeframe of the observation.

• Rescue experiments: An attractive means of determining if a perturbation effect is direct is to attempt a rescue of a knockout phenotype by introduction of the mRNA encoded by the presumed target gene. But, for a variety of reasons which cannot be elaborated here, global rescue experiments of this kind (i.e., injection of the rescue mRNA into the egg) rarely work well, if at all. A very powerful alternative is to do the experiment in a few blastomeres and transplant them to an otherwise normal recipient egg. Like clonal analysis, this avoids the supernumerary effects of gross developmental disturbance and of regulative readjustment. For example, a total rescue of the inactivation of the *pmar1 cis*-regulatory system in micromeres that occurs if the β-catenin/Tcf input is blocked was obtained in the following experiment (Oliveri *et al.*, 2003): *cadherin* mRNA, or *cadherin* mRNA + *pmar1* mRNA, were injected into donor eggs. Micromeres were then removed from these eggs and used to replace the micromeres of a normal recipient embryo. Embryos containing only the *cadherin* donor micromeres produce no skeleton; embryos containing micromeres which bear both *cadherin* and *pmar1* mRNAs generate normal skeleton. The *pmar1* gene is a direct target of the β-catenin/Tcf input blocked by Cadherin, and the experiment demonstrates conclusively that the product of this gene rescues skeletogenic functions in micromere descendants and also all other micromere functions.

B. Tests of *cis*-Regulatory Predictions

At one level, a GRN consists of a matrix of predicted *cis*-regulatory inputs, predicted logic functions executed in response to these inputs (for review of *cis*-regulatory logic processing, see Davidson, 2001), and predicted space/time regulatory outputs. This is the level at which the GRN directly addresses the genomic regulatory source code for development (Davidson *et al.*, 2003), ultimately in terms of As, Ts, Gs, and Cs. It is also the level at which the GRN can be subjected to direct experimental verification or, if need be, correction. By determining that the predicted inputs exist and are required in the relevant *cis*-regulatory elements, the GRN architecture may be confirmed, and issues of direct vs indirect linkages can be settled. Figure 5 describes the test process, with reference to a canonical

A) *cis*-Regulatory a prediction:

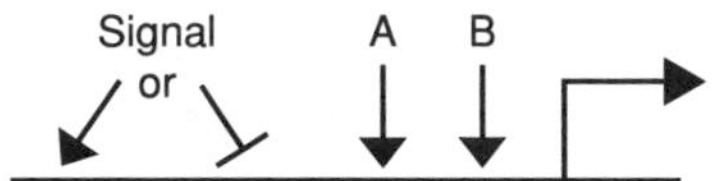

Perturbation analysis indicates a signal-mediated gate that represses the element (bar) except in cells where the signal is received. There, the two positive regulators (arrowheads) are permitted to activate the gene.

B) Analysis Steps:

1. Find *cis*-regulatory element.
2. Determine whether it responds to perturbation of signals and to A and B inputs, as does the endogenous gene.
3. Mutate the presumed target site and test for perturbation response: if the sites responsible are the predicted sites, QED.
4. If not, determine by mutation which sites mediate the response, and correct GRN architecture.
5. Determine by *cis*-regulatory response to mutations what logic functions the tested sites mediate.

Fig. 5 Stages of analysis of a canonical *cis*-regulatory element predicted at a GRN node.

GRN node (but one quite similar to several encountered in real life in the endomesoderm GRN).

The first step is to isolate the relevant *cis*-regulatory element. This is a fragment of DNA that, when associated with a reporter gene and introduced into an embryo, drives expression in the same pattern as described by the endogenous gene at that node of the GRN. We have been 100% successful in identifying desired *cis*-regulatory elements by means of interspecific sequence comparison. The elements sought are always among the conserved sequence patches and are identified by cloning them out and testing their activity in gene transfer experiments (Brown *et al.*, 2002; Yuh *et al.*, 2002; see Chapter 30 of Cameron *et al.*, this volume).

The second step is to determine whether the *cis*-regulatory element has the right stuff, that is, to assess its spatial response when it is injected into an embryo in which one of its presumed inputs has been perturbed. Figure 5 indicates that an α-MASO would depress expansion of the endogenous gene, so it must also depress expression of the reporter construct. Similarly, interruption of the signal input will also depress expression and ectopic expression of the signal would generate ectopic *cis*-regulatory construct expression whenever A and B inputs are also present.

The third step is to mutate the relevant sites, and determine whether the perturbation response of the expression construct is lost. For example, in the Fig. 5 case, mutation of the sites occupied by the transcription factor mediating the signal response should also cause ectopic expression wherever A and B inputs exist (Yuh *et al.*, 2004).

Suppose, in this example, that mutation of the site for the A input decreases response to <10% of control, and so does mutation of the site for B input. It would then be clear that the element is an "and" logic processor (Step 5 of Fig. 5): It is barely active unless A and B inputs are both present. The implication would be that the overlap in space of these two inputs is what determines where the gene is expressed, subject to the spatial and temporal constraint imposed by the signal-mediated gate.

IV. Summary

It may safely be predicted that GRN analysis will become increasingly important. It will come to underlie the causal study of development, the major effort underway to understand the regulatory code built into animal genomes and also the evolution of these genomes.

Partly by serendipity, sea urchin embryos turn out to be a superb experimental material for GRN analysis. Their natural properties have, in turn, influenced the predilections of those who work on them, and between them and us, so to speak, this is now a developmental system of which we are rapidly gaining an unusually complete understanding. The causal linkages that control development of the whole embryo will be revealed, leading all the way from the heritable genomic regulatory code to the events of embryology. The fundamental experimental operation is the perturbation analysis: Here is where causality permeates the exploration. We have in this chapter summarized in some detail the requirements for perturbation GRN analysis in sea urchin embryos. But that is not all, nor is it enough to enable the assembly of a GRN: What is required is the combined application of elegant computational methods, of gene regulation molecular biology, of genomic sequence data, and of experimental embryology. As the results crystallize together, we can begin to see how far this powerful combination of methods and ideas is going to carry us.

Acknowledgments

P. O. was supported by the Camilla Chandler Frost Fellowship. Research was supported by NIH grant HD37105, by the Office of Science (BER), U. S. Department of Energy, grant DE-FG02-03ER63584, and by the Caltech Beckman Institute.

References

Amore, G., Yavrouian, R. G., Peterson, K. J., Ransick, A., McClay, D. R., and Davidson, E. H. (2003). Spdeadringer, a sea urchin embryo gene required separately in skeletogenic and oral ectoderm gene regulatory networks. *Dev. Biol.* **261,** 55–81.

Angerer, L. M., Oleksyn, D. W., Levine, A. M., Li, X., Klein, W. H., and Angerer, R. C. (2001). Sea urchin goosecoid function links fate specification along the animal–vegetal and oral–aboral embryonic axes. *Development* **128,** 4393–4404.

Arnone, M. I., Bogarad, L. D., Collazo, A., Kirchhamer, C. V., Cameron, R. A., Rast, J. P., Gregorians, A., and Davidson, E. H. (1997). Green Fluorescent Protein in the sea urchin: New experimental approaches to transcriptional regulatory analysis in embryos and larvae. *Development* **124,** 4649–4659.

Brown, C. T., Rust, A. G., Clarke, P. J., Pan, Z., Schilstra, M. J., De Buysscher, T., Griffin, G., Wold, B. J., Cameron, R. A., Davidson, E. H., and Bolouri, H. (2002). New computational approaches for analysis of *cis*-regulatory networks. *Dev. Biol.* **246,** 86–102.

Calestani, C., Rast, J. P., and Davidson, E. H. (2003). Isolation of pigment cell specific genes in the sea urchin embryo by differential macroarray screening. *Development* **130,** 4587–4596.

Cavalieri, V., Spinelli, G., and Di Bernardo, M. (2003). Impairing Otp homeodomain function in oral ectoderm cells affects skeletogenesis in sea urchin embryos. *Dev. Biol.* **262,** 107–118.

Davidson, E. H. (1986). "Gene Activity in Early Development." Academic Press, Orlando, FL.

Davidson, E. H. (2001). "Genomic Regulatory Systems. Development and Evolution." Academic Press, San Diego, CA.

Davidson, E. H., Rast, J. P., Oliveri, P., Ransick, A., Calestani, C., Yuh, C. H., Minokawa, T., Amore, G., Hinman, V., Arenas-Mena, C., Otim, O., Brown, C. T., Livi, C. B., Lee, P. Y., Revilla, R., Rust, A. G., Pan, Z., Schilstra, M. J., Clarke, P. J., Arnone, M. I., Rowen, L., Cameron, R. A., McClay, D. R., Hood, L., and Bolouri, H. (2002a). A genomic regulatory network for development. *Science* **295,** 1669–1678.

Davidson, E. H., Rast, J. P., Oliveri, P., Ransick, A., Calestani, C., Yuh, C. H., Minokawa, T., Amore, G., Hinman, V., Arenas-Mena, C., Otim, O., Brown, C. T., Livi, C. B., Lee, P. Y., Revilla, R., Schilstra, M. J., Clarke, P. J., Rust, A. G., Pan, Z., Arnone, M. I., Rowen, L., Cameron, R. A., McClay, D. R., Hood, L., and Bolouri, H. (2002b). A provisional regulatory gene network for specification of endomesoderm in the sea urchin embryo. *Dev. Biol.* **246,** 162–190.

Davidson, E. H., McClay, D. R., and Hood, L. (2003). Regulatory gene networks and the properties of the developmental process. *Proc. Natl. Acad. Sci. USA* **100,** 1475–1480.

Ettensohn, C. A., Illies, M. R., Oliveri, P., and De Jong, D. L. (2003). Alx1, a member of the Cart1/Alx3/Alx4 subfamily of Paired-class homeodomain proteins, is an essential component of the gene network controlling skeletogenic fate specification in the sea urchin embryo. *Development* **130,** 2917–2928.

Heasman, J. (2002). Morpholino oligos: Making sense of antisense? *Dev. Biol.* **243,** 209–214.

Howard, E. W., Newman, L. A., Oleksyn, D. W., Angerer, R. C., and Angerer, L. M. (2001). SpKrl: A direct target of beta-catenin regulation required for endoderm differentiation in sea urchin embryos. *Development* **128,** 365–375.

Kenny, A. P., Oleksyn, D. W., Newman, L. A., Angerer, R. C., and Angerer, L. M. (2003). Tight regulation of SpSoxB factors is required for patterning and morphogenesis in sea urchin embryos. *Dev. Biol.* **261,** 412–425.

Kirchhamer, C. V., and Davidson, E. H. (1996). Spatial and temporal information processing in the sea urchin embryo: Modular and intramodular organization of the CyIIIa gene *cis*-regulatory system. *Development* **122,** 333–348.

Lee, J. J., and Costlow, N. A. (1987). A molecular titration assay to measure transcript prevalence levels. *Methods Enzymol.* **152,** 633–648.

Lee, Y. H., Huang, G. M., Cameron, R. A., Graham, G., Davidson, E. H., Hood, L., and Britten, R. J. (1999). EST analysis of gene expression in early cleavage-stage sea urchin embryos. *Development* **126,** 3857–3867.

Li, X., Wikramanayake, A. H., and Klein, W. H. (1999). Requirement of SpOtx in cell fate decisions in the sea urchin embryo and possible role as a mediator of beta-catenin signaling. *Dev. Biol.* **212,** 425–439.

Logan, C. Y., Miller, J., Ferkowicz, M., and McClay, D. (1999). Nuclear beta-catenin is required to specify vegetal cell fates in the sea urchin embryo. *Development* **126,** 345–357.

Minokawa, T., Rast, J. P., Arenas-Mena, C., Franco, C. B., and Davidson, E. H. (2004). Expression patterns of four different regulatory genes that function during sea urchin development. *Gene Exp. Patterns* **4,** 449–456.

Muller, P. Y., Janovjak, H., Miserez, A. R., and Dobbie, Z. (2002). Processing of gene expression data generated by quantitative real-time RT-PCR. *Biotechniques* **32,** 1372–1379.

Nemer, M., Rondinelli, E., Infante, D., and Infante, A. A. (1991). Polyubiquitin RNA characteristics and conditional induction in sea urchin embryos. *Dev Biol.* **145,** 255–265.

Oliveri, P., Carrick, D. M., and Davidson, E. H. (2002). A regulatory gene network that directs micromere specification in the sea urchin embryo. *Dev. Biol.* **246,** 209–228.

Oliveri, P., Davidson, E. H., and McClay, D. R. (2003). Activation of pmar1 controls specification of micromeres in the sea urchin embryo. *Dev. Biol.* **258,** 32–43.

Poustka, A. J., Herwig, R., Krause, A., Hennig, S., Meier-Ewert, S., and Lehach, H. (1999). Toward the gene catalogue of sea urchin development: The construction and analysis of an unfertilized egg cDNA library highly normalized by oligonucleotide fingerprinting. *Genomics* **59,** 122–133.

Ransick, A., Rast, J. P., Minokawa, T., Calestani, C., and Davidson, E. H. (2002). New early zygotic regulators expressed in endomesoderm of sea urchin embryos discovered by differential array hybridization. *Dev. Biol.* **246,** 132–147.

Rast, J. P., Amore, G., Calestani, C., Livi, C. B., Ransick, A., and Davidson, E. H. (2000). Recovery of developmentally defined gene sets from high-density cDNA macroarrays. *Dev. Biol.* **228,** 270–286.

Rast, J. P., Cameron, R. A., Poustka, A. J., and Davidson, E. H. (2002). Brachyury target genes in the early sea urchin embryo isolated by differential macroarray screening. *Dev. Biol.* **246,** 191–208.

Schulze, A., and Downward, J. (2001). Navigating gene expression using microarrays—A technology review. *Nature Cell Biol.* **3,** E190–195.

Sherwood, D. R., and McClay, D. R. (1999). LvNotch signaling mediates secondary mesenchyme specification in the sea urchin embryo. *Development* **126,** 1703–1713.

Summerton, J., and Weller, D. (1997). Morpholino antisense oligomers: Design, preparation, and properties. *Antisense Nuc. Acid Drug Dev.* **7,** 187.

Wang, D. G., Kirchhamer, C. V., Britten, R. J., and Davidson, E. H. (1995). SpZ12-1, a negative regulator required for spatial control of the territory-specific CyIIIa gene in the sea urchin embryo. *Development* **121,** 1111–1122.

Yuh, C.-H., Bolouri, H., and Davidson, E. H. (2001a). *Cis*-regulatory logic in the endo16 gene: Switching from a specification to a differentiation mode of control. *Development* **128,** 617–629.

Yuh, C.-H., Li, X., Davidson, E. H., and Klein, W. H. (2001b). Correct expression of spec2a in the sea urchin embryo requires both Otx and other *cis*-regulatory elements. *Dev. Biol.* **232,** 424–438.

Yuh, C.-H., Brown, C. T., Livi, C. B., Rowen, L., Clarke, P. J., and Davidson, E. H. (2002). Patchy interspecific sequence similarities efficiently identify positive *cis*-regulatory elements in the sea urchin. *Dev. Biol.* **246,** 148–161.

Yuh, C.-H., Dorman, E. R., Howard, M. L., and Davidson, E. H. (2004). An *otx cis*-regulatory module: A key node in the sea urchin endomesoderm gene regulatory network. *Dev. Biol.* **269,** 536–551.

PART VI

Echinoderm Eggs and Embryos in the Teaching Lab

CHAPTER 33

Sea Urchin Gametes in the Teaching Laboratory: Good Experiments and Good Experiences

David Epel,[*] Victor D. Vacquier,[†] Margaret Peeler,[‡] Pam Miller,[§] and Chris Patton[*]

[*]Hopkins Marine Station of Stanford University
Pacific Grove, California 93950

[†]Division of Marine Biology
Scripps Institution of Oceanography
La Jolla, California 92093

[‡]Department of Biology
Susquehanna University
Selinsgrove, Pennsylvania 17870

[§]Seaside High School
Seaside, California 93955

0091-679X/04 $35.00

Overview

Sea urchin gametes are exceptional for demonstrating cell and developmental phenomena and have long been classic material in college classrooms and, more recently, in high school laboratories. Here, we present salient information on using them in the teaching laboratory along with representative exercises which demonstrate many of the phenomena of early development. A major focus is using the microscope as a tool, allowing the student to obtain meaningful results without having to learn complex and time-consuming new techniques for each lab session. Emphasized are labs which can be completed in one to two days. Also emphasized is the ability to use this material for inquiry-based teaching, in which the student can ask questions and quickly obtain answers to these questions. The outcome is a highly satisfying experience for both student and instructor.

I. Introduction

Sea urchins have provided classical material for laboratory experimentation in the classroom, in part, because of their rich heritage and history and, in part, because of the ease with which urchin embryos can be fertilized and grown in the lab. Another advantage, which will be stressed in this chapter, is the use of this material for inquiry-based science. This approach allows the student to ask specific questions regarding embryonic development and obtain answers in the classroom setting. Inquiry-based teaching allows the student to experience the scientific process and, hopefully, to use this knowledge to better appreciate science as encountered in the real world. In ideal cases, it can open up to the student the possibilities of science as a lifetime career.

Sea urchin material provides such inquiry-based science education because the answers to such inquiries can often be obtained simply and easily in the course of single laboratory exercises. Many experiments have endpoints that can be scored microscopically, such as successful/unsuccessful fertilization or as effects of various experimental procedures on cell division or normal development. The relative technical simplicity of working with embryos means that a student does not need to spend most of the laboratory time learning a new technique; rather, he or she can simply observe early development with the microscope, ask questions about what is observed, and quickly obtain answers. Based on these answers, students can then ask another question, obtain an answer, and ask a different question. This satisfying iterative process has made sea urchin labs a favorite in many universities and even high schools, where initial experiments can lead to student-directed research projects. Indeed, the best-liked introductory biology lab at Stanford is the sea urchin lab.

Good sources of information on using sea urchins for instruction can be found on several Internet sites. The most comprehensive is "Sea Urchin Embryology" developed and maintained by Chris Patton of the Epel lab at http://www.stanford.edu/group/Urchin/index.html. Almost 300 pages long, it provides information, diagrams, videos, animations, and overheads on how to obtain sea urchins for classroom use, suggested lesson plans, suggestions for advanced experiments, and numerous videos and animations useful in the classroom setting. Spanish versions of the major parts of the site are also provided. The site maintained by Eric Davidson's lab at CalTech (http://sugp.caltech.edu:7000/) also provides current research information on the sea urchin genome, which is expected to be complete in 2003/2004.

II. Obtaining Adult Urchins and Gametes for the Classroom

A. Maintaining Adult Urchins

A major problem in using sea urchins is obtaining and maintaining adult sea urchins. Also of concern are issues of availability of material on a year-round basis. These issues are covered in Chapter 1 of this volume. Short-term maintenance of the urchins can be easily achieved in saltwater aquaria (see chapter 1 and Web site). A single order of urchins often can be used to support labs over the course of several weeks. Long-term maintenance is more problematic unless one has extensive experience with these aquaria. Teachers should also be sensitive to ethical and environmental concerns surrounding the treatment of this biological material.

B. Storing Eggs Useful for Classroom Experiments; an Alternative to Maintaining Aquaria

A simple protocol we (D. E., C. P.) have developed which eliminates problems in maintaining urchins in saltwater aquaria for long periods of time is to shed and store the gametes for later classroom use upon delivery of the adult urchins. The principle is based on our finding that "death" of sea urchin eggs after spawning results from bacterial action once the eggs are shed. To avoid this problem, the eggs are stored in simple antibiotic solutions that maintain the gametes in a fertilizable condition for at least one week, allowing use for many different classroom exercises. Simple storage of sperm at low temperatures will similarly preserve these gametes for 4 to 5 days.

Urchins are spawned as described in Chapter 1 and in the following text. The sperm are not spawned into seawater but are collected as a "dry" paste and stored in closed small tubes (conical "Eppendorf-type" are ideal) at 0 °C on ice in the refrigerator. They are good for 4 to 7 days under these storage conditions.

Eggs are spawned into Millipore-filtered seawater (SW) or Millipore-filtered artificial SW (do not use autoclaved seawater). The eggs settle quickly in the seawater and the supernatant is removed and the eggs resuspended in fresh filtered seawater. This settling of eggs, removing supernatant seawater, and resuspension is repeated three times in this sterile seawater. It is imperative to remove as much of the supernatant seawater as possible, since this contains the original contaminating bacteria. In a laboratory setting, this can be accomplished most easily with a water aspirator; in a classroom setting, a turkey baster is handy for removal of the supernatants.

After the third wash, the eggs are then resuspended in seawater containing a mixture of two antibiotics, sulfamethoxazole (final concentration at 200 ug/ml and trimethoprim (final concentration of 10 ug/ml). These concentrations are almost at the limits of solubility in seawater and getting them into solution is time-consuming. Continuous stirring and periodic adjustment of the pH are required until the antibiotics are dissolved (~2 h). At the end of the dissolution procedure, the antibiotic solution is passed through a Nalgene bacterial filter. Using this mixture of sulfamethoxazole and trimethoprim will permit storage for at least a week with *Strongylocentrotus purpuratus* eggs and up to three weeks with *Lytechinus pictus* eggs. If difficult to obtain both antibiotics, sulfamethoxazole alone will work although the eggs will not last as long. Although other antibiotics are not as effective, ampicillin will suffice for several days. More detailed protocols are described at http://www.stanford.edu/group/Urchin/protocol.html and in a paper in preparation (Epel *et al.*, in preparation).

Optimal storage is obtained under dilute conditions, 0.1 to 0.2% v/v%, as a monolayer in Falcon-type tissue culture flasks. Alternatives are baking dishes or other glass or plastic containers that allow the eggs to be stored in a quasi-monolayer condition. Storage temperature for *L. pictus* and *S. purpuratus* is 10 to 15 °C; lower temperatures lead to lysis of the eggs. We have not tried these procedures on eggs from other species but assume that similar principles will apply, albeit the storage temperatures should be adjusted to the ambient temperatures of the adult sea urchins.

III. Some Guidelines for Using Sea Urchin Gametes in the Classroom

Before describing specific experiments, we present several tips on handling gametes that will make the exercises more successful. These are based on our years of experience in the classroom and will allow you to avoid the most common pitfalls that students will experience.

A. Storage of Diluted Sperm during Classroom Exercises

You will need to use dilute sperm suspensions to fertilize eggs, but a problem in classroom use is that the diluted sperm quickly lose their ability to fertilize eggs when they are diluted in seawater. We have found that *S. purpuratus* sperm at a 1% concentration in seawater stored on ice will retain their fertilizability for at least 12 h. Similar procedures might work with sperm from other urchin species.

B. Temperature

Eggs of *S. purpuratus* develop in nature at temperatures of 12 to 15 °C, so development of this species should be followed at 12 to 17 °C. If this is not possible, one can fertilize them at room temperature; fertilization and cleavage stages will proceed, although later development will probably not be successful. Development is too slow in a refrigerator at 4 °C, unless you can adjust the refrigerator to a higher temperature. An option used in our local high school is to keep the culture vessels in a Styrofoam chest in which the water is adjusted with ice to 15 °C or to float the embryos in bottles in a cooled aquarium. Worst case: in the absence of below-ambient temperature equipment, try to keep the cultures in the coolest part of the room. Other species such as *Arbacia punctulata, Lytechinus pictus*, and *L. variegatus* develop well at room temperature, and have the added advantage of rapid development.

C. Need to Wash Eggs

During prolonged storage, substances are released from the eggs or the jelly layer of the eggs that can interfere with fertilization. If eggs have been shed more than an hour earlier, it is a good idea to wash the eggs before adding sperm. Since eggs are heavier than seawater, they will settle by gravity in the beaker and the supernatant can be removed and fresh seawater added.

Sometimes, settling is slowed by the presence of the jelly layer around the eggs; to speed up the washing, one can use low-speed centrifugation. The eggs are delicate, however, so be gentle when adding fresh seawater to the eggs or when resuspending the eggs in seawater.

D. Inadvertent Fertilization

It is very easy for students to accidentally fertilize the entire batch of eggs that is to be used by the class through contaminating the eggs with sperm from the sperm pipet. To avoid this, let the students know of this problem and, most importantly, label the pipettes being used for sperm or eggs.

E. Polyspermy

Polyspermy will result in abnormal development and is to be avoided (unless you wish to have a lab around this···which can be very interesting). To avoid polyspermy, emphasize to the students the need to fertilize with a very dilute sperm suspension. Details about sperm dilution will be given.

F. Overcrowding

Overcrowding can affect development. Proper development requires good oxygenation, which can be achieved by gentle stirring, as with a rocker-type stirrer. Since these may not be available in a classroom setting, keeping the eggs as a monolayer on the bottom of a beaker or other type of glass vessel will work just fine.

G. Egg Transparency

Transparency varies with the different species. Ideal, transparent eggs, which allow easy visualization of the nucleus and mitotic apparatus, are those of the genus *Lytechinus*. The eggs of *Arbacia* and *Strongylocentrotus* are relatively opaque so such cytological detail will be more challenging to see in the classroom. However, *S. purpuratus* has many advantages for classroom use, such as a prolonged reproductive season and robustness in the laboratory. Later developmental stages are easily seen with the microscope in all species of sea urchins.

IV. Basic Introductory Labs

Here we present a series of laboratory exercises that we have used in our classrooms at Stanford, University of California at San Diego, Susquehanna University, high schools on the Monterey Peninsula, and also in various international courses in which we have been involved over the years. These exercises provide specific examples that can be used as described or as ideas or scaffolds for use in specific situations. They start out with simple microscopic observations with light or phase microscopes that are essential introductory labs. We follow these with suggestions for a variety of experiments that could be used in the aforementioned inquiry-based science approach, some of which will require more specialized equipment as well as time beyond a normal laboratory period.

Most of the lab directions and experiments presented are the basis for handouts that we have used in our various classes. Instructors should feel free to use the text for their classes, as presented or modified for their particular settings. There are also excellent student exercises, handouts, and transparency texts available on the sea urchin embryology website (http://www.stanford.edu/group/Urchin/protocol.html).

A. Basic Protocol for *In Vitro* Fertilization and Observation of Early Development

This laboratory is a good introduction to events of fertilization and early embryonic development and works well with all sea urchin species we have used. The level of instruction written here is for introductory college labs and advanced high school instruction. It can also be adapted for upper division college labs.

1. Spawning Gametes

This part of the exercise is great if you have lots of sea urchins; it allows the student to see the adult urchin and understand how spawning takes place. The teacher should be sensitive to the ethical concerns of sacrificing the animals. This concern can be avoided if a seawater aquarium is available, since the animals can be placed back into the aquaria after spawning and will generally survive (but requires that the animals be kept moist while spawning and be placed back into the aquarium as soon as possible). Alternatively, the instructor can provide gametes to the students, either by inducing spawning just before the lab or using gametes obtained through the long-term egg and sperm storage procedure described earlier.

The sexes are separate and the gametes are spawned into seawater with fertilization and embryonic development taking place external to the adult body. Gravid animals spawn when injected with 0.5 M KCl [the K^+ ions depolarize a muscle which surrounds the outer surface of the ovary or testis and the eggs or sperm are forcibly extruded into the seawater through five gonopores on the aboral surface (the opposite surface from the mouth)]. To begin, the needle on the syringe of KCl is carefully inserted into the soft tissue around the mouth (Aristotle's lantern). When the needle comes to rest on the inner surface of the test, the syringe is depressed to release approximately 1.0 ml of the solution. This is repeated in two different areas, followed by gentle shaking of the animals. After injecting an animal, the hypodermic needle should be rinsed in tap water to kill any sperm that might have stuck to the needle so as to avoid unwanted fertilization of eggs (e.g., the next animal injected might be a female).

Within 1 to 5 min of injection, the gametes appear on the topside. The only way to distinguish males from females in most species is to see what comes out—eggs are orange, red, or yellow, depending on species, and sperm is creamy white and thick. Students may have trouble distinguishing the two initially until they see both.

Females are placed upside down on a beaker of seawater and the eggs settle to the bottom. It may take up to 10 min to spawn completely. Once the eggs are spawned, they should be "washed" through several changes of seawater, allowing them to settle by gravity to the bottom of the beaker after each addition of seawater. Eggs should be fertilized within a few hours of spawning, unless prepared for long-term storage as has been described.

Sperm are stored "dry," so males are placed upside down on an empty beaker. After 10 min, the sperm should be removed from the urchin surface with a Pasteur pipet and transferred to a small Eppendorf-type tube for storage in the refrigerator or on ice. Stored on ice, the sperm cells remain capable of fertilizing eggs for 4 to 5 days.

2. Jelly Removal

Sometimes, one would like to remove the jelly coat to facilitate handling the eggs (the jelly coat causes the eggs to be buoyant and impedes settling). This can be done by straining through Nitex mesh or cheesecloth; if using Nitex, use a mesh size that is about the diameter of the eggs (75 micron for *S. purpuratus* and 120 micron for *Lytechinus*). To do this, a suspension of eggs is poured through several layers of cheesecloth or one layer of Nitex into a new beaker. Use additional seawater to rinse the cheesecloth to remove any remaining eggs. Then wash the eggs several times to get rid of any jelly and broken eggs.

B. Observation of the Gametes

1. Eggs

The eggs are between 80 and 120 μm in diameter, depending on the species and 2 to 4 million eggs are present in one ml of packed eggs. Sea urchin eggs (and eggs of all species of the Echinoderm class "*Echinodea*") are unusual in that, unlike all other animal eggs (except for coelenterates), the mature eggs (ova) are post-meiotic and are arrested in the interphase of the first mitotic division (G1). Each egg contains a single haploid pronucleus (female pronucleus) with an intact nuclear envelope. The eggs are metabolically dormant and synthesize protein at very low rates. When they are fertilized, their metabolism is markedly activated, with large increases in protein synthesis (particularly the synthesis of cyclins necessary for cell division), appearance of new transport systems, alterations in the cytoskeleton, and initiation of DNA synthesis (Epel, 1997). These explosive changes in sea urchin eggs at fertilization have been well studied as a model for similar changes after fertilization in all animal eggs. Additionally, fertilized sea urchin eggs provide natural synchronously dividing eukaryotic cells for the first two divisions and present an exceptional model for understanding cell division in all eukaryotic cells (Sluder *et al.*, 1999).

To observe, drops of egg suspension are placed on a slide and viewed using the low power (10×) objective. The eggs may be dark in color due to yellow, purple, or red pigment granules (species-dependent). Students should focus up and down to see the round egg pronucleus (about 1/20th of an egg's diameter), which is most apparent in transparent eggs, such as those of *Lytechinus*. Students should note the egg pronucleus in an otherwise homogeneous, but granular, cytoplasm and observe that the nucleus is not in the center of the egg but is eccentrically placed on one side (as some eggs will be viewed from the ends or poles, this eccentricity will not be observed in all eggs; the instructor can use this as a means of demonstrating how the microscope image varies, depending on orientation of the cells). Eggs can also be viewed under higher (40×) power. Before going to higher power, students should add a "footed" coverslip by placing several small pieces of broken coverslip around the drop of eggs, then placing an intact coverslip on this support. This prevents the coverslip from crushing the eggs while protecting the objective lens from corrosive saltwater.

Occasionally, immature oocytes are present in the egg solution. These are easily identified because they contain a large germinal vesicle (pre-meiotic oocyte nucleus) inside of which is a round, extremely dark nucleolus. (If a large percentage of the ooctyes are immature, a new female should be spawned.)

The eggs are surrounded by transparent jelly layers, about one-half an egg diameter in thickness. This jelly layer cannot be seen because of its transparency but can be inferred since the cytoplasmic surfaces of the eggs will not touch each other and the eggs will be evenly spaced on the slide. The jelly coat can be directly visualized by adding a few drops of calligraphy ink (such as Sumi ink, which can be purchased at art stores) suspended in seawater to the slide. The ink particles cannot penetrate the viscous jelly coat, which then appears as an inkfree clear zone around each egg (see Fig. 1). The sperm must swim through this jelly coat, and there are a plethora of substances that affect sperm motility, cause the acrosome reaction, and induce chemotaxis (Mengerink and Vacquier, 2001). A procedure to isolate the egg jelly to study these sperm-altering substances will be described, after the fertilization protocol.

2. Sperm

One ml of undiluted semen contains 4×10^{10} sperm cells and the haploid amount of DNA in a single sperm cell is 0.85 picograms (*S. purpuratus*). The sperm cell (spermatozoan) is composed of a tail (flagellum) about 50 μm in length and a head 3 μm in length (approximate dimensions vary with species). Between the sperm flagellum and the head, and connected to the sperm head, is a midpiece made of a single mitochondrion.

A phase contrast microscope with a 100× oil immersion objective is needed to see details of the sperm morphology. The stored semen should be diluted 1:1000 in

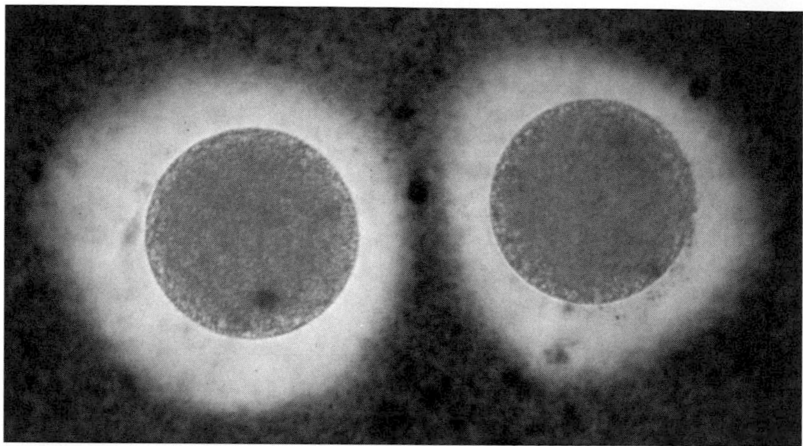

Fig. 1 A jelly layer surrounds the unfertilized sea urchin egg. The jelly layer is normally not visible because it is transparent. Here it is made visible by suspending the eggs in a thin suspension of Sumi ink (Japanese calligraphy ink). The jelly layer excludes the particles of ink and the layer then stands out as a clear structure against the black background of the ink suspension.

seawater. Then, a drop of the sperm suspension is placed on a slide and covered by a coverslip. Excess seawater from the slide can be drained by turning the slide on edge on a tissue, then flipping the slide over, coverslip down, on filter paper or Kimwipes. A slight pressure is applied by thumb for 1 s. The slide can then be turned over and viewed under high power with oil. Many of the sperm will have undergone a spontaneous acrosome reaction and are now stuck to the glass by their acrosomal process (this acrosome reaction is necessary for fertilization). Structures that should be discerniable are the flagellum ($50 \times 0.2\ \mu m$) and the head ($3 \times 1\ \mu m$). The head contains a single, giant, donut-shaped mitochondrion which comprises the sperm midpiece. The conical nucleus makes up most of the volume of the sperm head. At the most anterior region of the nucleus is the acrosome granule, with a clear space, the nuclear fossa, separating the granule from the nucleus. This will be visible only in non-acrosome reacted sperm.

If a phase contrast microscope is not available, sperm can be observed under bright field. At 40×, details of sperm structure are not readily discernible, but with a fresh sperm dilution, the beating of the flagella is easily seen.

C. Fertilization

To fertilize eggs in small batches, one medicine dropper/pipette full of eggs from the beaker into which the female was spawned is added to a clean beaker with approximately 30 ml of seawater. The eggs will have settled so the students must be careful to pipette from the bottom of the beaker. The more eggs that are transferred, the more easily the fertilized eggs for observation will be found.

A fresh sperm dilution (1:1000) is made as has been described. Generally, 1 ml of this diluted sperm added to the 30 mls of eggs will be sufficient for fertilization without inducing polyspermy. The sperm are added and the beaker is swirled to mix the gametes. After 2 min, during which time the cortical granule reaction occurs and the fertilization envelope is elevated and hardened, a drop of eggs from the bottom of the beaker is placed on a slide and observed with the 10× objective. Students should count the number of fertilized and unfertilized eggs on the slide to determine a percentage of fertilization. It is readily possible to count 100 eggs or more, provided sufficient eggs were transferred to the beaker initially, allowing students to obtain quantitative data. In most cases, a percentage of 90% or better represents successful fertilization. If the percentage is less than 90%, a fresh sperm dilution can be made and the eggs can be refertilized, as has been described. Eggs fertilized in the first batch will be protected from polyspermy because of the presence of the fertilization envelope.

Once students are confident that they can distinguish fertilized eggs from unfertilized eggs, they can attempt to observe the elevation of the fertilization envelope directly. In order to do this, a drop of unfertilized eggs is placed on a slide and the student focuses on a small cluster of eggs at 10× magnification with no coverslip (emphasize not to use higher power without a coverslip!). A drop of diluted sperm is added and students watch for several minutes to directly observe fertilization (an increased dilution factor should be used here; if there are too many sperm, it will be difficult to see the elevation). Students can time the elevation in several eggs by repeating the procedure with fresh eggs and sperm.

It should be possible to see sperm attachment to the egg surface. In 30 to 60 s (depending on the species), the elevation of the vitelline layer (VL) begins, followed by its transformation into the fertilization envelope (FE). The elevation of the VL initiates at the site of sperm–egg fusion, looks like a blister as it begins, and sweeps around the egg as a circular wave (Fig. 2). The flagellum of the fertilizing sperm may be visible, appearing straight and motionless, sticking out of the blister.

The VL to FE transformation is a result of the cortical reaction which involves the exocytosis of thousands of membrane-enclosed cortical granules which are attached to the inside surface of the egg plasma membrane (see Fig. 2). After the cortical reaction completes, a clear gelatinous layer, the hyaline layer, begins to become apparent on the surface of the egg. The hyalin protein that makes up the hyaline layer also originates from the cortical granules and forms a calcium-dependent gel that is necessary for adhesion of the blastomeres during development. The layer can be easily seen in *S. purpuratus* embryos in 5 to 10 min after fertilization. It forms more slowly in *L. pictus* embryos and should be apparent by 20 min after fertilization (Matese *et al.*, 1997).

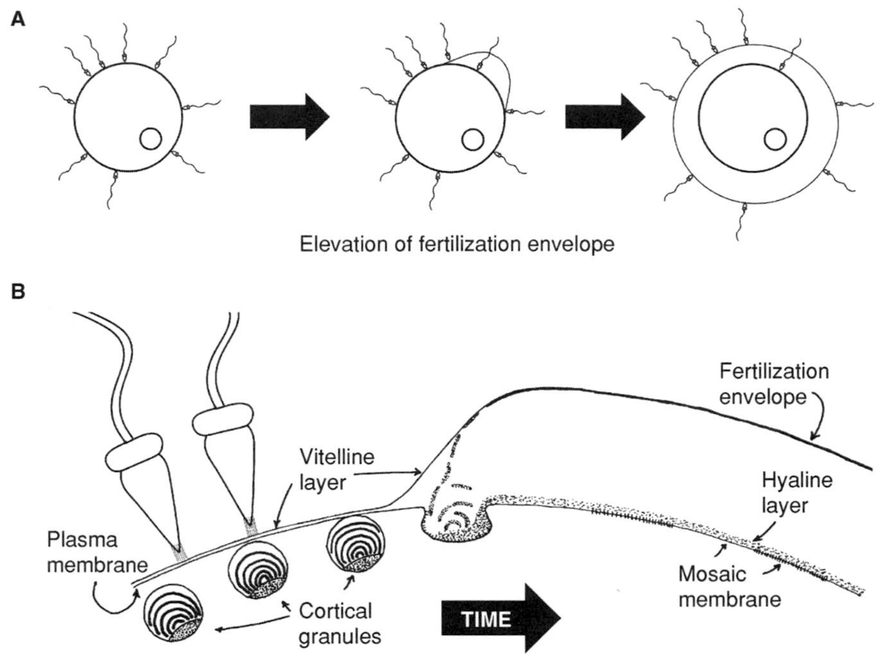

Fig. 2 A cartoon illustrating the elevation of the fertilization envelope. Part A depicts the elevation of the envelope as seen in an intact egg. As seen, the elevation begins as a blister on the surface, at the site of sperm-egg fusion. This then travels or propagates around the egg over the next 30 s, resulting in the elevation of the envelope. Part B depicts the mechanism of elevation, which involves the fusion or exocytosis of cortical granules with the plasma membrane of the egg. This fusion releases the granule content, which provides the structural materials for the membrane and its hardening over the next few minutes.

D. Observing Post-Fertilization Events

The transparent qualities of sea urchin eggs allow easy observation of the movement of the pronuclei, cell division, and later stages of development. To do this, a larger batch of eggs should be fertilized and resuspended in a dish with a large surface area-to-volume ratio so that the eggs will receive sufficient oxygenation. For class purposes, this could be in baking dishes, custard dishes, or other dishes. Embryos should be added to the dishes until a sparse monolayer forms. If available, one can also increase oxygenation by agitating the eggs in the dishes on a rocking platform.

1. Visualizing Nuclear and Mitotic Events—Light Microscope

Pronuclear fusion and mitosis are most easily seen with transparent eggs, such as those of *Lytechinus*. With these eggs, you will be able to observe with low-power microscopic magnification that the originally eccentric pronucleus moves to the center of the egg at about 20 to 30 min after insemination and that the nuclear envelope disappears as the egg enters late prophase. One can see these stages with much more detail using higher magnification (40×). Before going to higher power, a "footed" coverslip should be added, as described previously.

In transparent eggs, one can usually see directly the nuclear changes and formation of the sperm pronucleus and formation of mitotic apparatus. In the more opaque eggs, one will have to compress the eggs slightly. This can be done by withdrawing water from under the coverslip. With the footed coverslip, this is easily done by wicking off excess seawater by placing a paper towel or piece of filter paper next to one edge of the coverslip. The broken piece of coverslip will prevent the eggs from being crushed and lysed. With this compression, one should be able to see the nuclear movements and the fibrous mitotic apparatus. Timing-wise, the mitotic events will begin at approximately 50 to 90 min after insemination, depending on the temperature and species. Cytokinesis will begin about 90 to 120 min after insemination, and the successive divisions will then occur about every 30 to 60 min (again, times will vary with species and temperature).

2. Visualizing Nuclear Events—Fluorescence Microscope

If a fluorescent microscope is available, the dramatic events taking place in the nucleus during sperm–egg fusion and mitosis can easily be visualized.

Eggs are incubated in seawater containing 1 uM Hoechst 33342 for 10 min (Hinkley *et al.*, 1986). The eggs are then washed 2 to 3 times in seawater. This stain labels the eggs, so that if the eggs are used for subsequent fertilization, one should be able to visualize the fusing sperm, the movements of sperm and egg through the cytoplasm, the fusion of the two nuclei in the egg center, and the formation of the chromosomes during metaphase and their separation during anaphase of the first mitotic division. This will require high-power magnification (40×) and some compression of the eggs, so the footed coverslip technique will be necessary.

E. Later Developmental Stages

Depending on time, students should be able to observe the initial cleavage stages of development during the lab period. They may then be asked to return periodically to the lab during the next few days to observe later important developmental stages of embryonic development. Alternatively, batches of embryos fertilized by the instructor prior to the lab period can be made available for observation. Once the embryos hatch and begin to swim, it becomes more difficult for students to find and follow them under the microscope. Drops of embryos can be placed on slides that have been coated with protamine sulfate or poly(L)lysine (Mazia *et al.*, 1975), which provides a charged surface on the slide to which the embryos will adhere and become immobilized for easier visualization.

V. Inquiry-Based Experiments Using the Basic Fertilization Protocol

Fertilization and cleavage stage cell division are convenient systems for inquiry-based education since they occur in the timeframes of the typical classroom laboratory. Next, we describe some possible experiments that students can do which can lead to questions about the molecules and mechanisms involved in fertilization and early development. Many of these depend primarily on microscopic observation and the students should realize that to gain the imagination needed to be a good experimental cell biologist, a scientist must observe the behavior of living cells with his or her own eyes. Students will often look and then stop observing. It is important for them to realize that one cannot spend too much time observing living cells with a microscope. As Daniel Mazia stated about microscopy, "important discoveries continue to travel to us on beams of light." These beams provide a wonderful vehicle to learn and we hope these exercises will achieve this goal.

A. Chemotaxis

There are many examples in cell biology and development where cells sense a chemical gradient and move toward it or away from it. Slime mold amoebae move up gradients of cAMP to aggregate into the "cellular slug" stage. In mammals, white blood cells move to wounds and infections. Neural crest cells migrate in the vertebrate body to form many parts of the body. This directed movement is often a form of chemotaxis, or chemoattraction; few examples of the phenomenon are known in molecular detail in eukaryotic cells.

When treated with soluble egg jelly, sea urchin sperm cells show a change in swimming behavior that is some kind of an unexplained chemotaxis (Eisenbach, 2004; Ward *et al.*, 1985). It is easy to demonstrate this phenomenon using isolated egg jelly and sperm. To isolate egg jelly, the coat is solubilized by exposing the eggs to seawater at pH 5.0 for 2 min. This is done by putting a pH electrode in the egg suspension and gently stirring the eggs by hand while slowly dripping in 0.1 N HCl until the pH reaches 5.0. After 2 min, the pH is quickly readjusted to pH 7.8 by adding 1:100 1 M HEPES buffer, pH 8.0. The eggs are allowed to settle and the supernatant seawater, which contains the dissolved egg jelly, is removed and centrifuged to remove egg fragments. The isolated egg jelly solution then can be stored on ice or frozen for later use.

To observe chemotaxis, sperm are diluted about 1:100 with seawater. Then, 0.5 ml of diluted sperm is placed in a 12×75 mm glass tube and 0.5 ml of the egg jelly solution is added. Within 10 s, sperm begin to cluster, and they form clotlike aggregates which can be seen with the eye at the bottom of the tube. (The tube should be held up to the light to observe this.) People used to think the cells were sticking to each other in a true agglutination reaction. However, careful observation under the microscope demonstrates that the sperm are freely moving very rapidly, as bees do in a swarm (Collins, 1976). It is as if there is something in the center of each ball of swarming sperm that is attracting the cells. In 2 to 5 min, the clusters of sperm dissociate and the sperm swim freely in a once-again homogeneous suspension.

The swarming behavior of sperm cells by egg jelly can also be observed microscopically on a slide. Four drops of egg jelly solution are placed on a slide and two drops of freshly diluted sperm (1:100) are added. The clusters of sperm are irregular in shape for a few seconds before they become perfectly spherical swarms. In 2 to 5 min, the attraction is lost, the swarms spontaneously disperse, and the cells are evenly distributed in the field of view. Students can time these events of clustering and dispersion.

There are many investigative questions students can then begin to ask about this system; for example, will the same sperm swarm again if fresh egg jelly is added? Is the reaction species-specific? What is the effect of temperature on the reversal of swarms? (The lipids in the sperm membrane are frozen at about 7 °C.) How do proteases, which would digest sperm membrane receptors, affect the ability of the sperm to cluster? A solution of 0.2 mg/ml pronase in seawater makes a good test solution for this question—two drops of this solution are placed on a slide, and two drops of sperm suspension are added and mixed in with a pipette tip. After 10 min, four drops of egg jelly solution are added and the ability of the sperm to swarm is determined by comparing the response to a no-protease control.

B. Formation of the Fertilization Envelope

There are numerous examples of cells secreting glycoproteins that "harden" (intermolecular covalent crosslinking) to form tough structures that protect the cell and the organism. The epidermal cells of invertebrates such as insects and crustacea secrete an exoskeleton that hardens. Mussels secrete attachment threads of protein that are transformed to extremely resistant filaments. Reproductive stages of many animals are encased in hardened coats, such as insect egg chorions or the resistant cysts of protozoa and flatworms. One mechanism of hardening of such secreted proteins is by the formation of intermolecular crosslinks between adjacent proteins, which then are rendered extremely resistant to enzymatic and chemical attack. The hardening of the fertilization envelope of the sea urchin embryo occurs within a few minutes of sperm addition, and makes an excellent model system for asking questions about this hardening process (Kay and Shapiro, 1985).

1. Background on Transformation of the Fertilization Envelope

The egg is covered by a thin extracellular matrix called the vitelline layer or VL [it also contains the sperm binding (bindin) receptors]. The VL is about 30 nm thick and is made of a loose meshwork of glycoproteinaceous fibers that are intimately bonded to the cell membrane. In fact, you can't isolate the VL without destroying the egg.

Directly under the cell membrane are about 15,000 cortical granules (CG) that contain 6% of the egg protein. (Many other species, including mammalian eggs, also have CGs). At the point of sperm fusion the CGs begin to fuse with the egg membrane and release their contents extracellularly under the VL. The exocytosis of the CG sweeps around the egg as a circular wave, taking 30 s to complete. This is evidenced as the elevation of the VL and its transformation into the FE (see Fig. 2).

The mechanism of VL elevation and transformation to the hardened FE results from the interaction of the cortical granule components with the VL. First, a trypsinlike-protease, released from CGs, detaches the VL from the cell membrane. Simultaneously, two other proteins, one structural and called proteoliaisin (PLN), and the other an enzyme called ovoperoxidase, come out of the CG bound to each other in a 1:1 stoichiometry. This dimeric complex assembles in paracrystalline order on both the inner and outer surface of the elevating VL. The VL thus acts as a template for these proteins, which is part of FE assembly (Correa and Carrol, 1997; Weideman *et al.*, 1987).

The secreted proteins, aligned in an orderly array (paracrystalline) on the VL template, are now crosslinked together by action of the ovoperoxidase. This protein functions as both a structural component and an enzyme in FE formation and hardening. The ovoperoxidase uses H_2O_2 (hydrogen peroxide) to catalyze the formation of di- and tri-tyrosine crosslinks among adjacent molecules. These crosslinks are resistant to hydrolysis in boiling 6N HCl; if you hydrolyze purified FEs in HCl to break all peptide bonds, and then separate the amino acids by HPLC, you get a large peak of di-tyrosine and a smaller peak of tri-tyrosine (Shapiro, 1991).

But, where does the H_2O_2 come from that is the substrate used by the ovoperoxidase in the crosslinking reaction? If seawater had that much H_2O_2 in it, we ocean swimmers would all be blondes. The answer is that the egg has a cytoplasmic enzyme called ovo-oxidase that uses O_2 and nicotinamide adenine dinucleotide phosphate (NADPH) to catalyze the reaction: $\text{NADPH} + H^+ + O_2 \rightarrow \text{NADP} + H_2O_2$. A protein kinase C is physically associated with the oxidase and appears to activate it at fertilization. (A fascinating question is why the H_2O_2 is not toxic. The answer is that the eggs possess novel compounds that neutralize the effects of the peroxide and associated free radicals produced by the H_2O_2. These compounds—1-methyl-4-mercaptohistidines—are found in most marine invertebrate eggs at concentrations of 3 to 5 mM and are called ovothiols [Shapiro, 1991].)

The FE protects the embryo from physical damage and microbial attack and persists as a tough protective micro-incubation chamber until about the 600-cell blastula stage. At that time, the blastula secretes a hatching enzyme, a metalloendoprotease, that dissolves the FE. The embryo at this time has also synthesized and assembled flagella on the outer cells and can now swim in the plankton as an independent planktonic organism. In two days, it becomes a feeding larva and then spends 30 days in the plankton growing and continuing its development until metamorphosis.

2. A Simple Demonstration of the Assembly and Hardening of the FE

A drop of unfertilized eggs is placed on a slide, then one drop of 5% sodium dodecyl sulfate (SDS) is added. What should happen is complete solubilization of the cells; nothing is left but a yellow liquid. This is then repeated with fertilized eggs which have had elevated FEs for at least 5 min. What the student should see is that everything dissolves except the FE, leaving beautiful clean FEs on the slide that are resistant to SDS. In fact, one can boil the FEs in 5% SDS and they still do not dissolve.

3. Experiments Demonstrating the Mechanism of Assembly and Hardening of the Fertilization Envelope (FE)

a. Use of Inhibitors to Examine the Role of the Trypsinlike Cortical Granule Protease that Detaches the Vitelline Layer from the Egg Cell Membrane

Drops of concentrated unfertilized sea urchin eggs are placed in a series of tubes containing 1 ml seawater plus soybean trypsin inhibitor (SBTI) at concentrations of 2.0, 1.0, 0.5, 0.25, 0.12, and 0.06 mg/ml seawater. This protein binds tightly to trypsinlike enzymes with a 1:1 stoichiometry and completely blocks proteolytic activity. A seawater control (no SBTI) should also be used.

A few drops of the SBTI egg suspension are then placed on a clean slide. One drop of dilute sperm suspension is added and the eggs are observed to determine what happens to the elevation of the VL and its transformation into the FE. Students can ask questions regarding the effect of the different concentrations of SBTI as well as observe differences in sperm-binding behavior as compared to the no-SBTI control (see Vacquier *et al.*, 1972).

b. Preventing Di-tyrosine Crosslinking of the FE by Blocking the Egg Ovoperoxidase

Para-aminobenzoate (PABA) is an effective nontoxic inhibitor of the ovoperoxidase (the enzyme that forms the intermolecular di-tyrosine crosslinks). One ml of PABA dissolved in seawater at concentrations of 100, 50, 25, 12, and 6, 3, and 1.5 mM is placed in test tubes. Then 1 drop of concentrated unfertilized eggs is added to each tube. Three drops of each egg suspension are then placed on a slide and fertilized with a drop of diluted sperm. A PABA concentration-dependent effect is clearly seen on the elevation of the FE. When ovoperoxidase-mediated crosslinking is inhibited, then the FE becomes vulnerable to chemical agents which disrupt its molecular structure.

This can be demonstrated by repeating the above with a concentration of PABA that gives clear visual evidence that the FE is being affected. At 3 min after fertilization, 2 to 3 drops of 0.01 M DTT (dithiothreitol) in seawater (pH 9.1) are added. This reagent will reduce disulfide (S-S) bonds that help stabilize the proteins of the FE. Additionally, 2 drops of 0.1 mg/ml pronase in seawater can be added to degrade the peptide bonds of the FE proteins. Since the ovoperoxidase was blocked by PABA, no hardening of the FE occurred and the DTT or pronase should destroy the soft FE, leaving a "naked" egg. [A seawater (no PABA) control should be run in parallel.]

c. Demonstrating that Disulfide Bonds Stabilize the Elevated FE after Tyrosine Crosslinking Has Hardened It

As has been noted, the FE is hardened by intermolecular di-tyrosine crosslinks. However, disulfide (S-S) bonds between cysteine residues are extremely important to the stability of the VL and also the completely elevated, hardened FE (Epel *et al.*, 1970). This can be demonstrated as follows. Two drops of egg suspension on a slide are fertilized with dilute sperm and allowed to sit undisturbed at least 5 min to allow complete crosslinking of the FE to occur (do not use a coverslip). Two drops of 0.01 M dithiothreitol (DTT pH 9.0) in seawater is then added. What you should see is that addition of DTT causes a dramatic increase in diameter of the FE due to breakage of S-S bonds and relaxation of the protein structure. Blowing air across the slide for several seconds will cause the oxidation of the DTT, the reformation of S-S bonds, and the shrinkage of the FE to its normal size. The cycle of expansion and contraction of the FE by breakage and reformation of S-S bonds should be repeatable on the same eggs by adding more DTT, followed by blowing.

Students can then ask a variety of questions about this process, such as, is the hardened FE susceptible to proteolytic destruction when the S-S bonds are reduced to SH? To test, 2 drops of normal fertilized eggs with hardened FEs are mixed on a slide with 2 drops of a 0.1 mg/ml solution of pronase in seawater. This control is then compared to fertilized eggs pretreated with DTT to expand the FEs. Protease effects will be time dependent, so the control and DTT-treated FEs must be observed for at least 5 to 10 min.

d. Demonstrating that Transglutaminase Activity is Also Involved in the Hardening of the FE

Transglutaminases are enzymes that crosslink proteins together by linking the epsilon amino group of lysine in one protein to the carboxyl group of aspartic or glutamic acids in another protein. If transglutaminase activity is involved in FE hardening, fertilization in the presence of transglutaminase inhibitors will yield soft FEs which should be destroyed by pronase and DTT (see, e.g., Battaglia and Shapiro, 1988).

Primary amine inhibitors of transglutaminase activity are cadaverine, putrescine, and glycine ethyl ester (GEE). This experiment to test their effects is set up just like the PABA experiment already described. Seawater buffered with 10 mM HEPES, pH 7.8, is used to prepare GEE at the following concentrations: 100, 50, 25, 12, 6, 3, and 0 mg/ml. One ml of each concentration of GEE is placed into a series of tubes, then 1 drop of a 10% egg suspension is added. After 2 min, 3 drops of egg suspension are placed on a slide and fertilized by mixing in less than one drop of diluted sperm suspension using a pipette tip. Observe the formation of the FE. Does it look larger and thinner than the control? Additional questions can be asked, such as whether the FE formed in GEE is resistant to DTT and pronase.

C. The Role of Calcium Ions in the Fertilization Process

Calcium is required both for the acrosome reaction of sperm and for the activation of the egg, including triggering exocytosis of the cortical granules. There are a number of questions that students can ask about the role of calcium in the fertilization process, such as confirming that calcium is required for fertilization, determining where in the process calcium is required, and determining minimum levels of calcium required (see, e.g., Darszon *et al.*, 2001).

1. Is Calcium Required for Fertilization?

Five ml of calcium-free seawater (CaFSW) is placed in a tube and one drop of a 10% egg suspension is added. The eggs will settle through the CaFSW (or can be pelleted with very gentle centrifugation). The supernatant is removed with an aspirator and the eggs are resuspended in a fresh one ml of CaFSW. This procedure is repeated two more times to ensure that all thc calcium is removed from the suspension. Then two drops of the suspension is placed on a slide and fertilized with small drop of a 1:100 dilution of sperm in CaFSW. In normal seawater, the FE elevates and hardens by 3 min after adding sperm. Note that, in this experiment, even 5 min after adding sperm, the FE is not present. This is because, normally, when sperm contact egg jelly, two calcium channels open in the sperm membrane, allowing calcium ions to come into the cell and trigger the acrosome reaction. Without having undergone the acrosome reaction, the sperm cannot bind to or fuse with the egg; thus, fertilization is inhibited. To demonstrate that calcium is the limiting factor, a small drop of 0.34 M $CaCl_2$ buffered with 10 mM HEPES at pH 7.8 can be added. This should result in the elevation of FE within about 2 min.

Students then can ask questions about how much calcium is required for acrosome activation by mixing CaFSW with normal seawater to create a range of calcium concentrations. Drops of eggs washed into CaFSW can then be added to these varying concentrations and fertilized by sperm diluted in the corresponding concentration to determine the level of calcium required for the acrosome reaction to occur and permit sperm binding and fertilization to occur (see also Collins and Epel, 1977).

2. The Role of Calcium Ions in Cortical Granule Exocytosis

A unifying feature of exocytosis, including the cortical granule reaction of sea urchin fertilization, is that it is mediated by a transient increase in free calcium ions. For example, in the unfertilized egg, the free calcium ion concentration (not the total calcium) is about 100 nM. Fusion with a sperm causes the release of calcium from membranous channels that ramify throughout the cytoplasm and the free calcium ion concentration goes up to between 1 and 3 μM and this increase then triggers exocytosis of the CG, most likely by a conserved pathway involving SNARE proteins. This exocytosis and its calcium requirement can be easily studied with sea urchin egg cortical granule "lawns," which provide a model system for studying the mechanism of exocytosis (see review by Wessel *et al.*, 2001).

a. Isolation of Cortical Granule Lawns

The basis of the isolation is that eggs are adhered to a charged plastic dish and then sheared away, leaving a layer of plasma membrane with intact CG on the surface of the dish (see Vacquier, 1975).

A solution of one mg/ml protamine sulfate is used to coat a small circular area (2 cm) on the bottom of 5-cm-diameter tissue culture dish. It is best to circle the specific area with a marking pen on the bottoms and lids. After 1 min, the protamine sulfate is washed away with a jet of tap water, and the dish is then air dried. The dish can also be positively charged by applying a 2-cm-diameter circle of 1% alcian blue in water and then washing the dish with water and allowing it to air dry. The protamine remains stuck to the dish surface as a thin film of positively charged protein; as the surface of the egg is negative in charge, the eggs will electrostatically bind the dish surface and flatten out as hemispheres. Alcian blue works by binding to negatively charged polysaccharides on the cell surface.

The jelly coat of the eggs must first be removed by the low pH method previously described. The eggs are then washed three times in fresh seawater by settling or by gentle centrifugation. A few drops of concentrated egg suspension (10%) are placed in the circle and allowed to settle for 2 min. This results in the eggs sticking to the dish surface.

To make the CG lawns, the dishes are filled with cortical granule isolation medium (CGIM), which is calcium-free artificial seawater (CaFSW) with 10 mM EGTA to chelate the calcium ions, and adjusted to pH 8.0. After swirling for about 2 min to chelate all the calcium ions, the dish is tipped on its side and the eggs are squirted with a jet of CGIM from a squirt bottle. Sufficient force is used so that the egg cytoplasm is sheared away, leaving a perfectly clean preparation of CG bound to the exposed inner surface of the egg plasma membrane. All the excess liquid should be removed by flicking. The cortical granules can be observed by immediately dropping on a coverslip and observing with 40× objective of an inverted phase contrast microscope. (If the coverslip is floating, the excess liquid can be removed with a pipette.) The CGs are visible as a population of very regular granules about 1 μm in diameter.

b. Calcium and Exocytosis

To determine the effects of calcium, one student can observe the CG lawn through the microscope and have a another student add one drop of seawater to the edge of the coverslip. The CG will explode when the calcium ions hit them. The wave of breakdown of the CGs can be followed by moving along with the seawater flow, using the stage micrometer and the low power objective. The explosion of the CG takes less than a second. This is the only available preparation of exocytotic vesicles that can be isolated in seconds in an intact and highly pure form that responds to μM concentrations of free calcium ions.

3. Examining the Role of Calcium in Egg Activation with the Use of Calcium Ionophores

Ionophores are low molecular weight natural products which dissolve in the plasma membrane or intracellular membranes of cells and make the membrane permeable to specific ions. For example, A23187 and ionomycin allow the movement of calcium and hydrogen ions across membranes from a region of higher concentration to one of lower concentration.

Ionophores can be used to ask questions about the role of calcium in egg activation without the use of sperm to fertilize the eggs. The principle of these experiments is to raise the calcium concentration in cells by adding ionophore to eggs. A stock solution of 2 mg/ml A23187, dissolved in 100% DMSO, can be used. Ten ul of this solution is added to 0.5 ml of seawater in a tube and mixed well. Then 0.5 ml of 2% vol/vol egg suspension is added, and a drop is immediately put on a slide and observed. Diluting the ionophore stock in this manner results in a 1% final concentration of DMSO and 38 μM A23187. Eggs should be observed quickly as the elevation of the FE occurs rapidly (see Steinhardt and Epel, 1974).

Students can then ask questions regarding differences between fertilization and ionophore activation. For example, they should be able to observe that, in sperm-fertilized eggs, the CG exocytosis sweeps around the egg cell as an expanding wave, beginning at the point of successful sperm interaction with the egg. In A23187-induced CG exocytosis, all the CG appear to fuse at the same instant and there is no wave of FE elevation; the FE elevates from the entire egg surface all in one instant.

Students can also use this system to confirm that the calcium responsible for the CG reaction is released from internal cytoplasmic stores. To test this, the eggs can be activated with the ionophore in calcium-free seawater. The ionophore will permeabilize the ER membranes that surround the internal calcium stores in addition to permeabilizing the plasma membrane. Note that a problem in seeing whether the eggs are activated is that the fertilization envelope will be very thin in calcium-free seawater, so careful observation, perhaps with phase microscopy, may be necessary. There will also be a time delay relative to the ability of the ionophore to stimulate CG exocyotosis in eggs in normal seawater, where there is a vast excess of calcium (10 mM).

D. Consequences of Parthenogenesis

Activating eggs, as has been described with ionophore, is a form of artificial parthenogenesis or initiating development without sperm. Students can ask questions about the ability of development to proceed in the absence of sperm by observing what happens to the eggs activated with ionophore. Questions that can be asked include does the egg pronucleus center as it does in normal fertilization? Does the nuclear envelope break down? Does a mitotic apparatus form? Does the egg divide? These experiments will enable students to focus on the role of sperm in fertilization, such as the contribution of the centrosome used to assemble the mitotic spindle (see Schatten, 1994).

E. Consequences of Polyspermy

The role of the sperm can also be addressed by intentionally inducing polyspermy. This can be accomplished by addition of an excess of sperm or by adding excess sperm in the presence of soybean trypsin inhibitor (1 mg/ml). The consequences of the extra centrosomes for cell division are the most obvious, with the embryos going from one cell to multiple cell numbers at the first division. Students can also look at the nature of the mitotic apparatus under polyspermic conditions, such as observing multiple mitotic spindles instead of the usual bipolar spindle. This experiment thus demonstrates that the sperm brings in the centrosome and also that multiple centrosomes results in multiple divisions. Students can also observe the consequences on later development; the embryos will be abnormal since the chromosomes are distributed unequally to the daughter cells (see Schatten, 1994). Figure 3 depicts a polyspermic egg containing three sperm nuclei as observed under the phase contrast microscope. The nuclei were made visible by fixing the eggs in a 3:1 alcohol:glacial acetic acid mixture and then resuspending the eggs in 45% glacial acetic acid.

F. The Role of Protein Synthesis in Cell Division

A fundamental discovery made with sea urchins was that the synthesis of a specific protein after fertilization is required for cell division. This was first demonstrated by examining the consequences of inhibiting protein synthesis at various times after insemination. It was observed that applying these inhibitors shortly after fertilization prevented the later cell division (Wilt *et al.*, 1967). Most puzzling, however, was the finding that application of these inhibitors just before prophase had no effect on the cell division. This suggested that a critical protein was necessary for the cell division, but was made early in the cycle. This protein either acted early in the cell cycle or was already all synthesized by prophase, since inhibiting protein synthesis at prophase or later had no effect on that cell division. The answer to this question came from experiments by Tim Hunt and his colleagues, who showed that sea urchin eggs made a protein after fertilization, that this protein began to be degraded as the cell entered mitosis and it was resynthesized for the next cycle. This synthesis and degradation was cyclic and related to the cell division and Hunt named this protein "cyclin" (Evans *et al.*, 1983). Subsequent work revealed that this periodic synthesis and degradation was a normal part of the cell cycle in all organisms (and led to Tim Hunt's receiving a Nobel Prize for this important discovery).

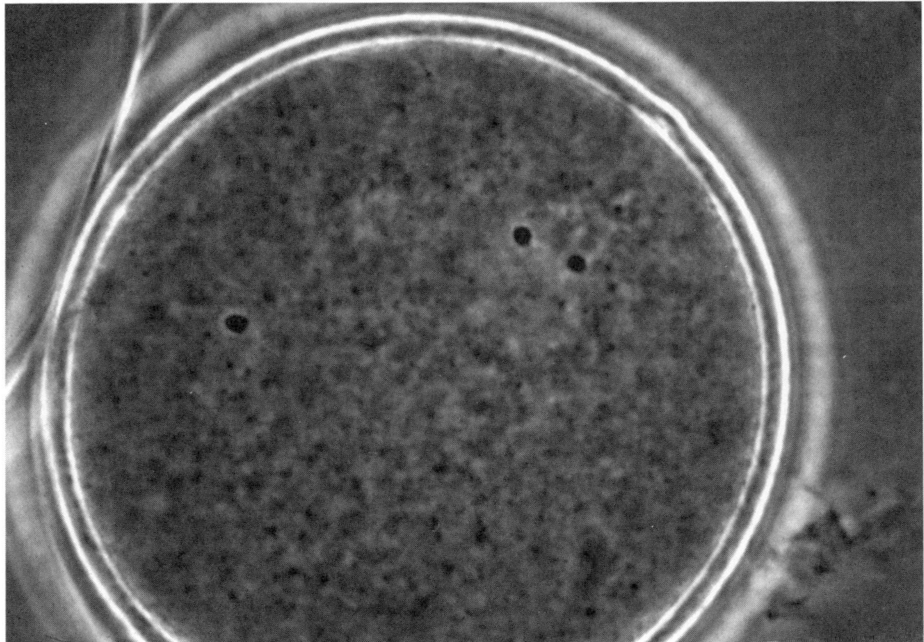

Fig. 3 Phase contrast micrograph of a polyspermic egg. Note the 3 sperm pronuclei, which are visible as dark spots within the egg. These eggs will not develop normally since each sperm brought in a centrosome, which will now direct the egg to form several mitotic apparatuses. These will compete for the chromosomes and there will be an unequal segregation of the chromosomes, which will upset the normal genetic regulation. The centrosomes will also each form a pole of a mitotic apparatus and the consequence will be that the cell will divide from one cell to 4 cells, corresponding to the number of centrosomes brought in by the sperm.

Students can demonstrate this essential protein synthesis in class. The experiment involves fertilizing eggs and then adding 100 uM emetine (a potent protein synthesis inhibitor) to the fertilized embryos at 10 to 15 min intervals after fertilization. The result will be that the cells will be prevented from entering mitosis until a specific time (in *S. pupuratus*, this is about 40 to 50 min after fertilization). After this time, the embryos will divide at the same time as the controls, but the next cell cycle will be inhibited. This concentration completely knocks out protein synthesis and indicates that synthesis of a protein is essential for cell division, but that this protein is completely synthesized long before the actual division takes place.

VI. Inquiry-Based Experiments in Later Development: Experiments on Differential Gene Expression

Sea urchin embryos are also excellent model systems for demonstrating differential gene expression during development. The expression of a number of cell-type or stage-specific genes can be followed by immunolocalization of the expressed protein using antibodies.

A. Immunolocalization of Cell Type Specific Proteins

This type of experiment utilizes antibodies generated against proteins that are known to be expressed in a cell-type and stage-specific manner to determine when and where during embryonic development expression of the protein occurs. Students can either work outside of the scheduled lab time to grow and fix embryos of the appropriate stage, or the instructor can provide fixed embryos and have the students do only the staining. (For upper-level students, these experiments work well as independent study units, with groups of students starting cultures of embryos, monitoring developmental progress, and fixing batches of embryos as they reach the appropriate developmental stage.)

1. Basic Protocol

Additional details regarding immunolocalization and a list of available antibodies can be found in Chapter 15 (Venuti *et al.*) of this volume. This method is a basic protocol that can easily be performed by undergraduates.

One of the easiest fixation methods that works well for staining embryos with many different antibodies is cold methanol. Embryos of the appropriate stages are collected by gentle centrifugation. At least 0.1 ml of gently packed embryos should be obtained in order to prevent excessive loss during the washing steps. The seawater supernatant is removed by aspiration and replaced by 10 mls of ice-cold methanol. The embryos are gently dispersed into the methanol by inverting the tube several times. The tube is placed on ice for 20 min, during which time the embryos settle into a loose pellet. At this point, they become relatively brittle and sticky and must be handled carefully—washing must be done by allowing them to settle by gravity rather than by centrifugation. The methanol is replaced by seawater, and this wash is repeated two more times to ensure that all the methanol is removed. Following fixation, embryos can be stored at 4 °C for a week or more; the addition of a few drops of sodium azide solution will slow their degradation. An alternative fixation is to use 3.7% formaldehyde in seawater for 10 min on ice, followed by three washes in seawater.

To stain the embryos, an aliquot of 100 ul from the pellet of embryos at each of the various stages to be tested is placed in a 1.5 ml microcentrifuge tube. As much of the seawater supernatant as possible is removed, and replaced by 100 ul of antibody solution. Monoclonal supernatants are generally used undiluted for this procedure. The tube is gently mixed and the embryos are incubated in the antibody solution for 1 h at room temperature. The antibody solution is carefully removed with a micropipette and the tube is filled with seawater. The embryos are allowed to settle (both the incubations and washes involve a lot of dead time, so you may wish to have additional activities for these long waiting periods). An additional seawater wash is done to remove any unbound antibody.

The second seawater wash is removed and 100 ul of the secondary antibody, conjugated to a fluorochrome, such as fluoroscein or Texas red, is then added at the appropriate dilution as directed. The tube is incubated for 30 to 60 min, then the embryos are washed with seawater as has been described. (If a fluorescence microscope is not available, an alternative method for localization is to use a secondary antibody coupled to an enzyme such as alkaline phosphate. A substrate is then added to form a colored precipitate at the site of antibody binding, which can then be visualized with a normal light microscope.)

After the last wash, embryos are allowed to settle to the bottom of the tube and then removed with a pipette tip and mounted on a slide in a small drop of mounting medium containing an antibleaching compound. A coverslip is added, and the edges are sealed with fingernail polish. Once dry, the slide can be viewed immediately or stored. We often have students complete the staining in one laboratory session and then come in at a scheduled time during the following week to do the microscopic analysis of their data, including image capture and analysis.

2. Experimental Questions

A number of inquiry-based labs can be designed based on this basic protocol. Antibodies to known germ layer-specific proteins can be used to ask questions about differentiation of cells and the regulation of gene expression both temporally and spatially. For example, monoclonal antibody 6a9 stains primary mesenchyme cells (PMCs) following their ingression and works well on most species of urchin. Students can fix embryos at key developmental stages, for example, unfertilized eggs, early cleavage, blastula stage, mesenchyme blastula, and various timepoints during gastrulation and later larval development. Students can then stain the stages with 6a9 to determine at what point in development the PMCs begin to express the protein recognized by 6a9. A complete list of useful antibodies and their species specificity is found in the Venuti chapter [Chapter 15 (see also Ettensohn, 1990)].

This tool can also be used as a way of determining the effect of various treatments known to perturb development (lithium chloride, nickel chloride, actinomycin D, etc.) on gene expression. It allows students to monitor specific molecular changes along with observing changes in morphological phenotype. Students can also try agents of their own choosing in similar types of experiments.

VII. Experiments Using Morphology as an Endpoint

In terms of inquiry-based education, the later phases of development lend themselves to asking many questions about how development is affected by environmental factors. Some examples could be the effects of UV irradiation; effects of common pollutants, such as detergents; personal care products, which have recently been shown to be present in the environment; hormone analogs; chemical pollutants, such as insecticides and herbicides. The student can pick a putative environmental agent and then observe the effects of this agent on later development. These experiments make excellent independent study projects, as different students can choose different agents.

VIII. Epilogue

The previously described exercises represent the tip of the iceberg of experiments that can be done with sea urchin embryos. Not covered here are experiments directly demonstrating new gene action during later development. This is an area of great research activity and intellectual ferment, but it has been omitted from this chapter since the laboratory study of such changes requires sophisticated equipment not available in most teaching laboratories. However, some experiments on the genetic changes in development can be done via computer analysis. As the sea urchin genome becomes available, many studies can be done with computer analysis of genes in the sea urchin genome and, for example, comparison with similar genes in other organisms or with similar genes in different species of sea urchins. The reader is referred to the sea urchin genome site http://sugp.caltech.edu:7000/.

References

Battaglia, D. E., and Shapiro, B. M. (1988). Hierarchies of protein cross-linking in the extracellular matrix: Involvement of an egg surface transglutaminase in early stages of fertilization envelope assembly. *J. Cell Biol.* **107,** 2447–2454.

Collins, F. D. (1976). A reevaluation of the fertilizin hypothesis of sperm agglutination and the description of a novel form of sperm adhesion. *Dev. Biol.* **49,** 381–394.

Collins, F. D., and Epel, D. (1977). The role of calcium ions in the acrosome reaction of sea urchin sperm: Regulation of exocytosis. *Exp. Cell Res.* **106,** 211–222.

Correa, L. M., and Carroll, E. J., Jr. (1997). Characterization of the vitelline envelope of the sea urchin *Strongylocentrotus purpuratus. Dev. Growth Differ.* **39,** 69–85.

Darszon, A., Beltran, C., Felix, R., Nishigaki, T., and Trevino, C. L. (2001). Ion transport in sperm signaling. *Dev. Biol.* **240,** 1–14.

Eisenbach, M., (2004). Towards understanding the molecular mechanism of sperm chemotaxis. *J. Gen. Physiol.* **124,** 105–108.

Epel, D., Weaver, A. M., and Mazia, D. (1970). Methods for removal of the vitelline membrane of sea urchin eggs. I. Use of dithiothreitol (Cleland's Reagent). *Exp. Cell Res.* **61,** 64–68.

Epel, D. (1997). Activation of sperm and egg during fertilization. *In* "Handbook of Physiology, Section 14: Cell Physiology" (J. F. Hoffman and J. J. Jamieson, eds.), pp. 859–884. Oxford Press, New York.

Ettensohn, C. A. (1990). The regulation of primary mesenchyme cell patterning. *Dev. Biol.* **140,** 261–271.

Evans, T., Rosenthal, E. T., Youngblom, J., Distel, D., and Hunt, T. (1983). Cyclin: A protein specified by maternal mRNA in sea urchin eggs that is destroyed at each cleavage division. *Cell* **33,** 389–396.

Hinkley, R. E., Wright, B. D., and Lynn, J. W. (1986). Rapid visual detection of sperm–egg fusion using the DNA-specific fluorochrome Hoechst 33342. *Dev. Biol.* **118,** 148–154.

Kay, E. S., and Shapiro, B. M. (1985). The formation of the fertilization membrane of the sea urchin egg. *In* "Biology of Fertilization" (C. B. Metz and A. Monroy, eds.), Vol. 3, pp. 45–81. Academic Press, San Diego, CA.

Matese, H. C., Black, S., and McClay, D. R. (1997). Regulated exocytosis and sequential construction of the extracellular matrix surrounding the sea urchin zygote. *Devel. Biol.* **186,** 16–26.

Mengerink, K. J., and Vacquier, V. D. (2001). Glycobiology of sperm–egg interactions in deuterostomes. *Glycobiology* **11,** 37R–43R.

Mazia, D., Schatten, G., and Sale, W. (1975). Adhesion of cells to surfaces coated with polylysine. Applications to electron microscopy. *J. Cell Biol.* **66,** 198–200.

Schatten, G. (1994). The centrosome and its mode of inheritance: The reduction of the centrosome during gametogenesis and its restoration during fertilization. *Dev. Biol.* **165,** 299–335.

Shapiro, B. M. (1991). The control of oxidant stress at fertilization. *Science* **252,** 533–536.

Sluder, G., Miller, F. J., and Hinchcliffe, E. H. (1999). Using sea urchin gametes for the study of mitosis. *Meth. Cell Biol.* **61,** 439–472.

Steinhardt, R. A., and Epel, D. (1974). Activation of sea urchin eggs by a calcium ionophore. *Proc. Nat. Acad. Sci. USA* **71,** 1915–1919.

Vacquier, V. D., Epel, D., and Douglas, L. (1972). Sea urchin eggs release protease activity at fertilization. *Nature* **237,** 34–36.

Vacquier, V. D. (1975). The isolation of intact cortical granules from sea urchin eggs: Calcium ions trigger granule discharge. *Dev. Biol.* **43,** 62–74.

Ward, G. E., Brokaw, C. J., Garbers, D. L., and Vacquier, V. D. (1985). Chemotaxis of Arbacia punctulata spermatozoa to resact, a peptide from the egg jelly layer. *J. Cell Biol.* **101,** 2324–2329.

Wessel, G. M., Brooks, J. M., Green, E., Haley, S., Wong, J., Zaydfudim, V., and Conner, S. (2001). The biology of cortical granules. *Int. Rev. Cytol.* **209,** 117–206.

Weidman, P. J., Teller, D. C., and Shapiro, B. M. (1987). Purification and characterization of proteoliaisin, a coordinating protein in fertilization envelope assembly. *J. Biol. Chem.* **262,** 15076–15084.

Wilt, F. H., Sakai, H., and Mazia, D. (1967). Old and new protein in the formation of the mitotic apparatus in cleaving sea urchin eggs. *J. Mol. Biol.* **27,** 1–7.

APPENDIX

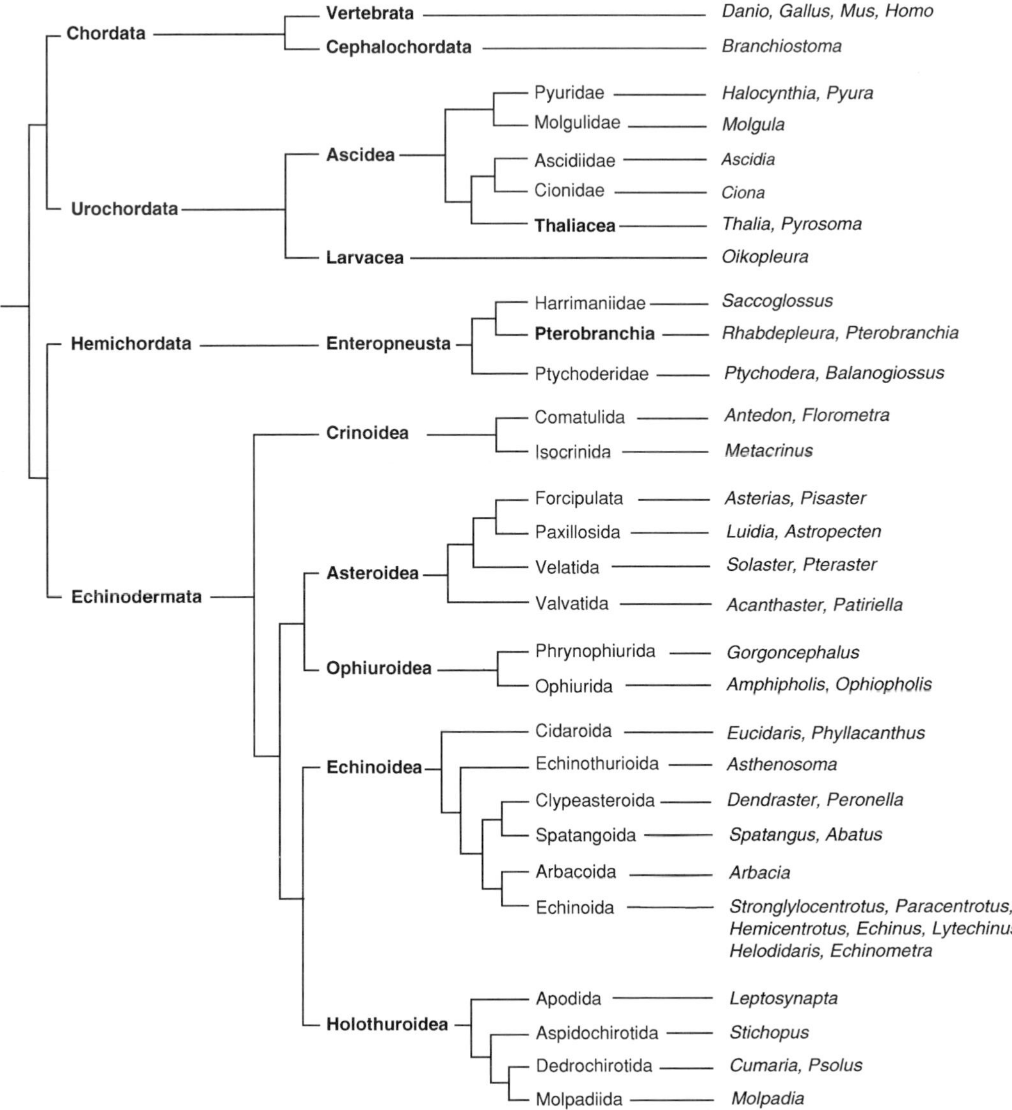

Outline of deuterostome phylogenetic relationships. The major lineages of deuterostomes are shown, with phyla in the left column, major subdivision within each phylum next (sub-phyla or classes), and representative smaller groups (usually orders or families); the right column lists representative genera

METHODS IN CELL BIOLOGY, VOL. 74

0091-679X/04 $35.00

Websites that may be of interest to researchers working with basal deuterostomes.

General Resources

- Echinoderm links:
- *http://www.calacademy.org/research/izg/echinoderm/echilink.htm*, an extensive listing of websites of interest to echinoderm biologists.

- The echinoderm newsletter:
- *http://www.nmnh.si.edu/iz/echinoderm/* and *http://www.nrm.se/ev/echinoderms/echinonews.html.en*.
- The newsletter contains information concerning meetings and conferences, publications of interest to echinoderm biologists, titles of doctoral theses on echinoderms, and research interests and addresses of echinoderm biologists.

Experimental Procedures:

- Fluorescent WMISH methods from D. Kosman: *http://www.biology.ucsd.edu/~davek/*.
- The ABI site *www.appliedbiosystems.com* contains an excellent primer entitled "DNA/RNA Real Time Quantitative PCR"

---

from each group, with an emphasis on genera represented in the literature of developmental biology. Note that some traditional sub-phyla and classes are likely to be paraphyletic (meaning that they encompass other groups of the same taxonomic rank). Examples are the Thaliacea, which is embedded within the Ascidea, and the Pterobranchia, which is embedded within the Enteropneusta. For larger groups, such as the Vertebrata, Ascidea, and all of the echinoderm classes only representative subgroups are shown; these contain the taxa most commonly employed in developmental studies. Common names for the major groups and characteristics for the less familiar groups, are as follows (top to bottom): Vertebrata = backboned animals, including fishes, amphibians, reptiles, birds, and mammals; Cephalochordata = lancets, also known as amphioxus; Ascidea = sea squirts, also known as tunicates (sessile, filter-feeding adults; can be solitary or colonial); Larvacea = appendicularians (diminutive, pelagic animals that construct a mucus "house" used for filter feeding); Thaliacea = salps (pelagic, solitary or colonial animals); Enteropneusta = acorn worms (soft, worm-like animals; many build tubes in the sediment); Pterobranchia = pterobranchs (small, sessile, filter-feeders); Crinoidea = sea lilies and feather stars; Asteroidea = sea stars; Ophiuroidea = basket stars, snake stars, and brittle stars; Echinoidea = sea urchins, sand dollars, and sea biscuits; Holothuroidea = sea cucumbers. The phylogenetic relationships shown are based on: Littlewood, D. T. J., and Smith, A. B. (1995). *Phil. Trans. Roy. Soc. B* **347,** 213–234; Smith, A. B., Littlewood, D. T. J., and Wray, G. A. (1995). *Phil. Trans. Roy. Soc. B* **349,** 11–18; Cameron, C. B., Garey, J. R., and Swalla, B. J. (2000). *Proc. Natl. Acad. Sci. USA* **97,** 4469–4474; Swalla, B. J., Cameron, C. B., Corley, L. S., and Garey, J. R. (2000). *Syst. Biol.* **49,** 122–134; Knott, K. E., and Wray, G. A. (2000). *Amer. Zool.* **40,** 382–392; Kerr, A. M., and Kim, J. H. (2001). *Zool. J. Linn. Soc. Lond.* **133,** 63–81.

- *http://docs.appliedbiosystems.com/pebiodocs/00777904.pdf.* Methods for measuring relative changes in RNA levels by DNA microarray hybridization can be found at *http://www.affymetrix.com.*
- The Gene Tools web site contains an extensive bibliography of work employing morpholinos and information concerning their design: *www.gene-tools.com.*
- Detailed protocols for long-term storage of sea urchin eggs are described at *http://www.stanford.edu/group/Urchin/protocol.html.*
- The on-line manual for microinjection and additional occyte handling methods are available from Laurinda Jaffe and Mark Terasaki *http://155.37.3.143/panda/injection/index.html.*
- The Axon guide for Electrophysiology and Biophysics Laboratory Techniques (*http://www.axon.com/mr_Axon_Guide.html*) is helpful in setting up the proper acquisition and analysis environment for ion channel work.

Genome Resources

The *C. intestinalis* genome assembly is accessible at *http://jgi.doe.gov/ciona.* Many unassembled sequences also have been made public through NCBI Megablast site (*http://www.ncbi.nlm.nih.gov/blast/tracemb.html*).

Other genome resources of two related species, *Ciona intestinalis* and *Ciona savignyi*, can (Dehal *et al.*, 2002) be found at *http://www-genome.wi.mit.edu/annotation/ciona/background.html*).

A number of EST studies have been published for *Halocynthia roretzi* (MAJEST: ***M***aboya ***G***ene ***E***xpression patterns and ***S***equence ***T***ags; *http://www.genome.ad.jp/magest/*).

All information concerning the *Ciona intestinalis* cDNA projects is accessible at *http://ghost.zool.kyoto-u.ac.jp/indexr1.html.*

The cDNA projects of ascidians include analyses of Expressed Sequence Tags (ESTs), cDNA sequence information, and spatial expression profiles of developmental genes. Besides small scale analyses from various ascidian species, a large-scale analysis was first aimed at maternal transcripts of *Halocynthia roretzi* (Kawashima *et al.*, 2000; Makabe *et al.*, 2001). This project, called "MAGEST", includes ~40,000 EST data and spatial expression profiles of ~1200 maternally expressed genes (Makabe, personal communication; Makabe *et al.*, 2001). All its current information is available at the project's website (*http://www.genome.ad.jp/magest/*).

The sea urchin genome page contains links to genome sequence and other genome resources (*http://sugp.caltech.edu/*). A large-scale provisional GRN for embryonic endomesoderm specification has been developed for sea urchin (*Strongylocentrotus purpuratus*) embryos. For it's current version see *http://www.its.caltech.edu/~mirsky/endomes.htm.*

A number of DNA resources have been generated for amphioxus and made freely available. The mitochondrial genomes of *B. floridae, B. lanceolatum and B. belcheri* have been sequenced (Boore *et al.*, 1999; Spruyt *et al.*, 1998). These sequences are available at *http://megasun.bch.umontreal.ca/ogmp/projects/other/mt_list.html*. In addition, gridded cDNA libraries of late neurula (26 hr) and adults, as well as two cosmid libraries of *B. floridae* are available at cost from the RZPD in Berlin [*http://www.RZPD.de*]. The 26-hr library has been subject to EST analysis (G. Panopoulou, unpublished). A genome library in pCYPAC7 has been constructed by Chris Amemiya, Virginia Mason Research Center Benaroya Research Institute, Seattle, WA 98101 USA. This library was made from 6 individuals and has an average insert size of 90 kb. It has approximately 5-fold coverage of the genome and is freely available [*http://www.vmresearch.org/*]. A large-insert BAC library of *B. floridae* made from a single ripe male has recently been constructed by the laboratory of P. de Jong, CHORI, Oakland, CA USA and will be available at cost from the BACPAC resource center at CHORI [*http://www.chori.org/bacpac/*].

Educational Resources

The most comprehensive is "Sea Urchin Embryology" developed and maintained by Chris Patton of the Epel lab at *http://www.stanford.edu/group/Urchin/index.html*. Almost 300 pages long, it provides information, diagrams, videos, animations and overheads on how to obtain sea urchins for classroom use, suggested lesson plans, suggestions for advanced experiments and numerous videos and animations useful in the classroom setting. Spanish versions of the major parts of the site are also provided.

There are also excellent student exercises, handouts and transparency texts available on the sea urchin embryology website (*http://www.stanford.edu/group/Urchin/protocol.html*).

The Society for Developmental biology web site [http://sdb.bio.purdue.edu/] contains several links to educational and research web sites of relevance to investigators of basal deuterostomes.

The green sea urchin web page containing general resources on echinoderm gametogenesis is at *http://zoology.unh.edu/faculty/walker/urchin/gametogenesis.html*.

Miscellaneous Resources

Stocks of most algae species and strains for feeding larvae of basal deuterostomes can be obtained from UTEX (*www.bio.utexas.edu/research/utex*), Provasoli-Guillard National Center for Culture of Marine Phytoplankton (CCMP) *http://ccmp.bigelow.org* or commercial sources (e.g., Carolina Biological Supply Co., *www.carolina.com*) in the US. A useful source in Japan is the National Research Institute of Aquaculture, Fisheries Research Agency (*www.nria.affrc.go.jp*). In Europe, these can be obtained from the Scandinavian Culture center for algae and protozoa. Botanical Inst. Univ. Copenhagen. Øster Far Farimagsgade 2D, DK-1353 Copenhagen K, Denmark; Curator at SCCAP: Niels H. Larsen *http://www.sccap.bot.ku.dk*.

(2) Culture Collection of Marine Micro-Algae, Instituto de Ciencias marinas de Andalucia (ICMAN) Cadiz Spain, *http://www.icman.csic.es/servic/servCmicroalgas_en.htm.*

In Australia (1) CSIRO collection of microalgae. Orders and enquiries to: Ms Cathy Johnston, CSIRO Marine Research, GPO Box 1538, Hobart, Tasmania 7001, Australia, Telephone: +61 (3) 6232 5316 (international); (03) 6232 5316 (Australia); Facsimile: +61 (3) 6232 5000 (international); (03) 6232 5000 (Australia); Email: *microalgae@marine.csiro.au.*

The animal collection staff of the Marine Resources Center at the Marine Biological Laboratory (*www.mbl.edu*) has extensive experience in both collecting and shipping a variety of basal deuterostomes.

Several different lineage-specific monoclonal antibodies are currently available from various sea urchin labs and the Developmental Studies Hybridoma Bank (*http://www.uiowa.edu/~dshbwww/*).

Formulae for General Solutions

Note that several different formulations are available for some of the artificial seawaters. Prepare the artificial seawater at room temperature using deionized H_2O. Add components to a volume of H_2O that is less than the final, desired volume and allow them to dissolve completely. Then, chill or warm the artificial seawater to the desired temperature and only then adjust the pH to 8.0 with NaOH followed by a final adjustment of the volume with H_2O. All recipes are for 1 liter of solution unless otherwise noted.

Artificial Sea Water (ASW)

28.3 g NaCl

0.77 g KCl

5.41 g $MgCl_2$ $6H_2O$

3.42 g $MgSO_4$ or 7.13 g $MgSO_4$ $7H_2O$

0.2 g $NaHCO_3$

1.56 g $CaCl_2$ $2H_2O$ (*add last*)

pH to 8.2, salinity in the range of 34-36 ppt

Calcium-Free Sea Water (CFSW)

26.5 g NaCl

0.7 g KCl

5.81 g $MgSO_4$ or 11.9 g $MgSO_4$ $7H_2O$

0.5 g NaHCO3

pH to 8.2, salinity in the range of 34–36 ppt

Calcium/Magnesium Free Sea Water (CMFSW)

31 g NaCl
0.8 g KCl
0.2 g $NaHCO_3$
1.6 g Na_2SO_4
pH to 8.2, salinity in the range of 34–36 ppt

Hyalin Extraction Media (HEM)

17.5 g NaCl
0.75 g KCl
1.25 g $MgSO_4$ or 2.5 g $MgSO_4$ $7H_2O$
22.5 g Glycine
1.21 g Tris
0.76 g EGTA
pH to 8.2, salinity in the range of 34–36 ppt

Low Sodium-Choline Substituted Seawater (Jaffe, 1980; Schuel and Schuel, 1981)

67.2 g Choline Cl[1,2]
0.76 g KCl
7.2 g $MgSO_4$ $7H_2O$
5.6 g $MgCl_2$ $6H_2O$
0.2 g $NaHCO_3$
1.6 g $CaCl_2$ $2H_2O$

Sodium-Free Seawater (NaFASW; Keller *et al.*, 1999)[2]

84.7 g Glycerol
0.745 g KCL
1.62 g $CaCl_2$ $2H_2O$
5.29 g $MgCl_2$ $6H_2O$
7.15 g $MgSO_4$ $7H_2O$
0.2 g $KHCO_3$

From Foltz *et al.*, Chapter 3.

[1]Choline Chloride has a limited shelf life. It normally is a white crystal and hydroscopic; if discolored, do not use. It can be purchased from SIGMA, catalog #C-1879. For the choline substituted seawater, adjust the pH to 8.0 using 1 M NaOH before adding the $CaCl_2$. Then add the $CaCl_2$, bring the solution up to volume, and readjust pH to 8.0 using 1 M NaOH.

[2]For both the choline-substituted and sodium-free ASW, keep track of how much NaOH is used so that the total amount of sodium being added can be calculated. Alternatively, use NH_4OH to adjust the pH.

Dissociation medium (DM; after Harkey and Whiteley, 1980)

1.0 M glycine
100 microM EDTA
adjusted to pH 8 with NaOH

See Chapter 12. By Fred Wilt.

Home-made anti-fade mounting medium (Elvanol):

0.2 M Tris buffer, pH 6.5 (12 ml)
Polyvinyl alcohol (2.4 g; average MW 30–70 × 10^3)
Glycerol (6.0 g)
H_2O (6.0 ml)
DABCO (1,4-Diazabicyclo[2.2.2]octane; Sigma cat # D2522) (0.8 g) – anti-fade reagent

From Wessel *et al.*, Chapter 5.

Sea Urchin Suppliers

See Table 1 for breeding seasons to determine when particular species are gravid. Adult sea urchins can be obtained from the following suppliers:

Marinus Scientific
11771 St. Mark St.
Garden Grove, CA, 92845
Tel. 714-901-9700 (*info@marinusscientific.com*)

Pacific BioMarine Laboratories, Inc.
PO Box 536
Venice, CA 90291
Tel. 310-677-1056

Marine Biological Laboratory Resources Center
7 MBL Street
Woods Hole, MA, 02543-1015
Tel. 508-289-7700

Carolina Biological Supply Co.
2700 York Road
Burlington, NC 27215
Tel. 800-334-5551

Beaufort Biological
135 Duke Marine Lab Road
Beaufort, NC 28516
Tel. 252-504-7567, fax 252-504-7648

Sue Decker
141400 SW 22nd Place
Davie, FL 33325
Tel. 954-424-2620

Gulf Specimen Marine Laboratories, Inc.
222 Clark Ave., PO Box 237
Panacea, FL 32346
Tel. 850-984-5297 (*gspecimen@sprintmail.com, www.gulfspecimen.org*).

Charles Hollahan
tidalflux@yahoo.com

Ascidian Suppliers

MBL Woods Hole, MA	FHL Friday Harbor, WA	Stazione Zoologica Naples, Italy	Station Biologique Roscoff, France	Japan	Hopkins Marine Station, Pacific Grove, CA
Ciona intestinalis	Ascidia columbiana A. paratropa	Ciona intestinalis	Ciona intestinalis Phallusia mammillata	Ciona savignyi Ciona intestinalis	Ciona intestinalis Ascidia ceratodes
Styela clava	Boltenia villosa		Styela clava	Halocynthia roretzi Styela plicata Herdmania pallida	
Botryllus schlosseri Botrylloides violaceus	Botrylloides violaceus	Botryllus schlosseri	Botryllus schlosseri	Botrylloides violaceus Botryllus schlosseri	B. violaceus Botryllus
Molgula manhattensis M. provisionalis M. citrina	Molgula pugetiensis		Molgula oculata M. occulta M. citrina (echinosiphonica) M. socialis Phallusia mammillata	Molgula tectiformis	

Table I

Summary of species information relating to collection of echinoids (sea urchins) and their breeding season. The reproductive seasons have been compiled for sea urchins from different populations. In different geographical areas breeding seasons can therefore vary slightly but will be within the reported range. Further information on British species can be found at http://www.marlin.ac.uk/. See Strathmann (1987a) and Giese *et al.* (1991) for a more extensive list of species

Species	Size	Depth	Substrate	Distribution	Reproductive season
Anthocidaris crassispina	<7 cm	0–70 m	rock	Japan, Southern China	June–August
Arbacia punctulata	<5.6 cm	0–230 m	rock, shell	Cape Cod-Florida, Texas, Yucatan, Cuba, Jamaica, West Indies	May–September
Echinus esculentus	15–16 cm	0–40 m	rock	British Isles	February–June
Hemicentrotus pulcherrimus	<5 cm	0–40 m	rock, coarse gravel	Japan, Northern China, Korea	January–March
Lytechinus variegatus	<7.6 cm	0–55 m	sand, gravel	North Carolina-Florida, Bahamas, West Indies	December–July
Loxechinus albus	<7.5 cm	0–15 m	rock, gravel sand, seagrass	Chile, Peru	July–September
Paracentrotus lividus	<7 cm	<3 m	rock	Mediterranean, western Atlantic	April–May
Psammechinus miliaris	<5.5 cm	0–10 m	coarse gravel	Britain & Ireland	February–November
Pseudocentrotus depressus	<8 cm	0–50 m	rock	Japan	October–December
Strongylocentrotus droebachiensis	<8.3 cm	0–1160 m	rock	Arctic-New Jersey, Alaska-Puget Sound	January–April
Strongylocentrotus fransciscanus	<12.7 cm	0–91 m	rock	Alaska-Baja California	February–August
Strongylocentrotus intermedius	<8 cm	0–40 m	rock	Pacific coasts of Asia and Siberia	August–October
Strongylocentrotus nudus	<8 cm	0–180 m	rock	Pacific coasts of Asia and Siberia	September–November
Strongylocentrotus *purpuratus*	<10.2 cm	0–160 m	rock	Alaska-Baja California	November–June
Tripneustes gratilla	<12 cm	0–30 m	hard substrate	Circumtropical extending to tropical areas in Indian and West Pacific Ocean	July–September (in Japan)

Table II

Summary of Species Information Relating to collection of asteroids (sea stars) and their reproductive season. The reproductive seasons have been compiled for sea stars from different populations. In different geographical areas breeding seasons can therefore vary slightly but will be within the reported range. Further information on British species can be found at http://www.marlin.ac.uk/

Species	Size	Depth	Substrate	Distribution	Reproductive season
Asterias amurensis	<40 cm	5–200 m	sand	South of Alaskan Peninsula, British Columbia, Canada, Japan, invading Australia & Tasmania	February–April (in Japan)
Asterias forbesi	<13 cm	0–49 m	rock, gravel, sand	Gulf of Maine-Texas (South of Cape Cod)	May–July
Asterias rubens	10–30 cm	30–100 m	gravel rock		February–April
Asterias vulgaris	<20 cm	0–349 m	rock, gravel	Labrador-Cape Hatteras (North of Cape Cod)	May–July
Asterias miniata	<10.2 cm	0–293 m	rock	Alaska-Baja California	January–November
Asterina pectinifera	<20 cm	0–100 m	rock, gravel, sand	Japan	May–July

From Walker, Chapter 2.

Table III

Brittlestars	Breeding season
Ophioderma brevispinum	May–August
Ophiopholis aculeata	April–September[a]
Ophiothrix angulata	March–September
Ophiothrix fragilis	June–September

[a]This species of brittle star has been reported to spawn in the lab January–April, July and October–November (Hendler, 1991; Strathmann and Rumrill, 1987).

Sea cucumbers	Breeding season
Psolus chitonoides	mid March–mid May
Psolus californicus	mid May–mid July

From Foltz *et al.*, Chapter 3.

Table IV
Culture Temperatures and Developmental Time Courses for Selected Echinoderm Species. See Strathmann (1987a) and Giese *et al.* (1991) for a More Extensive List of Species

Time after insemination	Developmental stage
S. purpuratus raised at 12 °C (see Strathmann, 1987b)	
2–2.5 hours	1st cleavage
4 hours	2nd cleavage
5.5 hours	3rd cleavage
6.5 hours	4th cleavage
27 hours	Hatching of blastula from FE
30 hours	Gastrulation
3 days	Early prism larva
4 days	Early pluteus larva
40–80 days	Metamorphosis
L. variegatus raised at 25 °C (Schoenwolf, 1995)	
1 hour	1st cleavage
1.5 hours	2nd cleavage
2 hours 40 min	32 cell stage
7.5 hours	Hatching of blastula from FE
11 hours	Gastrulation

(*continues*)

Table IV (*continued*)

Time after insemination	Developmental stage
15 hours	Early prism larva
20 hours	Early pluteus larva
L. pictus raised at 18 °C (Sea Urchin Embryology website)	
1.5 hours	1st cleavage
2.5 hours	2nd cleavage
24 hours	Blastula
2 days	Gastrula
5 days	Pluteus
A. miniata raised at 15 °C (see Strathmann, 1987c)	
2.5 hours	1st cleavage
9 hours	Blastula
18 hours	Hatching of blastula from FE
2 days	Gastrulation is complete
3–4 days	Bipinnaria larva
1 month	Brachiolaria larva
2 months	Metamorphosis
2 months and 1 day	Juvenile starfish
Ophiopholis aculeata raised at 8 °C (see Hendler, 1991; Strathmann and Rumrill, 1987)	
3 hours	1st cleavage
24–36 hours	Blastula
1.5–2 days	Gastrula
3 days	2-Arm stage
10.5 days	6-Arm stage
17 days	8-Arm stage (ophiopluteus)
30 days	Incipient metamorphosis
83–216 days	Complete metamorphosis
Ophiothrix angulata raised at 24–27 °C (see Hendler, 1991)	
0.5 days	Gastrula
1 day	2-Arm stage
2.5 days	6-Arm stage
7 days	8-Arm stage
>19 days	Complete metamorphosis
Ophiothrix fragilis (temperature not specified; see Hendler, 1991)	
24 hours	Blastula
1.5 days	Gastrula
1.5 – 2 days	2-Arm stage
3 days	4-Arm stage
4 days	6-Arm stage
7–10 days	8-Arm stage
16 days	Incipient metamorphosis
26 days	Complete metamorphosis
P. chitonoides raised at 11 °C (see McEuen, 1987)	
2 hours	1st cleavage
8 hours	32 cell stage

(*continues*)

Table IV (*continued*)

Time after insemination	Developmental stage
18 hours	Blastula
40 hours	Gastrula
60–70 hours	Early Larva
80–90 hours	Early Doliolaria
11–12 days	Settlement
26–32 days	Ossicle plates form
P. californicus raised at 11 °C (see McEuen, 1987)	
2.75 hours	1st cleavage
10.75 hours	32 cell stage
19 hours	Blastula
27 hours	Gastrula
3.25–4.25 days	Early Larva
52 days	Early Doliolaria
60–61 days	Settlement
61–65 days	Ossicle plates form

From Foltz.

Table V
Time Table for *Ciona intestinalis* Development (18 °C)

Stage	Time after fertilization
2-cell	60 min.
4-cell	90 min.
8-cell	120 min.
16-cell	160 min.
32-cell	210 min.
64-cell	250 min.
Early gastrula	5 hr.
Neurula	7 hr.
Early tailbud	9 hr.
Otolith pigmentation	12 hr.
Ocellus pigmentation	15 hr.
Tadpole	18 hr.

From Swalla, Chapter 6.

Table VI
Schedule of Development for *Branchiostoma floridae* (Amphioxus)

Stage	Time at 25 °C	Time at 30 °C
Insemination	0 h	
Fertilization envelope begins to elevate	30 sec	
Second polar body	10 min	
Fertilization envelope completes elevation	20 min	
2-cell	45 min	30 min
4-cell	60 min	50 min
8-cell	90 min	60 min
16-cell	2 hr	75 min
32-cell	2.25 hr	90 min
64-cell	2.5	105 min
128-cell	3 hr	2 hr
Blastula	4 hr	2.5 hr
Onset of gastrulation	4.5 hr	3.5 hrs
Capped-shaped gastrula	5 hr	4 hr
Cupped-shaped gastrula	6 hr	4.5 hr
Late gastrula/early neurula	6.5 hr	5 hr
Neurula, ciliated begins rotating inside fertilization envelope	8.5 hr	6 hr
Hatching	9.5 hr	6.5 hr
Anterior somites visible	10.5 hrs	7.5 hr
Late neurula; Onset of muscular movement	20 hrs	12 hr
Mouth and first gill slit	30–32 hr	23–24 hr
Anus open	36 hr	28 hr
2 gill slits	3 da	36 hr
3 gill slits	6 da	3 da
4 gill slits	15 da	7 da
6 gill slits	23 da	14 da
8 gill slits	29 da	19 da
11 gill slits	37 da	25 da
Metamorphosis	41–49 da	26–32 da

From Holland, Chapter 9.

Cell Lineage Chart Through Tenth Cleavage in the Sea Urchin *Lytechinus variegatus*

Cleavage Number

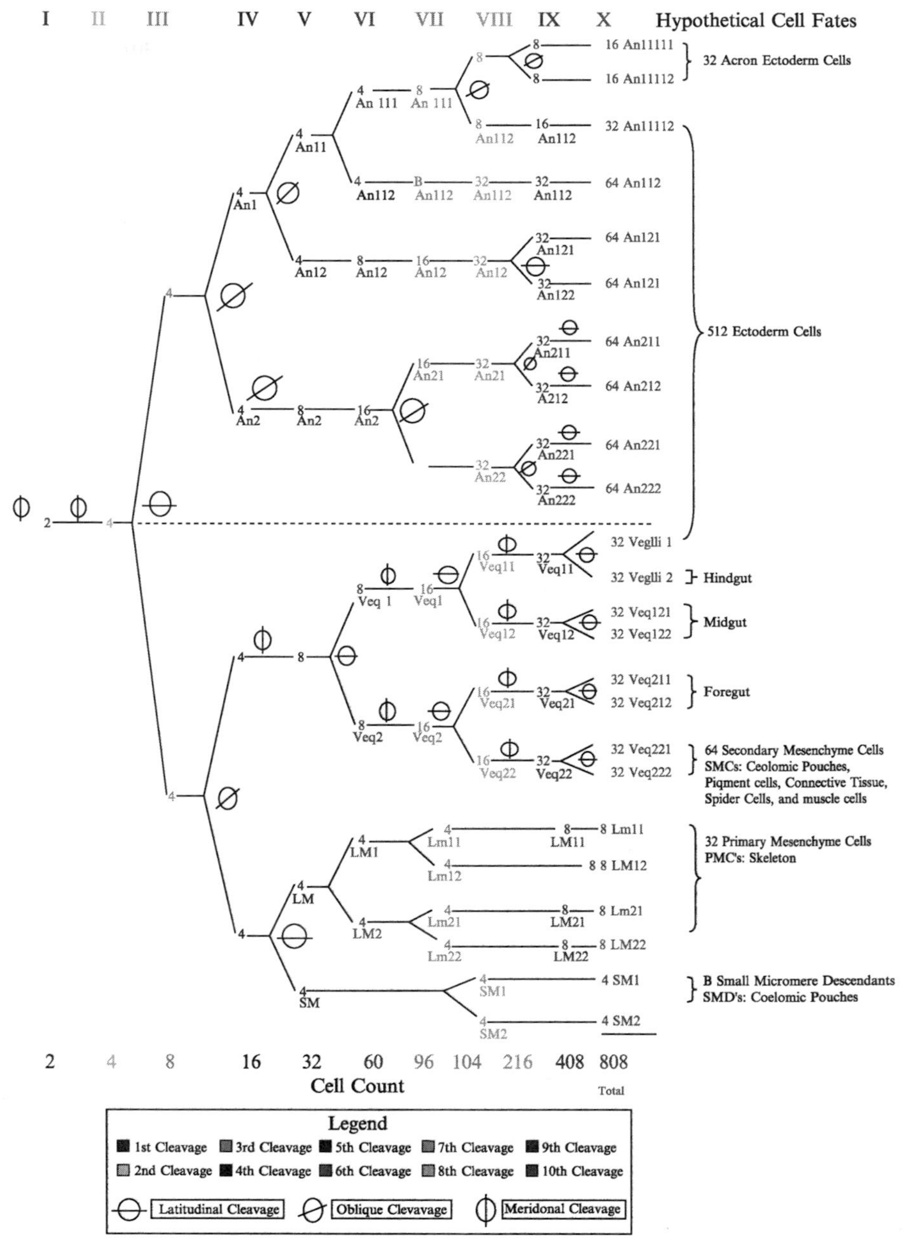

J. B. Morrill and Lauren Marcus, 2003. (See Color Insert.)

INDEX

G

H

M

Q

R

U

V

W

X

Y

VOLUMES IN SERIES

Founding Series Editor
DAVID M. PRESCOTT

Volume 1 (1964)
Methods in Cell Physiology
Edited by David M. Prescott

Volume 2 (1966)
Methods in Cell Physiology
Edited by David M. Prescott

Volume 3 (1968)
Methods in Cell Physiology
Edited by David M. Prescott

Volume 4 (1970)
Methods in Cell Physiology
Edited by David M. Prescott

Volume 5 (1972)
Methods in Cell Physiology
Edited by David M. Prescott

Volume 6 (1973)
Methods in Cell Physiology
Edited by David M. Prescott

Volume 7 (1973)
Methods in Cell Biology
Edited by David M. Prescott

Volume 8 (1974)
Methods in Cell Biology
Edited by David M. Prescott

Volume 9 (1975)
Methods in Cell Biology
Edited by David M. Prescott

Advisory Board Chairman
KEITH R. PORTER

Volume 21A (1980)
Normal Human Tissue and Cell Culture, Part A: Respiratory, Cardiovascular, and Integumentary Systems
Edited by Curtis C. Harris, Benjamin F. Trump, and Gary D. Stoner

Volume 21B (1980)
Normal Human Tissue and Cell Culture, Part B: Endocrine, Urogenital, and Gastrointestinal Systems
Edited by Curtis C. Harris, Benjamin F. Trump, and Gray D. Stoner

Volume 22 (1981)
Three-Dimensional Ultrastructure in Biology
Edited by James N. Turner

Volume 23 (1981)
Basic Mechanisms of Cellular Secretion
Edited by Arthur R. Hand and Constance Oliver

Volume 24 (1982)
The Cytoskeleton, Part A: Cytoskeletal Proteins, Isolation and Characterization
Edited by Leslie Wilson

Volume 25 (1982)
The Cytoskeleton, Part B: Biological Systems and *in Vitro* Models
Edited by Leslie Wilson

Volume 26 (1982)
Prenatal Diagnosis: Cell Biological Approaches
Edited by Samuel A. Latt and Gretchen J. Darlington

Series Editor
LESLIE WILSON

Volume 27 (1986)
Echinoderm Gametes and Embryos
Edited by Thomas E. Schroeder

Volume 28 (1987)
***Dictyostelium discoideum:* Molecular Approaches to Cell Biology**
Edited by James A. Spudich

Volume 29 (1989)
Fluorescence Microscopy of Living Cells in Culture, Part A: Fluorescent Analogs, Labeling Cells, and Basic Microscopy
Edited by Yu-Li Wang and D. Lansing Taylor

Volume 30 (1989)
Fluorescence Microscopy of Living Cells in Culture, Part B: Quantitative Fluorescence Microscopy—Imaging and Spectroscopy
Edited by D. Lansing Taylor and Yu-Li Wang

Volume 31 (1989)
Vesicular Transport, Part A
Edited by Alan M. Tartakoff

Volume 32 (1989)
Vesicular Transport, Part B
Edited by Alan M. Tartakoff

Volume 33 (1990)
Flow Cytometry
Edited by Zbigniew Darzynkiewicz and Harry A. Crissman

Volume 34 (1991)
Vectorial Transport of Proteins into and across Membranes
Edited by Alan M. Tartakoff

Selected from Volumes 31, 32, and 34 (1991)
Laboratory Methods for Vesicular and Vectorial Transport
Edited by Alan M. Tartakoff

Volume 35 (1991)
Functional Organization of the Nucleus: A Laboratory Guide
Edited by Barbara A. Hamkalo and Sarah C. R. Elgin

Volume 36 (1991)
Xenopus laevis: **Practical Uses in Cell and Molecular Biology**
Edited by Brian K. Kay and H. Benjamin Peng

Series Editors
LESLIE WILSON AND PAUL MATSUDAIRA

Volume 37 (1993)
Antibodies in Cell Biology
Edited by David J. Asai

Volume 38 (1993)
Cell Biological Applications of Confocal Microscopy
Edited by Brian Matsumoto

Volume 39 (1993)
Motility Assays for Motor Proteins
Edited by Jonathan M. Scholey

Volume 40 (1994)
A Practical Guide to the Study of Calcium in Living Cells
Edited by Richard Nuccitelli

Volume 41 (1994)
Flow Cytometry, Second Edition, Part A
Edited by Zbigniew Darzynkiewicz, J. Paul Robinson, and Harry A. Crissman

Volume 42 (1994)
Flow Cytometry, Second Edition, Part B
Edited by Zbigniew Darzynkiewicz, J. Paul Robinson, and Harry A. Crissman

Volume 43 (1994)
Protein Expression in Animal Cells
Edited by Michael G. Roth

Volume 44 (1994)
***Drosophila melanogaster:* Practical Uses in Cell and Molecular Biology**
Edited by Lawrence S. B. Goldstein and Eric A. Fyrberg

Volume 45 (1994)
Microbes as Tools for Cell Biology
Edited by David G. Russell

Volume 46 (1995) (in preparation)
Cell Death
Edited by Lawrence M. Schwartz and Barbara A. Osborne

Volume 47 (1995)
Cilia and Flagella
Edited by William Dentler and George Witman

Volume 48 (1995)
***Caenorhabditis elegans:* Modern Biological Analysis of an Organism**
Edited by Henry F. Epstein and Diane C. Shakes

Volume 49 (1995)
Methods in Plant Cell Biology, Part A
Edited by David W. Galbraith, Hans J. Bohnert, and Don P. Bourque

Chapter 1, Fig. 2 Representative adult invertebrate deuterostomes. (A) Sea urchin (*Lytechinus variegatus*, Phylum Echinodermata). (B) Starfish (*Linkia lavigata*, Phylum Echinodermata). (C) Ascidians (*Boltenia villosa* and *Styela gibbsii*, Subphylum Urochordata). Photograph courtesy of Swalla. (D) Hemichordate (*Saccoglossus kowalevskii*, Phylum Hemichordata). Photograph courtesy of Lowe. (E) Amphioxus (*Branchiostoma floridae*, Subphylum Cephalochordata). Photograph courtesy of Holland.

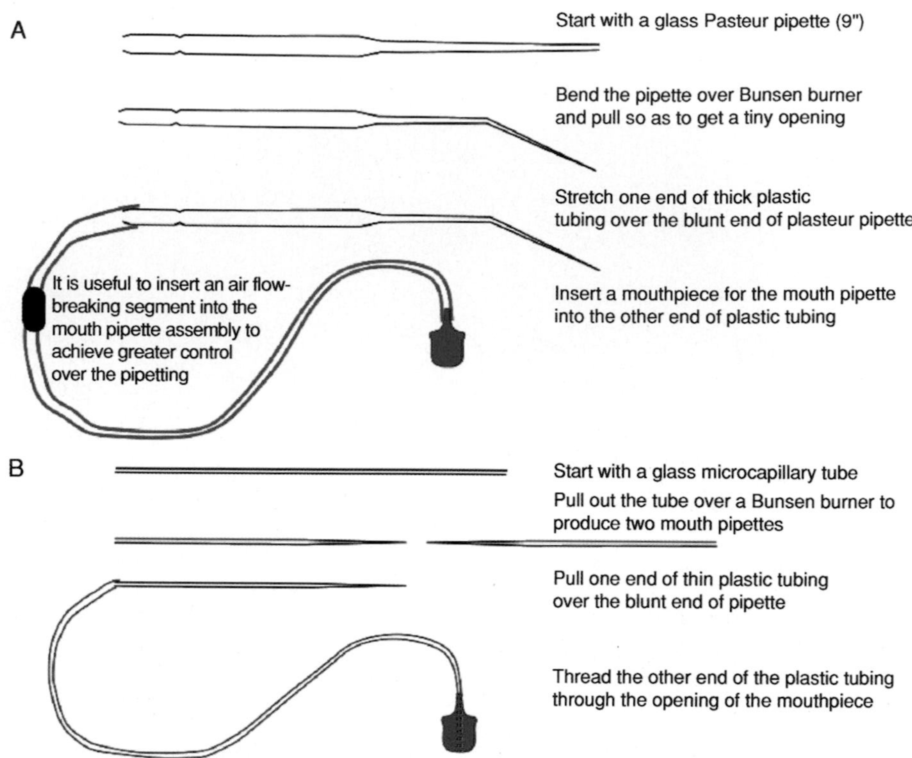
A
Start with a glass Pasteur pipette (9")
Bend the pipette over Bunsen burner
and pull so as to get a tiny opening
Stretch one end of thick plastic
tubing over the blunt end of plasteur pipette
It is useful to insert an air flow-
breaking segment into the
mouth pipette assembly to
achieve greater control
over the pipetting
Insert a mouthpiece for the mouth pipette
into the other end of plastic tubing
B
Start with a glass microcapillary tube
Pull out the tube over a Bunsen burner to
produce two mouth pipettes
Pull one end of thin plastic tubing
over the blunt end of pipette
Thread the other end of the plastic tubing
through the opening of the mouthpiece

Chapter 5, Fig. 2 Construction of mouth pipette. (A) Starting from a 9″ glass Pasteur pipette. (B) Starting from a glass microcapillary tube.

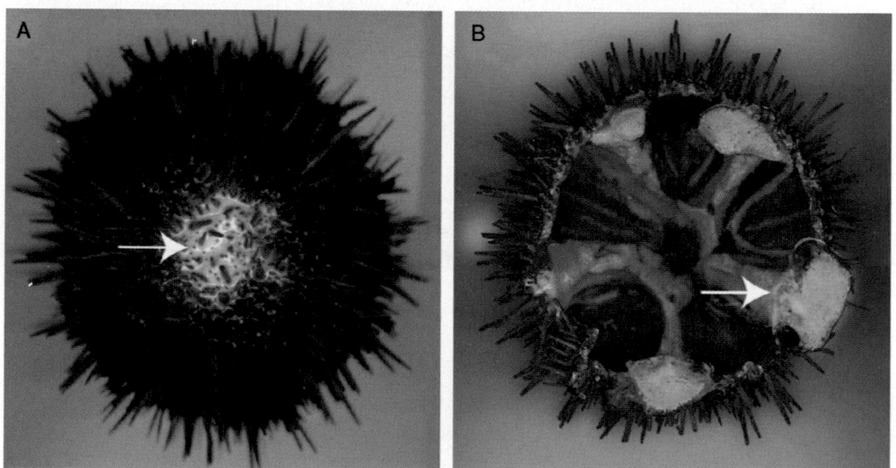
A
B

Chapter 5, Fig. 4 Panel A shows the top side of a female *Strongylocentrotus purpuratus* sea urchin shedding eggs (arrow) from its five gonadopores. Panel B shows a bottom-side view of this same urchin that has been cut away along the periphery of the test surrounding the mouth (Aristotle's lantern) to show the five ovaries (arrow) that fill the body cavity.

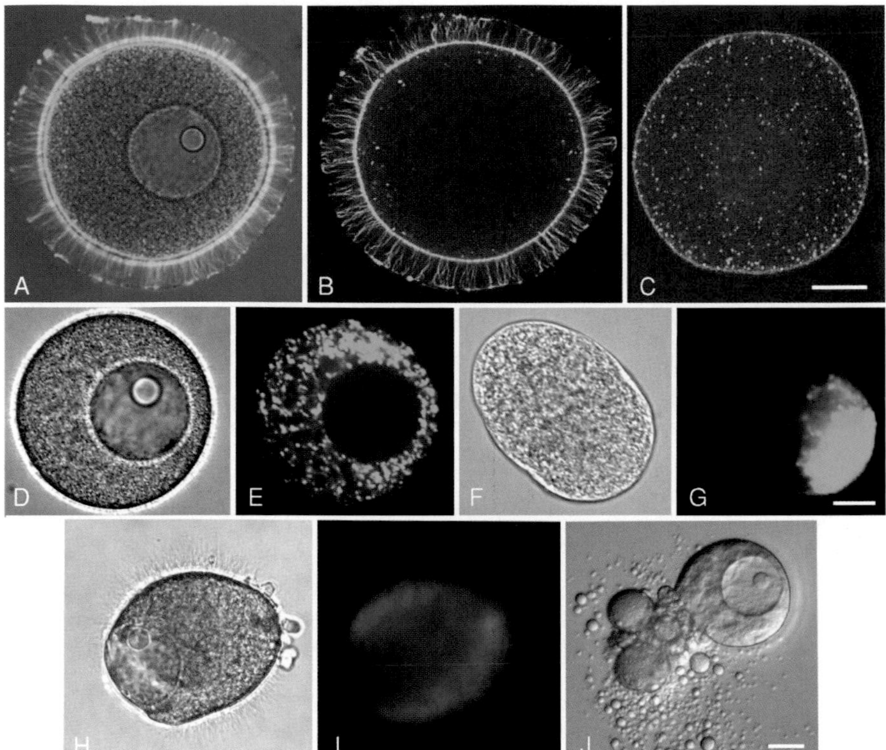

Chapter 5, Fig. 7 Oocyte labeling and manipulation. (A, B) Oocyte labeled with FM1-43 to delineate the plasma membrane and fine membrane extensions emanating from the surface. (A) is FM1-43 overlain with DIC (B) is fluorescence alone. (C) An oocyte fixed and labeled with antibodies to a cortical granule content protein, hyalin (green) and phalloidin to label the dense actin ring at the cortex (red). (D–G) Oocytes that have endocytically packaged the major yolk protein MYP that was fluorescently labeled. (E) shows the distribution of the newly packaged and fluorescent MYP in yolk platelets. (G) shows an oocyte as in (D) and (E) that was centrifuged to stratify its organelles, including the yolk platelets. These are readily seen concentrating to the centrifugal or heavy end of the elongated oocyte by virtue of its endocytosed, fluorochrome-labeled MYP. (H–J) Oocyte apoptosis. (H) and (I) shows an oocyte stimulated to apoptose by staurosporine and then labeled for caspase activity with FITC-VAD-fmk. (J) shows the characteristic blebbing phenotype of an advanced apoptotic oocyte. Each bar represents 25 microns.

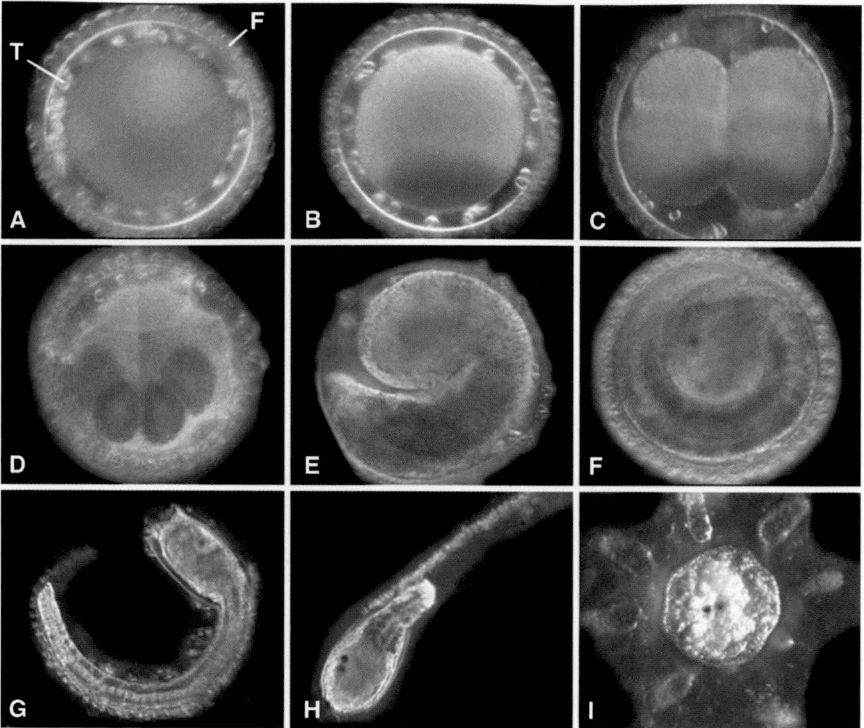

Chapter 6, Fig. 2 *Boltenia villosa* development. In *Boltenia villosa*, the myoplasm is colored a dark orange by pigment granules in the egg as in other pyurids and some styelids. (A) An unfertilized egg. The orange myoplasm is seen throughout the cortex of the egg. The test cells (T) float around the egg inside of the thick chorion. Attached to the chorion are a number of follicle cells (F). Most ascidian eggs have both test cells and follicle cells. (B) After fertilization, the myoplasm contracts toward the vegetal pole (bottom); this location will become the dorsal side of the embryo. The myoplasm marks where gastrulation will begin at the vegetal pole. Next, the myoplasm moves to the future posterior of the embryo (not shown). (C) At the 4-cell stage, the polarity of the embryos is visible by the bright orange myoplasm at the posterior (anterior is up). (D) A 16-cell embryo shows the distinct bilateral cleavage in an ascidian embryo. The plane of bilateral symmetry is in the center, anterior is up. At this stage, there are 4 myoplasm-containing cells, seen by the orange blastomeres in this vegetal view of the embryo. (E) After gastrulation, the tailbud embryo has a white head and the orange muscle cells surround the notochord in the posterior. (F) The tadpole larva just before hatching has undergone extensive convergence and extension, so now the tail wraps around the head and the white notochord cells are visible in the center of the tail. (G) A freshly hatched tadpole, with test cells still clinging to it. Note the palps (adhesive papillae) at the anterior of the larva. (H) After the larva swims for a period of time, the tail retracts during the process of metamorphosis. (I) In *Boltenia villosa*, the metamorphosing juvenile flattens down and makes a number of ectodermal ampullae that radiate out from the larva.

Chapter 7, Fig. 2 (A) Adult *Ciona intestinalis* cultured in a 10-cm petri dish. (B) Typical "cage" used for holding *C. intestinalis*. (C) Cages used for holding *Ciona savignyi*.

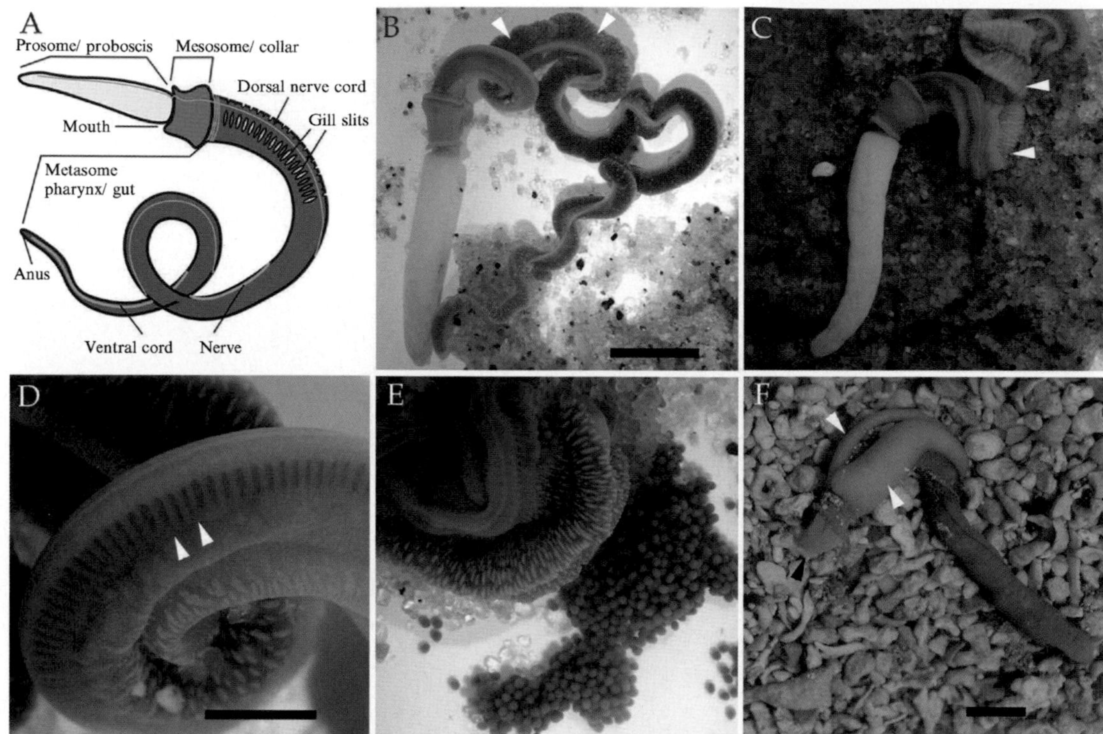

Chapter 8, Fig. 1 Adult enteropneust hemichordates, *Saccoglossus kowalevskii* and *Ptychodera flava*. (A) Model of an acorn worm adult based on *S. kowalevskii* outlining some of the characteristic features of the adult morphology. (B) Adult *S. kowalevskii* female. White arrowheads indicate the position of the green ovary. Black scale bar = 3 cm. (C) Adult *S. kowalevskii* male. White arrowheads indicate the position of the orange/white testes. (D) High magnification of branchial region of metasome on *S. kowalevskii*. Lateral view. White arrowheads indicate the position of two of the multiple pairs of gill slits on the dorsal side of the worm. Scale bar = 0.5 cm. (E) Spawning female of *S. kowalevskii*. Oocytes are green/blue-colored and are spawned along the length of the gonad. (F) Adult *P. flava*. Black arrowhead indicates the position of the proboscis. White arrowheads indicate the position of the genital wings. Scale bar = 1 cm.

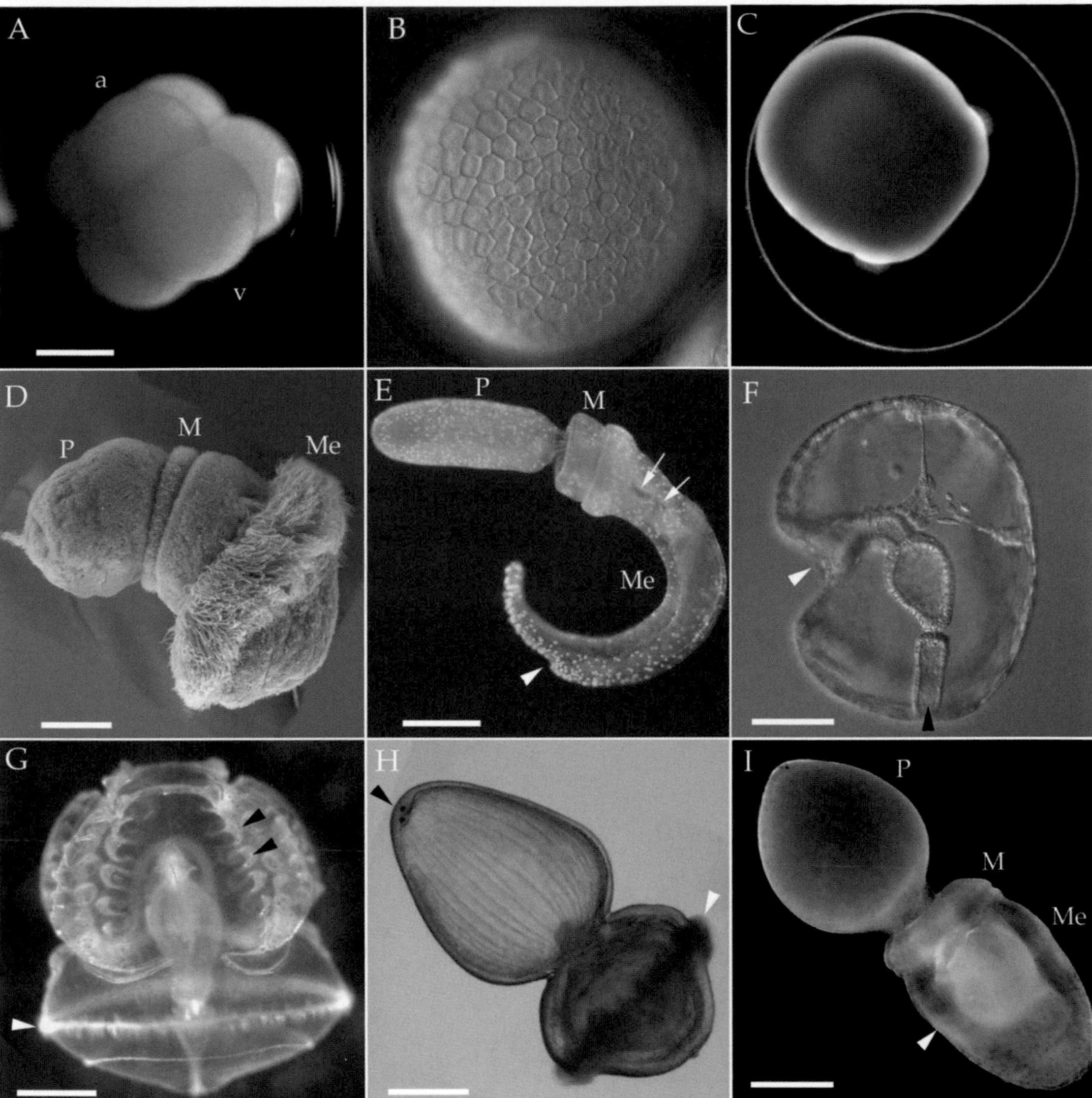

Chapter 8, Fig. 2 Embryological, larval, and juvenile stages of *S. kowalevskii* and *P. flava*. A through E reflect stages of *S. kowalevskii* development. F through I reflect stages of *P. flava* development. (A) Early cleavage stage within the vitelline envelope. Animal blastomeres are slightly smaller than vegetal ones. a = animal pole, v = vegetal pole. Scale bar = 100 μm. (B) Late blastula. (C) Post-gastrula with prominent telotroch at the posterior end of the embryo swimming within the vitelline envelope. (D) SEM micrograph of midstage embryonic development (day 4), lateral view. The tuft of cilia at the far left of the panel is the apical tuft. The ectoderm is evenly ciliated except for the dense telotroch at the posterior of the embryo. Scale bar = 75 μm. (E) Late juvenile 14 days old, lateral view. All three body regions clearly visible. White arrowhead indicates the position of the anus and, posterior to that, the ventral post-anal tail. White arrows indicate the position of the first two gill slits. Scale bar = 150 μm. (F) Lateral view, DIC micrograph of 3-day-old tornaria larva. White arrowhead indicates the larval mouth and black arrowhead, the anus. Scale bar = 50 μm. (G) Krohn stage larva. Black arrowheads indicate the folds, or epaulets, in the dorsal ciliated bands, and white arrowhead marks the telotroch. Scale bar = 300 μm. (H) Larva during transformation with clearly formed proboscis. The black arrowhead marks the position of the eyespots, and white arrowhead, the telotroch. Scale bar = 300 μm. (I) Juvenile worm one week after the initiation of transformation. White arrowhead indicates the band that persists after the cilia of the telotroch are shed upon settlement. Scale bar = 500 μm. P = prosome, M = mesosome, and Me = metasome.

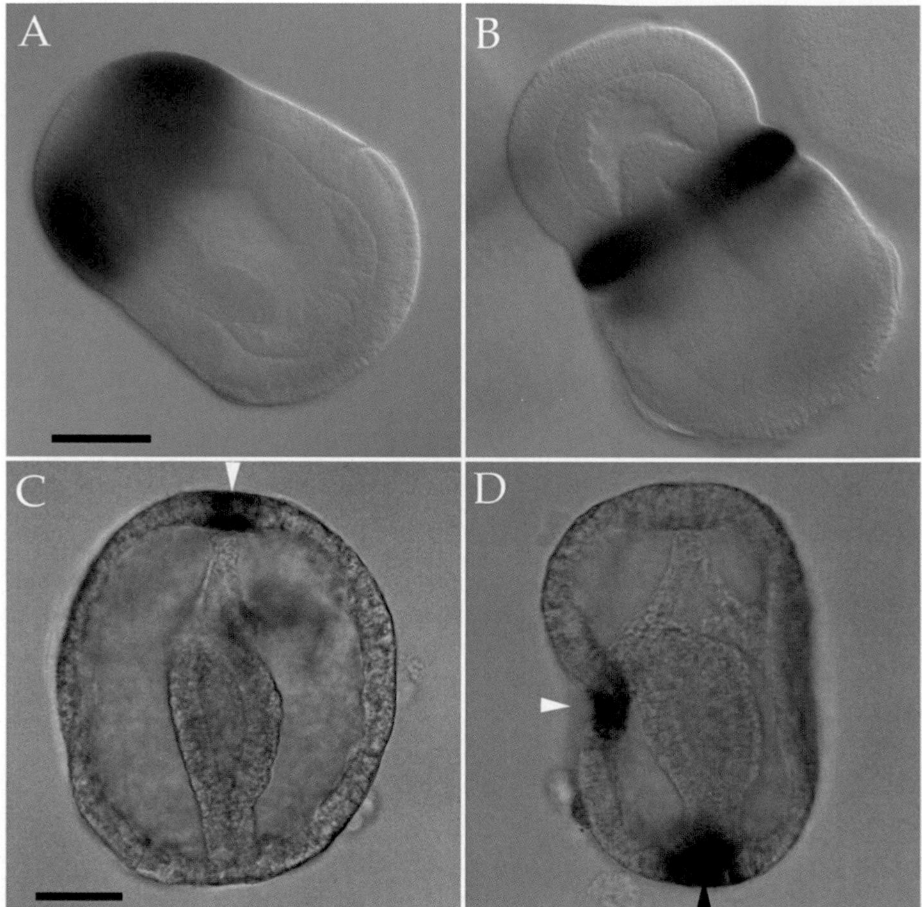

Chapter 8, Fig. 3 Examples of patterns of expression in hemichordate embryos of genes implicated in specifying vertebrate structures. Whole mount *in situ* hybridization in *S. kowalevskii* and *P. flava*. A and B show *S. kowalevskii*, C and D show *P. flava*. (A) Retinal homeobox expression in the developing proboscis ectoderm of a 2-day-old early embryo. Optical frontal section of a cleared specimen. Scale bar = 150 μm. (B) Expression of BarH, a neural determining gene, in the mesosome ectoderm of a 3-day-old early embryo. Frontal optical section of a cleared embryo. (C) Expression of the forebrain gene, T-brain in a 3-day-old tornaria larva at the apical organ, as marked by a white arrowhead. Scale bar = 50 μm. (D) Expression of Brachyury, a gene important in notochord development, in a 2-day-old tornaria larva. White arrowhead indicates the position of the mouth and black, the position of the anus.

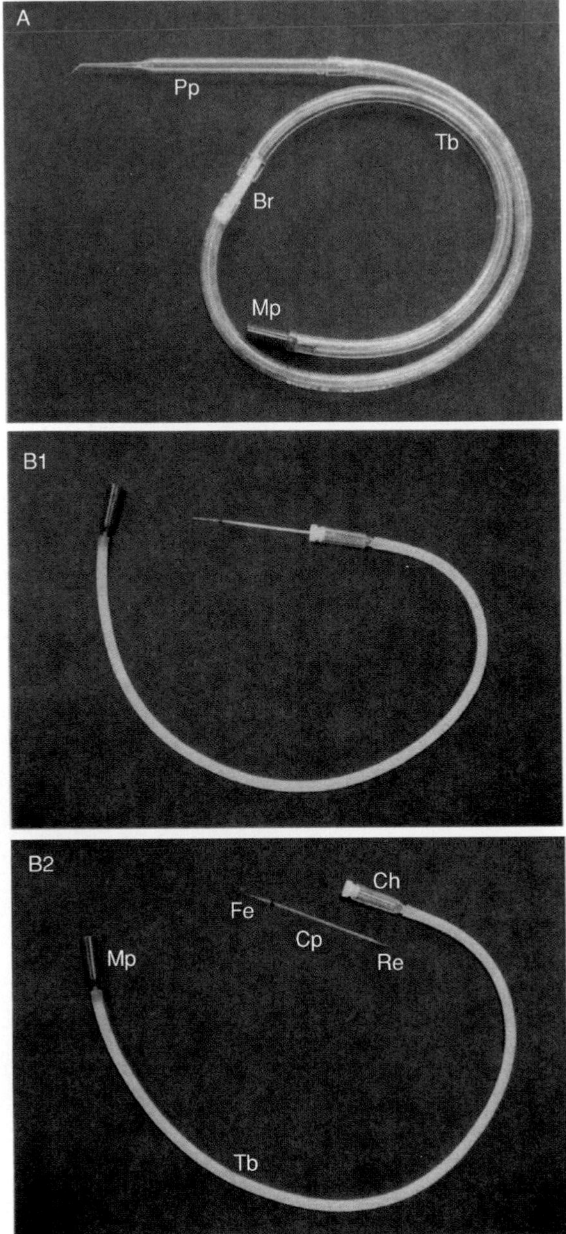

Chapter 11, Fig. 1 Mouth pipettes. (A) A mouth pipette composed of a pulled Pasteur pipette, a braking tube, a mouthpiece, and tubing made of rubber or plastic. (B) A mouth pipette composed of a pulled glass capillary, a capillary holder, a mouthpiece, and tubing. A completed pipette (B1) and a pipette in which the glass capillary is detached from the capillary holder (B2) are shown separately. The rear end of the glass capillary is pulled to make the open end thinner. The thin rear end works as a brake for fine control. Pp, Pasteur pipette; Br, braking tube; Mp, mouthpiece; Tb, tubing; Ch, capillary holder; Cp, capillary; Fe, front end of the capillary; Re, rear end of the capillary.

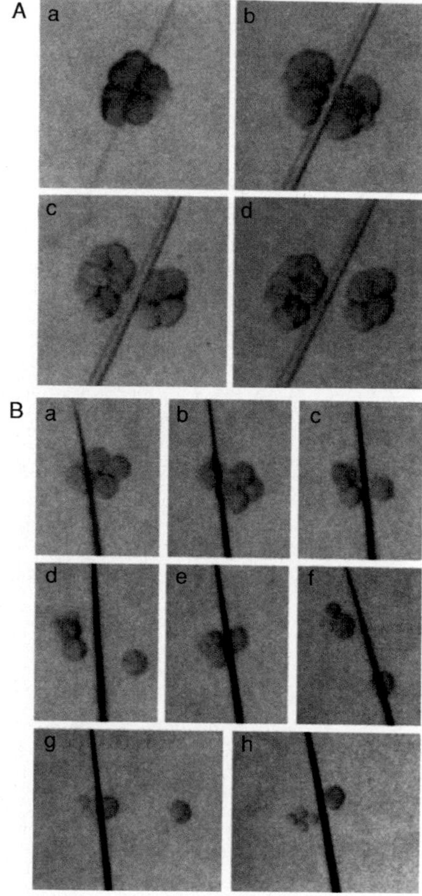

Chapter 11, Fig. 3 Sequence photographs on the processes for isolation of an animal cap and a vegetal half (A) and of a quartet of micromeres (B) from a 16-cell stage embryo (Photos by Dr. S. Amemiya).

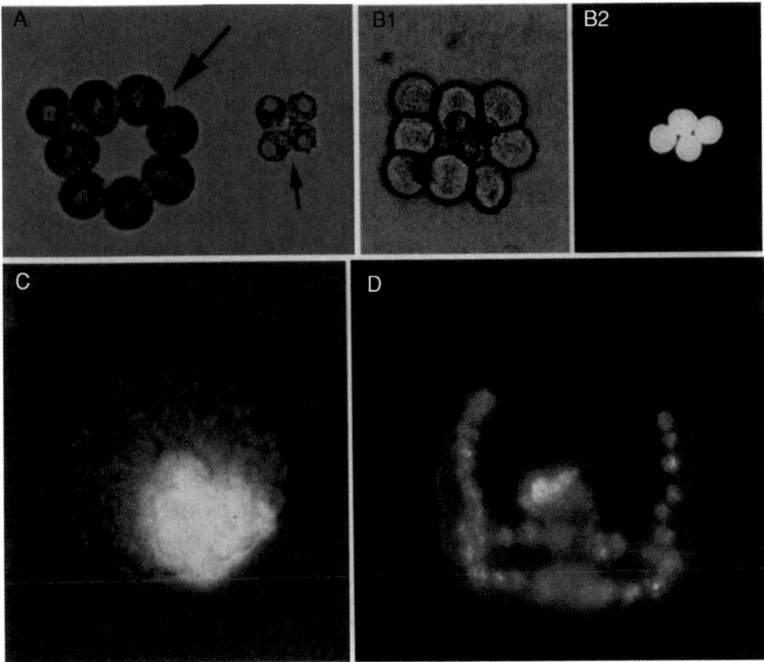

Chapter 11, Fig. 4 Development of a chimeric embryo derived from an animal cap recombined with a quartet of rhodaminated micromeres isolated from 16-cell stage embryos. (A) An animal cap (large arrow) and a quartet of micromeres (small arrow) just after isolation. (B) An animal cap recombined with a quartet of rhodaminated micromeres just after recombination viewed under a light microscope (B1) and an epifluorescence microscope (B2). (C) A morula stage embryo derived from an animal cap recombined with a quartet of rhodaminated micromeres. (D) A middle gastrula derived from an animal cap recombined with a quartet of rhodaminated micromeres (Photos by Dr. S. Amemiya).

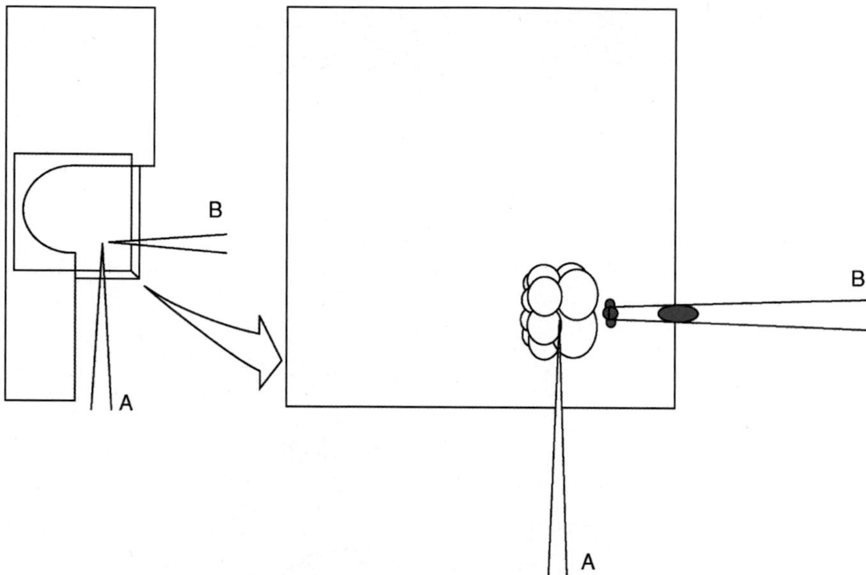

Chapter 11, Fig. 7 Modified Kiehart chamber. An aluminum slide of the shape illustrated is milled so two coverslips are separated by 1 mm. The coverslips form the dissection chamber by sealing the coverslips to the slide with Vaseline. Two micromanipulators are mounted to the dissection microscope to hold glass needles that approach the Kiehart chamber as illustrated. Needle A is unbroken and serves as either a knife or a "thumb" to maneuver the embryos. Needle B is broken with a blunt opening of a diameter and this needle is attached to a mouth pipette at the other end. When the suction needle is introduced into the chamber, a small amount of seawater backfills the needle by capillary action. Once the flow stops, one has very sensitive control over the suction, using this needle as a micro-mouth pipette. To the right of the Kiehart chamber is a blowup of the corner of the chamber where the surgeries are conducted. Shown is an embryo at an approximate magnification as seen during the surgery (Illustration produced by Dr. D. McClay).

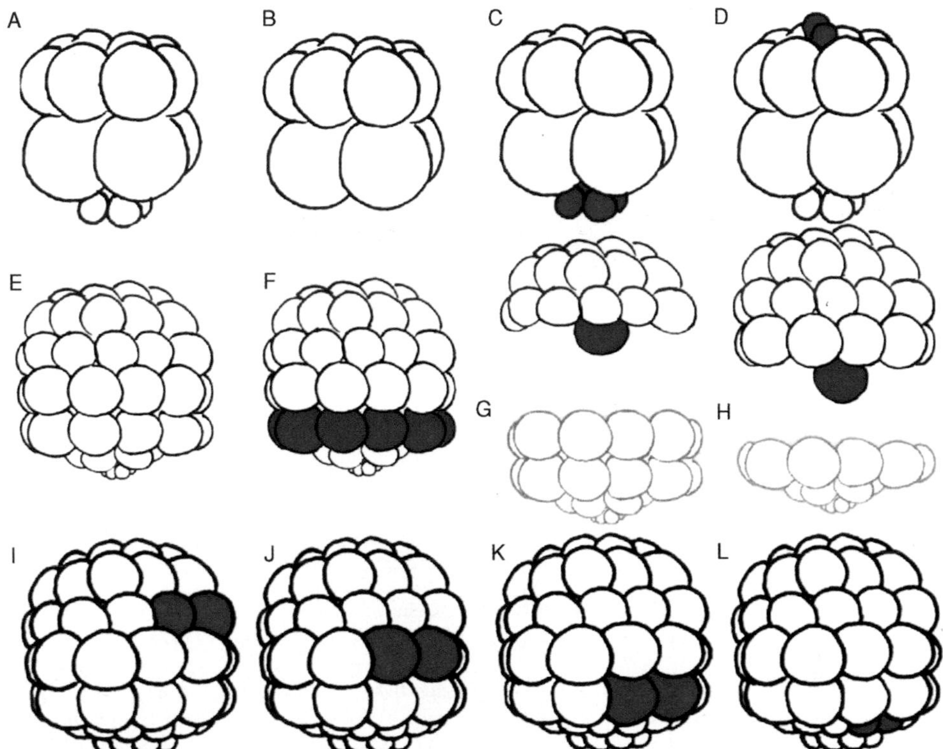

Chapter 11, Fig. 8 Hörstadius transplants and Mosaic analysis. Donor cells from embryos injected with RNA, a morpholino, or other perturbation are coinjected with rhodamine dextran. Examples of transplant combinations are shown and are described in the text. (A–H) Hörstadius transplants. (I–L) Mosaic analysis (Illustration produced by Dr. D. McClay).

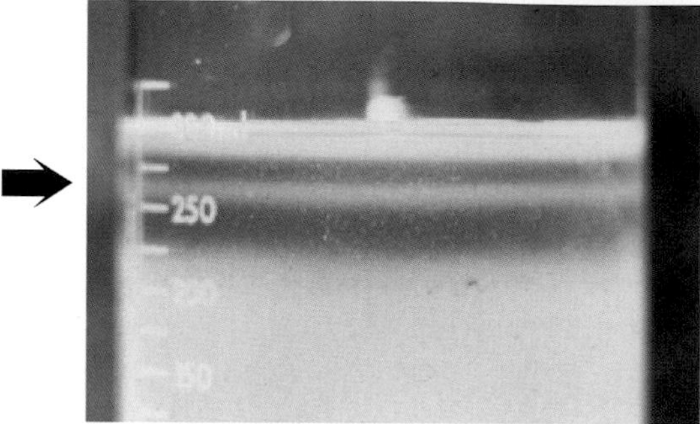

Chapter 12, Fig. 1 Sedimentation of blastomeres of 16-cell stage embryos at 1× g. Dissociated embryos were layered over the linear sucrose gradient, as discussed in the text. After 30 min of sedimentation, a clear micromere layer can be seen just above the 250 ml mark on the beaker (arrow). The illumination of the beaker was from the side and the background was black paper.

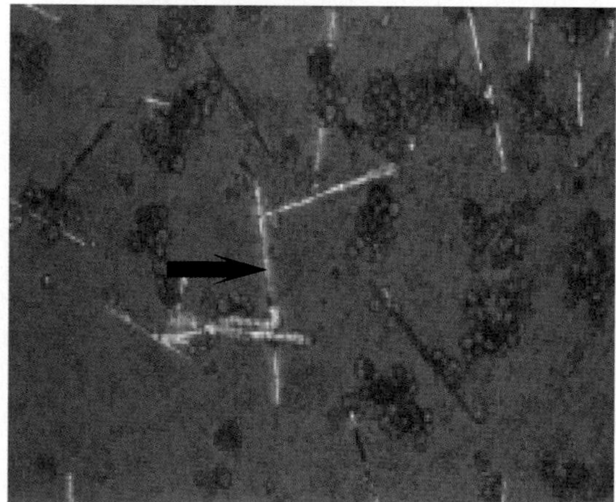

Chapter 12, Fig. 2 The appearance of a micromere culture after 72 h. Micromeres were cultured on coverslips, as discussed in the text. The coverslips were inverted and observed at a magnification of 200× using differential interference contrast microscopy. An arrow points to a spicule.

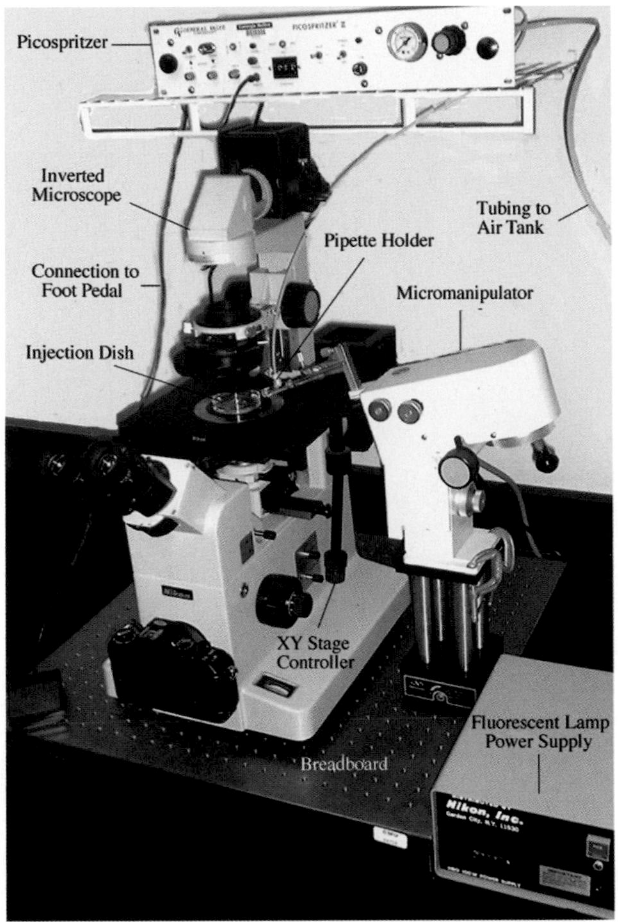

Chapter 13, Fig. 1 The microinjection apparatus. An open plastic injection dish with a row of immobilized eggs adhering to its surface is placed on the stage of an inverted, fixed-stage microscope. The microscope is equipped for transmitted light and epifluorescence optics. A freestanding micromanipulator is positioned on the right side of the microscope and used to move the microinjection pipette. The pipette is connected to a picospritzer, which is attached to a tank of compressed air. The picospritzer delivers controlled, timed pulses of pressurized air when activated by a foot pedal (note that the foot pedal normally rests on the floor, not on the tabletop as shown in the photograph). The XY stage controller of the microscope is used to move the eggs relative to the injection pipette. To control vibration, the entire apparatus is placed on a heavy metal breadboard that rests on natural rubber foam pads or other vibration-absorbing material.

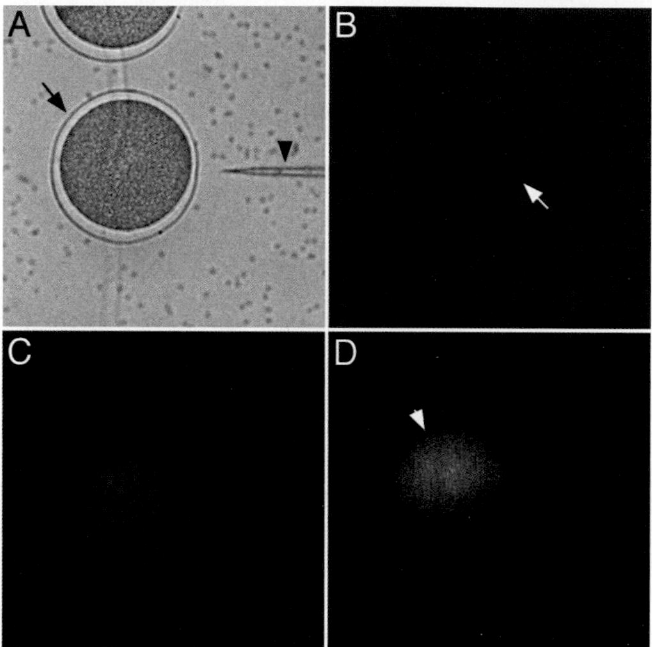

Chapter 13, Fig. 6 Microinjection of rhodamine dextran. A fertilized egg is attached to a protamine sulfate-coated injection dish. (A) Just prior to injection (transmitted light only). The egg has been fertilized and the fertilization envelope (arrow) has elevated. Numerous sperm are attached to the surface of the dish. The injection pipette is indicated by an arrowhead. (B) Penetration of the egg (combined transmitted light/epifluorescence). The tip of the injection pipette has pierced the fertilization envelope and is indenting the plasma membrane of the egg (arrow), immediately before puncturing it. (C, D) Delivery of injection solution (C—combined transmitted light/epifluorescence, D—epifluorescence only). A pulse of pressure has been delivered with the picospritzer and fluorescent injection solution has been expelled into the cytoplasm of the egg (arrowhead, D).

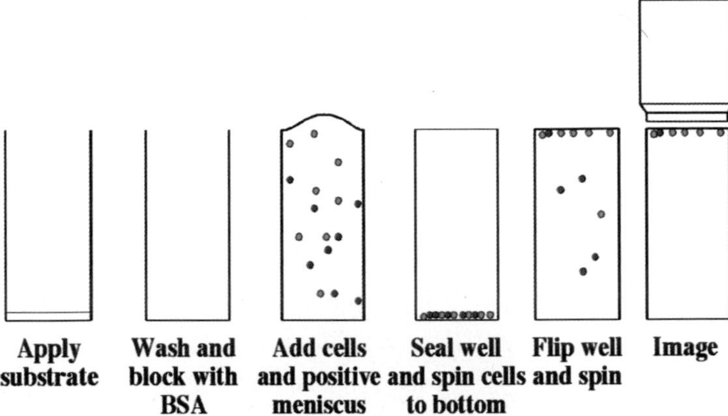
Apply substrate
Wash and block with BSA
Add cells and positive meniscus
Seal well and spin cells to bottom
Flip well and spin
Image

Chapter 14, Fig. 1 Quantitative adhesion assay. This assay modification allows one to measure strength of adhesion of small numbers of cells. The chambers represent the steps in the treatment of one well in a 384-well plate. First, the substrate is added and then background blocked with BSA. Then cells are added. In the experiment depicted, cells of two colors are added to the same well, which allows a direct comparison between these two cell types. A positive meniscus is left at the top so when the well is sealed, no air bubbles remain trapped in the well. The cells are spun to the bottom of the well. The plate is flipped and centrifuged (a dislodgement force) at a speed selected to challenge the adhesive avidity of the cells. In the diagram, the green cells are more adherent than the red cells. To see that difference, the plates are transferred to a microscope stage and the well bottoms imaged by red and then green fluorescence. The adherent cells then are counted from the images.

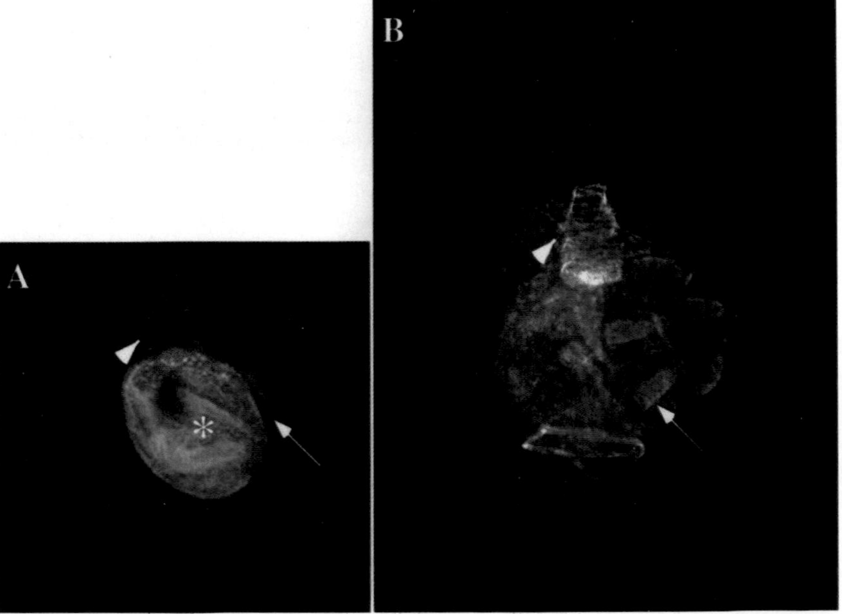
A
B

Chapter 15, Fig. 4 Phalloidin staining of larval stage embryos. Larvae were collected and labeled with phalloidin, as detailed in the text. An early rudiment (arrow) can be seen on one side of the 3-week-old larva in (A) that was also immunostained with Endo1 monoclonal antibody (*). A fully developed rudiment (arrow) containing well-defined tube feet is detected in the 6-week-old larva (B) by phalloidin staining. Arrowhead points to circumesophageal muscles.

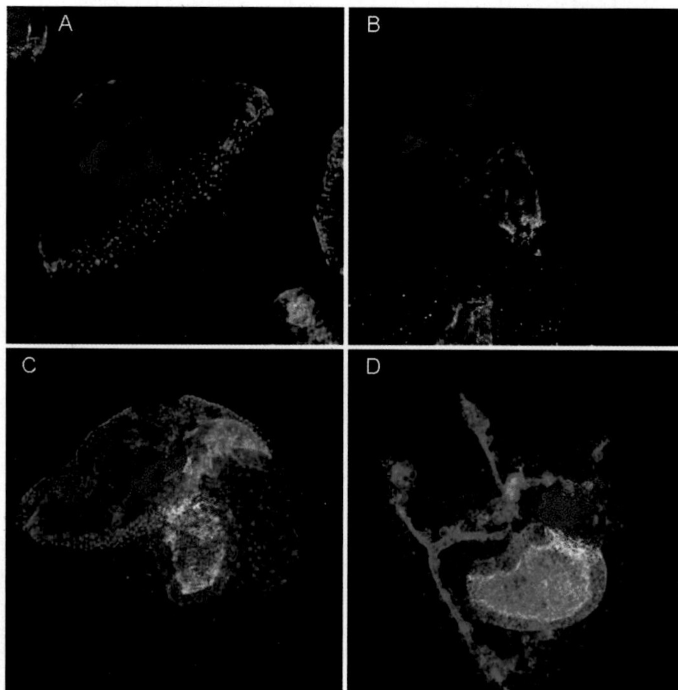

Chapter 15, Fig. 5 Double and triple immunofluorescent labeling of embryos. (A and B) For double labeling, embryos were simultaneously incubated with (A) MHC (polyclonal) and CBA (monoclonal) or (B) MHC (polyclonal) and Endo1 (monoclonal) primary antibodies, followed by simultaneous incubation in goat anti-mouse FITC and goat anti-rabbit TRITC secondary antibodies. The muscles cells appear red, and the ciliary band (A) or gut (B) are green. (C and D) For triple staining, embryos were incubated in MHC (polyclonal) and Endo1 (monoclonal) simultaneously, followed by simultaneous incubation in goat anti-mouse Cascade Blue and goat anti-rabbit TRITC secondary antibodies. After a second blocking step in 5% NGS/PBST, embryos were incubated with CBA (C) or 6A9 (D) monoclonal antibodies, followed by a final incubation in FITC-conjugated goat anti-mouse secondary antibody. Muscle cells appear red, the guts blue, and the ciliary band (C) and primary mesenchyme cells (D) are green. All embryos were fixed in MeOH. Images were obtained on a BioRad Radiance 2000 Confocal Microscope.

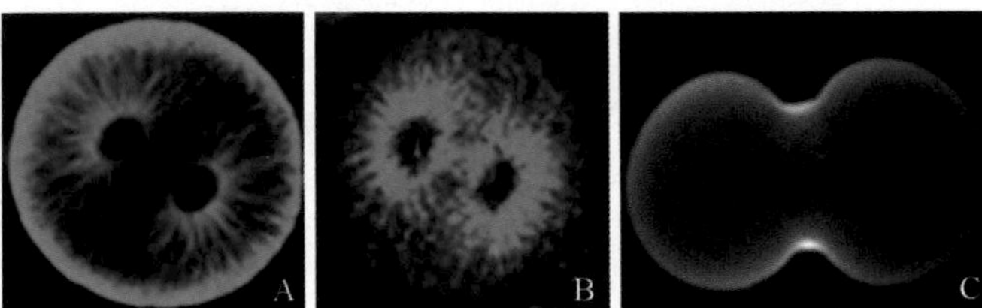

Chapter 16, Fig. 1 Common techniques for fluorescent labeling of fixed cells. (A) Tubulin immunofluorescence in an early anaphase *S. purperatus* egg. The cells were fixed in a large batch by the free fix method in section II.A.1.a. After fixation, and washing by hand centrifuging 3X through PBS, a 1 ml aliquot of cells was taken and brought to 3% with BSA. Primary antibody (E7; Developmental Studies Hybridoma Bank) was added at a dilution of 1:50, and the cells were incubated on a rotator for 2 h. The sample was again washed 3X in PBS, and resuspended in PBS with 3% BSA. Secondary antibody (Fluorescein-conjugated anti-mouse; Chemicon) was added at a dilution of 1:200. All post-fix incubations and washes were performed at room temperature. The cells were incubated on a rotator for 1 h, washed 3X, and imaged on a Leica confocal microscope using an argon/krypton laser for illumination. (Image: Laila Strickland) (B) Tubulin immunofluorescence in a metaphase *S. purperatus* egg. The cell was fixed on a glass coverslip by the method in Section II.A.1.d.i. The antibodies and dilutions used were the same as in (A). Incubations were performed at room temperature in a small tissue culture dish on a gentle rocker. Images were acquired as in (A). The punctate appearance of the microtubules and higher level of background fluorescence is characteristic of immunofluorescence in cells that have been fixed in this way. (Image: Laila Strickland) (C) Filipin-staining of sterol-rich membrane domains (lipid rafts) at the cleavage furrow of an *S. drobachiensis* egg. Lipid rafts are insoluble in 0.1% ice-cold Triton-X 100 and can be visualized after fixation and solubilization, by incubating with filipin for 30 min. (Image: Michelle Ng).

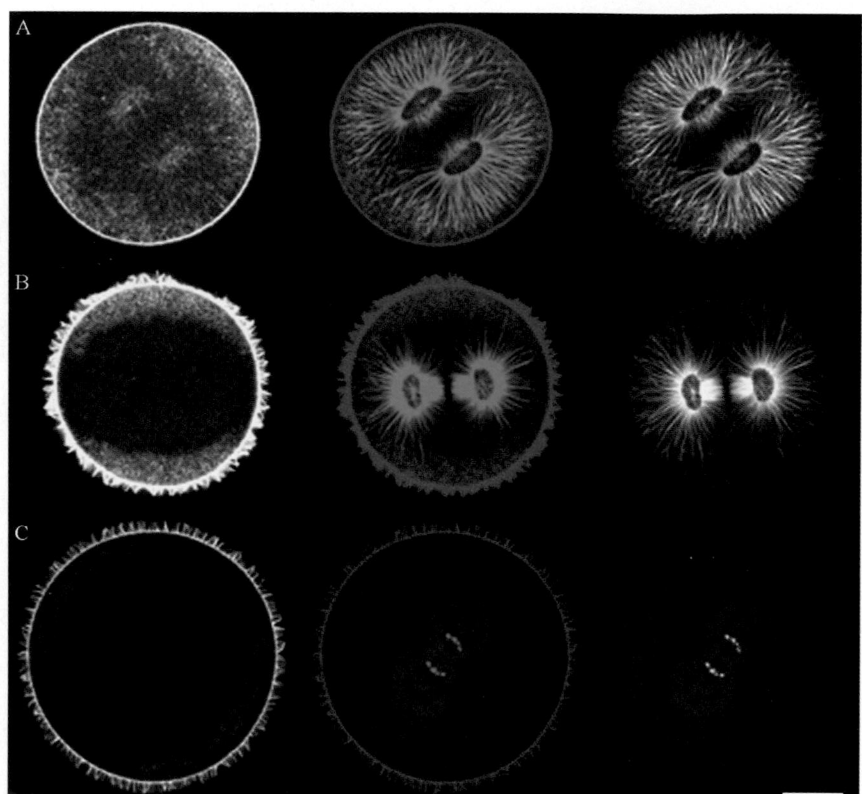

Chapter 16, Fig. 4 Simultaneous preservation and staining of actin filaments and microtubules. (A) Top row: A sand dollar embryo stained for F-actin with phalloidin (left) and tubulin by immunofluorescence (right). In the overlay (center), actin staining is shown in red and tubulin staining in green. Fixation and staining were performed by the protocols in Section IV. (B) One blastomere from a two-cell sea urchin embryo, fixed and stained as in (A). (C) A sea urchin embryo stained for F-actin with phalloidin (left) and for DNA with propidium iodide (right). In the overlay (center), actin staining is shown in red and nucleic acid staining is shown in green.

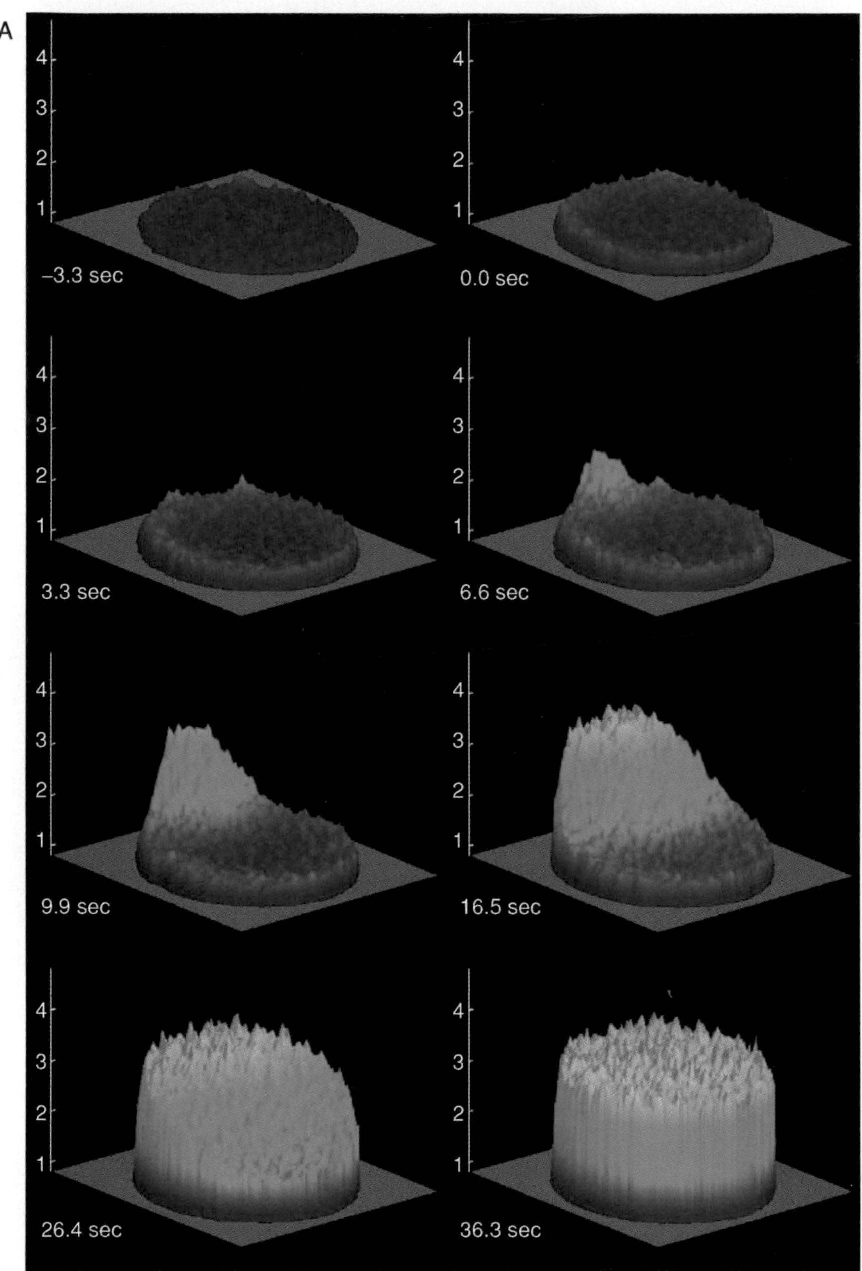

Chapter 18, Fig. 1 (*Continued*)

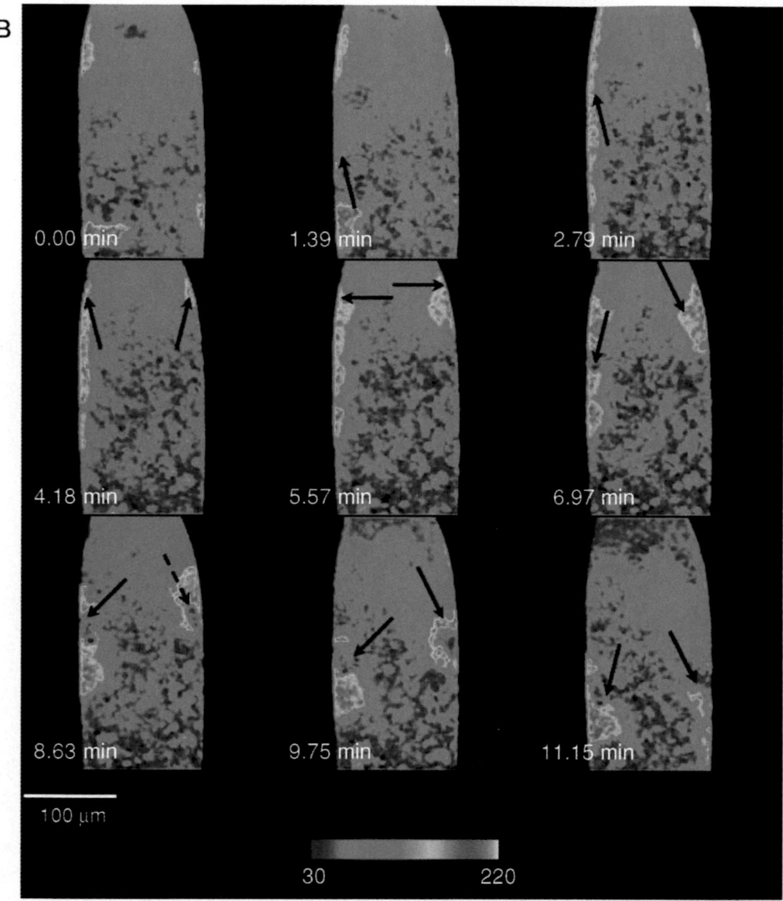

Chapter 18, Fig. 1 Calcium imaging using dextran-linked fluorescent calcium indicators. (A) The fertilization calcium wave in a sea urchin egg (L. pictus) imaged using calcium green dextran introduced by pressure microinjection (10 μM in the egg). Pseudoratio images are shown. Images were captured using a confocal microscope every 3.3 s. The time of fertilization ($t = 0$) is known from the rapid (ms) calcium influx that occurs when sperm and egg fuse; this can be seen as a small peripheral calcium increase. The pseudoratio series was obtained by dividing each image pixel by pixel by a resting image obtained at $t = -6.6$. The topographical representation of the data shows the ratio on the z-axis in a single equatorial confocal section. The first indication of the fertilization calcium wave initiation can be seen in the image at 6.6 s. Thereafter, the wave spreads across the egg. Martin Wilding's data. (B) Traveling calcium waves in a gastrulating *Drosophila* embryo (Bownes stage 9). In circumstances in which cells or their contents move during an experiment, as is the case here, pseudoratio imaging is misleading or impossible. An alternative, as here, is to pair calcium green dextran with rhodamine dextran, a dye indifferent to calcium. The assumption is that the dyes attached to the large dextran molecule will co-distribute in the cell, a supposition largely borne out by experiment. The successive images show calcium waves traveling intercellularly in the embryo. They first progress toward the anterior, then reverse and travel posteriorly. Their function is unknown. Jaime-Ann Tweedie's data.

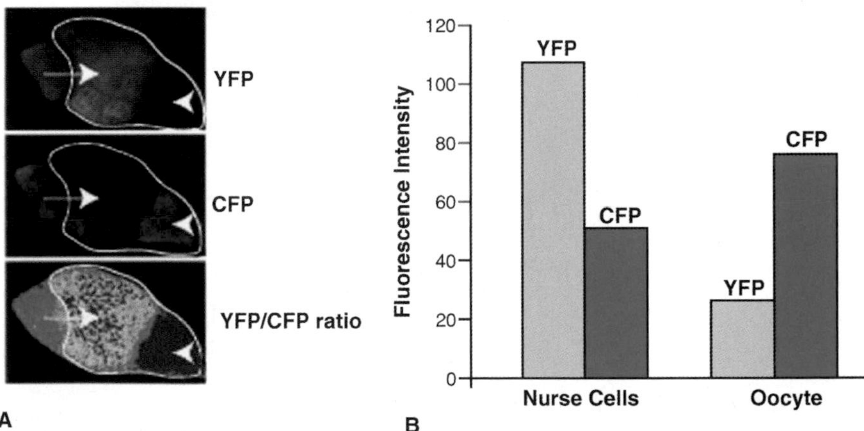

Chapter 18, Fig. 2 Expression of the transgenic chameleon indicator in *Drosophila* ovary. Transgenic flies expressing the chameleon indicator were constructed by expressing the chameleon transgene under the control of promoter sequences that permit germ line expression when crossed with a suitable Gal4 enhancer trap line. A single ovariole is shown (A). The nurse cells (arrows) show a strong YFP signal, while the oocyte shows a predominant CFP signal. Relative intensities are shown in B. The YFP/CFP ratio is 2.13 in the nurse cells and 0.34 in the ovary, indication that calcium concentrations are substantially higher in the nurse cells than in the oocyte. Jun-Yong Huang's data.

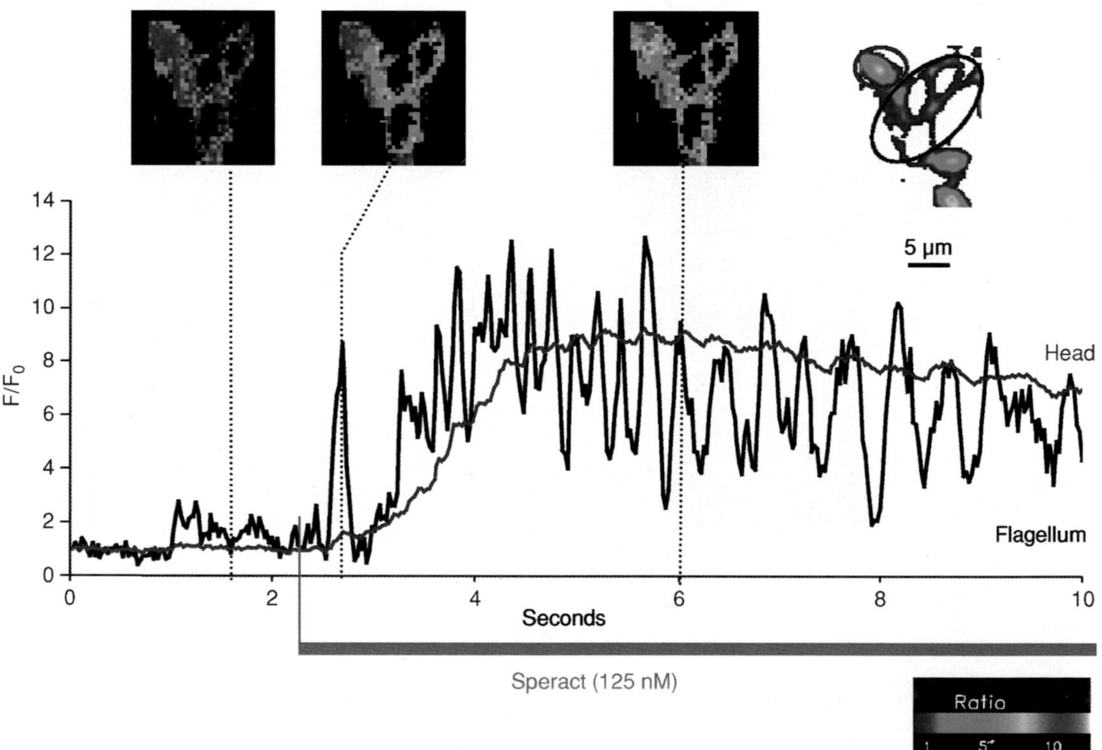

Chapter 22, Fig. 3 Complex, spatially discrete patterns of $[Ca^{2+}]$i increase as measured by the single-cell imaging techniques outlined in Section IV.E. Graph shows average per-pixel ratio increase (against value at time 0) of fluo-4 fluorescence in head and flagellum of a single *S. purpuratus* sperm following application of 125 nM speract. Regions of interest are as shown in the image at top right. Images above graph are ratio images against the frame immediately preceding speract addition. Images collected at 40 frames/s with a frame exposure time of 25 ms.

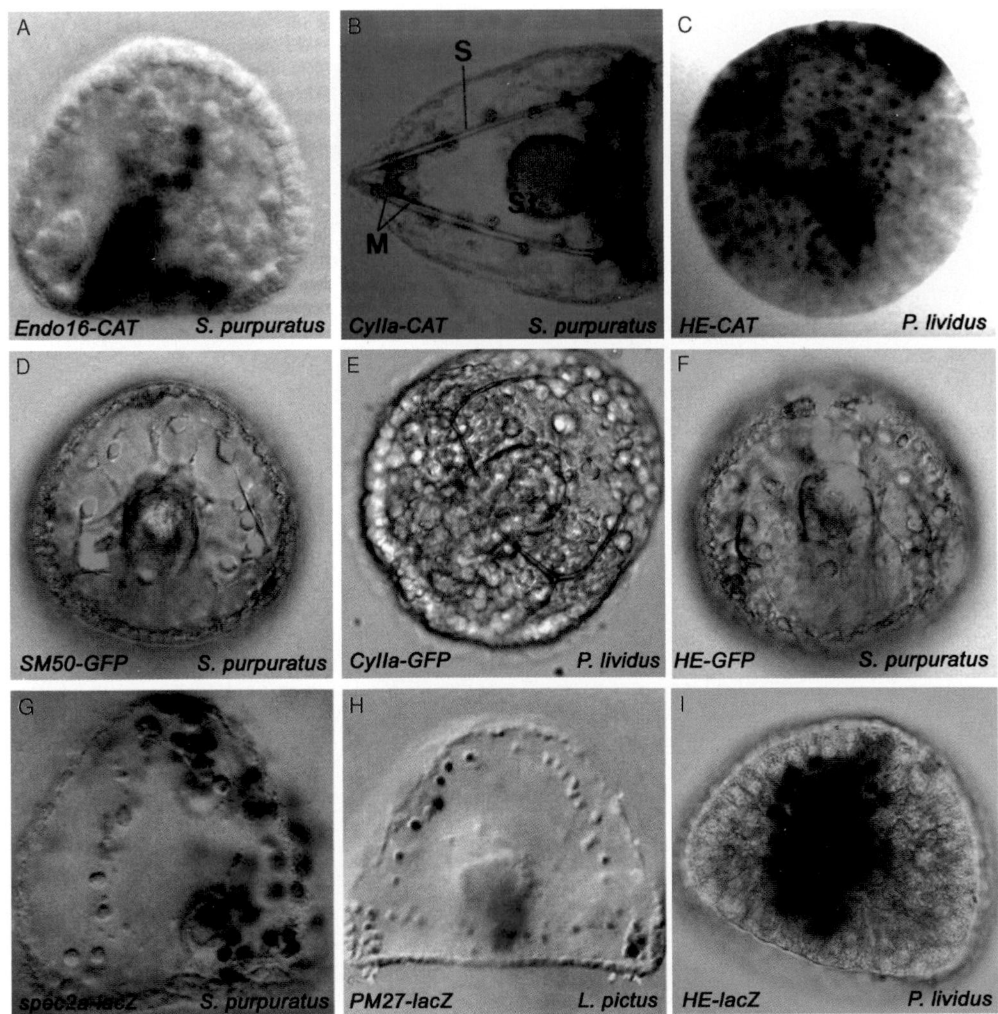

Chapter 25, Fig. 2 (*Continued*)

Chapter 25, Fig. 2 Visualization of expression of various reporter genes in transgenic sea urchin embryos. The construct and the sea urchin species injected are indicated at the bottom left and right corner of each panel, respectively. (A–C) Expression of the CAT reporter gene detected with different methods: (A) Whole mount *in situ* hybridization to detect CAT mRNA in the gut of a *S. purpuratus* gastrula driven by the *Endo16* regulatory region [adapted with permission from Yuh *et al.* (1994)]. (B) *CyIIa-CAT* expression detected in skeletogenic cells and stomach of a *S. purpuratus* pluteus using a rabbit polyclonal anti-CAT primary antibody and (FITC)-labeled goat anti-rabbit secondary antibody. Composite false color video images of bright field (blue) and fluorescent (red) exposures are shown superimposed [adapted with permission from Zeller *et al.* (1992)]. (C) *P. lividus* blastula showing ectodermal expression of *HE-CAT* visualized by immunolabeling (large spots; CAT protein diffuses in the cytoplasm of the expressing cell) using a sheep anti-CAT polyclonal Ig coupled to digoxigenin and a sheep alkaline-phosphatase-conjugated anti-digoxigenin Ig. The domain of expression of the endogenous HE gene in this embryo has been simultaneously visualized by immunolabeling of the gene product (small single spot/cell; the hatching enzyme concentrates in a subcellular compartment between the nucleus and the apical surface of the cell) using a rabbit anti-HE primary antibody and a goat alkaline-phosphatase-conjugated anti-(rabbit IgG) Ig [adapted with permission from Ghiglione *et al.* (1997)]. (D–F) GFP expression in different domains of gastrula stage embryos. GFP fluorescence is shown in green in these digital false color images superimposed to bright-field (gray) exposures obtained with DIC. The images in (D) and (F) are *S. purpuratus* specimen courtesy of P. Oliveri and show expression of the *SM 50* gene in skeletogenic cells and of the *SpHE* gene in ectoderm and gut cells, respectively. The *P. lividus* shown in (E) expresses GFP in skeletogenic cells driven by a regulatory element of the *CyIIa* gene of *S. purpuratus* (M. I. Arnone, unpublished). (G–I) Expression of the *lacZ* gene driven by the promoters of the indicated genes: (G) *Spec2a* in aboral ectoderm cells of *S. purpuratus* embryo at the prism stage [adapted with permission from Gan *et al.* (1990)]. (H) *PM27* in PMCs of *L. pictus* at the prism stage [adapted with permission from Klueg *et al.* (1997)]. (I) *HE* in ectoderm cells of *P. lividus* embryo at the gastrula stage (C. Ghiglione and C. Gache, unpublished).

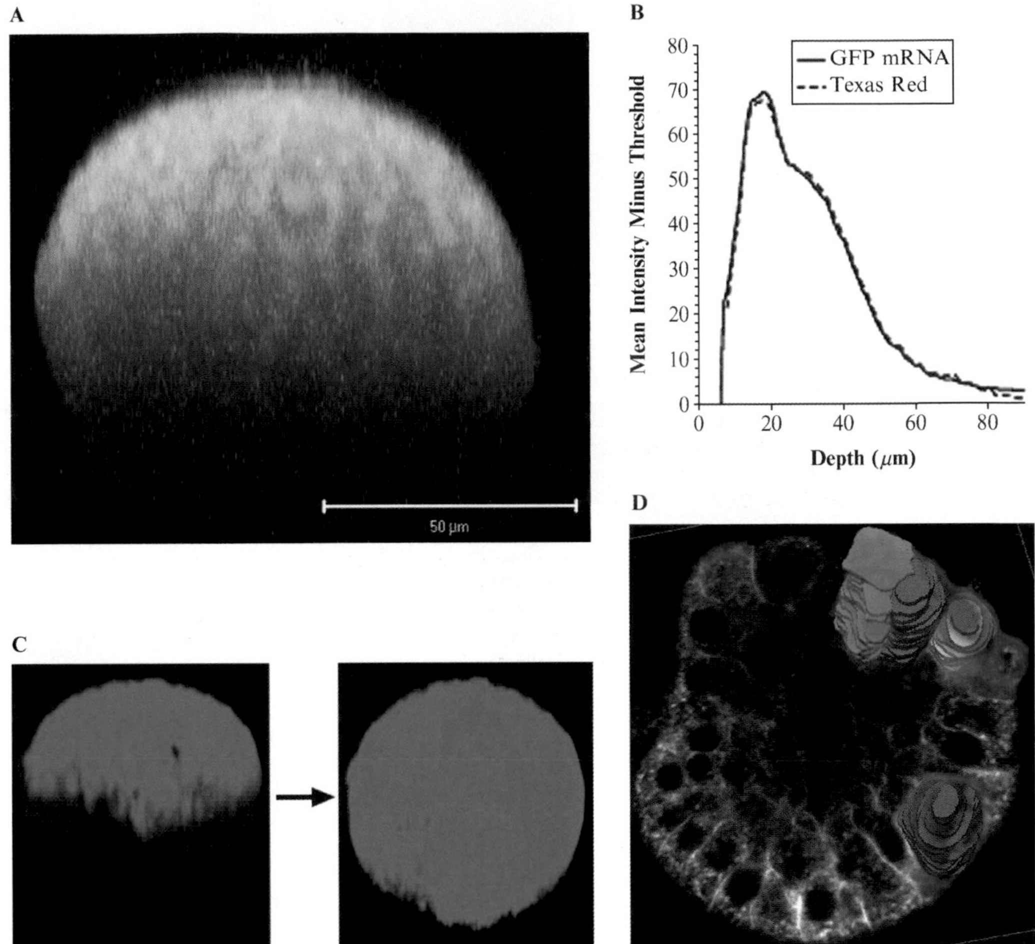

Chapter 25, Fig. 3 (A) Maximum intensity projection of confocal laser scanning microscopy (CLSM) fluorescence intensity (TR, red; GFP, green) in a 18 hpf *S. purpuratus* embryo. The embryo is coinjected with TR and *Krox-GFP* mRNA, and uniformly labeled. Image reconstruction shows marked loss of signal with increasing depth (imaging performed from the top.) (B) Mean intensity vs depth for both GFP and TR channels. The mean intensities are virtually superimposable, deviating only slightly at the brightest and dimmest regions of the embryo. This indicates that freely diffusing TR-dextran can serve as an internal fluorescent standard for quantitative measurements of GFP. (C) CLSM 3-dimensional image reconstructions of a *Krox-GFP* (mRNA) injected sea urchin embryo. The image on the left has not been depth corrected. After image processing, the GFP fluorescence (green) is restored to its proper intensity in the bottom half of the embryo (right image). Quantifying fluorescence within cell boundaries in a sea urchin blastula. Boundaries in a single plane (white) are indicated by the membrane dye FM 4-64 (Molecular Probes). The colorful stacks mark the entire boundary of each 3-dimensional cell, and are regions in which GFP has been quantified on a cell-by-cell basis (M. Souren and I. J. Dmochowski, unpublished).

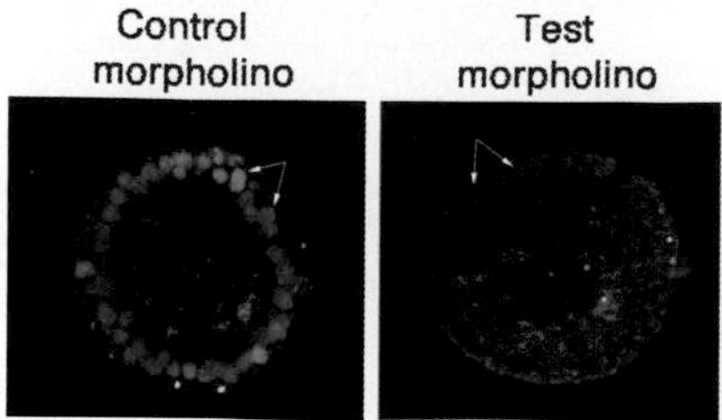

Chapter 28, Fig. 1 A morpholino knockdown of SpSoxB1 to undetectable levels in sea urchin blastulae. Embryos were stained with a polyclonal antibody to SpSoxB1 (green) and a monoclonal antibody, 6e10 (red), that specifically recognizes primary mesenchyme cells.

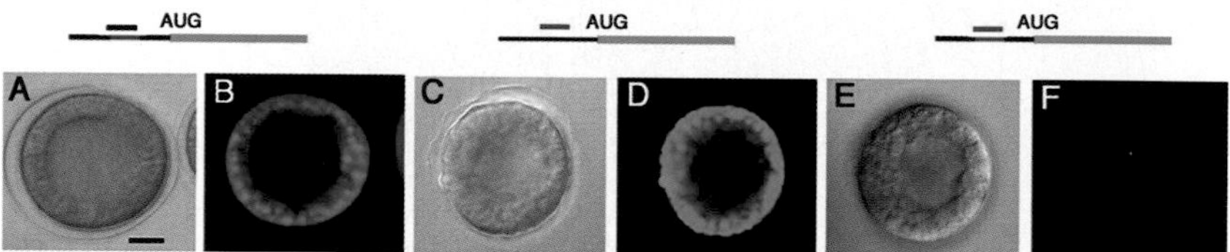

Chapter 28, Fig. 2 Demonstration that a morpholino can efficiently block translation *in vivo*. Early blastulae derived from eggs injected with (A, B) GFP mRNA containing the morpholino target sequence (red) and a control morpholino (black bar); (C, D) GFP mRNA lacking the morpholino target sequence and the test morpholino (yellow bar); or (E, F) GFP mRNA containing the morpholino target sequence and the test morpholino (yellow bars). (Figure from Howard *et al.*, 2001).

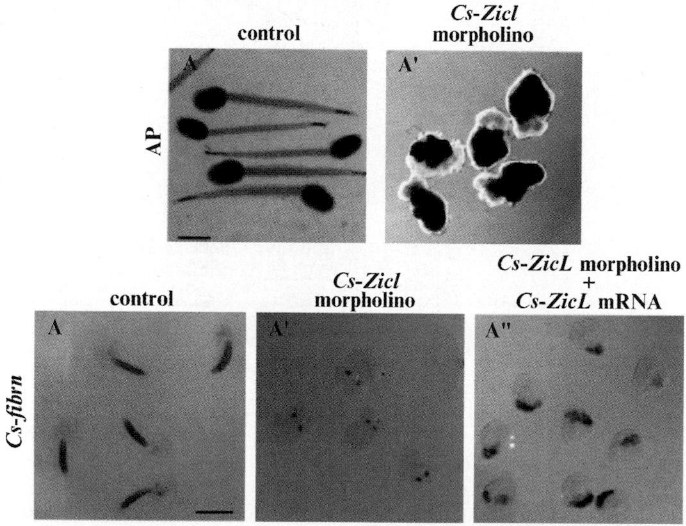

Chapter 28, Fig. 4 Demonstration of a specific morpholino-mediated knockdown in *Ciona*. Top panels demonstrate a dramatic loss-of-tail phenotype upon introduction of a Cs-ZicL morpholino. Embryos are stained for alkaline phosphatase (AP), which is expressed in the endoderm. Bottom panels show the reduction in number of cells expressing Cs-fibrn, a fibrinogen-like protein, in the presence of the morpholino (*cf.* A and A′) and a significant recovery of activity when CsZicL mRNA is co-injected (*cf.* A′ and A″). (Figures taken from Imai *et al.*, 2002.)

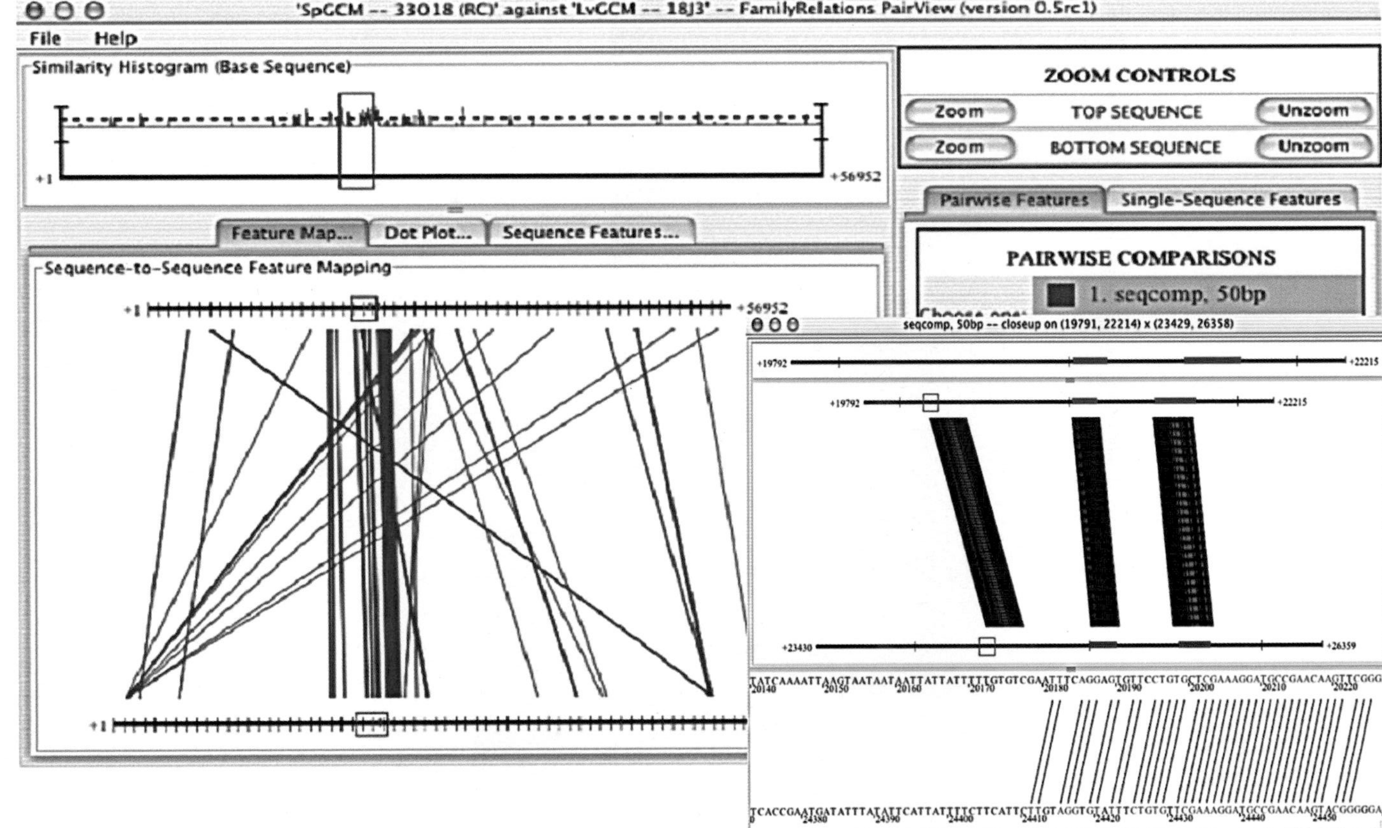

Chapter 30, Fig. 3 A Family Relations pairwise analysis of two BACs containing *gcm*, one sequence from *S. purpuratus* (top sequence) and the other from *L. variegatus*. The large panel shows an overall view of the pairwise analysis, with lines drawn between the two BACs where similarities exist above a threshold of 80% between any two 50 bp windows. The inset panel shows a zoomed view where matches between nucleotides are graphed. In both panels, the exons are plotted in red.

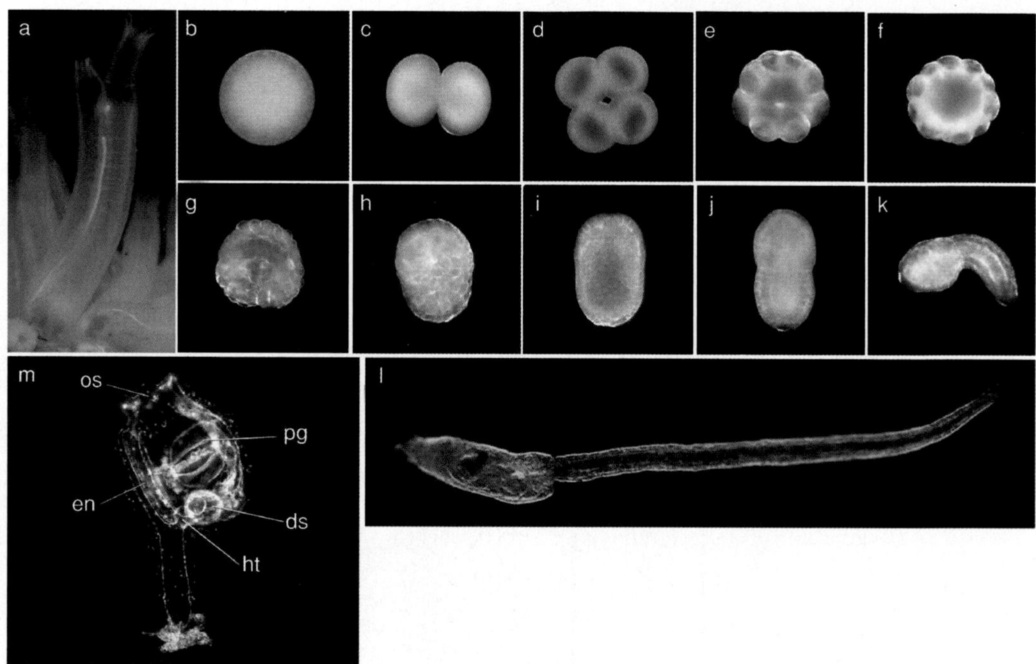

Chapter 31, Fig. 1 The sea squirt *Ciona intestinalis.* (a) Adults are filter feeders with incurrent and outcurrent siphons. The white duct is the sperm duct, while the orange duct paralleling it is the egg duct. (b–l) Embryogenesis. (b) Fertilized egg, (c) 2-cell embryo, (d) 4-cell embryo, (e) 16-cell embryo, (f) 32-cell embryo, (g) gastrula (about 150 cells), (h) neurula, (i–k) tailbud embryos, and (l) tadpole larva. Embryos were dechorionated to show their outer morphology clearly. (m) A juvenile a few days after metamorphosis, showing the internal structure. ds, digestive system; en, endostyle; ht, heart; os, neuronal complex; and pg, pharyngeal gill.

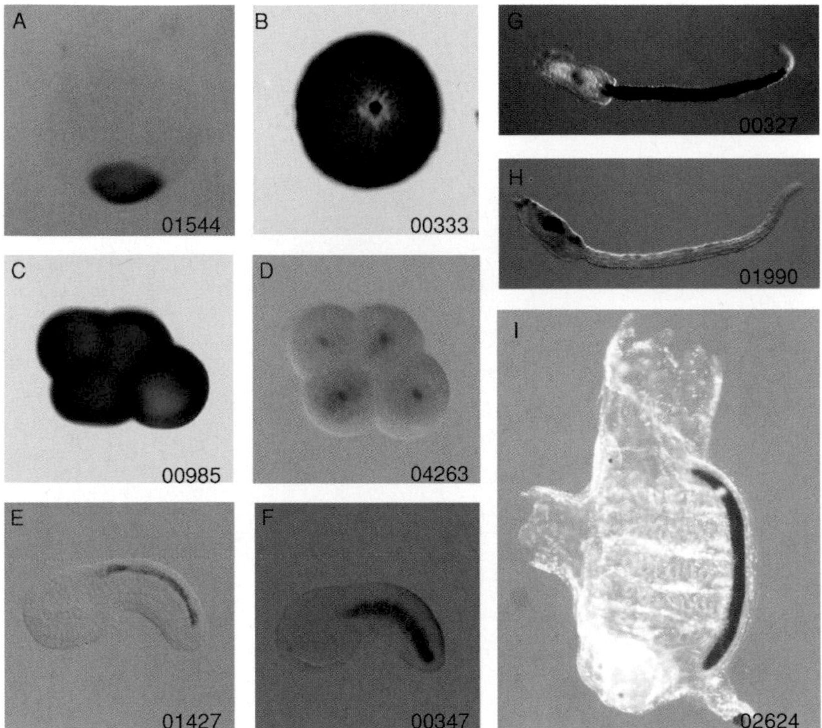

Chapter 31, Fig. 2 Whole-mount *in situ* hybridization showing the expression profiles of genes in *Ciona intestinalis*. Gene IDs are shown at the right bottom corner of each photograph. (A, B) Two examples of the distribution of maternal transcripts in fertilized eggs. (C, D) Examples of the distribution of (C) a maternal transcript and (D) a zygotic transcript in 8-cell embryos. (E) Expression of a nerve cord-specific gene and (F) notochord-specific gene in tailbud embryos. (G) Expression of a muscle-specific gene and (H) neuron-specific gene in larvae. (I) Expression of an endostyle-specific gene in a young adult.

Cell Lineage Chart Through Tenth Cleavage in the Sea Urchin *Lytechinus variegatus*

Cleavage Number

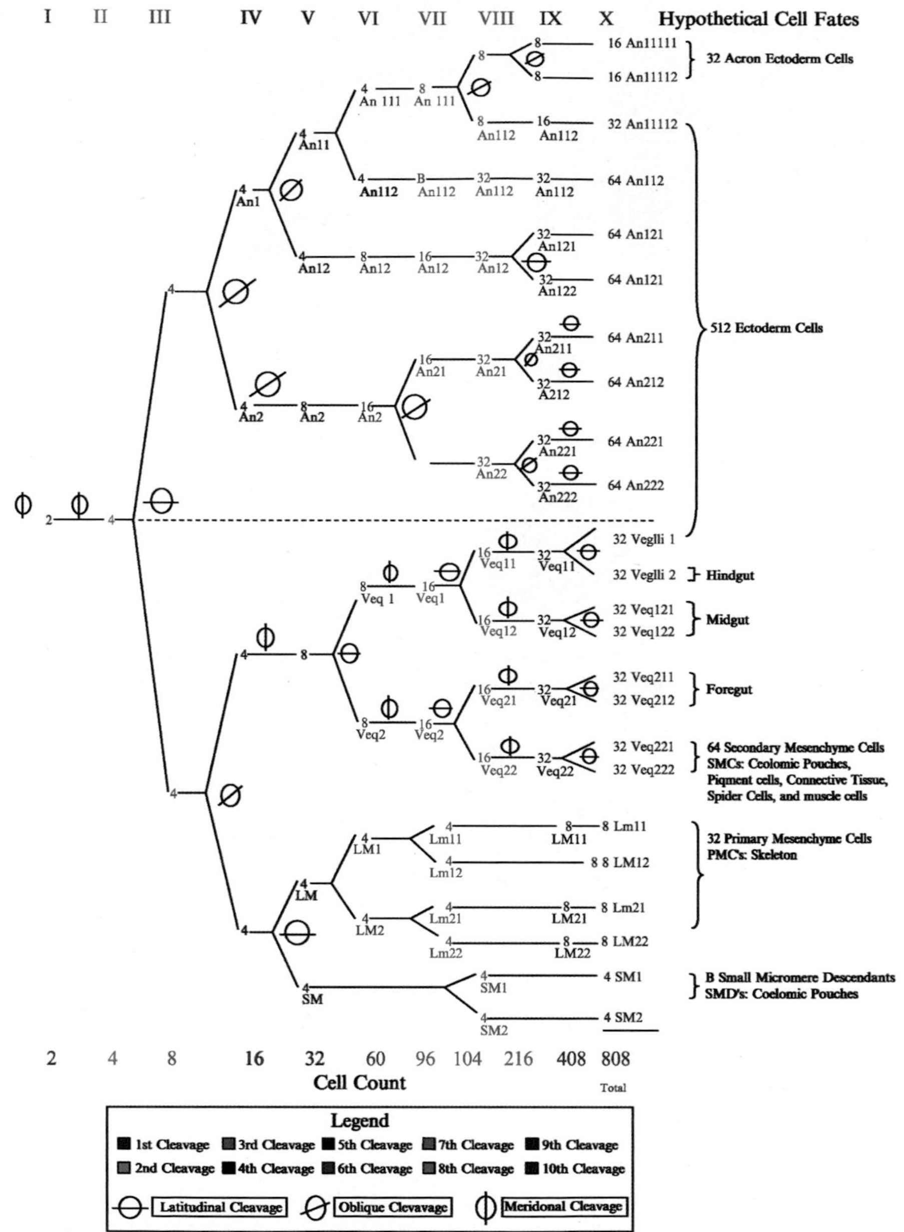

Appendix J. B. Morrill and Lauren Marcus, 2003.